Automotive Fuels Reference Book

Automotive Fuels Reference Book

Fourth Edition

BY PAUL RICHARDS AND JIM BARKER

SAE INTERNATIONAL

Warrendale, Pennsylvania, USA

400 Commonwealth Drive
Warrendale, PA 15096-0001 USA
E-mail: CustomerService@sae.org
Phone: 877-606-7323 (inside USA and Canada)
            724-776-4970 (outside USA)
Fax: 724-776-0790

Library of Congress Catalog Number 2023948255
http://dx.doi.org/10.4271/9781468605792

ISBN-Print 978-1-4686-0578-5
ISBN-PDF   978-1-4686-0579-2
ISBN-epub   978-1-4686-0580-8

To purchase bulk quantities, please contact: SAE Customer Service

E-mail: CustomerService@sae.org
Phone: 877-606-7323 (inside USA and Canada)
            724-776-4970 (outside USA)
Fax: 724-776-0790

Visit the SAE International Bookstore at books.sae.org

Publisher
Sherry Dickinson Nigam

Product Manager
Amanda Zeidan

Production and Manufacturing Associate
Brandon Joy

To Bernard and the memory of Rita.

—Paul

To the memory of Joseph and Edith.

—Jim

Contents

CHAPTER 6

Storage, Distribution, and Handling of Gasoline and Diesel Fuel   137

CHAPTER 7

Positive Ignition Engine Combustion Process   157

CHAPTER 10

Influence of Gasoline Composition on Stability, Gum Formation, and Engine Deposits   261

CHAPTER 11

Gasoline Additives   285

CHAPTER 13

Influence of Gasoline Characteristics on Emissions   323

CHAPTER 16

Diesel Engine Design and Influence of Fuel Characteristics   391

CHAPTER 19

Diesel Fuel Additives   499

CHAPTER 20

Other Diesel Specification and Nonspecification Properties   523

CHAPTER 21

Influence of Diesel Fuel Characteristics on Emissions   539

Preface to First Edition

n this book we have tried to include those aspects of automotive fuels that are most likely to be needed by automobile engineers, oil technologists, commercial and academic research establishments, fuel additive manufacturers, and personnel from governmental agencies dealing with the control and specification of these fuels. Because such a wide range of people of varying disciplines are likely to be interested, we have included, as Appendix 1, a brief outline of fuel chemistry for those readers who are unfamiliar with the chemical symbols and equations used, even though these have been kept to a minimum. A glossary of terms has also been included, although most terms will have been explained at some point in the text.

Automotive fuels are of interest worldwide and for this reason this book has been written to cover, as far as possible, the global situation rather than the position in any single country. Thus, although leaded fuels are now of limited interest in many countries, their use will continue in some parts of the world well into the next century and so we thought it necessary to include a full discussion on such fuels.

By far the most common of the currently used automotive fuels are gasoline and diesel, so it is inevitable that most of the book is taken up by these fuels. Alternative fuels such as liquefied petroleum gas (LPG) or ethanol are, however, very important in some countries and there are indications that other alternative fuels could become important in the future. In particular, methanol looks as if it eventually could become the fuel of the future and is discussed in some detail, along with ethanol, in Chapter 19 *(now discussed in Chapter 5 of the Fourth Edition)*. A chapter on racing fuels has also been added because this is a subject of some interest on which comparatively little has been written.

Without government intervention, the use of alternative fuels in diesel engines appears to be a fairly remote possibility. Environmental pressures are leading to ever-tighter legislation against diesel exhaust emissions, but the major variations in levels of particulates and gases are associated with the engine rather than the fuel. However, although the costs of modifications to engines, vehicles and distribution systems are substantial, methanol is being evaluated as a "clean" alternative diesel fuel by several bus fleet operators in the US, Japan, and Canada.

Because vehicle emission regulations are indirectly related to automotive fuels, we have incorporated as an Appendix a summary of current emissions legislation as compiled by CONCAWE (the oil companies' European Organization for Environmental and Health Protection) and included with their kind permission *(this is now discussed in Chapters 13 and 21 of the Fourth Edition)*. Legislation is changing at a rapid rate and so we have generally restricted our comments in the text to the general principles governing the formation of these emissions and particularly how they are influenced by fuel quality.

Chapter 15 *(Chapter 17 in the Fourth Edition)*, which covers diesel fuel low-temperature characteristics, includes a section on the measurement of low-temperature performance. Direct comparison of results obtained by different organizations is often difficult because of variations in test techniques. This measurement procedure, which includes testing in the field as well as in climatic chambers, provides guidance on vehicle instrumentation, cooling phases, driving patterns, etc. We hope that it will be accepted as the basis for an industry standard so that future test data will be more readily comparable with each other.

Regarding units, we have tried to stick to the SI system although there are some instances where alternative units are used. In these cases, the units are widely accepted and commonly used.

Finally, we would like to acknowledge with grateful thanks the assistance of many friends and colleagues who read various parts of the text and who improved the book with their helpful comments. In addition, we would like to thank Lucas Diesel Systems for their assistance with the chapters concerning diesel fuels. We are extremely grateful to the Esso/Exxon Chemicals Research Centre at Abingdon in the UK, who made available to us the use of their Information Centre, and we thank particularly Miss Andrea Strafford and her colleagues who carried out for us many computer-based literature searches and who provided copies of papers, and without whose help we would never have been able to complete this task.

Preface to Second Edition

The pace at which changes are taking place in the domain of automotive fuels has necessitated a number of revisions to bring the Handbook up to date.

Although the first edition was published as recently as October 1990, there has been a significant change of emphasis with the wider introduction of reformulated fuels for both spark-ignition and compression-ignition engines, and in the types of additives for primary and secondary treatment of diesel fuels.

The other major area of change—and growth—is in environmental legislation against noxious and undesirable exhaust emissions from road vehicles. We are grateful to CONCAWE, the oil companies' European organization for environment, health, and safety, for permission to incorporate their latest update on the situation in the appendices.

A new chapter has been contributed by Mr. C. S. Weaver to give more attention to the growing interest in the use of liquefied and compressed natural gas (LNG and CNG) and LPG as more environmentally friendly automotive fuels.

In view of concerns about automotive fuels and emissions, a section giving the health and environmental effects of gasolines and diesel fuels has been added.

The opportunity has been taken to increase the usefulness of the Handbook (and to resolve some uncertainties about units commonly used in the petroleum industry) by including Tables of Conversion and Heating Values. Inevitably, it has also been necessary to extend the list of abbreviations and acronyms which proliferate, often without explanation, in many technical publications.

Restructuring in Europe is also bringing in new names and abbreviations. At the end of 1993, the Single Europe Act came into effect, replacing the European Community (EC) with the European Union (EU). In May 1994, approval was given by the European Parliament to applications by four member countries of the European Free Trade Association (EFTA) to join the EU. The application by Norway was withdrawn after a referendum rejected joining but the other three countries, Austria, Finland, and Sweden, formally became members on January 1, 1995. Further enlargement and designation changes may occur if Eastern European countries are admitted but efforts will be made to minimize confusion in a situation which is still evolving.

The authors are appreciative of the advice and guidance provided by Don Goodsell, Ryozo Kato, former Esso/Exxon colleagues and other contributors during the preparation of this revision of the Handbook.

Keith Owen and Trevor Coley
January 1995

Preface to Third Edition

In the first and second editions of this book, the authors tried to include those aspects of automotive fuels that are most likely to be needed by automobile engineers, oil technologists, commercial and academic research establishments, fuel additive manufacturers, and personnel from governmental agencies dealing with the control and specification of these fuels. In writing this third edition I have tried to keep faithful to the goals and standards of the original authors, Keith Owen and Trevor Coley. I have also tried to follow the layout of their original concept.

Gasoline and diesel fuel continue to be the most commonly used automotive fuels, although alternative fuels such as liquefied and compressed natural gas (LNG and CNG) and liquefied petroleum gas (LPG) are covered. In the previous editions the vast majority of gasoline and diesel fuels were produced from crude oil with only a small proportion being made by the gas-to-liquids process, either from gasified coal or directly from natural gas, or from other fossil sources. The proportion of fuel being derived from these other fossil sources has now increased and as a result it has also been given its own chapter. Fuels such as ethanol and plant oil derivatives that were once thought of as alternative are now widely used blending components to increase the renewable component of gasolines and diesel fuels. A new chapter has thus been included discussing the manufacture of gasoline and diesel fuel from renewable sources. The use of renewable fuels will inevitably grow, with increasing research aimed at developing more energy and land efficient alternatives.

Another area of change since the second edition has been the general growth in popularity of the diesel passenger car (the US is probably an exception to that), and increasingly stringent regulation of diesel exhaust emissions. While the stability, deposit forming tendency, and influence of fuel characteristics on emissions were covered in the previous edition, I believe their growing importance now warrants separate chapters to mirror those for gasoline that were previously included. In respect of emissions legislation and fuel effects, the previous edition of this book included appendices that were taken from CONCAWE report 4/94. This report has since been updated every few years and has grown in stature. Due to the increased volume of that work and its ready availability via the Internet, it is no longer thought necessary to include the full sections of that report in this book. The latest version is listed as further reading at the end of the appropriate chapters. The website can be checked for later updates at www.CONCAWE.org

Automotive fuels are of interest worldwide, and for this reason this book has been written to cover, as far as possible, the many situations that exist throughout the globe. While leaded gasoline is sold to the populace in very few countries, it is still formulated for motorsport and aviation use. Thus, it is still of interest to the fuel technologist to retain the information on lead antiknock additives. Similarly, although carbureted cars are now part of the legacy fleet, they are still an important part of the fleet in some countries, and so how fuel characteristics affect these vehicles is thus also of interest. Similarly high-pressure common-rail fuel injection equipment is becoming de rigueur for diesel engines, but the older style pump-line-nozzle systems are still to be found. I have therefore retained much of what Keith and Trevor wrote regarding engine technology while adding what is new, although I have tried to restrict discussion to passenger car technology and performance.

I would now like to acknowledge with grateful thanks the assistance of many friends and colleagues throughout the various industries who have supplied material, reviewed my work, and provided advice. They are listed alphabetically by affiliation: Jerry Burton of Advanced Fuel Solutions (Chester) Ltd., Richard Jones at BP International Limited, Les Wolf at BP Products North America, Inc., Roger Hutcheson of Cameron Associates, Ken Rose at CONCAWE, Jan Tucker at the Coordinating Research Council, Inc., Diane Lance at Coryton Advanced Fuels Ltd., Nobuyasu Ohshio at Comso Oil Co., Ltd., Markus Paule at Daimler AG, Philip J Dingle at Delphi Diesel Systems, Bruce Beaver of Duquesne University, Nigel Elliott at ExxonMobil Research & Engineering, Steven Szwabowski at Ford Motor Company, John Maddox at Infineum UK Ltd., Trevor Russell and Jim Barker at Innospec Ltd., Naoki Shimazaki of Isuzu Advanced Engineering Center, Ltd., Wendy Clark at the National Renewable Energy Laboratory (NREL), Scott Curran of Oak Ridge National Laboratory, Ray Hooley of the Ray Hooley Collection, John Dec at Sandia National Laboratories, Paul Schaberg at Sasol Technology (Pty) Ltd., Nicki Welding and Roger Cracknell at Shell International Limited, Richard S. Jackson at Tesco Petrol Filling Station, Jan Czerwinski at the University of Applied Sciences Biel-Bienne, Yoshiyuki Kidoguchi at The University of Tokushima, Phil Easdown of www.vintagegarage.co.uk, Nigel Mathewson (no affiliation), plus Colette Wright, Martha Swiss, and Heather Slater at SAE International. Finally, I'd like to express my deep gratitude to my wife, Julie, for her support and patience during the many months of this project.

Paul Richards
March 2013

Preface to Fourth Edition

In the first three editions of this book, the authors tried to include those aspects of automotive fuels that are most likely to be needed by automobile engineers, oil technologists, commercial and academic research establishments, fuel additive manufacturers, and personnel from governmental agencies dealing with the control and specification of these fuels. In writing this fourth edition we have tried to keep faithful to the goals and standards of the original authors Keith Owen and Trevor Coley. We have also tried to follow the layout of their original concept as carried over to the third edition.

Gasoline and diesel fuel continue to be the most commonly used automotive fuels, although alternative fuels such as liquefied and compressed natural gas (LNG and CNG) and liquefied petroleum gas (LPG) are covered. In the previous editions the vast majority of gasoline and diesel fuels were produced from crude oil with only a small proportion being made by the gas-to-liquids process, either from gasified coal or directly from natural gas, or from other fossil sources. The proportion of fuel being derived from these other fossil sources has now increased and as a result has also been given its own chapter. Fuels such as ethanol and plant oil derivatives that were once thought of as alternative are now widely used blending components to increase the renewable component of gasolines and diesel fuels. Since the third edition we have also seen the advent of what we have termed Power-to-Liquid fuels, commonly referred to as e-fuels, that use renewables electricity to produce the components of liquid fuels. The chapter discussing the manufacture of gasoline and diesel fuel from renewable sources has thus been expanded to include e-fuels. The use of renewable fuels will inevitably grow with increasing research aimed at developing more energy and land efficient alternatives.

Another area of change since the third edition has inevitably been ever-tighter emissions regulation. Whilst the stability, deposit forming tendency and influence of fuel characteristics on emissions were covered in the previous edition and thus the separate chapters created for the third edition for gasoline and diesel have been maintained.

Automotive fuels are of interest worldwide and for this reason this book has been written to cover, as far as possible, the many situations that exist throughout the globe. Whilst leaded gasoline is no longer sold for use on public roads, it is still formulated for motorsport and aviation use. Thus, it is still of interest to the fuel technologist to retain the information on lead antiknock additives but due to this limited interest has been moved to an appendix. Similarly, although carbureted cars are now part of the legacy fleet, they are still an important part of the fleet in some countries and how fuel characteristics affect these vehicles is thus also of interest. Similarly direct injection diesel engines with high-pressure common-rail fuel injection equipment are becoming de rigueur for diesel engines, but the older indirect injection (IDI) engine technology with pump-line-nozzle injection systems is still to be found. We have therefore retained much of what Keith and Trevor wrote regarding engine technology whilst adding what is new, although we have tried to restrict discussion to passenger car technology and performance.

We would now like to acknowledge with grateful thanks the assistance of many friends and colleagues throughout the various industries who have supplied material, reviewed work, and provided advice. This includes those that contributed to the third edition whose invaluable contributions have been carried forward to this fourth edition. They are listed alphabetically by affiliation: Peter Kunz at APL Automobil-Prüftechnik Landau GmbH, Les Wolf at BP Products North America, Inc., Ben Lampertz and Steve Tapp at Coryton Advanced Fuels Ltd, Steven Szwabowski at Ford Motor Company, Joe Ciaravino and Shailesh Lopes at General

Motors, Sybille Riepe at H2 Mobility, María Jesús Sievers at HIF Global, SSV Ramakumar and Sandeep Sharma at Indian Oil, Avinash Kumar Agarwal at the Indian Institute of Technology Kanpur, John Maddox at Infineum UK Ltd., Trevor Russell and Simon Mulqueen at Innospec Ltd., Robert McCormick at the National Renewable Energy Laboratory (NREL), Liping Mian at Neste, Amal Agha and Kevin Cowie at Phillips 66, Ray Hooley of the Ray Hooley Collection, John Dec at Sandia National Laboratories, Paul Schaberg at Sasol Technology (Pty) Ltd., Nicki Welding and Roger Cracknell at Shell International Limited, Leonid Tartakovsky at Technion – Israel Institute of Technology, Jan Czerwinski at the University of Applied Sciences Biel-Bienne, Felix Leach at the University of Oxford, Yoshiyuki Kidoguchi at The University of Tokushima, Andrea Strzelec at the University of Wisconsin-Madison, Phil Easdown of www.vintagegarage.co.uk, plus Colette Wright, Sherry Nigam and Amanda Zeidan at SAE International. Finally, we would like to express our deep gratitude to our wives Julie and Linda for their support and patience throughout the months of this project.

Paul Richards and Jim Barker
August 2023

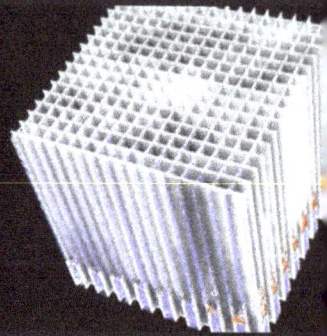

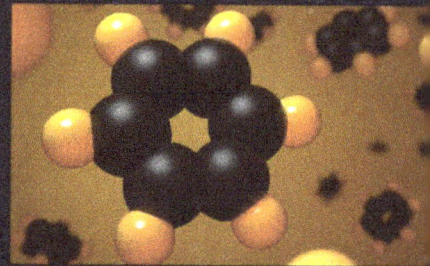

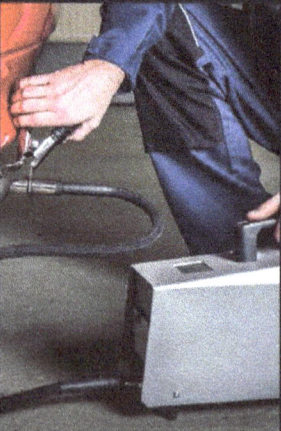

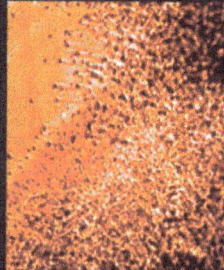

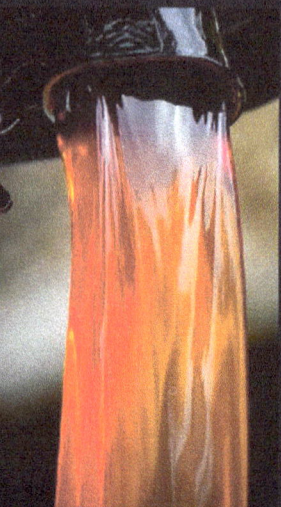

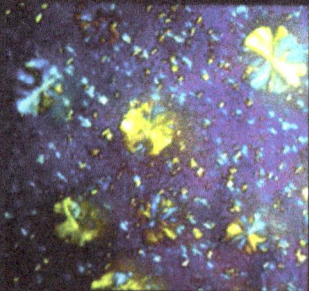

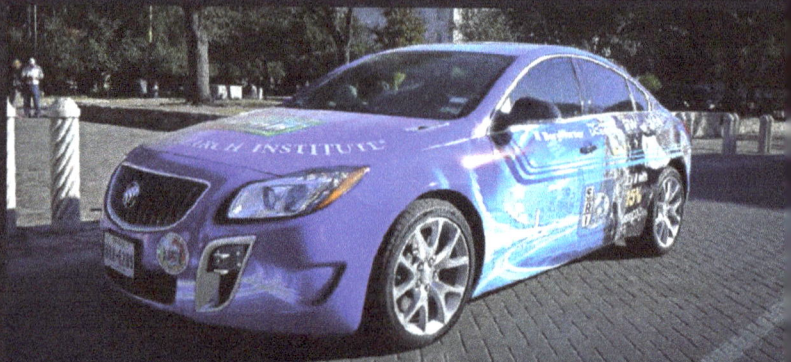

1

Introduction to Automotive Fuels and Their Specification

The origins of automotive transport can be debated, but by the start of the twentieth century, the power source for personal automotive transport was becoming a competition between electric vehicles (EV) and internal combustion (IC) engine-powered vehicles. Due to the limitations of the EV, the IC engine became a more popular form of motive power and fossil fuel derivatives became the dominant form of fuel.

During the early days of the petroleum industry when crude oil was processed primarily for kerosene, gasoline was a waste product and was separated from the oil used in lamps because of its tendency to vaporize and cause the lamps to explode. Later it was found that this ability to vaporize at reasonably low temperatures made possible the development of an internal combustion engine which would operate economically and satisfactorily on gasoline.

In the early development of the automobile, the average owner was satisfied if the engine would run continuously and start without too much difficulty. Nearly all engines were provided with priming cups so that fuel could be poured directly into the engine cylinder for starting. The motorist who was rash enough to drive his car in cold weather frequently provided himself with a can of ether mixed with a volatile gasoline for this purpose. The accelerating energy required to start the car moving was provided by using a heavy engine flywheel, speeding up the engine, and letting the clutch in. The engines themselves had practically no accelerating ability. Fuel economy in miles per gallon was not discussed because the distances traveled were short and the speeds low. Quoting from the 1892 catalog of a car manufacturer: "Speeds: The cars have three speeds; a small speed, an intermediate speed and a high speed. The high speed is generally adjusted to eleven m.p.h."

From this small beginning, motor cars, highways, and fuels developed together until gasoline, no longer a by-product, became the most important constituent of crude oil. As the demand for gasoline increased, the manufacturers endeavored to increase the available supply by using more and more of the heavier and less volatile portions of the crude [1.1].

These words were written over 85 years ago, and since then, distillate fuels have also become a significant product of the petroleum industry, and the demands of the automotive industry have also changed significantly. Automotive fuels, gasoline and diesel fuel, are now probably the most important products manufactured by oil refining companies, because such a large proportion (between 60 and 75%) of the crude oil run in a refinery is converted into gasoline and distillate fuels. Figure 1.1 [1.2, 1.3] shows this situation over the last 20 years of the 20th century and the first 20 years of the 21st century.

FIGURE 1.1   Refinery yields from a barrel of crude oil.

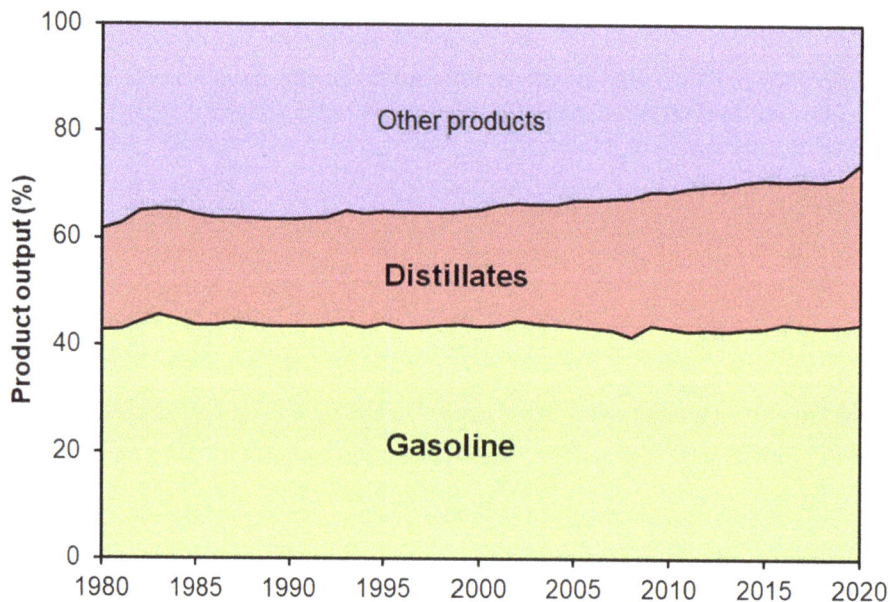

© SAE International.

However, the split between gasoline and diesel can vary significantly from country to country and with time. Figure 1.2 [1.4] shows the US product supplied since 1945.

FIGURE 1.2   US product supplied.

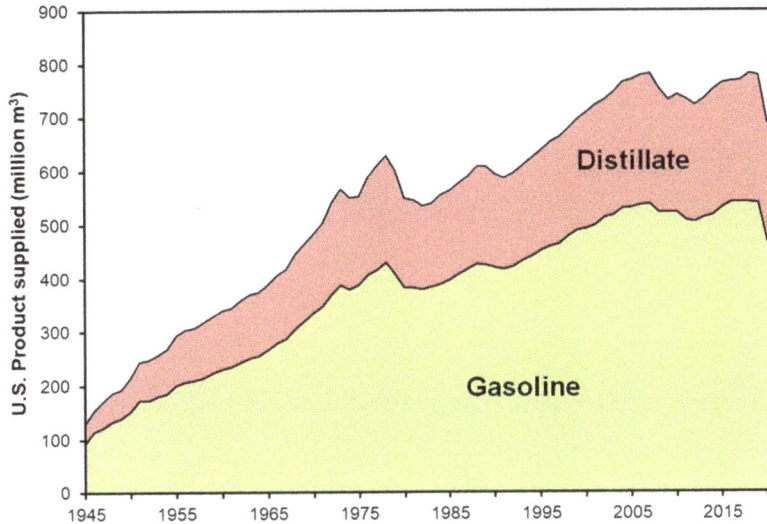

This figure clearly shows a dip in demand around 2020 due to the global COVID-19 pandemic caused by the SARS-CoV-2 virus [1.5]. Although the supply of product has increased more than fivefold over the illustrated time frame, the mix of product has been within quite a narrow range. Since 1950, the ratio of gasoline to diesel has remained within the range of 2 to 2.5. However, the situation in Europe has been somewhat different. There are two factors contributing to this situation: first, fiscal policy means that the cost of fuel in Europe is far higher than in the United States (US); second, in order to meet tighter fuel economy standards, the motor manufacturers were making diesel-powered vehicles more price-competitive. Figure 1.3 [1.6] shows the fuel usage in the 27 European Union countries from 1990 to 2010; it is clear that gasoline usage was diminishing while diesel fuel usage was generally increasing. Here, the ratio of gasoline to diesel has dropped from 1.34 in 1990 to 0.47 in 2010.

FIGURE 1.3   European fuel usage.

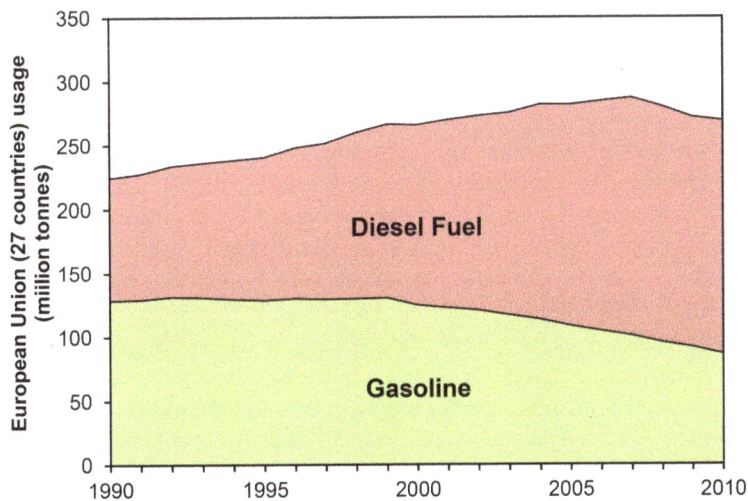

However, all that changed between 2010 and 2020 due to what has become known as "Dieselgate" [1.7–1.9]. "Dieselgate" shook the automotive world and triggered a mass migration away from diesel vehicles in Europe. In 2015, diesel vehicles accounted for 51.5% of sales, with gasoline vehicles trailing behind at 44.2%. Just three years later, in 2018, diesel vehicle sales had plummeted to 35.9%, while gasoline sales had climbed to 56.7%. Figure 1.4 [1.4] shows the steadily falling gasoline consumption until around this time when a reversal occurs and gasoline consumption starts to rise.

FIGURE 1.4   European gasoline consumption before and after "dieselgate."

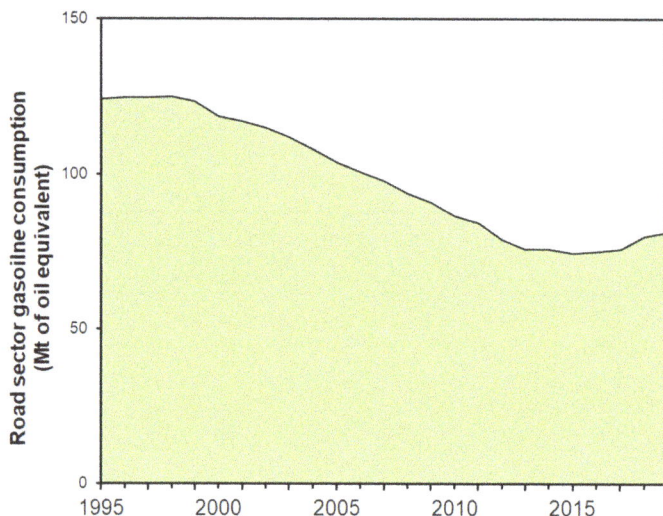

© SAE International.

A bias in demand can affect the fundamental nature of products produced and, to some extent, the national or regional specifications that have evolved to reflect that situation. US demand has favored maximizing gasoline production; in order to satisfy this demand, the gasoline distillation range has traditionally been wider than that in Europe, simply to transfer as many hydrocarbons as possible into the gasoline "pool." This has had an impact on US diesel fuel, which tends to have lower ignition quality than European product, reflecting the more aromatic nature of the streams available from the gasoline-dominated cracking processes. Such differences in demand therefore predetermine some basic fuel characteristics, and it is difficult to visualize a situation in which a single worldwide specification could become a practical or economic reality. However, the engine manufacturers associations of the US, Europe, and Japan have produced a document to outline the fuel quality needs of motor vehicles and to strive for harmonized fuel quality and specifications for the differing vehicle technologies around the world, which is known as the Worldwide Fuel Charter and some relevant details are given in Appendix 2. To meet the specification would normally require the inclusion of deposit control additive and in the US leading automotive and heavy-duty equipment manufacturers have developed a recognized premier fuel performance standard, in 2004 for gasoline and 2017 for diesel fuel. These standards are included in Appendix 3 and the fuels are marketed under the Top Tier™ branding. One of the aspects clearly highlighted within the Worldwide Fuel Charter is the reduced sulfur levels required for the satisfactory performance of emissions control equipment with Category 4, and upward, having a maximum sulfur content of 10 mg/kg, for both gasoline and diesel fuel. Figure 1.5 [1.4] shows the change in the sulfur content of the supplied US diesel fuel.

FIGURE 1.5   Sulfur content of US-supplied diesel fuel.

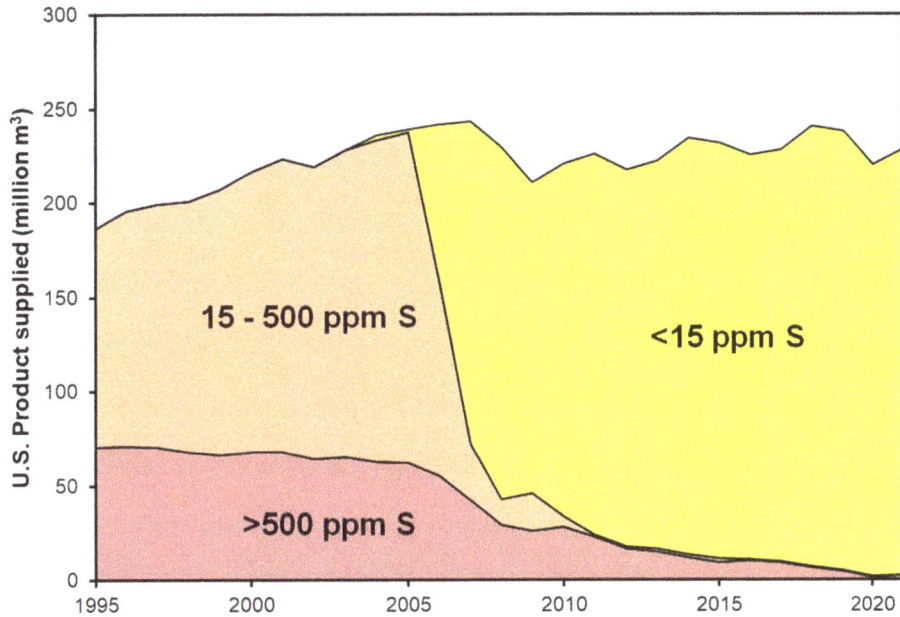

© SAE International.

This chart clearly shows how regulatory changes can significantly affect the way a refinery is operated, which in turn has an effect on the ability of the refiner to meet the product demand split from the crude supply, and hence the product price.

Of course, we can no longer overlook the burgeoning economies that will have a significant impact on the supply and demand for gasoline and diesel. Figure 1.6 [1.4] shows the growth in Chinese demand since 1975.

FIGURE 1.6   Chinese road sector fuel consumption.

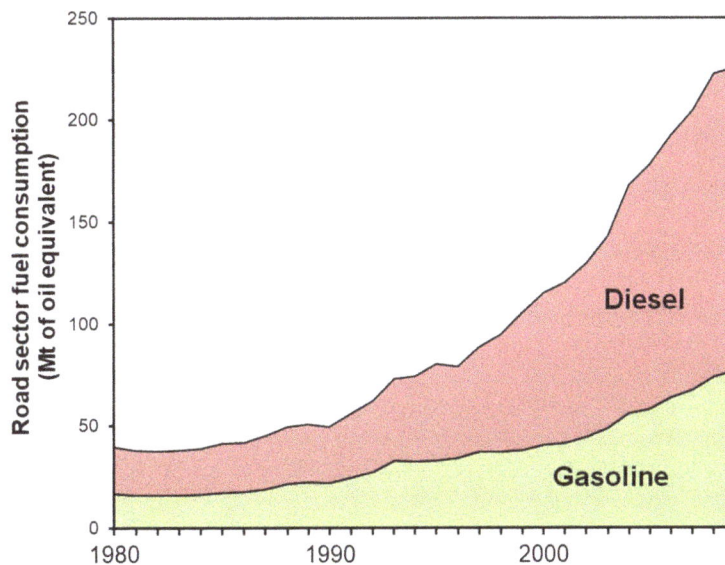

© SAE International.

Because of the high volumes of diesel and gasoline involved and the competitive marketing situation in most countries, these products determine, to a very large extent, the profit or loss situation for a refining company. The high fixed costs involved in manufacture and distribution mean that changes in market share can have a dramatic influence on profitability. For this reason, it is important for personnel involved with the production and marketing of these fuels to understand the quality aspects that can influence market share. Of course, there are many other factors that affect the profitability of producing and marketing these fuels that are outside the control of the refiner, such as the price of crude oil feedstock, interest rates, and general demand trends, but here we are concerned mainly with quality aspects of the finished fuels that can, to some extent, dictate the performance of the vehicle and which are within the control of the fuel manufacturer.

Against this background, the main aims of an automotive fuel technologist are to ensure that the product being manufactured meets the required legal specifications and that any in-house specifications imposed by the manufacturing or marketing company are realistic and worthwhile. National or other legally enforceable specifications represent the minimum quality that must be supplied, and it is implicit that engine designers should ensure that their vehicles will run satisfactorily on such a quality.

In order for vehicle and engine manufacturers to ensure that their products meet legislative requirements in terms of emissions and power and fuel economy and that they perform well on the range of fuels marketed, special legislative and reference fuels are available. Legislative fuels are used by the motor industry for certification or homologation and generally represent the average quality found in the marketplace. These are typically defined by the legislator (e.g., the European Commission, the US Environmental Protection Agency, and so on). Reference fuels can be used for engine development, and these represent qualities close to the worst case that may be marketed for particular properties, such as volatility or octane. Frequently, these fuels are specified by joint industry bodies [1.10].

Internal or in-house specifications are more severe in many respects than national specifications and usually include additional controls. They are used to define the quality a particular fuel company, refiner, or blender is producing. Differences from the national or standard specification exist in order to meet a competitive situation, capture a greater market share, meet a seasonal need, or overcome an existing field problem. Such specifications will include controls on the type and concentrations of additives to be used and the dates and levels at which seasonal specifications are to be introduced.

Close cooperation is important between fuel producers and motor manufacturers in the face of changes and regulatory pressures occurring in both industries. Joint committees have been set up in most countries to ensure that new developments, whether in legislation or engines or fuels, can be accommodated. In this respect, fuel quality and engine design are a chicken and egg situation. Designers of engines need to know what fuel will be available in a particular market, but equally, oil companies need to know the fuel requirements of vehicles on the road in order to maximize the number of vehicles, and hence customers, that they satisfy. Governmental bodies also need to be aware of the limited scope for modifying fuel properties and the influence such changes can have on vehicle performance, since they are involved in setting national or regional fuel specifications.

One other group with a close interest in fuel quality comprises the fuel additive manufacturers. Additives can reduce manufacturing costs, enhance quality, and provide benefits to the consumer that are recognizable—(although sometimes a little help is needed from advertisements) and so they can significantly influence market share and profitability.

Gasoline and diesel fuel qualities have never been static, as is clear from the history of these fuels. There have been many different pressures causing the changes. Sometimes it has been the cost of crude oil or the need to protect our environment. At other times, improvements in refinery processes have given rise to the potential to improve, and from time-to-time changes in vehicle design have led to the need to modify fuel composition. Occasionally, the changes are for purely commercial reasons in that they reduce manufacturing costs or increase sales potential.

Although gasoline and diesel fuel, derived from crude oil, are by far the most important automotive fuels, when considering the world as a whole, there are other niche fuels that are currently used quite extensively in specific locations. Examples are the use of liquefied petroleum gas (LPG), which is used extensively in Japanese taxi fleets, and compressed natural gas (CNG), which is used as fuel for taxis and city buses in heavily congested areas such as Delhi, India. These are discussed in greater detail in Chapter 4. Due to legislation resulting from concern about the environment, other fuels such as ethanol and fatty acid methyl esters (FAME), derived from vegetable oils or animal fats, which were until recently considered as alternative fuels, are now common blending components for gasoline and diesel, respectively, along with processed vegetable oils that can be used in both gasoline and diesel fuel. These fuels are discussed in more detail in Chapter 5. As an imbalance of demand for the different fuels can lead to significant movement of finished fuel between regions, legislation on fuel composition can also result in the movement of these "alternative fuel blending components." Figure 1.7 shows the consumption of ethanol in the US, along with the import or export quantities of this component.

FIGURE 1.7 US ethanol consumption and import/export volumes [1.3].

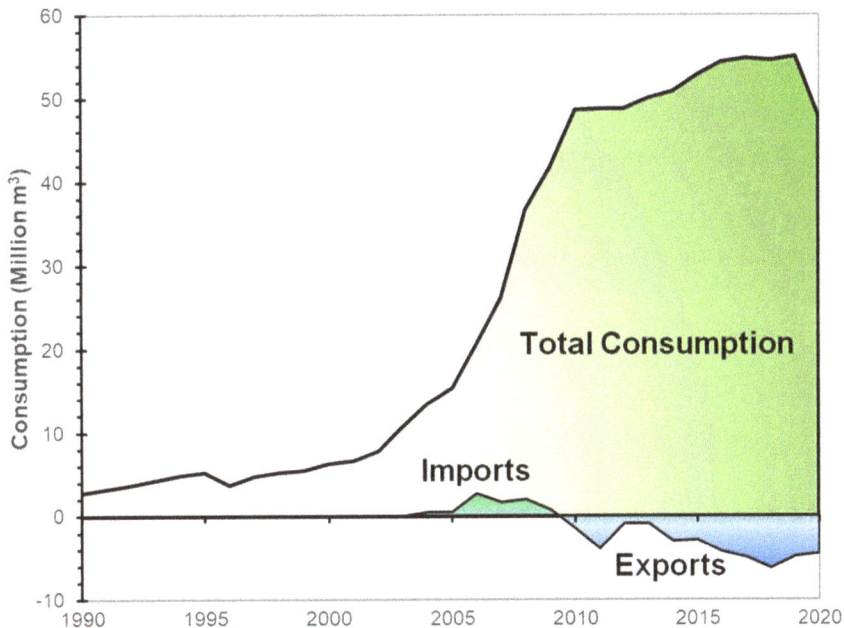

© SAE International.

Figure 1.8 shows the corresponding chart for biodiesel or FAME. From this chart, it is clear that the US was at times producing far more biodiesel than it was using.

FIGURE 1.8 US biodiesel consumption and import/export volumes [1.3].

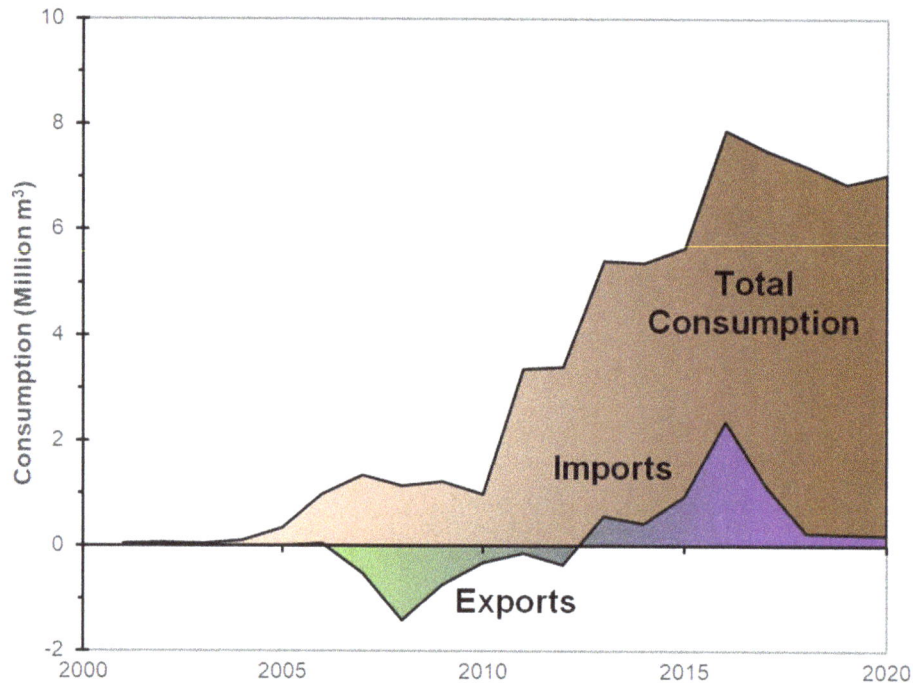

Currently much research is being conducted on fuel cells as a replacement for the IC engine. Hydrogen is the favored fuel for such devices, owing to the absence of carbon and hence carbon dioxide emissions; however, methanol has been proposed as an alternative. However, if hydrogen can be produced without producing CO_2 emissions, e.g., by electrolysis using renewably sourced electricity, then this offers another route to hydrocarbon fuels, by power-to-liquids technology as discussed in Chapter 5.

References

1.1. Robertson, B.J., "An Investigation of Motor Gasolines," Bulletin of the University of Minnesota, 1939.

1.2. Davis, S.C., Diegel, S.W., and Boundy, R.G., *Transportation Energy Data Book: Edition 31* (Oak Ridge, TN: Oak Ridge National Laboratory for the U.S. Department of Energy, 2012).

1.3. Davis, S.C. and Boundy, R.G., *Transportation Energy Data Book: Edition 40* (Oak Ridge, TN: Oak Ridge National Laboratory for the U.S. Department of Energy, 2022).

1.4. U.S. Energy Information Administration, "U.S. Product Supplied for Crude Oil and Petroleum Products," U.S. Department of Energy, Washington, DC, accessed June 2022, http://www.eia.gov/dnav/pet/pet_cons_psup_dc_nus_mbbl_m.htm.

1.5. Ciotti, M., Ciccozzi, M., Terrinoni, A., Jiang, W.C. et al., "The COVID-19 Pandemic," *Critical Reviews in Clinical Laboratory Sciences* 57, no. 6 (2020): 365-388.

1.6. Eurostat, "Complete Energy Balances [NRG_BAL_C__custom_2067018]," Luxembourg, accessed June 2022, www.ec.europa.eu/eurostat.

1.7. Bovens, L., "The Ethics of Dieselgate," *Midwest Studies in Philosophy* 40 (2016): 262-283.

1.8.   Mujkic, E. and Klingner, D., "Dieselgate: How Hubris and Bad Leadership Caused the Biggest Scandal in Automotive History," *Public Integrity* 21, no. 4 (2019): 365-377.

1.9.   Boretti, A., "The Future of the Internal Combustion Engine after 'Diesel-Gate'," SAE Technical Paper 2017-28-1933, 2017, doi:https://doi.org/10.4271/2017-28-1933.

1.10. Pearson J.K. and Hawkins, M.J., "Fuels for the Automotive Industry," in *International Conference on Petroleum Based Fuels and Automotive Applications*, Paper C304/86, Institution of Mechanical Engineers, London, UK, November 1986.

Further Reading

Asmus, T., Bragg, G., Cowan, C.W., Gibbs, L.M. et al., *The Automobile: A Century of Progress* (Warrendale, PA: SAE, 1997).

Eckermann, E., *World History of the Automobile* (Warrendale, PA: SAE, 2001).

Hogarty, T.F., "The History of Gasoline Marketing as a Guide to Marketing Theory," in *Proceedings of the Conference on Historical Analysis and Research in Marketing*, vol. 1, 73-78, Michigan State University, East Lansing, MI, June 1983.

Jackman, F.A., "The History of Gasoline," in Hancock, E.G. (Ed.), *Technology of Gasoline* (Oxford, UK: Society of Chemical Industry, 1985), 2-19.

Senecal, K. and Leach, F., *Racing toward Zero: The Untold Story of Driving Green* (Warrendale, PA: SAE, 2021).

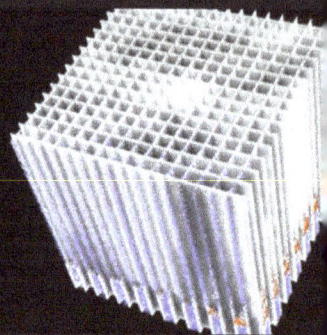

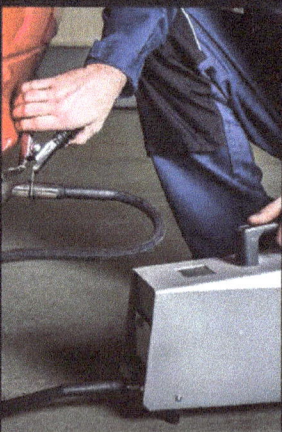

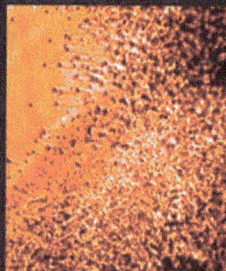

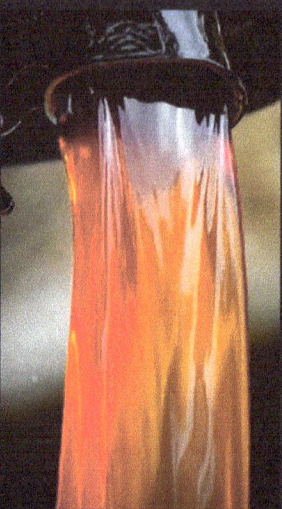

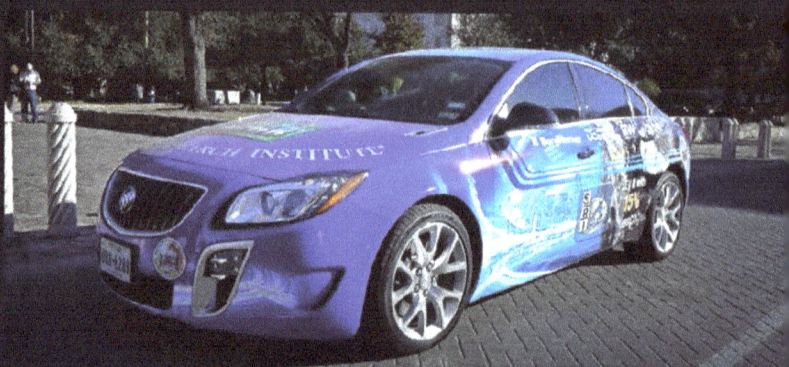

2

A History of Gasoline and Diesel Fuel Development

2.1. Gasoline

2.1.1. The Evolution of the Gasoline Engine

The first practical working IC engines appeared in the middle of the 19th century. A British engineer James Robson built a double-acting, atmospheric, spark-ignited gas engine in 1857, and three were in service before 1860, the year in which a Belgian, Jean-Joseph Étienne Lenoir, patented his own design of a double-acting, spark-ignited gas engine. The ignition was brought about by an induction coil, a vibrating contactor, and a rotating distributor. Not all gas engines of the time used an electrical spark to initiate combustion; other examples of positive ignition of the trapped gas were the introduction of a flame via an "igniting valve," as in William Barnett's patent of 1838 [2.1].

Although earlier engines built or designed at the beginning of that century had not been successful, a number of significant and innovative concepts were introduced.

An English patent granted in 1791 to John Barber had all the elements of a modern gas turbine, and Barber was the first to consider the use of manufactured coal-gas as fuel [2.2]. Robert Street's 1794 English patent is the first record of a mixture of gaseous fuel (vaporized turpentine) and air being burned in a reciprocating IC engine.

The Swiss engineer Isaac de Rivaz (1752–1828) [2.2] is credited with building the first vehicle powered by an IC engine using his 1813 design of a spark-ignited atmospheric gas engine. Its limited range of travel prompted Rivaz to put forward the need for a network of gas generators to enable the vehicle's leather gas storage bag to be refueled at regular intervals, about every 3 km!

The earliest carburetor for liquid fuel (alcohol or turpentine) for an IC engine was the "preparation vessel" of a portable version of the gas engine patented in 1826 by American engineer Samuel Morey (1762–1843) [2.2]. Morey also visualized the application of the IC engine to road and rail transport.

These early engines could be classified as either atmospheric or noncompression. An atmospheric engine uses the rapid combustion of the mixture to propel a piston with such force that it would travel far enough along the cylinder to produce a vacuum behind it. Atmospheric pressure would then slowly push the piston back. This return stroke was in fact the power stroke. The power output was low, but due to the high expansion ratio obtainable, the indicated thermal efficiency was relatively high. For production atmospheric engines, this was as high as 16%. The noncompression engine was similar in concept to the double-acting steam engines that were prevalent at the time, and produced power on every stroke.

The absence of compression of the fuel-air mixture, the need for which was indicated by Carnot [2.2, 2.3] in 1824, was a serious limitation. The answer to this was the four-stroke cycle: a suction stroke as the piston moves down to draw in an explosive charge; an upward stroke to compress the charge; a power stroke as the piston is forced down by the burning gases; and an upward exhaust stroke to expel the burned gases. Although some claim that French Engineer Alphonse Eugène Beau de Rochas (1815–1893) or Italian engineers Eugenio Barsanti (1821–1864) and Felice Matteucci (1808–1887) pre-dated Dr. Nikolaus August Otto (1832–1891) with the idea, none of them produced working examples. Otto successfully applied the concept of the "Otto cycle" in 1876, and the term "Otto engine" is still often used to describe the four-stroke spark ignition engine.

In 1864, together with Eugen Langen (1833–1895), Otto founded the world's first engine factory, originally known as N. A. Otto & Cie. Five years later, this duo was joined by Ludwig August Roosen-Runge, and the company moved to Deutz and became Langen, Otto & Roosen. Three years later, the company was renamed as Gasmotoren–Fabrik Deutz AG and following various commercial transformations became part of Klockner-Humboldt-Deutz. Their museum includes the first atmospheric gas engine developed by Otto and Langen and one of the first four-stroke gas engines built in 1878—both are still operable!

Exploitation of the early stationary Otto engines was limited by the availability of a gas supply, but the use of the more convenient liquid petroleum spirit as fuel made the engines applicable for transport purposes. Early pioneers in Germany were Gottlieb Daimler (1834–1900) and Karl Benz (1844–1929), who built their first single-cylinder engined motor car in 1885 (and whose names are still eminent in the world of motoring). Interest soon spread to other European countries and the US.

Since that time, more than a century ago, development of the gasoline engine has progressed steadily, from the modest single-cylinder unit of the 1880s with its rudimentary carburetor and ignition device to the powerful multicylindered, turbocharged models of today, with their sophisticated management systems for the introduction of fuel, the initiation of the ignition, and the control of emissions. The next section describes how the fuel itself has evolved to meet the progressively changing needs of these positive ignition engines and, more recently, to help preserve global resources and the environment in which we live.

2.1.2. Gasoline Development

Since the first horseless carriages appeared on the road at the end of the 19th century, fuels, vehicles, and the infrastructure have evolved together. Figure 2.1 shows an Anglo American Oil Co. fuel tanker at the end of the 19th century or the start of the 20th century, in Weston-super-Mare, UK; the tanker is still horse drawn! The Pratt's Motor Spirit would then be decanted into 2 gal (9.1 L) cans for sale to the motorist.

Courtesy of Nigel Muir Mathewson.

FIGURE 2.1   Horse-Drawn Fuel Tanker, Weston-super-Mare, UK.

In the beginning, the only fuels available were the lighter fractions from the distillation of crude oils and shale oils. These boiled within the range of about 50 to 200°C, much the same as today's gasolines, but with a very poor resistance to autoignition. As a result, engines of the time had to have low compression ratios to help them run without detonation, which produced a characteristic knocking or pinging sound and could cause severe damage to the engine.

The 1914–1918 war changed both the demand and the quality requirements of gasoline, or motor spirit, as it was generally called at that time. The relatively high power–weight ratio of the IC engine was quickly appreciated in terms of its potential for warfare, particularly as the Wright Brothers had already made their first manned flight in 1903. However, the restriction of having to use low compression ratios had become very apparent since the only way of achieving improved fuel quality at that time was to use crude oils rich in aromatics, such as those from Borneo or the Dutch East Indies rather than the more paraffinic crudes from Pennsylvania or Oklahoma.

The importance of the composition of gasoline relative to its performance had become very clear by the end of the war when it became desirable to improve mechanical efficiency by increasing the compression ratio. Work then started both in the US with Midgely and Boyd at the General Motors Research laboratory and in Great Britain by Ricardo to establish the factors that prevented fuels from burning smoothly without detonating in an engine. It was soon established by Ricardo that the hydrocarbons with most resistance to autoignition were aromatics and that those with the least resistance were the normal paraffins. The use of unrefined benzole (a mixture of benzene, toluene, and xylenes produced by the distillation of coal tar) in motor spirit was found to greatly improve performance at higher compression ratios. The benefits of alcohols as fuel components that would prevent detonation were also discovered at that time.

Midgely, in parallel work, had developed a similar test engine to the one used by Ricardo, but with the important addition of a device known as the "bouncing pin," shown in Figure 2.2 [2.4] for detecting the onset of detonation. This engine was ultimately adopted internationally as the standard for measuring the resistance to detonation, known as the antiknock quality of fuel, and was called the CFR engine, after the Cooperative Fuel Research Committee formed in 1921.

FIGURE 2.2 Midgely bouncing-pin pressure-element and knock indicator.

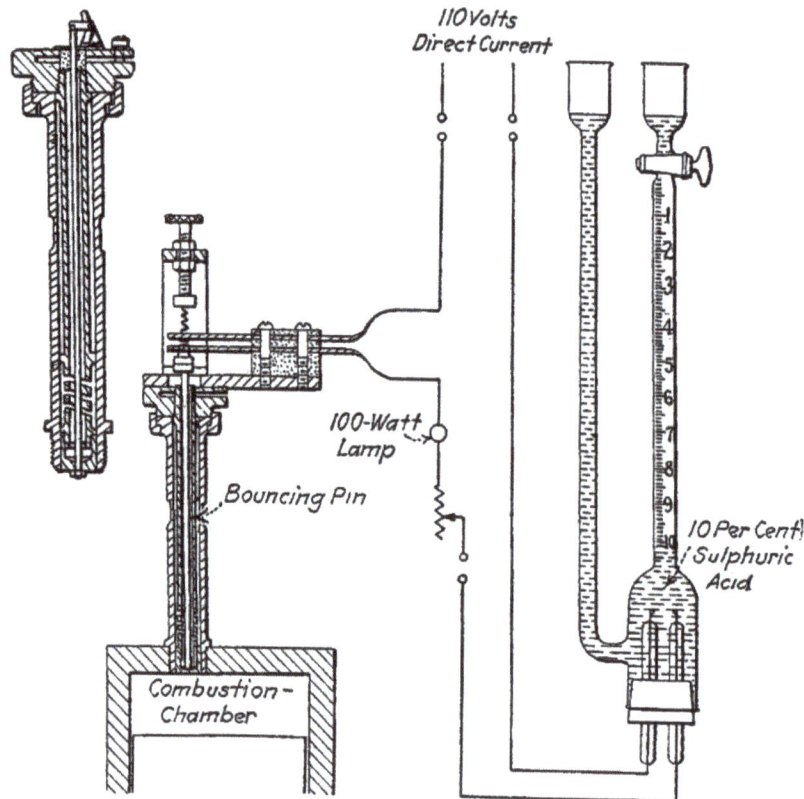

© SAE International.

One of Midgely's findings using this engine was that branch chain paraffins, unlike straight chain paraffins, had excellent resistance to knock. This encouraged his team to undertake a large-scale systematic investigation into chemical additives that would suppress knock.

They screened thousands of compounds and, at the end of 1921, selected tetra-ethyl-lead (TEL) as having the greatest effectiveness with the best potential for commercial development. Problems concerning lead-oxide deposits in the combustion chamber were overcome by combining with the lead compound some ethyl halide, which acts as a scavenger by volatilizing the lead in the combustion chamber. Around this time, the first gasoline dispenser pumps started to appear, selling unleaded or leaded fuels. Figure 2.3 shows one of these early pumps selling Pratt's Motor Spirit; the leaded fuel was sold as Pratt's Ethyl.

FIGURE 2.3   The first gasoline pumps appeared soon after World War I.

Courtesy of Exxon Mobil Corporation.

Since the late 1920s, lead compounds were the most important method of achieving required octane levels, and it is only since the 1970s and 1980s that the use of lead has diminished. This was due initially to environmental concerns, but primarily due to its deleterious effect on catalytic exhaust after-treatment. It is interesting that concerns regarding the toxicity of lead were widespread when it was first introduced, and an investigation was started by the US surgeon general of the Public Health Service in May 1925 to find out if there was any public health danger by the use of gasoline containing TEL. The sales of gasoline containing the additive were suspended until the results of this investigation were known. In January 1926, the surgeon general decided that there were no good grounds for prohibiting the use of lead anti-knock compounds, provided certain safety regulations were observed, and the sale of leaded gasolines restarted in the summer of 1926. Even so, the method of incorporating lead into gasoline was rather questionable. Dunstan and Card [2.5], writing about the work of Midgely and Boyd in 1924, gave the following description:

The antiknock is sold through the medium of specially fitted petrol pumps. These pumps are normal except that a small apparatus, called the ethylizer, is attached. If ethylized gasoline is required, the apparatus can be put in connection with the pump by operating a small lever, and every gallon of gasoline delivered will automatically have added to it the requisite quantity of lead tetraethyl and ethylene dibromide. Fresh supplies of the ethyl are supplied in sealed metal bottles containing about a quart, which are inverted and dropped in the top of the ethylizer, a pointed tube puncturing the seal and allowing the antiknock to run out as required.

In Great Britain, the sale of gasoline containing the lead antiknock compound started in 1928, but its use was severely hampered by adverse press comments. Eventually, a government inquiry was held, which cleared the product from a public health viewpoint. These concerns delayed the general acceptance of lead not only in the UK but also in many other countries, for several years.

In 1929, the octane scale was established in which two hydrocarbons were selected as references: one that tended to knock in an engine under almost all conditions (n-heptane) and the other having a much higher knock resistance than any known gasoline at that time (iso-octane). The antiknock performance of gasoline is compared with a blend of these two compounds, which had been given arbitrary values of 0 and 100, respectively. Two different engine operating conditions are commonly used, leading to the determination of either the research octane number (RON) or the motor octane number (MON). This scale is still the standard method for defining the antiknock quality of all marketed gasolines throughout the world (it is discussed in more detail in Chapter 7). Of course, it is possible to produce gasoline that has an octane rating greater than 100; this prompted further discussion on how best to extend the rating scale [2.6].

Considerable advances were also made in refining processes during this period between the wars. It had been found as early as 1861 that heating the heavier fractions of crude oil caused so-called "cracking reactions" to take place, leading to the formation of lighter hydrocarbons. This could be done using the Rittman process, and by 1916 there were seven US refineries using this process [2.7], although the quantity of cracked components was not sufficient to make a difference to the general gasoline pool until the end of the 1920s. Though this can increase the yield of gasoline, it is at the expense of the straight run distillate that would otherwise go to diesel fuel [2.8]. Cracking gives rise to the formation of olefinic compounds that, although they have relatively good antiknock properties, are quite easily oxidized to form gums. These gums form during storage, causing gels and sludges in the bottom of tanks, and also during engine operation, giving rise to deposits and causing piston ring and valve sticking. Additives or additional processing were found to be necessary to overcome these difficulties. Another problem was that these cracked streams often had an extremely unpleasant odor.

Cracking processes had become necessary in order to increase the yield of gasoline from crude oil beyond that obtainable from simple distillation. A number of variants of the first cracking process soon became available, all of which involved thermally cracking the larger molecules in the crude oil to smaller molecules. Continuous fractionation was also developed during this period, both for atmospheric and vacuum distillation. During the 1930s, processes to convert the gases formed during cracking into gasoline components were introduced, including polymerization and alkylation.

These process improvements not only improved the yield of gasoline from crude oil (which was very necessary since demand had risen dramatically) but also the quality. By the mid-1930s octane values were around 70, enough to allow compression ratios to increase to about 5.5:1.

World War II created an enormous demand for gasoline. Fighter planes needed 100 octane fuels in order to achieve the required power output, and this was affected by using such processes as alkylation and by adding high levels of lead. There was consequently a severe shortage of high-octane components, and octane values for gasoline for home use had to be reduced. However, during this period rapid progress was made in refining technology, and continuous catalytic cracking and reforming processes were developed as well as catalytic desulfurization.

After the war, the demand for high-octane components for aviation use subsided, and they were released back to the domestic gasoline markets, allowing a steady improvement in octane quality. There was a period of rapid growth in demand for gasoline, during which time brand names reappeared on the European market. From about 1950, octane values started to increase very rapidly and had reached about 95 RON by 1955. This was higher than necessary for the vehicles of that time but allowed engine designers to increase compression ratios in their new designs.

Brand images and advertising became a feature of the 1950s and 1960s. This is epitomized by the "put a tiger in your tank" slogan adopted by Esso Petroleum (depicted in the photograph shown in Figure 2.4).

FIGURE 2.4   Esso brand image establishment from the 1960s.

Courtesy of Exxon Mobil Corporation.

This period also saw the introduction of additives to overcome such problems as ice formation in the carburetor, fouling of spark plugs, and deposits in carburetors and on valves [2.9]. National specifications for gasoline were introduced toward the end of this period. Since then, national specifications have been changing and becoming more harmonized. Meeting these specifications has meant that the gasoline products have become more similar. However, global or national brand image is still an important factor for maintaining or gaining market share. In India, at the end of 2020 and the start of 2021, Indian Oil launched a 100 Octane fuel [2.10], which was aimed at owners of "performance" vehicles (see Figure 2.5).

FIGURE 2.5   Advert for premium gasoline.

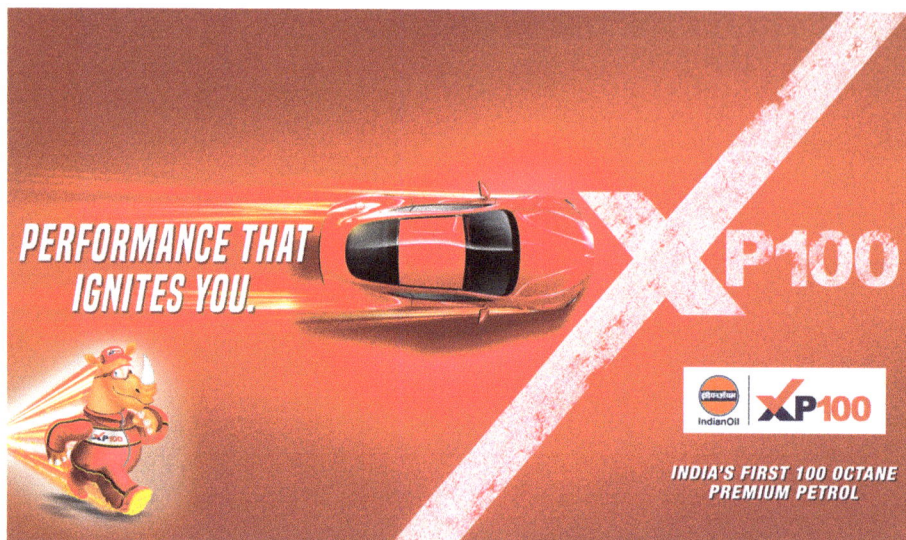

Another factor influencing gasoline development was concern for the environment. As early as the late 1960s, air quality deficiencies had been recognized for some time in the Los Angeles basin, and the blame for this was placed largely on exhaust emissions from vehicles. Similar problems had been recognized in other major cities where there was a high concentration of automobiles and where long periods of sunshine were the norm. Exhaust emission regulations were introduced in California in the 1960s, and by the 1970s, controls had been put in place for the entire country and in Japan.

The first emission controls on vehicles involved engine modifications that resulted in the sacrifice of performance. However, the introduction of catalytic converters in these countries gave better control of emissions such as carbon monoxide, unburned hydrocarbons, and nitrogen oxides and overcame any performance debits. Such vehicles require unleaded gasoline because lead poisons the precious metal catalyst used in these converters. This led to considerable changes in refining operations in order to achieve adequate octane levels without the use of lead. Later, there were concerns about the toxicity of lead itself and the effect it might be having on the health of people having to breathe in the exhaust fumes from vehicles for several hours a day. These worries caused lead levels to be reduced in stages in many of those countries, particularly in Europe, where catalytic converters had not yet been considered necessary as a method of improving exhaust emissions.

The introduction of unleaded and low-lead gasolines caused dramatic changes in gasoline composition and initially in Europe a real shortage of high-octane blend components. Oxygenated compounds such as alcohols and ethers were used to make up this deficiency and these, in turn, gave rise to a whole series of new problems.

In 1973, the Arab/Israeli (Yom Kippur) war occurred, and the Arab nations nationalized their oil reserves. These were the first of a series of Middle East oil crises that drastically increased the price of crude oil and led to a strong emphasis on fuel economy and to enormous changes in both the refining and automotive industries. They included the Iranian revolution in 1977 and the Iran/Iraq war in 1980, after which the price of crude gradually reduced, apart from a brief period following the invasion of Kuwait by Iraq in 1990. However, the high gasoline prices had already depressed sales, and the consequent spare capacity in most refineries led to a rationalization of the oil refining industry and the closure of many smaller and less efficient refineries.

The high price of crude oil had also caused balance of payments problems in many countries without domestic sources of crude oil, and this intensified the search for new oil fields and made some countries such as Brazil look for alternatives to petroleum-based fuels. Ethanol production from the fermentation of sugar cane became a viable alternative to gasoline in Brazil and necessitated the production of engines modified for satisfactory operation on this fuel.

In addition, there was a dramatic effect on product demand and on the relative volumes of different products needed from a refinery. The price of fuel oil had increased so much that alternative fuels such as coal became more economic for large energy users such as power stations and steel works. This gave rise to an imbalance in product demand with an excess of fuel oil relative to gasoline, particularly in Europe. To redress the balance, a massive increase in cracking processes took place. The changes in chemical composition of the resulting gasolines had an influence on the performance of the gasolines (as discussed in later chapters), and the high prices put a new emphasis on fuel economy. The pressure on fuel economy relaxed somewhat in the late 1980s and early 1990s, since oil prices had stabilized to some extent by then, apart from the short period of the Gulf War.

By 1989, because air quality, particularly in California, was still unsatisfactory in spite of very severe legislation to restrict noxious emissions from vehicles, a "new" type of gasoline was marketed by one oil company—Atlantic Richfield. It was called reformulated gasoline and was marketed on the basis that it helped reduce emissions from vehicles, particularly older vehicles not fitted with emission control devices. This gasoline was significantly different from the other gasolines of that time because it had a low vapor pressure (to minimize evaporative emissions), low benzene content (to reduce its toxicity), low aromatics (to reduce benzene exhaust emissions), a fixed level of oxygenated components (to reduce HC and CO emissions), and zero lead. Other oil companies followed suit, and by the early 1990s reformulated gasolines were mandatory in many parts of the US and were beginning to appear in Europe and other countries.

Toward the very end of the 20th century, the price of crude oil again began to rise due to increasing demand from developing countries. The price of crude oil rose by a factor of almost six between the years 2000 and 2008, when the financial crisis caused the price to fall to less than twice the 2000 level. The early 21st century also saw increasing concerns over global temperatures and claims that this was due to increasing emissions of so-called greenhouse gases, particularly carbon dioxide. This, along with continued unrest in the Middle East, has promoted calls for the production of fuels from renewable sources, with many countries now mandating that fuel should contain a percentage that is derived from renewable sources [2.11, 2.12]. For gasoline, this has predominantly been achieved by the inclusion of 5–10% ethanol, which can be used in a large proportion of the existing fleet with no modification. By 2016, blends of 10% ethanol (E10) had replaced nearly all pure gasoline (E0) sold in the US [2.13]. Blends of approximately 85% ethanol, known as E85, are available for specially designed vehicles. In Europe, the uptake of ethanol blends was not as rapid, but European Directive 2009/28/EC [2.11] set a mandatory 10% minimum target for the share of biofuels in gasoline by 2020. There is also renewed interest in methanol/gasoline blend, particularly M15 (15% methanol in gasoline), which has been shown to reduce not only CO_2 emissions but also other regulated emissions [2.14]. Israel began trials of M15 back in 2012 [2.15] and more recently India has begun a commercial trial of M15 fuel [2.16].

The increased use of catalytic after-treatment systems has necessitated the cessation of the use of lead antiknock additives in most markets. Concerns over catalyst poisoning and general health effects of all metallic additives coupled with the increasing use of high-octane oxygenated blending components will clearly lead to the eventual total elimination of metallic antiknock additives. The widespread introduction of national or regional specifications for automotive fuels means that while the oil companies obviously have to meet these requirements, they no longer need to ensure that their fuels satisfy virtually every vehicle on the road in a particular market. Thus, there is very little need for the oil industry to conduct tests on very large numbers of cars to find the octane and volatility qualities that satisfy them, and much of the satisfaction work described in Chapters 9 and 11 has been abandoned. The onus is now on the motor manufacturers to ensure that their engines operate satisfactorily on the specified fuels.

2.2. Diesel Fuel

2.2.1. The Evolution of the Diesel Engine

Herbert Akroyd Stuart was born in 1864, and after completing his technical training in London, he worked in his father's tinplate factory in Bletchley, England. In 1885, his interest in fuel combustion was aroused by a small explosion that occurred when some paraffin oil vaporized and ignited after accidentally spilling into a pot of hot metal. This incident led to Akroyd Stuart's development of the vaporizer engine. In 1890, Herbert Akroyd Stuart and Charles Richard Binney were granted a patent for "Improvements in Engines Operated by the Explosion of Mixtures of Combustible Vapour of Gas and Air" [2.17]. The patent describes an engine where liquid fuel is injected into compressed air within the combustion chamber. A vaporizer (the subject of the patent) heats the incoming fuel, allowing it to mix with the compressed air, the result of which is that "automatic ignition takes place." Figure 2.6 [2.17] shows a side elevation of the engine taken from the patent.

FIGURE 2.6   Side elevation of the engine from patent No. 7146.

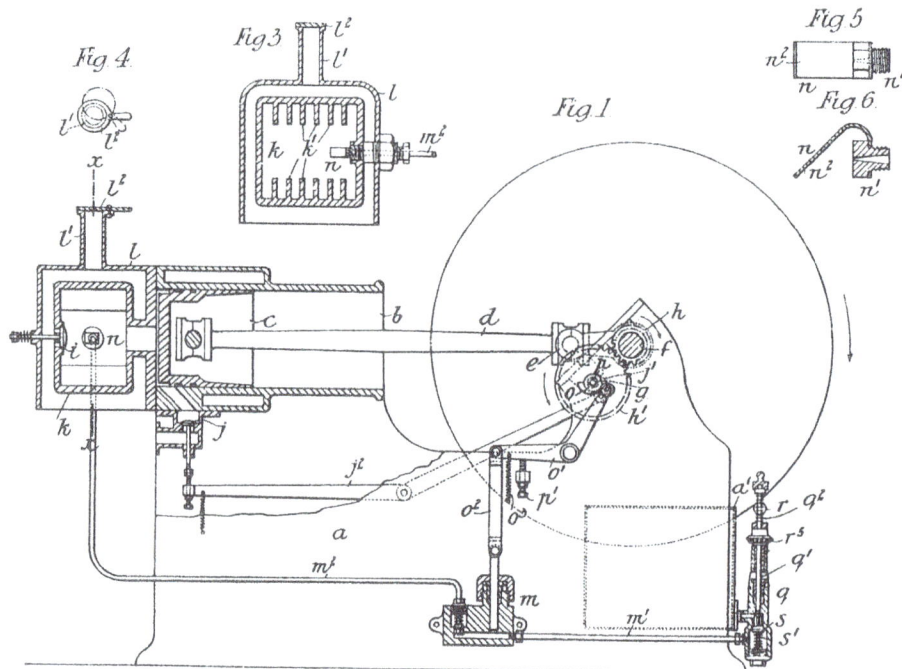

© SAE International.

By drawing only air into the compression chamber, rather than a mixture of hydrocarbon vapor and air, there was no risk of premature ignition of an explosive mixture. A working example of their invention was exhibited in 1891 at the Royal Agricultural Show in Doncaster, England. This was the first oil engine to run without a spark for ignition, relying instead on an externally heated vaporizer in the cylinder head to start the engine. External heating was only needed for starting, as fuel combustion maintained the vaporizer at a high temperature while the engine was running. The compression ratio was believed to be about 2:1. Figure 2.7 shows an early example of this type of engine produced by Richard Hornsby & Sons and sold as the Hornsby-Akroyd oil engine, presumably to distinguish it from the steam engines that were their usual product.

FIGURE 2.7   Hornsby-Akroyd engine No. 176 dispatched from the factory on October 24, 1892.

Rudolf Diesel (1858–1913) was born in Paris of German parents and studied engineering at Augsburg and Munich Technical College. After completing his doctorate, he went to Sulzer Bros., in Switzerland, where he studied thermodynamic principles, particularly as expounded by Sadi Carnot [2.3], and developed his theories on improving the efficiency of the IC engine. He was convinced that the significant heat loss during the power stroke was the major source of inefficiency. His 1892 patent entitled "A Process for Producing Motive Work from the Combustion of Fuel" [2.18] clearly outlines his idea that the fuel should be added and burned gradually so as not to raise the temperature of the compressed gas and hence not lead to additional heat loss; instead, all of the energy released by the fuel is converted into driving the piston. Diesel illustrated this in his patent, by means of pressure-volume diagrams, which are reproduced in Figure 2.8 [2.18].

FIGURE 2.8   Pressure-volume diagrams for Diesel's 1892 patent.

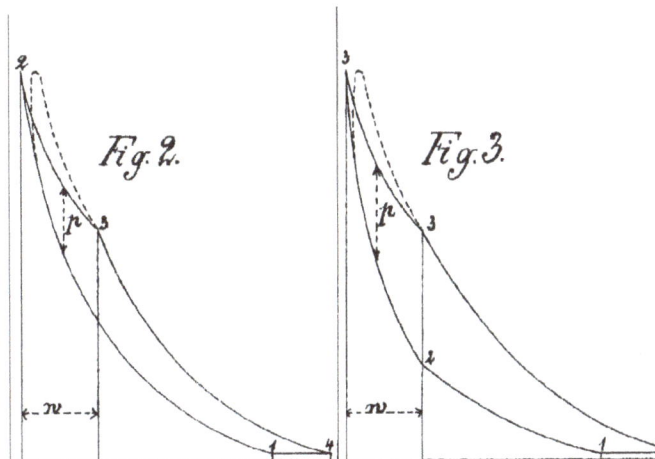

In this patent, Diesel stated that the engine could be run on any sort of fuel. However, at the time, Diesel obviously intended his engine to run on pulverized coal. In the patent Diesel describes a mechanism for delivering this pulverized coal into the combustion chamber. This is discussed further in Chapter 16.

A paper on Diesel's theories, published in 1893, received wide acceptance among German engineers, and experimental work was undertaken by both Krupp and Maschinenbau Actiengesellschaft Nürnberg (M.A.N.). The first prototype engine, made in the same year, was tested after a period of motoring. The purpose of running without power was to smooth high spots off the bearings, but it showed the need to reduce the piston diameter to avoid seizure. Changes were also made to the sealing devices to minimize leakage and ensure adequate compression. Initial work also demonstrated that the proposed mechanism for delivering dry pulverized coal had significant sealing problems, and Diesel proposed an air blast injection system that could also be used with liquid fuels. The first firing tests were performed with such a system utilizing benzin (gasoline) as the fuel; they very clearly demonstrated that automatic combustion had been achieved. The ignition was so violent that the pressure-indicating mechanism was blown off the cylinder and only narrowly missed hitting Dr. Diesel! The indicator diagram of this first firing on August 10, 1893, carrying the brief handwritten comment—"1st explosion (benzin)"—is held in the M.A.N. archives [2.19].

Since that time, around the turn of the century, what has become known as the diesel engine, the fuel injection equipment, and the fuel itself have been evolving in parallel with the spark ignition engine.

Progress was slow through World War I, but the pace increased afterward. Early use of the heavy, slow-revving diesel engine was mainly in industrial, marine, and railway installations, with only limited activity in the areas of road and air transport. However, shortages in the availability of gasoline in Germany after 1918 stimulated the development of diesel engines in that country, notably by Daimler-Benz and M.A.N., with the application to commercial road vehicles particularly in mind. Interest spread quickly to other European countries, stimulating the development of diesel engines specifically for road transport. By the mid-1930s, there were significant numbers of diesel-powered trucks and buses in service, although—despite successful efforts by racing and rally enthusiasts—there was only limited acceptance of the diesel engine as a car engine.

By the time war broke out in 1939, automotive diesel was well established in Europe, but except for the German Army, it had only a limited role in the military sphere. The German Air Force also had a few bomber aircraft powered by Junkers Jumo diesel engines. However, due to logistics problems of dealing with more than one type of fuel, most of the military vehicles used by the allied forces had gasoline engines. Although diesel engines were widely used in rail and sea transport, progress on automotive diesel engines virtually stopped until after 1945. It was then, after the war, that the importance of its better fuel economy was more generally recognized, bringing rapid growth in diesel-powered vehicles in the European commercial road transport sector.

Most of the diesel engines in commercial vehicles are direct injection (DI) engines, generally larger than 2.5 L, in which the fuel is injected directly into the cylinder and burned in a combustion chamber defined by the cylinder head and the piston (similar to what Diesel had proposed in his 1892 patent). Smaller-capacity engines of a size suitable for passenger cars were developed around the indirect injection (IDI) system, where the fuel is injected and ignited in a pre-chamber connected to the working cylinder by a narrow passageway through which the burning gases expand to force down the piston (similar to the Stuart and Binney patent of 1890). The aim of the designers of these later pre-chamber engines was to achieve steady and progressive burning of the fuel by means of a fairly high temperature in the pre-chamber and turbulent air movement to give good mixing with the fuel spray. Early pre-chamber designs include the Acro system and the Lanova air cell, both developed by the German engineer Franz Lang. A number of diesel engine manufacturers in Europe and the US built engines with Lanova combustion chambers but did not continue to use the system. The pre-chamber type of engine was adopted by the newly formed Daimler-Benz company, which was supplying engines for buses and trucks. An important advance in pre-chamber design was made in Britain with the development of the Ricardo Comet combustion chamber design of H. R. (Sir Harry) Ricardo. The Ricardo Comet head, patented in 1931, gave greater flexibility of operation and freedom from smoke than previous designs and was adopted by a number of British diesel engine builders.

Over the final quarter of the 20th century, Europe and Japan have seen a significant growth in the number of passenger cars fitted with diesel engines, most of which had IDI systems. Pre-chamber engines have a lower efficiency than DI engines because of higher thermal and pumping losses, but they are not as critical of fuel quality. While initially used mainly for taxis, where high annual mileages and lower running costs justified the higher initial outlay, the economy and performance of the modern small IDI diesel engine established it firmly in the private car sector. As discussed above, there are weaknesses in the indirect injection approach. Combustion chambers have larger surface-to-volume ratios than those of DI engines and therefore higher compression ratios, typically around 22:1, and are used to compensate for the greater heat losses during compression and to ensure an adequate ignition temperature. Even so, most IDI engines are equipped with heater plugs or glow plugs in the pre-chamber to facilitate starting from cold. The combination of this surface-to-volume ratio deficiency plus the heat loss through the pre-chamber throat endows the IDI engine with poorer thermal efficiency and lower specific power than an equivalent DI engine. For example, the fuel consumption of an IDI power unit is generally as much as 15% higher than a comparable DI diesel engine. More importantly, the gasoline engine has been reasserting its fuel economy credentials, particularly with the launch of a number of DI models. These competing developments led to the search for a way to emulate the fuel economy of the DI diesel truck engine without its inherent limitations (combustion noise and restricted speed and torque ranges). DI was clearly the way forward, and a combination of combustion chamber design, turbocharging, and two-stage injection provided the development route for the high-speed DI (HSDI) engine. Virtually all light-duty diesel engines are now of the DI configuration.

In the US, the availability of low-priced gasoline reduced the incentive to adopt the more economical diesel engine for road transportation of heavy goods and passengers. Although acceptance of the automotive diesel was slower than in Europe, production of diesels for off-highway use in marine, railroad, and earth-moving equipment was started as early as the 1920s by US companies such as Cummins, General Motors, and Caterpillar. American engine makers designed their own fuel injection systems rather than relying on the unit pump type systems designed by Robert Bosch in Germany. These US systems, usually a combined pump and injector assembly for each cylinder, operated by a pushrod and rocker from a camshaft in the engine.

Since the end of the war, in 1945, the automotive diesel has found its place in the US, powering buses and long-distance freight transport (the leading engine suppliers being GMC/Detroit Diesel, Cummins, and Mack). Cummins is also well established as a supplier of truck engines in several countries around the world, where manufacturing plants have been set up. For US passenger cars, however, the gasoline engine is by far the preferred option.

2.2.2. Diesel Fuel Development

In the years immediately following the exploitation of petroleum resources by drilling (initiated by Colonel Drake at Titusville, Pennsylvania, in 1859), kerosene lamp oil was the most valuable petroleum fraction. Surplus gasoline was disposed of by burning, surplus heavy residue was dumped into pits, and the "middle distillate" was used to enrich a town's gas, which explains why it is often still referred to as gas oil. Only with the invention of the diesel engine was a specific role found for the middle distillate fraction.

Fuel quality requirements of the early, heavy and slow-revving diesel engines were not very stringent, but, with engine design improvements to increase the power–weight ratio, more refinement of the fuel became necessary.

Elimination of fuels having high viscosities and high levels of hard combustion residues resulted in benefits in engine behavior. The tightening up of fuel quality also enabled engine speeds to be increased, with consequential improvements in output, efficiency, and reliability. These engines were the forerunners of the modern, high-speed diesel engine.

As a result of these improvements, the next fuel characteristic to come under scrutiny was the propensity of the fuel to autoignite—its ignition quality. Fuels that were satisfactory in slow-speed engines operating at a few hundred revolutions per minute had problems of poor startability and combustion noise under certain load conditions in engines designed to run at speeds of 2000 rev/min and higher.

The ignition quality of a fuel is obviously a function of the chemical composition, and experience had indicated that this was related to the density and aromatic content of the fuel. The density of petroleum products was traditionally expressed by its American Petroleum Institute (API) gravity, and the aniline point was an indication of the aromatic content of a fuel. The ignition quality of a diesel fuel was thus originally expressed by the calculated Diesel Index, which is a function of the fuel's API gravity and aniline point:

$$\text{Diesel Index} = \frac{\left(\text{API gravity}\right)\left(\text{Aniline point}\right)}{100}$$

Crude oils and their distillates are mixtures of hydrocarbon types: paraffins, naphthenes, and aromatics, in various proportions. Of the three, the paraffinic types of fuel have the best engine startability characteristics but the poorest cold properties because of the waxes, which come out of solution as the fuel temperature decreases. The opposite is true for aromatic fuels. The aniline point of a fuel is the lowest temperature at which equal quantities of fuel and aniline will go into solution. Aromatics have a higher density and better solvent power than paraffins, so a fuel which is high in aromatics will have a lower aniline point than a predominantly paraffinic fuel. Consequently, a highly paraffinic fuel will have a high aniline point and a high API gravity (low density), giving a high calculated Diesel Index to indicate good startability. This empirical approach, however, was not sufficiently discriminating, and before long, the CFR Cetane Engine test [2.20] was adopted as the standard procedure to define ignition quality.

Unfortunately, the CFR engine test method has poor precision, and so another calculated value, the cetane index [2.21], was introduced as a means of estimating the cetane number of a diesel fuel from its API gravity and mid-boiling point. The formula was revised from time to time, as fuels evolved, to maintain its predictive validity. A new four-variable cetane index method [2.22] has now generally replaced the older method [2.21] in many diesel specifications, using the 10, 50, and 90% recovery temperatures instead of the mid-boiling point. Further details are given in Chapter 15, Section 15.3. However, attempts have been made to produce other tools to predict the cetane number of fuels. These include the use of proton nuclear magnetic resonance spectroscopy [2.23, 2.24] and various constant volume combustion chambers [2.25, 2.26]. The most widely used of these are the automated Ignition Quality Tester (IQT™) [2.27] and the Fuel Ignition Tester (FIT™) [2.28], and the Advanced Fuel Ignition Delay Analyser (AFIDA) [2.29] which have been developed to determine a derived cetane number (DCN) for conventional and nonconventional fuels. Recently, there have been questions regarding the calibration of these methods with only the AFIDA calibration for ICN directly related to the primary reference fuels as used in the CFR engine. There are thus questions over whether some new advanced high-efficiency compression ignition engines will have their fuel performance fully described by these measurements [2.30].

As crude oils differ in their hydrocarbon composition, the ignition quality of diesel fuels depends heavily on the type of crude oil from which they are distilled. During the 1930s, when European interest in diesel-powered vehicles was starting to develop, most of the fuel production in Europe came from Middle East crudes, which were paraffinic and of good ignition quality. In other parts of the world, where different crude types were more common, some selectivity was necessary to ensure production of good quality diesel fuel.

Demand for automotive diesel fuel in Europe rose steadily through the 1950s and 1960s as highway networks were extended to cope with increased travel by commercial operators and also by private motorists who were beginning to move to diesel-engined passenger cars because of their better fuel economy.

Ignition quality during that period was generally satisfactory, and although additives to improve cetane number had been developed, there were no pressing technical or economic incentives for their use. Specifications were tightened to limit viscosity range, sulfur, and contaminants (water, sediment, and ash), while volatility had been established more as a means of product classification than of quality. The dominant fuel characteristic from both the refining and end-use viewpoint had now become its low-temperature performance, specified by cloud point or pour point. Cloud point and pour point are laboratory tests giving the temperature at which waxes start to separate and produce a cloudy appearance as the fuel is cooled and the lowest temperature at which the fuel can be poured. These tests are described in Chapter 17, Section 17.1. Seasonal grades were marketed because lowering the cloud point during winter months reduced the amount of diesel produced.

Very severe weather conditions over northern Europe during the 1962–1963 winters had created havoc with the operations of diesel-powered trucks and buses. Exposure to long periods of unusually low temperatures had resulted in wax plugging of fuel lines and filters, causing problems of starting and running diesel vehicles. Stalled vehicles were abandoned at the roadside when drivers were unable to clear the fuel systems and restart their engines. Similar problems had been experienced in previous years from time to time but never to such an extent, largely because in those earlier days the diesel vehicle population had been much smaller.

A committee comprising members of the European Motor and Petroleum Industries was set up to investigate the difficulties that had been encountered and to recommend preventive measures to minimize future occurrences.

The committee identified design features of some vehicles that made them more sensitive to waxing problems. Recommendations were made for vehicle manufacturers to review their fuel system designs and to carry out corrective modifications where necessary. Fuel suppliers were also asked to ensure that the quality of diesel fuel marketed would be adequate for winter conditions. Meteorological records giving long-term average values were used to help establish realistic low-temperature specification levels for winter diesel fuel.

Since that time, diesel fuel quality in Europe and elsewhere has changed in a number of ways. A major influencing factor was the nationalization of oil reserves by the Middle Eastern nations and the ensuing changes in crude oil price and governmental policies relating to the use of nonrenewable petroleum resources.

An important change soon after the severe winter problems of 1962–1963 was the introduction of cold flow improver additives as a routine treatment of winter-grade diesel fuels. These additives enabled refiners to meet winter quality standards with little or no reduction in the quantity of diesel produced. During the 1980s, environmental concerns saw the introduction of diesel fuels with reduced sulfur contents and lower back-end distillation. There was a concern that severe hydrotreatment would conflict with lubricity requirements by removing or inhibiting the natural lubricants in the fuel. Field failures were encountered in Canada, and experience in Sweden [2.31] confirmed the risk when strict environmentally driven fuel specifications were introduced in 1991. Lubricity additives are now widely used in automotive diesel fuels. Biodiesel fuel, which is discussed in Chapter 5, Section 5.4, usually possesses good lubricity characteristics and may negate the need for other lubricity additives.

The current trend is for greater use of additives in automotive diesel. Cold flow additives are often used in summer fuels as well as in winter grades, whereas ignition improvers are sometimes needed to adjust the cetane value of fuels prepared from certain crude types or in blends containing cracked gas oils. These additives are now regarded as standard diesel fuel blend constituents, available to help refiners produce fuels of the required quality in the quantities needed by the market. They are normally added at the refinery to ensure that the fuel conforms to the appropriate specification before release for sale.

Additionally, multifunctional treatments may be introduced into on-grade products to provide extra benefits that the fuel marketer can exploit. In Europe, the biggest diesel market, advertising campaigns are used to promote the concept of a premium quality product. Claims, which will depend on the additive combination used, can include better low-temperature performance; easier starting; protection against rust, corrosion, and injector fouling; reduced exhaust smoke and noise; lower foaming tendency, enabling faster refueling with less likelihood of spillage; and a more acceptable odor. The environmental benefits of the fuel can also form advertising claims, as evidenced by this early 2020s advert from the Indian Oil Company (Figure 2.9).

FIGURE 2.9   Advert for Indian Oil diesel fuel.

As with gasoline, fuel retailers must comply with current legislation not only in regard to the quality of the fuel but also in terms of the content from renewable sources [2.11, 2.12]. For diesel fuel, this obligation is currently met by the inclusion of fatty acid methyl esters (FAME). These esters are predominantly derived from plant oils or animal fats and are discussed further in Chapter 5. However, there is currently a significant move toward Hydrogenated Vegetable Oil (HVO) as an alternative to such esters. Such fuels derived from renewable sources can be used neatly in existing engines and infrastructure. This is discussed further in Chapter 5.

The quality standards have to take into account several factors: legal definitions; sulfur content; permitted levels of water, sediment, and acidity; ash and carbon residues from combustion; safety standards; climatic factors; and environmental considerations. All of these controls on quality are additional to the basic demands of the engine builder for a fuel that performs well in service. The relevance and importance of the various aspects of diesel fuel quality will be covered later in this book.

References

2.1. Clerk, D., *The Gas and Oil Engine* (New York: John Wiley & Sons, 1896).

2.2. Cummins, C.L. Jr., *Internal Fire: The Internal-Combustion Engine 1673–1900* (Warrendale, PA: SAE, Inc., 1989).

2.3. Carnot, N.L.S., "Réflexions sur la puissance motrice du feu et sur les machines propres à développer cette puissance," Chez Bachelier, Libraire, Paris, 1824.

2.4. Cummings, H.K., "Methods of Measuring the Antiknock Value of Fuels," SAE Technical Paper 270003, 1927, doi:https://doi.org/10.4271/270003.

2.5. Dunstan, A.E. and Card, S., "The Work of Midgely and Boyd," Volume XVIII, Part 1: 3–12, Report of the Empire Motor Fuels Committee, Embodying Other Allied Researches, Institution of Automobile Engineers, 1924.

2.6. Brooks, D.B., "Development of REFERENCE FUEL SCALES for KNOCK RATING—Report of Coordinating Fuel Research Committee of Coordinating Research Council, Inc.," SAE Technical Paper 460230, 1946, doi:https://doi.org/10.4271/460230.

2.7. Foljambe, E.S., "The Automobile Fuel Situation," SAE Technical Paper 160016, 1916, doi:https://doi.org/10.4271/160016.

2.8. Perriguey, W.G., "Automotive Fuels," SAE Technical Paper 480052, 1948, doi:https://doi.org/10.4271/480052.

2.9. Gibbs, L.M., "Gasoline Additives - When and Why," SAE Technical Paper 902104, 1990, doi:https://doi.org/10.4271/902104.

2.10. India Today, "Indian Oil Launches First-Ever XP100 100 Octane Petrol in India," India Today, accessed July 3, 2022, https://www.indiatoday.in/auto/latest-auto-news/story/indian-oil-launches-first-ever-xp100-100-octane-petrol-in-india-1753777-2020-12-28.

2.11. "Directive 2009/28/EC of the European Parliament and of the Council on the Promotion of the Use of Energy from Renewable Sources," Official Journal of the European Union, April 23, 2009.

2.12. US Environmental Protection Agency, "Regulation of Fuels and Fuel Additives: 2012 Renewable Fuel Standards," *Federal Register* 77, no. 5 (2012): 1319-1358; Rules and Regulations, Environmental Protection Agency, 40 CFR Part 80.

2.13. Johnson, C., Moriarty, K., Alleman, T., and Santini, D., "History of Ethanol Fuel Adoption in the United States: Policy, Economics, and Logistics," No. NREL/TP-5400-76260, National Renewable Energy Lab. (NREL), Golden, CO, 2021.

2.14. Goldwine, G., Sher, E., and Sher, D., "Comparison of Regulated and Unregulated Emissions and Fuel Economy of SI Engines with Three Fuels: RON95, M15, and E10," *SAE Int. J. Fuels Lubr.* 12, no. 3 (2019): 189-209, doi:https://doi.org/10.4271/04-12-03-0013.

2.15. IEA, "IEA—Advanced Motor Fuels Annual Report 2015," accessed May 2023, https://www.iea-amf.org/app/webroot/files/file/Annual%20Reports/IEA-AMF%202015%20Annual%20Report.pdf.

2.16. Chemical Industry Digest, "Indian Oil Launches Methanol Blended Petrol on Pilot Basis in Assam," accessed May 2023, https://chemindigest.com/indian-oil-launches-methanol-blended-petrol-on-pilot-basis-in-assam/.

2.17. Stuart, H.A., and Binney, C.R., Improvements in engines operated by the explosion of mixtures of combustible vapour of gas and air. G.B. Patent No. 7146, 1890.

2.18. Diesel, R., A process for producing motive work from the combustion of fuel. G.B. Patent No. 7241, 1892.

2.19. Cummins, L.C., *Diesel's Engine: From Conception to 1918* (Wilsonville, OR: Carrot Press, 1993).

2.20. ASTM International, "Standard Test Method for Cetane Number of Diesel Fuel Oil," ASTM D613-23, ASTM International, 2023.

2.21. ASTM International, "Standard Test Method for Calculated Cetane Index of Distillate Fuels," ASTM D976-21e1, ASTM International, 2023.

2.22. ASTM International, "Standard Test Method for Calculated Cetane Index by Four Variable Equation," ASTM D4737-21, ASTM International, 2021.

2.23. Glavinecvski, B., Gülder, Ö.L., and Gardner, L., "Cetane Number Estimation of Diesel Fuels from Carbon Type Structural Composition," SAE Technical Paper 841341, 1984, doi:https://doi.org/10.4271/841341.

2.24. Gülder, Ö.L., Glavinecvski, B., and Burton, G.F., "Ignition Quality Rating Methods for Diesel Fuels-A Critical Appraisal," SAE Technical Paper 852080, 1985, doi:https://doi.org/10.4271/852080.

2.25. Hurn, R. and Hughes, K., "Combustion Characteristics of Diesel Fuels as Measured in a Constant-Volume Bomb-A Report of the Coordinating Research Council, Inc.," SAE Technical Paper 520210, 1952, doi:https://doi.org/10.4271/520210.

2.26. Datschefski, G. and Rickeard, D.J., "Diesel Fuel Ignition Quality Measurement by a Constant Volume Combustion Test," SAE Technical Paper 932743, 1993, doi:https://doi.org/10.4271/932743.

2.27. ASTM International, "Standard Test Method for Determination of Derived Cetane Number (DCN) of Diesel Fuel Oils—Ignition Delay and Combustion Delay Using a Constant Volume Combustion Chamber Method," ASTM D7668-17, ASTM International, 2017.

2.28. ASTM International, "Standard Test Method for Determination of Derived Cetane Number (DCN) of Diesel Fuel Oils—Fixed Range Injection Period, Constant Volume Combustion Chamber Method," ASTM D7170-19, ASTM International, 2019, Withdrawn 2019.

2.29. ASTM International, "Standard Test Method for Determination of Indicated Cetane Number (ICN) of Diesel Fuel Oils Using a Constant Volume Combustion Chamber—Reference Fuels Calibration Method," ASTM D8183-22, ASTM International, 2022.

2.30. Abel, R.C., Luecke, J., Ratcliff, M.A., and Zigler, B.T., "Comparing Cetane Number Measurement Methods," in *Internal Combustion Engine Division Fall Technical Conference*, Vol. 84034, American Society of Mechanical Engineers, 2020, V001T02A009.

2.31. Tucker, R., Stradling, R., Wolveridge, P., Rivers, K. et al., "The Lubricity of Deeply Hydrogenated Diesel Fuels—The Swedish Experience," SAE Technical Paper 942016, 1994, doi:https://doi.org/10.4271/942016.

Further Reading

Banks, F.R., "The Influence of Tetraethyl Lead on Engine Design and Performance," *J. Inst. Pet.* 35 (1949): 264-292.

Bartholomew, E., "Four Decades of Engine Fuel Technology Forecast Future Advances," SAE Technical Paper 660771, 1966, doi:https://doi.org/10.4271/660771.

Berwick, I.D.G., "Key Developments in a Century of Road Transport Fuels," in *Conference on Petroleum-Based Fuels and Automotive Applications*, Paper C318/86, Institution of Mechanical Engineers, London, UK, November 1986.

Bryant, L., "The Origin of the Four-Stroke Cycle," *Technology and Culture* 8, no. 2 (1967): 178-198.

Hardenberg, H.O., "Samuel Morey and His Atmospheric Engine," SAE SP-922, Society of Automotive Engineers, Inc., Warrendale, PA, 1992.

Jackman, F.A., "The History of Gasoline," in Hancock, E.G. (Ed.), *Technology of Gasoline* (Oxford, UK: Society of Chemical Industry, 1985), 2-19.

Knight, P. and Wright, G., *A-Z of British Stationary Engines A-K.* Vol. 1 (Maidstone, England: Kelsey Publishing Ltd., 1996).

Knight, P. and Wright, G., *A-Z of British Stationary Engines L-Z.* Vol. 2 (Maidstone, England: Kelsey Publishing Ltd., 1996).

Leffler, W.L., *Petroleum Refining in Nontechnical Language* (Tulsa, OK: PennWell Books, 2008).

Raymond, L., "Today's Fuels and Lubricants and How They Got That Way," SAE Technical Paper 801341, 1980, doi:https://doi.org/10.4271/801341.

Vassiliou, M.S., *Historical Dictionary of the Petroleum Industry* (Lanham, MD: Scarecrow Press, 2009).

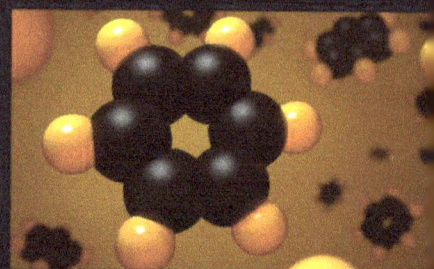

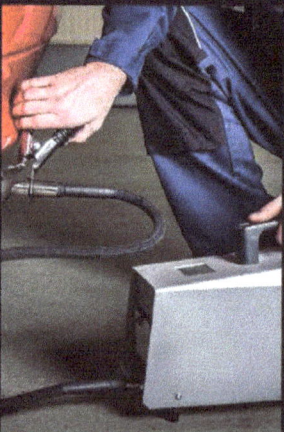

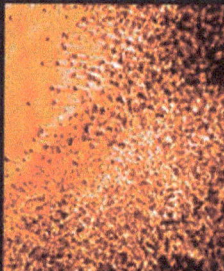

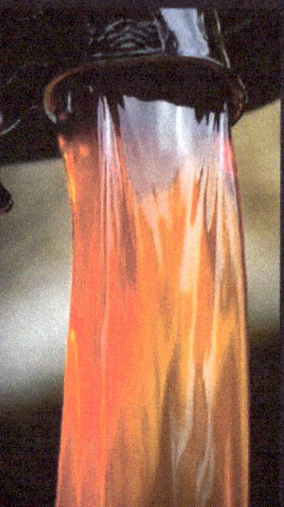

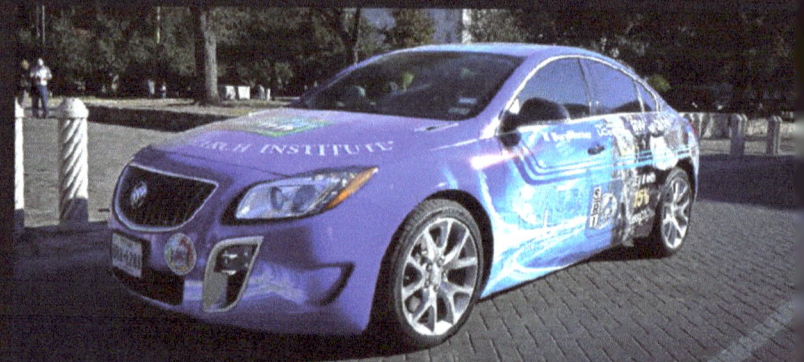

3

Manufacture of Gasoline and Diesel Fuel from Crude Oil

3.1. Introduction

As discussed in the previous chapter, the refining industry has been undergoing a transition due to changes in product demand and in the quality of the gasoline and diesel blend components required. This has arisen partly because of increases in the price of crude oil and the requirement for vehicles to meet legislated targets for emissions and fuel economy and also because of the requirement to incorporate increasing quantities of renewable components. This chapter discusses the effect that these changes have had on refining operations; it describes the most important processes for manufacturing fuels and the characteristics of each component in terms of how it blends into a finished gasoline or diesel fuel.

It must be appreciated that gasolines and diesel fuels can differ quite widely one from another in composition, even if they are of the same grade. This is because refineries are rarely the same in their general configuration and processes and also because the crude oils that are processed can be very different and can even vary from day to day. In addition, the operating conditions on the plants themselves have to be modified according to seasonal and other factors influencing product demand and product quality, and this changes the chemical composition of the finished fuel. Traditionally, gasolines and diesel fuels were refined from crude oil as the input material, but recent changes have seen the source of input to the traditional oil refineries expand. To the traditional petroleum-based "sweet" and "sour" crudes, the use of tar sand-derived synthetic crude and biomass-derived liquids have been added. Thus, the complexity of many refineries has grown and co-processing has become more common. This allows the processing of waste bio-feedstocks in existing refineries, along with intermediate products such as vacuum gas oil to produce hydrocarbon fuels such as renewable diesel [3.1–3.4].

3.2. Crude Oil

Crude oil is the liquid part of the naturally occurring organic material composed mostly of hydrocarbons that is trapped geologically in underground reservoirs. It is by no means uniform and varies in density, chemical composition, boiling range, and so forth, from oil field to oil field and also with time from any given oil field. Table 3.1 [3.5] shows the yields of a number of fractions when four different crude oils are split into different boiling ranges by a distillation process.

TABLE 3.1 Yield (wt %) of main products from crude oil by distillation.

Crude type	Arabian Light	Nigerian	Brent	Maya
LPG	0.7	0.6	2.1	1.0
Naphtha	17.8	12.9	17.8	11.7
Gas oil/Kerosene	33.1	47.2	35.5	23.1
Residue	48.4	39.3	44.6	64.2

© SAE International.

In this table, naphtha represents the percentage of hydrocarbons boiling in the gasoline range, and gas oil/kerosene is the hydrocarbon boiling roughly in the diesel fuel range (but including also jet fuel and kerosene). It can be seen that there are large differences in the percentages of the various streams that can be obtained from each crude by simple distillation. Synthetic crude is produced from heavy oil sources such as tar sands [3.6] or shale oil. A heavy oil upgrader is used to process these heavy products before they are then sent to the refinery.

3.3. Influence of Product Demand Pattern on Processing

If there were a free choice of which crude oils to purchase and if the product demand pattern could be met by selecting appropriate crudes, then a relatively simple refinery is all that would be necessary. Such refineries were commonplace in Europe in the 1960s and 1970s, up to the time that the crude oil price increases of 1973 and later had taken effect. Until then, there had been a large demand for fuel oil (shown as residue in Table 3.1), but the availability of alternative cheaper fuels such as coal changed this. The simple refineries of the time had no way to process this excess residue.

These refineries consisted basically of distillation units with processes for upgrading the octane quality of the naphtha and for removing malodorous sulfur compounds. They were known as hydroskimming refineries, and the layout of a typical one is shown in Figure 3.1.

FIGURE 3.1   Simplified flow diagram of a hydroskimming refinery.

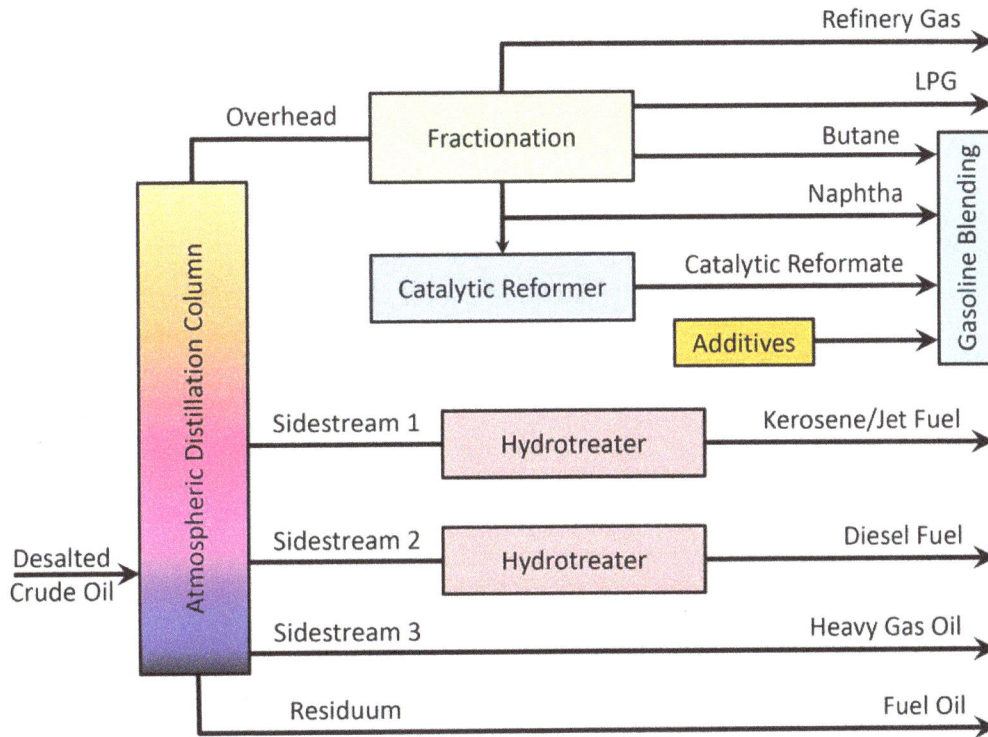

It can be seen that this type of refinery produces a gasoline consisting of just a few refined components, and that the diesel fuel is even simpler, with only one refined component. Fuel for the refinery is normally any stream that is in excess of market requirements and which cannot command a higher price than the cheapest purchased petroleum fuel.

Nowadays, however, refineries generally need to be much more complex in order to be able to run any crude oil that happens to be available and to meet any reasonable change in product demand. In addition, the range of products that are made is often much greater than with a hydroskimming refinery, particularly if petrochemical processing to produce such materials as ethylene (for polyethylene manufacture) and butadiene (for synthetic rubber manufacture) is also part of the refinery operations.

In this chapter, only processes relevant to the manufacture of gasoline and diesel fuel will be discussed, and others will be mentioned only to show how they fit into the general refinery scheme. Figure 3.2 shows the layout of a complex refinery that has all the major processes used in gasoline and diesel manufacture.

FIGURE 3.2 Simplified flow diagram of a complex refinery.

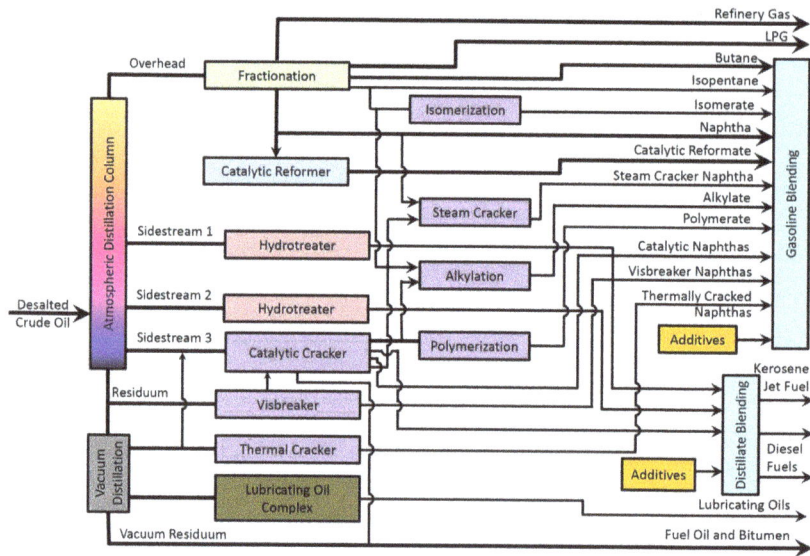

© SAE International.

This is a hypothetical refinery plan only, designed to show the feedstocks for each type of process that could be used and to indicate where in the general processing scheme each process fits. It is perhaps worth mentioning that heat integration is an extremely important aspect of refinery layout. It refers to making maximum use of all the heat energy available by, for example, using hot product streams that need to be cooled down before they go to tankage to partially heat other streams prior to putting them through a furnace on a distillation or other unit.

Processes such as cracking, which convert heavier streams to lighter streams, are known as conversion processes, and refineries able to carry out such operations are sometimes called conversion refineries.

Refineries can normally cope with the changes in product demand that occur as a result of seasonal or other short-term, but predictable, influences. However, when massive changes occur, as was the case when fuel oil demand in Europe fell away dramatically due to oil price rise, the ability of individual units to produce the required product pattern may not be adequate. In such instances, there will often be a shortfall while existing units are expanded or new units installed, and this has to be made up in some way. Purchases of suitable streams from other refineries with spare capacity are one way out, and another, which is still in use to meet the demand for unleaded gasoline in many countries, is to use oxygenated blendstocks. Only two types of oxygenated compounds, alcohols and ethers, are used to any significant extent in gasoline as components (i.e., at concentrations greater than 1 or 2%), and of these the most important are the following:

Alcohols:	Methanol (MeOH)
	Ethanol (EtOH)
	Butanol or butyl-alcohol
	Iso-propanol (IPA)
	Tertiary-butanol (TBA)
	Mixed C1 to C5 alcohols
Ethers:	Methyl-tertiary-butyl-ether (MTBE)
	Tertiary-amyl-methyl-ether (TAME)
	Ethyl-tertiary-butyl-ether (ETBE)
	Mixed ethers

Because of its poor solubility in gasoline when water is present, methanol is used with a co-solvent such as TBA and, in fact, all gasolines containing alcohols require careful handling to avoid or minimize water contact. Ethers tend to be relatively trouble-free as gasoline blend components. However, recent concerns over the leaching of MTBE into ground water have led to its prohibition in certain US states. The manufacture of the oxygenates commonly used in motor gasoline is described in Section 3.9. There are two major problems that effectively limit the amount of oxygenate that can be used in gasolines marketed in areas where the vehicle population is designed for conventional hydrocarbon gasolines. These are the effect that the oxygen content of the compound has on the engine's air-to-fuel ratio and the adverse effect on vehicle fuel system materials. These will be discussed in more detail later.

3.4. Distillation

The crude oil delivered to the refinery will usually contain water, suspended solids, inorganic salts, and water-soluble trace metals. These impurities can cause corrosion and fouling of equipment and can poison the catalysts used in various processing units. These impurities are thus removed before the oil passes to the refinery, via a process usually referred to as desalting. The process consists of removing the impurities by the use of one or more of the following techniques: electrostatic separation, chemical desalting, or the least commonly used, physical filtering. All of these processes require the crude oil to be heated in order to reduce its viscosity and surface tension.

After desalting, the oil passes to the atmospheric distillation unit. This process is used in all refineries, and its aim is to separate the crude oil into different boiling range fractions, each of which may be a product in its own right, a blend component or feed for a further processing step.

Crude oil contains many thousands of different hydrocarbons, each of which has its own boiling point. The lightest are usually gases at normal ambient temperature but can remain dissolved in the heavier liquid hydrocarbons unless the temperature is raised. The heaviest hydrocarbons are solids at ambient temperature but are able to stay in solution except at low temperatures.

Distillation is not intended to chemically change the crude oil in any way but to split it up into fractions, each of which has its own boiling range. In practice, the distillation process is not able to effect perfect separation, as illustrated in Figure 3.3, and there is always a small amount of lighter and heavier materials present in any fraction, depending on the efficiency of the distillation tower.

FIGURE 3.3 Distillation curve for sidestreams 1 and 2 showing imperfect separation.

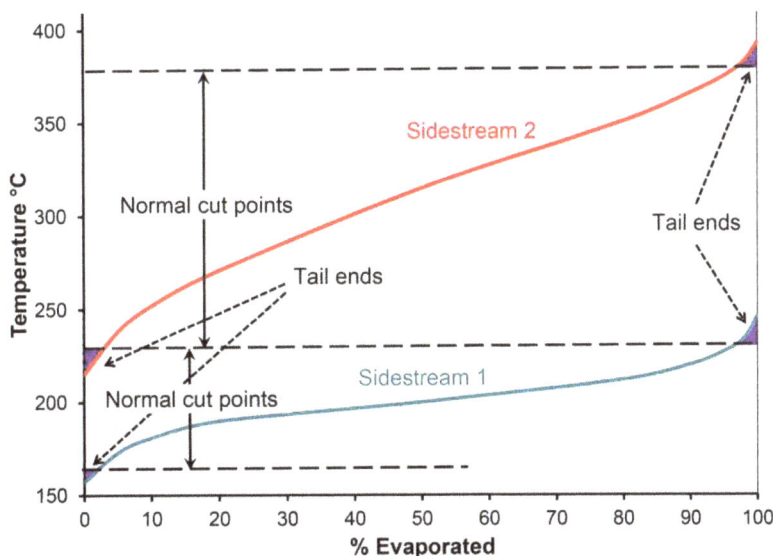

Due to the elevated temperatures, the presence of air, and metals in the feed, there are also some reactions that can occur especially in the heavier fractions. Figure 3.4 illustrates the primary distillation process that is carried out in a unit that is often referred to as a pipestill.

FIGURE 3.4 Flow through a pipestill.

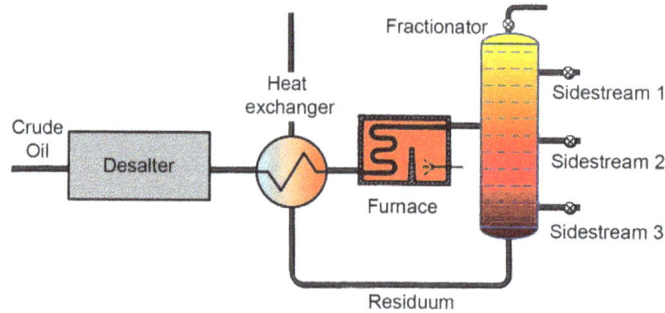

The desalted crude oil is preheated by one or more of the refinery streams through heat exchangers and is then further heated in a furnace to about 400°C. The resulting mixture of liquid and vapor is then put into a fractionating tower at atmospheric pressure, the internal construction of which is designed to affect the maximum separation possible between compounds of different boiling points. Gases such as ethane, propane, and butane are part of the overhead stream, and then successive sidestreams will contain progressively higher boiling range groups of hydrocarbons. The residue (fuel oil) stays liquid and flows out of the bottom of the tower.

In a complex refinery, as shown in Figure 3.2, the residue from the atmospheric tower would be further fractionated under vacuum to produce other streams, which might then in turn be fed back into a cracking unit or be used for lubricating oil manufacture, bitumen manufacture, or heavy fuel oil. Figure 3.5 shows an atmospheric distillation unit on the left and a vacuum distillation alongside it on the right.

FIGURE 3.5 Atmospheric and vacuum distillation units.

3.5. Cracking Processes

There are two general types of cracking process—thermal and catalytic—and each proceeds by quite different mechanisms [3.7]. *Thermal cracking* is believed to take place through the creation of hydrocarbon-free radicals by homolytic fission, for example:

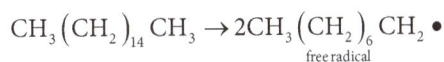

$$CH_3\left(CH_2\right)_{14}CH_3 \rightarrow 2CH_3\underset{\text{free radical}}{\left(CH_2\right)_6 CH_2 \bullet}$$

These free radicals are extremely reactive and undergo various reactions to form lower molecular weight hydrocarbons and free radicals, as for example:

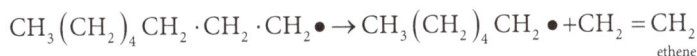

$$CH_3\left(CH_2\right)_4 CH_2 \cdot CH_2 \cdot CH_2 \bullet \rightarrow CH_3\left(CH_2\right)_4 CH_2 \bullet + \underset{\text{ethene}}{CH_2 = CH_2}$$

or

$$CH_3(CH_2)_4 CH\bullet \rightarrow CH_3(CH_2)_2CH_2\bullet \;+\; CH_2{=}CH.CH_2.CH_3$$
$$\hspace{0.5cm}\vert \hspace{4cm}\text{n-butene}$$
$$CH_3CH_2$$

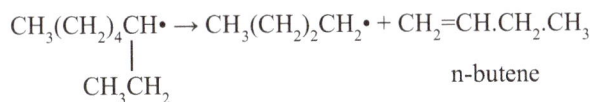

Eventually the chain reactions can be terminated by two free radicals reacting with each other to form a new paraffinic molecule.

Dehydrogenation of olefins can also occur under severe thermal cracking conditions to form dienes:

$$CH_2 = CHCH_2CH_3 \rightarrow CH_2 = CH - CH = CH_2 + H_2$$

Naphthenes may undergo ring cleavage or may dehydrogenate to form aromatics, which are the most stable class of hydrocarbons to further cracking.

Catalytic cracking, unlike thermal cracking, proceeds by the formation of carbonium ions, for example:

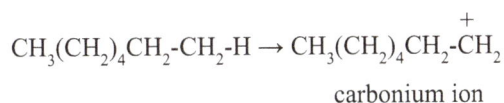

$$CH_3(CH_2)_4CH_2\text{-}CH_2\text{-}H \rightarrow CH_3(CH_2)_4CH_2\text{-}\overset{+}{C}H_2$$
$$\text{carbonium ion}$$

Once formed, carbonium ions rapidly undergo a number of subsequent reactions, such as isomerization, to give secondary or tertiary carbonium ions:

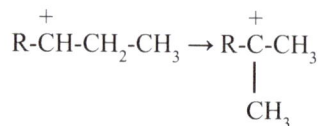

$$\overset{+}{\text{R-CH-CH}_2\text{-CH}_3} \rightarrow \text{R-}\overset{+}{\underset{\underset{\text{CH}_3}{|}}{\text{C}}}\text{-CH}_3$$

Cracking at the C-C bond in the β position relative to the positively charged carbon atom also occurs to form olefins:

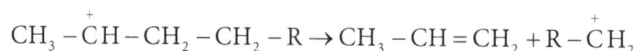

$$\text{CH}_3 - \overset{+}{\text{CH}} - \text{CH}_2 - \text{CH}_2 - \text{R} \rightarrow \text{CH}_3 - \text{CH} = \text{CH}_2 + \text{R} - \overset{+}{\text{CH}}_2$$

The olefins themselves then undergo further reactions to form branch chain and aromatic compounds. The reactions are terminated when the carbonium ion is destroyed by donating a proton to an acid site on the catalyst or by abstracting a hydride ion from the catalyst.

The yields and the characteristics of the finished product can therefore be quite different for thermally cracked naphthas (and this includes naphthas from visbreakers and cokers as well as thermal crackers) as compared to catalytic crackers. Thermally cracked naphthas tend to contain lower aromatics levels and fewer branch chain compounds than catalytically cracked naphthas but more lighter olefins and dienes.

It is the olefin and aromatics content that are important in cracked streams because they have a large effect in determining the behavior of the gasoline or diesel fuel into which they are blended. Olefins are very reactive compounds that are easily oxidized and have better octane characteristics than the paraffinic streams from which they are derived, although their cetane number values are somewhat worse. Aromatic compounds have excellent octane properties but have very poor cetane numbers.

The yields obtained depend on, among many other factors, the hydrocarbon makeup of the feed. Paraffins are the easiest hydrocarbons to crack and aromatics the most difficult, with naphthenic compounds being intermediate.

Individual cracking processes are described next.

3.5.1. Thermal Cracking

Conventional thermal cracking is now virtually obsolete for the production of motor gasoline since there are many other superior cracking processes available. The process requires the feed to be heated to around 500°C at a pressure of up to 7 MPa when it is passed into a soaking chamber where the cracking reactions take place. The cracked products are fractionated, and the yield of cracked naphtha obtained can be as much as 60% by volume.

The product is relatively unstable and requires the use of antioxidants and/or other treatments to prevent gum formation in storage or during use. It has a relatively poor research octane number (RON) and motor octane number (MON) quality. The gas oils from both this process and visbreaking (see the next section) are characterized by low cetane numbers, thermal instability due to the presence of diolefins, and high sulfur contents.

3.5.2. **Visbreaking**

Visbreaking is a low-severity form of thermal cracking originally developed to reduce the viscosity (break the viscosity, hence visbreaker) of residual fuel oils, thereby minimizing the amount of higher-value distillate needed to achieve the required viscosity. The process is similar to conventional thermal cracking using temperatures up to 500°C and pressures up to 35 MPa. Figure 3.6 shows a visbreaker scheme as proposed by Standard Oil Development Company in their 1951 patent [3.8].

FIGURE 3.6   Visbreaker schematic flow diagram.

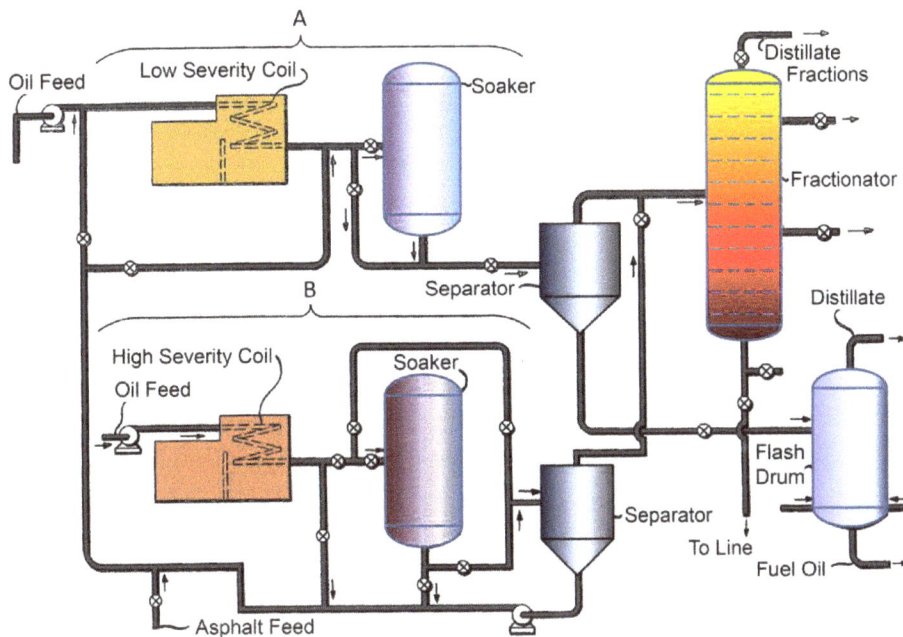

This particular system utilizes a two-stage visbreaker system: the first high-severity system (B) uses more severe conditions (495 to 515°C and 2 to 7 MPa), whereas the lower severity system uses more conventional conditions.

Nowadays, visbreaking is most often used as a relatively cheap process for converting some of the excess fuel oil into the more profitable distillate. Yields of visbreaker naphtha for gasoline blending vary from about 4 to 8%. For diesel fuel manufacture, the yield of gas oil varies from about 12 to 25%, using atmospheric residuum as feed.

3.5.3. **Coking**

This is a severe form of thermal cracking designed to convert residual products such as fuel oils into gas, naphtha, heating oil, gas oil, and coke. It is particularly valuable as a process when the market for residual fuel oils is restricted. The coke produced can be used as a fuel or as a feed to a coke gasifier to produce hydrogen. It can also be marketed for other uses such as electrode manufacture. Two major forms of coking process are in use, namely delayed coking and fluid coking.

Delayed coking is a semicontinuous process in which the heated charge is subjected to a long residence time in a soaking drum to allow all the cracking reactions to go to completion. Coke is deposited in the drum; there are two of these so that while one is on-stream the other can be cleaned. The cracked products are fractionated, and heavy distillates are often recycled with feed. Figure 3.7 shows coker units at a US refiner.

FIGURE 3.7 Coker units at a US refinery.

© SAE International.

Fluid coking is a continuous process that converts vacuum residuum to gaseous and liquid products using a fluidized solids technique. Fluidization involves subjecting small particles, in this case of coke, to a stream of gas of an appropriate velocity so that the suspended particles behave as if they were a liquid. Coking of the residuum is achieved by spraying it into a fluidized bed of hot, fine coke particles. The use of a fluidized bed allows the coking reactions to be carried out at higher temperatures and with shorter residence times than can be employed in delayed coking. Two vessels are used in fluid coking, a reactor and a burner. A portion of the coke is burned to provide heat, and this heat is transferred to the reactor by the circulation of hot particles of coke.

Flexicoking is an extension of the fluid coking process, which is able to process very heavy, high-sulfur residues. The coke formed, which contains most of the sulfur and heavy metals from the feed, is converted to a low-calorie gas suitable for use in process burners.

Coker naphthas suffer from the same drawbacks as the previously described thermally cracked streams. The gas oil product is of rather poor quality, suffering a high sulfur content, thermal instability (due to the presence of diolefins), and a high aromatics content. Indeed, all thermally cracked gas oils require upgrading before blending to diesel fuel.

3.5.4. Catalytic Cracking

Catalytic cracking is the most important and widely used process for converting heavy refinery streams to lighter products, and it is the most popular method of increasing the ratio of light to heavy products from crude oil. The input to the catalytic cracker is also a point for inclusion of renewable components into the feed [3.1]. The catalytic cracker has superseded conventional thermal cracking because of its higher yields, better product quality (particularly for gasoline), and superior economics. Figure 3.8 shows an early catalytic cracking system where the gasoline yield was enhanced by careful fractionation of the feedstock [3.9].

FIGURE 3.8   Catalytic cracker flow schematic.

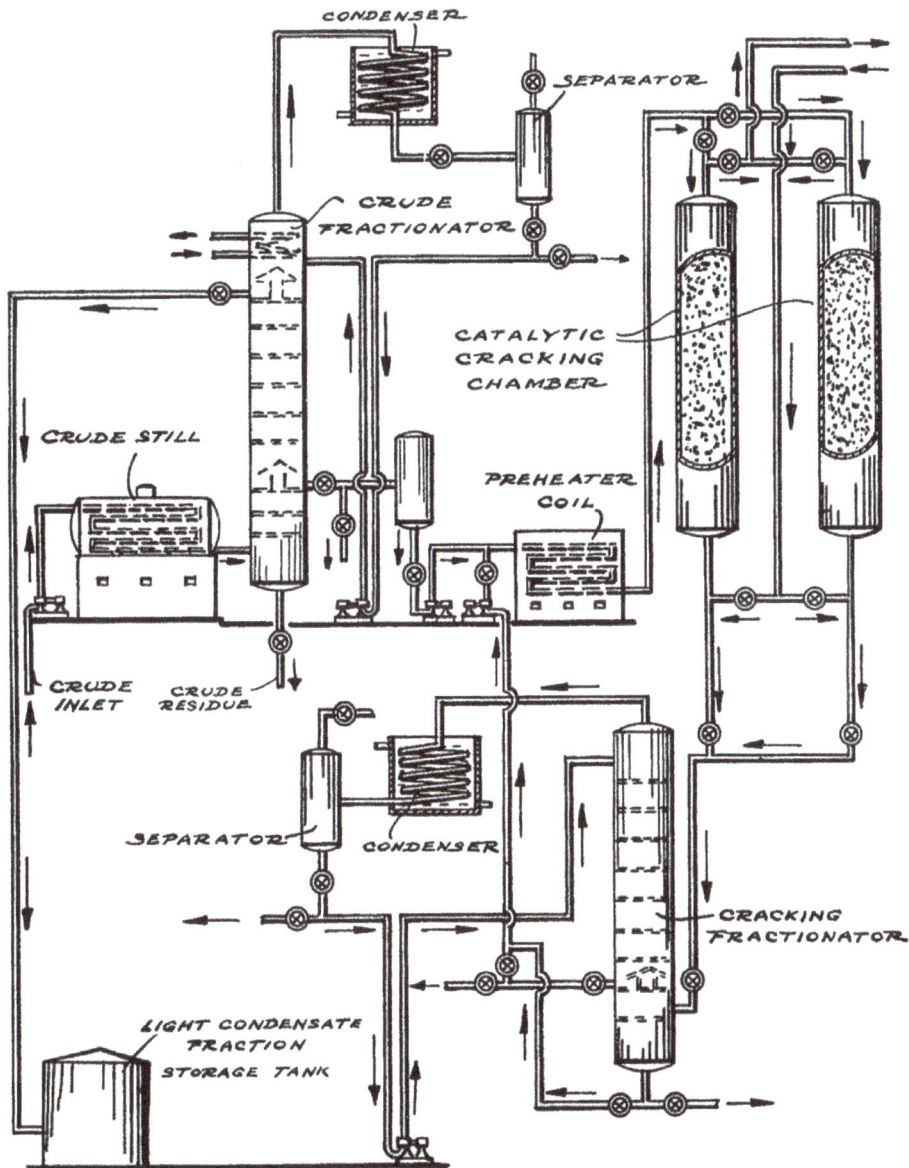

Most catalytic cracking processes use a fluidized bed of catalyst into which feed is introduced. Fluidization has the advantage that the catalyst can be made to flow through pipes and from one vessel to another. The gas velocity is quite critical; at low velocities the catalyst will flow down pipes and at higher velocities it can be made to flow upward. This allows the catalyst to be moved readily between the reactor and the regenerator. Fluidized beds also allow high mass and heat transfer rates between the catalyst surface and the gas phase to give fast reaction rates.

The catalyst itself is a form of aluminum silicate known as zeolite, which has a high activity and which suppresses the formation of light olefins.

Figure 3.9 shows a simplified schematic of one possible layout of a fluidized catalytic cracker (FCC) based on a Shell patent filed in 1945 [3.10].

FIGURE 3.9 Schematic of a fluid catalytic cracker.

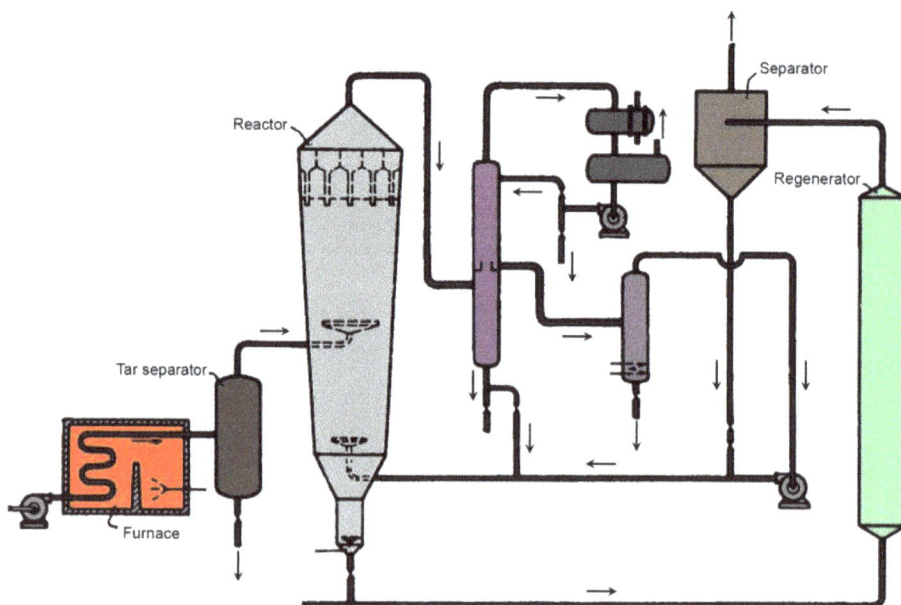

The catalyst is moved continuously from the reactor, where carbon is deposited on it during the reactions, to the regenerator, where the carbon is burned off in a stream of hot air. As the catalyst is recycled back to the reactor, it carries with it the heat needed to vaporize and crack the feedstock, which is usually a heavy gas oil from the atmospheric or vacuum distillation towers. The vapor and the catalyst flow together into the reactor where the cracking takes place, although with today's high-activity catalysts, much of the cracking can take place in the transfer lines. Figure 3.10 is a photograph of a fluid catalytic cracker unit.

FIGURE 3.10 Fluid catalytic cracker unit.

The cracked oil vapors pass to fractionating towers in which the new, smaller molecules are separated from the heavier products. The products from such a unit would be gas, catalytic naphthas, cycle oils, and residue.

Developments in catalyst technology and in regeneration techniques allow the use of a mixture of residue and heavy distillate as feed. Without these developments the catalyst would become deactivated very rapidly when residuum is present.

Highest conversion is obtained with paraffinic feedstock, resulting in C_5/220°C final boiling point naphtha yield of around 40% w/w on feed. The naphtha is usually split into two or three fractions to take advantage of aromatics/olefins distribution and improve the economics of subsequent processing. Typical octane number qualities for FCC naphthas are as follows:

Naphtha cut	RON	MON
Light	94.5	81.4
Medium	92.5	79.3
Heavy	90.3	77.6

© SAE International.

FCC gas oil is characterized by its high aromatics content, resulting in very high density and very low cetane number. Cetane numbers below 20 and densities above 900 kg/m³ are not uncommon, and as a consequence, it is very difficult to blend even small quantities of this stream into finished diesel fuels.

3.5.5. Hydrocracking

The mechanism of hydrocracking is similar to that of catalytic cracking, but with hydrogenation superimposed. Carbonium ions are formed, which crack, and the fragments are rapidly hydrogenated under the high hydrogen partial pressures used in the process. The rapid hydrogenation prevents coke deposition on the catalyst as occurs in catalytic cracking so that long runs without catalyst regeneration are achieved.

A dual-function catalyst is needed, which performs satisfactorily for both cracking and hydrogenation reactions. This is achieved by the use of a silica alumina base to promote the cracking reactions, on which is dispersed metals such as molybdenum, tungsten, cobalt, nickel, and so forth, to promote hydrogenation reactions.

Hydrocracking is used mainly to produce low boiling fractions from feedstocks such as heavy gas oils, waxy distillates, and even residues. The heavy feedstocks contain polycyclic aromatic compounds, and these are quickly partially hydrogenated followed by splitting of the rings to form substituted monocyclic hydrocarbons. The side chains are then removed to form iso-paraffins. The degree of cracking depends greatly on the feedstock, and in general, the heavier the feed, the more middle distillate is produced.

The hydrogen requirement of hydrocrackers is extremely high so that unless there is plenty available from such units as catalytic reformers, it is necessary to have a hydrogen production plant as part of the hydrocracker complex.

The yield and product quality achieved depend on the feed to the unit and the severity of operation. The product boiling in the gasoline range is usually further processed by passing it through a catalytic reforming process. This is because hydrocracked naphtha is predominantly paraffinic and thus exhibits relatively low RON and MON values. The units are often run to maximize production of middle distillates as they produce high cetane number gas oils with a very low sulfur content.

3.5.5.1. Catalytic Distillate Dewaxing

Distillates with a high wax content can make it difficult for the refinery planner to meet low-temperature quality specifications. While refrigerative dewaxing is a normal procedure for lowering the pour point of lubricant base stocks, it would be impractical and uneconomical for diesel fuel components because of the large amounts of product (and wax) involved.

Catalytic dewaxing is a process previously used for jet fuel [3.11] but later developed for waxy middle distillate streams to improve their low-temperature properties, an example of which is the Mobil distillate dewaxing (MDDW) process, which involves selective cracking of wax inside the pores of a ZSM-5 catalyst [3.12]. It is a type of mild hydrocracking in which a very low pour point gas oil can be produced from a waxy straight-run feedstock. Figure 3.11 shows a flow plan of the process [3.13].

FIGURE 3.11 Flow plan of catalytic distillate dewaxing process.

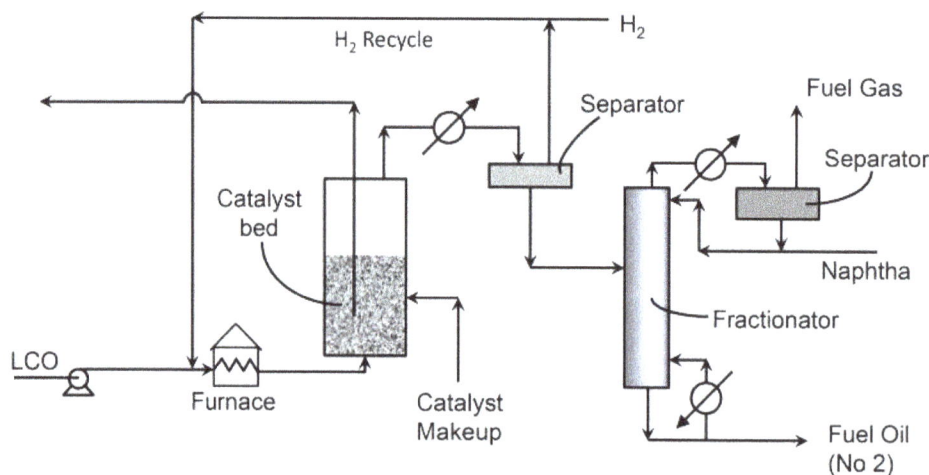

A shape-selective catalyst preferentially hydrocracks the normal and near-normal paraffins in the waxy feed, producing a dewaxed gas oil and small proportion of lighter components. The effectiveness of the process is evident from results obtained with two feed streams of different wax contents.

A light Arabian heavy atmospheric gas oil with a pour point of 18°C cracked in the MDDW unit yielded 85% volume of dewaxed product with a −18°C pour point. From a more waxy Libyan heavy gas oil with a 33°C pour point, a 65% yield of 0°C pour point distillate was obtained.

The process uses a lot of hydrogen and can be very costly, particularly if hydrogen availability is limited and a new plant has to be installed. Although distillate dewaxing is justified in some refining situations, only a few units are in operation around the world.

3.5.6. Steam Cracking

Steam cracking produces ethylene and other low boiling olefins used as chemical intermediates, particularly for the production of polythene. A by-product from this process used in gasoline blending is steam-cracked naphtha or, as it is often called, pyrolysis gasoline or pygas.

The process does not employ a catalyst and involves cracking naphtha or gas oil at temperatures in the range of 750 to 900°C in the presence of steam. The steam reduces the partial pressure of the feedstock, thereby increasing the yield of gaseous rather than liquid products. It also minimizes the formation of coke. Other feeds such as ethane or LPG can be used, but these produce very little steam-cracked naphtha for gasoline blending.

Steam-cracked naphtha must be pretreated before it is used in gasoline because it contains dienes and other very reactive olefins that reduce storage stability and can lead to excessive deposit formation in engines. The pretreatment can be a hydrogenation step or some other process such as heat soaking that removes the gum-forming materials.

Many steam-cracked naphthas contain high levels of benzene and other aromatics, depending on the feed and the operating conditions.

These aromatics are sometimes removed for use as chemical intermediates, but if they are not, care must be taken that the maximum allowable level of benzene in gasoline is not exceeded when this component is used.

3.6. Catalytic Reforming

Catalytic reforming is an extremely important, widely used process for increasing the antiknock properties of naphtha. There are many variations developed by different companies such as Platforming (Universal Oil Products, UOP), Powerforming (Exxon), and Ultraforming (Amoco). An example of a catalytic reformer is shown in Figure 3.12.

FIGURE 3.12 A catalytic reformer.

All of these systems use a catalyst in which platinum, often together with rhenium or other precious metals, is deposited on a support of alumina. Others, such as the RZ-Platforming™ system (UOP) and the Aromax™ (Chevron Phillips Chemical Company), utilize Pt/L-zeolite catalyst. The Pt catalyst is readily poisoned by such materials as arsenic, copper, and lead and its activity is reduced by the presence of sulfur and nitrogen compounds. For this reason, the feed to a catalytic reformer is hydrodesulfurized (see Section 3.8.3) to remove these undesirables.

The main route for improving the antiknock properties of naphtha in catalytic reforming is the formation of aromatics by the dehydrogenation of naphthenes:

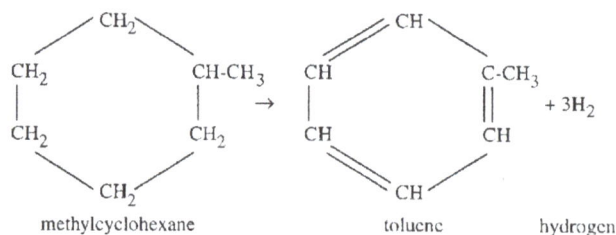

It can be seen that the process is a net producer of hydrogen, which can be used for hydrotreating and for hydrocracking.

Other reactions that take place are isomerization, in which straight chain paraffins are rearranged into the higher octane number quality branched hydrocarbons; dehydrocyclization, in which paraffins are cyclized to naphthenes and then dehydrogenated to aromatics; hydrocracking to form lower, smaller paraffinic molecules from larger ones; and dealkylation, in which the side chains from higher aromatics are removed to form lower aromatics and paraffins. Most of these reactions simply involve rearranging the carbon atoms in the molecule so that the boiling range of the product is not vastly different from that of the feed.

Small amounts of polycyclic aromatic hydrocarbons (PAH) are also formed, particularly when the units are operated at high severities, and these are undesirable in a gasoline because of their tendency to lay down combustion chamber deposits and also because they tend to be carcinogenic. Catalytic reformates are sometimes rerun by passing them through a flash tower to remove these very high boiling materials.

The process itself can be semicontinuous or continuous. In the semicontinuous form, long on-stream runs are achieved followed by a period when the catalyst, which is contained in several reactors arranged in series, is regenerated by carefully burning off the coke. There are two forms of continuous reforming: one involves the use of a swing reactor that is used in place of each of the other reactors in turn so that they can be regenerated without disrupting the run; the other has the reactors stacked one on top of the other, and the catalyst flows through them by gravity and is then lifted to a regenerator and recycled.

3.7. Alkylation, Isomerization, and Polymerization

3.7.1. Alkylation

Alkylation is the name given to a process for producing a high-octane number gasoline component (alkylate) by combining light olefins with iso-butane in the presence of a strongly acidic catalyst. Mineral acids such as sulfuric and hydrofluoric acid were commonly used, but these have now been largely replaced by solid acid catalysts. It was an extremely important process during World War II when it was used to manufacture high-octane number aviation gasoline for military use.

The olefins used in the feed are usually derived from catalytic cracking units and are normally a mixture of propenes and butenes. The reactions that take place are complex, although basically, the iso-paraffins formed have the same number of carbon atoms as the sum of those in the iso-butane and olefin reactants. An example of this type of reaction can be represented simplistically as

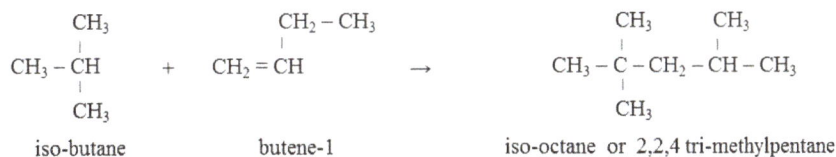

iso-butane   butene-1   iso-octane or 2,2,4 tri-methylpentane

In the preceding example, the product is iso-octane (2,2,4-trimethylpentane), which has, by definition, an RON and a MON of 100. Alkylation produces a mixture of high-octane number branch chain paraffins with a low sensitivity (the difference between RON and MON, see Chapter 6) and can be a valuable component when MON is a limiting specification point. Figure 3.13 shows a sulfuric acid alkylation unit at a US refinery.

FIGURE 3.13   Alkylation unit.

© SAE International.

The feeds required are very specific. Iso-butane is obtained by super-fractionation of virgin, fluid catalytic cracked, or reformer LPG streams, whereas the n- and iso- C_3 and/or C_4 olefins come from the FCC light ends recovery unit. Thus, any alkylate contribution to the total gasoline pool is limited by feedstock availability.

3.7.2. Isomerization

This is a process for converting straight chain paraffins to branch chain and may be used to provide iso-butane feed for the alkylation process or simply to convert the relatively low-octane number quality of straight paraffins to the more valuable branch chain molecules. Thus, n-pentane with a RON of 62 can be converted to iso-pentane having a RON of 92.

In principle, the process involves contacting the hydrocarbons with the catalyst (platinum on a zeolite base) and separating any unchanged straight paraffins for recycling through the unit. Feedstocks must be essentially sulfur free to avoid catalyst poisoning. The product is clean burning and has good RON and MON qualities. Figure 3.14 shows an isomerization unit (to the left of the picture) at a UK refinery.

FIGURE 3.14 Isomerization unit.

3.7.3. Polymerization

In this process light olefins such as propene and butenes are reacted together to give heavier olefins, which have a good octane quality and do not increase unduly the vapor pressure of the gasoline. An example of the type of reaction that takes place is as follows:

$$\underset{\text{butane}}{CH_3CH_2CH=CH_2} + \underset{\text{propene}}{CH_3CH=CH_2} \rightarrow \underset{\text{heptene}}{CH_3(CH_2)_4CH=CH_2}$$

The catalyst most commonly used is phosphoric acid on keiselguhr.

The product is almost 100% olefinic and has a relatively poor MON compared with RON (high sensitivity). The process is now seldom used for gasoline because the product composition tends to be variable and because the feed is more valuable in other processes such as alkylation or higher olefin synthesis.

3.8. Finishing Processes

Gasoline and diesel fuel component streams produced by the preceding processes are often unsuitable for immediate use for a number of reasons, including objectionable odor or instability. For this reason, they are subjected to secondary treatments, the most important of which from a gasoline and diesel fuel viewpoint are caustic washing, UOP Merox treating, and hydrodesulfurization. Other processes in this category are copper chloride sweetening [3.14, 3.15, 3.16], inhibitor sweetening [3.17, 3.18], acid treatment [3.19], and so on, but these are generally no longer used and will not be discussed here.

3.8.1. Caustic Washing

Washing with caustic soda or caustic potash will remove a number of undesirable contaminants such as hydrogen sulfide, mercaptans (thiols), cresylic acids, and naphthenic acids, as well as acidic materials carried over from processes such as alkylation that use an acid catalyst. Caustic washing is mainly used as a final cleanup operation to remove the last traces of contaminant. It is sometimes followed by a water wash to ensure that there is no carry-over of caustic soda into the finished product.

Caustic washing is carried out by mixing together the hydrocarbon stream and aqueous caustic soda and then allowing the mixture to settle into two layers in a horizontal drum so that each layer can be drawn off separately. Because of the environmental problems caused by the disposal of spent caustic soda, processes in which the caustic is regenerated, such as the Merox Process, are generally preferred.

3.8.2. Merox Treating

This process, developed by Universal Oil Products (UOP), catalytically oxidizes the evil-smelling mercaptans to nonodorous disulfides, which can either be left in solution (Merox Sweetening) or removed altogether (Merox Extraction). The basic reactions are the following:

1. Extraction of mercaptans from the oil phase by means of aqueous caustic soda:

$$RSH + NaOH \rightarrow NaSR + H_2O$$

2. Regeneration of the caustic and oxidation of the mercaptides by air blowing in the presence of a catalyst:

$$2NaSR + \tfrac{1}{2}O_2 + H_2O \rightarrow RSSR + 2NaOH$$

The disulfides formed are insoluble in caustic soda and so are allowed to re-dissolve in the hydrocarbon in the sweetening process. But in the extraction process, the sodium mercaptide solution is separated from the hydrocarbon phase before air blowing so that the separated disulfides can be removed prior to the return of the regenerated caustic for reuse. The catalyst consists of a metal chelate and can be supplied either in a form that is soluble in caustic soda or as a solid supported on a carrier.

3.8.3. Hydrodesulfurization

Naphtha, kerosene, and gas oils, which contain high molecular weight sulfur compounds in addition to hydrogen sulfide and light mercaptans, can be desulfurized by hydrogen treatment. The process also improves stability by saturating olefinic compounds, particularly the more reactive ones, and by removing nitrogen- and oxygen-containing compounds.

Examples of the reactions that take place are the following:

$$RSH + H_2 \rightarrow RH + H_2S$$

$$RSSR' + 3H_2 \rightarrow RH + R'H + 2H_2S$$

The catalyst used contains cobalt and molybdenum on an alumina base and can be regenerated in situ or by removing it for regeneration offsite. The hydrogen sulfide formed is normally converted into elemental sulfur in a sulfur recovery plant using the Claus process.

There are many variations on the hydrodesulfurization process developed by different companies such as Hydrofining (Exxon Research and Engineering) and Ultrafining (Standard Oil Co. [Indiana]), but the basic principles are the same and are as outlined above. One of the challenges facing refiners has been how to selectively desulfurize FCC naphthas without saturating olefins and losing octane number quality. Fortunately, a number of proprietary processes, including SCANfining and CDHydro, have been developed, which reduced RON loss from around 5 RON to about 1.5 RON.

3.8.3.1. Sources of Hydrogen

As noted in Section 3.6, the catalytic reforming process is a valuable source of hydrogen. However, the increasing demand for hydrogen means that this will rarely be sufficient. Additional hydrogen can be obtained via a number of alternative routes. These typically include the following:

- Recovering hydrogen from refinery fuel gas

- Steam reforming using fuel gas or naphtha

- Gasifying coke from the coking unit

- Importing hydrogen from outside the refinery, also now as over the fence (OTF) supply

Hydrogen can be recovered from flue gases by what is generally termed hydrogen purification processes. The three major hydrogen purification processes are pressure swing adsorption (PSA), selective permeation using membranes, and cryogenic separation.

The PSA process is a technique by which the non-hydrogen components, considered as impurities, of the gas stream are removed in adsorbent beds. PSA units are based on the ability of the adsorbents to adsorb more impurities at high gas-phase partial pressure than at low partial pressure. The pressure is then swung to desorb the impurities, which are purged from the system. Membrane systems rely on the fact that different molecules will permeate the membranes to a different degree and at different rates. Cryogenic separation relies on the fact that different components of the flue gas will condense at different temperatures. The latter is usually the most cost-intensive system and is thus used less frequently.

Steam reforming and the gasification of coke are partial oxidation processes where the oxygen in the steam in the absence of molecular oxygen acts as the oxidant leaving hydrogen as the more valuable product. This is indicated by the following two equations:

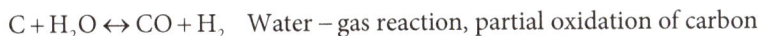

$$CH_4 + H_2O \leftrightarrow CO_2 + 3H_2 \quad \text{Steam reforming of methane}$$

$$C + H_2O \leftrightarrow CO + H_2 \quad \text{Water} - \text{gas reaction, partial oxidation of carbon}$$

3.9. Oxygenated Gasoline Components

As a result of the phase-out of lead antiknock additives, there have been difficulties in many parts of the world in meeting both octane number quality and the required volume of gasoline. These have been overcome to some extent by the use of oxygenated blend components. Work has been conducted on many oxygenates including alcohols, ethers, furan derivatives, substituted phenols, aliphatic amines and polyamines, various amide structures, and Mannich base phenols [3.20]. However, the most commonly used are alcohols and ethers. The lower alcohols are the most commonly used, although due to poor miscibility these are often used in combination with the higher alcohols such as iso-propyl alcohol and tertiary-butyl alcohol [3.21]. However, recently the use of butanol is receiving increased attention [3.22–3.25]. The most commonly used ethers are MTBE, ETBE, TAME, or mixtures of ethers. However, due to concerns regarding the leaching of MTBE into ground water [3.26–3.28], in 1999 California proposed the phase-out of MTBE, and other US states have since set in motion plans to either ban or severely restricted the use of MTBE. The behavior of these components in gasoline is discussed in later chapters; this section covers the most important processes used in the manufacture of these materials from fossil fuels.

3.9.1. Alcohols

Methanol (CH_3OH) was widely used, together with one or more higher alcohols, within certain restrictions whenever the price made it economical to do so. This applied in the 1970s, but (as will be discussed later) methanol is probably the least attractive alcohol for use as a gasoline blend component. It is still widely used as a racing fuel but is rarely employed in commercial gasoline. China is a notable exception, where the use of methanol is growing [3.29].

Virtually all methanol is manufactured from natural gas, although it can be produced from any carbonaceous raw material such as coal or wood that can be converted to synthesis gas (carbon monoxide and hydrogen). The reactions are the following:

$$CH_4 + \tfrac{1}{2}O_2 \rightarrow CO + 2H_2$$

$$CO + 2H_2 \rightarrow CH_3OH$$

Highly selective catalysts are used to maximize the yield of methanol, although less selective catalysts will give higher alcohols such as ethanol (C_2H_5OH), butanol (C_4H_9OH), and so forth:

$$2CO + 4H_2 \rightarrow C_2H_5OH + H_2O$$

$$4CO + 8H_2 \rightarrow C_4H_9OH + 3H_2O$$

Mixtures of alcohols can also be obtained from synthesis gas using the above route and have been used as gasoline components, because the higher alcohols present reduce the water sensitivity of the methanol. The process developed by Dow/UCC uses one of a family of active and stable molybdenum-sulfide catalysts, which have a good selectivity to the production of alcohols. The water produced is removed in this process by the use of molecular sieves.

Ethanol is most commonly produced by hydration of ethylene or by fermentation of biomass, as discussed in Chapter 5. Most of the ethanol used industrially is made synthetically by mixing ethylene with steam at 60–70 atmospheres and about 300°C over a phosphoric-acid catalyst supported on diatomaceous earth. The reaction is

$$CH_2 = CH_2 + H_2O \rightarrow CH_3CH_2OH$$

Due to the drive toward renewable fuel, many jurisdictions, including the US, India, Indonesia, the European Union, and the UK, have mandated the inclusion of bioethanol in gasoline [3.30]. This has to a degree eliminated the need to use petroleum-derived ethanol as a blending component in these areas. There are some countries in which there is some resistance to the widespread inclusion of ethanol, severely limiting its concentration in some areas [3.31, 3.32].

TBA (t-butanol, $(CH_3)_3COH$), which was used as a co-solvent for methanol, is most often made by controlled oxidation of iso-butane to TBA and tertiary-butyl-hydro-peroxide (TBHP). The hydro-peroxide is then reacted with propene to give propylene oxide and additional butyl-alcohol. The reactions are the following:

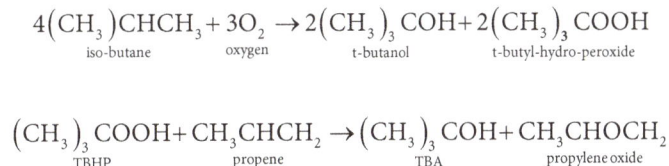

$$\underset{\text{iso-butane}}{4(CH_3)CHCH_3} + \underset{\text{oxygen}}{3O_2} \rightarrow \underset{\text{t-butanol}}{2(CH_3)_3COH} + \underset{\text{t-butyl-hydro-peroxide}}{2(CH_3)_3COOH}$$

$$\underset{\text{TBHP}}{(CH_3)_3COOH} + \underset{\text{propene}}{CH_3CHCH_2} \rightarrow \underset{\text{TBA}}{(CH_3)_3COH} + \underset{\text{propylene oxide}}{CH_3CHOCH_2}$$

Although the TBA produced by this route tends to be somewhat impure, it is satisfactory for use in gasoline without further refining.

3.9.2. Ethers

The most important ether used in gasoline blending is MTBE, although others such as TAME and ETBE are receiving more and more attention. However, it should be noted that environmental concerns are now limiting the use of MTBE; this is discussed further in Chapter 6, Section 6.7.1.

MTBE is manufactured by reacting methanol with iso-butylene:

$$\underset{\text{iso-butylene}}{(CH_3)_2C=CH_2} + \underset{\text{methanol}}{CH_3OH} \rightarrow \underset{\text{MTBE}}{(CH_3)_3COCH_3}$$

Different processes for manufacturing MTBE vary only in the route that is used to make the iso-butylene. Thus, the Arco process dehydrates TBA, whereas the Houdry process dehydrogenates iso-butane.

TAME is produced commercially in a similar way to MTBE in that methanol is reacted with an iso-amylene such as 2-methyl-2-butene or 2-methyl-1-butane:

$$\underset{\text{2-methyl-2-butene}}{(CH_3)_2C=CHCH_3} + \underset{\text{methanol}}{CH_3OH} \rightarrow \underset{\text{TAME}}{(CH_3)_2C_2H_5COCH_3}$$

ETBE is again similar to MTBE, except that ethanol is used instead of methanol.

Mixed ethers have also been produced for use as a gasoline blend component, and here, instead of using relatively pure olefins such as tertiary butylene or iso-amylene, a mixture of C_4 to C_7 olefins is reacted with methanol.

3.10. Gasoline Blending

Gasoline blending, when carried out on a routine basis with a relatively small number of blend components, provides few difficulties once the blending data for the components in question have been determined.

However, in a complex refinery or when untried components are purchased, it can require a number of trial blends to establish blending behavior. One reason is that the blend does not always behave as expected in terms of how the calculated quality compares with actual measured quality, particularly if there are plant changes or new components being used. The main causes of these discrepancies are the following:

- Many of the important specification parameters such as octane and vapor pressure do not blend linearly.

- The octane behavior of a given component is modified by the nature of other components in the blend, that is, factors other than RON and MON, such as hydrocarbon composition, influence the way in which a component blends.

- The blender is trying to meet several specification points at once and any move to meet one specification point may well result in another being put out of grade.

There are other factors that conspire to make the life of a gasoline blender difficult:

- Restrictions in the maximum level allowed for some components or additives, such as oxygenates, anti-knock additives, sulfur, and so on.

- Tankage limitations—a refiner has to find a home for all of the product streams made, and if all of a product stream cannot be used, it may necessitate cutting back in plant throughput to avoid overfilling tanks. This, in turn, may mean a deficiency in other streams and an increase in costs.

In addition, many specification points are legally enforced so that the penalties in terms of adverse publicity and fines if out-of-grade product is put on the market can be very expensive. Because test procedures often have a poor precision, this has to be taken into account when setting "internal" specification standards so as to be sure that any sample picked up by a controlling body has a very high probability (usually at least 95%) of meeting the required specifications.

The cost of "giving away" quality, that is, the cost of making a higher quality than is required by the specification, can be very high indeed, particularly for important points such as octane and vapor pressure. This is not to say that all refiners manufacture as close to the legal specifications as they can; many will have their own "internal" specifications that can be significantly better than the minimum legal limits. They do this in order to make their product more attractive to customers so that they can increase, or at least maintain, their share of the gasoline market.

Fungibility, the interchangeability of a gasoline between suppliers and the compatibility of gasolines from different sources, is another problem and is important for two reasons: First, a refiner may wish to sell to or purchase from another manufacturer either to make up a deficiency in manufacturing capacity or to minimize distribution costs. In these cases, the material must meet the specification of the purchasing company, which may be different from the internal specification of the supplying company. Second, if a customer purchases gasoline from two different sources and they are not compatible, then the resulting mixture in the tank may not perform satisfactorily even though both batches of gasoline meet the required specifications. An example of this latter effect is when two gasolines, one containing methanol and the other not, are mixed together, the vapor pressure of the mixture may be higher than that of either of the two constituents. If both were on the limit of the vapor pressure specification, then it is possible that the mixture would be out of specification.

With the widespread use of ethanol as a finished gasoline component and a desire to keep significant volumes of ethanol from entering the distribution pipelines, there are cases of refiners producing blendstock for oxygenated blends (BOB) grades of fuel. The BOB grade fuel is such that when the predefined percentage of ethanol is added to the fuel, it will meet all the required specifications for the finished gasoline. The ethanol can then be added downstream, for example, at the distribution terminal.

3.10.1. Blending Operations

Blending can be either continuous or batch-wise, although it is always necessary to check each finished tank of gasoline prior to shipment or sale to ensure that it is satisfactory and meets the required specification limits. Batches of gasoline are normally blended by line mixing through a manifold to ensure that the blend in the tank is homogeneous and not layered, as can happen when components are added to the tank one by one. It is important to check for layering by taking samples from the top, middle, and bottom of a tank, and if they are identical in terms of density (or some other simple-to-measure characteristic) to then combine these samples for final testing of the tank. If layering is found, some method of mixing the tank contents must be used such as recirculation through a pump. Great care has to be taken in the handling of samples used for measuring vapor pressure, because it is very easy to lose light ends, the more volatile components, from them.

Continuous analyzers are frequently used for measuring octane quality so that the overall quality going into a rundown tank from a processing plant or into a blend tank can be accurately assessed by integrating the individual results. Such instruments have, as an internal standard, a gasoline or stream having an accurately known octane level close to that of the product. The standard is run through the instrument automatically at regular intervals and compared with the test result, so that the blender has confidence in the data being produced. Another approach relies on spectroscopy, [near infrared (NIR), mid infrared (MIR), nuclear magnetic resonance (NMR) and ultraviolet (UV)] [3.33] to build complex mathematical models. Alternatively, a number of different analyzers, contained in "analyzer house," measuring different parameters may be used. These instruments can allow a refiner to operate closer to the specification limit than if they had to rely on individual CFR ratings [3.34], and this can represent a very real cost saving.

The blender has to identify which specification points are critical, that is, which points are most likely to be difficult to meet in terms of cost and the available component streams. For many refiners, this will be octane quality or Reid vapor pressure, and it will be necessary to concentrate particularly on these points. Many specification limits will always be satisfactory, whatever the blending method, and these are only monitored on an occasional basis.

3.10.2. Blending Calculations

Blend calculations are rarely made by hand but rely on computational models in order to optimize the gasoline pool within the restraints imposed by the specification, the volumes required, the need to produce other products, tankage limitations, and so on. Techniques such as linear programming (LP) [3.35] are widely used, but due to the nonlinear nature of many of the gasoline properties, others have tried nonlinear programming (NLP) or to linearize the nonlinear models and use sequential linear programming (SLP) [3.36]. Linear programming plus bias [3.37] and nonlinear programming plus bias, for example the Ethyl RT-70 model [3.38], and the Stewart model [3.39], give better results. Sophisticated software packages are now available that control and optimize blends by manipulating the set points of a digital blender, using feedback from online analyzer measurements [3.40, 3.41].

The main difficulty in achieving an accurate prediction of the properties of a gasoline blend is that most of the important parameters, such as RON, MON, and vapor pressure, do not blend linearly, and the way in which they blend will vary according to the other components in the pool. A number of different ways of overcoming this problem have been described [3.42], as summarized in the next section.

3.10.3. **Octane Blending**

The most widely used procedures for octane blending are the following:

1. *The Blending Bonus Approach.* The properties of a blend can be calculated simply by making a weighted average using measured values for each of the components, as shown below for a three-component blend:

$$P_B = V_1 P_1 + V_2 P_2 + V_3 P_3$$

where

P_B is the property (such as RON, MON, and so on) of the blend
P_1, P_2, and P_3 are the properties of the components
V_1, V_2, and V_3 are the volume fractions

However, this approach will usually give a significant error, in that it will overestimate the MON and underestimate the RON. Nevertheless, if all the blends are fairly similar in composition, then the error is usually quite constant, and an equation can be used in which this under- or overestimation is corrected:

$$P_B = V_1 P_1 + V_2 P_2 + V_3 P_3 + BONUS$$

The value of the bonus, which is the expected deviation from linear blending, is determined by measuring the RON of a series of trial blends. For more universal application, it can be calculated from a knowledge of the olefin and aromatic contents of each component. One such equation for determining the value of the bonus in a specific gasoline pool is given below:

$$BONUS = 0.43(\bar{R}) - 0.004(\bar{O}^2) + 0.004(\bar{A}^2) - 0.01(\bar{R}^2) - 0.005(\bar{O}\bar{A}) - 55.18$$

where
$\bar{R}$ is the volumetric average RON
$\bar{O}$ and $\bar{A}$ are the average olefin and aromatic contents

Because this method uses compositional data on each component, it takes into account changes in blending bonus when larger than normal changes in blend composition occur.

A variation of this method [3.43] uses an excess octane number that can be positive or negative and represents the deviations from the ideal blending behavior. It is claimed to require little data, to produce simpler and more accurate correlations, and is highly suited to use in a computer-based LP-optimized blending model.

2. *The Component Blending Value Approach.* Here blending numbers that do blend linearly are used for each component, so that the blending equation becomes the following:

$$P_B = V_1 B_1 + V_2 B_2 + V_3 B_3$$

where B_1, B_2, and B_3 are the linear blending numbers for each component. These blending values are usually obtained by determining the octane quality of a carefully designed set of fuels blended from all the components available and on which the octane levels have been determined.

3. *Blending Interaction Coefficient Approach.* In this procedure, the effect of different components on blending behavior is taken into account by assuming that each blend component will interact on the other components, as shown by the following equation:

$$P_B = V_1 P_1 + V_2 P_2 + V_3 P_3 + V_1 V_2 IC_{12} + V_1 V_3 IC_{13} + V_2 V_3 IC_{23}$$

where

IC_{12} is the interaction coefficient between components 1 and 2
IC_{13} is the interaction coefficient between components 1 and 3, and so on

For refineries producing leaded aviation gasolines, the octane values at the lead level of the final blend must be used. Lead response curves (see Appendix 6), in which octane quality (RON or MON) is plotted against lead concentration, are used, although since lead levels are usually very restricted, most operators will operate at the maximum level allowed in order to minimize manufacturing costs. The response to lead is nonlinear, with the greatest octane benefit occurring with the first increment of lead and with progressively smaller benefits as the lead concentration increases.

It should also be noted that the error when using the blending bonus approach is much less than with unleaded gasolines. However, with low-lead gasolines the blending bonus approach will overestimate both the RON and the MON of the blend.

3.10.4. Reid Vapor Pressure (RVP) Blending

Deviations from linear blending mainly occur when light hydrocarbon components such as butane or iso-pentane are used and also when alcohols such as methanol or ethanol are blend components.

Two methods are used for hydrocarbon blends. One widely used procedure is to blend by molar proportion, and the other is to use determined linear blending values, as for octane number. For example, although the determined RVP of butane is 0.45 MPa, it blends as if it is actually about 0.55 MPa, depending, as always, on what other components are present.

The effect of alcohols is more difficult to take into account, although the interactive coefficient approach has been found to give good results. However, in countries such as the US and Europe, where renewable fuels legislation encourages the use of renewable ethanol, then the effect of the maximum allowable concentration of ethanol will be factored in to yield a revised target for the refinery stream blending. The ethanol will then be blended with any other additive package required to meet specifications, particularly corrosion inhibitors.

3.10.5. ASTM Distillation Blending

It is better to use the "percent evaporated at a given temperature" rather than the "temperature at which a given amount evaporates" for blending calculations. The "levels of percent evaporated at given temperatures" (e.g., % evaporated at 100°C) blend linearly, unless alcohols are present or unless the component boiling ranges are very different. With alcohols, interaction models give good prediction.

There are, however, no simple equations that enable initial boiling points, final boiling points, or the temperatures for any distillation level (e.g., temperature, °C for 50% evaporated) of blends to be predicted.

3.11. Diesel Fuel Blending

Diesel fuels are normally blended from at least two refinery streams. The operation becomes more complicated as the number of components increases because, as with gasoline, some of the most important fuel characteristics do not blend in a linear manner. In a simple hydroskimming refinery, blending two or possibly three components is a relatively straightforward procedure once the blending parameters of the individual components are known. It is a more complex operation with a conversion refinery having as many as seven or eight possible blend components that can differ significantly in key characteristics. Processing a variety of crude oil types may also contribute to the complexity.

The planner responsible for blending will be restricted by the scheduled, long-term pattern of refinery operation as regards crude types, throughput, and cut points of the various streams. However, within these limitations, the planner has scope to respond to short-term, market-driven requirements. In many countries, it is still the case that the dominating factor is the demand for high value gasoline, followed by the jet fuel/kerosene market, with diesel and distillate heating oil next in line. A refinery with a big market for jet fuel will try to minimize the amount of kerosene in the diesel blend and, if possible, use an alternative stream such as light cracked gas oil to lower the diesel cloud point.

The constraints on blend composition are the specifications to be met, the volume of product needed by the market, and the quality and quantity of available component streams. The relative values of those streams are also pertinent because the aim is to make the blend that will be most profitable for the refinery.

Refinery planners cannot formulate a diesel fuel blend in isolation. For logistic and economic reasons, other products the refinery is making also need to be taken into consideration because the same stream may have to be shared among several different products. The refiner has to find an outlet for all the product streams, and tankage capacity provides only limited possibilities for stockpiling if all of a particular stream cannot be used. Cutting back on crude throughput would be a last resort, as it results in shortages of other products and increased operating costs.

As with gasoline, many specification points are legally enforceable, and failure to comply can be very expensive for a company if it results in adverse publicity as well as heavy fines for marketing off-grade product. As test procedures used to measure fuel properties often have poor precision, this has to be taken into account to ensure that any sample picked up by a controlling body has at least a 95% probability of meeting the required specification points. For this reason, refiners usually work to limits that are slightly tighter than those required by the specifications.

Many other aspects of diesel blending are similar to those of gasoline blending, such as compatibility of diesel fuels from different sources and the need for fuels to be fungible (or interchangeable) for deals between suppliers. These are important for the same reasons as given in Section 3.10 on gasoline blending.

3.11.1. Diesel Blending Operations

As with gasoline, blending can be either continuous or in batches, but it is always necessary to check each finished tank of fuel prior to shipment to ensure that it is satisfactory and meets all the required specification limits. Blends of diesel fuel are normally prepared by simultaneous transfer of streams from the various component tanks into the finished product tank by pumping them through a line containing a mixing section. Where multiple stream blending is not possible, component streams are added one by one to the tank, and the contents are blended by recirculation around an external system that transfers fuel from the bottom of the tank to the top. Components will settle into different density layers if mixing has been inadequate, so it is important to check for layering by sampling at various levels in the tank. Pumping round is continued until top, middle, and bottom samples show the product to be homogeneous.

Some refineries are equipped with on-line analyzers; this enables continuous monitoring of a specification property by drawing samples either from the rundown line to intermediate storage or from the blending line to the finished product tank. On-line analyzers are available to measure the cloud point and cold filter plugging point of diesel fuels. The cold filter plugging point (CFPP) is described in Chapter 17. An optional capability that may be incorporated with a stream analyzer for cloud point or CFPP is a closed-loop feedback system to adjust the blend proportions or the treat rate of flow improver additive to maintain a consistent low-temperature quality.

3.11.2. Diesel Blend Calculations

Before the blender can start to formulate a blend to meet the quality criteria in the quantity needed for the market, the following information is needed:

- The identity of the most critical/difficult specification points to be met.
- The quantity of product to be made.
- The qualities and quantities of the available component streams.

For a diesel fuel most specifications include the following parameters:

- Cetane number—for ease of ignition and emissions control.
- Density—for emissions control and good fuel economy.
- Sulfur—to minimize engine wear and degradation of emissions control systems and to meet legislation.
- Flash point—for safe handling and storage; also a legal requirement.
- Volatility—for complete combustion, but also a legal requirement.
- Viscosity—to meet injection pump requirements and aid atomization.
- Cold flow characteristics—to meet seasonal specifications.

The relevance of these various fuel properties will be discussed in later chapters.

Instead of imposing a minimum or maximum limit for viscosity and density, some specifications define a tolerance band for one or both of these properties to avoid extreme values causing wide variations in engine performance and emissions.

Blend calculations to optimize the diesel formulation within the specification restraints are usually made by linear programming rather than by hand. However, as some of the important fuel parameters such as cloud point and viscosity do not blend linearly, they must be converted to linear blending numbers, or blending indices, before the program can be run.

For cloud point blending, component cloud points are used in the following formula to calculate their cloud point blending indices. From these, the index for the blend is obtained by summation of the product of volume and index for each component [3.44]. Back substitution of the blend index in the formula then gives the cloud point of the blend.

$$\text{CBI} = 7.710e^{(0.071\text{CL})} + 0.0556\text{CL}$$

where
   CBI is the cloud point blending index
   CL is the cloud point, °C

An example of a cloud point blending curve is included as Figure 3.15.

FIGURE 3.15 Cloud point blending curve.

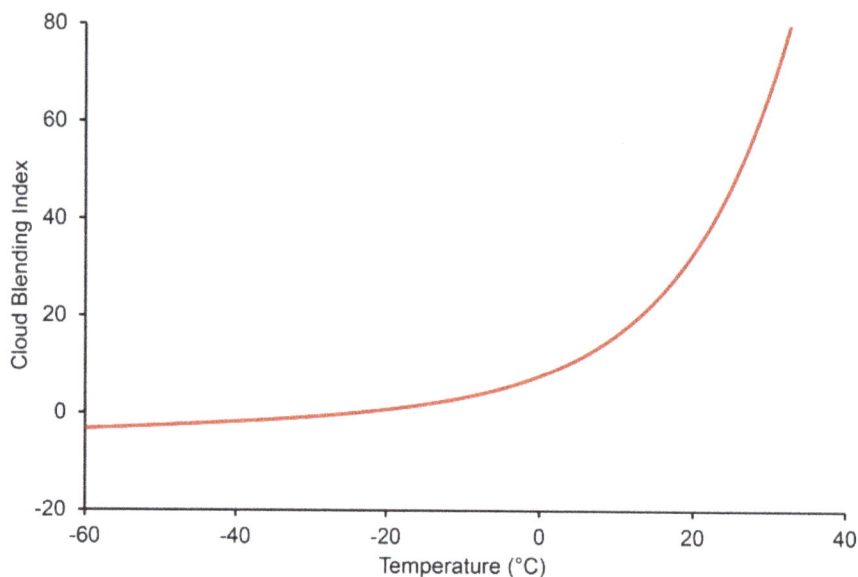

© SAE International.

A similar procedure is followed for viscosity blending, with component viscosities converted into viscosity blending numbers by means of the Refutas blending chart, which uses scales devised to express the temperature-viscosity relationship for Newtonian fluids as a straight line:

$$VBN_1 = 14.534 \log\left[\log\left(CS_1 + 0.8\right)\right] + 10.975$$

where
    VBN_1 is the viscosity blending number for component 1
    CS_1 is the viscosity, cSt, for component 1

Component VBNs are summated by weight and the resulting blend VBN converted to blend viscosity by back substitution in the formula.

If additives such as flow improvers or cetane improvers are to be used as a blend component to enable specification targets to be met, the planner will need to take account of other relationships to ascertain the responsiveness of the blend, because the result of additive treatment is also a nonlinear function.

Response to a flow improver additive, for example, is very dependent on the distillation characteristics of the blend, particularly at the heavy end, as well as the wax content [3.45]. One way of stimulating response is to incorporate a small amount of a heavier stream (see Chapter 17, Section 17.2.7). While this can often enable the CFPP target to be attained, it may have an unfavorable effect on cloud point and perhaps other quality criteria. This can be corrected by using more light component material to bring the blend back on specification, but if a catalytically cracked gas oil is used for this purpose, the cetane number may become too low.

Diesel ignition improver additives are used to increase the cetane number and do not normally involve blend reformulation to limit the improvement permitted by additive treatment, unless cetane index is also specified, as in the current European standard.

A formula for flash point blending relates the flash point to the vapor pressure of the blend components [3.46]. There is also a formula for color in which light absorbance is the basis for the blending index table [3.46].

References

3.1. Bezergianni, S., Dimitriadis, A., Kikhtyanin, O., and Kubička, D., "Refinery Co-Processing of Renewable Feeds," *Progress in Energy and Combustion Science* 68 (2018): 29-64.

3.2. van Dyk, S., Su, J., Mcmillan, J.D., and Saddler, J., "Potential Synergies of Drop-In Biofuel Production with Further Co-Processing at Oil Refineries," *Biofuels, Bioproducts and Biorefining* 13, no. 3 (2019): 760-775.

3.3. Han, X., Wang, H., Zeng, Y., and Liu, J., "Advancing the Application of Bio-Oils by Co-Processing with Petroleum Intermediates: A Review," *Energy Conversion and Management* 10 (2021): 100069.

3.4. Yáñez, É., Meerman, H., Ramírez, A., Castillo, É. et al., "Assessing Bio-Oil Co-Processing Routes as CO_2 Mitigation Strategies in Oil Refineries," *Biofuels, Bioproducts and Biorefining* 15, no. 1 (2021): 305-333.

3.5. van Paasen, C.W.C. "Changing Refining Practice to Meet Gasoline and Diesel Demand and Specification Requirement," in *Institution of Mechanical Engineers Symposium on Petroleum Based Fuels*, Paper No. C316186, London, UK, 1986.

3.6. Yui, S., "Producing Quality Synthetic Crude Oil from Canadian Oil Sands Bitumen," *Journal of the Japan Petroleum Institute* 51, no. 1 (2008): 1-13.

3.7. Hobson, G.D. (Eds), *Modern Petroleum Technology*, 5th ed. (Chichester, UK: John Wiley & Sons, 1984).

3.8. Jennings, J.M. and Pappas, G.F., Hydrocarbon conversion. U.S. Patent 2,687,986, 1954.

3.9. Sweeney, W.J., Catalytic cracking of hydrocarbon oil. U.S. Patent 2,246,959, 1941.

3.10. Brackenbury, J.M., Catalytic cracking of hydrocarbon oils. U.S. Patent 2,243,277, 1947.

3.11. Egan, C.J., Combined hydrocracking and catalytic dewaxing process. U.S. Patent 3,681,232, 1972.

3.12. Donnelly, S.P. and Green, J.R., "Catalytic Dewaxing Process Improved," *Oil and Gas Journal* 78 (1980): 77-82.

3.13. Demmel, E.J., Nace, D.M., Owen, H., and Rosinski, E.J., Combination operation to maximize fuel oil product of low pour. U.S. Patent 3,891,540, 1975.

3.14. Bolt, J.A. and Shoemaker, B.H., Copper chloride sweetening. U.S. Patent 2,332,048, 1941.

3.15. Messmore, H.E. and Mason, J.M., Method of copper sweetening. U.S. Patent 2,378,092, 1943.

3.16. Brown, R.H., Fairchild, W.P., and Kawahara, F.K., Copper chloride sweetening. U.S. Patent 2,848,373, 1958.

3.17. Keller, J.L., Inhibitor sweetening of straight-run heating oils containing added olefins with a phenylene-diamine, alkali and oxygen. U.S. Patent 2,793,171, 1957.

3.18. Sorg, L.V., Naphtha sweetening with phenylene-diamine followed by alkali. U.S. Patent 2,766,181, 1956.

3.19. Stark, D.D. and Edwards, T.O. Jr., Acid treatment of oils. U.S. Patent 2,052,852, 1936.

3.20. Gouli, S., Stournas, S., and Lois, E., "Antiknock Performance of Gasoline Substitutes and their Effects on Gasoline Properties," SAE Technical Paper 981367, 1998, doi:https://doi.org/10.4271/981367.

3.21. Gibbs, L., "How Gasoline Has Changed," SAE Technical Paper 932828, 1993, doi:https://doi.org/10.4271/932828.

3.22. Bata, R., Elrod, A., and Lewandowskia, T., "Butanol as a Blending Agent with Gasoline for I. C. Engines," SAE Technical Paper 890434, 1989, doi:https://doi.org/10.4271/890434.

3.23. Cairns, A., Stansfield, P., Fraser, N., Blaxill, H. et al., "A Study of Gasoline-Alcohol Blended Fuels in an Advanced Turbocharged DISI Engine," *SAE Int. J. Fuels Lubr.* 2, no. 1 (2009): 41-57, doi:https://doi.org/10.4271/2009-01-0138.

3.24. Williams, J., Goodfellow, C., Lance, D., Ota, A. et al., "Impact of Butanol and Other Bio-Components on the Thermal Efficiency of Prototype and Conventional Engines," SAE Technical Paper 2009-01-1908, 2009, doi:https://doi.org/10.4271/2009-01-1908.

3.25. Stansfield, P., Bisordi, A., OudeNijeweme, D., Williams, J. et al., "The Performance of a Modern Vehicle on a Variety of Alcohol-Gasoline Fuel Blends," *SAE Int. J. Fuels Lubr.* 5, no. 2 (2012): 813-822, doi:https://doi.org/10.4271/2012-01-1272.

3.26. Williams, P.R.D., "MTBE in California Drinking Water: An Analysis of Patterns and Trends," *Journal of Environmental Forensics* 2, no. 1 (2001): 75-85.

3.27. Hartley, W.R., Englande, A.J. Jr., and Harrington, D.J., "Health Risk Assessment of Groundwater Contaminated with Methyl Tertiary Butyl Ether (MTBE)," *Water Science and Technology* 39, no. 10-11 (1999): 305-310.

3.28. Cooney, C.M., "California Struggles with Presence of MTBE in Public Drinking Water Wells," *Environ. Sci. Technol.* 31, no. 6 (1997): 269.

3.29. Yang, C.-J. and Jackson, R.B., "China's Growing Methanol Economy and Its Implications for Energy and the Environment," *Energy Policy* 41 (2012): 878-884.

3.30. Abel, R.C., Coney, K., Johnson, C., Thornton, M.J. et al., "Global Ethanol-Blended-Fuel Vehicle Compatibility Study," No. NREL/TP-5400-81252, National Renewable Energy Lab. (NREL), Golden, CO, 2021.

3.31. Aguilar-Rivera, N., Michel-Cuello, C., Cervantes-Niño, J.J., Gómez-Merino, F.C. et al., "Effects of Public Policies on the Sustainability of the Biofuels Value Chain," in *Sustainable Biofuels*, Ed. Ray, R.C. (London, UK: Academic Press, 2021), 345-379.

3.32. "Regulator Program Resolution No. A/024/2020 that Modifies the Official Mexican Standard NOM-016-CRE-2016, Quality Specifications for Petroleum Products." Comisión Reguladora de Energía (CRE) [Energy Regulatory Commission (Mexico)].

3.33. Martínez, E., Huertas, S., Ménez, H., Peña, J.L. et al., "Comparison of Chemometric Techniques Applied to Near Infrared Spectra for a Gasoline Blending Control," *Journal of Near Infrared Spectroscopy* 16, no. 3 (2008): 297-303.

3.34. Chung, H., "Applications of Near-Infrared Spectroscopy in Refineries and Important Issues to Address," *Applied Spectroscopy Reviews* 42, no. 3 (2007): 251-285.

3.35. Diaz, A. and Barsamian, J.A., "Meet Changing Fuel Requirements with Online Blend Optimization," *Hydrocarbon Processing* 75, no. 2 (1996): 71-76.

3.36. Ramsey, J.R. Jr. and Truesdale, P.B., "Blend Optimization Integrated into Refinery-Wide Strategy," *Oil and Gas Journal* 88, no. 12 (1990): 40-44.

3.37. Forbes, J.F. and Marlin, T.E., "Model Accuracy for Economic Optimizing Controllers: The Bias Update Case," *Ind. Eng. Chem. Res.* 33, no. 8 (1994): 1919-1929.

3.38. Healy, W.C. Jr., "A New Approach to Blending Octanes," in *Proceedings of the 24th Meeting API Refining Division*, Vol. 39, New York, 1959.

3.39. Stewart, W.E., "Predict Octanes for Gasoline Blends," *Petroleum Refiner* 38, no. 12 (1959): 135-139.

3.40. White, J. and Kemp, H.V.D., "Optimized Gasoline Blending at Lindsey Oil Refinery," *Petroleum Review* 46 (1992): 317-319.

3.41. Espinosa, A., Sanchez, M., Osta, S., Boniface, C. et al., "On-Line NIR Analysis and Advanced Control Improve Gasoline Blending," *Oil and Gas Journal* 92, no. 42 (1994): 49-56.

3.42. Searle, G., "Gasoline Blending," College of Petroleum and Energy Studies Manual for Gasoline Technology Course RF5, 1993.

3.43. Muller, A., "New Method Produces Accurate Octane Blending Values," *Oil and Gas Journal* 90 (1992): 80.

3.44.   Auckland, M.H.T. and Charnock, D.J., "The Development of Linear Blending Indices for Petroleum Properties," *Journal of the Institute of Petroleum* 55, no. 545 (1969): 322-329.

3.45.   Zielinski, J., Rossi, F., and Stevens, A., "Wax and Flow in Diesel Fuels," SAE Technical Paper 841352, 1984, doi:https://doi.org/10.4271/841352.

3.46.   Butler, R.M., Cooke, G.M., Lukk, G.G., and Jameson, B.G., "Prediction of Flash Points of Middle Distillates," *Industrial and Engineering Chemistry* 48, no. 4 (1956): 808-812.

Further Reading

CONCAWE, "Oil Refining in the EU in 2020, with Perspectives to 2030," Report No. 1/13R, CONCAWE, 2013.

Hsu, C.S. and Robinson, P.R. (Eds), *Practical Advances in Petroleum Processing.* Vol. 1 (New York: Springer, 2006), 23-34.

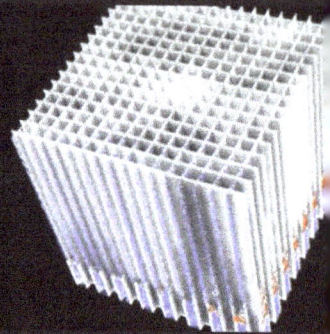

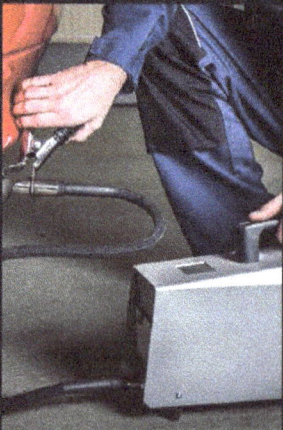

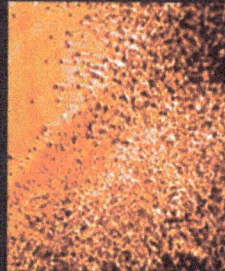

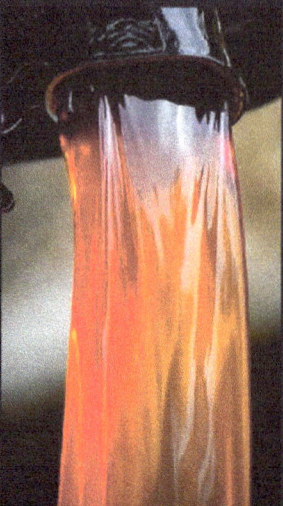

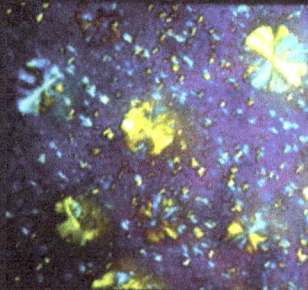

4

Manufacture of Gasoline and Diesel Fuel from Non-Crude Oil Fossil Sources

4.1. Introduction

As noted in Chapter 2, the early development of the IC engine relied on gas as a fuel. At the time, most of this gas was generated (i.e., synthetic gas—syn-gas) from coal or wood due to the lack of availability of liquid fuels. A question asked at the time was "can we obtain this gas without having to mine the coal?" In 1910, A. G. Betts patented the idea of generating this syn-gas directly underground [4.1]; in his patent, he wrote:

> *Hitherto coal has been dug and brought to the surface for utilization at the expense of much sweat and loss of life. The object of my invention is to obviate these sacrifices and also to provide for the utilization of coal seams too thin or too poor for commercial use by present methods.*

The limited potential range of an automobile relying on gas at atmospheric or near atmospheric pressure was, however, a serious disadvantage. This resulted in a good deal of research in converting gaseous and solid fuels into liquid fuels. The discovery of plentiful reserves of crude oil steered engine and refinery development over most of the last century. The location of these oil reserves was not—and still is not—necessarily geographically or politically in the right location for their use. This resulted in the development of technology to utilize other fossil reserves that can compete with crude oil, when and where necessary. This chapter gives some insight into these technologies and the possibilities they hold.

4.2. Coal to Liquids

Coal is by far the most abundant of known fossil fuels. Coal had long been a source of heat, and its mining required more and more power. In the 17th century, this was literally horse power. This led to the development of the first commercially successful atmospheric-steam engine, developed by Thomas Newcomen approximately 1712 [4.2].

However, it was a hundred years before this technology was implemented to produce a self-propelled vehicle in the form of a traction engine. It was not until the end of the 19th century when the technology produced anything that would be recognized as an automobile. By this time, Otto and Diesel engines powered by liquid fuels were vying as the preferred power plant. Although the energy density of coal was greater than that of the competing liquid fuels, the inefficiencies of the external combustion (steam) engine and the need to carry not just the fuel but also the water to make the steam consigned the steam engine to an also-ran.

Interestingly, when Rudolf Diesel drew up his patent [4.3] in 1892, he chose to exemplify his design by describing an engine that would burn pulverized coal. Although diesel never succeeded in producing a working model of this design, other German engineers did pursue the idea. By the end of World War II, 19 pulverized coal-powered engines had been developed. These spanned the power range from about 7 to 450 kW, but these were all for stationary application [4.4].

At the start of the 20th century, the main source of fuel for the automobile was gasoline, and by then it was already predominant made from petroleum. In 1927, A. C. Fieldner, chief chemist, Bureau of Mines, Pittsburg, US, wrote [4.5],

> *The possibility of future shortages of petroleum fuel suitable for automotive engines, however, and of the production of substitutes to avoid such a contingency is receiving considerable attention in America and Europe.*

Around this time, Germany was also becoming more dependent on gasoline and diesel fuel and becoming concerned about its availability. Germany had a plentiful supply of coal and was developing ways of converting this into liquid fuels. The French scientist Pierre Eugène Marcellin Berthelot had succeeded in hydrogenating coal for the first time back in 1869 [4.6]. He had combined finely powdered coal with a saturated solution of hydrogen iodide, heating the reactants to 275°C under atmospheric pressure. Because of the high cost of the hydrogen iodide solution, this had no potential industrial value, and the process received little attention until the first decades of the twentieth century when chemists began to extensively study the hydrogenation process. Two factors contributed to the eventual success of the hydrogenation process: the introduction of metallic catalysts and the use of high pressure.

Friedrich Bergius's invention of high-pressure coal hydrogenation occurred during his experiments on the water gas reaction. Bergius believed that at pressures in excess of 20 MPa, he could run the reaction at temperatures of 300–600°C, rather than 900°C. This would shift the equilibrium point to favor the production of carbon dioxide and hydrogen, eliminating entirely the formation of carbon monoxide, and thus the problem of separating it from hydrogen, the hydrogen could then be reacted directly with the coal, or other carbonaceous material, to form liquid product [4.7].

$$nC + (n+1)H_2 \rightarrow C_nH_{2n+2}$$

The step shown in the above equation is now commonly referred to as direct coal liquefaction (DCL).

Prior to the high-pressure process devised by Bergius, syn-gas would normally contain CO and H_2, and as was documented by Paul Sabatier and Jean-Baptiste Sanderens [4.8] in 1902, in the presence of a nickel catalyst, these two gases could be reacted to form methane.

From 1915 to 1921, Professor Franz Fischer, working at the Kaiser Wilhelm-Institut für Kohlenforschung Mülheim-Rühr in Germany, published six volumes of Gesammelte Abhandlungen zur Kenntnis der Kohle (collected papers to the attention of coal) and subsequently Die Umwandlung der Kohle in Öle [4.9]. In 1923, Franz Fischer and Hans Tropsch reported the use of alkalized Fe catalysts to produce liquid hydrocarbons rich in oxygenated compounds; at the time they termed this the synthol process [4.10]. They later patented the idea for producing paraffins other than methane [4.11]. This process has subsequently become known as the Fischer–Tropsch (F-T) process. At this point, a cobalt/thorium/manganese catalyst [4.12] on a kieselguhr (silicious diatomatous earth) carrier had become the standard catalyst [4.13] for this atmospheric pressure process. In 1936, Fischer and Helmut Pichler [4.14] proposed a medium pressure process with pressures "above 4 atmospheres" in claim one and from "5 to 20 atmospheres" in claims two and three.

There were thus two easily distinguished routes to the conversion of coal to liquid hydrocarbons, namely DCL and indirect coal liquefaction (ICL), where the coal is first gasified and then the gas is converted to liquid; these will be discussed in Sections 4.2.1 and 4.2.2.

In addition to a resurgence of interest in converting coal to hydrocarbons that can be separated into gasoline and diesel fuels there is also an increase of interest in coal rich, oil poor countries such as Australia, China, India, and Indonesia in converting coal into chemicals; some of which can then be converted to gasoline or diesel substitutes or blending components [4.15, 4.16].

4.2.1. Direct Coal Liquefaction (DCL)

Using the Bergius high-pressure technique to bias the water gas shift reaction in favor of hydrogen and CO_2 production, coal can be used to produce hydrogen, which can then be used for coal liquefaction. Initially the CO_2 was considered as a waste gas but is now considered as a pollutant and much work is being conducted into what is termed carbon capture and storage (CCS) [4.17–4.19]. Even with a ready supply of hydrogen, simply combining this gaseous hydrogen with solid coal, even if pulverized, does not work very well, combining the coal with a solvent makes the process a lot more effective. The coal solution, slurry, or coal tar, is then reacted with hydrogen in the presence of a catalyst; it is effectively hydrocracked to form a product similar to partially refined crude oil. This product can then be processed in a similar manner to a product from crude oil; see Chapter 3, Section 3.5.5.

The coal liquefaction process is very complex, but it is generally believed to include the following fundamental steps [4.20].

1. Coal dissolution, i.e., disintegration or dispersion of the larger molecules.
2. Reduction of the molecular size, which will allow access to catalyst pores.
3. Removal of the heteroatoms, such as N, O, and S.
4. Hydrogenation of the aromatic rings and ring opening of the saturated rings.
5. Reduction of molecular size to 5–20 carbon atoms.
6. Increase in the atomic H/C ratio to about 2:1.

The optimum temperature and heating rate are coal type dependent.

This liquefaction can be accomplished via a single reactor, a single-stage process, or by a series of reactors and a two-stage process, although more stages have been proposed [4.21, 4.22]. These different approaches are outlined below.

4.2.1.1. Single-Stage DCL

A single-stage DCL process produces distillates via a single primary reactor. In the US, pilot plants were built to develop these processes during the 1960s–1980s. These included H-coal, solvent-refined coal (SRC), and

Exxon donor solvent (EDS). The H-coal process was developed by Hydrogen Research Inc. [4.23, 4.24] (that later became HRI Inc.) and was based on a novel catalytic reactor in which the catalyst was ebullated in the liquid phase; this is illustrated in Figure 4.1.

FIGURE 4.1   Schematic of a single-stage DCL plant.

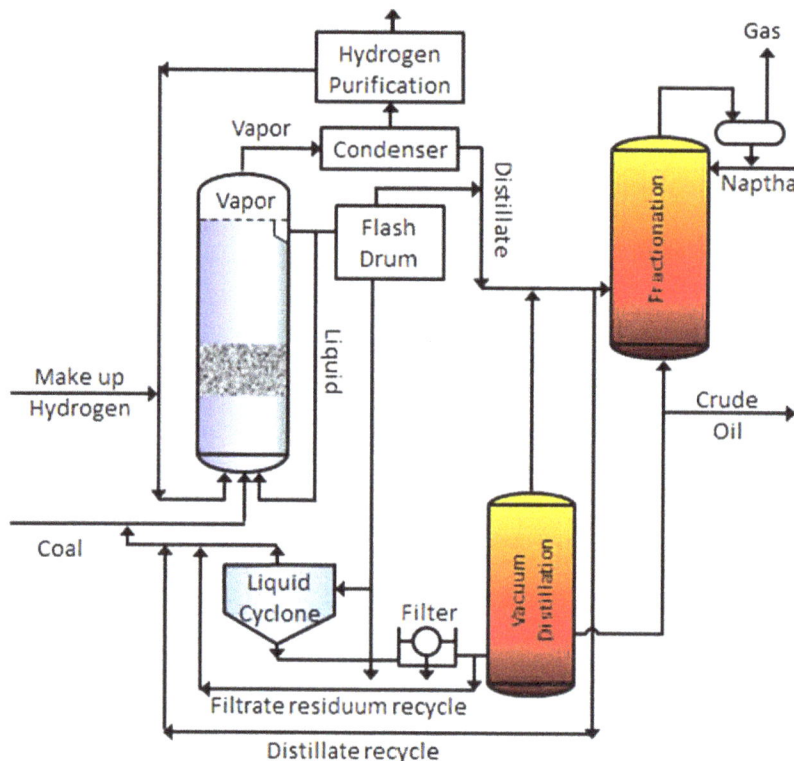

Both the SRC [4.25, 4.26] and EDS [4.27] processes used iron-containing ores in the coal as the catalyst and the solvent as the hydrogen donor. The EDS process used a catalytic stage to hydrotreat part of the recycled solvent in a second reactor. However, the primary coal liquefaction unit did not directly use a catalyst.

Funding for these projects was reduced in the early 1980s and ended by the end of the twentieth century. Work continued on the development of DCL in China. In 2002, China Shenhua Coal Liquefaction Co. Ltd. began site preparation for a DCL plant (SH-1) in Inner Mongolia. The technology of the SH-1 plant was based on the HTI process modified using technology developed in Japan (NEDOL) and Germany (Kohleoel), and their own innovations.

4.2.1.2. Two-Stage DCL

A two-stage DCL process produces products using two distinct reactor stages in series. These two stages operate under two different sets of conditions. The primary function of the first stage is the dissolution of the coal. This stage operates with a low activity catalyst, usually iron-based, or no catalyst at all. The second reactor stage is used to hydrotreat the output of the first stage that is of similar average composition to the original coal. The second stage requires a catalyst of higher activity. There may be some de-ashing between the two stages. A schematic of a two-stage liquefaction process is shown in Figure 4.2 [4.28].

FIGURE 4.2   Schematic of a two-stage DCL plant.

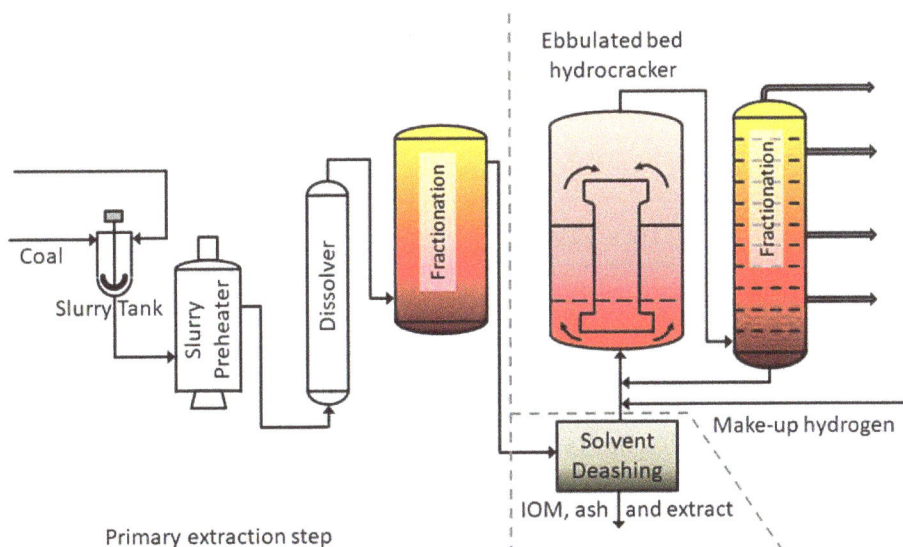

© SAE International.

From 1978 to 1983, three DCL processes were developed by the Japanese companies Nippon Kokan, Sumitomo Metals Industries and Mitsubishi Heavy Industries. This work was managed by the New Energy and Industrial Technology Development Organization (NEDO). In 1983, the features of all these processes were combined to produce the NEDOL process. This process used a fine powder iron-based liquefaction catalyst with a Ni-Mo-Al$_2$O$_3$-based hydrogenation catalyst. The liquefaction reactors would typically operate at 15 to 20 MPa at a temperature between 430 and 465°C. The re-hydrogenation reactors would typically operate at a lower pressure of 10 to 15 MPa and a temperature between 320 and 400°C.

4.2.2. Indirect Coal Liquefaction (ICL)

As noted at the start of this chapter, in 1910, Anson G. Betts proposed generating syn-gas directly underground [4.1]; this is known as underground coal gasification (UCG). In 1912, Sir William Ramsay, who discovered noble gases and received the Nobel Prize in Chemistry in 1904, began work on the idea of UCG near Tursdale colliery, just south of the city of Durham in the UK. Test shafts were sunk that apparently followed the scheme proposed in the Betts patent as shown in Figure 4.3 [4.29]. Unfortunately, the work was never completed due to the outbreak of World War I in 1914 and Sir William's death in 1916.

FIGURE 4.3 Proposed UCG layout according to A. G. Betts.

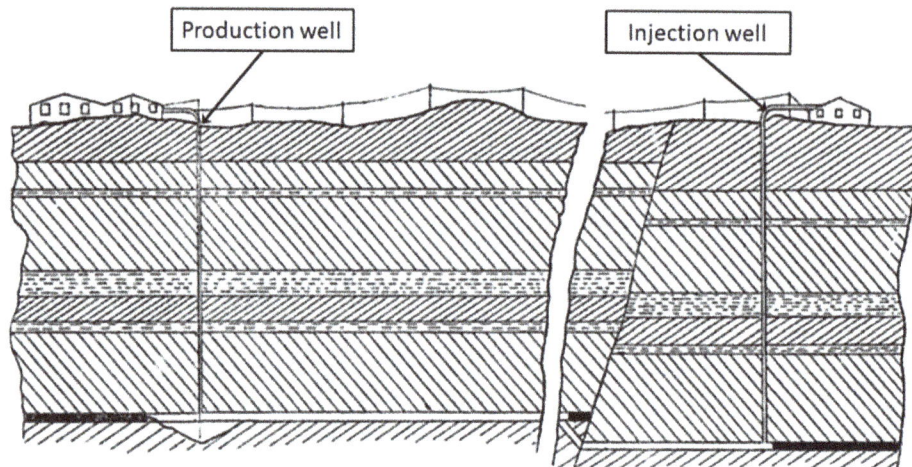

In 1930, further trials were undertaken in Ukraine and Russia, then part of the former USSR. From the late 1940s into the 1950s work was also performed in the US and the UK [4.29]. Of the commercial projects that were begun in the former USSR countries, the only one that still operates is in Angren, Uzbekistan, which utilizes the syn-gas produced to generate electricity. In China, over the last quarter of a century, there have been a number of commercial UCG projects for chemicals production and with other projects under development. There are also developmental projects in countries including in Australia, Canada, India, New Zealand, South Africa and the US [4.30]. At present, however, the vast majority of coal for gasification and conversion to liquids is still mined and then converted.

By gasification of coal followed by use of the F-T process, it was possible to convert coal-to-liquid (CTL) fuels and other chemical feedstocks, i.e., indirect coal liquefaction (ICL).

During World War II, Germany operated five low-pressure plants, two medium-pressure plants and two plants that included both low- and medium-pressure units [4.13]. About 46% of the output of these plants was gasoline and 23% was diesel fuel. However, the total output of these plants was only about 9% of the total domestic production [4.13]. The majority of the production, just over 60%, was from hydrogenation of coal or tar [4.31]. At the end of the war, there were also three small plants in operation in Japan [4.32], with experimental facilities in the US, Great Britain, Spain, Australia, and Canada [4.33]. Just prior to the war, Japan had built a CTL plant in Jinzhou, China; this plant was moth-balled after the war but was brought back into operation between 1951 and 1957 [4.34]. A couple of years later, China found large oil reserves in northeastern China and the Jinzhou plant began to lose their significance and suspended their operation in 1967 [4.35].

After the war, there was interest and some development work in the US. By 1949, Standard Oil Development Company and Pittsburgh Consolidation Coal Company had erected a pilot plant, and Koppers Company and Gulf Oil Corporation were cooperating on research projects [4.36]. Hydrocarbon Research Inc. developed their version of the F-T process that became known as the Hydrocol process [4.37], which used natural gas as a feed and is discussed further in Section 4.4. This technology was originally trialed in New York, but a commercial plant was constructed in Brownsville, Texas. However, the plant only ran from 1951 to 1957. The major development work was done in South Africa with the development of a new company named Sasol (Suid-Afrikaanse Steenkool-, Olie- en Gasmaatskappy, *South African Coal, Oil and Gas Company*) [4.38], which has had commercial plants running since 1955.

A typical early CTL plant is shown schematically in Figure 4.4.

FIGURE 4.4   Schematic of a typical early CTL plant.

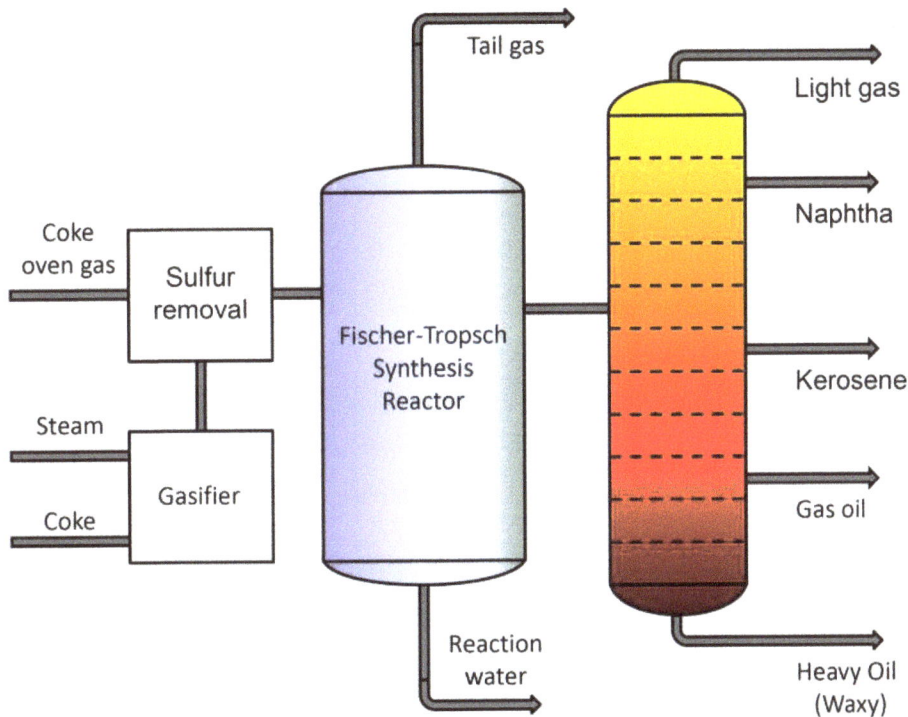

Due to shortages of oil supply during the 1970s, there was renewed interest in the DCL route to producing liquid fuels from coal.

4.2.2.1. Syn-Gas Production

The chemistry of coal gasification is complex but is usually described by eight reaction equations. The rate of these different reactions, and hence which reactions dominate, will depend on the prevailing conditions, temperature, pressure, and the feedstock used. The overall process can be considered as a partial oxidation process, with only a minor portion of the carbon being completely oxidized to carbon dioxide (CO_2). The major reactions within the gasification process are those involving carbon, CO, CO_2, H_2, water (steam) and methane, as follows:

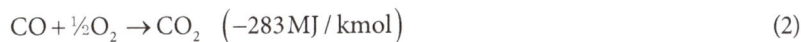

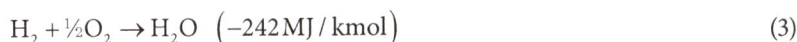

$$C + \tfrac{1}{2}O_2 \rightarrow CO \ \left(-111\,MJ\,/\,kmol\right) \tag{1}$$

$$CO + \tfrac{1}{2}O_2 \rightarrow CO_2 \ \left(-283\,MJ\,/\,kmol\right) \tag{2}$$

$$H_2 + \tfrac{1}{2}O_2 \rightarrow H_2O \ \left(-242\,MJ\,/\,kmol\right) \tag{3}$$

$$C + H_2O \leftrightarrow CO + H_2 \quad \left(+131\,MJ\,/\,kmol\right) \quad \text{Water–gas reaction} \tag{4}$$

$$C + CO_2 \leftrightarrow 2CO \quad \left(+172\,MJ\,/\,kmol\right) \quad \text{Boudouard reaction} \tag{5}$$

$$C + 2H_2 \leftrightarrow CH_4 \quad \left(-75\,MJ\,/\,kmol\right) \quad \text{Methanation reaction} \tag{6}$$

$$CO + H_2O \leftrightarrow CO_2 + H_2 \quad \left(-41\,MJ\,/\,kmol\right) \quad \text{Water–gas-shift reaction} \tag{7}$$

$$CH_4 + H_2O \leftrightarrow CO_2 + 3H_2 \quad \left(+206\,MJ\,/\,kmol\right) \quad \text{Steam-methane-reforming reaction} \tag{8}$$

The first three reactions are combustion reactions essentially carried out to completion under normal gasification operating conditions. These oxidation reactions provide most of the energy required to drive the endothermic gasification reactions. The three heterogeneous reactions (Reactions 4 through 6) can be reduced to two homogeneous gas-phase reactions of water-gas-shift and steam-methane-reforming (Reactions 7 and 8), which collectively play a key role in determining the final equilibrium syn-gas composition. Figure 4.5 shows the coal gasifier at the Sasol Secunda facility in South Africa.

FIGURE 4.5   Coal gasifier at the Sasol Secunda facility.

Source: Sasol Limited.

As noted previously, the temperature of the coal gasifier has an effect on the syn-gas composition. If the temperature of the gasifier is low, then this will favor the production of CH_4 but will provide an H_2/CO ratio that is preferable as a feed to the F-T reactor. With low-temperature gasifiers, the CH_4 produced will have to be reprocessed to produce a feed for the F-T reactors, or it can be burned to produce a source of heat. High-temperature gasifiers produce a syn-gas with a lower CH_4 content, but the H_2/CO ratio tends to be too low. This can be increased by feeding the syn-gas through a water-gas-shift reactor.

Under the sub-stoichiometric reducing conditions of gasification, most of the fuel's sulfur is converted to hydrogen sulfide (H_2S) and (to a lesser degree) carbonyl sulfide (COS). If air is also used as a feed gas, then some of the nitrogen (there may also be some coal bound nitrogen) in the feed may be converted to ammonia (NH_3) and a small amount of hydrogen cyanide (HCN). Any chlorine in the feed is primarily converted to hydrogen chloride (HCl). These compounds along with other trace elements associated with both organic and inorganic components of the feedstock, such as mercury and other heavy metals that do not accumulate with the ash, must be removed from the syn-gas prior to the F-T reactor stages; otherwise, they will poison the F-T catalyst.

Various physical solvents can be used to scrub the raw syn-gas of NH_3, H_2S, tars, and CO_2 to the required levels. The Rectisol® process [4.39] uses refrigerated-methanol, the Selexol™ [4.40] uses di-methyl-ether of poly-ethylene-glycol (DEPG), and the Purisol® process [4.41] uses n-methyl-2-pyrrolidone. These techniques are best suited to high-pressure systems, and the Fluor Solvent™ process [4.42] using propylene carbonate is suitable for medium pressure. Chemical absorption techniques are more suited to low-pressure systems; these include amine scrubber processes, which commonly use mono-ethanol-amine (MEA), di-ethanol-amine (DEA) or methyl-di-ethanol-amine (MDEA). Fixed-bed reactors containing ZnO are also used for removing sulfur (known as sulfur polishing) [4.43].

4.2.2.2. F-T Synthesis

The nature of the liquid hydrocarbons produced, often referred to as synthetic crude oil or syn-crude, is dependent upon the F-T process and in turn determines the amount and type of refining required to produce the desired end products. Product distributions are influenced by temperature, feed gas composition (i.e., the H_2/CO ratio), pressure, catalyst type, and catalyst composition.

The low-temperature Fischer-Tropsch (LTFT) process, operating at 180–250°C, favors the production of heavier waxy hydrocarbons, and the high-temperature Fischer–Tropsch (HTFT) process, operating at 300–350°C, favors the production of lighter hydrocarbons and higher yields of low molecular weight olefin. Temperatures are usually kept below 400°C to minimize CH_4 production. Without further refining, the LTFT process thus produces relatively low yields of both gasoline and diesel-type products, whereas the HTFT process can produce reasonably high yields of gasoline-type product. Table 4.1 shows a comparison of the yields from two types of processes. The oxygenates produced are alcohols, aldehydes, acids, and ketones, which generally come off with water.

TABLE 4.1 Comparison of LTFT and HTFT percentage outputs.

Product	LTFT	HTFT
CH_4	4	7
C_2 to C_4 paraffins	4	6
C_2 to C_4 olefins	4	24
Gasoline	18	36
Middle distillates	19	12
Heavy oils and waxes	48	9
Oxygenates	3	6

LTFT reactors initially used a fixed-bed reactor design, which was superseded by the multi-tubular fixed-bed reactor design. This design forms the basis of the Sasol arge process and the Shell middle distillate synthesis (SMDS) process, described in Section 4.4. The multi-tubular fixed-bed design is shown schematically in Figure 4.6, along with other designs that are discussed briefly following.

FIGURE 4.6 Schematic of different F-T reactors.

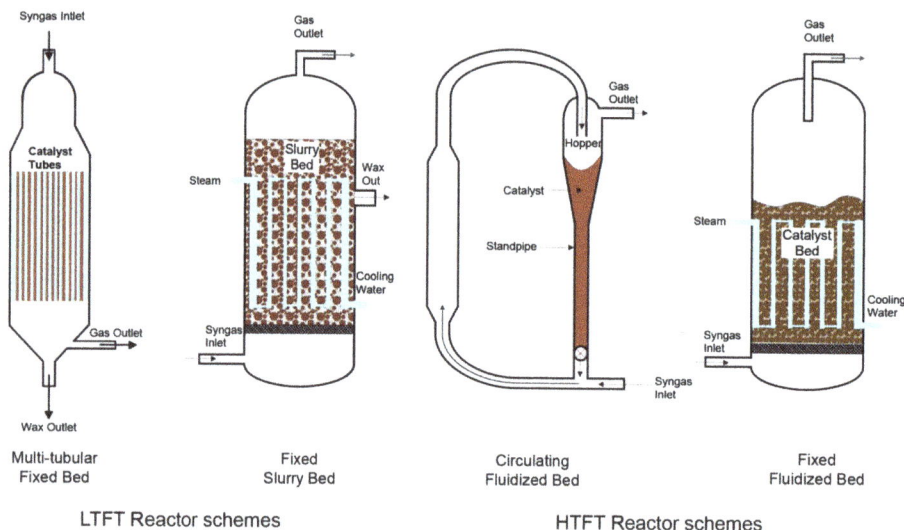

LTFT Reactor schemes   HTFT Reactor schemes

© SAE International.

As can be seen from Figure 4.6, in the multi-tubular design, the feed gases enter at the top of the reactor and pass down through the catalyst, which is packed into the tubes that are surrounded by water to carry away the heat produced. The products are collected from the bottom of the reactor. The advantage of these reactors is that they are easy to operate and can be operated over a wide range of temperatures. It is clearly easy to separate the products from the catalyst as they flow down under gravity. The disadvantage is economic: they are expensive to construct, and replacing the catalyst is labor intensive. Due to the compromise that has to be made regarding catalyst particle size, the pressure drop across the catalyst tubes is high, which has a knock-on effect to the cost of pumping the gases through the reactor. These disadvantages are largely overcome by the use of a slurry bed reactor. Here the catalyst is suspended in a fluid and the reactant gases are bubbled through the catalyst slurry; hence they are often referred to as bubble column slurry reactors.

For the HTFT process, circulating fluidized bed reactors were developed. These are known as synthol reactors. An advantage of this type of reactor is that the catalyst can be topped up or replaced while the reactor is in operation, which reduces costly down-time. The down side is that finer catalyst particles must be used and an effective system must be employed to ensure that catalyst fines do not get passed with the products, cyclones, and oil scrubbers used. There are also regions of high fluid velocities where the catalyst particles can cause abrasive wear and erosion. To overcome this later point and to allow larger reactors to be made, the fixed fluidized bed reactor has been developed. Figure 4.7 shows the advanced synthol reactor at the Sasol Secunda facility.

FIGURE 4.7 Advanced synthol reactor at the Sasol Secunda facility.

Source: Sasol Limited.

The basic chemistry of the F-T synthesis can be considered as a polymerization of methylene units [4.44]. The initial step is thought to be CO adsorption on the catalyst surface, CO dissociation, followed by hydrogenation to yield surface methyl species, as summarized in the following equation:

$$CO + 2H_2 \rightarrow -CH_{2-} + H_2O$$

Chain growth from the addition of methylene into the metal-alkyl bond, chain termination, and finally desorption from the catalyst surface—the equations for the production of the different species can be summarized as shown:

1. $CO + 3H_2 \rightarrow CH_4 + H_2O$ Methanation reaction
2. $nCO + (2n+1)H_2 \rightarrow C_nH_{2n+2} + nH_2O$ Paraffins synthesis
3. $nCO + 2nH_2 \rightarrow C_nH_{2n} + nH_2O$ Olefins synthesis
4. $nCO + 2nH_2 \rightarrow C_nH_{2n+1}OH + (n-1)H_2O$ Alcohols synthesis

The composition of the synthetic crude oil is thus, understandably, different from petroleum crude oil. The syn-crude is predominantly paraffinic with quantities of olefins and lower carbon number alcohols but contains no sulfur or nitrogen compounds. Petroleum crude on the other hand does not contain olefins and has significant amounts of sulfur and nitrogen compounds. The alcohols in petroleum crude, although found in very low concentrations, tend to be higher carbon number than those in syn-crude. The other major difference, which is apparent from the preceding equations, is that the F-T process produces water as well as the syn-crude; the water content of syn-crude and petroleum crude are thus also significantly different. These differences dictate that the bias of downstream processes will be somewhat different from those of a petroleum refinery.

Over the years, much of the focus has been on catalyst development, specifically, to improve catalyst lifetimes, activity, and selectivity. Generally, cobalt-based catalysts are used for LTFT process while iron-based catalysts are used for the HTFT process. The catalysts primarily lose activity as a result of poisoning, thermal degradation, and fouling.

4.2.2.3. Product Upgrading

As with crude oil, the primary processing unit for the output of the F-T reactors is atmospheric distillation (described in Chapter 3). A vacuum distillation unit will probably also be included to avoid temperatures above approximately 300°C. This is because the output of the F-T reactor will contain olefins and oxygenates that are susceptible to thermal decomposition above this temperature. Waxy residue from the reactor can also be fed through the vacuum distillation unit.

4.2.2.3.1. Hydrocracking. Hydrocracking is used to reduce the high carbon number F-T waxes to lower carbon number products that can be blended into the fuel pool. Unlike most petroleum feeds to hydrocrackers, F-T feeds are free of sulfur, which allows the use of noble metal catalysts. With F-T feeds, lower temperatures and pressures can usually be used when compared with petroleum feeds. This process can be used to produce highly isomerized products with good cold flow characteristics and high cetane numbers [4.45]. The product from the hydrocracker can make up almost two-thirds of the diesel pool [4.38].

4.2.2.3.2. Oligomerization. The F-T synthesis process produces low carbon number olefins as shown in Table 4.1. While these compounds have high octane numbers and are therefore desirable for inclusion in gasoline, they have high vapor pressures, which limit the quantity that can be blended into the gasoline. Oligomerization allows lower carbon number olefins to be converted into higher carbon number liquid compounds.

The process is often carried out over a solid phosphoric acid catalyst [4.46], but other catalysts such as acidic resin [4.47], zeolite [4.48], and organometallic catalysts [4.49] can be used.

4.2.2.3.3. Catalytic Reforming. As in petroleum refining, catalytic reforming can be used to increase the octane number of the gasoline blending pool. However, unlike the situation in petroleum refineries (see Section 3.6) that favors the use of Pt/Al_2O_3-based catalysts, the low naphthenes and aromatic content of syn-crude make a Pt/L-zeolite catalyst more attractive [4.50]. The absence of sulfur in the syn-gas eliminates the issues around catalyst poisoning.

4.2.2.3.4. Alcohol Dehydration. Alcohol dehydration can also be useful for upgrading syn-crude. The alcohols that are produced by the F-T process can be dehydrated to produce olefins [4.51], which are processed along with the rest of the syn-crude, or they can be partially dehydrated to produce ethers as high cetane diesel pool additive.

4.3. Gaseous Fuels

At the end of 2020, the ratio of proven gas reserves to annual production was slightly lower than for oil (including tar-sands, shale oil, etc.), at 48.8 years for gas as against 53.5 years for oil, compared with 139 years for coal [4.52]. The largest known reserves of natural gas are in the Russian Federation, Iran, Saudi Arabia, Turkmenistan, China, Qatar and the US [4.52]. Large quantities of additional gas are available in less accessible resources such as tight sands, coal seams, and geopressurized brines. The more accessible sources of gas can now be included with the liquid petroleum gas (LPG) that has been used as vehicle fuel for about 100 years [4.53]. These fuels, whether naturally sourced or petroleum derived, are predominantly methane with other gaseous components as discussed later.

These fuels are abundant and relatively inexpensive in many parts of the world, and when used as engine fuel they produce far lower emission levels than a diesel engine without after-treatment. These fuels have played a limited but important role as road transport fuels in countries including Russia, Argentina, Italy, the Netherlands, Canada, New Zealand, and the US. Initially, the major motivation for using these fuels was economic [4.54]: the low cost of natural gas (NG) and LPG compared with gasoline or diesel fuel made their use attractive in certain applications such as taxicabs, where the fuel savings were sufficient to offset the higher cost of on-board storage and compression/dispensing systems. In recent years, attention has focused increasingly on the environmental benefits of gaseous fuels [4.55, 4.56]. Recent advances in the technology for gaseous-fueled vehicles and engines, new technologies and international standardization for compressed natural gas (CNG) storage cylinders, and the production of new, factory-manufactured gaseous-fueled vehicles in a number of countries have all combined to boost the visibility and market potential of these fuels. However, the widespread fitment of diesel particulate filters (DPF) to diesel engines has greatly reduced the particulate emissions from diesel engines and hence eliminated one of the big environmental benefits of using gaseous fuels [4.57, 4.58]. These fuels cannot be used as a direct replacement for gasoline or diesel fuel, changes to engine technology to take advantage of these fuels is discussed in Chapters 8 and 16.

4.3.1. Gaseous Fuel Supply

Natural gas has been used for thousands of years. People have constructed religious sites around lighted natural gas seeping through porous rock fissures. The Chinese drilled, using bamboo pipes, for gas deposits in 211 BC [4.59]. In 1825, in Fredonia, NY in the US, the first commercial gas well was hand-dug in to shale by William Hart [4.60, 4.61]. Oil shale fuel is discussed further in Section 4.7. Since then, there has been an enormous expansion in the demand for and recovery of gas supplies.

4.3.1.1. Natural Gas

Today, most major urban centers and many minor ones in industrial countries are served by a large network of high-pressure natural gas pipelines. These are connected through "city gate" valves to urban gas distribution networks operating at moderate to low pressures, which transport gas directly to the point of use, including automotive gas filling stations. To be moved through these pipelines the gas must be processed to pipeline quality standards. Alternatively, it can be compressed and transported in large banks of cylinders.

To be used in vehicles, NG must be compressed (CNG) or liquefied for on-board storage. This liquefied gas is then referred to as liquefied natural gas (LNG), as opposed to liquid petroleum gas (LPG), which is discussed later.

Natural gas, as recovered, contains varying amounts of non-methane hydrocarbons, water vapor, sulfur compounds, predominantly H_2S, but may include mercaptans (RSH, where R is a hydrocarbon chain), carbonyl sulfide (COS), and carbon disulfide (CS_2), which is a neurotoxin in high concentrations [4.62]. The NG will usually also contain CO_2, N_2, He, Ar, and other trace gases. Trace amounts of mercury may also be present, which would be detrimental due its corrosive properties towards things like aluminum alloys, and of course due to its toxicity. In most cases, it is necessary to upgrade the gas to meet pipeline specifications, this is accomplished in a gas processing plant before it is delivered to the distribution network. Acid gases such as H_2S and CO_2 can be removed by means of chemical solvents, physical solvents, membranes, or cryogenic fractionation [4.63]. Water can be removed through condensation, absorption, or adsorption. Where high levels of He are found it is extracted as a valuable by-product. Excess amounts of other inert gases such as Ar and N_2 are also removed in processing. The acid gases are removed to prevent corrosion to the pipeline network.

Pipeline-quality natural gas is a mixture of several different gases. The primary constituent is methane, which typically makes up 80 to 99% of the total. The remainder is primarily ethane and some remaining inert gases such as N_2 and CO_2, with smaller amounts of propane, butanes, and higher hydrocarbons. The mix of minor constituents varies considerably from place to place and from time to time, depending on the source and processing of the gas. In order to ensure consistent combustion behavior, major natural gas pipelines generally impose specifications on the composition of the gas they will accept for transport. These specifications typically limit the percentage of propane, butane, and higher hydrocarbons, the volumetric heating value, and the Wobbe Index (see Section 4.3.2.2).

Although pipeline gas generally exhibits a limited range of composition and properties, natural gas found in distribution systems may exhibit greater variability. In some cases, distribution systems in gas-producing areas receive gas directly from the well, with minimal processing. The resulting gas may be rich in non-methane hydrocarbons, inert gases, or both. Because the gas pipelines also deliver gas for domestic use and once processed the NG is practically odor free, the addition of an odorant is required by most regulations to make the presence of the gas more detectable to the consumer. This odorant is usually trace amounts of a mercaptan (thiol). The standard requirement is that a user will be able to smell the gas when the gas concentration reaches 1% in air.

4.3.1.2. LPG

LPG from refineries includes light hydrocarbons originally dissolved in the crude oil and separated during the distillation process, as well as those produced in the process of "cracking" heavy hydrocarbons to lighter products. Refinery LPG often contains significant quantities of olefinic compounds (propenes and butenes) produced in the cracking process. This is shown in Chapter 3.

LPG comes from two sources, the first is from the production well, the second is from the refinery. The crude oil and gas from a production well will naturally include some of the three- and four-carbon hydrocarbons mentioned above. Many of these lighter hydrocarbons will have to be removed to "stabilize" the crude oil, reduce its vapor pressure, before storage or transportation. Refining the crude oil will also produce LPG as shown in Chapter 3. LPG can also contain the other contaminants mentioned for NG, and these can be removed as they are from NG.

The composition of commercial LPG varies greatly from one country to another. In the US, automotive LPG is generally more than 80% propane, with small amounts of ethane and butanes, and up to 10% propene. In the US automotive LPG is covered by the ASTM D1835 [4.64] specification. In Europe automotive LPG is covered by the EN 589 standard [4.65]. Countries having relatively cold climates tend to use a high percentage of propane and propene in order to provide adequate vapor pressure in winter, while warmer countries such as Italy use mostly butane and butenes. LPG composition may also vary between summer and winter, with a higher percentage of propane and propenes in the winter months.

4.3.2. Gaseous Fuel Composition and Properties

4.3.2.1. Gaseous Fuel Components

The major sources of commercial CNG, LNG and LPG are natural gas processing and petroleum refining. As found in the earth, NG often contains excess propane and butanes that must be removed to prevent their condensing in high-pressure pipelines and to control variation in gas properties.

The properties of the main hydrocarbon constituents of NG and LPG are summarized in Table 4.2.

TABLE 4.2 Properties of the main hydrocarbon fuel gases.

	Methane	Ethane	Propane	Propene	n-Butane	iso-Butane	Butenes
Energy content (LHV) (MJ/kg)	50.01	47.48	46.35	45.78	45.74	45.59	45.32
Liquid density (kg/L)	0.466	0.572	0.501	0.519	0.601	0.549	0.607
Liquid energy density (MJ/L)	23.30	27.16	23.22	23.76	27.49	25.03	27.51
Gas energy density (MJ/m³)	32.6	58.4	84.4	79.4	111.4	110.4	113.0
Gas specific gravity (at 25°C)	0.55	1.05	1.55	1.47	2.07	2.06	1.93
Boiling point (°C)	–164	–89	–42	–47	–0.5	–12	–6.3 to 3.7
Research octane no.	>127	—	109	—	—	—	—
Motor octane no.	122	101	96	84	89	97	77
Wobbe index (MJ/m³)	50.66	65.11	74.54	71.97	85.46	84.71	81.27

© SAE International.

4.3.2.2. Wobbe Index and Fuel Metering

The Wobbe index, also referred to as the Wobbe number, is an important parameter for gaseous fuels. The Wobbe index of a gaseous fuel is determined by its composition. The value of the Wobbe index, W, is calculated as

$$W = H / \sqrt{\rho}$$

where

H is the volumetric heating value of the gas

ρ is the specific gravity

Since specific gravity is dimensionless, the Wobbe index has the same units as H, that is, MJ per standard cubic meter in SI and Btu per standard cubic foot in the imperial units commonly used in the international gas industry. Wobbe indices can be calculated from gas composition data or heating value and density measurements.

The Wobbe index of a gas is proportional to the heating value of the quantity of gas that will flow subsonically through an orifice in response to a given pressure drop. Since virtually all gaseous fuel metering systems are based on orifices, a change in the Wobbe index of the fuel (other things being equal) will result in a nearly proportional change in the rate of energy flow and thus in the air-to-fuel ratio. Departures from strict proportionality may occur in fuel systems using choked (sonic) flow or because of changes in the H:C ratio of the fuel. Even in these cases, however, the Wobbe index provides a good indicator of the change in air-to-fuel ratio.

The effect of variations in the Wobbe index for gaseous-fuel vehicles is similar to the effect of varying the fuel's volumetric energy content in gasoline vehicles. A lower Wobbe index results in a leaner air-to-fuel ratio, and a higher Wobbe index gives a richer mixture. Depending on the fuel metering technology, variations in the Wobbe index may affect engine performance and emissions. Modern, stoichiometric spark-ignition engines with closed-loop control of the air-to-fuel ratio are able to compensate for reasonable variations in the Wobbe index, just as they compensate for variations in gasoline energy content due to refining differences or use of alcohol blends. For engine control systems without air-to-fuel ratio feedback, such as those used in heavy-duty, lean-burn engines, variations in fuel composition can present a significant problem possibly resulting either in poor engine performance (due to a mixture that is too lean) or engine damage due to overheating (with the mixture too rich). Gas used for automotive applications may be covered by additional standards to pipeline gas [4.66, 4.67].

Because changes in air-to-fuel ratio affect combustion and efficiency in many gas-burning appliances as well as engines, natural gas pipelines and distribution utilities have long striven to maintain close control of the Wobbe index of the gas they deliver. Pure methane has a Wobbe index of 50.67 MJ/m^3. Increasing concentrations of higher hydrocarbons such as ethane and propane increase the Wobbe index, whereas increasing concentrations of inert gases lower it. In practice, these two effects are used to cancel each other out, so as to maintain the Wobbe index of natural gas in the pipeline close to the nominal specification. In the US, the Wobbe index is set by ASTM D8080 at between 46 and 53 MJ/m^3.

The three- and four-carbon species in LPG differ in volumetric energy content, so that a change in LPG composition can affect the air-to-fuel ratio in engines and other combustion devices operating on LPG. Although seldom used in reference to LPG, the Wobbe index is equally applicable to assessing the effect of varying fuel composition on air-to-fuel ratio. The Wobbe indices and other properties of the main constituent gases of LPG were shown in Table 4.2. As the table shows, the indices for propane and propene are nearly identical so that, from an air-to-fuel ratio standpoint, these gases are virtually interchangeable. The same is true for normal- and iso-butane. The index for butenes reflects a mixture of several butene isomers and lies between those for propane and butanes.

The effect of variations in the Wobbe index for LPG vehicles is similar to the effect of variations in gasoline liquid density for gasoline vehicles. The levels of variation in gasoline density that are considered acceptable can therefore serve as a guide to acceptable levels of variation in the Wobbe index.

4.3.2.3. Propensity to Knock: Octane and Methane Numbers

As is the case with liquid gasoline, the degree of resistance to engine knock is an important property of gaseous fuels. This tendency was measured in several different ways. The knock resistance of gaseous fuels was reported in terms of the familiar research octane number (RON) and motor octane number (MON) used with gasoline. However, the RON and MON test methods are intended for liquid fuels and are not well adapted for measuring the knock resistance of gaseous fuel. ASTM defined a standard, ASTM D2623 [4.68], for measuring the MON of LPG mixtures, but this method was withdrawn in 1989. It was withdrawn because the octane numbers could be accurately determined from the MON of a compositionally weighted linear sum of the MONs of its pure constituents, and that the compositional limits of LPG standard ASTM D1835 [4.64] were sufficiently restrictive [4.69]. The ASTM D2623 test method used a retrofitted gaseous fuel carburetor, in conjunction with a control valve, to regulate the fuel flow, and a flow meter and vaporizing unit. A piece of work in Australia demonstrated that the CFR engine, using modern hardware, allowed both the RON and MON of any arbitrary LPG mixture to be determined, and produce similar results to the withdrawn Motor (LP) test method [4.70]. In order to determine the knock resistance of natural gas blends, a separate methane number scale has been created [4.71]. In this scale, the reference fuels are mixtures of methane and hydrogen. Pure methane has a methane number of 100, and pure hydrogen has a methane number of 0. To define the relationship between MON and methane number (MN), workers at Southwest Research Institute [4.72] extended the ASTM MON method for LPG to a number of typical natural gas blends, as well as samples of pure methane, ethane, and propane, and methane-propane blends. It was found that MON and methane number are closely correlated. The best-fit relationships were found to be

$$MON = 0.679 \times MN + 72.32$$

$$MN = 1.445 \times MON - 103.42$$

with R^2 in each case greater than 0.95.

ASTM has now introduced a method for measuring the calculated methane number [4.73]. Because of the excellent knock resistance of natural gas, engines designed specifically for natural gas fuel can use higher compression ratios than gasoline engines, with a consequent improvement in efficiency and power output. Typical compression ratios for natural gas engines range from 10:1 (for large engines) to 13:1. The knock-resistance of natural gas also permits supercharging with much higher boost pressures than gasoline engines, enabling these engines to attain torque levels comparable to those of modern heavy-duty diesel engines. The antiknock performance of natural gas is best for pure methane or methane/inert gas mixtures and declines somewhat with increasing concentrations of non-methane hydrocarbons. This effect is not usually significant for the typical range of pipeline gas composition but may become important in high-compression engines burning unprocessed gas or propane-air mixtures.

Of the hydrocarbons commonly included in LPG, propane has good antiknock properties compared with gasoline. The antiknock performance of the other LPG constituents is markedly inferior to that of propane, raising the possibility that an engine optimized for use on high-octane propane might suffer damage from knock if operated on LPG containing significant quantities of propylene, butanes, or butenes.

4.3.3. Gaseous Fuels Dispensing and Storage

Natural gas may be stored onboard a vehicle either as a compressed gas in high-pressure cylinders or as a cryogenic liquid. From the engine's standpoint, CNG and LNG/LPG are essentially interchangeable; it is only the on-board storage medium that is different. LNG/LPG is stored and dispensed at ambient temperatures, as a liquid under pressure.

4.3.3.1. CNG Dispensing and Storage

Handling of CNG is similar to that of any high-pressure gas. Piping and connections must be strong and gas-tight. Refueling is accomplished by connecting a manifold on the vehicle to a high-pressure gas line using a positive-lock connection and then admitting the gas to the vehicle tanks.

CNG refueling systems can be divided into slow-fill and fast-fill designs. The main components of a typical "fast fill" station are a compressor to boost the gas pressure from the pressure in the distribution pipe to about 34.5 MPa; a bank of storage vessels (often called a cascade) to store the high-pressure gas; and a dispenser, which often resembles a common gasoline pump from the outside but is very different within. The compressor is driven by an electric motor or natural gas engine and requires considerable power. The cost of the compressor and motor constitute a large part of the overall costs of a CNG refueling system. These costs are typically much higher than those of refueling systems for liquid fuels. The complete package also includes controls and safety devices, and filters to eliminate oil and particulate matter from the compressed gas. Generally, the compressor, motor, controls, and auxiliaries are packaged and sold as a single pre-engineered unit. The package may also include the cascade storage and the dispenser, or these may be sold separately.

"Fast fill" CNG systems are designed to refuel one or two vehicles at a time, but to refuel each vehicle very quickly, within about 5 minutes. This is achieved by using the compressor to pump high-pressure gas into the cascade storage, which serves as a "buffer." The cascade storage vessels are divided into several groups that can be connected independently to the refueling connector. To refill a vehicle, the cascade storage units are connected in sequence to the vehicle's fuel intake and allowed to equalize pressure. The group having the lowest pressure is connected first, then the next lowest, and so forth until the storage pressure onboard the vehicle reaches the desired level. This arrangement makes it possible to use a smaller compressor than would be required to achieve the same refueling time by pumping directly from the compressor to the vehicle.

Because of the rapid filling characteristic of fast-fill systems, the gas in the cylinders has no time to lose heat to the environment. At the end of the fill, the gas in the vehicle cylinders will be considerably warmer than the ambient temperature. This makes it difficult to obtain a complete refueling by fast fill: although the tank pressure may be at the nominal full level after refueling, the pressure will drop as the gas in the cylinders cools to ambient temperature.

A "slow fill" CNG fueling system is designed to refuel vehicles such as buses that can be parked overnight. All of the vehicles are connected in parallel to the compressor, eliminating the need for cascade storage but requiring a large number of high-pressure hoses and connectors. This arrangement makes it possible to achieve good compressor utilization, since the compressor can run continuously all night. For even more effective compressor utilization, it is also possible to design a "hybrid" system that can be used for fast fill during the day and slow fill at night.

The current standard working pressure for compressed natural gas cylinders is 20 MPa. There has been a recent push to move to 25 MPa working pressures for CNG tanks to provide better range (using equivalent fuel storage system volume). The Institute of Gas Technology (IGT) is currently studying potential methods of safely increasing current design 20 MPa compressor output up to 25 MPa. Because most current systems operate at 20 MPa, a complete move to 25 MPa systems would be a disadvantage to those people who have already installed the lower pressure systems.

The size and weight of CNG cylinders are often cited as major drawbacks of natural gas use in vehicles. About 0.9 m³ of natural gas, weighing 0.65 kg, the national average [4.74], are required to provide the same energy content (lower heating value) as 1 L of gasoline. With conventional steel cylinders, the weight of the cylinder required to contain 0.9 m³ of gas is about 4.55 kg, for a total weight of fuel and storage of 5.2 kg to store the equivalent of 1 L of gasoline. For comparison, the weight of the fuel and tank for gasoline would be about 0.93 kg/L; for diesel fuel it would be about 1.02 kg/L. However, as shown in Figure 4.8, recent developments in high-strength composite materials have made it possible to reduce the weight of CNG cylinders substantially.

FIGURE 4.8 Comparison of fuel and storage system weights for CNG and gasoline.

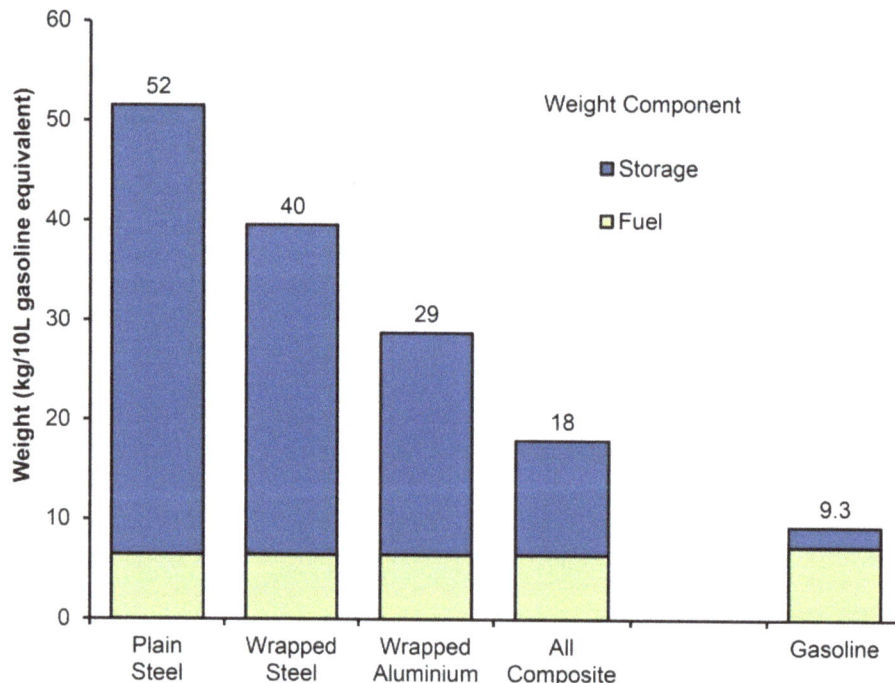

With present fiber-wrapped steel tanks, the tank weight is reduced by about 30%, so that the total weight of gas and tank would be about 4.0 kg/L of gasoline equivalent. Fiber-wrapped aluminum cylinders weigh about 2.9 kg/L of gasoline equivalent. All-composite cylinders weigh even less at about 1.8 kg/L of gasoline equivalent. This is still higher than gasoline (or diesel) fuel storage, but these newer tanks are significantly lighter than the plain steel tanks of the past.

4.3.3.2. LNG Dispensing and Storage

LNG is stored in double-wall, vacuum-insulated cryogenic (−160°C or less) containers (commonly known as dewars) and then vaporized, usually by engine heat, to produce engine-usable natural gas. A sophisticated fuel tank storage/delivery system is required to allow pumping of LNG to and from the fuel system while maintaining cryogenic temperatures over long periods of time. Current design LNG tanks are of double-walled construction: an inner vessel made of stainless steel surrounded by an insulating material and an outer casing made of high-strength carbon steel. The space between the inner and outer shells is vacuum evacuated to 0.013 Pa. For applications requiring multiple LNG tanks, vacuum-jacketed piping (double-walled pipe with vacuum separation) is used to interconnect the tanks.

The quality of LNG can vary greatly, and the quality needed by the user may also vary. The liquefaction process generally removes all of the minor natural gas constituents except ethane and nitrogen. Typical merchant LNG is about 87 to 92% methane, with most of the remainder being liquid ethane. With additional processing, the ethane and nitrogen components can be removed as well, yielding a product that is greater than 99% pure methane. Since methane has substantially better antiknock properties than ethane, the use of pure liquid methane may be desirable for certain high-performance engines, such as those used in locomotives.

An important concern with the use of liquid methane/ethane mixtures is the possibility of changes in fuel composition during handling and processing. This is known as "aging," "weathering," or "enrichment." The liquid ethane has a higher boiling point than methane. Therefore, at every stage in processing where evaporation can take place—central storage, trucking, transfer, refueling station storage, and final fueling—methane boils off, leaving the ethane behind. At the end of processing, a higher concentration of ethane exists than did at the beginning. The Wobbe index of the fuel thus becomes progressively higher while the knock-resistance becomes lower. This may be a significant problem for high-compression engines or those with very high cylinder pressure levels, due to the increased potential for knock.

A number of LNG refueling system designs exist. They can be divided into two categories: those with and those without vapor return. It appears that the industry is leaning toward single-hose, no-vapor-return refueling systems. Tank depressurizing is automatic, with integrated hoses and nozzles. Both fueling hoses and dispensing units must be designed according to the following needs: accurate flow measurement, accurate pressure regulation, sufficient hose support (hoses are heavier than gasoline or diesel hoses), leak detection, fire detection, frost and ice avoidance (especially in high-humidity areas), and protection of refueling personnel. The lack of a standard LNG nozzle/receptacle design is presently a significant barrier to the commercialization of LNG technology.

The newest design LNG tanks are filled from the top, providing a vapor-only interface at the nozzle connection. This practice is one of design convenience and safety. LNG entering the tank at the vapor interface cools the vapor present in the tank, condensing some of it back into liquid form. This eliminates the need for a vapor return line. If the fuel is pumped into the liquid interface at the bottom of the tank, it will not cool the overhead vapor but rather will compress it and increase the internal tank pressure. Older LNG tank designs using bottom fill require a vapor return system to maintain proper internal tank pressure during refueling.

Nearly all LNG engines use the fuel in gaseous form. To accomplish this, a "vaporizer" borrows engine heat (usually through heat exchange with the engine coolant) to help expand the cold liquid into its gaseous form. In some applications, the LNG is vaporized to a gas at close to ambient conditions and then pumped to the required pressure with a compressor. In other applications, the fuel is pumped as a liquid to high pressure and then vaporized. The latter requires much less pumping power because liquids are much less compressible than gases. On some designs, a CNG buffer, or expansion tank, gives additional space for sudden expansion of cold LNG when the fuel demand from the engine is suddenly reduced or interrupted. This buffer is also used to store natural gas for start-up of the engine. When the engine is restarted, there is enough gas in the buffer system to start the engine running and get the LNG vaporizer system working and producing gas for normal operation.

4.3.3.3. LPG Dispensing and Storage

LPG is stored on the vehicle as a liquid under pressure. LPG tanks must be designed to contain an internal pressure of 1.6 to 1.7 MPa. They are generally cylindrical, with rounded ends, and are much stronger than tanks used for storing gasoline or diesel fuel, albeit much less so than those used for CNG. LPG can be pumped from one tank to another like any liquid, but the need to maintain pressure requires a gas-tight seal. Except for the need for a standardized, gas-tight connection, LPG used as vehicle fuel can be dispensed in much the same way as gasoline or diesel fuel. To ensure that some vapor space is always available for expansion, LPG tanks used in automotive service must never be filled more than 80% full. Automatic fill limiters are incorporated in the tanks to ensure that this does not occur.

4.4. Gas-to-Liquids

The discovery of large quantities of natural gas allowed a modification of the Fischer-Tropsch process (discussed in the previous section) to produce liquid fuels from these light gases. This gives rise to the term gas-to-liquids (GTL) technology. The discovery of this abundant resource also spurred the development of technology to allow the transportation of this gas and the direct combustion of this gas within the engine; as discussed in the previous section.

The use of natural gas as opposed to coal as a feedstock for syn-gas production has the effect of producing lower CO_2 emissions. The equations for the stoichiometric production of syn-gas from coal and methane are given below [4.75]:

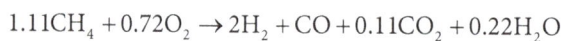

$$1.78CH_{0.5} + 0.5O_2 + 1.56H_2O \rightarrow 2H_2 + CO + 0.78CO_2$$

$$1.11CH_4 + 0.72O_2 \rightarrow 2H_2 + CO + 0.11CO_2 + 0.22H_2O$$

The quality of the feed gas determines the first stage of a GTL plant. Natural gas, as extracted, will normally contain some liquids. These are first separated from the gas in a natural gas liquids recovery unit. The condensate from this unit can then be used to provide a naphtha feed to the refinery and a straight run diesel blending fraction. The gas from the liquids recovery section is then converted to syn-gas.

The syn-gas production is usually accomplished by steam reforming and an autothermal reformer. The steam reformer produces an excess of hydrogen for the F-T synthesis, as shown below:

$$CH_4 + H_2O \rightarrow CO + 3H_2$$

This is effectively the reverse of the methanation reaction that takes place in the F-T synthesis.

This hydrogen-rich gas along with some of the basic natural gas feedstock is then fed to the autothermal reactor, where the methane is partially oxidized to produce CO and water, which is subject also to the water-gas shift reaction. The syn-gas can then be fed to LTFT and HTFT synthesizers as described in the previous section. However, due to the presence of butanes in the distillate from the liquids recovery section and the abundance of butenes from the HTFT process, an alkylation process can be included to produce high octane number gasoline components.

An alternative approach was taken by Shell in the early 1980s. In this case, the LTFT synthesis process was tailored to produce heavier hydrocarbons, which were then selectively cracked to produce the components in the fuel range desired [4.76]. This has become known as the SMDS process. The SMDS process is used as the Shell plant at Bintulu, Malaysia (shown in Figure 4.9), and the Qatar Petroleum/Shell Pearl plant at Las Raffan, Qatar (shown in Figure 4.10).

FIGURE 4.9   The Shell Bintulu plant in Malaysia.

Source: Photographic Services, Shell International Ltd.

Source: Photographic Services, Shell International Ltd.

FIGURE 4.10 The Qatar Petroleum/Shell Pearl plant at Las Raffan, Qatar.

A multi-tubular fixed-bed reactor is used for the F-T synthesis with a "Co-impregnation catalyst" [4.77]; this is referred to as the heavy paraffin synthesis (HPS) process [4.78], as opposed to the Sasol LTFT slurry phase distillate (SPD) process [4.79]. The Sasol SPD technology is used at the Qatar Petroleum/Sasol Oryx plant at Las Raffan, Qatar. Note: a "Co-impregnation catalyst," according to [4.77] is a catalyst containing 10–40 parts by weight of Co and 0.25–5 parts by weight of Zr, Ti, or Cr per 100 parts by weight of silica and prepared by impregnating a silica carrier with one or more aqueous solutions of salts of Co and Zr, Ti or Cr, and then drying and calcining at 350–700°C.

The subsequent steps comprise heavy paraffin conversion and distillation sections to produce naphtha, kerosene, gas oil, paraffins, and various grades of wax. Lighter grades are fed back for conversion into hydrogen while the long-chain waxy paraffins are either left intact or broken into smaller molecules, depending on market needs. A simplified schematic of the Bintulu plant is shown in Figure 4.11.

FIGURE 4.11 Simplified schematic of the Bintulu GTL plant.

A number of other GTL processes exist (e.g., the Exxon/Mobil AGC-21 technology), and there is considerable worldwide interest in exploiting natural gas fields to produce "synthetic" fuels. A disadvantage of the process is that it produces substantial quantities of CO_2, and this is still viewed as a major development challenge. Product upgrading follows the same basics steps as outlined for a CTL plant.

4.5. Methanol to Fuel

As described in Section 3.9.1, methanol can easily be synthesized from syn-gas using an appropriate catalyst. The Methanol to Gasoline (MTG) or Methanol to Fuel (MTF) process was developed in the 1970s by scientists from Mobil Oil Corporation [4.80, 4.81]. The process can be controlled to produce only olefins instead of gasoline, the so-called methanol-to-olefin process (MTO). The oil crisis of 1973 and 1978 initiated the development of a commercial MTG process to provide an alternative route from coal or natural gas to high-octane gasoline, as well as olefins and other chemicals [4.82]. A first commercial plant in New Zealand, utilized methanol produced from natural gas over a fixed bed catalyst [4.83, 4.84]. In Germany further developments resulted in a plant using the fluidized-bed MTO process for producing high quality gasoline (MTG operation), but the process conditions could be varied to produce olefins (MTO operation), which could be converted to gasoline and diesel fuel [4.85].

More detailed investigation of the processes suggested that the main reaction steps are:

$$2CH_3OH \leftrightarrow CH_3OCH_3 + H_2O \quad \text{Dimethylether formation}$$

$$CH_3OCH_3 \rightarrow \text{Light olefins} + H_2O \quad \text{Olefins synthesis}$$

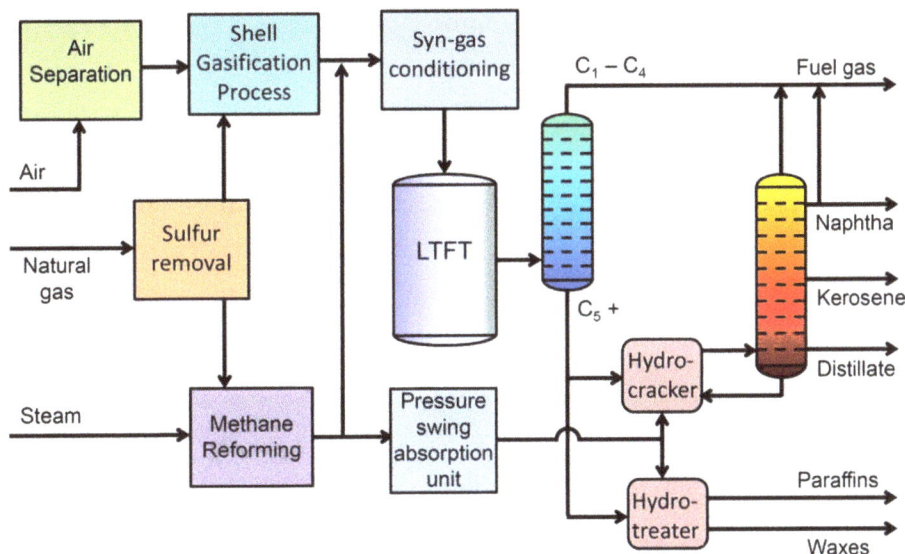

$$\text{Light olefins} + CH_3OCH_3 \rightarrow \text{Heavy olefins} + H_2O$$

$$\text{Heavy olefins} \rightarrow \text{Aromatics} + \text{Paraffins}$$

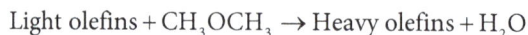

As can been seen from the reactions above, methanol is first dehydrated to dimethylether (DME). The equilibrium mixture of methanol, DME and water is then converted to light olefins (C_2–C_4). The reactions can then proceed to form a mixture of higher olefins, n/iso-paraffins, aromatics and naphthenes. The catalyst used for this early work was the synthetic zeolite ZSM-5 but much work was performed to tune the reaction with different catalysts; Union Carbide developed a process to convert methanol to olefins using a molecular sieve silicoaluminophosphate (SAPO) catalyst [4.86, 4.87].

Due to easing of oil prices around this time it became unattractive to produce olefins via this route rather than directly from crude oil and the MTG process was not commercially developed at the time. Now, due to concerns over the environment and the production of bio-methanol and e-fuels the MTG process is again being considered as a route to "carbon neutral" gasoline as discussed in Section 5.6.

4.6. Oil Sands Fuel

Oil sands, also known as tar sands, are a mixture of sand, clay, water, and bitumen. This mixture can be mined and processed to extract the bitumen, which is then refined to produce oil. Like oil, oil sands can be found in many countries worldwide; currently, the largest know deposits have been found in Canada and Venezuela. In the first quarter of the last century, investigations began into methods of recovering the bitumen from these deposits with small-scale plants being built in the 1930s [4.88]. Commercial production of oil from oil sands began in Canada in 1967 [4.89]. Canada has the largest industry to produce oil from oil sands; Venezuela has only small, but is increasing, production capacity. There is also fierce debate about large-scale production within Utah.

The bitumen in oil sands cannot be pumped from the ground in the way that crude oil can. Crude oil has a viscosity in the range of 1 to 10 mPas, while heavy crudes have a viscosity of about 50 to 100 mPas at reservoir conditions, and the viscosity of crude bitumen ranges from 500 to 5000 Pa·s [4.90]. The oil sands must therefore be mined by strip mining or from open pits in a manner similar to open cast coal mining, or by techniques to increase the temperature [4.91–4.94] of the bitumen or to lower its viscosity [4.95, 4.96]. However, the oil sands do not break like rock or flow like sand. The terrain is also not like rock, which means that mining equipment must exert very low pressure on the ground, otherwise there is danger of it sinking. Once mined, the oil sands must be transported to the processing plant in the same way that coal would be transported. This further complicates the process of recovering the material.

Once the oil sands have been recovered, the bitumen must be separated from the sand, clay, and water that make up the oil sands. The bitumen that is separated out may also require additional upgrading before it is suitable as input material to a conventional oil refinery. If it is to be transported by pipeline it will require dilution to make it pumpable. These processes are discussed in the following sections.

4.6.1. Oil Sands Extraction and Processing

Oil sands that are close to the surface can be recovered by open pit mining techniques, although due to the climatic conditions in Canada, this in itself can be problematic [4.97]. This involves removing the overburden, the non-bitumous rock above the tar-sands deposits, and then removing the tar-sands for transportation to the processing plant. Figure 4.12 shows an example of an open pit mining operation.

FIGURE 4.12   Open pit mining of oil sands.

In the picture a Bucyrus (now part of Caterpillar Inc.) electric rope shovel loads a Caterpillar 797B truck. Each shovel load is approximately 100 tons, and three or four shovelfuls will fill a truck. These machines have wider tracks specifically to reduce ground pressure for oil shale applications.

4.6.1.1. Surface Mining

In a typical oil extraction process, the oil sands are fed to a tumbler or agitator where a hot aqueous caustic solution is added to the oil sands. Additional heat can be supplied by the addition of steam. The agitation of the resultant mix causes the bitumen to be separated from the solids and to form a bitumen froth. This slurry is then screened to remove larger solids and passes to a separator where the sand and sediment settle to the bottom and the bitumen froth rises to the top. The viscous sludge left in the middle is known as middlings, which consists of the finer solids (clay) and probably also some finer particles of bitumen and is often passed to a second separator. The solids from the separator, known as tailings, are passed to a tailings pond where over time there is further settling out and the water can be recycled. A typical oil extraction process is shown in Figure 4.13.

The bitumen froth from the separators is skimmed off and passed to the froth treatment section. The bitumen froth is basically an aerated emulsion of small bitumen particles that also contains some fine solids from the clay. In the froth breaker, the froth is heated and broken to allow the air to escape. Naphtha may be added to reduce the density of the bitumen before it passes to a centrifuge where the water and hydrocarbons are separated. Figure 4.13 shows a two-stage centrifuge arrangement, but more are often used. The resultant hydrocarbon mix, which could contain about 5% water, is not suitable for passing to the refinery and is usually fed to a coker, which drives off the naphtha and drops out the asphaltenes. In a variation of this process [4.98] the naphtha or other light hydrocarbon (kerosene) is added with the caustic and then recovered after the centrifuge for recycling.

FIGURE 4.13 Oil sands water extraction process.

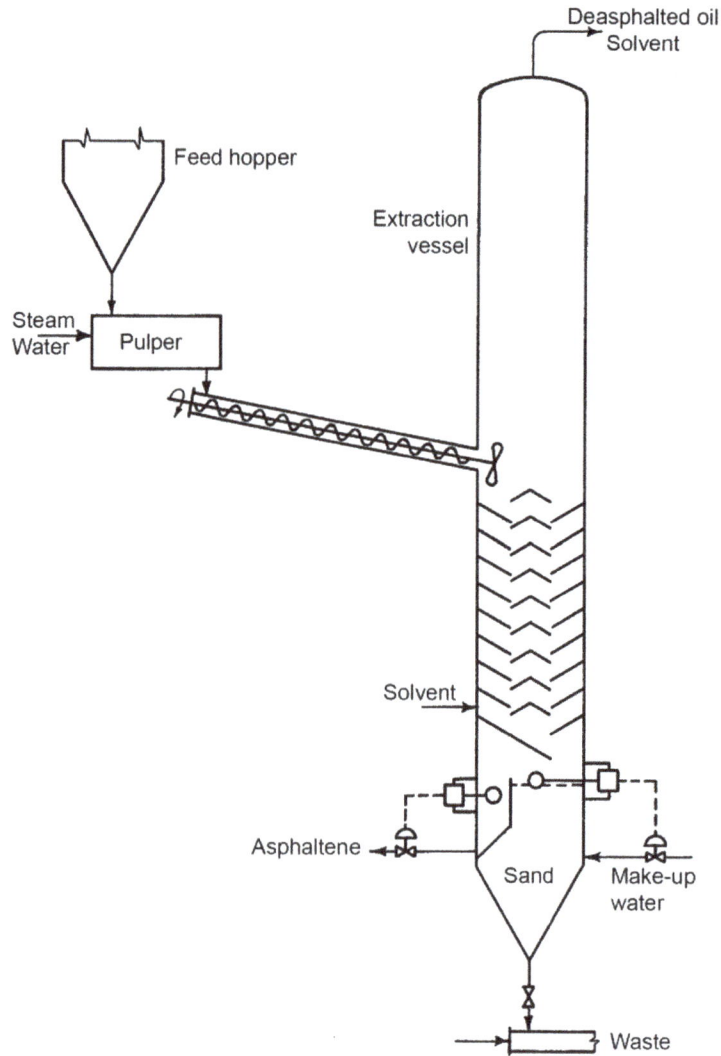

In a modification of this process, the oil sands and the hot caustic solution are mixed close to the mine site, and the resultant mix flows down a large pipe to the extraction plant. This is known as the hydro-transport system. The turbulent flow along the pipe, which can be aided by air injection along the pipe, is sufficient to aerate the mixture and form the bitumen froth [4.99]. This eliminates the need for the agitator at the oil extraction plant. A variation of this method uses sodium bicarbonate and/or potassium bicarbonate (with or without a source of calcium ions and/or magnesium ions) instead of the sodium hydroxide (caustic) [4.100].

Another method of separating the bitumen from the sand while also separating the asphaltenes from the bitumen is the reverse flow solvent method [4.101]. This process is illustrated in Figure 4.14: the oil sands are fed in through the feed hopper, steam and water are added, and the mixture is pulped and fed into the extraction vessel, where it flows down over the baffles.

FIGURE 4.14 Reverse flow solvent separation process.

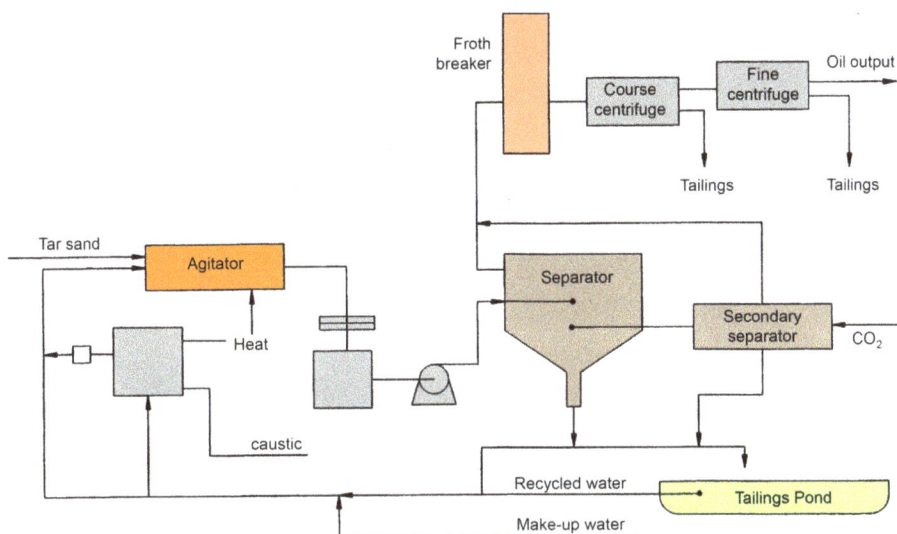

A low molecular weight paraffin such as propane is introduced toward the bottom of the extraction vessel and passes up through the downward-flowing sand mix. The solvent carries the de-asphalted oil to the top of the vessel, where it is extracted; the solvent is separated off and recycled, leaving the dasphalted oil to be used in a conventional refinery. The wet sand sits in the bottom of the vessel before it is drawn off as waste. The asphaltene-rich layer sits above the wet sand, flows over the weir, and is drawn off.

4.6.1.2. Subterranean Recovery

Proposals have been made to mine the oil sands that are too deep for strip mining [4.102], but these have not been widely adopted and are not suitable for mining deep deposits. An alternative approach to this problem has been to heat the oil sands to the point at which they can be pumped out of the ground, transported to the processing facility, and then processed as described in the previous section. There are a number of ways of heating the in situ oil sands.

Steam can also be used as a source of heat. In the cyclic steam stimulation (CSS) technique [4.103, 4.104, 4.105], steam is injected into a well bore, left to soak for weeks or months, and the hot oil is pumped out of the same bore hole. The process can then be repeated until the value of the bitumen recovered falls below the cost of recovering it. The steam is injected at a temperature of between 300 and 340°C at a pressure of up to 13.5 MPa. A more efficient method is the steam-assisted gravity drainage (SAGD) method [4.106, 4.107]. In this method, the steam is injected through one bore hole and recovered from a second bore hole that is usually about 5 m below the injection bore. The heat from the steam causes the viscosity of the bitumen to fall, and it then flows under gravity to be extracted from the recovery well. The SAGD method will typically recover about twice as much bitumen as the CSS method, using about half as much steam to recover each unit of bitumen. Solvent can be included with the steam [4.108] which further reduces the viscosity of the bitumen. Electrical heaters can be used to raise the temperature of the oil sands, either in conjunction with steam [4.109] or alone [4.110]. Figure 4.15 shows a set of pumps recovering material from a subterranean seam of oil sands.

FIGURE 4.15 Subterranean recovery of oil sands.

Source: Photographic Services, Shell International Ltd.

4.6.1.3. In Situ Processing

In situ processing refers to recovering oil that is chemically different from the bitumen that is present in the deposit. Usually this means changing the chemistry to make the recovered product more like the desired product (more like conventional oil).

As noted previously, heaters may be used to heat the oil sands and thus reduce the viscosity of the bitumen. However, if the temperature is raised to a high enough level, then this will lead to visbreaking of the trapped bitumen [4.111].

In another approach, two wells are drilled. The temperature around one of these bore holes, the injection bore, is raised by a bore heater. The temperature is raised to the combustion temperature of the bitumen, and an oxidant (oxygen or air) is injected, either through the injection bore, direct or forward combustion [4.112], or through the recovery bore, reverse combustion [4.113]. The ensuing combustion raises the temperature of the bitumen, causing it to flow, but also brings about chemical changes. Reverse combustion is considered advantageous, as the lower viscosity oil behind the flame front is immediately available at the recovery well, whereas with the forward combustion method the flame front must pass through to the recovery bore before the oil can be recovered.

4.7. Oil Shale Fuel

Shale is a hard, slate-like sedimentary rock; oil shale is shale that also contains a solid hydrocarbon material known as kerogen. Kerogen is the organic material that may, under the right conditions, go on to form bitumen as part of the petroleum generation process [4.114]. Shale oil is therefore intrinsically different to oil sands. The kerogen must be heated to yield oil by pyrolysis. The kerogen can be released and pyrolyzed through heating the shale to in excess of 500°C in the absence of oxygen, a process called retorting. An alternative to mining the shale and then retorting is a process referred to as in situ retorting, which involves heating the oil shale while it is still underground. The resultant liquid is then pumped out.

Scotland was the global pioneer of the modern oil industry and, for a few decades in the second half of the 19th century, was the leading oil-producer in the world.

So wrote Barbara Harvie as the opening to her paper for the Estonian Academy Publishers journal "Oil Shale" [4.115]. Following independence from Russia in 1918, and after failed attempts to develop a shale oil industry based on Scottish practices, the 1920s allowed shale oil to secure energy independence for Estonia. Recovery of hydrocarbons from oil shale has also taken place in Australia, Brazil, Canada, China, France, South Africa, Spain, Sweden, Switzerland, and the US. But shale oil is present in many countries throughout the world. The largest known reserves are in the US, in the Green River Formation, which lies beneath part of Colorado, Utah, and Wyoming. Other important deposits are located in North Dakota and Montana, and these have been producing oil for some years [4.116]. It has been estimated that recovery of shale oils could help the US become the second largest oil producer in the world after Saudi Arabia [4.117]. There have also been significant discoveries of shale oil resources in several basins in China, including the Junggar, Ordos, and Songliao basins. The technically recoverable hydrocarbon resources were estimated at approximately 30 to 60×10^8 tons [4.118].

Although the reserves of oil trapped in shale are thought to be considerable, there is no significant shale oil industry, primarily due to the economics of extracting the oil. However, political and economic drivers may alter this situation, and a brief outline of the processes required are included in the following sections.

4.7.1. Oil Shale Mining and Processing

Oil shale can be mined in the same way as coal, either through open cat, strip mining, or through deep pit mining. The shale must then be transported to the location of the retort and must be reduced to a manageable granular size, probably using a crusher. The crushed shale is then passed to the retort, where the temperature is raised to a sufficient level to pyrolyze the kerogen, which is driven off as a gas. The remaining shale must eventually be disposed of, preferably by returning it to the mine. However, due to the thermal process, the shale will usually occupy a greater volume than that originally mined. The nature of the kerogen means that the shale oils contain high levels of sulfur and nitrogen, which will poison refinery catalysts. These impurities must be removed before the oils can be processed by normal oil refinery techniques. Various techniques have been proposed for removing sulfur compounds [4.119–4.121], nitrogen [4.122, 4.123], and arsenic that can be present [4.124, 4.125], or preventing its liberation into the shale gas [4.126].

4.7.2. In Situ Retorting

In situ retorting of oil shale is also utilized. In this case the retorting process often follows the same rationale as in situ processing of oil sands. However, due to the nature of the shale compared with the sand, it may also be necessary to fracture the shale to allow the oil to flow through it [4.127], a process that has become known as fracking. Injection and recovery wells are used. Air is injected and an ignition source is utilized to allow the burning and pyrolysis of the oil, which flows out through the shale to the recovery wells. It can be difficult to control the flame front and the flow of oil, leaving areas unheated and some oil unrecovered. This can limit the degree of oil recovery.

Modified in situ can also be used. This involves mining some shale from above or below the main shale deposit. The main shale deposit is then heated by igniting the top of the deposit and recovering the oil from ahead of or from beneath the burning zone. Modified in situ retorting can improve degree of recovery by heating a larger proportion of the shale and allowing an improved flow of gas and oil through the rock formation.

In situ retorting has greater potential to cause environmental problems. Because in situ retorting relies on fluid flows through the shale, there is the potential for these fluid flows to carry oil and residuals into the ground water.

References

4.1.   Betts, A.G., Method of utilizing buried coal. U.S. Patent 947,608, 1910.

4.2.   Kerker, M., "Science and the Steam Engine," *Technology and Culture* 2, no. 4 (1961): 381-390.

4.3.   Diesel, R., A process for producing motive work from the combustion of fuel. GB Patent 7241, 1892.

4.4.   Soehngen, E.E., "Development of Coal-Burning Diesel Engines in Germany. A State-of-the-Art Review," Technical Report No. FE-WAPO/3387-1, DOE Contract No.: PO-WA-76-3387, 1976.

4.5.   Fieldner, A., "Production of Gasoline Substitutes from Coal," SAE Technical Paper 270010, 1927, doi:https://doi.org/10.4271/270010.

4.6.   Donath, E.E., "Coal-Hydrogenation Vapor-Phase Catalysts," in *Advances in Catalysis*, Vol. 8, Eds. Frankenburg, W.G., Komarewsky, V.I., and Rideal, E.K. (New York: Academic Press, 1956), 239-292.

4.7.   Bergius, F. and Billwiller, J., Process for producing liquid or soluble organic combinations from hard coal and the like. U.S. Patent 1,251,954, 1918.

4.8.   Sabatier, P. and Sendenes, J.-B., "Nouvelles syntheses du methane," *Comptes Rendus Hebdomadaires des Séances de L'Académie des Sciences* 134 (1902): 514-517.

4.9.   Fischer, F., *The Conversion of Coal into Oils*, translated by Lessing, R. (London, UK: Ernest Benn, Ltd., 1925).

4.10.   Fischer, F. and Tropsch, H., "Über die Herstellung synthetischer Ölgemische (Synthol) durch Aufbau aus Kohleoxyd und Wasserstoff," *Brennstoff-Chemie* 4 (1923): 276-285.

4.11.   Fisher, F. and Tropshe, H., Process for the production of paraffin hydrocarbons with more than one carbon atom. GB Patent 255,818, 1927.

4.12.   Heckel, H. and Roelen, O., Catalyst preparation. U.S. Patent 2,219,042, Priority date 1937, Granted 1940.

4.13.   U.S. Naval Technical Mission in Europe, "The Synthesis of Hydrocarbons and Chemicals from CO and H_2," Technical Report No. 248-45, 1945.

4.14.   Fischer, F. and Pichler, H., Process for preparing solid paraffins. U.S. Patent 2,206,500, Priority date 1936, Granted 1940.

4.15.   National Bureau of Statistics, *The 13th Five-Year Plan of Coal Deep Processing Industry Demonstration* (Beijing, China: National Bureau of Statistics, 2017).

4.16.   Xie, K., Li, W., and Zhao, W., "Coal Chemical Industry and Its Sustainable Development in China," *Energy* 35, no. 11 (2010): 4349-4355.

4.17.   Anderson, S. and Newell, R. "Prospects for Carbon Capture and Storage Technologies." Annu. Rev. Environ. Resour., 29, pp.109-142. 2004.

4.18.   Van Alphen, K., Noothout, P.M., Hekkert, M.P., and Turkenburg, W.C., "Evaluating the Development of Carbon Capture and Storage Technologies in the United States," *Renewable and Sustainable Energy Reviews* 14, no. 3 (2010): 971-986.

4.19.   Su, H. and Fletcher, J.J., "Carbon Capture and Storage in China: Options for the Shenhua Direct Coal Liquefaction Plant," International Association for Energy Economics, Cleveland, OH, 2014.

4.20.   Vasireddy, S., Morreale, B., Cugini, A., Song, C. et al., "Clean Liquid Fuels from Direct Coal Liquefaction: Chemistry, Catalysis, Technological Status and Challenges," *Energy & Environmental Science* 4, no. 2 (2011): 311-345.

4.21.   Comolli, A.G., Johanson, E.S., Lee, L.K., and Stalzer, R.H., "Catalytic Multi-Stage Liquefaction of Coal," First Quarterly Report, No. DOE/PC/92147-1, December 8–31, 1992, Hydrocarbon Research, Inc., Princeton, NJ, 1993.

4.22.   Kuehler, C.W., Three-stage coal liquefaction process. U.S. Patent 4,331,531, 1982.

4.23.   Stotler, H.H., H-coal process: slurry oil recycle system. U.S. Patent 3,962,070, 1976.

4.24.   Kydd, P.H., Chervenak, M.C., and DeVaux, G.R., H-coal process and plant design. U.S. Patent 4,400,263, 1983.

4.25.   Bull, W.C., Wright, C.H., and Pastor, G.R., Solvent refined coal process including recycle of coal minerals. U.S. Patent 3,884,794, 1975.

4.26.   Hinderliter, C.R. and Perrussel, R.E., Solvent refined coal process with retention of coal minerals. U.S. Patent **3,884,796,** 1975.

4.27.   Maa, P.S., Hydrogen donor solvent coal liquefaction process. U.S. Patent 4,022,680, 1977.

4.28.   Neuworth, M.B., Two stage liquefaction of coal. U.S. Patent 4,298,451, 1981.

4.29.   Klimenko, A.Y., "Early Ideas in Underground Coal Gasification and Their Evolution," *Energies* 2, no. 2 (2009): **456-476.**

4.30.   Burton, E.A., Upadhye, R., and Friedmann, S.J., "Best Practices in Underground Coal Gasification," No. LLNL-TR-225331, Lawrence Livermore National Lab. (LLNL), Livermore, CA, 2019.

4.31.   U.S. Naval Technical Mission in Europe, "The Production of Synthetic Fuels by the Hydrogenation of Solid and Liquid Carbonaceous Materials," Technical Report No. 217-45, 1945.

4.32.   Rutherford, R.W., "Oil from Coal in Japan," *Colliery Eng.* 21 (1944): 40-42.

4.33.   Storch, H.H., Anderson, R.B., Hofer, L.J.E., Hawk, C.O. et al., "Synthetic Liquid Fuels from Hydrogenation of Carbon Monoxide," Technical Paper 709, Bureau of Mines, Washington, DC, 1948.

4.34.   Yang, Y., Xu, J., Liu, Z., Guo, Q. et al., "Progress in Coal Chemical Technologies of China," *Reviews in Chemical Engineering* 36, no. 1 (2020): 21-66.

4.35.   Rong, F. and Victor, D.G., "Coal Liquefaction Policy in China: Explaining the Policy Reversal Since 2006," *Energy Policy* 39, no. 12 (2011): 8175-8184.

4.36.   Schneider, R.B., "Recent Developments in Petroleum Chemistry," *The Analysts Journal* 5, no. 2 (1949): 17-19.

4.37.   Keith, P.C., "Gasoline from Natural Gas," *Oil and Gas Journal* 45, no. 6 (1946): 102-112.

4.38.   Leckel, D., "Diesel Production from Fischer-Tropsch: The Past, the Present, and New Concepts," *Energy & Fuels* 23, no. 5 (2009): 2342-2358.

4.39.   Fleming, D.K. and Primack, H.S., "Purification Processes for Coal Gasification," *Coal Process. Technol.* 3 (1977): 66-78.

4.40.   Raney, D.R., "Remove Carbon Dioxide with Selexol," *Hydrocarbon Process* 55, no. 4 (1976): 73-75.

4.41.   Hochgesand, G., "Rectisol and Purisol," *Ind. Eng. Chem.* 62, no. 7 (1970): 37-43.

4.42.   Kohl, A.L. and Buckingham, P.A., "The Fluor Solvent CO_2-Removal Process," *Oil Gas Journal* 58 (1960): 146-155.

4.43.   Anderson, G.L. and Berry, F.O., "Development of a Hot Gas Cleanup System," in *Proceedings of the 7th Annual Gasification and Gas Stream Cleanup Systems Meeting*, Morgantown, WV, Vol. 2, 1987.

4.44.   Fischer, F. and Tropsch, H., "Synthesis of Petroleum at Atmospheric Pressure from Gasification Products of Coal," *Brennstoff-Chem.* 7 (1926): 97-104.

4.45.   Leckel, D., "Low-Pressure Hydrocracking of Coal-Derived Fischer–Tropsch Waxes to Diesel," *Energy Fuels* 21, no. 3 (2007): 1425-1431.

4.46.   Ipatieff, V.N., Corson, B.B., and Egloff, G., "Polymerization, a New Source of Gasoline," *Ind. Eng. Chem.* 27, no. 9 (1935): 1077-1081.

4.47.   Marchionna, M., Di Girolamo, M., and Paterini, R., "Light Olefins Dimerization to High Quality Gasoline Components," *Catalysis Today* 65, no. 2-4 (2001): 397-403.

4.48.   Quann, R.J., Green, L.A., Tabak, S.A., and Krambeck, F.J., "Chemistry of Olefin Oligomerization over ZSM-5 Catalyst," *Ind. Eng. Chem. Res.* 27, no. 4 (1988): 565-570.

4.49.   Chauvin, Y., Commereuc, D., Gaillard, J., Leger, G. et al., Catalyst composition and its use for oligomerizing olefins. U.S. Patent 4,283,305, 1981.

4.50.   Hughes, T.R., Jacobson, R.L., and Tamm, P.W., "Catalytic Processes for Octane Enhancement by Increasing the Aromatics Content of Gasoline," *Studies in Surface Science and Catalysis* 38 (1988): 317-333.

4.51.   Nel, R.J.J. and de Klerk, A., "Fischer–Tropsch Aqueous Phase Refining by Catalytic Alcohol Dehydration," *Ind. Eng. Chem. Res.* 46, no. 11 (2007): 3558-3565.

4.52.   The Energy Institute, "The Energy Institute Statistical Review of World Energy™, 72nd Edition," London, 2023.

4.53. Browne, A., "LPG Fuels for Automotive Engines," SAE Technical Paper 510091, 1951, doi:https://doi.org/10.4271/510091.

4.54. Horiak, E., "High Economy of Hercules Spark Ignition Engines for Heavy Duty - Operating on LPG," SAE Technical Paper 590338, 1959, doi:https://doi.org/10.4271/590338.

4.55. Meyer, R., Cole, J., Kienzle, E., and Wells, A., "Development of a CNGg Engine," SAE Technical Paper 910881, 1991, doi:https://doi.org/10.4271/910881.

4.56. Conti, L., Ferrera, M., Garlaseo, R., Volpi, E. et al., "Rationale of Dedicated Low Emitting CNG Cars," SAE Technical Paper 932763, 1993, doi:https://doi.org/10.4271/932763.

4.57. Pucher, E. and Mueller, J., "Real-Life Emission Measurement of Light Duty Trucks with CNG, Diesel and Gasoline Engines," SAE Technical Paper 2005-01-3445, 2005, doi:https://doi.org/10.4271/2005-01-3445.

4.58. Schreiber, D., Forss, A.M., Mohr, M., and Dimopoulos, P., "Particle Characterisation of Modern CNG, Gasoline and Diesel Passenger Cars," SAE Technical Paper 2007-24-0123, 2007, doi:https://doi.org/10.4271/2007-24-0123.

4.59. Meyerhoff, A.A. and Willums, J.O., "Petroleum in the People's Republic of China, 1949-1979," in *Energy Resources of the Pacific Region*, Ed. Halbouty, M.T. (Tulsa, OK: American Association of Petroleum Geologists, 1981).

4.60. Charles, H.H. and Page, J.H., "Shale-Gas Industry of Eastern Kansas," *AAPG Bulletin* 13, no. 4 (1929): 367-381.

4.61. Chew, K.J., "The Future of Oil: Unconventional Fossil Fuels," *Philosophical Transactions of the Royal Society A: Mathematical, Physical and Engineering Sciences* 372, no. 2006 (2014): 20120324.

4.62. Gottfried, M.R., Graham, D.G., Morgan, M., Casey, H.W. et al., "The Morphology of Carbon Disulfide Neurotoxicity," *Neurotoxicology* 6, no. 4 (1985): 89-96.

4.63. Dalrymple, D.A., Trofe, T.W., and Evans, J.M., "An Overview of Liquid Redox Sulfur Recovery," *Chem. Eng. Prog.* 85, no. 3 (1989): 43.

4.64. ASTM International, "ASTM D1835-20 Standard Specification for Liquefied Petroleum (LP) Gases," ASTM International, 2022.

4.65. European Committee for Standardization, "EN 589 Automotive Fuels—LPG—Requirements and Test Methods," European Committee for Standardization, Brussels, 2022.

4.66. ASTM International, "ASTM D8080-21 Standard Specification for Compressed Natural Gas (CNG) and Liquefied Natural Gas (LNG) Used as a Motor Vehicle Fuel," ASTM International, 2021.

4.67. European Committee for Standardization, "UNE EN 16723-2:2018 Natural Gas and Biomethane for Use in Transport and Biomethane for Injection in the Natural Gas Network-Part 2: Automotive Fuels Specification," European Committee for Standardization, Brussels, 2018.

4.68. ASTM International, "ASTM D2623 Methods for Knock Characteristics of Liquefied Petroleum (LP) Gases by the Motor (LP) Method (Withdrawn 1989)," ASTM International, 1988.

4.69. Falkiner, R.J., "Liquefied Petroleum Gas," in *Fuels and Lubricants Handbook*, Eds. Totten, G.E., Westbrook, S.R., and Shah, R.J. (West Conshohocken, PA: ASTM International, 2003), 31-59.

4.70. Morganti, K., Foong, T., Brear, M., Da Silva, G. et al., "Design and Analysis of a Modified CFR Engine for the Octane Rating of Liquefied Petroleum Gases (LPG)," *SAE Int. J. Fuels Lubr.* 7, no. 1 (2014): 283-300, doi:https://doi.org/10.4271/2014-01-1474.

4.71. Callahan, T.J., Ryan, T.W. III, and King, S.R., "Engine Knock Rating of Natural Gases—Methane Number," ASME Paper No. 93-ICE-18, 1993.

4.72. Kubesh, J., King, S.R., and Liss, W.E., "Effect of Gas Composition on Octane Number of Natural Gas Fuels," SAE Technical Paper 922359, 1992, doi:https://doi.org/10.4271/922359.

4.73. ASTM International, "ASTM D8221-18ae1 Standard Practice for Determining the Calculated Methane Number (MNC) of Gaseous Fuels Used in Internal Combustion Engines," ASTM International, 2019.

4.74. Liss, W.E., Thrasher, W.H., Steinmetz, G.F., Chowdiah, P. et al., "Variability of Natural Gas Composition in Select Major Metropolitan Areas of the United States," GRI-92/0123, Gas Research Institute, Chicago, IL, March 1992.

4.75. Dry, M.E., "Practical and Theoretical Aspects of Catalytic Fischer-Tropsch Process," *Applied Catalysis A: General* 138, no. 2 (1996): 319-344.

4.76. Eilers, J., Posthuma, S.A., and Sie, S.T., "The Shell Middle Distillate Synthesis Process (SMDS)," *Catalysis Letters* 7, no. 1-4 (1990): 253-269.

4.77. Bijwaard, H.M.J., Boersma, M.A.M., and Sie, S.T., Process for the preparation of middle distillates. U.S. Patent 4,385,193, 1983.

4.78. Sie, S.T., Senden, M.M.G., and Van Wechem, H.M.H., "Conversion of Natural Gas to Transportation Fuels via the Shell Middle Distillate Synthesis Process (SMDS)," *Catalysis Today* 8, no. 3 (1991): 371-394.

4.79. Espinoza, R.L., Steynberg, A.P., Jager, B., and Vosloo, A.C., "Low Temperature Fischer–Tropsch Synthesis from a Sasol Perspective," *Applied Catalysis A: General* 186, no. 1–2 (1999): 13-26.

4.80. Chang, C.D., Silvestri, A.J., and Smith, R.L., Production of gasoline hydrocarbons. U.S. Patent 3,928,483, 1975.

4.81. Kuo, J.C., Conversion of methanol to gasoline components. U.S. Patent 3,931,349, 1976.

4.82. Chang, C.D. and Silvestri, A.J., "MTG: Origin, Evolution Operation," *CHEMTECH* 17, no. 10 (1987): 624-631.

4.83. Haggin, J., "First Methanol-to-Gasoline Plant Nears Startup in New Zealand," *Chem. Eng. News* 63, no. 12 (1985): 39-41.

4.84. Maiden, C.J., "The New Zealand Gas-to-Gasoline Project," in *Studies in Surface Science and Catalysis*, Vol. 36, Eds. Bibby, D.M., Chang, C.D., Howe, R.F., and Yurchak, S. (Amsterdam, the Netherlands: Elsevier, 1988), 1-16.

4.85. Grimmer, H.R., Thiagarajan, N., and Nitschke, E., "Conversion of Methanol to Liquid Fuels by the Fluid Bed Mobil Process (A Commercial Concept)," in *Studies in Surface Science and Catalysis*, Vol. 36, Eds. Bibby, D.M., Chang, C.D., Howe, R.F., and Yurchak, S. (Amsterdam, the Netherlands: Elsevier, 1988), 273-291.

4.86. Kaiser, S.W., Production of light olefins. U.S. Patent 4,499,327, 1985.

4.87. Kaiser, S.W., Production of hydrocarbons with aluminophosphate molecular sieves. U.S. Patent 4,524,234, 1985.

4.88. Harrison, T., "Excavating Tar Sands in Canada," SAE Technical Paper 750723, 1975, doi:https://doi.org/10.4271/750723.

4.89. Innes, E.D. and Fear, J.V.D., "Canada's First Commercial Tar Sand Development," in *Proceedings of the 7th World Petroleum Congress*, Mexico, Vol. 3, Elsevier Publishing Co., 633, 1967.

4.90. Evans, R., "Alberta's Oil Sands-Present and Future," SAE Technical Paper 770671, 1977, doi:https://doi.org/10.4271/770671.

4.91. Allen, C.J., Thermal recovery of viscous oil. U.S. Patent 3,964,546, 1976.

4.92. Redford, D.A. and Creighton, S.M., Thermal recovery of hydrocarbon from tar sands. U.S. Patent 4,006,778, 1977.

4.93. Cram, P.J. and Redford, D.A., Thermal recovery of hydrocarbon from tar sands. U.S. Patent 4,046,195, 1977.

4.94. Coskuner, G., In situ thermal process for recovering oil from oil sands. U.S. 2009/0288827, 2009.

4.95. Kelly, J.T. and Poettmann, F.H., Situ recovery of oil from tar sands using oil-external micellar dispersions. U.S. Patent 3,800,873, 1974.

4.96. Cram, P.J. and Pachovsky, R.A., Recovery of petroleum from viscous petroleum-containing formations including tar sands. U.S. Patent 4,133,382, 1979.

4.97. Allen, A., "Equipment Operations in the Athabasca Tar Sands," SAE Technical Paper 730426, 1973, doi:https://doi.org/10.4271/730426.

4.98. Humphreys, R.D., Tar sands extraction process. U.S. Patent 5,985,138, 1999.

4.99. Cymerman, G.J., Leung, A.H.S., and Maciejewski, W.B., Pipeline conditioning process for mined oil-sand. U.S. Patent 5,264,118, 1993.

4.100. Humphreys, R.D., Tar sands extraction process. U.S. Patent 5,626,743, 1997.

4.101. Lowman, M.C. Jr. and Fisch, E., Recovery of oil from Tar sands. U.S. Patent 2,871,180, 1959.

4.102. Drake, R.D., Kobler, M.H., and Watson, J.D., Method and system for mining hydrocarbon-containing materials. U.S. Patent 6,554,368, 2003.

4.103. Satter, A. and Craig, F.F., Method for conducting cyclic steam injection in recovery of hydrocarbons. U.S. Patent 3,434,544, 1969.

4.104. Hutchinson, S.O., Method of improving steam-assisted oil recovery. U.S. Patent 3,583,488, 1971.

4.105. Butler, R.M., Method for oil recovery using a horizontal well with indirect heating. U.S. Patent 4,085,803, 1978.

4.106. Butler, R.M., Method for continuously producing viscous hydrocarbons by gravity drainage while injecting heated fluids. U.S. Patent 4,344,485, 1982.

4.107. Sanchez, J.M., Method for reducing startup time during steam assisted gravity drainage process in parallel horizontal wells. U.S. Patent 5,215,146, 1993.

4.108. Hocking, G., Enhanced hydrocarbon recovery by vaporizing solvents in oil sand formations. U.S. Patent 2007/0199711, 2007.

4.109. Perkins, T.K., Production of bitumen from tar sand formation. U.S. Patent 3,958,636, 1976.

4.110. Gill, W.G., Electrical method and apparatus for the recovery of oil. U.S. Patent 3,642,066, 1972.

4.111. Karanikas, J.M., Colmenares, T.R., Zhang, E., Marino, M. et al., Heating Tar Sans formations to visbreaking temperatures. U.S. Patent 2008/0283246, 2008.

4.112. Strange, L.K., Forward in situ combustion method for recovering viscous hydrocarbons. U.S. Patent 3,399,721, 1968.

4.113. Morse, R.A., Oil recovery by underground combustion. U.S. Patent 2,793,696, 1957.

4.114. Ishiwatari, R., Ishiwatari, M., Kaplan, I.R., and Rohrback, B., "Thermal Alteration of Young Kerogen in Relation to Petroleum Genesis," *Nature* 264 (1976): 347-349.

4.115. Harvie, B.A., "Historical Review Paper the Shale-Oil Industry in Scotland 1858-1962. I: Geology and History," *Oil Shale* 27, no. 4 (2010): 354.

4.116. Maugeri, L., "The Shale Oil Boom: A U.S. Phenomenon," Discussion Paper 2013-05, Belfer Center for Science and International Affairs, Harvard Kennedy School, Cambridge, MA, June 2013.

4.117. Maugeri, L., "Oil: The Next Revolution," Discussion Paper 2012-10, Belfer Center for Science and International Affairs, Harvard Kennedy School, Cambridge, MA, June 2012.

4.118. Caineng, Z., Zhi, Y., Jingwei, C., Rukai, Z. et al., "Formation Mechanism, Geological Characteristics and Development Strategy of Nonmarine Shale Oil in China," *Petroleum Exploration and Development* 40, no. 1 (2013): 15-27.

4.119. Compton, L.E., Removal of sulfur from shale oil. U.S. Patent 4,218,309, 1980.

4.120. Ridley, R.D. and Cha, C.Y., Two-stage removal of sulfur dioxide from process gas using treated oil shale. U.S. Patent 4,140,181, 1979.

4.121. Ridley, R.D., Removing sulfur dioxide from gas streams with retorted oil shale. U.S. Patent 4,125,157, 1978.

4.122. Tucker, S., Process for removing nitrogen compounds from hydrocarbon oil. U.S. Patent 3,004,913, 1961.

4.123. Compton L.E., Process for renoving nitrogen from shale oil using pyrrole polymerization. U.S. Patent 4,274,934, 1981.

4.124. Myers, G.A. and Wunderlich, D.K., Shale oil treatment. U.S. Patent 3,804,750, 1974.

4.125. Young, D.A., Process for removing arsenic from hydrocarbons. U.S. Patent 4,046,674, 1977.

4.126. Young, D.A., Oil shale retorting process. U.S. Patent 4,127,469, 1978.

4.127. Britton, M.W., Martin, W.L., McDaniel, J.D., and Wahl, H.A., Fracture preheat oil recovery process. U.S. Patent 4,265,310, 1981.

Further Reading

Blinderman, M. and Klimenko, A., *Underground Coal Gasification and Combustion* (Duxford, UK: Woodhead Publishing, 2017).

Hu, J. and Burns, A., "Index Predicts Cloud, Pour and Flash Points of Distillate Fuel Blends," *The Oil and Gas Journal* 68 (1970): 45-66.

Hutcheson, R.C. and van Paasen, C.W.C. "Diesel Fuel Quality into the Next Century," in *Institute of Petroleum Symposium on the European Auto Diesel Challenge*, London, UK, April 1990.

Rutherford, R.W., "Oil from Coal in Japan," *Colliery Eng.* 21 (1944): 40-42.

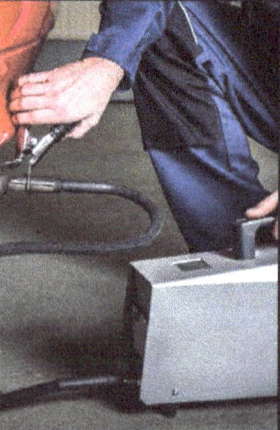

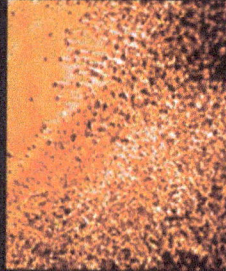

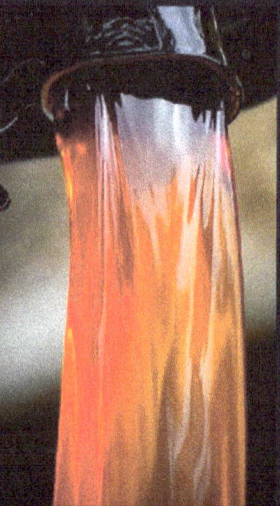

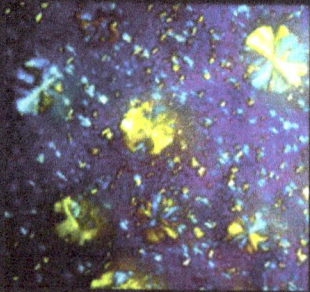

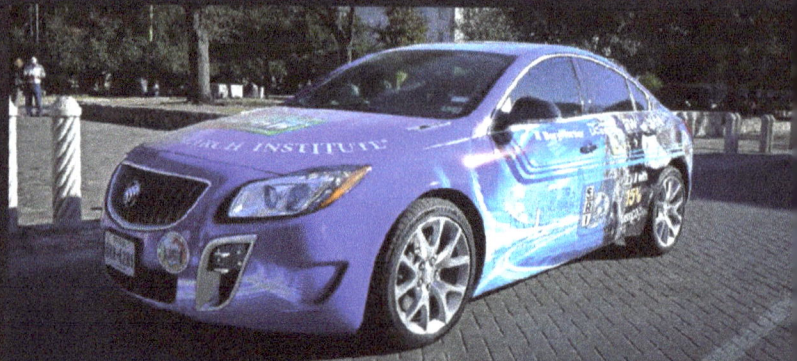

Manufacture of Gasoline and Diesel Fuel from Renewable Sources

5.1. Introduction

The previous chapters have discussed the production of gasoline and diesel fuel from fossil sources, coal, sand and tar deposits, crude oil, and natural gas. However, gasoline and diesel fuel, as we would know them today, did not exist at the inception of the automotive industry. As noted in Chapter 2, the forebears of today's gasoline engines were developed to run on coal gas, while Rudolf Diesel envisaged an engine running on pulverized coal. These early engines were of necessity fairly flexible in their appetite for fuel, Diesel's patent [5.1] explicitly stating,

Every kind of fuel in any state of aggregation is suitable for carrying out the process.

In 1900, a diesel engine was famously (although not at the time) demonstrated running on groundnut (*Arachis hypogaea*) oil, as Diesel himself later wrote [5.2],

At the Paris Exhibition in 1900 there was shown by the Otto Company a small Diesel engine, which, at the request of the French Government, ran on Arachis (earth-nut or pea-nut) oil, and worked so smoothly that only very few people were aware of it. The engine was constructed for using mineral oil, and was then worked on vegetable oil without any alterations being made. The French Government at the time thought of testing the applicability to power production of the Arachide, or ground-nut, which grows in considerable quantities in their African colonies...

This finding was confirmed 80 years later by investigators at the University of Michigan [5.3]. In 1907, Thomas L. White wrote [5.4],

In considering the possibilities of alcohol as a fuel for automobile motors, it is impossible to avoid alluding, however briefly, to the economic conditions which must eventually determine its use as a fuel at all, and this independently of all technical considerations. Gasoline is the by-product of a geographically limited and monopolistically controlled industry, and there are reasons to believe that the available supply is more than mortgaged by a worldwide and growing demand. Alcohol is, one might say, the product of the four seasons. It can be manufactured from any vegetable substance which contains sugar or some material like starch, which is easily convertible into sugar. As to available supply, it can be and will be produced in unlimited quantities at a steadily diminishing cost. From corn-cobs it has already been experimentally prepared at five cents a gallon, and there seems little reason to doubt that if the technical problems connected with its use can only be solved, and if, also, what is important at the present moment, its use can be reduced to current practice; if, in other words, some means can be devised of economically burning this fuel in the thousands of automobile motors in existence to-day, there is little doubt that the demand so created will be satisfactorily met, both as to quality, quantity and price

Some years later, Henry Ford told *The New York Times* [5.5],

The fuel of the future is going to come from fruit like that sumac out by the road, or from apples, weeds, sawdust—almost anything. There is fuel in every bit of vegetable matter that can be fermented. There's enough alcohol in one year's yield of an acre of potatoes to drive the machinery necessary to cultivate the fields for a hundred years.

The farming industries of the time clearly held competing sway with the oil industry! However, it was clear from these early years that the use of bio-derived fuels was clearly of great interest.

The emergence of the oil refining industry and the supply of cheap oil consigned these renewable fuels to niche markets, despite early concerns of imminent exhaustion of petroleum supplies [5.6, 5.7]. As early as the end of the first quarter of the last century, there were concerns over security of supply. In 1922, Professor C. A. Norman wrote [5.8],

The best estimates that can be procured are made by Eugene Stebinger, of the United States Geological Survey, who is an expert on foreign oil fields. There is a small supply in Canada; in Mexico about 4,500,000,000 bbl.; in Northern South America, including Peru, 5,700,000,000 bbl.; and in Southern South America about 3,500,000,000 bbl. That includes about all the supplies of oil in this hemisphere. We find oil in quantities of, roughly, 5,500,000,000 bbl. available in Persia and Mesopotamia, 5,500,000,000 bbl. in Russia and 3,000,000,000 bbl. in the East Indies. Can we get this oil if we want it? Here we are faced with a most startling situation. It appears that while we have been exhausting our natural resources at a terrific rate, Great Britain has put oil resources under her commercial and in some cases her political sway, all over the globe. She controls at present, directly and indirectly, 75 percent of the world's future oil supply. The oil in Persia is controlled by an agreement with the Persian Government. The oil in Mesopotamia is controlled jointly by France and England. Holland shuts out all nationals except the Dutch from the exploitation of oil in Java. In other places British capital simply is in control. We must compete with Great Britain to obtain oil from the countries on this side of the globe.

The US was not the only country that was concerned about scarcity of crude oil supply and considered the use of alcohols to extend the gasoline pool. In the 1920s, Germany produced a fuel comprising 25% bio-alcohol (from fermented potatoes) and 75% gasoline that was marketed under the name Monopolin, and in Czechoslovakia, a 50% alcohol blend was marketed as Dynacol. In the UK, by the mid-1930s, a 10% alcohol blend was launched under the name Koolmotor [5.9] and Cleveland Discol fuel [5.10] was marketed by Cleveland Petroleum Company, which later became Esso Petroleum Company Ltd. Interestingly, this fuel was marketed based on its volatility, which aided winter starting, whereas later use of ethanol blend generated concerns due to the volatility of the ethanol, leading to an increase in evaporative emissions [5.11–5.13].

World War II brought about a drastic reduction in the free flow of petroleum products, which caused a flurry of activity in the production of fuels from domestic sources. China, for example, installed cracking plants [5.14]. The crude oil yield was 70% by volume of the Tung oil, 25% of which was usable as gasoline.

After World War II, the geopolitical situation changed, and concerns about the supply of crude oil appeared to abate until the drastic rise in the price of crude oil in the early 1970s caused a renewed appraisal of the benefits of producing fuel from indigenous sources. The use of coal is discussed in the previous chapter, but not all countries have plentiful supplies of coal and attention was turned to the use of bio sources. One of the most successful examples of the time was the 1975 Brazilian ProAlcool program that was set up with the aim of substituting gasoline with ethanol derived from biomass. Suggested sources were sugarcane (*Saccharum officinarum*), cassava (*Manihot esculenta*), and sorghum (*Sorghum bicolor*) [5.15]. All three suggested feedstocks have a high accessible glucose (sugar) content. In tropical and subtropical countries such as Brazil, sugar cane is the most important feedstock for producing ethanol, whereas in countries with a more moderate climate, such as Europe, sugar beets are more commonly used, and research suggests that sorghum would be preferable feedstock in more arid countries [5.16]. Sugar cane therefore became the feedstock of choice for the ProAlcool program. However, although ethanol had been widely used in low concentration, engine and vehicle technology changes were necessary to allow the use of the high concentrations envisaged in the ProAlcool program. By 1979, vehicles were being produced that would run on 100% hydrated ethanol [5.17]. The Brazilian market offered two types of oxygenated fuels: alcohol or E100, which was 100% hydrated ethanol, and gasohol or E22, which was a mixture of gasoline and anhydrous ethanol [5.18]. Other countries also began to use ethanol in either neat or blended form, including the US, the Philippines, South-Africa, and Kenya [5.19].

Latterly growing concern about global temperatures and the belief that these are linked to anthropogenic CO_2 emissions [5.20, 5.21] has prompted renewed interest, and in many countries, legislation to promote the use of bio-derived fuel [5.22, 5.23]. The following sections describe briefly the different technologies employed to convert renewable resources into automotive fuels.

5.2. Minimally Processed Vegetable Oil

The concept of using vegetable oils as an automotive fuel is not new. As noted, the Otto Company exhibited an engine running on groundnut oil at the *Exposition Universelle* of 1900 in Paris. Although interest in vegetable oils as fuel waned during the following years, the farming community remained interested because of their ready access to such sources.

There are no records of how long these early engines could run on vegetable oil, but it very soon became apparent that one of the major problems of trying to operate diesel engines on vegetable oil was the propensity of such fuels to form carbonaceous deposits in the engine, particularly on the fuel injectors and the piston rings, leading to piston ring sticking and scuffing of the cylinder liners [5.24]. The vegetable oil would also pass the piston rings and mix with the lubricant, where it would polymerize resulting in failure of the lubricant. The use of neat vegetable oils also had a tendency to plug fuel filters, with reports of filters clogging and reducing fuel flow every 30 hours of operation [5.25].

The neat unrefined vegetable oils, which are predominantly triglycerides, will typically also contain free fatty acids, phospholipids, sterols, water, and other impurities. These additional components will affect the performance of these oils as a substitute, or extender, for diesel fuel, and as discussed later will affect their use and performance as a feedstock for biodiesel production [5.26]. The gums are soluble in dehydrated vegetable oils, but small quantities of water will cause them to be precipitated. The oils may thus be degummed by adding about 3% of water (with or without a degumming agent such as phosphoric acid) and mixing at 65°C for between 30 and 60 minutes. The impurities tend to migrate to the water, which is removed from the mixture after it has been centrifuged [5.27].

In the early 1980s, durability testing was performed in Brazil using fully refined soybean oil [5.28]. Five large-engine (1.75 L per cylinder or larger) off-road vehicles accumulated over 10,000 hours of operation between them, without major problems. However, inspection of the engines revealed higher ring and liner wear than would be acceptable with a US diesel fuel. All of these engines were of the indirect injection (IDI) type. A German study [5.29] using rapeseed oil also concluded that large cylinder capacity IDI engines were able to operate satisfactorily on neat rapeseed oil but direct injection (DI) engines and smaller IDI engines were not suited to operate on rapeseed oil. In the US, using sunflower oil, it was confirmed that DI engines were not suited to this type of fuel [5.30]. In all these cases, the reasons given for the incompatibility of fuel and engine operation were deposit build-up, ring sticking, and oil contamination. Researchers in Japan also concluded that DI engines could be operated satisfactorily only for short periods on neat rapeseed oil and palm oil [5.31]. Operation for a longer period of time resulted in carbon deposit build-up and piston ring sticking. However, they did investigate possible routes to overcome these problems; these included increasing the fuel temperature to over 200°C, blending diesel fuel or ethanol into the vegetable oil, or converting the vegetable oils into methyl esters! A US study also found that by diluting the vegetable oil with diesel fuel, satisfactory operation could be achieved [5.32]. However, in the US study, the blend ratio was only 20% rapeseed oil, whereas the Japanese had concluded that 75% vegetable oil was acceptable.

Work in the US confirmed that IDI type engines could run satisfactorily on vegetable oils such as sunflower oil [5.33] and soybean oil [5.28].

Subsequent work has been conducted to investigate whether fuel additives can be used to overcome some of these problems. Work was performed using a barium base smoke suppressant (which would act as an oxidation catalyst on carbon-based deposits) in combination with an ashless dispersant [5.34]. However, later concerns about increased toxicity of exhaust emissions when using this additive [5.35] and a ban on metal fuel additives in many countries renders this approach impractical. More recently, other researchers found good performance using only an ashless fuel additive [5.36].

Due to the difficulties of using minimally processed vegetable oils and the relative ease of producing transesterified vegetable oils (biodiesel), which have fewer problems, there is presently no large-scale commercial production of these oils as a fuel or diesel fuel extender; although it has been suggested that there may be examples of blending of cooking oil into diesel fuel to illegally avoid road fuel tax [5.37]. As will be discussed later these vegetable oils can be hydrotreated to produce good hydrocarbon fuels.

5.3. Bio-Ethanol

All plants consist of the following polysaccharides: cellulose, hemicellulose, lignin, and pectin. The proportion of these constituents varies within the plant and from species to species.

Put simply, cellulose is a polymer of glucose joined by beta-linkages. The linear cellulose molecules are typically arranged as a parallel array with extensive hydrogen bonding between them. The result is a highly ordered, crystalline material that is generally resistant to chemical attack. Hemicellulose is a polymer composed of several sugars, predominantly xylose, a five-carbon sugar as opposed to glucose, which is a six-carbon sugar. Hemicellulose has a more branched structure and is thus essentially noncrystalline and more chemically reactive [5.38].

The process for producing ethanol from the plant material will usually include the following steps: some form of pretreatment to prepare the biomass for saccharification, a saccharification process, which is breaking down the polysaccharides to their monomers (sugars); a fermentation step to convert the sugars into alcohol; and finally, a distillation step to separate the alcohol. The pretreatment may be as simple as grinding, milling, or chopping the biomass, to chemical and/or biological processes to break down the plant structure to make it more amenable to saccharification. The feedstock will determine how many of these steps are required and the details of each step: sugars will not need the same degree of pretreatment and will not need saccharification, starchy crops will need saccharification as well as fermentation, and cellulosic feedstocks will need the greatest amount of pretreatment before they can undergo saccharification and fermentation.

Common sugar crops used as feedstocks are sugar cane, sugar beets, and, to a much lesser extent, sweet sorghum [5.39]. Common starchy feedstocks include corn (maize), wheat, and cassava.

The fermentation step produces CO_2, which can easily be captured and forms a useful co-product for the process.

5.3.1. Ethanol from Sugar Crops

The sugarcane, when harvested, consists of stalk, tops, and leaves. Most of the sugars are contained in the stalks, which are the main source of input to the ethanol production process, and therefore most of the leaves and tops (sugarcane trash) are left in the fields as mulch. The sugarcane stalks delivered for processing contain the sugars (12–17%), rich in sucrose (about 90% of all the sugars) [5.40], plus unwanted water, fiber, and other impurities. The first stage of processing is to remove any physical contaminants, and the cane is then pressed to squeeze out the sugars from the remaining fibrous material, known as the bagasse. The bagasse can be burned to produce heat for the production processes or supplied for paper-making, and so on. The sugar is then fermented (the yeast *Saccharomyces cerevisiae* is frequently used for this stage); the sugar undergoes initial hydrolysis to break the sucrose down into smaller molecules of glucose. The enzyme *invertase*, found in the yeast, catalyzes this process.

$$C_{12}H_{22}O_{11} + H_2O + Invertase \rightarrow 2C_6H_{12}O_6$$

The enzyme complex zymase, derived from the yeast, then catalyzes the conversion of the glucose into alcohol and CO_2.

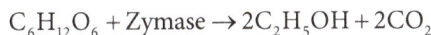

$$C_6H_{12}O_6 + Zymase \rightarrow 2C_2H_5OH + 2CO_2$$

Sugar beets (*Beta vulgaris*) are processed in a similar fashion but require a greater net energy input to the processing plant. This is because there is no appreciable equivalent of the bagasse produced by sugar cane; thus, the energy that can be recovered by burning the bagasse is not available when processing beets. However, financially, there are other by-products from processing beets, namely the tops and the pulp can be used as animal feed. The beets delivered to the processing plant are first chopped into thin chips called cossettes. The cossettes are washed and then pressed to remove the remaining water and sugar. This sucrose-rich liquor is then cleaned of impurities and passed to the fermentation step, as described for the sugar cane.

Traditionally the fermentation process has been conducted as a batch process, similar to the production of wines. However, the development of continuous fermentation processes has resulted in more cost-effective production [5.41, 5.42]. As might be imagined, for a fermentation process the feedstock is constantly (or quasi-continuously) fed into the reactor while an equal flow of the fermented product is drawn off to maintain the volume within the reactor [5.43]. There are two basic types of continuous reactors: the continuous stirred tank reactor (CSTR) or the plug flow reactor (PFR). As implied by the name, in a CSTR, the mixture is stirred, leading to an almost homogenous mixture that is of the same composition as that drawn off. Because the mixture is nearly homogeneous, it is inevitable that the product drawn off contains some unfermented product. Higher yields are therefore obtained by cascading a number of CSTRs [5.44]. In a PFR, the reactants are pumped through a pipe, and the reaction proceeds as the mixture travels slowly through the reactor.

5.3.2. Ethanol from Grain Crops

Sugars can also be extracted from starchy crops, including grains such as corn and wheat, and from tubers such as potatoes and cassava. Starch molecules are made up of long chains of glucose molecules. Breaking down the starch molecules to form sugars (saccharification) requires the additional step of hydrolysis. This converts the starch into sugar by means of an enzymatic process before fermentation of the sugar into ethanol. This additional step inevitably adds to the cost and energy requirement of the production process.

The initial step in the production of ethanol from grain can be divided into two basic categories: dry mill and wet mill. In the dry mill process, the entire grain is ground up and then subsequently mixed with water to form a mash. In many dry mills, the "spent mash," often called distillers dried grains with solubles (DDGS), is a valuable coproduct that can be sold as a high protein animal feed. For the wet mill process, the grain is steeped in water and a dilute acid to allow it to be separated into its four basic components: germ, fiber, gluten, and starch. The starch is separated for subsequent hydrolysis to produce sugars; the germ can be used to produce oil while the fiber and the gluten can be recovered to produce gluten meal.

The subsequent steps of hydrolysis and fermentation can again be subdivided into two approaches. An initial step where α-amylase is used to attack the starch, usually at a temperature between 90 and 180°C with a holding time of about 30 minutes, is common to both approaches. The resultant liquefied starch then undergoes saccharification to convert it to glucose, often using glucoamylase. The glucose then undergoes fermentation to ethanol as previously described for the sugar-to-ethanol process. To reduce the energy cost of heating the broth to such high temperatures, noncooking and low-temperature cooking fermentation systems have been developed [5.45].

In the second approach, the saccharification and fermentation processes can be combined in one reaction vessel, known as simultaneous saccharification and fermentation (SSF) [5.46]. The advantage of the SSF is that it will tend to convert the sugars into ethanol as soon as they are formed, thus reducing the accumulation of glucose in the broth. Because the sugars are much more inhibitory of the conversion process than ethanol, higher reaction rates and yields can be achieved when compared with the sequential processes. The disadvantage of SSF is that the optimum conditions for hydrolysis and fermentation are different; this makes the control and optimization of process more difficult [5.47]. The SSF approach also requires a greater amount of enzyme. The SSF process is considered more appropriate for the dry mill approach [5.48]. The dry mill approach with SSF is shown schematically in Figure 5.1 alongside the wet mill approach with sequential hydrolysis and fermentation.

FIGURE 5.1   Schematic comparison of dry mill and wet mill approaches.

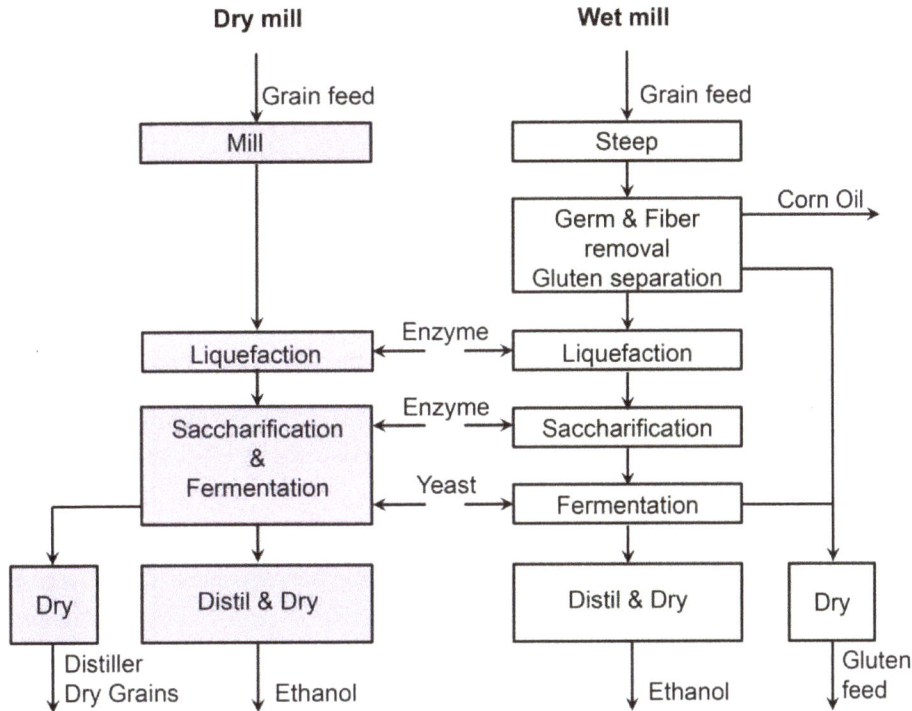

After fermentation, the ethanol is distilled and filtered to produce the fuel grade ethanol in a similar manner to that used for the production of ethanol from sugar crops.

5.3.3. Lignocellulosic Ethanol

The production of ethanol from lignocellulosic feedstocks is preferred over the manufacture from sugar or starch crops as it eliminates any potential conflict between the use of the feedstock for food or fuel. Manufacturing ethanol from lignocellulosic material also allows for the use of wastes such as wood chippings, corn stover, corn cobs, or straw or specifically grown crops such as switch grass (*Panicum virgatum*) and miscanthus. For a long time, there has been interest in using these wastes to produce useful products [5.49], but as yet they have not proved economically competitive. Other routes for the use of such materials to provide renewable fuel components are also being considered [5.50, 5.51].

The production of ethanol from lignocellulosic material differs from the production from starch crops only in the pretreatment processes employed. The cellulous is intimately bound with the lignin and cannot be readily hydrolyzed and must therefore undergo a greater degree of pretreatment before it can be passed to the saccharification stage of the production process. A variety of different pretreatment techniques are being developed, including acid hydrolysis, alkaline hydrolysis, steam explosion (auto-hydrolysis), ammonia fiber explosion (AFEX), organic solvent pretreatments, and others.

Acid hydrolysis can be further subdivided into concentrated acid and dilute acid hydrolysis. Concentrated acid hydrolysis uses highly concentrated sulfuric acid or hydrochloric acid to dissolve the cellulose. The concentration of the acid is then reduced and the mixture is cooked for about 3 hours at 100 to 120°C to covert the cellulous oligomers to the glucose monomers. This process produces high yields of glucose but presents manufacturing problems due to the high concentration of the acid. The dilute acid hydrolysis approach uses a much more dilute acid but requires higher cooking temperatures, typically 180 to 240°C. The cooking times are considerably shorter, typically minutes, thus allowing continuous processing. However, the glucose yields are typically only half of those from the concentrated acid hydrolysis approach.

Alkaline hydrolysis follows a similar route, but using alkali instead of acid to separate the lignin from the cellulose. Various alkalis including H_2O_2 [5.52, 5.53], NaOH [5.54], and $Ca(OH)_2$ [5.55] have been proposed. The dilute alkali is thought to cause swelling of the lignocellulosic material, leading to a decrease in the degree of polymerization and crystallinity, making the cellulose more accessible for enzymatic attach in subsequent processes. Typical conditions for alkaline hydrolysis include a temperature of between 100 and 150°C and may include an oxidative step [5.56].

The steam explosion technique utilizes rapid vaporization of water to bring about explosive decomposition of the cellulosic material. The mechanically reduced feedstock is subjected to saturated steam, at a pressure of 1.7 to 6.9, before the pressure is rapidly reduced to atmospheric pressure [5.57]. Steam explosion can be combined with dilute acid to enhance the process.

The AFEX process is similar to the steam explosion technique, except that the lignocellulosic material is impregnated with ammonia rather than steam [5.58, 5.59]. The biomass is steeped in liquid ammonia under high pressure before the pressure is released to bring about the explosive decomposition. Pressures of 1.4 to 3.5 MPa and a temperature of 60 to 200°C are typically used. The AFEX process is not as effective for materials with a very high lignin content.

Organic solvent pretreatments include the use of acetone [5.60] and alcohols [5.61] to enhance the separation of the cellulose from the lignin.

5.4. Biodiesel Fuel

Biodiesel is recognized as a term to represent mono-alkyl esters of long-chain fatty acids, which are used as a replacement for or a blending component for fossil diesel fuel. The predominant mono-alkyl ester, presently in use, is for economic reasons methanol based, with the possible exception of Brazil, where ethanol is sometimes used [5.62, 5.63]. The predominant biodiesel is fatty acid methyl ester (FAME). Biodiesel is now commonly used as a blending component to help meet the various legislative requirements to include a percentage of the fuel makeup from renewable sources. FAME is used in preference to some of the other possible alternatives because of the relatively straightforward and inexpensive processing methods and equipment. The fatty acids, which are usually bound as triglycerides, that are used to make the FAME, can be derived from vegetable or animal sources. A great many oil-bearing crops have been identified, which are pressed to extract the oil; extraction, or further extraction from the pressed oil, with hexane is also commonly used. The biodiesel is produced by chemically reacting the plant oils or animal fat with an alcohol (methanol to make methyl esters) in the presence of a catalyst. The product of this reaction is a mixture of methyl esters, dependent upon the starting products, and glycerol, which may be a valuable product in its own right.

Again, the notion of using vegetable oils as a fuel source is not new and the idea of transesterification of vegetable oils to enhance their suitability as a fuel also goes back more than 70 years. In 1937, Charles George Chavanne was granted a Belgian patent for a "Method for transformation of vegetable oils for use as fuels" [5.64] in which he illustrates the transesterification of palm oil with ethanol in the presence of an acid catalyst.

The most important issue during biodiesel production is the completeness of the transesterification reaction. If the feedstock is high in free fatty acids (FFAs), as opposed to the fatty acids being bound in the triglyceride molecule, then these are usually reacted in an acid pretreatment, as shown:

$$R\text{-}COOH + CH_3OH \rightarrow R\text{-}COOCH_3 + H_2O$$

Fatty Acid   Methanol   Methyl-ester   Water

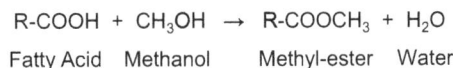

where R is a long-chain hydrocarbon.

The basic chemical reactions that occur during the transesterification process are indicated below, where R_1, R_2, and R_3 are different long-chain hydrocarbons, usually containing from 8 to 24 carbon atoms.

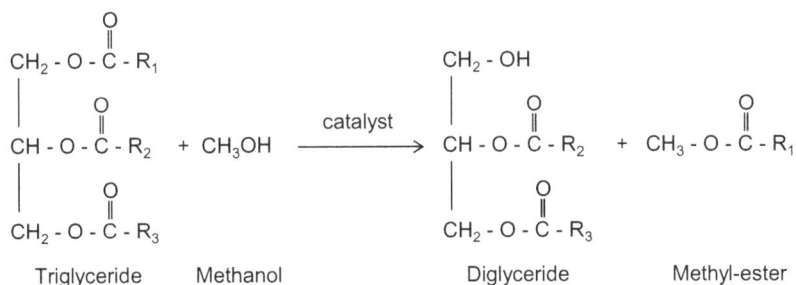

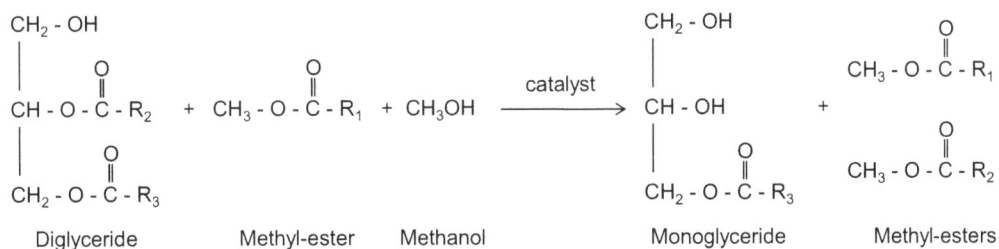

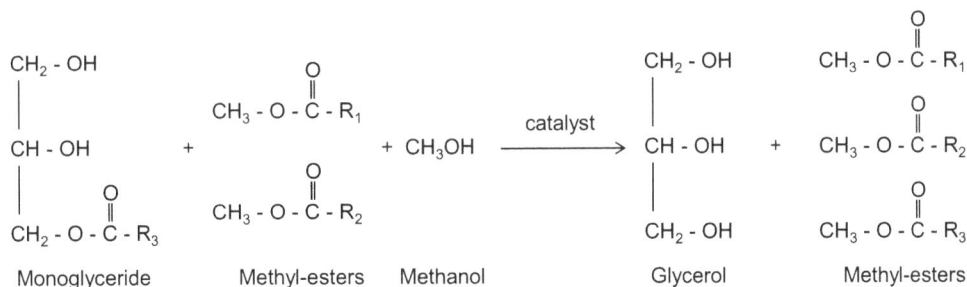

As seen above, diglycerides and monoglycerides are formed as intermediate products during the reaction. The reaction rates for the different equations shown above and hence the concentration of the various compounds, over time, will be dependent upon the concentration of the reactants, the type and quantity of catalyst, the reaction conditions (pressure and temperature), and the presence of any impurities. It is also clear from the equations that the feedstocks for the transesterification process must contain a minimal amount of water, otherwise the water will react with the methyl esters formed to form FFAs, as shown below.

$$R\text{-COOCH}_3 + H_2O \rightarrow R\text{-COOH} + CH_3OH$$

Methyl-ester   Water   Fatty Acid   Methanol

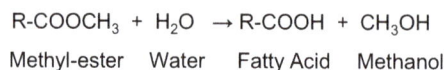

Once the transesterification process has been completed, the FAME must be separated from the glycol. The methyl esters and the glycerol are not miscible, so once the reactions are complete the methyl esters can easily be separated, either by allowing the products to rest in a settling tank or by centrifuging. Once separated from the glycerol, the biodiesel is further processed to remove any unreacted product and catalyst. The excess alcohol is removed by a vacuum flash process or a falling film evaporator. The most commonly used catalysts are alkali metal hydroxides, usually sodium or potassium. The catalyst is then neutralized by the addition of acid, which also has the effect of breaking down any metal soaps that have formed. The soaps are broken down to yield fatty acid and a metal salt. The biodiesel is then water washed to remove the salts, and any remaining catalyst, soap, salts, alcohol, and free glycerol. Any FFA produced will remain in the biodiesel. A basic biodiesel production plant is show schematically in Figure 5.2.

FIGURE 5.2 Schematic of basic biodiesel production plant.

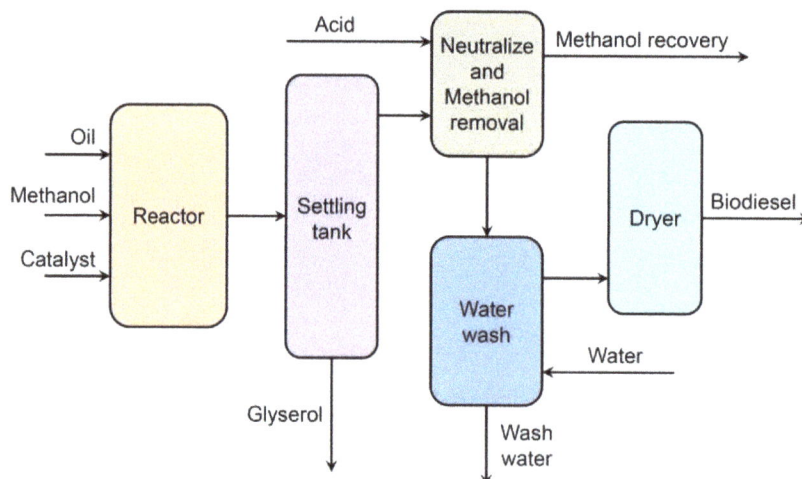

The composition of the resulting FAME is clearly dependent upon the feedstock, and with such a wide choice of feedstock there can be significant differences in the resultant FAME. The different methyl esters are usually classified according to the number of carbon atoms in the fatty acid chain (i.e., the total number of carbon atoms in the methyl ester molecule, excluding those originating from the alky group of the alcohol) and the degree of saturation of that chain. Thus, caprylic acid, which can be written as $CH_3(CH_2)_6COOH$ and forms the methyl ester $CH_3(CH_2)_6COOCH_3$, is classified as C8:0, the 8 indicating the number of carbon atoms in the parent fatty acid and the 0 that there are no double bonds in the molecule, meaning that it is saturated. The methyl ester of Linoleic acid, which can be written as $CH_3(CH_2)_4CH=CHCH_2CH=CH(CH_2)_7COOCH_3$, is thus classified as C18:2 to indicate that there are two double bonds in the molecule, and it is polyunsaturated. The composition of the FAME produced from a number of different feedstocks is given in Appendix 4.

FAME produced from soybeans (*Glycine max*), which is the predominant feedstock for FAME production in the US, is predominantly C18:2, with the second major component being C18:1. This implies that it is going to be more prone to oxidation due to the preponderance of polyunsaturated and monounsaturated molecules, the polyunsaturated molecules being the more important in this respect. Rapeseed (*Brassica napus*, of which canola is a cultivar) oil is the main source of FAME within Europe; *Brassica napus* is illustrated in Figure 5.3.

FIGURE 5.3   The familiar sight of a bright yellow field of *Brassica napus* in bloom (upper left), the plant in bloom (upper right), the ripening seed pods (lower left), and a seed pod showing the mature seeds (lower right).

© SAE International.

Rapeseed oil is predominantly C18:1, with the second major component being C18:2. Thus the rape methyl ester (RME), as it is commonly known, will be slightly less prone to oxidation than the soy methyl ester (SME). Palm oil (*Elaeis guineensis*), which is commonly used is Asia, contains roughly equal parts of C16:0 and C18:1, while coconut (*Cocos nucifera*), which is again commonly used in Asia, contains predominantly C12:0 and C14:0, with typically less than 10% unsaturated molecules. The latter is thus likely to be more oxidatively stable than the other FAMEs. However, because the coconut methyl ester (CME) is predominantly composed of straight chain molecules, it is likely to have a higher melting point and thus poorer cold flow properties. These characteristics of different FAMEs will be discussed in more detail in later chapters.

The feedstock not only determines the composition, and hence the characteristics, of the finished biodiesel, it also determines the optimum conditions and catalyst for the production process. The commonest production processes use homogenous base catalysts, such as sodium hydroxide or potassium hydroxide, because of their ready availability and low cost. However, if the quantity of FFA in the feedstock is too high, then the FFA will react with the base catalyst to produce unacceptable levels of soap. As noted above, this is usually countered by an acid pretreatment. The base catalyst acts by dissociating to form hydroxyl ions, which then react with the methanol to form methoxide ions, as shown below:

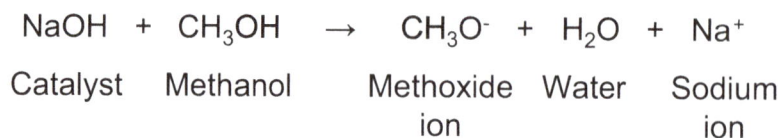

$$NaOH \; + \; CH_3OH \quad \rightarrow \quad CH_3O^- \; + \; H_2O \; + \; Na^+$$

| Catalyst | Methanol | | Methoxide ion | Water | Sodium ion |

The methoxide ions, which are strong nucleophiles, then attack the carbonyl moiety in glyceride molecules to form alkyl esters. As the active species is the methoxide, sodium or potassium methoxide are sometimes purchased and used as the catalyst. Acid catalysts have been proposed [5.65–5.67], but the homogeneous acid-catalyzed reaction is orders of magnitude slower than the homogeneous base-catalyzed reaction [5.68].

The choice of sodium or potassium hydroxide can be optimized according to the feedstock; for example, sodium hydroxide is favored for the processing of canola (*Brassica napus*) [5.69], sunflower (*Helianthus annuus*) [5.70], and waste cooking oils [5.71], while potassium hydroxide was favored for the processing of Jatropha (*Jatropha curcas*) [5.72], Karanja (*Pongamia pinnata*) [5.73], and Mahwa (*Madhuca*) [5.74], economics can change this situation. Reactions are usually performed at atmospheric pressure and a temperature between 55 and 65°C, although the optimization of the esterification, with potassium hydroxide, of mustard seed oil (*Brassica carinata*) and jojoba (*Simmondsia chinensis*), was found to give a 97 and 83.5% yield, respectively, at a temperature of only 25°C [5.75, 5.76]. Reaction times tend to be between 1 and 2 hours, but again there are exceptions, with a 99% yield reported for jatropha after only 24 minutes [5.72] and a 98% yield for Mahwa [5.74].

Impurities in the biodiesel can significantly affect engine performance. Metal compounds, either from the esterification catalyst or the purification process, can adversely affect exhaust aftertreatment devices [5.77–5.79] and saturated monoglycerides, and sterol glucosides can have a deleterious effect on the cold flow performance of biodiesel blends, as discussed in Chapter 17.

There have been numerous studies to investigate the performance of neat biodiesel and biodiesel/fossil diesel blends, which will be discussed in later chapters. It is clear that some of the reported problems are due to poor quality biodiesel. It is therefore imperative that the biodiesel must meet standards that are currently in force, for example BS EN 14214 [5.80], IS:15607 [5.81], and ASTM D6751 [5.82]. There is no equivalent standard in Japan as B100 is not allowed. Standards are set for B5 fuel and following testing to investigate acid generation in FAME/diesel blends and the effect on corrosion [5.83]. This work concluded that besides a limit in TAN there should also be a limit of 30 ppm on the sum total of formic, acetic, and propionic acids.

5.5. Co-Processing Feedstocks

To increase the renewable content of gasoline and diesel fuel requires the incorporation of bio-derived feedstock. These bio-feedstocks must be processed in some way before they can be incorporated in to gasoline and diesel fuel, conversion to alcohols or esters, as discussed above, is now a proven route. Another approach is to process these bio-feedstocks to make them more compatible with refined or partially refined crude oil and to incorporate them into the refinery process as co-processing feedstocks. This processing usually includes deoxygenation and some hydrogenation to increase the effective hydrogen/carbon ratio to closer to that of the fuel it is to replace. The bio-derived starting product is usually starchy or cellulosic materials [5.84, 5.85] or oils [5.86] and fats [5.87], such as may be used to produce bio-ethanol or biodiesel as discussed above. There are three approaches to the processing of these products: oleochemical processes, thermochemical processes, and biochemical processes, each is discussed below.

5.5.1. Oleochemical Route

Oleochemical processes take vegetable oils and fats, as are used for the production of biodiesel, but instead of esterification, use of catalytic deoxygenation/decarboxylation/decarbonylation [5.88, 5.89] and/or hydroprocessing to catalytically remove the oxygen from the fatty acid chains present in the lipid feedstock. As illustrated in Section 5.4 and Appendix 4 most of the fatty acid chains contain 8 to 24 carbon atoms which is similar to diesel fuel range. Because the feedstock contains a relatively low oxygen content the process requires relatively low additional hydrogen input. The hydroprocessing step results in these fuels being called hydrotreated esters and fatty acids (HEFA). This technology is well developed and is discussed in Section 5.6 as Hydrotreated Vegetable Oil (HVO). Due to the difficulties of incorporating bio products into aviation fuel via other routes HEFA has seen noticeable uptake in this field [5.90, 5.91]. This technology can, and is, being used to produce a similar product that can then be co-processed in an oil refinery.

5.5.2. Thermochemical Route

Thermochemical processing of biomass produces three basic products, bio-oil, synthesis gas and char. The process conditions can be optimized to influence the ratio of these three products. There are two main routes to fuel production which can be categorized as gasification and pyrolysis.

Gasification involves oxidizing the biomass in a reduced oxygen environment, similar to the process of indirect coal liquefaction outlined in Section 4.2.3. The syn-gas that this process produces can then be processed by the F-T process or other routes.

Pyrolysis is thermal decomposition brought about in the absence of oxygen. A lower process temperature and a longer hot vapor residence time favors the production of high carbon/hydrogen ratios, such as coke or charcoal. Higher process temperatures and long vapor residence times favors higher hydrogen/carbon ratios, i.e., gases, and moderate temperatures and short hot vapor residence times favor the production liquids. The choice of process conditions obviously determines the proportions of products produced. Fast pyrolysis, with temperature of about 500°C and very short hot vapor residence times can maximize liquid yield. The bio-oils produced can be used as co-feeds to a fluidic catalyst cracker unit [5.92].

5.5.3. Biochemical Route

This is a far less mature route that could be described as advanced fermentation, which can result in relatively pure molecular streams, such as higher alcohols [5.93]. These metabolic processes are, at present, far more energy intensive and not as efficient as conventional sugar-to-ethanol fermentation processes.

5.6. Hydrotreated Vegetable Oil

As discussed in Section 5.4, vegetable oils and animal fats, triglycerides, are a renewable source of carbon-based fuels. The triglycerides are oxygenated compounds which, using the relatively straightforward esterification processes described previously, are broken down into smaller molecules that have physical properties that are more akin to the diesel fuel that they are intended to replace. However, they remain as oxygenated compounds. This has some advantages in that the bound oxygen in these compounds can aid oxidation within the combustion process but has the disadvantage that it reduces the energy content of the fuel and thus the customer perception of fuel economy. An alternative approach to esterification is hydrogenation and deoxygenation of the triglyceride. This approach results in paraffins rather than esters. These paraffins are then directly compatible with fossil-derived diesel fuel chemistry and can be processed, if necessary, using the same downstream techniques that are used for paraffinic streams derived from fossil sources, in particular those derived via the Fischer-Tropsch process. The downside is that the equipment necessary to hydrogenate the triglycerides is more complex and expensive than that for the esterification process. The overall chemistry of the process is illustrated by the following equation.

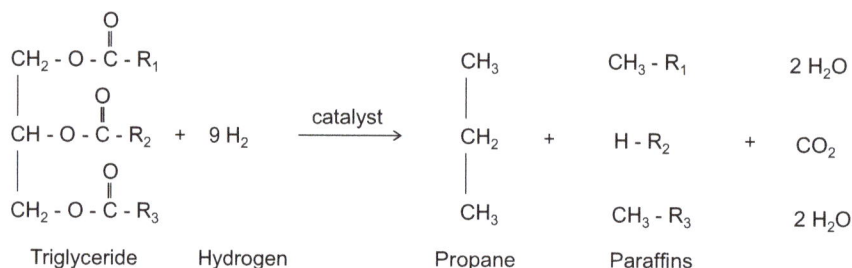

$$
\begin{array}{c}
CH_2 - O - \overset{\displaystyle O}{\overset{\|}{C}} - R_1 \\[4pt]
CH - O - \overset{\displaystyle O}{\overset{\|}{C}} - R_2 \quad + \quad 9\,H_2 \xrightarrow{\;\text{catalyst}\;} \\[4pt]
CH_2 - O - \overset{\displaystyle O}{\overset{\|}{C}} - R_3
\end{array}
\qquad
\begin{array}{c}
CH_3 \\[4pt]
CH_2 \quad + \\[4pt]
CH_3
\end{array}
\qquad
\begin{array}{c}
CH_3 - R_1 \\[4pt]
H - R_2 \quad + \\[4pt]
CH_3 - R_3
\end{array}
\qquad
\begin{array}{c}
2\,H_2O \\[4pt]
CO_2 \\[4pt]
2\,H_2O
\end{array}
$$

Triglyceride Hydrogen   Propane   Paraffins

The paraffins produced from hydrogenation and deoxygenation of most triglycerides are straight chain paraffins which tend to have high cetane numbers but poor cold flow properties. Including an isomerization step increases the number of branched paraffins which have better cold flow properties. A simplified schematic of the process is illustrated in Figure 5.4 [5.94].

FIGURE 5.4 Schematic of a simplified triglyceride hydrotreatment plant.

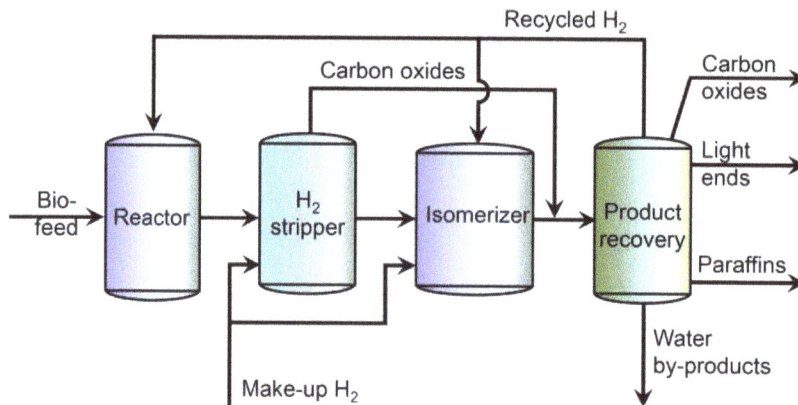

© SAE International.

The hydrogenation and deoxygenation reactions occur over a catalyst bed, typically nickel/molybdenum or cobalt/molybdenum, with the isomerization taking place over a platinum catalyst in combination with a zeolite or other molecular sieve such as SAPO-11 [5.95]. The reaction temperature is usually in the range of 200 to 300°C at a pressure from about 1.38 to 4.83 MPa.

The Honeywell/Universal Oil Products (UOP) process [5.94] specifically allows for up to 20 ppm (w/w) of phosphorous in the feedstock, unless it is in the form of phospholipids that have been found to cause gums within the reactor, or a pre-heater. More than one reactor may be employed, and by careful control of the temperature and pressure in each reactor, the hydrogenation and deoxygenation reactions can be, to some degree, separated.

As discussed in the previous section, the different feedstocks contain differing proportions of different fatty acids, which when hydrogenated and deoxygenated will yield different carbon number paraffins. As shown above in the simplified chemical equation, the reactions also produce propane as opposed to the glycerol produced by the esterification process. Some of the output of the hydrotreatment process can thus be directed to gasoline streams. The output from the hydrogenation/deoxygenation reactors tends to be n-paraffins, which have poor cold flow characteristics; the isomerization unit produces more branched paraffins. However, besides the added cost of converting the n-paraffins, if the isomerization unit is driven too hard it will produce unwanted cracking and reduce the yield of product within the diesel fuel boiling range. Particularly in Europe, which is deficient in diesel fuel, the process tends to be tuned to favor diesel fuel production. The world's first commercial plant for producing fuel from HVO was built by Neste at Porvoo, Finland. The plant started production in 2007 using the NEXBTL™ (an acronym for "next generation biomass-to-liquid") technology [5.96], nowadays manufacturing products marketed worldwide as Neste MY Renewable Diesel™, Neste MY Sustainable Aviation Fuel™ and Neste RE™. The second NEXBTL processing unit started production in 2009 at Porvoo, Finland. Neste has setup two dedicated facilities to produce the above-mentioned renewable and circular solutions (including renewable diesel, sustainable aviation fuel and renewable feedstock for polymer and chemicals, etc.) one in Singapore, and one in Rotterdam, the Netherlands, in 2010 and 2011, respectively. The Singapore refinery has a production capacity of 2,600,000 tonnes/year and the Rotterdam refinery has a capacity of 1,400,000 tonnes/year, with a planned expansion to 2,700,000 tonnes/year of renewable diesel using the NEXBTL technology. The Singapore facility is shown as Figure 5.5.

FIGURE 5.5   Neste's refinery in Singapore producing exclusively renewable and circular solutions using NEXBTL technology.

© Neste.

Hydrotreated vegetable oil is highly paraffinic, and like other synthetic diesel fuels, such as Fischer-Tropsch fuels, it lacks the low levels of polar compound that are found in fossil diesel fuel. These fuels must therefore be treated with a lubricity additive to provide the necessary degree of lubrication for diesel fuel injection equipment. Because these fuels also tend to have quite a narrow cut of carbon number paraffins, their cold flow behavior can differ markedly from that of fossil diesel fuels. There is less of a spread between cloud point and cold filter plugging point. It is therefore imperative to match the right fuel additives to the product stream. Purely paraffinic fuels can affect elastomers differently than other fuels, and this must be considered if switching from conventional fossil diesel fuel to a purely paraffinic supply. Because the hydrotreated vegetable oil has a density of 780 ± 5 kg/m^3 at 15°C [5.97], which is below normal diesel fuel density specifications, the bio-derived fuel can be blended with fossil diesel fuel, and a process has been patented for producing hydrocarbons with a relatively high cetane rating being highly effective as diesel fuel ignition improvers [5.98].

A different process has been patented by UOP and Eni to produce renewable diesel. The process is known by the name Ecofining™ [5.99, 5.100] and uses a two-stage deoxygenation and hydrogenation process. This is shown schematically is Figure 5.6.

FIGURE 5.6 Schematic of a the Ecofining™ process.

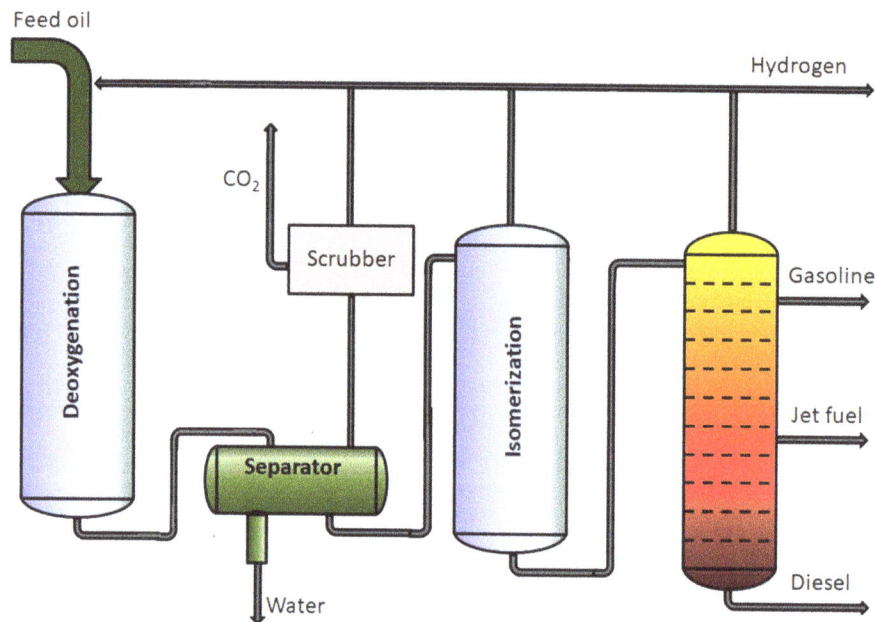

© SAE International.

In 2014, Eni had converted the oil refinery at Porto Marghera near Venice, Italy, into a biorefinery [5.101] producing about 360,000 tonnes/year. Diamond Green Diesel are also using the Ecofining process at their Norco, Louisiana facility in the US. In San Francisco, California the Phillips 66 refinery, known as Rodeo Renewed, process pretreated vegetable oil feedstocks to produces over 500,000 tonnes/year of renewable diesel. This facility is shown as Figure 5.7. Total production of renewable diesel fuel is over 5 million tonnes/year.

FIGURE 5.7   Phillips 66 San Francisco Refinery in Rodeo, California.

A variation of the vegetable oil hydrotreatment process is to co-feed the hydrogenation reactor with a mixture of vegetable oil (or other fatty acid or triglyceride product) and fossil-derived hydrocarbons [5.102, 5.103] (the fossil hydrocarbon stream is hydrotreated in the same reactor as the triglyceride or other bio-derived feedstock). A further variation allows for the production of a range of hydrocarbons that are useful for both gasoline and diesel fuel [5.104]. For this combined production process, a combined feed composed of glycerides and/or FFAs in combination with polyols, such as sucrose, sorbitol (a sugar alcohol found in fruit) and other sugar alcohols, glycols, and glycerol.

5.7. Power-to-Liquids

For consistency with Section 4.2, where we discussed Coal-to-Liquids, and Section 4.3 where we discussed Gas-to-Liquids, we are calling this Power-to-Liquids. These fuels are also referred to as: PtL, P2L, electro-fuels, e-fuels, PtX (Power-to-X), PtG (Power-to-Gas) and Powerfuels. These fuels have the same basis; the combination of hydrogen, produced by the electrolysis of water using renewably produced electricity, and

CO_2 captured from the atmosphere or a concentrated source such as exhaust from industrial processes, for example, fermentation to produce ethanol, again using renewably generated power; so that they can be considered as CO_2 neutral or even CO_2 negative, "negative emissions" technologies (NETs). This process clearly eliminates the use of hydrocarbon starting points and thus any potential conflict between food and fuel.

Renewable electricity, by its very nature, is not dependable or constant. A typical PtL facility would need the ability to store the hydrogen it produced and would probably have its own power source, such as wind turbines, solar panels, geothermal, or hydro-electricity. The electricity produced by this renewable source is then used to power an electrolyzer to separate water into hydrogen and oxygen, the hydrogen is then stored under pressure. This hydrogen is referred to as green hydrogen and is the basis of a future hydrogen economy [5.105, 5.106]. The amount of hydrogen storage required would depend on the variability of the electricity generation and the demands of the downstream plant. A separate piece of plant would produce CO_2, either as part of the process of CO_2 removal from another facility, part of a carbon capture and storage (CCS) process, or from the atmosphere by a process known as direct air capture (DAC). This hydrogen and carbon are then combined to produce a renewably sourced hydrocarbon, either directly using the F-T process [5.107] or via methanol production [5.108–5.110]. The different aspects of a PtL plant will be briefly discussed in the following subsections. The first commercial, fully integrated, plant of this nature was constructed in the Magallanes Region of Chile by HIF Global. A view of the plant is shown in Figure 5.8.

FIGURE 5.8 Haru Oni facility under construction.

HIF Global.

The output of this demonstration plant will be 130,000 L/year of synthetic. Capacity is planned to be scaled up in future e-Fuels projects in Chile, US and Australia, to up to 700 million L/year by 2027.

5.7.1. Hydrogen Generation

Electrolysis of water for hydrogen production is now a mature technology, but work is ongoing to improve the efficiency of the different processes, and possibly to find new processes. The process is designed to dissociated water into hydrogen (H_2) and oxygen (O_2) under the influence of electricity. Water electrolysis can be classified according to the electrolyte, operating conditions, and ionic agents (OH⁻, H⁺, O₂⁻) used. Three common electrolysis methods are

- Proton Exchange Membrane (PEM) electrolysis
- Solid oxide electrolysis (SOE)
- Alkaline water electrolysis (AWE)

These processes are shown schematically in Figure 5.9.

FIGURE 5.9   Three different approaches to water electrolysis for hydrogen production.

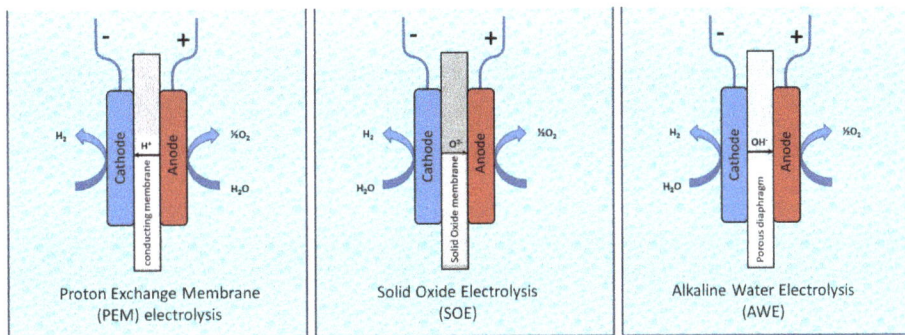

5.7.1.1. PEM Water Electrolysis

The idea of PEM water electrolysis was possibly first suggested by William T. Grubb in the late 1950s [5.111, 5.112]. A PEM fuel cell employs similar technology, but working in reverse, to produce electricity. A solid membrane, such as sulfonated resins are used as a proton conductor [5.113, 5.114]. Current is conducted via a coated bipolar plate which not only serves as a current carrier but also to conduct heat and distribute the reacting agents within the electrolyzer. Because the bipolar plates are exposed to a highly oxidizing atmosphere, titanium parts are widely used in current PEM water electrolysis cells to avoid corrosion [5.115], although work is ongoing to replace the titanium with a less expensive alternative [5.116]. The overall reaction consists of two electrochemical half reactions, often referred to as the oxygen evolution reaction (OER) at the anode,

$$H_2O \rightarrow \tfrac{1}{2}O_2 + 2H^+$$

and the hydrogen evolution reaction (HER) at the cathode,

$$2H^+ + 2e^- \rightarrow H_2$$

giving the overall reaction as

$$H_2O \rightarrow H_2 + \tfrac{1}{2}O_2$$

The reaction kinetics of the OER is normally much slower than the kinetics of the HER, which can limit the overall efficiency of the process. The kinetics of both reactions can be influenced by electrocatalyst and can be different for anode and cathode.

At the anode the OER catalysts commonly used include oxides of iridium and ruthenium, with ruthenium oxides showing better performance than iridium oxides, although IrO_2 is probably more stable than RuO_2 [5.117]. Other noble metal, and composite-based catalysts, used for the OER catalysts are Ir, Pt, Ag, IrO_2/Pt, and nanoporous gold/IrO_2. The stability of IrO_2 and RuO_2 can be improved through combining with less expensive oxides such as SnO_2, Co_3O_4, MnO_2, and Ta_2O_5 [5.118–5.120].

At the cathode, to catalyze the HER the noble metals platinum and palladium are known to be effective [5.121] but very expensive. Transition metal carbides have been explored to replace the noble metal but they can also be used as support material to reduce noble metal loading [5.122–5.127]. PEM water electrolysis technology is used to provide the hydrogen in the Haru Oni facility shown in Figure 5.8.

5.7.1.2. Solid Oxide Electrolysis (SOE)

The SOE was proposed in the late 1970s/early 1980s [5.128, 5.129]. SOE offers major advantages over alternative electrolysis technologies. Electrolyzing steam, rather than water, means that the high operating temperatures greatly improve the reaction kinetics, SOE systems can be thermally integrated with downstream chemical syntheses, such as the production, e.g., methanol, SOE technology uses abundant raw materials, such as nickel, zirconia and steel, with precious metals. SOE can be used for splitting water and carbon dioxide (CO_2), or simultaneously splitting H_2O and CO_2, co-electrolysis [5.130, 5.131]. The resulting hydrogen (H_2), carbon monoxide (CO) [5.132] or syn-gas [5.133–5.135] can be used as feedstock for fuel synthesis, and the chemical industry. The basic principle of an SOE unit, for water electrolysis, is as shown in the center panel of Figure 5.9. A solid O^{2-} conductor is sandwiched between an OER anode and a HER cathode.

The overall reaction consists of two electrochemical half reactions, with electrons, instead of protons, flowing through the membrane, to give the OER at the anode,

$$O^{2-} \rightarrow \frac{1}{2}O_2 + 2e^-$$

and the HER at the cathode,

$$H_2O + 2e^- \rightarrow H_2 + O^{2-}$$

giving the overall reaction as

$$H_2O \rightarrow H_2 + \frac{1}{2}O_2$$

In most cases, the solid oxide membrane comprises either yttria-stabilized zirconia (YSZ)—a solid solution of few mole-% yttria (Y_2O_3) in zirconia (ZrO_2)—or a composite of metallic Ni and YSZ. Noble metal catalysts have been proposed for the SOE anode material but these are too expensive to be considered commercially. The anodes are thus usually ceramic or ceramic composites [5.136, 5.137], but the SOE anode material can delaminate from the electrolyte [5.138–5.140] and perovskites have been used to reduce this tendency [5.141].

5.7.1.3. Alkaline Water Electrolysis (AWE)

Alkaline water electrolysis (AWE) is probably the oldest, and most well-established technology for hydrogen production. The concept of electrolysis of water probably dates back to 1789 and is attributed to the Dutch scientists A. P. van Troostwijk and J.R. Deiman [5.142] who used an electrostatic generator which then discharged electricity through two gold wires in a tube of water. Pure water is a relatively poor conductor of electricity, the process is improved if an alkaline solution is used. In most commercial elecrolyzers, a weak solution of KOH or NaOH is used. At the cathode the alkaline molecules (KOH or NaOH) are reduced to hydroxyl ions (OH⁻) and protons which combine to form molecular hydrogen (H$_2$). The hydroxyl ions migrate through the porous diaphragm, under the influence of the electrical current, where they combine to form ½ molecule of oxygen (O$_2$) and one molecule of water (H$_2$O). The overall reaction can again be considered as two electrochemical half reactions, with hydroxyl ions, flowing through the diaphragm, to give the OER at the anode,

$$2OH^- \rightarrow H_2O + \tfrac{1}{2}O_2 + 2e^-$$

and the HER at the cathode,

$$2H_2O + 2e^- \rightarrow H_2 + 2OH^-$$

giving the overall reaction as

$$H_2O \rightarrow H_2 + \tfrac{1}{2}O_2$$

AWE was first commercialized using asbestos as the porous diaphragm. Asbestos is a fibrous silicate with many varieties, the main one is crisotila which because of its alkaline structure cannot be used in acidic solutions and in basic solutions it is also liable to corrosion at higher temperatures. The adverse health effects of asbestos have also led to the development metal oxide diaphragms [5.143, 5.144]. One downside of the traditional AWE system is that the diaphragm is permeable and thus allows some limited cross-over of the gases produced at the electrodes. Oxygen diffusion to the cathode reacts with the hydrogen being produced, to form water, thus reducing the efficiency. In addition, there are safety aspects associated with hydrogen diffusion to the anode [5.145]. Anion exchange membranes have been developed for water electrolysis [5.146, 5.147]. Anion exchange membrane water electrolysis (AEMWE) is thus very similar to proton exchange membrane water electrolysis (PEMWE); the membrane does not allow permeation of the hydrogen produced. The primary difference being that charge is transferred across the membrane by hydroxide ions, as opposed to protons in PEM systems. Many of these membranes have now been commercialized [5.148].

5.7.2. Carbon Capture

Due to concerns around global warming and greenhouse gas emissions, mainly CO_2, and efforts to reduce the concentration of CO_2 in the atmosphere, there are a number of nonexclusive options which can be split into two basic groups; CO_2 emissions reduction where CO_2 emissions are significantly reduced or captured at sources such, e.g., flue gases from fossil fueled power plants, or by capturing CO_2 directly from the atmosphere. These two approaches are discussed briefly in the following sections.

5.7.2.1. Carbon Capture at Source

Due to concerns over the effects of high CO_2 emissions from such industries as power generation, cement production, iron & steel production, chemicals production, aluminum and nonferrous metals production and pulp, paper and printing, etc., there has been growing research into ways of capturing the CO_2 at source and sequestrating this CO_2, carbon capture and sequestration, or CCS [5.149, 5.150]. But more recently effort has been directed toward carbon capture and utilization (CCU) or carbon capture utilization and storage (CCUS) [5.151–5.155].

For carbon capture at source the most mature technology utilizes amine solvents [5.149, 5.156, 5.157] with the general trend for increasing loading capacity of the amines being, primary amines, hindered amines, secondary amines, tertiary amines and diamines [5.158]. The first CCS plant to become operational was the SaskPower Integrated Carbon Capture and Storage (ICCS) Demonstration Project in Saskatchewan, Canada. The CO_2 recovered from the flue gas in this facility is compressed and transferred through a pipeline and injected into an oil field to provide enhanced oil recovery (EOR) [5.159]. Many more facilities have been built with the primary aim of removing CO_2 from the flue gas before it reached the atmosphere. A nonexhaustive list of technologies and facilities include; the KM CDR Process®, jointly developed by Mitsubishi Heavy Industries (MHI) and Kansai Electric Power Co. Inc. [5.160] which is employed in facilities in the US, the Middle East, the Far East, including Japan and the Indian Sub-continent, other commercial processes are the Econamine FG Plus^{SM} technology [5.161, 5.162] from Fluor Corporation, Cansolv® Technology [5.163, 5.164] from Shell, the Advanced Carbon Capture™ system from Aker, the Chilled Ammonia Process (CAP) from Alstom, and amino acid-based system, PostCap™, from Siemens. Major research effort is being directed toward improving the solvents used.

Oxy-fuel combustion can also be used as part of the CCSU technologies. Oxy-fuel combustion is where the fuel is combusted in a combination of oxygen and recycled flue gas, instead of air, thereby increasing the concentration of CO_2 in the flue gas. The process can be refined by separating the oxygen and nitrogen present in the pre-combustion air. The concentration of CO_2 in the flue gas, after oxy-combustion, can be as high as 99% [5.165, 5.166].

5.7.2.2. CO_2 Direct Air Capture

Direct air capture of CO_2 collects the CO_2 present in the atmosphere, currently estimated at about 400 ppm [5.167], and then utilizes this CO_2 or sequesters it [5.168–5.170]. DAC technologies can be classified as adsorption/desorption or separation. The adsorption/desorption technique can then be subdivided into Liquid DAC (L-DAC) and Solid DAC (S-DAC).

For an L-DAC system the ambient air is passed through an aqueous alkaline solution, e.g., KOH or NaOH, the CO_2 reacts with the basic solution to produce a carbonate, with the alkaline solution being regenerated in a separate process requiring higher temperatures, up to 900°C, [5.171, 5.172] and thus higher energy input. However, because the sorbent is liquid, the two processes can be physically connected and the whole process can thus be configured to run continuously. These aqueous alkaline sorbents have the advantage of ready availability and relatively fast adsorption kinetics, but they are highly corrosive and the regeneration is energy intensive.

For an S-DAC system the ambient air is passed over a support structure for the sorbent, the support can be made from zeolites [5.173], silica materials [5.174], porous organic polymers [5.175], metal–organic frameworks (MOFs) [5.176], or carbon nanotubes [5.177], the sorbent itself is commonly an amine [5.178]. The regeneration takes place at a lower temperature than for L-DAC systems which thus requires a lower energy input. The sorbent must be regenerated in a separate process which necessitates operating as a batch process so that at least two units are operating, one adopting and one desorbing.

An emerging technology which could also be classed as adsorption/desorption is electro-swing adsorption. Electro-swing adsorption is based on an electrochemical cell where a porous solid electrode adsorbs CO_2 when negatively charged and releases it when positive charged. This porous electrode is sandwiched between two electrically connected parts of the opposite polarity electrode of the opposite polarity [5.179, 5.180]. Another possible technique that is attracting much attention is membrane separation. The use of membranes offers greater energy efficient because sorbent regeneration is not required. The membrane material, permeability, selectivity can be configured to enhance performance [5.181]. Membranes may be either organic or inorganic, with inorganic membranes able to operate at high temperatures. At present commercialization is hampered by the high cost of the membranes. Polymeric membranes have been used to capture CO_2 for natural gas sweetening [5.182].

5.7.3. Hydrocarbon Generation

The previous sections illustrate how hydrogen can be produced from water using electrolysis and CO_2 can be captured from the air or directly from its production source. The following sections will look at different routes from hydrogen and carbon dioxide, to what is considered as automotive fuel. Unfortunately, CO_2 is a fully oxidized molecule, making it rather thermodynamically stable and chemically inert. The activation of CO_2 and its hydrogenation to hydrocarbons, or alcohols, is therefore not straight forward. The following sections will outline different routes to achieve this.

5.7.3.1. Production of Methane from H_2 and CO_2

At the end of the 20th century the French scientists Paul Sabatier and Jean-Baptiste Senderens were experimenting with converting unsaturated organic molecules into corresponding saturated compounds by passing the vapor of these organic molecules and hydrogen over hot, finely divided, nickel. The hydrogenation of CO_2 became known as the Sabatier process and won Sabatier the Nobel Prize in Chemistry in 1912 [5.183, 5.184]. This is probably most technologically advanced conversion process combining H_2 plus CO_2 to produce methane, also known as synthetic methane or synthetic natural gas (SNG). The overall reaction can be shown by the following equation:

$$4H_2 + CO_2 \rightarrow CH_4 + 2H_2O$$

The most widely accepted mechanism of the CO_2 methanation reaction is the combination of an endothermic reversed water gas shift (RWGS) reaction and an exothermic CO methanation, as shown, respectively, in the following two equations.

$$H_2 + CO_2 \rightarrow CO + H_2O$$

$$3H_2 + CO \rightarrow CH_4 + H_2O$$

The overall reaction is exothermic making it possible to use the energy generated in the methane production to provide heat for regenerating a DAC sorbate.

The methanation reaction requires a catalyst that is elective toward methane. Typical commercial catalysts are nickel-based using a metal oxide support. Oxides such as Al_2O_3, TiO_2, ZrO_2, and CeO_2 are commonly used. Ni- and Ru-based catalysts offer very good selectivity toward methane, other metal catalysts of such as Pd, Pt, Rh, Mo, Re, and Au can also simultaneously produce methanol and unwanted CO [5.185]. Methane produced directly from H_2 and CO_2 is effectively a pure form of natural gas and can be converted to fuel in the same way as natural gas, as discussed in Section 4.4.

5.7.3.2. Production of Syn-Gas from H_2 and CO_2

As noted in Section 5.7.1.2 water and CO_2 can be co-electrolyzed to produce syn-gas. This syn-gas can then be converted using the Fischer-Tropsch (F-T) process (Section 4.2). Alternatively, the CO_2 can be reduced to CO in a hydrogen atmosphere resulting in the production of syn-gas [5.186–5.188], that can then be converted to fuel using the F-T process.

5.7.3.3. Production of Methanol from H_2 and CO_2

Hydrogenation of CO_2 to methanol follows the simple equation set out below:

$$CO_2 + 3H_2 \rightarrow CH_3OH + H_2O$$

Catalyst can obviously be used to promote this reaction [5.189–5.193]. However, there is also the possibility of some of the CO_2 undergoing the RWGS reaction as shown below:

$$CO_2 + H_2 \leftrightarrow CO + H_2O$$

It has been found that by using the CAMERE process the reaction temperatures can be reduced and the methanol yield increased [5.194]. The CAMERE name derives from the description <u>ca</u>rbon dioxide hydrogenation to form <u>me</u>thanol via a <u>re</u>verse-water-gas-shift reaction (RWGS) reaction by capitalizing the letters underlined. The process involves placing two reactors in series with the first producing CO by the RWGS reaction, water is then removed and the second reactor hydrogenates the CO to methanol. Methanol can then be converted to higher hydrocarbons via the methanol to gasoline process described in Section 4.5.

5.7.3.4. Direct Conversion of H_2 and CO_2 to Higher Hydrocarbons

Because of the extreme inertness of CO_2 and the high activation energy of the C–C coupling reaction, as well as many competing reactions, this favors the formation of C_1 products including methane and methanol, mentioned above, and methanoic acid also known as formic acid. Thus, there is a lot of ongoing work to develop routes to go directly from H_2 and CO_2 to higher carbon number hydrocarbons, often referred to as C_{2+}, and in particular C_{5+} hydrocarbons [5.195]. These are the hydrocarbons that dominate the gasoline range. Due to the stability of CO_2 the hydrogenation to higher carbon number products is very dependent on catalyst development. These tend to be Fe-based, often in combination with zeolites such as HZSM-5 and SAPO-34 [5.196, 5.197] with other metals such as cerium and indium being considered [5.109, 5.198]. The combination of cobalt and manganese has also been reported to give good selectivity [5.199]. The Fe-based catalysts have been found to give better selectivity when used in combination with the oxides of other metals such as K, Mn, Cu and Ce that reduce the unwanted methane formation [5.200]. The catalyst support is also key to the outcomes [5.201].

References

5.1. Diesel, R., A process for producing motive work from the combustion of fuel. UK Patent 7,241, 1892.

5.2. Diesel, R., "The Diesel Oil-Engine and Its Industrial Importance Particularly for Great Britain," *Proc. Inst. Mech. Eng.* 82, no. 1 (1912): 179-280.

5.3. Forgiel, R. and Varde, K., "Experimental Investigation of Vegetable Oils Utilization in a Direct Injection Diesel Engine," **SAE** Technical Paper 811214, 1981, doi:https://doi.org/10.4271/811214.

5.4. White, T., "Alcohol as a Fuel for the Automobile Motor," SAE Technical Paper 070002, 1907, doi:https://doi.org/10.4271/070002.

5.5. Ford, H. "Ford Predicts Fuel from Vegetation," *New York Times*, September 20, 1925, 25.

5.6. Foljambe, E., "The Automobile Fuel Situation," SAE Technical Paper 160016, 1916, doi:https://doi.org/10.4271/160016.

5.7. Dean, E., "Fuel for Automotive Apparatus," SAE Technical Paper 180007, 1918, doi:https://doi.org/10.4271/180007.

5.8.   Norman, C., "Internal-Combustion Engine Fuels," SAE Technical Paper 220005, 1922, doi:https://doi.org/10.4271/220005.

5.9.   Rosa, L.Z. and Ribeiro, S.K., "Avoiding Emissions of Carbon Dioxide through the Use of Fuels Derived from Sugar Cane," *Ambio* 27, no. 6 (1998): 465-470.

5.10.   Chaudhari, A.B., Dandi, N.D., Vadnere, N.C. Patil, U.K. et al., "Bioethanol: A Critical Appraisal," in *Microorganisms in Sustainable Agriculture and Biotechnology*, Eds. Satyanarayana, T., Johri, B.N., and Prakash, A. (2012), 793-824.

5.11.   Austin, T. and Rubenstein, G., "Gasohol: Technical, Economic, or Political Panacea?" SAE Technical Paper 800891, 1980, doi:https://doi.org/10.4271/800891.

5.12.   Lang, J. and Black, F., "Impact of Gasohol on Automobile Evaporative and Tailpipe Emissions," SAE Technical Paper 810438, 1981, doi:https://doi.org/10.4271/810438.

5.13.   Gibbs, L. and Gilbert, B., "Contra Costa County's One-Year Experience with Gasohol," SAE Technical Paper 810440, 1980, doi:https://doi.org/10.4271/810440.

5.14.   Chang, C.-C. and Wan, S.-W., "China's Motor Fuels from Tung Oil," *Industrial Engineering Chemistry* 39, no. 12 (1947): 1543-1548.

5.15.   Rosillo-Calle, F. and Cortez, L.A.B., "Towards Proalcool II—Review of the Brazilian Bioethanol Programme," *Biomass and Bioenergy* 14, no. 2 (1998): 115-124.

5.16.   Almodares, A. and Hadi, M.R., "Production of Bioethanol from Sweet Sorghum: A Review," *African Journal of Agricultural Research* 4, no. 9 (2009): 772-780.

5.17.   Kremer, F. and Fachetti, A., "Alcohol as Automotive Fuel—Brazilian Experience," SAE Technical Paper 2000-01-1965, 2000, doi:https://doi.org/10.4271/2000-01-1965.

5.18.   Rodrigues, A. and Alto, A., "Cold Start System Development for Flex Fuel Vehicles," SAE Technical Paper 2007-01-2595, 2007, doi:https://doi.org/10.4271/2007-01-2595.

5.19.   Bustani, M., "The Effect of Blending Alcohol in Gasoline on Physical and Chemical Properties," SAE Technical Paper 830956, 1983, doi:https://doi.org/10.4271/830956.

5.20.   Broecker, W.S., "Unpleasant Surprises in the Greenhouse?" *Nature* 328 (1987): 123-126.

5.21.   Intergovernmental Panel on Climate Change (IPCC), in: Houghton, J.T. et al. (Eds), *Climate Change: The IPCC Scientific Assessment*, (Cambridge, UK: Cambridge University Press, 1990).

5.22.   "Directive 2009/28/EC of the European Parliament and of the Council on the Promotion of the Use of Energy from Renewable Sources," April 23, 2009.

5.23.   U.S. Environmental Protection Agency, "40 CFR Part 80, Regulation of Fuels and Fuel Additives: 2012 Renewable Fuel Standards," *Federal Register* 77, no. 5 (2012): 1320-1358, Rules and Regulations.

5.24.   Babu, A. and Devaradjane, G., "Vegetable Oils and Their Derivatives as Fuels for CI Engines: An Overview," SAE Technical Paper 2003-01-0767, 2003, doi:https://doi.org/10.4271/2003-01-0767.

5.25.   Niemi, S., Hätönen, T., and Laiho, V., "Results from a Durability Test of a Mustard Seed Oil Driven Tractor Engine," SAE Technical Paper 982528, 1998, doi:https://doi.org/10.4271/982528.

5.26.   Maa, F. and Hanna, M.A., "Biodiesel Production: A Review," *Bioresource Technology* 70, no. 1 (1999): 1-15.

5.27.   Wiedermann, L.H., "Degumming, Refining and Bleaching Soybean Oil," *Journal of the American Oil Chemists' Society* 58, no. 3 (1981): 159-166.

5.28.   Suda, K., "Vegetable Oil or Diesel Fuel—A Flexible Option," SAE Technical Paper 840004, 1984, doi:https://doi.org/10.4271/840004.

5.29.   Hemmerlein, N., Korte, V., Richter, H., and Schröder, G., "Performance, Exhaust Emissions and Durability of Modern Diesel Engines Running on Rapeseed Oil," SAE Technical Paper 910848, 1991, doi:https://doi.org/10.4271/910848.

5.30.   Fuls, J., Hawkins, C.S., and Hugo, F.J.C., "Tractor Engine Performance on Sun Flower Oil Fuel," *Journal of Agricultural Engineering Research* 30 (1984): 29-35.

5.31. **Murayama, T.,** Oh, Y., Miyamoto, N., Chikahisa, T. et al., "Low Carbon Flower Buildup, Low Smoke, and Efficient Diesel Operation with Vegetable Oils by Conversion to Mono-Esters and Blending with Diesel Oil or Alcohols," SAE Technical Paper 841161, 1984, doi:https://doi.org/10.4271/841161.

5.32. **Fishinger, M.,** Engelman, H., and Guenther, D., "Service Trial of Waste Vegetable Oil as a Diesel Fuel Supplement," SAE Technical Paper 811215, 1981, doi:https://doi.org/10.4271/811215.

5.33. **Hawkins, C.,** Fuls, J., and Hugo, F., "Engine Durability Tests with Sunflower Oil in an Indirect Injection Diesel Engine," SAE Technical Paper 831357, 1983, doi:https://doi.org/10.4271/831357.

5.34. **Ziejewski, M.,** Kaufman, K., and Tupa, R., "Laboratory Endurance Testing of a 25/75 Sunflower Oil-Diesel Fuel Blend Treated with Fuel Additives," SAE Technical Paper 840236, 1984, doi:https://doi.org/10.4271/840236.

5.35. **Draper, W.,** Phillips, J., and Zeller, H., "Impact of a Barium Fuel Additive on the Mutagenicity and Polycyclic Aromatic Hydrocarbon Content of Diesel Exhaust Particulate Emissions," SAE Technical Paper 881651, 1988, doi:https://doi.org/10.4271/881651.

5.36. **Li, H.,** Lea-Langton, A., Biller, P., Andrews, G. et al., "Effect of Multifunctional Fuel Additive Package on Fuel Injector Deposit, Combustion and Emissions Using Pure Rape Seed Oil for a DI Diesel," *SAE Int. J. Fuels Lubr.* 2, no. 2 (2010): 54-65, doi:https://doi.org/10.4271/2009-01-2642.

5.37. **Lance, D.** and Andersson, J., "Emissions Performance of Pure Vegetable Oil in Two European Light Duty Vehicles," SAE Technical Paper 2004-01-1881, 2004, doi:https://doi.org/10.4271/2004-01-1881.

5.38. **Lynd, L.R.,** "Overview and Evaluation of Fuel Ethanol from Cellulosic Biomass: Technology, Economics, the Environment, and Policy," *Annual Review of Energy and the Environment* 21 (1996): 403-465.

5.39. **Aggarwal, N.K.,** Nigam, P., Singh, D., and Yadav, B.S., "Process Optimization for the Production of Sugar for the Bioethanol Industry from Sorghum, a Non-Conventional Source of Starch," *World Journal of Microbiology and Biotechnology* 17, no. 4 (2001): 411-415.

5.40. **Limtong, S.,** Sringiew, C., and Yongmanitchai, W., "Production of Fuel Ethanol at High Temperature from Sugarcane Juice by a Newly Isolated *Kluyveromyces marxianus*," *Bioresource Technology* 98, no. 17 (2007): 3367-3374.

5.41. **Tyagi, R.D.** and Ghose, T.K., "Studies on Immobilized *Saccharomyces cerevisiae*. I. Analysis of Continuous Rapid Ethanol Fermentation in Immobilized Cell Reactor," *Biotechnology and Bioengineering* 24, no. 4 (1982): 781-795.

5.42. **Nagashima, M.,** Azuma, M., Noguchi, S., Inuzuka, K. et al., "Continuous Ethanol Fermentation Using Immobilized Yeast Cells," *Biotechnology and Bioengineering* 26, no. 8 (1984): 992-997.

5.43. **Boeckeler, B.C.,** Fermenting method. U.S. Patent 2,440,925, 1944.

5.44. **Zanin, G.M.,** Santana, C.C., Bon, E.P.S., Giordano, R.C.L. et al., "Brazilian Bioethanol Program," *Applied Biochemistry and Biotechnology* 84-86 (2000): 1147-1161.

5.45. **Shigechia, H.,** Fujitab, Y., Koha, J., Uedac, M. et al., "Energy-Saving Direct Ethanol Production from Low-Temperature-Cooked Corn Starch Using a Cell-Surface Engineered Yeast Strain Co-Displaying Glucoamylase and α-Amylase," *Biochemical Engineering Journal* 18, no. 2 (2004): 149-153.

5.46. **Gauss, W.F.,** Suzuki, S., and Takagi, M., Manufacture of alcohol from cellulosic materials using plural ferments. U.S. Patent 3,990,944, 1976.

5.47. **Claassen, P.A.M.,** van Lier, J.B., López Contreras, A.M., van Niel, E.W.J. et al., "Utilisation of Biomass for the Supply of Energy Carriers," *Applied Microbiology and Biotechnology* 52, no. 6 (1999): 741-755.

5.48. **Cardona, C.A.** and Sánchez, Ó.J., "Fuel Ethanol Production: Process Design Trends and Integration Opportunities," *Bioresource Technology* 98, no. 12 (2007): 2415-2457.

5.49. **Leonard, R.H.** and Hajny, G.J., "Fermentation of Wood Sugars to Ethyl Alcohol," *Industrial & Engineering Chemistry* 37, no. 4 (1945): 390-395.

5.50. **Brosse, N.,** Dufour, A., Meng, X., Sun, Q. et al., "Miscanthus: A Fast-Growing Crop for Biofuels and Chemicals Production," *Biofuels, Bioproducts and Biorefining* 6, no. 5 (2012): 580-598.

5.51. Eschenbacher, A., Myrstad, T., Bech, N., Thi, H.D. et al., "Fluid Catalytic Co-Processing of Bio-Oils with Petroleum Intermediates: Comparison of Vapour Phase Low Pressure Hydrotreating and Catalytic Cracking as Pretreatment," *Fuel* 302 (2021): 121198.

5.52. Gould, J.M., "Alkaline Peroxide Delignification of Agricultural Residues to Enhance Enzymatic Saccharification," *Biotechnology and Bioengineering* 26, no. 1 (1984): 46-52.

5.53. Gould, J.M., Alkaline peroxide treatment of agricultural byproducts. U.S. Patent 4,806,475, 1989.

5.54. Elshafei, A.M., Vega, J.L., Klasson, K.T., Clausen, E.C. et al., "The Saccharification of Corn Stover by Cellulase from *Penicillium funiculosum*," *Bioresource Technology* 35, no. 1 (1991): 73-80.

5.55. Chang, V.S., Nagwani, M., Kim, C.-H., and Holtzapple, M.T., "Oxidative Lime Pretreatment of High-Lignin Biomass: Poplar Wood and Newspaper," *Applied Biochemistry and Biotechnology* 94, no. 1 (2001): 1-28.

5.56. Wingerson, R.C., Method of treating lignocellulosic biomass to produce cellulose. U.S. Patent 6,419,788, 2002.

5.57. Foody, P., Method for increasing the accessibility of cellulose in lignocellulosic materials, particularly hardwoods agricultural residues and the like. U.S. Patent 4,461,648, 1984.

5.58. O'Connor, J.J., Exploding of ammonia impregnated wood chips. U.S. Patent 3,707,436, 1972.

5.59. Dale, B.E., Method of increasing the reactivity and digestibility of celulose with ammonia. U.S. Patent 4,600,590, 1986.

5.60. Pazner, L. and Chang, P.-C., High efficiency organosolv saccharification process. U.S. Patent 4,470,851, 1984.

5.61. Schläpfer, P. and Silberman, H.C., Process for the saccharification of cellulose and cellulosic materials. U.S. Patent 2,959,500, 1960.

5.62. Pousa, G.P.A.G., Santos, A.L.F., and Suarez, P.A.Z., "History and Policy of Biodiesel in Brazil," *Energy Policy* 35, no. 11 (2007): 5393-5398.

5.63. Hoekman, S., Gertler, A., Broch, A., Robbins, C. et al., "Biodistillate Transportation Fuels 1. Production and Properties," *SAE Int. J. Fuels Lubr.* 2, no. 2 (2010): 185-232, doi:https://doi.org/10.4271/2009-01-2766.

5.64. Chavanne, C.G., Procédé de transformation d'huiles végétales en vue de leur utilisation comme carburants. Belgian Patent 422877, August 1937.

5.65. Lotero, E., Liu, Y., Lopez, D.E., Suwannakarn, K. et al., "Synthesis of Biodiesel via Acid Catalysis," *Industrial Engineering Chemistry Res.* 44, no. 14 (2005): 5353-5363.

5.66. Kulkarni, M.G., Gopinath, R., Meher, L.C., and Dalai, A.K., "Solid Acid Catalyzed Biodiesel Production by Simultaneous Esterification and Transesterification," *Green Chem.* 8 (2006): 1056-1062.

5.67. Shu, Q., Gao, J., Nawaz, Z., Liao, Y. et al., "Synthesis of Biodiesel from Waste Vegetable Oil with Large Amounts of Free Fatty Acids Using a Carbon-Based Solid Acid Catalyst," *Applied Energy* 87, no. 8 (2010): 2589-2596.

5.68. Formo, M.W., "Ester Reactions of Fatty Materials," *Journal of the American Oil Chemists' Society* 31, no. 11 (1954): 548-559.

5.69. Leung, D.Y.C. and Guo, Y., "Transesterification of Neat and Used Frying Oil: Optimization for Biodiesel Production," *Fuel Processing Technology* 87, no. 10 (2006): 883-890.

5.70. Rashid, U., Anwar, F., Moser, B.R., and Ashraf, S., "Production of Sunflower Oil Methyl Esters by Optimized Alkali-Catalyzed Methanolysis," *Biomass and Bioenergy* 32, no. 12 (2008): 1202-1205.

5.71. Meng, X., Chen, G., and Wang, Y., "Biodiesel Production from Waste Cooking Oil via Alkali Catalyst and Its Engine Test," *Fuel Processing Technology* 89, no. 9 (2008): 851-857.

5.72. Tiwari, A.K., Kumar, A., and Raheman, H., "Biodiesel Production from Jatropha Oil (*Jatropha curcas*) with High Free Fatty Acids: An Optimized Process," *Biomass and Bioenergy* 31, no. 8 (2007): 569-575.

5.73. Meher, L.C., Dharmagadda, V.S.S., and Naik, S.N., "Optimization of Alkali-Catalyzed Transesterification of *Pongamia pinnata* Oil for Production of Biodiesel," *Bioresource Technology* 97, no. 12 (2006): 1392-1397.

5.74. Ghadge, S.V. and Raheman, H., "Process Optimization for Biodiesel Production from Mahua (*Madhuca indica* L.) Oil Using Response Surface Methodology," *Bioresource Technology* 97, no. 3 (2006): 379-384.

5.75. **Vicente, G.,** Martinez, M., and Aracil, J., "Optimization of *Brassica carinata* Oil Methanolysis for Biodiesel Production," *Journal of the American Oil Chemists' Society* 82, no. 12 (2005): 899-904.

5.76. **Bouaid, A.,** Bajo, L., Martinez, M., and Aracil, J., "Optimization of Biodiesel Production from Jojoba Oil," *Process Safety and Environmental Protection* 85, no. 5 (2007): 378-382.

5.77. **Williams, A.,** McCormick, R., Luecke, J., Brezny, R. et al., "Impact of Biodiesel Impurities on the Performance and Durability of DOC, DPF and SCR Technologies," *SAE Int. J. Fuels Lubr.* 4, no. 1 (2011): 110-124, doi:https://doi.org/10.4271/2011-01-1136.

5.78. **Lance, M.,** Wereszczak, A., Toops, T., Ancimer, R. et al., "Evaluation of Fuel-Borne Sodium Effects on a DOC-DPF-SCR Heavy-Duty Engine Emission Control System: Simulation of Full-Useful Life," *SAE Int. J. Fuels Lubr.* 9, no. 3 (2016): 683-694, doi:https://doi.org/10.4271/13-02-02-0009.

5.79. **Kamp, C.** and Bagi, S., "Perspectives on Current and Future Requirements of Advanced Analytical and Characterization Methods in the Automotive Emissions Control Industry," *SAE J. STEEP* 2, no. 2 (2021): 141-160, doi:https://doi.org/10.4271/13-02-02-0009.

5.80. **British Standards Institute,** "Liquid Petroleum Products. Fatty Acid Methyl Esters (FAME) for Use in Diesel Engines and Heating Applications," BS EN 14214:2012+A2:2019, British Standards Institute, 2021.

5.81. **Bureau of Indian Standards,** "Biodiesel (B-100)-Fatty Acid Methyl Esters (FAME)-Specification," IS 15607: 2022, Bureau of Indian Standards, 2022.

5.82. **ASTM International,** "Standard Specification for Biodiesel Fuel Blend Stock (B100) for Middle Distillate Fuels," ASTM D6751–23a, ASTM International, 2023.

5.83. **Tsuchiya, T.,** Shiotani, H., Goto, S., Sugiyama, G. et al., "Japanese Standards for Diesel Fuel Containing 5% FAME: Investigation of Acid Generation in FAME Blended Diesel Fuels and Its Impact on Corrosion," SAE Technical Paper 2006-01-3303, 2006, doi:https://doi.org/10.4271/2006-01-3303.

5.84. **de Rezende Pinho, A.,** de Almeida, M.B., Mendes, F.L., Casavechia, L.C. et al., "Fast Pyrolysis Oil from Pinewood Chips Co-Processing with Vacuum Gas Oil in an FCC Unit for Second Generation Fuel Production," *Fuel* 188 (2017): 462-473.

5.85. **Magrini, K.,** Olstad, J., Peterson, B., Jackson, R. et al., "Feedstock and Catalyst Impact on Bio-Oil Production and FCC Co-Processing to Fuels," *Biomass and Bioenergy* 163 (2022): 106502.

5.86. **Lappas, A.A.,** Bezergianni, S., and Vasalos, I.A., "Production of Biofuels via Co-Processing in Conventional Refining Processes," *Catalysis Today* 145, no. 1-2 (2009): 55-62.

5.87. **de Paz Carmona, H.,** Vráblík, A., Herrador, J.M.H., Velvarská, R. et al., "Animal Fats as a Suitable Feedstock for Co-Processing with Atmospheric Gas Oil," *Sustainable Energy & Fuels* 5, no. 19 (2021): 4955-4964.

5.88. **Santillan-Jimenez, E.** and Crocker, M., "Catalytic Deoxygenation of Fatty Acids and Their Derivatives to Hydrocarbon Fuels via Decarboxylation/Decarbonylation," *Journal of Chemical Technology & Biotechnology* 87, no. 8 (2012): 1041-1050.

5.89. **Cheah, K.W.,** Yusup, S., Loy, A.C.M., How, B.S. et al., "Recent Advances in the Catalytic Deoxygenation of Plant Oils and Prototypical Fatty Acid Models Compounds: Catalysis, Process, and Kinetics," *Molecular Catalysis*, (2021): 111469.

5.90. **Monteiro, R.R.,** dos Santos, I.A., Arcanjo, M.R., Cavalcante, C.L. Jr. et al., "Production of Jet Biofuels by Catalytic Hydroprocessing of Esters and Fatty Acids: A Review," *Catalysts* 12, no. 2 (2022): 237.

5.91. **Vardon, D.R.,** Sherbacow, B.J., Guan, K., Heyne, J.S. et al., "Realizing 'Net-Zero-Carbon' Sustainable Aviation Fuel," *Joule* 6, no. 1 (2022): 16-21.

5.92. **Talmadge, M.,** Kinchin, C., Chum, H.L., de Rezende Pinho, A. et al., "Techno-Economic Analysis for Co-Processing Fast Pyrolysis Liquid with Vacuum Gasoil in FCC Units for Second-Generation Biofuel Production," *Fuel* 293 (2021): 119960.

5.93. **Zhao, J.,** Lu, C., Chen, C.C. and Yang, S.T., "Biological Production of Butanol and Higher Alcohols," in *Bioprocessing Technologies in Biorefinery for Sustainable Production of Fuels, Chemicals, and Polymers*, (2013), 235-262.

5.94. **Marker, T.L.,** Sabatino, L.M.F., and Baldiraghi, F., Production of diesel fuel from renewable feedstocks containing phosphorous. U.S. Patent 2009/0321311 A1, 2009.

5.95. **Aalto, P.,** Piirainen, O., and Kiiski, U., Keskitisleen valmistus (middle distillate production). Finnish Patent FI 100248, 1997.

5.96.   Rantanen, L., Linnaila, R., Aakko, P., and Harju, T., "NExBTL—Biodiesel Fuel of the Second Generation," SAE Technical Paper 2005-01-3771, 2005, doi:https://doi.org/10.4271/2005-01-3771.

5.97.   Kuronen, M., Mikkonen, S., Aakko, P., and Murtonen, T., "Hydrotreated Vegetable Oil as Fuel for Heavy Duty Diesel Engines," SAE Technical Paper 2007-01-4031, 2007, doi:https://doi.org/10.4271/2007-01-4031.

5.98.   Craig, W.K. and Soveran, D.W., Production of hydrocarbons with a relatively high cetane rating. U.S. Patent 4,992,605, 1991.

5.99.   Brandvold, T.A. and McCall, M.J., Production of blended fuel from renewable feedstocks. U.S. Patent 8,329,967, 2012.

5.100. Bianchi, D. and Spadavecchia, F., "Substainable Feedstock for Hydrogenated Vegetable Oil (HVO) Based Biorefineries," in *DGMK Conference 'Challenges for Petrochemicals and Fuels: Integration of Value Chains and Energy Transition'*, Berlin, Germany, 2018.

5.101. Zhang, B. and Seddon, D., "Commercial Hydroprocessing Processes for Biofeedstock," in *Hydroprocessing Catalysts and Processes: The Challenges for Biofuels Production*, 2018), 207-223.

5.102. Yao, J., Sughrue, E.L. II, Cross, J.B., Kimble, J.B. et al., Process for converting triglycerides to hydrocarbons. U.S. Patent 7,550,634 B2, 2009.

5.103. Verdier, S., Alkilde, O.F., Chopra, R., Gabrielsen, J. et al., *Hydroprocessing of Renewable Feedstocks-Challenges and Solutions* (Copenhagen, Denmark: Haldor Topsoe A/S, 2019).

5.104. Yao, J. and Sughrue, I.E.L., Co-production of renewable diesel and renewable gasoline. U.S. Patent 8,809,607 B2, 2014.

5.105. Maack, M.H. and Skulason, J.B., "Implementing the Hydrogen Economy," *Journal of Cleaner Production* 14, no. 1 (2006): 52-64.

5.106. Tu, K.J., "Prospects of a Hydrogen Economy with Chinese Characteristics," *Études de l'Ifri* 52, no. 3 (2020): 62.

5.107. Zang, G., Sun, P., Elgowainy, A.A., Bafana, A. et al., "Performance and Cost Analysis of Liquid Fuel Production from H_2 and CO_2 Based on the Fischer-Tropsch Process," *Journal of CO2 Utilization* 46 (2021): 101459.

5.108. Ganesh, I., "Conversion of Carbon Dioxide into Methanol—A Potential Liquid Fuel: Fundamental Challenges and Opportunities: A Review," *Renewable and Sustainable Energy Reviews* 31 (2014): 221-257.

5.109. Gao, P., Li, S., Bu, X., Dang, S. et al., "Direct Conversion of CO_2 into Liquid Fuels with High Selectivity over a Bifunctional Catalyst," *Nature Chemistry* 9, no. 10 (2017): 1019-1024.

5.110. Alsayegh, S.O., Varjian, R., Alsalik, Y., Katsiev, K. et al., "Methanol Production Using Ultrahigh Concentrated Solar Cells: Hybrid Electrolysis and CO_2 Capture," *ACS Energy Letters* 5, no. 2 (2020): 540-544.

5.111. Grubb, W.T., "Batteries with Solid Ion Exchange Electrolytes: I. Secondary Cells Employing Metal Electrodes," *Journal of the Electrochemical Society* 106, no. 4 (1959): 275.

5.112. Grubb, W.T. and Niedrach, L.W., "Batteries with Solid Ion-Exchange Membrane Electrolytes: II. Low-Temperature Hydrogen-Oxygen Fuel Cells," *Journal of the Electrochemical Society* 107, no. 2 (1960): 131.

5.113. Yoshio, T. and Maomi, S., Sulfonic cation exchange resins prepared in the presence of plasticizer and polymer. U.S. Patent 2,891,014, 1959.

5.114. Fox, D.W. and Popkin, S., Sulfonated polyphenylene ether cation exchange resin. U.S. Patent 3,259,592, 1966.

5.115. Wakayama, H. and Yamazaki, K., "Low-Cost Bipolar Plates of Ti_4O_7-Coated Ti for Water Electrolysis with Polymer Electrolyte Membranes," *ACS Omega* 6, no. 6 (2021): 4161-4166.

5.116. Kellenberger, A., Vaszilcsin, N., Duca, D., Dan, M.L. et al., "Towards Replacing Titanium with Copper in the Bipolar Plates for Proton Exchange Membrane Water Electrolysis," *Materials* 15, no. 5 (2022): 1628.

5.117. Muradov, N.Z. and Veziroğlu, T.N., "From Hydrocarbon to Hydrogen–Carbon to Hydrogen Economy," *International Journal of Hydrogen Energy* 30, no. 3 (2005): 225-237.

5.118. Fernandez, J.L., Gennero De Chialvo, M.R., and Chialvo, A.C., "Preparation and Electrochemical Characterization of Ti/Ru_x $Mn_{1-x}O_2$ Electrodes," *Journal of Applied Electrochemistry* 32, no. 5 (2002): 513-520.

5.119. Marshall, A.T., Sunde, S., Tsypkin, M., and Tunold, R., "Performance of a PEM Water Electrolysis Cell Using $Ir_xRu_yTa_zO_2$ Electrocatalysts for the Oxygen Evolution Electrode," *International Journal of Hydrogen Energy* 32, no. 13 (2007): 2320-2324.

5.120. Wu, X., Tayal, J., Basu, S., and Scott, K., "Nano-Crystalline $Ru_xSn_{1-x}O_2$ Powder Catalysts for Oxygen Evolution Reaction in Proton Exchange Membrane Water Electrolysers," *International Journal of Hydrogen Energy* 36, no. 22 (2011): 14796-14804.

5.121. Grigoriev, S.A., Millet, P., and Fateev, V.N., "Evaluation of Carbon-Supported Pt and Pd Nanoparticles for the Hydrogen Evolution Reaction in PEM Water Electrolysers," *Journal of Power Sources* 177, no. 2 (2008): 281-285.

5.122. Esposito, D.V., Hunt, S.T., Kimmel, Y.C., and Chen, J.G., "A New Class of Electrocatalysts for Hydrogen Production from Water Electrolysis: Metal Monolayers Supported on Low-Cost Transition Metal Carbides," *Journal of the American Chemical Society* 134, no. 6 (2012): 3025-3033.

5.123. Esposito, D.V., Hunt, S.T., Stottlemyer, A.L., Dobson, K.D. et al., "Low-Cost Hydrogen-Evolution Catalysts Based on Monolayer Platinum on Tungsten Monocarbide Substrates," *Angewandte Chemie International Edition* 49, no. 51 (2010): 9859-9862.

5.124. Kelly, T.G., Hunt, S.T., Esposito, D.V., and Chen, J.G., "Monolayer Palladium Supported on Molybdenum and Tungsten Carbide Substrates as Low-Cost Hydrogen Evolution Reaction (HER) Electrocatalysts," *International Journal of Hydrogen Energy* 38, no. 14 (2013): 5638-5644.

5.125. Meyer, S., Nikiforov, A.V., Petrushina, I.M., Köhler, K. et al., "Transition Metal Carbides (WC, Mo_2C, TaC, NbC) as Potential Electrocatalysts for the Hydrogen Evolution Reaction (HER) at Medium Temperatures," *International Journal of Hydrogen Energy* 40, no. 7 (2015): 2905-2911.

5.126. Tang, C., Sun, A., Xu, Y., Wu, Z. et al., "High Specific Surface Area Mo2C Nanoparticles as an Efficient Electrocatalyst for Hydrogen Evolution," *Journal of Power Sources* 296 (2015): 18-22.

5.127. Ramakrishna, S.U.B., Reddy, D.S., Kumar, S.S., and Himabindu, V., "Nitrogen Doped CNTs Supported Palladium Electrocatalyst for Hydrogen Evolution Reaction in PEM Water Electrolyser," *International Journal of Hydrogen Energy* 41, no. 45 (2016): 20447-20454.

5.128. Blum, P. and Viguie, J.C., Cell for electrolysis of steam at high temperature. U.S. Patent 3,993,653, 1976.

5.129. Dönitz, W. and Erdle, E., "High-Temperature Electrolysis of Water Vapor—Status of Development and Perspectives for Application," *International Journal of Hydrogen Energy* 10, no. 5 (1985): 291-295.

5.130. Graves, C., Ebbesen, S.D., and Mogensen, M., "Co-electrolysis of CO_2 and H_2O in Solid Oxide Cells: Performance and Durability," *Solid State Ionics* 192, no. 1 (2011): 398-403.

5.131. Ebbesen, S.D., Knibbe, R., and Mogensen, M., "Co-electrolysis of Steam and Carbon Dioxide in Solid Oxide Cells," *Journal of the Electrochemical Society* 159, no. 8 (2012): F482.

5.132. Zhan, Z. and Zhao, L., "Electrochemical Reduction of CO_2 in Solid Oxide Electrolysis Cells," *Journal of Power Sources* 195, no. 21 (2010): 7250-7254.

5.133. Chen, X., Guan, C., Xiao, G., Du, X. et al., "Syngas Production by High Temperature Steam/CO_2 Coelectrolysis Using Solid Oxide Electrolysis Cells," *Faraday Discussions* 182 (2015): 341-351.

5.134. Zheng, Y., Wang, J., Yu, B., Zhang, W. et al., "A Review of High Temperature Co-electrolysis of H_2O and CO_2 to Produce Sustainable Fuels Using Solid Oxide Electrolysis Cells (SOECs): Advanced Materials and Technology," *Chemical Society Reviews* 46, no. 5 (2017): 1427-1463.

5.135. Song, Y., Zhang, X., Xie, K., Wang, G. et al., "High-Temperature CO_2 Electrolysis in Solid Oxide Electrolysis Cells: Developments, Challenges, and Prospects," *Advanced Materials* 31, no. 50 (2019): 1902033.

5.136. Li, S. and Xie, K., "Composite Oxygen Electrode Based on LSCF and BSCF for Steam Electrolysis in a Proton-Conducting Solid Oxide Electrolyzer," *Journal of The Electrochemical Society* 160, no. 2 (2013): F224.

5.137. Gan, Y., Zhang, J., Li, Y., Li, S. et al., "Composite Oxygen Electrode Based on LSCM for Steam Electrolysis in a Proton Conducting Solid Oxide Electrolyzer," *Journal of The Electrochemical Society* 159, no. 11 (2012): F763.

5.138. Kaiser, A., Monreal, E., Koch, A., and Stolten, D., "Reactions at the Interface $La_{0.5}Ca_{0.5}MnO_3$-YSZ/Al_2O_3 under Anodic Current," *Ionics* 2, no. 3 (1996): 184-189.

5.139. Momma, A., Kato, T., Kaga, Y., and Nagata, S., "Polarization Behavior of High Temperature Solid Oxide Electrolysis Cells (SOEC)," *Journal of the Ceramic Society of Japan* 105, no. 1221 (1997): 369-373.

5.140. Keane, M., Mahapatra, M.K., Verma, A., and Singh, P., "LSM–YSZ Interactions and Anode Delamination in Solid Oxide Electrolysis Cells," *International Journal of Hydrogen Energy* 37, no. 22 (2012): 16776-16785.

5.141. Yang, C., Coffin, A., and Chen, F., "High Temperature Solid Oxide Electrolysis Cell Employing Porous Structured (La0.75Sr0.25)0.95MnO$_3$ with Enhanced Oxygen Electrode Performance," *International Journal of Hydrogen Energy* 35, no. 8 (2010): 3221-3226.

5.142. van Troostwijk, A.P. and Deiman, J.R., "Sur une manière de décomposer l'Eau en Air inflammable et en Air vital," *Obs. Phys.* 35 (1789): 369-378.

5.143. Divisek, J. and Mergel, J., Diaphragms for alkaline water electrolysis and method for production of the same as well as utilization thereof. U.S. Patent 4,394,244, 1983.

5.144. Darlington, W.B. and DuBois, D.W., Electrolyte permeable diaphragm including a polymeric metal oxide. U.S. Patent 4,680,101, 1987.

5.145. In Lee, H., Dung, D.T., Kim, J., Pak, J.H. et al., "The Synthesis of a Zirfon-Type Porous Separator with Reduced Gas Crossover for Alkaline Electrolyzer," *International Journal of Energy Research* 44, no. 3 (2020): 1875-1885.

5.146. Konovalova, A., Kim, H., Kim, S., Lim, A. et al., "Blend Membranes of Polybenzimidazole and an Anion Exchange Ionomer (FAA3) for Alkaline Water Electrolysis: Improved Alkaline Stability and Conductivity," *Journal of Membrane Science* 564 (2018): 653-662.

5.147. Kim, H.J., Taekyung, L.E.E., Lee, S.Y., Park, H.Y. et al., Polybenzimidazole-based electrolyte membrane for alkaline water electrolysis and water electrolysis device comprising the same. U.S. Patent 17/589,466, 2022.

5.148. Henkensmeier, D., Najibah, M., Harms, C., Žitka, J. et al., "Overview: State-of-the Art Commercial Membranes for Anion Exchange Membrane Water Electrolysis," *Journal of Electrochemical Energy Conversion and Storage* 18, no. 2 (2021): 024001.

5.149. Figueroa, J.D., Fout, T., Plasynski, S., McIlvried, H. et al., "Advances in CO$_2$ Capture Technology—The US Department of Energy's Carbon Sequestration Program," *International Journal of Greenhouse Gas Control* 2, no. 1 (2008): 9-20.

5.150. Kapetaki, Z. and Scowcroft, J., "Overview of Carbon Capture and Storage (CCS) Demonstration Project Business Models: Risks and Enablers on the Two Sides of the Atlantic," *Energy Procedia* 114 (2017): 6623-6630.

5.151. Pérez-Fortes, M., Bocin-Dumitriu, A., and Tzimas, E., "CO$_2$ Utilization Pathways: Techno-Economic Assessment and Market Opportunities," *Energy Procedia* 63 (2014): 7968-7975.

5.152. Wang, M. and Oko, E., "Special Issue on Carbon Capture in the Context of Carbon Capture, Utilisation and Storage (CCUS)," *International Journal of Coal Science & Technology* 4, no. 1 (2017): 1-4.

5.153. Jung, S.H., Lee, S.H., Min, J., Lee, M.H. et al., "Analysis of the State of the Art of International Policies and Projects on CCU for Climate Change Mitigation with a Focus on the Cases in Korea," *Sustainability* 13, no. 1 (2020): 19.

5.154. Vishal, V., Chandra, D., Singh, U., and Verma, Y., "Understanding Initial Opportunities and Key Challenges for CCUS Deployment in India at Scale," *Resources, Conservation and Recycling* 175 (2021): 105829.

5.155. Zhang, L., Song, Y., Shi, J., Shen, Q. et al., "Frontiers of CO$_2$ Capture and Utilization (CCU) towards Carbon Neutrality," *Advances in Atmospheric Sciences* 39, no. 8 (2022): 1252-1270.

5.156. Wu, X., Yu, Y., Qin, Z., and Zhang, Z., "The Advances of Post-Combustion CO$_2$ Capture with Chemical Solvents: Review and Guidelines," *Energy Procedia* 63 (2014): 1339-1346.

5.157. Reiter, G. and Lindorfer, J., "Evaluating CO$_2$ Sources for Power-to-Gas Applications—A Case Study for Austria," *Journal of CO2 Utilization* 10 (2015): 40-49.

5.158. Liang, Z.H., Rongwong, W., Liu, H., Fu, K. et al., "Recent Progress and New Developments in Post-Combustion Carbon-Capture Technology with Amine Based Solvents," *International Journal of Greenhouse Gas Control* 40 (2015): 26-54.

5.159. Hirata, T., Kishimoto, S., Inui, M., Tsujiuchi, T. et al., "MHI's Commercial Experiences with CO$_2$ Capture and Recent R&D Activities," *Mitsubishi Heavy Industries Technical Review* 55, no. 1 (2018): 32.

5.160. Mimura, T., Shimojo, S., Suda, T., Iijima, M. et al., "Research and Development on Energy Saving Technology for Flue Gas Carbon Dioxide Recovery and Steam System in Power Plant," *Energy Conversion and Management* 36, no. 6-9 (1995): 397-400.

5.161. Reddy, S., Scherffius, J., Freguia, S., and Roberts, C., "Fluor's Econamine FG Plus SM Technology," in *Proceedings of the 2nd Annual Conference on Carbon Sequestration*, Washington, DC, May 5-8, 2003.

5.162. Reddy, S., Yonkoski, J., Rode, H., Irons, R. et al., "Fluor's Econamine FG PlusSM Completes Test Program at Uniper's Wilhelmshaven Coal Power Plant," *Energy Procedia* 114 (2017): 5816-5825.

5.163. Singh, A. and Stéphenne, K., "Shell Cansolv CO_2 Capture Technology: Achievement from First Commercial Plant," *Energy Procedia* 63 (2014): 1678-1685.

5.164. Cotton, A., Gray, L., and Maas, W., "Learnings from the Shell Peterhead CCS Project front End Engineering Design," *Energy Procedia* 114 (2017): 5663-5670.

5.165. Romano, M.C., "Ultra-High CO_2 Capture Efficiency in CFB Oxyfuel Power Plants by Calcium Looping Process for CO_2 Recovery from Purification Units Vent Gas," *International Journal of Greenhouse Gas Control* 18 (2013): 57-67.

5.166. Font-Palma, C., Errey, O., Corden, C., Chalmers, H. et al., "Integrated Oxyfuel Power Plant with Improved CO_2 Separation and Compression Technology for EOR Application," *Process Safety and Environmental Protection* 103 (2016): 455-465.

5.167. Rae, J.W., Zhang, Y.G., Liu, X., Foster, G.L. et al., "Atmospheric CO_2 over the Past 66 Million Years from Marine Archives," *Annual Review of Earth and Planetary Sciences* 49 (2021): 609-641.

5.168. Tsuji, T., Sorai, M., Shiga, M., Fujikawa, S. et al., "Geological Storage of CO_2–N_2–O_2 Mixtures Produced by Membrane-Based Direct Air Capture (DAC)," *Greenhouse Gases: Science and Technology* 11, no. 4 (2021): 610-618.

5.169. Ragipani, R., Sreenivasan, K., Anex, R.P., Zhai, H. et al., "Direct Air Capture and Sequestration of CO_2 by Accelerated Indirect Aqueous Mineral Carbonation under Ambient Conditions," *ACS Sustainable Chemistry & Engineering* 10, no. 24 (2022): 7852-7861.

5.170. Castro-Muñoz, R., Ahmad, M.Z., Malankowska, M., and Coronas, J., "A New Relevant Membrane Application: CO_2 Direct Air Capture (DAC)," *Chemical Engineering Journal* 446 (2022): 137047.

5.171. Seipp, C.A., Williams, N.J., Kidder, M.K., and Custelcean, R., "CO_2 Capture from Ambient Air by Crystallization with a Guanidine Sorbent," *Angewandte Chemie International Edition* 129, no. 4 (2017): 1062-1065.

5.172. Brethomé, F.M., Williams, N.J., Seipp, C.A., Kidder, M.K. et al., "Direct Air Capture of CO_2 via Aqueous-Phase Absorption and Crystalline-Phase Release Using Concentrated Solar Power," *Nature Energy* 3, no. 7 (2018): 553-559.

5.173. Liu, S., Chen, Y., Yue, B., Wang, C. et al., "Regulating Extra-Framework Cations in Faujasite Zeolites for Capture of Trace Carbon Dioxide," *Chemistry–A European Journal* 28, no. 50 (2022): e202201659.

5.174. Anyanwu, J.T., Wang, Y., and Yang, R.T., "Amine-Grafted Silica Gels for CO_2 Capture Including Direct Air Capture," *Industrial & Engineering Chemistry Research* 59, no. 15 (2019): 7072-7079.

5.175. Sekizkardes, A.K., Wang, P., Hoffman, J., Budhathoki, S. et al., "Amine-Functionalized Porous Organic Polymers for Carbon Dioxide Capture," *Materials Advances* 3, no. 17 (2022): 6668-6686.

5.176. Shekhah, O., Belmabkhout, Y., Chen, Z., Guillerm, V. et al., "Made-to-Order Metal-Organic Frameworks for Trace Carbon Dioxide Removal and Air Capture," *Nature Communications* 5, no. 1 (2014): 1-7.

5.177. Rim, G., Feric, T.G., Moore, T., and Park, A.H.A., "Solvent Impregnated Polymers Loaded with Liquid-Like Nanoparticle Organic Hybrid Materials for Enhanced Kinetics of Direct air Capture and Point Source CO_2 Capture," *Advanced Functional Materials* 31, no. 21 (2021): 2010047.

5.178. Rochelle, G.T., "Amine Scrubbing for CO_2 Capture," *Science* 325, no. 5948 (2009): 1652-1654.

5.179. Voskian, S. and Hatton, T.A., "Faradaic Electro-Swing Reactive Adsorption for CO_2 Capture," *Energy & Environmental Science* 12, no. 12 (2019): 3530-3547.

5.180. Liu, Y., Chow, C.M., Phillips, K.R., Wang, M. et al., "Electrochemically Mediated Gating Membrane with Dynamically Controllable Gas Transport," *Science Advances* 6, no. 42 (2020): eabc1741.

5.181. Kárászová, M., Zach, B., Petrusová, Z., Červenka, V. et al., "Post-Combustion Carbon Capture by Membrane Separation, Review," *Separation and Purification Technology* 238 (2020): 116448.

5.182. Kentish, S.E., "Polymeric Membranes for Natural Gas Processing," in *Advanced Membrane Science and Technology for Sustainable Energy and Environmental Applications*, Eds. Basile, A. and Nunes, S.P. (Cambridge, UK: Woodhead Publishing, 2011), 339-360.

5.183. Sabatier, P. and Senderens, J.B., "New Synthesis of Methane," *Comptes Rendus Hebd. Des. Seances Acad. Des. Scrences* 134 (1902): 514-516.

5.184. Sabatier, P., "The Method of Direct Hydrogenation by Catalysis," Nobel Lectures, Chemistry, 1912.

5.185. Garbarino, G., Bellotti, D., Riani, P., Magistri, L. et al., "Methanation of Carbon Dioxide on Ru/Al_2O_3 and Ni/Al_2O_3 Catalysts at Atmospheric Pressure: Catalysts Activation, Behaviour and Stability," *International Journal of Hydrogen Energy* 40, no. 30 (2015): 9171-9182.

5.186. Lu, Q., Rosen, J., Zhou, Y., Hutchings, G.S. et al., "A Selective and Efficient Electrocatalyst for Carbon Dioxide Reduction," *Nature Communications* 5, no. 1 (2014): 1-6.

5.187. Wang, W., Wang, S., Ma, X., and Gong, J., "Recent Advances in Catalytic Hydrogenation of Carbon Dioxide," *Chemical Society Reviews* 40, no. 7 (2011): 3703-3727.

5.188. Porosoff, M.D., Yang, X., Boscoboinik, J.A., and Chen, J.G., "Molybdenum Carbide as Alternative Catalysts to Precious Metals for Highly Selective Reduction of CO_2 to CO," *Angewandte Chemie International Edition* 53, no. 26 (2014): 6705-6709.

5.189. Erdöhelyi, A., Pásztor, M., and Solymosi, F., "Catalytic Hydrogenation of CO_2 over Supported Palladium," *Journal of Catalysis* 98, no. 1 (1986): 166-177.

5.190. Saito, M., Fujitani, T., Takeuchi, M., and Watanabe, T., "Development of Copper/Zinc Oxide-Based Multicomponent Catalysts for Methanol Synthesis from Carbon Dioxide and Hydrogen," *Applied Catalysis A: General* 138, no. 2 (1996): 311-318.

5.191. Melian-Cabrera, I., López Granados, M., and Fierro, J.L.G., "Effect of Pd on Cu-Zn Catalysts for the Hydrogenation of CO_2 to Methanol: Stabilization of Cu Metal Against CO_2 Oxidation," *Catalysis Letters* 79, no. 1 (2002): 165-170.

5.192. Studt, F., Sharafutdinov, I., Abild-Pedersen, F., Elkjær, C.F. et al., "Discovery of a Ni-Ga Catalyst for Carbon Dioxide Reduction to Methanol," *Nature Chemistry* 6, no. 4 (2014): 320-324.

5.193. Martin, O., Martín, A.J., Mondelli, C., Mitchell, S. et al., "Indium Oxide as a Superior Catalyst for Methanol Synthesis by CO_2 Hydrogenation," *Angewandte Chemie International Edition* 55, no. 21 (2016): 6261-6265.

5.194. Joo, O.S., Jung, K.D., Moon, I., Rozovskii, A.Y. et al., "Carbon Dioxide Hydrogenation to form Methanol via a Reverse-Water-Gas-Shift Reaction (the CAMERE Process)," *Industrial & Engineering Chemistry Research* 38, no. 5 (1999): 1808-1812.

5.195. Wei, J., Ge, Q., Yao, R., Wen, Z. et al., "Directly Converting CO_2 into a Gasoline Fuel," *Nature Communications* 8, no. 1 (2017): 1-9.

5.196. Khan, M.K., Butolia, P., Jo, H., Irshad, M. et al., "Selective Conversion of Carbon Dioxide into Liquid Hydrocarbons and Long-Chain α-Olefins over Fe-Amorphous AlO_X Bifunctional Catalysts," *ACS Catalysis* 10, no. 18 (2020): 10325-10338.

5.197. Wang, X.X., Duan, Y.H., Zhang, J.F., and Tan, Y.S., "Catalytic Conversion of CO_2 into High Value-Added Hydrocarbons over Tandem Catalyst," *Journal of Fuel Chemistry and Technology* 50, no. 5 (2022): 538-563.

5.198. Ghasemi, M., Mohammadi, M., and Sedighi, M., "Sustainable Production of Light Olefins from Greenhouse Gas CO_2 over SAPO-34 Supported Modified Cerium Oxide," *Microporous and Mesoporous Materials* 297 (2020): 110029.

5.199. He, Z., Cui, M., Qian, Q., Zhang, J. et al., "Synthesis of Liquid Fuel via Direct Hydrogenation of CO_2," *Proc. Natl. Acad. Sci. USA* 116 (2019): 12654-12659.

5.200. Albrecht, M., Rodemerck, U., Schneider, M., Bröring, M. et al., "Unexpectedly Efficient CO_2 Hydrogenation to Higher Hydrocarbons over Non-doped Fe_2O_3," *Applied Catalysis B: Environmental* 204 (2017): 119-126.

5.201. Owen, R.E., Plucinski, P., Mattia, D., Torrente-Murciano, L. et al., "Effect of Support of Co-Na-Mo Catalysts on the Direct Conversion of CO_2 to Hydrocarbons," *Journal of CO2 Utilization* 16 (2016): 97-103.

Further Reading

Agarwal, A.K. and Valera, H., *Greener and Scalable E-Fuels for Decarbonization of Transport* (Singapore: Springer Singapore Pte. Limited, 2021), ISBN:978-981-16-8344-2.

Bajpai, P., *Developments in Bioethanol* (Singapore: Springer Nature, 2020).

Knothe, G., Van Gerpen, J.H., and Krahl, J., *The Biodiesel Handbook* (Boca Raton, FL: Taylor & Francis Ltd., 2005).

Rackley, S.A., *Carbon Capture and Storage*, 2nd ed. (Amsterdam, the Netherlands: Butterworth-Heinemann, 2017), ISBN:978-0128120422.

Scibioh, M.A. and Viswanathan, B., *Carbon Dioxide to Chemicals and Fuels* (Amsterdam, the Netherlands: Elsevier, 2018).

Senecal, K. and Leach, F., *Racing toward Zero: The Untold Story of Driving Green* (Warrendale, PA: SAE International, 2021), ISBN:978-1-4686-0146-6.

Tremel, A., *Electricity-Based Fuels*. Vol. 941 (Cham, Switzerland: Springer International Publishing, 2018), ISBN:978-3-319-72458-4.

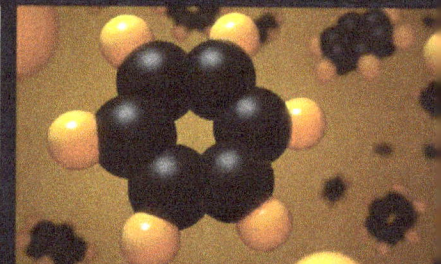

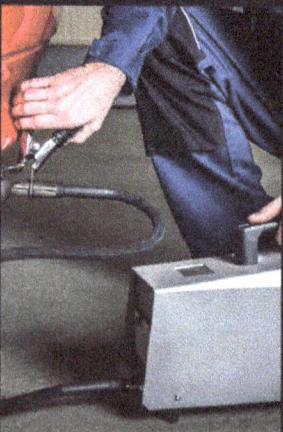

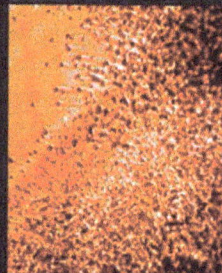

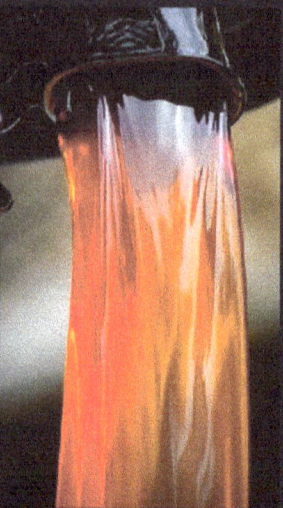

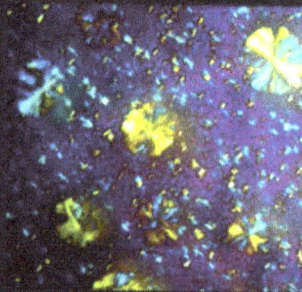

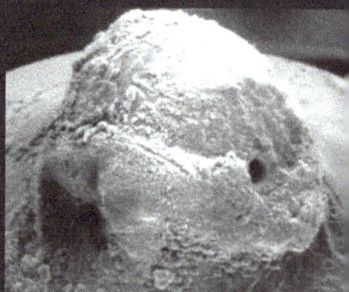

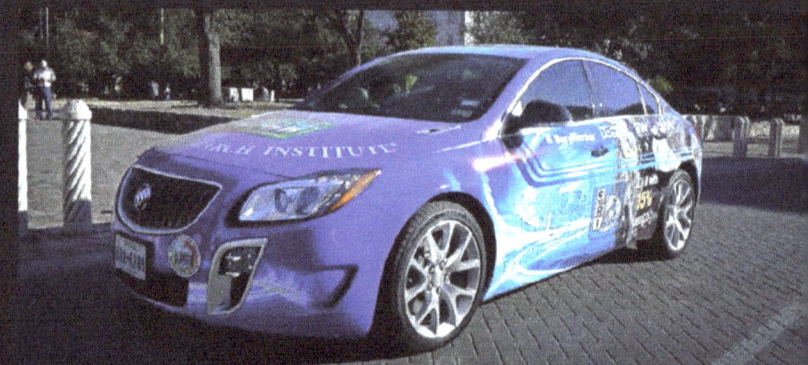

6

Storage, Distribution, and Handling of Gasoline and Diesel Fuel

6.1. Introduction

The most important aspect of the storage, distribution, and handling of any fuel is that it is a chemical product and should be treated accordingly. This is covered by the United Nations Globally Harmonized System of Classification and Labelling of Chemicals (GHS) [6.1]. Fuels should therefore be accompanied by a safety data sheet (SDS) detailing such aspects of storage and handling as physical properties, hazard classification, first aid measures, firefighting, toxicological, and ecological advice. Examples of material safety data sheets (MSDS) for a European gasoline, both with and without an ethanol content, and a European diesel fuel, again with and without an oxygenate content, are included in Appendix 5. For general comments regarding health and safety aspects, see Sections 6.2 to 6.4.

Beyond the health and safety aspects of storage and handling, a number of factors can influence the quality and performance of gasoline and diesel fuels after they leave the refinery and before they reach the customer's tank. At an oil terminal, they can be reblended, mixed with other batches of the same sort of material, treated with various additives, accidentally contaminated with other products, and subjected to electrostatic discharges during pumping operations. Once they have left the terminal, they can be contaminated with water; held for long periods in contact with air; subjected to severe mixing; brought into contact with metals and elastomeric materials, which they can attack; and they can be subjected to a range of temperature and humidity conditions.

Most of these factors can modify product quality; great efforts are therefore made by the fuel producers, distributors, and retailers to minimize the adverse effect of these factors.

6.2. Safety Considerations for Storage and Handling

6.2.1. Flash Point

To minimize the likelihood of an accidental fire while petroleum fuels are being handled and transferred into and out of storage containers, it is important that the appropriate codes of practice be carried out. Typical regulations include restrictions on smoking and define the type of electrical equipment that may be used. Flammable liquids are usually classified according to their flash points, for example, in the UK gasoline with a flash point below −21°C (actually it is normally below −40°C and too low to measure by standard methods) is given an Energy Institute (Institute of Petroleum [IP]) classification of Class I, and diesel fuel with a minimum flash point of 56°C is in Class III. The flash point is the temperature to which the fuel must be heated under specified conditions to produce a vapor-air mixture that will ignite when a test flame is applied.

A minimum flash point of around 56°C is typical of diesel fuel in many countries, although other flash point levels may apply where climatic or other considerations prevail. In the US, for example, the permitted minimum flash point for No. 1D grade diesel fuel is 38°C, and in Canada it is a minimum of 40°C.

6.2.2. Electrical Conductivity

Although conductivity is not a specification requirement for either gasoline or diesel fuel, it is occasionally measured because there is a potential safety risk due to the buildup of a static electricity charge during bulk handling at high pumping rates. Grounding leads are normally used to conduct away any charge when large quantities of fuel are pumped into or out of storage tanks. Aviation kerosene (jet fuel) must be treated with a static, a dissipator additive, and it is now becoming routine to treat automotive fuels with such additives.

Gasoline tends to be more conductive than diesel fuel, which, in turn, is more conductive than kerosene, and there is little risk of an excessive charge building up when refueling vehicles at a service station. However, faster pumping rates are used when loading and offloading road and rail fuel tankers, and an antistatic additive (see Chapters 11 and 19) can be used in these cases, even though the equipment is almost always grounded.

6.3. Health and Environmental Effects of Gasoline

The currently available data on the health, safety, and environmental properties of gasoline have been collated by CONCAWE [6.2] (the European oil companies' organization for environmental, health, and safety issues), and the following is a very brief review of their findings. The references given by CONCAWE in their report have not been included in this summary.

6.3.1. Health Aspects

The International Agency for Research on Cancer (IARC) has reviewed the carcinogenic risks from gasoline by looking at occupations where gasoline exposure may have occurred, such as service station attendants and car mechanics. IARC allocated gasoline an overall classification of Group 2B (i.e., possibly carcinogenic to humans), based on what was accepted as inadequate and limited evidence in experimental animals plus other evidence including the presence of benzene and 1,3-butadiene in gasoline.

6.3.1.1. Inhalation

High concentrations of gasoline vapor can accumulate in confined spaces since it is heavier than air, and it can present both a safety and a health hazard. Short and infrequent exposures are unlikely to result in a health risk. Exposure to gasoline vapor concentrations in the range of 500 to 1000 mg/m^3 can cause irritation of the upper respiratory tract, and if continued will give rise to a narcotic effect with symptoms such as headache, dizziness, nausea, and mental confusion with eventual loss of consciousness. These central nervous system effects can occur rapidly with a sudden loss of consciousness even after a brief exposure to very high concentrations. Cardiac irregularities have also been reported after exposure to high vapor concentrations.

6.3.1.2. Ingestion

The taste and smell of gasoline will usually limit ingestion to a small amount. In adults, ingestion is usually the result of siphoning attempts, and in children, it results from drinking from unlabeled or wrongly labeled containers. Gasoline is of moderate to low oral toxicity for adults, but for children ingestion of even small quantities may prove dangerous or even fatal.

Spontaneous vomiting is common after ingestion, with the likely consequence of aspiration of liquid gasoline into the lungs. This is the principal hazard, and no attempt should be made to induce vomiting.

6.3.1.3. Aspiration

Aspiration of even small amounts of liquid gasoline into the lungs, either directly or as a consequence of vomiting, can have very serious results. Irritation of the respiratory tract may lead rapidly to difficulty in breathing and development of a potentially fatal chemical pneumonitis.

6.3.1.4. Skin Contact

Repeated or prolonged skin contact can result in drying, cracking, and possible dermatitis. In rare cases, an individual may become sensitized to the dyes used in some gasolines. Repeated contact may also make the skin more susceptible to irritation and penetration by other materials.

6.3.1.5. Eye Contact

Eye contact of liquid gasoline may cause moderate to severe irritation and conjunctivitis. This effect is transient, and permanent injury is unlikely to result.

6.3.2. Exposure Limits

Because of the complex and variable composition of gasoline, there is no widely accepted exposure limit that is generally applicable. In the US, the American Conference of Governmental Industrial Hygienists (ACGIH) recommendation is an 8-hour time-weighted average threshold limit value (TLV-TWA) of 300 mg/m^3 and a short-term exposure limit threshold limit value (TLV-STEL) of 500 mg/m^3 over a period of 15 minutes. These limits are based on the typical vapor composition of gasoline in the US. Individual countries may have their own standards based on the composition of the gasoline, particularly its aromatic contents, such as benzene and toluene.

6.3.3. Ecotoxicity

The environmental effect of gasoline spills is mainly due to the water solubility of its components. Each of the several hundred hydrocarbon components in gasoline will have its own very low solubility, and of these, aromatic hydrocarbons have the greatest solubility and therefore represent the major part of the water-soluble fraction. The monoaromatic hydrocarbons are of most concern, and of these, benzene, toluene, ethylbenzene, and xylene are the most important, with naphthalene and methylnaphthalene as the most important di-aromatics. When oxygenated compounds are also present, since these are much more soluble in water than hydrocarbons (with the lower alcohols being completely miscible), such spills can represent a much more severe environmental problem.

Following a spillage, the more volatile components are rapidly lost by evaporation and, to a lesser extent, by dissolution into water. Local environmental conditions such as temperature, wind, wave action, soil type, and so forth, together with photo-oxidation, biodegradation, and adsorption onto suspended material, all contribute to the weathering of the remains of the spilled gasoline.

Microorganisms present in sediments and in the water are capable of degrading gasoline components, and the time to reduce the fuel concentration to 50% of the initial amount has been reported as between 1.2 and 2.7 days in sand, loam, or clay soils. Nutrient addition and inoculation with bacterial isolates enhance biodegradable losses.

Gasoline exhibits some short-term toxicity to freshwater and marine organisms. The components that are most prominent in the water-soluble fraction and cause aquatic toxicity are also highly volatile and can be readily biodegraded by microorganisms. Considering these factors, spilled gasoline, unless it contains alcohols, is unlikely to remain in water in sufficient quantities to cause aquatic effects and as a result presents a minimal overall risk to the environment. The situation when gasoline contains alcohol is discussed in Section 6.7.1.

6.3.4. Disposal

It is seldom necessary to dispose of large quantities of gasoline, but when it is, such as with residues from tank cleaning, it should be done by incineration. Materials that have been highly contaminated with gasoline should also be incinerated, but less contaminated materials may be acceptable for authorized landfill sites. Contaminated soil may be treated by land farming.

6.4. Health and Environmental Effects of Diesel Fuel

The currently available data on the health, safety, and environmental properties of diesel fuel has been collated by CONCAWE [6.3], and the following is a very brief review of their findings. The references given by CONCAWE in their report have not been included in this summary.

6.4.1. Health Aspects

Diesel fuels are complex and variable mixtures of hydrocarbons, predominantly of the carbon number range C11 to C25 and boiling over the temperature interval 150 to 450°C. The generic chemical composition of diesel fuels depends on the nature of the crude oils from which they are derived and the refinery processes that they have undergone.

6.4.1.1. Inhalation

Under normal conditions of storage and use, the vapor pressure of diesel fuels is too low for significant concentrations of vapor to develop. However, where temperatures are high and ventilation poor, vapor inhalation may result in health effects such as central nervous and respiratory system depression with eventual loss of consciousness. In some cases, a mist may be generated; at concentrations well above 5 mg/m^3, this mist could irritate the mucous membranes of the upper respiratory tract.

6.4.1.2. Ingestion

Ingestion of diesel fuel is an unlikely event in normal use. The taste and smell of diesel fuel will usually limit ingestion to small amounts. Although diesel fuels are of low acute oral toxicity, spontaneous vomiting may occur, with the associated risks of aspiration of diesel fuel into the lungs. Ingestion may also give rise to irritation of the mouth, throat, and gastrointestinal tract.

6.4.1.3. Aspiration

Aspiration of diesel fuel into the lungs, either directly or as a consequence of vomiting following ingestion, may result in damage to lung tissue. Breathing difficulties may arise, and a potentially fatal chemical pneumonitis may follow.

6.4.1.4. Skin Contact

In common with other low viscosity hydrocarbons, diesel fuels will remove natural fat from the skin; repeated or prolonged exposure can result in drying and cracking, irritation, and dermatitis. Some individuals may be especially susceptible to these effects. Excessive exposure under conditions of poor personal hygiene may also lead to oil acne and folliculitis, and with some products development of warty growths may occur, and these may become malignant subsequently.

6.4.1.5. Eye Contact

Accidental eye contact with liquid diesel fuel may cause mild, transient stinging and/or redness. Exposure to high concentrations of mist or vapor may also cause slight eye irritation.

6.4.2. Exposure Limits

Legislative limits for inhalation exposure have not been established. However, the guidance provided by the UK for an occupational exposure limit for mixed exposures may help in setting an appropriate exposure limit. In any case it is advisable to reduce the exposure to mist or vapor of diesel fuels to the lowest practicable level.

6.4.3. Ecotoxicity

Spillages of diesel fuel may penetrate the soil, causing ground water contamination, and it may be harmful to aquatic organisms. Films of fuel on water surfaces may cause physical damage to organisms and also impair oxygen transfer. The product is inherently biodegradable, and there is no evidence to show that bioaccumulation will occur.

6.4.4. Disposal

Because diesel fuels are primarily used as fuel, disposal of large quantities is seldom necessary. When disposal is necessary, for example from spillages or tank cleaning, this can be done by combustion. Alternatively, re-distillation with other fuel oils is possible. Contaminated material and oil/water mixtures could be shipped to refineries or other treatment plants for separation and reutilization.

6.5. Influences on Product Quality during Distribution

6.5.1. Sea Transport

For strong economic reasons, shipping companies prefer not to move their vessels loaded only with ballast. There is thus a high probability that refined products, such as gasoline and diesel fuel, can be carried in tanks that have previously been used to carry other materials, such as crude oil. There is thus a risk of contamination of the refined product if the tanks are not adequately cleaned. On a seagoing vessel, there is also the possibility of some sea water ingress and the formation of water condensation. The presence of water can give rise to haze formation, especially when it is mixed in with the fuel by the centrifugal pumps during discharge. When this occurs, it can require the use of dehazer additives to avoid excessive holdup in storage tanks while the product clears.

Evaporation losses can also take place but are not usually large enough to have a significant effect on product quality. When light products such as gasoline are transported long distances, the cargo can undergo considerable changes in temperature that can accentuate evaporative losses.

For safety reasons, oil-containing tanks will be blanketed with an inert gas (less than 5% oxygen). However, this can be generated by a combustor, which burns hydrocarbon fuel to consume the oxygen. This will inevitably generate water, which must be removed by filters or equivalent devices. Some ships now use onboard nitrogen generators.

6.5.2. Pipeline

This method of transporting fuel is growing because it is safe, cost-effective, and energy conserving. However, it does have potential difficulties because most pipelines will carry several products.

Contamination between one product and the next in a pipeline does occur to some extent since there is no physical barrier between them, although they are always put through in a predetermined sequence to minimize the effects of contamination. A typical sequence might be gasoline followed by kerosene or aviation fuel, diesel fuel, kerosene, and then gasoline again. However, the potential contamination of jet fuel by ethanol/gasoline blends is a serious concern due of the affinity of ethanol for water (see Section 6.7.2), and the potential for water can be carried through the pipeline system. Lead anti-knock compounds are no longer used in motor gasoline (mogas), only in aviation gasoline (avgas) and these fuels are kept separate so there is no chance of contamination. As the regulated level of sulfur in given fuel classes is reduced, there is a need to ensure that low-sulfur fuels are not contaminated by high-sulfur fuels; for example, low-sulfur fuel for highway use must not be contaminated by diesel fuel for off-highway use, which often has a high-sulfur limit. The rate of flow through the pipeline is kept above a certain minimum speed (depending on pipe diameter) to minimize mixing between products.

The interface between one product and the next could be estimated by knowledge of the pipeline capacity, flow rate, and switching cycle. However, to guarantee switching before interface product arrives at the switching valve, this will usually lead to large interface product volumes. It is therefore better to use an instrument or instruments to detect changes in the product in the pipeline. Traditionally, density was used as the metric for determining a change in product. Optical techniques, viscosity sensors, and online process analyzers may also be used. Even laser technology can be applied to the problem [6.4]. The interface material is segregated and blended back at a later stage into a product where the product specification will not be compromised.

Apart from interface contamination, there are other minor possibilities for product quality problems during pipeline transportation. Some additives or components used in gasoline and diesel fuel can loosen rust and other deposits from the sides of the pipe and hold it in suspension, causing filter blockage. Anticorrosion additives (see Chapters 11 and 19) are often added to keep the inside of the pipeline free from rust since it can restrict the maximum flow rate through the pipeline. The anticorrosion additives themselves cause no problems provided they are used at appropriate levels and that other fuel properties meet specification. However, some pipeline corrosion inhibitors, such as sodium nitrite, introduce metals into the fuel system, which is actively discouraged. Sodium has also been associated with interactions with other fuel additives leading to diesel fuel injector deposit formation [6.5]. In addition, ethanol/gasoline blends should be avoided, and are banned in some jurisdiction, due to ethanol's affinity to water, see Section 6.7.2. Any pipeline used for the shipment of ethanol/gasoline blends must be completely dehydrated prior to shipment and more frequent cleaning and inspection should be carried out. Corrosion inhibitors should be selected that are compatible with gasoline/ethanol blends. Similarly, biodiesel has an affinity to water and can thus become contaminated by water, which can cause corrosion in the pipeline systems.

Drag-reducing agents can also be used in products flowing through pipelines. All fluids are subject to resistance or drag as they flow through a pipe; this causes a pressure drop along the pipe. In general, the drag increases as the flow rate increases; as the flow rate increases beyond a critical value, the flow becomes turbulent, and the rate of increase of the drag also increases. This effectively puts an economic limit on the volume flow rate of the fluid through a given size of pipe. The flow through the pipe could therefore be increased by the use of an additive that increased the flow rate at which the flow became turbulent. An early example of such an additive for gasoline was a mixture of aluminum soap of coconut oil acid, aluminum naphthenate, and aluminum oleate (otherwise known as napalm) [6.6]. Of course, such additives had to be removed at the delivery end of the pipeline before the gasoline became saleable and usable. The drag can also be reduced without inducing thickening of the fluid and additives such as isobutylene resins [6.7] and poly-alpha-olefins [6.8]. Drag-reducing additives were first used commercially in 1979, in the 1.2-m-diameter Trans-Alaska pipeline [6.9].

Besides these additives incorporated in the fuels for transportation through pipelines, other complex mixtures are used as fuel additives to enhance the performance of the finished fuels (see Chapters 11 and 19). Many of these additives are surface active compounds that can be adsorbed onto the surface of the pipeline and desorbed into other products going through the pipeline. When this occurs, problems such as haze stabilization and additive loss can ensue.

6.5.3. Road and Rail

Rail is often used to transport fuel to terminals and via road is almost invariably the final method of transportation used to reach the filling station. Contamination problems are quite rare since dedicated tankers are usually used. The biggest potential problem is in ensuring that human error does not result in the wrong materials being put into the filling station tanks.

The use of switch-loading in road tankers is now becoming fairly common. This means that a tanker compartment can be switched from holding either gasoline or diesel to the other fuel. Switch-loading makes the conductivity of the fuel an important factor. With gasoline, which is more volatile, the head space is usually too fuel rich to allow combustion, whereas for diesel fuel the head space is too lean due to the low volatility of the diesel fuel. However, putting diesel into a tank that has some gasoline vapor can result in a mixture within the flammable range. This makes preventing sparks of paramount importance.

6.6. Influences on Product Quality during Storage

Quality can improve in the tanks used to hold the product at various stages in its transportation from the refinery to the filling station if the fuel is held long enough for the particles of dirt and rust that might have become entrained to separate out. Water, which can originate from processing steps or be picked up during transportation, may also separate out at this stage and settle to the bottom of the tank.

Initial tests for identifying the presence of water and sediment are ASTM D4176 [6.10] and ASTM D1796 [6.11]. ASTM D4176 is the appropriate test for the presence of water and sediment in gasoline and involves a sample being swirled in a clean glass jar and examined for visual sediment or water droplets. It is claimed that an experienced tester can detect as little as 40 mg/kg of free water in the fuel. For diesel fuel, the centrifuge method ASTM D1796 can be used. This test is a constituent of the ASTM D975 [6.12] specification for diesel fuel oils in the US. A 50 mL sample of the fuel is mixed with an equal volume of water-saturated solvent and centrifuged to concentrate the water and sediment at the bottom of conical centrifuge tubes. This test method is not suitable for fuels containing alcohol; the ASTM D6304 [6.13] method will give a quantitative assessment of water content from 20 to 25,000 mg/kg, but the presence of sulfur compounds can interfere with the results.

Some diesel specifications impose separate limits for water and sediment, and others limit the combined amount of water and sediment. Typical maximum levels are 0.05% for water and sediment together, 0.05% for water alone, and 0.01% for sediment alone. In the European specification for automotive diesel fuel, EN 590 [6.14], sets a maximum limit value of 200 mg/kg (0.02%) for water and 24 mg/kg for particulate matter.

Oxidation stability problems may occur with excessive storage times, as is sometimes required for strategic fuel stocks. This is particularly true with the inclusion of oxygenated components within the fuel. For storage at high ambient temperatures, inert gas blanketing (often nitrogen) is sometimes used to minimize oxidation of the fuel.

6.6.1. Water Contamination in Tankage

Water can be introduced into the gasoline while it is in tankage because many tanks have floating roofs, and rainwater can find its way past the seals into the fuel. Water is also sometimes used to push fuel out of the tank; such a fuel is, therefore, always saturated with water. A drop in temperature of the product may also cause dissolved water to separate out as a haze that will eventually clear unless certain types of surfactant additives are present, which can stabilize such hazes.

Most storage tanks have a layer of water on the bottom that should be kept to a minimum by frequently drawing it off. The product draw-off point is usually well above any water layer so that there should be little chance of contamination from tank water bottoms. However, if the water layer is allowed to build up excessively, or if a tank is not given enough time to allow the water to settle out, then free water can be present in the product being discharged. In cases where the amount of tankage is restricted, gasolines or diesel fuels may be pumped out of a tank at the same time as fresh product is being pumped in, and in these cases, particularly if the tank level is low, it is possible for the water bottoms to be stirred up and contaminate the product.

6.6.2. Microbiological Contamination

Bacterial and fungal growths can sometimes lead to operational problems such as cloudy fuel and filter blocking if storage tanks are not cleaned regularly. Gasoline is generally less susceptible to this problem than diesel fuel. The organisms, which can be aerobic or anaerobic, live in water bottoms of tanks and feed on the fuel at the interface. Anaerobic, or sulfate-reducing, bacteria derive their oxygen from sulfates, and this gives rise to a slimy sulfide deposit and hydrogen sulfide gas. Hydrogen sulfide can cause corrosion problems as well as give an unpleasant odor. Both the water and oil phases become darker when this occurs. Biocides [6.15] can be used to kill these organisms (see Chapters 11 and 19), but cleaning and good housekeeping are also necessary to remove the by-products of their activity and to minimize the likelihood of further problems.

6.6.3. Sludge in Tankage

Other materials that may separate out in tankage are gums from gasoline, and wax from diesel fuel. These can contribute to a layer of sludge at the bottom of the tank.

Wax settling from diesel fuels can occur when the fuel has cooled to below its cloud point as a result of unexpectedly low temperatures or long storage in small, exposed tanks. Although settled wax usually re-dissolves when the tank is refilled or as the temperature rises, its removal is advisable if the amount of wax remaining in a fuel could put a new batch of fuel out of specification. Additives to overcome wax separation problems are discussed in Chapter 19. The inclusion of biodiesel can cause additional problems as precipitate can form at temperatures higher than the cloud point. This precipitate is also more resistant to re-dissolving [6.16]. This is discussed in more detail in Section 6.7.5.

The sludge layer in gasoline tanks can also build up to a considerable level, and it is necessary from time to time to take tanks out of service to remove this material before it becomes a potential contaminant. Where leaded aviation gasoline is still stored, lead compound degradation products can accumulate as sludge and require specialist attention. The gums present in the sludge can cause severe deposit problems in engines.

6.6.4. Evaporative Losses

A significant amount of evaporation is possible with very volatile gasolines. This can change the performance of a gasoline and increase its costs, as well as have an adverse environmental effect. The amount of evaporative loss from a tank will depend on whether it has a fixed or a floating roof. In fixed roof tanks, which operate with a space above the liquid, vapor loss can occur during tank filling and in tank breathing as the temperature changes. The losses are reduced to some extent by the provision of pressure/vacuum (P/V) valves that allow some positive or negative pressure to develop before vapor expulsion or air induction can occur. Better control of evaporative emissions is possible with the use of internal floating roofs. With floating roof tanks, losses can occur from the vapor seals between the roof and the tank shell, from various fittings in the roof for taking samples, and from the wet sides as the tank level drops during discharge. Improved tank seals can minimize these losses. Control of evaporative emissions during manufacture and distribution is known as Stage 1 recovery and includes the installation of vapor recovery facilities at terminals and depots.

Evaporative emissions are also of concern during the refueling of vehicles. Vapor recovery equipment to minimize these losses can be installed in service stations, the so-called Stage 2 controls, required to do so.

6.6.5. Oxidation

Both gasoline and diesel fuel can oxidize during storage, giving rise to the formation of gums and gum precursors that can cause deposit formation in engines and seriously influence their performance. Biodiesel is particularly prone to oxidation; this is discussed in Section 6.7.4. Hydrocarbon fuels containing olefinic components, arising mainly from cracking operations, are the most susceptible to gum formation and therefore may need some special processing or the use of antioxidants (see Chapters 11 and 19). When gums form, they initially remain in solution, but as the amount increases, they begin to separate out to give cloudy fuels and, in extreme cases, blocked lines and filters, and can result in high sludge levels in tankage.

The mechanism by which the oxidation of hydrocarbons progresses occurs in several stages, as follows:

1. Chain initiation involving the generation of free radicals:

$$R - H \rightarrow R \bullet + \left(H \bullet \right)$$

2. Chain propagation:
   Once a hydrocarbon free radical (R•) has been formed, it can combine with oxygen to form a peroxide radical (R-O-O•), which in turn can react with another hydrocarbon molecule, thereby generating other hydrocarbon free radicals and a hydroperoxide (R-O-OH):

$$R \bullet + O_2 \rightarrow R - O - O \bullet$$

$$R - O - O \bullet + R'H \rightarrow R - O - OH + R' \bullet$$

The oxidation process is therefore self-perpetuating. The free radicals can also give rise to polymerization as well as oxidation reactions to form high-molecular-weight materials. These can deposit in the fuel system.

3. Chain termination:

The chain reaction is only terminated, in the absence of an antioxidant, by reactions that lead to non–free radical products:

$$R \bullet + R \bullet \rightarrow R - R$$

$$R - O - O \bullet + R \bullet \rightarrow R - O - O - R$$

The chain breaking or terminating function of antioxidants is thought to proceed by the donation of a hydrogen atom from the reactive center of the antioxidant to the peroxy radical. The activity is then sufficiently stabilized by resonance to discontinue chain propagation [6.17, 6.18].

The presence of copper and certain other metals dissolved in the gasoline will actively catalyze the oxidation process. The use of metal deactivating additives (see Chapter 11, Section 11.1) can overcome this problem.

6.7. Considerations with Oxygenated Blends

Fuels containing alcohols, and methanol in particular, can cause problems unless care is taken during blending, storage, and distribution. There are also concerns about the stability of fatty acid methyl esters (biodiesel). These points are discussed in more detail in the following sections.

6.7.1. Environmental Aspects of Fuels Containing MTBE

Methyl tertiary butyl ether (MTBE) is readily soluble in water and has a pungent odor and taste that can render water undrinkable at concentrations that do not pose any risk to human health. Leakage of gasoline from underground storage tanks or spillage at service stations can lead to the contamination of groundwater as the MTBE permeates through the soil. These very low levels of contamination have occurred in California and to a limited extent in other US states.

The concern in California reached such a level that on March 25, 1999, the governor of California announced a three-year, eight-month phase-out of MTBE from Californian gasoline. He additionally called upon California's federal delegation to establish a forum for the removal of MTBE without violating the Federal Clean Air Act. The appropriate state regulatory agencies were ordered to devise and carry out a plan to begin the immediate phase-out of MTBE from California gasoline, with 100% removal to be achieved no later than December 31, 2002. On December 9, 1999, the California Air Resources Board (CARB) approved a new set of gasoline rules banning MTBE while "preserving all the air-quality benefits obtained from the state's cleaner-burning gasoline program." The new rules, known officially as the Phase 3 gasoline regulation, prohibit the formulation of gasoline with MTBE after December 31, 2002. This was subsequently delayed by one year, with the prohibition coming into effect after December 2003.

On September 14, 1999, the EPA Office of Mobile Services director made a statement to the US House of Representatives Subcommittee on Energy and Environment "to put the issues surrounding the use of MTBE and ethanol in perspective." Neither the Clean Air Act nor the Environmental Protection Agency require the use of MTBE in reformulated gasoline (RFG). The Clean Air Act Amendments of 1990 required that RFG contain 2.0% m/m minimum oxygen but does not specify which oxygenate to use. Both ethanol and MTBE are employed in the current RFG program, with fuel suppliers at that time choosing to use MTBE in about 80% of RFG. The EPA's position is that oxygenates help to reduce emissions of smog precursors and air toxics by diluting or displacing gasoline components such as benzene, olefins, aromatics, and sulfur and by altering distillation characteristics. In addition, because oxygenates increase octane, refiners have chosen to add them to gasoline since the late-1970s. Because oxygenates contribute up to 11% of the volume of RFG, they can extend the gasoline supply through displacement of other gasoline components.

Despite the air-quality aspects of oxygenates in RFG, the EPA admitted that there was growing concern about contamination of drinking water by MTBE in Santa Monica, California, and several other areas in the state, as well as in Maine and other states. For the most part, levels detected in drinking water were quite low. In June 1999, 3.7% of California's drinking water systems sampled detected MTBE. Most of those detections were below the state's secondary standard (or taste and odor action level) of 5 ppb.

The US Geological Survey reported that about 3% of groundwater wells in RFG program areas had detections of MTBE at or above 5 ppb. MTBE detections at higher concentrations in groundwater, such as those experienced in Santa Monica, result primarily from leaking underground fuel storage tanks and possibly from spills from distribution facilities. These leaks are unacceptable, regardless of whether or not MTBE is present in the gasoline. However, the presence of MTBE at these sites suggests the need for improved early warning systems for underground storage tank leaks. The agency's underground storage tank program is expected to substantially reduce future leaks of all fuels and additives, including MTBE, from underground fuel storage tanks. Underground tanks were required to be upgraded, closed, or replaced to meet these requirements by December 1998.

In response to concerns associated with the use of oxygenates in gasoline, the administrator established a blue-ribbon panel of leading experts from public health and scientific communities, water utilities, environmental groups, industry, and local and state governments. The objective was to assess issues posed by the use of oxygenates in gasoline. The panel presented its recommendations to the Clean Air Act Advisory Committee in late July 1999:

- Enhance water protection and monitoring
- Prevent leaks through improvement of existing programs
- Remediate existing contamination
- Amend the Clean Air Act to remove the requirement that federal reformulated gas contain 2% m/m oxygen
- Maintain current air benefits (no environmental backsliding)
- Reduce the use of MTBE
- Accelerate research on MTBE substitutes

The EPA aims to address the panel's recommendations to the extent possible within the agency's current administrative authority. This will include strengthening underground storage tank and drinking water protection programs and, where possible, providing more flexibility to states and refiners as they move to decrease the use of MTBE in gasoline.

The EPA intends "to provide a targeted legislative solution that maintains our air-quality gains and allows for the reduction of MTBE, while preserving the important role of renewable fuels like ethanol." Furthermore,

As MTBE use is reduced, ethanol is clearly the most likely substitute for MTBE and is already used to meet the oxygen standard in Milwaukee and Chicago. We believe it is likely that substantial amounts of ethanol will continue to be used in the federal RFG program. EPA is working closely with all fuel providers to assure a smooth transition to the second phase of the RFG program.

The situation is somewhat different in Europe. Draft Fuels Directive: COM(2001) 241 final of July 2001 recognized US concerns but pointed out that consumption in Europe had been lower. This is because the use of oxygenates in gasoline was not mandated but permitted subject to a maximum content. Although MTBE had been found in groundwater in certain member states, the Commission is of the opinion that such contamination was not widespread across the community. Moreover, a study confirmed that such contamination was unlikely if standards governing the construction and operation of underground storage tanks at service stations were robustly enforced. Accordingly, the commission did not propose any amendment to Directive 98/70/EC in respect of the MTBE content of gasoline. This did not affect the prerogative of member states to request stricter environmental specifications.

6.7.2. Water Sensitivity of Alcohol Blends

Both methanol and ethanol are completely miscible in water and so are partially extracted from gasoline if it is contacted with water. This is one reason for avoiding adding ethanol at the refinery before transferring via the pipeline system. This results in blendstock for oxygenated blends (BOB) grades being made and piped to terminals (as mentioned in Chapter 3, Section 3.10). In the presence of water, the gasoline/alcohol blend can separate into two layers: an alcohol-rich aqueous layer and a gasoline-rich upper layer. This has a number of undesirable effects: first, it depletes the gasoline of alcohol and so it may not meet the octane specifications. Second, the lower phase will have increased in volume due to the alcohol and have a lower density so that it will be more readily entrained and carried over into the motorist's tank. If this happens, it can stop a vehicle completely.

The water tolerance of a gasoline/alcohol blend (the amount of water that a blend can tolerate before it breaks into two phases) depends on temperature, the hydrocarbon composition of the gasoline, the type and concentration of the alcohol, and the presence of cosolvents such as tertiary butanol (TBA). Figure 6.1 [6.19] shows the effect of alcohol concentration and temperature for methanol and ethanol in an unleaded gasoline having an aromatics content of 26% by volume. No cosolvent is present.

FIGURE 6.1   Water tolerance of methanol and ethanol blends in gasoline with 26% volume aromatic content.

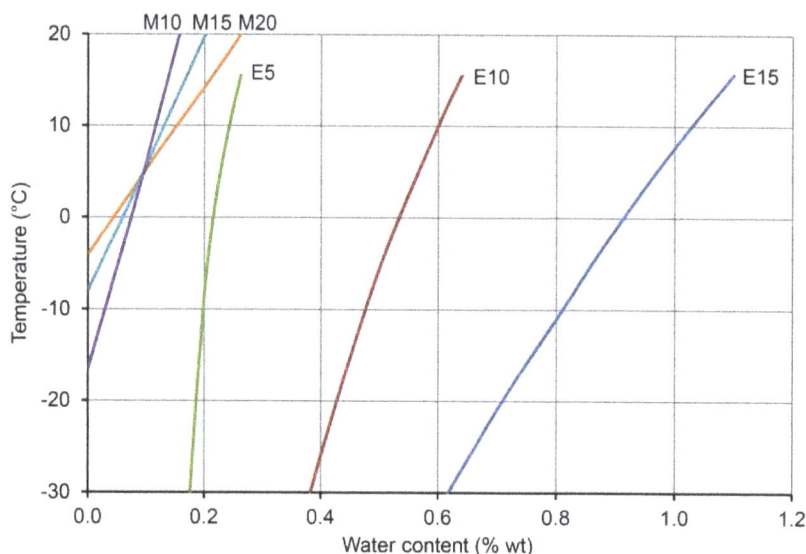

© SAE International.

It can be seen that ethanol/gasoline blends have a much greater water tolerance than methanol/gasoline blends. It is also seen that water tolerance tends to increase with methanol content at temperatures above approximately 2°C, and to decrease with increasing methanol concentration below this temperature [6.19].

Figure 6.2 shows how water tolerance varies with aromatic content and methanol-ethanol ratio in blends containing 10% alcohol [6.19]. Figure 6.3 illustrates how the use of a higher alcohol acting as cosolvent increases the water tolerance [6.19]. Water tolerance curves for TBA and methanol mixtures, which have been widely used in practice, are shown in Figure 6.4 [6.20].

FIGURE 6.2   Water tolerance of 10% volume alcohol blends in gasolines of different aromatic content.

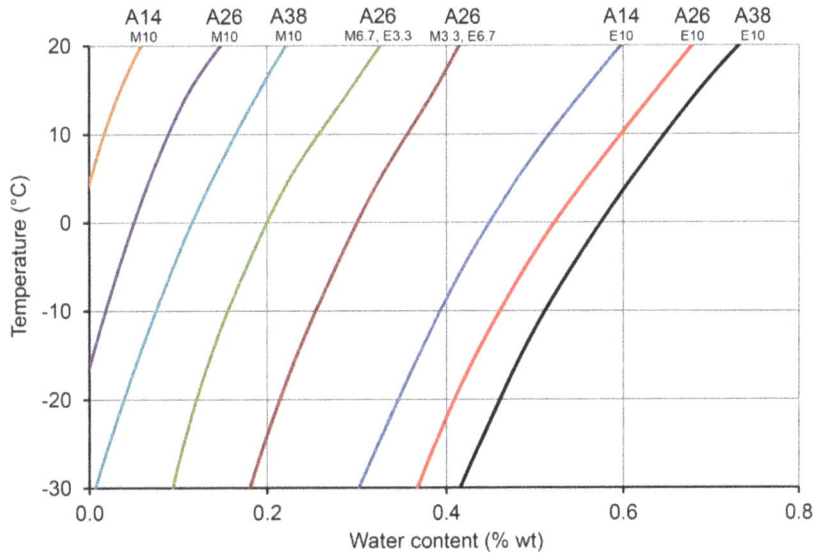

© SAE International.

FIGURE 6.3   Effect of cosolvents on water tolerance of methanol and ethanol blends in gasoline with 26% volume aromatic content.

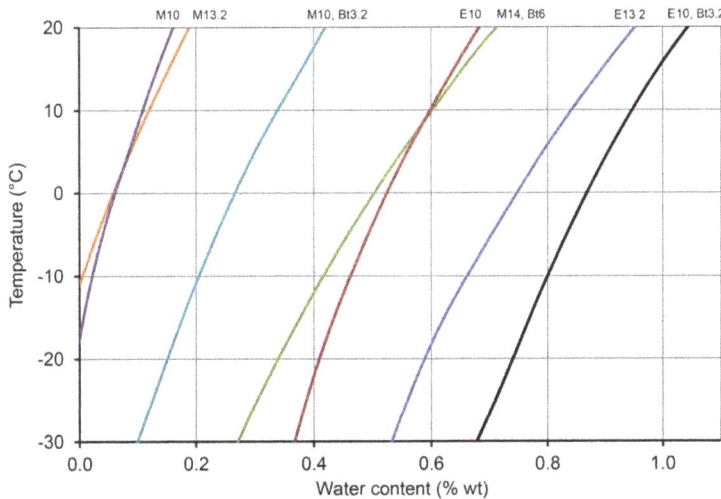

© SAE International.

FIGURE 6.4 Effect of TBA on water tolerance of fuels containing up to 5% volume methanol.

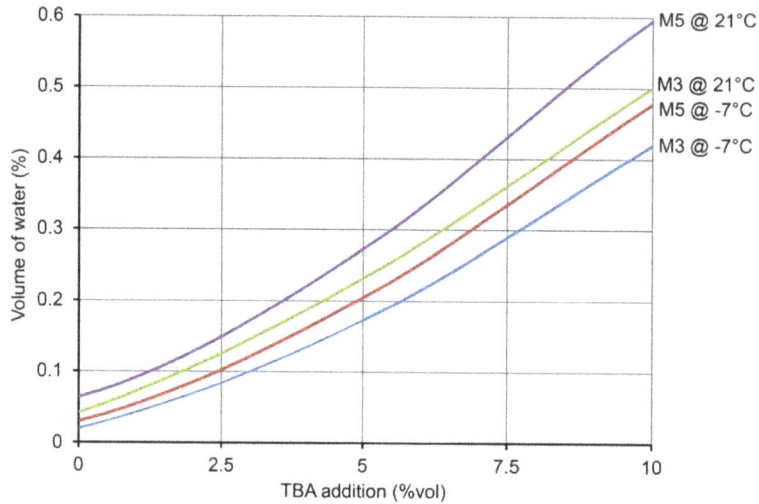

The use of certain surfactant additives to prevent or reduce the tendency for phase separation in alcohol mixtures with gasoline has been investigated [6.21] and found to significantly reduce the temperature at which phase separation occurs.

Water bottoms growth is a feature of alcohol blends in storage tanks. The water in the bottom of a gasoline tank is normally left behind every time gasoline is discharged, so that if alcohol is present in the gasoline, the level of water bottoms will increase as alcohol enters the water phase from successive batches of gasoline. Eventually, the aqueous phase will contain so much alcohol that it becomes soluble in the gasoline so that the water bottoms will gradually disappear. Figure 6.5 shows the percentage volume increase in the water bottoms for successive passes as a 3% methanol blend in gasoline with 26% aromatic content. The tests were carried out with different cosolvents, MTBE and a gasoline grade TBA, and an initial water bottom volume of 0.5% of the gasoline volume [6.20].

FIGURE 6.5 Effect of cosolvents on water bottom growth for gasoline containing 3% methanol.

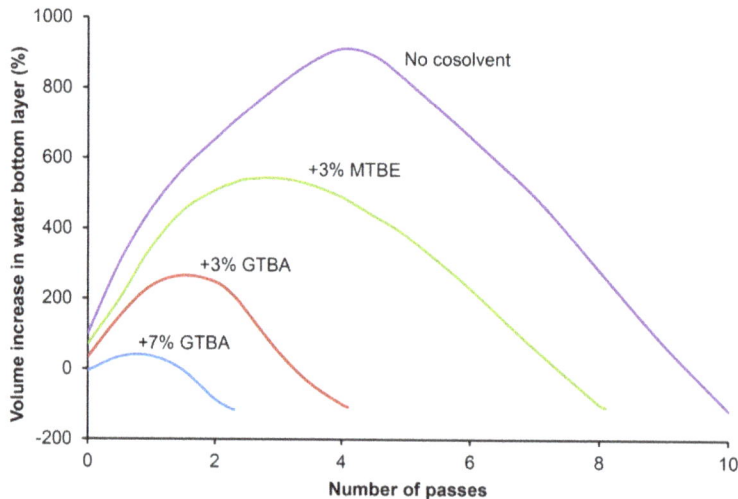

Once the storage tank has been "dried out" in this way, no further losses of methanol will occur, but care must be taken when this drying-out process is taking place to avoid discharging some of the aqueous phase at the same time as the gasoline.

Haziness is another form of phase separation in which separated water droplets are in suspension. If water is forced out of solution for any reason, such as a reduction in ambient temperature or the addition during blending of a component having a low aromatics content, it will usually appear as a haze that will gradually disperse as the droplets agglomerate and sink to the bottom of the tank. If surfactant additives are present, such hazes can be very stable and require the use of demulsifiers to clear them (see Section 11.2.6).

Only a small amount of water usually accumulates in a vehicle's fuel tank, and this is much less of a problem when alcohol is present because the alcohol will eventually dry out the tank completely. Sometimes alcohol is added to a vehicle's fuel tank solely for this purpose. Gross amounts of water in the fuel tank can cause problems if alcohol fuels are used because the increase in water bottoms may cause the engine to take in a blend of methanol and water.

6.7.3. Safety and Fire Protection for Alcohol Blends

Underground leaks of gasoline/alcohol blends are particularly undesirable because the alcohol will separate from the gasoline and dissolve in groundwater. These blends cannot be separated from the water using conventional gasoline techniques, such as gravity separation. There is evidence, however, that the alcohol is biodegraded [6.22].

In general, the handling of blends containing alcohols or ethers from a safety standpoint is no different from that of a conventional gasoline. Care should be taken to minimize skin contact and the breathing of vapors and to avoid ingestion. Any spills should be cleaned up promptly and work areas should be well ventilated [6.23].

All types of gasoline/alcohol or gasoline/ether fires can be extinguished with dry chemical, carbon dioxide fire extinguishers. Suitable firefighting foams are also available [6.24].

6.7.4. Oxidation Stability of Biodiesel

Biodiesel is prone to oxidation, and to what extent depends not only on the quality of the finished biodiesel but also on the feedstock from which it was made. Vegetable oils contain naturally occurring antioxidants, such as tocopherols, sterols, and tocotrienols (vitamin E) [6.25, 6.26]. The biodiesel production process will determine whether any of the naturally occurring antioxidants within some plant oils will still exist in the finished product. The oxidation stability of the biodiesel will also depend upon whether antioxidant additives have been used. If the biodiesel is blended with petroleum diesel, then this will increase the oxidation stability of the blend compared with that of the neat biodiesel, but it can have an antagonistic effect on deposit formation (as will be discussed in Chapter 18).

In a pure biodiesel, the fatty acid alkyl chains have different levels of unsaturation, normally in non-conjugated and *cis* configurations. For the C18 groups, the predominant groups in soybean methyl ester and rapeseed methyl ester, this ranges from no double bonds (fully saturated) to three double bonds (polyunsaturated): one double bond for oleic acid chain; two double bonds for linoleic acid chains, which are far more reactive than the oleic chains; and three double bonds for the linolenic acid chains, which in turn are more reactive than the linoleic chains [6.27, 6.28]. The polyunsaturated methyl esters are more vulnerable to oxidation than monounsaturated esters because they contain more allylic methylene configurations [6.29]. There is no direct correlation with the total number of double bonds [6.30] but with the total number of *bis*-allylic sites (the methylene moiety between two double bonds). The *bis*-allylic site is the most vulnerable position for the formation of the free radicals required to initiate the oxidation process. This is illustrated by the sites 11 and 14 in Figure 6.6 [6.31].

FIGURE 6.6 Basic chemistry for oxygen incorporation and dimerization of polyunsaturated fatty acid chains.

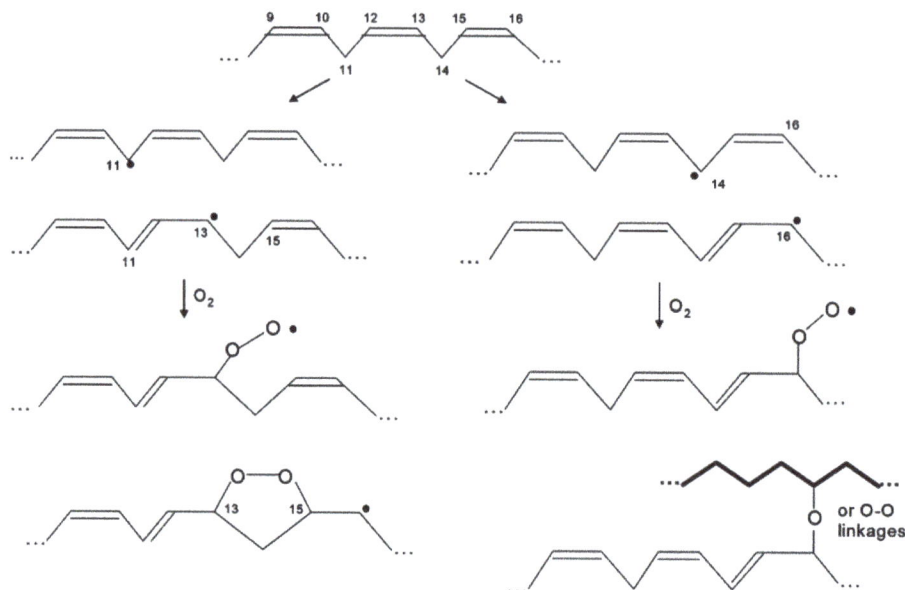

As shown in Figure 6.6, the radicals formed at the *bis*-allylic sites will then readily isomerize to form a more stable conjugated structure, which can react directly with oxygen to form a peroxide. As shown to the left of Figure 6.6, the peroxide species can form a cyclic intermediate that will cleave to form acids, alcohols, aldehydes, and ketones [6.32, 6.33], or as shown to the right of Figure 6.6, the peroxide can react with another fatty acid chain to form a dimer [6.33]. The process of peroxide formation at the *bis*-allylic sites can result in oligomerization even at ambient temperature but is accelerated at higher temperatures [6.34]. The presence of metals will also accelerate the oxidation process [6.35], copper in particular being a strong catalyst for these reactions. Although the quality of neat biodiesel (B100) is regulated, as discussed in Chapter 5, Section 5.4, in use biodiesel blends are subject to water contamination and this may affect the efficacy of additives used to allow the B100 to meet specification. It has been found that more polar antioxidant additives that are more effective for B100 can be less effective in wet biodiesel blends [6.36].

6.7.5. Low-Temperature Operability of Biodiesel

As noted in Section 6.6.3, diesel fuel contains wax, which can begin to crystalize at low temperatures. These wax crystals will then re-dissolve as the temperature rises. Biodiesel and biodiesel blends are also sometimes prone to precipitate formation at temperatures above the cloud point or the cold filter plugging point; these precipitates have become known as "above the cloud point precipitates" [6.37], and these deposits do not readily re-dissolve in the fuel. These above the cloud points have been attributed to the presence of mono-glycerides [6.38, 6.39] and steryl glucosides [6.40, 6.41] within the biodioesel, particularly saturated mono-glycerides [6.42]. Figure 6.7 shows crystals formed in a biodiesel blend, including a cold flow improver additive, of 5% palm methyl ester (PME) that had been stored at 10°C for less than 7 days [6.38]. These were found to be approximately 95% mono-glycerides. Compliance with ASTM D7501 [6.43] should reduce these problems.

FIGURE 6.7   Microscope image of crystals formed at 10°C in a biodiesel blend with 5% PME.

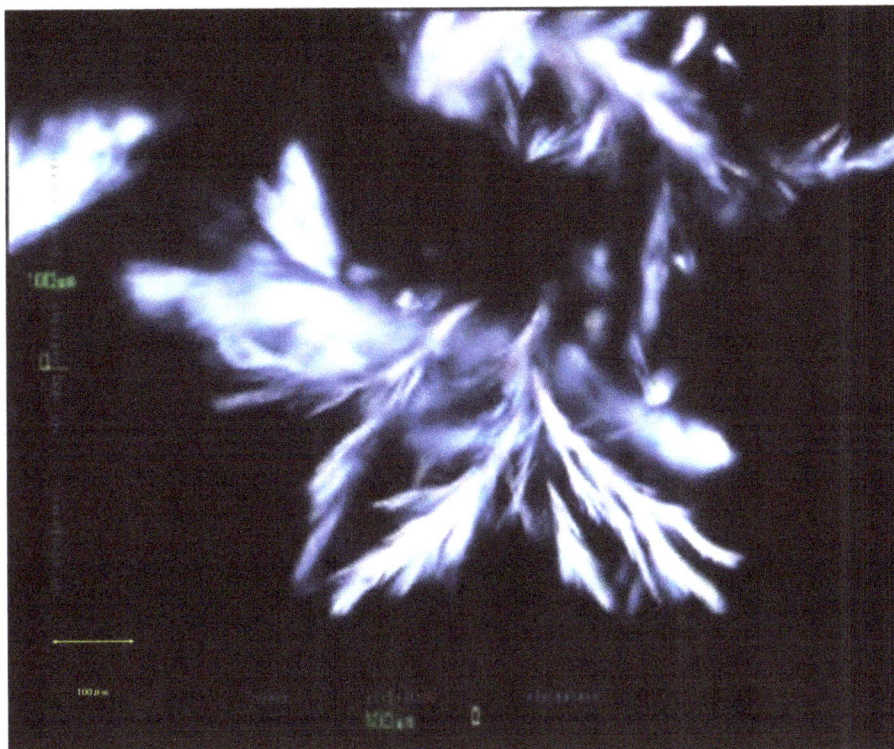

These deposits, when formed in bulk storage, can lead to problems with filters and pumps in the distribution system. When the deposits form in the vehicle tank, they can lead to issues with vehicle operability. These aspects are discussed further in Chapters 17 and 18, however, these problems should not occur if the biodiesel complies with the latest standards, as discussed in Chapter 5, Section 5.4.

References

6.1.   United Nations, "Globally Harmonized System of Classification and Labelling of Chemicals (GHS) Fourth Edition," United Nations ST/SG/AC.10/30/Rev. 4, 2011.

6.2.   CONCAWE, "Gasolines," CONCAWE 0, CONCAWE, Brussels, July 1992.

6.3.   CONCAWE, "Gas Oils (Diesel Fuels/Heating Oils)," CONCAWE Product Dossier No. 95/107, CONCAWE, Brussels, September 1996.

6.4.   Gamble, H.A., Robbins, J.C., Mackay, G.I., and Schiff, H.I., Development of a compact Raman spectrometer for detecting product interfaces in a flow path. U.S. Patent 6,734,963, 2004.

6.5.   Lacey, P., Gail, S., Kientz, J., Milovanovic, N. et al., "Internal Fuel Injector Deposits," *SAE Int. J. Fuels Lubr.* 5, no. 1 (2012): 132-145, doi:https://doi.org/10.4271/2011-01-1925.

6.6.   Mysels, K.J., Flow of thickened fluids. U.S. Patent 2,492,173, 1949.

6.7.   White, J.L. and Gibson, D.L., Method of reducing friction loss in flowing hydrocarbon liquids. U.S. Patent 3,215,154, 1965.

6.8.　Cutler, J.D. and McClaffin, G.G., Method of friction loss reduction in oleaginous fluids flowing through conduits. U.S. Patent 3,692,676, 1972.

6.9.　Burger, E.D., "Flow Increase in the Trans Alaska Pipeline through Use of a Polymeric Drag-Reducing Additive," *Journal of Petroleum Technology* 34, no. 2 (1982): 377-386.

6.10.　ASTM International, "Standard Test Method for Free Water and Particulate Contamination in Distillate Fuels (Visual Inspection Procedures)," ASTM D4176-21a, ASTM International, 2021.

6.11.　ASTM International, "Standard Test Method for Water and Sediment in Fuel Oils by the Centrifuge Method (Laboratory Procedure)," ASTM D1796-11 (2016), ASTM International, 2018.

6.12.　ASTM International, "Standard Specification for Diesel Fuel Oils," ASTM D975-11b, ASTM International, 2013.

6.13.　ASTM International, "Standard Test Method for Determination of Water in Petroleum Products, Lubricating Oils, and Additives by Coulometric Karl Fischer Titration," ASTM D6304-20, ASTM International, 2021.

6.14.　BSI, "Automotive Fuels. Diesel. Requirements and Test Methods," BS EN 590:2009+A1:2010, 2009.

6.15.　Coley, T. "Diesel Fuel Additives Influencing Flow and Storage Properties," in *Gasoline and Diesel Fuel Additives*, Ed. Owen, K. (Chichester, UK: John Wiley & Sons, 1989).

6.16.　Tang, H., Salley, S.O., and Ng, K.Y.S., "Fuel Properties and Precipitate Formation at Low Temperature in Soy-, Cottonseed-, and Poultry Fat-Based Biodiesel Blends," *Fuel* 87, no. 13–14 (2008): 3006-3017.

6.17.　Polss, P., "What Additives Do for Gasoline," *Hydrocarbon Processing* 52, no. 2 (1973): 61-68.

6.18.　Giles, H.N., Bowden, J.N., and Stavinoha, L.L., "Overview on Assessment of Crude Oil and Refined Product Quality during Long Term Storage," U.S. Department of Energy and U.S. Army Fuels and Lubricants Laboratory, Washington, DC, June 1985.

6.19.　Keller, J.L., "Methanol and Ethanol Fuels for Modem Cars," 44th Refinery Mid-year Mtg./Session on Fossil Fuels in the 1980's, Reprint No. 08-79, May 15, 1979.

6.20.　ARCO Chemical Europe Inc, "Technical Bulletin, GTBA, Gasoline Grade Tertiary Butyl Alcohol," Atlantic Richfield Company, La Palma, CA, 1986.

6.21.　Smith, E. and Jordan, D., "The Use of Surfactants in Preventing Phase Separation of Alcohol Petroleum Fuel Mixtures," SAE Technical Paper 830385 (1983), doi:https://doi.org/10.4271/830385.

6.22.　Novak, J.T., Goldsmith, C.D., Benoit R.E., and O'Brien, J.H., "Biodegradation of Methanol and Tertiary Butyl Alcohol in Subsurface Systems," in *International Seminar on Degradation, Retention and Dispersion of Pollutants in Groundwater*, Copenhagen, Denmark, 1984.

6.23.　American Petroleum Institute, "Storage and Handling of Gasoline-Methanol/Co-Solvent Blends at Distribution Terminals and Service Stations," API Recommended Practice 1627, August 1986.

6.24.　National Fire Protection Association, "Standard for Low-, Medium-, and High-Expansion Foam," NFPA 11, 2010.

6.25.　Tarandjiiska, R.B., Marekov, I.N., Nikolova-Damyanova, B.M., and Amidzhin, B.S., "Determination of Triacylglycerol Classes and Molecular Species in Seed Oils with High Content of Linoleic and Linolenic Fatty Acids," *Journal of the Science of Food and Agriculture* 72, no. 4 (1996): 403-410.

6.26.　Schwartz, H., Ollilainen, V., Piironen, V., and Lampi, A.-M., "Tocopherol, Tocotrienol and Plant Sterol Contents of Vegetable Oils and Industrial Fats," *Journal of Food Composition and Analysis* 21, no. 2 (2008): 152-161.

6.27.　Miyashita, K. and Takagi, T., "Study on the Oxidative Rate and Prooxidant Activity of Free Fatty Acids," *Journal of the American Oil Chemists' Society* 63, no. 10 (1986): 1380-1384.

6.28.　Cosgrove, J.P., Church, D.F., and Pryor, W.A., "The Kinetics of the Autoxidation of Polyunsaturated Fatty Acids," *Lipids* 22, no. 5 (1987): 299-304.

6.29.　Tang, H., De Guzman, R.C., Salley, S.O., and Ng, K.Y.S., "The Oxidative Stability of Biodiesel: Effects of FAME Composition and Antioxidant," *Lipid Technology* 20, no. 11 (2008): 249-252.

6.30.　Knothe, G., "Structure Indices in FA Chemistry," *Journal of the American Oil Chemists" Society* 79, no. 9 (2002): 847-854.

6.31. Fang, H. and McCormick, R., "Spectroscopic Study of Biodiesel Degradation Pathways," SAE Technical Paper 2006-01-3300 (2006), doi:https://doi.org/10.4271/2006-01-3300.

6.32. McCormick, R.L. and Westbrook, S.R., "Empirical Study of the Stability of Biodiesel and Biodiesel Blends," NREL/TP-540-41619, 2007.

6.33. Ogawa, T., Kajiya, S., Kosaka, S., Tajima, I. et al., "Analysis of Oxidative Deterioration of Biodiesel Fuel," *SAE Int. J. Fuels Lubr.* 1, no. 1 (2009): 1571-1583, doi:https://doi.org/10.4271/2008-01-2502.

6.34. McCormick, R.L. and Westbrook, S.R., "Storage Stability of Biodiesel and Biodiesel Blends," *Energy Fuels* 24, no. 1 (2010): 690-698.

6.35. Shiotani, H. and Goto, S., "Studies of Fuel Properties and Oxidation Stability of Biodiesel Fuel," SAE Technical Paper 2007-01-0073 (2007), doi:https://doi.org/10.4271/2007-01-0073.

6.36. Christensen, E.D. and McCormick, R.L., "Water Contamination Impacts on Biodiesel Antioxidants and Storage Stability," *Energy & Fuels* 37, no. 7 (2023): 5179-5188.

6.37. Jolly, L., Strojek, W., Bunting, W., Sakata, I. et al., "A Study of Mixed-FAME and Trace Component Effects on the Filter Blocking Propensity of FAME and FAME Blends," SAE Technical Paper 2010-01-2116 (2010), doi:https://doi.org/10.4271/2010-01-2116.

6.38. Ohshio, N., Saito, K., Kobayashi, S., and Tanaka, S., "Storage Stability of FAME Blended Diesel Fuels," SAE Technical Paper 2008-01-2505 (2008), doi:https://doi.org/10.4271/2008-01-2505.

6.39. Tang, H., De Guzman, R.C., Salley, S.O., and Ng, K.Y.S., "Formation of Insolubles in Palm Oil-, Yellow Grease-, and Soybean Oil-Based Biodiesel Blends after Cold Soaking at 4°C," *Journal of the American Oil Chemists' Society* 85, no. 12 (2008): 1173-1182.

6.40. Moreau, R.A., Scott, K.M., and Haas, M.J., "The Identification and Quantification of Steryl Glucosides in Precipitates from Commercial Biodiesel," *Journal of the American Oil Chemists' Society* 85, no. 8 (2008): 761-770.

6.41. Bondioli, P., Cortesi, N., and Mariani, C., "Identification and Quantification of Steryl Glucosides in Biodiesel," *European Journal of Lipid Science and Technology* 110, no. 2 (2008): 120-126.

6.42. Jolly, L., Kitano, K., Sakata, I., Strojek, W. et al., "A Study of Mixed-FAME and Trace Component Effects on the Filter Blocking Propensity of FAME and FAME Blends," SAE Technical Paper 2010-01-2116 (2010), doi:https://doi.org/10.4271/2010-01-2116.

6.43. ASTM International, "Standard Test Method for Determination of Fuel Filter Blocking Potential of Biodiesel Fuel Blendstock (B100) by Cold Soak Filtration Test (CSFT)," ASTM D7501-22, ASTM International, 2022.

Further Reading

Chesneau, H.L. and Dorris, M.M., "Distillate Fuel—Contamination, Storage and Handling," American Society for Testing & Materials, 1988.

CONCAWE, "Guidelines for Handling and Blending FAME," Report No. 9/09, CONCAWE, 1999.

Groysman, A., *Corrosion in Systems for Storage and Transportation of Petroleum Products and Biofuels: Identification, Monitoring and Solutions* (Berlin, Heidelberg: Springer Science & Business Media, 2014).

World Health Organization, "Occupational Exposures in Petroleum Refining; Crude Oil and Major Petroleum Fuels," in *IARC Monographs on the Evaluation of Carcinogenic Risks to Humans*, Vol. 45 (Lyon, France: World Health Organization, International Agency for Research on Cancer, 1998).

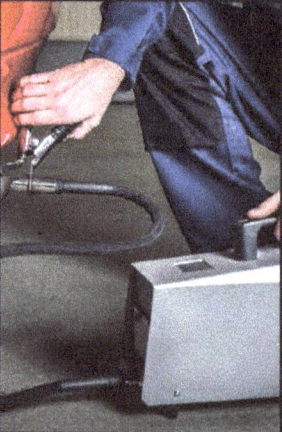

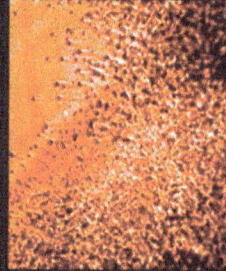

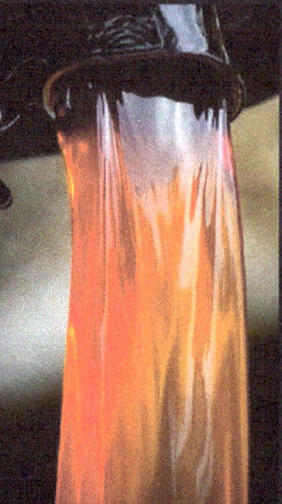

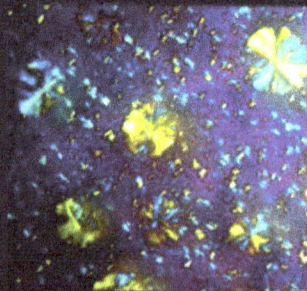

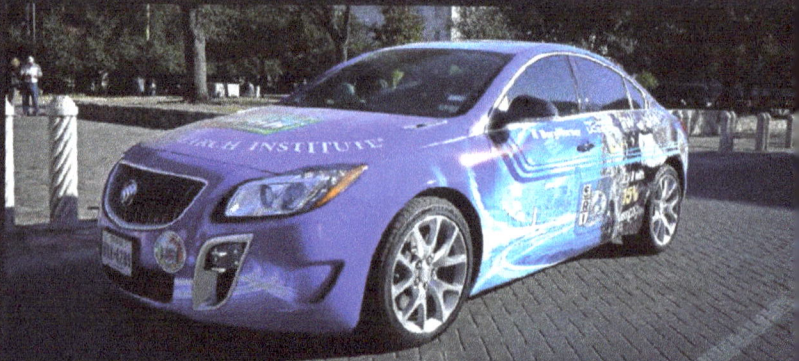

7

Positive Ignition Engine Combustion Process

The automotive transport sector still relies, to a very large extent, on the internal combustion engine to provide motive power. The vast majority of these engines are of the reciprocating type, with a few rotary engines, such as those of the Wankel type [7.1]. However, whether they are of the reciprocating or rotary design, they all operate by intermittent combustion as opposed to steady-state burning of the fuel, as would be the case in external combustion engines such as steam engines or gas turbines. Intermittent combustion occurs under complex and continually varying conditions. The combustion process in such engines can be divided into two categories: positive ignition, where the combustible mixture of fuel and air is ignited by a positively controlled source such as a spark, and compression ignition, where the temperature and pressure generated by compression provide the conditions for spontaneous ignition. In this chapter, we will consider the positive ignition engine combustion process. The fuel for such engines should be such that it does not readily auto-ignite due to temperature and pressure but will burn well once positively ignited. This requirement is traditionally met by such fuels as gasoline and gaseous fuels such as liquefied petroleum gas (LPG), liquefied natural gas (LNG), compressed natural gas (CNG), and hydrogen. The combustion efficiency of such engines is not only determined by the engine design but is also very sensitive to fuel characteristics and its quality. The fuel quality requirement of a particular engine design is strongly dependent on operating conditions.

7.1. Normal Combustion

7.1.1. Mixture Requirements

The amount of air required to combust a known hydrocarbon fuel completely to carbon dioxide and water can readily be calculated (the stoichiometric ratio). For example, heptane, a hydrocarbon boiling in about the middle of the gasoline range, under perfect combustion conditions undergoes oxidation as follows:

$$C_7H_{16} + 11O_2 \rightarrow 7CO_2 + 8H_2O$$

In this example, one volume of heptane requires 11 volumes of oxygen or 52.6 volumes of air. On a weight basis this is equivalent to an air-to-fuel ratio of 15:1. One part of gasoline, because it differs from heptane in its carbon-hydrogen ratio, requires about 14.5 parts by weight of air for complete combustion; the exact stoichiometric amount will clearly depend on the composition of the fuel. For a pure hydrocarbon gasoline and a homogeneous mixture, generally speaking, if the air-to-fuel ratio is less than about 7:1 it will be too rich to ignite, and if it is more than about 20:1 in a conventional engine it will be too weak. Lean burn engines have been manufactured that will operate satisfactorily on air-to-fuel ratios leaner than 20:1 [7.2].

To control the amount of energy released from the fuel and hence the power generated by the engine, the amount of fuel must be regulated. Because of this range of air-to-fuel ratios that can be ignited, the amount of air must also be controlled. Thus, to reduce the amount of power produced, the amount of air used must be reduced, and was traditionally done by throttling the intake. This leads to significant losses, pumping losses, in trying to draw the air through the throttled intake. A great deal of research was carried out early in the final quarter of the 20th century to develop engines that promoted mixture stratification so that control of the power output was more dependent on fuel flow than air flow [7.3–7.7]. These engines had to ensure that the mixture around the source of ignition was still within the limits of ignitability. The requirement to meet ever-tightening emissions limits led to the development of the three-way catalyst (TWC), which required that the overall mixture strength was very close to stoichiometric [7.8]; this impeded development of such stratified charge engines. However, with the necessity to improve fuel efficiency, much work is being done to improve catalytic systems that will allow satisfactory emissions control with an overall lean mixture [7.8–7.11]. These aspects of engine design will be discussed in more detail in Chapter 8. The combustion process remains basically the same.

Toward the end of the 20th century and into the 21st century, work has been performed on controlling the amount of air, or air/fuel mixture, entering the combustion chamber by controlling the opening times and the lift of the inlet valves, known as variable valve actuation (VVA) [7.12–7.15] and this is discussed further in Chapter 23, Section 23.4.1.

7.1.2. The Combustion Process

When a fuel is mixed with oxygen, so-called "preflame" reactions will commence as soon as the fuel mixes with the air, which may be even before the mixture has reached the combustion chamber, and will continue after ignition until all the fuel has been consumed by the advancing flame front [7.16, 7.17]. The extent of these reactions will depend on a number of factors, such as the chemical composition of the fuel and "air" mixture and the temperature and pressure of this mixture. The nature of the reactions will determine whether the fuel will burn "normally," that is, whether it will combust smoothly and efficiently or whether it will give rise to some form of "abnormal" combustion such as preignition, knock or misfire.

Normal combustion occurs when a flame front that, after being initiated by the positive ignition source—usually a spark—moves smoothly but in a somewhat irregular fashion, across the combustion chamber until combustion is complete. The irregularities along the flame front are due to small-scale turbulence and incomplete mixing [7.18]. The general shape of the flame front is influenced by larger-scale turbulence and possible stratification of the charge. The overall pressure change in the chamber during this process is shown in Figure 7.1. It will be seen that the pressure begins to rise as the mixture is compressed and then rises rapidly after ignition due to the temperature increase and the formation of the combustion gases, mainly carbon dioxide and water vapor. Maximum pressure occurs soon after the piston has reached the top of its travel, top dead center (TDC).

FIGURE 7.1   Cylinder pressure trace during normal combustion.

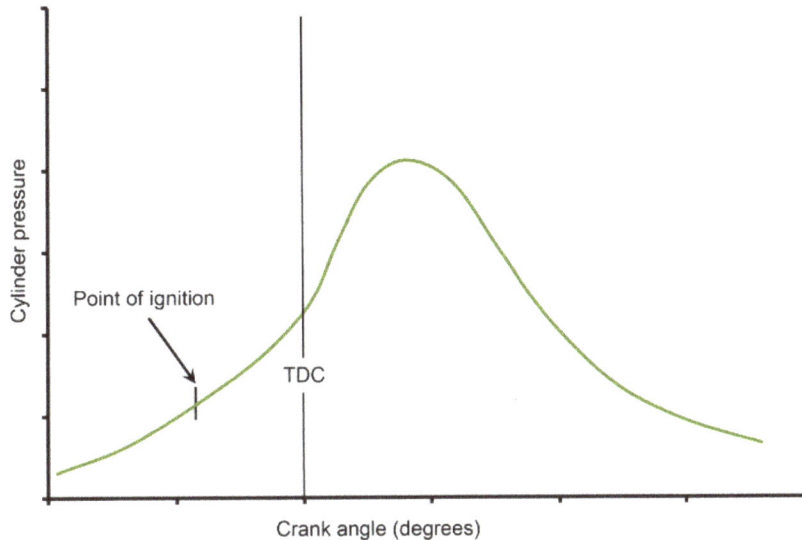

The position, within the cycle, of the peak pressure is determined by the timing of the ignition event and the rate of the chemical reactions, in relation to the speed of the engine. The greater the rate of chemical reaction, the sooner all of the energy will be released after the ignition. The faster the engine is running, the shorter the cycle time; a fixed time for the combustion processes to occur will therefore represent a larger part of the engine cycle. Ignoring heat losses, the maximum efficiency will be achieved if all of the energy is released very soon after TDC. However, as the combustion process requires a finite period of time, ignition usually occurs slightly before TDC. This will be discussed further in the following chapter.

Even with normal combustion, all spark ignition engines show variations in the maximum cylinder pressure and rate of pressure rise from cycle to cycle (this is known as cyclic dispersion) in spite of strict control of the running conditions [7.19]. This is believed to be due to variations in turbulence between cycles, which cause variations in flame speeds across the combustion chamber. If reductions in cyclic dispersion are achieved, significant benefits in terms of improved fuel consumption can be obtained and the knock resistance qualities of the fuel are also lowered [7.20]. Cylinder to cylinder variations also occur, due to variations in manufacturing tolerances but primarily due to variations in the proportions of air and fuel trapped within each cylinder. Again, there is significant potential for improvements in overall efficiency by reducing these variations. This will be highlighted in Chapter 8 with the discussion of the changes in fuel system technology.

7.2. Spark Knock

Spark knock, so called because it can be influenced by the timing of the spark, is one of the most important forms of abnormal combustion; it determines, to some extent, the thermal efficiency that can be achieved in an engine. The higher the engine's compression ratio is, the better the thermal efficiency is, but the greater the tendency for spark knock to occur and hence the higher the fuel octane quality that is required. Igniting the air-to-fuel mixture earlier in the compression stroke means that the preflame reactions begin earlier, and some of the fuel energy is released earlier in the cycle and at a lower cylinder temperature and pressure. The temperature and pressure of the end gas are thus reduced and the tendency for this gas to auto-ignite is thus also reduced. However, it should be clear that releasing some of the fuel energy during the compression stroke is counterproductive and will reduce the engines efficiency. Similarly, releasing the energy too late in the cycle will also reduce the efficiency. There is thus an optimum timing of the ignition to get the maximum efficiency out of the engine, usually referred to as the minimum for best torque (MBT) ignition advance.

Even vehicles operating on a fuel for which they have been designed will sometimes knock, which may be due to a number of factors, such as excessive deposit formation in the combustion chamber, over-advancement of the ignition timing, particularly severe driving conditions, or a combination of several factors during manufacture in which the production tolerances all conspire to increase the knocking tendency. It can also be due to the fuel being of a poorer quality than specified.

7.2.1. How Spark Knock Occurs

It is believed that knock occurs as a result of uncontrolled spontaneous auto-ignition of the fuel–air mixture ahead of the flame front, the end gas [7.21–7.23]. The sequence of events when knock occurs is shown diagrammatically in Figure 7.2.

FIGURE 7.2 Sequence of events for normal (left) and knocking (right) combustion cycles.

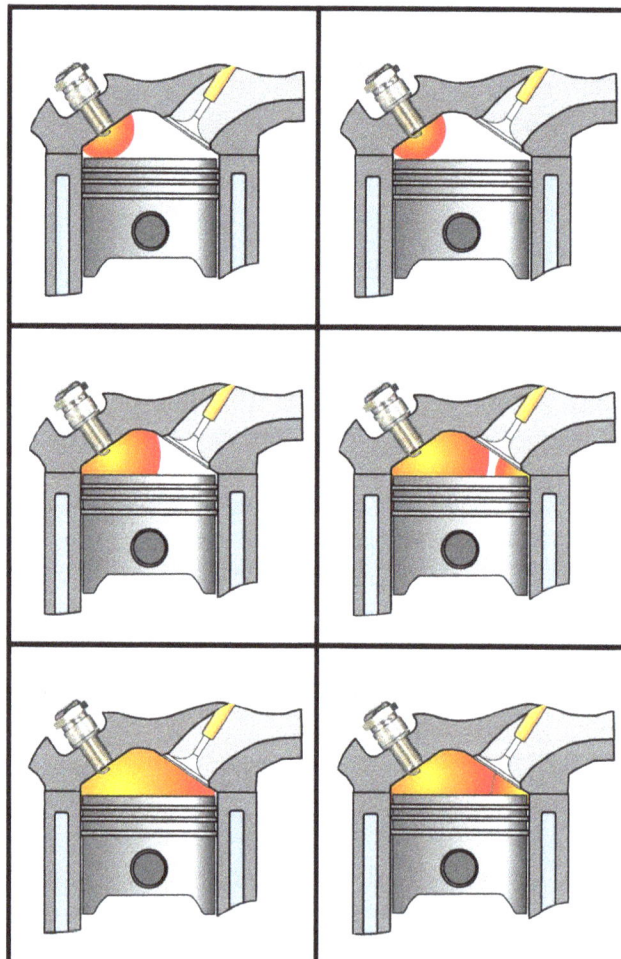

© SAE International.

The left side of the figure shows normal combustion, and the right side of the figure shows the occurrence of knock. As the flame propagates from the spark plug, the top two images in Figure 7.2, the temperature and pressure of the unburned gas ahead of the flame front are raised due to heat from the flame front itself and the increase in pressure from the expanding burning gases. Because of this, preflame reactions will take place at an increasing rate and eventually may reach the point when the mixture will auto-ignite, the center right image in Figure 7.2. In normal combustion, this stage is never reached because there is insufficient time for the preflame reactions to progress to the point of auto-ignition. In the knocking situation the flame front from the auto-ignition source propagates until it reaches the flame front from the spark plug (lower right image in Figure 7.2). This sets up the shock waves that are heard as knock.

The shock wave then passes through the combustion chamber in a complex manner determined by the conditions in the chamber. This shock wave generates vibrations with a frequency content dependent upon the cylinder dimensions and the temperature conditions within the chamber, which in turn affect the speed of sound [7.24]. For most automotive engines this gives rise to the classic knock frequency of between 5 and 18 kHz [7.25]. This is the characteristic pinking sound associated with knock. Figure 7.3 shows a typical cylinder pressure trace during a knocking cycle, superimposed on the normal pressure trace as shown in Figure 7.1.

FIGURE 7.3 Cylinder pressure trace during knock.

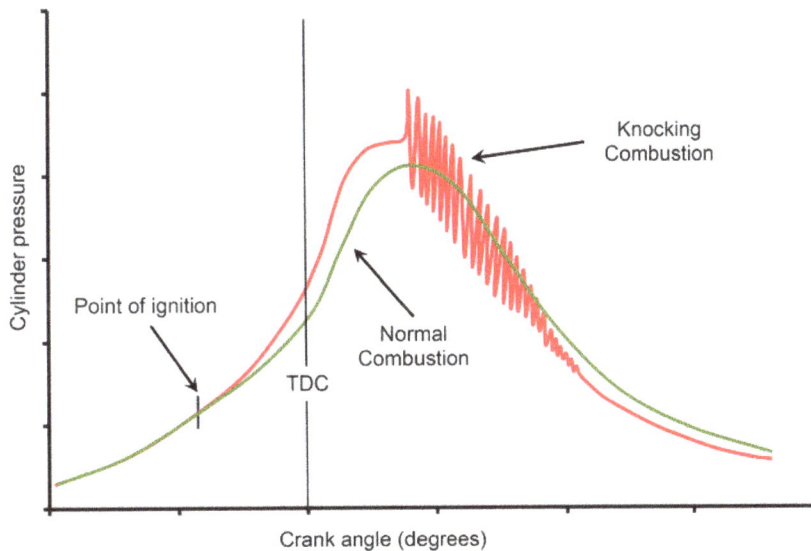

© SAE International.

As engine speed is increased, the time for preflame reactions to take place is reduced and so the tendency to knock decreases, although normally the ignition timing is advanced as engine speed increases. Conversely, increasing engine speed tends to increase the combustion chamber temperature, which increases the likelihood of knock. Knock during acceleration at wide-open throttle from a low engine speed is of such short duration that it does not usually cause damage and, unless it is very severe, will not cause discernible loss of power [7.26]. High constant-speed knock, on the other hand, can cause loss of power and severe damage, usually to the cylinder head gasket, the spark plug electrodes, and the pistons [7.27, 7.28]. In extreme cases, knock can lead to preignition (see Section 7.8.1) and/or so-called "runaway knock," where the knock intensity gets progressively higher until catastrophic engine damage occurs. This is because the temperature in the chamber is raised during heavy knock, and this in turn increases the tendency for more knock to occur, and so on.

Knock continues to be one of the most important aspects of engine design and gasoline technology, and many aspects of it are still not completely understood and are under investigation [7.29–7.33].

7.3. Measurement of Gasoline Antiknock Quality

7.3.1. Research and Motor Octane Number

Prior to 1929, fuels were rated using a Ricardo engine, in which the compression ratio could be varied between 2.7:1 and 7.9:1, as shown in Figure 7.4 [7.34].

FIGURE 7.4   Ricardo variable-compression research engine.

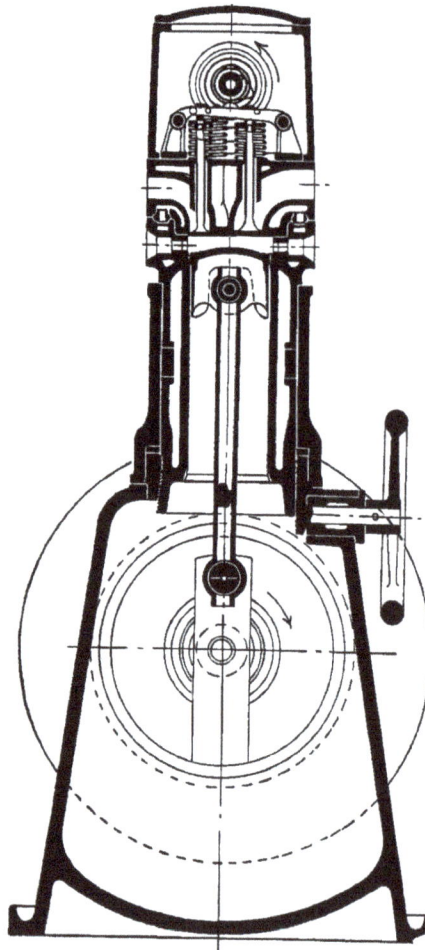

Each fuel was run in this engine at various air-to-fuel ratios and ignition timings to obtain conditions for maximum power, and the highest compression ratio was then established beyond which knock and power loss occurred. Fuels were assigned values in terms of highest useful compression ratio (HUCR), a scale that was later standardized against toluene. The scale was nonlinear—an example is shown in Figure 7.5 [7.34].

FIGURE 7.5 Antiknock value of blended fuels.

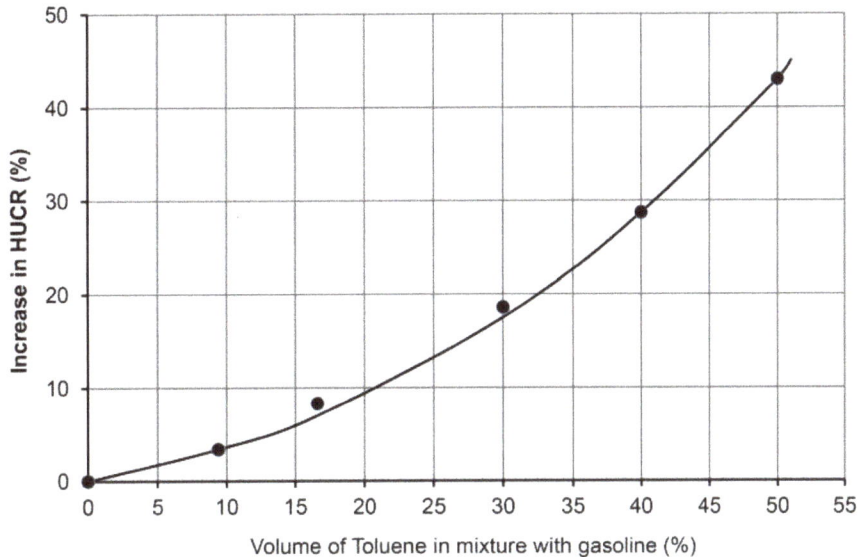

The octane scale proposed by Graham Edgar in 1926 [7.35] was established in 1929 and has been used ever since. In this scale, two pure hydrocarbons were chosen, one with a high resistance to knock and the other with a low resistance to knock. Iso-octane (2,2,4-trimethylpentane) was chosen as the compound with a high resistance to knock and was arbitrarily assigned a value of 100; n-heptane was chosen as the other compound with a low resistance to knock and was assigned a value of 0. These two compounds had the advantage of being similar in characteristics other than their resistance to knock. This is shown in Table 7.1.

TABLE 7.1 Properties of iso-octane and n-heptane.

Property	Iso-octane	n-Heptane
Density (kg/m³ at 25°C)	0.69	0.68
Boiling point (°C)	99–100	98–99
Vapor pressure (kPa at 25°C)	5.5	5.3
Flammability limits (% in air)	1–6	1–7

The octane number of a fuel is the volume percentage of iso-octane in a blend with n-heptane that shows the same antiknock performance as the test fuel when tested in a standard engine under standard conditions. For fuels having octane values above 100, mixtures of iso-octane and tetraethyl lead are used, and the relationship between octane number and the lead content of iso-octane is published in the standard test method [7.36, 7.37].

The test engine used for determining octane values of fuels was developed in the US by the Waukesha Company under the direction of the Cooperative Fuel Research Committee (CFR) and is a single-cylinder, variable-compression ratio engine known as a CFR engine. It was found in a series of tests in the early 1930s that it was not possible to correlate the performance of cars on the road with just one type of octane number because engine designs and driving conditions were continually changing, and this modified the response to the fuels [7.38]. Nowadays, it is usual to define the resistance to knock or "octane quality" using at least two octane parameters, and sometimes three. These are research octane number (RON) [7.36], motor octane number (MON) [7.37], and a number concerned with the distribution of research octane quality through the boiling range of the gasoline, which is discussed later.

The test procedures for RON and MON are carried out under the conditions summarized in Table 7.2.

TABLE 7.2 Test conditions for the research and motor test procedures.

Test	Research	Motor
CRC designation	F-1	F-2
ASTM method	D2699	D2700
Engine speed (rpm)	600	900
Intake air temperature (°C)	Depends on barometric pressure	38
Mixture temperature (°C)	Not specified	149
Coolant temperature (°C)	100	100
Ignition advance (deg. BTDC)	13	Linked to compression ratio

© SAE International.

Both procedures are very similar and involve adjusting the compression ratio to obtain a standard knock intensity as indicated by a knock meter, using the fuel under test. The air-to-fuel ratio is then adjusted to give maximum knock. This is done by raising or lowering the carburetor bowl, which adjusts the fuel flow rate; the air flow rate is ostensibly constant as the engine speed is constant and the air inlet is not throttled. The compression ratio is then readjusted to give a midscale reading on the knock meter. A blend of iso-octane and n-heptane, known as primary reference fuels (PRF), of approximately the same octane number as the test fuel is placed in a second carburetor bowl, which is then switched through to the engine and the height again adjusted to give maximum knock. The reading on the knock meter is noted, provided it is within a certain range, and a second PRF differing by two octane numbers from the first is tested. This is continued until the reading for the test fuel is bracketed by the readings for two PRFs, which differ by no more than two octane numbers. The octane quality of the test fuel is then obtained by interpolation and is usually reported to the nearest 0.1.

7.3.2. Road Octane Number

The RON test was thought to correlate best with low-speed, relatively mild driving conditions, whereas the MON relates to high-speed, high-severity conditions, although later research has questioned this view. Clearly, vehicles operating on the road will spend most of the time operating at a severity other than these two levels. Empirical correlations that permit calculation of automotive antiknock performance or road octane number are based on the general equation:

$$\text{Road ON} = k_1 \, \text{RON} + k_2 \, \text{MON} + k_3$$

where k_1, k_2, and k_3 are empirically determined constants for a given vehicle parc. In the US, both k_1 and k_2 are set to 0.5, and k_3 is set to 0 to arrive at the antiknock index:

$$\text{Antiknock index} = \frac{1}{2}\left(\text{RON} + \text{MON}\right)$$

In the US, fuel is specified and sold according to the antiknock index [7.39], whereas in many other jurisdictions, RON and MON are usually specified separately, and gasoline is often designated for identification or marketing purposes only by the RON value.

Most gasolines have a higher RON than MON, and the difference between these two ratings is called the "sensitivity." For fuels of the same RON, a high-sensitivity gasoline will have a lower MON than a low-sensitivity gasoline. It represents the sensitivity of the fuel to changes in the severity of engine operating conditions in terms of antiknock performance. This is illustrated in Figure 7.6 [7.40] in which octane number is plotted against engine severity. The location of the research and motor test procedures in terms of their relative severities is shown, and many engines will operate at a severity somewhere between these two positions for most of the time. However, there are conditions when some engine designs may be more severe than the motor method (as with some two-stroke engines) and others that may be less severe than the research method (as with many modern engines technologies, such as direct fuel injection and pressure charging). Three fuels are shown in Figure 7.6: Fuel A having a high RON (97) and a low sensitivity (S = 3), Fuel B having a somewhat lower RON (96) and a high sensitivity (S = 12), and Fuel C having a much lower RON (84) but a very low sensitivity (S = 1). It can be seen that if the vehicle happens to be operating at a severity corresponding to X, then Fuel B performs better than Fuel A, even though it has a lower RON and MON. Similarly, if the vehicle is operating at a severity corresponding to Y, Fuel C is better than Fuel B even though, again, it has a lower RON and MON than Fuel B. Of course, at the severity levels that occur with most vehicles for most of the time, Fuel A is better than Fuel B, which is better than Fuel C.

FIGURE 7.6   Octane quality and engine severity.

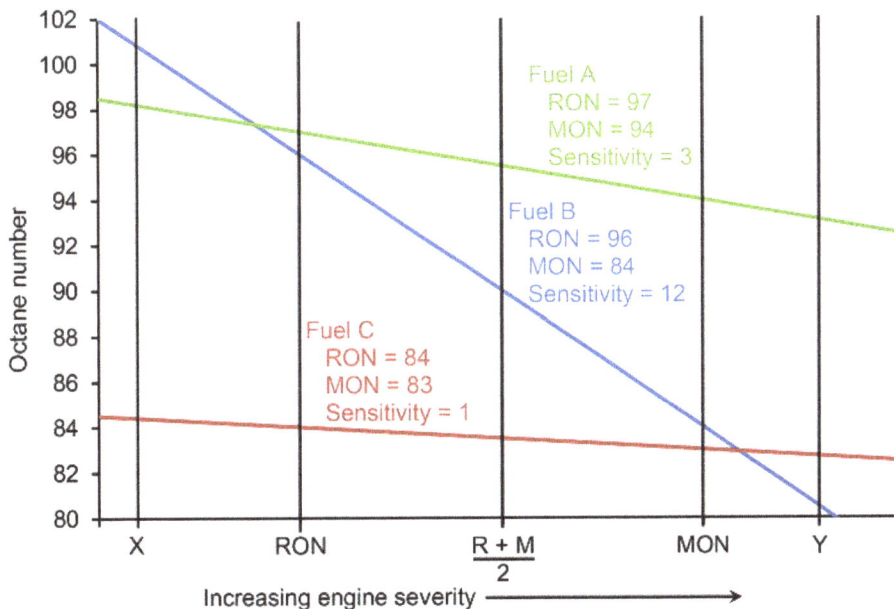

A third octane parameter that gives a measure of the distribution of RON quality through the boiling range of the gasoline has been widely used in Europe, although its importance is declining as engine designs improve and become more sophisticated. It is possible to make two fuels that have the same RON, but the way that the octane quality is distributed throughout the boiling range can be quite different, as shown in Figure 7.7.

FIGURE 7.7 Distribution of octane quality through the boiling range of two gasolines.

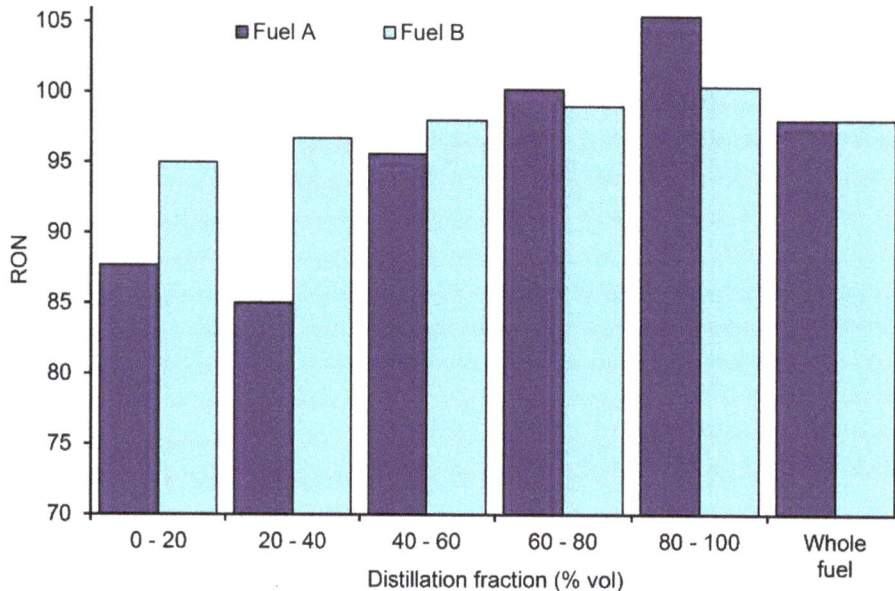

The RON levels of the lighter parts of Fuel A are much lower than the corresponding portions of Fuel B, and the RON levels of the heavier fractions are higher than those of Fuel B. In Fuel B, all the components have similar octane qualities, whereas Fuel A is composed of light components of relatively poor octane quality and heavier components of higher octane quality. In practice, most gasolines have a lower octane quality in the front, or low boiling end, than in the back, or high boiling end.

The importance of octane distribution in the fuel is that during full throttle accelerations in some vehicles, large amounts of liquid fuel are forced into the inlet manifold and initially only the lighter parts are swept on into the combustion chamber by the air stream. If these have a low octane quality, transient knock may occur until the heavier ends catch up as the acceleration continues. Although this type of transient knock does no damage to the engine because it does not last long enough for combustion chamber temperatures to rise very much, the noise can be disturbing to the driver. The increasing use of fuel injection and more sophisticated fuel intake systems have reduced this problem of fuel segregation in the inlet manifold. This is particularly so with the increased prevalence of direct injection.

The distribution of octane quality within the fuel can be measured by a number of different tests, but probably the most well-known involves simply distilling off the portion of fuel boiling up to 100°C and measuring its RON. This value is usually designated as the R100°C of the fuel. The difference between the RON of the whole fuel and the R100°C is abbreviated to ΔR100°C. Other tests that have been used to define front-end octane quality include measuring the RON of the first 75% to distill off (R75%) and the use of a modified CFR engine with a manifold, which segregates some of the heavier ends before they reach the cylinder [7.41].

Other methods of measuring octane quality by correlating laboratory test data with the CFR engine results have been devised, mainly because the engine test is slow and expensive. They are often based on composition [7.42, 7.43] or on infrared spectra of the fuel [7.44–7.47], and there are numerous variations of these and other methods for predicting octane quality without going through the expensive and time-consuming engine test procedure. Many of the procedures can be operated continuously and have been developed purely for process control.

7.3.3. Octane Index and Modern Engines

It has been suggested that the accepted criteria, as described above, of specifying the octane quality in terms of the antiknock index is not relevant to the modern automotive engine [7.48]. It is suggested that the octane quality or antiknock quality of the fuel is better defined as an octane index given by

$$OI = RON - K\,S$$

where
   K is a constant
   S is the sensitivity, RON – MON
   This is in effect the road octane number given above:

$$Road\,ON = k_1\,RON + k_2\,MON + k_3$$

where k_2 now equals K and k_1 equals (1 – K) or (1 – k_2), k_3 is again equal to zero.

If K = 0.5, then the OI is the same as the antiknock index. However, it has been found that in most of the cases considered, K was negative [7.49–7.53]. Thus, for a given RON, a fuel with the higher sensitivity (lower MON value) had better antiknock quality and gave better performance. Even when K was not negative, it had only a small (<0.2) positive value so that MON contributed much less to fuel antiknock quality than has generally been assumed. The empirical results show that the K value is different at different operating conditions and for different engines. In general, K decreases and even becomes negative as the engine becomes more prone to knock.

One approach to understanding auto-ignition and knock is to work with the auto-ignition delay time, τ, which depends on pressure (p) and temperature (T) for a given fuel and mixture strength and has to be measured in separate experiments, for example, in a shock tube. If sufficient data exist for a given fuel, τ can be expressed as a function of pressure and temperature. However, such detailed information on τ does not really exist for fuels of different chemistries. Also, the evolution of the pressure pulse following auto-ignition is determined by its interaction with the temperature gradient in the end gas [7.31], and this can also affect knock intensities.

If two fuels of equal RON are considered, an olefinic or aromatic fuel with a high sensitivity, and a paraffinic fuel with a low sensitivity, then the high-sensitivity fuel, by definition, will have a lower MON. Classically, octane sensitivity is viewed from the basis that paraffinic behavior is normal. This is understandable as two paraffins, iso-octane and n-heptane, define the octane scale. From this perspective, sensitive fuels are considered to be "de-rated" or behave abnormally under MON rating conditions. An alternative approach [7.29], and one which makes understanding the chemistry of sensitivity somewhat easier, is to consider the behavior of sensitive fuels to be normal. On this basis, the nonsensitive paraffins are "super rated" at MON conditions. Paraffins exhibit what is classically called negative temperature coefficient (NTC) behavior.

The term NTC derives from low-temperature auto-ignition studies conducted in static reactors. In these experiments, fuel and air (or some other oxidizing medium) are admitted into a heated reaction vessel, and the time required for auto-ignition is determined as a function of initial reactor temperature and pressure. Figure 7.8 presents the results of such a study of propane.

FIGURE 7.8 Auto-ignition times as a function of temperature for propane in a static reactor.

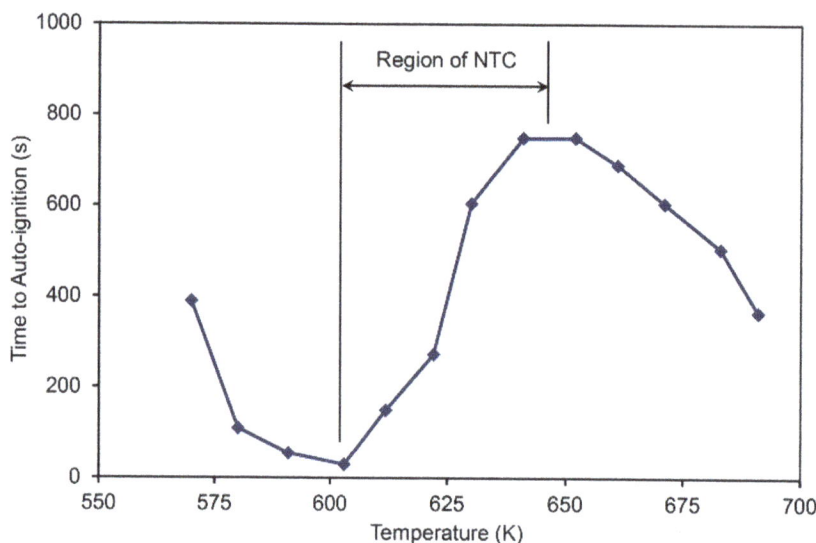

At temperatures below 600 K, there is a monotonic decrease in induction time as temperature is increased. This behavior is consistent with Arrhenius kinetics, which dictates that reaction rates increase (hence induction times decrease) with temperature. However, in the temperature range 600 to 650 K the reverse is true. This abnormal behavior where induction times increase (reaction rates decrease) is termed the NTC for obvious reasons. Above 650 K, Arrhenius behavior is again observed. Paraffin chemistry is dominated by this NTC. This has been demonstrated [7.29] to be more pronounced at lower pressures, higher temperatures, and shorter reaction times (i.e., higher engine speeds). By contrast, olefin and aromatic chemistries do not exhibit NTC behavior. The result is that paraffins have enhanced antiknock qualities under MON rating conditions than at RON rating conditions when compared to olefins and aromatics [7.29].

It has been found [7.54] that the peak cylinder pressure for the MON test is lower than that for the RON test. This contradicts the perceived wisdom in that the MON test conditions are usually regarded as more severe, that is, higher engine speed. The RON test therefore has higher peak pressures but lower temperatures when compared with the MON test, for a given primary reference fuel rating condition. The trend for increased specific power output forces in-cylinder conditions toward higher mean effective pressure but lower temperatures to avoid knock. The engine's octane requirement therefore moves farther from that indicated by the MON test and closer to that indicated by the RON test, even to the point where it moves to the other side of the RON value and hence K becomes negative.

7.3.4. Influence of Chemical Structure on Octane Quality

The chemical structure of hydrocarbons has a great influence on the octane quality, as illustrated in Figures 7.9 and 7.10. Figure 7.9 shows the effects of the structure of hexane on the RON and MON value. Figure 7.10 shows how the bond structure affects the RON and MON values. From Figure 7.9, it is clear that there is a significant difference between the octane numbers of the straight chain paraffin and those with some branching. The position of the branch does not make that much difference—for instance, between the 2-methyl pentane and the 3-methyl pentane. The addition of a second branch again provides a noticeable improvement in the RON and MON values, and here there is a slight difference brought about by the position of this second branch. Branched chain paraffinic hydrocarbons are present in the products from isomerization and alkylation processes (i.e., in isomerate and alkylate). The sensitivity of all these alkanes is fairly small. Maintaining the straight chain structure but introducing a double bond again makes a significant, in fact slightly larger, difference to the RON value. There is a significant difference to the MON value but the sensitivity is increased (see Figure 7.10). The refinery cracking processes give rise to olefinic compounds, and so cracked streams tend to have high sensitivities.

FIGURE 7.9 Influence of chemical structure on octane quality of C_6 alkanes.

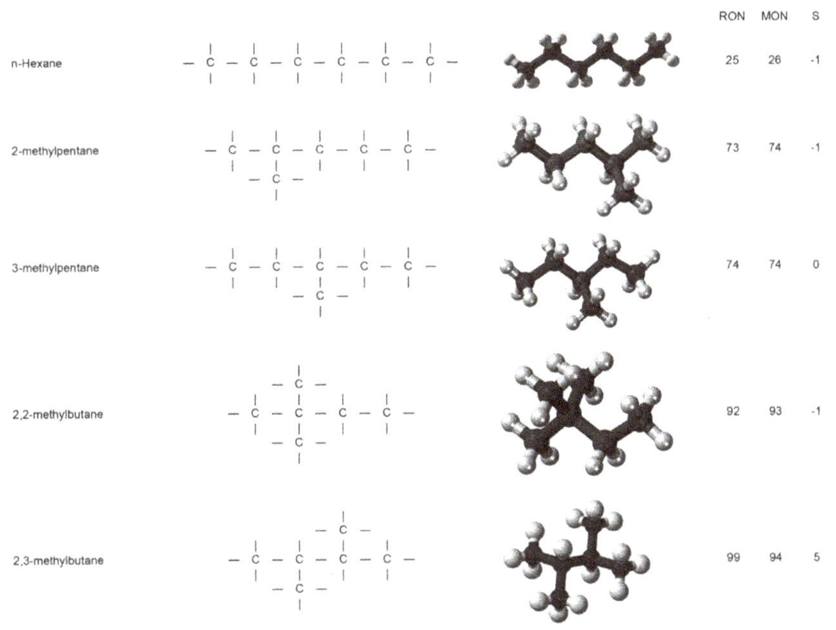

	RON	MON	S
n-Hexane	25	26	-1
2-methylpentane	73	74	-1
3-methylpentane	74	74	0
2,2-methylbutane	92	93	-1
2,3-methylbutane	99	94	5

FIGURE 7.10 Influence of chemical structure on octane quality of C_6 compounds.

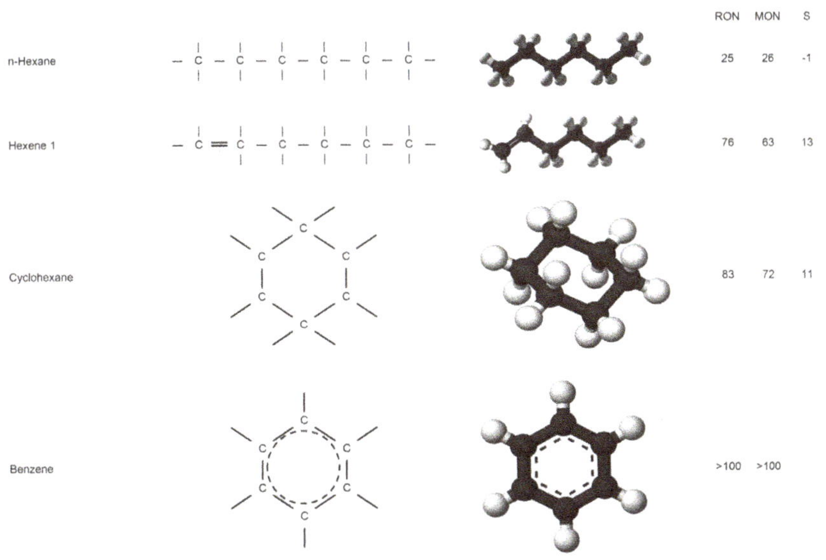

	RON	MON	S
n-Hexane	25	26	-1
Hexene 1	76	63	13
Cyclohexane	83	72	11
Benzene	>100	>100	

Saturated cyclic compounds (i.e., naphthenes) such as cyclohexane can be seen to have better octane levels than the corresponding straight chain compound, in this case hexane, but with an intermediate sensitivity. Aromatic compounds such as benzene have still higher octane values but with a relatively high sensitivity. Aromatics are formed in the catalytic reforming process and, to a lesser extent, in cracking processes.

7.4. Antiknock Additives

7.4.1. Lead Alkyls

Lead alkyls were the most widely used antiknock additives as they represented the most economical way to achieve required octane number levels. However, due to environmental concerns and the adverse effects of lead on catalyst the use of lead antiknock additives is no longer allowed in motor gasoline for on-road use (pump gasoline). It is still allowed in specialist fuel for motor-sport (see Chapter 14) and for aviation gasoline, a fuller discussion of such additives can be found in Appendix 6. As noted above, it is also used in the RON [7.36] and MON [7.37] test procedures in order to achieve reference fuels of octane numbers greater than 100, i.e., pure iso-octane.

7.4.2. MMT—Methylcyclopentadienyl Manganese Tricarbonyl

Although over the years there have been many investigations into possible organometallic antiknock compounds to replace lead alkyls [7.55], the nearest to a successful alternative that has been found is MMT. This compound is more effective than lead on a metal concentration basis, as shown in Figure 7.11 [7.56], and was commercialized in 1958 as a supplement to lead [7.57]. It also has some benefits in preventing valve-seat recession (VSR) in vehicles that were designed to run on leaded fuel and have soft valve seats.

FIGURE 7.11 Effectiveness of organometallic antiknock additives.

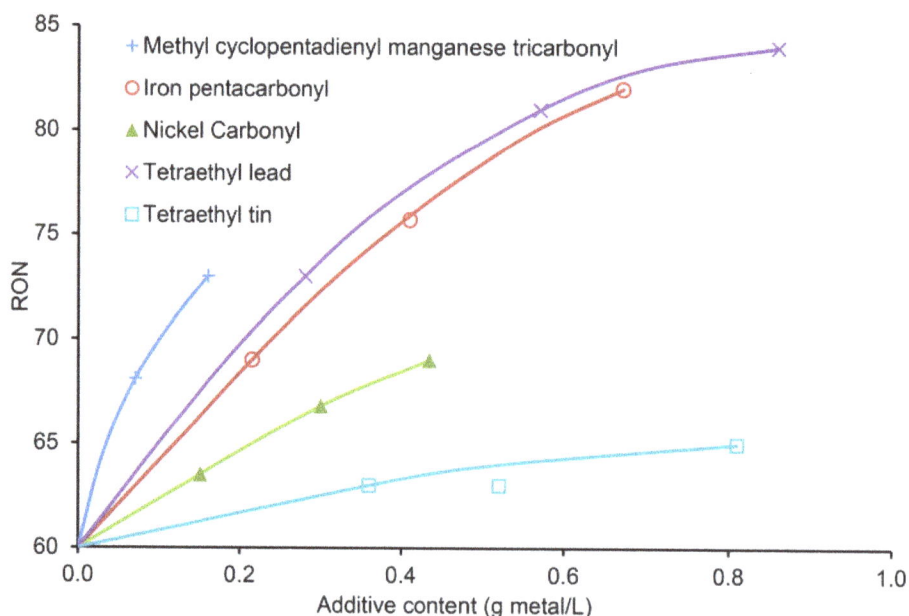

© SAE International.

MMT has been used in unleaded gasoline for catalyst-equipped vehicles at manganese concentrations up to 0.018 g Mn/L. At this concentration level it can provide up to three octane numbers at low cost, and it provides blending flexibility to the refiner where its use is permitted. MMT is a manganese-based organo-metallic compound and cannot be used in countries where there are restrictions on the use of such compounds in gasoline. In the US and Canada, it is only permitted at concentrations of up to 1/32 g/US gal, equivalent to 8¼ mg/L, but is prohibited in some states, in the EU the limit is 2 mg/L.

It can only be used at relatively low concentrations because of reported problems with fuel instability, deposit buildup in engines, lack of response at higher concentrations, and its adverse effect on hydrocarbon emissions from catalyst-controlled cars [7.56, 7.58–7.61]. Other data contradict these observations; vehicle studies have demonstrated that gasoline containing MMT does not harm emission control systems or the operation of on-board diagnostics (OBD) system [7.62, 7.63]. Although manganese from the combustion of MMT is found on the catalyst, the catalysts display similar if not better performance and contain lower levels of phosphorus and zinc compared with catalysts from vehicles using gasoline without the additive [7.64–7.66]. However, many motor manufacturers oppose the use of MMT. The major objection presented in the Worldwide Fuel Charter is that the additive adversely affects catalysts, spark plugs, and OBD systems, which could lead to higher tailpipe emissions [7.67–7.72]. More recently, concern has moved to the effects of MMT and other metallic fuel additives on low-speed preignition (LSPI), or stochastic preignition (SPI), with reports that increasing levels of MMT increase the level of LSPI resulting in an increase in particulate number (PN) emissions [7.73].

7.4.3. Other Metallic Antiknocks

Figure 7.11 shows the relative effectiveness of a number of alternative metallic antiknocks in addition to lead and MMT. The only ones to have been used commercially contain iron and were used during the 1930s in both the US and Europe either as pentacarbonyl or as ferrocene (dicyclopentadienyl iron). Ferrocene has been reassessed as an antiknock more recently [7.74–7.76] at low concentrations (15 to 30 ppm), and, although it increases RON by up to 1.5 units and MON by up to 0.9 units at 30 ppm, it is not clear whether the drawbacks of iron compounds (namely excessive engine wear and spark plug fouling) have been overcome at the low concentrations suggested. In one set of tests [7.76] which indicated that ferrocene may also give some improvement to emissions and fuel economy, one pair of cars fitted with a TWC ran for 85,000 km without any reported problems, and another pair completed their scheduled 20,000 km also apparently without problems.

Although other metals have been shown to have antiknock effectiveness, they have generally failed to be commercialized for a number of reasons including combustion chamber and spark plug deposits, toxicity, wear, and cost-effectiveness. As noted in the previous section, the Worldwide Fuel Charter notes that

Ash-forming additives and contaminants must be avoided, and gasoline that contains these additives or contaminants cannot be considered good quality fuel.

7.4.4. Organic Antiknocks

Most of the undesirable characteristics of metallic antiknocks arise because they leave a deposit on combustion. Organic antiknocks, being ashless, have always been of considerable interest, and in fact the first anti-knocks to be identified were the aromatic amines. Of the readily available aromatic amines, n-methyl-aniline (NMA) is one of the most effective, although many others have been investigated [7.77]. It requires about 1% by volume of NMA to give an activity similar to that of 0.1 g lead/L, and so it is much less cost-effective than lead. More recently [7.78], a comprehensive investigation was carried out in which a wide range of potential ashless antiknocks were evaluated. Four groups of compounds showed up as having comparable activity to N-methylaniline: aromatic compounds containing nitrogen; aromatic compounds containing oxygen; iodine and aliphatic iodine compounds; and selenium compounds. The best of these are compared in Table 7.3 in terms of their relative effectiveness with NMA.

TABLE 7.3 Relative effectiveness of organic antiknocks.

Compound	Relative effectiveness (weight basis, NMA = 1)
Best amine	1.1
Best hydrazine	0.9
Best N-nitrosamine	1.5
Best phenol	0.5
Best formate	0.5
Best oxalate	0.4
Iodine	1.1
Best selenium compound	2.8

© SAE International.

It was concluded from the study that none of the compounds tested were as cost-effective as lead alkyls or further processing. However, with the increasing use of downsized, pressure-charged engines and increased awareness of LSPI there have also been concerns raised over the use of anilines and aniline derivatives as organic antiknock additives [7.79].

Some types of ashless compound have proved to be valuable in terms of antiknock performance and have been used widely since lead was phased out. These are the oxygenated compounds, such as alcohols and ethers. They are used at concentrations of several percent and so could be classed more as blend components than antiknock additives. They are, however, very important in helping refiners to meet both the volumes required and the octane quality of unleaded gasolines.

7.4.5. Oxygenated Blending Components

It is not possible to give absolute values for the octane quality of the various oxygenates because the way they blend into gasoline depends on the other components that are present. However, Table 7.4 summarizes a range of octane blending data that have been obtained with different gasoline blends and demonstrates that these components can be valuable in helping to meet required octane levels, particularly in an unleaded situation.

TABLE 7.4 Approximate octane blending values.

Compound	Blending RON	Blending MON
Methanol	127–136	99–104
Ethanol	120–135	100–106
Tert. butanol	104–110	90–98
Methanol/TBA 50/50	115–123	96–104
MTBE	115–123	98–105
TAME	111–116	98–103
ETBE	110–119	95–104

© SAE International.

It will be seen that the sensitivities (RON minus MON) of the oxygenates are high, and this means that in a situation where MON is a critical specification point, it may be difficult to meet MON specifications using these materials without going over the top of the RON specification, that is, without giving quality away. In this situation, high-cost, restricted availability blending components with low sensitivities such as alkylate or isomerate may be needed.

The standard tests for RON and MON (i.e., ASTM D2699 for RON and ASTM D2700 for MON) are equally applicable to fuels containing oxygenates. However, the high latent heats of vaporization of the alcohols make the RON, and particularly the MON, method unsuitable at very high alcohol concentrations.

The high octane blending values of oxygenates make them attractive to refiners, and it is important to establish if these octane benefits are seen by vehicles on the road. There has been some question as to whether the conventional octane parameters of RON and MON give a reliable guide to road octane performance when some oxygenates, particularly ethanol, are used in gasoline [7.80, 7.81]. However, other work has shown a satisfactory correlation [7.82].

A CRC program [7.82] in the US showed that in regular-grade fuels at full throttle all the oxygenates tested, which included both alcohols and ethers, improved road octane performance relative to hydrocarbon blending components, although no trend was readily apparent in premium fuels. High-speed road octane performance with some oxygenates, particularly methanol, has been shown to be relatively poor [7.83]. MTBE does not appear to show this deterioration in performance at high speed [7.84].

A comprehensive study in Europe using a wide range of different oxygenates in gasolines blended to constant RON and MON levels showed the following [7.85]:

- At low olefin levels (10%), accelerating knock performance of all oxygenate-containing fuels was better than that of the hydrocarbon-only fuel of the same RON and MON. For most oxygenates, when used at levels near the maximum oxygen content allowable, there was a reduction in octane requirement of around 0.5 numbers. However, at high olefin levels (20%), the reverse was true.

- At constant speed, the oxygenated fuels were generally inferior to the corresponding hydrocarbon fuel, increasing octane requirements by an average of about one number. Ethanol and methanol give the biggest increase, and MTBE and mixtures of ethers appear to have only a marginal effect on octane requirement.

It is only possible to generalize on the effect of oxygenates because, as with all octane work, the actual effect will depend on the vehicle, the composition of the fuel, and the method of test.

7.5. Octane Blending

The constraints imposed on gasoline blenders are many and diverse. First, they must meet all the octane specifications, usually RON, MON, and R100°C. Because octane numbers, as measured, do not blend linearly and because their blending behavior is dependent on the other components that are present in a blend, it can be difficult to predict the octane quality of a finished gasoline. If blending is carried out at a refinery rather than using only purchased components, maximum use must be made of all the components produced, particularly if their alternative uses are limited and/or much less profitable. Whatever the source of the components, each blend must be made at the lowest possible cost. Octane quality is only one of several specification points that must be met with minimum quality giveaway, and often, correcting a blend for one parameter will put it out of specification for another. Tankage restrictions and the relatively poor precision of many of the tests also increase the difficulties, although the use of continuous analyzers, particularly for octane quality, helps in this respect.

Octane blending is discussed in more detail in Chapter 3, Section 3.10.3, and covers the several methodologies that are available to help blenders in their task. Many of them require lengthy calculations and detailed input due to the interactions that occur between components. The blending characteristics of each component can change with time due to different crude oils being run, changes in plant conditions, changes in blend composition, the use of oxygenated components, and so forth. Appendix 7 summarizes the physical properties of some of the hydrocarbons that can be present in gasoline.

7.6. Octane Requirements of Vehicles and Engines

Vehicles differ widely in the way they respond to octane parameters and in the level of octane quality they require to be clear of knock. It is important to both the fuel and the motor industry to know what the octane requirements of vehicles are under both normal and severe driving conditions so that fuels can be made available to satisfy essentially all cars in a given population regardless of driving conditions. It is also important that vehicles are not produced requiring higher octane number fuels than are available.

The measurement of octane requirement is difficult because the test is rather imprecise and the results are influenced by weather conditions and by the way the engine/vehicle has been driven immediately prior to the test. The situation is further complicated because vehicle manufacturing tolerances can give rise to large variations between different vehicles of the same model. A spread of seven octane numbers has been observed [7.40] which means that in order to define with reasonable accuracy the octane requirement of a vehicle, it is necessary to test at least 10 examples. Manufacturing tolerances of compression ratio account for much of this spread of requirements, and it has been suggested [7.86] that reducing the manufacturing tolerances on compression ratio alone would enable vehicles to make better use of available octane quality and give an overall benefit of about 2% in fuel consumption.

Clearly it would be extremely expensive for an individual company to test enough vehicles to be able to estimate the octane requirement distribution of the vehicles in any one country, and for this reason such testing is normally carried out cooperatively. In the US such tests were carried out by the Coordinating Research Council (CRC), in Europe by the Cooperative Octane Requirement Committee (CORC), in Japan by the Japanese Petroleum Institute (JPI), and in Australia by the Australian Cooperative Octane Requirement Council (ACORC). Each of these groups prepared its own full boiling range (FBR) reference fuels and used standard test procedures so that all the data obtained were compatible within a geographical area. As regional or national specifications for gasoline have increased in importance, interest in octane requirement distribution has waned. In Europe for example, the unleaded gasoline market is dominated by the 95 RON grade. The European specification (EN 228) does not include higher octane grades, although the increase in the marketing of premium grades does mean that higher octane number fuels are quite widely available. As a consequence, CORC was disbanded in the mid-1990s. Nevertheless, some European car models are optimized on 98 RON gasoline and only run satisfactorily on 95 RON fuels because they are fitted with knock sensors as discussed in the following section.

The reference fuels used for the work described above consist of a number of fuel series in which the RON is increased in one-number increments from a level below that of the lowest commercial grade available up to as high as possible without deviating too far from commercial reality; typically, each series varied from about 88 RON to a little over 100 RON. The number of fuel series necessary depended on how the vehicles in the area under consideration responded to octane number parameters. In Europe, for many cars, R100°C was important as well as RON and MON, and this meant that at least three series in which these parameters were varied independently of each other were required. In the US, R100°C was not found to be an important variable and thus only two fuel series were used. Where R100°C and sensitivity were important, it was preferable to use fuel series having two levels of sensitivity (i.e., RON – MON) and two levels of ΔR100°C (i.e., RON – R100°C). Vehicles were also tested using PRFs, which are blends of n-heptane (RON and MON = 0) and iso-octane (RON and MON = 100).

The test procedures used [7.87, 7.88] varied somewhat from country to country but all involve accelerations from a low speed, at full and at part throttle, and finding at which octane number on each fuel series trace knock occurs. Trace knock is also called borderline knock and is usually defined as the lowest level of knock that can be detected by ear, although some laboratories use instrumentation to detect knock and relate the readings back to the trace knock level. In addition, tests were also frequently carried out at constant speeds and wide-open throttle to simulate travel on high-speed roads under various conditions of load. In these situations, the use of instrumentation will greatly assist in the detection of knock. These latter tests do not include fuels in which R100°C is varied, since this parameter is not considered important at constant speed. Constant-speed tests are almost invariably carried out using a chassis dynamometer (rolling road) to allow more accurate control and to allow higher speeds than may be legal on public roads.

7.6.1. Vehicles with Knock Sensor Systems

Knock sensors [7.89, 7.90] are now fitted to many engines in order to protect them from damage and to enable higher compression ratios to be used to achieve an improvement in fuel economy. They are usually attached to the cylinder head or block, and when they detect vibration at the frequencies that are characteristic of knock, they activate a mechanism which reduces the octane requirement of the vehicle. This is most often done by retarding the ignition timing although other methods are available, such as using the turbocharger waste gate to control cylinder pressure in turbocharged vehicles. The ignition timing (or other mechanism) is then gradually relaxed back to the default setting for optimum power and economy until driving conditions put the engine back into incipient knock again.

Figure 7.12 is a simplified illustration of what happens to knock intensity in a vehicle fitted with a knock sensor when the octane number of the fuel is gradually lowered. With high octane numbers the vehicle is clear of knock. As the octane quality is gradually reduced, knock will eventually start and will gradually increase in intensity until it reaches a level at which the knock sensor system determines that action is required. At this stage the ignition timing begins to be retarded—or some other method of reducing the octane requirement of the engine is activated—and this will hold the knock intensity at this maximum acceptable level as set by the engine manufacturer. If the octane number of the fuel is reduced still further, the mechanism for reducing knock intensity continues to function until it reaches the limit of its operation. At this stage, knock level will then increase rapidly from the level set by the engine manufacturer up to medium or heavy knock.

FIGURE 7.12 Effect of a knock sensor on knock intensity as octane quality is lowered.

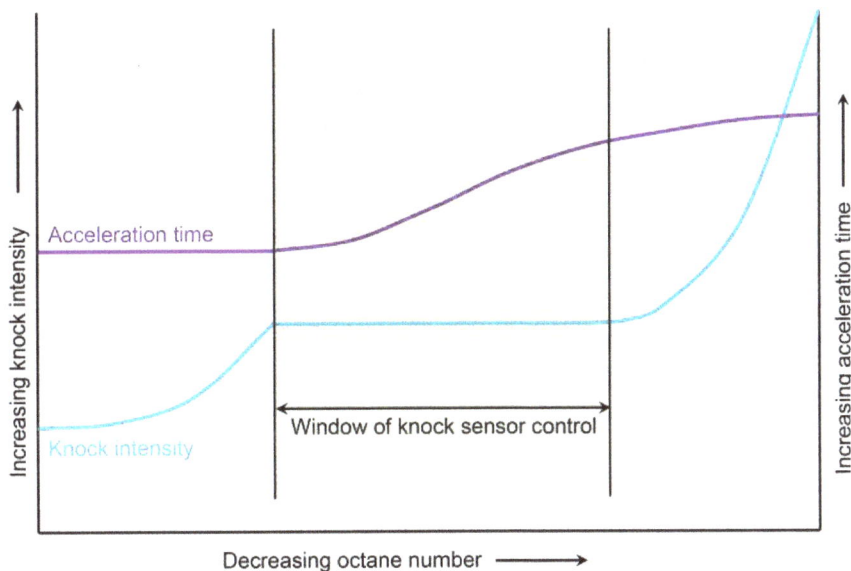

Unfortunately, as the timing of a vehicle is retarded, so is its performance, and Figure 7.12 also shows what happens to acceleration time while the knock sensor system is functioning. This worsening of acceleration performance can be very noticeable [7.91] and can be used as a criterion for determining octane requirement, if desired, in this type of vehicle. The knock sensor system should hold the engine at or just below the trace knock level, the traditional methods of determining octane requirement, as described above. This would thus result in the octane requirement being measured as the value at the end of the plateau where the knock sensor system is operating, just before the knock intensity starts increasing rapidly, although the vehicle acceleration performance is likely to be severely curtailed at this point.

Some vehicles with complex engine management systems have a learning curve that adapts to the way the vehicle is normally driven so that engine conditions can be optimized. It is thus possible that the same engine may exhibit different octane requirements dependent upon previous driving history and the learned engine management strategy. To get consistent results, it is thus necessary to "precondition" such a vehicle in a standard manner before measuring its octane requirement.

7.6.2. Data Analysis

After tests have been carried out cooperatively on a large number of vehicles, the results are pooled so that each participating organization can carry out its own data analysis if it wishes, and, in addition, some general analysis of the results is usually carried out, which may be published [7.92, 7.93]. A number of studies have been carried out on methods of analyzing such data [7.40, 7.94, 7.95], and these can involve preparing satisfaction curves as in Figure 7.13 and developing equations that relate how a car population responds to the various octane parameters. Such equations might be of the form:

$$\text{Road octane number} = a\,\text{RON} + b\,\text{MON} + c\,\text{R100°C} + d$$

where road octane number is the octane requirement in terms of PRFs and a, b, c, and d are constants for any given car population and driving mode.

FIGURE 7.13 Examples of vehicle satisfaction curves.

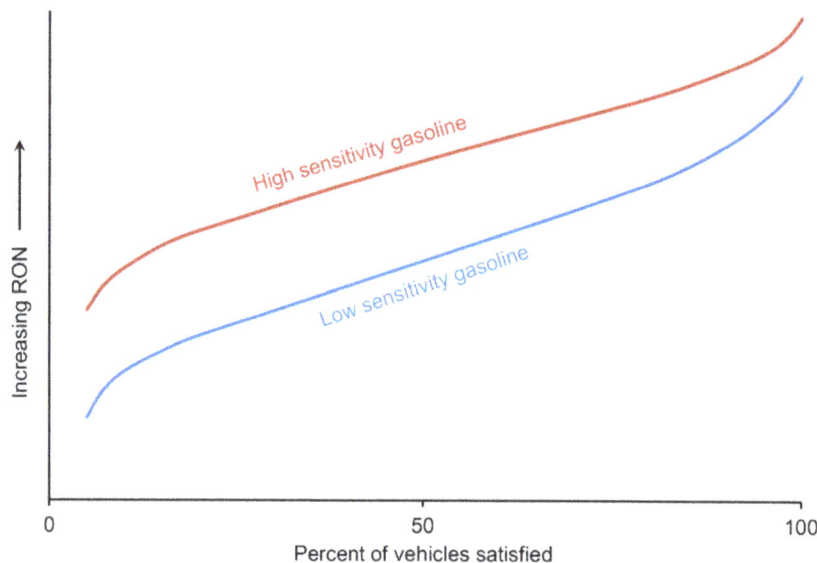

These correlations allow fuel manufacturers to predict the percentages of vehicles satisfied by a gasoline in any given car population and to design a gasoline to maximize the satisfaction level. However, what they really need to know is how many customers are satisfied rather than how many vehicles are technically satisfied, that is, are free of knock when driven by technical raters using standard driving procedures. A considerable amount of work has been carried out on this problem [7.96–7.103], and all agree that customers do not hear or recognize knock as readily as technical raters and in addition do not drive in as severe a manner as required by the test procedure and so are less likely to provoke knock in the engine. In some of the tests it has been suggested that customers are less sensitive than technical raters by as much as five octane numbers on average, although there are large differences between customers. Continuous improvement in the sound insulation of the engine bay and cabin also makes knock less perceptible to the driver. The methods of analysis of the data are complex [7.102] but the procedures available do enable reasonable estimates of customer satisfaction levels to be made.

7.6.3. Octane Rating of Fuels Using Vehicles or Engines

Octane requirements are valuable in establishing the fuel quality required to satisfy a vehicle or population of vehicles. It is sometimes useful to be able to compare the performance of a number of different fuels in a vehicle on the road in much the same way as is done using a CFR engine. The octane numbers thus obtained are also called road octane numbers because they are related to the performance of PRFs and are carried out on the road. To carry out such fuel ratings, the ignition timing of the vehicle is adjusted to find a setting that gives trace knock for a particular fuel and driving mode, whereas when measuring octane requirements, the fuel quality is varied to find an octane number at which trace knock occurs.

Two standard procedures were developed and commonly used to rate fuels in vehicles: the Modified Uniontown procedure and the Modified Borderline procedure [7.103]. The Modified Uniontown procedure was developed from an earlier procedure, the Uniontown method, which involved assessing the knock intensity of different fuels at a fixed ignition setting and comparing the results with those from a series of standard reference fuels. This procedure fell out of favor because knock above the trace level can remove deposits and thereby lower octane requirements. It is also more subjective to compare different knock severities than to compare a series of ignition timing readings all at trace knock levels.

The Modified Uniontown procedure requires the basic ignition timing control of the vehicle to be modified so that it can be readily adjusted and the new timing measured. Full throttle accelerations are made from a low speed using PRF, and the timing adjusted so that a trace level of knock is detected during some part of the acceleration. These results are plotted to give a curve relating ignition timing with PRF octane value at the trace knock level, as shown in Figure 7.14. The test fuel is then run and the trace knock ignition timing recorded, reading up from this point on the *x*-axis to the curve, and then reading across to the *y*-axis, determines the road octane number. The procedure is rapid and simple but does not give much information about how the octane rating changes over the speed range.

FIGURE 7.14   Modified Uniontown method for determining road octane number.

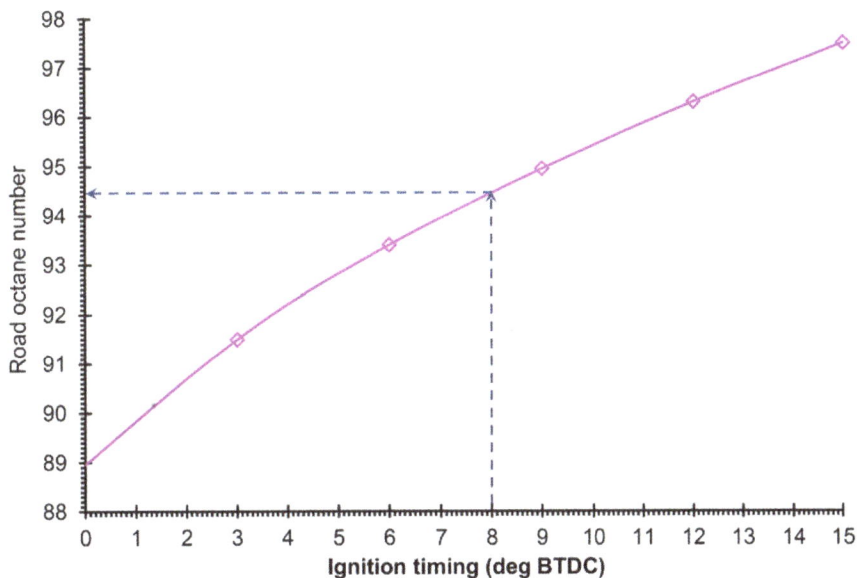

The Modified Borderline knock procedure is the second method. Again, the automatic ignition timing control is disabled and arrangements made to adjust it manually during an acceleration. Accelerations are made as before, but this time the spark advance is adjusted continually in order to maintain a trace knock level throughout the acceleration. The data obtained can be handled in a number of ways. One method is to construct a series of curves for different PRFs and then to read off these values against the test fuel trace, as shown in Figure 7.15. Here it can be seen that Fuel A has a higher road octane number than Fuel B at low speeds and a lower value at high speeds.

FIGURE 7.15 Modified Borderline method for determining road octane number.

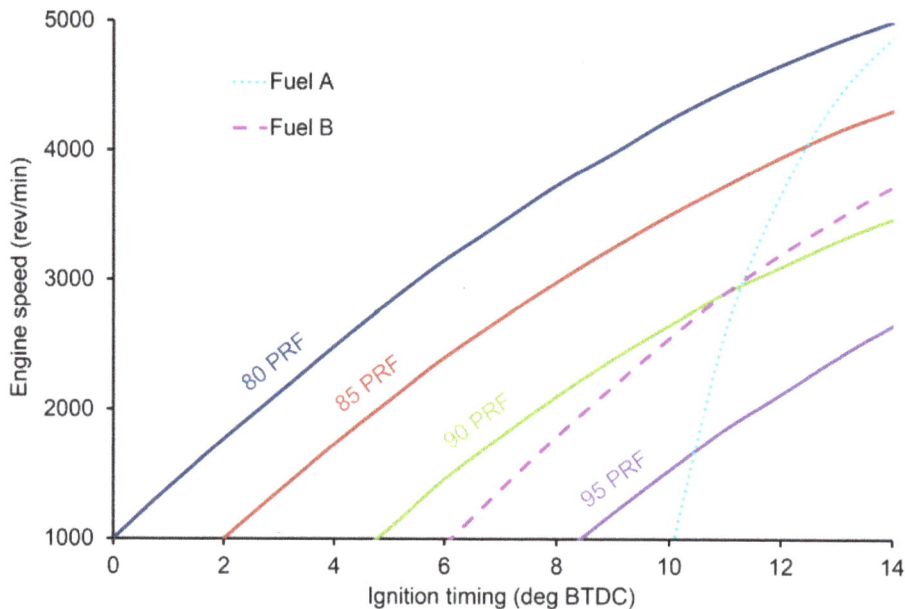

With increasing levels of sophistication of engine management systems, it becomes increasingly difficult to bypass them and to change the ignition timing manually; therefore, complex electronic equipment tailored to a specific car is needed for this purpose.

7.6.4. Engine and Other Factors That Influence Octane Requirements

Of the factors that influence the octane requirement of an engine, probably the most important is compression ratio. However, it is not possible to predict the octane requirement of a vehicle directly from its compression ratio because there are a large number of other variables that influence it [7.104]. Nevertheless, for a given engine, a unit increase in compression ratio will increase octane requirement by anything from three to six units depending on other design features.

The second most important factor is ignition timing; a change in ignition timing by advancing it two degrees will typically increase octane requirement by between a half and one unit. The selection of the optimum ignition timing to use throughout the speed range is extremely important. If the timing is too advanced in the low-speed range, then knock may occur during acceleration. If it is too advanced at higher speeds, then knock, although present, may not be heard because of wind and engine noises, and engine damage can result.

As engine speed increases, knocking tendency at a fixed ignition timing will normally decrease because less time is available for preflame reactions to take place. However, ignition timing is advanced as engine speed increases in order to ensure that maximum power is obtained over the whole speed range. This means that high-speed operation can be as critical as low-speed operation. On some models, as discussed in Section 7.6.1, knock sensors are fitted that activate a mechanism to reduce octane requirement immediately when the engine begins to knock. Then, if no further knock occurs, the engine is allowed to gradually revert back to normal [7.90]. In many such vehicles, there can also be a benefit in that improved acceleration performance can be obtained with fuels having octane levels above the nominal octane requirement of the vehicle. The acceleration performance of vehicles without knock sensors is largely unaffected by fuel octane quality.

Other vehicle design or operating factors that influence knock are load, mixture strength, and charge temperature. Increasing the load increases the engine temperature and the end gas pressure and, hence, increases the tendency to knock. With regard to mixture strength, knocking tendency is at a maximum at an air-to-fuel ratio of about 13.5:1, but this varies with the characteristics of the fuel. Charge temperature is also important because the higher the charge temperature, the higher the tendency for knock, as would be expected from the consideration of knock as an end gas explosion.

The effect of engine variables on knock means that because of manufacturing tolerances, apparently identical cars of the same model can differ significantly in octane requirement. When this is coupled with variations in deposit laydown in the combustion chamber and errors in measuring octane requirements, it can be seen that it is difficult to establish the octane requirement characteristics of a car population with any precision.

External factors such as the ambient temperature, pressure, and humidity can also influence the octane requirements of cars, and these can be taken into account when setting octane specification for specific locations and for different seasons of the year [7.105]. On average, octane requirements increase with increasing ambient temperature by 0.097 MON/°C and decrease with increasing specific humidity by 0.25 MON/g of water/kg of dry air [7.39].

Atmospheric pressure also influences octane requirements; a reduction lowers the octane requirement because the mixture density is reduced. Atmospheric pressure changes are not taken into account in seasonal octane specifications because they are not predictably dependent on the time of year. However, atmospheric pressure does reduce with increasing altitude, and this is often taken into consideration. The extent of the effect varies considerably with vehicle design, and previous ASTM specifications allowed reductions of 1.0 to 1.5 octane numbers per 300 m [7.106]. This is now a nonmandatory appendix to the current procedure [7.39]. However, the current level of sophistication of electronic engine management systems tends to minimize this altitude effect, and tests have shown [7.107] that with such vehicles the reduction in octane requirement is only about 0.2 octane numbers per 300 m.

7.7. Octane Requirement Increase (ORI)

Combustion chamber deposits significantly increase octane requirements, and it is normal to measure the octane requirement of vehicles only after they have accumulated several thousands of kilometers, when, on leaded fuels at least, most of the increase in octane requirement will have taken place, as shown in Figure 7.16 [7.40]. The increase in octane requirement is usually between two and ten numbers from a new engine [7.108], and this needs to be taken into account by engine designers to ensure that vehicles operate satisfactorily on the fuels that are available.

FIGURE 7.16 Typical curves of the increase in octane requirement with distance traveled under normal everyday driving conditions.

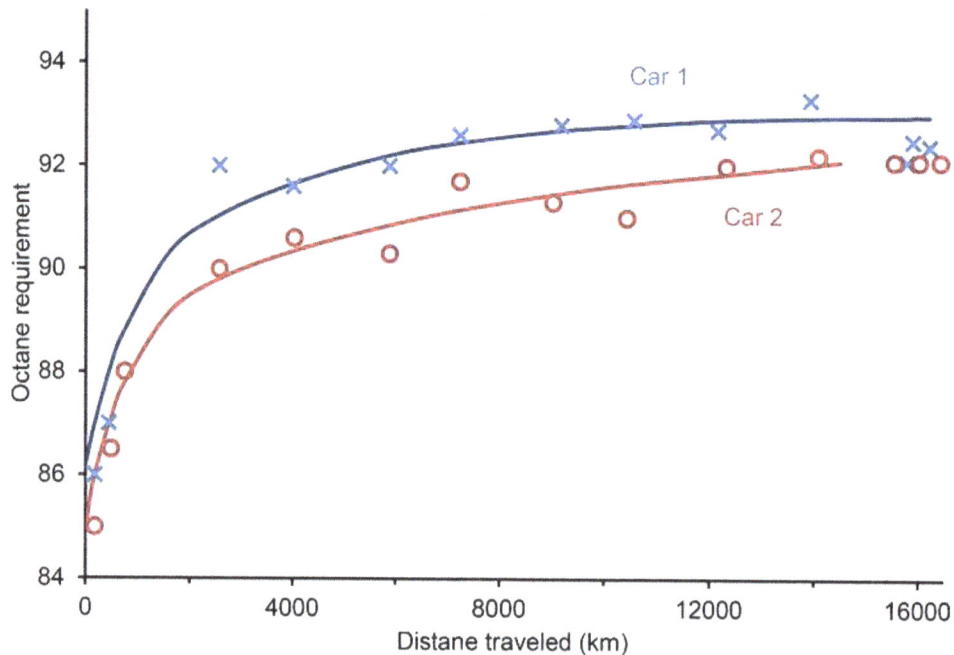

There were conflicting views as to whether leaded fuels show different ORIs and stabilization times than unleaded. Several investigators have found [7.109, 7.110] that there was no significant difference between leaded and unleaded ORIs, whereas other data [7.111–7.113] suggested that unleaded fuel gave higher ORIs than leaded. Although the curves shown in Figure 7.16 suggest that the ORI has plateaued by about 16,000 km, other work has suggested that the ORI is still increasing beyond 24,000 km [7.113, 7.114]. The vehicles in this survey showed average ORI levels of about 5 RON units and 3 MON units and there was an indication that the higher the initial requirement of the vehicle, the lower the ORI.

The measurement of ORI is difficult and the test is rather imprecise. This is not surprising since there are a number of variables that are either difficult or impossible to control. These arise because the octane requirement measurements are taken at intervals of 1500 km or more, and during this time the weather conditions may have changed and the rater's appraisal of knock intensity may have varied. In addition, changes to the engine may have occurred, and there may have been variations in the driving cycle over the distance accumulation because it is difficult to maintain a constant test cycle on the road. A distance accumulation chassis dynamometer with computer control of the driving cycle would obviously produce better repeatability.

City-type driving involving a great deal of stopping and starting tends to lay down more deposits than high-speed driving. If, during the octane requirement test, knock is allowed to increase above the borderline level, deposits may be removed and the octane requirement reduced. Similarly, if a vehicle is subjected to a high-speed run immediately prior to carrying out the test, the requirement may be lower since some of the deposits may have burned off.

Deposits outside the combustion chamber, particularly ridge-type deposits in the inlet ports, have also been claimed to increase octane requirements [7.115, 7.116], although other work has suggested that combustion chamber deposits account for virtually all of the ORI in cars [7.109]. It seems likely that engine design plays an important part in determining whether ridge deposits in the inlet ports will influence octane requirements.

The nature of the combustion chamber deposits has been investigated [7.117], and it has been shown that they are partly carbonaceous and partly inorganic. They are derived from the combustion of the fuel and from any lubricant that finds its way into the chamber via the piston rings or valve guides. Thus, any metals present will have come from either or both of these sources. The organic part arises from incomplete combustion, and some components or additives are known to contribute more than others to ORI [7.118]. Aromatic compounds, and particularly those with condensed ring structures, are prone to give rise to deposits [7.119]. Heavy lubricants and especially brightstocks also contribute disproportionately to ORI [7.109, 7.120], as can some multifunctional fuel additives [7.121].

The most important factor in causing ORI is the thermal insulation effect of the deposits [7.117], which increases the temperature in the combustion chamber. It has been estimated [7.109] that this accounts for 90% of the ORI. The remaining 10% is due to the increase in compression ratio caused by the presence of the deposits and the chemical nature of the deposits that themselves may catalyze those reactions known to favor detonation.

7.8. Other Abnormal Combustion Phenomena

7.8.1. Preignition

Preignition can be defined as the initiation of a flame front at any point in the mixture before the flame front, initiated by the controlled positive ignition, would normally reach that point. This definition encompasses phenomena that have previously been described by such terms as wild ping, surface ignition, and hot spot ignition [7.122].

Surface ignition is the ignition of the charge by hot deposits or surfaces and can occur before or after the spark ignition. Mild surface ignition gives rise to rough operation of the engine and can be accompanied by a noise very similar to spark knock and is often referred to as "ping" [7.122]. It can be distinguished from spark knock in that it is not controllable by adjustment of the ignition timing. This form of abnormal combustion should not be confused with knock, which is the auto-ignition of the end gas after it has been heated and compressed by the advancing flame front.

Clearly, if the surface ignition occurs only fractionally before the spark, and at a single point, the effect will be the same as advancing the ignition. If, however, surface ignition occurs some time before the spark, then negative work is obtained, resulting in very high temperatures. In extreme cases the high temperatures can promote yet more surface ignition, and when this occurs it is often called runaway surface ignition. This is a violently unstable condition in which so much heat is produced that damage such as a hole burned in the piston crown can occur in a very short time [7.123]. The rapid release of energy before it is intended can give rise to a hard sound similar to that of an old diesel engine and is usually referred to as "rumble." Figure 7.17 shows the pressure traces for normal combustion, a knocking cycle (as shown in Figure 7.3), and a cycle with preignition [7.124].

FIGURE 7.17 Pressure traces for normal combustion, knock, and preignition cycles.

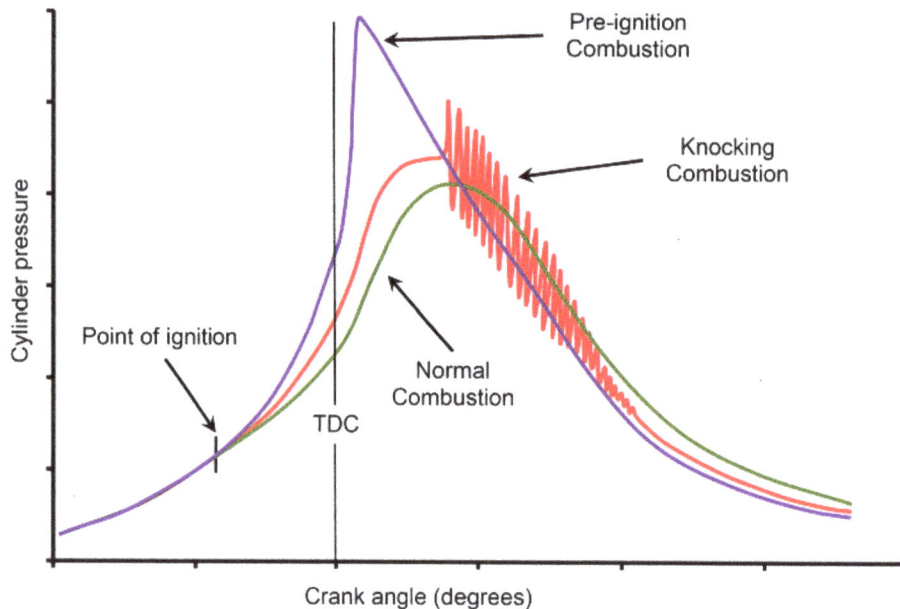

If an engine is operating to its design specification, that is, the correct mixture strength and ignition timing, then the main cause of surface ignition is the buildup of carbonaceous deposits within the combustion chamber. This type of preignition is usually referred to as deposit-induced ignition and has a somewhat random nature. Once it has started it may die away as the carbonaceous deposits are burned off, or it may increase in severity. Test work has shown that the combustion chamber surface temperature is the dominant factor in deposit formation [7.125], and hence driving mode is correspondingly important. Deposits on the exhaust valves are particularly likely to cause preignition problems, as the face of the exhaust valve is more remote from engine coolant than other parts of the combustion chamber. Phosphorus and boron additives have been used to reduce the glowing tendency of deposits but are rarely used nowadays because they adversely affect exhaust aftertreatment catalysts.

As engine design tends toward higher compression temperatures and pressures (higher compression ratios and pressure charging), then the auto-ignition of oil and fuel droplets also becomes a possible source of preignition [7.126, 7.127]. These are more likely to arise from hydrocarbon accumulation within the top land piston crevices volume [7.128, 7.129] rather than poor atomization and mixing. The widespread introduction of direct injection, along with the aforementioned increase in compression ratios and pressure charging has made this phenomenon more of an issue and it has now become known as stochastic preignition (SPI) or low-speed preignition (LSPI) and this can lead to what is known as super-knock [7.130] and mega-knock [7.131].

Engines can be checked for their susceptibility to preignition. This is done by first running them under a low-speed, low-load cycle to lay down deposits and then changing to a high-speed, high-load operation to raise the temperature of the deposits. The onset of preignition can be detected in a variety of ways, such as fall in power output, increase in peak pressure, increase in temperature, presence of a flame front prior to the spark by using an ionization detector [7.132], displacement of peak pressure curves [7.133], and so on. A fast response is required so that steps can be taken to avoid damaging the engine. The test can be made more severe by incorporating a small amount of a critical lubricant in the gasoline. Stochastic preignition is by its very nature a lot more difficult to monitor and therefore a lot more difficult to "design-out."

7.8.1.1. Fuel Quality Effects on Preignition

Fuel quality can be an influence on engine susceptibility to preignition, and several different aspects of the fuel quality are important. These are its deposit-forming tendency, its resistance to ignition by a hot source, and its ability to heat up a potential ignition source.

7.8.1.2. Deposits and Preignition

Turning first to the deposit-forming tendency of fuels, this can be due to all or any of the following:

- The presence of heavy, nonvolatile hydrocarbons in the gasoline. These may arise during manufacture from poor fractionation and/or from high-severity catalytic reformer operation. In the latter case, polynuclear aromatics are formed, which are particularly difficult to combust completely and so readily form carbonaceous deposits. It is best to remove these heavy compounds at the refining stage by re-running the reformate through a distillation tower. With the move toward plug-in electric hybrid vehicle designs, where the battery can be charged from an electrical supply and the vehicle can be operated upon the battery alone, there is the possibility that fuel can be stored in the vehicle tank for extended periods, which could lead to the loss of less-volatile components. This could alter the fuels distillation characteristics.

- Deposits derived from metals. The deliberate addition of metals to fuel is not permitted in many jurisdictions and greatly limited in others, metal derived deposits are thus predominantly derived from lubricants.

- Additives used in gasoline can give rise to carbonaceous deposits. Mineral oils are sometimes added for a number of reasons, and these can be deposit-forming. Brightstocks (heavy lube oils derived from residuum) are particularly prone to deposit formation [7.109, 7.132]. Other additives that have the potential to cause problems are any metal-containing compounds and some polymeric surfactant additives [7.134].

It is inevitable that some engine lubricant can find its way into the combustion chamber, and this can cause deposits, which can be especially bad from a preignition point of view [7.132, 7.135]. The metals that are often present are one factor, but so also are some of the polymeric additives such as viscosity index (VI) improvers.

7.8.1.3. Inherent Fuel Resistance to Preignition

The second important factor is the tendency for the fuel to be ignited by a hot source. This preignition resistance has been evaluated for a wide range of compounds using an electrically heated hot spot to find the relative ease with which they can be ignited [7.132, 7.136–7.138]. A rating scale has been used [7.136] in which iso-octane (having a high resistance to preignition) is given an arbitrary value of 100 and cyclohexane (low resistance to preignition) is given a value of zero. Ratings outside this range are determined by extrapolation.

Ratings obtained using this scale show that hydrocarbons, even of the same chemical type, vary enormously in their preignition resistance, as, for example:

- Aromatics: Benzene (26) is poor but toluene (93) is good.

- Naphthenes: Cyclopentane (70) is good but cyclohexane (0) is poor.

- Olefins: Di-isobutylene (64) is good but hexene-2 (−26) is poor.

- Paraffins: All are good and mainly in the range 50 to 100.

- Alcohols: Methyl alcohol (<0) is poor but isopropyl alcohol (62) is good.

Similar rankings are obtained using hot spot temperature to rate the fuel as shown in Figure 7.18 [7.132].

FIGURE 7.18   Hot spot ignition temperature data obtained using a single-cylinder engine at 4000 rev/min and equivalence ratio of 1.10.

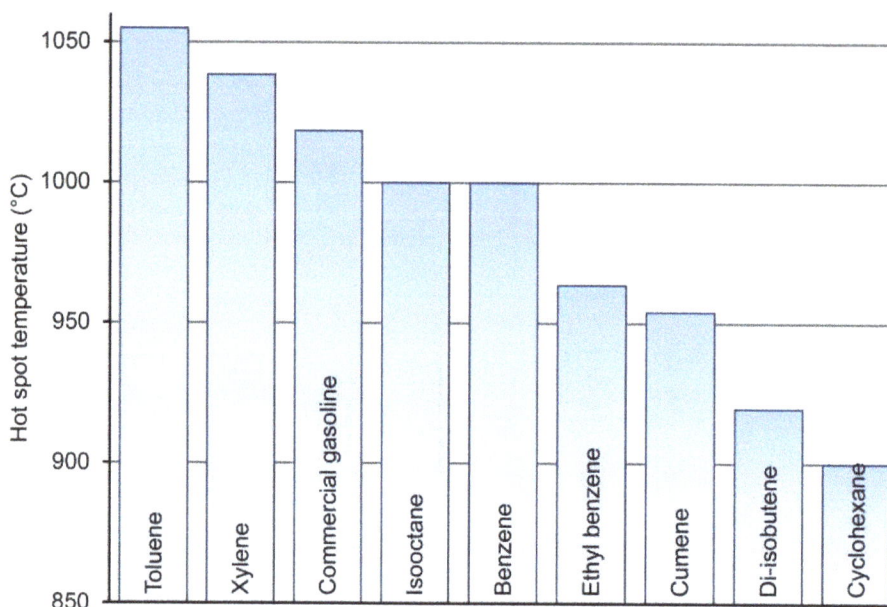

The final fuel factor that influences preignition is the ability of the fuel to heat up a potential hot spot. This was found to be partially correlated with flame speed [7.132], the higher the flame speed, the higher the hot spot temperature. The temperature of a metal filament in the combustion chamber of an engine running at 4000 rev/min and full throttle was measured using different pure hydrocarbon fuels; the data obtained are shown in Table 7.5.

TABLE 7.5   Deposit heating tendency of different fuels.

Hydrocarbon	Temperature attained by hot spot (°C)
Iso-octane	705
Xylene	750
Toluene	770
Cyclohexane	790
Cumene	800
Ethylbenzene	800
Benzene	835

It has been shown [7.132] that a combination of the resistance to hot spot ignition and the heating ability of the fuel gives excellent agreement with the preignition tendency for a given ignition source.

Hot spot ignition is where a part of the combustion chamber itself becomes hotter than was intended by the engine design team. An obvious candidate for this is the spark plug, which can get very hot if it has the wrong heat range [7.139–7.142]. The temperature that a plug reaches depends on how far the heat has to travel from the center electrode to the cooler outer shell and to the cylinder head. A long path will result in a hotter plug that is more prone to preignition but is less likely to be fouled by carbonaceous deposits. As engine designs move toward smaller capacities with higher compression ratios and pressure charging to improve low-load fuel economy while still maintaining high power outputs, there is a commensurate increase in the range of temperatures with which the spark plug must cope. The use of colder plugs, to cope with the peak powers, increases the tendency for deposit formation [7.143]. Deposits formed on the spark plug can in itself lead to preignition or misfire. The exhaust valve can also be the ignition source, so engine design factors are extremely important in determining whether or not an engine is likely to be prone to preignition problems.

Engine operating conditions also influence preignition because they affect combustion chamber temperatures. High-speed and high-load driving obviously results in high temperatures, although low-speed stop/start operation tends to maximize deposits. Advancing the ignition timing will increase temperatures, but changes in air-to-fuel ratio have little effect. If an engine is knocking, this can increase combustion chamber temperatures and lead to preignition. For this reason, it can sometimes be difficult to establish whether the primary cause of an engine failure is knock or preignition.

7.8.2. Misfire

Misfire, unless it is due to mechanical or electrical problems, can be influenced by gasoline quality. The main causes are the following:

1. Air-to-fuel ratio is outside the combustible limit.
   With most vehicles this type of failure is likely to be due to the mixture being too weak to burn, particularly when many vehicles operate under very lean conditions. This can often happen in cold weather during accelerations when the excess liquid fuel forced into the intake manifold does not volatilize sufficiently to give an ignitable mixture. Increasing the volatility of the fuel will normally overcome this malfunction. Driveability problems such as hesitations and stumbles can result until the mixture strength moves into the right range. Oxygenated gasoline components can aggravate this situation because they effectively lean the mixture still further, depending on the oxygen content of the compound being used. Most vehicles are now fitted with a TWC and an oxygen sensor to provide feedback to control the mixture at stoichiometric. Unless there is a failure of this sensor or a problem with the fueling system that prevents the necessary adjustment, these problems do not occur with modern vehicles.

   Too rich mixtures can also give rise to misfiring and can be due to incorrect fuel system calibration such as carburetor settings on older vehicles or injector dribble on newer vehicles; certain types of hot weather driveability problems aggravated by high fuel volatility as discussed in Chapter 9, Section 9.4; or extended cold start enrichment, possibly caused by a malfunctioning coolant temperature sensor. In older vehicles, during idle, deposit or ice formation in the carburetor can lead to rough idle and even stalling due to misfire. The deposits prevent enough air from flowing past the closed throttle plate to give an ignitable mixture. Additives are available that can prevent deposit and ice formation.

2. Deposit formation on the spark plug electrodes.
   Deposits may simply bridge the gap so that there is no spark at all, or they may provide an insulation layer that makes it more difficult to achieve a spark of the required energy to ignite the mixture. The deposits may be largely carbonaceous due to the mixture being too rich or because of heavy, nonvolatile compounds in the gasoline, or they may be inorganic arising from any lubricant that has found its way into the combustion chamber. In areas where metallic antiknock additives are still used then these will inevitably form inorganic residues that can foul spark plugs.

   A problem sometimes experienced by vehicle manufacturers is that vehicles coming off the production line frequently need to be moved for short distances before they are delivered to the dealer. Almost every time the vehicle is moved it will involve starting from cold and will require the use of the choke. This can cause so much cold fouling that misfire can occur and the manufacturer may need to fit new plugs prior to delivery to the customer. The problem can be overcome or alleviated by the use of a special factory-fill gasoline [7.144] containing, somewhat surprisingly, a lube oil that can be anything from a light spindle oil to a brightstock. The amount used will depend on the viscosity of the oil, but 5% of a light mineral oil or 0.5% of a brightstock have been found to be effective in eliminating misfire. Other materials have also been found to be effective [7.144], such as a polybutene or 5% methanol; the latter presumably works because it leans the mixture. It is important to appreciate that these additives should only be used in factory-fill gasolines and are not necessarily suitable for normal operation requiring much less frequent use of the choke.

3. Erosion of the plug gap or damage to the plug electrodes.
   Preignition can cause mechanical damage to the electrodes and thus give rise to misfire.

7.8.3. Run-On

This is also sometimes called after-running and refers to the tendency for an engine to continue to fire after the ignition has been switched off. It is most frequently due to the auto-ignition of the fuel–air mixture that is drawn into the combustion chamber before the engine stops rotating. Surface ignition can also cause run-on, but this is much less common and is characterized by the engine firing on every cycle and maintaining its speed rather than the very erratic firing typical of run-on caused by auto-ignition.

A detailed investigation of this phenomenon [7.145] has shown that during run-on, preflame reactions take place in the combustion chamber after the ignition has been switched off, which result in spontaneous ignition of the charge. For these reactions to proceed to the stage of auto-ignition, the charge has to be hot, it has to be under pressure, and it has to have sufficient time. The combustion chamber will be hot after switching off, and because engines continue to rotate to some extent, at least one of the cylinders will be under compression. If the engine speed is too high, there will be insufficient time for the preflame reactions to reach the auto-ignition point, and if the temperature is not high enough, they will not progress fast enough. If the temperature and speed fall within the critical region for spontaneous ignition, the engine will re-fire. This may result in the temperature and speed increasing out of the critical range, so that the engine may not fire again for one or two cycles, that is, until conditions are once again in the critical region. This gives rise to the very uneven type of running that is characteristic of run-on. Its duration is limited by the overall cooling of the engine.

Run-on is less of a problem nowadays because fuel cut-off valves are frequently fitted. Too high an idle speed setting and inadequate cooling of the combustion chamber are mechanical factors that worsen run-on.

The fuel factor that has the greatest influence on run-on is its resistance to spontaneous ignition, and this is its research octane quality [7.146]. There are many examples of vehicles requiring a higher RON to overcome run-on than needed to avoid knock.

References

7.1. Wankel, F., Rotary piston machines. U.S. Patent 2,988,008, 1961.

7.2. Germane, G., Wood, C., and Hess, C., "Lean Combustion in Spark-Ignited Internal Combustion Engines—A Review," SAE Technical Paper 831694 (1983), doi:https://doi.org/10.4271/831694.

7.3. Haslett, R.A., Monaghan, M.L., and McFadden, J.J., "Stratified Charge Engines," SAE Technical Paper 760755 (1976), doi:https://doi.org/10.4271/760755.

7.4. Quader, A.A., "The Axially-Stratified-Charge Engine," SAE Technical Paper 820131 (1982), doi:https://doi.org/10.4271/820131.

7.5.   Frank, R.M. and Heywood, J.B., "Combustion Characterization in a Direct-Injection Stratified-Charge Engine and Implications on Hydrocarbon Emissions," SAE Technical Paper 892058 (1989), doi:https://doi.org/10.4271/892058.

7.6.   Kono, S., "Development of the Stratified Charge and Stable Combustion Method in DI Gasoline Engines," SAE Technical Paper 950688 (1995), doi:https://doi.org/10.4271/950688.

7.7.   Brehob, D.D., Fleming, J.E., Haghgooie, M., and Stein, R.A., "Stratified-Charge Engine Fuel Economy and Emission Characteristics," SAE Technical Paper 982704 (1998), doi:https://doi.org/10.4271/982704.

7.8.   Thompson, C.E., Mooney, J.J., Keith, C.D., and Mannion W.A., Polyfunctional catalysts. U.S. Patent 4,157,316, 1977.

7.9.   Kemmler, R., Waltner, A., Schön, C., and Godwin, S., "Current Status and Prospects for Gasoline Engine Emission Control Technology—Paving the Way for Minimal Emissions," SAE Technical Paper 2000-01-0856 (2000), doi:https://doi.org/10.4271/2000-01-0856.

7.10. Tamura, Y., Kikuchi, S., Okada, K., Koga, K. et al., "Development of Advanced Emission-Control Technologies for Gasoline Direct-Injection Engines," SAE Technical Paper 2001-01-0254 (2001), doi:https://doi.org/10.4271/2000-01-0856.

7.11. Krebs, R., Pott, E., Kreuzer, T.P., Göbel, U. et al., "Exhaust Gas Aftertreatment of Volkswagen FSI Fuel Stratified Injection Engines," SAE Technical Paper 2002-01-0346 (2002), doi:https://doi.org/10.4271/2002-01-0346.

7.12. Hara, S., Kumagai, K., and Matsumoto, Y., "Application of a Valve Lift and Timing Control System to an Automotive Engine," SAE Technical Paper 890681 (1989), doi:https://doi.org/10.4271/890681.

7.13. Theobald, M., Lequesne, B., and Henry, R., "Control of Engine Load via Electromagnetic Valve Actuators," SAE Technical Paper 940816 (1994), doi:https://doi.org/10.4271/940816.

7.14. Turner, C.W., Babbitt, G.R., Balton, C.S., Raimao, M.A. et al., "Design and Control of a Two-Stage Electro-Hydraulic Valve Actuation System," SAE Technical Paper 2004-01-1265 (2004), doi:https://doi.org/10.4271/2004-01-1265.

7.15. Trajkovic, S., Milosavljevic, A., Tunestål, P., and Johansson, B., "FPGA Controlled Pneumatic Variable Valve Actuation," SAE Technical Paper 2006-01-0041 (2006), doi:https://doi.org/10.4271/2006-01-0041.

7.16. Haskell, W. and Bame, J., "Engine Knock - An End-Gas Explosion," SAE Technical Paper 650506 (1965), doi:https://doi.org/10.4271/650506.

7.17. Taylor, C.F., *The Internal Combustion Engine in Theory and Practice, Volume 2: Combustion, Fuels, Materials, Design* (Cambridge, MA: The MIT Press, 1968).

7.18. Abraham, J., Williams, F.A., and Bracco, F.V., "A Discussion of Turbulent Flame Structure in Premixed Charges," SAE Technical Paper 850345 (1985), doi:https://doi.org/10.4271/850345.

7.19. Patterson, D., "Cylinder Pressure Variations, A Fundamental Combustion Problem," SAE Technical Paper 660129 (1966), doi:https://doi.org/10.4271/660129.

7.20. Lyon, D., "Knock and Cyclic Dispersion in a Spark Ignition Engine," in *Institution of Mechanical Engineers Conference on Petroleum Based Fuels and Automotive Applications*, Paper C307/86, London, UK, November 1986.

7.21. Kettering, C.F., "More Efficient Utilization of Fuel," *The Journal of the Society of Automotive Engineers* 4 (1919): 263-269.

7.22. Ricardo, H., "Recent Research Work on the Internal-Combustion Engine," SAE Technical Paper 220001 (1922), doi:https://doi.org/10.4271/220001.

7.23. Livengood, J.C., Taylor, C.F., and Wu, P.C., "Measurement of Gas Temperature in an Engine by the Velocity of Sound Method," SAE Technical Paper 580064 (1958), doi:https://doi.org/10.4271/580064.

7.24. Scholl, D., Russ, S., and Stockhausen, W., "Detection of Spark Knock Oscillations: Dependence on Combustion Temperature," SAE Technical Paper 970038 (1997), doi:https://doi.org/10.4271/970038.

7.25. Borg, J.M., Saikalis, G., Oho, S., and Cheok, K.C., "Knock Signal Analysis Using the Discrete Wavelet Transform," SAE Technical Paper 2006-01-0226 (2006), doi:https://doi.org/10.4271/2006-01-0226.

7.26. Addicott, J.L., "First Paper: Some Considerations on Detecting and Preventing Spark Knock in the Car Engine," *Proceedings of the Institution of Mechanical Engineers: Automobile Division* 182, no. 1 (1967): 71-83.

7.27. Sezzi, F., Cornetti, G., Arrigoni, V., Vicenzetto, F. et al., "Possible Mechanisms of Piston Failure due to Detonation and Preignition," Paper Presented at in *Assoziazione Tecnica Dell'Automobile Congress*, Turin, June 1969.

7.28.   Betts, W.E. "Knock and Engine Damage," in *International Symposium on the Performance and Evaluation of Automotive Fuels and Lubricants*, Paper EF/2/2 CEC, Rome, Italy, June 1981.

7.29.   Leppard, W., "The Chemical Origin of Fuel Octane Sensitivity," SAE Technical Paper 902137 (1990), doi:https://doi.org/10.4271/902137.

7.30.   Westbrook, C., Pitz, W., and Leppard, W., "The Autoignition Chemistry of Paraffinic Fuels and Pro-Knock and Anti-Knock Additives: A Detailed Chemical Kinetic Study," SAE Technical Paper 912314 (1991), doi:https://doi.org/10.4271/912314.

7.31.   Konig, G. and Sheppard, C., "End Gas Autoignition and Knock in a Spark Ignition Engine," SAE Technical Paper 902135 (1990), doi:https://doi.org/10.4271/902135.

7.32.   Leppard, W., "The Autoignition Chemistries of Primary Reference Fuels, Olefin/Paraffin Binary Mixtures, and Non-Linear Octane Blending," SAE Technical Paper 922325 (1992), doi:https://doi.org/10.4271/922325.

7.33.   Xiaofeng, G., Stone, R., Hudson, C., and Bradbury, I., "The Detection and Quantification of Knock in Spark Ignition Engines," SAE Technical Paper 932759 (1993), doi:https://doi.org/10.4271/932759.

7.34.   Cummings, H., "Methods of Measuring the Antiknock Value of Fuels," SAE Technical Paper 270003 (1927), doi:https://doi.org/10.4271/270003.

7.35.   Edgar, G., "Measurement of Knock Characteristics of Gasoline in Terms of a Standard Fuel," *Industrial and Engineering Chemistry* 19, no. 1 (1927): 145-146.

7.36.   ASTM International, "Standard Test Method for Research Octane Number of Spark-Ignition Engine Fuel," ASTM D2699-23, ASTM International, 2023.

7.37.   ASTM International, "Standard Test Method for Motor Octane Number of Spark-Ignition Engine Fuel," ASTM D2700-23, ASTM International, 2023.

7.38.   Veal, C.B., Best, H.W., Campbell, J.M., and Holaday, W.M., "Antiknock Research Coordinates Laboratory and Road Tests," SAE Technical Paper 330015 (1933), doi:https://doi.org/10.4271/330015.

7.39.   ASTM International, "Standard Specification for Automotive Spark-Ignition Engine Fuel," ASTM D4814-22, ASTM International, 2022.

7.40.   Bell, A., "The Relationship between Octane Quality and Octane Requirement," SAE Technical Paper 750935 (1975), doi:https://doi.org/10.4271/750935.

7.41.   ASTM International, "Method of Test for Knock Characteristics of Motor Fuels by the Distribution Octane Number (DON)," ASTM D2886-86, ASTM International, 1986 (Method Withdrawn 1989).

7.42.   Protić-Lovasić, G., Jambrec, N., Deur-Siftar, D., and Prostenik, M.V., "Determination of Catalytic Reformed Gasoline Octane Number by High Resolution Gas Chromatography," *Fuel* 69, no. 4 (1990): 525-528.

7.43.   Ghosh, P., Hickey, K.J., and Jaffe, S.B., "Development of a Detailed Gasoline Composition-Based Octane Model," *Industrial & Engineering Chemistry Research* 45, no. 1 (2006): 337-345.

7.44.   Finch, P., "Near-Infrared Online Analysis of Octane Number Testing," *Measurement and Control* 27, no. 4 (1994): 115-118.

7.45.   Fodor, G.E., Kohl, K.B., and Mason, R.L., "Analysis of Gasolines by FT-IR Spectroscopy," *Analytical Chemistry* 68, no. 1 (1996): 23-30.

7.46.   Özdemir, D., "Determination of Octane Number of Gasoline Using Near Infrared Spectroscopy and Genetic Multivariate Calibration Methods," *Petroleum Science and Technology* 23, no. 9-10 (2005): 1139-1152.

7.47.   Brudzewski, K., Kesik, A., Kołodziejczyk, K., Zborowska, U. et al., "Gasoline Quality Prediction Using Gas Chromatography and FTIR Spectroscopy: An Artificial Intelligence Approach," *Fuel* 85, no. 4 (2006): 553-558.

7.48.   Kalghatgi, G.T., "Fuel Anti-Knock Quality–Part I. Engine Studies," SAE Technical Paper 2001-01-3584 (2001), doi:https://doi.org/10.4271/2001-01-3584.

7.49.   Kalghatgi, G.T., "Fuel Anti-Knock Quality–Part II. Vehicle Studies–How Relevant is Motor Octane Number (MON) in Modern Engines?" SAE Technical Paper 2001-01-3585 (2001).

7.50. Mittal, V. and Heywood, J., "The Relevance of Fuel RON and MON to Knock Onset in Modern SI Engines," SAE Technical Paper 2008-01-2414 (2008), doi:https://doi.org/10.4271/2008-01-2414.

7.51. Amer, A., Babiker, H., Chang, J., Kalghatgi, G. et al., "Fuel Effects on Knock in a Highly Boosted Direct Injection Spark Ignition Engine," *SAE Int. J. Fuels Lubr.* 5, no. 3 (2012): 1048-1065, doi:https://doi.org/10.4271/2012-01-1634.

7.52. Orlebar, C., Joedicke, A., and Studzinski, W., "The Effects of Octane, Sensitivity and K on the Performance and Fuel Economy of a Direct Injection Spark Ignition Vehicle," SAE Technical Paper 2014-01-1216 (2014), doi:https://doi.org/10.4271/2014-01-1216.

7.53. Prakash, A., Cracknell, R., Natarajan, V., Doyle, D. et al., "Understanding the Octane Appetite of Modern Vehicles," *SAE Int. J. Fuels Lubr.* 9, no. 2 (2016): 345-357, doi:https://doi.org/10.4271/2016-01-0834.

7.54. Bradley, D. and Morley, C., "Autoignition in Spark Ignition Engines," in Pilling, M.J. (ed.), *Low-Temperature Combustion and Autoignition*, Vol. 35 (Amsterdam, the Netherlands: Elsevier, 1997), Chapter 7, 661-760.

7.55. Whitcombe, R.M., *Non-lead Antiknock Agents for Motor Fuels* (Park Ridge, NJ: Noyes Data Corporation, 1975).

7.56. Russell, T.J., "Motor Gasoline Antiknock Additives," Associated Octel Company Report 87/10, 1987.

7.57. Bailie, J.D., Michalski, G.W., and Unzelman, G.H., "The MMT Outlook, 1977," in *API Refining Department 42nd Midyear Meeting, Product Quality and Conservation*, Chicago, IL, 1977.

7.58. Lenane, D., "Effect of a Fuel Additive on Emission Control Systems," SAE Technical Paper 902097 (1990), doi:https://doi.org/10.4271/902097.

7.59. Otto, K. and Sulak, R.L., "Effects of Mn Deposits from MMT on Automobile Catalysts in the Absence and Presence of Other Fuel Additives," *Environmental Science and Technology* 12 (1978): 181-184.

7.60. Hurley, R.G., Hansen, L.A., Guttridge, D.L., Gandhi, H.S. et al., "The Effect on Emissions and Emission Component Durability by the Fuel Additive Methylcyclopentadienyl Manganese Tricarbonyl (MMT)," SAE Technical Paper 912437 (1991), doi:https://doi.org/10.4271/912437.

7.61. Hammerle, R.H., Korniski, T.J., Weir, J.E., Chladek, E. et al., "Particulate Emissions from Current Model Vehicles Using Gasoline with Methylcyclopentadienyl Manganese Tricarbonyl," SAE Technical Paper 912436 (1991), doi:https://doi.org/10.4271/912436.

7.62. Lenane, D.L., Fort, B.F., Ter Haar, G.L., Lynam, D.R. et al., "Emission Results from a 48-Car Test Evaluation of MMT Performance Additive," *Science for the Total Environment* 146, no. 147 (1994): 245-251.

7.63. Roos, J.W., Scull, H.M., Dykes, K.L., Hotchkiss, A.R. et al., "Evaluation of On-Board Diagnostic Systems and the Impact of Gasoline Containing MMT," SAE Technical Paper 972849 (1997), doi:https://doi.org/10.4271/972849.

7.64. Williamson, W.B., Gandhi, H.S., and Weaver, E.E., "Effects of Fuel Additive MMT on Contaminant Retention and Catalyst Performance," SAE Technical Paper 821193 (1982), doi:https://doi.org/10.4271/821193.

7.65. Aradi, A.A., Roos, J.W., Fort, B.F., Lee, T.E. et al., "The Physical and Chemical Effect of Manganese Oxides on Automobile Catalytic Converters," SAE Technical Paper 940747 (1994), doi:https://doi.org/10.4271/940747.

7.66. Roos, J.W., Lenane, D.L., Dykes, K.L., Scull, H.M. et al., "A Systems Approach to Improved Exhaust Catalyst Durability: The Role of the MMT Fuel Additive," SAE Technical Paper 2000-01-1880 (2000), doi:https://doi.org/10.4271/2000-01-1880.

7.67. Hubbard, C.P., Hepburn, J.S., and Gandhi, H.S., "The Effect of MMT on the OBD-II Catalyst Efficiency Monitor," SAE Technical Paper 932855 (1993), doi:https://doi.org/10.4271/932855.

7.68. Hurley, R.G., Hansen, L.A., Guttridge, D.L., Hammerle, R.H. et al., "The Effect of Mileage on Emissions and Emission Component Durability by the Fuel Additive Methylcyclopentadiencyl Manganese Tricarbonyl (MMT)," SAE Technical Paper 920730 (1992), doi:https://doi.org/10.4271/920730.

7.69. Benson, J.D. and Dana, G., "The Impact of MMT Gasoline Additive on Exhaust Emissions and Fuel Economy of Low Emission Vehicles (LEV)," SAE Technical Paper 2002-01-2894 (2002), doi:https://doi.org/10.4271/2002-01-2894.

7.70. Shimizu, C. and Ohtaka, Y., "Parametric Analysis of Catalytic Converter Plugging Caused by Manganese-Based Gasoline Additives," SAE Technical Paper 2007-01-1070 (2007), doi:https://doi.org/10.4271/2007-01-1070.

7.71. **Hoekman, S.** and Broch, A., "MMT Effects on Gasoline Vehicles: A Literature Review," *SAE Int. J. Fuels Lubr.* 9, no. 1 (2016): 322-343, doi:https://doi.org/10.4271/2016-01-9073.

7.72. **Zhang, M.,** Ning, T., Sun, P., Zhang, D. et al., "Poisoning Mechanisms of Mn-Containing Additives on the Performance of TiO_2 Based Lambda Oxygen Sensor," *Sensors and Actuators B: Chemical* 267 (2018): 565-569.

7.73. **Nakajo, T.** and Matsuura, K., "Effect of Properties and Additives of Gasoline on Low-Speed Pre-Ignition in Turbocharged Engines," SAE Technical Paper 2022-01-1077 (2022), doi:https://doi.org/10.4271/2022-01-1077.

7.74. **Toma, S.,** Elecko, P., Salisova, M., and Vesely, V., "Ferrocene Derivatives as Gasoline Additives," *Acta Fac. Rerum Nat. Univ. Comenianae, Form. Prot. Nat.* 7 (1981): 18792.

7.75. **Wilke, G.,** "Veba Develops Octane Improver," *Eur. Chem. News* 46, no. 1229 (1986): 32.

7.76. **Schug, K.P.,** Guttmann, H.-J., Preuss, A.W., and Schädlich, K., "Effects of Ferrocene as a Gasoline Additive on Exhaust Emissions and Fuel Consumption of Catalyst Equipped Vehicles," SAE Technical Paper 900154 (1990), doi:https://doi.org/10.4271/900154.

7.77. **Brown, J.E.,** Markley, F.X., and Shapiro, H., "Mechanism of Aromatic Amine Antiknock Action," *Industrial and Engineering Chemistry* 47, no. 10 (1965): 2140.

7.78. **Mackinven, R.,** "A Search for an Ashless Replacement for Lead in Gasoline," Jahrestagung, DGMK, West Germany, 1974.

7.79. **Chapman, E.,** Geng, P., Konzack, A., Heppes, S. et al., "Update on Gasoline Fuel Property and Gasoline Additives Impacts on Stochastic Preignition with Review of Global Market Gasoline Quality," SAE Technical Paper 2022-01-1071 (2022), doi:https://doi.org/10.4271/2022-01-1071.

7.80. **Histon, P.D.** and Roles, R.T., "The Road Antiknock and Preignition Characteristics of Gasoline Containing Oxygenates," in *5th Alcohol Fuel Symposium*, Auckland, New Zealand, 1982.

7.81. **Campbell, K.** and Russell, T.J., "The Effect on Gasoline Quality of Adding Oxygenates," Associated Octel Publication OP82/1, April 1982.

7.82. **Coordinating** Research Council, "Fuel Rating Program—Road Octane Performance of Oxygenates in 1982 Model Cars," CRC Report No. 541, July 1985.

7.83. **Brinkman, N.D.,** Gallapoulos, N.E., and Jackson, M.W., "Exhaust Emissions, Fuel Economy and Drivability of Vehicles Fueled with Alcohol-Gasoline Blends," SAE Technical Paper 750120 (1977), doi:https://doi.org/10.4271/750120.

7.84. **Tanaguch, B.** and Johnson, R.T., "MTBE for Octane Improvement," *Chemtech* 9, no. 8 (1979): 502-510.

7.85. **Lang, G.J.** and Palmer, F.H., "The Use of Oxygenates in Motor Gasolines," in Owen, K. (ed), *Gasoline and Diesel Fuel Additives* (New York: John Wiley & Sons, 1989), 147-152.

7.86. **Betts, W.E.,** "Improved Fuel Economy by Better Utilization of Available Octane Quality," SAE Technical Paper 790940 (1979), doi:https://doi.org/10.4271/790940.

7.87. **CRC,** "1983 CRC Octane Number Requirement Survey," Attachment 2, CRC Report No. 539, 1983.

7.88. **Cooperative** Octane Requirement Committee, "Summary of the Technique for Determining Road Octane Requirements under Accelerating Conditions," 1977.

7.89. **Dues, S.M.,** Adams, J.M., and Shinkle, G.A., "Combustion Knock Sensing: Sensor Selection and Application Issues," SAE Technical Paper 900488 (1990), doi:https://doi.org/10.4271/900488.

7.90. **Decker, H.** and Gruber, H.-U., "Knock Control of Gasoline Engines—A Comparison of Solutions and Tendencies, with Special Reference to Future European Emission Legislation," SAE Technical Paper 850298 (1985), doi:https://doi.org/10.4271/850298.

7.91. **McNally, M.J.,** Callison, J.C., Evans, B., Graham, J.P. et al., "The Effects of Gasoline Octane Quality on Vehicle Acceleration Performance—A CRC Study," SAE Technical Paper 912394 (1991), doi:https://doi.org/10.4271/912394.

7.92. **CRC Report,** "A 1986 CRC Octane Number Requirement Survey," 1987.

7.93. **The Subcommittee** of Octane Number Requirement Survey, Gasoline Section, Products Division of the Japanese Petroleum Institute, "Octane Number Requirement Survey of 1985 Japanese Passenger Cars," *Sekiyu Gakkaishi* 29, no. 6 (1986).

7.94. **Brinegar, C.S.** and Miller, R.R., "Statistical Estimation of the Gasoline Octane Number Requirement of New Model Automobiles," *Technometrics* 2, no. 1 (1960).

7.95. Ingamells, J.C. and Jones, E.R., "Developing Road Octane Correlations from Octane Requirement Surveys," SAE Technical Paper 810492 (1981), doi:https://doi.org/10.4271/810492.

7.96. Corner, E.S., Hochhauser, A.M., and Shannon, H.F., "Technical versus Customer Knock Satisfaction—Two Decades," SAE Technical Paper 780322 (1978), doi:https://doi.org/10.4271/780322.

7.97. Jones, E.R. and Ingamells, J.C., "Predicting Customer Octane Satisfaction," SAE Technical Paper 810493 (1981), doi:https://doi.org/10.4271/810493.

7.98. Bettoney, W.E., Rogers, J.D., Benson, J.D., Keller, B.D. et al., "Knock Perception—A 1975 Customer/Rater Study by CRC," SAE Technical Paper 780321 (1978), doi:https://doi.org/10.4271/780321.

7.99. Becker, R.F. and Taylor, M.A., "The Results of the European Interindustry Study on Driver Reaction to Knock," in *International Symposium on Knocking of Internal Combustion Engines*, Wolfsburg, Germany, 1981.

7.100. Bettoney, W.E., Rogers, J.D., Stanke, G.W., and Taniguchi, B.Y., "Customer versus Rater Octane Number Requirements—A 1978 CRC Study," SAE Technical Paper 801355 (1980), doi:https://doi.org/10.4271/801355.

7.101. Uihlein, J.P., Biller, W.F., Bonés, C.J., Bouffard, R.A. et al., "CRC Customer Versus Rater Octane Number Requirement Program," SAE Technical Paper 932673 (1993), doi:https://doi.org/10.4271/932673.

7.102. Rodriguez, R.N. and Taniguchi, B.Y., "A New Statistical Model for Predicting Customer Octane Satisfaction Using Trained-Rater Observations," SAE Technical Paper 801356 (1980), doi:https://doi.org/10.4271/801356.

7.103. Coordinating Research Council, "Road Rating Techniques," CRC-259, 1/51, 1951.

7.104. Betts, W.E., Gozzelino, R., Poullot B., and Williams, D., "Knock and Engine Trends," in *CEC Second International Symposium on the Performance of Automotive Fuels and Lubricants*, Paper EF2, Wolfsburg, Germany, June 1985.

7.105. Keller, B.D., Steury, J.H., and Wagner, T.O., "Seasonal Octane Specifications," SAE Technical Paper 780668 (1978), doi:https://doi.org/10.4271/780668.

7.106. ASTM International, "Standard Specification for Automotive Gasoline," ASTM D439-85a, ASTM International, 1985 (Withdrawn 1990).

7.107. Callison, J.C., "Octane Number Requirements of Vehicles at High Altitude," SAE Technical Paper 872160 (1987), doi:https://doi.org/10.4271/872160.

7.108. Callison, J.C., Wusz, T., and Biller, W.F., "Coordinating Research Council Trends in Octane Number Requirement Increase," SAE Technical Paper 892036 (1989), doi:https://doi.org/10.4271/892036.

7.109. Benson, J.D., "Some Factors Which Affect Octane Requirement Increase," SAE Technical Paper 750933 (1975), doi:https://doi.org/10.4271/750933.

7.110. Cooperative Octane Requirement Committee, "The Effect of Removing Lead Alkyl Antiknock from European Motor Gasolines," October 1972.

7.111. Forster E.J. and Stinson, L.E., "Effect of Leaded versus Unleaded Fuels on Stabilized Octane Requirements," in *National Fuel and Lubricants Meeting of the National Petroleum Refiners Association*, New York, 1970.

7.112. Niles, H.T., McConnell, R.J., Roberts, M.A., and Saillant, R., "Establishment of ORI Characteristics as a Function of Selected Fuels and Engine Families," SAE Technical Paper 750451 (1975), doi:https://doi.org/10.4271/750451.

7.113. CRC Report, "Octane Requirement Increase of 1982 Model Cars," Report No. 540, September 1984.

7.114. CRC Report, "Octane Requirement Increase of 1984 Model Cars," Report No. 549, October 1986.

7.115. Alquist, H.E., Holman, G.E., and Wimmer, D.B., "Some Observations of Factors Affecting ORI," SAE Technical Paper 750932 (1975), doi:https://doi.org/10.4271/750932.

7.116. Graiff, L.B., "Some New Aspects of Deposit Effects on Engine Octane Requirement Increase and Fuel Economy," SAE Technical Paper 790938 (1979), doi:https://doi.org/10.4271/790938.

7.117. Ebert, L.B. (Eds), *Chemistry of Engine Combustion Chamber Deposits* (New York: Plenum Press, 1985).

7.118. Megnin, M.K. and Furman, J.B., "Gasoline Effects on Octane Requirement Increase and Combustion Chamber Deposits," SAE Technical Paper 922258 (1992), doi:https://doi.org/10.4271/922258.

7.119.   **Choate, P.J.** and Edwards, J.C., "Relationship between Combustion Chamber Deposits, Fuel Composition, and Combustion Chamber Deposit Structure," SAE Technical Paper 932812 (1993), doi:https://doi.org/10.4271/932812.

7.120.   **Barber, P.A.,** Lonstrup, T.F., and Tunkel, N., "The Role of Lubricant Additives in Controlling Abnormal Combustion (ORI)," SAE Technical Paper 750449 (1975), doi:https://doi.org/10.4271/750449.

7.121.   **Schreyer, P.,** Starke, K.W., Thomas, J., and Crema, S., "Effect of Multifunctional Fuel Additives on Octane Number Requirement of Internal Combustion Engines," SAE Technical Paper 932813 (1993), doi:https://doi.org/10.4271/932813.

7.122.   **Winch, R.F.** and Mayes, F.M., "A Method for Identifying Preignition," SAE Technical Paper 530246 (1953), doi:https://doi.org/10.4271/530246.

7.123.   **Perry, R.H.** and Lowther, H.V., "Knock-Knock: Spark Knock, Wild Ping, or Rumble?" SAE Technical Paper 590019 (1959), doi:https://doi.org/10.4271/590019.

7.124.   **Pless, L.G.,** "Surface Ignition and Rumble in Engines: A Literature Review," SAE Technical Paper 650391 (1965), doi:https://doi.org/10.4271/650391.

7.125.   **Cheng, S.S.** and Kim, C., "Effect of Engine Operating Parameters on Combustion Chamber Deposits," SAE Technical Paper 902108 (1990), doi:https://doi.org/10.4271/902108.

7.126.   **Dahnz, C.,** Han, K.-M., Spicher, U., Magar, M. et al., "Investigations on Pre-Ignition in Highly Supercharged SI Engines," *SAE Int. J. Engines* 3, no. 1 (2010): 214-224, doi:https://doi.org/10.4271/2010-01-0355.

7.127.   **Zahdeh, A.,** Rothenberger, P., Nguyen, W., Anbarasu, M. et al., "Fundamental Approach to Investigate Pre-Ignition in Boosted SI Engines," *SAE Int. J. Engines* 4, no. 1 (2011): 246-273, doi:https://doi.org/10.4271/2011-01-0340.

7.128.   **Amann, M.,** Mehta, D., and Alger, T., "Engine Operating Condition and Gasoline Fuel Composition Effects on Low-Speed Pre-Ignition in High-Performance Spark Ignited Gasoline Engines," *SAE Int. J. Engines* 4, no. 1 (2011): 274-285, doi:https://doi.org/10.4271/2011-01-0342.

7.129.   **Amann, M.,** Alger, T., Westmoreland, B., and Rothmaier, A., "The Effects of Piston Crevices and Injection Strategy on Low-Speed Pre-Ignition in Boosted SI Engines," *SAE Int. J. Engines* 5, no. 3 (2012): 1216-1228, doi:https://doi.org/10.4271/2012-01-1148.

7.130.   **Kalghatgi, G.T.** and Bradley, D., "Pre-Ignition and 'Super-Knock' in Turbo-Charged Spark-Ignition Engines," *International Journal of Engine Research* 13, no. 4 (2012): 399-414.

7.131.   **Spicher, U.** and Palaveev, S., "Pre-Ignition and Knocking Combustion in Spark Ignition Engines with Direct Injection," JSAE Technical Paper 20105390, 2010.

7.132.   **Guibet, J.C.** and Duval, A., "New Aspects of Preignition in European Automotive Engines," SAE Technical Paper 720114 (1972), doi:https://doi.org/10.4271/720114.

7.133.   **Arrigoni, V.,** Cornetti, G., Gaetani, B. and Ghezzi, P. "Quantitative systems for measuring knock." Proceedings of the Institution of Mechanical Engineers, 186(1), pp.575-583, 1972.

7.134.   **Sabina, J.R.,** Mikita, J.J., and Cambell, M.H., "Pre-ignition in Automotive Engines," Presented at A.P.I. Division of Refining, Mid-Year Meeting, New York, May 1953.

7.135.   **Marciante, A.** and Chiampo, P., "The Influence of Lubricating Oil Ash on Surface Ignition Phenomena," SAE Technical Paper 700458 (1970), doi:https://doi.org/10.4271/700458.

7.136.   **Downs, D.** and Pigneguy, J.H., "An Experimental Investigation into Preignition in the Spark Ignition Engine," *Proceedings of the Institution of Mechanical Engineers: Automobile Division* 4, no. 1: 125-149.

7.137.   **Downs, D.** and Theobald, F.B., "The Effect of Fuel Characteristics and Engine Operating Conditions on Preignition," *Proceedings of the Institution of Mechanical Engineers: Automobile Division* 178, no. 1 (1963): 89-108.

7.138.   **Massa, V.F.,** "A Study of the Normal and Abnormal Combustion Behavior of Gasolines," SAE Technical Paper 610403 (1961), doi:https://doi.org/10.4271/610403.

7.139.   **Sparrow, S.W.,** "Preignition and Spark-Plugs," SAE Technical Paper 200015 (1920), doi:https://doi.org/10.4271/200015.

7.140.   **Ward, R.M.,** "Spark Plugs," SAE Technical Paper 460016 (1946), doi:https://doi.org/10.4271/460016.

7.141. Igashira, T., Kawai, H., Yoshinaga, T., and Nakamura, N., "Evaluation of a Spark Plug by the Preignition Simulator," SAE Technical Paper 880692 (1988), doi:https://doi.org/10.4271/880692.

7.142. Shimanokami, Y., Ishikawa, M., Matsubara, Y., and Suzuki, T., "Development of a Wide Range Spark Plug," SAE Technical Paper 2006-01-0406 (2006), doi:https://doi.org/10.4271/2006-01-0406.

7.143. Mogi, K., Ohno, E., and Nakamura, N., "Spark Plug Fouling: Behavior and Countermeasure," SAE Technical Paper 922093 (1992), doi:https://doi.org/10.4271/922093.

7.144. Owen, K., "Comments at Champion Ignition and Engine Performance Conference," Paragraphs 111 and 121, Brussels, 1984.

7.145. Affleck, W.S., Bright, P.E., and Ellison, R.J., "Run-On in Gasoline Engines," *Institution of Mechanical Engineers, Automobile Division Proceedings* 183, no. 1 (1968): 21-51.

7.146. Ingamells, J.C., "Effect of Gasoline Octane Quality and Hydrocarbon Composition on After-Run," SAE Technical Paper 790939 (1979), doi:https://doi.org/10.4271/790939.

Further Reading

Corrigan, D.J. and Fontanesi, S., "Knock: A Century of Research," *SAE Int. J. Engines* 15, no. 1 (2022): 57-127, doi:https://doi.org/10.4271/03-15-01-0004.

Kalghatgi, G., *Fuel/Engine Interactions* (Warrendale, PA: SAE International, 2014).

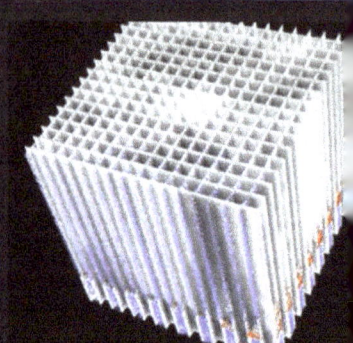

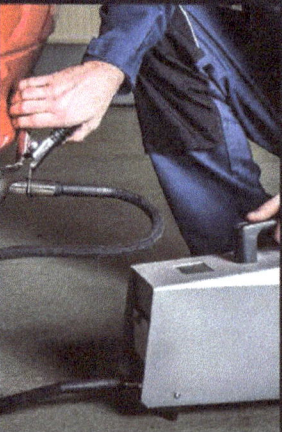

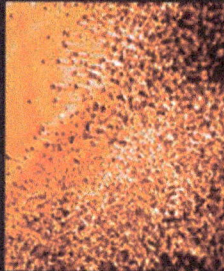

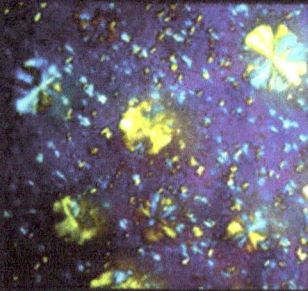

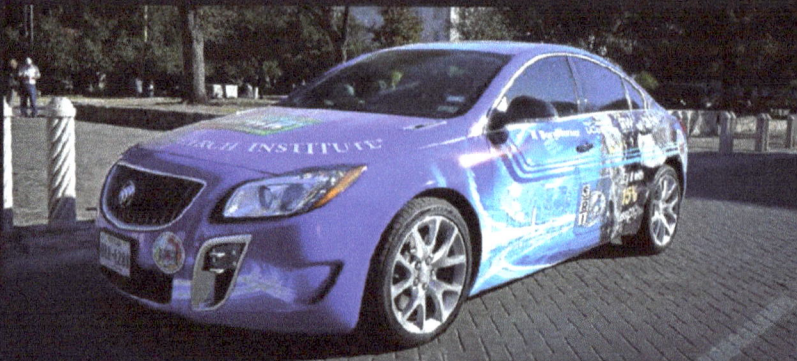

8

Gasoline Engine Design and Influence of Fuel Characteristics

8.1. Introduction

The effect of fuel quality on vehicle performance is relatively small compared with the influence of vehicle design, although fuel quality can have dramatic effects under certain conditions. However, the engine designer must work to try and ensure that the engine design is commensurate with the specification and quality of fuel that will be available to the customer. For example, as discussed in the previous chapter, the octane requirement of an engine will be of significant importance to the customer, as knock not only can be an irritant but also can give rise to catastrophic engine damage. It is therefore pointless to design an engine that will give outstanding performance when run on a 100-research octane number (RON) primary reference fuels (PRF) if the engine management system will limit this to mediocre performance when fueled with the commercially available gasolines, or worse, in the absence of engine management, it could lead to heavy knock and engine damage.

Pressures to improve air quality have led to a whole range of design modifications and innovations by the motor industry, which includes the use of catalysts to minimize the main pollutants from vehicle tailpipes and the use of vapor adsorption systems that prevent the loss of light hydrocarbons to the atmosphere by evaporation. In addition, complex electronic control devices have been developed that can ensure the vehicle is operating as close to optimum or design conditions at all times. The fuel industry, in turn, has made important changes to facilitate some of these advances. These have included removing lead antiknock additives and significantly reducing fuel sulfur content, both of which can poison catalysts; introducing fuel additive packages to minimize deposit formation and thus ensure that critical equipment remains within design specification; and changing gasoline specifications to avoid excessive evaporative losses and to minimize undesirable exhaust emissions.

Vehicle fuel economy has always been a factor in customer choice. The relative importance of this factor varies with the price of fuel. The periodic sharp rises in the price of crude oil that have occurred in the last quarter of the 20th century and into the 21st century draw customer attention to the importance of fuel economy. More recently, concern over global temperatures and the belief that this is due to anthropogenic CO_2 emissions has added to the pressure on the vehicle designer to reduce vehicle fuel consumption. While there are many things that can be done to reduce vehicle fuel consumption, such as reducing weight and improving aerodynamics, there will always be a place in this equation for increasing the efficiency of the engine. However, as noted, the engine and vehicle design must be commensurate with the available fuel. The fuel itself also has a minor role to play in that the base fuel composition can affect the fuel consumption, and the use of fuel additives can in themselves affect fuel economy but they can also help to prevent the formation of performance robbing deposits. Better fuels allow the engine designer to develop better, more efficient engines. Better fuels come at a cost, however, both financial and in terms of energy utilization. The two industries must therefore work together to target optimum use of resources.

From the fuel technologists' point of view, the vehicle consists of a fuel storage area and delivery system, namely the fuel tank and fuel pump, a fuel metering device, the fuel consuming unit (i.e., the engine), and finally an exhaust aftertreatment system to minimize unwanted products that may be emitted from the tailpipe. These systems will be discussed in more detail in the following sections.

8.2. The Gasoline Engine

This is only a brief guide to the positive ignition internal combustion engine, but it highlights those aspects where the interplay of engine design and fuel characteristics are important factors in determining engine and vehicle performance, and hence directing fuel specification.

As noted, in the early development of these engines, gaseous fuels were employed as the energy source. The energy density of liquid fuels favored their use and the gas engine evolved into the gasoline engine. During time periods when the availability of these liquid fuels was curtailed, during World War II and into the 1950s, and during the 1970s, there was a resurgence in the use of gaseous fueled vehicles, and in the 21st century although gaseous fuels are still used in a number of situations, the majority of automotive positive ignition engines utilize liquid fuels, which we have been classifying as gasoline. On-vehicle storage of gaseous fuels and gasoline are obviously going to be different, and the delivery of the fuel to the combustion chamber will also be different, but the basic concept remains the same and will be discussed in more detail below.

In virtually all automotive applications of the positive ignition engine, the ignition source is a spark; in this chapter we will therefore be discussing the spark ignition gasoline engine, unless otherwise stated.

8.2.1. Otto Cycle

In this section, we will consider the most common type of spark ignition engine, using a reciprocating piston and crank assembly to convert the energy released from the fuel into power. The general principle of operation of the four-stroke internal combustion engine (also known as the Otto engine, after its German inventor Nikolaus August Otto) is described briefly. The four strokes are illustrated in Figure 8.1.

FIGURE 8.1   Schematic representation of the four-stroke Otto cycle.

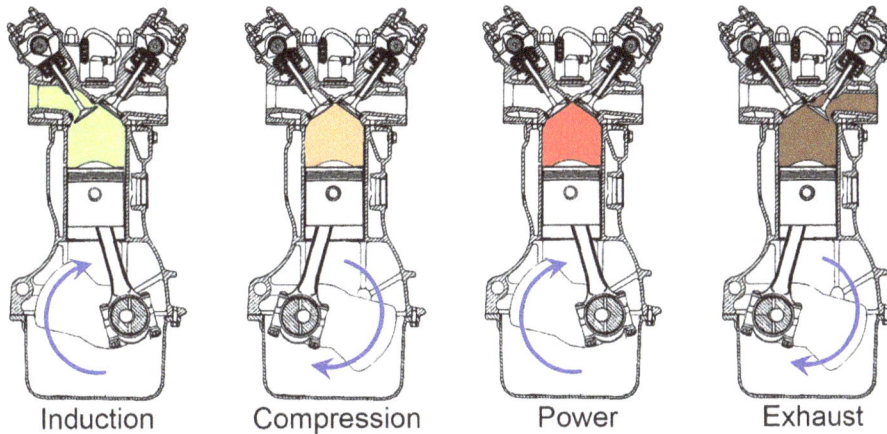

Induction   Compression   Power   Exhaust

The first stroke is the induction stroke: either air or a mixture of air and fuel is induced into the combustion chamber through the open inlet valve, to the left of the cross sections shown in Figure 8.1. The downward motion of the piston ensures a lower pressure within the combustion chamber than in the inlet manifold. Toward the bottom of the piston stroke the inlet valve closes, trapping the incoming air or mixture within the cylinder. The subsequent upward motion of the piston, the second stroke, compresses this air or mixture. If only air is drawn into the cylinder through the inlet valve, then fuel must be injected directly into the cylinder during one of these two strokes. This will be discussed later under direct injection. Around the top of this stroke the spark is initiated, leading to the ignition of the fuel air mixture. Hopefully, as discussed in the previous chapter, the ensuing combustion will progress smoothly, burning all the available fuel without abnormal combustion such as knock. The rapid burning of the mixture causes a rapid rise in temperature and pressure; it is this pressure that forces the piston down and produces power, the power stroke. The momentum and the power from other cylinders then force the piston back up, the exhaust stroke. The exhaust valve will be open at around the time the piston reaches the bottom of the power stroke, and the burned gas will thus be forced out of the cylinder during this exhaust stroke. As the piston reaches the top of the stroke, the exhaust valves will close, the inlet valves will open, and the whole cycle will be repeated.

The ideal situation is that at the start of the induction stroke, the inlet valve will be fully open and the incoming air or mixture completely fills the cylinder. The compression stroke and the power stroke, or expansion stroke, will have the same length of piston travel, that is, the same compression ratio and expansion ratio. The valves will be completely closed throughout both of these strokes. The exhaust valve will be fully open at the beginning of the exhaust stroke and will close at the end of the stroke, just as the inlet valve opens. This is the essence of the Otto cycle. In practice, it takes a finite time for the valves to open and close, and the incoming and exhausted gases have a finite mass and thus inertia. The valves must therefore start to open before the start of the relevant stroke and finish closing after the end of that stroke. Using a simple camshaft arrangement as shown in Figure 8.1 restricts the freedom to alter the timing of the valve events. This has implications for determining the optimum operation of the engine, but such detailed discussion of engine design is beyond the scope of this book. It is worth noting that over the years tremendous advances have been made in this area, such as variable valve timing [8.1–8.5], variable valve lift [8.6, 8.7], and work to achieve completely flexible valve events [8.8–8.11]. Some of the implications of this are discussed in subsequent sections.

The first and the last of the four strokes described can be considered as the air exchange part of the cycle. If this part of the cycle can be condensed into a much smaller part of the whole cycle, then the power stroke becomes a more significant portion of the cycle, and a higher specific power output can be achieved. This is the objective of a two-stroke cycle. The air exchange process is condensed into a fraction of the cycle toward the bottom of the piston travel. Inlet and exhaust valves, or ports, are opened simultaneously and pressurized incoming air or mixture is used to force the residual gases out of the exhaust.

If a porting arrangement is used instead of a poppet valve, then a two-stroke engine can have the advantages of light weight, compactness, lower manufacturing costs, and higher specific power than a comparable four-stroke engine. However, during the gas exchange process that takes place during only the end of the power stroke and the start of the compression stroke, there is ample opportunity for some of the incoming air and fuel mixture to pass straight through to the exhaust. This is clearly not good for economy or emissions; this was a serious limitation until the commercialization of direct injection. In simple low-cost, two-stroke designs, the downward stroke of the piston is used to generate a positive pressure for the incoming air. This has the added drawback of mingling the incoming mixture with the lubrication of the crankshaft. This was usually achieved by mixing lubricant with the fuel. This again had an adverse effect on emissions and could lead to spark plug fouling. These factors have made the commercial use of two-stroke engines for automotive applications virtually obsolete since about 1970.

8.2.2. The Atkinson Cycle

Shortly after Nikolaus Otto had developed his concept of the four-stroke spark ignition engine, English inventor James Atkinson was granted a patent in 1885 for a gas engine that had different compression and expansion ratios [8.12]. His initial patent was for an opposed piston arrangement with inlet and exhaust ports in the cylinder liner and clever mechanical linkage, as illustrated in Figure 8.2, which also shows the operation of the four-stroke cycle.

FIGURE 8.2   Mechanical linkage and method of operation for Atkinson opposed piston engine.

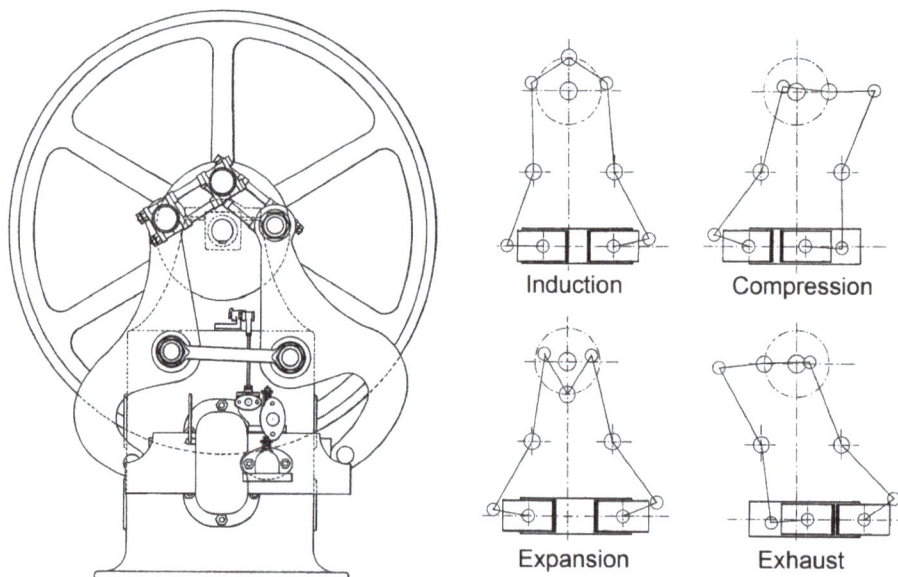

However, it will be noted that these four strokes are achieved within a single revolution of the crankshaft as opposed to the two revolutions of the four-stroke Otto cycle. The following year, Atkinson was granted a patent for a single piston arrangement [8.13]. Similar, but simpler, as there is only one piston, the linkage is shown in Figure 8.3.

FIGURE 8.3 Single piston Atkinson engine.

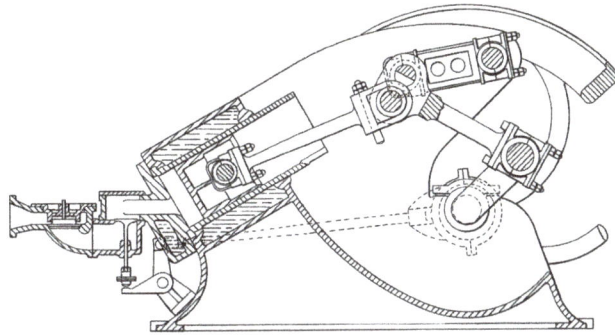

By having a larger expansion stroke than the compression stroke, it is possible to achieve a higher thermal efficiency than with the Otto cycle. However, the disadvantage is that for a given cylinder capacity it is not possible to admit as much mixture due to the shorter compression stroke. The Atkinson engine therefore has a lower potential power density than a similar-sized Otto engine. During the many years of development of the gasoline engine, the lack of achievable power density, the added cost, and bulk of the additional linkage have outweighed the thermal efficiency gains. The engines proposed by Atkinson also suffered from the same part load problem as did the Otto engines, that is, to significantly reduce the load the intake had to be throttled in order to reduce the fuel input. There was thus no fundamental difference in the fuel requirements of the two types of engines for a given compression ratio.

8.2.3. The Miller Cycle

An alternative approach to achieving a different compression and expansion ratio is to use valve timing to regulate the amount of air trapped within the cylinder. This has become known as the Miller cycle, after Ralph Miller, who proposed the idea [8.14]. In 1956, Miller applied for a patent to cover the concept using an addition valve in the cylinder head in order to control the trapped gas [8.15]. More recent efforts to implement the Miller cycle have relied on the ability of the newer variable valve actuation systems to change the timing of the inlet valve closure. Closing the valve early [8.16] means that the cylinder does not have time to fill, whereas closing the valve late means that some of the gas that has entered the cylinder is pushed out again by the rising piston [8.17, 8.18]. An engine can thus be designed such that at full load or wide-open throttle (WOT) the engine operates as a conventional Otto engine, but for lower power outputs, late [8.19] or early [8.20] inlet valve closing is used to regulate the amount of air or mixture entering the cylinder. This not only reduces the power output without introducing the added pumping losses encountered with the use of a throttle, but it also reduces the effective compression ratio. This can reduce the required octane rating of the fuel [8.21], although others have suggested that it can increase knocking tendency at lower engine speeds [8.22]. From a fuel requirement perspective, the other difference is that by eliminating the throttle, this significantly reduces the vacuum that will be generated within the intake manifold at part load conditions. If a fuel metering system other than direct injection is being employed, then this will reduce the volatility of the fuel under these conditions. However, this will be no worse than at the WOT conditions.

8.3. Vehicle Fuel Systems

8.3.1. The Fuel Metering System

As outlined in the previous chapter, the spark ignition engine requires an air-to-fuel ratio in the region of the spark plug that is within the range of about 7:1 and 20:1. For an engine equipped with a three-way catalyst (TWC), the required overall mixture strength should be very close to stoichiometric. Over the years different approaches have been employed to meter the required amount of fuel to the corresponding flow of air. These will now be described briefly, highlighting their relevance to fuel requirements.

8.3.1.1. Carburetors

A pump supplies fuel directly to the carburetor, which then provides a mixture of atomized fuel and air in the required ratio, depending on the engine operating conditions. Figure 8.4 illustrates two basic types of carburetors: the fixed venturi or fixed jet type and the variable venturi or constant depression type.

FIGURE 8.4   Fixed venturi and variable venturi type carburetors.

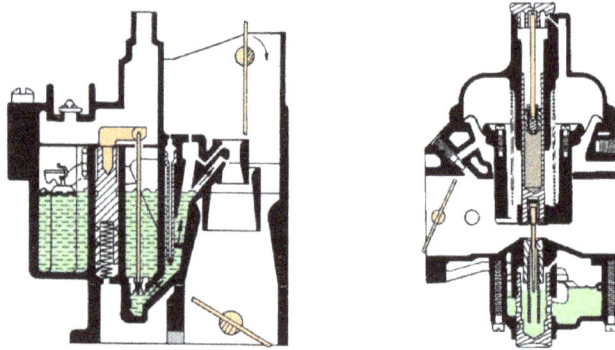

© SAE International.

    As air flows through the venturi, it produces a partial vacuum, which causes the fuel jet to deliver a fine spray of gasoline into the air stream. The throttle or butterfly valve controls the air passage and, hence, the amount of gasoline entering the air stream. The evaporative cooling of the air, by the vaporizing fuel, can lead to ice formation on the throttle valve; this is discussed further in Chapter 9, Section 9.7.

    Between the fuel supply pump and the fuel jet is a float bowl that maintains a constant head of fuel. For open-loop carburetors the float bowl has a vent to atmosphere or to a carbon canister to control evaporative emissions. In an open-loop carburetor, when no feedback is required to maintain a stoichiometric mixture for a TWC, it is possible to produce air-to-fuel ratio variations to suit different driving conditions. For example, lean air-to-fuel ratios at cruise conditions will deliver better fuel economy but slightly rich air-to-fuel ratios are required for maximum power. Very rich mixtures required for cold starting can be achieved by an additional throttle valve or choke upstream of the carburetor. These could be manually operated or controlled by a metallic strip or another system that reacts to ambient or coolant temperature. In an open-loop carburetor, because the fuel is metered solely by the air flow and the size of the orifice through which the fuel flows, there will be variations of air-to-fuel ratio as a result of changes in the fuel's viscosity and density. Carburetor design can be made to accommodate variations due to temperature variation, but they cannot be made to allow for variations in the gasoline specification.

During idle operation the throttle is virtually closed so that very little air can get by the butterfly valve, and it is necessary to have a special system to allow the required amount of mixture at the correct air-to-fuel ratio to get to the combustion chamber. This involves a fairly complex series of passages in the carburetor through which the air-fuel mixture passes until it emerges just upstream of the throttle valve, where it is mixed with the small amount of air getting past to give the air-to-fuel ratio required for smooth idle performance. Deposits from the fuel and from gases and particulates pulsing back from the combustion chamber can build up on the throttle plate and the throttle body and restrict the air going past. This is only of concern at low throttle openings, as occurs during idle, when the mixture going forward may be too rich for smooth running and gives rise to uneven firing and stalling. Some fuel compositions accentuate this problem, but additives are available that can minimize deposit formation; this is discussed in Chapter 11, Section 11.3.3.1. There are many different designs of carburetors that became extremely sophisticated before being supplanted by fuel injection systems. They vary in the extent to which they are subject to deposit formation and in their sensitivity to such deposits in terms of performance.

The use of an inlet manifold with only a single carburetor could lead to significant mal-distribution between different engine cylinders; this was not only bad for fuel efficiency but it meant that not all of the engines cylinders were producing their peak power when required. Where maximum power was a requisite, such as in racing, the use of multiple carburetors was a way around this problem. The attainment of increased specific power and economy was also one of the driving forces in the development of fuel injection [8.23, 8.24].

The widespread adaption of the TWC and the necessity to control the air-to-fuel ratio to stoichiometric proportions necessitated some means of providing closed loop control of these carburetors. Various approaches have been suggested [8.25], including controlling the pressure above the fuel in the float chamber, which effectively allowed control of the fluid head provided to the metering jet, and electronically controlling the pressure above the needle holder in a variable venturi type carburetor [8.26]. An alternative approach was to use vacuum-operated control rods to regulate the fuel flow to the main jet circuit. An example of this is illustrated in Figure 8.5 [8.27]. An alternative to controlling the air-to-fuel ratio leaving the carburetor was to provide a controlled air bleed downstream of the carburetor [8.28].

FIGURE 8.5   Closed loop carburetor mechanism.

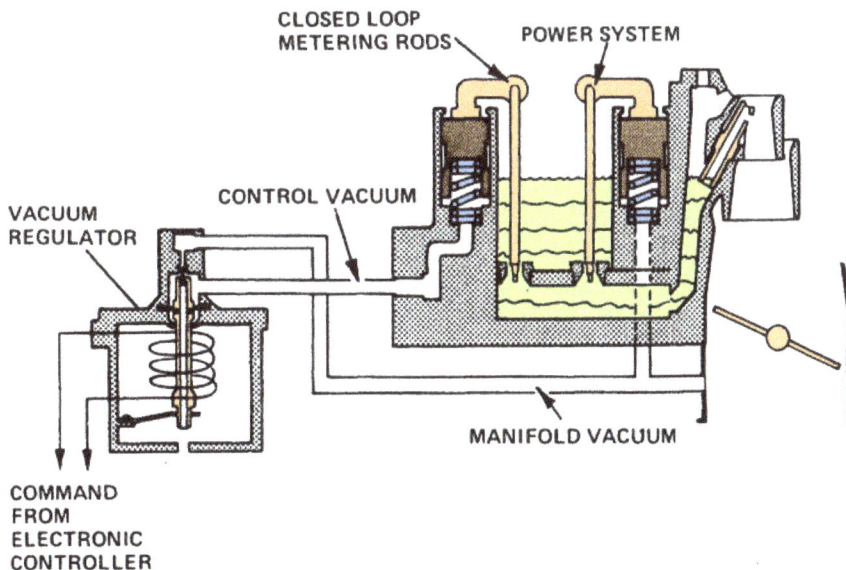

The position of the carburetor is also very important. If it is positioned so that excessive heat soak-back can occur from the engine after it is switched off, then a number of hot-fuel handling problems can occur, as described in Chapter 9, Section 9.4.2. The distance of the carburetor from the engine and the manifold arrangement can also affect the responsiveness of the engine. It is known that some of the fuel will be transported to the engine as a liquid film on the manifold wall [8.29], and if the carburetor is not designed to compensate for this, it can affect the transient performance [8.30, 8.31].

8.3.1.2. Throttle Body Fuel Injection

Carburetors had served the automotive industry well for over a century and had been adapted to allow closed loop control to comply with tightening emissions regulation. Fuel injection had become quite widely used on more exotic engines as a replacement for multiple carburetors where high-power outputs were required. It therefore seemed a logical step to combine the technology of fuel injection with the cost advantages of a single carburetor feeding multiple cylinders. This could be achieved through the use of a throttle body injector (TBI), in effect replacing the carburetors venturi and fuel jet with an electronic fuel injector. Of course, multibarrel carburetors could be replaced with multiinjector TBI assemblies. Figure 8.6 shows a single-injector TBI arrangement [8.32] and a twin injector TBI arrangement [8.33].

FIGURE 8.6 Single-injector (left) and twin injector (right) TBI arrangements.

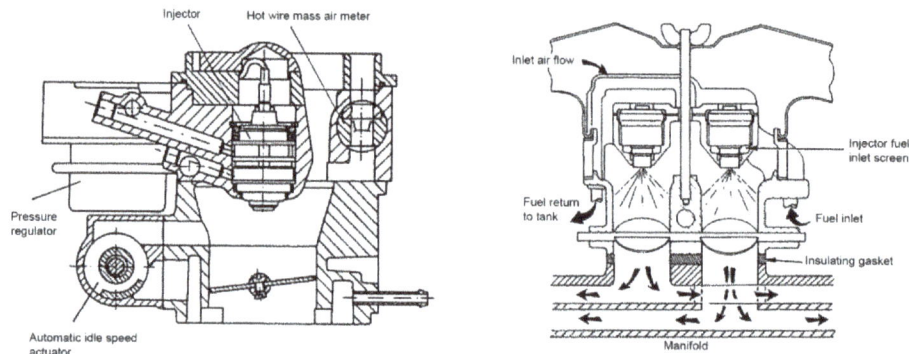

© SAE International.

The TBI approach still retains some of the carburetor characteristics of poor cylinder-to-cylinder matching of air-to-fuel ratio, and the injector is subject to fouling. This later point can be addressed with good quality fuel and additive package. Careful design of the TBI system can minimize the mal-distribution problem to some extent [8.34] but not at all speed load conditions. This favors the use of multipoint injection (MPI), where the injectors are usually placed as close as possible to the inlet port of each cylinder, as discussed in the following section.

8.3.1.3. Port Fuel Injection

By using multiple injectors and positioning them as close to the inlet port of each cylinder, many of the response and distribution issues associated with carburetors, and the TBI systems that were used to replace them, can be overcome. Such port fuel injection (PFI) systems were developed in the middle of the last century [8.24] but only became popular in the final quarter of the century. They provided a means of meeting the demands of increased power output relative to engine size, lower specific fuel consumption, higher torque at low engine speeds, lower exhaust emissions, and improved hot and cold weather driveability performance.

There are two types of injectors—mechanical and electronic. In a mechanical injection system, a valve opened against a spring as a result of fuel pressure. This produced a continuous fuel spray that was usually directed toward the back of the intake valve. As it was unlikely that all of this fuel would be vaporized during the roughly three-quarters of the cycle when the inlet valve was closed, liquid fuel inevitably would accumulate on the back of the intake valve. This could lead to a significant buildup of deposits on the back of the intake valve. The quantity of fuel flowing through the injector is determined by the fuel pressure, which in turn is determined by the air flowing into the inlet engine. Figure 8.7 [8.32] shows a schematic of such a system.

FIGURE 8.7   Bosch K-Jetronic fuel injection system.

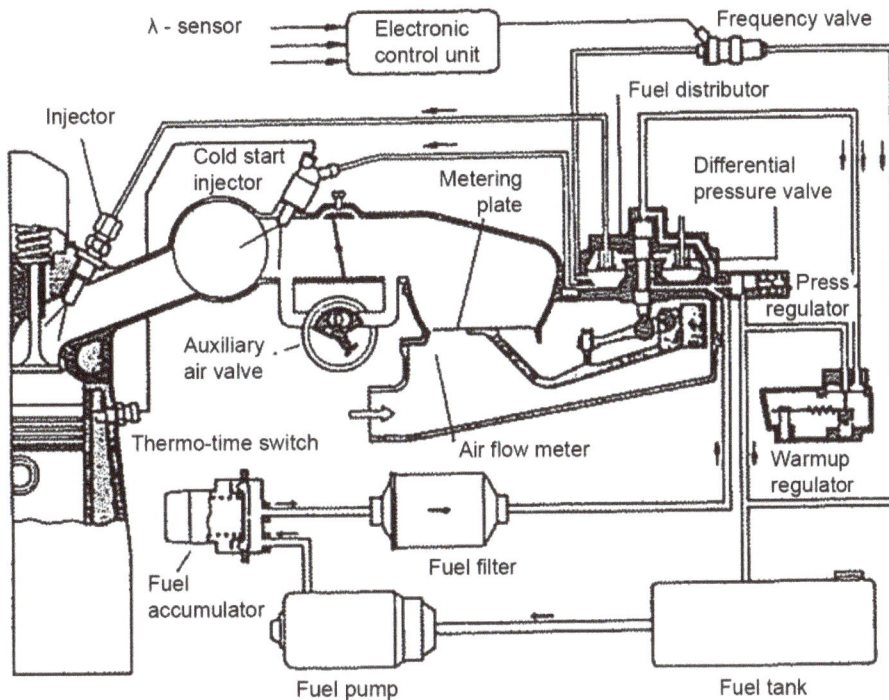

This type of system is fitted to the Mercedes-Benz M102E engine that is still widely used as an intake valve deposit test, but the injector itself is relatively free of problems.

The electronic systems use electromagnetically actuated injectors that spray fuel into the inlet ports of the engine on a pulsed basis. These pulses can be timed to coincide with the engine cycles. The injection pulse can therefore be timed so that it starts as the inlet valve opens and ends when the required amount of fuel has been delivered. As the intake stroke, in a four-stroke engine, only occurs every other revolution of the engine, the control unit must have some means of knowing which revolution this is. A simpler system will fire the injector every revolution of the engine, and some fuel will be "stored" in the manifold until the intake stroke. Good atomization in the spray is vital to ensure low hydrocarbon emissions and good driveability [8.35]. Because injection is not continuous, there can be a tendency for deposits to occur on the pintle due to any residual gasoline present being subjected to heat from the engine. These deposits can affect the spray pattern and upset vehicle driveability. The problem is most acute in stop/start driving conditions, which maximize the opportunities for heat soak-back into the injectors. It has been largely overcome by the use of appropriate additives in the fuel and, in new models, by improved injector design. Figure 8.8 compares the construction of typical mechanical and electronic type injectors.

FIGURE 8.8   A typical mechanical (left) and electronic (right) port fuel injector.

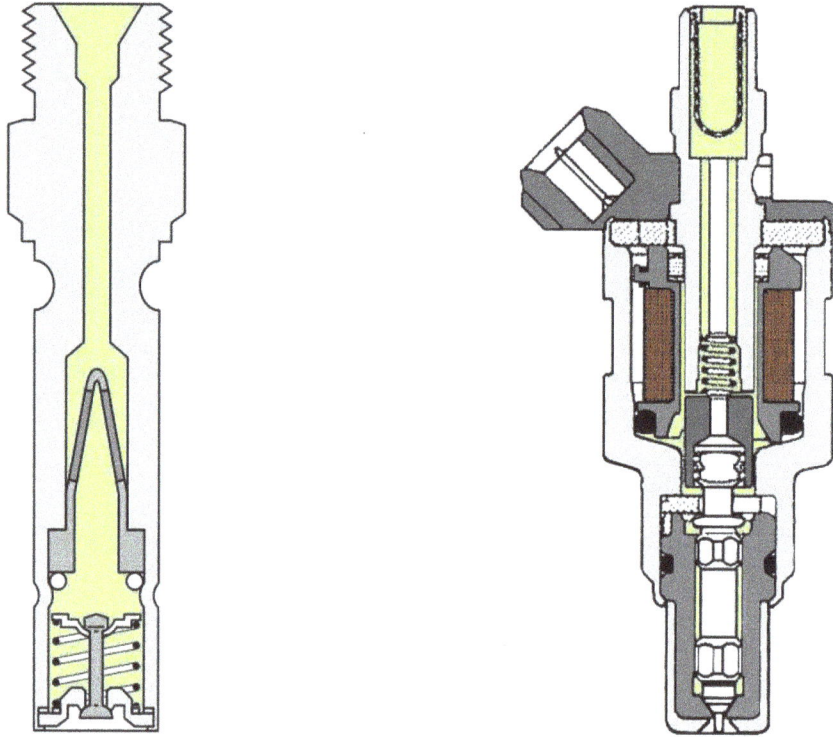

© SAE International.

Three other areas of the fuel system can be susceptible to fuel quality: the intake manifold, the intake ports, and the intake valves. The likelihood of different gasoline qualities forming deposits in these areas is discussed in Chapter 10, Section 10.2.3, but such deposits can cause driveability problems. The intake manifold is probably less sensitive, except that, if a hot spot is used to warm up the charge and help vaporization, particularly on TBI systems, deposits are more likely to form on the hot spot and will increase warm-up time.

Deposits on the intake valves and in the ports will interfere with the breathing of the engine and adversely affect cold starting and acceleration performance. In extreme cases, engine damage can occur as a result of the valve being held off the seat. Such damage is sometimes caused by hot gases escaping past the open valve that can burn away part of the seat, or, in some designs, by the open valve being struck by the rising piston.

8.3.1.4. Direct Fuel Injection

The previous three sections are in effect a chronology of automotive fuel metering systems. The changes have largely been driven by the need to meet legislative emissions limits. When technology has been developed that provides better performance but at a price, it has not generally been applied until it is necessary to meet legislation. For example, multiple carburetors were available on racing cars and top-end vehicles for the general public. This was replaced by PFI systems for top-end vehicles. Legislation drove the adaption of PFI to produce TBI for mass production; this then led to PFI for mass production. This is also true of direct fuel injection.

Direct injection spark ignition (DISI) engines were developed in the second quarter of the 20th century to meet the requirements of the aeronautical industry [8.36] but were also soon implemented in the automotive industry in a very limited number of applications where high specific power was required. They used diesel fuel injection equipment, which was limited in speed and flexibility. As other systems developed, the advantages of these early DISI engines were overtaken. The late 1960s saw a resurgence of interest in DISI engines, particularly for direct injection stratified charge (DISC) engines that held out the promise of far superior fuel economy by dispensing with the wasteful throttling process and injecting the required quantity of fuel directly into an air-filled cylinder. This led to the development of systems such as the MAN-FM [8.37], the Ford programmed combustion control (PROCO) system [8.38–8.40], and the Texaco controlled combustion system (TCCS) [8.41, 8.42]. The widespread adoption of the TWC and the requirement for overall stoichiometry curtailed the application of this technology. The advances in fuel injection equipment and electronic controls have once more made these concepts attractive.

8.3.1.4.1. Direct Injection Strategies. By injecting the fuel during the induction stroke and/or very early in the compression stroke, a relatively homogeneous mixture can be achieved, and the combustion process is as with a manifold metered fuel delivery system. Overall stoichiometry can be maintained, and a TWC will still reduce all the regulated emissions. However, the benefits of more precise fuel delivery are still achieved, and there is often the added advantage that evaporative cooling of the charge is brought about by the evaporation of the fuel spray. This can reduce the end gas temperature and reduce octane requirement. Conversely, for a given octane number fuel, a higher compression ratio can be used that will improve efficiency. If injection occurs during the intake stroke, then the charge cooling can also increase volumetric efficiency, the cooler air is denser, and a greater mass is therefore inducted. As noted, major efficiency and economy benefits can be achieved if the engine can be operated unthrottled, and ignition can still be achieved by stratifying the charge so that a readily ignitable mixture occurs around the spark plug but with an overall lean mixture. This will leave excess oxygen in the exhaust and requires a revised exhaust aftertreatment regime, which will be discussed later in this chapter.

There are many ways to bring about such charge stratification, but they can usually be categorized under the following headings: spray-guided, wall-guided, and air-guided. These approaches are shown schematically in Figure 8.9 [8.43].

FIGURE 8.9   Categories for charge stratification in DISI engines.

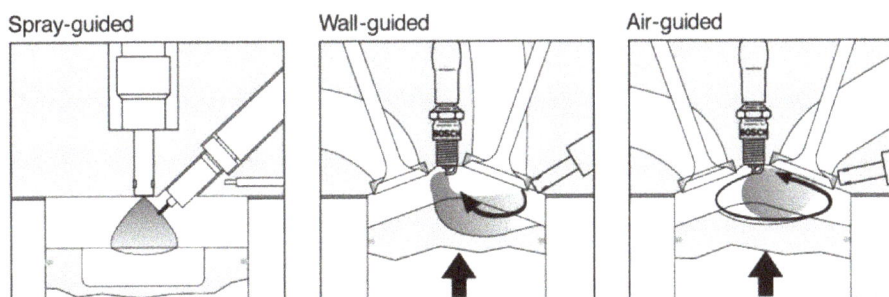

From Figure 8.9 it is clear that in a spray-guided configuration the fuel injector is close to the spark plug (a narrow spacing arrangement). This has implications for the engine designer but also from the fuel technologist perspective. The narrow spacing means that liquid fuel is more likely to impinge on the spark plug; this can lead to spark plug fouling, particularly with fuels containing high boiling components. The positioning of the injector close to the spark plug and hence the center of the flame kernel usually results in a higher injector tip temperature. This will increase the tendency for deposits to form on the injector. This design is also more sensitive to variations in fuel spray that may result from injector deposits. There is also a greater possibility of liquid fuel impinging on the piston crown, leading to deposit formation.

For the wall-guided and spray-guided configurations, the injector is farther from the spark plug, that is, a wide spacing. This arrangement usually provides the engine designer with more flexibility in terms of engine geometry, but the remoteness of the injector implies less stratification as the fuel has more time to mix and evaporate during the transport to the spark plug. From the fuel technologist's perspective, it means a low injector tip temperature and less chance of spark plug fouling. On the other hand, there is a greater chance that liquid film will impinge on the cylinder wall, leading to migration of fuel into the lubricating oil and the accompanying problems.

8.3.1.4.2. Direct Injection Injector Technology. Early applications of direct injection to gasoline engines used injector technology from diesel engine systems. This meant pintle type injectors with a single spray hole or multihole injectors. Lower fuel injection pressures resulted in poor atomization of the fuel and wall or piston impingement. By modifying the geometry of the pintle, the fuel spray could be made into a hollow cone type spray. This hollow cone effect could be enhanced by imparting rotary momentum to the fuel just before it left the nozzle. This is known as a swirl injector [8.44, 8.45]. A single hole can be modified into a slit to produce an almost two-dimensional spray of a fan shape. Figure 8.10 shows photographs of the spray produced by two swirl type injectors, with different degrees of swirl, and a slit type injector [8.46] and a multihole nozzle [8.47].

FIGURE 8.10 Effect of injector design on spray formation.

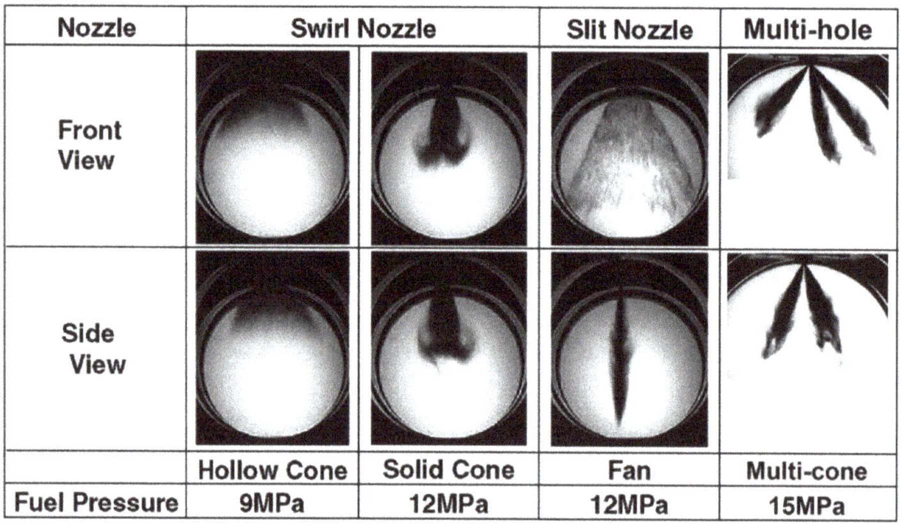

Nozzle	Swirl Nozzle		Slit Nozzle	Multi-hole
Front View				
Side View				
	Hollow Cone	Solid Cone	Fan	Multi-cone
Fuel Pressure	9MPa	12MPa	12MPa	15MPa

© SAE International.

The injector activation technology followed that of the PFI injectors in using a solenoid. For tighter control of smaller quantities of fuel and smaller injection-to-injection variation, piezo injectors have also been developed [8.48, 8.49]. Work on closed loop control of a solenoid inject has also shown similarly low levels of variation [8.50].

The design of the injector tip can obviously be used to help determine the direction of the spray in order to help in accommodating the injector within the cylinder head. Multihole nozzles have been developed, and these produce a number of fuel sprays. The sprays have a much narrower cone angle than the swirl type injectors, but the cone angle is less sensitive to combustion chamber pressure than the swirl type injector [8.51–8.54].

8.3.1.5. Gaseous Fuel Systems

With the desire to move to a carbon free economy there has naturally been a lot of interest in using hydrogen as an automotive fuel, this is still under development and is discussed further in Chapter 23, Section 23.5.1. As discussed in Chapter 4, Section 4.3, there has been commercial application of gaseous fuels for over half a century. Discussions in this section will therefore concentrate on existing gaseous fuel systems. Natural gas occupies considerably more volume than the amount of gasoline having the same energy. As a result, the volumetric energy content of a stoichiometric natural gas/air mixture is less than that of gasoline-air mixture. In addition, natural gas does not benefit from the practice of "power enrichment"—the best power output from natural gas engines occurs at essentially the stoichiometric air-to-fuel ratio. When a gasoline engine is converted to natural gas, the combination of these two effects typically results in a loss in maximum brake mean effective pressure (BMEP) and power output of approximately 10%. For LPG, these effects are smaller, and the power loss is typically only a few percent. In dedicated engines, the reduction in power output with gaseous fuel can be compensated for by increasing the compression ratio, thus increasing the amount of useful work extracted from a given amount of fuel input. Further increases in BMEP and power output can be achieved by turbo-charging.

Gaseous fuel vehicles require precise control of the air-to-fuel ratio to minimize emissions while maintaining good performance and fuel economy. Until about 1990, nearly all gaseous fuel metering systems relied on mechanical principles, analogous to the mechanical carburetors used in gasoline engines until the early 1980s [8.55]. Although these mechanical systems can be designed to give good engine performance and efficiency, they are susceptible to fuel metering errors due to wear, drift, changes in elastomer properties, changes in fuel and air temperature, changes in fuel properties, and other causes. The effects of fuel composition on metering are discussed in Chapter 4, Section 4.3.2. The old mechanical systems are thus unable to meet the requirement of modern three-way catalytic converter systems for very precise control of the air-to-fuel ratio. Thus, gaseous fuel metering systems have followed gasoline system evolution, from mechanical systems to full-authority digital electronic fuel injection. The fuel metering and engine control systems installed on natural gas vehicles produced by, Chrysler [8.56] and others are essentially identical to the multipoint sequential fuel injection systems installed on production gasoline vehicles, except for details of the fuel rail and injectors.

As engine design and fuel injection technology moves toward direct injection (DI) this offers the possibility to remove the volumetric efficiency losses due to the fact that gaseous fuel displaces a corresponding amount of fresh intake air. DI also offers the possibilities of stratified charge operation, thus reducing the pumping losses at part loads [8.57]. This also opens up the possibilities of dual-fueling where a DI system is used to supply one fuel while a port injection system supplies a different fuel [8.58, 8.59]. The fuel system itself is very similar in architecture to a gasoline system. The CNG is stored in pressurized tanks, at about 200–250 bar, equipped with shut-off valves to isolate the tanks when the engine is shut-off, or in the case of collision. The gas from these tanks is reduced in pressure before passing to the fuel rail. Due to the pressure drop across this regulator, and commensurate temperature drop, it may be necessary to provide some heating of this unit, again shut of valves may be included for safety reasons. Alternatively, the pressure may be reduced in stages using two regulators. A schematic of a manifold injection system is shown in Figure 8.11 [8.60].

FIGURE 8.11 Multipoint CNG fuel system layout.

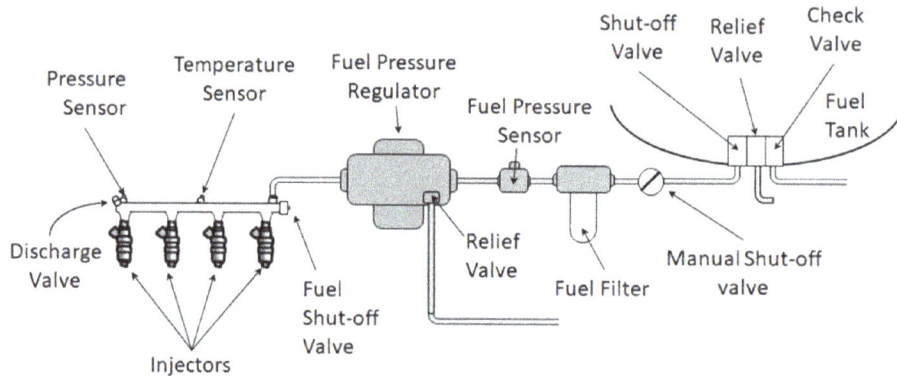

8.3.2. The Fuel Tank and Pump

For vehicle packaging considerations, the fuel tank is usually situated at the opposite end of the vehicle to the engine. It should be manufactured from material that is not prone to be attacked by the gasoline. Plastic fuel tanks are generally satisfactory, although problems of permeability have been encountered with some of them when fuels containing alcohols, and particularly methanol, have been used. Terne plate, which is a steel sheet coated with a lead/tin alloy, is also generally satisfactory, provided that there are no pinholes for corrosion to start by attacking the underlying steel. A small amount of water is not uncommon in a fuel tank since it can separate from the fuel during a decrease in temperature. Alcohol-containing fuels are prone to adsorb water, and this can antagonize their propensity to attack terne plates, with the extent of the attack depending on the exact fuel composition. Therefore, it is important to ensure that the correct terne plate grade is used to minimize the risk of attack. Other coatings on steel are used by different manufacturers, and it is important that they are fully evaluated using as wide a range of fuels as possible before being introduced.

Most vehicle fuel tanks have a vent system to allow air to escape during filling, and in countries where there are restrictions on evaporative emissions, the vent pipe (and other parts of the fuel system where it is possible for vapors to escape) is connected to a charcoal canister to prevent vapors being released to the atmosphere. Clearly the vapor pressure of the gasoline will be a factor in determining the size of canister required. The canister is regenerated by sucking intake air through it when the engine is running so that the vapors are recovered and burned and the canister is regenerated. The filler cap often incorporates a two-way relief valve that allows a slight pressure or vacuum to develop, and there is frequently a rollover valve that prevents fuel from discharging if the vehicle rolls over in an accident. Some countries require that gasoline vapor escaping from the tank during filling be controlled. Special filling nozzles are used in these cases that provide a seal at the filler pipe and a vapor return line that takes the vapor back to the storage tank.

The vehicle fuel tank will also incorporate a strainer/filter in the bottom of the tank near the outlet line. Particles of dirt and rust that may be introduced with the fuel or that may appear due to corrosion of the tank are held here to prevent problems later in the system.

The fuel pump delivers fuel from the tank to the carburetor or fuel injectors; its siting and design are critical to the performance of the vehicle in hot weather. Mechanical fuel pumps are usually mounted on the side of the cylinder block and operate by raising and lowering a diaphragm that draws fuel in and then forces it out through a series of valves. Many vehicles are fitted with a vapor return line from the fuel pump to the tank in order to return any vapors and excess fuel to the tank. Pumping stops if vapor forms in the pump, causing "vapor lock," which can halt a car completely. Heat can be transmitted to the pump body from the cylinder block, and this, together with the fact that the pump operates on the principle of using a partial vacuum to draw in the fuel, means that this type of pump is somewhat prone to fuel vaporization problems, resulting in poor hot weather driveability. This is discussed in Chapter 9, Section 9.4.2. The fuel lines from the tank to the pump are also subject to vapor formation unless they are carefully sited because they are on the suction side of the pump and may be subjected to heat from the engine or exhaust system.

Mechanical pumps will usually only provide low-pressure fuel feed necessary to supply carburetor float bowl. Fuel injection systems require higher pressures, and electric fuel pumps are used to satisfy these requirements. Electric fuel pumps are usually sited in the fuel tank itself so they can utilize the thermal capacity of the fuel as a coolant. This also minimizes the suction leg of the system. As the fuel from the pump to the fuel metering device is under higher pressure than on a carbureted engine and the suction leg is short, these pumps and the associated fuel systems are much less susceptible to vapor lock issues.

8.4. Ignition Systems

For the air-fuel mixture to ignite in the combustion chamber and provide the maximum amount of usable work requires an ignition source of adequate energy at the right time in the cycle, the air-to-fuel ratio to be in the burnable range, and for the ignition source to be positioned so that it occurs within a combustible mixture of air and fuel. Most of these requirements are determined by engine design and are outside the influence of the fuel. The combustion process and the fuel requirements are discussed in some detail in Chapter 7. The minimum electrical energy required to ignite a mixture of gasoline and air is when the mixture is in stoichiometric proportions. Going on either side of stoichiometric requires more and more energy until the limits of burnability are reached. The fuel can influence ignition because some components are able to burn over a wider range of air-to-fuel ratio than others. Additives are available that help a spark and overcome to some degree the problems of spark plug fouling; this is discussed in Chapter 11, Section 11.4.

The timing of the spark is vital because if it occurs too soon (if the timing is over-advanced), knock can occur, causing damage to the engine and loss of power. This can be avoided by the use of higher-octane quality fuels, but suitable fuels are not always readily available. If the spark is fired later than the optimum point (minimum for best torque [MBT]), that is, if it is retarded, then loss of power can result. In some engines there is a knock sensor fitted [8.61] that detects knock and puts into effect some action to prevent it and avoid engine damage. Usually this means retarding the ignition, but other actions are possible. Because knock is influenced by driving conditions, it will often only occur for the time that the engine is being driven under knock-critical modes. Of course, there will be no knock at all if the fuel has a higher antiknock value than required by the engine. The knock sensor system is designed so that after a few cycles without knock, the ignition is gradually advanced back to the optimum setting until the next knocking cycle. Quite complex systems for retarding and re-advancing the ignition timing have been developed to minimize the time that the engine is running in a nonoptimized state. From a fuel quality viewpoint, this can mean that the engine will function on a wide range of octane quality fuels without damage, although lower quality fuels will give lower power outputs and hence reduced acceleration performance.

All of the above assumes that the mixture is homogeneous, stoichiometric and there is a single point of ignition, a spark. Multiple sparks, from multiple spark plugs, will allow multiple flame fronts and a faster combustion event. However, as discussed previously, attempts were made to stratify the charge to allow these conditions to prevail locally while maintaining an overall lean condition to aid efficiency. The need to maintain an overall stoichiometric charge to allow a TWC to function meant that these systems did not progress. With increasing demands for fuel efficiency and reduced emissions there is a resurgence of interest in using weaker mixtures, a higher overall air/fuel ratio along with improved aftertreatment technology that is more effective under these conditions. An early example of how to achieve charge stratification was published over a century ago [8.62], which used an auxiliary intake valve to feed a prechamber, the main aim with this design, as with the Ricardo comet diesel prechamber, was to induce greater turbulence in the main chamber. To greatly simplify the idea, there have been a number of designs where the prechamber is simply a small part of the main chamber connected by a small passage or throat. An early example is the Toyota Turbulence Generating Pot (TGP) design illustrated in Figure 8.12 [8.63, 8.64].

FIGURE 8.12   Turbulence Generating Pot (TGP).

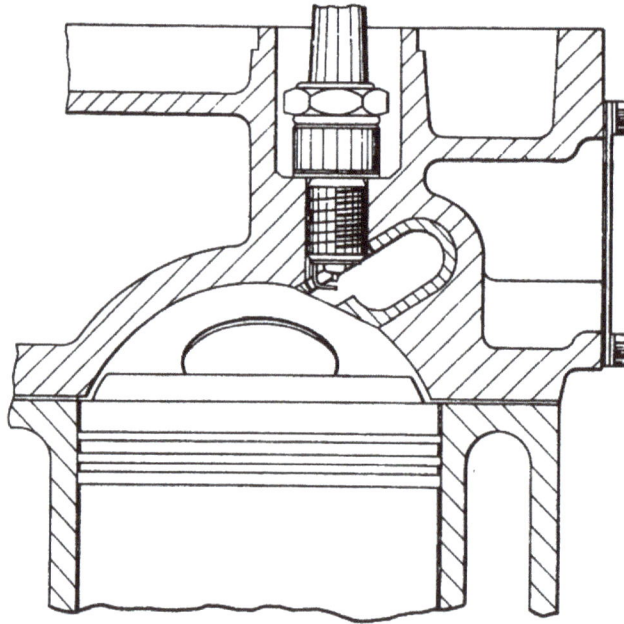

Adapted from "Development of Toyota lean burn engine." SAE Transactions, pp.2358-2373, 1976 © SAE International.

Similar systems had been proposed by Volkswagen, termed the "prechamber ignition" (PCI) system [8.65], the Ford Torch ignition systems [8.66], the GM Torch Jet Spark Plug [8.67], the Delphi Directed Jet Spark Plug [8.68], and latterly the Mahle Jet Ignition® (MJI) system [8.69, 8.70], as shown in Figure 8.13.

FIGURE 8.13   Mahle Jet Ignition® (MJI) system arrangement.

Reprinted from "Knock Mitigation Benefits Achieved through the Application of Passive MAHLE Jet Ignition Enabling Increased Output under Stoichiometric operation." SAE Technical Paper 2021-01-0477, 2021 © SAE International.

It had been discovered that the jet of gas from the throat of such arrangements contained partially combusted products, which can enhance flame propagation in fuel lean mixtures, either stoichiometrically lean or stoichiometric mixtures diluted with EGR [8.71–8.73]. The turbulent jet of radicals entering the main chamber not only causes turbulence but also provides multiple ignition sources.

8.5. Combustion and Exhaust Emission Control Systems

The combustion of gasoline is discussed in Chapter 7; in this section, we discuss engine design factors that influence the fuel quality required. Many of the design features introduced to control exhaust emissions also impact gasoline quality requirements; these are discussed in more detail in Chapter 13.

The most important aspects of combustion are that the fuel should burn smoothly and efficiently, without knocking, and without giving rise to unacceptable levels of toxic exhaust gases. The exhaust gases that are of most concern, and for which there is control legislation, are carbon monoxide (CO), unburned hydrocarbons (HC), total oxides of nitrogen (NO_X) and particulate matter, particularly when measured as particulate number (PN). In some areas there are also limits on non-methane hydrocarbon (NMHC) emissions or non-methane organic gases (NMOG). There are other pollutants such as benzene and polycyclic aromatic hydrocarbons (PAHs) for which there are also concerns. More recently, with the increasing use of DI technology, particularly where charge stratification is used, there is also growing interest in the number of ultra-fine particles that are emitted from the tailpipe.

8.5.1. Combustion Chamber Configuration

Considering first the combustion chamber itself, there are two major factors that are important, and these are the compression ratio and the geometry. Compression ratio is the ratio of the volume of the chamber when the piston is at bottom dead center (BDC) to when it is at top dead center (TDC), and it is a measure of how much the incoming charge is compressed in the cylinders. It is important because the higher the compression ratio, the better the thermal efficiency and hence the better the fuel economy and power for a given-sized engine. Higher compression ratios mean not only higher pressures at the end of the compression stroke but also higher temperatures. Unfortunately, there is thus an increasing tendency for the fuel to self-ignite (detonate) with increasing compression ratio. The use of pressure charging, using turbochargers or superchargers, to increase specific power output from an engine also effectively increases the compression ratio and so increases the detonation tendency.

8.5.1.1. Combustion Chamber Shape

From consideration of the combustion event, the ideal combustion chamber shape would be spherical with the spark plug electrodes in the center. However, this would result in a very long reach spark plug that is likely to get hot and become a source of surface ignition, as discussed in the previous chapter. To achieve such a combustion chamber shape and high compression ratios would require a long stroke and small diameter cylinder. More vertically compact combustion chamber shapes are therefore necessary. The engine designer is also interested in maximizing the area of the intake and exhaust valves, particularly the intake valve, to maximize the amount of air that can be drawn into the engine. This allows a greater amount of fuel to be burned, thus maximizing the specific output of the engine, and allowing a smaller, more economical engine to be used while still achieving the same maximum power; this is now commonly referred to as downsizing. When only one intake and one exhaust valve were used per cylinder, this led to the development of so-called hemispherical combustion chamber designs such as that in the famous Jaguar XK engine that won the Le-Mans 24-hour race in 1951, 1953, 1955, 1956, and 1957 [8.74] and is depicted in Figure 8.1. To achieve the required compression ratio, an almost corresponding dome was necessary on the top of the piston. A more popular arrangement had the inlet and exhaust valve in the same plane, which made it easier to operate them from a single camshaft, but this did result in a smaller valve area. Figure 8.14 shows the elevation and plan of a hemispherical and wedge-shaped combustion chamber; it can be seen that for the same cylinder bore the hemispherical chamber gives a greater valve area.

FIGURE 8.14 Different combustion chamber geometries.

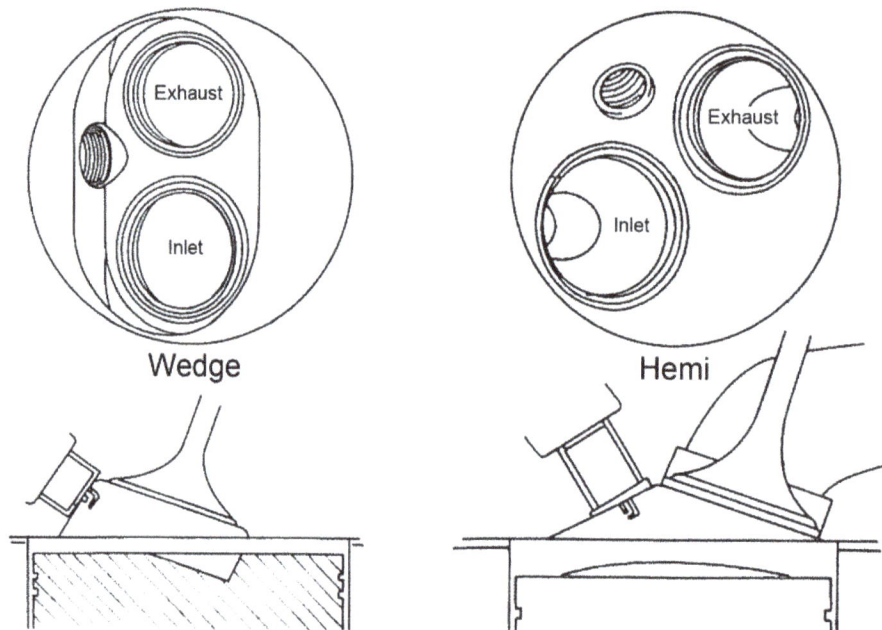

8.5.1.2. Valve Layouts

As the quest for increased power and efficiency continues, different valve configurations have been investigated to maximize the valve area, moving from two valves per cylinder to three [8.75], four [8.76–8.79], and five [8.80, 8.81]. Four valves per cylinder is probably the commonest arrangement with a pent roof combustion chamber. The shape influences three factors that are important for both detonation tendency and exhaust gas quality: turbulence, squish, and quench.

Turbulence ensures good mixing of the air and fuel and reduces the time for a flame front to sweep through the mixture, thereby allowing less time for detonation to occur. Squish is the term used for the squeezing of the mixture away from the end gas region; this can cause further turbulence. Quench is the cooling effect of the sides of the combustion chamber, which can make the mixture in the vicinity of the walls too cool to detonate. The squish region of the combustion chamber is often also the quench area since heat is readily conducted away, thereby reducing the tendency for the end gas to explode. Quenching also has the effect of preventing the thin layer of air-to-fuel mixture next to the walls of the combustion chamber from completely burning so that some unburned hydrocarbons are present in the exhaust gases, this thin layer of air and fuel is also likely to contain lubricant.

For low hydrocarbon emissions, a compact combustion chamber is required with a small surface-to-volume ratio to minimize quench effects; a spherical chamber would be ideal, but as discussed, there are practical limitations. The spark plug should be located centrally so that the flame front has a relatively short distance to travel and hence allows complete combustion without detonation. There also needs to be turbulence. The hemispherical combustion chamber meets a number of these requirements in that it has a low surface-to-volume ratio and the spark plug is located near the center of the dome. However, there are no squish or quench areas as in the wedge type of chamber, and the lack of turbulence prevents good mixing within the chamber; good preparation within the manifold is therefore important to avoid poor exhaust emissions.

The use of a very small squish zone clearance, below about 1.0 mm, helps to minimize hydrocarbon emissions. However, if high levels of deposits build up on the top of the piston, contact can be made with the cylinder head in the squish zone [8.82] when the piston is at TDC, this is discussed in Chapter 10, Section 10.2.4.2. This mechanical contact will give rise to a knocking noise that has been called "carbon knock" or "carbon rap," among other names, and is totally unrelated to combustion knock. This phenomenon is most prevalent on start-up when the engine is cold and due to thermal contraction of the cylinder block the squish clearance is at a minimum; it usually disappears after the engine has warmed up. Excessive carbon buildup causing this problem may result from high treats of certain types of additives [8.83] or from the presence of too high a level of deposit-forming, high boiling point compounds in the gasoline.

8.5.2. The Effect of Air-to-Fuel Ratio

Mixture strength has an important influence on exhaust emissions, as shown in Figure 8.15. Here the symbol λ represents the air-to-fuel ratio, and a value of 1.0 represents the stoichiometric ratio; lower values refer to mixtures richer than stoichiometric and higher values to leaner than stoichiometric.

FIGURE 8.15   Influence of air-to-fuel ratio on exhaust emissions, power, and specific fuel consumption.

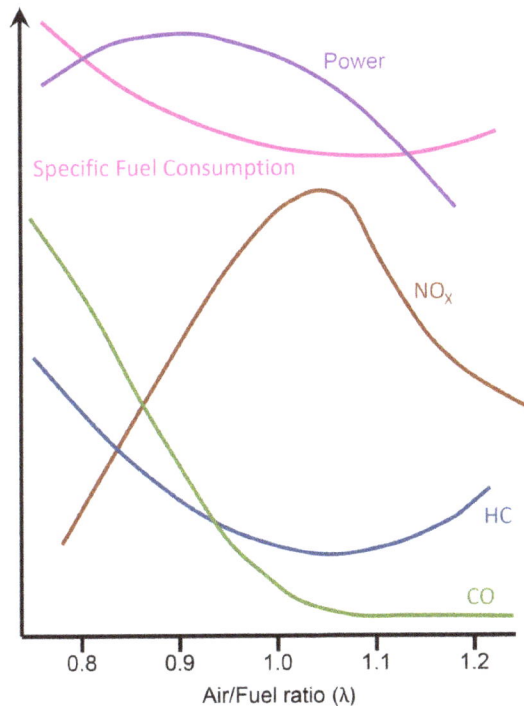

From inspection of the chemistry CO and HC will be produced when the mixture is rich, when there is not enough oxygen to completely combust the fuel. These pollutants would fall to zero at $\lambda = 1$ and remain there at values greater than 1. Oxides of nitrogen are formed as a result of the very high temperatures reached as a result of the combustion process. In reality, due to the kinetics of the combustion process, the possibility of incomplete vaporization and mixing of the fuel, quenching, and so forth, this ideal is not totally met. Figure 8.13 shows that CO and HC emissions are high when λ levels are low, although nitrogen-oxide emissions are low. This is due to a lack of oxygen to completely combust all of the fuel: as some fuel is not being burned, the specific fuel consumption is higher under these conditions. Maximum torque is obtained when the mixture strength is about 0.9λ, when the mixture is slightly rich. This maximizes the air utilization with minimal fuel wastage. Where emissions control strategies allow, the air-to-fuel ratio is usually adjusted to this value when maximum power is required, that is, when running at WOT.

At stoichiometric conditions the CO and HC levels drop to a low level as does the specific fuel consumption, and as λ values increase these parameters continue to fall slightly, until the hydrocarbon emissions start to increase as the misfire limit is approached; driveability will also deteriorate from this point. Charge stratification as discussed in Section 8.3.1.4 allows operation to very low overall λ values. But because the high temperatures still occur in the rich areas, oxides of nitrogen will still be produced and will require control.

Nitrogen oxides reach a maximum just lean of stoichiometric where maximum fuel utilization occurs, and there is a slight excess of oxygen. As the air-to-fuel ratio increases, there is less fuel available, less heat is generated, and obviously less power is generated.

Gasoline volatility and composition, especially the presence of oxygenates, can have an effect on emissions and on driveability, as is discussed in Chapter 9, and these become particularly important when vehicles are designed to operate near the lean limit of combustion.

8.5.3. Exhaust Aftertreatment Systems

There is a trade-off between the different regulated emissions, power, and fuel economy for any given air-to-fuel ratio. How the engine designer weights these different parameters is determined to a large extent by the legislative emissions targets and the aftertreatment technologies available.

Initial investigations into meeting proposed targets had centered on using exhaust gas recirculation (EGR) to reduce NO_X emissions [8.84, 8.85] and secondary air injection [8.86–8.88] to provide an oxidizing atmosphere in the exhaust for HC and CO reduction. EGR can be achieved either by means of adjusting the valve timing or by feeding exhaust gas back through valves into the inlet manifold or elsewhere in the intake system. The use of EGR was reported to result in significant power loss [8.89], but careful tuning could reduce this [8.90]. Electronic control and engine management were still in their infancy at this time, and it was difficult to control the amount of EGR to match the engine operating conditions. Developments on catalytic systems meant that they soon became the accepted approach. The different catalytic systems are outlined briefly in the following section.

8.5.3.1. Oxidation Catalysts

Proposals had been made as early as the 1950s for a *"catalytic apparatus to render nonpoisonous exhaust gases from internal combustion engines"* [8.91, 8.92]. Because lead antiknock additives were still widely used, there was interest in developing catalytic systems that were tolerant of the lead emissions that were found in the exhaust [8.93]. Due to the financial implications at that time of removing lead antiknock additives from fuel, there was also research into a means of preventing the lead emissions from reaching the atmosphere or catalytic systems [8.94, 8.95]. This ultimately proved difficult and resulted in the phase-out of lead antiknock additives.

Many catalyst formulations have been developed [8.96–8.99] and alternative catalyst constructions were also developed [8.100–8.103]. Monolithic catalysts tended to exhibit faster warm-up characteristics [8.104] and have ultimately proved to be the favored approach.

8.5.3.2. Three-Way Catalysts

The development of TWCs, which simultaneously oxidize the HC and CO to CO_2 and water and reduce the NO_X back to N_2, meant that all three regulated pollutants could be controlled with far greater efficiency than the previous systems. As mentioned in the previous chapter, these catalytic systems are prone to poisoning and plugging by metallic fuel additive. They are also susceptible to poisoning by sulfur compounds, and strenuous efforts are being made by the fuel industry to reduce sulfur levels. The only significant drawback to TWCs is the requirement for the engine to operate at a stoichiometric air-to-fuel ratio. This ratio is often nonideal from the standpoint of power or economy.

8.5.3.3. Lean NO$_X$ Traps

A TWC requires the engine to operate at a stoichiometric air-to-fuel ratio; for maximum economy, it is advantageous to operate the engine with a slightly lean mixture, when the mixture is homogeneous. Even greater economies can be achieved if the engine operates with a stratified charge. Under these conditions the HC and CO emissions are generally low and a TWC will still oxidize the HC and CO to reduce their emission to acceptable levels. Unfortunately, due to the excess oxygen, the reduction reaction that takes place in the TWC when the engine is operating at stoichiometric will not occur. Some additional device is necessary to reduce the NO_X emissions.

Attention was initially focused on the development of reduction catalysts that would use hydrocarbons, such as unburned fuel to selectively reduce the NO_X emissions. Examples such as copper zeolite catalysts [8.105–8.107], copper-silver zeolites [8.108], and other transition metal zeolites [8.109] have been investigated. Some of these have been commercialized [8.110]. However, their application has been limited by low NO_X conversion efficiencies, a narrow operating temperature window, and poor thermal durability. A more promising solution is the so-called lean-NO_X trap (LNT) or NO_X storage reduction (NSR) catalyst [8.111, 8.112]. This technology relies upon the different NO_X constituents, mainly NO, being converted to NO_2, which then reacts at the catalyst surface to form a nitrate. The nitrogen is thus "trapped" on the catalyst surface. Due to the subsequent introduction of a hydrocarbon spike, or high temperatures, the nitrate decomposes to N_2. The mechanism is illustrated in Figure 8.16 [8.113].

FIGURE 8.16   Reaction scheme for NSR catalysts.

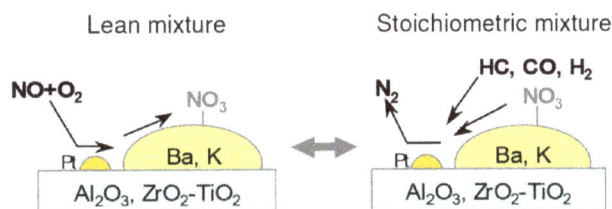

The LNT would therefore need to be placed downstream of a catalyst that would oxidize the NO to NO_2; this catalyst could be a TWC that would also reduce NO_X when the engine was operating stoichiometrically. Periodic regimes of rich operation would be required to regenerate the NSR catalyst, and this would require careful integration into the engine management system. A completely integrated DI gasoline engine and aftertreatment system may look like that shown in Figure 8.17 [8.114].

FIGURE 8.17   Exhaust aftertreatment system for DI gasoline engine.

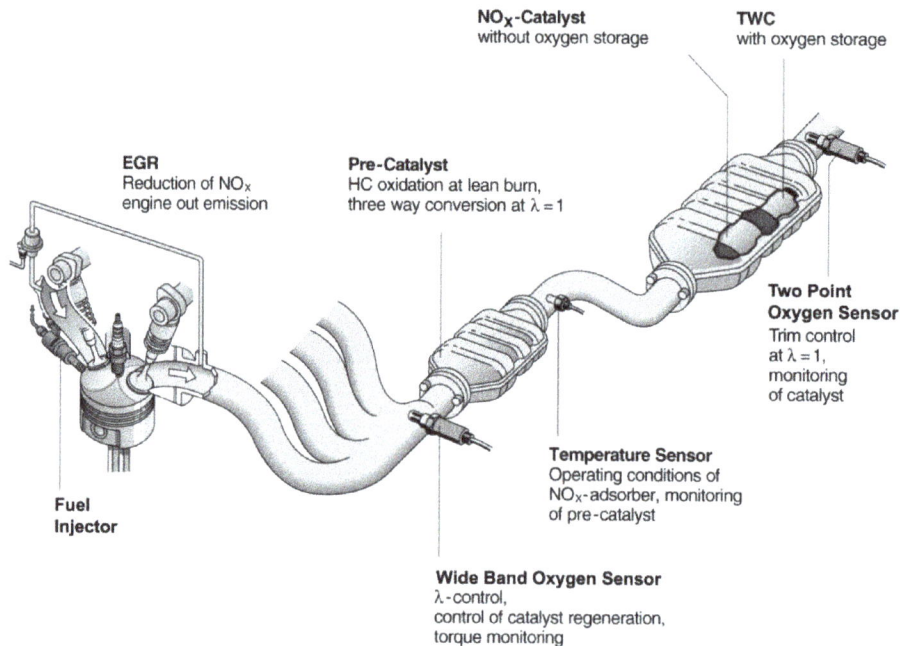

Unfortunately, the catalytic oxidation of NO to NO_2 will also oxidize SO_2 to SO_3, and under oxygen-rich conditions the SO_X will adsorb more strongly than the NO_2. The adsorbed SO_X does not completely desorb during the rich spike, and a dedicated de-sulfation cycle is required to purge the NSR catalyst [8.115]. The de-sulfation process requires raising the temperature of the catalyst to typically over 700°C. If this is not properly controlled it can give rise to undesirable sulfide emissions, such as H_2S [8.116].

Another approach that has recently been investigated utilizes the fact that ammonia can be generated over a TWC [8.117] during the rich excursions that are required to regenerate an NSR. This ammonia can be stored on a downstream SCR catalyst and used to reduce NO_X when the engine is operating lean [8.118, 8.119].

8.5.3.4. Particulate Filters

Gasoline engines, like diesel engines, will produce soot if fuel is burned with a deficiency of oxygen. Old vehicles running rich during cold starts would produce sooty exhaust pipes. With improved fuel control and automatic fast warm-up systems, soot emissions from gasoline vehicles ceased to be an issue. However, with increasing concern over the carcinogenic effects of diesel soot and efforts to decrease not just the mass but the number of diesel particulate emissions, attention has turned to DI gasoline engines. When these engines are operating in a stratified charge regime, the fuel is injected late in the compression stroke. The time available for evaporation and mixing of the fuel is therefore limited. As a consequence, it is possible that some of this fuel will burn in a rich regime and may produce soot particles. If there is insufficient time and/or temperature to ensure that these particles are completely oxidized in the remainder of the expansion stroke or in the exhaust stroke, then some of these particles may make the journey to the tailpipe. Work has shown that this is in fact the case. Figure 8.18 [8.120] shows the result of work, looking at both the mass and the number of particles emitted from different engine technologies.

FIGURE 8.18   Particulate mass and number emissions for different engine technologies.

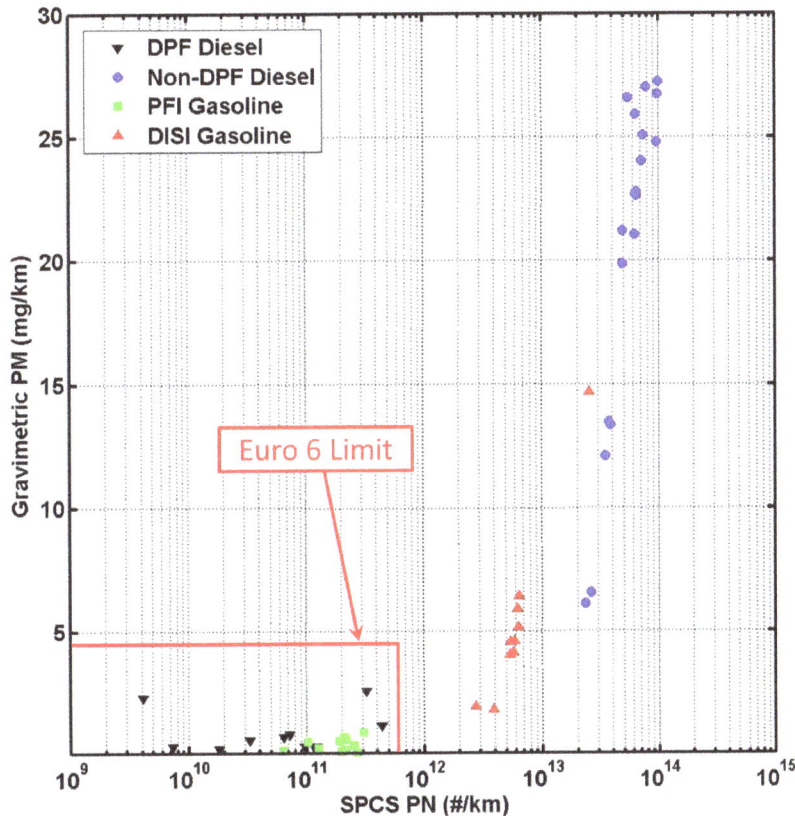

From this figure it is clear that most DISI engine vehicles, of that time, would fail to meet current limits for both particle mass and number. Systems were therefore developed to trap these very fine particles that are being produced in DISI engines.

Due to the need to reduce diesel particulate emissions, the filter technology has been well developed. The technology cannot be simply transferred to gasoline engines due to the nature of the particulates formed. Particulate matter generated by a diesel engine tends to be larger and this helps to form a layer of particles on the wall of the filter, a soot cake layer. This soot cake can boost the filtration efficiency of a DPF from as low as 50% to over 99% [8.121]. The characteristics of the filter substrate, wall thickness and porosity, must therefore be modified to suit a gasoline engine [8.122].

As with diesel particulate filters at the start of the century, there is now a desire to integrate the function of the GPF with other aftertreatment devices by the use of catalytic coatings on the filter substrate [8.123– 8.126].

GPFs also face the same challenges as DPFs in that the accumulated soot needs to be burned out [8.127– 8.129], although this is less of a problem because a DISI engine tends to run hotter with regular excess oxygen events (fuel shut-off during over-run) which promotes soot burn-out. The other issue facing both GPFs and DPFs is ash accumulation from the metals in the lubricating oil [8.130–8.132].

Lubricating oil is not only an issue when considering ash accumulation in a GPF it must also be considered when considering the nature of the emitted particles that lead to this ash. With increasing awareness of gasoline characteristics and engine and fuel injection technology that cause particulate emission the observed emissions are being reduced. The use of lower carbon/hydrogen ratio fuels, to reduce carbon emissions will obviously reduce the formation of carbonaceous particulates generated from the fuel, but as fuel generated particulates diminish the lubricating oil generated particulates become more important. This has been clearly demonstrated by using hydrogen as a carbon-free fuel and observing particulate emissions [8.133], and it has been shown that particulate number (PN) emissions can be increased due to lubricant vapor entrainment.

References

8.1.   Gray, C., "A Review of Variable Engine Valve Timing," SAE Technical Paper 880386, 1988, doi:https://doi.org/10.4271/880386.

8.2.   Dresner, T. and Barkan, P., "A Review of Variable Valve Timing Benefits and Modes of Operation," SAE Technical Paper 891676, 1989, doi:https://doi.org/10.4271/891676.

8.3.   Titolo, A., "The Variable Valve Timing System—Application on a V8 Engine," SAE Technical Paper 910009, 1991, doi:https://doi.org/10.4271/910009.

8.4.   Kramer, U. and Phlips, P., "Phasing Strategy for an Engine with Twin Variable Cam Timing," SAE Technical Paper 2002-01-1101, 2002, doi:https://doi.org/10.4271/2002-01-1101.

8.5.   Hattori, M., Inoue, T., Mashiki, Z., Takenaka, A. et al., "Development of Variable Valve Timing System Controlled by Electric Motor," *SAE Int. J. Engines* 1, no. 1 (2009): 985-990, doi:https://doi.org/10.4271/2008-01-1358.

8.6.   Ha, K., Han, D., and Kim, W., "Development of Continuously Variable Valve Lift Engine," SAE Technical Paper 2010-01-1187, 2010, doi:https://doi.org/10.4271/2010-01-1187.

8.7.   Fernandez, H., Kazour, Y., Knauf, M., Sinnamon, J. et al., "Development of Continuously Variable Valve Lift Mechanism for Improved Fuel Economy," SAE Technical Paper 2012-01-0163, 2012, doi:https://doi.org/10.4271/2012-01-0163.

8.8.   Ahmad, T. and Theobald, M., "A Survey of Variable-Valve-Actuation Technology," SAE Technical Paper 891674, 1989, doi:https://doi.org/10.4271/891674.

8.9.   Gould, L., Richeson, W., and Erickson, F., "Performance Evaluation of a Camless Engine Using Valve Actuators with Programmable Timing," SAE Technical Paper 910450, 1991, doi:https://doi.org/10.4271/910450.

8.10. Schechter, M. and Levin, M., "Camless Engine," SAE Technical Paper 960581, 1996, doi:https://doi.org/10.4271/960581.

8.11. Postrioti, L., Battistoni, M., Foschini, L., and Flora, R., "Application of a Fully Flexible Electro-Hydraulic Camless System to a Research SI Engine," SAE Technical Paper 2009-24-0076, 2009, doi:https://doi.org/10.4271/2009-24-0076.

8.12. Atkinson, J., Gas engine. G.B. Patent No. 2,712, 1985.

8.13. Atkinson, J., Gas engine. G.B. Patent No. 3,522, 1986.

8.14. Miller, R.H., "Supercharging and Internal Cooling Cycle for High Output," *ASME Transactions* 69, no. 4 (1947): 453-464.

8.15. Miller, R., Supercharged engine. U.S. Patent 2,817,322, 1957.

8.16. Tuttle, J.H., "Controlling Engine Load by Means of Early Intake-Valve Closing," SAE Technical Paper 820408, 1982, doi:https://doi.org/10.4271/820408.

8.17. Tuttle, J.H., "Controlling Engine Load by Means of Late Intake-Valve Closing," SAE Technical Paper 800794, 1980, doi:https://doi.org/10.4271/800794.

8.18.   Anderson, M., Assanis, D., and Filipi, Z., "First and Second Law Analyses of a Naturally-Aspirated, Miller Cycle, SI Engine with Late Intake Valve Closure," SAE Technical Paper 980889, 1998, doi:https://doi.org/10.4271/980889.

8.19.   Goto, T., Hatamura, K., Takizawa, S., Hayama, N. et al., "Development of V6 Miller Cycle Gasoline Engine," SAE Technical Paper 940198, 1994, doi:https://doi.org/10.4271/940198.

8.20.   Urata, Y., Umiyama, H., Shimizu, K., Fujiyoshi, Y. et al., "A Study of Vehicle Equipped with Non-Throttling S.I. Engine with Early Intake Valve Closing Mechanism," SAE Technical Paper 930820, 1993, doi:https://doi.org/10.4271/930820.

8.21.   Hitomi, M., Sasaki, J., Hatamura, K., and Yano, Y., "Mechanism of Improving Fuel Efficiency by Miller Cycle and Its Future Prospect," SAE Technical Paper 950974, 1995, doi:https://doi.org/10.4271/950974.

8.22.   Rabia, S.M. and Korah, N.S., "Knocking Phenomena in a Gasoline Engine with Late-Intake Valve Closing," SAE Technical Paper 920381, 1992, doi:https://doi.org/10.4271/920381.

8.23.   Taylor, C.F., Taylor, E.S., and Williams, G.L., "Fuel Injection with Spark Ignition in an Otto-Cycle Engine," SAE Technical Paper 310005, 1931, doi:https://doi.org/10.4271/310005.

8.24.   Miller, S.E., "Automotive Gasoline Injection," SAE Technical Paper 560321, 1956, doi:https://doi.org/10.4271/560321.

8.25.   Masaki, K., Aono, S., Akaeda, M., and Minami, H., "Development of the Nissan Electronically Controlled Carburetor System," SAE Technical Paper 780204, 1978, doi:https://doi.org/10.4271/780204.

8.26.   Nakamura, N., Itoh, T., and Morino, T., "Development of a New Variable Venturi Carburetor," SAE Technical Paper 830617, 1983, doi:https://doi.org/10.4271/830617.

8.27.   Spilski, R.A. and Creps, W.D., "Closed Loop Carburetor Emission Control System," SAE Technical Paper 750371, 1975, doi:https://doi.org/10.4271/750371.

8.28.   Aono, S., Hosaka, A., and Inoue, M., "An Electronic Carburetor Controller," SAE Technical Paper 790743, 1979, doi:https://doi.org/10.4271/790743.

8.29.   Kay, I.W., "Manifold Fuel Film Effects in an SI Engine," SAE Technical Paper 780944, 1978, doi:https://doi.org/10.4271/780944.

8.30.   Tanaka, M. and Durbin, E.J., "Transient Response of a Carburetor Engine," SAE Technical Paper 770046, 1977, doi:https://doi.org/10.4271/770046.

8.31.   Winterbone, D.E. and Richards, P., "A Study of Petrol Engine Dynamics," in *Paper 83/WA/DSC-41, Proceedings ASME Winter Annual Meeting*, Boston, MA, November 1983.

8.32.   Manger, H., "Electronic Fuel Injection," SAE Technical Paper 820903, 1982, doi:https://doi.org/10.4271/820903.

8.33.   Bowler, L.L., "Throttle Body Fuel Injection (TBI)—An Integrated Engine Control System," SAE Technical Paper 800164, 1980, doi:https://doi.org/10.4271/800164.

8.34.   Takeda, K., Shiozawa, K., Oishi, K., and Inoue, T., "Toyota Central Injection (Ci) System for Lean Combustion and High Transient Response," SAE Technical Paper 851675, 1985, doi:https://doi.org/10.4271/851675.

8.35.   Senda, J., Nishikori, T., Tsukamoto, T., and Fujimoto, H., "Atomization of Spray under Low-Pressure Field from Pintle Type Gasoline Injector," SAE Technical Paper 920382, 1992, doi:https://doi.org/10.4271/920382.

8.36.   Marshall, J.T., "Direct Injection for Wright Eighteen Cylinder Engine," SAE Technical Paper 480018, 1948, doi:https://doi.org/10.4271/480018.

8.37.   Meurer, J.S. and Urlaub, A.C., "Development and Operational Results of the MAN FM Combustion System," SAE Technical Paper 690255, 1969, doi:https://doi.org/10.4271/690255.

8.38.   Bishop, I.N. and Simko, A., "A New Concept of Stratified Charge Combustion—The Ford Combustion Process (FCP)," SAE Technical Paper 680041, 1968, doi:https://doi.org/10.4271/680041.

8.39.   Simko, A., Choma, M.M., and Repko, L.L., "Exhaust Emission Control by the Ford Programmed Combustion Process–PROCO," SAE Technical Paper 720052, 1972, doi:https://doi.org/10.4271/720052.

8.40.   Scussel, A.J., Simko, A.O., and Wade, W.R., "The Ford PROCO Engine Update," SAE Technical Paper 780699, 1978, doi:https://doi.org/10.4271/780699.

8.41. Mitchell, E., Cobb, J.M., and Frost, R.A., "Design and Evaluation of a Stratified Charge Multifuel Military Engine," SAE Technical Paper 680042, 1968, doi:https://doi.org/10.4271/680042.

8.42. Alperstein, M., Schafer, G.H., and Villforth, F.J. III, "Texaco's Stratified Charge Engine—Multifuel Efficient, Clean, and Practical," SAE Technical Paper 740563, 1974, doi:https://doi.org/10.4271/740563.

8.43. Preussner, C., Döring, C., Fehler, S., and Kampmann, S., "GDI: Interaction between Mixture Preparation, Combustion System and Injector Performance," SAE Technical Paper 980498, 1998, doi:https://doi.org/10.4271/980498.

8.44. Yamauchi, T. and Wakisaka, T., "Computation of the Hollow-Cone Sprays from a High-Pressure Swirl Injector for a Gasoline Direct-Injection SI Engine," SAE Technical Paper 962016, 1996, doi:https://doi.org/10.4271/962016.

8.45. Jang, C., Choi, S., Bae, C., Kim, J. et al., "Performance of Prototype High Pressure Swirl Injector Nozzles for Gasoline Direct Injection," SAE Technical Paper 1999-01-3654, 1999, doi:https://doi.org/10.4271/1999-01-3654.

8.46. Takeda, K., Sugimoto, T., Tsuchiya, T., Ogawa, M. et al., "Slit Nozzle Injector for a New Concept of Direct Injection SI Gasoline Engine," SAE Technical Paper 2000-01-1902, 2000, doi:https://doi.org/10.4271/2000-01-1902.

8.47. Serras-Pereira, J., Aleiferis, P.G., Richardson, D., and Wallace, S., "Mixture Preparation and Combustion Variability in a Spray-Guided DISI Engine," SAE Technical Paper 2007-01-4033, 2007, doi:https://doi.org/10.4271/2007-01-4033.

8.48. Skogsberg, M., Dahlander, P., and Denbratt, I., "Spray Shape and Atomization Quality of an Outward-Opening Piezo Gasoline DI Injector," SAE Technical Paper 2007-01-1409, 2007, doi:https://doi.org/10.4271/2007-01-1409.

8.49. Smith, J., Szekely, G. Jr., Solomon, A., and Parrish, S., "A Comparison of Spray-Guided Stratified-Charge Combustion Performance with Outwardly-Opening Piezo and Multi-Hole Solenoid Injectors," *SAE Int. J. Engines* 4, no. 1 (2011): 1481-1497, doi:https://doi.org/10.4271/2011-01-1217.

8.50. Skiba, S. and Melbert, J., "Dosing Performance of Piezo Injectors and Sensorless Closed-Loop Controlled Solenoid Injectors for Gasoline Direct Injection," *SAE Int. J. Engines* 5, no. 2 (2012): 330-335, doi:https://doi.org/10.4271/2012-01-0394.

8.51. Stach, T., Schlerfer, J., and Vorbach, M., "New Generation Multi-hole Fuel Injector for Direct-Injection SI Engines—Optimization of Spray Characteristics by Means of Adapted Injector Layout and Multiple Injection," SAE Technical Paper 2007-01-1404, 2007, doi:https://doi.org/10.4271/2007-01-1404.

8.52. Yamashita, H., Seto, M., Ota, N., Murakami, Y. et al., "Spray Guided DISI Using Side Mounted Multi-Hole Injector," SAE Technical Paper 2007-01-1413, 2007, doi:https://doi.org/10.4271/2007-01-1413.

8.53. Mitroglou, N., Nouri, J.M., Yan, Y., Gavaises, M. et al., "Spray Structure Generated by Multi-Hole Injectors for Gasoline Direct-Injection Engines," SAE Technical Paper 2007-01-1417, 2007, doi:https://doi.org/10.4271/2007-01-1417.

8.54. Tanaka, M., Mizobuchi, T., Yamamoto, N., and Shigenaga, M., "Development of Multi-hole Nozzle Injector for Spray-Guided DISI Engine," SAE Technical Paper 2011-01-1888, 2011, doi:https://doi.org/10.4271/2011-01-1888.

8.55. Klimstra, J., "Carburetors for Gaseous Fuels: On Air-to-Fuel Ratio, Homogeneity, and Flow Restriction," SAE Paper 892141, 1989, doi:https://doi.org/10.4271/892141.

8.56. Geiss, R., Burkmyre, W., and Lanigan, J., "Technical Highlights of the Dodge Compressed Natural Gas Ram Van/Wagon," SAE Technical Paper 921551, 1992, doi:https://doi.org/10.4271/921551.

8.57. Chen, H., He, J., and Zhong, X., "Engine Combustion and Emission Fuelled with Natural Gas: A Review," *Journal of the Energy Institute* 92, no. 4 (2019): 1123-1136.

8.58. Singh, E., Morganti, K., and Dibble, R., "Dual-Fuel Operation of Gasoline and Natural Gas in a Turbocharged Engine," *Fuel* 237 (2019): 694-706.

8.59. Wallner, T., Pamminger, M., Scarcelli, R., Powell, C. et al., "Performance, Fuel Economy, and Economic Assessment of a Combustion Concept Employing In-Cylinder Gasoline/Natural Gas Blending for Light-Duty Vehicle Applications," *SAE Int. J. Engines* 12, no. 3 (2019): 271-289, doi:https://doi.org/10.4271/03-12-03-0019.

8.60. Kato, K., Igarashi, K., Masuda, M., Otsubo, K. et al., "Development of Engine for Natural Gas Vehicle," SAE Technical Paper 1999-01-0574, 1999, doi:https://doi.org/10.4271/1999-01-0574.

8.61. Dues, S.M., Adams, J.M., and Shinkle, G.A., "Combustion Knock Sensing: Sensor Selection and Application Issues," SAE Technical Paper 900488, 1990, doi:https://doi.org/10.4271/900488.

8.62.   Ricardo, H.R., "Recent Research Work on the Internal-Combustion Engine," SAE Technical Paper 220001, 1922, doi:https://doi.org/10.4271/760757.

8.63.   Noguchi, M., Sanda, S., and Nakamura, N., "Development of Toyota Lean Burn Engine," SAE Technical Paper 760757, 1976, doi:https://doi.org/10.4271/760757.

8.64.   Sanda, S. and Nakamura, N., Internal combustion engine provided with pre-combustion chamber. U.S. Patent 4,048,973, 1977.

8.65.   Brandstetter, W., Decker, G., and Reichel, K., "The Water-Cooled Volkswagen PCI-Stratified Charge Engine," SAE Technical Paper 750869, 1975, doi:https://doi.org/10.4271/750869.

8.66.   Adams, T.G., "Torch Ignition for Combustion Control of Lean Mixtures," SAE Technical Paper 790440, 1979, doi:https://doi.org/10.4271/790440.

8.67.   Durling, H.E., Johnston, R.P., and Polikarpus, K.K., Torch jet spark plug. U.S. Patent 5,421,300, 1995.

8.68.   Durling, H.E. and Ralph, J.G., Directed jet spark plug. U.S. Patent 6,213,085, 2001.

8.69.   Bunce, M., Blaxill, H., and Cooper, A., "Development of Both Active and Passive Pre-Chamber Jet Ignition Multi-Cylinder Demonstrator Engines," in *28th Aachen Colloquium Automobile and Engine Technology*, Aachen, Germany, 907-942, 2019.

8.70.   Cooper, A., Harrington, A., Bassett, M., and Pates, D., "Knock Mitigation Benefits Achieved through the Application of Passive MAHLE Jet Ignition Enabling Increased Output under Stoichiometric Operation," SAE Technical Paper 2021-01-0477, 2021, doi:https://doi.org/10.4271/2021-01-0477.

8.71.   Murase, E. and Hanada, K., "Enhancement of Combustion by Injection of Radicals," SAE Technical Paper 2000-01-0194, 2000, doi:https://doi.org/10.4271/2000-01-0194.

8.72.   Boretti, A., "Progress of Direct Injection and Jet Ignition in Throttle-Controlled Engines," *SAE Int. J. Adv. & Curr. Prac. in Mobility* 1, no. 1 (2019): 61-66, doi:https://doi.org/10.4271/2019-26-0045.

8.73.   Millo, F., Rolando, L., Piano, A., Sementa, P. et al., "Experimental and Numerical Investigation of a Passive Pre-Chamber Jet Ignition Single-Cylinder Engine," SAE Technical Paper 2021-24-0010, 2021, doi:https://doi.org/10.4271/2021-24-0010.

8.74.   Hassan, W.T.F., "The New Jaguar 12-Cyl Engine," SAE Technical Paper 720163, 1972, doi:https://doi.org/10.4271/720163.

8.75.   Okuno, T., Mase, Y., and Naritomi, T., "Development of a New 12 Valve 4 Cylinder Engine," SAE Technical Paper 881776, 1988, doi:https://doi.org/10.4271/881776.

8.76.   Gruden, D., Richter, H., and Wurster, W., "Combustion Chamber Investigations at Porsche-Paving the Way for the 4-Valve Engine," SAE Technical Paper 845005, 1984, doi:https://doi.org/10.4271/845005.

8.77.   Kimbara, Y., Konishi, M., Kitagawa, K., Watanabe, H. et al., "Innovative Toyota Standard Engine Equipped with 4-Valve," SAE Technical Paper 870352, 1987, doi:https://doi.org/10.4271/870352.

8.78.   Larsson, T., Bergstrom, K., Hinderman, T., Hauptmann, L. et al., "The Volvo 3-Litre 6-Cylinder Engine with 4-Valve Technology," SAE Technical Paper 901715, 1990, doi:https://doi.org/10.4271/901715.

8.79.   Regueiro, A., "DaimlerChrysler's New 1.6L, Multi-Valve 4-Cylinder Engine Series," SAE Technical Paper 2001-01-0330, 2001, doi:https://doi.org/10.4271/2001-01-0330.

8.80.   Sykes, R., "Tickford Five Valve Per Cylinder Technology for Optimised Performance and Combustion," SAE Technical Paper 950815, 1995, doi:https://doi.org/10.4271/950815.

8.81.   Pfalzgraf, B., Fitzen, M., Siebler, J., and Erdmann, H.-D., "First ULEV Turbo Gasoline Engine—The Audi 1.8 l 125 kW 5-Valve Turbo," SAE Technical Paper 2001-01-1350, 2001, doi:https://doi.org/10.4271/2001-01-1350.

8.82.   Moore, S., "Combustion Chamber Deposit Interference Effects in Late Model Vehicles," SAE Technical Paper 940385, 1994, doi:https://doi.org/10.4271/940385.

8.83.   Schreyer, P., Starke, K.W., Thomas, J., and Crema, S., "Effect of Multifunctional Fuel Additives on Octane Number Requirement of Internal Combustion Engines," SAE Technical Paper 932813, 1993, doi:https://doi.org/10.4271/932813.

8.84.   Newhall, H.K., "Control of Nitrogen Oxides by Exhaust Recirculation—A Preliminary Theoretical Study," SAE Technical Paper 670495, 1967, doi:https://doi.org/10.4271/670495.

8.85. Deeter, W.F., Daigh, H.D., and Wallin, O.W., "An Approach for Controlling Vehicle Emissions," SAE Technical Paper 680400, 1968, doi:https://doi.org/10.4271/680400.

8.86. Brownson, D.A., Johnson, R.S., and Candelise, A., "A Progress Report on ManAirOx-Manifold Air Oxidation of Exhaust Gas," SAE Technical Paper 620403, 1962, doi:https://doi.org/10.4271/620403.

8.87. Brownson, D.A. and Stebar, R.F., "Factors Influencing the Effectiveness of Air Injection in Reducing Exhaust Emissions," SAE Technical Paper 650526, 1965, doi:https://doi.org/10.4271/650526.

8.88. Steinhagen, W.K., Niepoth, G.W., and Mick, S.H., "Design and Development of the General Motors Air Injection Reactor System," SAE Technical Paper 660106, 1966, doi:https://doi.org/10.4271/660106.

8.89. Benson, J.D., "Reduction of Nitrogen Oxides in Automobile Exhaust," SAE Technical Paper 690019, 1969, doi:https://doi.org/10.4271/690019.

8.90. Tanaka, K., Akutagawa, M., Ito, K., Higashi, Y. et al., "Toyo Kogyo Status Report on Low Emission Concept Vehicles," SAE Technical Paper 720486, 1972, doi:https://doi.org/10.4271/720486.

8.91. Houndry, E.J., Catalytic converter for exhaust gases. U.S. Patent 2,674,521, 1954.

8.92. Houndry, E.J., Catalytic structure and composition. U.S. Patent 2,742,437, 1956.

8.93. Jagel, K.I. and Dwyer, F.G., "HC/CO Oxidation Catalysts for Vehicle Exhaust Emission Control," SAE Technical Paper 710290, 1971, doi:https://doi.org/10.4271/710290.

8.94. Cantwell, E.N., Jacobs, E.S., Kunz, W.G., and Liberi, V.E., "Control of Particulate Lead Emissions from Automobiles," SAE Technical Paper 720672, 1972, doi:https://doi.org/10.4271/720672.

8.95. Treuhaft, M.B. and Wisnewski, J.P., "Trapping of Lead Particulates in Automotive Exhaust," SAE Technical Paper 770059, 1977, doi:https://doi.org/10.4271/770059.

8.96. Keith, C.D., Kenah, P.M., and Bair, D.L., Method of preparing an oxidation catalyst. U.S. Patent 3,331,787, 1967.

8.97. Yonehara, K., Ohhata, T., and Hara, H., Oxidation catalyst for purifying exhaust gases of internal combustion engines. U.S. Patent 3,839,224, 1974.

8.98. Hegedus L., and Summers, J.C., Pellet-type oxidation catalyst. U.S. Patent 4,051,073, 1977.

8.99. Wheelock, K.S., Selective automotive exhaust catalysts and a process for their preparation. U.S. Patent 4,171,289, 1979.

8.100. Keith, C.D., Schreuders, T., and Cunningham, C.E., "Apparatus for purifying exhaust gases of an internal combustion engine. U.S. Patent 3,441,381, 1969.

8.101. Keith, C.D., and Schreuders, T., Catalyst cartridge. U.S. Patent 3,441,382, 1969.

8.102. Pereira, C.J., Kubsh, J.E. and Hegedus, L., Monolith washcoat having optimum pore structure and optimum method of desgning the washcoat. U.S. Patent 4,771,029, 1988.

8.103. Pereira, C.J., Kubsh, J.E., and Hegedus, L., Process for treating automotive exhaust gases using monolith wascoat having optimum pore structure. U.S. Patent 4,859,433, 1989.

8.104. Snyder, P.W., Stover, W.A., and Lassen, H.G., "Status Report on HC/CO Oxidation Catalysts for Exhaust Emission Control," SAE Technical Paper 720479, 1972, doi:https://doi.org/10.4271/720479.

8.105. Held, W., König, A., Richter, T., and Puppe, L., "Catalytic NOx Reduction in Net Oxidizing Exhaust Gas," SAE Technical Paper 900496, 1990, doi:https://doi.org/10.4271/900496.

8.106. Ozawa, M., Suzuki, S., and Toda, H., "High-Temperature Automotive Catalytic Nitrogen Oxide Reduction over Copper-Lanthanum-Alumina," *Journal of the American Ceramic Society* 80, no. 8 (1997): 1957-1964.

8.107. Yahiro, H. and Iwamoto, M., "Copper Ion-Exchanged Zeolite Catalysts in deNOx Reaction," *Applied Catalysis A: General* 222, no. 1–2 (2001): 163-181.

8.108. Kharas, K.C.C., Copper-silver zeolite catalysts in exhaust gas treatment. U.S. Patent 5,968,466, 1999.

8.109. Bhore, N.A., Dwyer, F.G., Marler, D.O., and McWilliams, J.P., Method for reducing automotive NOX emissions in lean burn internal combustion engine exhaust using a transition metal-containing zeolite catalyst which is in-situ crystallized. U.S. Patent 5,254,322, 1993.

8.110. Takami, A., Takemoto, T., Iwakuni, H., Saito, F. et al., "Development of Lean Burn Catalyst," SAE Technical Paper 950746, 1995, doi:https://doi.org/10.4271/950746.

8.111. Miyoshi, N., Matsumoto, S., Katoh, K., Tanaka, T. et al., "Development of New Concept Three-Way Catalyst for Automotive Lean-Burn Engines," SAE Technical Paper 950809, 1995, doi:https://doi.org/10.4271/950809.

8.112. Hachisuka, I., Hirata, H., Ikeda, Y., and Matsumoto, S., "Deactivation Mechanism of NOX Storage-Reduction Catalyst and Improvement of Its Performance," SAE Technical Paper 2000-01-1196, 2000, doi:https://doi.org/10.4271/2000-01-1196.

8.113. Yazawa, Y., Watanabe, M., Takeuchi, M., Imagawa, H. et al., "Improvement of NOx Storage-Reduction Catalyst," SAE Technical Paper 2007-01-1056, 2007, doi:https://doi.org/10.4271/2007-01-1056.

8.114. Küsell, M., Moser, W., and Philipp, M., "Motronic MED7 for Gasoline Direct Injection Engines: Engine Management System and Calibration Procedures," SAE Technical Paper 1999-01-1284, 1999, doi:https://doi.org/10.4271/1999-01-1284.

8.115. Collier, T., Brogan, M., Retman, P., and Bye, R., "Development of a Rapid Sulfation Technique and Fundamental Investigations into Desulfation Process," SAE Technical Paper 2003-01-1162, 2003, doi:https://doi.org/10.4271/2003-01-1162.

8.116. Asik, J.R., Dobson, D.A., and Meyer, G.M., "Suppression of Sulfide Emission During Lean NOx Trap Desulfation," SAE Technical Paper 2001-01-1299, 2001, doi:https://doi.org/10.4271/2001-01-1299.

8.117. Kim, C.H., Perry, K., Viola, M., Li, W. et al., "Three-Way Catalyst Design for Urealess Passive Ammonia SCR: Lean-Burn SIDI Aftertreatment System," SAE Technical Paper 2011-01-0306, 2011, doi:https://doi.org/10.4271/2011-01-0306.

8.118. Li, W., Perry, K., Narayanaswamy, K., Kim, C.H. et al., "Passive Ammonia SCR System for Lean-Burn SIDI Engines," *SAE Int. J. Fuels Lubr.* 3, no. 1 (2010): 99-106, doi:https://doi.org/10.4271/2010-01-0366.

8.119. Guralp, O., Qi, G., Li, W., and Najt, P., "Experimental Study of NOx Reduction by Passive Ammonia-SCR for Stoichiometric SIDI Engines," SAE Technical Paper 2011-01-0307, 2011, doi:https://doi.org/10.4271/2011-01-0307.

8.120. Braisher, M., Stone, R., and Price, P., "Particle Number Emissions from a Range of European Vehicles," SAE Technical Paper 2010-01-0786, 2010, doi:https://doi.org/10.4271/2010-01-0786.

8.121. Surenahalli, H.S., Backhaus, J., Liu, Q., Dabhoiwala, R. et al., "Experimental Study of Impact of Ash and Soot on Tail Pipe Particle Number," SAE Technical Paper 2019-01-0976, 2019, doi:https://doi.org/10.4271/2019-01-0976.

8.122. Araji, F., Matida, E., and Huang, X., "Effects of Porosity, Wall Thickness, and Length on the Filtration Efficiency of Gasoline Particulate Filters," SAE Technical Paper 2022-01-5010, 2022, doi:https://doi.org/10.4271/2022-01-5010.

8.123. Richter, J., Klingmann, R., Spiess, S., and Wong, K.-F., "Application of Catalyzed Gasoline Particulate Filters to GDI Vehicles," *SAE Int. J. Engines* 5, no. 3 (2012): 1361-1370, doi:https://doi.org/10.4271/2012-01-1244.

8.124. Weiwei, G., Ball, D., Yang, C., Meng, X. et al., "Advantages of Coated Gasoline Particulate Filters in the CC2 Position for China 6B," SAE Technical Paper 2021-01-0587, 2021, doi:https://doi.org/10.4271/2021-01-0587.

8.125. Qiao, D., Pan, X., He, J., Chen, J. et al., "Novel Pt/Rh Coated Gasoline Particulate Filter Development and Its Application," SAE Technical Paper 2022-01-0546, 2022, doi:https://doi.org/10.4271/2022-01-0546.

8.126. Ota, Y., Takahasi, H., and Maekawa, R., "Development of Coated Gasoline Particulate Filter Design Method Combining Simulation and Multi-Objective Optimization," *SAE Int. J. Adv. & Curr. Prac. in Mobility* 4, no. 1 (2022): 204-210, doi:https://doi.org/10.4271/2021-01-0838.

8.127. Van Nieustadt, M.J., Ruhland, H.H., Springer, M.K., Lorenz, T. et al., System and method for regenerating a particulate filter for a direct injection engine. U.S. Patent 2011/072794 A1, 2011.

8.128. Sorensen, C.M. Jr., Process and devices for regenerating gasoline particulate filters. U.S. Patent 2011/0120090 A1, 2011.

8.129. Harada, K., Yamada H., and Takami, A., Particulate filter regenerating system. U.S. Patent 8,051,646 B2, 2011.

8.130. Custer, N., Kamp, C., Sappok, A., Pakko, J. et al., "Lubricant-Derived Ash Impact on Gasoline Particulate Filter Performance," *SAE Int. J. Engines* 9, no. 3 (2016): 1604-1614, doi:https://doi.org/10.4271/2016-01-0942.

8.131. Masumitsu, N., Otsuka, S., Fujikura, R., Imai, Y. et al., "Analysis of the Pressure Drop Increase Mechanism by Ash Accumulated of Coated GPF," SAE Technical Paper 2019-01-0981, 2019, doi:https://doi.org/10.4271/2019-01-0981.

8.132. Caillaud, S., Courtois, O., Delvigne, T., and Hennebert, B., "Contribution of Lubricant Additives to Ash Generation on a Close Coupled GPF," *SAE Int. J. Adv. & Curr. Prac. in Mobility* 3, no. 1 (2021): 477-484, doi:https://doi.org/10.4271/2020-01-2162.

8.133. Thawko, A. and Tartakovsky, L., "The Mechanism of Particle Formation in Non-Premixed Hydrogen Combustion in a Direct-Injection Internal Combustion Engine," *Fuel* 327 (2022): 125187.

Further Reading

Boger, T. and Cutler, W., *Reducing Particulate Emissions in Gasoline Engines* (Warrendale, PA: SAE International, 2018).

Ferrari, A. and Pizzo, P., *Injection Technologies and Mixture Formation Strategies for Spark-Ignition and Dual-Fuel Engines* (Warrendale, PA: SAE International, 2022).

Kasab, J. and Strzelec, A., *Automotive Emissions Regulations and Exhaust Aftertreatment Systems* (Warrendale, PA: SAE International, 2020).

Mondt, J.R., *Cleaner Cars—The History and Technology of Emission Control Since the 1960s* (Warrendale, PA: SAE International, 2000).

Stone, R., *Introduction to Internal Combustion Engines*, 3rd ed. (Warrendale, PA: SAE International, 1999).

Zhao, F. (Eds), *Technologies for Near-Zero-Emission Gasoline-Powered Vehicles* (Warrendale, PA: SAE International, 2006).

Zhao, F., Harrington, D.L., and Lai, M.-C.D., *Automotive Gasoline Direct-Injection Engines* (Warrendale, PA: SAE International, 2002).

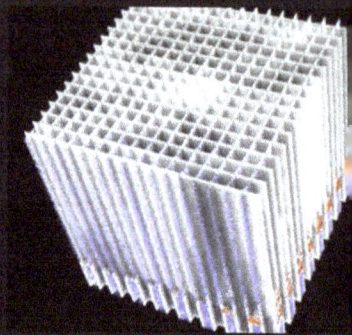

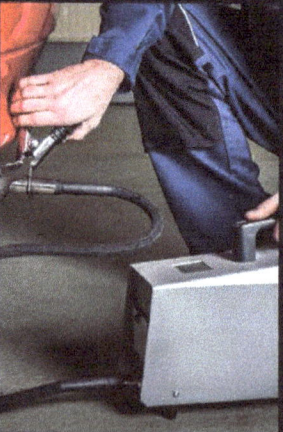

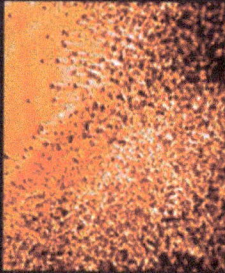

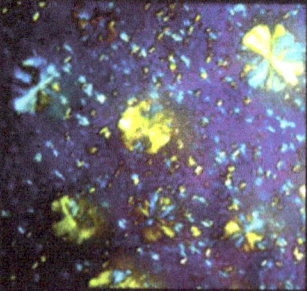

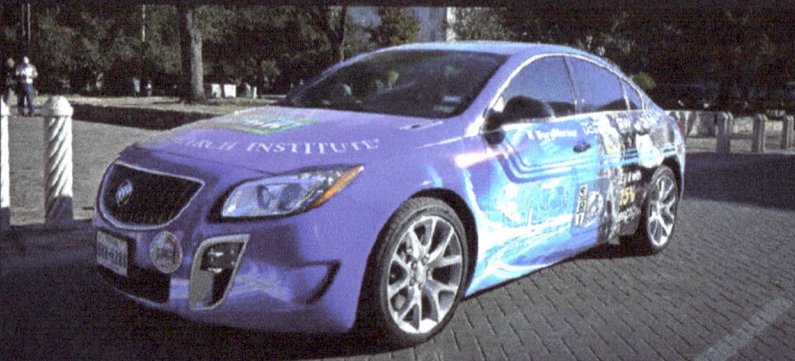

9

Gasoline Volatility

From Chapter 7, it is clear that a positive ignition engine, a gasoline engine, requires a well-mixed combination of air and fuel vapor around the ignition source, the spark plug. To get to this situation, the liquid fuel must be turned into vapor. Users have long recognized that the requirements for a satisfactory fuel are as follows [9.1]:

1. That it shall permit ready starting of a cold engine.
2. That it shall permit smooth and reliable operation of an engine at varied and rapidly changing loads and speeds.
3. That it shall not contain substances such as free acid, which by corrosion or otherwise can injure an engine.
4. That there shall be a minimum tendency for either solid or liquid residues to accumulate in the engine.

The latter two points are discussed elsewhere in this book. Regarding the first two points, this has been expressed [9.2] in terms of the fuel characteristics as "*The distillation curve is the best index of a fuel's suitability.*" This chapter looks at the distillation curve and other aspects of fuel volatility and how they affect the performance of the engine.

Gasoline is a mixture of many different compounds, each having its own boiling point and vapor-forming characteristics. Thus, when distilled using a simple still with little fractionation, gasolines show a boiling range covering a temperature spread of around 170°C from the initial boiling point (IBP) to the final boiling point (FBP). The latter value is usually specified in national specifications. The temperature range over which the gasoline distills will depend on the composition of the gasoline and the fractionation efficiency of the distillation column. A very efficient column may be able to separate individual compounds if they have moderate differences in boiling point and if each is present in a reasonable amount. Such a column would give a distillation curve looking something like Figure 9.1 for a mixture containing equal amounts of five different components, each having a different boiling point.

FIGURE 9.1 Good fractionation can separate compounds differing in boiling point.

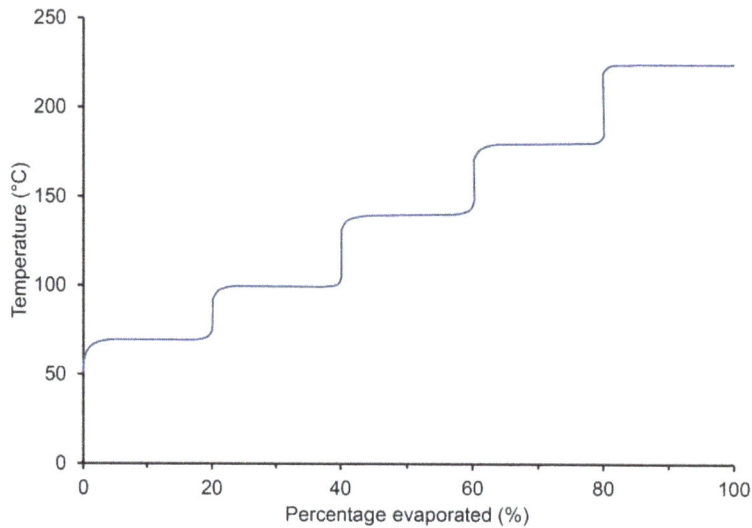

Gasolines usually contain so many readily identifiable compounds (up to about 400) that the distillation curve is quite smooth, even when a high degree of fractionation is used. Figure 9.2 shows gasoline distillation curves obtained at two different levels of fractionation efficiency; curve A is the sort of curve that might be obtained with good fractionation, whereas curve B is the same gasoline but with very little fractionation using the standard ASTM D86 [9.3] test.

FIGURE 9.2 Gasoline distillation curves.

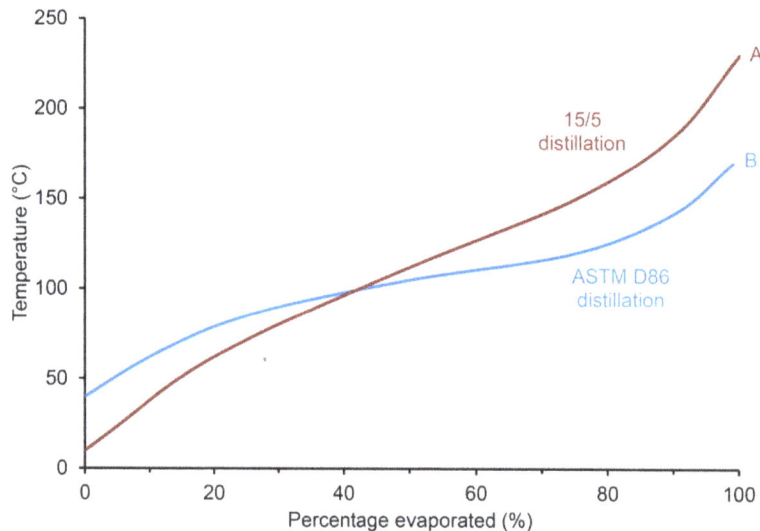

The temperature at which the mixture begins to distill is much lower with a high degree of fractionation, and the final temperature is much higher, showing that individual compounds or groups of similar boiling compounds are separated much more readily in this case. Of course, if even better separation was required, then techniques such as gas/liquid chromatography (GLC) could be used.

However, it is not necessary to be concerned with high-efficiency fractionation when considering the behavior of a gasoline in an engine, because the type of evaporation that occurs represents only a coarse separation of different boiling range materials. This is not to say that the distillation curve is unimportant; it is very important and correlates with different performance aspects of different vehicles, including its driveability. Driveability can be defined as the response of the vehicle to the throttle; a vehicle with good driveability will accelerate smoothly without stumbling or hesitating, will idle evenly, and will cruise without surging.

Figure 9.3 shows a gasoline distillation curve obtained using a low fractionation efficiency still and indicates the aspects of vehicle performance that different parts of it influence.

FIGURE 9.3 Gasoline volatility is a compromise.

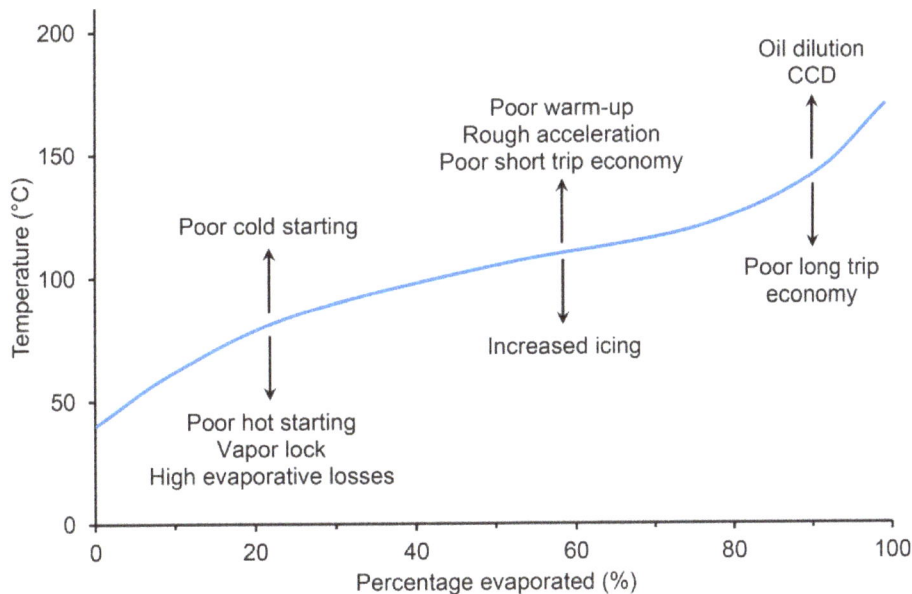

If the distillation curve is displaced downwards, the gasoline becomes more volatile and vice versa. The front end, that is, the compounds in the gasoline having boiling points up to about 70°C, is the first to be distilled over, and this controls the ease of starting and the likelihood of hot weather problems such as vapor lock occurring. The midrange largely controls the way that the vehicle drives in cold weather, and particularly the time for the engine to warm up. It also influences to some extent the tendency for ice to form in a carburetor or TBI system during cool, humid weather. The back end contains all the heavier, high boiling point compounds. These compounds usually have a higher carbon-to-hydrogen ratio and thus have a higher energy content and are important in improving fuel economy when the engine is fully warmed up. However, some of these heavier compounds may find their way past the pistons into the crankcase and dilute the crankcase oil. They are also not as readily combusted as the lighter components and may give rise to combustion chamber deposits.

Figure 9.3 shows that if the gasoline is made more volatile, one set of problems can occur, and if the gasoline is made less volatile, another set of difficulties is possible. Weather conditions, particularly ambient temperature, influence the choice of volatility required for satisfactory operation. Altitude also has a small effect because atmospheric pressure affects the rate of evaporation of gasoline. Vehicles themselves vary enormously in the way that they respond to gasoline volatility. Some vehicles are extremely tolerant to changes, whereas others can give severe driveability problems if the gasoline volatility is not closely matched to the weather conditions prevailing. The most important vehicle design aspect in this respect is the proximity of the fuel system to hot engine parts. It is necessary to avoid excessive vaporization during hot weather and yet to make sure that there is enough heat present during cold weather to adequately vaporize the gasoline.

The above is generally true but with the wider application of direct injection technology and the wider awareness of low-speed preignition (LSPI) or super-knock, as discussed in Chapter 7, Section 7.8.1, there has been increased focus on the causes of this preignition phenomenon. While some research has focused on deposit particles within the combustion chamber [9.4–9.6] others have focused on lubricating oil passing the piston ring pack and being ejected from the crevice around the top land of the piston [9.7–9.9]. Research in this latter area has suggested that fuel volatility, and in particular the back-end volatility, can contribute to wall-wetting, which can increase the probability of LSPI [9.10–9.13]. These considerations may influence the setting of volatility specifications in the future, but currently the setting of volatility specifications is a compromise that is influenced by weather conditions, geographical location, and the characteristics of the vehicle population including the large legacy fleet.

9.1. Measurement of Gasoline Volatility

The tests most commonly used to define gasoline volatility are vapor pressure, the Reid vapor pressure (RVP), the dry vapor pressure equivalent (DVPE), the ASTM Distillation test, and the vapor–liquid ratio. Each of these is discussed in more detail in the following sections.

9.1.1. Vapor Pressure

Vapor pressure is a measure of volatility and is defined as the pressure at which gasoline liquid and its vapor are in equilibrium at 37.8°C. This was traditionally determined under standard conditions using a bomb in a water bath (the Reid method, described in Section 9.1.2). However, oxygenates blends create measurement problems due to their affinity to the water-saturated air employed in the test procedure. Other methods are therefore generally used.

9.1.2. Reid Vapor Pressure

This is the vapor pressure obtained under standard conditions, using an air-to-liquid ratio of between 3.8:1 and 4.2:1 and a temperature of 37.8°C. It is determined by ASTM Procedure D323 [9.14], which involves filling a metal chamber with a chilled sample and connecting it to an air chamber that is in turn connected to a pressure gauge, as shown in Figure 9.4.

FIGURE 9.4 Reid vapor pressure apparatus.

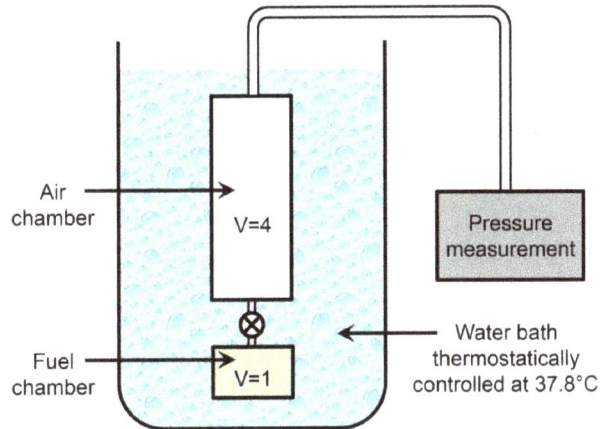

The apparatus is immersed in a water bath at 37.8°C and is shaken periodically until a constant pressure is obtained; this is the RVP. The value of vapor pressure determined by the Reid method differs from the true vapor pressure as a result of some small sample vaporization and the presence of water vapor and air in the confined space. The procedure is not suitable for gasolines containing oxygenated compounds other than methyl tertiary butyl ether (MTBE).

9.1.3. Dry Vapor Pressure and DVPE

The procedure used for measuring the vapor pressure of a gasoline blend containing alcohols can affect the result. The standard RVP procedure (ASTM D323) measures the vapor pressure in the presence of water-saturated air, and this gives rise to a lower value than if the dry procedure (ASTM D4953) is used. Some fuel specifications will allow the "wet" procedure to be used, and some require the use of a "dry" procedure. The differences that can occur between the two procedures are shown in Figure 9.5 [9.15].

FIGURE 9.5 Vapor pressure of gasoline-methanol blends depending on different methods and different base gasolines.

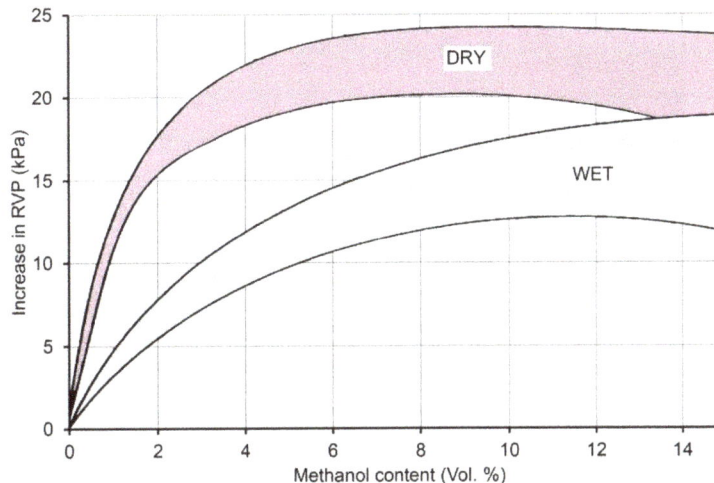

Due to the fact that the Reid method tends to underestimate the vapor pressure of alcohol-containing fuels, new methods have been developed [9.16, 9.17] that ensure that the sample is not exposed to water or water vapor during testing. The dry method is a modification of the Reid method and uses the same apparatus but differs in the sample handling to ensure water does not contaminate the measurement.

The test method ASTM D5190 uses an automated instrument, available form Southwest Research Institute, San Antonio, TX, to determine a vapor pressure for the sample. The sample is calibrated using n-hexane and n-pentane to zero and span the pressure transducer. However, the instrument is known to have a bias relative to the dry method. The measured value of vapor pressure is then used to determine a DVPE. The equation relating the measured value to DVPE is stated in the procedure and is based on an interlaboratory correlation program relating the automatic method to the dry method.

The test mini method, like the automatic method, relies on introducing the sample into a chamber that can be expanded to five times its original capacity to produce the 4:1 ratio of liquid to vapor. This is in contrast to the Reid method, which connects two chambers. The sample size for the mini method is only 1 to 10 mL. Again, the result obtained is not a true vapor pressure, and DVPE must be calculated using the equation given in the procedure. The calculations for DVPE are different for the automatic method and the mini method.

9.1.4. Distillation by ASTM D86

In this test [9.3], a 100 mL sample of gasoline is distilled in a standard apparatus under specified conditions of heat input and coolant temperature so that the distillation rate is strictly controlled. The basis of the apparatus is shown schematically in Figure 9.6, although automatic distillation units that conform to this method are available from a variety of different suppliers.

FIGURE 9.6   ASTM D86 distillation apparatus.

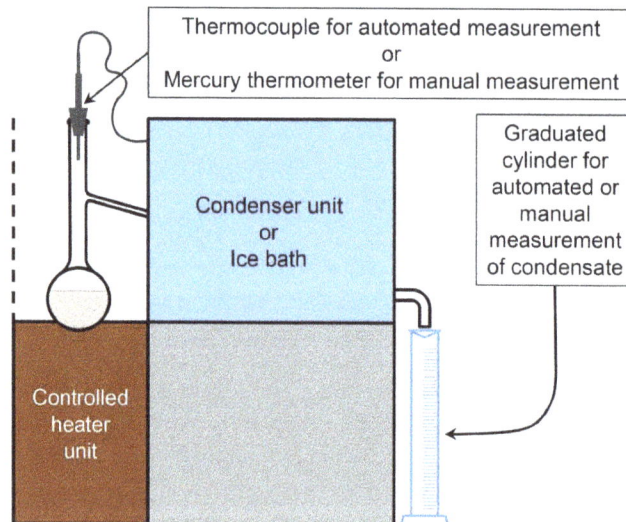

In the automated units the distillation rate is controlled by varying the heat input based on feedback signals from the distillate in the receiver. The fractionation efficiency is equivalent to about one theoretical plate, and a curve such as that shown in Figure 9.7 is obtained.

FIGURE 9.7 ASTM D86 distillation curve.

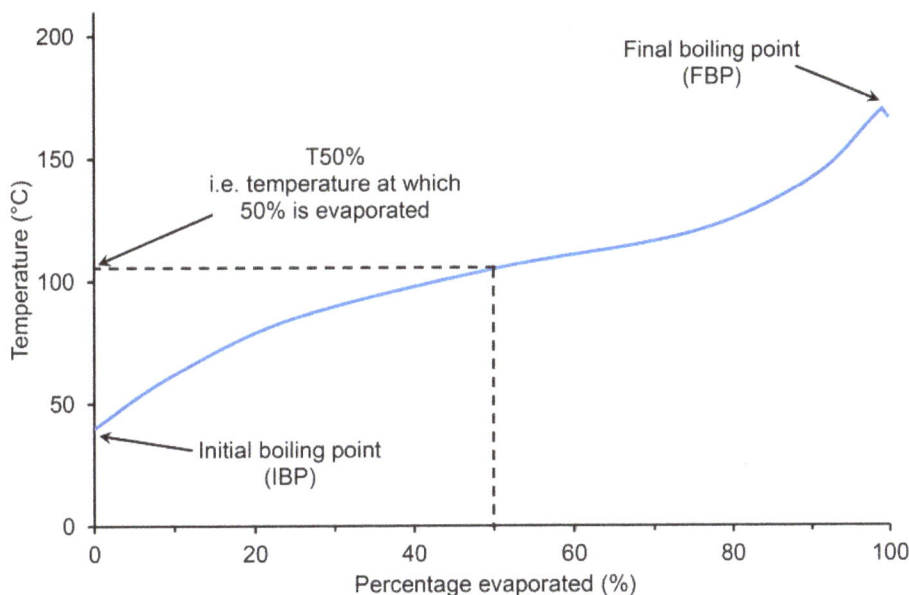

Some of the various terms used in relation to distillation data are illustrated in this figure, and others that are important when considering distillation data are the following:

- Percent distilled: the volume in milliliters of condensate in the receiver corresponding to a simultaneous temperature reading. Also called percent recovered.

- Percent residue: the volume of residue left in the flask when allowed to cool after the distillation is complete.

- Percent loss: 100 minus (maximum percent distilled plus percent residue). This represents mainly those very light hydrocarbons that are not condensed.

- Percent evaporated: the sum of the percent distilled and the percent loss.

Distillation data are often represented and specified by the temperature at which a given percentage of the gasoline is evaporated so that T10% (sometimes written T10, T_{10}, or $T_{10\%}$) is the temperature at which 10% of the gasoline is evaporated using ASTM D86. The data can also be represented by the percentage evaporated at a given temperature so that E70°C (E70, E_{70}, or $E_{70°C}$) represents the percentage evaporated at 70°C. It is considered preferable and more meaningful by many to use percentages evaporated rather than temperatures, particularly when carrying out blending calculations. It is important to remember that it is always percent evaporated rather than percent distilled or recovered when considering specifications or comparing gasolines.

The distillation characteristics of a gasoline are not always similar to that shown in Figure 9.6 because the shape is dependent on the blend composition. Figure 9.8 illustrates two other types of distillation curve that can be found and compares them with a more conventional curve.

FIGURE 9.8 Atypical gasoline distillation curves.

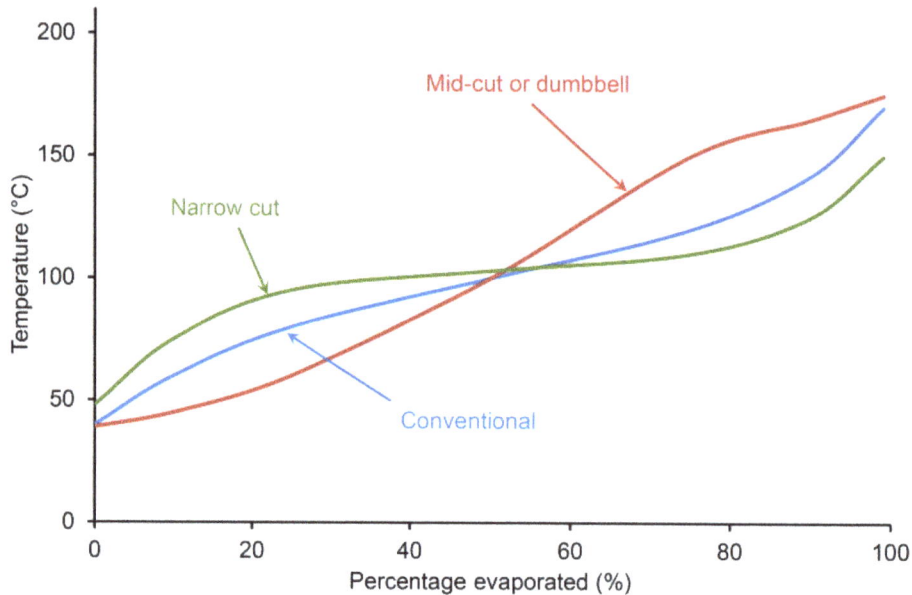

Gasolines containing only components that boil within the same narrow range are known as narrow-cut blends and show a fairly flat plateau for a large part of the curve. Blends consisting of light and heavy components with very little material boiling in the intermediate temperature range are known as dumbbell or mid-cut blends, and these show a steep rise in temperature once the light materials have been distilled until the heavy components start distilling. Other types of distillation curve are possible, such as one with a low front-end volatility and a high back-end volatility, and this is sometimes known as a front-cut gasoline. The driveability performance of fuels with these atypical distillation curves can be quite different from that of conventional gasolines.

9.1.5. Vapor–Liquid Ratio

The RVP test involves an arbitrary 4:1 ratio of vapor-to-liquid at a given temperature, but when a gasoline is subjected to the levels of temperature found in the fuel system of a vehicle, it can be important to know what ratio of vapor-to-liquid is likely to form. The amount of vapor produced can have a direct influence on the performance of the vehicle, particularly under conditions of high ambient temperature.

A test to determine the vapor–liquid ratio (V/L) [9.18] was in place until 2008. The test measured the volume of vapor formed at atmospheric pressure from a given volume of gasoline at a specified test temperature. The test was carried out by introducing a measured volume of liquid fuel that had been chilled to about 0°C through a rubber septum into a glycerin-filled burette. The charged burette was then placed in a temperature-controlled water bath and the volume of vapor in equilibrium with liquid fuel was determined, that is, the temperature was set and the V/L was determined. Due to the fact that gasoline-oxygenate blends may be soluble in glycerin, the method specified mercury as an alternative containing liquid. For health and safety reasons the alternative method was no longer used and the method has been withdrawn.

The V/L at a given temperature can now be inferred from the determination of the temperature to achieve a given V/L using ASTM D5188. This method determines the temperature to which the given volume of gasoline must be raised to achieve a vapor pressure of one atmosphere (101.3 kPa) in an evacuated chamber of given volume, meaning, the V/L is set and the temperature is determined.

A good correlation between V/L and a combination of RVP and certain distillation points has been shown [9.19]. Procedures have been established in some fuel specifications [9.10] for estimating temperature-V/L values from these parameters. There is also a procedure for correcting the estimated V/L values for ethanol blends [9.20].

9.1.6. Effect of Oxygenated Blending Components

The presence of alcohols can have a dramatic effect on the vapor pressure of a gasoline to which it is added, as illustrated in Figure 9.9 [9.21]. Note Oxinol™ 50 was an Area Chemical Company's product containing gasoline-grade t-butanol (GTBA) and a corrosion inhibitor.

FIGURE 9.9   Effect of oxygenates concentration on blend vapor pressure.

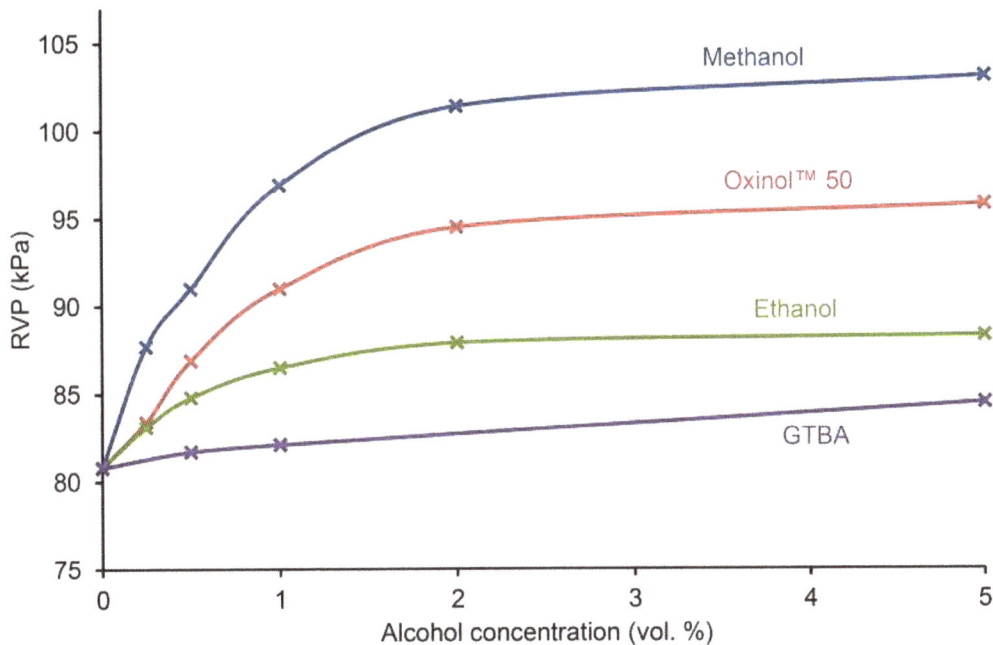

© SAE International.

The effect on vapor pressure is generally greater with the lower the carbon number of the alcohol; thus, blending significant quantities of methanol into the gasoline pool may require other lighter components, such as butane, to be backed out in order to meet the relevant vapor pressure specification. This can adversely affect the refinery economics. It can also lead to atypical distillation curves.

It is apparent from Figure 9.9 that gasolines and alcohols blend to form nonideal solutions, that is, they do not obey Raoult's law. This is also true of oxygenated fuels. If two hydrocarbon fuels are mixed, then the RVP of commingled blend can be easily estimated from the RVPs of each of the constituents because they form a nearly ideal solution. The RVP of hydrocarbon fuel blend is equal to the volume-weighted average of RVPs of each of the constituents [9.22]. If one or both of the fuels contain ethanol, the vapor pressure of the blend can be higher than that estimated by volume-weighted averages. Mathematical equations and models have been developed to estimate the vapor pressure of adding alcohol and of commingling oxygenated gasoline blends [9.22–9.24]. The fact that fuels do not blend as ideal solutions can have implications for vehicle performance. For example, if a vehicle's tank is part full of a nonoxygenated gasoline and it is refueled with an oxygenated gasoline with the same vapor pressure, then the vapor pressure of the mixture can be higher than that of the two different gasolines. This is shown in Figure 9.10, where two 63 kPa fuels, one nonoxygenated (E0) and the other containing 10% ethanol (E10), are commingled.

FIGURE 9.10   Effect on vapor pressure of commingling an E0 and E10 fuel.

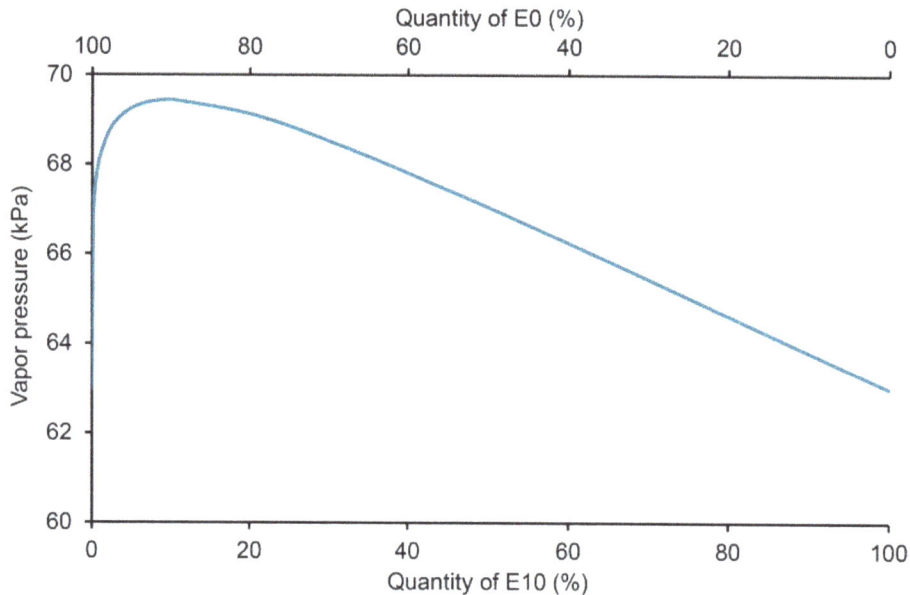

© SAE International.

Figure 9.11 shows the influence on the ASTM D86 distillation curve of adding various oxygenates to a summer-grade gasoline [9.25]; it can be seen that the curve can be highly distorted by the presence of alcohols. Figure 9.12 shows the influence on a winter-grade gasoline [9.25].

FIGURE 9.11 Effects of oxygenates on the distillation curve of a summer-grade gasoline.

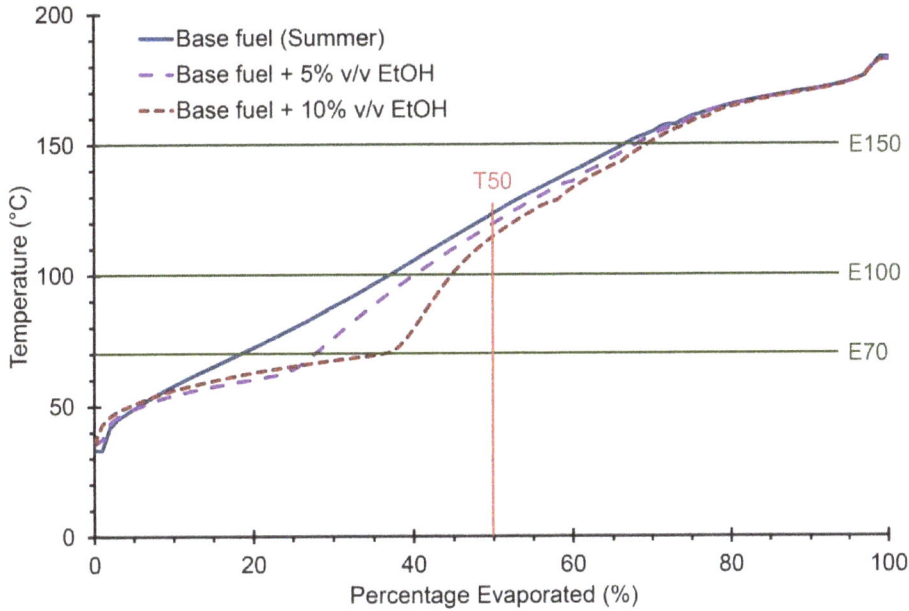

FIGURE 9.12 Effects of oxygenates on the distillation curve of a winter-grade gasoline.

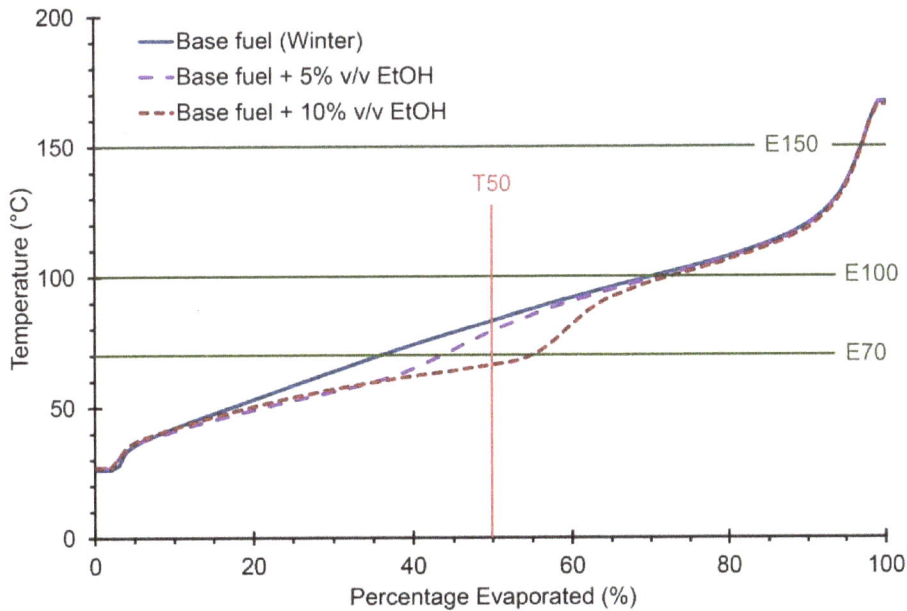

© SAE International.

As the molecular weight of the alcohol increases, the effect becomes less and the temperature range of the section of the distillation curve that is influenced moves up the curve.

Ethers do not have the same effect of changing just a part of the distillation curve, but they can make the whole gasoline lighter or heavier depending on the volatility of the base gasoline and the specific ether being used [9.26–9.28]. An example of the effect of MTBE on the distillation curve is also shown in Figure 9.11.

The vapor–liquid ratio is increased by the use of alcohols as would be expected from a consideration of the effect on RVP and the distillation curve. The use of ethers tends to have a far less significant effect [9.26].

9.2. Cold Starting

For a spark ignition engine to start, the air-to-fuel ratio of the gaseous mixture in the combustion chamber must be within the ignitable range; in general, it must be between about 7:1 and 20:1 by weight, although some engines may be able to start with air-to-fuel ratios outside these limits. When the engine is cold, it may be difficult to achieve even the leanest ignitable air-to-fuel ratio because the fuel may not vaporize sufficiently, and under these conditions the mixture is richened to bring it within the ignitable range by the use of a choke or other enrichment device with carbureted vehicles, or with additional fuel injection or by increasing the injection pulse with fuel-injected vehicles. However, unvolatilized fuel can have a negative impact on cold start hydrocarbon emissions.

There are many other reasons for a vehicle to fail to start easily [9.29]. An adequate cranking speed is important because higher speeds favor good atomization of the fuel in carbureted vehicles and provide some warming of the engine due to the frequent compressions. Battery condition is also important because of its effect on cranking speed, and it is unfortunate that battery power output drops as the temperature reduces. The lubricant viscosity at low temperature also influences cranking speed and so affects cold starting. Finally, the ignition system and ignition timing are significant factors because it is important to have a spark of sufficient power at the optimum time to ignite the mixture. Thus, the fuel has a comparatively small effect on cold starting compared with the mechanical and electrical factors mentioned, except at low temperatures.

9.2.1. Different Fuel System Technologies

In spite of its minor role, fuel volatility becomes increasingly important. Its decrease lowers the ambient temperature below about −10°C [9.29]. For fuel-injected vehicles and many later carbureted vehicles, it is the low boiling components in a gasoline that are most relevant in cold starting, and these are characterized by such parameters as the vapor pressure, the temperature at which 10% evaporates (T10), the amount that evaporates at 70°C (E70°C), or a combination of some of these terms. In practice, these parameters are often highly intercorrelated so that a high E70°C gasoline will have a low T10, and so it is of little consequence which of these is chosen in a specification to ensure adequate starting under low ambient temperature conditions.

9.2.1.1. Carbureted Engines

There are some carbureted vehicles, however, in which heat soak-back occurs from the hot engine into the carburetor after the engine has been switched off. This can cause the lighter portion of the gasoline in the float bowl to evaporate so that when the engine is being restarted after it has cooled down, the gasoline going forward initially will be depleted of volatile components and may not ignite. For this reason, in such vehicles, E100°C may provide a better predictor of cold start performance than the terms relating to the front end only.

It also explains why some highly volatile fuels containing perhaps excessive amounts of butane may show poor cold-starting characteristics in some older vehicles.

9.2.1.2. Port Fuel-Injected Engines

Port fuel-injected engines tend to utilize one of two injection strategies: valve-closed injection or valve-open injection. The former injects fuel when the intake valve or valves are closed, which allows time for the fuel to volatilize before the intake valve opens. This tends to result in a more homogeneous mixture entering the cylinder and hence gives better combustion. The valve-open injection strategy gives better control over the amount of fuel entering the cylinder on each intake stroke but means that the majority of the fuel enters the cylinder as liquid droplets. This gives less time for the fuel to evaporate before the spark is fired [9.30, 9.31]. However, under cold start conditions, there is little volatilization of the fuel injected while the intake valve is closed, and this fuel enters the cylinder as large droplets as the inlet valve begins to open. With open-valve injection, the better atomization out of the injector results in a smaller droplet size distribution entering the combustion chamber [9.32]. A higher volatility fuel reduces the size of the droplet [9.33].

9.2.1.3. Direct Injection Engines

With modern direct injection engines, it is necessary to ensure a rapid rise in fuel rail pressure to enable rapid and reliable starting of the engine. Heat soak can lead to vapor formation in the fuel system, which will give unreliable fuel rail pressure rise [9.34]. When injection does occur, the volatility of the fuel will affect the mixture preparation and hence the viability of ignition during the first few cycles of the starting procedure. High-volatility fuel components will vaporize more quickly, resulting in smaller fuel droplets. Low-volatility fuel components vaporize more slowly, resulting in larger fuel droplets; these larger droplets will penetrate farther into the combustion chamber and are more likely to impinge on the combustion cylinder walls. The presence of a low-volatility fuel component can significantly degrade engine performance during the initial transient phase of cold starting, increasing the number of cycles required to attain stable combustion [9.35]. This can also significantly affect the cold start emissions and oil dilution, discussed later in this chapter.

9.2.2. Relevant Specifications

Specification levels to ensure satisfactory cold start performance will clearly depend on the climatic conditions in the geographical area under consideration, as well as the vehicle population itself. For example, the minimum vapor pressure, expressed as DVPE, in some European countries can vary from a winter minimum of 70 kPa to a summer minimum of 45 kPa, while in warmer climates the minimum winter DVPE is set at only 50 kPa. Too high a front-end volatility, however, will increase the risk of driveability problems caused by excessive vaporization in the fuel system. To prevent this, upper limits on volatility parameters are set in national specifications.

In many studies, cold starting performance is included in cold weather driveability, where the overall cold weather performance is related to a driveability index, discussed in the next section. These driveability indices do not usually include the vapor pressure but rely primarily on specific parameters from the distillation curve.

9.3. Cold Weather Driveability

The driveability of a vehicle has been defined as the degree to which a vehicle starts readily, idles evenly, drives smoothly when cruising and accelerating, and generally responds well to the throttle inputs from the driver. It is most critical when the vehicle is warming up. It was well known that vehicle driveability deteriorated as ambient temperature decreased. However, in today's market, drivers expect their vehicles to perform well during all conditions, including very cold weather. Emissions regulations also require rapid warm-up with consistent combustion to minimize unburned hydrocarbon emissions.

Driveability malfunctions are caused by variations in mixture strength giving an air-fuel mixture outside the ignitable range in one or more cylinders for a few cycles. When a single-point injection or a carbureted engine is cold and the ambient temperature is low, a large proportion of the fuel traveling to the cylinder can be present in the inlet manifold as a liquid film [9.36, 9.37] rather than as a vapor. It is this lack of vaporization that gives rise to a hesitation before a burnable mixture reaches the cylinders at the start of acceleration. Under these circumstances, too rich a mixture can also occur during the acceleration as the excess liquid "catches up." An uneven idle and "surging" during cruise in carbureted or single-point injection vehicles may be caused by maldistribution of the fuel between cylinders, and this can be another reason for a stumble during an acceleration.

The ease with which a gasoline vaporizes in the engine determines the cold weather driveability of the fuel/vehicle combination. Fuel volatility and ambient temperature are obviously both important factors, but even more so is fuel system design. The use of multipoint fuel injectors rather than carburetors has a positive effect on driveability, as also do heating systems for the inlet manifold and good air-to-fuel ratio control during warm-up.

9.3.1. Cold Weather Driveability Test Procedures

A number of procedures have been developed to quantitatively measure cold weather driveability, such as the Coordinating Research Council (CRC) procedure in the US [9.38] and the Coordinating European Council for the development of performance tests for fuels, lubricants, and other fluids (CEC) code of practice in Europe [9.39]. With many vehicles having adaptive engine management systems, it is necessary to carry out preliminary tests to precondition the adaptive memory while the vehicle is warming up.

9.3.1.1. US Test Procedures

The CRC procedure (E-28-94) is described briefly below. More detail of the procedure is given in various CRC reports [9.40–9.42]. The test involves evaluating the starting, idle, and warm-up performance of the vehicle.

Up to six attempts are made to start the engine. Once it is started it is allowed to idle for 30 seconds with the gear selector in park. The gear selector is then moved to drive, and after 5 seconds of idle with the gear selector in drive, a light throttle acceleration is made from 0 to 15 mph (24.14 km/h) at a constant throttle opening and at a predetermined manifold vacuum. The vehicle is then braked to a standstill and allowed to idle for 3 seconds; the 0 to 15 mph (0 to 24.14 km/h) acceleration is then repeated. The vehicle is then allowed to cruise at 15 mph (24.14 km/h) to the 0.1 mile (0.16 km) distance before being braked an allowed to idle for 3 seconds. The vehicle then undergoes a wide-open throttle (WOT) acceleration to 20 mph (32.19 km/h) before being braked to 10 mph (16.09 km/h) and cruises at this speed until 0.2 mile (0.32 km) accumulated distance. Moderate braking is again applied to bring the vehicle to stop, where it is allowed to idle for 3 seconds. The 0 to 15 mph (0 to 24.14 km/h) accelerations are then repeated, but after the second acceleration the vehicle is allowed to cruise at 10 mph (16.09 km/h) until 0.3 miles (0.48 km) accumulated distance. A 10 to 20 mph (16.09 to 32.19 km/h) light acceleration is then performed, followed by light braking to take the vehicle to the 0.4 mile (0.64 km) accumulated distance, again followed by a 3-second idle. A moderate acceleration from 0 to 20 mph (0 to 32.19 km/h) and moderate braking takes the vehicle to the 0.5 mile (0.80 km) accumulated distance, at which point it idles for 5 seconds before repeating the whole process from the first acceleration. Light and moderate accelerations are defined by a manifold vacuum at the start of the acceleration. This vacuum will then change during the acceleration as the engine speed changes.

An additional sequence may be included for low ambient temperatures. This immediately follows the testing outlined and includes a constant manifold vacuum acceleration from 0 to 45 mph (72.42 km/h) and deceleration to 25 mph (40.23 km/h) within a total distance of 0.4 miles (0.64 km); this is followed by a 25 to 35 mph (40.23 to 56.33 km/h) acceleration at the set throttle position, just above that which causes a downshift, until a total accumulated distance of 1 mile (1.61 km). The vehicle is then braked to a standstill and allowed to idle for 30 seconds with the gear selector still in drive. During these maneuvers the severity of the various malfunctions are recorded: stall, idle roughness, backfire, hesitation, stumble, and surge, each of which is given a severity rating. This information is then used to determine a total weighted demerits (TWD) value, which are used as an indication of driveability.

9.3.1.2. European Procedures

The CEC procedure, which is no longer supported by CEC, used a set drive cycle that involves a cold start, an idle period, and part-throttle accelerations, a 55 km/h cruise and a WOT acceleration mode, as shown in Figure 9.13 [9.39].

FIGURE 9.13   CEC cold weather driveability test cycle.

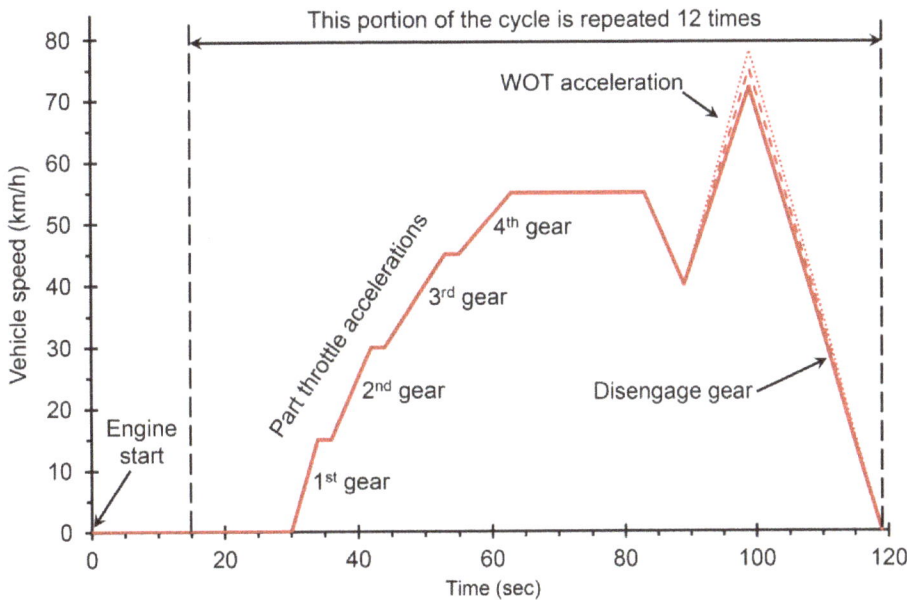

The WOT acceleration is over a set time rather than a set speed; the acceleration rate and the final speed increase as the engine warms up. Tests can be carried out on the road or on a chassis dynamometer in a climate-controlled chamber. Numerical ratings are assigned to each part of the test procedures so that an overall demerit rating can be calculated for each vehicle/fuel/temperature combination. The change in speed during the first 5 seconds of the WOT acceleration for each cycle is used to calculate the warm up time. The warm-up time is defined as the time from starting the engine to achieving an acceleration, with no cold start enrichment such as a choke, of 70% of the fully warmed-up acceleration on the standardization fuel. This is shown graphically in Figure 9.14 [9.39].

FIGURE 9.14 Warm-up time curve.

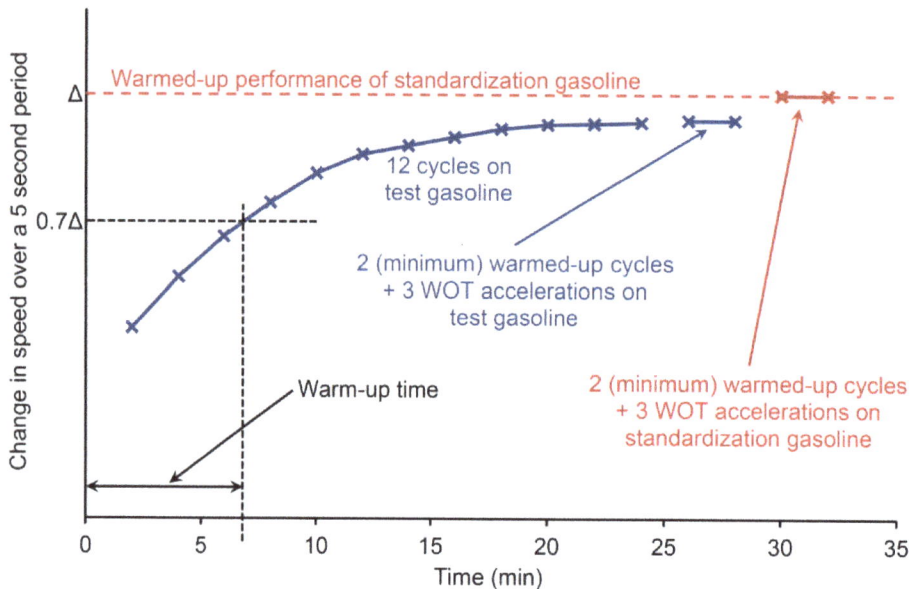

Electronically controlled fuel-injected engines tended to show fewer driveability problems using this procedure and a revised procedure was proposed [9.43] that consisted of an increased proportion of part-throttle accelerations and cruises compared to the old CEC procedure, which had a strong emphasis on WOT accelerations.

The above methods all rely on a trained rater driving the vehicle and subjectively assessing the performance. Attempts have also been made to assess fuel performance using a given engine on a dynamometer [9.44]. Besides removing the variability inherent in the driver, assessment was performed by analysis of cylinder pressure data obtained from a cylinder pressure transducer in each cylinder. Another possible advantage of using an engine on a test bench is that the test bed cooling system can be used to bring about a rapid cool-down of the engine to allow for rapid turnaround between different fuel tests. Unfortunately, a test bench method does not allow for rapid assessment of different engine/vehicle configurations.

9.3.2. Relevant Fuel Parameters

The fuel parameters that influence cold weather driveability are not simple and can vary widely from one vehicle to another. As engine technology changes, the importance of different parameters will also change [9.45]. With the introduction of hybrid electric vehicles (HEV), where the internal combustion engine can be replaced or supplemented by an electric motor under start-up and low-speed running, there are question marks over the applicability of existing cold start test procedures. Some preliminary work has suggested that HEVs do show a response to fuel characteristics [9.46], and that based on the limited data set, there is a similar trend to conventional spark ignition engine vehicles, and that ethanol does degrade performance.

9.3.2.1. US Driveability Index

Over the years, attempts have been made to determine a fuel parameter or parameters that can be included in a fuel specification that will correlate with driveability. Based on work in the US, an overall driveability index (DI) was developed [9.47] where

$$DI = 0.5T10 + T50 + 0.5T90$$

Another driveability equation developed for some US vehicles [9.48] also uses the same parameters of T10, T50, and T90, but this equation included a temperature term resulting in a driveability number for a given ambient temperature.

As engine technology developed, the importance of the T90 diminished and the equation was subsequently adjusted to the following [9.49]:

$$DI = 1.5T10 + 3T50 + T90$$

This version has been shown to correlate well with driver satisfaction both for hydrocarbon-only gasolines and also when MTBE is present—although at the same DI, the hydrocarbon fuels had a higher satisfaction level than gasolines containing MTBE [9.50]. For fuels representing typical commercial gasolines, vapor pressure and DI are usually intercorrelated so that meeting low vapor pressure levels in order to minimize evaporative emissions can lead to poor cold weather driveability [9.51]. With the wider use of ethanol as a blending component, and the changes in vehicle technology, further studies have resulted in the DI once more being adjusted to take account of the presence of oxygenates [9.40, 9.52–9.54]. The revised equation for the new DI (NDI) is given as

$$NDI = 1.5 * T10 + 3 * T50 + T90 - 1.33 * \left(\%EtOH \right)$$

where the T10, T50, and T90 values are measured in °C.

A value of DI that will give acceptable driveability will depend upon the climatic conditions in the same way as other measures of gasoline volatility. Therefore, the US Standard Specification for Automotive Spark-Ignition Engine Fuel, ASTM D4614 [9.20] specifies a maximum DI for each volatility class. Maximum permitted values range from 569 for the highest volatility class to 597 for the lowest volatility class of gasoline.

9.3.2.2. European and Japanese Indices

In Europe and Japan, a single distillation parameter such as T50 or the E100°C point has been used as a DI. The distillation terms T50 and E100°C are highly intercorrelated for most gasolines and so are equally valid, and both have been widely used. When such single predictors are used, it is assumed that the whole of the distillation curve has a conventional shape and that, for example, the T10 and the T90 points have the normal relationship to the T50 point. However, testing using fuels with atypical distillation characteristics [9.55] has shown that, in general, the further the fuel deviates from the conventional, the worse the cold weather driveability. Thus, when significant quantities of single component blend stocks are added, the distillation curve becomes atypical and a single distillation parameter is no longer enough to describe the fuels performance, that is, when MTBE is blended into the gasoline [9.56]. The reason was that the MTBE vaporized very readily and so gave a lean mixture under transient operation such as during an acceleration, resulting in a deterioration in driveability. To overcome this, a new DI has been suggested for fuels containing MTBE [9.57]:

$$DI = T50 + M/2$$

where M is the %vol. MTBE in the blend.

A high level of involatiles such as heavy aromatics will increase the T90 point, and this will cause poor driveability and also increase hydrocarbon exhaust emissions [9.57–9.59].

Test work in France using some 25 vehicles and four fuel series [9.60] indicated that the best overall correlation with driveability demerits was obtained with an expression containing two variables: RVP + 22*E100°C (where RVP is in mbars).

Two similar variables, RVP and T50%, were used in a study conducted in the US by CRC at intermediate temperatures of −1 to 13.5°C (30 to 56°F) [9.61]. Under these conditions, fuels containing MTBE behaved very similarly to hydrocarbon-only fuels, but fuels containing ethanol gave a somewhat poorer driveability performance at the same volatility level.

The relationship between these technical assessments of vehicle driveability and customer perception of driveability malfunctions has also been studied [9.50–9.63]. It was shown that there is a good correlation between the technical and customer assessments and that the test procedure is, therefore, a good indicator of consumer response to cold weather driveability.

An exercise in Japan determined a new DI given by

$$NDI = 4*A_1 + 3*A_2 + 2*A_3 - 1*A_4 - 4*A_5$$

where $A_1 = E_{70°C}$, $A_2 = E_{100°C} - E_{70°C}$, $A_3 = E_{130°C} - E_{100°C}$, $A_4 = E_{160°C} - E_{130°C}$, and $A_5 = 100 - E_{160°C}$.

This was found to be suitable for hydrocarbon fuels and oxygenated fuels containing MTBE, ETBE, and ethanol [9.64].

9.4. Hot Weather Driveability

When gasoline vaporizes prematurely in the fuel system (upstream of the injectors or carburetor jets), driveability problems can also ensue. The likelihood of such problems occurring will depend, as with cold weather driveability, on engine/vehicle design, ambient temperature, and pressure, driving mode, and fuel volatility.

Fuel system design is perhaps the most important factor because although many vehicles appear to be virtually free of problems in hot weather even with relatively volatile fuels, others can exhibit very severe problems indeed. Thus, vehicles with port fuel injectors are much less susceptible to driveability problems at high and intermediate temperatures than carbureted or throttle body injected vehicles [9.61, 9.65]. The position of the fuel pump(s) is extremely important. If at least one is submerged in the fuel tank, then the gasoline being delivered to the injectors or carburetor will be under pressure and much less likely to vaporize prematurely. If the pump or the fuel lines are positioned so that they can be heated by the engine, this will increase the possibility of undesirable vaporization. The trend toward low drag coefficients and hence smaller front grill apertures can lead to reduced cooling air flowing into the engine compartment, which may make these vehicles more susceptible to hot weather driveability problems. Equipment in the engine compartment that is likely to be hot, such as turbochargers and close-coupled catalysts, will also give rise to high under-hood temperatures and increase vaporization problems. Even the fuel in the tank can get hot due to fuel recirculation and to the fact that the exhaust system is often positioned close by. The situation is worse when the engine is shut down shortly after a prolonged period of high duty operation, when cooling air flow is reduced to a minimum. Vapor formed in the fuel system will tend have a deleterious effect on attempts to restart the engine.

Ambient temperature is obviously important in causing hot weather driveability malfunctions. High-altitude driving is also a problem, not only because of the reduced pressure but also because it often means that the engine compartment is very hot due to the high-load, low-speed driving mode used when climbing, even though ambient temperature tends to drop with altitude. Both ambient temperature and altitude are considered when setting gasoline volatility specifications [9.30].

9.4.1. Hot Weather Driveability Testing

Driving mode is clearly important, because this determines how hot the engine compartment and the fuel system can get. The methods of test for hot weather driveability [9.66] usually involve three phases. The first requires the vehicle to be driven at a relatively high speed to get it fully warmed up; the second allows time for the heat of the engine to soak back into the fuel system, and this is achieved by switching off the engine or by allowing it to idle for 15 minutes or so; and the third measures the ease with which the engine restarts, the idle quality, the smoothness of the acceleration, and the time to reach a given speed. These data are recorded and compared with results when a low-volatility reference fuel is used. Other test procedures have been used [9.65, 9.67], and as with testing for cold weather driveability, for vehicles fitted with engine management systems it can be necessary to carry out preliminary runs to precondition the adaptive memory of the system.

9.4.2. Hot Weather Driveability Problems

The type of hot weather driveability problems that occur can be classified as vapor lock, fuel weathering, and for older vehicles carburetor percolation or foaming.

9.4.2.1. Fuel Weathering

Fuel weathering is when over time, the fuel loses the more volatile components, the light ends, resulting in a fuel with a lower vapor pressure; this is exacerbated by higher temperatures. As engine design moves to higher pressures in the fuel delivery system, from carburetors to manifold fuel injection to direct injection, there is a concomitant increase in the temperatures experienced by the fuel within the fuel system. With fuel injection systems there is some fuel returned to the fuel tank, which will increase the temperature of the fuel in the tank. This can lead to a very significant drop in vapor pressure of the remaining fuel [9.68], particularly for high-volatility fuels [9.69].

9.4.2.2. Vapor Lock

Vapor lock is probably the main field problem that occurs, and this also causes poor restarting and other malfunctions, such as hesitation and stumbles during accelerations. It can be so severe that it can cause the engine to cut out completely. These malfunctions are due to the mixture being too lean as a result of excessive vapor formation, preventing the fuel pump from operating satisfactorily so that the fuel in the float bowl is not replenished [9.70].

9.4.2.3. Carburetor Percolation

Carburetor percolation occurs when fuel in the float bowl starts to boil, either during or following a hot soak, and forces excess fuel into the inlet manifold via the carburetor vent (in vehicles with an internally vented system) or through the carburetor jet system [9.71]. This gives rise to an over-rich mixture that can cause difficulty in restarting the engine and persistent stalling.

9.4.2.4. Carburetor Foaming

Carburetor foaming [9.72, 9.73] is caused by the rapid boiling of fuel as it enters a hot carburetor, thereby generating foam that causes the float to sink. This allows more and more fuel to enter the bowl so that it fills with foam, thereby blocking the air vent and causing an increase in pressure inside the bowl. This pressure increase, in turn, forces excess fuel through the metering jet and vent so that the vehicle suffers similar malfunctions to those caused by vapor lock even though they are caused by excessive enrichment of the mixture rather than the over leaning that occurs with vapor lock. In cases where the vent goes directly to atmosphere, fuel can be forced out into the engine compartment and cause a fire hazard.

9.4.3. Hot Weather Fuel Parameter Specification

It is generally accepted that it is the front-end volatility of a gasoline that determines the extent to which hot weather driveability problems are likely, and a number of volatility terms have been investigated to correlate with field and laboratory experience [9.74, 9.75]. The gasoline temperature to obtain a V/L ratio of 20 is widely used to control front-end volatility [9.75–9.77] and is the basis of the front-end volatility specifications of ASTM D439 (now superseded by ASTM D4814) [9.10]. This US specification defines minimum quality standards in different geographic regions, depending on their climatic conditions, to protect consumers against hot weather driveability and other problems.

The alternative to using V/L ratio is to use a combination of RVP and an ASTM D86 distillation point such as E70°C. This is sometimes preferred because, as discussed earlier, such expressions are highly correlated with V/L ratio [9.9] and are easier to measure. The general expression

$$RVP + n\,E70°C$$

where n is a constant, has been shown to give a good correlation with TV/L = 20, and hence with hot weather driveability performance [9.78, 9.79]. In the US, this expression has been called the vapor lock index (VLI) [9.80] or front-end volatility index (FEVI) when a value of n = 9 has been used (this relates to the use of RVP in mbar and E70°C in %vol. (if RVP is measured in psi, n is 0.13) [9.75]. For European and Japanese vehicles, the value of n has been shown [9.78] to give the best correlation with vehicle performance when it lies between about 4 and 7 (RVP in mbar and E70°C in %vol.). This is consistent with values of 5 [9.81] and 9.2 [9.60] that have also been obtained. In general, the expression RVP + 0.7*E70°C (where RVP is measured in kPa) has been widely used for European and Japanese vehicles and has been called the flexible volatility index, flexivolatility index, or simply the VLI. RVP and T50% have also been used to correlate with hot driveability [9.65], but since T50 is itself highly intercorrelated with E70°C for most marketed hydrocarbon fuels, this does not necessarily mean that T50% is a better predictor of hot driveability than E70°C.

The FEVIs discussed have been correlated with consumer response and found to be satisfactory, although it has been shown that motorists are generally less critical than technical raters for most driveability malfunctions [9.82].

A later review of front-end volatility indices [9.83] led to the conclusion that for European markets at least, a summer VLI specification is no longer necessary. It must be remembered that it was taking into account the proposed increase in minimum levels of E70 and E100, in addition to the introduction of an E150 minimum limit. In this work, a database was used containing many hundreds of vehicles tested over many years, and this represented a wide selection of engine technologies (carbureted, single-point, and multipoint injection). The data include an assessment of vehicles' sensitivity to a wide range of fuel volatility at ambient temperatures representative of the European market. Using vehicle population and ambient temperature data for individual markets allowed technical satisfaction levels to be established, which, in turn, permitted total customer satisfaction levels to be determined for the year 2000. The conclusion that a summer VLI specification was no longer necessary also extended, with some important exceptions, to other seasons. The transition periods, between summer and winter, and between winter and summer, were identified as critical for some markets, and these findings are reflected in the current European gasoline specification [9.84] where minimum VLI levels are specified for different regions.

The presence of oxygenates such as MTBE in the fuel can improve hot weather driveability [9.65] as compared with a hydrocarbon fuel of the same front-end volatility since such problems are usually caused by the air-to-fuel mixture being too rich.

Data on the driveability characteristics of different vehicle models are usually obtained cooperatively within the oil and other industries, both in the US and Europe, using standard test procedures and fuels. The raw data obtained are then available to all participants in the test program. A volatility data bank can be built up and used to establish performance levels for individual vehicle models or whole vehicle populations for a range of fuel volatilities and ambient temperatures. The data can also be used to estimate levels of FEVI to satisfy different vehicle populations at various ambient temperatures [9.85], as shown in Figure 9.15. In this figure, FEVI is given by

$$FEVI = RVP + 0.7E70°C$$

where RVP is measured in kPa.

FIGURE 9.15   Hot weather driveability performance of European vehicle populations, 80% technical satisfaction.

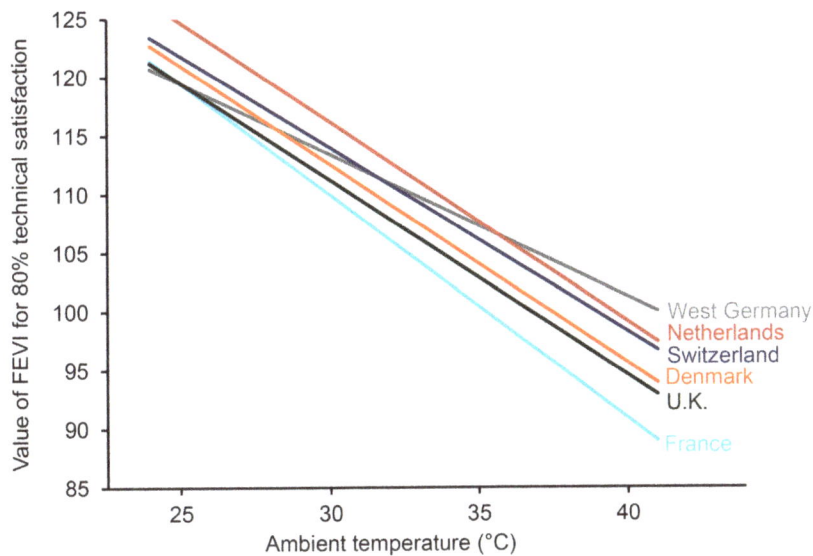

9.5. Evaporative Emissions from Vehicles

Atmospheric hydrocarbon emissions contribute to the formation of photochemical smog, which is a major problem in certain cities such as Los Angeles and Santiago, due to their geography, and Beijing, Delhi, and Tokyo, due to traffic density; in Europe the problem is much less severe. The presence of volatile toxic compounds such as benzene in gasoline [9.86, 9.87] is another reason for wishing to minimize such vapors in the atmosphere. Evaporative loss control devices are capable of reducing evaporative emissions by over 90%. In Western Europe in 1986, it was estimated that about half of the volatile organic compounds (VOC) in the atmosphere came from man-made sources and that of these, exhaust hydrocarbons contributed about 25%, evaporative losses from vehicles about 10%, and evaporative losses during refueling about 2% [9.88]. A 2001–2003 study in two cities in the Los Angeles basin showed that at the two sites, 71 and 80% of the VOCs were attributable to mobile sources [9.89]. This was broken down as 34% evaporative losses, 24% tailpipe emissions, and 13% unburned gasoline at the first site, and 31% evaporative losses, 22% tailpipe emissions, and 27% unburned gasoline at the second site.

Exhaust emission legislation has become progressively more severe in most countries, and the use of, or intention to introduce, evaporative emission legislation is becoming more widespread.

Evaporation losses from a vehicle can occur from several different sources on the vehicle unless control systems have been fitted. For a moving or parked vehicle, without evaporative emission controls, the most important of these are breathing losses from the fuel tank; on vehicles with carburetors there is also significant vapor loss from the carburetor bowl. Vehicle design features that have an important influence on the potential for evaporative losses are the fuel system, the presence of heat shields and insulation, the shape and liquid surface area of the fuel tank, and the fuel tank venting system.

There is also the potential for significant vapor release during refueling, and this must also be addressed.

It is usual to divide vehicle evaporative losses into three categories:

1. Running losses. These are the losses that occur while a vehicle is being driven normally.
2. Diurnal losses. These losses occur when the vehicle is stationary for an extended period with the engine switched off and are due to the normal temperature changes that occur over a 24-hour period. Most of these emissions occur whenever there is an increase in temperature that causes an expansion of the vapor in the fuel tank.
3. Hot-soak losses. These occur when a fully warmed-up vehicle is left to stand and are due to engine heat soaking back into the fuel tank and the remainder of the fuel system.

9.5.1. Measuring Evaporative Emissions

The measurement of evaporative losses from vehicles has been carried out by a variety of different methods, the two most important of which are the adsorption of vapors from different parts of the fuel system in canisters containing activated charcoal and the measurement of the hydrocarbon concentration within a sealed chamber in which the vehicle is contained. The most commonly used of these methods is the latter and is known as the SHED (sealed housing for evaporative determination) test procedure [9.90]. In the US, the diurnal and hot-soak emissions are measured and the sum of these two is used for federal certification of the vehicle. The running losses are of less concern because a purge system is fitted that draws vapors back into the engine when the vehicle is operating. The diurnal portion of the test originally simulated the temperature rise that occurs over 24 hours by raising the temperature of the fuel in the tank from 16 to 29°C in 1 hour while the vehicle is in the SHED. However, new regulations replaced the 1-hour diurnal simulation with a 72-hour test in which the entire vehicle experiences three successive 24-hour diurnal temperature cycles [9.91]. The hot-soak part of the test takes place after the vehicle has undergone a standard warm-up procedure on a chassis dynamometer, when it is pushed into the SHED and the hydrocarbon concentration measured after a given time. From a knowledge of the volume of the SHED, the total hydrocarbons evaporated during both portions of the test can be calculated. The SHED procedure has a disadvantage in that it picks up all hydrocarbon emissions and not just those from the fuel system. With a new vehicle, a surprisingly large amount of hydrocarbons can come from the tires, upholstery, undersealant, and so forth, and it is usual to steam-clean and weather vehicles before testing if fuel system hydrocarbon loss is the prime concern.

A comprehensive study of evaporative emissions was carried out by the US Bureau of Mines in the late 1960s to early 1970s [9.92–9.94]. Emissions from a wide range of vehicles at several ambient temperatures were measured using fuels of varying volatility. The test procedure involved measurement of the losses from particular points in the fuel system by condensing the vapors and weighing, since the SHED procedure had then not yet been developed. RVP was used to categorize the fuels, and it was found, for example, that at 21°C with a fuel of 70 kPa, total evaporative emissions varied from 8 to 40 g/test, depending on the vehicle. A 1978 study [9.95] showed that the relationship between evaporative losses and RVP is nonlinear, with losses increasing disproportionately as RVP is increased. Subsequent test work has confirmed the importance of RVP [9.96].

9.5.2. Reducing Evaporative Emissions

Controls were introduced in 1970–1971 in the US that required only relatively simple modifications to enable vehicles to meet the initial emission standards of 6 g/test. However, as the limits were progressively tightened, it became necessary to introduce more sophisticated methods of control. The technique that has been universally adopted to meet the current limits employs an evaporative emission canister(s) containing activated carbon sorbents that trap and store fuel vapors. When the vehicle is started, intake vacuum draws air through the canister, and adsorbed vapors are removed from the sorbent and into the intake manifold and then the combustion chamber. This type of system is successful in reducing evaporative emissions by about 95% from uncontrolled levels [9.97] and allows vehicles to meet a 2 g/test limit.

Evaporative emission limits are also imposed in Europe, Japan, and Australia. Studies have been carried out in Europe to determine typical emission levels from a number of European and Asian vehicles, to find the effect of various fuel and vehicle parameters on them using a slightly modified SHED procedure, and, in a later test, to determine the effects of temperature on them [9.98–9.100]. The work confirmed that vehicle and fuel system design have the greatest effect on evaporative emissions and that carbon canisters are very effective in controlling these emissions. It also showed that

- The emissions from the 10 vehicles tested varied from 9 to 25 g/test on a winter-type fuel and from 4 to 16 g/test on a lower volatility summer-grade fuel.

- RVP is the only significant fuel parameter influencing these emissions. It has a relatively small effect—a reduction of 10 kPa reduced evaporative emissions by 23%.

- The cycle used to warm up the vehicle is important in its influence on total evaporative losses.

- The presence of oxygenated compounds in a gasoline did not influence evaporative losses.

- Diurnal losses are significant, but are readily controlled by carbon canisters.

- Hot-soak losses and running losses from uncontrolled vehicles increase progressively with ambient temperature and RVP. A change of 1°C in ambient air temperature was found to have the same effect on evaporative emissions as a 3.8 kPa change in RVP.

In a test using mainly US vehicles, reducing RVP by 0.7 kPa reduced diurnal emissions in nonoxygenated fuels and ethanol-containing fuels by 46.4 and 53.7%, respectively, but there was an unexplained zero effect with MTBE-containing fuels [9.101].

Devices to control refueling evaporative emissions both on the vehicle (on-board refueling vapor recovery (ORVR)) [9.102–9.106] and at the filling station [9.107, 9.108] have been developed and compared [9.109]. These systems can reduce refueling emissions by 98%. In certain areas of the US vapor recovery systems at the filling station have been mandatory since 1994, but as more vehicles become fitted with ORVR systems the use of the filling station systems is thought to be no longer necessary.

9.6. Influence of Fuel Volatility on Exhaust Emissions

A considerable amount of test work has been carried out to establish the effect of both fuel composition and fuel volatility on exhaust emissions. It is generally agreed that their influence is small relative to the effect of vehicle technology [9.110]. This topic will be discussed more fully in Chapter 13, Section 13.5.

9.7. Intake System Icing

Ice formation in the intake system became apparent on early carbureted engines where the throttle plate was downstream of the fuel metering. Evaporative cooling brought about by vaporization of the fuel would exacerbate the ice formation as a result of cold, humid air [9.111–9.113]. With fuel-injected engines, the fuel tends to be injected downstream of the throttle plate, which lessens the problem. However, the problem still exists even with direct injection engines [9.114], although this is clearly beyond the influence of the fuel characteristics.

Ice formation and deposition occurs when the water in the intake air condenses and then freezes. Throttle icing can be classified as soak icing and run icing. Soak icing is where ice forms on the stationary throttle plate and bore walls after the vehicle has been shut down after an extended drive. Run icing occurs while the vehicle is in operation and was the more prevalent form on older carbureted and TBI engines.

Intake system icing should not be confused with the freezing of any extraneous free water that may have found its way into the fuel system, causing a blockage of fuel lines. This occurs only when the ambient temperature is well below freezing and when severe contamination of the fuel with water has occurred. It is not influenced by fuel volatility, although the presence of ethanol can influence water pickup, but ethanol is known as a good anti-icing additive; it can also be minimized by certain types of anti-icing additives.

9.7.1. Throttle Icing in Carbureted and Throttle Body Injected Engines

Throttle icing on these older types of engines tends to occur only in cool, humid weather, when the ambient temperature is roughly in the range −5 to +12°C and the relative humidity is above about 80%. At lower temperatures and at low humidity levels there is not enough water in the air to cause problems, and at higher ambient temperatures the temperature depression caused by the evaporating gasoline is not enough to freeze any condensed water. It is made more severe by increasing fuel volatility because this, in turn, increases the temperature drop that occurs as the gasoline evaporates. It is usually accepted that the mid-fill volatility as defined, for example, by the E100°C figure, is the best single fuel volatility parameter for controlling icing. It is, however, more usual and more economical to control intake system icing by the use of additives rather than by adjusting volatility. This is discussed in Chapter 11, Section 11.3.2.

The effect of ice deposition in a carburetor is to restrict the air or mixture flow. The area where this ice deposition occurs varies with different carburetors, and because of this it can have quite different effects on driveability. Ice formation on the throttle plate will only have a significant effect on air-to-fuel ratio at very low throttle openings or when the throttle is closed completely, such as when the engine is idling. This type of icing, known as idle icing, is most common in city driving during the period that the vehicle is warming up and tends to cause the engine to stall whenever it idles because the mixture is too rich. Fuel consumption is also increased during this period of idling, and so are the CO and hydrocarbon exhaust emissions, particularly for noncatalyst vehicles. Figure 9.16 shows throttle plate icing at idle and also at the light load condition where there is maximum evaporative cooling affect from the fuel [9.115].

FIGURE 9.16   Throttle plate icing at idle (left) and light load (right).

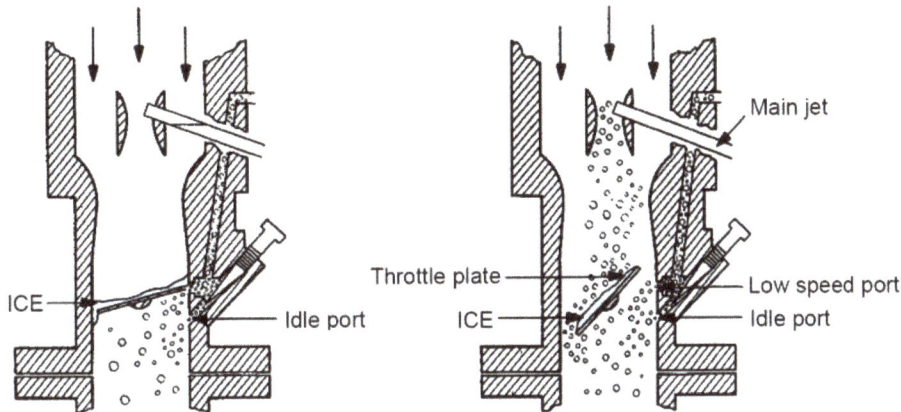

Due to evaporative cooling, ice can also form in the venturi area [9.116, 9.117], and this tends to cause power loss during cruise conditions, again because of restrictions in the air flow. The drop in power can be so great that the vehicle can come to a complete halt. When this happens, it can be quite puzzling, because if the motorist waits for a little time, perhaps while looking for a fault, heat soak-back from the engine will cause the ice to melt and the vehicle will operate quite normally again on restarting. This type of icing is known as cruise icing and will also have an adverse effect on fuel economy and exhaust emissions.

Standard test procedures were developed in Europe for evaluating the icing characteristics of a vehicle or of a gasoline or gasoline/additive combination [9.118]; because carbureted vehicles are now rare in Europe, the procedure is no longer supported. The test procedures covered idle icing, cruise icing, and an extended idle test meant to simulate the conditions that occur when vehicles are left idling unattended on cold mornings in order to allow them to warm up before the owner gets in and drives away. The tests can be carried out on the road or on a track, but it is more precise to use a chassis dynamometer in a temperature- and humidity-controlled climate chamber. In the idle icing test, a standard cycle is used that is repeated a dozen or so times and which consists of a full throttle acceleration to 50 km/h, a cruise at this speed for 30 seconds, and then an idle period of 30 seconds. Only the idle mode is assessed and demerits applied that relate to the smoothness of the idle and the occurrence of stalls. The cruise icing test is usually carried out with a fully warmed-up vehicle, unlike the idle icing test, and involves running the vehicle at a constant throttle position at a speed in the range 70 to 120 km/h. If ice forms, the speed will gradually drop and the vehicle may, in severe cases, completely stall. The extent of the speed loss is used as a measure of the icing that has occurred. When assessing gasolines or anti-icing additives it is necessary to select a test vehicle that is known to be susceptible to icing problems and to use a relatively volatile base fuel free of oxygenates and anti-icing additives.

There were also many in-house tests for evaluating anti-icing additives that were used mainly by fuel additive manufacturers [9.117]. They almost always involved an engine on a test bed, fed with inlet air having a controlled humidity and temperature. Fuel additives were tested under idling conditions and the number of stalls, or the time for the first stall is taken as a measure of the degree of icing that has occurred.

9.7.2. Throttle Icing in Multipoint Fuel-Injected Engines

Throttle plate icing can still occur on fuel-injected engines. As noted, soak icing is more prevalent in these engines than run icing and is beyond the control of fuel characteristics. In these engines, soak icing can occur when the moisture in the air in the intake system condenses out and freezes on the throttle plate as it cools to below freezing point in cold ambient conditions.

The absolute humidity of the incoming air can be increased by the high water content of exhaust gas recirculation and by the water content of gases from the crankcase that are drawn into the intake by positive crankcase ventilation (PCV) systems. In cold climates, repeated stop start driving patterns can introduce a significant amount of unburned fuel and water to the oil pan [9.119], which as the engine warms up on a longer journey can be vaporized and drawn into the intake via the PCV system. These factors can raise the absolute humidity of the air in the intake system above the 100% relative humidity level for the ambient temperature, and thus as the engine cools the air cools and the water condenses out.

9.8. Oil Dilution and Combustion Chamber Deposits

The least volatile part of the gasoline will contain the molecules having the highest heat content and the highest density. This portion of the gasoline will be important, therefore, in contributing to fuel economy once the vehicle has warmed up. However, because this part of the gasoline is less volatile, it may not be fully vaporized by the time it gets to the combustion chamber, particularly when starting and driving away on a cold day with the cold start enrichment strategy in place. Under these circumstances liquid fuel can flow down the bores past the piston rings and into the lubricant in the crankcase [9.120, 9.121]. This can have two effects: it can wash lubricant off the bores and increase bore wear, and it can dilute the lubricant itself and increase wear in general. It is not uncommon for as much as 10% of the lubricant to consist of gasoline heavy ends after starting and driving the first few miles [9.122], and in cold weather it can take several miles for the oil to heat up sufficiently to evaporate off these gasoline components. Because the inclusion of ethanol in the fuel can significantly alter the distillation characteristics, this can have a detrimental effect on oil dilution [9.123].

Another potential problem caused by the heavy ends of the gasoline is that they frequently contain compounds that are difficult to combust. Some of the heavier aromatic compounds that arise from catalytic reforming fall into this category, and particularly the polycyclic aromatics. In the combustion chamber these compounds may only partially burn and give rise to carbonaceous deposits that can gradually increase the octane requirement of the engine [9.124, 9.125]. They can also cause spark plug fouling and misfire. In addition, these partially burned hydrocarbons can find their way into the lubricant, where they increase sludge levels, and they can also appear in the fuel intake system due to valve overlap or exhaust gas recirculation and thereby increase inlet system deposits. Deposit control additives can also markedly affect CCDs and this is discussed in more detail in Chapter 11, Section 11.3.3.5.

The back end of the gasoline is usually controlled by the FBP plus either the 90% distillation temperature or the percent evaporated at a temperature such as 180°C. FBPs in excess of about 220°C are usually not allowed by gasoline specifications, and values well below this are preferred in cold climates.

9.9. Fuel Economy and Gasoline Volatility

Gasoline volatility has a rather complex influence on fuel economy [9.126]. Most of the elements have already been mentioned, and this section attempts only to try to pull all these effects together.

For short runs in cold weather, high front-end and mid-fill volatilities are important because a more volatile fuel will ensure stable combustion for starting and the warm-up period. This can shorten the warm-up period and thus reduce the necessary fuel enrichment leading to improved fuel economy. However, as the volatility and the density, and hence calorific value of the fuel, tend to be inversely related, a high-volatility fuel will tend to have less energy content and hence deliver poorer economy over a longer distance. High volatility can also lead to icing problems which reduce economy. Without proper control of vapor loss, this can also lead to significantly reduced efficiency.

Long-trip economy is provided by a high heat content in the gasoline; this means a high density and therefore a low volatility. As mentioned in Section 9.8, most of the energy in a gasoline is contained in the heavy ends and so setting back-end volatility specifications is a compromise between good long-trip fuel economy and increased crankcase lubricant dilution/combustion chamber deposits with all the attendant problems that these bring.

In an analysis of fuel economy data obtained using the federal test procedure in a study primarily designed to look at fuel effects on exhaust emissions [9.127], the following conclusions were drawn.

Volumetric fuel economy (VOLFE) is lowered when:

- Aromatics are reduced. A reduction from 45 to 20% lowered VOLFE by about 3%.

- Oxygenates are added. The addition of 2.7 wt% oxygen lowered VOLFE by 2.3% in current vehicles and by 1.6% in older vehicles.

- T90% is reduced from 182 to 138°C. This lowered VOLFE by about 1.5%.

- Olefins are reduced from 20 to 5%. This lowered VOLFE by 0.2 to 0.6%.

References

9.1.   Dean, E.W., "Fuel for Automotive Apparatus," SAE Technical Paper 180007, 1918, doi:https://doi.org/10.4271/180007.

9.2.   Kettering, C.F., "More Efficient Utilization of Fuel," SAE Technical Paper 190010, 1919, doi:https://doi.org/10.4271/190010.

9.3.   ASTM International, "Standard Test Method for Distillation of Petroleum Products at Atmospheric Pressure," ASTM D86-23, ASTM International, 2023.

9.4.   Chapman, E., Davis, R., Studzinski, W., and Geng, P., "Fuel Octane and Volatility Effects on the Stochastic Pre-Ignition Behavior of a 2.0L Gasoline Turbocharged DI Engine," *SAE Int. J. Fuels Lubr.* 7, no. 2 (2014): 379-389, doi:https://doi.org/10.4271/2014-01-1226.

9.5.   Wang, Z., Qi, Y., Liu, H., Long, Y. et al., "Experimental Study on Pre-Ignition and Super-Knock in Gasoline Engine Combustion with Carbon Particle at Elevated Temperatures and Pressures," SAE Technical Paper 2015-01-0752, 2015, doi:https://doi.org/10.4271/2015-01-0752.

9.6.   Ben Amara, A., Tahtouh, T., Ubrich, E., Starck, L. et al., "Critical Analysis of PM Index and Other Fuel Indices: Impact of Gasoline Fuel Volatility and Chemical Composition," SAE Technical Paper 2018-01-1741, 2018, doi:https://doi.org/10.4271/2018-01-1741.

9.7.   Takeuchi, K., Fujimoto, K., Hirano, S., and Yamashita, M., "Investigation of Engine Oil Effect on Abnormal Combustion in Turbocharged Direct Injection - Spark Ignition Engines," *SAE Int. J. Fuels Lubr.* 5, no. 3 (2012): 1017-1024, doi:https://doi.org/10.4271/2012-01-1615.

9.8.   Kocsis, M., Briggs, T., and Anderson, G., "The Impact of Lubricant Volatility, Viscosity and Detergent Chemistry on Low Speed Pre-Ignition Behavior," *SAE Int. J. Engines* 10, no. 3 (2017): 1019-1035, doi:https://doi.org/10.4271/2017-01-0685.

9.9.   Teng, H., Miao, R., Cao, L., Luo, X. et al., "Characteristics of Abnormal Combustion in the Scavenging Zone for a Highly-Boosted Gasoline Direct Injection Engine," SAE Technical Paper 2017-01-1721, 2017, doi:https://doi.org/10.4271/2017-01-1721.

9.10. Dahnz, C., Han, K., Spicher, U., Magar, M. et al., "Investigations on Pre-Ignition in Highly Supercharged SI Engines," *SAE Int. J. Engines* 3, no. 1 (2010): 214-224, doi:https://doi.org/10.4271/2010-01-0355.

9.11. **Zahdeh, A.,** Rothenberger, P., Nguyen, W., Anbarasu, M. et al., "Fundamental Approach to Investigate Pre-Ignition in Boosted SI Engines," *SAE Int. J. Engines* 4, no. 1 (2011): 246-273, doi:https://doi.org/10.4271/2011-01-0340.

9.12. **Palaveev, S.,** Magar, M., Kubach, H., Schiessl, R. et al., "Premature Flame Initiation in a Turbocharged DISI Engine - Numerical and Experimental Investigations," *SAE Int. J. Engines* 6, no. 1 (2013): 54-66, doi:https://doi.org/10.4271/2013-01-0252.

9.13. **Kocsis, M.,** Anderson, G., and Briggs, T., "Real Fuel Effects on Low Speed Pre-Ignition," SAE Technical Paper 2018-01-1456, 2018, doi:https://doi.org/10.4271/2018-01-1456.

9.14. **ASTM International,** "Standard Test Method for Vapor Pressure of Petroleum Products (Reid Method)," ASTM D323-20a, ASTM International, 2020.

9.15. **Menrad, H.** and Nierhauve, B., "Engine and Vehicle Concepts for Methanol-Gasoline Blends," SAE Technical Paper 831686, 1983, doi:https://doi.org/10.4271/831686.

9.16. **ASTM International,** "Standard Test Method for Vapor Pressure of Gasoline and Gasoline-Oxygenate Blends (Dry Method)," ASTM D4953-20, ASTM International, 2020.

9.17. **ASTM International,** "Standard Test Method for Vapor Pressure of Petroleum Products (Mini Method)," ASTM D5191-22, ASTM International, 2022.

9.18. **ASTM International,** "Standard Test Method for Vapor-Liquid Ratio of Spark-Ignition Engine Fuels," ASTM D2533-99, ASTM International, 2010 (Withdrawn 2008).

9.19. **Jenkins, G.I.,** "Control of the Front-End Volatility of Motor Gasolines. Calculation of Vapor/Liquid Ratios from the Reid Vapor Pressure and Distillation Test," *Journal of Institute of Petroleum* 54 (1968): 80-86.

9.20. **ASTM International,** "Standard Specification for Automotive Spark-Ignition Engine Fuel," ASTM D4814-22, ASTM International, 2022.

9.21. **Furey, R.L.,** "Volatility Characteristics of Gasoline-Alcohol and Gasoline-Ether Fuel Blends," SAE Technical Paper 852116, 1985, doi:https://doi.org/10.4271/852116.

9.22. **Furey, R.L.** and Perry, K.L., "Vapor Pressures of Mixtures of Gasolines and Gasoline-Alcohol Blends," SAE Technical Paper 861557, 1986, doi:https://doi.org/10.4271/861557.

9.23. **Caffrey, P.E.** and Machiele, P.A., "In-Use Volatility Impact of Commingling Ethanol and Non-Ethanol Fuels," SAE Technical Paper 940765, 1994, doi:https://doi.org/10.4271/940765.

9.24. **Reddy, S.R.,** "A Model for Estimating Vapor Pressures of Commingled Ethanol Fuels," SAE Technical Paper 2007-01-4006, 2007, doi:https://doi.org/10.4271/2007-01-4006.

9.25. **CONCAWE,** "Ethanol/Petrol Blends: Volatility Characterisation in the Range 5-25 vol% Ethanol (BEP525) Final Report," 2009.

9.26. **Furey, R.L.** and Perry, K.L., "Volatility Characteristics of Blends of Gasoline with Ethyl Tertiary-Butyl Ether (ETBE)," SAE Technical Paper 901114, 1990, doi:https://doi.org/10.4271/901114.

9.27. **Shiblom, C.M.,** Schoonveld, G.A., Riley, R.K., and Pahl, R.H., "Use of Ethyl-t-Butyl Ether (ETBE) as a Gasoline Blending Component," SAE Technical Paper 902132, 1990, doi:https://doi.org/10.4271/902132.

9.28. **Kivi, J.,** Niemi, A., Nylund, N.-O., Kytö, M. et al., "Use of MTBE and ETBE as Gasoline Reformulation Components," SAE Technical Paper 922379, 1992, doi:https://doi.org/10.4271/922379.

9.29. **Nakajima, Y.,** Saito, T., Takagi, Y., Katoh, K. et al., "The Influence of Fuel Characteristics on Vaporization in the S.I. Engine Cylinder During Cranking at Low Temperature," SAE Technical Paper 780612, 1978, doi:https://doi.org/10.4271/780612.

9.30. **Kim, H.,** Yoon, S., Lai, M.-C., Quelhas, S. et al., "Correlating Port Fuel Injection to Wetted Fuel Footprints on Combustion Chamber Walls and UBHC in Engine Start Processes," SAE Technical Paper 2003-01-3240, 2003, doi:https://doi.org/10.4271/2003-01-3240.

9.31. **Luan, Y.,** Henein, N.A., and Tagomori, M.K., "Port-Fuel-Injection Gasoline Engine Cold Start Fuel Calibration," SAE Technical Paper 2006-01-1052, 2006, doi:https://doi.org/10.4271/2006-01-1052.

9.32. **Meyer, R.** and Heywood, J.B., "Liquid Fuel Transport Mechanisms into the Cylinder of a Firing Port-Injected SI Engine during Start Up," SAE Technical Paper 970865, 1997, doi:https://doi.org/10.4271/970865.

9.33. Meyer, R. and Heywood, J.B., "Effect of Engine and Fuel Variables on Liquid Fuel Transport into the Cylinder in Port-Injected SI Engines," SAE Technical Paper 1999-01-0563, 1999, doi:https://doi.org/10.4271/1999-01-0563.

9.34. Burke, D., Foti, D., Haller, J., and Fedor, W.J., "Fuel Rail Pressure Rise during Cold Start of a Gasoline Direct Injection Engine," SAE Technical Paper 2012-01-0393, 2012, doi:https://doi.org/10.4271/2012-01-0393.

9.35. Tong, K., Quay, B.D., Zello, J.V., and Santavicca, D.A., "Fuel Volatility Effects on Mixture Preparation and Performance in a GDI Engine During Cold Start," SAE Technical Paper 2001-01-3650, 2001, doi:https://doi.org/10.4271/2001-01-3650.

9.36. Kay, I.W., "Manifold Fuel Film Effects in an S.I. Engine," SAE Technical Paper 780944, 1978, doi:https://doi.org/10.4271/780944.

9.37. Servati, H.B. and Yuen, W.W., "Deposition of Fuel Droplets in Horizontal Intake Manifolds and the Behavior of Fuel Film Flow on Its Walls," SAE Technical Paper 840239, 1984, doi:https://doi.org/10.4271/840239.

9.38. Benson, J.D., Bigley, H.A., and Keller, J.L., "Passenger Car Driveability in Cool Weather," SAE Technical Paper 710138, 1971, doi:https://doi.org/10.4271/710138.

9.39. Falk, K., "The Development of a European Cold Weather Driveability Test Procedure for Motor Vehicles with Spark Ignition Engines," SAE Technical Paper 831754, 1983, doi:https://doi.org/10.4271/831754.

9.40. Coordinating Research Council, "1995-97 CRC Study of Fuel Volatility Effects on Cold-Start and Warmup Driveability with Hydrocarbon, MTBE, and Ethanol Gasolines: Phase 3, Cold Temperature," CRC Report No. 605, Coordinating Research Council, Inc., 1998.

9.41. Coordinating Research Council, "2003 CRC Intermediate-Temperature Volatility Program," CRC Report No. 638, Coordinating Research Council, Inc., 2004.

9.42. Coordinating Research Council, "2008 CRC Cold-Start and Warmup E85 and E15/E29 Driveability Program–Final Report," CRC Report No. 652, Coordinating Research Council, Inc., 2008.

9.43. Stephenson, T. and Paesler, H., "The Development of a Cold Weather Driveability Test Cycle for Fuel Injected Vehicles," SAE Technical Paper 961220, 1996, doi:https://doi.org/10.4271/961220.

9.44. Arters, D.C., Schiferl, E.A., and Szappanos, G., "Effects of Gasoline Driveability Index, Ethanol and Intake Valve Deposits on Engine Performance in a Dynamometer-Based Cold Start and Warmup Procedure," SAE Technical Paper 2002-01-1639, 2002, doi:https://doi.org/10.4271/2002-01-1639.

9.45. Horowitz, A.H. and Stawnychy, M.S., "Cold Weather Driveability Performance of 1975-1981 Model Cars," SAE Technical Paper 821203, 1982, doi:https://doi.org/10.4271/821203.

9.46. Thornton, M., Jorgensen, S., Evans, B., and Wright, K., "Cold-Start and Warmup Driveability Performance of Hybrid Electric Vehicles Using Oxygenated Fuels," SAE Technical Paper 2003-01-3196, 2003, doi:https://doi.org/10.4271/2003-01-3196.

9.47. Brinkman, N.D., Steinke, E.D., Villforth, F.J., and Williams, W.C., "Effect of Fuel Volatility on Driveability at Low and intermediate Ambient Temperatures," SAE Technical Paper 830593, 1983, doi:https://doi.org/10.4271/830593.

9.48. Barker, D.A. and Dunn, M.R., "Driveability Number - A Gasoline Volatility Parameter Related to Cold Start Passenger Car Performance," SAE Technical Paper 831756, 1983, doi:https://doi.org/10.4271/831756.

9.49. Barker, D.A., Gibbs, L.M., and Steinke, E.D., "The Development and Proposed Implementation of the ASTM Driveability Index for Motor Gasoline," SAE Technical Paper 881668, 1988, doi:https://doi.org/10.4271/881668.

9.50. Buczynsky, A., "Effects of Driveability Index and MTBE on Driver Satisfaction at Intermediate Ambient Temperatures," SAE Technical Paper 932671, 1993, doi:https://doi.org/10.4271/932671.

9.51. Abramo, L., Baxter, C.E., Costello, P.J., and Kouhl, F.A., "Effect of Volatility Changes on Vehicle Cold-Start Driveability," SAE Technical Paper 892088, 1989, doi:https://doi.org/10.4271/892088.

9.52. Coordinating Research Council, "1995-97 CRC Study of Fuel Volatility Effects on Cold-Start and Warmup Driveability with Hydrocarbon, MTBE, and Ethanol Gasolines: Phase 1, Intermediate Temperature," CRC Report No. 599, Coordinating Research Council, Inc., 1997.

9.53. Coordinating Research Council, "1995-97 CRC Study of Fuel Volatility Effects on Cold-Start and Warmup Driveability with Hydrocarbon, MTBE, and Ethanol Gasolines: Phase 2, Warm Temperature," CRC Report No. 603, Coordinating Research Council, Inc., 1998.

9.54. Jorgensen, S., Eng, K., Evans, B., McNally, M. et al., "Evaluation of New Volatility Indices for Modern Fuels," SAE Technical Paper 1999-01-1549, 1999, doi:https://doi.org/10.4271/1999-01-1549.

9.55. Baudino, J.H. and Copeland, L.C., "Atypical Fuel Volatility Effects on Driveability, Emissions, and Fuel Economy of Stratified Charge and Conventionally Powered Vehicles," SAE Technical Paper 780610, 1978, doi:https://doi.org/10.4271/780610.

9.56. Tomita, M., Okada, M., Katayama, H., and Nakada, M., "Effect of Gasoline Quality on Throttle Response of Engines during Warm-Up," SAE Technical Paper 900163, 1990, doi:https://doi.org/10.4271/900163.

9.57. Oda, K., Hosono, K., Isoda, T., Aihara, H. et al., "Effect of Gasoline Composition on Engine Performance," SAE Technical Paper 930375, 1993, doi:https://doi.org/10.4271/930375.

9.58. Gething, J.A., "Distillation Adjustment: An Innovative Step to Gasoline Reformulation," SAE Technical Paper 910382, 1991, doi:https://doi.org/10.4271/910382.

9.59. Quader, A.A., Sloane, T.M., Sinkevitch, R.M., and Olson, K.L., "Why Gasoline 90% Distillation Temperature Affects Emissions with Port Fuel Injection and Premixed Charge," SAE Technical Paper 912430, 1991, doi:https://doi.org/10.4271/912430.

9.60. Le Breton, M.D., "Hot and Cold Fuel Volatility Indexes of French Cars: A Cooperative Study by the GFC Volatility Group," SAE Technical Paper 841386, 1984, doi:https://doi.org/10.4271/841386.

9.61. Graham, J.P., Evans, B., Reuter, R.M., and Steury, J.H., "Effect of Volatility on Intermediate-Temperature Driveability with Hydrocarbon-Only and Oxygenated Gasolines," SAE Technical Paper 912432, 1991, doi:https://doi.org/10.4271/912432.

9.62. Robinson, J.E., "Influence of Fuel Volatility on Customer Perception of Cold Start Driveability," SAE Technical Paper 801351, 1980, doi:https://doi.org/10.4271/801351.

9.63. Pearson, J.K., Reders, K.H., and Tertois, V.M., "The Correlation of Consumer and Chassis Dynamometer Cold Weather Drivability," in *Paper EF4, CEC Symposium*, Wolfsburg, Germany, 1985.

9.64. Shibata, G., Omata, T., Isoda, T., Hosono, K. et al., "The Development of Driveability Index and the Effects of Gasoline Volatility on Engine Performance," SAE Technical Paper 952521, 1995, doi:https://doi.org/10.4271/952521.

9.65. Jorgensen, S.W. and Reuter, R.M., "Hot-Start Driveability of Low T50 Fuels," SAE Technical Paper 932672, 1993, doi:https://doi.org/10.4271/932672.

9.66. Coordinating Research Council, "2009-2010 CRC/ASTM Hot-Fuel-Handling Program (for Classes D-4 and E-5 Gasoline) CRC Project No. CM-138-09-2 Final Report," CRC Report No. 658, 2010.

9.67. CEC, "CEC Hot Weather Drivability Code of Practice for Use on Road, Track and Vehicle Dynamometer for Vehicles with Spark Ignition Engines," Ref. CEC M09T84, 1984.

9.68. Brownlow, A.D., Brunner, J.K., and Welstand, J.S., "Changes in Reid Vapor Pressure of Gasoline in Vehicle Tanks as the Gasoline is Used," SAE Technical Paper 892090, 1989, doi:https://doi.org/10.4271/892090.

9.69. Lavoie, G.A. and Smith, C.S., "Vapor Pressure Equations for Characterizing Automotive Fuel Behavior under Hot Fuel Handling Conditions," SAE Technical Paper 971650, 1997, doi:https://doi.org/10.4271/971650.

9.70. Pearson, J.K., Caddock, B.D., and Orman, P.L., "A Computer Model for the Prediction of Vapor Lock in the Fuel Pump of a Carbureted Engine," SAE Technical Paper 821201, 1982, doi:https://doi.org/10.4271/821201.

9.71. Riley, R.K., "Vapor Lock in Late Model Cars," SAE Technical Paper 831707, 1983, doi:https://doi.org/10.4271/831707.

9.72. Tertois, V.M. and Caddock, B.D., "Carbureter Foaming and Its Influence on the Hot Weather Performance of Motor Vehicles," SAE Technical Paper 821202, 1982, doi:https://doi.org/10.4271/821202.

9.73. Clarke, L.J., "The Causes and Control of Carburetor Foaming," SAE Technical Paper 841400, 1984, doi:https://doi.org/10.4271/841400.

9.74. Morrison, E.D., Ebersole, G.D., and Elder, H.J., "Laboratory Expressions for Motor Fuel Volatility and Their Significance in Terms of Performance," SAE Technical Paper 650859, 1965, doi:https://doi.org/10.4271/650859.

9.75. Coordinating Research Council, "Evaluation of Expressions for Fuel Volatility," CRC Report No. 403, Coordinating Research Council, Inc., 1967.

9.76. Becker, R.F., Ciardiello, U., Fitch, F.B., and Smith, C.N., "Hot Weather Volatility Requirements of European Passenger Cars," SAE Technical Paper 780651, 1978, doi:https://doi.org/10.4271/780651.

9.77. Morgan, C.R. and Smith, C.N., "Fuel Volatility Effects on Driveability of Vehicles Equipped with Current and Advanced Fuel Management Systems," SAE Technical Paper 780611, 1978, doi:https://doi.org/10.4271/780611.

9.78. Clarke, P.J., "Front-End Volatility Requirements of Late Model Cars at Intermediate Ambient Temperatures," SAE Technical Paper 830595, 1983, doi:https://doi.org/10.4271/830595.

9.79. Caddock, B.D., Davies, P.T., Evans, A.W., and Barker, R.F., "The Hot-Fuel Handling Performance of European and Japanese Cars," SAE Technical Paper 780653, 1978, doi:https://doi.org/10.4271/780653.

9.80. Clarke, P.J., "The Effect of Gasoline Volatility on Emissions and Driveability," SAE Technical Paper 710136, 1971, doi:https://doi.org/10.4271/710136.

9.81. Palmer, F.H., "The Development of CEC Driveability Test Procedures for European Vehicles and Fuels," SAE Technical Paper 811230, 1981, doi:https://doi.org/10.4271/811230.

9.82. Palmer, F.H., "Hot Weather Drivability—Does the CEC CF24 Test Method Reflect Motorists Requirements?" in *Paper EF3, CEC Symposium*, Wolfsburg, Germany, 1985.

9.83. CONCAWE, "Proposal for Revision of Volatility Classes in EN 228 Specification in Light of EU Fuels Directive," CONCAWE Report No. 99/51, 1999.

9.84. CEN, "Automotive Fuels. Unleaded Petrol. Requirements and Test Methods," EN 228:2012+A1:2017, 2017.

9.85. Jones, J.M., Pearson, J.K., and McArragher, J.S., "The Setting of European Gasoline Volatility Levels to Control Hot-Weather Driveability," SAE Technical Paper 852118, 1985, doi:https://doi.org/10.4271/852118.

9.86. Santos-Mello, R. and Cavalcante, B., "Cytogenetic Studies on Gas Station Attendants," *Mutation Research/Genetic Toxicology* 280, no. 4 (1992): 285-290.

9.87. Maltoni, C., Conti, B., Perino, G., and Di Maio, V., "Further Evidence of Benzene Carcinogenicity," *Annals of the New York Academy of Sciences* 534 (1988): 412-426.

9.88. CONCAWE, "Position Paper on Gasoline Vehicle Evaporative Emissions," CONCAWE, 1987.

9.89. Brown, S.G., Frankel, A., and Hafner, H.R., "Source Apportionment of VOCs in the Los Angeles Area Using Positive Matrix Factorization," *Atmospheric Environment* 41, no. 2 (2007): 227-237.

9.90. U.S. Code of Federal Regulations, "Title 40: Protection of Environment; Part 86: Control of Emissions from New and In-Use Highway Vehicles and Engines: Certification and Test Procedures; Subpart A—General Provisions for Emission Regulations for 1977 and Later Model Year New Light-Duty Vehicles, Light-Duty Trucks and Heavy-Duty Engines, and for 1985 and Later Model Year New Gasoline Fueled, Natural Gas-Fueled, Liquefied Petroleum Gas-Fueled and Methanol-Fueled Heavy-Duty Vehicles, § 86.098-28 Compliance with Emission Standards," 2010, Commission Regulation (EC) No. 692/2008, Annex VI, Determination of Evaporative Emissions (Type 4 Test), 2008.

9.91. Dudek, W. and Fisher, D., "Multiple Diurnal Evaporative Emissions Determinations with a Naturally Controlled Variable Volume Enclosure," SAE Technical Paper 932674, 1993, doi:https://doi.org/10.4271/932674.

9.92. Hum, R.W. and Eccleston, B.H., "Evaporative Losses from Automobiles: Fuel and Fuel System Influences," Report C150/71, Institute of Mechanical Engineers, London, UK, 1971.

9.93. Eccleston, B.H., Noble, B.F., and Hum, R.W., "Influence of Volatile Fuel Components on Vehicle Emissions," Report No. 7291, U.S. Bureau of Mines, Washington, DC, 1970.

9.94. Eccleston, B.H. and Hum, R.W., "Effect of Fuel Front-End and Mid-Range Volatility on Automobile Emissions," Report No. 7707, U.S. Bureau of Mines, Washington, DC, 1972.

9.95. American Petroleum Institute, "A Study of Factors Influencing the Evaporative Emissions from In-Use Automobiles," Publication No. 4406, American Petroleum Institute, 1985.

9.96. Koehl, W.J., Benson, J.D., Burns, V., Gorse, R.A. et al., "Effects of Gasoline Composition and Properties on Vehicle Emissions: A Review of Prior Studies—Auto/Oil Air Quality Improvement Research Program," SAE Technical Paper 912321, 1991, doi:https://doi.org/10.4271/912321.

9.97. Heinen, C.M., "We've Done the Job—What's It Worth?" SAE Technical Paper 801357, 1980, doi:https://doi.org/10.4271/801357.

9.98. McArragher, J.S., Betts, W.E., Brandt, J., Kiessfing, D. et al., "Evaporative Emissions from Modern European Vehicles and Their Control," SAE Technical Paper 880315, 1988, doi:https://doi.org/10.4271/880315.

9.99. CONCAWE, "An Investigation into Evaporative Hydrocarbon Emissions from European Vehicles," Report No. 87/60, CONCAWE, 1987.

9.100. CONCAWE, "The Effects of Temperature and Fuel Volatility on Vehicle Evaporative Emissions," Report No. 90/51, CONCAWE, 1990.

9.101. Reuter, R.M., Benson, J.D., Burns, V.R., Gorse, R.A. et al., "Effects of Oxygenated Fuels and RVP on Automotive Emissions— Auto/Oil Air Quality Improvement Program," SAE Technical Paper 920326, 1992, doi:https://doi.org/10.4271/920326.

9.102. Koehl, W.J., Lloyd, D.W., and McCabe, L.J., "Vehicle Onboard Control of Refueling Emissions—System Demonstration on a 1985 Vehicle," SAE Technical Paper 861551, 1986, doi:https://doi.org/10.4271/861551.

9.103. Braddock, J.N., Gabele, P.A., and Lemmons, T.J., "Factors Influencing the Composition and Quantity of Passenger Car Refueling Emissions—Part I," SAE Technical Paper 861558, 1986, doi:https://doi.org/10.4271/861558.

9.104. Musser, G.S. and Shannon, H.F., "Onboard Control of Refueling Emissions," SAE Technical Paper 861560, 1986, doi:https://doi.org/10.4271/861560.

9.105. Musser, G.S., Shannon, H.F., and Hochhauser, A.M., "Improved Design of Onboard Control of Refueling Emissions," SAE Technical Paper 900155, 1990, doi:https://doi.org/10.4271/900155.

9.106. Johnson, P.J., Khami, R.J., Bauman, J.E., Goebel, T.D. et al., "Carbon Canister Development for Enhanced Evaporative Emissions and On-Board Refueling," SAE Technical Paper 970312, 1997, doi:https://doi.org/10.4271/970312.

9.107. Weidenaar, B.E., Grimes, H.J., and Jewell, R.G., "Vapor Recovery Nozzle Development and Field Testing," SAE Technical Paper 741038, 1974, doi:https://doi.org/10.4271/741038.

9.108. Austin, T.C. and Rubenstein, G.S., "A Comparison of Refueling Emissions Control with Onboard and Stage II Systems," SAE Technical Paper 851204, 1985, doi:https://doi.org/10.4271/851204.

9.109. Graboski, M.S., Mowery, D.L., and McClelland, J., "Microenvironmental Exposure Analysis Evaluation of the Toxicity of Conventional and Oxygenated Motor Fuels," SAE Technical Paper 982535, 1998, doi:https://doi.org/10.4271/982535.

9.110. Jeffrey, J.G. and Elliott, N.G., "Gasoline Composition Effects in a Range of European Vehicle Technologies," SAE Technical Paper 932680, 1993, doi:https://doi.org/10.4271/932680.

9.111. Sanders, R., "Carburetor Icing," SAE Technical Paper 380067, 1938, doi:https://doi.org/10.4271/380067.

9.112. Eltinge, L., Gray, D.S., Oblad, S.R., and Kay, R.E., "Gasolines, Cars and Carburetor Icing," SAE Technical Paper 620025, 1962, doi:https://doi.org/10.4271/620025.

9.113. Jackson, H.R. and Wallin, O.W., "Carburetor Icing Tendencies of Late Model Cars," SAE Technical Paper 690209, 1969, doi:https://doi.org/10.4271/690209.

9.114. Galante-Fox, J.M., Jarvis, D.E., Garrick, R.D., and Chen, A.J., "Throttle Icing: Understanding the Icing Mechanism and Effects of Various Throttle Features," SAE Technical Paper 2008-01-0439, 2008, doi:https://doi.org/10.4271/2008-01-0439.

9.115. Kung, J.F., Haworth, J.P., and Hickok, J.E., "A New Look at Motor Gasoline Quality—Carburetor Icing Tendency," SAE Technical Paper 510005, 1951, doi:https://doi.org/10.4271/510005.

9.116. Owen, K. and Landells, R.G.M., "Precombustion Fuel Additives," in *Critical Reports on Applied Chemistry*, Vol. 25, Gasoline and Diesel Fuel Additives, Ed. Owen, K. (New York: John Wiley & Sons, 1989).

9.117. Tupa, R.C. and Dorer, C.J., "Gasoline and Diesel Fuel Additives for Performance/Distribution Quality–II," SAE Technical Paper 861179, 1986, doi:https://doi.org/10.4271/861179.

9.118. Coordinating European Council, "Intake System Icing Procedures for Use on Road, Track or Vehicle Dynamometer with Spark Ignition Vehicles," Tentative Test Method No. CEC M10T87, Coordinating European Council, 1987.

9.119. Schwartz, S.E., "Observations Through a Transparent Oil Pan during Cold-Start, Short-Trip Service," SAE Technical Paper 912387, 1991, doi:https://doi.org/10.4271/912387.

9.120. Hallock, E.F., "Crankcase-Oil Dilution," SAE Technical Paper 230035, 1923, doi:https://doi.org/10.4271/230035.

9.121. Shayler, P.J., Davies, M.T., and Scarisbrick, A., "Audit of Fuel Utilisation during the Warm-Up of SI Engines," SAE Technical Paper 971656, 1997, doi:https://doi.org/10.4271/971656.

9.122. Sagawa, T., Fujimoto, H., and Nakamura, K., "Study of Fuel Dilution in Direct-Injection and Multipoint Injection Gasoline Engines," SAE Technical Paper 2002-01-1647, 2002, doi:https://doi.org/10.4271/2002-01-1647.

9.123. Sehr, A., Thewes, M., Müther, M., Brassat, A. et al., "Analysis of the Effect of Bio-Fuels on the Combustion in a Downsized DI SI Engine," *SAE Int. J. Fuels Lubr.* 5, no. 1 (2012): 274-288, doi:https://doi.org/10.4271/2011-01-1991.

9.124. Choate, P.J. and Edwards, J.C., "Relationship between Combustion Chamber Deposits, Fuel Composition, and Combustion Chamber Deposit Structure," SAE Technical Paper 932812, 1993, doi:https://doi.org/10.4271/932812.

9.125. Richardson, C.E., Fischer, J.L., and Pawczuk, G., "Evaluation of the Effect of Fuel Composition and Gasoline Additives on Combustion Chamber Deposits," SAE Technical Paper 962012, 1996, doi:https://doi.org/10.4271/962012.

9.126. Foringer, D.E., "Gasoline Properties Affecting Fuel Economy," SAE Technical Paper 650427, 1965, doi:https://doi.org/10.4271/650427.

9.127. Hochhauser, A.M., Benson, J.D., Burns, V.R., Gorse, R.A. et al., "Fuel Composition Effects on Automotive Fuel Economy—Auto/Oil Air Quality Improvement Research Program," SAE Technical Paper 930138, 1993, doi:https://doi.org/10.4271/930138.

Further Reading

Montemayor, R.G. (Eds), *Distillation and Vapor Pressure Measurement in Petroleum Products* (West Conshohocken, PA: ASTM International, 2008).

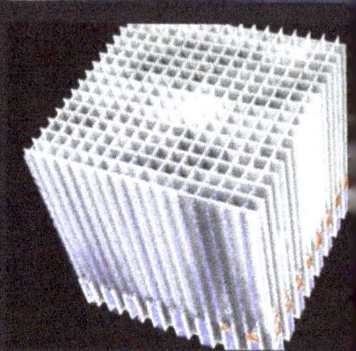

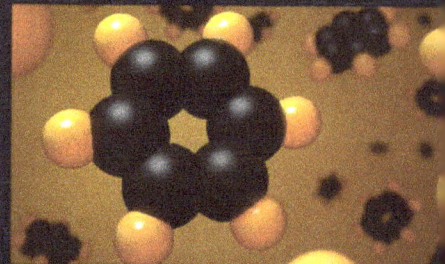

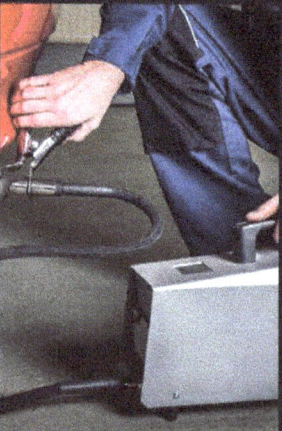

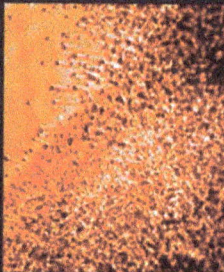

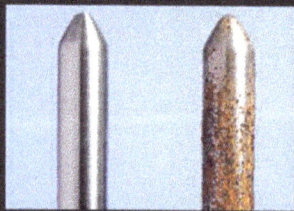

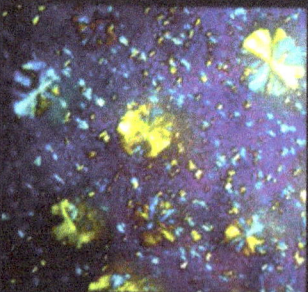

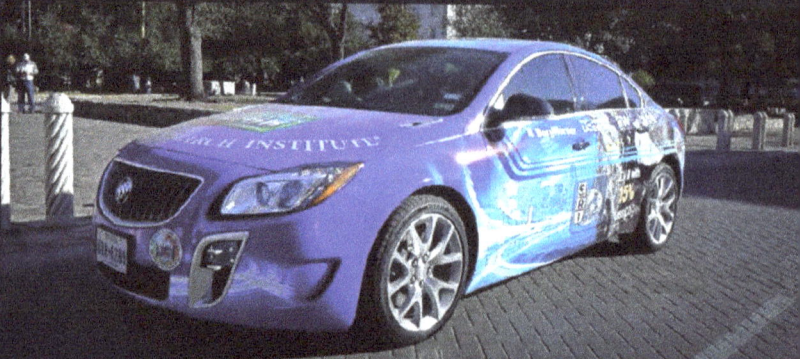

10

Influence of Gasoline Composition on Stability, Gum Formation, and Engine Deposits

10.1. The Influence of Gasoline Composition on Stability

The chemical structure of the different hydrocarbons present in gasoline is very important in terms of the stability of that gasoline. This is exemplified by the publication of two patents for deposit-forming gasoline reference fuels [10.1, 10.2] The tendency of specific chemical groups to autoxidize has been found to be as follows, in order of decreasing reactivity: di-olefins, aromatic olefins, aliphatic olefins, alkyl aromatics, naphthenes, and paraffins [10.3]. Olefinic compounds are much less stable to oxidation than aromatic or paraffinic compounds, so when considering using gasoline components that are olefinic in nature, precautions must be taken to ensure that they do not give rise to undesirable oxidation products. These precautions normally consist of using antioxidant fuel additives and/or using additional refinery processing such as hydrogen treatment or heat soaking to improve the stability. Under these conditions, and provided that no metals are present that could catalyze hydrocarbon oxidation, long-term storage of gasoline can be successfully carried out [10.4].

It is mainly the cracking processes such as catalytic cracking, thermal cracking, coking, steam cracking, and visbreaking that give rise to olefinic gasoline blend components. In particular, thermal cracking processes (visbreaking, coking, and thermal cracking) can give rise to di-olefinic compounds in addition to mono-olefins, and these are extremely susceptible to oxidation and formation of gummy compounds. The presence of some sulfur [10.5, 10.6] and nitrogen compounds [10.5–10.9] will also contribute to the autoxidation of gasoline, as will the use of alcohols as blend components [10.10]. Gasoline ethanol blends are now mandatory in some geographical areas and have shown an increase in water content and gum formation [10.11, 10.12] while other studies were conversely claiming a dilution effect [10.13, 10.14].

The oxidation of hydrocarbons, prior to combustion, proceeds via the formation of peroxide and hydroperoxide radicals, which eventually result in the formation of gums and acidic products. The gums are only slightly soluble in gasoline and, if more and more of them are allowed to form, perhaps because of insufficient oxidation inhibitor being present, they will come out of solution and settle as a sludge on the bottom of the storage tank or the vehicle fuel tank. If they are present to an excessive degree in the finished gasoline, they will deposit in the intake system of the engine, as discussed later in this chapter. Certain metals, notably copper, catalyze the oxidation process and must be avoided in fuel systems, or, if present in the fuel, must be deactivated by the use of appropriate fuel additives [10.15].

The hydroperoxides formed during the oxidation of hydrocarbons are pro-knocks. If present, they will reduce the octane quality of the gasoline [10.16]. Where lead antiknocks are still used, the antiknock performance can also be reduced if there is insufficient antioxidant present to prevent any tetra-ethyl-lead (TEL) from degrading due to oxidation. If this occurs, a white deposit of lead oxide and lead carbonate can form and block filters, and the octane quality can also be reduced depending on the extent of loss of the TEL.

10.1.1. Measurement of Stability

There are a number of different laboratory tests that are used to establish the oxidation stability of gasoline. Some of them are rather lengthy and so are unsuitable for routine control of gasoline stability, and none of them can be said to give a very precise idea of the resistance of gasoline to oxidation during storage. The reason for this is that storage conditions are extremely variable. Some storage tanks have fixed roofs, whereas others have floating roofs that reduce contact with air; ambient temperature conditions will vary according to geographical location; and the nature of the gasoline itself varies, with some blends containing high levels of olefins.

The most commonly used tests are discussed briefly in the following sections.

10.1.1.1. Measurement of Oxidation Stability by Induction Period Method

For the induction period method [10.17], a sample of gasoline is oxidized in a bomb initially filled with oxygen at 689 kPa. The bomb is held at a temperature of 100°C and the pressure is recorded. The pressure will begin to fall slowly but will reach a point where the rate of decay increases rapidly, until it begins to fall at a rate of at least 13.8 kPa over a 15-minute period. This is known as the break point. The induction period (sometimes called breakdown time) is the elapsed time from placing the bomb in the bath at 100°C up to the break point.

It is generally accepted that an induction period of 240 minutes corresponds to a storage life of at least six months. The full duration of the test is 960 minutes, and most good gasolines will exceed this. However, this test has been shown to have a poor correlation with gum deposition under more realistic conditions [10.18].

10.1.1.2. Automated Measurement of Induction Period

From the previous section, it is clear that it would normally take 16 hours to measure the induction period. In response to requests from the US gasoline industry a faster method has therefore been developed [10.19] that has been found to correlate well with the full induction period method. For this automated method, a small fuel sample, 5 mL, is again sealed in a bomb and charged with oxygen under pressure at room temperature, with a fixed amount of oxygen. A slightly higher pressure of 700 kPa is used and a higher temperature of 140°C. As the sample heats up, the pressure will rise; the pressure in the bomb is automatically monitored and the point at which the pressure drops by 10% of the maximum pressure attained is used as the break point. The higher temperature and pressure produce a higher oxidation rate, and testing is usually completed within 1 hour but can take 4 hours.

The measured induction period can be used to as an indication of oxidation and storage stability. Further, it has been used to evaluate the deposit-forming tendencies of gasoline in conjunction with an IFP energies nouvelles (IFPEN) in-house autoclave method, in conjunction with both the M102E IVD engine test and the direct injection VW EA1112 engine test. The RSSOT showed some correlation with M102E IVD deposit mass but none with the VW EAA1112 engine test [10.20].

10.1.1.3. Measurement of Gum Content by Jet Evaporation

The existent gum or solvent-washed gum is measured by the gum content in fuels by jet evaporation [10.21] test that is used for aviation gasoline and aircraft turbine fuel as well as for motor gasoline. For aviation turbine fuels, large quantities of gum indicating the presence of higher boiling components or particulate matter tend to reflect poor handling within the distribution system. For motor gasoline the existent gum test is only intended as a guide to the propensity of the gasoline to result in induction system deposits and valve sticking, rather than as a measure of the fuel's storage stability.

When testing gasoline, a measured volume is evaporated under controlled conditions of temperature in a stream of air. The residue is weighed, extracted with n-heptane, and then reweighed. The existent gum is the heptane insoluble part of the residue. The total residue, before extraction, is known as the unwashed gum.

The unwashed gum will give an indication of the amount of nonvolatile components present in the gasoline, but it has been shown not to correlate with combustion chamber deposits (CCDs) [10.22]. For a finished gasoline, this will include many of the fuel additives that are now included in the gasoline, plus any existent gum and any contaminants that may be present as a result of poor handling practices. Because of the mild conditions of the test, only a small amount of gum will generally be formed; the existent gum test is thus really only a measure of the amount of gum already present in the fuel. However, a high value may indicate that there has already been some oxidation of the gasoline, and unless it is a severely aged gasoline it may indicate some instability and that more gum is likely to be formed.

10.1.1.4. Measurement of Oxidation Stability by Potential Residue Method

The potential residue method [10.23] is primarily intended for use with and in the specification of aviation fuels, although it can also be used for motor gasolines. A sample of gasoline is oxidized in a bomb using the same conditions as in the induction period test, but for the full 960-minute duration of the test method. Any precipitate is filtered off and the unwashed gum is determined as described above. This is recorded as the potential gum. A shortened version of this procedure with an aging time of 300 minutes can be used and features in the aviation gasoline specification [10.24].

The potential gum represents the overall potential of the fuel to form gums and deposits, and so is some indication of the storage stability of the fuel. However, the oxidation conditions used are much more severe than that occurs in practice, and the correlation with storage stability is thus poor. It is probably true to say, however, that a low potential gum represents a stable fuel, although the reverse may not necessarily be true.

10.1.1.5. Measurement of Long-Term Stability

In this test, fuels are stored in the dark at 43°C and the existent gums are then determined. A modified Arrhenius equation can be used to predict the storage stability at lower temperatures. The results of previous investigations [10.25–10.27] show that aging at this temperature for 32 weeks was equivalent of aging for 5.7 years at 27°C. However, this work was conducted on the gasolines available at the time, which were all leaded gasolines. No work has been performed to investigate the validity of this work with oxygenated fuels.

10.2. Deposit Formation in Engines Due to Gasoline Oxidation

Deposit formation in the engine due to lack of fuel stability can have a severe influence on vehicle driveability performance [10.28–10.31] and can cause piston ring and valve sticking [10.32, 10.33], although piston ring sticking is usually due to oil deposits with only a minor contribution from the heavy ends of the fuel. The nature and amount of the deposits laid down in different parts of the engine will depend very much on a variety of factors, including temperature, driving mode, contact with air, and so on. In addition, foreign matter such as sand or rust can find its way into the fuel and may attach itself to the sticky gums formed by oxidation, thereby increasing the volume and adverse effects of the deposits.

The most potent fuel factor that influences deposit formation in the engine intake system is the presence of appropriate additives designed to keep the system clean and/or to clean up existing deposits. Different types of additives may be needed, depending on the position in the fuel system where they form; these are discussed in Chapter 11. There are additives such as some antioxidants that can worsen intake valve deposits (IVDs) if used at too high a concentration [10.34].

Regarding other fuel factors that cause intake system deposits, none of the preceding laboratory tests are entirely satisfactory for predicting deposit-forming tendencies. However, there are engine tests that will give an indication of the relative deposit-forming tendency of different gasolines and the effectiveness of additives in overcoming them. These are discussed briefly later in the chapter.

It is primarily the olefin content and the types of olefins in the base fuel that are thought to have the greatest influence [10.34, 10.35]. The most reactive of the olefins and, in particular, the heavier conjugated di-olefins are generally regarded as the most likely to cause deposit problems [10.36]. However, it is not always easy to identify precisely which olefins are present in a gasoline, and so total olefin content, coupled with satisfactory levels for existent gum and induction period, is sometimes taken as a guide to likely stability in an engine. The ASTM Bromine number test [10.37] gives a measure of the olefinic content, but it is not very reliable in predicting deposit-forming tendency because some olefins, and particularly the important di-olefins, are not fully determined in the test. Diene number is another test that has been used [10.38], in which the gasoline is refluxed with a known amount of maleic anhydride and the amount of unreacted maleic anhydride is determined. The maleic anhydride combines with conjugated dienes to form cyclic compounds by the Diels-Alder reaction, and so the test gives a measure of the amount of this type of di-olefin present. However, it does not determine the presence of other reactive di- and mono-olefins and so is not entirely satisfactory. Recent work with supercritical chromatography (SFC) and mass spectrometry using positive ion atmospheric pressure photo-ionization (APPI) has shown significant promise [10.39]

The existent gum test [10.21] is a measure only of the amount of gum actually present in a gasoline, not of how much can form under the conditions in an engine, and so is only of limited value in predicting the deposit-forming tendency of a gasoline. The potential residue test [10.23] measures the total gum that can be formed in a fuel when it is oxidized under prescribed conditions in a bomb filled with oxygen, but this can be much more severe than occurs in an engine and does not correlate well with field experience [10.18].

10.2.1. Deposit Formation in the Fuel Tank and Fuel Lines

The antioxidant fuel additives that are used to enhance storage stability are most effective when the fuel is liquid rather than vaporized, since they are relatively nonvolatile. There would normally be sufficient additive present to fully protect the fuel in its liquid state, and so problems of oxidation in the tank or fuel lines are relatively rare. However, too much of certain antioxidants can sometimes contribute to intake system deposits [10.40, 10.41].

The fuel in a vehicle's tank is normally no more than a few weeks old from the time it was blended, although it may sometimes have been stored for considerable periods. However, there have been cases during periods of economic recession when cars have been stockpiled for well over a year prior to being sold. These vehicles would have needed to have some gasoline in their tanks in order to move them from the production line to the storage area, and so in these cases there is a considerable potential for problems due to gasoline oxidation. When such long storage times are predicted, it would be normal for a vehicle manufacturer to use a special "factory fill" gasoline containing additional oxidation inhibitors and other additives. With the increasing acceptance of plug-in hybrid electric vehicles (PHEV), there is also the possibility that such vehicles may be used predominantly for short journeys, recharged overnight, and thus not use the internal combustion engine for prolonged periods of time.

The oxidation that occurs in the vehicle's tank is the same as in any storage tank, with gums being formed that will come out of the solution, block filters, and deposit in various parts of the fuel intake system. Even though the fuel is not normally held for long periods, in use it can be heated to temperatures as high as 50°C in the tank due to recirculation and heat transfer from the exhaust system. The tank will generally be under a slight positive pressure, and there may be metals present that can catalyze oxidation reactions. In addition, vehicle movement will agitate the tank contents and mix them with air, further increasing the likelihood of oxidation.

Gasolines that have undergone oxidation on storage and contain hydroperoxides are sometimes known as "sour" gasolines (although this term is also applied to petroleum streams containing mercaptans). The application of SFC-MS [10.39] has been shown to be capable of identifying problem peroxides in gasoline.

These have been shown [10.42, 10.43] to have an adverse effect on certain elastomeric materials used in parts of the fuel system, such as the fuel hose. Sour gasolines have been shown to affect the tensile strength, hardness, and volume swell of such materials as nitrile rubbers and epichlorohydrin copolymers. There has been concern that such attack could result in the hoses becoming permeable and contributing to increased hydrocarbon emissions [10.44]. The deliberate inclusion of oxygenated component, for example, alcohols, will also affect the permeability of elastomers, and it has been shown that the effect can persist for some time after changing to a nonoxygenated fuel [10.45].

10.2.2. Deposit Formation in Fuel Injectors and Carburetors

The fuel is vaporized and mixed with air in the intake port, or in the carburetor venturi of such engines, so this is the time when oxidation really starts. Pre-flame oxidation reactions occur (discussed in Chapter 7), and some already partially oxidized hydrocarbons may polymerize to form gums that will stick to surfaces so that deposits will build up. These deposits can occur in the fuel injectors or carburetors, intake manifolds, and the intake valves and ports, depending on the temperature regime in each part of the system. Oxidation accelerated by heat soak-back from the engine when it is switched off can be a major cause of deposit formation. For this reason, driving that involves a great deal of stopping and starting is the most critical [10.46, 10.47]. Vehicles with a large heat capacity in the engine compartment such as those making use of turbochargers and close coupled catalysts, for example, can give rise to high under-hood temperatures and may be more susceptible to this type of deposit formation. The use of electric cooling fans that continue to operate after the vehicle is switched off may often be beneficial in avoiding or minimizing these deposits, particularly if the air stream is directed at critical components.

Although the build-up of deposits probably starts with gums and nonvolatiles sticking to the metal surfaces, it is increased by particulates and other materials from a number of possible sources. These include exhaust gases arising from other traffic and other airborne particles which penetrate the air filter, combustion gases blown back into the fuel inlet system due to valve overlap, exhaust gas recirculation (EGR) [10.48], and blow-by gases via the positive crankcase ventilation (PCV) system [10.49–10.51]. Additives are available that can largely overcome these deposit problems.

In a carburetor, deposits can form on the walls of the throttle body, on the throttle plate, in the idle air circuit, and in metering orifices and jets, as shown in Figure 10.1 [10.46]. Many carbureted engines are open loop controlled, and so these deposits will cause the air-to-fuel ratio to move away from the optimum setting as selected by the vehicle manufacturer and may thus cause driveability malfunctions, a decrease in fuel economy, and for noncatalyst vehicles, an increase in exhaust emissions [10.52].

FIGURE 10.1 Location of deposit formation in a carburetor.

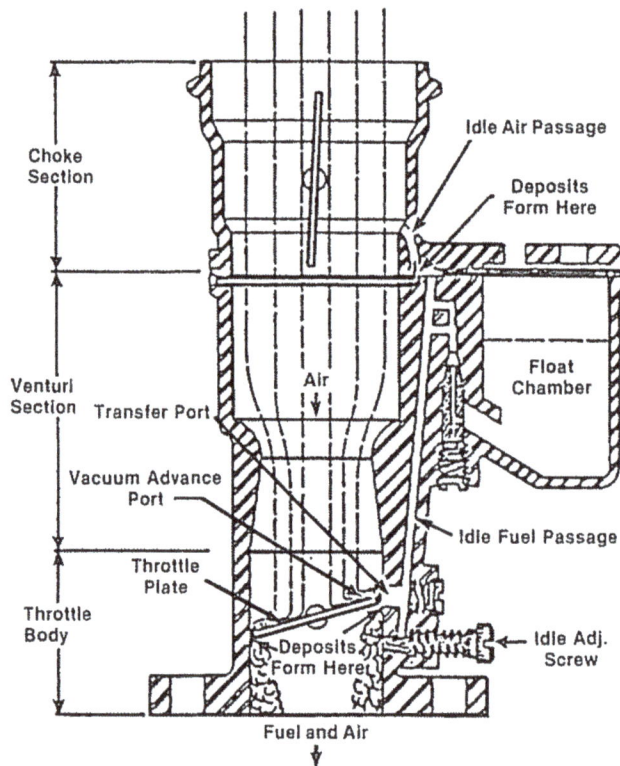

A great deal of work was performed in the US from the early 1970s to investigate the driving conditions that promoted carburetor deposits, with the aim of developing a test method to assess fuel and additive performance [10.53]. Techniques had been developed for assessing the deposits as they built up in field trials [10.52], but there was no recognized test procedure for forming the deposits. Due to the efforts within the CRC group, over 750 tests were run, and a tentative procedure was developed that used a removable aluminum sleeve inserted into the carburetor barrel that was both weighed and visually rated.

The proposed procedure used an in-line six-cylinder engine fitted with a single-barrel carburetor. The engine ran a cycle alternating between 3 minutes at an idle speed of 700 rev/min and 7 minutes at a cruise speed of 2000 rev/min. The load conditions were set by the fuel flow at each speed. The test ran for 20 hours. To increase test severity, the piston ring gaps were opened to increase blow-by, which was then routed back into the air intake, and during the cruise condition full EGR was applied.

However, by the time the procedure had been developed to this stage, port fuel injection (PFI) engines were already making in-roads into the market, and the need for a carburetor cleanliness test was diminishing, and so the proposed method was never taken beyond this point.

In Japan, work was performed using a similar cycle, but they found little deposit formation in their compound carburetor [10.34]. Their work found that a constant operating point of 2000 rev/min with a manifold depression of 26.7 kPa and 20% EGR produced more deposits and would discriminate between an untreated fuel and an additive treated fuel.

In Europe a test method had been developed [10.54] that relied on increasing the blow-by, by removing one of the piston rings, which was routed back into the air intake. This procedure used Renault type 810-26, in-line four-cylinder engine with a single-barrel carburetor, but the assessment was purely visual as there was no removable insert that could be weighed. The engine ran for 2 minutes at 800 rev/min and 8 minutes at 1800 rev/min. The engine ran for 6 hours, then stood for 18 hours before running for a further 6 hours. As with the case in the US, this procedure never progressed beyond the tentative stage, although it was widely recognized and accepted as an indication of performance.

Fuel injectors can be particularly susceptible to deposit formation if they inject into the ports rather than into the inlet manifold, due to the higher temperatures that may be achieved as a result of hot soaking. There is an increased problem if the injectors are electronically controlled so that they inject only at the appropriate time rather than continuously [10.5, 10.35, 10.47, 10.55–10.57]. PFI was widely adopted because of the benefits in terms of fuel economy, exhaust emissions, and overall vehicle performance [10.58]. The deposits form in the pintle area, as shown in Figure 10.2 [10.46], and can reduce fuel flow rate and alter spray pattern. The fuel factors that increase injector deposit formation are di-olefins, olefins, and sulfur content [10.59]. Unwashed gum from heavy aromatics has no influence, and there is no difference in deposit-forming tendency between leaded and unleaded gasolines.

FIGURE 10.2 Deposits form in the pintle area of port fuel injectors.

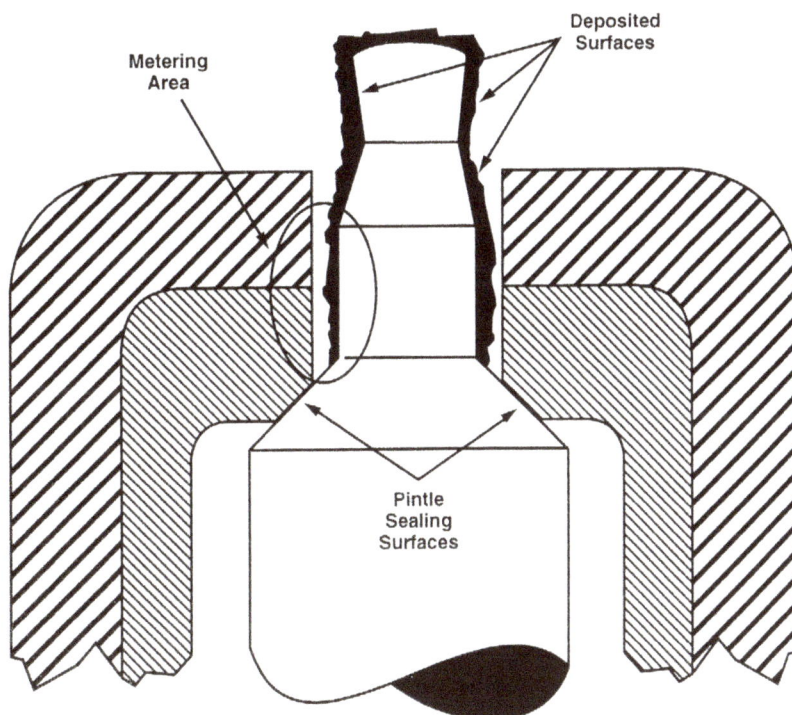

© SAE International.

The recognized PFI deposit-forming tests are described briefly next.

10.2.2.1. Chrysler 2.2

As a result of customer complaints during the mid-1980s, the CRC set up the Automotive Fuel Injector Deposit Group [10.60, 10.61], which led to the development of this procedure. The test procedure [10.62] uses a 2.2 L turbo-charged engine with an in-line four-cylinder configuration. The vehicle can be run on the road or a chassis dynamometer. The engine operates to a two-mode cycle, commonly referred to as the 15/45 cycle. This consists of running the vehicle at 88 km/h for 15 minutes and then shutting the engine off and allowing it to soak for 45 minutes. The normal test duration is 16,100 km. A laboratory bench flow apparatus is used to determine the fuel flow through the injectors, and the results are reported as a percentage loss of flow rate.

10.2.2.2. Other Vehicle-Based Injector Deposit Test

In Europe, a similar test method was employed using the Peugeot XU5 engine. The test was developed by the Institut Français du Pétrole (the French Institute of Petroleum), now known as IFP energies nouvelles, and was designated as the IFP-TAE-1-87 Peugeot 205 GTI test. The engine was a naturally aspirated in-line four-cylinder engine, and the test procedure relied on long engine-off periods to allow heat soak-back from the engines to "cook" any residual fuel in and on the injector nozzle. The test method never received any official Europe-wide approval, although it was widely used; it is now essentially obsolete.

10.2.2.3. CRC Injector Deposit Bench Test

Although the vehicle test procedure ASTM D5598 [10.62] had been developed to replicate field problems and allow for the development of fuels to avoid these problems, it was recognized that such a test was both costly and time consuming. It had also become necessary to assess injector fouling as part of the fuel certification process. A quicker and less expensive method was therefore developed that tried to replicate the vehicle within the laboratory [10.63]. For this test procedure [10.64] four fuel injectors are placed in a heated aluminum block that is heated to 160°C. The block is allowed to cool for 5 minutes before the injectors are set to pulse for 15 seconds before being turned off for the remainder of a 1-hour cycle. Fuel flow through the injectors is measured after 22 and 44 cycles, and the results are reported as percentage loss of flow. It has been concluded that 33 cycles in the bench test rig is equivalent, in terms of deposit build-up, to 3219 km in the vehicle test [10.65]. The bench test has been shown to correlate with the vehicle test but is more sever [10.66].

10.2.3. Deposit Formation in the Inlet Manifold, Ports, and on Valves

For a period of carbureted engine evolution, a "hot spot" or early fuel evaporation (EFE) plate was incorporated in the inlet manifold to help fuel vaporization. This area was particularly prone to deposit formation, and the insulating and fuel absorbing properties of these deposits would interfere with the emission control. They will increase warm-up time and so worsen fuel economy, and they may give driveability problems. Work was undertaken by the CRC in the mid-1970s with the aim of developing an engine test procedure for assessing such deposits [10.67]. The engine test was based on a 1974 5.7 L Chevrolet V8 engine fitted with a four-barrel carburetor. The EFE had already been phased out from this engine. The changing engine technology resulted in the emphasis moving to IVDs, and the test procedure was biased toward this, although intake manifold deposits were reported for the ensuing test.

The deposits that form in the intake port have been claimed to contribute to octane requirement increase (ORI) [10.68, 10.69], and the combination of port and valve deposits will influence the breathing of the engine and so will affect power, emissions, and driveability [10.30, 10.70]. The mechanism by which IVDs adversely affect driveability and exhaust emissions is claimed to be due to two factors [10.71].

1. C9 and C10 hydrocarbons are held up as liquids by the deposits, the amount increasing with increasing valve deposits, so that during accelerations the air-to-fuel ratio becomes momentarily lean, possibly to the extent of causing a misfire.

2. The heavy fuel absorbed on the valves is released after the acceleration is completed and again upsets the air-to-fuel ratio, influencing both driveability and emissions.

Removing deposits from valves has been shown to reduce both CO and NO_X emissions in specific engine types [10.72].

Valve deposits can build up to such an extent that they interfere with the closing of the valve so that valve burning can result. The nature of the deposit varies with the fuel and the valve temperature so that sometimes they are soft and sticky and in other cases they are hard and brittle [10.70]. A number of investigations into the nature of valve deposits have been carried out [10.73–10.75], and there is a consistency of views that heavy aromatic compounds from reformate and heavy polar compounds formed by oxidation are largely responsible for deposit formation. The presence of alcohol in the fuel appears to increase the deposit rate on intake valves, and such blends may need additional additive treatment to overcome the higher deposit levels [10.76, 10.77]. It has also been shown that corrosion inhibitors used to protect the fuel delivery infrastructure can increase IVDs [10.78]. The concentration of these additives may become greater than intended as a result of splash blending ethanol that has been treated into gasoline that has also been treated.

Where the deposits are brittle, pieces can break off and hold the valve open so that loss of power or even valve burning can occur. Pieces of hard valve deposit can also pass into the combustion chamber and become wedged in the ring pack or get lodged between the piston crown and valves or cylinder head. If sticky deposits extend up the valve stem, they can become extremely viscous in cold weather while the vehicle is standing idle and can cause failure to start from cold also and engine damage [10.32].

As with all aspects relating to the interaction of fuel and engine factors, it is the latter which have the greatest effect, so that some vehicles are entirely free from deposit problems and others can suffer very quickly. The lubricant can play a part, particularly with respect to valve deposits, because the valve stems are lubricated by oil flowing down them onto the valve undersides. The contribution of various additives in the lubricant has been assessed [10.79–10.82], and it has been shown that increasing amounts of viscosity index improver additive in the oil increases valve deposits [10.79, 10.83]. As noted earlier, none of the standard laboratory tests for fuel stability will adequately correlate with the deposit-forming tendency of fuels within the engine; there are no tests to specifically look at the deposits within the manifold, but there are a number of engine tests that have been developed, and widely recognized, to assess inlet valve deposits. These are discussed next, along with rig tests that have been developed to simulate the problem.

In direct injection engines, IVD can still be an issue, as shown in Figure 10.3 [10.84], but as fuel is not injected into the manifold or onto the back of the valves, in such engines the IVDs are found to be dependent on the lubricating oil and the operating cycle but not the fuel.

FIGURE 10.3   IVDs from a direct injection engine.

10.2.3.1. Opel Kadett Test

This test procedure [10.85] uses a 1.2 L engine fitted with twin down-draft carburetors that was originally fitted in the Opel Kadett car. This was a four-cylinder engine with push rod-operated valves that made disassembly a straight forward operation. The engine had a cast iron cylinder head. Because the engine was equipped with twin carburetors, it was possible, with only slight modification to the inlet manifold, to operate the engine as two twin-cylinder engines operating to exactly the same cycle. A different fuel was supplied to each carburetor, usually a base fuel to one carburetor and an additive fuel to the other carburetor. The engine is run on a test bench according to the 4.5-minute cycle, shown in Table 10.1.

TABLE 10.1 Opel Kadett test cycle.

Stage	Time (seconds)	Speed (rev/min)	Torque (Nm)
1	30	1200	2
2	60	3000	35
3	60	1300	29
4	120	1850	32

© SAE International.

The cycle is repeated to give a total test running time of 40 hours. The induction system was then removed and the cylinder head removed so that the intake valves could be removed for visual rating and to determine the weight of deposits.

As it was known that the lubricating oil that passed down the valve guide would affect inlet valve deposits, the later version of the procedure was modified to incorporate a specially modified cylinder head that was drilled to allow a controlled feed of lubricating oil to the valve guides. The valve stem seals were replaced with specified seals to restrict the flow of the normal lubricating oil, and a syringe pump was used to supply the controlled oil feed to the valve stems.

Due to the lack of availability of spare parts, this procedure is no longer supported by CEC.

10.2.3.2. Mercedes-Benz M102E Test

This test procedure was developed as a replacement for the Opel Kadett procedure as it was considered that carburetors would become obsolete and would be replaced with fuel injection. The test procedure [10.86] uses a 2.3 L four-cylinder, in-line engine that was widely used in Mercedes-Benz vehicles from the 1980s. The engine has an aluminum alloy cylinder head and is fitted with a Bosch KE-Jetronic electromechanical fuel injection system that continuously sprays fuel toward the inlet valve.

The engine is again run to a 4.5-minute cycle as shown in Table 10.2. The cycle is repeated 800 times to give a total running time of 60 hours. The test cycle differs from that used for the Opel Kadett in that successive stages are at a higher speed, and there is thus only one deceleration at the end of the cycle as opposed to the two decelerations in the carbureted engine cycle.

TABLE 10.2 Mercedes-Benz M102E test cycle.

Stage	Time (seconds)	Speed (rev/min)	Torque (Nm)
1	30	800	0–2
2	60	1300	30
3	120	1850	33
4	60	3000	35

© SAE International.

Because the KE-Jetronic fuel system is constantly delivering fuel (except during over-run conditions), it is constantly wetting the valve tulip except for where the valve stem obstructs the spray. At the higher engine speeds, the intake valve will slowly rotate due to the action of the valve spring as the valve open and closes. This means that the "shadow" produced by the valve stem will slowly move around the valve tulip. To make the test more severe, later versions of the test procedure required that the inlet valves be pegged to prevent this rotation. A valve from such a test where the fuel additive package does not keep the valve completely clean but allows some deposit build-up in this "shadow" area is shown by the valve in the center of Figure 10.4.

FIGURE 10.4 Intake valves from different CEC F-05-A-93 tests.

Source: APL Automobil-Prüftechnik Landau GmbH.

This valve has a deposit weight of 44 mg, which would satisfy the Worldwide Fuel Charter requirements for a Category 2 fuel of <50 mg per valve. The figure also shows, on the left, a valve from a test on untreated fuel, with a weight of 456 mg, and on the right, a valve where the additive package is more effective and does not allow any deposit to build up. This valve has a deposit weight if 0.5 mg, which would satisfy the Worldwide Fuel Charter requirements for a Category 3 to 6 fuel of <30 mg per valve.

Although this test was developed in the late 1980s and early 1990s, it is still widely used and features in the Worldwide Fuel Charter to specified inlet valve cleanliness but recently due to an unacceptable interlaboratory variation in IVD results, with market fuels, the working group agreed to change the status for the test procedure from surveillance group (SG) to test development group (TDG) until the issues are resolved [10.87].

10.2.3.3. BMW 318i Test

The BMW 318i test procedure was developed at Southwest Research Institute [10.29] and later adopted as an ASTM standard test procedure [10.88]. The test is conducted on the public road, although it can be adapted for a distance accumulation dynamometer, using 1985 BMW 318i vehicles that are powered by a 1.8 L in-line four-cylinder engine with PFI. The driving cycle is designed to give a mix of driving that can be broken down by distance as 10% city driving (64 km), 20% suburban driving (128 km), and 70% highway driving (450 km). On a time basis, this equates to approximately 15% city, 46% suburban, and 39% highway. The vehicle is run for a total of 16,093 km with an intermediate assessment at 8,046 km. The assessment is by visual rating and weighing. A valve weight of <90 mg/valve would satisfy the Worldwide Fuel Charter requirements for a Category 2 fuel but must be <50 mg/valve to meet the requirements for a Category 3 to Category 6 fuels.

10.2.3.4. Mercedes-Benz M111 Test

In the mid-1990s a new European test was developed with the intention of replacing the M102E test procedure. The test procedure was initially intended to be used to assess both IVDs and CCDs [10.89] but never managed to achieve the repeatability or reproducibility necessary to be approved for the determination of CCDs. The new test procedure [10.90] was approved only for the determination of IVDs. The test uses a Mercedes-Benz M111 engine, a four-cylinder naturally aspirated engine that differs from the M102E in that it has four valves per cylinder and sequential electronic fuel injection.

The test cycle differed from that used for the M102E engine and is shown in Table 10.3.

TABLE 10.3 Mercedes-Benz M111 test cycle.

Stage	Time (seconds)	Speed (rev/min)	Torque (Nm)
1	30	800	0-2
2	60	1350	40
3	120	1800	40
4	60	3000	40

© SAE International.

Because the M111 engine has two inlet valves per cylinder, the results are usually reported on a per cylinder basis rather than a per valve basis; this makes the results more comparable to the results for other test procedures that use engines with only a single inlet valve per cylinder. As with the M102E test procedure, the latest version of the test method requires the use of pegged inlet valves to increase the severity of the test. The upper part of Figure 10.5 shows a pair of valves from one cylinder of a test with untreated fuel while the lower part of the figure shows two valves from a test with additive treated fuel.

FIGURE 10.5 Untreated (above) and treated (below) valves from a M111 test.

© SAE International.

10.2.3.5. **Ford 2.3 Test**

The Ford 2.3 L engine test [10.91] was developed [10.92–10.95] to provide an engine test bench method of assessing IVDs that would be comparable to the BMW 318i method that was performed using a vehicle. The method utilizes a 2.3 L Ford engine, an in-line four-cylinder engine with a cast iron cylinder head. The engine is naturally aspirated with two valves per cylinder. The engine is run to a two-stage cycle, the first stage being 2000 rev/min with no load and the second stage 2800 rev/min with a nominal 65% throttle opening; the test conditions are specified in terms of intake manifold pressure. The first stage is 4-minute duration, and the second stage is 8 minutes; there is a 30-second linear ramp between each stage. The total test running time is 100 hours. The inlet air conditions are tightly specified for pressure and humidity as well as the temperature, which is normally specified for the European test procedures. It has been concluded that the Ford 2.3 L test correlates well with the BMW 318i test and that the Ford engine test is slightly more severe than the BMW [10.96].

10.2.3.6. Other Engine Tests

There are many other engine tests, but most of them are in-house procedures developed by different oil or additive companies using engines known to suffer from intake system deposit problems or which have been modified to exaggerate the problems. These have been mainly developed to assess the effectiveness of additives. Universities and other research institutions have also used the variations of the above standard procedures, for example, substituting an engine of similar technology.

Another test method that used a far higher duty cycle was based on a 1.05 L in-line four-cylinder VW engine [10.97]. The higher speed and load operating points in this test procedure, which includes a time at full power, made it interesting because it gave a different perspective on performance to the other test, which covered only light to medium duty.

10.2.3.7. Bench Simulator Rigs

To simulate inlet valve deposit formation without the complication of engine testing, it has been proposed [10.98] that droplets of fuel are dropped onto an inclined aluminum bar and flow down the bar. The high end of the bar, where the fuel drops, is maintained at a temperature of 400°C, and by natural heat loss along the bar the lower end is 120°C. As the fuel flows down the bar, it begins to evaporate and decompose; however, this is not merely another way of determining the distillation characteristics as the deposits will tend not to adhere to the bar if the fuel contains an effective additive package. The performance of the fuel can be determined by how far down the bar the deposit formation begins. It was also shown that the chemical structure of the deposits formed on the simulator was similar in nature to those formed on inlet valves.

A more complex rig that is closer to an actual engine has been proposed and used. The rig is shown schematically in Figure 10.6 [10.99].

FIGURE 10.6   Schematic of inlet valve deposit simulator rig.

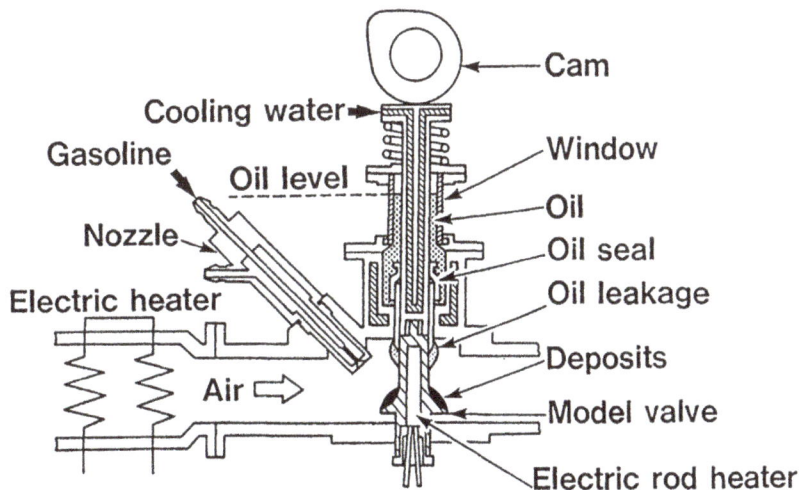

© SAE International.

The rig allows independent control of the model valve temperature, the oil leakage rate, and the fuel flow rate. The ability to independently control these parameters allows investigation of how different engine operating conditions influence the deposit build-up mechanism. The simulator has also been used to demonstrate how the inclusion of ethanol and MTBE tend to increase deposit formation [10.100].

Another example of a rig that sprays fuel onto the back of a simulated inlet valve is the Intake Valve Deposit Apparatus (IVDA) that was developed and commercialized by Southwest Research Institute [10.101]. The IVDA uses a simulated intake manifold, consisting of a fuel injector, air-intake runner, and intake valve specimen. The intake valve specimen replicates that used in the BMW engine discussed in Section 10.2.2.3 but is designed with a hollow valve stem that incorporates an electrical element capable of producing a uniform temperature of up to 316°C on the valve surface. A heat exchanger is used to pre-heat bottled air that is then passed over the valve specimen at a rate of 0.28 m^3/min. The gas stream is maintained at 205°C and has the provision for the addition of small quantities of NO$_X$ and/or SO$_2$ to simulate recirculated exhaust gases. At the end of the test the valve specimen is washed and weighed to determine deposit mass. Good correlation has been found between the IVDA and both the BMW and Ford test procedures [10.101], when the IVDA was operated at a valve specimen temperature of 218°C for 20 hours.

10.2.3.8. Valve Stick Test

This test [10.102] was developed using the Volkswagen 1.9 L water-cooled "boxer" engine (the term "boxer" being used to describe horizontally opposed piston engines). Being a boxer engine, the cylinder axis is horizontal, and as such the valve stems are closer to horizontal than vertical. On this particular engine the valve stem is angled slightly downward from the valve tulip; liquid on the valve stem will thus have a tendency to flow down the valve stem toward the valve guide. As the test cycle involves a short period of engine operation followed by an engine shutdown and cool down, fuel residues that form on the valve will thus have a tendency to accumulate around the valve guide.

The rate of cool-down is crucial as to the behavior of these residues: if the cool-down rate is too high then the residues will freeze on the valve head and stem and will not flow down toward the guide; if the cool-down rate is too low then the residues will not accumulate. As the deposits cool, they become more viscous; if there is too great an accumulation of these deposits around the valve guide, they will interfere with the operation of the valves and may prevent proper valve closing and thus reduce compression pressure and the engine's ability to start. The starting ability of the engine is assessed by replacing the spark plugs with pressure transducers and recording the cylinder pressure during engine cranking. Although it is possible to try and imply graduations of performance from the pressure traces, this is basically a pass/fail test.

10.2.4. Combustion Chamber Deposits

For engines where the fuel is metered upstream of the intake valve—carbureted, throttle body injection (TBI) and PFI engines—one of the most important effects of CCDs is that they promote ORI. However, they also influence exhaust emissions and lead to carbon knock, which will be discussed later, and are prone to flaking [10.103, 10.104]. In addition, they can cause surface ignition and, if they are on the spark plug, misfire. Each of these aspects is of significance—ORI, because increases in octane requirement can give rise to catastrophic failure of the engine, as can surface ignition; emissions, because of ever more stringent emissions regulation; and carbon knock, because, although it does not appear to cause any serious damage, it can give real concern to the driver because of the noise it makes.

The amount and nature of CCDs depend on the fuel, the lubricant, engine design, and driving mode [10.105, 10.106].

Regarding the fuel, with leaded gasoline, regardless of the lead concentration, the deposits contain up to 70% of lead by weight, being present mainly as the halide, oxyhalide, and sulfate; the remainder of the deposit being mostly carbonaceous. Of course, if there are metals present in the gasoline other than lead, such as manganese from the antiknock methylcyclopentadienyl manganese tricarbonyl (MMT), these will also appear in the deposits. If phosphorus is also in the fuel, it will modify the nature of the metallic deposits.

For unleaded gasolines, the deposit-forming tendency increases with the boiling point of the heaviest part of the fuel [10.107–10.110], but with aromatics giving the highest amount of deposits, olefins an intermediate level, and paraffins the least. The presence of certain types of multifunctional additive can increase combustion chamber deposits [10.111, 10.112] as can the presence of certain carrier additives [10.113], although it has also been found that synthetic carrier oils do not contribute to CCDs [10.114, 10.115].

Engine conditions and driving mode also play a large part in deposit formation. It has been found [10.116] that coolant temperature has the greatest effect and that air-to-fuel ratio is also important. Compression ratio and intake air temperature have only a very small influence on deposits.

The lubricating oil also contributes to CCDs [10.113], and the amount depends on oil consumption [10.69], particularly on the piston top [10.117], and on the volatility [10.113, 10.118] and sulfated ash content [10.119].

The effect on emissions is to increase HC from the exhaust by a mechanism involving the absorption and desorption of partially burned fuel on the deposits. The formation of NO_X is also increased because of the thermal insulation effect of the deposits [10.120].

10.2.4.1. Deposit Formation in Direct Injection Gasoline Engines

Direct injection gasoline engines, which are now commonplace and continue to grow in market-share because of fuel economy requirements, present a new set of challenges. The fuel injector is located within the combustion chamber, and deposit formation on the injector can be considered as part of the CCDs. At the beginning of this century, a large project was begun to develop a global test procedure to allow the evaluation of IVD and CCDs, including fuel injectors, with different fuels [10.121–10.123]. A test procedure was developed, but it was agreed that it needed further development and has not yet become an approved procedure, although it is still widely used. Some of the findings from this work are incorporated in the following sections.

10.2.4.1.1. Direct Injection Injector Deposits. In a direct injection gasoline engine, the fuel injector is located in a more severe environment. Not only are the average temperatures and pressures higher, but the injector is the subject of direct exposure to the flame front and combustion gases. As the injector is mounted in the engine block, it will also experience higher soak temperatures. These factors make the injectors in a direct injection engine more prone to deposit formation. There are two types of injector configuration, spray-guided injection with a centrally located injector, and wall guided where the injector id mounted toward the intake side of the chamber, as discussed in Chapter 8, Section 8.3.1.4.1. These strategies are sometimes used in conjunction.

As with PFIs, deposits will restrict fuel flow and alter the spray characteristics. While the lambda control system on stoichiometric PFI engines will compensate for constricted fuel flow, at all but full throttle conditions, a direct injection engine designed to run fuel lean at part throttle conditions may run excessively lean with a severely restricted injector flow. If a wide-band oxygen sensor is used for feedback control, then this problem may be reduced [10.124, 10.125].

It must also be remembered that the spray pattern is crucial to mixture preparation when late injection is used for charge stratification. Spray distortion due to deposits can thus affect the mixing process and stratification, which in turn will impact on driveability, fuel economy, and engine-out emissions. Figure 10.7 shows a clean in-service multihole injector (left) and a fouled (right) multihole injector [10.126].

FIGURE 10.7 Direct injection spark ignition fuel injectors, showing clean in-service (left) and fouled (right) injectors.

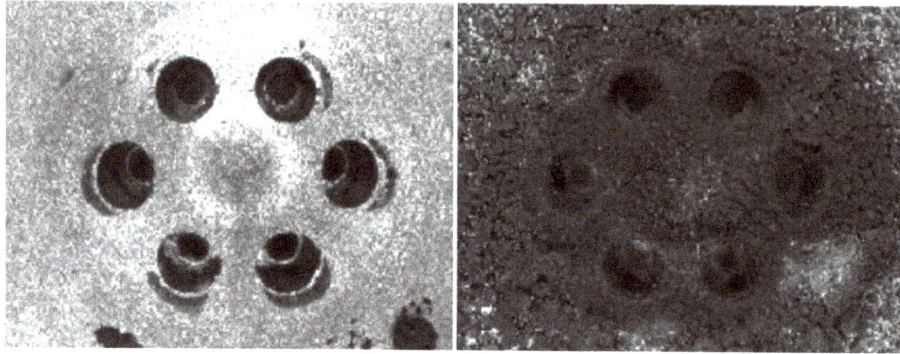

© SAE International.

Injector deposit build-up will also affect the "fuel trim," used to adjust the fuel injection map, which can eventually result in engine warning lights being triggered, customer complaints. In Europe, this led to the CEC setting up a test development group (TDG-F-113 – DISI test) to develop a test method [10.127]. The test is quite widely used but due to poor precision is not fully approved [10.87] and is not included in the Worldwide Fuel Charter.

Tests using an engine fitted with side-mounted, solenoid-operated multihole injectors, the standard engine calibration, including closed-loop lambda control, showed that after 55 hours of deposit accumulation the injector pulse width increased by an average of 23.5%, indicating significant flow restrictions [10.128]. A flow restriction of 25% would probably cause a fault code to be triggered. At a part load condition, this level of fouling was also associated with a fuel economy deterioration of about 2%, a doubling of engine-out CO emissions and a large increase in HC emissions. At full load, the fouling resulted in increased fuel consumption; additionally, a reduction in power and an increase in the number of pre-ignition events were observed. Other work has shown increases in increased driveability demerits and PM emissions [10.129].

The injector deposit problem appears to be worse in direct injection spark ignition (DISI) engines than in PFI engines [10.84, 10.126], but this is controllable. A recent study using a traditional PFI engine deposit control additive (DCA) was shown to be ineffective in dealing with DISI deposits, while DISI specific DCAs not only provided reduced deposit build-up (keep-clean), but also deposit clean up and significant particle emission reduction [10.39, 10.129]. The distillation characteristics of the fuel appear to play a major role. It has been suggested [10.130] that if the area around the injector orifice clearance space is allowed to become "dry," the deposits will begin to adhere to this region and deposit growth will ensue. If the T90 remains below the injector tip temperature, then the deposit formation is much reduced [10.130, 10.131], however, other work has found that fuels with a T90 above 180°C reduced the injector fouling tendency [10.132].

10.2.4.1.2. Direct Injection Chamber Deposits. Deposit formation within the combustion chamber of a direct injection engine can significantly change the mixing process that is critical to such engines, especially with late injection. With direct injection engines, most of the CCD in the piston cavity is derived from fuel [10.133]. This area is crucial to mixture preparation. It was considered that fuel properties have larger effects on CCD formation in such engines than in PFI engines. It has also been found that the composition of inorganic portions of the CCDs in a direct injection engine have the same chemical composition as those in PFI engines.

Fuel volatility is probably more important to the optimum performance of DISI engines than it is to PFI engines. Fuel volatility has a major influence on mixture formation [10.134] and thus on emissions [10.135].

However, the CCDs will also adversely affect emissions; CCDs will increase both HC and CO emissions and will reduce fuel economy [10.84]. Deposit formation within the combustion chamber is influenced by fuel characteristics in a similar manner to that in a port-fueled engine, but because more liquid fuel is initially delivered into the combustion chamber then the DISI engine is more prone to deposit formation [10.84] and that the fuel characteristics have a greater effect in a DISI engine [10.135].

10.2.4.2. Combustion Chamber Deposit Interference

The phenomenon of combustion chamber deposit interference (CCDI), which is also called deposit interference noise, carbon rap, and carbon knock, occurs when deposits build up on the top of the piston to such an extent that there is mechanical contact, resulting in a knocking or rapping noise. The problem is most acute on engines with small squish clearances. While the engine is warming up, the squish clearance is diminished by the rates of thermal contraction as the engine cools down. The problem came to prominence in the early 1990s in the US and coincided with a general increase in unwashed gum level due to a higher use of fuel additives to overcome IVDs. Apparently, the same engines in use prior to 1991 had not experienced these problems, and there had been no problems with similar engines in Japan where the unwashed gum specification was a maximum of 20 mg/100 mL [10.136].

Test work has shown that the factors that are important in CCD formation are also important for carbon knock—engine design, heavy aromatics in the fuel, certain types of additives, lubricant type, and consumption, and so on.

Although no damage is caused by carbon rap, it is of concern because the noise is upsetting to drivers, and also because tight squish clearances of 1 mm or less are becoming more common since they help to reduce HC emissions. When the engine is cold, the piston can tilt slightly toward the lag side, and this effectively reduces the squish clearance.

References

10.1. Hemberger, Y.V., Strunk, J., Krueger-Venus, J., Gross, J.-H. et al., Gasoline fuel composition. World Patent WO 2020/254518, 2020.

10.2. Studzinski, W.M. and Cummings, J.M., Reference fuel composition. U.S. Patent 8,764,854, 2014.

10.3. Lundberg, W.O. (Eds), *Autoxidation and Antioxidants*. Vol. 2 (New York: Interscience Publishers, 1961).

10.4. Giles, H.N., Bowden, J.N., and Stavinoha, L.L., "Overview on Assessment of Crude Oil and Refined Product Quality During Long Term Storage," Report DOE/FE0048, United States Department of Energy and U.S. Army Fuels and Lubricants Research Laboratory, Washington, DC, 1985.

10.5. Taniguchi, B.Y., Peyla, R.J., Parsons, G.M., Hoekman, S.K. et al., "Injector Deposits—The Tip of Intake System Deposit Problems," SAE Technical Paper 861534, 1986, doi:https://doi.org/10.4271/861534.

10.6. Nagpal, J.M., Joshi, G.C., and Aswal, D.S., "Gum Formation Tendencies of Olefinic Structures in Gasoline and Synergisyic Effects of Sulphur Compounds," in *5th International Conference on Stability and Handling of Liquid Fuels*, Rotterdam, the Netherlands, 1994.

10.7. Schwartz, F.G., Whisman, M.L., Allbright, C.S., and Ward, C.C., "Storage Stability of Gasoline Fundamentals of Gum Formation Including a Discussion of Radiotracer Techniques," U.S. Bureau of Mines Bulletin 626, 1964.

10.8. Dimitroff, E. and Johnston, A.A., "Mechanism of Induction System Deposit Formation," SAE Technical Paper 660784, 1966, doi:https://doi.org/10.4271/660784.

10.9. Gibbs, L.M. and Richardson, C.E., "Carburetor Deposits and Their Control," SAE Technical Paper 790202, 1979, doi:https://doi.org/10.4271/790202.

10.10. Por, N., "Stability Properties of Gasoline Alcohol Blends," in *3rd International Conference on Stability and Handling of Liquid Fuels*, London, UK, 1988.

10.11. D'Ornellas, C., "The Effect of Ethanol on Gasoline Oxidation Stability," SAE Technical Paper 2001-01-3582, 2001, doi:https://doi.org/10.4271/2001-01-3582.

10.12. Jęczmionek, Ł., Danek, B., Pałuchowska, M., and Krasodomski, W., "Changes in the Quality of E15–E25 Gasoline during Short-Term Storage up to Four Months," *Energy & Fuels* 31, no. 1 (2017): 504-513.

10.13. Pereira, R.C. and Pasa, V.M., "Effect of Mono-Olefins and Diolefins on the Stability of Automotive Gasoline," *Fuel* 85, no. 12-13 (2006): 1860-1865.

10.14. Pradelle, F., Leal Braga, S., Fonseca de Aguiar Martins, A.R., Turkovics, F. et al., "Modeling of Unwashed and Washed Gum Content in Brazilian Gasoline–Ethanol Blends during Prolonged Storage: Application of a Doehlert Matrix," *Energy & Fuels* 30, no. 8 (2016): 6381-6394.

10.15. Polsse, P., "What Additives Do for Gasoline," *Hydrocarbon Processing* 52, no. 2 (1973): 61-68.

10.16. Pitz, W.J., Westbrook, C.K., and Leppard, W.R., "Autoignition Chemistry of N-Butane in a Motored Engine: A Comparison of Experimental and Modeling Results," SAE Technical Paper 881605, 1988, doi:https://doi.org/10.4271/881605.

10.17. ASTM International, "Standard Test Method for Oxidation Stability of Gasoline (Induction Period Method)," ASTM D525-12a(2019), ASTM International, 2019.

10.18. Morris, D.L., Bowden, J.N., and Stavinoha, L.L., "Evaluation of Motor Gasoline Stability," in *3rd International Conference on Stability and Handling of Liquid Fuels*, London, UK, 1988.

10.19. ASTM International, "Standard Test Method for Oxidation Stability of Spark Ignition Fuel—Rapid Small Scale Oxidation Test (RSSOT)," ASTM D7525-14(2019)e1, ASTM International, 2019.

10.20. Alves-Fortunato, M., Baroni, A., Neocel, L., Chardin, M. et al., "Gasoline Oxidation Stability: Deposit Formation Tendencies Evaluated by PetroOxy and Autoclave Methods and GDI/PFI Engine Tests," *Energy & Fuels* 35, no. 22 (2021): 18430-18440.

10.21. ASTM International, "Standard Test Method for Gum Content in Fuels by Jet Evaporation," ASTM D381-22, ASTM International, 2022.

10.22. Kalghatgi, G.T., Sutkowski, A., Pace, S., Schwahn, H. et al., "ASTM Unwashed Gum and the Propensity of a Fuel to Form Combustion Chamber Deposits," SAE Technical Paper 2000-01-2026, 2000, doi:https://doi.org/10.4271/2000-01-2026.

10.23. ASTM International, "Standard Test Method for Oxidation Stability of Aviation Fuels (Potential Residue Method)," ASTM D873-22, ASTM International, 2022.

10.24. ASTM International, "Standard Specification for Aviation Gasolines," ASTM D910-21, ASTM International, 2022.

10.25. Schwartz, F.G., Allbright, C.S., and Ward, C.C., "Storage Stability of Gasoline: Oven Test for Prediction of Gasoline Storage Stability," Report of Investigations 7179, U.S. Bureau of Mines, Washington, DC, 1968.

10.26. Schwartz, F.C., Allbright, C.S., and Ward, C.C., "Prediction of Gasoline Storage Stability," SAE Technical Paper 690760, 1969, doi:https://doi.org/10.4271/690760.

10.27. Bowden, J.N., and Brinkman, D.W., "Stability Survey of Hydrocarbon Fuels," Report BETC/17784, U.S. Department of Energy, Washington, DC, 1979.

10.28. Amberg, G.H. and Craig, W.S., "Gasoline Detergents Control Intake System Deposits," SAE Technical Paper 620253, 1962, doi:https://doi.org/10.4271/620253.

10.29. Gething, J.A., "Performance-Robbing Aspects of Intake Valve and Port Deposits," SAE Technical Paper 872116, 1987, doi:https://doi.org/10.4271/872116.

10.30. Graham, J.P. and Evans, B., "Effects of Intake Valve Deposits on Driveability," SAE Technical Paper 922220, 1992, doi:https://doi.org/10.4271/922220.

10.31. Arters, D.C., Schiferl, E.A., and Szappanos, G., "Effects of Gasoline Driveability Index, Ethanol and Intake Valve Deposits on Engine Performance in a Dynamometer-Based Cold Start and Warmup Procedure," SAE Technical Paper 2002-01-1639, 2002, doi:https://doi.org/10.4271/2002-01-1639.

10.32. Mikkonen, S., Karlsson, R., and Kivi, J., "Intake Valve Sticking in Some Carburetor Engines," SAE Technical Paper 881643, 1988, doi:https://doi.org/10.4271/881643.

10.33. Megnin, M.K., Cobb, J.M., and Eng, K.D., "Development of a Gasoline Additive Screening Test for Intake Valve Stickiness and Deposit Levels," SAE Technical Paper 892121, 1989, doi:https://doi.org/10.4271/892121.

10.34. Nishizaki, T., Maeda, Y., Date, K., and Maeda, T., "The Effects of Fuel Composition and Fuel Additives on Intake System Detergency of Japanese Automobile Engine," SAE Technical Paper 790203, 1979, doi:https://doi.org/10.4271/790203.

10.35. Benson, J.D. and Yaccarino, P.A., "The Effects of Fuel Composition and Additives on Multiport Fuel Injector Deposits," SAE Technical Paper 861533, 1986, doi:https://doi.org/10.4271/861533.

10.36. Hilden, D.L., "The Relationship of Gasoline Diolefin Content to Deposits in Multiport Fuel Injectors," SAE Technical Paper 881642, 1988, doi:https://doi.org/10.4271/881642.

10.37. ASTM International, "Standard Test Method for Bromine Numbers of Petroleum Distillates and Commercial Aliphatic Olefins by Electrometric Titration," ASTM D1159-07(2017), ASTM International, 2018.

10.38. Universal Oil Products, "Diene Value by Maleic Anhydride Addition Reaction," UOP Method 326-82, Universal Oil Products Inc., 1965.

10.39. Reid, J., Mulqueen, S., Langley, G., Wilmot, E. et al., "The Investigation of the Structure and Origins of Gasoline Direct Injection (GDI) Deposits," SAE Technical Paper 2019-01-2356, 2019, doi:https://doi.org/10.4271/2019-01-2356.

10.40. Sweeney, W.J., Kunc, J.F., and Morris, W.E., "Aircraft Engine Induction System Deposits," SAE Technical Paper 460122, 1946, doi:https://doi.org/10.4271/460122.

10.41. Albright, R.E., Nelson, F.L., and Raymond, L., "Effect of Additives on Gasoline Engine Deposits," *Industrial & Engineering Chemistry* 41, no. 5 (1949): 897-902.

10.42. Nersasian, A., "Effect of 'Sour' Gasoline on Fuel Hose Rubber Materials," SAE Technical Paper 790659, 1979, doi:https://doi.org/10.4271/790659.

10.43. Orloff, G.A., Pellegrini da Silva, J.A., and Gentil, A.A., "Effects of Oxidated and Oxygenated Fuels on Rubber Hoses," SAE Technical Paper 921501, 1992, doi:https://doi.org/10.4271/921501.

10.44. MacLachlan, J.D., "Automotive Fuel Permeation Resistance—A Comparison of Elastomeric Materials," SAE Technical Paper 790657, 1979, doi:https://doi.org/10.4271/790657.

10.45. Stevens, R.D. and Fuller, R.E., "Fuel Permeation Rates of Elastomers after Changing Fuel," SAE Technical Paper 970307, 1997, doi:https://doi.org/10.4271/970307.

10.46. Tupa, R.C. and Dorer, C.J., "Gasoline and Diesel Fuel Additives for Performance/Distribution Quality – II," SAE Technical Paper 861179, 1986, doi:https://doi.org/10.4271/861179.

10.47. Tupa, R.C. and Koehler, D.E., "Gasoline Port Fuel Injectors - Keep Glean/Clean Up with Additives," SAE Technical Paper 861536, 1986, doi:https://doi.org/10.4271/861536.

10.48. Gerber, A.F. and Smith, R., "Some Effects of Exhaust Gas Recirculation upon Automotive Engine Intake System Deposits and Crankcase Lubricant Performance," SAE Technical Paper 710142, 1971, doi:https://doi.org/10.4271/710142.

10.49. Pearce, A.F. and Shannon, H.F., "PCV—Problems, Cures, Variables," SAE Technical Paper 640260, 1964, doi:https://doi.org/10.4271/640260.

10.50. Tracy, C.B. and Frank, W.W., "Fuels, Lubricants, and Positive Crankcase Ventilation Systems," SAE Technical Paper 640261, 1964, doi:https://doi.org/10.4271/640261.

10.51. Pless, L.G., "Some Effects of Experimental Vehicle Emission Control Systems on Engine Deposits and Wear," SAE Technical Paper 710583, 1971, doi:https://doi.org/10.4271/710583.

10.52. Tandrup, E.L., "Evaluating Carburetor Detergent Performance," SAE Technical Paper 660782, 1966, doi:https://doi.org/10.4271/660782.

10.53. Coordinating Research Council, "Carburetor Cleanliness Test Procedure State-of-the-Art Summary Report: 1973–1981," CRC Report No. 529, 1983.

10.54. CEC, "Evaluation of Gasolines with Respect to Maintenance of Carburettor Cleanliness," CEC F-03-T-81, 1981.

10.55. Abramo, G.P., Horowitz, A.M., and Trewella, J.C., "Port Fuel Injector Cleanliness Studies," SAE Technical Paper 861535, 1986, doi:https://doi.org/10.4271/861535.

10.56. Lenane, D.L. and Stocky, T.P., "Gasoline Additives Solve Injector Deposit Problems," SAE Technical Paper 861537, 1986, doi:https://doi.org/10.4271/861537.

10.57. Tupa, R.C., "Port Fuel Injectors-Causes/Consequences/Cures," SAE Technical Paper 872113, 1987, doi:https://doi.org/10.4271/872113.

10.58. Greiner, M., Romann, P., and Steinbrenner, U., "BOSCH Fuel Injectors—New Developments," SAE Technical Paper 870124, 1987, doi:https://doi.org/10.4271/870124.

10.59. Shiratori, A. and Saitoh, K., "Fuel Property Requirements for Multiport Fuel Injector Deposit Cleanliness," SAE Technical Paper 912380, 1991, doi:https://doi.org/10.4271/912380.

10.60. Coordinating Research Council, "A Program to Evaluate a Vehicle Test Method for Port Fuel Injector Deposit-Forming Tendencies of Unleaded Base Gasolines," CRC Report No. 565, Coordinating Research Council, Inc., 1989.

10.61. Tupa, R.C., Taniguchi, B.Y., and Benson, J.D., "A Vehicle Test Technique for Studying Port Fuel Injector Deposits—A Coordinating Research Council Program," SAE Technical Paper 890213, 1989, doi:https://doi.org/10.4271/890213.

10.62. ASTM International, "Standard Test Method for Evaluating Unleaded Automotive Spark-Ignition Engine Fuel for Electronic Port Fuel Injector Fouling," ASTM D5598-20, ASTM International, 2020.

10.63. Richardson, C.B., Gyorog, D.A., and Beard, L.K., "A Laboratory Test for Fuel Injector Deposit Studies," SAE Technical Paper 892116, 1989, doi:https://doi.org/10.4271/892116.

10.64. ASTM International, "Standard Test Method for Evaluating Automotive Spark-Ignition Engine Fuel for Electronic Port Fuel Injector Fouling by Bench Procedure," ASTM D6421-20, ASTM International, 2020.

10.65. Coordinating Research Council, "A Program to Evaluate a Bench Scale Test Method to Determine the Deposit Forming Tendencies of Port Fuel Injectors," CRC Report No. 592, Coordinating Research Council, Inc., 1995.

10.66. Coordinating Research Council, "Port Fuel Injector Bench Test Method, Interlaboratory Study, and Vehicle Test Correlation," CRC Report No. 617, Coordinating Research Council, Inc., 1999.

10.67. Coordinating Research Council, "Intake Manifold Deposit Engine Dynamometer Test Procedure State-of-the-Art Summary Report: 1973–1978," Report No. 505, Coordinating Research Council Inc., 1979.

10.68. Alquist, H.E., Holman, G.E., and Wimmer, D.B., "Some Observations of Factors Affecting ORI," SAE Technical Paper 750932, 1975, doi:https://doi.org/10.4271/750932.

10.69. Graiff, L.B., "Some New Aspects of Deposit Effects on Engine Octane Requirement Increase and Fuel Economy," SAE Technical Paper 790938, 1979, doi:https://doi.org/10.4271/790938.

10.70. Bitting, B., Gschwendtner, F., Kohlhepp, W., Kothe, M. et al., "Intake Valve Deposits—Fuel Detergency Requirements Revisited," SAE Technical Paper 872117, 1987, doi:https://doi.org/10.4271/872117.

10.71. Shibata, G., Nagaishi, H., and Oda, K., "Effect of Intake Valve Deposits and Gasoline Composition on S.I. Engine Performance," SAE Technical Paper 922263, 1992, doi:https://doi.org/10.4271/922263.

10.72. Houser, K.R. and Crosby, T.A., "The Impact of Intake Valve Deposits on Exhaust Emissions," SAE Technical Paper 922259, 1992, doi:https://doi.org/10.4271/922259.

10.73. Bunting, B.G., "An Analysis of Intake Valve Deposits from Gasolines Containing Polycyclic Aromatics," SAE Technical Paper 912378, 1991, doi:https://doi.org/10.4271/912378.

10.74. Martin, P., McCarty, F., and Bustamante, D., "Mechanism of Deposit Formation: Deposit Tendency of Cracked Components by Boiling Range," SAE Technical Paper 922217, 1992, doi:https://doi.org/10.4271/922217.

10.75. Martin, P. and Bustamante, D., "Deposit Forming Tendency of Gasoline Polar Compounds," SAE Technical Paper 932742, 1993, doi:https://doi.org/10.4271/932742.

10.76. Shilbolm, C.M. and Schoonveld, G.A., "Effect on Intake Valve Deposits of Ethanol and Additives Common to the Available Ethanol Supply," SAE Technical Paper 902109, 1990, doi:https://doi.org/10.4271/902109.

10.77. Belincanta, J., Mello, E., and Sá, R.A.B., "Evaluating the Induction System Deposit Tendencies: Bench Technique," *SAE Int. J. Fuels Lubr.* 3, no. 2 (2010): 60-66, doi:https://doi.org/10.4271/2010-01-1467.

10.78. Chapman, E. and Cummings, J., "Effects of Fuel Corrosion Inhibitors on Powertrain Intake Valve Deposits," SAE Technical Paper 2011-01-0908, 2011, doi:https://doi.org/10.4271/2011-01-0908.

10.79. Bidwell, J.B. and Williams, R.K., "The New Look in Lubricating Oils," SAE Technical Paper 550258, 1955, doi:https://doi.org/10.4271/550258.

10.80. Mitsui, J., Akiyama, K., Ueda, F., Okada, M. et al., "Effect of Gasoline Engine Oil Components on Intake Valve Deposit," SAE Technical Paper 932792, 1993, doi:https://doi.org/10.4271/932792.

10.81. Cheng, S.-W., "The Effects of Engine Oils on Intake Valve Deposits and Combustion Chamber Deposits," SAE Technical Paper 932810, 1993, doi:https://doi.org/10.4271/932810.

10.82. Mendiratta, R.L. and Singh, D., "Effect of Base Oil and Additives on Combustion Chamber and Intake Valve Deposits Formation in IC Engine," SAE Technical Paper 2004-28-0089, 2004, doi:https://doi.org/10.4271/2004-28-0089.

10.83. Rogers, D.T. and Jonach, F.L., "Mechanism of Intake Valve Underside Deposit Formation," SAE Technical Paper 580083, 1958, doi:https://doi.org/10.4271/580083.

10.84. Arters, D.C., Bardasz, E.A., Schiferl, E.A., and Fisher, D.W., "A Comparison of Gasoline Direct Injection Part I—Fuel System Deposits and Vehicle Performance," SAE Technical Paper 1999-01-1498, 1999, doi:https://doi.org/10.4271/1999-01-1498.

10.85. CEC, "The Evaluation of Gasoline Engine Intake System Deposition," CEC F-04-A-87, 1987.

10.86. CEC, "Inlet Valve Cleanliness in the MB M102E Engine," CEC F-05-A-93, 2010.

10.87. CEC, " CEC Activity Report July – December 2022," https://cectests.org/assets/CEC-Activity-Report_Jul---Dec-2022.pdf. accessed June 2023.

10.88. ASTM International, "Standard Test Method for Vehicle Evaluation of Unleaded Automotive Spark-Ignition Engine Fuel for Intake Valve Deposit Formation," ASTM D5500-20a, ASTM International, 2020.

10.89. Daniel, K., Aarnink, T.J., Gairing, M., and Schmidt, H., "Combustion Chamber Deposits and Their Evaluation by a European Performance Test," SAE Technical Paper 2000-01-2023, 2000, doi:https://doi.org/10.4271/2000-01-2023.

10.90. CEC, "Deposit Forming Tendency on Intake Valves," CEC F-20-A-98, 2012.

10.91. ASTM International, "Standard Test Method for Dynamometer Evaluation of Unleaded Spark-Ignition Engine Fuel for Intake Valve Deposit Formation," ASTM D6201-19a, ASTM International, 2019.

10.92. Bannon, S.A., Avery, N.L., Bitting, W.H., Carlson, C.A. et al., "Coordinating Research Council Development of a CRC Intake Valve Deposit Test," SAE Technical Paper 940348, 1994, doi:https://doi.org/10.4271/940348.

10.93. Bannon, S.A., Ahmadi, M., Buckingham, J.P., Corkwell, K.C. et al., "Coordinating Research Council Ford 2.3L Intake Valve Deposit Test—Interlaboratory Study," SAE Technical Paper 961099, 1996, doi:https://doi.org/10.4271/961099.

10.94. Corkwell, K.C. and Firmstone, G.P., "Repeatability of Intake Valve Deposit Measurements in the CRC 2.3L Ford Intake Valve Deposit Dynamometer Test," SAE Technical Paper 962011, 1996, doi:https://doi.org/10.4271/962011.

10.95. Coordinating Research Council, "CRC Intake Valve Deposit Study," CRC Report No. 606, Coordinating Research Council, Inc., 1998.

10.96. Crosby, T.A., Ahmadi, M.R., Schiferl, E.A., Arters, D.C. et al., "A Statistical Review of Available Data Correlating the BMW and Ford Intake Valve Deposit Tests," SAE Technical Paper 981365, 1998, doi:https://doi.org/10.4271/981365.

10.97. Gohn, M.R., "Evaluating Gasoline Additives for Intake-Valve Cleanliness Using the Volkswagen Polo Engine," SAE Technical Paper 892118, 1989, doi:https://doi.org/10.4271/892118.

10.98. Daneshagari, P., "A Test Procedure for Identifying the Gasoline's Deposit Formation Tendency on Intake Valves," SAE Technical Paper 892120, 1989, doi:https://doi.org/10.4271/892120.

10.99. Nomura, Y., Ohsawa, K., Ishiguro, T., and Nakada, M., "Mechanism of Intake-Valve Deposit Formation Part 2: Simulation Tests," SAE Technical Paper 900152, 1990, doi:https://doi.org/10.4271/900152.

10.100. Ohsawa, K., Nomura, Y., Moritani, H., Okada, M. et al., "Mechanism of Intake Valve Deposit Formation Part III: Effects of Gasoline Quality," SAE Technical Paper 922265, 1992, doi:https://doi.org/10.4271/922265.

10.101. Lacey, P.I., Kohl, K.B., Stavinoha, L.L., and Estefan, R.M., "A Laboratory-Scale Test to Predict Intake Valve Deposits," SAE Technical Paper 972838, 1997, doi:https://doi.org/10.4271/972838.

10.102. CEC, "Assessment of the Inlet Valve Sticking Tendency of Gasoline Fuels (VW Waterboxer Gasoline Engine)," CEC F-16-96, 2012.

10.103. Kalghatgi, G.T. and Price, R.J., "Combustion Chamber Deposit Flaking," SAE Technical Paper 2000-01-2858, 2000, doi:https://doi.org/10.4271/2000-01-2858.

10.104. Kalghatgi, G.T., "Combustion Chamber Deposit Flaking Studies Using a Road Test Procedure," SAE Technical Paper 2002-01-2833, 2002, doi:https://doi.org/10.4271/2002-01-2833.

10.105. Kalghatgi, G.T., "Deposits in Gasoline Engines—A Literature Review," SAE Technical Paper 902105, 1990, doi:https://doi.org/10.4271/902105.

10.106. Cheng, S.-W., "The Impacts of Engine Operating Conditions and Fuel Compositions on the Formation of Combustion Chamber Deposits," SAE Technical Paper 2000-01-2025, 2000, doi:https://doi.org/10.4271/2000-01-2025.

10.107. Shore, L.B. and Ockert, K.F., "Combustion-Chamber Deposits—A Radiotracer Study," SAE Technical Paper 580030, 1958, doi:https://doi.org/10.4271/580030.

10.108. Kim, C., Cheng, S.-W., and Majorski, S.A., "Engine Combustion Chamber Deposits: Fuel Effects and Mechanisms of Formation," SAE Technical Paper 912379, 1991, doi:https://doi.org/10.4271/912379.

10.109. Megnin, M.K. and Furman, J.B., "Gasoline Effects on Octane Requirement Increase and Combustion Chamber Deposits," SAE Technical Paper 922258, 1992, doi:https://doi.org/10.4271/922258.

10.110. Choate, P.J. and Edwards, J.C., "Relationship between Combustion Chamber Deposits, Fuel Composition, and Combustion Chamber Deposit Structure," SAE Technical Paper 932812, 1993, doi:https://doi.org/10.4271/932812.

10.111. Schreyer, P., Starke, K.W., Thomas, J., and Crema, S., "Effect of Multifunctional Fuel Additives on Octane Number Requirement of Internal Combustion Engines," SAE Technical Paper 932813, 1993, doi:https://doi.org/10.4271/932813.

10.112. Jung, H.S., Kim, S.H., Lee, S.M., Park, S.I. et al., "The Effect of Combustion Chamber Deposits on Octane Requirement Increase in a Spark Ignition Engine," SAE Technical Paper 2008-01-1761, 2008, doi:https://doi.org/10.4271/2008-01-1761.

10.113. Benson, J.D., "Some Factors Which Affect Octane Requirement Increase," SAE Technical Paper 750933, 1975, doi:https://doi.org/10.4271/750933.

10.114. Jackson, M.M. and Pocinki, S.B., "Effects of Fuel and Additives on Combustion Chamber Deposits," SAE Technical Paper 941890, 1994, doi:https://doi.org/10.4271/941890.

10.115. Kelemen, S.R., Siskin, M., Most, W.J., Kwiatek, P.J. et al., "Combustion Chamber Deposits from Base Fuel and Commercial IVD Detergent Packages," SAE Technical Paper 982716, 1998, doi:https://doi.org/10.4271/982716.

10.116. Cheng, S.-W. and Kim, C., "Effect of Engine Operating Parameters on Engine Combustion Chamber Deposits," SAE Technical Paper 902108, 1990, doi:https://doi.org/10.4271/902108.

10.117. Keller, B.D., Meguerian, G.H., Tracy, C.B., and Smith, J.B., "ORI of Today's Vehicles," SAE Technical Paper 760195, 1976, doi:https://doi.org/10.4271/760195.

10.118. McNab, J.C., Moody, L.E., and Hakala, N.V., "Effect of Lubricant Composition on Combustion-Chamber Deposits," SAE Technical Paper 540237, 1954, doi:https://doi.org/10.4271/540237.

10.119. Marciante, A. and Chiampo, P., "Influence of Lubricating Oil Ash on the ORI of Engines Running on Unleaded Fuel," SAE Technical Paper 720945, 1972, doi:https://doi.org/10.4271/720945.

10.120. Studzinski, W.M., Liiva, P.M., Choate, P.J., Acker, W.P. et al., "A Computational and Experimental Study of Combustion Chamber Deposit Effects on NOx Emissions," SAE Technical Paper 932815, 1993, doi:https://doi.org/10.4271/932815.

10.121. Noma, K., Noda, T., Ashida, T., Kamioka, R. et al., "A Study of Injector Deposits, Combustion Chamber Deposits (CCD) and Intake Valve Deposits (IVD) in Direct Injection Spark Ignition (DISI) Engines," SAE Technical Paper 2002-01-2659, 2002, doi:https://doi.org/10.4271/2002-01-2659.

10.122. Noma, K., Noda, T., Isomura, H., Ashida, T. et al., "A Study of Injector Deposits, Combustion Chamber Deposits (CCD) and Intake Valve Deposits (IVD) in Direct Injection Spark Ignition (DISI) Engines II," SAE Technical Paper 2003-01-3162, 2003, doi:https://doi.org/10.4271/2003-01-3162.

10.123. Coordinating Research Council, "Report on the Work of the CEC/CRC/OACIS Task Force on the Development of a Test Method to Evaluate Fuel Quality with Respect to Injector Fouling in Direct Injection Gasoline Engines," CRC Project CM-136-00, 2007.

10.124. Maloney, P.J., "A Production Wide-Range AFR Sensor Response Diagnostic Algorithm for Direct-Injection Gasoline Application," SAE Technical Paper 2001-01-0558, 2001, doi:https://doi.org/10.4271/2001-01-0558.

10.125. Hackel, V., Schnabel, C., and Tiefenbach, A., "Wide Band Oxygen Sensor Electronic Control Unit (LambdaTronic)," SAE Technical Paper 2005-01-0061, 2005, doi:https://doi.org/10.4271/2005-01-0061.

10.126. Kuo, C.-H., Smocha, R., Loeper, P., Mukkada, N. et al., "Aftermarket Fuel Additives and Their Effects on GDI Injector Performance and Particulate Emissions," SAE Technical Paper 2022-01-1074, 2022, doi:https://doi.org/10.4271/2022-01-1074.

10.127. CEC, "CEC Activity Report January – July 2017," https://cectests.org/assets/presentations_publications/CEC-Activity-Report_Jan-Jul2017_FINAL.pdf. accessed June 2023.

10.128. Joedicke, A., Krueger-Venus, J., Bohr, P., Cracknell, R. et al., "Understanding the Effect of DISI Injector Deposits on Vehicle Performance," SAE Technical Paper 2012-01-0391, 2012, doi:https://doi.org/10.4271/2012-01-0391.

10.129. Arters, D.C. and Macduff, M.J., "The Effect on Vehicle Performance of Injector Deposits in a Direct Injection Gasoline Engine," SAE Technical Paper 2000-01-2021, 2000, doi:https://doi.org/10.4271/2000-01-2021.

10.130. Kinoshita, M., Saito, A., Matsushita, S., Shibata, H. et al., "A Method for Suppressing Formation of Deposits on Fuel Injector for Direct Injection Gasoline Engine," SAE Technical Paper 1999-01-3656, 1999, doi:https://doi.org/10.4271/1999-01-3656.

10.131. Aradi, A.A., Imoehl, B., Avery, N.L., Wells, P.P. et al., "The Effect of Fuel Composition and Engine Operating Parameters on Injector Deposits in a High-Pressure Direct Injection Gasoline (DIG) Research Engine," SAE Technical Paper 1999-01-3690, 1999, doi:https://doi.org/10.4271/1999-01-3690.

10.132. Zhang, W., Ma, X., Xinhui, L., Shuai, S. et al., "Impact of Fuel Properties on GDI Injector Deposit Formation and Particulate Matter Emissions," SAE Technical Paper 2020-01-0388, 2020, doi:https://doi.org/10.4271/2020-01-0388.

10.133. Parsinejad, F. and Biggs, W., "Direct Injection Spark Ignition Engine Deposit Analysis: Combustion Chamber and Intake Valve Deposits," SAE Technical Paper 2011-01-2110, 2011, doi:https://doi.org/10.4271/2011-01-2110.

10.134. Lindgren, R., Skogsberg, M., Sandquist, H., and Denbratt, I., "The Influence of Injector Deposits on Mixture Formation in a DISC SI Engine," SAE Technical Paper 2003-01-1771, 2003, doi:https://doi.org/10.4271/2003-01-1771.

10.135. Takei, Y., Kinugasa, Y., Okada, M., Tanaka, T. et al., "Fuel Property Requirement for Advanced Technology Engines," SAE Technical Paper 2000-01-2019, 2000, doi:https://doi.org/10.4271/2000-01-2019.

10.136. Richardson, C.E., Fischer, J.L., and Pawczuk, G., "Evaluation of the Effect of Fuel Composition and Gasoline Additives on Combustion Chamber Deposits," SAE Technical Paper 962012, 1996, doi:https://doi.org/10.4271/962012.

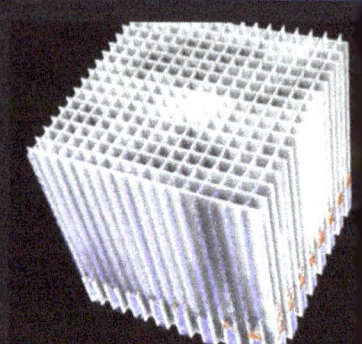

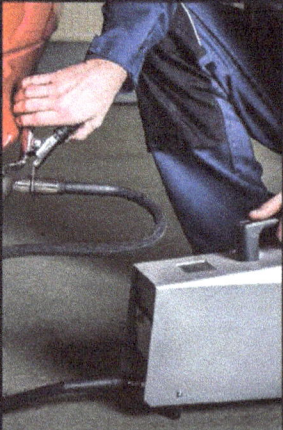

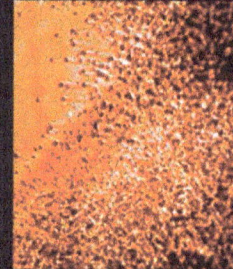

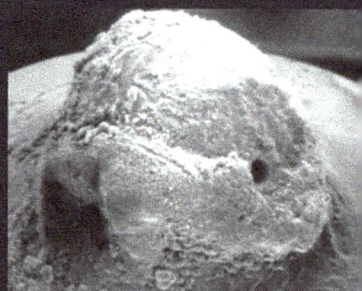

11

Gasoline Additives

To be competitive in today's market it is necessary to provide high quality at low prices. This is particularly true in the refining and marketing of liquid hydrocarbon fuels for internal combustion engines.

Additives play an important role in the over-all petroleum picture. They help the refiner maintain the flexibility required to meet the various demands created by the production of many types of products. In addition, many additives impart new and useful properties frequently unobtainable by processing. Additives also supplement refining processes as a means of producing high quality product at low cost to the consumer. [11.1]

These words were written just over 50 years ago, yet the sentiment is as true today as it was then. However, the role of these additives has changed with history because of changes in engine technology and legislation. Increasingly stringent exhaust emission regulations require the use of exhaust aftertreatment. This restricts the use of certain fuel additives and also limits the quantities of certain elements (which occur naturally in the feedstock for fuel production) being allowed into the finished fuel. Emission limits apply to the full working life of a vehicle, not just when it is first produced. It is therefore essential that vehicles maintain their design calibration for long periods. Deposit build-up or anything else that might erode this calibration must be minimized.

Fuels themselves may be less stable because the contribution of desulfurization and cracking processes in gasoline manufacture has expanded, bringing with it higher olefin contents. This started in many countries with the increase in crude price in the 1970s, which changed the demand pattern. This made it important to increase the yield of gasoline from crude. Although prices can still be rather volatile, it seems likely that the relatively high olefin contents associated with cracked components are here to stay. However, lighter olefins may be restricted in some gasolines to avoid the presence of highly reactive hydrocarbons in evaporative emissions. The price increases of crude oil also emphasized the importance of fuel economy, and this brought with it many changes, including the use of smaller engines operating at higher ratings.

Another change in fuel composition that has increased the need for additives is the use of oxygenated components. Initially this was to achieve the required gasoline volumes, to meet octane specifications, and to reduce exhaust emissions of CO and HC. More recently, this has been to meet renewable fuels directives and the mandated use of ethanol in jurisdictions such as Europe and the US.

Finally, there has been an increasing awareness of the value of additives in product differentiation. It can be difficult and expensive to modify the octane number, volatility, or other parameters controlled during the manufacturing process in order to achieve a performance quality advantage over competitive products. This is particularly true when exchange agreements to minimize transportation costs mean that many different fuel marketing companies may be using the same base gasoline. However, additives can always be introduced into specific brands or grades to provide very real marketing benefits. There are many examples of valuable increases in market share of gasolines because of the use of additives supported by advertising campaigns.

A variety of units are used for indicating additive treat rates. In Europe, the term ppm is often used, but as to whether this is by volume per volume, mass per mass, or even mass per volume is often omitted, leading to confusion. In the US, it is usual to specify pounds per thousand barrels (ptb). In this case, the treat rate on a mass per mass or volume per volume basis will vary with the density of the fuel being treated. To avoid confusion, it is advantageous to always use mg/kg; this is equal to ppm on a mass per mass basis, and 1 ptb is approximately equivalent to 4 mg/kg, depending on the density of the gasoline. For a 0.72 kg/L gasoline the conversion is 1 ptb equals 3.96 mg/kg, while for a 0.78 kg/L gasoline 1 ptb is equal to 3.66 mg/kg.

Gasoline additives can be classified in many different ways [11.2]; in the following sections, they will be classified and discussed according to their primary function within the engine or the fuel chain. Some of these additives may be added at the refinery to ensure that the fuel meets specifications; others may be added at the terminal as part of a marketing performance package. Additionally, there are aftermarket additives provided to the consumer, by motor manufacturers, additive producers, and through other retail outlets. The inclusion of ethanol in gasoline in many parts of the world has increased the use of additives to safeguard legacy fleets.

11.1. Additives to Improve Oxidation Stability

11.1.1. Antioxidants

Antioxidants (also called oxidation inhibitors) function by terminating the free radical chain reactions involved in hydrocarbon oxidation (see Chapter 6, Section 6.6.5). The products of oxidation are gums that can cause a number of problems during storage and in use in an engine, for example, catalytically cracked gasoline olefins [11.3], as discussed in Chapter 10. The additive requirement is even more stringent during long-term storage, such as military use, or enforced reduction in throughput such as during the pandemic in 2020/2021.

The type and amount of antioxidant to use will depend on such factors as the gasoline composition and the storage conditions. It is difficult to predict the optimum additive type and concentration, and to assess this, plant trials are best carried out using the components that will actually be in the gasoline blend. For cracked stocks, it is important that the antioxidant is injected into the stream at the earliest possible time, and preferably in the rundown line from the process unit. Tests such as induction period and existent gum are used to establish that gasoline has satisfactory storage stability. These tests are described in Chapter 10, Section 10.1.1.

There are two main types of antioxidants in use: the aromatic di-amines and the alkylphenols. Aminophenols are also used to a smaller extent but have the disadvantage of being soluble in water and reactive to caustic soda and so can be lost to water bottoms in tankage.

Aromatic di-amines such as the para- or ortho-phenylenediamines are extremely active oxidation inhibitors and are usually used in the range 5 to 20 mg/kg. They are particularly useful in gasolines having high olefin contents. The compounds most frequently used have the following general structure:

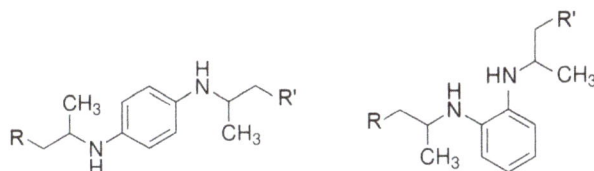

where R and R′ can be the same or different and are often sec-butyl, iso-propyl, 1,4-di-methyl-pentyl, or 1-methyl-pentyl. The compound N,N′-di-sec-butyl-p-phenylenediamine is usually considered essential and is thus widely used.

These aromatic di-amines are somewhat soluble in acidic tank water bottoms, and if these conditions exist, there could be a loss of additive on storage and hence diminished protection. Tank water bottoms are usually close to neutral or somewhat alkaline, but if upsets occur with such units as alkylation plants, then acidic water bottoms can result from un-neutralized process water separating out of the product.

Alkylphenols such as 2,6-di-tert-butylphenol are used mainly when the gasoline has a low olefin content. The most commonly used compounds have sterically hindered hydroxyl groups—for example, 2,6-di-tert-butyl-4-methylphenol or 2,6-di-tert-butyl-p-cresol (DBPC), also known as butylated hydroxytoluene (BHT), which has the structure shown below.

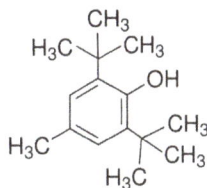

Quite often mixtures of various alkylphenols are used, and these can have the advantage of low freezing points and of being more cost-effective. They are used in the range of 5 to 100 mg/kg.

For gasoline compositions containing higher concentrations of olefins, mixtures of alkylphenols and phenylenediamines are frequently used and are claimed to be more effective than equivalent concentrations of either of the constituents alone. An approximate guide to the ratio of phenylenediamine type to alkylphenol type is given in Figure 11.1 [11.4].

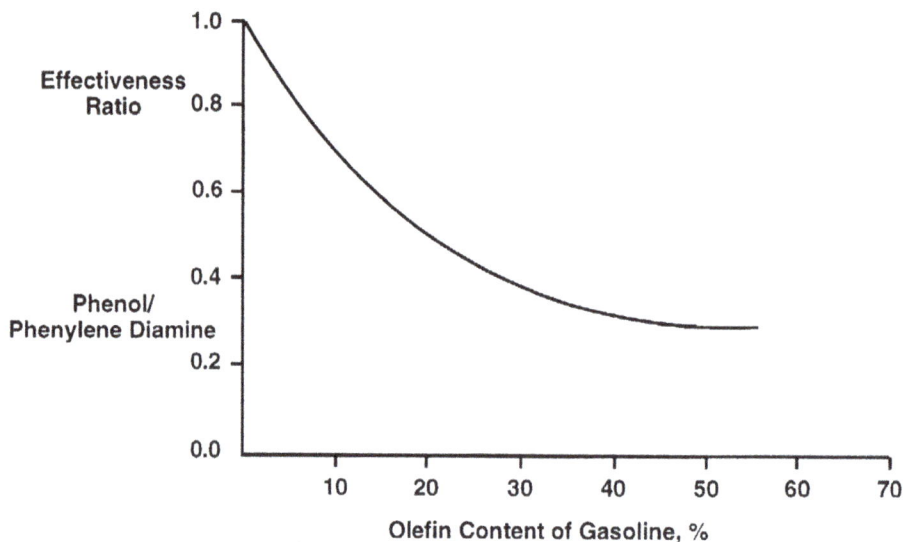

FIGURE 11.1 Effectiveness ratio of phenol/phenylenediamine antioxidants versus olefin content of gasoline.

Where tetra-ethyl-lead (TEL) is still used, antioxidants are also valuable in protecting the TEL from decomposition [11.5], and although this is rare, it can occur under certain circumstances. Some gasolines containing no cracked stocks may not have antioxidant added, but if they also contain lead, then it is always desirable to use a low level of antioxidant (i.e., up to about 10 mg/kg) to avoid lead decomposition products forming and blocking filters.

11.1.2. Metal Deactivators

Metal deactivators (MDAs) are used to prevent metals such as copper in a gasoline from catalyzing oxidation reactions. Although metals are present in crude oil, most of these remain in the heavier fractions rather than in the gasoline. However, metals may be picked up as dissolved metals by the action of acidic compounds, such as mercaptans in the gasoline attacking metals in the distribution and vehicle fuel systems.

Various compounds have been proposed as MDAs. These are usually chelating agents that form a complex with the metal ions that prevent the ions acting as a catalyst for the fuel oxidation processes. Early examples included N,N'-disalicylidene-1,2-propanediamine, salicylaldoxime, n-salicylidine o-aminophenol, 2,2',4'-trihydroxyazobenzene [11.6] and 2-(2-hydroxyphenyl)benzoxazole [11.7] as shown:

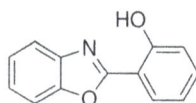

The most commonly used metal deactivator is probably N,N′-disalicylidene-1,2-propanediamine; it deactivates metals such as copper, iron, chrome, and nickel by forming a chelate with the structure shown in Figure 11.2 [11.4].

FIGURE 11.2 Copper chelate.

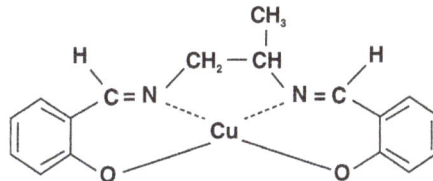

From this it can also be visualized how two molecules of 2-(2-hydroxyphenyl) benzoxazole would chelate the metal ion in a similar fashion but with a greater molecular weight.

Metal deactivators can be used at up to 12 mg/kg and must be added downstream of any caustic wash unit because they are soluble in aqueous sodium hydroxide.

11.2. Additives Used in Gasoline Distribution

11.2.1. Dyes and Markers

Dyes and markers are often needed for legal reasons; for example, a lower rate of tax is payable for off-road fuels that are dyed to prevent their illegal use for on-road vehicles. Dyes and markers can also be used to distinguish one product or brand from another. This is particularly important for different grades of aviation gasoline. Addition may be around 2–20 mg/kg.

Azo dyes are normally used where red or orange colors are needed and anthroquinone dyes when greens or blues are required. Concentrations are kept very low, less than 10 mg/kg to avoid deposit formation on evaporation and staining if they are spilled on the sides of light-colored vehicles. They are usually supplied as concentrated solutions.

Marker chemicals such as furfural or diphenylamine do not color the gasoline and can be used for security reasons, such as tracking stolen products. They are detected by adding appropriate reagents to the fuel, which produce colored anions.

11.2.2. Corrosion Inhibitors

Corrosion inhibitors are needed to protect pipelines and tanks from corrosion and hence to avoid rust being formed, which will suspend in the gasoline and cause filters to be blocked.

These additives are surfactant materials that have a polar group at one end and an oleophilic/hydrophobic group at the other. The polar group is attracted to a metal or other surface, and the other group repels water and provides an oily layer to prevent rust formation. This is illustrated in Figure 11.3 [11.1].

FIGURE 11.3 Schematic of how a corrosion inhibitor protects a metal surface.

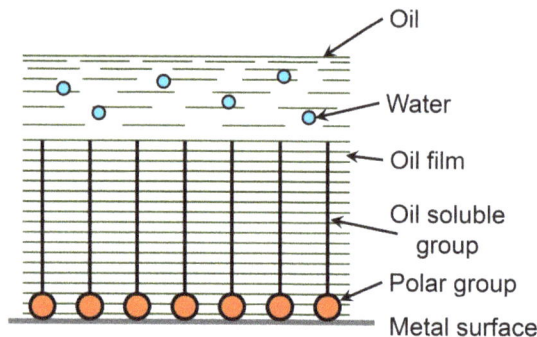

© SAE International.

They are used at low treat rates, often only 5 mg/kg, although treat levels as high as 20 mg/kg or more can be used in areas where severe corrosion is expected. In many cases, the amount injected is controlled so that only 1 mg/kg or so emerges at the end of the pipeline or distribution system. A wide range of chemical types are used as anticorrosion additives for distribution and storage, and these additives are also used to protect vehicle fuel systems (discussed further in Section 11.3.1). Typical examples are alkenyl succinic acids, esters, and amine salts [11.8], dimer acid and other carboxylic acids, and amine salts [11.9, 11.10], mixed alkyl ortho-phosphoric acids and amine salts [11.11], alkyl phosphonic acids and amine salts, aryl sulfonic acids and amine salts, Mannich amines [11.12], and carboxylic acid salts of Mannich amines [11.13]. Due to the adverse effects of phosphorous and sulfur on exhaust aftertreatment devices, additives containing these elements are now rarely used.

Test methods for gasolines containing corrosion inhibitor additives are discussed in Section 11.3.1.

11.2.3. Biocides

Tanks can accumulate water bottoms, as discussed in Chapter 6, Section 6.1. Biocides prevent microbial growth in the bottom of tanks that can cause suspended matter to accumulate in the water bottoms and in the gasoline. The microbial activity takes place mainly at the water/gasoline interface; this is discussed further in Chapter 6, Section 6.2.

The types of chemicals that have been used to overcome this problem include boron compounds, quaternary ammonium salts of salicylic acid, and glycol ethers. Typical treatment levels range from 135 to 1000 mg/kg in the fuel [11.4].

Microbial growth is less frequently encountered with leaded gasoline because the lead antiknock additives tend to inhibit microbial growth. This is likely to be a result of the halogenated compounds that are used as scavengers, providing a biocide effect.

11.2.4. Anti-Static Additives

Anti-static additives, or static dissipaters, are occasionally added to gasoline when it is being pumped at a high rate into a vessel or tank in which there is a danger of an explosive mixture being formed. Because gasolines are particularly volatile, they generally produce a vapor/air mixture above the liquid in a tank, which is too rich to explode.

The additives used increase the conductivity of the gasoline and prevent an electrostatic charge from building up and causing a spark. Nonmetallic additives were preferred for use in gasoline to avoid spark plug deposits. Chromium-containing material was previously used in jet fuel, although this has now ceased. Some of the other additives or components used in gasoline, particularly ethanol and other water-soluble oxygenates, may increase electrical conductivity and give some anti-static protection to the gasoline.

11.2.5. Drag Reducing Agents

Drag reducing agents (DRAs) can increase pipeline capacity (discussed in more detail in Chapter 6, Section 6.5.2). DRAs are high-molecular-weight polymers, for example, polyacrylamide suspended in the hydrocarbon solvent that shear readily and thus reduce drag [11.14–11.16]. Concentrations of up to 50 mg/kg have been used.

11.2.6. Demulsifiers and Dehazers

Demulsifiers and dehazers are occasionally added to gasoline if a water haze or emulsion has formed that will not readily clear by allowing it to stand. Stable hazes and emulsions can form when gasoline containing free water and a surfactant has been pumped or agitated in some way. Most gasoline surfactant additives are formulated to minimize this, but accidental overtreatment can occasionally cause problems. Hazes can form when there is a drop in fuel temperature and may be stabilized if a surfactant additive is present.

Demulsifiers and dehazing additives are often complex mixtures [11.4, 11.17] and function by promoting coalescence of the small droplets. They are themselves surfactants and so their use must be carefully controlled. Too much can make matters worse, and, more importantly, they can interfere with other surfactants that may be present and can reduce or eliminate their effectiveness. Treat rates ~10–500 mg/kg to break existing emulsions and 1–10 mg/kg as a preventative dose rate.

It is important to be sure that any surfactant additive used in a gasoline will not cause haze and emulsion problems and will not suspend rust and other particulates from the distribution system. Water sensitivity tests should be carried out using a range of water samples of different pH values. A number of test methods are used by the oil and additive industries, including the following:

1. The ASTM D1094 water tolerance test [11.18] is a simple test that involves shaking 80 mL of fuel with 20 mL of a phosphate buffer solution with a pH of 7, and then, after it stands for 5 minutes, rating the fuel, water, and interface for clarity and emulsions. It should be noted that if the gasoline contains water-soluble alcohols, then these can be pulled out of the gasoline and settle in the aqueous layer.

2. The ASTM D7451 water separation test [11.19] is similar to the D1094 water tolerance test that was developed for aviation fuels. The primary use of the water separation test is to evaluate new additive packages to ensure that fuel-water separability will not be compromised by the use of the additive package. As with the D1094 test, 80 mL of fuel is shaken with 20 mL of an aqueous solution. The D7451 procedure differs from the D1094 method in that it uses a specified mechanical shaker and allows the use of aqueous solutions of different pH value.

3. The 10 Cycle Multiple Contact test [11.20] gives a better simulation of field conditions than the ASTM D1094 test because the water bottoms are contacted with several batches of gasoline, as normally happens in practice. In this test, 100 mL of fuel is shaken with 10 mL of water for 5 minutes and then allowed to stand for 24 hours. It is then rated for clarity and emulsion formation. The gasoline is carefully poured off, leaving the water bottoms behind, and a fresh 100 mL of gasoline is poured in, shaken as before, and rated after 24 hours. This is repeated for a total of ten cycles. There is tendency for alcohols to be drawn out of the fuel and into the water bottom. The results give a good prediction of the performance of an additive in gasoline, provided that a range of different water bottoms are used varying in pH from about 4 to 9.

4. The Waring blender test uses a flame-proof laboratory blender to mix 475 mL of gasoline and 25 mL water for 4 minutes. The fuel and water phases are rated as before at set time intervals.

5. The particle suspension test evaluates the tendency for a fuel to suspend particulate matter such as rust or dirt. A small amount (0.1 g) of dry precipitated iron oxide is added to 100 mL of fuel, shaken, and then allowed to stand for a set period. Samples of gasoline are withdrawn periodically and filtered through a millipore filter to obtain a measure of the suspended matter. Sometimes a drop of water is added after shaking, followed by further shaking.

11.3. Additives Used to Protect Engines and Fuel Systems

11.3.1. Corrosion Inhibitors

Corrosion inhibitors are important not only to protect the fuel distribution system, as discussed previously, but also to prevent corrosion of the fuel system in a vehicle. The problem has long been recognized but became particularly acute in the late 1930s in new vehicles [11.21]. If corrosion occurs, particles of corrosion products can block fuel line filters, the small filters within fuel injectors or carburetor jets, on older vehicles. In severe cases, corrosion can lead to pin-holes in fuel tanks and lines leading to fuel leaks, an obvious safety hazard.

The same types of compounds used to combat corrosion in the distribution system (see Section 11.2.2) are also used to protect the fuel system of vehicles. However, there are issues associated with the use of acidic corrosion inhibitors in gasoline. Because of their acidic nature, they can interact with metals, metal oxides, and hydroxides to form soaps. These soaps are often insoluble in gasoline and can result in the premature blocking of filters. For example, refinery water bottoms are often alkaline, with pH values as high as 13. The metal involved is usually sodium as a result of caustic carryover from refinery operations. Iron oxides and hydroxides are also usually present. The underground tank fill pipes at service stations may be galvanized (zinc coated), and the zinc alloys have been found to react with alkylphosphoric acid type corrosion inhibitors. As noted in Section 11.2.2, the use of phosphorous and sulfur compounds has also been generally discontinued due to their adverse effects on exhaust aftertreatment systems.

Tests to evaluate the effectiveness of gasoline corrosion inhibitors fall into two categories: dynamic, to simulate use in vehicles or use in pipelines, and static, to simulate storage conditions.

The dynamic corrosion tests most frequently used are ASTM D665 [11.22], the colonial pipeline test [11.23], and the National Association of Corrosion Engineers (NACE) test [11.24]. These procedures are used for the assessment of the blended fuel; for the assessment of the anti-rusting performance of the additive, a variation of the ASTM D665 can be used with iso-octane as the hydrocarbon [11.25].

The ASTM D665 test was originally developed to evaluate the anti-rust properties of steam turbine oils and has been modified for use with gasoline. A cylindrical steel spindle is polished to a clearly defined procedure and is then immersed in a mixture of 300 mL of the treated gasoline and 30 mL water, which is stirred for a fixed period of time, often 24 hours, at a temperature in the range of ambient to 38°C. The lower temperatures are used when the flash point of the gasoline is a concern. At the end of this period, the spindle is visually rated for the degree of rust that has formed. The water used in this test can be deionized to simulate rainwater, it can be synthetic seawater, or it can be at different pH levels to simulate the different types of process water that can find their way into gasoline. The test may be shortened when there is concern over volatilization of the gasoline.

Another static corrosion test that can be used [11.4] consists of immersing a strip of carbon steel in a blend of 90 mL gasoline and 10 mL water in a stoppered glass bottle. After shaking, it is left to stand for a fixed period, typically about three weeks, with estimates of the degree of rusting being made at set time intervals, for example, after 24 hours, 48 hours, one week, two weeks, and three weeks.

11.3.2. Anti-Icing Additives

The mechanism by which ice can form in a carburetor or throttle body has been described in Chapter 9, Section 9.7, together with the effects of ice deposition and the test methods used to assess them. Two types of additives can be used to control icing: cryoscopic and surfactant.

Cryoscopic anti-icing additives function by depressing the freezing point of water and are usually alcohols or glycols. Materials that have been widely used in nonoxygenated base fuels are iso-propyl-alcohol, hexylene-glycol, and di-propylene-glycol (DPG), as well as a number of complex mixtures of various glycols and other water-soluble oxygenated compounds. The treat rates used will depend on the additive and the likelihood and severity of problems in the vehicle population under consideration. Different compounds will require different treat levels to give the same protection. For example, 2% of isopropyl alcohol will give about the same effect as 500 mg/kg of DPG. Where significant quantities of alcohols are blended into the fuel to meet octane requirements or renewable fuels limits, anti-icing requirements are usually satisfied by the base gasoline.

Surfactant anti-icing additives function in much the same way as corrosion inhibitors, and good corrosion inhibitors are often also good anti-icing additives. They form a monomolecular layer on metal surfaces and thus prevent ice crystals from attaching themselves to the throttle plate and other surfaces so that they are swept into the combustion chamber. They also prevent ice crystals from sticking to each other and thus inhibit the build-up of large crystals. Many types of surfactants are used, including amine or imidazoline salts of carboxylic or alkenyl succinic acids [11.4]. As with all surfactant additives, care must be taken to be sure that they do not stabilize hazes and emulsions, as discussed in Section 11.2.6.

11.3.3. Deposit Control Additives (DCAs)

11.3.3.1. Carburetor Cleanliness Additives

Carburetors have largely been replaced by fuel injection equipment, but it is still considered important for fuels to satisfy the legacy fleet. Carburetor cleanliness additives are surfactant products that function by the polar group, at one end of the molecule, attaching itself to a particle or a surface and the large nonpolar, oleophilic group at the other end dissolving in the fuel. The monomolecular film that is formed around any particle effectively solubilizes it by forming a micelle that prevents aggregation of the particles and allows the particulate matter to be carried into the combustion chamber with the fuel. Metal surfaces are protected against deposition in a similar way to that in which they are protected by surfactant corrosion inhibitors: the oleophilic group repels the deposit particle or the micelle in the same way it would repel a water droplet. This is illustrated in Figure 11.4 [11.26] for a molecule having a low-molecular-weight oleophilic group, which give very good surface protection but poorer dispersion.

FIGURE 11.4 Low-molecular-weight deposit control additive.

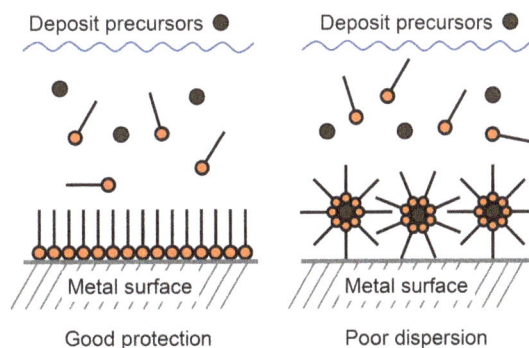

Figure 11.5 [11.26] shows the proposed mechanism for an additive with a high-molecular-weight oleophilic group; by contrast, this gives poorer surface protection but acts as a better dispersant.

FIGURE 11.5   High-molecular-weight deposit control additive.

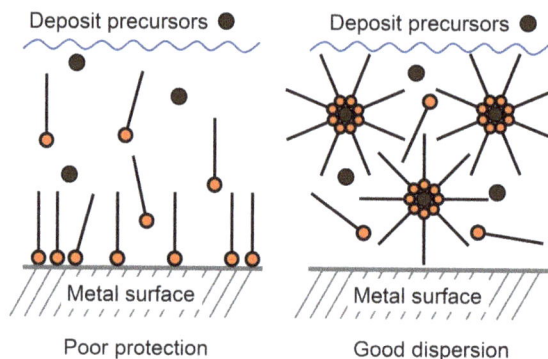

© SAE International.

These two figures are obviously a simplification, only to illustrate the mechanism; the actual structure of the DCA molecules is rarely as simple. Figure 11.6 shows (on the left) an amine carboxylate salt [11.27] that is a relatively low-molecular-weight DCA and (on the right) a modified succinimide molecule [11.28] having a far higher molecular weight.

FIGURE 11.6   Amine carboxylate salt (left) and modified succinimide (right) DCAs.

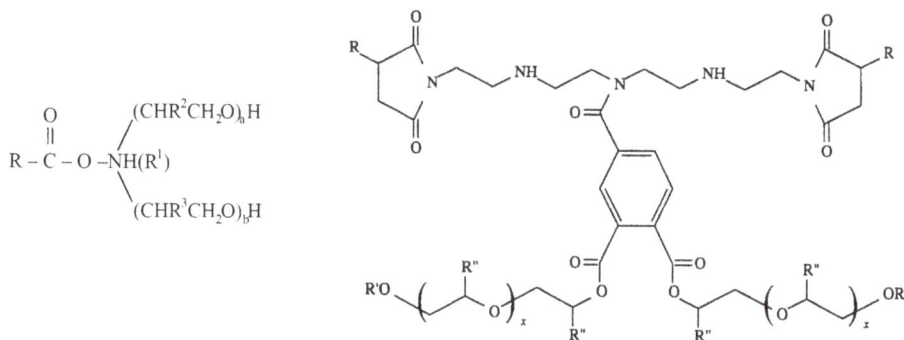

© SAE International.

For the amine carboxylate salt shown, R is a C_{11-21} saturated or unsaturated aliphatic hydrocarbyl, R^1 is C_{8-24} saturated or unsaturated aliphatic hydrocarbyl, each of R^2 and R^3 is selected from H and CH_3, each of a and b is 1-14, and the sum of a and b is 2-15. For the modified succinimide shown, R can be an alkyl or alkenyl group from C_6 to C_{100}, with polyisobutene being the preferred substituent; R' can be hydrogen, alkyl, aryl, alkaryl, or aryl alkyl from C_1 to C_{100}, with C_1 to C_{12} alkyl or alkyl substituted phenyls being the preferred substituents; R'' can be hydrogen, alkyl, aryl, alkaryl, or aryl alkyl from C_1 to C_{100}, with C_1 to C_{12} alkyl or alkyl substituted phenyls being the preferred substituents, with x being 5 to 100.

Some early carburetor cleanliness additives have included amino-amide salts of monocarboxylic acids [11.29], oxazoline borate [11.30], phosphated amino amides [11.31], copolymers of N-vinyl pyrrolidones and alkyl esters of acrylic and methacrylic acids [11.32], a combination of a fatty propylene diamine and an alkenyl-substituted propylene urea [11.33], alkyl amine phosphates [11.34], and fatty acid amides [11.35]. Polymeric dispersants such as alkenyl succinimides [11.36, 11.37], polybuteneamines [11.38–11.40], and polyetheramines [11.41, 11.42] are also widely used now. Treat rates of the earlier additives were usually in the range of 20 to 100 mg/kg; some of the more recent developments are not used solely to keep carburetors clean but also for other parts of the inlet system. For other applications they are usually used at higher treat rates and usually in combination with at least a fluidizer [11.38], but more typically as part of a completed additive package, as discussed in Section 11.6.

Surfactant additives are normally multifunctional so that although an additive might be primarily designed to be a carburetor cleanliness additive, it will often have other benefits as well, such as anti-icing and/or anticorrosion effectiveness.

The effectiveness of carburetor cleanliness additives is assessed by means of engine testing or road tests. A large number of engine tests have been developed, many of them in-house procedures that may be adaptions of other standard procedures used by additive companies to demonstrate the effectiveness of their products. An example of this is using a standard lubricating oil evaluation test [11.43] to assess the performance of a multifunctional additive package on carburetor cleanliness. The lubricant test used a 6.03 L V8 engine that ran 48 iterations of a 4-hour cycle. Figure 11.7 [11.6] shows the results of such a test, using an additive treat rate of 135 mg/kg. The untreated carburetor is on the left and the treated carburetor is on the right.

FIGURE 11.7 Untreated (left) and treated (right) carburetors from a Sequence V lubricating oil test.

© SAE International.

However, a number of procedures that have emerged over the years became widely accepted. These standard tests included the CRC Carburetor Detergency test [11.44] in the US and the CEC Renault 5 Carburetor Cleanliness test [11.45] in Europe, although, due to the scarcity of carbureted vehicles in the European fleet, this procedure is no longer supported.

11.3.3.2. Port Fuel Injector Anti-Fouling Additives

There are two main categories of fuel injector systems: multipoint injection (MPI) and throttle body injection (TBI). The use of fuel injection is now almost universal because it gives better control of air-to-fuel ratio than do carburetors; this is valuable in helping to meet emissions and fuel economy requirements. Electronic fuel injection also lends itself to computer implementation of closed loop control of air-to-fuel ratio. However, deposits can form in the pintle area of these injectors. These deposits will reduce the flow rate and modify the fuel spray patterns. Even low deposit levels can have a significant influence on vehicle driveability. This has already been discussed more fully in Chapter 10, Section 10.2.2.

These deposits became a serious problem in the US in 1985–1986, and a number of fuel additive solutions were developed that should overcame the problem. Fuel injector deposits are formed at much higher temperatures than carburetor deposits and are thus more difficult to control. Polymeric dispersants and amine-type additives (as mentioned in the previous section) have been used successfully to control these deposits [11.46–11.52]. Additive chemistries, treat rates, and multifunctional packages are being developed to address the problem of deposit formation in the inlet manifold, intake ports, and on inlet valves. These are discussed in the following section. Such additives are usually more than sufficient to control fuel injector deposits.

The deposits are formed by heat from the engine that evaporates any droplets of gasoline remaining in the pintle area of the injector when the engine is off. Therefore, stop/start type driving most likely causes these deposits to form. Test procedures to evaluate the effectiveness of gasoline additive formulations in overcoming this problem (or the susceptibility of vehicles to it) all involve test cycles in which the engine or vehicle is driven at a moderate speed for a short period of time and is then allowed to stand and soak with the engine off, usually for long enough for the engine to cool. This cycle is repeated until driveability malfunctions begin to occur. This is discussed in more detail in Chapter 10, Section 10.2.2.

11.3.3.3. Additives to Control Inlet Manifold, Inlet Valve, and Port Deposits

Deposits in this region can cause many difficulties, including driveability problems, increased exhaust emissions, and poor fuel consumption, as discussed in Chapters 8 and 9.

For carbureted engines and those using TBI systems, deposits tend to accumulate in the hot spot region because the higher temperature there favors the gum-forming reactions. Surfactant additives such as carburetor cleanliness additives or injector anti-fouling additives will usually provide adequate deposit control. However, to ensure adequate distribution and fluidity of the additive to control deposits on the intake valves and inlet ports will often require the addition of a high boiling point, thermally stable, oily material known as a fluidizer, carrier oil, or solvent oil. These can be a petroleum-based, lube oil base-stock, a synthetic oil, or a polymeric material such as a polybutene or a polyetheramine [11.53]. These carrier fluids may have surfactant properties of their own, or they may act as a thermally stable diluent for the primary DCA, which may have high viscosity. As the more volatile fuel components evaporate, the remaining droplets of DCA would therefore become less mobile and less effective at providing the protective and dispersant characteristics described in Section 11.3.3.1. These additives are used at relatively high concentrations, sometimes as much as 2000 mg/kg [11.39]. Problems of valve sticking have been associated with some DCAs [11.54, 11.55] in vehicles that are particularly prone to this type of problem. It is thus necessary to ensure that a fully formulated additive package is not going to result in such problems. This is discussed further in Section 11.6.

When ethanol is present in the gasoline, deposits within the intake system can be more difficult to control and may need up to 50% greater treat level to achieve adequate cleanliness [11.56]. Although reduced port deposits are claimed to reduce an engine's octane requirement increase (ORI) (see Chapter 7, Section 7.7), increased combustion chamber deposits can result from some DCAs [11.57], and this may contribute to increasing the ORI [11.58]. Because these additives are not very volatile, there is a strong possibility that some of them may find their way into the lubricating oil; care must also be taken to ensure that this does not cause the lubricant to thicken or form sludge, although it has been shown that fuel additives may be beneficial [11.59].

Test methods for assessing the propensity of fuels or fuel additive combinations to produce deposits in the induction system are discussed in Chapter 10, Section 10.2.3.

11.3.3.4. Additives for Direct Injection Injector Cleanliness

As discussed in Chapter 10, Section 10.4.2.1, the injectors in a direct injection engine are located in a more severe operating condition; this puts greater demand upon the DCA. Existing DCAs have been tested in direct injection engines, and a poly-iso-butylene-amine (PIBA) additive has demonstrated the injectors are kept clean, but it did not clean up fouled injectors at the treat rate that was used [11.60]. It has also been demonstrated that a PIBA additive performs better than a polyether amine. A Mannich DCA performed better than either of these other chemistries [11.61, 11.62]. Mannich DCAs can also be tailored to meet the more exacting requirements of direct injection engines [11.63].

As we have seen with the change from carbureted engines to engines with PFI systems, as engine technology changes, the demands placed on the DCA also change. With this change, old DCA technology had to be improved and new chemistries developed. So, it is with direct injection engine technologies: the existing chemistries have to be improved [11.63] and new products must be developed to specifically address the new demands of direct injection gasoline engines. These new products could be a combination of products, such as a succinimide and a Mannich [11.64], or variations on existing chemistries such as hydrocarbyl amines [11.65] and quaternary ammonium salts [11.66] that are specifically developed to meet these new demands. The increase in direct injection gasoline vehicles has shown new challenges for additive use [11.67] with work showing the reduction in exhaust particles, using a new generation of additives is indicative of this [11.68]. Work has also shown the importance of optimizing the concentration of additive to both control particulate emissions and deposits, although it should be emphasized that the study was limited to older technology additives [11.69].

11.3.3.5. Combustion Chamber Deposit Control

The occurrence of and deleterious effect of combustion chamber deposits (CCDs) has long been recognized [11.70]. Issues such as increased ORI and preignition that are primarily caused by CCDs have in the past been specifically targeted by additive solutions (as discussed in Sections 11.4.2 and 11.4.3).

The CCDs are thought to arise as a result of both fuel and lubricant. The fuel can either impact the combustion chamber walls directly as fuel droplets due to incomplete vaporization, or the higher boiling components that can include fuel additives condense on the combustion chamber walls. These condensates will, over time, lose volatile material and undergo chemical reactions causing the viscosity to increase. As these deposits grow in thickness, they produce an insulating effect that allows the deposit exposed to the combustion process to undergo thermal degradation, leading to a highly carbonaceous deposit [11.71].

Some carburetor cleanliness additives also claimed to control CCDs [11.72], but as engine technology moved toward fuel injection new additive chemistries were required to keep the fuel injectors clean. These new additives often required blending with fluidizers as discussed in Section 11.3.3.3. Some of these additives, and the fluidizers in particular, have been found to contribute to CCDs [11.73–11.75]. Fuel additives have therefore been designed with control of CCDs as a specific target. Additive chemistries include polyether hydroxyethylaminoethyl oxalamide [11.76], amine-esters [11.77], ester amines [11.78], polyoxyalkyene triamide alkoxylates [11.79], succinic acid derivatives [11.80], and polybutene amines or succinimides, in combination with a mineral carrier oil but specifically at high treat rates [11.81].

11.3.4. Factory Fill Additives

Factory fill additives are used by some vehicle manufacturers if it is likely that engines in their vehicles will not be run for a long period. Examples of this are if the vehicles are being shipped for long distances or if there is a possibility they may be stockpiled before being sold [11.82]. They consist of a blend of a number of additives such as antioxidants, metal deactivators, corrosion inhibitors, and biocides. However, some factory fill additive packages may be antagonistic toward some deposit control additive packages intended for continuous use, but this is rarely problematic because only a limited amount of factory fill fuel is loaded into the fuel tank.

11.4. Additives That Influence Combustion

11.4.1. Antiknock Additives

Antiknock additives are discussed in Chapter 7, Section 7.4. These additives can also help to overcome run-on and, in the case of TEL, valve seat recession.

11.4.2. Anti-ORI Additives

As discussed in Chapter 7, Section 7.7, the octane requirement of an engine can increase with use. This is known as ORI and is mainly due to deposit formation. Anti-ORI additives aim to reduce the ORI that occurs due to deposits in the combustion chamber and, to some extent, in the ports. The reduction of port deposits by the use of surfactant additives/fluidizers has been discussed in Section 11.3.3.3. These materials are claimed to reduce ORI by reducing these port deposits [11.83], although the presence of high levels of some fluidizers [11.84] and high-molecular-weight DCAs [11.58] are known to increase CCDs and increase ORI.

Aftermarket fuel additives such as polyetheramines can be used to reduce ORI. These aftermarket additives are usually added to the fuel intermittently and at high concentrations [11.85]. They can rapidly reduce octane requirement by one or two numbers but it will immediately start to increase again. The benefit is only obtained for a few thousand kilometers before a further additive treat is needed. This is shown in Figure 11.8 [11.85].

FIGURE 11.8   Intermittent additive treats can reduce ORI.

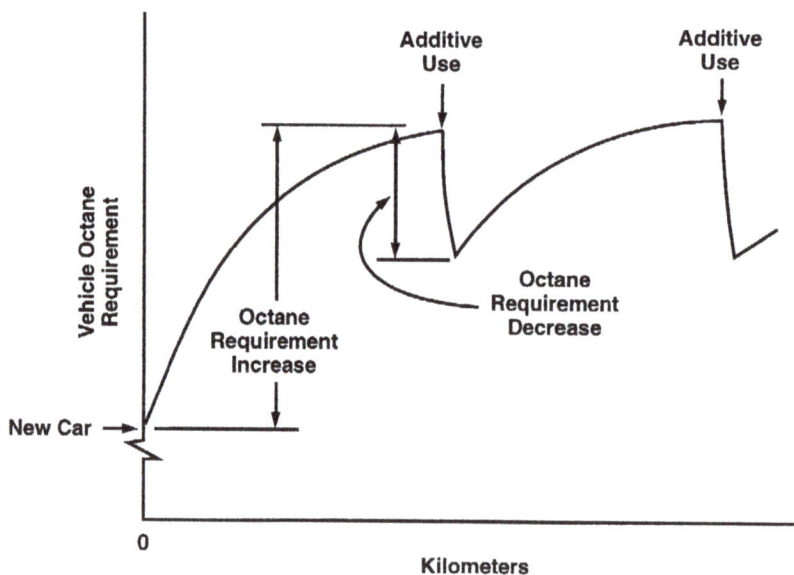

Other fuel additives have been proposed that will control ORI on a continuous use basis [11.86–11.91]. However, effective control in CCDs, as discussed in Section 11.3.3.5, should eliminate the problem.

11.4.3. Anti-Pre-Ignition and Anti-Misfire Additives

Anti-pre-ignition and anti-misfire additives have been used in leaded gasolines, particularly when lead levels were high. They consist of phosphorus compounds such as tricresyl phosphate, which converts the lead halides; oxyhalides; and oxides formed in the combustion chamber into lead phosphate. This has two effects: first, lead phosphate needs to be at a much higher temperature than the other lead compounds before it starts to catalyze the combustion of carbonaceous deposits and cause them to glow. Glowing deposits can be a source of preignition, so phosphorus compounds reduce the risk of preignition when using leaded gasolines. Second, lead phosphate has a much lower electrical conductivity at high temperatures than the halides, and so phosphorus additives are valuable in reducing misfire due to lead deposits on the plug conducting the charge away.

Phosphorus additives were used for both of these purposes at 0.2 to 0.5 of the theoretical amount required to combine with all the lead in the fuel. As the use of lead has declined with lower concentrations in the gasoline, the use of these additives has diminished. Phosphorus also has an adverse effect on the efficiency of catalytic converters for improving exhaust gas quality and is banned in countries where they are used.

11.4.4. Spark-Aider Additives

Spark-aider additives [11.92] are designed to improve the spark so that when the air-to-fuel ratio is in the borderline condition, such that it is almost too lean to burn, these additives will help prevent misfire and improve exhaust emissions. This is most likely to happen in cold weather with vehicles that are designed to operate near the lean limit, because during the warm-up period it is sometimes not possible to get enough fuel vaporization to achieve a readily combustible mixture.

The additive is based on an organic gasoline-soluble compound containing an alkali metal or alkaline metal [11.93]. Potassium was the favored metal [11.52] with a treat rate to give a level of potassium in the fuel of only a few milligrams/kilograms. This is claimed to give a layer of deposit on the spark plug electrodes, which has a low electron work function; this in turn increases the efficiency with which the spark energy is transferred to the gas by reducing the cathode fall in the glow discharge phase and also by promoting the transition from the glow phase to the arc phase during some part of the discharge [11.93–11.95]. The initial flame is therefore stronger, and there is a small increase in flame speed at the start of combustion.

The use of such additives would not be allowed in unleaded gasolines because the potassium is a known poison of catalytic converters. When used in leaded gasolines, there is a possibility that there could be an antagonistic reaction between the alkali metal or alkaline metal component and the halogen compounds used as lead scavengers. This could result in the formation of a viscous compound that in certain car models could cause valves to stick, with consequent engine damage.

11.4.5. Additives for Improving Fuel Distribution between Cylinders

In older engines, such as carbureted and TBI engines, the fuel is added to the incoming air before it is split and fed to individual cylinders. There are a few exceptions, such as racing engines using multiple carburetors. As the fuel is usually insufficiently vaporized and mixed with the air before this split, there is a very strong possibility that there will be a maldistribution of the fuel between the cylinders of a multicylinder engine. This will result in different air-to-fuel ratios in the different cylinders and hence in different amounts of power and emissions being produced by each of the cylinders. This can give rise to driveability problems and cause a deterioration in exhaust gas quality and fuel economy [11.96]. It is most likely to occur during cold weather with vehicles designed to operate on the lean side of stoichiometric.

Additives to reduce this problem appeared in the early 1970s, at a time when many US vehicles suffered from this fuel maldistribution problem. Some were purely hydrocarbon [11.97] or contained compounds composed only of carbon, hydrogen, and nitrogen [11.98]. Others were fluorinated hydrocarbons [11.99, 11.100] or contained silicone [11.101]. These additives function by forming a low surface energy coating on the internal surface of the inlet manifold so that instead of a layer of liquid fuel being spread over the manifold surface, it is in the form of small, discrete droplets that are much more readily entrained by the fuel-air stream [11.102]. As engine technology has moved toward multipoint and direct injection, the problem of maldistribution has subsided and with it the need for such additives.

11.4.6. Anti-Valve-Seat Recession Additives

Valve seat recession is a problem that can occur with engines that do not have hardened valve seats. When the gasoline engine was first developed, with simple cast iron cylinder heads, the operating regimes, particularly engine speeds, were low and did not put too much wear on the valve seats. With the widespread introduction of leaded fuels that (probably unknown at the time) laid down a lubricating layer of lead oxide and sulfate on the valve seat [11.103], engine technology evolved that produced higher engine speeds and higher stresses on the valve seats. High-performance engines had hardened valve seats of valve seat inserts. With the phase-out of leaded fuel, the beneficial effect of the lead was realized, and new engines were built with hardened valve seats. However, the legacy fleet had to be accommodated and thus requires a fuel additive. Without the additive, the exhaust valves will effectively grind their way through the cylinder head so that the valve tappet clearance can be taken up in just a few thousand miles and may result in poor seating of the exhaust valve and hence increased emissions [11.104, 11.105].

Lead is probably the best anti-valve-seat recession additive, but other materials are known to be at least partially effective. Phosphorus compounds such as tricresyl phosphate and other phosphorous compounds [11.106] reduce this type of wear, but they cannot be used in an unleaded gasoline that is also likely to be used in vehicles fitted with catalytic converters because the phosphorus has an adverse effect on catalysts. Other compounds have been proposed, usually containing alkali or alkaline earth metals [11.107–11.109]. Although these additives have been shown to be effective [11.110, 11.111], they cannot be included in commercial fuels in countries where there is a restriction on the use of additives containing metallic elements. They can be used as aftermarket additives by owners of older vehicle that still require valve seat protection. Organic additive has been proposed [11.112] and claim to have some effect [11.113].

11.5. Additives That Improve Lubricant Performance

A small amount of gasoline and hence gasoline additive finds its way into the lubricant, mainly by cylinder wall wetting during cold starts. There can also be wall wetting due to poor fuel atomization on a warmed-up engine. For engines with purely mechanical fuel metering, such as a carburetor, some gasoline can be drawn into the cylinder after switching off the ignition. This is because the engine will continue to rotate for one or two cycles, drawing fuel into the cylinder but not igniting it. Fuel additives can thus be used to directly assist the lubrication of the piston ring to bore interface on initial start-up. These are known as upper cylinder lubricants and are intended to enhance lubrication in this area, thus reducing frictional losses. Other additives rely on this wall wetting to transfer additive into the lubricant. This will enhance the lubricant performance or replenishing lubricant additives, such as dispersants, that might become depleted over the life of the lubricant.

This class of additive was used in the 1950s and 1960s but is of limited use with today's highly developed lubricants, although there is some renewed interest because of a persistent "black sludge" problem in certain vehicles (see Section 11.5.4).

11.5.1. Upper Cylinder Lubricants

Upper cylinder lubricants were one of the first gasoline additives to be used and consisted of light mineral oils that were used at concentrations of up to 0.5 % volume. They were valuable because they provided immediate lubrication to the ring and bore area after starting up. The early lubricants tended to drain away from the bores on switching off, and it took some time before they reached them again on restarting. This caused considerable wear, particularly for vehicles engaged in mainly stop-start operation. Many modern lubricating oils contain components that promote the retention of an oil film on the metal surfaces; this combined with the fact that these additives would contribute to hydrocarbon emissions made the use of such additives rare.

11.5.2. Anti-Wear Additives

Halogenated compounds—such as are used as lead scavengers or that are present in the atmosphere, sulfur compounds, and some alcohols in the gasoline—will form corrosive gases during combustion. These gases will condense on the bores when the engine is switched off and begins to cool down. A coating of rust is formed on the metal surfaces, and when the engine is restarted, this rust is swept away, exposing a clean metal surface for further corrosion to occur on when the engine is again switched off [11.114]. Additives have been developed [11.115–11.117] and used commercially that prevent corrosive wear in the cylinder bores and on the rings [11.118]. These additives are surface active agents that form a lubricating film on the cylinder walls. The elimination of lead scavengers, the significant reduction in sulfur levels, the use of ethanol rather than methanol, and the use of improved modern lubricants will reduce the need for such additives. Due to fuel economy considerations, there is a growing interest in using fuel additives to reduce engine friction. This is discussed in the next section.

11.5.3. Friction Modifiers

Upper cylinder lubricants (discussed in Section 11.5.1) were designed to reduce wear when the engine was cold; this wear was brought about by friction between the piston rings and the cylinder bore. Reducing this friction reduced the wear, but it also reduced the amount of energy that had to be expended to overcome this friction; fuel economy under these start-up conditions would thus be improved. However, as noted previously, the advances in lubricant technology have to a large extent eliminated the need for and the benefits of this type of additive.

Increasing pressures to improve engine efficiency have again focused attention on whether fuel additives can be used to reduce the friction in this vital area. The frictional losses in this one area of the engine can account for 65% of the total frictional losses [11.119]. A major determinant of the frictional losses in this region are the viscosity of the lubricant, or more correctly the combination of fuel and lubricant that exist in this region [11.120]. Due to lubricant degradation, the viscosity increases during the lifetime of the lubricant [11.121]. One way to improve fuel economy is to include an additive within the fuel that will migrate to the lubricant to either reduce the rate of degradation of the lubricant additive package or simply to replenish the additives included in the lubricant [11.122]. An alternative approach is to formulate an additive, which in itself will reduce the friction coefficient at the piston ring to cylinder bore interface [11.123].

This latter approach is receiving the greatest attention because the inclusion of such additives can provide an immediate reduction in fuel economy as opposed to merely offsetting a gradual deterioration as the lubricant ages. Some of the additive will accumulate in the lubricating oil and will thus give an additional longer-term benefit [11.123]. Popular chemistries for this purpose are surface active materials that include amines, amides, and esters [11.124–11.131].

11.5.4. Anti-Sludge Additives

Low-temperature sludge formed in gasoline engine lubricants was initially a result of fuel-derived oxidation products [11.132, 11.133]. The use of fuel additives to improve the stability of the fuel eased the problem.

A black (or hot) sludge problem reappeared during the 1980s in both Europe and the US when some vehicles were run under high-speed, high-temperature conditions [11.134]. Tests have shown that the cause can be due to both the fuel and the lubricant and also to fuel/lubricant interaction effects [11.59, 11.135]. The fuel parameters that are important are fuel end point [11.136], the presence of high olefin content [11.137], and the presence of certain fuel additives [11.138]. Standard test procedures have been developed within the lubricants industry to assess the sludge-forming tendency of lubricants. It is advantageous when developing fuel additives to run these tests to ensure that the additives will not interact with the lubricant to promote sludge, even at higher than the recommended fuel additive treat rates [11.139].

11.6. Multifunctional Additive Packages

The preceding sections of this chapter clearly show that there are many facets of gasoline performance that are more readily achieved by the use of fuel additives than by added refining. Some of these issues can be addressed by a single additive; for example, a DCA that will keep a port fuel injector clean should also keep a carburetor clean, but not necessarily vice versa. A DCA, being surface active, will probably also act as a corrosion inhibitor. However, to meet all the requirements of a modern high-quality gasoline, it is necessary to include more than one additive component. Sometimes these components are synergistic—for example, an anti-valve seat recession additive in combination with an octane enhancing additive has been found to be synergistic in protecting the valve seat [11.111]. Other combinations are also shown to produce desirable outcomes, for example, fuel economy benefits from mixtures of two or more products of the type, a polyetheramine, an aliphatic hydrocarbon-substituted amine, a Mannich reaction product formed by reacting an aliphatic hydrocarbon-substituted phenol and an aldehyde and an amine [11.128].

As noted, the active components of DCA often need a fluidizer to aid their functionality or to ensure that they do not produce undesirable side effects such as valve sticking. Conversely, it must be ensured that such additional components as the fluidizer do not contribute to other issues such as CCDs. A solvent may also be required to control the viscosity of the additive to allow easy addition to and mixing with the fuel.

Clearly, the producer of a finished fuel cannot simply buy individual additives off the shelf, add them to the base gasoline, and hope that the finished fuel achieves all the required performance attributes. It is thus desirable that the fuel producer buys and adds a single fuel additive package that contains all the required components to give the desired quality fuel.

Over 50 years ago, it was stated

> *A gasoline additive must meet many stringent requirements by passing many tests before it can become a commercial product. The evaluation of additives is extremely complex because of the great variability of engines, equipment, fuels, lubricants, and operating conditions. Moreover, the evaluation of "side effects" on engines and equipment may be more costly and time-consuming than tests for effectiveness. [11.140]*

This, as with the quote at the start of the chapter, is still true today, possibly more so in light of the finer tolerances to which engines are designed and manufactured, increased emission control, and increasing pressure to increase engine efficiency. Testing these additive packages is no longer a case of only running laboratory tests or even engine tests to look at various aspects such as corrosion resistance or carburetor cleanliness. Testing these additive packages must involve looking at all the intended benefits, stability, engine cleanliness, and so on, as well as the possible side effects, including valve sticking and sludge formation. It is also important to consider the influence on vehicle emissions and emission control devices, and whether the performance can be achieved in a range of vehicles in the real world [11.139–11.144].

References

11.1.   Collings, H.E. and Squerciati, E.C., "Additives for Liquid Hydrocarbon Fuels," SAE Technical Paper **600324, 1960,** doi:https://doi.org/10.4271/600324.

11.2.   Gibbs, L., "Gasoline Additives - When and Why," SAE Technical Paper 902104, 1990, doi:https://doi.org/10.4271/902104.

11.3.   D'Ornellas, C.V. and de Oliveira, E.J., "Gasoline Oxidation Stability—Influence of Composition and Antioxidant Response," in *Proceedings of 4th International Colloquium Fuels*, Esslingen, Germany, January 15–16, 2003.

11.4.   Tupa, R.C. and Dorer, C.J., "Gasoline and Diesel Fuel Additives for Performance/Distribution Quality–II," SAE Technical Paper 861179, 1986, doi:https://doi.org/10.4271/861179.

11.5.   Schwartz, F.G., Allbright, C.S., and Ward, C.C., "Prediction of Gasoline Storage Stability," SAE Technical Paper 690760, 1969, doi:https://doi.org/10.4271/690760.

11.6.   Asseff, P.A., "Multifunctional Gasoline Additives Reduce Engine Deposits," SAE Technical Paper 660543, 1966, doi:https://doi.org/10.4271/660543.

11.7.   Chenicek, J.A., Metal deactivator. U.S. Patent 2,754,216, 1956.

11.8.   Phillips, T.A., Corrosion inhibitor for liquid fuels. U.S. Patent 4,737,159, 1988.

11.9.   Garth, B.H. and Schmidt, F.H., Corrosion inhibitor composition. U.S. Patent 4,214,876, 1980.

11.10. Perilstein, W.L., Corrosion inhibitors for alcohol-based fuels. U.S. Patent 4,426,208, 1984.

11.11. Moore, F.W. and Bailey, B.S., Motor fuel composition. U.S. Patent 3,502,451, 1970.

11.12. Miller, C.O. and Dorer, C.J., Acylated polyamine composition. U.S. Patent 3,240,575, 1966.

11.13. Dorer, C.J. Jr. and Miller, C.O., Fuel compositions containing esters and ester-type dispersants. U.S. Patent 3,427,141, 1969.

11.14. White, J.L. and Gibson, D.L., Method of reducing friction loss in flowing hydrocarbon liquids. U.S. Patent 3,215,154, 1965.

11.15. Culter, J.D. and McClaffin, G.G., Method of friction loss reduction in oleaginous fluids flowing through conduits. U.S. Patent 3,692,676, 1972.

11.16. Coordinating Research Council, in *Proceedings of the CRC Workshop on Pipeline Drag Reducing Agents and Their Impact on Fuel Product Quality*, Houston, TX, November 29–30, 1988.

11.17. Johnston, A.A. and Dimitroff, E., "A Bench Technique for Evaluating the Induction System Deposit Tendencies of Motor Gasolines," SAE Technical Paper 660783, 1966, doi:https://doi.org/10.4271/660783.

11.18. ASTM International, "Standard Test Method for Water Reaction of Aviation Fuels," ASTM D1094-07(2019), ASTM International, 2022.

11.19. ASTM International, "Standard Test Method for Water Separation Properties of Light and Middle Distillate, and Compression and Spark Ignition Fuels," ASTM D7451-21, ASTM International, 2021.

11.20. Sheahan, T.J., Dorer, C.J., and Miller, C.O., "Detergent-Dispersant Fuel Performance and Handling," SAE Technical Paper 690516, 1969, doi:https://doi.org/10.4271/690516.

11.21. Bolt, J.A., "Storage Stability of Gasolines in Automotive Fuel Systems (Report of Gasoline Additives Groups of CFR Motor Fuels Div. Coordinating Research Council)," SAE Technical Paper 470121, 1947, doi:https://doi.org/10.4271/470121.

11.22. ASTM International, "Standard Test Method for Rust-Preventing Characteristics of Inhibited Mineral Oil in the Presence of Water," ASTM D665-19, ASTM International, 2020.

11.23. Colonial Pipeline Company, "Pipeline Rust Test (Modified ASTM D665)," 1967.

11.24. National Association of Corrosion Engineers, "NACE Standard TM0172-2001," 2001.

11.25. U.S. Department of Defense, Performance Specification, "Inhibitor, Corrosion/Lubricity Improver, Fuel Soluble (NATO S-1747)," MIL-PRF-25017H w/Amendment 1, 2011.

11.26. Udelhofen, J.H. and Zahalka, T.L., "Gasoline Additive Requirements for Today's Smaller Engines," SAE Technical Paper 881644, 1988, doi:https://doi.org/10.4271/881644.

11.27. Polss, P., Gasoline. U.S. Patent 3,873,278, 1975.

11.28. Abramo, G.P., Blain, D.A., and Cardis, A.B., Modified succinimides as dispersants and detergents and lubricant and fuel compositions containing same. U.S. Patent 5,486,301, 1996.

11.29. Lindstrom, E.G. and Barusch, M.R., Amino amide salts of organic monocarboxylic acids as additives for reducing carburetor deposits. U.S. Patent 2,922,707, 1960.

11.30. De Gray, R.J. and Belden, S.H., Gasoline containing borated oxazolines. U.S. Patent 2,965,459, 1960.

11.31. Lindstrom, E.G. and Barusch, M.R., Gasoline composition containing phosphated amino amides. U.S. Patent 2,974,022, 1961.

11.32. Michaels, A.E., Nostrand, E.D., and Tutwiler, T.S., Gasoline containing polymeric additive agents. U.S. Patent 3,058,818, 1962.

11.33. Malec, R.E., Motor fuel. U.S. Patent 3,139,330, 1964.

11.34. Kautsky, G.J., Gasoline fuel containing alky orthophospahtes of n-aminoalkyl-subsituted 2-amino-alkane detergents. U.S. Patent 3,389,980, 1968.

11.35. Schlicht, R.C., Levin, M.D., Herbstman, S., and Sung, R.L., Gasoline composition containing reaction products of fatty acid esters and amines as carburetor detergents. U.S. Patent 4,729,769, 1988.

11.36. Meyer, G.R. and Lyons, W.R. Jr., Hydrocarbon fuel detergent. U.S. Patent 4,863,487, 1989.

11.37. Cherpeck, R.E., Fuel additive compositions containing polyisobutenyl succinimides. U.S. Patent 5,393,309, 1995.

11.38. Alquist, H.E. and Ebersole, G.D., Fuel additive package containing polybutene amine and lubricating oil. U.S. Patent 4,022,589, 1977.

11.39. Kummer, R., Franz, D., and Rath, H.P., Polybutyl- and polyisobutylamines, their preparation, and fuel compositions containing these. U.S. Patent 4,832,702, 1989.

11.40. Dever, J.L., Menon, M.C., Phillips, S.D., and Baldwin, L.J., Process for the production of fuel additives from chlorinated polybutenes. U.S. Patent 5,346,965, 1994.

11.41. Rath, H.P., Mach, H., Oppenläender, K., Schoenleben, W. et al., Gasoline-engine fuels containing polyetheramines or polyetheramine derivatives. U.S. Patent 5,112,364, 1992.

11.42. Oppenläender, K., Günther, W., Henne, A., Menger, V. et al., Polyetheramine-containing fuels for gasoline engines. U.S. Patent 5,660,601, 1997.

11.43. Gagliardi, J.C., Ghannam, F.E., Gasvoda, R.F., and Potter, R.I., "Engine Tests for Evaluating Crankcase Oils in Stop-and-Go MS Service," SAE Technical Paper 600117, 1960, doi:https://doi.org/10.4271/600117.

11.44. Malakar, J.J., Retzloff, J.B., and Gibbs, L.M., "Throttle Body Deposits—The CRC Carburetor Cleanliness Test Procedure," SAE Technical Paper 831708, 1983, doi:https://doi.org/10.4271/831708.

11.45. CEC, "Evaluation of Gasolines with Respect to Maintenance of Carburetor Cleanliness," CEC F-03-T-81, 1981.

11.46. Taniguchi, B.Y., Peyla, R.J., Parsons, G.M., Hoekman, S.K. et al., "Injector Deposits - The Tip of Intake System Deposit Problems," SAE Technical Paper 861534, 1986, doi:https://doi.org/10.4271/861534.

11.47. Benson, J.D. and Yaccarino, P.A., "The Effects of Fuel Composition and Additives on Multiport Fuel Injector Deposits," SAE Technical Paper 861533, 1986, doi:https://doi.org/10.4271/861533.

11.48. Abramo, G.P., Horowitz, A.M., and Trewella, J.C., "Port Fuel Injector Cleanliness Studies," SAE Technical Paper 861535, 1986, doi:https://doi.org/10.4271/861535.

11.49. Tupa, R.C. and Koehler, D.E., "Gasoline Port Fuel Injectors—Keep Glean/Clean Up with Additives," SAE Technical Paper 861536, 1986, doi:https://doi.org/10.4271/861536.

11.50. Tupa, R.C., "Port Fuel Injectors—Causes/Consequences/Cures," SAE Technical Paper 872113, 1987, doi:https://doi.org/10.4271/872113.

11.51. Bitting, B., Gschwendtner, F., Kohlhepp, W., Kothe, M. et al., "Intake Valve Deposits—Fuel Detergency Requirements Revisited," SAE Technical Paper 872117, 1987, doi:https://doi.org/10.4271/872117.

11.52. Spink, C.D., Barraud, P.G., and Morris, G.E.L., "A Critical Road Test Evaluation of Two High-Performance Gasoline Additive Packages in a Fleet of Modern European and Japanese Vehicles," SAE Technical Paper 912393, 1991, doi:https://doi.org/10.4271/912393.

11.53. Sung, R.L., Daly, D.T., and Hayden, T.E., "A Novel Gasoline Additive Package Removes Induction System Deposits and Reduces Engine Octane Requirement Increase," SAE Technical Paper 891298, 1989, doi:https://doi.org/10.4271/891298.

11.54. Mikkonen, S., Karlsson, R., and Kivi, J., "Intake Valve Sticking in Some Carburetor Engines," SAE Technical Paper 881643, 1988, doi:https://doi.org/10.4271/881643.

11.55. Megnin, M.K., Cobb, J.M., and Eng, K.D., "Development of a Gasoline Additive Screening Test for Intake Valve Stickiness and Deposit Levels," SAE Technical Paper 892121, 1989, doi:https://doi.org/10.4271/892121.

11.56. Shilbolm, C.M. and Schoonveld, G.A., "Effect on Intake Valve Deposits of Ethanol and Additives Common to the Available Ethanol Supply," SAE Technical Paper 902109, 1990, doi:https://doi.org/10.4271/902109.

11.57. Richardson, C.E., Fischer, J.L., and Pawczuk, G., "Evaluation of the Effect of Fuel Composition and Gasoline Additives on Combustion Chamber Deposits," SAE Technical Paper 962012, 1996, doi:https://doi.org/10.4271/962012.

11.58. Schreyer, P., Starke, K.W., Thomas, J., and Crema, S., "Effect of Multifunctional Fuel Additives on Octane Number Requirement of Internal Combustion Engines," SAE Technical Paper 932813, 1993, doi:https://doi.org/10.4271/932813.

11.59. Galliard, I.R. and Lillywhite, J.R.F., "Field Trial to Investigate the Effect of Fuel Composition and Fuel-Lubricant Interaction on Sludge Formation in Gasoline Engines," SAE Technical Paper 922218, 1992, doi:https://doi.org/10.4271/922218.

11.60. Sandquist, H., Denbratt, I., Owrang, F., and Olsson, J., "Influence of Fuel Parameters on Deposit Formation and Emissions in a Direct Injection Stratified Charge SI Engine," SAE Technical Paper 2001-01-2028, 2001, doi:https://doi.org/10.4271/2001-01-2028.

11.61. Aradi, A.A., Colucci, W.J., Scull, H.M., and Openshaw, M.J., "A Study of Fuel Additives for Direct Injection Gasoline (DIG) Injector Deposit Control," SAE Technical Paper 2000-01-2020, 2000, doi:https://doi.org/10.4271/2000-01-2020.

11.62. China, P. and Rivere, J.-P., "Development of a Direct Injection Spark Ignition Engine Test for Injector Fouling," SAE Technical Paper 2003-01-2006, 2003, doi:https://doi.org/10.4271/2003-01-2006.

11.63. Aradi, A.A., Evans, J., Miller, K., and Hotchkiss, A., "Direct Injection Gasoline (DIG) Injector Deposit Control with Additives," SAE Technical Paper 2003-01-2024, 2003, doi:https://doi.org/10.4271/2003-01-2024.

11.64. Colucci, W.J., Loper, J.T., and Aradi, A.A., Fuels compositions and methods for using same. U.S. Patent 7,491,248, 2009.

11.65. Graupner, O., Mundt, M., Schütze, A., Louis, J.J.J. et al., Gasoline additive. U.S. Patent 7,901,470, 2011.

11.66. Burgess, V., Reid, J., and Mulqueen, S., Composition, method and use. World Patent WO 2011/141731, 2011.

11.67. Maricq, M.M., "Engine, Aftertreatment, Fuel Quality and Non-Tailpipe Achievements to Lower Gasoline Vehicle PM Emissions: Literature Review and Future Prospects," *Science of The Total Environment* 866 (2023): 161225.

11.68. Reid, J., Mulqueen, S., Langley, G., Wilmot, E. et al., "The Investigation of the Structure and Origins of Gasoline Direct Injection (GDI) Deposits," SAE Technical Paper 2019-01-2356, 2019, doi:https://doi.org/10.4271/2019-01-2356.

11.69. Monroe, R., Studzinski, W., Parsons, J., La, C. et al., "Engine Particulate Emissions as a Function of Gasoline Deposit Control Additive," *SAE Int. J. Fuels Lubr.* 14, no. 1 (2021): 3-11, doi:https://doi.org/10.4271/04-14-01-0001.

11.70. Tongberg, C.O., Hakala, N.V., Moody, L.E., and Patberg, J.B., "Gasoline Additives," SAE Technical Paper 540022, 1954, doi:https://doi.org/10.4271/540022.

11.71. Daly, D.T., Bannon, S.A., Fog, D.A., and Harold, S.M., "Mechanism of Combustion Chamber Deposit Formation," SAE Technical Paper 941889, 1994, doi:https://doi.org/10.4271/941889.

11.72. Machleder, W.H. and Bollinger, J.M., Multipurpose fuel additive. U.S. Patent 4,048,081, 1977.

11.73. Keller, C.T. and Corkwell, K.C., "Honda Generators Used to Evaluate Fuels and Additive Effects on Combustion Chamber Deposits," SAE Technical Paper 940347, 1994, doi:https://doi.org/10.4271/940347.

11.74. **Nagao, M.**, Kaneko, T., Omata, T., Iwamoto, S. et al., "Mechanism of Combustion Chamber Deposit Interference and Effects of Gasoline Additives on CCD Formation," SAE Technical Paper 950741, 1995, doi:https://doi.org/10.4271/950741.

11.75. **Crema, S.**, Schreyer, P., and Miin, T., "Effect of Thermal Stability of Detergents and Carrier Fluids on the Formation of Combustion Chamber Deposits," SAE Technical Paper 961097, 1996, doi:https://doi.org/10.4271/961097.

11.76. **Su, W.-Y.**, Herbstman, S., Zimmerman, R.L., and Cuscurida, M., Polyether hydroxyethylaminoethyl oxalamide motor fuel detergent additives. U.S. Patent 5,285,267, 1994.

11.77. **Loper, J.T.**, Amine ester-containing additives and methods of making and using same. U.S. Patent 5,597,390, 1997.

11.78. **Kanakia, M.D.**, Franklin, R., Steichen, D., and Gadberry, J.F., Fuel additive compositions for simultaneously reducing intake valve and combustion chamber deposits. U.S. Patent 6,013,115, 2000.

11.79. **Lin, J.-J.**, Su, I.-F., Lin, K.-H., Ho, Y.-S. et al., Star-like poly(oxyalkyene) triamide alkoxylates gasoline additive and the method for producing the same. U.S. Patent 6,270,540, 2001.

11.80. **Keleman, S.R.**, Siskin, M., Rose, K.D., Schwahn, H. et al., Gasoline additives for reducing the amount of internal combustion engine, intake valve deposits and combustion chamber deposits. U.S. Patent 7,226,489, 2007.

11.81. **Forde, R.M.**, Ahmadi, M.R., and Cherpeck, R.E., Method and composition for reduction of combustion chamber deposits. U.S. Patent 6,136,051, 2000.

11.82. **Wolf, L.R.**, Fuel stability additive. U.S. Patent 2003/0196372, 2003.

11.83. **Graiff, L.B.**, "Some New Aspects of Deposit Effects on Engine Octane Requirement Increase and Fuel Economy," SAE Technical Paper 790938, 1979, doi:https://doi.org/10.4271/790938.

11.84. **Benson, J.D.**, "Some Factors Which Affect Octane Requirement Increase," SAE Technical Paper 750933, 1975, doi:https://doi.org/10.4271/750933.

11.85. **Bert, J.A.**, Gething, J.A., Hansel, T.J., Newhall, H.K. et al., "A Gasoline Additive Concentrate Removes Combustion Chamber Deposits and Reduces Vehicle Octane Requirement," SAE Technical Paper 831709, 1983, doi:https://doi.org/10.4271/831709.

11.86. **Feldman, N.**, Method and fuel composition for reducing octane requirement increase. U.S. Patent 4,787,916, 1988.

11.87. **Sung, R.L.**, Behrens, M.D., Caggiano, M.A., Knifton, J.F. et al., Reaction product additive and ORI-inhibited motor fuel composition. U.S. Patent 4,810,261, 1989.

11.88. **Powers, W.J. III**, Matthews, L.A., Leong, M.F., and Erikson, R.W. Jr., Process for producing ORI control additives. U.S. Patent 5,147,414, 1992.

11.89. **Garapon, J.**, Bregent, R., Touet, R., Mulard, P. et al., Polynitrogen compound having two terminal cycles of the imide type, their preparation and uses. U.S. Patent 5,234,476, 1993.

11.90. **Mulard, P.**, Labruyere, Y., Forestiere, A., and Bregent, R., Additive formulation for fuels incorporating ester function products and a detergent-dispersant. U.S. Patent 5,433,755, 1995.

11.91. **Delhomme, H.**, Gaillard, J., Mulard, P., and Eber, D., Gasoline additive containing alkoxylated imidazo_oxazoles. U.S. Patent 5,472,457, 1995.

11.92. **van Es, C.**, Miles, R., Kalghatgi, G.T., McArragher, J.S. et al., Gasoline composition. U.S. Patent 4,765,800, 1988.

11.93. **Kalghati, G.T.**, "Improvements in Early Flame Development in a Spark Ignition Engine Brought about by 'Spark-Aider' Fuel Additives," *Combustion Science and Technology* 52, no. 4-6 (1987): 427-446.

11.94. **Kalghatgi, G.T.**, "Spark Ignition, Early Flame Development and Cyclic Variation in I.C. Engines," SAE Technical Paper 870163, 1987, doi:https://doi.org/10.4271/870163.

11.95. **Kalghati, G.T.**, "Effect of a Spark Aider Fuel Additive on the Misfire Characteristics of a Spark Ignition Engine," *Combustion Science and Technology* 62, no. 1-3 (1988): 1-19.

11.96. **Yu, H.T.C.**, "Fuel Distribution Studies—A New Look at an Old Problem," SAE Technical Paper 630485, 1963, doi:https://doi.org/10.4271/630485.

11.97. **Zimmerman, A.A.**, Furlong, L.E., and Shannon, H.F., Composition for improving air-fuel distribution in internal combustion engines. U.S. Patent 3,773,184, 1973.

11.98. Zimmerman, A.A., Furlong, L.E., and Shannon, H.F., Method and composition for optimizing air-fuel ratio distribution in internal combustion engines. U.S. Patent 3,707,362, 1972.

11.99. Furlong, L.E., Zimmerman, A.A., and Shannon, H.F., Enhancing the operation of a gasoline engine. U.S. Patent 3,791,066, 1974.

11.100. Laity, J.L., Gasoline composition. U.S. Patent 4,039,301, 1977.

11.101. Zimmerman, A.A., Furlong, L.E., and Shannon, H.F., Air-fuel ratio distribution to enhance gasoline engine operation. U.S. Patent 3,762,891, 1973.

11.102. Zimmerman, A.A., Furlong, L.E., and Shannon, H.F., "Improved Fuel Distribution—A New Role for Gasoline Additives," SAE Technical Paper 720082, 1972, doi:https://doi.org/10.4271/720082.

11.103. Godfrey, D. and Courtney, R.L., "Investigation of the Mechanism of Exhaust Valve Seat Wear in Engines Run on Unleaded Gasoline," SAE Technical Paper 710356, 1971, doi:https://doi.org/10.4271/710356.

11.104. Felt, A.E. and Kerley, R.V., "Engines and Effects of Lead-Free Gasoline," SAE Technical Paper 710367, 1971, doi:https://doi.org/10.4271/710367.

11.105. Kent, W.L. and Finnigan, F.T., "The Effect of Some Fuel and Operating Parameters on Exhaust Valve Seat Wear," SAE Technical Paper 710673, 1971, doi:https://doi.org/10.4271/710673.

11.106. Croudace, M.C. and Howland, W.W., Gasoline fuel composition. U.S. Patent 4,720,288, 1988.

11.107. Graham, J.P., Ryer, J., and Ketz, S.J., Gasoline composition containing a sodium additive. U.S. Patent 3,955,938, 1976.

11.108. Johnston, T.E. and Dorer, C.J. Jr., Fuel composition for lessening valve seat recession. U.S. Patent 4,659,338, 1987.

11.109. Crawford, J., Kikabhai, T., McCleary, D.B., and Pearce, A., Fuel composition containing an additive for reducing valve seat recession. U.S. Patent 5,090,966, 1992.

11.110. Akarapanjavit, N. and Boonchanta, P., "Valve Seat Recession and Protection Due to Lead Phase Out in Thailand," SAE Technical Paper 962029, 1996, doi:https://doi.org/10.4271/962029.

11.111. Mulqueen, S., Vincent, M.W., and Mulcare, C., "Development of a High Performance Anti-Wear Additive Providing Protection Against Valve Seat Recession Combined with Octane Enhancement in Treated Fuel," SAE Technical Paper 2000-01-2016, 2000, doi:https://doi.org/10.4271/2000-01-2016.

11.112. Nelson, M.L. and Nelson, O.L. Jr., Fuel conditioner. U.S. Patent 4,753,661, 1988.

11.113. Nelson, O.L., Larson, J.E., Fein, R.S., Fuller, D.D. et al., "A Broad-Spectrum, Non-Metallic Additive for Gasoline and Diesel Fuels: Performance in Gasoline Engines," SAE Technical Paper 890214, 1989, doi:https://doi.org/10.4271/890214.

11.114. Jackson, H., "Why Does Your Car Wear Out," SAE Technical Paper 570230, 1957, doi:https://doi.org/10.4271/570230.

11.115. Annable, W.G., Gasoline compositions. U.S. Patent 3,707,361, 1972.

11.116. Washecheck, P.H., Liu, A.T.C., and Kennedy, E.F., Methanol fuel and methanol fuel additives. U.S. Patent 4,375,360, 1983.

11.117. Craig, R.C., Panzer, J., Wisotsky, M.J., and Beltzer, M., Anti-wear additives in alkanol fuel. U.S. Patent 4,609,376, 1986.

11.118. Hudnall, J.R., Gerard, P.L., Hood, C.B., Burns, I.D.M. et al., "New Gasoline Formulations Provide Protection Against Corrosive Engine Wear," SAE Technical Paper 690514, 1969, doi:https://doi.org/10.4271/690514.

11.119. McGeehan, J.A., "A Literature Review of the Effects of Piston and Ring Friction and Lubricating Oil Viscosity on Fuel Economy," SAE Technical Paper 780673, 1978, doi:https://doi.org/10.4271/780673.

11.120. Smith, O., Priest, M., Taylor, R.I., Price, R. et al., "Simulated Fuel Dilution and Friction Modifier Effects on Piston Ring Friction," *Proceedings of the Institution of Mechanical Engineers, Part J: Journal of Engineering Tribology* 220, no. 3 (2006): 181-189.

11.121. Rappaport, S.T., Ferner, M.D., Hecker, L.S., and Tierney, T.B., "Evaluation of API/ILSAC GF-4 Oil Life in Today's U.S. Fleet," SAE Technical Paper 2008-01-1740, 2008, doi:https://doi.org/10.4271/2008-01-1740.

11.122. Rappaport, S., Nattrass, S., Smith, S., Brewer, M. et al., "Management of Lubricant Fuel Economy Performance over Time through Fuel Additives," SAE Technical Paper 2012-01-1270, 2012, doi:https://doi.org/10.4271/2012-01-1270.

11.123. Hayden, T.E., Ropes, C.A., and Rawdon, M.G., "The Performance of a Gasoline Friction Modifier Fuel Additive," SAE Technical Paper 2001-01-1961, 2001, doi:https://doi.org/10.4271/2001-01-1961.

11.124. Culotta, A.M., Fuel economy additives. U.S. Patent 5,632,785, 1997.

11.125. Oumar-Mahamat, H. and Carey, J.T., Friction reducing additives for fuels and lubricants. U.S. Patent 5,858,029, 1999.

11.126. Fuentes-Afflick, P.A. and Gething, J.A., Fuel composition containing an amine compound and ester. U.S. Patent 6,203,584, 2001.

11.127. DeRosa, T.F., Kaufman, B.J., DeBlase, F.J., Hayden, T.E. et al., Fuel additive composition for improving delivery of fricton modifier. U.S. Patent 6,743,266, 2004.

11.128. Jackson, M.M. and Corkwell, K.C., Gasoline additive concentrate composition and fuel composition and method thereof. U.S. Patent 7,195,654, 2007.

11.129. Aradi, A.A., Malfer, D.J., Schwab, S.D., and Colucci, W.J., Friction modifier alkoxyamine salts of carboxylic acids as additives for fuel compositions and methods of use thereof. U.S. Patent 7,435,272, 2008.

11.130. Kaufman, B.J., Ketcham, J.R., and Acker, W.P., Fuel additives and fuel compositions and methods for making and using the same. U.S. Patent 2010/0132253, 2010.

11.131. Cooney, A., Ross, A., and Burgess, V., Improvements in or relating to additives for fuels and lubricants. World Patent WO 2012/076896, 2012.

11.132. Rogers, D.T., Rice, W.W., and Jonach, F.L., "Mechanism of Engine Sludge Formation and Additive Action," SAE Technical Paper 560067, 1956, doi:https://doi.org/10.4271/560067.

11.133. Dimitroff, E. and Quillian, R.D., "Low Temperature Engine Sludge—What?—Where?—How?" SAE Technical Paper 650255, 1965, doi:https://doi.org/10.4271/650255.

11.134. Lillywhite, J.R.F., Sant, P., and Saville, S.B., "Sludge Formation: Investigation of Sludge Formation in Gasoline Engines," *Industrial Lubrication and Tribology* 42, no. 1 (1990): 4-10.

11.135. Graf, R.T., Copan, W.G., Kornbrekke, R.E., and Murphy, J.P., "Sludge Formation in Engine Testing and Field Service," SAE Technical Paper 881580, 1988, doi:https://doi.org/10.4271/881580.

11.136. Cracknell, R.F. and Head, R.A., "Influence of Fuel Properties on Lubricant Oxidative Stability: Part 1—Engine Tests," SAE Technical Paper 2005-01-3839, 2005, doi:https://doi.org/10.4271/2005-01-3839.

11.137. Kawamura, M., Moritani, H., Nakada, M., and Oohori, M., "Sludge Formation and Engine Oil Dispersancy Evaluation with a Laboratory Scale Sludge Simulator," SAE Technical Paper 892105, 1989, doi:https://doi.org/10.4271/892105.

11.138. Lewis, R.A., Newhall, H.K., Peyla, R.J., Voss, D.A. et al., "A New Concept in Engine Deposit Control Additives for Unleaded Gasolines," SAE Technical Paper 830938, 1983, doi:https://doi.org/10.4271/830938.

11.139. Papachristos, M.J., Vincent, M.W., Burton, J., Cooney, A.J. et al., "Fully Synthetic Gasoline Additive Packages to Meet the Needs of the 90s," SAE Technical Paper 932809, 1993, doi:https://doi.org/10.4271/932809.

11.140. Felt, A.E. and Sumner, H.C., "Gasoline Additives—A Review for Engineers," SAE Technical Paper 600323, 1960, doi:https://doi.org/10.4271/600323.

11.141. Corkwell, K.C. and Megnin, M.K., "Fleet Test Evaluation of Gasoline Additives for Intake Valve and Combustion Chamber Deposit Clean Up," SAE Technical Paper 950742, 1995, doi:https://doi.org/10.4271/950742.

11.142. Haury, E.J. and Graham, J., "A Fleet Test Evaluation of the Effect of a Unique Gasoline Additive on Octane Requirement Emissions," SAE Technical Paper 961098, 1996, doi:https://doi.org/10.4271/961098.

11.143. Spink, C.D., McDonald, C.R., Morris, G.E.L., and Stephenson, T., "A Critical Road Test Evaluation of a High-Performance Gasoline Additive Package in a Fleet of Modern European and Asian Vehicles," SAE Technical Paper 2004-01-2027, 2004, doi:https://doi.org/10.4271/2004-01-2027.

11.144. Spink, C.D., Coleman, A.T., Nelson, E.C., Wakefield, S. et al., "Multi-Vehicle Evaluation of Gasoline Additive Packages: A Fourth Generation Protocol for the Assessment of Intake System Deposit Removal," *SAE Int. J. Fuels Lubr.* 2, no. 2 (2010): 27-37, doi:https://doi.org/10.4271/2009-01-2635.

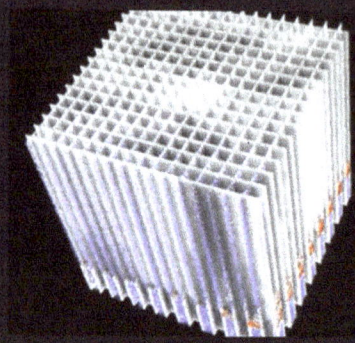

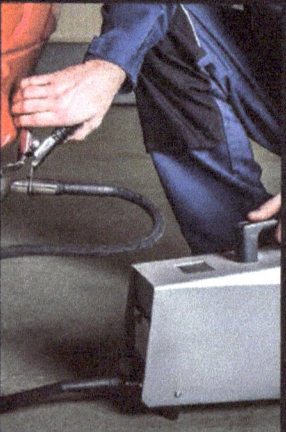

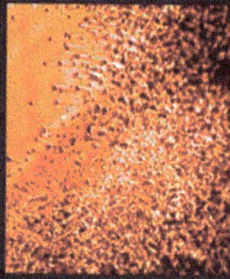

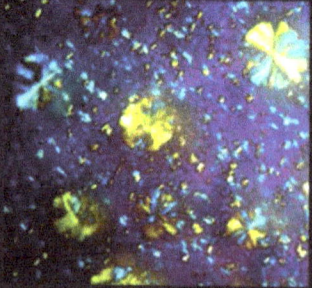

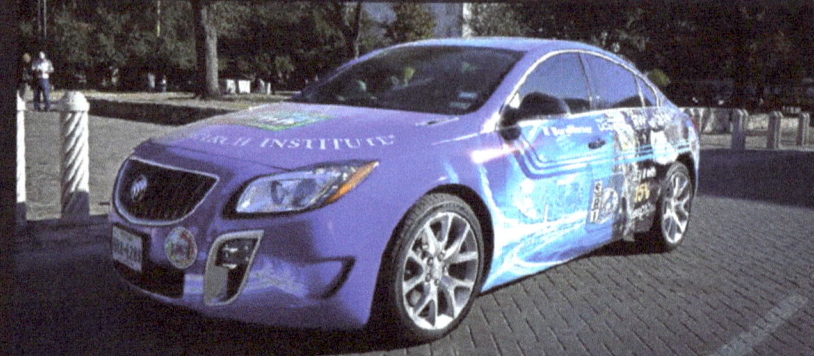

12

Other Gasoline Specification and Non-Specification Properties

Two of the major gasoline properties that affect the performance of a gasoline engine are octane quality and volatility; these have been discussed in Chapters 7 and 9, respectively. Ensuring that the gasoline quality remains as the fuel producer intended and also that it does not produce unwanted degradation products within the engine has been discussed in Chapters 10 and 11. How fuel composition influences exhaust emissions will be dealt with in Chapter 13. This chapter will consider other gasoline properties that can influence vehicle performance and customer perception, and how these properties are measured.

12.1. Density

The density of the fuel, often expressed as relative density or specific gravity, is the ratio of the weight of a given volume of gasoline to the weight of the same volume of water, when both are at a temperature of 15°C and at a pressure of 101.325 kPa. To a small extent, the older term "degrees API" is also used and is based on an arbitrary hydrometer scale that is related to specific gravity, as follows:

$$\text{Degrees API} = \frac{141.5}{\text{SG}} - 131.5$$

where SG is the specific gravity that is the ratio of the density of the gasoline to the density of water when both are measured at 15°C.

Fuel is usually sold by units of volume; because of this, motorists usually measure fuel economy or fuel consumption on a volumetric basis. A denser fuel means that motorists are actually getting more fuel, by mass, every time they fill their tank. Density tends to correlate well with the amount of carbon in the fuel, and this tends to correlate with the energy content of the fuel. For hydrocarbon components, generally speaking, aromatic compounds have the highest density, with the di-olefins and olefins being intermediate and the paraffins having the lowest density when compounds having the same number of carbon atoms are compared. Also, a high carbon number compound will have a higher density than a low carbon number compound of the same type. However, as discussed in Chapter 9, there is generally an inverse relationship between density and volatility for the different hydrocarbon types. There is clearly a trade-off between these inter-related properties.

Most gasolines have a relative density between about 0.72 and 0.78, equivalent to an API range of 65 to 50. It is important in long-trip fuel economy (see Chapter 9, Section 9.9) because it is a guide to the energy content of a gasoline (see Section 12.2). Measurement is usually by a hydrometer method [12.1, 12.2] or by an oscillating U-tube method [12.3, 12.4].

The stoichiometric air-to-fuel ratio is easily determined on a mass basis; fuel metering is usually performed on a volumetric basis. Therefore, when setting the original engine calibrations, a fuel of a known density must be used. If a fuel of a different density is used, the engine will receive a different air-to-fuel ratio than intended. This can affect driveability and the exhaust emissions. For vehicles equipped with closed loop feedback control systems, such changes in density will be compensated for.

For engines employing an open-loop fuel injection system, this compensation will not apply. For carbureted engines, there is some built-in compensation. When considering the flow of fuel through a carburetor jet, the coefficient of discharge for normal hydrocarbon fuels is virtually constant above a critical value of the Reynolds number. Under these conditions, the mass flow is a function of the density of the fuel so that increasing density increases mass flow and lowers the air-to-fuel ratio of the mixture. However, for a denser fuel, the float in the float bowl will sit higher and is thus shut of the fuel sooner, leading to a lower level of fuel in the float bowl. The flow through the carburetor jet is determined by the difference in pressure between the carburetor throat and the head of fuel in the float bowl, so there will be some reduction in volume flow.

12.2. Heat of Combustion

The heat of combustion is also referred to as the heating value or calorific value of the fuel. The gross heat of combustion is the heat released by the combustion of a unit mass of fuel in a constant volume bomb with substantially all the water condensed to the liquid state. The net heat of combustion is the heat released by the combustion of a unit mass of fuel at a constant pressure of 1 atmosphere (101.325 kPa) with the water remaining in the vapor state. It is the net or lower heat of combustion that is normally used when considering automotive fuels because exhaust gases leave the combustion chamber at a high temperature, carrying the uncondensed water vapor with them. It is usually expressed gravimetrically in terms of MJ/kg of fuel or on a volumetric basis as MJ/L.

The heat of combustion is clearly important because it is a measure of the energy content of the gasoline and will influence the fuel economy that can be achieved. Figure 12.1 illustrates the net heating value of a range of pure hydrocarbon categories that may be found in gasoline.

FIGURE 12.1 Correlation between net heating value and specific gravity for a range of hydrocarbon components.

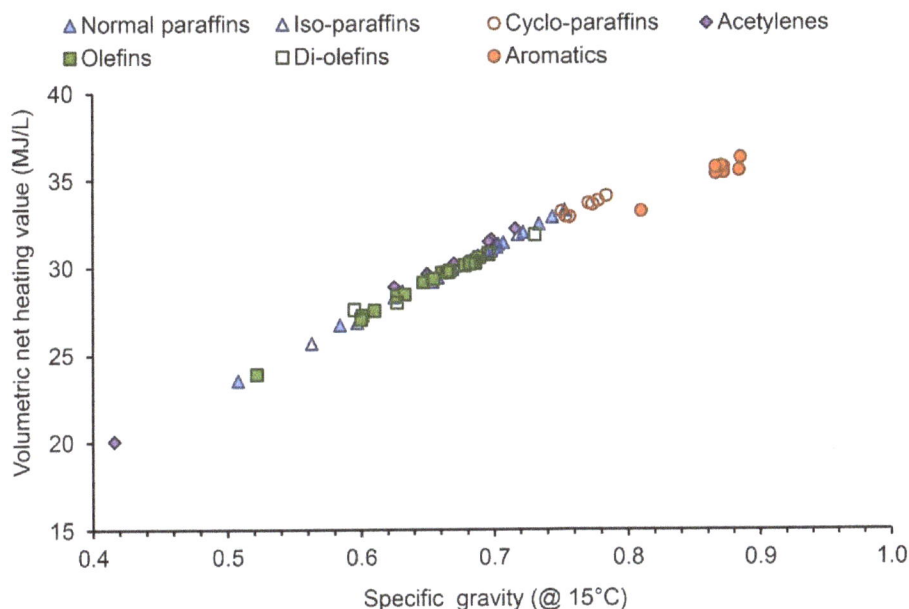

© SAE International.

It is clear that there is a very strong linear relationship between the density of the hydrocarbon and its volumetric energy content. The aromatics, although they have a higher density than a cyclo-paraffin of the same carbon number, have a lower energy content than might be expected.

Heat of combustion of gasoline can be measured using ASTM D240 [12.5], but ASTM D4809 [12.6] is preferred, as it is more repeatable. Both of these methods involve combusting a known mass of sample in a bomb calorimeter and measuring the temperature increase. If only a rough value is required, it is often more convenient to measure the density of the fuel and relate it to the heat content because there is a direct relationship between both parameters. Gasolines, which do not contain oxygenates, typically have a net heating value of between 32 and 33 MJ/L.

12.3. Composition

A number of compositional restrictions can be applied to gasoline for environmental or other reasons. As discussed in Chapter 10, the composition of the gasoline can affect its stability and the tendency for it to form deposits. To an extent this can be corrected by the use of appropriate additives as discussed in Chapter 11. However, the composition of the gasoline will also affect the composition of the exhaust emissions and the efficacy of emissions control equipment, which cannot readily be influenced by the use of fuel additives. The important compositional characteristics are discussed briefly in the following sections.

12.3.1. Hydrocarbon Composition

One of the first and most commonly restricted individual hydrocarbon species is benzene. Benzene is thought to pose a serious health risk and is possibly a human carcinogen [12.7]. Maximum allowable levels of 5%, 3%,

1%, or even 0.8% have been imposed on motor gasolines in many countries to minimize both evaporative and exhaust emissions of this compound. The benzene content is usually determined by gas chromatography using a method such as ASTM D3606 [12.8], but other methods are available, although often less repeatable [12.9].

Although benzene is usually limited because of its toxicity, the presence of other aromatic compounds can result in the production of benzene during the combustion process [12.10] or in the exhaust aftertreatment system [12.11]. The presence of aromatics can also lead to the production of polynuclear aromatic hydrocarbon (PNA), also known as polycyclic aromatic hydrocarbons (PAH), many of which are also carcinogenic.

There are strong links between the hydrocarbon composition of gasoline and the tailpipe emissions produced; this will be discussed further in the next chapter. The composition of the gasoline has also been linked to its propensity to form deposits within the induction system and the engine itself. It is the aromatic content that is believed to cause problems, and it is thus desirable to limit the aromatic content to reduce the deposit-forming tendency of the fuel. The Worldwide Fuel Charter calls for a maximum aromatic content of 40% for a category 2 fuel but for a maximum of 35% for all other categories of fuel. Knowledge of how the fuel composition affects emissions and deposit-forming tendency is useful for formulating reference fuels. For example, a fuel that is prone to producing deposits is useful as a severe reference fuel for engine tests for assessing the efficacy of fuel additives to overcome such problems.

Apart from the legislative and desirable restrictions mentioned, most gasoline specifications only require that paraffins, olefins, naphthenes, and aromatics (PONA) are measured and give only a loose control over composition. This is because individual hydrocarbons within any of these categories can differ in the way they affect performance and the way they interact with other species. Sometimes, as an extra control, specific active components such as dienes are also specified.

PONA analysis can be carried out according to ASTM procedure D6730 [12.12], which will also determine the oxygenated compounds methanol, ethanol, tertiary-butanol, methyl-tertiary-butyl-ether (MTBE), ethyl-tertiary-butyl-ether (ETBE), and tertiary-amyl-methyl-ether (TAME), as long as they are present at below 30% of the gasoline.

A frequently specified procedure for compositional analysis is the fluorescent indicator adsorption (FIA) method [12.13]. If oxygenates are present in the fuel, then they are best removed before carrying out this test. The fuel sample is introduced into a special glass adsorption column packed with activated silica gel, a small layer of which contains a mixture of fluorescent dyes. Alcohol is used to desorb the sample and move it through the column. The hydrocarbons are thus separated into aromatics, olefins, and saturates (paraffins) according to their tendency to be adsorbed. Their position is identified along the column by the fluorescent dyes visible under ultraviolet light that separate selectively with the hydrocarbon types. This method is simple but does not give a reliable separation into different types. Thus, some di-olefins, aromatics with olefinic side chains, and sulfur, nitrogen, and oxygenated compounds all appear in the aromatic band. Due to the arbitrary separation involved in the FIA test, it is difficult to get results from this test to agree with the ASTM D6730 [12.12] or other gas chromatography (GC) methods. There may also be issues due to the florescent dye used.

The compositional analysis by FIA is also useful for determining the hydrogen-to-carbon (C/H) ratio of the fuel. This parameter is required for calculating fuel consumption from emission measurements during regulated tests procedures; this is known as fuel consumption by carbon balance. The C/H ratio is given by the following equation [12.14]:

$$\frac{C}{H} = 0.527 * Sat + 0.610 * Olef + 0.1039 \, Arom$$

where Sat, Olef, and Arom represent the % volume values, from the FIA analysis, for saturates, olefins and aromatics, respectively. This is often referred to as the Sirtori equation.

12.3.2. Elemental Composition

In many jurisdictions it is not allowed to add any metals to the fuel. However, the gasoline may contain traces of metals, derived from the crude oil or from refinery catalysts, that are carried over during the refining process. The Worldwide Fuel Charter calls for no detectable metals to be present in any category of gasoline. Other elements that may act as poisons to automotive catalysts are also controlled. These are discussed next.

12.3.2.1. Lead

The use of lead antiknock compounds is no longer allowed in any motor gasolines. However, there is still a maximum lead content specified for all gasolines. The lead content is important because it may cause deposits in automotive emission control equipment; it is also a known poison of precious metal catalysts. Where low levels of lead are expected, 0.0025 to 0.025 g Pb/L then an atomic absorption spectroscopy method is used [12.15] and for higher lead levels, 0.026 to 1.3 g Pb/L, a method using iodine monochloride is employed [12.16]. An X-ray spectrometry method can be used to cover the full range of 0.0026 to 1.32 g Pb/L [12.17].

12.3.2.2. Manganese

Where the manganese antiknock additive methylcyclopentadienyl manganese tricarbonyl (MMT) (see Chapter 7, Section 7.4) is allowed to be used in gasoline, its maximum concentration is specified. The test procedure ASTM D3831 [12.18] uses atomic absorption and is suitable for estimating manganese concentrations in the range 0.25 to 40 mg Mn/L.

12.3.2.3. Phosphorus

Phosphorus is another element that will adversely affect automotive catalytic converters; therefore, the maximum level is controlled in countries where this type of vehicle is present. The phosphorus can be present as an additive since phosphorus-containing additives have been used as anti-misfire and anti-pre-ignition agents and as surfactants. These are discussed in Chapter 11, Section 11.4.3. It can be determined in gasoline by the ASTM method D3231 [12.19], which is applicable for the determination of phosphorus in the range from 0.2 to 40 mg P/L. The Worldwide Fuel Charter calls for no detectable phosphorous in all remaining gasoline categories.

12.3.2.4. Sulfur

Due to normal refining procedures, the amount of sulfur present in gasoline is generally below 0.1% wt, although some specifications allow up to 0.2% wt. The maximum sulfur level in gasoline has been low compared with most other hydrocarbon fuels. Despite this, it has been recognized that sulfur not only contributes directly to tailpipe emissions but that it will also impair the functioning of emission control equipment [12.20–12.22]. It has also been shown that fuel sulfur levels can influence benzene emissions [12.23]. The magnitude of the effect is dependent upon the fuel composition. There is thus a trend toward minimizing the permitted level of sulfur in the gasoline. Most of the US now has a limit of 15 mg/kg. Japan and the EU have introduced, with some derogations, so-called "zero" sulfur gasoline, where the maximum allowable level of sulfur is 10 mg/kg. This limit agrees with the maximum set forth in the Worldwide Fuel Charter for a category 4 gasoline. Most other countries are therefore planning a move to such zero sulfur gasolines.

Apart from its effect on catalysts, there are a number of reasons for controlling the level of sulfur present. These include its contribution to the odor of the gasoline, its corrosivity, and its general adverse environmental effect in that sulfur dioxide, a poisonous gas, and sulfuric acid, which leads to acid rain, will be emitted in the exhaust gases [12.24]. In addition to these environmentally deleterious gases, the malodorous hydrogen sulfide and other sulfides can be formed over oxidation catalysts [12.25, 12.26]. The effect of sulfur on emissions is discussed in more detail in Chapter 13, Section 13.3. It has also been blamed as a factor in increasing the deposit-forming tendency of a gasoline [12.27–12.29] and for its adverse effect on fuel system elastomers.

Sulfur can be present in a number of forms. There can be very small amounts of mercaptans (thiols) present, which are particularly undesirable because of the smell they give to the gasoline, although the various sweetening processes used during manufacture are designed to remove these compounds. It can also be present as elemental sulfur, sulfides, disulfides, polysulfides, thiophenes, and in many partially oxidized forms such as sulfoxides.

Mercaptans can be determined qualitatively by the doctor test [12.30], which is sometimes still specified even though it is an old test because it is simple to carry out and does not require expensive or complex equipment. A sample of gasoline is shaken up with sodium plumbite (Na_2PbO_2), and if the mixture slowly darkens, it indicates the presence of both mercaptans and elemental sulfur. If it gives an immediate black precipitate, it indicates that hydrogen sulfide is present, which would be highly undesirable and unusual and would necessitate washing a fresh sample with caustic soda solution and then starting again. If the mixture does not darken at all, a small quantity of flowers of sulfur is added, and if, after re-shaking, the mix darkens, the presence of mercaptans is indicated. This test will indicate the presence of about 5 to 10 mg/kg of mercaptan, and the results are reported as a pass or fail. Mercaptans can be measured quantitatively by a number of methods, including ASTM D3227 [12.31], which involves a potentiometric titration with silver nitrate.

Total sulfur in gasoline is usually determined by burning a known amount in a wick lamp and absorbing the combustion gases in hydrogen peroxide so that the sulfur dioxide is converted to sulfuric acid, the amount of which is then determined by titration with standard caustic soda solution. The lamp method, ASTM D1266 [12.32], uses a special sulfate analysis procedure and permits the determination of sulfur in concentrations as low as 5 mg/kg. Other procedures that can be used are an oxidative microcoulometry method, ASTM D3120 [12.33], which will allow determinations to as low as 3 mg/kg, or alternatively an x-ray spectrographic method, ASTM D2622 [12.34], which uses a far more expensive piece of equipment but will again allow determination down to 3 mg/kg.

12.3.3. Oxygenates

As discussed in Chapter 5, Section 5.1, the use of oxygenated compounds, such as alcohols, started in the 1920s when the high-octane quality of methanol and ethanol made them extremely valuable blend components at a time when only relatively low-octane components were available from refinery processing. The use of petroleum-derived oxygenates expanded considerably in the final quarter of the 20th century for a number of reasons. As discussed in Chapter 3, Section 3.9, they allowed refiners to extend the gasoline pool, they provided a means of overcoming an octane shortage due to the phase-out of lead antiknock compounds, and from an environmental standpoint they produced gasolines that reduced exhaust emissions of CO and HC.

Only two types of oxygenated compounds, namely alcohols and ethers, are used to any significant extent as gasoline components (i.e., at concentrations greater than 1 or 2%). The most important alcohols are methanol (MeOH), ethanol (EtOH), butanol (also known as butyl-alcohol), iso-propanol (IPA), tertiary-butanol (TBA), and mixtures of C1 to C5 alcohols. The most important ethers are MTBE, TAME, ETBE, di-iso-propyl-ether (DIPE), and mixed ethers.

Because of its poor solubility in gasoline when water is present, methanol is used with a cosolvent such as TBA. In fact, all gasolines containing alcohols require careful handling to avoid or minimize water contact. Ethers tend to be relatively trouble-free as gasoline blend components.

Specifications can include limits on individual oxygenates, usually specified as % volume, or on total oxygen content, expressed as % mass. The total amount of oxygen in the gasoline can be determined using various instruments as specified in ASTM method D5622 [12.35], and individual oxygenates can be determined by various methods, including ASTM D5986 [12.36] and ASTM D5599 [12.37], which will both make determinations from 0.1 to 20% mass of the individual oxygenates. ASTM method D4815 [12.38] is more portable but will not reliably resolve individual oxygenates present at below 0.2% mass.

12.3.4. Water

Both dissolved and free water can be present in gasoline; free water is undesirable because it can freeze and block lines, promote corrosion, worsen intake system icing, and cause emulsion and haze formation under appropriate conditions. Dissolved water in gasoline cannot usually be avoided because the components are almost always contacted with aqueous solutions during manufacture, and storage tanks invariably contain free water at the bottom. Haze formation can occur during a drop in temperature of the gasoline due to some of the water coming out of solution. Such hazes usually clear rapidly as the droplets coalesce and sink into the tank water bottoms, and this separation can be helped by the use of dehazing additives. These are discussed in Chapter 11, Section 11.2.6. The amount of water that will dissolve will depend on the composition of the gasoline. Aromatics will increase water solubility, and Figure 12.2 shows a fairly typical curve of how water solubility varies with temperature. The presence of certain oxygenated components in the gasoline can substantially increase the amount of water that will dissolve in a gasoline.

FIGURE 12.2   Water solubility against temperature.

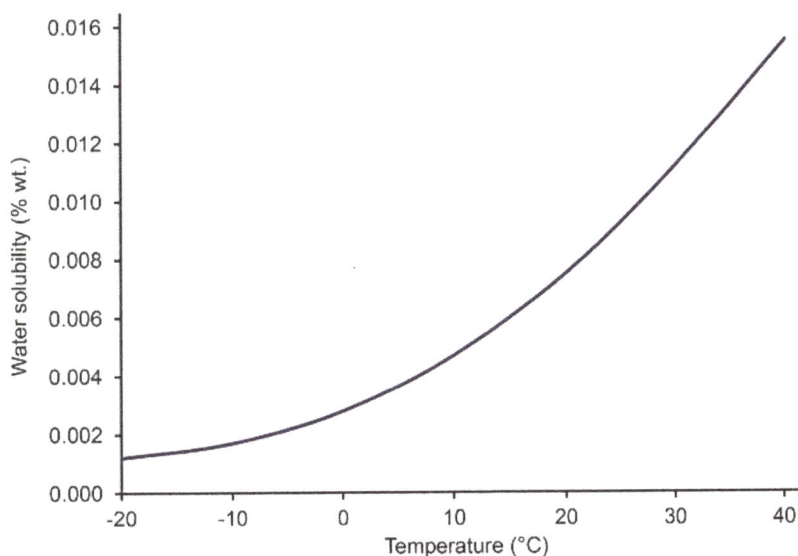

© SAE International.

12.4. Flash Point

The flash point is the lowest temperature of a sample at which application of an ignition source causes the vapor of the sample to ignite under the specified conditions of the test. For gasoline, it is so low (normally below −40°C) that it cannot be determined by any standard methods.

12.5. Surface Tension

Surface tension has an effect on flow through carburetor jets because increased venturi suction is required to overcome the surface tension at the discharge nozzle [12.39]. Increasing surface tension will reduce mass flow and air-to-fuel ratio. It increases as the temperature is reduced (as does density and viscosity). Fuel injection characteristics may be affected as atomization of the fuel is also influenced by surface tension. The droplet size distribution and the mean droplet size may thus be changed, although it has been shown [12.40, 12.41] that the practical effects are small.

Measurement can be made by a variety of methods, including the maximum bubble pressure method [12.42, 12.43] using commercial tensiometers or by methods given in ASTM D1331 [12.44]. A typical value of surface tension for gasoline is 20×10^{-3} N/m at 15°C, but this will decrease as temperature increases.

12.6. Viscosity

The fuels viscosity can also influence the flow through metering orifices because the Reynolds number is an inverse function of viscosity. In a carburetor, the coefficient of discharge is fairly constant above a critical value of the Reynolds number [12.45]. Because for most vehicles the Reynolds number is usually above this critical value, normal variations in viscosity have a relatively small but significant influence on flow. However, if the fuel viscosity is above the normal range, the Reynolds number may fall below the critical level so that non-turbulent flow occurs. Under these circumstances, increasing fuel viscosity will then decrease fuel flow and so the air-to-fuel ratio will increase. Thickening agents have been proposed that will effectively lean off the mixture strength in this way for an improvement in fuel economy [12.46]. With fuel injection systems, the viscosity will again influence droplet size and mean droplet diameter with increasing viscosity increasing droplet size. Increasing fuel viscosity will increase droplet size, although the practical variations that would be found in gasoline would not make an appreciable difference [12.41, 12.44]. The closed loop control system will maintain the required air-to-fuel ratio, and any variation in viscosity would therefore not affect engine performance.

Viscosity is measured by ASTM D445 [12.47], in which the flow under gravity through a calibrated capillary tube is measured. A typical range for gasoline is 0.5 to 0.6 centistokes.

12.7. Conductivity

The conductivity of a fuel is important in determining whether a charge of static electricity will build up during pumping operations. A high-conductivity fuel will dissipate an electrical charge more rapidly than a low-conductivity fuel and reduce the risk of an electrical discharge causing a fire. Explosions or fires will only occur if such a discharge occurs in the presence of an air-fuel mixture that is within the combustible range. The conductivity affects the rate at which charged ions are removed and is measured in picosiemen/meter, one unit of which is equivalent to one conductivity unit (CU). The time taken for the charge to fall to half its value (half time value) and the time to drop to approximately 37% of its value (relaxation time) are sometimes used to indicate whether or not static charges are likely to be a hazard.

Values of conductivity for gasoline are difficult to quote because they depend on the hydrocarbon composition and whether there are oxygenates and/or surfactants present. A pure dry hydrocarbon-only gasoline will have a conductivity of about 1 CU, but this can increase very rapidly with compositional changes and the use of certain additives (as discussed in Chapter 11, Section 11.2.4). A Standard test methods for determining conductivity is ASTM D4308 [12.48].

12.8. Corrosivity

Corrosion is a problem not only because of the damage it does to equipment but also because dissolved metals such as copper can catalyze oxidation reactions and give rise to excessive deposit formation. In addition, the products of corrosion can block filters and orifices in a vehicle's fuel system and increase wear rates. The extent of corrosion will depend on the water content of a gasoline, the type and quantity of oxygenates present, the level and type of sulfur compounds present, and most importantly on the efficacy of corrosion inhibitor additives that are present (as discussed in Chapter 11, Sections 11.2.2 and 11.3.1).

The corrosivity of gasoline to copper-containing parts of a fuel system is evaluated using the copper strip corrosion test [12.49]. A polished copper strip is immersed in a sample of the gasoline at fixed temperature and for a fixed period of time (usually 3 hours with the gasoline at a temperature of 50°C) and the degree of corrosion assessed by comparing the sample with standard ASTM colors representing various degrees of tarnish.

Corrosion of steel is assessed by using a modified ASTM D665 [12.50] test procedure or the National Association of Corrosion Engineers (NACE) procedure [12.51].

12.9. **Freezing Point**

This is also not very relevant for motor gasolines, although a maximum temperature of –60°C is usually specified for aviation gasolines using ASTM Method D2386 [12.52]. It is the temperature at which crystals of hydrocarbons formed on cooling the fuel just disappear when the temperature is allowed to rise. Motor gasoline typically has a freezing point of about –70°C, depending on its composition.

12.10. **Appearance**

Gasoline should always be clear and bright, without any suspended or other particulate matter and with no apparent free water. Particulate matter can block filters, fine passages, and metering orifices; water can block lines at ambient temperatures below freezing, increase intake system icing, and worsen corrosion. The color of gasoline is specified on a global basis for aviation gasolines where the gasoline is dyed different colors for easy identification of grade. However, certain local regulations specify a color for either grade identification or for taxation purposes.

The standard procedures for evaluating the appearance of gasoline are contained in ASTM D4176 [12.53] in which Method A is for field use at ambient temperature and Method B is for use in the laboratory at a controlled temperature of 25°C. In both cases a 500 mL sample of the fuel is swirled in a clean, glass jar and examined for visual sediment or water droplets just below the vortex formed by the swirling. Any sign of haziness or dullness is also noted. The results are recorded as a pass or a fail.

If free water and/or sediment are evident, it may be worthwhile to determine them quantitatively by centrifuging, as in ASTM D2709 [12.54], with particulate contamination determined by a filtration test such as ASTM D2276 [12.55].

References

12.1. ASTM International, "Standard Test Method for Density, Relative Density, or API Gravity of Crude Petroleum and Liquid Petroleum Products by Hydrometer Method," ASTM D1298-12b(2017)e1, ASTM International, 2023.

12.2. International Organization for Standardization, "Crude Petroleum and Liquid Petroleum Products—Laboratory Determination of Density—Hydrometer Method," ISO 3675:1998, International Organization for Standardization, 1998.

12.3. Energy Institute, "Crude Petroleum and Petroleum Products—Determination of Density—Oscillating U-Tube Method," IP 365:1997 (R2004), Energy Institute, 2004.

12.4. ASTM International, "Standard Test Method for Density, Relative Density, and API Gravity of Liquids by Digital Density Meter," ASTM D4052-22, ASTM International, 2022.

12.5. ASTM International, "Standard Test Method for Heat of Combustion of Liquid Hydrocarbon Fuels by Bomb Calorimeter," ASTM D240-19, ASTM International, 2019.

12.6. ASTM International, "Standard Test Method for Heat of Combustion of Liquid Hydrocarbon Fuels by Bomb Calorimeter (Precision Method)," ASTM D4809-18, ASTM International, 2018.

12.7. Environmental Protection Agency, "Health Effects of Benzene: A Review," EPA 560/5-76-003, Environmental Protection Agency, 1976.

12.8.   ASTM International, "Standard Test Method for Determination of Benzene and Toluene in Finished Motor and Aviation Gasoline by Gas Chromatography," ASTM D3606-22, ASTM International, 2022.

12.9.   Pauls, R.E., Weight, G.J., and Munowitz, P.S., "A Comparison of Methods to Determine Benzene in Gasoline Boiling Range Material," *Journal of Chromatographic Science* 30, no. 1 (1992): 32-39.

12.10.   Marshall, W.F. and Gurney, M.D., "Effect of Gasoline Composition on Emissions of Aromatic Hydrocarbons," SAE Technical Paper 892076, 1989, doi:https://doi.org/10.4271/892076.

12.11.   Nagel, H., Frey, R., Hartgerink, C., Rikeit, H.-E. et al., "On-Line Analysis of Individual Aromatic Hydrocarbons in Automotive Exhaust: Dealkylation of the Aromatic Hydrocarbons in the Catalytic Converter," SAE Technical Paper 971606, 1997, doi:https://doi.org/10.4271/971606.

12.12.   ASTM International, "Standard Test Method for Determination of Individual Components in Spark Ignition Engine Fuels by 100-Metre Capillary (with Precolumn) High-Resolution Gas Chromatography," ASTM D6730-22, ASTM International, 2022.

12.13.   ASTM International, "Standard Test Method for Hydrocarbon Types in Liquid Petroleum Products by Fluorescent Indicator Adsorption," ASTM D1319-20a, ASTM International, 2020.

12.14.   Sirtori, S., Garibaldi, P., and Vicenzetto, F.A., "Prediction of the Combustion Properties of Gasolines from the Analysis of Their Composition," SAE Technical Paper 741058, 1974, doi:https://doi.org/10.4271/741058.

12.15.   ASTM International, "Standard Test Method for Lead in Gasoline by Atomic Absorption Spectroscopy," ASTM D3237-22, ASTM International, 2022.

12.16.   International Organization for Standardization, "Petroleum Products—Determination of Lead Content of Gasoline—Iodine Monochloride Method," ISO 3830:1993, International Organization for Standardization, 1996.

12.17.   ASTM International, "Standard Test Methods for Lead in Gasoline by X-Ray Spectroscopy," ASTM D5059-21, ASTM International, 2021.

12.18.   ASTM International, "Standard Test Method for Manganese in Gasoline by Atomic Absorption Spectroscopy," ASTM D3831-22, ASTM International, 2022.

12.19.   ASTM International, "Standard Test Method for Phosphorus in Gasoline," ASTM D3231-18, ASTM International, 2018.

12.20.   Benson, J.D., Burns, V., Koehl, W.J., Gorse, R.A. et al., "Effects of Gasoline Sulfur Level on Mass Exhaust Emissions—Auto/ Oil Air Quality Improvement Research Program," SAE Technical Paper 912323, 1991, doi:https://doi.org/10.4271/912323.

12.21.   Summers, J.C., Skowron, J.F., Williamson, W.B., and Mitchell, K.I., "Fuel Sulfur Effects on Automotive Catalyst Performance," SAE Technical Paper 920558, 1992, doi:https://doi.org/10.4271/920558.

12.22.   Tauster, S., Rabinowitz, H., and Heck, R., "Understanding Sulfur Interaction Key to OBD of Low Emission Vehicles," SAE Technical Paper 2000-01-2929, 2000, doi:https://doi.org/10.4271/2000-01-2929.

12.23.   Akimoto, J., Kaneko, T., Ichikawa, T., Hamatani, K. et al., "The Effects of Sulfur on Emissions from a S.I. Engine," SAE Technical Paper 961219, 1996, doi:https://doi.org/10.4271/961219.

12.24.   Hammerle, R.H. and Mikkor, M., "Some Phenomena Which Control Sulfuric Acid Emission from Automotive Catalysts," SAE Technical Paper 750097, 1975, doi:https://doi.org/10.4271/750097.

12.25.   Barnes, G.J. and Summers, J.C., "Hydrogen Sulfide Formation Over Automotive Oxidation Catalysts," SAE Technical Paper 750093, 1975, doi:https://doi.org/10.4271/750093.

12.26.   Cadle, S.H. and Mulawa, P.A., "Sulfide Emissions from Catalyst-Equipped Cars," SAE Technical Paper 780200, 1978, doi:https://doi.org/10.4271/780200.

12.27.   Gibbs, L.M. and Richardson, C.E., "Carburetor Deposits and Their Control," SAE Technical Paper 790202, 1979, doi:https://doi.org/10.4271/790202.

12.28.   Tseregounis, S.I., "Effects of Sulfur Chemistry on Deposits Derived from a Gasoline Oxidized at 100°C," SAE Technical Paper 902106, 1990, doi:https://doi.org/10.4271/790202.

12.29.   Martin, P. and Bustamante, D., "Deposit Forming Tendency of Gasoline Polar Compounds," SAE Technical Paper 932742, 1993, doi:https://doi.org/10.4271/790202.

12.30.   ASTM International, "Standard Test Method for Qualitative Analysis for Active Sulfur Species in Fuels and Solvents (Doctor Test)," ASTM D4952-12(2017), ASTM International, 2017.

12.31.   ASTM International, "Standard Test Method for (Thiol Mercaptan) Sulfur in Gasoline, Kerosine, Aviation Turbine, and Distillate Fuels (Potentiometric Method)," ASTM D3227-16, ASTM International, 2016.

12.32. ASTM International, "Standard Test Method for Sulfur in Petroleum Products (Lamp Method)," ASTM D1266-18, ASTM International, 2018.

12.33. ASTM International, "Standard Test Method for Trace Quantities of Sulfur in Light Liquid Petroleum Hydrocarbons by Oxidative Microcoulometry," ASTM D3120-08(2019), ASTM International, 2019.

12.34. ASTM International, "Standard Test Method for Sulfur in Petroleum Products by Wavelength Dispersive X-Ray Fluorescence Spectrometry," ASTM D2622-21, ASTM International, 2022.

12.35. ASTM International, "Standard Test Methods for Determination of Total Oxygen in Gasoline and Methanol Fuels by Reductive Pyrolysis," ASTM D5622-17, ASTM International, 2017.

12.36. ASTM International, "Standard Test Method for Determination of Oxygenates, Benzene, Toluene, C8-C12 Aromatics and Total Aromatics in Finished Gasoline by Gas Chromatography/Fourier Transform Infrared Spectroscopy," ASTM D5986-96(2019), ASTM International, 2019.

12.37. ASTM International, "Standard Test Method for Determination of Oxygenates in Gasoline by Gas Chromatography and Oxygen Selective Flame Ionization Detection," ASTM D5599-22, ASTM International, 2022.

12.38. ASTM International, "Standard Test Method for Determination of MTBE, ETBE, TAME, DIPE, Tertiary-Amyl Alcohol and C1 to C4 Alcohols in Gasoline by Gas Chromatography," ASTM D4815-22, ASTM International, 2022.

12.39. Bolt, J.A., Derezinski, S.J., and Harrington, D.L., "Influence of Fuel Properties on Metering in Carburetors," SAE Technical Paper 710207, 1971, doi:https://doi.org/10.4271/790202.

12.40. Williams, P. and Beckwith, P., "Correlation between the Liquid Fuel Properties Density, Viscosity and Surface Tension and the Drop Sizes Produced by an SI Engine Pintle-Type Port Fuel Injector," SAE Technical Paper 941864, 1994, doi:https://doi.org/10.4271/941864.

12.41. Williams, P. and Beckwith, P., "The Effect of Fuel Composition and Manifold Conditions upon Spray Formation from an SI Engine Pintle Injector," SAE Technical Paper 941865, 1994, doi:https://doi.org/10.4271/941865.

12.42. Mysels, K.J., "The Maximum Bubble Pressure Method of Measuring Surface Tension, Revisited," *Colloids and Surfaces* 43, no. 2 (1990): 241-262.

12.43. Levi, O., Freud, R., and Sher, E., "Dynamic Surface Tension of Gasoline and Alcohol Fuel Blends," *Atomization and Sprays* 27, no. 1 (2017): 1-5.

12.44. ASTM International, "Standard Test Method for Dynamic Surface Tension by the Fast-Bubble Technique (Withdrawn 2016)," ASTM D3825-09, ASTM International, 2016.

12.45. Lichtarowicz, A., Duggins, R.K., and Markland, E., "Discharge Coefficients for Incompressible Non-Cavitating Flow through Long Orifices," *Journal of Mechanical Engineering Science* 7, no. 2 (1965): 210-219.

12.46. Reuter, R.M. and Eckert, G.W., Method of operating an internal combustion engine and motor fuel therefor. U.S. Patent 3,164,138, 1965.

12.47. ASTM International, "Standard Test Method for Kinematic Viscosity of Transparent and Opaque Liquids (and Calculation of Dynamic Viscosity)," ASTM D445-21e2, ASTM International, 2022.

12.48. ASTM International, "Standard Test Method for Electrical Conductivity of Liquid Hydrocarbons by Precision Meter," ASTM D4308-21, ASTM International, 2022.

12.49. ASTM International, "Standard Test Method for Corrosiveness to Copper from Petroleum Products by Copper Strip Test," ASTM D130-19, ASTM International, 2019.

12.50. ASTM International, "Standard Test Method for Rust-Preventing Characteristics of Inhibited Mineral Oil in the Presence of Water," ASTM D665-19, ASTM International, 2020.

12.51. National Association of Corrosion Engineers, "NACE Standard TM0172-2001," 2001.

12.52. ASTM International, "Standard Test Method for Freezing Point of Aviation Fuels," ASTM D2386-19, ASTM International, 2019.

12.53. ASTM International, "Standard Test Method for Free Water and Particulate Contamination in Distillate Fuels (Visual Inspection Procedures)," ASTM D4176-22, ASTM International, 2022.

12.54. ASTM International, "Standard Test Method for Water and Sediment in Middle Distillate Fuels by Centrifuge," ASTM D2709-22, ASTM International, 2022.

12.55. ASTM International, "Standard Test Method for Particulate Contaminant in Aviation Fuel by Line Sampling," ASTM D2276-22, ASTM International, 2022.

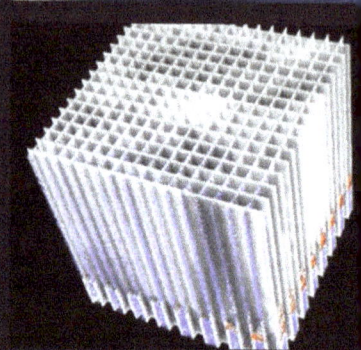

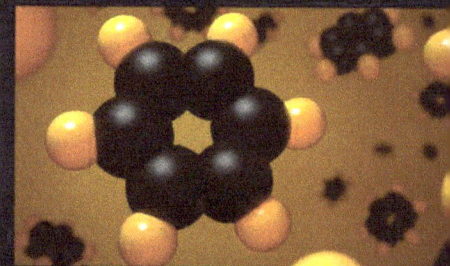

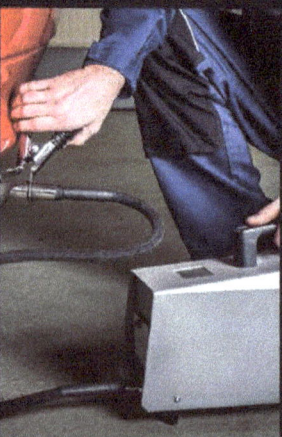

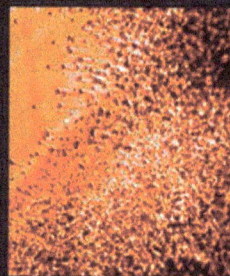

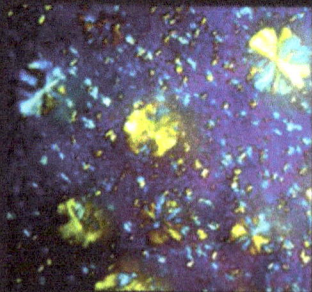

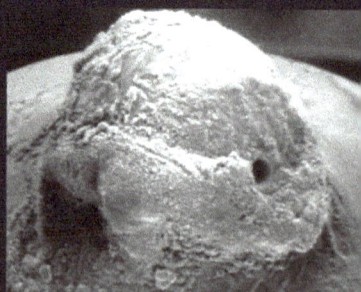

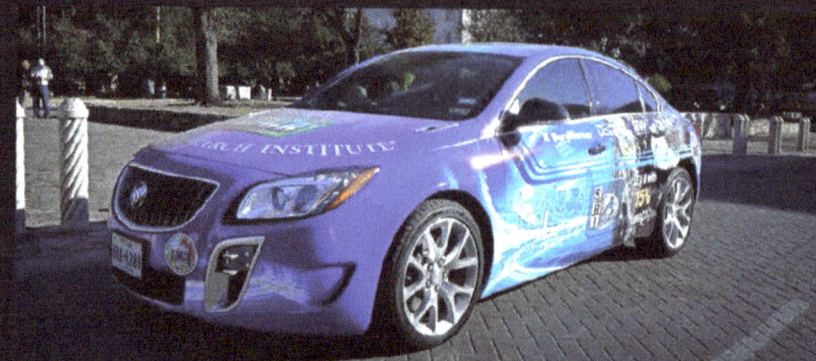

Influence of Gasoline Characteristics on Emissions

The topographical and climatic conditions that made the Los Angeles basin such an attractive place to live proved to have unwanted side-effects. A key feature of both the Los Angeles basin and Santiago, Chile, is a persistent thermal inversion [13.1]. This inversion traps pollutants formed with the basin, forming the now infamous photochemical smog. In 1947, the smog problem became so great that the community forced state action and formed the first countywide Air Pollution Control District [13.2]. The rapidly increasing personal mobility of the 1950s brought with it increasing automotive emissions and health concerns [13.3, 13.4]. The oxides of nitrogen present in automotive exhaust emissions were soon identified as ozone precursors [13.5], which acted as a strong oxidant for the 922 tons of hydrocarbons that automotive exhaust emissions at the time were added to the atmosphere every day [13.6]. It was estimated that this would have to be curtailed by about 60% to return to the relatively smog-free conditions that existed before 1945 [13.7].

Concerns not only among the populace but also within the industry led to the establishment of the Inter-Industry Emission Control Program (IIEC) in 1967. This was initially set up in the US by the Ford Motor Company and Mobil Oil Corporation. By mid-1968, the list of participants had grown to include Amoco Oil Company, Atlantic Richfield Company, Fiat, S.p.A., Ford Motor Company, Marathon Oil Company, Mitsubishi Motors Corp., Mobil Oil Corp., Nissan Motor Co., Ltd., The Standard Oil Co. (Ohio), Sun Oil Company, and Toyo Kogyo Co., Ltd. "Know how" agreements were also executed with Volkswagenwerk A.G. and Toyota Motor Co., Ltd. The aim of the IIEC was to accelerate the development of a virtually emission-free, gasoline-powered vehicle by combining expertise from both the petroleum and automotive industries. The program was soon divided into two periods, known as IIEC-1 and IIEC-2, as its goals were broadened to include energy conservation [13.8]. This was the first program of its kind, bringing together the petroleum and automotive industries on a global basis. The basis of emission-control technology was born of this program's cooperation.

Twenty years after the formation of the Air Pollution Control District, the then California Governor, Ronald Regan, signed into law the Mulford-Carrell Air Resources Act, which brought together the Bureau of Air Sanitation and the Motor Vehicle Pollution Control Board to form the California Air Resources Board (CARB), with the responsibility for the control of emissions from mobile sources [13.9]. The CARB received waivers from the federal government to establish its own emission standards for motor vehicles and has since become a leader in setting such standards.

It is interesting to note that one of the unexpected benefits of automotive emission control has been the reduction in fatalities from motor vehicle exhaust gas in the US. The decrease in accidents and avoided suicides together with the projected cancer risk avoidance show that, in terms of mortality, the unanticipated benefits could outweigh the intended benefits [13.10].

Although the initial focus of emission control was to eliminate the obvious pollution that was becoming known as a health hazard, attention is now focusing on the toxicity of the pollutants emitted by vehicles, many of which are carcinogenic or mutagenic.

13.1. Development of Emission Legislation and Fuel Quality Regulations

Los Angeles was not the only place to suffer from pollution episodes that could be related to road transport [13.11], but the strength of an organization such as the CARB gave impetus to the formation of control measures. The following subsections will briefly outline how automotive emission legislation and fuel quality standards have evolved globally in over half a century of development.

13.1.1. Development in the US

In the US the original legislated national automotive exhaust emission limits were set by the Clean Air Act (CAA) of 1968. This was updated in 1970 to become the Clean Air Act Amendments (CAAA), which required a 90% reduction from the then current levels of exhaust emission. The aim was that this should be achieved by 1975–1976, but following discussions between the industry and the Environmental Protection Agency (EPA) implementation were delayed to give time for the required technology to be developed. Interim standards were set for 1975, which for most vehicle technologies could only be met by the use of exhaust aftertreatment in the form of oxidation catalysts. This required the first major change to fuel quality regulation in that unleaded fuel was required for such vehicles, which necessitated changes to refinery practices in order to maintain the octane quality the customers demanded.

A further revision to the CAAA, in 1977, incorporated National Ambient Air Quality Standards (NAAQS) that were formulated by the EPA. The NAAQS set the level of air cleanliness required throughout the US, with areas that failed to meet these standards being declared nonattainment areas. Such areas were required to develop state implementation plans (SIPs) to bring the area into compliance. The SIPs cover all emission sources and must be approved. The EPA is empowered to develop a federal implementation plan (FIP) if the SIP is not approved. This set a firm target for achieving the 90% reduction in volatile organic compounds (VOC) and CO emissions for 1980 and 1981, respectively. It was considered that the technology to meet the 90% NO_X reduction was not yet available, and a target of 75% NO_X reduction was included for 1981. These targets were met by the adoption of three-way catalysts (TWCs), so-called because they controlled all the three regulated pollutants, HC, CO, and NO_X.

These new limits could only be met by the use of exhaust aftertreatment, so the new regulations incorporated a requirement that the aftertreatment system be effective over the "useful life" of the vehicle, which was initially set at 80,000 km.

An update to the CAAA was signed into law in 1990. Following discussions between the EPA and the fuel and motor industries, detailed rules were drawn up to implement the legislation. The most important of these were the Tier One exhaust emission limits, to be followed by Tier Two limits from 2004, for light-duty vehicles (LDVs), evaporative emission test procedures and limits, plus the rules for reformulated gasoline; Tier 3 was to follow 10 years later. However, the state of California believed that stricter measures were required and implemented its own limits. During the 1990s, the 12 eastern states that comprised the Ozone Transport Region (OTR) adopted some elements of the California legislation. To avoid the complexity of different emission limits applying in different federal states, a voluntary agreement was developed that was known as the National Low Emissions Vehicles (NLEV) program. The first NLEVs reached consumers in New England in 1999 and the rest of the country in 2001. These vehicles operated with a NO_X standard that was 50% below the Tier One standards. The NLEV agreement also called for a 17% NO_X reduction, from Tier One, for light-duty trucks. The tightening of the emission standards can be seen visually in Figure 13.1. The dashed lines to the left of the chart show estimated average emissions prior to legislation while the solid lines show the legislated levels up to the introduction of the Tier Two limits. Note that the US limits are set in mixed units of g/mile but have been converted to g/km for comparison with other regions.

FIGURE 13.1   Evolution of US emission limits until 1994.

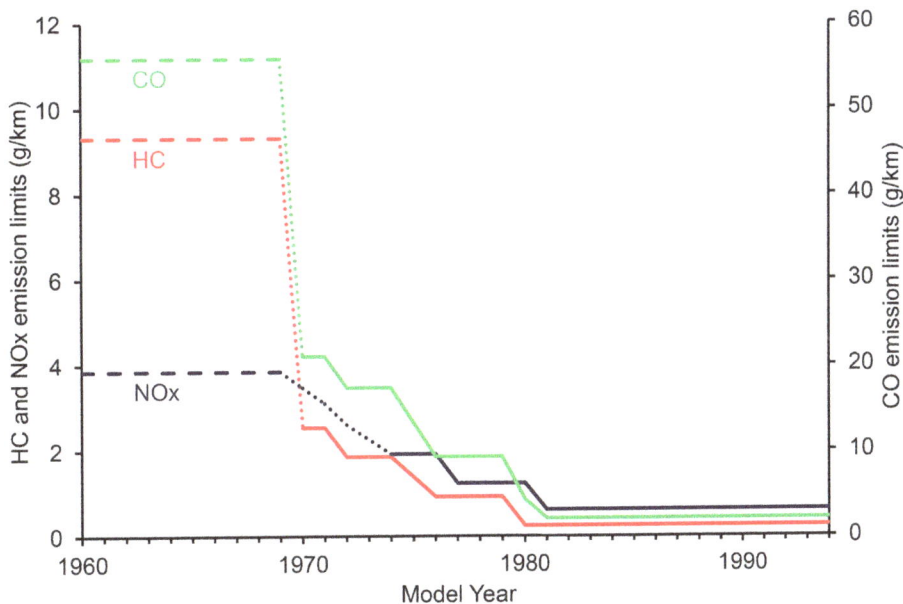

In 1998, as required by the CAAA of 1990, the EPA issued the Tier Two Report to Congress. The report contained strong justification for tighter tailpipe emission standards beginning in 2004. The EPA also determined gasoline sulfur reductions necessary to enable the full performance of the necessary emission-control devices. The new Tier Two standards would be phased in from 2004 to 2008. The Tier Two regulations also introduced the idea of fleet averaging and the "bin" structure. The bins are a series of slightly more severe limits as the bin number decreases, until bin 1 is zero emissions. The limits for each bin are shown in Table 13.1.

TABLE 13.1 Tier two emission limits converted to g/km.

Bin no.	Emission limits (g/km)				
	NO$_X$	NMOG	CO	HCHO	PM
11	0.559	0.174	4.536	0.020	0.075
10	0.373	0.097 (0.143)	2.610 (3.977)	0.011 (0.017)	0.050
9	0.186	0.056 (0.112)	2.610	0.011	0.037
8	0.124	0.078 (0.097)	2.610	0.011	0.012
7	0.093	0.056	2.610	0.011	0.012
6	0.062	0.056	2.610	0.011	0.006
5	0.043	0.056	2.610	0.011	0.006
4	0.025	0.043	1.305	0.007	0.006
3	0.019	0.034	1.305	0.007	0.006
2	0.012	0.006	1.305	0.002	0.006
1	0.000	0.000	0.000	0.000	0.000

© SAE International.

In Table 13.1, where a second, higher value is given, this applied only to heavy light-duty trucks (HLDTs) until 2008, when this relaxed limit expired. Bin 11 was only applicable to medium-duty passenger vehicles (MDPVs) and along with bins 9 and 10 were deleted in 2006, except for HLDTs, for which they expired in 2008. These emission limits apply equally to gasoline and diesel-powered vehicles.

Under the bins system, the motor manufacturer selects a bin to which they will produce a family of vehicles. A test group must meet all of the limits within that bin. The manufacturer must also ensure that the corporate average emissions meet the 0.043 g/km NO$_X$ limit put in place to ensure air quality targets should be met. The number of vehicles produced in each family and complying with each bin must be controlled to meet this corporate average.

In 2014 the Tier 3 standards came into law, to be phased-in from 2017 through 2025. Tier 3 followed a similar structure to Tier 2 with the idea of fleet averaging and the bin structure. For Tier 3 the limits were set in terms of NO$_X$ + NMOG rather than for the individual pollutants and the bins were named according to the NO$_X$ + NMOG value expressed in terms of mg/mile. The Tier 3 Bin 160 and Bin 30 are equivalent to the old Tier 2 Bin 5 and Tiers 2 Bin 2, respectively, shown in Table 13.2, all values have been converted to SI units of mg/km.

TABLE 13.2 Tier 3 emission limits converted to mg/km.

Bin no.	Emission limits (mg/km)			
	NO$_X$ + NMOG	CO	HCHO	PM
Bin 160	99.4	2609.8	2.5	1.9
Bin 125	77.7	1304.9	2.5	1.9
Bin 70	43.5	1056.3	2.5	1.9
Bin 50	31.1	1056.3	2.5	1.9
Bin 30	18.6	621.4	2.5	1.9
Bin 20	12.4	621.4	2.5	1.9
Bin 0	0.0	0.0	0.0	0.0

© SAE International.

Vehicles had to comply with these limits for a vehicle life of 150,000 miles (241,402 km) to demonstrate the durability of the emission-control system. The emission testing was performed with E10 fuel and a maximum gasoline sulfur content of 10 ppm.

13.1.2. Development in Europe

Emission regulations in Europe were formulated in the 1970s and 1980s by the United Nations Economic Commission for Europe (UN-ECE) through its technical advisory body, the Group Rapporteurs Pollution and Energy (GRPE). Its role was to produce model standards that may be adopted by member nations, but it has no power to enforce compliance. In its early years, the European Union generally adopted regulations that were technically identical with the ECE equivalents. This position has changed over time with the Motor Vehicle Emissions Group (MVEG), an expert group of the European Commission, gradually assuming a major role in formulating automotive emission standards. The GRPE is now unlikely to adopt any proposal which has not been put forward by the MVEG.

The European Union (EU) publishes its standards as directives that have the force of law within EU member states, under the provisions of the Treaty of Rome. The EU countries may not prohibit the marketing of vehicles that comply with the provisions of the directives but may prohibit vehicles that do not comply. With the introduction of the Consolidated Emissions Directive in June 1991 (91/441/EEC), implementation became mandatory for all EU member states and was no longer left to the discretion of individual national governments. With the later Directive 93/59/EEC, these became known as the Euro 1 limits. The EU Council of Ministers adopted Directive 94/12/EC and 96/69/EC, applying more stringent emission limits from 1996 that are known as the Euro 2 limits. Contrary to earlier standards, no conformity of production allowance was permitted over and above the type approval limits.

As a result of Directive 94/12/EC, the European Commission set up the European Programme on Emissions, Fuels and Engines (EPEFE). This initiative involved legislators, the European Parliament, academia, consumer groups, the fuel industry, and automotive manufacturers. This program resulted in the submission of proposals for the further regulation of vehicle emissions to take effect during the period 2000/2005/2010. The proposals also included the introduction of requirements for onboard diagnostic systems, more rigorous periodic inspections, a tightening of evaporative emission limits, and, importantly, constraints on fuel quality specifications.

A common position was reached regarding the first step, and this became EU Directive 98/69/EC, which set the Euro 3 and Euro 4 limits that came into force from 2000 and 2005, respectively. The regulations were updated by Directive 2002/80/EC, which included new specifications for reference fuels for testing LDVs. These new regulations had separate limits for gasoline- and diesel-powered vehicles.

Further reductions were considered necessary, and Regulation (EC) 715/2007 set forth a further two levels of reduction, known as Euro 5 and Euro 6. This was later amended by Commission Regulation (EC) No. 692/2008. The evolution of the limits for the criteria pollutants is illustrated in Table 13.3.

TABLE 13.3 Evolution of EU emission limits.

| | Year | Emission limit (g/km) | | | | | | | PN (#/km) |
		CO	HC	NMHC	HC + NO$_x$	NO$_x$	PM	NH$_3$	
Euro 1	1992	2.72	—	—	0.97	—	0.140	—	—
Euro 2	1996	2.2	—	—	0.50	—	—	—	—
Euro 3	2000	2.3	0.20	—	—	0.15	—	—	—
Euro 4	2005	1.0	0.10	—	—	0.08	—	—	—
Euro 5	2009	1.0	0.10	0.068	—	0.06	0.005	—	—
Euro 6	2014	1.0	0.10	0.068	—	0.06	0.005	—	6.0 E^{11}
Euro 7*	2025	0.5	0.10	0.068	—	0.06	0.0045	0.02	6.0 E^{11}

* Proposal yet to be ratified by the European Parliament and the Council.

The Euro 1 limits applied equally to gasoline- and diesel-powered vehicles, so the values stated in the table apply only to gasoline-powered vehicles. The situation regarding diesel-powered vehicles is discussed in Chapter 21. The Euro 5 limits reintroduced a particulate mass limit for gasoline-powered vehicles, and for the Euro 6 standard there is an additional requirement for gasoline-powered vehicles using direct injection technology to emit less than $6*10^{11}$ particles per km. As in the US, the test fuel is E10 with a maximum sulfur content of 10 mg/kg. The proposed Euro 7 limits are technology and fuel neutral, thus while they appear little changed for gasoline vehicles it does represent a tightening of the emission limits for diesel vehicles, as is explained in Chapter 21. Besides introducing a limit on ammonia emissions, it also includes limits on PM emissions from brakes; this is vehicle related matter and outside the scope of this book. The Euro 7 proposal is meeting a lot of opposition within the industry [13.12, 13.13] and have yet to be ratified.

13.1.3. Development in Japan

In Japan, the first controls on vehicle emissions came into place in 1966. These early regulations simply limited the CO emissions. The first long-term plan was put forward by the Ministry of Transport (MOT) in 1970 [13.14]. These proposals limited the emissions of the criteria pollutants (i.e., CO, HC, and NO_X) with separate limits for hot start and cold start cycles. However, in 1971 the Central Council for Environmental Pollution Control (CCEPC) recommended more stringent exhaust emission standards. The limits that took effect in 1975 required the use of catalysts on gasoline vehicles. These limits were further tightened in 1978, by additional NO_X reductions. Subsequent revisions to the test procedures effectively made these limits harder to achieve.

In December 1989, the CCEPC recommended new emission limits with both short- and long-term targets. The aim was to set the most stringent standards that were technologically feasible and to apply the same standards for both gasoline- and diesel-fueled vehicles. Based on this proposal, the MOT revised the emission regulations in May 1991 to meet the short-term limits. The major changes to apply to gasoline passenger vehicles were a revision of the test cycles and measurement modes. The 10-mode light-duty test cycle was modified to include a high-speed section. This now referred to as the 10–15 mode cycle and applies to both gasoline and diesel engine LDVs. Emission measurement results are now expressed as g/kWh.

A joint MOT/Ministry of International Trade and Industry (MITI)/Environmental Agency (EA) proposal to further reduce NO_X in urban areas took effect from December 1993. Toward the end of 1997, the Central Environment Council (CEC) issued a report, "Future Policy for Motor Vehicle Exhaust Emissions Reduction (Second Report)," and the EA embarked on a revision of the existing regulations for motor vehicles to take effect between 2000 and 2002, referred to as New Short-Term Regulations. Amendments include an extension of durability requirements, the introduction of onboard diagnostics, and the adoption of a sealed housing for evaporative determination (SHED) procedure for evaporative emission control.

In addition to the emission legislation, the CEC recommended new long-term regulations for fuels. The proposed fuel quality requirements included a reduction in the gasoline sulfur limit to less than 50 mg/kg by the end of 2004. This was further reduced to 10 mg/kg by 2007, this is now the standard set by JIS K 2202.

Further amendments, New Long-Term Regulations, phase in by 2010, and the Post New Long-Term Regulations, which were phased in by 2010. This has been followed by the Japan 2018 Target Emission Regulations which are illustrated in Table 13.4, testing is performed according to the World-Harmonized Light-Duty Transient Cycle (WLTC).

TABLE 13.4 Japanese future criteria pollutant emission standards.

Test	Emission limit (g/km)			
	CO	NMHC	NO$_X$	PM
Passenger cars	1.15	0.10	0.05	0.005
Vehicles with GVW ≤ 1700 kg	1.15	0.10	0.05	0.005
Vehicles with GVW ≤ 3500 kg	2.25	0.15	0.07	0.007

13.1.4. Development in the Rest of the World

Other areas across the globe did not initially have the same traffic densities as the three major areas discussed above. This has now changed, particularly in economies such as Brazil, China, India, and the Russian Federation. These countries have had to address the issues of increasing automotive-generated air pollution. The rate of growth in traffic density in some of the big cities in these economies far outweighs what has taken place in the US, Europe, and Japan. The introduction and tightening of emission legislation in these countries is thus often far more rapid than has been experienced by the vanguard. The approach taken by other countries is briefly outlined below.

Motor vehicle emission control became effective in Canada from 1997. In 2002, under the Canadian Environmental Protection Act, 1999, the Canadian government passed new on-road vehicle and engine emission regulations. These new regulations established standards and procedures in line with those of the US. Central and South American countries also tend to follow the US procedures and limits, although different countries are at different stages of the development of these standards.

Australia and New Zealand follow European procedures and are broadly in line with the EU limits.

China and India have procedures equivalent to the EU procedures, but implementation of the limit values lags Europe and is also different for the country as a whole and for the largest metropolitan areas. Other Asian countries are adopting the EU procedures and regulations, although the timing is different from country to country. While these countries are currently some years behind Europe, many have a timetable for bringing their regulations in to line with Europe. The exception is Taiwan, which is following the US procedures, with limits slightly behind the US stages.

Central and Eastern European countries are adopting the EU regulations. The Russian Federation currently has its own limits. These are less severe than the EU limits.

South Africa, the Middle East, and other African countries follow EU procedures, although different countries are at different stages of the reduction steps that have taken place in Europe. Many of the African countries that have no indigenous vehicle manufacturing capability rely on the fact that their imported vehicles comply with relevant standards of the country of manufacture when the vehicles were first put into production. It is thus anticipated that most countries outside of the Americas and Japan will eventually adopt the Euro 7 regulations.

13.2. The Introduction of Reformulated Gasolines

Reformulated gasolines (RFG), as they are generally called in the US, are designed to reduce both exhaust and evaporative emissions from vehicles. In Europe and elsewhere, these fuels were sometimes termed clean gasolines. They were first developed by the Atlantic Richfield Company (ARCO), who later became part of BP and then part of Tesoro Corp [13.15]. In September 1989, ARCO started marketing an emission-control gasoline, EC-1, in Southern California. It was intended as a fuel for vehicles not equipped with catalytic exhaust aftertreatment, of which there were an estimated 1.2 million in the South Coast Air Basin of Southern California, representing at that time about 15% of the cars and trucks in the area but causing more than 30% of the vehicular air pollution.

13.2.1. The Adoption of RFG in the US

This EC-1 gasoline was unleaded, although it was designed to replace leaded regular gasoline and had the same octane specification as the gasoline it replaced. Its effect in reducing emissions was achieved by several different routes, including reduced vapor pressure to minimize evaporative emissions; lower aromatic, benzene, and olefin content to reduce smog-forming hydrocarbons and benzene exhaust emissions; the use of oxygenates to improve CO and HC exhaust emissions; and a reduced sulfur content.

Tests carried out on ARCO's EC-1 [13.16], based on a comparison with regular leaded gasoline that was the norm at that time, with a 20-car fleet test indicated statistically significant reductions in regulated and other emissions, as shown in Table 13.5.

TABLE 13.5 Effect of EC-1 in reducing vehicle emissions.

	Approximate reduction (%)
Exhaust hydrocarbons	5
Evaporative emissions	22
Carbon monoxide	10
NO$_X$	6
Exhaust benzene emissions	43

© SAE International.

Soon after the introduction of EC-1, many other US companies followed suit with their own versions of RFG, most of which were marketed only in restricted US areas. There were considerable variations between the specifications that each company imposed on itself.

As noted in Section 13.1.1, the CAAA of 1990 included a requirement for RFG; this was to be sold in over 40 ozone and carbon monoxide nonattainment areas starting in November 1992. The primary targets for these fuels were the ten cities that had the worst record of noncompliance with the US NAAQS set by the 1970 CAA. This act set the major requirements for RFG, but it was left to the EPA to define the detailed specifications. The EPA announced its final rule for the program on December 15, 1993. It has been implemented in two phases: Phase I of the program began in 1995, and Phase II commenced in 2000.

There was customer resistance to RFG in some areas, particularly those containing MTBE. Groundwater contamination with MTBE raised questions regarding its suitability as a gasoline component, and as discussed in Chapter 3, Section 3.9, the use of MTBE has now been severely limited.

The EPA introduced a ruling allowing refiners to add up to 10% v/v of ethanol in the summer months rather than the statutory 7.7% v/v limit allowed in RFG. As the ethanol has a higher volatility than the gasoline, and it is usually splash blended at the terminal; the refiners will be required to produce a lower volatility blend stock because the rule does not provide a waiver on Reid vapor pressure (RVP). Furthermore, the ruling only applies to RFG blended by the simple model (which expired in 1998). Blending under the complex model is NO$_X$ rather than ethanol limited. In July 1997, the EPA proposed a minor relaxation in that refiners would be required to meet average NO$_X$ reduction targets rather than a per-gallon NO$_X$ reduction standard.

The implications of all these regulations, in terms of administrative burden alone, have not been estimated. However, a vast array of records must be kept accurately, because missing data, any misclassification of gasoline, inconsistency between blending and shipping data, or other record-keeping problems constitute a violation of the regulations. Such violations carry a very high financial penalty.

The first phase of the RFG program, from 1995 through 1999, required average reductions of smog-forming VOC and air toxics (benzene, formaldehyde, acetaldehyde, and 1,3-butadiene) of 17% each, and NO$_X$ emissions were reduced by 1.5%. The second phase of the RFG program is claimed to achieve even greater average benefits: a 27% reduction in VOCs, a 22% reduction in air toxics, and a 7% reduction in NO$_X$ emissions. The EPA suggests that this is equivalent to taking more than 16 million vehicles off the road.

13.2.2. RFG Specifications

When they were first introduced, all RFG had to contain a minimum of 2.0% by mass of oxygen, a maximum of 1.0% by volume of benzene. The fuel sulfur and olefins contents, and the T90 value, were limited to a refiners' 1990 average. In addition, RFGs had to meet certain VOCs, air toxics, and NO_X emission performance requirements. Again, these were judged against qualities produced in 1990. Emission performance had to be calculated on the basis of empirical models.

From 1995 to1997, refiners were allowed to use a simple model to certify their RFGs. The simple model was designed to reduce VOC emissions by limiting RVP (deemed to be equivalent to a 15% VOC reduction) and total air toxics. The latter were calculated from benzene, 1,3-butadiene, polycyclic organic compounds, formaldehyde, and acetaldehyde emissions using formulae devised by the US EPA.

From January 1998, refiners have been required to use a complex model for certification. The complex model is a set of equations correlating a gasoline's properties to its emission characteristics. Refiners could also have used this complex model for the first three years, which would have given them more flexibility in meeting the requirements. Refiners can comply with the standards either on a batch basis or on a quarterly average basis. Average limits are more severe overall but have more latitude on a batch basis. The complex emission model uses formulae to calculate total VOCs, air toxics, and NO_X reductions.

The major advantage of changing fuel characteristics is that it has an immediate beneficial effect, especially on the emissions of noncontrolled and hence the most polluting vehicles in the vehicle population. In addition, emitted hydrocarbons from these fuels will have lower photochemical reactivities. It can take many years before the older vehicles disappear completely and are replaced by newer designs, and so it can take a long time for improvements in air quality to take place if reliance is only placed on vehicle technology changes.

Even with vehicles that have emission controls, there are still advantages to using RFGs, although these benefits will diminish as control systems get more sophisticated. Many of the feedback controls on older vehicles do not operate under some driving regimes, for example, during the warm-up phase or under full-throttle accelerations. During these periods there is clearly a benefit in using RFG. In addition, the changes in fuel quality will reduce engine-out and evaporative emissions, thus reducing the burden on the emission-control equipment.

The RFG program persists, however, due to changes to all US gasoline the benefits of using RFG have now virtually all disappeared [13.17]. The convergence of fuel properties such as sulfur content, oxygen content, total aromatics, and benzene means there is little difference between RFG and other gasolines. The exception to this that RG still provides a predicted reduction in VOC emissions due to the lower RVP limits.

13.2.3. European Experience of RFGs

Some RFGs appeared in Europe, although new EU fuel specifications, introduced in 2005, made these products redundant.

Tests with RFGs all show significant benefits, although these will depend on the vehicles used and the test conditions [13.18–13.20]. The principle of diminishing returns seems to apply because little additional benefits seem to be gained by moving to extreme reformulation such as by having a very low aromatics level [13.20].

13.2.4. The Fuel and Emission Relationship Going Forward

Establishing unambiguous relationships between gasoline properties and vehicle emissions is difficult due to the inter-relationship of fuel properties and composition. For example, it is difficult to vary the aromatics content of gasoline without changing other characteristics, such as volatility, that may also influence emission performance. The increasing complexity of engine technology, aftertreatment, and control systems will alter the engine calibration as a result of changes in fuel characteristics and minimize tailpipe emissions to the point where they may be below background levels. Fundamental studies with research engines allow some insight but as yet do not give a true picture of what the effect of fuel compositional changes will have on the overall emission burden of the existing and future vehicle fleet.

It is now clearer than ever that the automotive industry and the fuel industry form complementary parts of a single system. Cooperative programs, such as the US Air Quality Improvement Program (AQIRP), the EPEFE, and the Japan Clean Air Program (JCAP), have assisted in understanding this concept, along with numerous joint programs between individual motor manufacturers and fuel companies.

The following sections draw these various programs together in an attempt to elucidate common conclusions. However, it should be remembered that these programs are very time and resource consuming and they cannot be undertaken with every evolutionary or revolutionary change in vehicle or fuel technology. These large programs can therefore easily become outdated, and smaller, less all-encompassing programs can add important new insight. Although the carefully designed fuel matrices (which are necessary to try and isolate a fuel property as an independent variable so that its influence on emission performance can be studied) can point toward the "ideal" fuel, these gasolines are not necessarily practically or commercially feasible products.

13.3. The Influence of Gasoline Sulfur Content on Emissions

In response to concerns over automotive emissions, the motor manufacturers investigated different approaches to reducing tailpipe emissions. One such approach was the fitment of catalytic reactors within the exhaust pipe. It was soon found that the accumulation of sulfur could impair the function of oxidation catalysts [13.21] and TWCs [13.22, 13.23]. This could happen in the short term, producing an immediate but reversible reduction in the efficacy of the aftertreatment system (as discussed in Section 13.3.1), or it could degrade the performance over a longer period (as discussed in Section 13.3.2).

It was also found that catalyst-equipped vehicles could produce higher emissions of sulfuric acid [13.24, 13.25]. The catalyst was promoting the oxidation of SO_2 to SO_3, which was then forming sulfuric acid. The oxidation of SO_2 to SO_3 within the catalyst was obviously temperature and hence operating condition dependent [13.26, 13.27].

There was strong pressure from the motor manufacturers for reducing the amount of sulfur in gasoline. However, to do this required investment and additional refinery operating expenses. This would consume additional energy, increase refinery CO_2 emissions, and negatively affect other fuel properties [13.28]. The use of catalytic exhaust aftertreatment has become the most widely used method of reducing tailpipe emissions. To enable this technology to function at its optimum, the maximum permitted level of sulfur in the gasoline has now been legislated to very low levels in many areas of the world. The following sections consider the impact of fuel sulfur on different vehicle technologies in terms of the tailpipe emissions, the durability of the emission-control technology, and the control and monitoring system.

13.3.1. The Effect of Sulfur on Tailpipe Emissions

Most TWC systems oxidize HC and CO emissions while also reducing NO_X emissions by rapidly switching from slightly rich of the stoichiometric air-to-fuel ratio and slightly lean of this ratio. This switching occurs about every second. Current TWC formulations contain one or more precious metals that are generally dispersed on alumina with the addition of base metal compounds, including the oxides of cerium (CeO_2), lanthanum (La_2O_3), barium (BaO), zirconium (ZrO_2), and nickel (NiO), as co-catalysts, promoters, and stabilizers. During the rich part of the cycle the NO_X in the exhaust gas or stored in the catalyst is reduced to N_2, and the HC and CO are converted by water-gas shift type reactions or by stored oxygen. During the lean part of the cycle there is sufficient oxygen to enable the catalytic oxidation of the HC and CO. There is also some storage of NO_X within the catalyst. Sulfur is known to inhibit the water-gas shift reactions that are important for controlling HC and CO when operating on the fuel-rich side of stoichiometric [13.29]. The SO_2 in the exhaust gas can also dissociate on noble metals to form adsorbed elemental sulfur that is then difficult to remove from the catalyst surface under rich conditions [13.30–13.33]. When the engine is operating on the fuel lean side of stoichiometric, sulfur can be stored due to interaction with the alumina [13.34]. The sulfur can also react with ciria to form $Ce_2^{III}(SO_4)_3$ and/or $Ce^{IV}(SO_4)_2$ [13.35]; this reaction can be stronger than that with alumina [13.36].

The influence of sulfur on emissions was therefore considered an important part of the cooperative research programs mentioned in Section 13.2.4. Differences in sulfur content can be achieved without affecting other fuel properties by doping a low sulfur fuel with di-tertiary-butyl disulfide [13.37, 13.38]. Early data from phase 1 of the AQIRP showed that reducing sulfur from a nominal 500 mg/kg to a nominal 50 mg/kg reduced the HC, CO, and NO_X emissions by 16, 13, and 9%, respectively [13.39]. The measured sulfur concentrations for this work were 466 and 49 mg/kg, so the actual reduction was slightly less than the tenfold decrease intended. The ten vehicles tested employed a selection of fuel metering systems: throttle body injection, port fuel injection, including two with sequential injection, and one carbureted engine vehicle. Beyond the headline results noted, which were achieved in typically less than 16 km, it was also found that engine-out emissions were not significantly affected by the concentration of sulfur in the fuel. Changes in tailpipe emissions were therefore due to deterioration in the conversion efficiency of the catalyst. There was no apparent carryover or memory effect as the emission levels returned to their original levels, whether the order of testing was from high-to-low-to-high or low-to-high-to-low sulfur levels.

These results were largely confirmed by a small part of a larger program in which there was the intention of producing a tenfold reduction in sulfur concentration [13.38]. In this case, a target of 300 and 30 mg/kg was used, with measured values of 298 and 25 mg/kg. This resulted in an almost twelvefold increase in sulfur concentration, from the low sulfur to the high sulfur, which increased the HC, CO, and NO_X emissions by 50, 17, and 59%, respectively. These changes were greater than previously observed but were for a greater change in sulfur concentration and, interestingly, also at a lower overall sulfur level.

A European study [13.40] demonstrated the possibility of fleet average reduction in HC, CO, and NO_X emissions of 8.6, 9.0, and 10.4%, respectively, by reducing the sulfur concentration from 382 to 18 mg/kg.

This provides an indication that the relationship between sulfur content of the gasoline and the emissions is not a linear relationship. This has been confirmed by further work as part of AQIRP [13.41] that also considered variations in sulfur content from 11 to 50 mg/kg.

As discussed in Chapter 8, Section 8.3.1.4, there is a potential for significant fuel economy benefits by operating a gasoline engine on the lean side of stoichiometry. Under these conditions there is excess oxygen, and the reduction reactions that would normally take place in the TWC to control the NO_X emissions will not occur. The motor manufacturers and aftertreatment technologists have therefore developed lean-NO_X trap (LNT) or NO_X storage reduction (NSR) catalyst technology [13.42–13.44]. As discussed in Chapter 8, Section 8.5.3.3, this technology relies upon NO_2 being converted to a nitrate on the catalyst surface. With an excess oxygen environment, any SO_2 in the exhaust will be converted to SO_3. Because the sulfate is more chemically stable than the nitrate, the SO_3 will adsorb more strongly than the NO_2. This poisons the catalyst by rendering the NO_X storage sites ineffective [13.45]. Figure 13.2 [13.46] shows how the NO_X conversion efficiency of an NSR is reduced over time with different levels of fuel sulfur.

FIGURE 13.2 Effect of fuel sulfur on NO$_x$ conversion efficiency of an NSR.

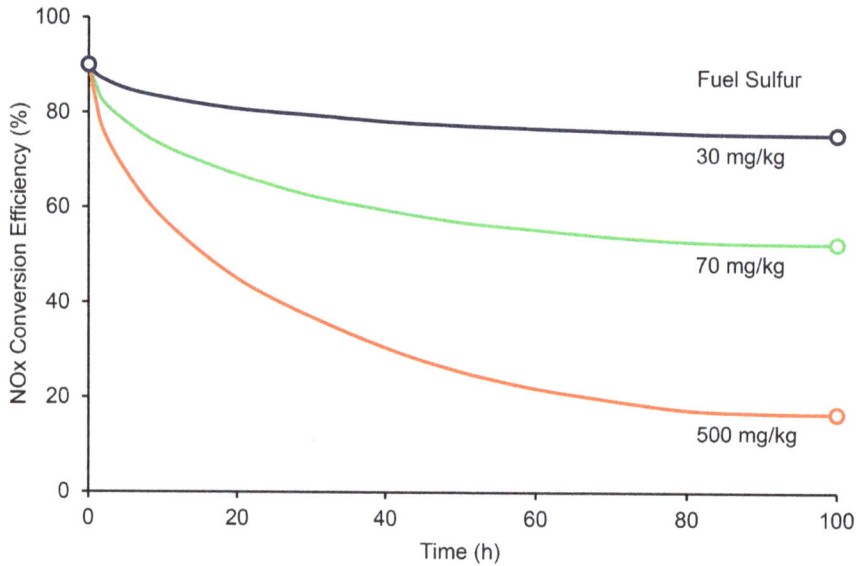

© SAE International.

Because the sulfate is more stable than the nitrate, it does not completely desorb during the stoichiometric/rich spike, and a dedicated desulfation cycle is required to purge the NSR catalyst [13.47]. The desulfation process requires raising the temperature of the catalyst to typically over 700°C. If this is not properly controlled, it can give rise to undesirable sulfide emissions such as H$_2$S [13.48]. The mechanism of hydrogen sulfide production is illustrated in Figure 13.3 [13.49].

FIGURE 13.3 Mechanism of hydrogen sulfide production from an NSR.

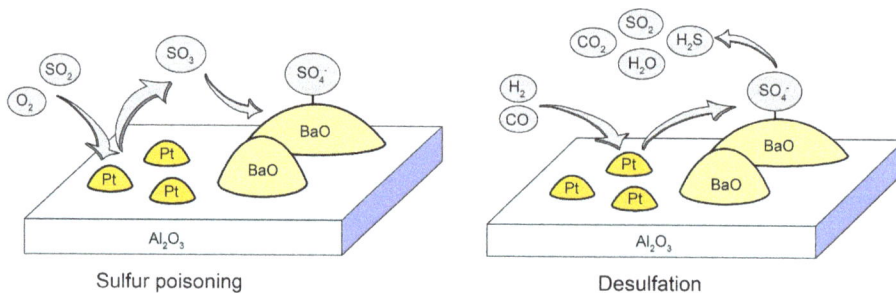

© SAE International.

Careful design of the NSR by incorporating the oxides of metals such as nickel [13.50] and titanium [13.46] can inhibit the formation of hydrogen sulfide and improve the sulfur desorption process, thus improving the long-term efficiency of the NSR.

13.3.2. The Effect of Sulfur on Durability of Aftertreatment Systems

The discussion in the previous section highlights that sulfur can inhibit the functioning of catalytic after-treatment. Some of these processes should be reversible under the necessary conditions. Whether there is any long-term or durability effects on the system depends on what the necessary conditions are and whether they are achieved under expected driving patterns.

Under low-temperature, lean conditions, SO_2 chemisorption occurs on platinum, palladium, and rhodium surfaces [13.51–13.53]. This chemisorption of sulfur will compete with the reactants in the exhaust gas for adsorption sites and will thus increase catalyst light-off temperatures. As temperature increases under lean conditions, the equilibrium concentration of SO_2 on these metal surfaces will decrease. The effect of sulfur poisoning will thus be less during warmed-up conditions.

As the operating cycle moves from lean to rich, the SO_2 dissociatively adsorbs, aided by the removal of adsorbed oxygen from the catalyst surface by the reductants H_2 and/or CO. For rhodium, the impact on catalyst performance is significantly less than for palladium and platinum [13.29]. This is because for rhodium, there is a lower equilibrium level for the adsorption of SO_2 and lower stability for the adsorbed sulfur species under reducing conditions. The adsorption of sulfur species onto platinum and palladium results in significant loss in catalyst performance due to a combination of physical blockage and electronic effects inhibiting the adsorption of reactants [13.54, 13.55]. For palladium this is accentuated by the ability of sulfur to migrate into the bulk palladium [13.56]. This increases the severity of sulfur poisoning, the difficulty of its removal, and hence the long-term effect. The adsorbed sulfur is removed from the platinum surface at temperatures in excess of 800°C, whereas for palladium, temperatures in excess of 850°C are required. This may be a result of sulfur migration back from the bulk palladium as the surface sulfur is removed.

Under lean conditions, the removal of sulfur from palladium will take place from about 650°C, whereas for platinum, temperatures in excess of 700°C are required. Removal of the sulfur is aided by its oxidation to SO_2 and possibly SO_3, that thermal desorb catalyst surface. The availability of oxygen will therefore influence the temperatures at which the desulfation occurs. Also under lean conditions, the oxygen storage and release capability and loss of catalytic activity for the water-gas shift are impaired by the formation of $Ce_2(SO_4)_3$, which does not decompose until temperatures in excess of 700°C. Under rich conditions, the $Ce_2(SO_4)_3$ will start to decomposition at temperatures above 600°C.

TWCs that have operated in the normal manner with rapid switching either side of stoichiometry but with high sulfur fuel will form $Ce_2(SO_4)_3$ and strongly adsorbed sulfur on the noble metal sites. The reversibility of the sulfur poisoning will depend upon whether the required conditions are achieved during normal operation.

Under the reducing atmosphere resulting from rich operation, it is very difficult to remove the sulfur from platinum and palladium if the temperature remains below 750°C. At temperatures above 600°C, these reducing conditions will readily decompose $Ce_2(SO_4)_3$.

Under oxidizing conditions resulting from lean operation, removal of sulfur from platinum and palladium requires temperatures between 650 and 750°C. A temperature above 700°C is required to decompose $Ce_2(SO_4)_3$ under these oxidizing conditions.

At stoichiometric conditions, temperatures above 750°C are required to remove the sulfur from the platinum and palladium sites and to decompose the $Ce_2(SO_4)_3$.

The effective reversal of sulfur poisoning requires high catalyst temperatures and rich/lean excursions to decompose/desorb sulfur species from the catalyst surface. The degree of reversibility and hence the long-term effect will vary significantly with catalyst formulation and its position in the exhaust system. This can explain the variability in the effect of sulfur on the emissions of different vehicles, catalyst technologies, and whether the catalyst is aged or relatively new.

As NSR catalysts are introduced, the high temperatures required for the desulfation cycle will help to produce the necessary conditions for desulfating the TWC. However, the widespread introduction of vehicles equipped with NSRs will tend to happen only with the widespread introduction of zero sulfur fuels.

13.3.3. The Effect of Sulfur on Onboard Diagnostics

As part of the drive to reduce automotive emissions, stricter inspection and maintenance programs were introduced and the concept of onboard diagnostics (OBD) was developed to monitor the health of the engine and its emission-control equipment [13.57]. Over the years, these systems have become more sophisticated [13.58].

The most widely used method of OBD of catalyst efficiency is currently via the dual exhaust gas oxygen (EGO) sensor method [13.59, 13.60]. These sensors are often heated and referred to as heated exhaust oxygen sensors (HEGO). With this technique, one oxygen sensor is placed upstream of a TWC and a second sensor is placed downstream. The upstream EGO is normally used to monitor the engine-out oxygen in order to modulate the air-to-fuel ratio slightly rich and slightly lean to enable the TWC to function correctly. The second, down-stream EGO is used to monitor the oxygen concentration after the TWC to assess the storage and release of oxygen and thus infer the correct functioning of the catalyst.

The perturbations in the air-to-fuel ratio about the stoichiometric point, resulting from the signal from the upstream, controlling EGO, are stored internally. Because the catalyst functions by storing and releasing oxygen, the perturbations in the oxygen concentration entering the catalyst will be dampened by the catalyst, resulting in less variation downstream. This is illustrated in Figure 13.4 [13.61], which shows the EGO signal voltage for the pre- and postcatalysts devises for a 97% HC conversion efficiency and an 82% HC conversion efficiency.

FIGURE 13.4   EGO signals for 97% (left) and 82% (right) HC conversion efficiency.

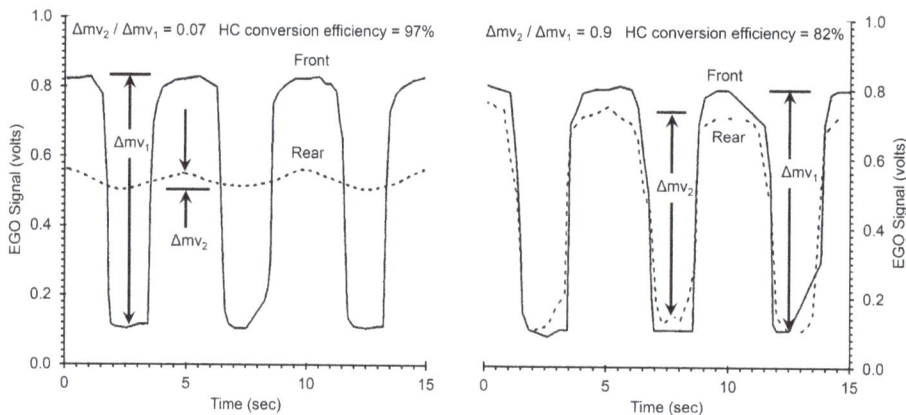

A comparison of the upstream and downstream signals provides a measurement of the oxygen storage capacity (OSC) of the catalyst. As shown in Figure 13.4, one way of calculating a metric is simply to divide the peak voltage of the downstream EGO by that of the upstream EGO. This is sometimes referred to as the rear HEGO/EGO index. As the storage capacity of the catalyst is reduced with age, the damping effect diminishes; the monitor signal will then become closer to the signal from the upstream control sensor, and this ratio will increase.

An alternative strategy switches the function of the two sensors. By using the downstream EGO as the controlling sensor, the frequency of air-to-fuel ratio perturbations is reduced because of the damping effect of the OSC. Comparison of the perturbation frequencies depending upon whether the upstream or downstream sensor is controlling will give an indication of OSC and hence the degradation of the catalyst.

With the assumption that a well-functioning catalyst with a high OSC will exhibit high conversion efficiency toward HC, it should be possible to predict catalyst HC conversion efficiency. However, the relationship is not simple. Figure 13.5 [13.62] shows some data on the relationship between the HC conversion efficiency and the rear HEGO/EGO index.

FIGURE 13.5   Relationship between rear HEGO/EGO index and HC conversion efficiency.

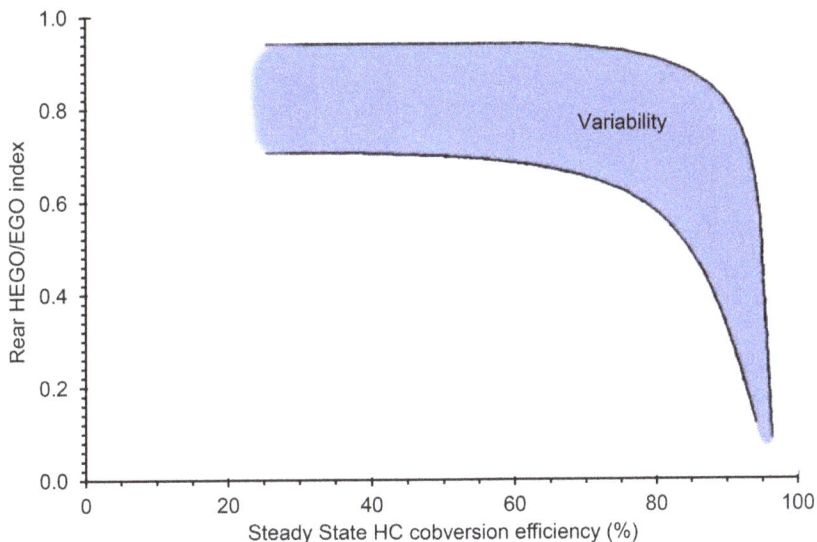

In general, the index does not discriminate between catalysts with HC conversion efficiencies below 80%, and the index falls sharply: the HC conversion efficiencies exceeds 80 to 90%. There is also a large degree of variability in this relationship between different systems. One of the main reasons for this problem is that the relationship between the OSC and the HC conversion efficiency is complex, nonlinear, and system specific [13.61–13.63]. This is illustrated in Figure 13.6, [13.62] where the OSC has been plotted against the HC conversion efficiency; both were measured at 500°C for 17 different catalysts [13.62]. The results have been superimposed on the plot of rear HEGO/EGO index against HC conversion efficiency.

FIGURE 13.6   Relationship between OSC and HC conversion efficiency.

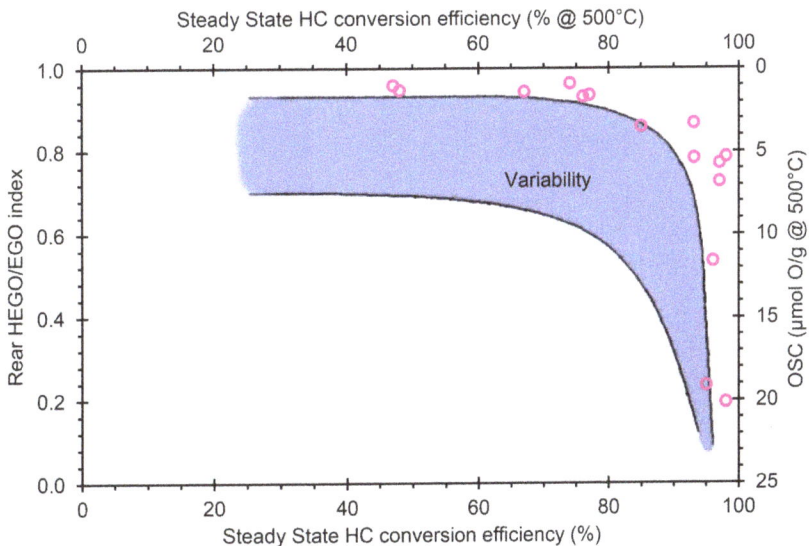

There are two mechanisms by which fuel sulfur can affect dual oxygen sensor catalyst monitoring systems. As discussed in Section 13.3.1, the sulfites and sulfates, formed from the sulfur in the fuel, can block the available oxygen storage sites. The degree to which this occurs can be influenced by the catalyst formulation [13.61, 13.64–13.67] as well as the sulfur concentration in the fuel. The second mechanism by which sulfur affects the OBD system is by the poisoning of the EGO [13.68–13.71]. Figure 13.7 [13.72] shows how the response of an EGO to switching from rich to lean to rich is reduced by a high sulfur concentration in the fuel. The rich to lean response is the most severely affected.

FIGURE 13.7   Effect of sulfur on EGO response.

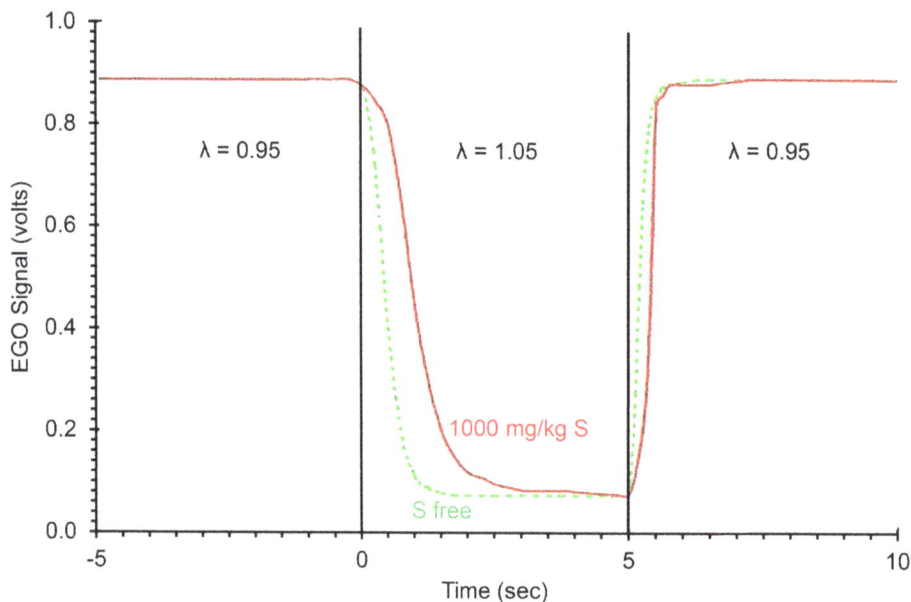

© SAE International.

However, the net response of a particular vehicle OBD system to the degree of sulfur poisoning will be a product of the magnitude of loss of OSC relative to the increase in the rear EGO response time.

13.4. The Influence of Gasoline Hydrocarbon Composition on Emissions

The previous sections have focused on the effect of fuel borne sulfur on emissions and emission-control devices. In many regions of the world, fuel sulfur is controlled to very low levels. While fuel producers have incurred considerable expense to achieve these targets, they have usually been achieved with relatively small impact on other fuel properties. The adverse effects of sulfur have therefore been largely eliminated in the regions where the prevailing technology is most susceptible to these effects.

The properties and performance of gasoline are intrinsically linked to its chemical composition and in broad terms to the hydrocarbon groups present. It is therefore much more difficult to vary a parameter such as aromatic content without influencing other properties. As noted, large-scale cooperative programs have been performed to assess the impact of varying properties such as the aromatic and olefin content of gasolines, on the emissions from a sample of the available technologies. As new technologies have evolved since those programs, smaller pieces of work have been performed to assess how these technologies respond to compositional changes in the gasoline [13.73–13.75].

The following sections look at the influence of certain gasoline component groups on vehicle emissions.

13.4.1. Aromatics

Heavy aromatics and other high molecular weight compounds have been linked to engine deposit formation, particularly combustion chamber deposits (discussed further in Chapter 10). These deposits can also act as a storage and release site for other high boiling point hydrocarbons, particularly polynuclear aromatic hydrocarbon (PNA), also referred to as polycyclic aromatic hydrocarbons (PAH) (many of which are carcinogenic) [13.76]. This mechanism can influence the total hydrocarbon emissions. Benzene is also a known carcinogen [13.77] and in many jurisdictions has been singled out for control within gasoline compositions. However, other aromatics are precursors to benzene production during the combustion process and hence benzene emissions [13.78].

As a hydrocarbon group, the aromatics within gasoline tend to have a higher boiling point than other hydrocarbon groups. This reduced volatility makes them more likely to pass through the engine unburned. Increased aromatic content can result in higher peak combustion temperatures that can result in increased NO_X emissions [13.79, 13.80]. However, aromatics are also very reactive in the reduction of NO_X over a TWC [13.81]. These effects can be self-cancelling to a greater or lesser extent. In addition, the inclusion of ethanol with its high octane number can lead to a reduction in the aromatic content of the blendstock for oxygenated blends (BOB) with work looking specifically at this aspect [13.82]. The following sections consider some of these aspects in greater detail.

13.4.1.1. Aromatics and Regulated Gaseous Emissions

As discussed in Chapter 8, Section 8.5.2, the air-to-fuel ratio is a major influence on emissions. As the mixture gets richer, the emissions of HC and CO increase. As the mixture reaches stoichiometry, the HC and CO emissions decrease, while NO_X emissions increase; as the mixture becomes leaner, the NO_X emissions then tend to decrease. Reducing the aromatic content of the fuel will tend to make it less dense and reduce the C/H ratio [13.83]. These parameters rather than the aromatic content per se can have a major effect on the emissions.

If an open-loop fuel metering system meters a controlled volume of fuel, then reducing the density will result in a lower mass of fuel being delivered, which will make the air-to-fuel ratio leaner. Reducing the C/H ratio also means that more oxygen is required to achieve stoichiometry. These two effects will combine to move the mixture strength toward higher lambda values. The result on emissions will depend upon the original engine calibration. However, if the other parameters could be kept constant, then reducing aromatics would still have the effect of reducing peak flame temperatures and would thus tend to reduce NO_X emissions [13.80, 13.84, 13.85]. While CO emissions are also related to peak temperatures, they are less dependent than NO_X emissions [13.86].

Changing the aromatic content of the fuel will have a profound effect on the aromatic composition of the HC emissions [13.78, 13.87]. This is discussed further in Section 13.4.1.2. In addition, reducing aromatics can also result in a reduction in the total mass of engine-out HC emissions [13.88, 13.89].

In a vehicle equipped with a TWC, the air-to-fuel ratio is controlled around stoichiometric using feedback from the EGO; any effects of composition on the stoichiometric air-to-fuel ratio will therefore tend to be minimized. However, even under these controlled conditions the effect of reducing aromatics will affect the peak flame temperature as noted. In addition, the greater tendency for CO_2 to dissociate at high temperatures, than H_2O, will lead to lower equilibrium concentrations of O_2 and O radicals. This, together with the lower flame temperature, will lower in-cylinder NO_X production, and hence engine-out emissions [13.90].

With an engine equipped with aftertreatment, the effect of the composition of the HC emissions must also be considered. Aromatics within the exhaust gas are far more reactive in their contribution to NO_X

conversion within the catalyst than low molecular weight alkanes, particularly methane [13.81]. Reducing the aromatics in the fuel will reduce the aromatics within the exhaust gas. This will reduce the conversion efficiency of the catalyst toward NO_X. The trade-off between reduced engine-out NO_X emissions and the reduced conversion efficiency of the catalyst will therefore vary from vehicle to vehicle, making the overall effect difficult to predict.

Results from the AQIRP showed that in older vehicles a reduction in aromatics increased HC but reduced NO_X emissions; the effect on CO was not significant. However, in the then current vehicles, reducing aromatics reduced HC and CO emissions with no significant effect on NO_X emissions [13.91].

In the EPEFE study it was found that reducing aromatic content reduced emissions of HC and CO but increased emissions of NO_X over the composite cycle [13.92].

In the JCAP it was found that for most test vehicles, reducing the fuel aromatic content reduced the HC and CO emissions but increased the NO_X emission for the 10–15 mode test procedure. However, for the 11-mode test procedure, the reduction in fuel aromatics content reduced the HC and NO_X emissions but increased the CO emissions [13.93]. The 11-mode test procedure includes a cold phase when the catalyst would not be fully up to temperature.

The JCAP Step II work included vehicles using direct injection engine technology [13.94]. In these vehicles, the air-to-fuel ratio is not controlled to stoichiometric and thus the inter-correlation between aromatic content and density will become relevant. This was apparent in that some of the results were indicative of a leaner mixture as the aromatic content was reduced. However, the directional trend in the effect on emissions varied between different vehicles.

A test program in Thailand showed that for a catalyst-equipped vehicle, reducing aromatics reduced both HC and NO_X emissions, with a reduction in CO emissions that was not statistically significant [13.95]. For a noncatalyst-equipped vehicle, there was a directional trend for reduced HC and NO_X emissions, but only the HC emissions reduction was significant. The directional trend for CO emissions was reversed, but the changes were again not statistically significant.

The overall conclusion from the combination of these programs is that the effect of fuel aromatic content on the regulated exhaust emissions is very dependent upon the engine and aftertreatment technology. These parameters will have a far greater effect on the emissions than fuel aromatic content. These studies did not consider particulate emissions as they were not regulated at the time. The effect of aromatics and other fuel components on particulate emissions is discussed in Section 13.4.3.

13.4.1.2. Aromatics and Air Toxins

One of the major air toxins related to gasoline vehicles is benzene. As discussed, the allowable amount of benzene in gasoline has been severely limited in many jurisdictions. Other PAH compounds are also known to be toxic. Data from animal studies indicate that several PAHs may induce a number of adverse effects, such as immunotoxicity, genotoxicity, carcinogenicity, and reproductive toxicity and may possibly also influence the development of atherosclerosis [13.96].

PAHs are formed during incomplete combustion or pyrolysis of organic material. They may therefore be present in the fuel, or they may be formed during the combustion process. Part of the small amount of PAH present in the fuel may evade the combustion process in the flame quench zones near the surfaces of the combustion chamber, crevice volumes, and by wetting of combustion chamber deposits. This can be particularly true with direct injection technology.

PAH synthesis involves bimolecular and higher order reactions of hydrocarbon fragments to form large compounds. These chemical reactions would be expected to occur only in regions that are rich in hydrocarbons, such as those mentioned previously. Thus, both PAH survival and PAH synthesis could take place at the walls and in deposits of the combustion chamber [13.97]. PAH that accumulates or are formed in these regions can then be emitted [13.76]. When a catalytic converter is fitted to the engine, there is also the possibility that benzene can be formed over the catalyst via dealkylation of alkyl-benzene groups [13.98, 13.99]. A modeling study has shown that increasing aromatics increased the emissions of the toxins, 1,3-butadiene, benzene, toluene, ethylbenzene, and xylene while increasing ethanol content lowered 1,3-butadiene emissions but increase emissions of acetaldehyde [13.100].

13.4.2. Olefins

Olefins are thermally unstable and may lead to gum formation and deposits in an engine's intake system. Their evaporation into the atmosphere as chemically reactive species contributes to ozone formation, and their combustion can form toxic dienes [13.101]. Because they are more reactive, they are more readily consumed during the combustion process. However, olefins are known to be formed in significant concentrations during the partial oxidation of paraffins, aromatics, and other olefins [13.102, 13.103]. The major route by which olefins affect emissions is therefore via evaporative emissions.

The effect on ozone-forming potential was demonstrated by the AQIRP. A modeling exercise based on the findings of AQIRP predicted that reducing total olefins from 20 to 5% would significantly decrease ozone-forming potential in three critical cities: Los Angeles, Dallas-Fort Worth, and New York [13.104]. The projected emissions for these hypothetical fuels were calculated using regressions models developed from AQIRP Phase I testing, a vapor headspace model, and other information. The effect of reducing olefin content was due primarily to a reduction in the reactivity of the evaporative, running loss, and refueling emissions. The C5 olefins had a larger effect than the C6+ olefins on evaporative-related emissions because of the higher volatility of the C5 olefins.

The effect of olefins on regulated engine-out emissions is small and outweighed by vehicle-to-vehicle variation. These differences tend to be reduced even further by aftertreatment devices that drastically reduce the overall mass of emissions.

There is no doubt that olefins increase the reactivity of evaporative emissions. However, modern onboard controls and the increasing use of vapor recovery systems at filling stations reduce any problem to minimal proportions.

13.4.3. Oxygenates

As mentioned in Chapter 3 the use of oxygenated compound has been of interest initially as octane number enhancers and latterly as a source of renewable carbon components; bio-ethanol now being mandated in much of today's gasoline. There have been issues with the use of some oxygenates, as discussed in Chapter 3, which has resulted in restrictions now in place, but as just noted bio-ethanol is now common. The inclusion of bio-ethanol and other oxygenates can also affect the emissions performance of the engines/vehicles that they are used in. The inclusion of oxygen in the fuel will obviously reduce the amount of hydrocarbon present thus altering the air/fuel ratio, when the fuel is metered in an open-loop fashion. As discussed in Chapter 8, Section 8.5.2, the air/fuel ratio can significantly alter the engine-out emissions; particularly reducing emissions of hydrocarbons and carbon monoxide, when moving to a slightly leaner mixture. The inclusion of oxygenates was thus initially seen as an emissions reduction strategy, leading to their inclusion in RFG as discussed above, and the early development of flexible fuel vehicles (FFV). Early FFVs were developed using methanol/gasoline fuels [13.105–13.107] but some ethanol/gasoline vehicles were also developed [13.108].

Adding ethanol into the fuel will tend to increase ethanol emissions, both from the tailpipe [13.109] and via evaporative emissions [13.110]. This effect is dependent upon the RVP of the finished gasoline, ambient temperature, vehicle technology, etc. [13.111]. Partial oxidation of alcohols can form aldehydes resulting in increased emissions of these compounds. With the inclusion of ethanol in gasoline there can be increased engine-out emissions of acetaldehyde [13.112–13.114] which can still persist even after a catalytic converter [13.115].

The shift in air/fuel ratio due to the inclusion of oxygenates also has a tendency to increase NO_X emissions [13.116] but as engine technology has improved the control of air/fuel ratio this trend tends to be eliminated [13.117]. As discussed in Chapter 8, Section 8.5.3 exhaust aftertreatment systems can also eliminate and potential increase in NO_X.

With modern DISI engine technology it is possible that the increased heat of vaporization of the oxygenates can lower the end of compression in-cylinder temperature resulting in reduced NO_X emissions [13.118] but this can also affect particulate emissions [13.119]. There are competing mechanisms here so overall the effect is difficult to predict. Ethanol has a higher heat of evaporation than gasoline, and required fuel to achieve stoichiometry is greater leading to a tendency for greater inhomogeneity of the mixture at the start of combustion [13.120]. This can lead to greater particulate formation. However, the increased fuel ethanol content can result in reduced aromatic content needed to meet specification, thus resulting in reduced particulate formation [13.121]. The inclusion of oxygenates in the fuel is accounted for in the particulate forming tendency of the fuel as discussed in the following section.

13.4.4. Fuel Composition and Particulate Emissions

As discussed in previous sections the fuel composition can affect the formation of gums and deposits within the engine, and what is emitted from it. With the wide adoption of DISI technology, and concern over the particulate emissions that can result, much work has been performed to consider the relationship between the gasoline composition and its particulate forming tendency. At the end of the first decade of this century, work was performed by Honda to correlate the fuel composition, as determined by detailed hydrocarbon analysis (DHA), with a value of the gasoline's potential to form PM emissions, this became known as the PM Index (PMI) [13.122]. Each component, from a DHA, is assessed for its double bond equivalent (DBE) value and vapor pressure, and a weighted average value is calculated to give the PMI. For this work the DHA was acquired using the ASTM D6729 procedure [13.123]. At that time ethanol was not a standard fuel component but since then a study was performed by the National Renewable Energy Laboratory [13.124], based on data for more modern engine technology [13.125], to update the previous PMI while also taking account of the more prevalent oxygenates. A number of other studies have been performed to produce different indices correlating fuel composition and the fuels potential to form PM emissions. This included work done by the CRC to update the PMI, which was termed the particulate matter emissions (PME) index [13.126]. Other examples include, the Wittmann/Menger (WM) Index [13.127] that uses the ASTM D6839 [13.128] method to assess the fuel composition, the particulate evaluation index (PEI) [13.129] that also uses the ASTM D6839 method to assess the fuel composition. Other DHA techniques have can also be used with appropriate correlations [13.130].

13.5. The Influence of Gasoline Distillation Characteristics on Emissions

As already discussed, it is virtually impossible to independently vary certain gasoline properties. This is obviously also true for the distillation characteristics. It is possible within certain limits to fix given points on the distillation curve while varying other parameters. This has been done in the major interindustry studies such as AQIRP, EPEFE, and JCAP. Different programs have considered the front-end volatility as measured by the vapor pressure, the mid-range volatility as characterized by the E100 value [13.92], or the T50 value [13.131], and the back-end volatility as measured by the T90 value [13.91]. These parameters were varied while trying to keep other characteristics constant. The effects of these three parameters are discussed briefly in the following sections. Overall, there are no wholly definitive data establishing which distillation characteristics really influence tailpipe emissions. Distillation effects on HC emissions are generally larger and more significant than those for CO and NO_X. The effect of distillation on NO_X emissions tends to be in the opposite direction to that for HC and CO. It appears that any effects on tailpipe emission are primarily due to the mid-range volatility.

13.5.1. Front-End Volatility

The major effect of front-end volatility is on the evaporative emissions of vehicles not equipped with evaporative emissions controls. This is discussed in more detail in Chapter 9, Section 9.5, but the conclusions were that the relationship between evaporative losses and vapor pressure is nonlinear, with losses increasing disproportionately as the vapor pressure is increased [13.132]. The vehicle and fuel system design have the greatest effect on evaporative emissions, and carbon canisters are very effective in controlling these emissions [13.133].

An investigation was conducted in the early 1990s into the effect of vapor pressure on regulated exhaust emissions using five European vehicles [13.134]. This showed that a reduction in vapor pressure gave overall reductions in CO and NO_X emissions but no significant effect on exhaust HC emissions. The results were vehicle dependent with only three of the vehicles showing a reduction in CO emission as a result of reducing vapor pressure. The other two vehicles showed no CO emissions sensitivity to vapor pressure. Only four of the vehicles showed NO_X emission sensitivity to vapor pressure changes. Although there was no overall effect of vapor pressure on exhaust HC emissions, there was a reduction of exhaust HC emissions with a reduction in vapor pressure during hotter phases of the testing. The other sensitivities were also greater when the vehicle was warmed up.

13.5.2. Mid-Range Volatility

Trying to formulate fuels to vary the mid-range volatility while keeping other parameters constant is extremely difficult. As discussed in Chapter 9, Section 9.3.2.1, in the US a driveability index (DI) has been developed to characterize the performance of vehicles in relation to the fuel's distillation characteristic. As part of a program [13.135] to investigate any relationship between HC emissions and driveability, three sets of fuels were blended with no oxygenates, approximately 10% ethanol, and approximately 15% MTBE, respectively. These fuels had different distillation characteristics and correspondingly different driveability indices. One of the conclusions of the work was that there was a correlation between DI and HC and CO emissions, but not for NO_X emissions. The largest effect was on HC emissions. Though there was poor correlation between the T10 and T90 values that made up the DI, there was a good correlation with the T50 value. This is illustrated in Figure 13.8 [13.135].

FIGURE 13.8 Relationship between T50 and HC emissions.

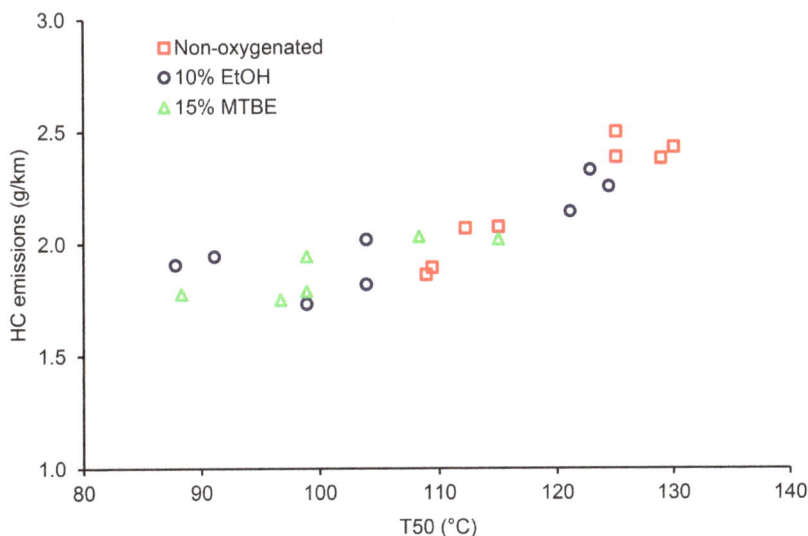

© SAE International.

Other work [13.104] has also indicated that there is a strong correlation between HC emissions and the T50 value.

In the EPEFE study [13.103], a nine-fuel matrix was blended to determine the independent effects of aromatics and mid-range distillation, as characterized by the E100 value. Back-end volatility was fixed by establishing a constant E150 target value. Increasing the E100 value, similar to reducing the T50 value, reduced the HC emissions. The size of the reduction increased as the aromatic content of the fuel increased. The effect of E100 on CO emissions was not as well defined. The modeling work suggested a minimum occurred close to the intermediate level of E100 used in the program; this was 50%. At E100 values above and below this level, CO emissions increased by around 6 to 10%. The work once more showed that NO_X emissions had the opposite tendency. The modeling work suggested that NO_X emissions would increase linearly with increasing E100.

13.5.3. Back-End Volatility

Phase 1 of AQIRP considered two fleets of vehicles with two levels of emissions control technology [13.91]. The older fleet was predominantly carbureted vehicles employing a less-sophisticated emissions control technology. This fleet showed a small but significant reduction in tailpipe HC emissions with a reduction in T90. These vehicles also showed a large increase in CO emissions but no statistically significant change in NO_X emissions

For the then current fleet, a larger and statistically significant reduction in HC emissions was observed with a small increase in NO_X emissions as a result of reducing T90 from 182 to 138°C. No statistically significant effect of T90 on CO emissions was found for these newer vehicles. This clearly demonstrates that vehicle and particularly emissions control technology have a far greater effect on emissions than do changes in fuel characteristics.

Further work provides some clarification of this point [13.136, 13.137]. Increasing the T90 value produced a modest but consistent increase in engine-out emissions during the parts of the test procedure when the vehicle was warmed up. During the cold start and the warm-up phases of the procedure, the increase in T90 value produced a far greater increase in engine-out HC emissions. This is almost certainly due to the less-volatile components of the fuel passing unburned through the combustion chamber. As the increased HC emissions are not primarily due to poorer combustion but lack of combustion, there is not a commensurate increase in CO emissions. Because the aftertreatment systems would not be functioning well during the cold start and warm-up phases, this increase in engine-out HC emissions would feed through to the tailpipe emissions. The aftertreatment would significantly reduce the HC emissions during the remainder of the test procedure, resulting in a low overall level of tailpipe HC emissions. The effect of the increased cold start phase emissions would therefore have a far larger impact for the newer vehicles.

Reducing cold start and warm-up HC emissions is a major challenge for the automotive industry [13.138–13.141]. As aftertreatment technology improves [13.140–13.143] to address this concern, the effect of gasoline distillation will become smaller.

References

13.1.   Robinson, E., "Some Air Pollution Aspects of the Los Angeles Temperature Inversion," *Bulletin of the American Meteorological Society* 33 (1952): 247-250.

13.2.   Faith, W.L., "Air Pollution Foundation—What It Is and What It Does," SAE Technical Paper 550279 (1955), doi:https://doi.org/10.4271/550279.

13.3.   Goldsmith, J.R. and Rogers, L.H., "Health Hazards of Automobile Exhaust," *Public Health Reports* 74, no. 6 (1959): 551-558.

13.4.   Schoettlin, C.E. and Landau, E., "Air Pollution and Asthmatic Attacks in the Los Angeles Area," *Public Health Reports* 76, no. 6 (1961): 545-551.

13.5.   Haagen-Smit, A.J. and Fox, M.M., "Automobile Exhaust and Ozone Formation," SAE Technical Paper 550277 (1955), doi:https://doi.org/10.4271/550277.

13.6.   Larson, G.P., Fischer, G.I., and Hamming, W.J., "Evaluating Sources of Air Pollution," *Industrial & Engineering Chemistry* 45, no. 5 (1953): 1070-1074.

13.7.   Hutchison, D.H. and Holden, F.R., "An Inventory of Automobile Gases," SAE Technical Paper 550278 (1955), doi:https://doi.org/10.4271/550278.

13.8.   McCabe, L.J. and Koehl, W.J., "The Inter-Industry Emission Control Program—Eleven Years of Progress in Automotive Emissions and Fuel Economy Research," SAE Technical Paper 780588 (1978), doi:https://doi.org/10.4271/780588.

13.9.   Lawson, D.R., Groblicki, P.J., Stedman, D.H., Bishop, G.A. et al., "Emissions from In-Use Motor Vehicles in Los Angeles: A Pilot Study of Remote Sensing and the Inspection and Maintenance Program," *Journal of the Air & Waste Management Association* 40, no. 8 (1990): 1096-1105.

13.10.   Shelef, M., "Unanticipated Benefits of Automotive Emission Control: Reduction in Fatalities by Motor Vehicle Exhaust Gas," SAE Technical Paper 922335 (1992), doi:https://doi.org/10.4271/922335.

13.11.   Wilkins, E.T., "Exhaust Gases from Motor Vehicles: Some Measurements of Carbon Monoxide in the Air of London," *Journal of the Royal Society of Health* 4: 677-687.

13.12.   Automotive News Europe, "EU, Germany Reach Deal to Allow e-Fuels after 2035 Law Ends Sales of CO2-Emitting Cars," March 25, 2023, accessed June 2023, https://europe.autonews.com/environmentemissions/eu-germany-reach-car-emissions-deal-includes-e-fuels.

13.13.   Automotive News Europe, "Europe's Euro 7 Emission Limits Face Pushback from Eight Countries," May 22, 2023, accessed June 5, 2023, https://europe.autonews.com/environmentemissions/eight-eu-nations-call-scrapping-euro-7-emission-rules.

13.14.   Miyazaki, T. and Nishimoto, T., "Motor Vehicle Emission Control Measures of Japan," SAE Technical Paper 922178 (1992), doi:https://doi.org/10.4271/922178.

13.15.   Clossey, T.J., DeJovine, J.M., McHugh, K.J., Paulsen, D.A. et al., "The EC-X Test Program—Reformulated Gasoline for Lower Vehicle Emissions," SAE Technical Paper 920798 (1992), doi:https://doi.org/10.4271/920798.

13.16.   Cohu, L.K., Rapp, L.A., and Segal, J.S., *EC-1 Emission Control Gasoline* (Anaheim, CA: ARCO Products Co., 1989).

13.17.   Hoekman, S.K., Leland, A., and Bishop, G., "Diminishing Benefits of Federal Reformulated Gasoline (RFG) Compared to Conventional Gasoline (CG)," *SAE Int. J. Fuels Lubr.* 12, no. 1 (2019): 5-28, doi:https://doi.org/10.4271/04-12-01-0001.

13.18.   Morgan, T.D.B., den Otter, G.J., Lange, W.W., Doyon, J. et al., "An Integrated Study of the Effects of Gasoline Composition on Exhaust Emissions Part I: Programme Outline and Results on Regulated Emissions," SAE Technical Paper 932678 (1993), doi:https://doi.org/10.4271/932678.

13.19.   Schoonveld, G.A. and Marshall, W.F., "The Total Effect of a Reformulated Gasoline on Vehicle Emissions by Technology (1973 to 1989)," SAE Technical Paper 910380 (1991), doi:https://doi.org/10.4271/910380.

13.20.   den Otter, G.J., Malpas, R.E., and Morgan, T.D.B., "Effect of Gasoline Reformulation on Exhaust Emissions in Current European Vehicles," SAE Technical Paper 930372 (1993), doi:https://doi.org/10.4271/930372.

13.21.   Hunter, J.E., "Studies of Catalyst Degradation in Automotive Emission Control Systems," SAE Technical Paper 720122 (1972), doi:https://doi.org/10.4271/720122.

13.22.   Furey, R.L. and Monroe, D.R., "Fuel Sulfur Effects on the Performance of Automotive Three-Way Catalysts during Vehicle Emissions Tests," SAE Technical Paper 811228 (1981), doi:https://doi.org/10.4271/811228.

13.23.   Williamson, W.B., Gandhi, H.S., Heyde, M.E., and Zawacki, G.A., "Deactivation of Three-Way Catalysts by Fuel Contaminants—Lead, Phosphorus and Sulfur," SAE Technical Paper 790942 (1979), doi:https://doi.org/10.4271/790942.

13.24.   Pierson, W.R., Hammerle, R.H., and Kummer, J.T., "Sulfuric Acid Aerosol Emissions from Catalyst-Equipped Engines," SAE Technical Paper 740287 (1974), doi:https://doi.org/10.4271/740287.

13.25.   Trayser, D.A., Blosser, E.R., Creswick, F.A., and Pierson, W.R., "Sulfuric Acid and Nitrate Emissions from Oxidation Catalysts," SAE Technical Paper 750091 (1975), doi:https://doi.org/10.4271/750091.

13.26.   Hammerle, R.H. and Mikkor, M., "Some Phenomena Which Control Sulfuric Acid Emission from Automotive Catalysts," SAE Technical Paper 750097 (1975), doi:https://doi.org/10.4271/750097.

13.27. Creswick, F.A., Blosser, E.R., Trayser, D.A., and Foster, J.F., "Sulfuric Acid Emissions from an Oxidation-Catalyst Equipped Vehicle," SAE Technical Paper 750411 (1975), doi:https://doi.org/10.4271/750411.

13.28. CRC, "Review of Prior Studies of Fuel Effects on Vehicle Emissions," CRC Report E-84, 2008.

13.29. Joy, G.C., Lester, G.R., and Molinaro, F.S., "The Influence of Sulfur Species on the Laboratory Performance of Automotive Three Component Control Catalysts," SAE Technical Paper 790943 (1979), doi:https://doi.org/10.4271/790943.

13.30. Köhler, U. and Wassmuth, H.-W., "Surface Reactions of Sulfur with Oxygen on Pt. (111)," *Surface Science* 117, no. 1-3 (1982): 668-675.

13.31. Köhler, U. and Wassmuth, H.-W., "SO_2 Adsorption and Desorption Kinetics on Pt. (111)," *Surface Science* 126, no. 1–3 (1983): 448-454.

13.32. Gutleben, H. and Bechtold, E., "Desorption of Sulfur from Pt(100)," *Surface Science* 191, no. 1–2 (1987): 157-173.

13.33. Burke, M.L. and Madix, R.J., "SO_2 Structure and Reactivity on Clean and Sulfur Modified Pd(100)," *Surface Science* 194, no. 1-2 (1988): 223-244.

13.34. Datta, A., Cavell, R.G., Tower, R.W., and George, Z.M., "Claus Catalysis. 1. Adsorption of Sulfur Dioxide on the Alumina Catalyst Studied by FTIR and EPR Spectroscopy," *Journal of Physical Chemistry* 89, no. 3 (1985): 443-449.

13.35. Henk, M.G., White, J.J., and Denison, G.W., "Sulfur Storage and Release from Automotive Catalysts," SAE Technical Paper 872134 (1987), doi:https://doi.org/10.4271/872134.

13.36. Diwell, A.F., Hallett, C., and Taylor, J.R., "The Impact of Sulphur Storage on Emissions from Three-Way Catalysts," SAE Technical Paper 872163 (1987), doi:https://doi.org/10.4271/872163.

13.37. Pahl, R.H. and McNally, M.J., "Fuel Blending and Analysis for the Auto/Oil Air Quality Improvement Research Program," SAE Technical Paper 902098 (1990), doi:https://doi.org/10.4271/902098.

13.38. Takei, Y., Hoshi, H., Kato, M., Okada, M. et al., "Effects of California Phase 2 Reformulated Gasoline Specifications on Exhaust Emission Reduction," SAE Technical Paper 922179 (1992), doi:https://doi.org/10.4271/922179.

13.39. Benson, J.D., Burns, V., Koehl, W.J., Gorse, R.A. et al., "Effects of Gasoline Sulfur Level on Mass Exhaust Emissions—Auto/Oil Air Quality Improvement Research Program," SAE Technical Paper 912323 (1991), doi:https://doi.org/10.4271/912323.

13.40. Petit, A., Jeffrey, J.G., Palmer, F.H., and Steinbrink, R., "European Programme on Emissions, Fuels and Engine Technologies (EPEFE)—Emissions from Gasoline Sulphur Study," SAE Technical Paper 961071 (1996), doi:https://doi.org/10.4271/961071.

13.41. Koehl, W.J., Benson, J.D., Burns, V.R., Gorse, R.A. et al., "Effects of Gasoline Sulfur Level on Exhaust Mass and Speciated Emissions: The Question of Linearity—Auto/Oil Air Quality Improvement Program," SAE Technical Paper 932727 (1993), doi:https://doi.org/10.4271/932727.

13.42. Miyoshi, N., Matsumoto, S., Katoh, K., Tanaka, T. et al., "Development of New Concept Three-Way Catalyst for Automotive Lean-Burn Engines," SAE Technical Paper 950809 (1995), doi:https://doi.org/10.4271/950809.

13.43. Brogan, M.S., Clark, A.D., and Brisley, R.J., "Recent Progress in NO_X Trap Technology," SAE Technical Paper 980933 (1998), doi:https://doi.org/10.4271/980933.

13.44. Maunula, T., Vakkilainen, A., Heikkinen, R., and Härkönen, M., "NO_X Storage and Reduction on Differentiated Chemistry Catalysts for Lean Gasoline Vehicles," SAE Technical Paper 2001-01-3665 (2001), doi:https://doi.org/10.4271/2001-01-3665.

13.45. Theis, J.R., Li, J.J., Ura, J.A., and Hurley, R.G., "The Desulfation Characteristics of Lean NO_X Traps," SAE Technical Paper 2002-01-0733 (2002), doi:https://doi.org/10.4271/2002-01-0733.

13.46. Hachisuka, I., Hirata, H., Ikeda, Y., and Matsumoto, S., "Deactivation Mechanism of NO_X Storage-Reduction Catalyst and Improvement of Its Performance," SAE Technical Paper 2000-01-1196 (2000), doi:https://doi.org/10.4271/2000-01-1196.

13.47. Guyon, M., Blejean, F., Bert, C., and Faou, L., "Impact of Sulfur on NO_X Trap Catalyst Activity - Study of the Regeneration Conditions," SAE Technical Paper 982607 (1998), doi:https://doi.org/10.4271/982607.

13.48. Ura, J.A., Goralski, C.T., Graham, G.W., McCabe, R.W. et al., "Laboratory Study of Lean NO_X Trap Desulfation Strategies," SAE Technical Paper 2005-01-1114 (2005), doi:https://doi.org/10.4271/2005-01-1114.

13.49. Asik, J.R., Dobson, D.A., and Meyer, G.M., "Suppression of Sulfide Emission during Lean NO_X Trap Desulfation," SAE Technical Paper 2001-01-1299 (2001), doi:https://doi.org/10.4271/2001-01-1299.

13.50. Elwart, S., Surnilla, G., Ura, J., and Theis, J., "H₂S Suppression during the Desulfation of a Lean NO$_X$ Trap with a Nickel-Containing Catalyst," SAE Technical Paper 2005-01-1116 (2005), doi:https://doi.org/10.4271/2005-01-1116.

13.51. Astegger, S. and Bechtold, E., "Adsorption of Sulfurdioxide and the Interaction of Coadsorbedoxygen and Sulfur on Pt(111)," *Surface Science* 122, no. 3 (1982): 491-504.

13.52. Burke, M.L. and Madix, R.J., "Formation of Adsorbed Sulfate from the Oxidation of Sulfur Dioxide on Palladium(100)," *The Journal of Physical Chemistry* 92, no. 7 (1988): 1974-1981.

13.53. Ku, R.C. and Wynblatt, P., "SO₂ Adsorption on Rh(110) and Pt(110) Surfaces," *Applications of Surface Science* 8, no. 3 (1981): 250-259.

13.54. Beck, D.D., Krueger, M.H., and Monroe, D.R., "The Impact of Sulfur on Three-Way Catalysts: Storage and Removal," SAE Technical Paper 910844 (1991), doi:https://doi.org/10.4271/910844.

13.55. Gandhi, H.S., Yao, H.C., Stepien, H.K., and Shelef, M., "Evaluation of Three-Way Catalysts - Part III Formation of NH3, Its Suppression by SO₂ and Re-Oxidation," SAE Technical Paper 780606 (1978), doi:https://doi.org/10.4271/780606.

13.56. Alfonso, D.R., "Initial Incorporation of Sulfur into the Pd(111) Surface: A Theoretical Study," *Surface Science* 600, no. 19 (2006): 4508-4516.

13.57. Calhoun, J.C., Blass, G.F., Zemke, B.E., and Evernham, T.W., "Changing Inspection and Maintenance Requirements: … A Result of New Emission Control Technology," SAE Technical Paper 790783 (1979), doi:https://doi.org/10.4271/790783.

13.58. Baltusis, P., "On Board Vehicle Diagnostics," SAE Technical Paper 2004-21-0009 (2004), doi:https://doi.org/10.4271/2004-21-0009.

13.59. Clemmens, W.B., Sabourin, M.A., and Rao, T., "Detection of Catalyst Performance Loss Using On-Board Diagnostics," SAE Technical Paper 900062 (1990), doi:https://doi.org/10.4271/900062.

13.60. Koupal, J.W., Sabourin, M.A., and Clemmens, W.B., "Detection of Catalyst Failure On-Vehicle Using the Dual Oxygen Sensor Method," SAE Technical Paper 910561 (1991), doi:https://doi.org/10.4271/910561.

13.61. Hepburn, J.S., Dobson, D.A., Hubbard, C.P., Guldberg, S.O. et al., "A Review of the Dual EGO Sensor Method for OBD-II Catalyst Efficiency Monitoring," SAE Technical Paper 942057 (1994), doi:https://doi.org/10.4271/942057.

13.62. Hepburn, J.S. and Gandhi, H.S., "The Relationship between Catalyst Hydrocarbon Conversion Efficiency and Oxygen Storage Capacity," SAE Technical Paper 920831 (1992), doi:https://doi.org/10.4271/920831.

13.63. Beck, D.D., Silvis, T.W., and Mahan, S.T., "Impact of Fuel Sulfur on OBD-II Catalyst Monitoring Using the Dual Oxygen Sensor Approach," SAE Technical Paper 941054 (1994), doi:https://doi.org/10.4271/941054.

13.64. Summers, J.C., Skowron, J.F., Williamson, W.B., and Mitchell, K.I., "Fuel Sulfur Effects on Automotive Catalyst Performance," SAE Technical Paper 920558 (1992), doi:https://doi.org/10.4271/920558.

13.65. Beckwith, P., Bennett, P.J., Goodfellow, C.L., Brisley, R.J. et al., "The Effect of Three-Way Catalyst Formulation on Sulphur Tolerance and Emissions from Gasoline Fuelled Vehicles," SAE Technical Paper 940310 (1994), doi:https://doi.org/10.4271/940310.

13.66. Rieck, J.S., Collins, N.R., and Moore, J.S., "OBD-II Performance of Three-Way Catalysts," SAE Technical Paper 980665 (1998), doi:https://doi.org/10.4271/980665.

13.67. Tauster, S., Rabinowitz, H., and Heck, R., "Understanding Sulfur Interaction Key to OBD of Low Emission Vehicles," SAE Technical Paper 2000-01-2929 (2000), doi:https://doi.org/10.4271/2000-01-2929.

13.68. Ohata, A., Ohashi, M., Nasu, M., and Inoue, T., "Model Based Air Fuel Ratio Control for Reducing Exhaust Gas Emissions," SAE Technical Paper 950075 (1995), doi:https://doi.org/10.4271/950075.

13.69. Carnevale, C., Coin, D., Secco, M., and Tubetti, P., "A/F Ratio Control with Sliding Mode Technique," SAE Technical Paper 950075 (1995), doi:https://doi.org/10.4271/950075.

13.70. Browning, L.H. and Moyer, C.B., "Sulfur Effects on California OBD-II Systems," SAE Technical Paper 952422 (1995), doi:https://doi.org/10.4271/952422.

13.71. Beck, D.D. and Short, W.A., "Vehicle Testing of the OBD-II Catalyst Monitor on a 2.2 L Corsica TLEV," SAE Technical Paper 952424 (1995), doi:https://doi.org/10.4271/952424.

13.72. Hepburn, J., Sweppy, M., and Zaghati, Z., "The Effect of Fuel Sulfur on the OBD-II Catalyst Monitor," SAE Technical Paper 972855 (1997), doi:https://doi.org/10.4271/972855.

13.73.   Hajbabaei, M., Karavalakis, G., Miller, J.W., Villela, M. et al., "Impact of Olefin Content on Criteria and Toxic Emissions from Modern Gasoline Vehicles," *Fuel* 107 (2013): 671-679.

13.74.   Karavalakis, G., Short, D., Vu, D., Russell, R. et al., "Evaluating the Effects of Aromatics Content in Gasoline on Gaseous and Particulate Matter Emissions from SI-PFI and SIDI Vehicles," *Environmental Science & Technology* 49, no. 11 (2015): 7021-7031.

13.75.   Peng, J., Hu, M., Du, Z., Wang, Y. et al., "Gasoline Aromatics: A Critical Determinant of Urban Secondary Organic Aerosol Formation," *Atmospheric Chemistry and Physics* 17, no. 17 (2017): 10743-10752.

13.76.   Gross, G.P., "The Effect of Fuel and Vehicle Variables on Polynuclear Aromatic Hydrocarbon and Phenol Emissions," SAE Technical Paper 720210 (1972), doi:https://doi.org/10.4271/720210.

13.77.   Environmental Protection Agency, "Health Effects of Benzene: A Review," EPA 560/5-76-003, Environmental Protection Agency, 1976.

13.78.   Marshall, W.F. and Gurney, M.D., "Effect of Gasoline Composition on Emissions of Aromatic Hydrocarbons," SAE Technical Paper 892076 (1989), doi:https://doi.org/10.4271/892076.

13.79.   Wimmer, D.B. and McReynolds, L.A., "Nitrogen Oxides and Engine Combustion," SAE Technical Paper 620561 (1962), doi:https://doi.org/10.4271/620561.

13.80.   Carr, R.C., Starkman, E.S., and Sawyer, R.F., "The Influence of Fuel Composition on Emissions of Carbon Monoxide and Oxides of Nitrogen," SAE Technical Paper 700470 (1970), doi:https://doi.org/10.4271/700470.

13.81.   van den Brink, P.J. and McDonald, C.R., "The Influence of the Fuel Hydrocarbon Composition on NO Conversion in 3-Way Catalysts: The NO_X/Aromatics Effect," SAE Technical Paper 952399 (1995), doi:https://doi.org/10.4271/952399.

13.82.   Clark, N.N., McKain, D.L. Jr., Klein, T., and Higgins, T.S., "Quantification of Gasoline-Ethanol Blend Emissions Effects," *Journal of the Air & Waste Management Association* 71, no. 1 (2021): 3-22.

13.83.   Bascunana, J.L. and Stahman, R.C., "Impact of Gasoline Characteristics on Fuel Economy," SAE Technical Paper 780628 (1978), doi:https://doi.org/10.4271/780628.

13.84.   Newhall, H.K., "Control of Nitrogen Oxides by Exhaust Recirculation - A Preliminary Theoretical Study," SAE Technical Paper 670495 (1967), doi:https://doi.org/10.4271/670495.

13.85.   Kataoka, K., Tsurusaki, M., and Kadota, T., "Effect of Fuel Properties on the Combustion Process and NO Emission in a Spark Ignition Engine," SAE Technical Paper 931940 (1993), doi:https://doi.org/10.4271/931940.

13.86.   D'Alleva, B.A. and Lovell, W.G., "Relation of Exhaust Gas Composition to Air-Fuel Ratio," SAE Technical Paper 360106 (1936), doi:https://doi.org/10.4271/360106.

13.87.   Ellis, J.C., "Gasolines for Low-Emission Vehicles," SAE Technical Paper 730616 (1973), doi:https://doi.org/10.4271/730616.

13.88.   Quader, A.A., "How Injector, Engine, and Fuel Variables Impact Smoke and Hydrocarbon Emissions with Port Fuel Injection," SAE Technical Paper 890623 (1989), doi:https://doi.org/10.4271/890623.

13.89.   Rapp, L.A., Benson, J.D., Burns, V.R., Gorse, R.A. et al., "Effects of Fuel Properties on Mass Exhaust Emissions during Various Modes of Vehicle Operation," SAE Technical Paper 932726 (1993), doi:https://doi.org/10.4271/932726.

13.90.   Le Jeune, A., McDonald, C.R., and Lee, R.G., "NO_X Aromatics Effects in Catalyst-Equipped Gasoline Vehicles," SAE Technical Paper 941869 (1994), doi:https://doi.org/10.4271/941869.

13.91.   Hochhauser, A.M., Benson, J.D., Burns, V., Gorse, R.A. et al., "The Effect of Aromatics, MTBE, Olefins and T90 on Mass Exhaust Emissions from Current and Older Vehicles—The Auto/Oil Air Quality Improvement Research Program," SAE Technical Paper 912322 (1991), doi:https://doi.org/10.4271/912322.

13.92.   Goodfellow, C.L., Gorse, R.A., Hawkins, M.J., and McArragher, J.S., "European Programme on Emissions, Fuels and Engine Technologies (EPEFE)—Gasoline Aromatics/E100 Study," SAE Technical Paper 961072 (1996), doi:https://doi.org/10.4271/961072.

13.93.   Hamasaki, M., Yamaguchi, M., and Hirose, K., "Japan Clean Air Program (JCAP) - Step 1 Study of Gasoline Vehicle and Fuel Influence on Emissions," SAE Technical Paper 2000-01-1972 (2000), doi:https://doi.org/10.4271/2000-01-1972.

13.94.   Saitoh, K. and Hamasaki, M., "Effects of Sulfur, Aromatics, T50, T90 and MTBE on Mass Exhaust Emissions from Vehicles with Advanced Technology—JCAP Gasoline WG STEP II Report," SAE Technical Paper 2003-01-1905 (2003), doi:https://doi.org/10.4271/2003-01-1905.

13.95.   Thummadetsak, T., Wuttimongkolchai, A., Tunyapisetsak, S., and Kimura, T., "Effect of Gasoline Compositions and Properties on Tailpipe Emissions of Currently Existing Vehicles in Thailand," SAE Technical Paper 1999-01-3570 (1999), doi:https://doi.org/10.4271/1999-01-3570.

13.96.   International Agency for Research on Cancer, "Polynuclear Aromatic Compounds. Part 1. Chemical, Environmental and Experimental Data," IARC Monographs on the Evaluation of the Carcinogenic Risk of Chemicals to Humans, Vol. 32, 1989.

13.97.   Laity, J.L., Malbin, M.D., Haskell, W.W., and Doty, W.I., "Mechanisms of Polynuclear Aromatic Hydrocarbon Emissions from Automotive Engines," SAE Technical Paper 730835 (1973), doi:https://doi.org/10.4271/730835.

13.98.   Takei, Y., Hoshi, H., Okada, M., and Abe, K., "Effect of Gasoline Components on Exhaust Hydrocarbon Components," SAE Technical Paper 932670 (1993), doi:https://doi.org/10.4271/932670.

13.99.   Nagel, H., Frey, R., Hartgerink, C., Rikeit, H.E. et al., "On-Line Analysis of Individual Aromatic Hydrocarbons in Automotive Exhaust: Dealkylation of the Aromatic Hydrocarbons in the Catalytic Converter," SAE Technical Paper 971606 (1997), doi:https://doi.org/10.4271/971606.

13.100.   Kazemiparkouhi, F., Karavalakis, G., Falconi, T.M.A., MacIntosh, D.L. et al., "Comprehensive US Database and Model for Ethanol Blend Effects on Air Toxics, Particle Number, and Black Carbon Tailpipe Emissions," *Atmospheric Environment: X* 16 (2022): 100185.

13.101.   Azev, V.S., Emel'yanov, V.E., and Turovskii, F.V., "Automotive Gasolines. Long-Term Requirements for Composition and Properties," *Chemistry and Technology of Fuels and Oils* 40, no. 5 (2004): 291-297.

13.102.   Dempster, N.M. and Shore, P.R., "An Investigation into the Production of Hydrocarbon Emissions from a Gasoline Engine Tested on Chemically Defined Fuels," SAE Technical Paper 900354 (1990), doi:https://doi.org/10.4271/900354.

13.103.   Pelz, N., Dempster, N.M., Hundleby, G.E., and Shore, P.R., "The Composition of Gasoline Engine Hydrocarbon Emissions—An Evaluation of Catalyst and Fuel Effects," SAE Technical Paper 902074 (1990), doi:https://doi.org/10.4271/902074.

13.104.   Schleyer, C.H., Koehl, W.J., Leppard, W.R., Dunker, A.M. et al., "Effect of Gasoline Olefin Composition on Predicted Ozone in 2005/2010—Auto/Oil Air Quality Improvement Research Program," SAE Technical Paper 940579 (1994), doi:https://doi.org/10.4271/940579.

13.105.   Menrad, H., Bernhardt, W., and Decker, G., "Methanol Vehicles of Volkswagen—A Contribution to Better Air Quality," SAE Technical Paper 881196 (1988), doi:https://doi.org/10.4271/881196.

13.106.   Nishide, H., Yata, T., Hirota, T., Fujiwara, H. et al., "Performance and Exhaust Emissions of Nissan FFV NX Coupe," SAE Technical Paper 920299 (1992), doi:https://doi.org/10.4271/920299.

13.107.   Hüttebräucker, D., "The Flexible Fuel Concept by Mercedes-Benz," SAE Technical Paper 921456 (1992), doi:https://doi.org/10.4271/921456.

13.108.   Cowart, J., Boruta, W., Dalton, J., Dona, R. et al., "Powertrain Development of the 1996 Ford Flexible Fuel Taurus," SAE Technical Paper 952751 (1995), doi:https://doi.org/10.4271/952751.

13.109.   Kar, K. and Cheng, W., "Speciated Engine-Out Organic Gas Emissions from a PFI-SI Engine Operating on Ethanol/Gasoline Mixtures," *SAE Int. J. Fuels Lubr.* 2, no. 2 (2010): 91-101, doi:https://doi.org/10.4271/2009-01-2673.

13.110.   Tanaka, H., Matsumoto, T., Funaki, R., Kato, T. et al., "Effects of Ethanol or ETBE Blending in Gasoline on Evaporative Emissions for Japanese In-Use Passenger Vehicles," SAE Technical Paper 2007-01-4005 (2007), doi:https://doi.org/10.4271/2007-01-4005.

13.111.   Yassine, M. and La Pan, M., "Impact of Ethanol Fuels on Regulated Tailpipe Emissions," SAE Technical Paper 2012-01-0872 (2012), doi:https://doi.org/10.4271/2012-01-0872.

13.112.   Stump, F.D., Knapp, K.T., and Ray, W.D., "Influence of Ethanol-Blended Fuels on the Emissions from Three Pre-1985 Light-Duty Passenger Vehicles," *Journal of the Air & Waste Management Association* 46, no. 12 (1996): 1149-1161.

13.113.   Winebrake, J.J. and Deaton, M.L., "Hazardous Air Pollution from Mobile Sources: A Comparison of Alternative Fuel and Reformulated Gasoline Vehicles," *Journal of the Air & Waste Management Association* 49, no. 5 (1999): 576-581.

13.114.   Anderson, L.G., "Ethanol Fuel Use in Brazil: Air Quality Impacts," *Energy & Environmental Science* 2, no. 10 (2009): 1015-1037.

13.115. Poulopoulos, S.G. and Philippopoulos, C.J., "The Effect of Adding Oxygenated Compounds to Gasoline on Automotive Exhaust Emissions," *J. Eng. Gas Turbines Power* 125, no. 1 (2003): 344-350.

13.116. Singh, P.K., Ramadhas, A.S., Mathai, R., and Sehgal, A.K., "Investigation on Combustion, Performance and Emissions of Automotive Engine Fueled with Ethanol Blended Gasoline," *SAE Int. J. Fuels Lubr.* 9, no. 1 (2016): 215-223, doi:https://doi.org/10.4271/2016-01-0886.

13.117. Karavalakis, G., Durbin, T.D., Shrivastava, M., Zheng, Z. et al., "Impacts of Ethanol Fuel Level on Emissions of Regulated and Unregulated Pollutants from a Fleet of Gasoline Light-Duty Vehicles," *Fuel* 93 (2012): 549-558.

13.118. Szybist, J.P., Youngquist, A.D., Barone, T.L., Storey, J.M. et al., "Ethanol Blends and Engine Operating Strategy Effects on Light-Duty Spark-Ignition Engine Particle Emissions," *Energy & Fuels* 25, no. 11 (2011): 4977-4985.

13.119. Karavalakis, G., Short, D., Vu, D., Villela, M. et al., "Regulated Emissions, Air Toxics, and Particle Emissions from SI-DI Light-Duty Vehicles Operating on Different ISO-Butanol and Ethanol Blends," *SAE Int. J. Fuels Lubr.* 7, no. 1 (2014): 183-199, doi:https://doi.org/10.4271/2014-01-1451.

13.120. Chen, L., Braisher, M., Crossley, A., Stone, R. et al., "The Influence of Ethanol Blends on Particulate Matter Emissions from Gasoline Direct Injection Engines," SAE Technical Paper 2010-01-0793 (2010), doi:https://doi.org/10.4271/2010-01-0793.

13.121. Ratcliff, M.A., Windom, B., Fioroni, G.M., John, P.S. et al., "Impact of Ethanol Blending into Gasoline on Aromatic Compound Evaporation and Particle Emissions from a Gasoline Direct Injection Engine," *Applied Energy* 250 (2019): 1618-1631.

13.122. Aikawa, K., Sakurai, T., and Jetter, J.J., "Development of a Predictive Model for Gasoline Vehicle Particulate Matter Emissions," *SAE Int. J. Fuels Lubr.* 3, no. 2 (2010): 610-622, doi:https://doi.org/10.4271/2010-01-2115.

13.123. ASTM International, "Standard Test Method for Determination of Individual Components in Spark Ignition Engine Fuels by 100 Metre Capillary High Resolution Gas Chromatography," ASTM D6729-20, ASTM International, 2020.

13.124. St. John, P.C., Kim, S., and McCormick, R.L., "Development of a Data-Derived Sooting Index Including Oxygen-Containing Fuel Components," *Energy & Fuels* 33, no. 10 (2019): 10290-10296.

13.125. Sobotowski, R., Butler, A., and Guerra, Z., "A Pilot Study of Fuel Impacts on PM Emissions from Light-Duty Gasoline Vehicles," *SAE Int. J. Fuels Lubr.* 8, no. 1 (2015): 214-233, doi:https://doi.org/10.4271/2015-01-9071.

13.126. Crawford, R. and Lyons, J., "Assessment of the Relative Accuracy of the PM Index and Related Methods," CRC Report No. RW-107-2, 2021.

13.127. Wittmann, J.H. and Menger, L., "Novel Index for Evaluation of Particle Formation Tendencies of Fuels with Different Chemical Compositions," *SAE Int. J. Fuels Lubr.* 10, no. 3 (2017): 690-697, doi:https://doi.org/10.4271/2017-01-9380.

13.128. ASTM International, "Standard Test Method for Hydrocarbon Types, Oxygenated Compounds, Benzene, and Toluene in Spark Ignition Engine Fuels by Multidimensional Gas Chromatography," ASTM D6839-21a, ASTM International, 2021.

13.129. Chapman, E., Winston-Galant, M., Geng, P., and Konzack, A., "Global Market Gasoline Range Fuel Review Using Fuel Particulate Emission Correlation Indices," SAE Technical Paper 2016-01-2251 (2016), doi:https://doi.org/10.4271/2016-01-2251.

13.130. Chapman, E., Salyers, J.T., Wispinski, D., Scussel, M. et al., "Comparison of the Particulate Matter Index and Particulate Evaluation Index Numbers Calculated by Detailed Hydrocarbon Analysis by Gas Chromatography (Enhanced ASTM D6730) and Vacuum Ultraviolet Paraffin, Isoparaffin, Olefin, Naphthene, and Aromatic Analysis (ASTM D8071)," SAE Technical Paper 2021-01-5070 (2021), doi:https://doi.org/10.4271/2021-01-5070.

13.131. Jessup, P.J., Croudace, M.C., and Wusz, T., "An Overview of Unocal's Low Emission Gasoline Research Program," SAE Technical Paper 920801 (1992), doi:https://doi.org/10.4271/920801.

13.132. Koehl, W.J., Benson, J.D., Burns, V., Gorse, R.A. et al., "Effects of Gasoline Composition and Properties on Vehicle Emissions: A Review of Prior Studies—Auto/Oil Air Quality Improvement Research Program," SAE Technical Paper 912321 (1991), doi:https://doi.org/10.4271/912321.

13.133. McArragher, J.S., Betts, W.E., Brandt, J., Kiessfing, D. et al., "Evaporative Emissions from Modern European Vehicles and Their Control," SAE Technical Paper 880315 (1988), doi:https://doi.org/10.4271/880315.

13.134. Bennett, P.J., Beckwith, P., Goodfellow, C.L., and Skårdalsmo, K., "The Effect of Gasoline RVP on Exhaust Emissions from Current European Vehicles," SAE Technical Paper 952526 (1995), doi:https://doi.org/10.4271/952526.

13.135. Jorgensen, S.W. and Benson, J.D., "A Correlation between Tailpipe Hydrocarbon Emissions and Driveability," SAE Technical Paper 962023 (1996), doi:https://doi.org/10.4271/962023.

13.136. Gething, J.A., "Distillation Adjustment: An Innovative Step to Gasoline Reformulation," SAE Technical Paper 910382 (1991), doi:https://doi.org/10.4271/910382.

13.137. Leppard, W.R., Benson, J.D., Burns, V.R., Gorse, R.A. et al., "How Heavy Hydrocarbons in the Fuel Affect Exhaust Mass Emissions: Modal Analysis—The Auto/Oil Air Quality Improvement Research Program," SAE Technical Paper 932724 (1993), doi:https://doi.org/10.4271/932724.

13.138. Nakayama, Y., Maruya, T., Oikawa, T., Fujiwara, M. et al., "Reduction of HC Emission from VTEC Engine during Cold-Start Condition," SAE Technical Paper 940481 (1994), doi:https://doi.org/10.4271/940481.

13.139. Fischer, H.C. and Brereton, G.J., "Fuel Injection Strategies to Minimize Cold-Start HC Emission," SAE Technical Paper 970040 (1997), doi:https://doi.org/10.4271/970040.

13.140. Samenfink, W., Albrodt, H., Frank, M., Gesk, M. et al., "Strategies to Reduce HC-Emissions during the Cold Starting of a Port Fuel Injected Gasoline Engine," SAE Technical Paper 2003-01-0627 (2003), doi:https://doi.org/10.4271/2003-01-0627.

13.141. Andrianov, D.I., Keynejad, F., Dingli, R., Voice, G. et al., "A Cold-Start Emissions Model of an Engine and Aftertreatment System for Optimisation Studies," SAE Technical Paper 2010-01-1274 (2010), doi:https://doi.org/10.4271/2010-01-1274.

13.142. Noda, N., Takahashi, A., and Mizuno, H., "In-line Hydrocarbon (HC) Adsorber System for Cold Start Emissions," SAE Technical Paper 970266 (1997), doi:https://doi.org/10.4271/970266.

13.143. Yamamoto, S., Matsushita, K., Etoh, S., and Takaya, M., "In-line Hydrocarbon (HC) Adsorber System for Reducing Cold-Start Emissions," SAE Technical Paper 2000-01-0892 (2000), doi:https://doi.org/10.4271/2000-01-0892.

13.144. Yamazaki, H., Endo, T., Ueno, M., and Sugaya, S., "Research on HC Adsorption Emission System," SAE Technical Paper 2004-01-1273 (2004), doi:https://doi.org/10.4271/2004-01-1273.

13.145. Zammit, M., Wuttke, J., Ravindran, P., and Aaltonen, S., "The Effects of Catalytic Converter Location and Palladium Loading on Tailpipe Emissions," SAE Technical Paper 2012-01-1247 (2012), doi:https://doi.org/10.4271/2012-01-1247.

Further Reading

CONCAWE, "Motor Vehicle Emission Regulations and Fuel Specifications: Part 1—2004/2005 Update," Report No. 5/06, CONCAWE, 2006.

CONCAWE, "Motor Vehicle Emission Regulations and Fuel Specifications: Appendix to Part 1 2004/2005 Update," Report No. 5/06, Appendix, CONCAWE, 2006.

CONCAWE, "Motor Vehicle Emission Regulations and Fuel Specifications: Part 2 Historic Review (1996-2005)," Report No. 6/06, CONCAWE, 2006.

CONCAWE, "Comparison of Particle Emissions from Advanced Vehicles Using DG TREN and PMP Measurement Protocols," Report No. 2/09, CONCAWE, 2009.

Cozzarini, C. and Lenz, H.P., *Emissions and Air Quality* (Warrendale, PA: SAE International, 1999).

Johnson, J.H., *Combustion & Emission Control for SI Engines—Modeling and Experimental Studies* (Warrendale, PA: SAE International, 2005a).

Johnson, J.H., *Emission Control and Fuel Economy for Port and Direct Injected SI Engines* (Warrendale, PA: SAE International, 2005b).

Kasab, J. and Strzelec, A., *Automotive Emissions Regulations and Exhaust Aftertreatment Systems* (Warrendale, PA: SAE International, 2020).

Kubsh, J.E., *Advanced Three-way Catalysts* (Warrendale, PA: SAE International, 2006).

Mondt, J.R., *Cleaner Cars—The History and Technology of Emission Control Since the 1960s* (Warrendale, PA: SAE International, 2000).

Pearson, J.K., *Improving Air Quality—Progress and Challenges for the Auto Industry* (Warrendale, PA: SAE International, 2001).

Zhao, F., *Technologies for Near-Zero-Emission Gasoline-Powered Vehicles* (Warrendale, PA: SAE International, 2006).

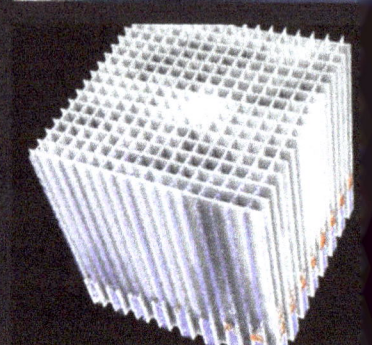

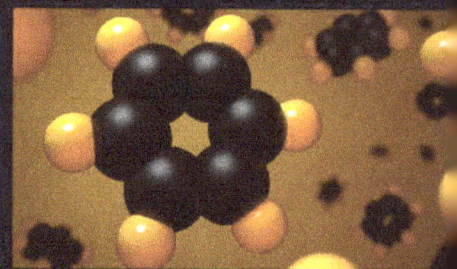

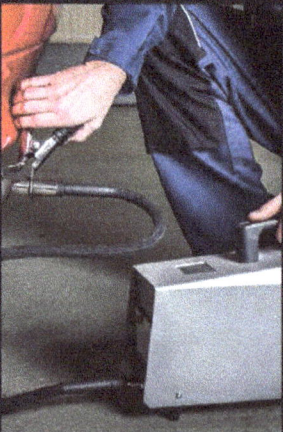

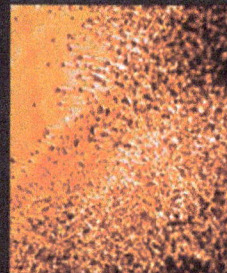

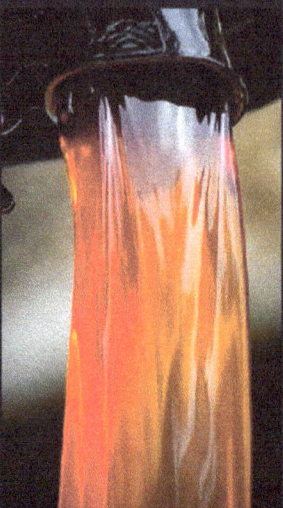

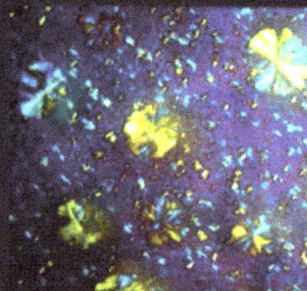

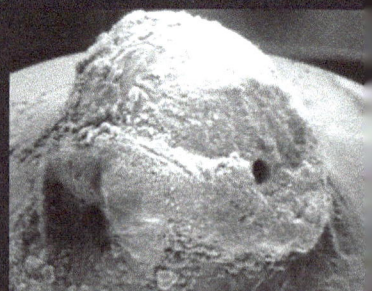

14

Racing Fuels

Almost as soon as motor vehicles were invented, people began to race them. The aim then, as now, was to go faster than the competition. This could involve covering a set distance in a shorter time or covering a greater distance in a set time. The distance can vary from a standing quarter mile (402.3 m) drag race to the 500-mile (804.7 km) Indianapolis 500 race, or over 5000 km that is now typically covered in a 24-hour endurance race. Drag racing in the US is administered by the National Hot Rod Association (NHRA), and the Indianapolis 500 race is administered by the Indy Racing League, LLC. The Le Mans 24-hour endurance race and the American Le Mans Series (ALMS) are run according to rules set by the Automobile Club de l'Ouest (ACO), which operates under the auspices of the Fédération Internationale de l'Automobile (FIA), which also sets the rules for the Formula One championship. Another major sanctioning body in the US is the National Association for Stock Car Auto Racing (NASCAR), which runs national series for both cars and trucks. The oldest and largest sanctioning US body is the International Motor Sport Association (IMSA). In Australia, the sanctioning controlling body is the Confederation of Australian Motor Sport (CAMS).

In the early years, motor racing was not as constrained by technical rules and regulations as it is today. This led to a vast variety of vehicles competing against one another. Over the years, this has resulted in a myriad of race formulae and even vehicle categories within the same event.

The mainstay of motor racing was the spark ignition engine. Though two-stroke [14.1], rotary [14.2], and even gas turbine [14.3] engines have been developed and shown success, the majority of cars are powered by four-stroke engines. Due to its superior fuel economy, attempts have been made to use the diesel engine as a power source for longer races, where the reduced number or lack of refueling stops outweigh any power disadvantage. In events that allow the use of diesel engines, such as the Le Mans 24-hour endurance race, diesel-powered vehicles, with appropriate aftertreatment [14.4], are now very successful.

Over the years, gasoline racing engines have employed the different means of fuel delivery systems described in Chapter 8, Section 8.3, including carburetors [14.5], port fuel injection [14.6], and direct fuel injection [14.7].

In most formulae of racing, there is now some form of control over both the vehicle and the fuel that is being used. The exact nature of these rules restricts the scope of every fuel developer because they not only control the characteristics of the final product, but they can also limit the sources of components and the ratios in which these components may be included in a fuel. For example, the international Formula One championship series the fuel specification is set by the FIA and the "2026 Formula 1 Power Unit Technical Regulations" [14.8] states

> *With regard to fuel, the detailed requirements of this Article are intended to ensure the use of Advanced Sustainable (AS) fuels comprising solely AS components, that are composed of certified compounds and refinery streams and fuel additives and to prohibit the use of specific power-boosting chemical compounds. The final, blended fuel must achieve a greenhouse gas (GHG) emissions savings, relative to fossil-derived gasoline, of at least that defined for the transport sector in the EU Renewable Energy Directive (RED), which was current on January 1st in the year prior to the relevant Formula One Championship.*

In other events the competitors must source their fuel from specified suppliers, for example, the NHRA publish a list accepted gasolines for given categories of racing [14.9]. However, in formulae where only basic rules or no rules apply, a race team who has the support of a competent fuel supplier can make the most of the freedom provided. The team can then maximize their performance utilizing the synergistic development of fuel, engine, and race demands. In such circumstances, it is only the costs that the supplier or the team is willing to incur that will be a limiting factor. Modern racing fuels can be tailored with absolute precision to give a competitor exactly the characteristics that are needed. The difference in performance between a well-formulated modern racing fuel and a more old-fashioned racing fuel, such as aviation gasoline (Avgas), clearly demonstrates this principle.

The development of racing fuel, especially for Formula One, is a closely guarded and secretive practice. The following section can therefore only give a broad outline from which the fuel technologist can work to formulate the ideal fuel for their application. As the vast majority of formulae still exclusively use gasoline engines, the discussion will be limited to fuel for these engines.

14.1. General Considerations

A primary objective in racing is to maximize engine power output consistent with any imposed restrictions on fuel quality and the ability of the engine to survive for the duration of the race or races. Power output is mainly increased by engine design features, but the use of specialized fuels to maximize power can play an important part in achieving higher performance than can be obtained by mechanical means alone.

In motorsport, the race regulations normally specify the permissible range of properties for the fuel. Within this specification, the challenge for fuel designers is to find the best compromise of properties for optimum performance. At the higher levels of the sport, this may stretch as far as formulating a fuel specifically for a particular event. In road racing, different circuits will have different proportions of acceleration to cruising, and this can influence the balance of fuel characteristics. However, in some series, such as the FIA Formula One championship, the same fuel must be used throughout. The race officials have the authority to compare any fuel sample with a reference supplied by the team at the start of the season. This is done by gas chromatography.

The factors that are important in a racing fuel are described in the following sections.

14.1.1. Safety

The fuel must be sufficiently stable to ensure that it does not generate an explosion hazard when being transported and handled under the normal use conditions. Its toxicity should be such that exposure of personnel handling the fuel is within the limits accepted for the components present in the fuel.

Pure low-molecular-weight aliphatic alcohols that are currently favored in motorsport (e.g., methanol and ethanol) will tend to burn with a colorless flame. While soluble metallic additives have been proposed [14.10] to make alcohol flames luminous, consideration of the deleterious effects of the metals usually precludes their use in a modern engine.

14.1.2. Volatility

The fuel must have a boiling range that allows it to be transported readily in the liquid form and yet volatilize satisfactorily in the engine. Depending upon the engine design, it must also be such that it avoids any vapor lock issues. A reduction in Reid vapor pressure to 48.3 kPa to avoid vapor lock problems was the innovative step of the first commercial racing gasoline. This gasoline was developed in 1955 by the Pure Oil Company (now part of Chevron Corporation) for NASCAR racing [14.11].

14.1.3. Resistance to Detonation and Pre-Ignition

These properties are vital because under high-speed, high-load racing conditions, these uncontrolled ignition phenomena can very rapidly destroy an engine. The relatively severe driving conditions used in racing may mean that motor octane number (MON) is generally more important than research octane number (RON) as a guide to the antiknock performance of the fuel, although for turbocharged or supercharged vehicles, the reverse may be true. However, under the rather extreme conditions in a racing engine, the RON and MON of a fuel may not have very much meaning in terms of predicting whether knock, or pre-ignition resulting from knock, will occur.

14.1.4. Flammability Limits

The flammability limits should be such that fuel-rich air-to-fuel ratios can be used. As combustion is never perfect, there is usually some unconsumed fuel or air. As the limit is usually on the amount of air that can pass through the engine, it is therefore important to fully utilize this air even if this requires wasting some fuel. Power is therefore maximized by running rich. This could be as much as 15% rich [14.6, 14.12].

14.1.5. Flame Speed

A fast burn rate is important in a racing engine because it helps to maintain good combustion under varying conditions such as speed, load, and air-to-fuel ratio. To ensure maximum thermal efficiency, the flame must have consumed all the available air as soon after top-dead-center of the pistons stroke as is possible [14.12]. As power is proportional to engine speed, the engine designer is always looking to develop higher engine speeds. In an engine operating at over 18,000 rpm or higher [14.13], the time available for complete combustion is becoming shorter and shorter. Branched olefins have been found to have a higher flame speed than alkanes. This is shown in Figure 14.1 [14.14].

FIGURE 14.1 Flame travel times for alkanes and branched olefins.

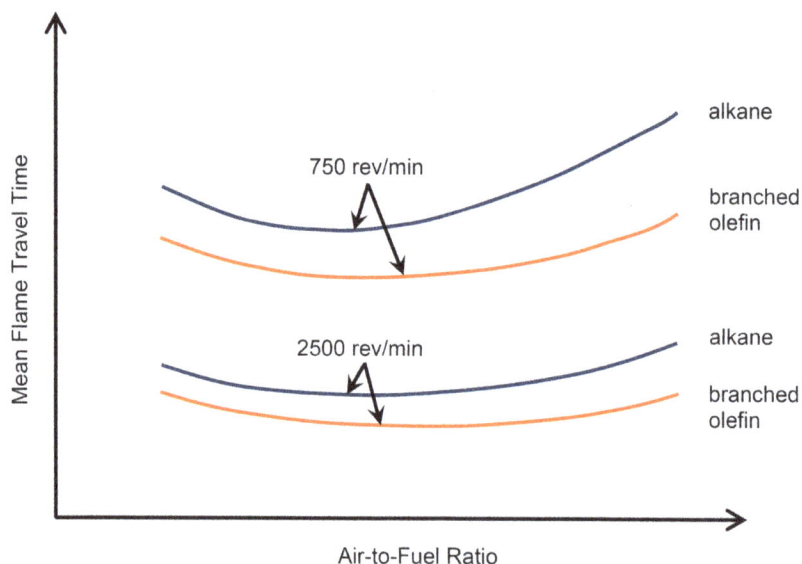

© SAE International.

14.1.6. Heat of Vaporization

A high heat of vaporization will increase the volumetric efficiency of an engine by cooling the intake charge as the fuel evaporates [14.15]. The increased mass of air in the cylinder allows for an increased mass of fuel and hence a higher energy density. However, as engine speed increases, the time available for fuel evaporation and charge cooling diminishes.

14.1.7. Density

There is often a fuel tank capacity limit applied to the vehicle design. Therefore, a higher density fuel allows a greater mass of fuel to be carried. This means that higher fuel consumption can be tolerated or fewer refueling stops required. This also means that more weight has to be carried. This trade-off must be considered with due regard to the fuel for the intended type of event.

14.1.8. Heating Value

The greater the energy content of the fuel, usually quoted as its heating value, the higher the possible energy output. However, as discussed in Section 14.1.10, the highest energy content does not necessarily result in the best fuel.

14.1.9. Stoichiometry

The lower the stoichiometric air-to-fuel ratio (or oxygen-to-fuel ratio, if a different oxidant than air is used), the more fuel can be burned for the amount of air (or oxidant) in the combustion chamber. The downside of this is a higher fuel consumption, meaning more fuel has to be carried.

14.1.10. Specific Energy

Clearly, the three parameters are very closely interrelated, and as explained next, there is usually a trade-off between these parameters.

  All racing engines are restricted to the amount of air that they can consume, either by an air inlet restrictor, their cubic capacity, or their speed, or any combination of these factors. Thus, assuming that the liquid fuel volume is negligible, the amount of fuel energy liberated per cycle depends not only on the fuel's unit energy content but also the useable air-to-fuel ratio. An important parameter that provides a method of comparing the heat released by different fuels in an engine is thus the specific energy (SE). The theoretical SE is calculated by dividing the heating value (usually the lower heating value) by the air-to-fuel ratio so that it represents the fuel energy delivered to the combustion chamber per unit mass of air inducted. Specific energy ratio is also used to compare fuels and is the ratio of the SE of the fuel to that of a standard reference fuel, usually iso-octane, since this is similar to a commercial gasoline. This is illustrated in Figure 14.2 [14.16], where iso-octane, methanol, and nitromethane at different equivalence ratios are compared to iso-octane at stoichiometry.

FIGURE 14.2   Theoretical SE ratio at different equivalence ratios.

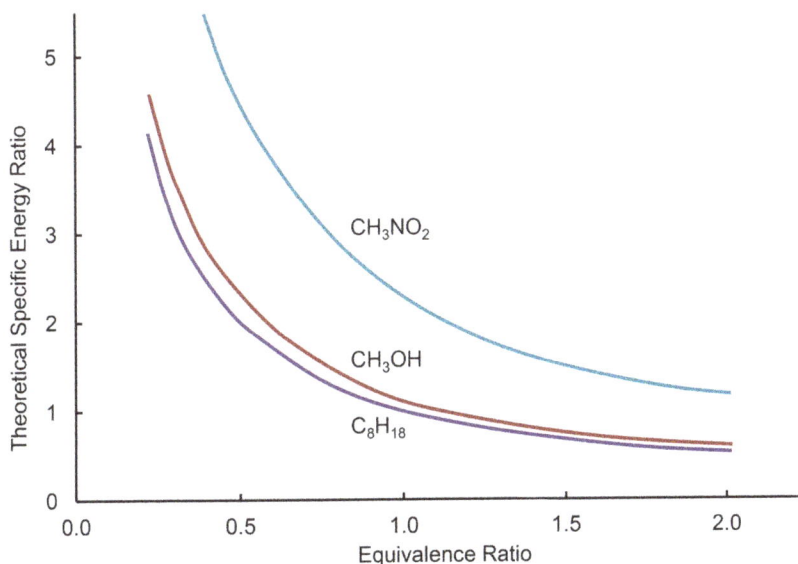

The theoretical values shown in Figure 14.2 assume complete combustion with no dissociation. A more realistic value of SE can be calculated, allowing for dissociation and heat losses from the cylinder walls. This is shown in Figure 14.3 [14.16].

FIGURE 14.3 SE ratio based on dissociation computations with heat loss at different equivalence ratios.

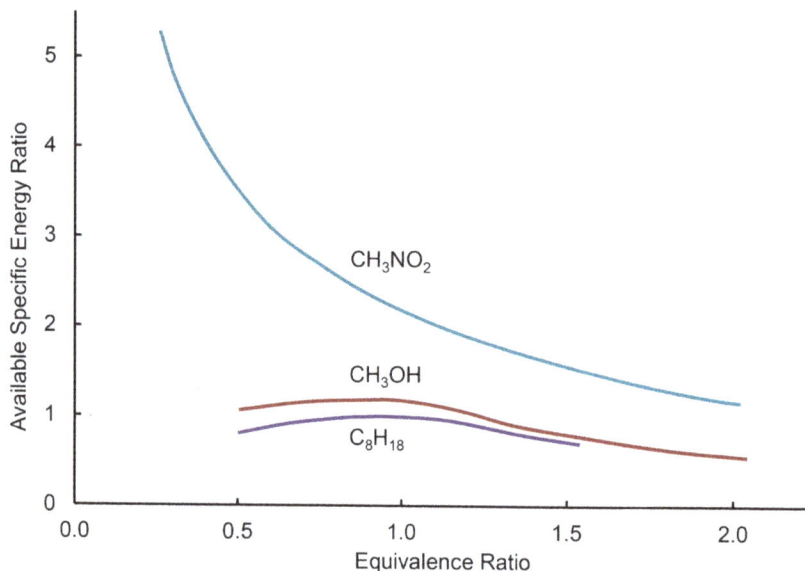

Bearing in mind that a racing engine usually operates rich of stoichiometry, it can be seen by comparing Figures 14.2 and 14.3 that the practical SE benefit of nitromethane is greater than may be expected from simple calculation. Table 14.1 [14.16] compares some characteristics of nitromethane and methanol to iso-octane. It can be seen that nitromethane has a theoretical SE of 2.3 times that of iso-octane and that methanol is only marginally better than iso-octane. The primary reason for this is the much lower stoichiometric air-to-fuel ratio of methanol and nitromethane. Thus, despite the fact that iso-octane has a higher energy content, it is usually a less-desirable racing fuel.

TABLE 14.1 Properties of some common racing fuels.

Property	Nitromethane	Methanol	Iso-octane
Formula	CH_3NO_2	CH_3OH	C_8H_{18}
Mol. wt.	61	32	114
Oxygen content (wt.%)	52.5	49.9	0
Stoichiometric air-to-fuel ratio	1.7:1	6.45:1	15.1:1
Lower heating value (MJ/kg)	11.3	19.9	44.3
Specific energy at stoichiometric	6.6	3.1	2.9
SE ratio	2.30	1.06	1.00
Heat of vaporization (MJ/kg)	0.56	1.17	0.27

© SAE International.

It is not possible to formulate a fuel that has the optimum level of all the characteristics listed, and so racing fuels must always be a compromise. The characteristics of individual fuels vary widely from one to another. It is therefore imperative that the engine and fuel are optimized as a combination. An engine that is designed for a specific fuel blend will not normally run efficiently on other blends.

14.2. Hydrocarbon Racing Fuels

Gasoline is the most commonly used racing fuel in view of its ready availability and low cost. Its greatest limitations are its relatively low specific energy and its tendency to knock at the high compression ratios necessary to maximize power output. The following sections look at how the racing fraternity and fuel producers have tried to overcome these shortcomings in the past and in the current climate of rules and regulation.

14.2.1. Historical Perspective

In the period between World War I and II, blends of the then-poor octane quality gasoline were used together with benzene and higher aromatics to increase the octane number of the fuel. Soon after World War I, General Motors patented an Avgas that became known as Hector Fuel. This fuel consisted of a 70:30 blend of cyclohexane and benzene [14.17], and this would have had an excellent octane quality and a high relative density. Blend components containing benzene would not be acceptable today because of the toxicity of this compound, and in any case, such blends that boil over a restricted temperature range do not always give good driveability.

Other proposed Avgas formulations have included the following:

- Up to 50% spiropentane [14.18], although 5 to 25% were suggested to be preferable.

- From 30 to 90% cyclopentane [14.19].

- From 20 to 60% of 1,1,2-rimethyl-cyclopropane [14.20].

- Up to 10% isopropyl benzene [14.21].

- From 2 to 7% of isomeric butyl benzenes [14.22].

Triptane has also been investigated as a possible Avgas component [14.23, 14.24] in view of its exceptionally high knock resistance, but it has the disadvantage of a high production cost.

14.2.2. Current Perspective

Nowadays, motor gasolines having high octane levels are not normally available commercially because of restrictions on the use of lead, although fuels with research octane levels of 100 or more and with motor octane levels of at least 88 have been marketed in the past. They were often highly leaded blends of butane, light catalytically cracked naphtha, and heavy catalytic reformate.

A fuel having a RON in the range of 110 to 120 and a MON from 105 to 110 has been suggested as being capable of satisfying most naturally aspirated piston engines operating under the most severe conditions likely to be encountered in racing [14.16]. Leaded blends of selected compounds would be needed to meet such a target. Specific hydrocarbons and, in particular, the highly branched paraffins such as iso-octane (2,2,4-trimethylpentane) and triptane (2,2,3-trimethylbutane), which have excellent antiknock properties, have been used in racing fuel blends to give acceptable antiknock performance. Table 14.2 summarizes the octane quality of some of the compounds that have been used in such blends [14.14].

TABLE 14.2 Octane qualities of some pure hydrocarbons.

Compound	Formula	RON	MON
Iso-octane	C_8H_{18}	100	100
Triptane	C_7H_{16}	112	101
Isodecane	$C_{10}H_{22}$	113	92
Cyclopentane	C_5H_{10}	101	85
Cyclohexane	C_6H_{12}	83	77
Toluene	C_7H_8	120	109
Xylene	C_8H_{10}	118	115

© SAE International.

Although cyclohexane appears to have a relatively low octane quality, it has, in fact, quite high blending values, depending on the other components present.

Racing gasolines blended from refinery streams usually consist of a large proportion of alkylate together with some highly aromatic stream such as catalytic reformate, and treated with the maximum amount of antiknock additive allowed.

Because of the toxicity of tetra-ethyl-lead (TEL), this antiknock additive is not generally available as a blending compound, so many racing fuel blenders have utilized Avgas as a blending source of lead antiknock [14.16]. The Avgas grades with relatively high lead levels are the most valuable for blending, such as 115/145 and 100/130. The first number refers to the antiknock rating performed by the standard method [14.25] and the second refers to the rating performed by the supercharge method [14.26]. The 100/130 grade contains a maximum of 0.85 g lead/L and the 115/145 a maximum of 1.28 g lead/L. A further grade designated as 100LL is less valuable as a blend component because it is a low-lead grade with a maximum content of 0.56 g lead/L.

Avgas consists mainly of alkylate and can be blended with aromatics such as toluene, xylene, or heavy catalytic reformate to give a satisfactory fuel for many racing applications [14.16]. Another reported practice is to blend it with a premium unleaded automotive gasoline containing a high proportion of alkylate and/or isomerate, since this will give a blend having a higher antiknock quality than either of the two components. This is because TEL gives a better octane response in paraffinic hydrocarbons such as alkylate than it does in aromatic-type blend components. Table 14.3 shows some of the racing fuels commercially available from some of the suppliers on the NHRA accepted gasoline list [14.9]. The table shows a low, intermediate, and high antiknock-index fuel, whether it is oxygenated or leaded, and what color it is.

TABLE 14.3 Examples of some available racing fuel specifications.

Supplier	Product	RON	MON	wt.% O_2	Leaded	S.G.	Color
Dragon Racing [14.27]	Tarragon 118	>120	116	0	Yes	0.710	Red
	Tarragon 114	118	112	0	Yes	0.716	Violet
	Tarragon 100	104	96.7	0	Yes	0.730	Red
Renegade [14.28]	PRO 120+	123	>117	0	Yes	0.701	Rose
	PRO 114+	119	111	0	Yes	0.705	Green
	PRO 110+	113	108	0	Yes	0.721	Purple
Rockett Brand® [14.29]	118L	>120	>115	0	Yes	0.701	Yellow
	114L	119	110	0	Yes	0.727	Orange
	100E	105	97	3.5	No	0.717	Blue
Sunoco [14.30]	SR18	120	116	0	Yes	0.704	Yellow
	HRC Plus	118	110	0	Yes	0.736	Orange
	260 GT	105	95	3.7	No	0.734	none
VP [14.31]	C25	>120	114	0	Yes	0.6985	Yellow
	C14	118	114	0	Yes	0.6930	Yellow
	C10	104	96	0	No	0.765	none

© SAE International.

For Formula One racing, the regulations do not allow the use of nitro-paraffins and the other highly effective power boosters discussed later. Because of this, and the intense competition in Formula One, a considerable amount of work has been carried out to develop hydrocarbon-based racing fuels that give enhanced performance under the exacting conditions of racing engines. In the turbo-charged era of the 1980s, synthetic fuels, made from a few pure hydrocarbons rather than refinery streams, could be designed to maximize power. One successful 1985 Formula One fuel for turbocharged engines consisted of just two chemicals, compared with around four hundred in a typical pump fuel. Another racing blend was reported to contain 84% toluene [14.6] in order to give good octane quality and high energy content.

The fuel regulations were changed in the early 1990s with the objective of limiting power and making Formula One racing blends similar to pump fuels. The fuel specifications were based on those for commercial fuels designed for normal road vehicles. In addition, only chemicals that appeared in commercial fuels were permitted, and the concentrations of each chemical could not exceed that present in a survey of pump gasolines. Components such as olefins that could increase flame speed [14.32] were restricted. This approach continued in line with environmental pressures for cleaner gasolines and with further restrictions on the amounts of benzene, total aromatics, olefins, oxygen, and sulfur. By the year 2000, Formula One fuels had to meet the exacting standards targeted for European-marketed gasolines of 2005, making Formula One a proving ground for the technology of future, ultra-clean pump fuels. As noted earlier the FIA is now striving to make Formula One racing fuels sustainable.

These regulations eliminated the potential for large performance gains (10% or more) in Formula One race fuels and increased the importance of exactly matching the fuel to the appetite of individual Formula One engine designs to obtain the last 2 to 3% of power.

It is essential to avoid knock in race engines because the risk of knock damage increases with engine speed. However, the very high speeds in Formula One (up to 18,000 rpm) limit the time for pre-flame reactions to develop. As a consequence, many Formula One engines are satisfied by lower octane gasolines than are needed in other levels of racing.

14.3. Alcohols as Racing Fuels

Straight methanol is well known as a racing fuel, having many advantages over hydrocarbon fuels. It is somewhat better than gasoline in terms of specific energy (see Table 14.1) because its low heating value is more than offset by the low stoichiometric air-to-fuel ratio. In addition, maximum power is obtained at the very rich air-to-fuel ratio of 4:1 so that under these conditions the SE is considerably higher than that of gasoline, where maximum power is obtained only at about 20% richer than stoichiometric. Methanol has excellent antiknock properties, allowing high compression ratios to be used—up to about 16:1 have been reported [14.33]. It also has a high heat of vaporization, giving good volumetric efficiency. Finally, it is readily available and relatively inexpensive.

The main problem is high fuel consumption, particularly when operating at fuel-rich conditions that makes it necessary to carry a large fuel supply. At an air-to-fuel ratio of 4:1, the engine is quite cool running, and, as one leans off, the engine will give better fuel economy but will run hotter and show some reduction in performance. Cold starting can be a problem with straight methanol because it has a low vapor pressure, and small amounts of hydrocarbons or ether are sometimes added for this reason. Because methanol boils at a single temperature, carburetion can be difficult and can result in poor driveability that can be overcome by blending in some gasoline or other hydrocarbons. Fuel injection can avoid or minimize these driveability problems.

Ethanol has also been used as a racing fuel and is intermediate in properties between methanol and gasoline, as can be seen from Table 14.4, where some of the important characteristics of methanol and ethanol are summarized and compared with those of gasoline. The air-to-fuel ratio for maximum power is approximately 7:1, and the relatively high heat of vaporization makes it a useful racing fuel. As one goes up the aliphatic alcohol series, the oxygen content reduces, and the closer one approaches the combustion characteristics of gasoline.

TABLE 14.4 Properties of methanol and ethanol.

	Methanol	Ethanol	Gasoline
Oxygen content (wt.%)	50.0	34.8	0
Boiling point (°C)	65	78	35–210
Lower heating value (MJ/kg)	19.9	26.8	42.7 approx.
Heat of vaporization (MJ/kg)	1.17	0.93	0.18 approx.
Stoichiometric air-to-fuel ratio	6.45:1	9.0:1	14.6:1 approx.
Specific energy	3.08	3.00	2.92 approx.
Blending RON	115–130	112–120	90–100
Blending MON	95–103	95–106	80–90

Alcohols have a poor lead response and so treatment with lead alkyls is not very useful.

Bio-ethanol has also been considered as a component for racing fuels because of its effect on environmental carbon balance [14.34].

14.4. Antiknock Components

Methanol and ethanol are high-octane blend components. They have both been widely used to upgrade gasoline for use as a racing fuel [14.33]. In the 1920s, European factory teams used blends of ethanol (E), benzene (B), and Avgas (G) in the ratios of E:B:G of between 20:20:60 to 80:10:10. These blends had a good water tolerance but were somewhat prone to pre-ignition. Methanol (M) took over to some extent in the 1930s due to its high heat of vaporization and ability to combust at lower air-to-fuel ratios, with M:E:B:G blends varying from 30:30:20:20 to 60:10:25:5. Three-component blends of M:E:G were also used and ranged from 20:60:20 to 80:10:10 [14.33].

Current commercial gasolines have restrictions on the amount of oxygenate that can be included because vehicles need to be able to run on these gasolines as well as on conventional gasolines. This restriction does not apply to racing vehicles, where the whole system can be tailored to a specific fuel.

It can be seen from Table 14.4 that alcohols are excellent blend components in terms of octane quality, particularly RON, and that they are all somewhat better than gasoline from a specific energy viewpoint. The actual blending numbers will depend on the other components present in the blend and the amount of alcohol used.

The high heat of vaporization of the alcohols, especially methanol, will cool the intake charge and improve volumetric efficiency, as it does when neat alcohol is used.

Another advantage of the use of methanol as a blend component is that, as discussed previously, maximum power is achieved at an air-to-fuel ratio of 4:1, some 40% richer than the stoichiometric air-to-fuel ratio. This tends to keep the engine cool and allows the possibility of leaning off under lighter load conditions to improve economy.

Disadvantages of the alcohols are that they increase the fuel's vapor pressure when blended with gasoline, causing hot driveability problems, and that they attack metals and elastomers in the fuel system. When the engine is not in use, it is advisable to drain off the fuel if it contains a high amount of alcohol, and especially methanol. Alcohols also absorb moisture from the atmosphere, and such water can cause phase separation.

Ethers such as MTBE or TAME have also been used as blend components in racing fuels in order to improve octane quality. They are both relatively trouble-free but do not have a significant cooling effect on the intake charge because of their relatively low heats of vaporization as compared with gasoline. Their characteristics are summarized in Table 14.5.

TABLE 14.5 Characteristics of MTBE and TAME as racing fuel blend components.

	MTBE	TAME
Oxygen content (wt.%)	18.2	15.7
Boiling point (°C)	55	86
Lower heat of combustion (MJ/kg)	35.1	37.7
Heat of vaporization (MJ/kg)	0.32	0.32
Stoichiometric air-to-fuel ratio	11.7:1	11.9:1
Blending RON (approx.)	115	111
Blending MON (approx.)	104	100

© SAE International.

Although water is not strictly an antiknock additive, it has been known for many years that its use will suppress knock [14.35]. It can be introduced into the engine in many different ways, such as by injection into the inlet manifold, ports, or directly into the cylinder. It can even be incorporated into the fuel in the form of an emulsion [14.36]. In addition to suppressing knock, water can also improve performance and substantially reduce exhaust emissions of nitrogen oxides [14.37]. However, it must be noted that if the water has vaporized before it reaches the combustion chamber, it will be displacing some of the inducted air and reducing the available oxygen for fuel combustion. This will reduce the possible energy release. This must be balanced against any gains that can be made by utilizing a more advance ignition timing or higher compression ratio to take advantage of the reduced knocking tendency.

Emissions of hydrocarbons are increased, but CO is only slightly affected. Adverse effects are that the lean limit of combustion is narrowed and that corrosion and wear are increased. The introduction of 40% wt. of water into a gasoline in the form of an emulsion increased the RON of a gasoline blend from 91.2 to 100 [14.37], although such an emulsion would be difficult to use because of its high viscosity and the difficulty of preventing the emulsion from breaking. Lower amounts injected separately will have a very real effect.

Finally, it is perhaps worth mentioning that nitrogen compounds such as N-methyl aniline and aniline are antiknock agents and have been used to boost the octane quality of gasolines. Rather large amounts, on the order of 1 or 2 vol.%, are needed to gain 1 or 2 octane numbers, and at this concentration they tend to increase the deposit-forming tendency of the gasoline. They are unpleasant and toxic materials that are relatively expensive, so they have not enjoyed widespread use.

14.5. Nitro-Paraffins as Racing Fuels

One of the limiting factors in increasing power output of the gasoline engine is volumetric efficiency. Hydrocarbon fuels require a rather narrow range of air-to-fuel ratio in order to burn efficiently and give maximum performance, so there is little scope for increasing this.

By using fuels that require less oxygen for complete combustion, it is possible to increase the available energy, provided, of course, that the heating value of the fuel is not so low that it more than offsets the gain due to the lower air-to-fuel ratio.

Table 14.1 showed that nitromethane has an SE ratio almost 2.3 times higher than iso-octane, so it is potentially an excellent fuel from this viewpoint. Nitromethane combusts in oxygen or air according to the following equation:

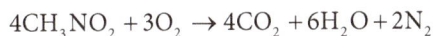

$$4CH_3NO_2 + 3O_2 \rightarrow 4CO_2 + 6H_2O + 2N_2$$

Thus, 7 moles of reactant give 12 moles of product, or, taking atmospheric nitrogen into account, 18.29 moles of reactant give 23.29 moles of product, a ratio of 1.27:1 as compared with 1.00:1 for a hydrocarbon fuel, such as iso-octane. In fact, nitromethane will "combust" in the complete absence of air, according to the following equation [14.38]:

$$4CH_3NO_2 \rightarrow CO_2 + 3CO + 3H_2O + 3H_2 + 2N_2$$

Nitromethane will therefore act as a monopropellant and will combust over a wide range of air-to-fuel ratios from 100% fuel to extremely lean fuel-air mixtures. Other nitro-paraffins can be used as engine fuels, including nitroethane, nitropropane, and 2,2 dinitropropane [14.38].

Table 14.1 also shows that the heat of vaporization for nitromethane is quite high, so that when the increased fuel charge resulting from the low air-to-fuel ratio is taken into account, a cooling effect about twice that of pure methanol is obtained. Of course, not all of this cooling effect is achieved in practice, mainly because of incomplete vaporization and heat transfer.

Nitro fuels are fast burning and particularly suited to engines running at high speeds where the slower-burning methanol cannot release all its energy at the peak of piston travel. However, because of severe engine stress, nitromethane is only used as a primary fuel in short duration runs such as in drag racing, where the engine is usually run at full power for only a few seconds at a time [14.39]. The tendency to produce severe knock and pre-ignition is usually offset by operating at very rich air-to-fuel ratios and by using low compression ratios and retarded spark timing.

In practice, in order to reduce peak flame temperatures, mixtures with methanol are used. The amount of nitromethane used in such a blend with methanol usually varies from 10 to 90%. Fifty percent of nitromethane in methanol was shown to increase the power output over that of pure methanol by almost 45%, although knock increased by about 20% [14.39]. Work has also been carried out using nitroethane or nitropropane as cosolvent instead of methanol, and it has been found that with these blends increasing nitromethane concentration increases power and reduces knock [14.39].

The addition of TEL or methylcyclopentadienyl manganese tricarbonyl (MMT) to nitromethane does not reduce significantly the tendency for the engine to knock. Nitromethane is highly corrosive to aluminum and magnesium, so it is important to drain and purge the fuel system to avoid damage.

14.6. Fuel Additives

14.6.1. Hydrazine

Hydrazine (N_2H_4) is a liquid boiling at 113°C and has been used as a fuel additive in nitromethane and methanol blends in which the fuel properties and composition are not restricted by regulations. It forms an explosive salt with nitromethane that requires only the oxygen in the nitromethane for it to combust. High rates of pressure rise are achieved in the cylinder with dramatic increases in power but with an increased risk of damage to the engine. Very little is needed, and less than 1% of hydrazine by volume gives a significant power increase [14.16]. The mixture is so unstable that it causes a severe safety hazard.

14.6.2. Antiknock Additives

The use and mode of action of the most common antiknock additives have already been discussed in Chapter 7, Section 7.4, and the benefits of lead antiknocks in racing fuels are referred to in Section 14.2. In particular, the toxicity of lead antiknock compounds means that they are not generally available except to refiners and blenders with the appropriate handling equipment. The response of different types of hydrocarbons to the addition of lead alkyls is variable, with the lowest response being found with aromatic components and the best response with paraffinic components such as alkylate and isomerate. It is partly for this reason that these two components appear so frequently in racing gasolines and partly because branch chain paraffins tend to have good motor octane levels and low sensitivities.

Other antiknock additives that have been used to a limited extent in racing gasolines include MMT [14.16] (see Chapter 6, Section 6.4). Although it is classed as a Class B poison according to US shipping regulations and is toxic by all exposure routes, the handling of MMT is much less restrictive than required for lead compounds because of its low vapor pressure and high thermal stability. In unleaded gasolines at a treat level of 0.07 g Mn/L, it is capable of giving octane gains of 2 to 3 units for MON and over 4 for RON, but such high treat levels are not acceptable for normal use on the road. At higher treat levels, spark plug fouling and valve wear problems have been reported [14.40]. It also has a synergistic effect when used with lead so that octane gains of up to one number are possible when it is added to leaded gasoline. Dicyclopentadienyl iron (ferrocene) has also been investigated as a way to enhance octane quality in racing fuels [14.41], although again there are issues with spark plug fouling.

14.6.3. Stabilizers

Racing fuels do not sit in vehicle fuel tanks for very long. However, commercial racing fuels have to be stable and remain constant for much greater lengths of time. For example, if a drag racing team buys a drum (205 L) of fuel, it can last for a number of meetings. Under the current FIA Formula One rules, a team must lodge a sample of fuel at the start of the season, which is then used as a reference for comparison with any samples that may be taken during the season. It is therefore important that racing fuels are stable, typically for a year or more.

Fuel stability additives are discussed in more detail in Chapter 11, Section 11.1. Clearly, as the composition of some racing fuels can be significantly different from normal road gasoline, it will thus be necessary to tailor the fuel additives packages accordingly.

14.6.4. Deposit Control Additives

It is not unusual for a race engine to be completely stripped and rebuilt after each race. This would allow the removal of deposits that could cause issues in road-going vehicles. However, engine deposits can accumulate in relatively short periods of time. In motor racing, the slight performance differences that result from these deposits can make the difference between winning and not winning. It has been shown that in an FIA Formula One engine, piston top deposits were responsible for a 13.2 kW reduction in power during the engine life span of 2200 km. Figure 14.4 [14.13] shows cylinder pressure data for such an engine and clearly showed total power loss during the life of the engine.

FIGURE 14.4   Cylinder pressure data from a Formula One engine.

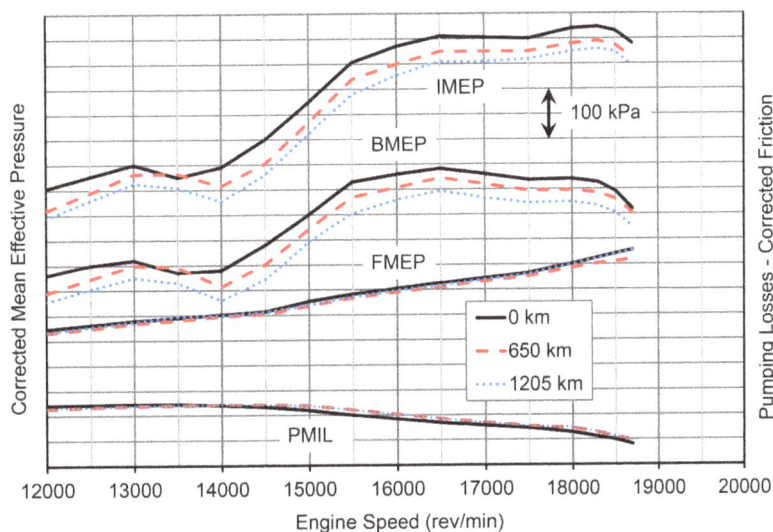

Deposit control additives are discussed in more detail in Chapter 11, Section 11.3.3. As noted in the previous section, racing fuels may be significantly different from road gasoline, and the additive formulation may have to be tailored accordingly.

14.6.5. Dyes and Markers

Racing fuels do not necessarily have to comply with legislation for road fuels. For example, some race regulations still allow the use of lead antiknock compounds. Some race fuels are therefore not road legal and are dyed to make it obvious. Different commercial producers can color their fuels according to their own criteria, as shown in Table 14.3. Fuel dyes and markers are discussed in Chapter 11, Section 11.2.1.

14.6.6. Static Dissipaters

When race refueling is required, it is usually advantageous to refuel in the shortest possible time. It is therefore necessary to have high fuel flow rates that can result in the build-up of static electricity. Thus, for safety reasons static dissipaters may be used to reduce the possibility of pit-stop fires. Static dissipaters are discussed in Chapter 11, Section 11.2.4.

14.7. Nitrous Oxide as an Oxidant for Racing Fuels

Nitrous oxide (N_2O) is a gas containing 36% oxygen by weight as compared with 23% for air. Nitrous oxide therefore has an attraction in that for the same intake pressure, there is an increased amount of oxygen in the combustion chamber. This allows more fuel to be burned, as it effectively reduces the stoichiometric air-to-fuel ratio, thus increasing the SE [14.16]. In tests during the 1940s, it was found that a nitrous oxide-to-air ratio of 0.1 resulted in an increase of about 14% in power output, and a ratio of 0.2 resulted in an increase of about 25% [14.42].

Nitrous oxide is stable and readily available but it is an anesthetic and must be handled with care. It is normally supplied in gas cylinders so that storage on racing vehicles limits its use.

It is usually used for short durations in view of the storage problems, the high combustion temperatures achieved, and the high fuel flow rates required. It has been calculated [14.16] that a gaseous stoichiometric mixture of iso-octane, air, and 5% nitrous oxide would produce a 9% increase in available specific energy.

References

14.1. Duesenberg, F.S., "Racing-Car Developments," SAE Technical Paper 260039 (1926), doi:https://doi.org/10.4271/260039.

14.2. Shimizu, R., Tadokoro, T., Nakanishi, T., and Funamoto, J., "Mazda 4-Rotor Rotary Engine for the Le Mans 24-Hour Endurance Race," SAE Technical Paper 920309 (1992), doi:https://doi.org/10.4271/920309.

14.3. Spear, P. and Penny, N., "The Development of the Rover—B.R.M," SAE Technical Paper 640100 (1964), doi:https://doi.org/10.4271/640100.

14.4. Li, C., Koelman, H., Ramanathan, R., Baretzky, U. et al., "Particulate Filter Design for High Performance Diesel Engine Application," *SAE Int. J. Fuels Lubr.* 1, no. 1 (2009): 1307-1312, doi:https://doi.org/10.4271/2008-01-1747.

14.5. Hassan, W.T.F., "The New Jaguar 12-Cyl Engine," SAE Technical Paper 720163 (1972), doi:https://doi.org/10.4271/720163.

14.6.   Otobe, Y., Goto, O., Miyano, H., Kawamoto, M. et al., "Honda Formula One Turbo-Charged V-6 1.5L Engine," SAE Technical Paper 890877 (1989), doi:https://doi.org/10.4271/890877.

14.7.   Baretzky, U., Andor, T., Diel, H., and Ullrich, W., "The Direct Injection System of the 2001 Audi Turbo V8 Le Mans Engines," SAE Technical Paper 2002-01-3357 (2002), doi:https://doi.org/10.4271/2002-01-3357.

14.8.   Federation Internationale de l'Automobile, "2026 Formula 1 Power Unit Technical Regulations," August 2022, accessed November 2022, https://www.fia.com/sites/default/files/fia_2026_formula_1_technical_regulations_pu_-_issue_1_-_2022-08-16.pdf.

14.9.   National Hot Rod Association, "The NHRA Technical Department Has Posted the Updated Approved Fuel List," December 2014, accessed November 2022, https://www.nhraracer.com/content/general.asp?articleid=61901&zoneid=132.

14.10. de Witt, C.J. and Reese, R., Alcohol compositions having luminous flames. U.S. Patent 4,536,188, 1985.

14.11. Wusz, T., "Gasoline for NASCAR Stock Car Racing: 1951-1994," SAE Technical Paper 942539 (1994), doi:https://doi.org/10.4271/942539.

14.12. Glover, A.R., Yasuoka, A., Galliard, I.R., and Matsumoto, Y., "Optimizing Formula 1 Engine Performance through Fuel and Combustion System Development," SAE Technical Paper 9302510 (1993), doi:https://doi.org/10.4271/942540.

14.13. Takeuchi, K., Pfeilmaier, P., Esaki, Y., and Choi, E., "Anti- Combustion Deposit Fuel Development for 2009 Toyota Formula One Racing Engine," SAE Technical Paper 2011-01-1983 (2011), doi:https://doi.org/10.4271/2011-01-1983.

14.14. Bryce, S.G., Lindsay, R., Galliard, I., and Glover, A.R., "Fuels Development for Formula One," SAE Technical Paper 942540 (1994), doi:https://doi.org/10.4271/942540.

14.15. Wyszynski, L.P., Stone, C.R., and Kalghatgi, G.T., "The Volumetric Efficiency of Direct and Port Injection Gasoline Engines with Different Fuels," SAE Technical Paper 2002-01-0839 (2002), doi:https://doi.org/10.4271/2002-01-0839.

14.16. Germane, G., "A Technical Review of Automotive Racing Fuels," SAE Technical Paper 852129 (1985), doi:https://doi.org/10.4271/852129.

14.17. Dickinson, H.C., Sparrow, S.W., and Gage, V.R., "Comparison of Hector Fuel with Export Aviation Gasoline," Issue 90 of Report, National Advisory Committee for Aeronautics, Washington, DC, 1920.

14.18. McCulloch, T.B., Aviation fuel. U.S. Patent 2,411,582, 1946.

14.19. Marschner, R.F., Superfuel. U.S. Patent 2,407,716, 1946.

14.20. Marschner, R.F., Aviation superfuel. U.S. Patent 2,407,717, 1946.

14.21. Schulze, W.A. and Alden, R.C., Fuel composition. U.S. Patent 2,409,156, 1946.

14.22. Schulze, W.A. and Alden, R.C., Fuel composition. U.S. Patent 2,409,157, 1946.

14.23. Branstetter, J.R., "Comparison of Knock-Limited Performance of Triptane with 23 Other Purified Hydrocarbons," Memorandum Report E5E15, National Advisory Committee for Aviation, 1945.

14.24. Meyer, C.L., "The Knock-Limited Performance of Fuel Blends Containing Spiropentane, Methylenecyclobutane, Di-tert-Butyl Ether, Methyl-tert-Butyl Ether, and Triptane," Restricted Bulletin, No. E6D22, National Advisory Committee for Aviation, Washington, DC, 1946.

14.25. ASTM International, "Standard Test Method for Motor Octane Number of Spark-Ignition Engine Fuel," ASTM D2700-23, ASTM International, 2023.

14.26. ASTM International, "Standard Test Method for Supercharge Rating of Spark-Ignition Aviation Gasoline," ASTM D909-22, ASTM International, 2022.

14.27. Dragon Racing Fuels, accessed November 2022, https://www.dragonracingfuels.com/.

14.28. Renegade Race Fuel & Lubricants, accessed November 2022, https://renegaderacefuel.com/.

14.29. Rockett Brand Racing Fuel, accessed November 2022, https://www.rockettbrand.com/.

14.30. Sunoco Race Fuels, accessed November 2022, https://www.sunocoracefuels.com/.

14.31. VP Racing Fuels, accessed November 2022, https://vpracingfuels.com/.

14.32. **Goodger,** E.M., *Hydrocarbon Fuels: Production, Properties and Performance of Liquids and Gases* (London, UK: The MacMillan Press Ltd., 1975), 112.

14.33. **Powell,** T., "Racing Experiences with Methanol and Ethanol-Based Motor-Fuel Blends," SAE Technical Paper 750124 (1975), doi:https://doi.org/10.4271/750124.

14.34. **Dwyer,** K., "Cost Effective Sustainable Fuels for Performance Vehicles," SAE Technical Paper 2008-01-2954 (2008), doi:https://doi.org/10.4271/2008-01-2954.

14.35. **Hopkinson,** B., "A New Method of Cooling Gas Engines," *Proceedings of the Institution of Mechanical Engineers* 85, no. 1 (1913): 679-715.

14.36. **Peters,** B. and Stebar, R., "Water-Gasoline Fuels—Their Effect on Spark Ignition Engine Emissions and Performance," SAE Technical Paper 760547 (1976), doi:https://doi.org/10.4271/760547.

14.37. **Harrington,** J., "Water Addition to Gasoline-Effect on Combustion, Emissions, Performance, and Knock," SAE Technical Paper 820314 (1982), doi:https://doi.org/10.4271/820314.

14.38. **Starkman,** E.S., "Nitroparaffins as Potential Engine Fuel," *Industrial and Engineering Chemistry* 51, no. 12 (1959): 1477-1480.

14.39. **Bush,** K.C., Germane, G.J., and Hess, G.L., "Improved Utilization of Nitromethane as an Internal Combustion Engine Fuel," SAE Technical Paper 852130 (1985), doi:https://doi.org/10.4271/852130.

14.40. **Ethyl Corporation,** "Information for the National Research Council Concerning Methylcyclopentadienyl Manganese Tricarbonyl," Ethyl Corporation, Richmond, VA, 1971.

14.41. **Leveque,** R., Marcusich, M., and Patriquin, G., "Unleaded Racing Gasoline Components and Blends in the 110 Octane Range," SAE Technical Paper 942541 (1994), doi:https://doi.org/10.4271/942541.

14.42. **Tauschek,** M.J., Corrington, L.C., and Huppert, M.C., "Nitrous Oxide Supercharging of an Aircraft-Engine Cylinder," Memorandum Report E5F26, National Advisory Committee for Aviation, Washington, DC, 1945.

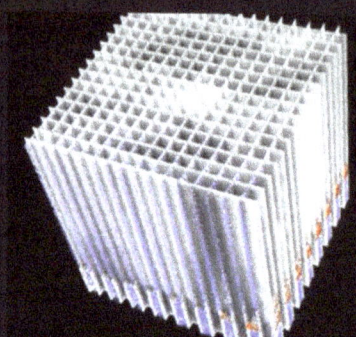

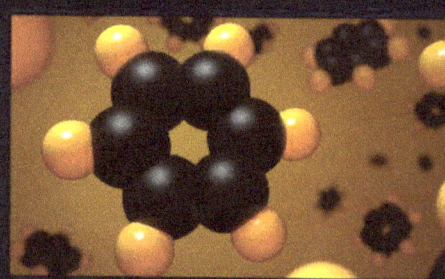

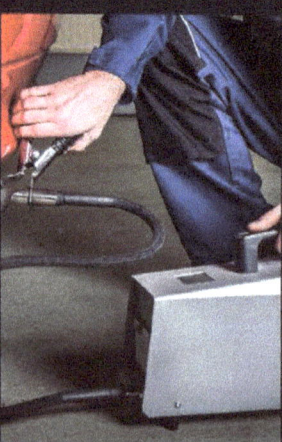

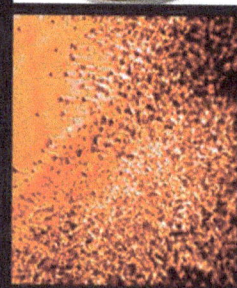

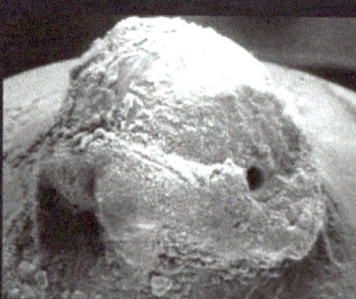

15

The Diesel Engine Combustion Process

As discussed in Chapter 2, Section 2.2.1, in 1890, Herbert Akroyd Stuart and Charles Richard Binney were granted a patent for "Improvements in Engines Operated by the Explosion of Mixtures of Combustible Vapour of Gas and Air" [15.1]. The engine illustrated in their patent (see Chapter 2, Figure 2.5) would probably be recognized today as a diesel engine. However, when Rudolf Diesel took out his patent in 1892, it was not an engine that he was patenting but a combustion process [15.2]. Diesel's proposal was that all of the fuel energy should be added so as not to raise the temperature of the compressed gas and hence not lead to additional heat loss. In reality, this ideal was never achieved by Diesel and is not achieved in the modern engines that bear his name.

The fundamental difference between the existing gas engines and the engine of Stuart and Binney and that proposed by Diesel was that they relied upon the temperature and pressure generated by the compression stroke to bring about combustion of the fuel. There was no positive ignition as with a spark-ignited engine. On this basis these, two types of engines were often classified as spark ignition (SI) and compression ignition (CI). More recently, this simple differentiation has been clouded by research into a type of engine that has some characteristics in common with both of these existing types.

Initially designated as active thermo-atmosphere combustion (ATAC) [15.3], the new engine type is variously referred to as controlled auto-ignition (CAI) [15.4], homogeneous charge compression ignition (HCCI) [15.5], premixed compression ignition (PCI) [15.6], premixed charge compression ignition (PCCI) [15.7], and partially premixed combustion (PPC) [15.8], which will be discussed in Chapter 22. This has recently led to an attempt to differentiate what we refer to as the diesel engine from the emerging engine technology by classifying the former as mixing controlled CI combustion and the latter as kinetically controlled CI combustion. However, this classification is potentially confusing, as the diesel engine does include some kinetically controlled combustion (discussed in the following section).

15.1. The Diesel Combustion Process

Diesel combustion is a complex, turbulent, three-dimensional, multiphase process that occurs in a high-temperature and high-pressure environment. [15.9]

The diesel combustion process differs from that in the gasoline engine in that the mixture preparation process and the combustion process can take place simultaneously in different parts of the cylinder. These two processes also take place consecutively as opposed to SI combustion, where the two processes can occur independently. For example, in a gasoline engine, mixture preparation can take place in the inlet manifold or during the induction stroke. For diesel combustion, the whole process can thus be considered on an overall basis, considering the engine cylinder as a whole, or on a discreet basis where individual packets of fuel are considered.

Considering the process on an overall basis, it can be viewed in terms of the cylinder pressure history. A rise in cylinder pressure is driven by the energy released from the fuel. Figure 15.1 shows a cylinder pressure against crank angle history.

FIGURE 15.1 Typical diesel cylinder pressure against crank angle trace.

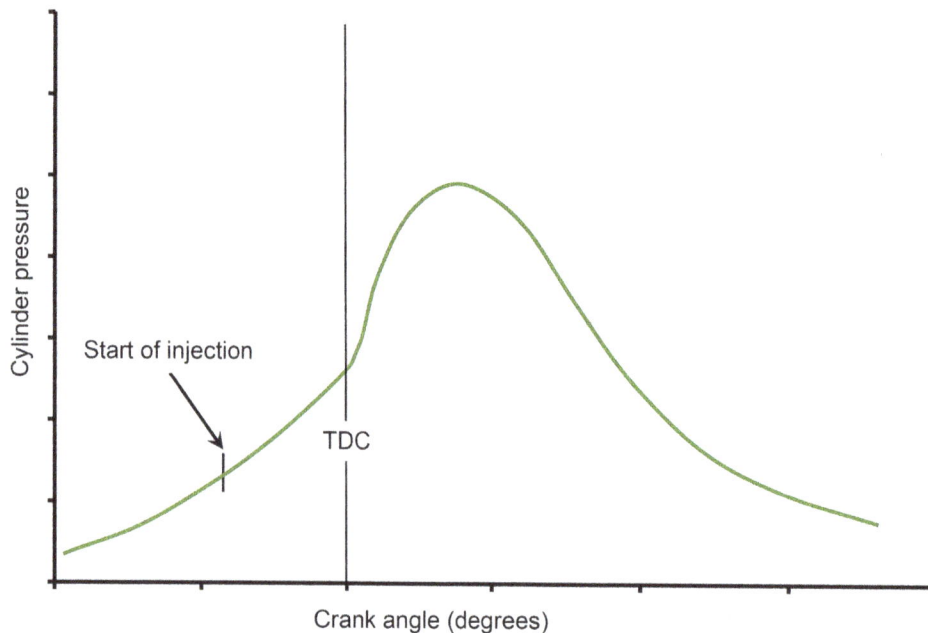

This can also be compared with a corresponding graph for normal SI combustion as shown in Chapter 7, Figure 7.1. The two graphs are not too dissimilar, and a thermodynamic analysis will show them to be far from Diesel's ideals. Figure 15.2 shows a typical heat release against crank angle plot that breaks the combustion and heat release process into four steps.

FIGURE 15.2   Typical heat release diagram showing phases of diesel combustion.

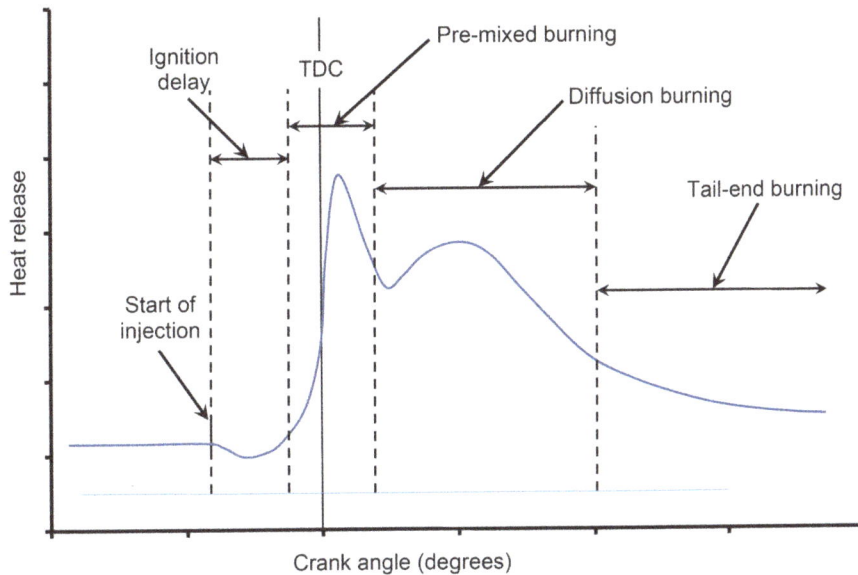

Notice that the heat release rate is initially negative. This is due to evaporating fuel taking energy out of the compressed gas as it begins to evaporate. The combustion process can then be broken down into the four stages shown in the figure: the ignition delay, premixed burning, diffusion burning, and tail-end burning.

As noted, the process could also be considered from the point of view of an individual fuel droplet that will experience both physical and chemical processes. However, there will be a great number of individual droplets that will be going through similar processes at different points in time and space within the combustion chamber. For engineers to model the combustion process, it is therefore necessary to consider these discrete droplets of fuel in order to arrive at the four stages of the combustion event.

The diesel combustion process begins with the oxidant and diluent being trapped within the cylinder and undergoing compression. The oxidant is usually oxygen and the diluent nitrogen—that is to say, fresh air—however, there may also be some residual exhaust gas left in the cylinder, or exhaust gas may be deliberately introduced as a further diluent. Many modern diesel engines, especially for automotive applications, differ from their predecessors in that they are pressure charged and often use exhaust gas recirculation (EGR) to control the emissions of oxides of nitrogen. How these design factors affect the combustion process will be discussed later in the chapter.

Unlike gasoline, combustion fuel is not present throughout most of the compression. There is thus no concern regarding chemical processes within the fuel at this stage and hence no concern regarding preignition. The fuel is only introduced late in the compressions stroke and must undergo vaporization and mixing in a very short space of time before combustion can take place. How this takes place and how it results in the four stages of the process noted will be described in the following sections.

15.1.1.   Ignition Delay Period

The ignition delay period is usually defined as the time from the start of fuel injection to the point at which perceptible heat release occurs. The actual definition depends upon the measuring techniques employed. The start of injection can be measured as the static injection timing; by measuring injector needle lift or if optical methods are available; and then by the first appearance of fuel out of the injector. The point of first heat release can also be defined in a number of different ways. A common example is when the heat release trace changes direction. This can be assessed visually or can be computed from when the derivative of the heat release changes from negative to positive. The heat release rate is initially negative because heat is taken out of the compressed gas to volatilize the incoming fuel. Alternatively, the onset of heat release can simply be defined as when the heat release itself becomes positive. If optical access is available, the onset of heat release can be assessed as when the first visible emissions occur [15.10, 15.11].

During the ignition delay period, both physical and chemical processes are taking place. It is claimed [15.12] that physical delay was initially doubted, but later photographic work clearly demonstrated that fuel underwent vaporization before ignition [15.13].

The aim of the fuel injection process is to atomize the fuel to produce as small a droplet as possible. Fuel injection systems are discussed in more detail in Chapter 16, Section 16.3.3. As the first droplets of fuel are injected (the start of injection), these droplets will rapidly begin to heat up as they enter the hot compressed air within the combustion chamber. As the droplets heat up, they will begin to vaporize [15.14–15.16]. The fuel vapor that is produced will mix with the air that is being entrained in the penetrating jet due to motion of the droplet away from the injector or due to turbulence within the combustion chamber.

Even at very low fuel flow rates, there will be many fuel droplets making up the fuel spray that will have very fuel-dense cores and more disperse droplets toward its periphery. See Figure 15.3 [15.17], which shows evaporating droplets in combustion gas at 427°C and with an injection pressure of 40 MPa.

FIGURE 15.3 Fuel droplets and vapor formation at 427°C, 40 MPa injection pressure.

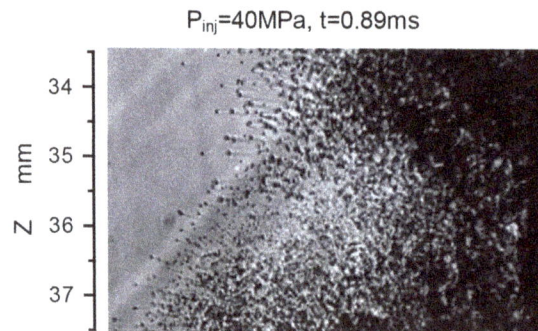

P_{inj}=40MPa, t=0.89ms

© SAE International.

Relatively large droplets are formed at this low injection pressure, and this delays the evaporation process. As a result, there are many droplets that are clearly visible moving out from the core of the spray. Figure 15.4 [15.17] shows a corresponding image with a higher injection pressure.

FIGURE 15.4 Fuel droplets and vapor formation at 427°C, 100 MPa injection pressure.

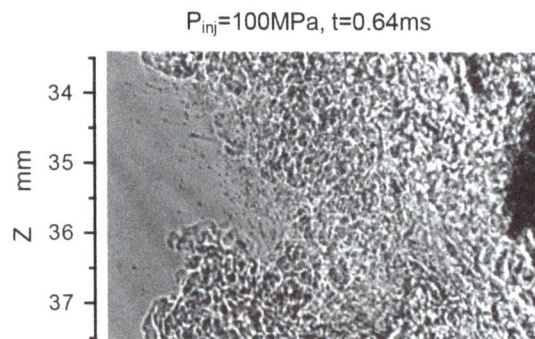

P_{inj}=100MPa, t=0.64ms

© SAE International.

In this picture, the higher fuel injection pressure produces smaller droplets, which vaporize more rapidly. Because of this, there is more vapor and fewer droplets visible in the picture.

In this more disperse region on the spray periphery, the mean distance between droplets is much greater than the mean droplet diameter. Therefore, the probability that droplets will collide with one another is very small [15.18]. Initially, it is valid to consider the fate of a single droplet, and a great deal of work has been performed to study the fate of single fuel droplets [15.17, 15.19–15.24]. Some computer models of diesel combustion are thus based on the behavior of single droplets as a starting point [15.25–15.27].

As the fuel droplet moves away from the injector, it slows down, and its surface temperature increases. The fuel vapor left in its wake will become progressively richer. At some point this mixture of air and vapor will contain an ignitable mixture. This can be considered the end of the physical delay for this droplet, and after the chemical delay ignition will begin to take place. At this point it would not be detectable, but because this droplet's wake will probably intersect with that of other droplets, a flame front would be established.

In parallel with the physical processes, the following chemical processes are also taking place: fuel molecules break up into smaller molecules and radicals, oxygen from the air reacts with these species, and when enough radicals are formed, the chemical reaction rate increases exponentially to become an auto-ignition site [15.18].

The length of the ignition delay period will therefore depend on both the physical and the chemical processes. The progression of the physical process is primarily determined by the mechanical characteristics of the engine and fuel injection equipment. The compression ratio, engine speed, cooling system, and injection timing will determine the temperature and pressure into which the fuel is injected. The inlet port geometry and combustion chamber design will determine the level of air turbulence within the combustion chamber. The fuel injection equipment will determine the characteristics of the fuel spray(s), droplet size, distribution, penetration, and so on. The presence or absence of EGR and the inlet and exhaust valve timing will influence the composition and temperature of the compressed air. The physical processes will also be influenced by the nature of the fuel. The fuel injection process and hence the fuel spray characteristics will be influenced by the fuel's physical characteristics, its density, viscosity, surface tension, and so forth. The volatility of the fuels will affect the rate of evaporation, which will in turn affect the mixing process.

The chemical process will also be influenced by the temperature and pressure but will be largely determined by the chemical composition of the fuel. To compare the performance of different fuels, a fixed set of engine and injection conditions must be used. As with gasoline, this has been done using a specific engine and operating condition and using specific pure chemicals as reference fuels. This is discussed in Section 15.2.

15.1.2. Premixed Burning Period

By the end of the ignition delay period, a quantity of fuel has vaporized. Much of this fuel has been mixed with air and will be within the flammability range for the hydrocarbon species present. Once auto-ignition has occurred, the flame front will spread rapidly through this mixture. This produces the high rate of heat release that can be seen in Figure 15.2. This will lead to a high rate of pressure rise within the engine cylinder.

In current diesel engines, the fuel velocity leaving the injector nozzle will be very high. This will promote a high degree of air entrainment into the fuel jet. The mixture within the limits of flammability will therefore be some distance away from the injection nozzle. The premixed burning zone will also be away from the nozzle tip.

Because the air and fuel that are being burned at this time are considered to be premixed, the reaction rate has been said to be controlled by the chemical kinetics. The premixed burning period is therefore sometimes referred to as the kinetic burning period. However, it has been shown [15.28] that the characteristic time for the chemical delay is far shorter than the characteristic time for mixing, so in reality this part of the combustion process is also diffusion controlled.

The amount of fuel that is available for combustion during the premixed phase will obviously dictate the magnitude of this spike in the heat release curve. If the quantity of heat released is too great, then the rate of pressure rise will also be too great. This can lead to the characteristic diesel knock sound that is most prevalent when an old diesel engine is started from cold. The amount of fuel available will be determined by the length of the ignition delay period relative to the rate of injection.

The ignition delay can be controlled by both the fuel properties and by engine design. As will be discussed in Chapter 16, Section 16.3.3, modern fuel injection equipment allows precise control of the injection characteristics, including multiple injections. This can be used to carefully regulate the amount of premixed burning. The end of the premixed burning period is usually defined by the local minima of the heat release diagram. This equates to the point at which the premixed fuel within the necessary flammability limits has been consumed.

15.1.3. Diffusion Burning Period

In Diesel's 1892 patent, he did not consider ignition delay and premixed burn as such. Diesel envisaged that the whole combustion process was controlled by the rate of fuel addition. In current diesel engines, with the possible exception of very light loads, the ignition delay and premixed burning phases will be completed before the end of the fuel injection. The rest of the combustion process will then follow Diesel's original conception.

Fuel is still being injected during the diffusion burning phase of the combustion process. However, the liquid fuel jet only penetrates a relatively short distance. Entrainment of hot in-cylinder gas promotes rapid vaporization of the liquid fuel and results in a rich but combustible mixture just downstream of the tip of the liquid jet.

Downstream of this point, the majority of the fuel in the combustion zone is in the vapor phase or as partially oxidized or pyrolyzed fuel particles. This will be discussed further in Section 15.3, considering the formation of soot.

15.1.4. Tail-End Burning Period

At the end of injection, the tail of the liquid jet moves away from the injector tip. Hot in-cylinder gases are still being entrained to feed the fuel-rich diffusion flame. Any remaining fuel vapor along with the partially oxidized fuel and soot particles will continue to undergo further oxidation.

As the piston is now in the expansion stroke and the majority of the fuel has been burned, the increasing cylinder volume will tend to reduce the pressure. The rate of heat release has now been reduced significantly, and heat transfer is limiting any further rise of in-cylinder temperature.

15.2. Fuel Properties Influencing Combustion

As discussed in the previous section, in a diesel engine the majority of the fuel is burned during the diffusion burning period. The temperature and pressure prevailing in the combustion chamber means that most fuels will burn readily as long as there is oxygen available. Because of the locally high temperature around the incoming fuel jet, the fuel will readily vaporize. This can be ensured by controlling the distillation characteristics of the fuel. One of the advantages of a diesel engine is that it runs with excess air due to the absence of any throttling of the intake air. The oxygen necessary for combustion is available, and it is incumbent upon the engine designer to try and ensure good mixing. The fuel characteristics have a lesser role in the diffusion burning part of the process. However, the fuel characteristics can significantly affect the emissions performance (this will be discussed in Chapter 21).

The main way in which fuel characteristics influence the diesel combustion process is on the ignition delay phase of the process. This in turn will affect the amount of fuel available for the premixed burning phase.

One of the fuel characteristics that will influence the ignition delay period is the temperature at which the fuel, or some fraction of it, will auto-ignite. Some of the earliest attempts to assess this were done almost a century ago [15.29] using a rapid compression machine with a glass-walled cylinder!

Early engine experiments indicated that the ignition quality of a fuel was related to its density and aromatic content [15.30]. The density of petroleum products was traditionally expressed by its American Petroleum Institute (API) gravity, and the aniline point was an indication of the aromatic content of a fuel. This led to the proposal to quantify the ignition quality of fuels using a Diesel Index (DI), which was a function of the fuel's API gravity and aniline point:

$$\text{Diesel Index} = \frac{\left(\text{API gravity}\right)\left(\text{Aniline point}\right)}{100}$$

This empirical approach, however, was not sufficiently discriminating and led to the development of the Cooperative Fuels Research (CFR) cetane engine test [15.31, 15.32]. This has now become the recognized method [15.33] of assessing diesel fuel ignition quality in the same way as the octane scale has become the standard for gasoline. Unfortunately, the CFR engine test method has poor precision, and so another calculated value, the Cetane Index [15.34], was introduced as a means of estimating the cetane number of a diesel fuel from properties that could be determined repeatably in the laboratory. The coefficients for the equation were determined to provide a good correlation with the engine test method. The formula has been revised from time to time, as fuels have evolved, to maintain its predictive validity. A more recent attempt to replace the engine test method with a more repeatable method has been the use of a constant volume combustion chamber [15.35, 15.36]. The output of these methods has been termed the derived cetane number (DCN), and these methods will be discussed in more detail next.

15.2.1. Ignition Quality

The ignition quality of a diesel fuel is a measure of how readily it will ignite. This can be assessed either by measuring the ignition delay time or by measuring some physical characteristics that will correlate with the ignition delay time. The most widely recognized method of performing these assessments is described in the following sections.

15.2.1.1. Measurement of Cetane Number

The most widely recognized measure of diesel fuel ignition quality is the cetane number (CN). The CN is determined by testing the fuels performance relative to two reference fuels in a CFR engine under specified standard conditions. The cetane scale is derived from two pure compounds, n-hexadecane (n-cetane) and α-methyl naphthalene (1-methyl naphthalene), which are assigned values of 100 and 0, respectively. Due to handling concerns around α-methyl naphthalene [15.37], it was soon replaced with 2,2,4,4,6,8,8-heptamethylnonane (iso-cetane), which has a CN of 15. These two fuels are the primary reference fuels, but for most tests a pair of secondary reference fuels is used. These fuels are designated as T and U fuels. These are reblended on a periodic basis and are available from Chevron Phillips Chemical Company, LP. The current batches are T-26 and U-19 and have CNs of 75.2 and 19.4, respectively. If CN determination is required outside of this range, then the primary reference fuels can be used either as primary reference fuel blends or as blends with the secondary reference fuels.

The CN of a fuel is determined by comparing its ignition delay under standard operating conditions with two reference fuel blends of known CN. The reference fuels are prepared by blending normal cetane (n-hexadecane), having by definition a value of 100, with heptamethyl nonane, a highly branched paraffin with an assigned value of 15.

In practice, reference fuel blends are made at five CN intervals, and the compression ratio of the engine is varied to give the same ignition delay period for the test fuel and bracketing pair of reference fuel blends, which differ by less than five CN. The CN of the test fuel is calculated by interpolation.

At the time of the development of the cetane scale, the chemical composition of the fuel had a greater relevance than the diesel engine technology. Unfortunately, the current engine technology is significantly different from the CFR engine technology, and diesel fuels are no longer simply a blend of petroleum-derived hydrocarbons. The ignition characteristics determined by this method do not always correlate well with the ignition delays in production diesel engines, particularly for some alternative fuels. Fuel characteristics such as viscosity can have a significant effect on the physical ignition delay. The fuel handling, temperature, injection pressure, injector design, combustion chamber design, and so forth can all influence the ignition delay, and the relative performance of widely different fuels can be different in the CFR engine than a modern high-pressure direct injection engine [15.38–15.41].

Another disadvantage of the standard CN determination method [15.33] is that it is slow, taking 1 to 2 hours to measure the cetane number of one sample, and it also requires a skilled operator. Attempts have been made to automate the process by fixing the compression ratio rather than the ignition delay time. Measurements of ignition delay time are taken and the cetane number determined by interpolation based on the ignition delay time for the bracketing reference fuels [15.42].

Critically, however, it is the poor precision of the method that is of most concern. Repeatability lies between 0.8 and 1.0 cetane numbers, while reproducibility is in the range of 2.8 to 4.8 cetane numbers.

15.2.1.2. Measurement of DCN

The idea of measuring the CN in a constant volume combustion chamber is not new [15.43].

A great deal of work has been conducted to assess the ignition quality of diesel fuels using a constant volume combustion chamber [15.44–15.47]. In the late 1990s, work was conducted to determine operating conditions for constant volume combustion chamber experiments to allow better correlation with the standard CFR engine text method [15.48–15.51]. This work on the Ignition Quality Tester (IQT™) has resulted in an ASTM standard method using that apparatus [15.36]. The CN is by definition a value determined from the CFR engine test. The constant volume combustion chamber procedure does not measure CN but ignition delay and combustion times. These results are then converted into a DCN by the software within the IQT™. The test method is suitable for ignition delays ranging from 2.8 to 6.5 ms and combustion delays ranging from 5.5 to 120 ms; these times equate to a DCN of 30 to 65. However, the precision stated in the method only covers the range of DCN from 35 to 60.

An alternative ASTM method uses a Fuel Ignition Tester (FIT™) that is now marketed by the same company that markets the CFR engine. This is an automated method that produces a DCN value.

Both of these systems comprise three main elements: the constant volume combustion chamber; a single-shot fuel injection system; and the software and hardware for data acquisition, analysis, and control. The time taken for a DCN determination is significantly shorter than for the CN determination using the engine test. The FIT™ and the IQT™ produce a result in approximately 20 or 30 minutes, respectively. They also require a smaller volume of sample at approximately 220 and 370 mL, respectively.

Both of the test methods are intended to produce results that equate to those produced by the CFR engine. An apparent discrepancy in the response of the IQT™ has been noted and investigated [15.52]. When the IQT™ is set up according to the procedure, the ignition delay of the n-heptane calibration reference fuel is 3.78 ms, which converts to a DCN of 52.5. The DCN is significantly different from the CN of 56 obtained using the CFR engine. This highlights that different fuels respond differently to changes in temperature and pressure during the ignition delay period. This is further evidence that a single measurement can be only indicative of relative fuel performance and that it cannot be used for predicting performance in a given engine.

15.2.1.3. Measurement of Cetane Index

Due to the difficulty of running the CFR engine for on-line and routine measurement of the ignition quality of fuel, it has always been desirable to be able to approximate the CN by some routine laboratory testing of physical parameters of the fuel. As a result, many different formulae and using different properties have been proposed and evaluated [15.53].

One of the first attempts to do this was by the calculation of a DI that was defined as a function of the fuel's API gravity and aniline point [15.30].

A higher DI was considered to be a good thing. In practice, conventional petroleum distillates that were highly paraffinic would tend to have a high DI. This would not necessarily ensure that they had a high cetane rating. Some testing to compare the DI with other methods of assessing diesel quality, including ignition delay in the CFR engine, showed that there was a directional trend but the correlation was not that good [15.31]. This is shown in Figure 15.5.

FIGURE 15.5 Correlation between CN and DI.

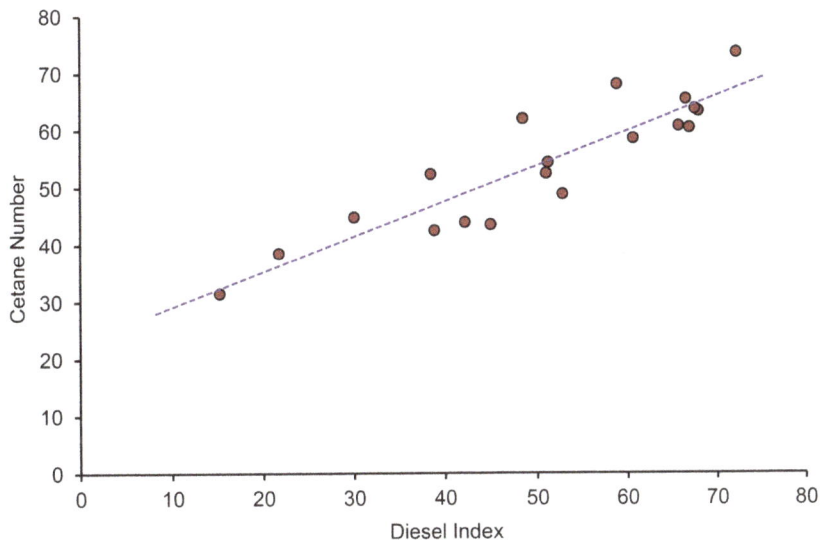

Further work was performed to determine a better predictor of CN. The calculated cetane index (CCI) was proposed, and this became an ASTM procedure, ASTM D976-66. The coefficients of the equation were later updated in ASTM D976-80 and this followed through into ASTM D976-06 and ASTM D976-21 [15.34]. The CCI is defined by the following equation:

$$CCI = 454.74 - 1641.416\,D + 774.74\,D^2 - 0.554\,T_{50} + 97.803\left(\log T_{50}\right)^2$$

where
CCI is the calculated cetane index
D is the density at 15°C (kg/L)
T_{50} is the mid-boiling temperature (°C)

The CCI is still included in the US diesel specification [15.54] because it is recognized by the US EPA as an alternative method to meet the US federal diesel aromatics limit for diesel fuels containing less than 500 mg/kg sulfur.

The CCI is clearly a correlation between observed CN and the physical properties of a batch of diesel fuels. Like the DI, it relied on only two physical properties. Attempts were made to incorporate other properties, such as the hydrogen content of the fuel [15.55], other points on the distillation curve [15.56], and hydrocarbon type, determined by proton nuclear magnetic resonance (^{1}H NMR) [15.57]. However, if the nature of a diesel fuel is significantly different to the fuels that were used to derive the equation, then the predicted value could be somewhat different from the true value [15.58]. This led to a further proposal that included the aniline point, three points from the distillation curve, the density, and the viscosity [15.59]. However, an ASTM Task Force set up in 1982 adopted a new CCI based on a four variables equation [15.60]. The new equation based on density and the 10%, 50%, and 90% distillation points is given:

$$
\begin{aligned}
\mathrm{CCI} = {} & 45.2 + 0.0892\left(T_{10} - 215\right) + 0.131\left(T_{50} - 260\right) \\
& + 0.0523\left(T_{90} - 310\right) + 0.901\,\mathrm{B}\left(T_{50} - 260\right) - 0.420\,\mathrm{B}\left(T_{90} - 310\right) \\
& + 0.0049\left(T_{10} - 215\right)^2 - 0.0049\left(T_{90} - 310\right)^2 + 107.0\,\mathrm{B} + 60.0\,\mathrm{B}^2
\end{aligned}
$$

where T_{10}, T_{50}, and T_{90} are the 10%, 50%, and 90% recovery temperatures in °C, corrected to standard barometric pressure, respectively; D is density at 15°C and

$$
\mathrm{B} = \mathrm{e}^{\left[-3.5(D - 0.85)\right]} - 1
$$

15.3. Emissions Characteristics of Diesel Combustion

Early in automotive history, the tailpipe emissions of gasoline engines were largely ignored because they did not pose a visible problem. The smog that they produced in specific locations such as Los Angeles was only attributed to automotive emissions following detailed research. This is discussed in Chapter 13. With the diesel engine, there has always been a potential visible emissions problem. On cold starting, diesel engines would traditionally emit a large cloud of white smoke. Under heavy load, they could then emit a plume of black smoke. These obnoxious emissions meant that the diesel engine underwent voluntary emissions control before any legislation was introduced. By limiting the maximum amount of fuel that could be delivered to the engine, the amount of black smoke could be limited. The maximum amount of power the engine would produce would also be limited, but if this was an issue then a larger engine could be used! However, it was very easy for the operator to override the fueling stop in order to attain more power. Los Angeles was again at the forefront of emissions control, and in 1946, the Smoke Control Hearing Board was set up under the Los Angeles County Air Pollution Control Board (the A.P.C.D.). Laws were set to limit the emissions of visible smoke, and the official patrol car of the A.P.C.D. was legally authorized to stop any highway vehicle violating the ordinance [15.61].

Clearly, more fuel could be introduced without hitting this smoke limit if more air could also be introduced. This was one of the reasons for the widespread adoption of pressure charging. Pressure charging also allowed for a significant increase in the specific power output of the engine (discussed in the next chapter).

With growing investigation of tailpipe emissions, it was found that the diesel engine produced low levels of carbon monoxide (CO) and unburned hydrocarbons (UHC). However, the diesel engine produced significant emission levels of oxides of nitrogen (NO_X). The black smoke was in fact a plume of soot particles; these could be assessed in a variety of ways including opacity, the reflectance of a sample collected on a filter paper, the mass collected on the filter paper, or the number of particles in a given volume of exhaust gas.

Engine manufacturers found that they could limit the emissions of NO_X, but this was usually at the expense of increased soot emissions. The soot emissions were measured as the mass of particulate matter (PM). The fact that limiting NO_X increased PM emissions, and visa-a-versa, is hence known as the NO_X – PM trade-off.

It also became apparent that the emissions characteristics could be significantly affected by the type and quality of the fuel. The fuel effects will be discussed further in Chapter 21. The remainder of this chapter will be devoted to a basic explanation of how the nature of the diesel combustion process produces these emissions.

15.3.1. Smoke and Particulates

The basic concept of the diesel combustion process that was briefly discussed in Section 15.1 will now be related to the formation of soot particles. Figure 15.6 [15.9] shows the ignition delay period, the premixed burn period, and the start of the diffusion burn period as a sequence of idealized schematics.

FIGURE 15.6 Schematic of the early stages of diesel combustion.

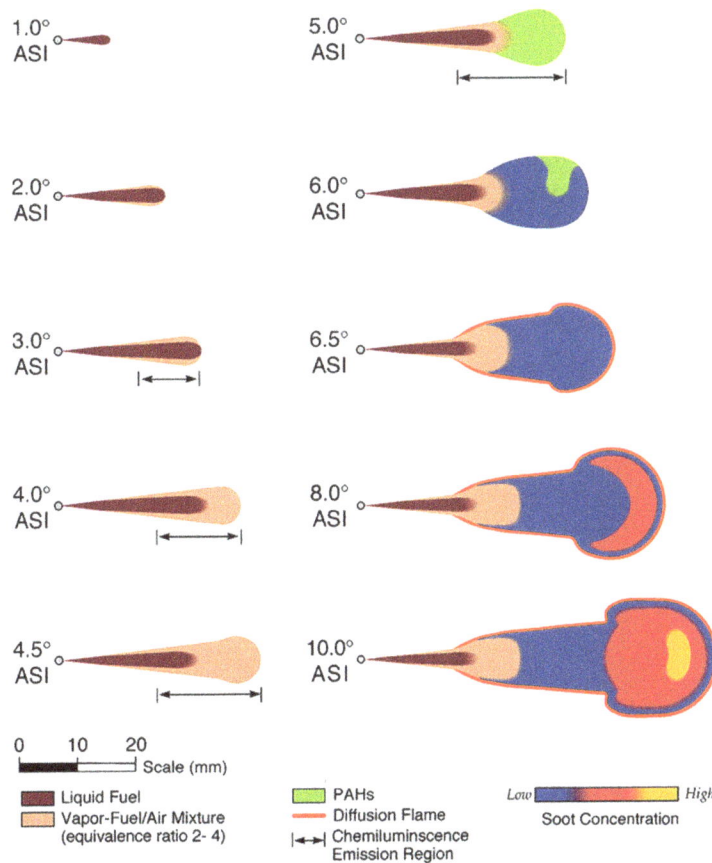

© SAE International.

The sequence starts shortly after the start of injection (ASI). The individual schematics are labeled according to their position in degrees crank angle ASI. These figures are only indicative and will vary according to engine, fuel injection, equipment, fuel, operating conditions, and so on. It also assumes that the combustion chamber is quiescent in the area of the fuel spray. The penetrating fuel jet draws hot gas from the combustion chamber into the fuel jet. This hot gas causes a rise in the fuel temperature, leading to evaporation. This in turn results in a sheath comprising a mixture of hot fuel vapor and air to form around the jet and particularly at the penetrating tip. This is shown in the sequence down the left-hand side of Figure 15.6. By about 5°ASI, the temperatures in the plume will have reached approximately 480°C.

At these temperatures, the preflame chemical reactions will begin to occur. As these reactions progress, additional hot air is entrained into the jet, increasing the temperatures to about 550°C. This increase in temperature will increase the rate of these reactions. If EGR is present, then this will reduce the amount of oxygen available for these reactions [15.62, 15.63]. Air movement, such as swirl, within the cylinder can also affect the mixing process that can lead to reduced soot emissions [15.64, 15.65].

One of the products of these initial reactions is small, partially burned fragments of hydrocarbons, usually comprising less than four carbon atoms. Many of these fragments are C_2H_2, C_2H_4, and C_3H_3. These species are generally believed to lead to the formation of the polycyclic aromatic hydrocarbons (PAHs) that are the building blocks for soot particulates. This is depicted in the schematic on the top right of Figure 15.6. It is at about this time that the premixed burn begins, consuming the well-mixed fuel vapor and establishing the diffusion flame around the spray plume. The combustion then enters the diffusion burn period, and the form of the spray structure stabilizes. This is depicted in the bottom schematics in Figure 15.6. A more detailed image of the stabilized structure is shown in Figure 15.7 [15.9].

FIGURE 15.7 Schematic of stabilized burning jet.

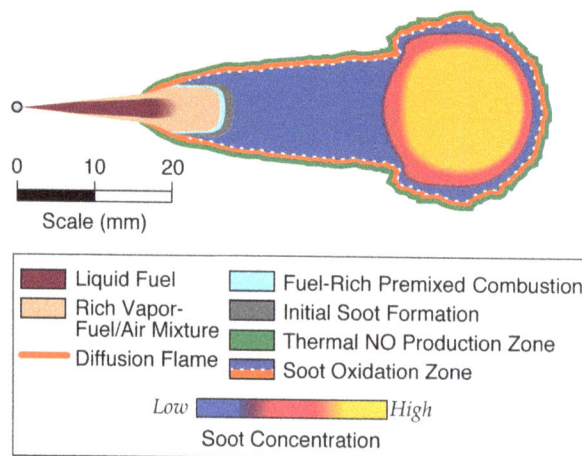

© SAE International.

The temperature reached by this partial oxidation of fuel is about 1600 K, and the reaction products will subsequently receive heat diffusing from the hot diffusion flame around the periphery of the reaction zone. This hot oxygen-depleted region has chemical constituents that are favorable for the formation and agglomeration of soot particles.

If the engine is pressure charged, then this will probably raise the temperature of the incoming air. The degree to which the temperature is raised will depend on the type and degree of intercooling, which is how much the air is cooled between the pressure charger and the engine. If EGR is used, then this can also raise the temperature of the incoming charge. However, the EGR will also reduce the oxygen availability because the EGR contains significant amounts of water vapor and carbon dioxide (CO_2) [15.66].

15.3.2. Oxides of Nitrogen

Diesel fuel should naturally contain very little nitrogen and as such the NO_X emissions are not believed to be formed by the combustion of the fuel, per se. However, the inclusion of organic nitrates as CN improves could increase the amount of fuel-bound nitrogen. By comparing the performance of nitrate and peroxide CN improver additives, it has been shown that nitrate-based additives do not contribute to NO_X emissions [15.67, 15.68]. The reduction in ignition delay, as a result of higher CN, can then lead to a reduction in NO_X emissions.

The NO_X emissions are therefore derived from the oxidation of the nitrogen in the incoming air. Most of the NO_X formed is nitric oxide (NO). This can occur as "thermal" NO via the extended Zeldovich mechanism [15.69], through the "prompt" NO formation mechanism [15.70] and via the formation of N_2O [15.71].

In the extended Zeldovich mechanism, NO is formed via the following reactions:

$$O + N_2 \leftrightarrow NO + N$$

$$N + O_2 \leftrightarrow NO + O$$

$$N + OH \leftrightarrow NO + H$$

In the prompt NO formation mechanism, nitrogen reacts with CH radicals via the following reaction:

$$CH + N_2 \leftrightarrow HCN + N$$

The products of this reaction then undergo rapid oxidation to form NO [15.72].

It is also thought that under the high pressures that can exist in a diesel engine, oxygen atoms can directly attack the N_2 molecules, resulting in the formation of nitrous oxide. Most of the N_2O is then converted to NO through further reactions in the presence of free radical oxygen and hydrogen. This is described by the following equations:

$$N_2O + O \leftrightarrow 2NO$$

$$N_2O + H \leftrightarrow NO + NH$$

The NH then undergoes further oxidation to produce more NO.

Figure 15.8 [15.73] shows the stabilized diffusion burn depicted in Figure 15.7 with additional notation to indicate the temperatures prevailing in the different regions.

FIGURE 15.8   Schematic of the burning process with the inclusion of temperature information.

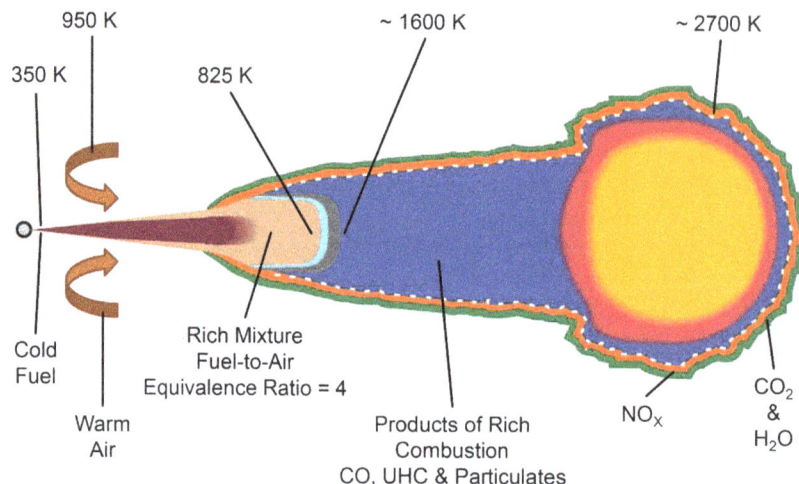

As discussed in Section 15.1, most of the premixed combustion occurs in a fuel-rich region with a fuel-to-air equivalence ratio of about 4. The adiabatic flame temperature in this fuel-rich region is about 1600 K. The kinetics suggests that there will be very little thermal NO produced at these temperatures. However, the remaining fuel burns on the jet periphery as a diffusion flame. At the jet periphery, the mixture strength will be a lot closer to stoichiometric and as indicated in Figure 15.8 the temperatures will be a lot higher. There will also be a greater availability of oxygen, and this will favor the production of thermal NO. These high temperatures and oxygen availability will also favor NO production after the end of injection.

Pressure charging and the use of EGR will again affect the processes described. If the engine is pressure charged, then this will probably raise the temperature of the incoming air. The degree to which the temperature is raised will depend on the type and degree of intercooling.

The presence of EGR can also affect the charge temperature and affect the NO_X formation process in other ways. The replacement of some of the oxygen with water vapor and CO_2 will increase the heat-absorbing capacity of the inlet charge, which in turn will slightly reduce the flame temperature. There will also be a slight reduction in NO_X emissions due to the dissociation of the CO_2. However, the major effect of EGR on NO_X formation is due to the dilution effect of replacing some of the oxygen [15.74, 15.75]. The flame temperature can be reduced by 50 to 100 K, depending on the operating condition, for 50% EGR [15.76].

References

15.1. Stuart, H.A. and Binney, C.R., Improvements in engines operated by the explosion of mixtures of combustible vapour of gas and air. G.B. Patent No. 7146, 1890.

15.2. Diesel, R., A process for producing motive work from the combustion of fuel. G.B. Patent No. 7241, 1892.

15.3. Onishi, S., Jo, S.H., Shoda, K., Jo, P.D. et al., "Active Thermo-Atmosphere Combustion (ATAC)—A New Combustion Process for Internal Combustion Engines," SAE Technical Paper 790501 (1979), doi:https://doi.org/10.4271/790501.

15.4. Lavy, J., Dabadie, J.-C., Angelberger, C., Duret, P. et al., "Innovative Ultra-low NOx Controlled Auto-Ignition Combustion Process for Gasoline Engines: The 4-SPACE Project," SAE Technical Paper 2000-01-1837 (2000), doi:https://doi.org/10.4271/2000-01-1837.

15.5. Thring, R.H., "Homogeneous-Charge Compression-Ignition (HCCI) Engines," SAE Technical Paper 892068 (1989), doi:https://doi.org/10.4271/892068.

15.6. Wåhlin, F., Cronhjort, A., Olofsson, U., and Ångström, H., "Effect of Injection Pressure and Engine Speed on Air/Fuel Mixing and Emissions in a Pre-Mixed Compression Ignited (PCI) Engine Using Diesel Fuel," SAE Technical Paper 2004-01-2989 (2004), doi:https://doi.org/10.4271/2004-01-2989.

15.7. Mohammadi, A., Kee, S., Ishiyama, T., Kakuta, T. et al., "Implementation of Ethanol Diesel Blend Fuels in PCCI Combustion," SAE Technical Paper 2005-01-3712 (2005), doi:https://doi.org/10.4271/2005-01-3712.

15.8. Lewander, C., Johansson, B., and Tunestal, P., "Extending the Operating Region of Multi-Cylinder Partially Premixed Combustion Using High Octane Number Fuel," SAE Technical Paper 2011-01-1394 (2011), doi:https://doi.org/10.4271/2011-01-1394.

15.9. Dec, J.E., "A Conceptual Model of DI Diesel Combustion Based on Laser-Sheet Imaging," SAE Technical Paper 970873 (1997), doi:https://doi.org/10.4271/970873.

15.10. Edwards, C.F., Siebers, D.L., and Hoskin, D.H., "A Study of the Autoignition Process of a Diesel Spray via High Speed Visualization," SAE Technical Paper 920108 (1992), doi:https://doi.org/10.4271/920108.

15.11. Dec, J.E. and Espey, C., "Ignition and Early Soot Formation in a DI Diesel Engine Using Multiple 2-D Imaging Diagnostics," SAE Technical Paper 950456 (1995), doi:https://doi.org/10.4271/950456.

15.12. Tausz, J. and Schulte, F., "Determination of Ignition Points of Liquid Fuels under Pressure," Technical Memorandum No. 299, National Advisory Committee for Aeronautics, Washington, DC, 1925.

15.13. Rothrock, A.M., "The N.A.C.A. Apparatus for Studying the Formation and Combustion of Fuel Sprays and the Results from Preliminary Tests," Technical Report No. 429, National Advisory Committee for Aeronautics, Washington, DC, 1932.

15.14. Yu, T.C., Uyehara, O.A., Myers, P.S., Collins, R.N. et al., "Physical and Chemical Ignition Delay in an Operating Diesel Engine Using the Hot-Motored Technique," SAE Technical Paper 560061 (1956), doi:https://doi.org/10.4271/560061.

15.15. El Wakil, M.M., Myers, P.S., and Uyehara, O.A., "Fuel Vaporization and Ignition Las in Diesel Combustion," SAE Technical Paper 560063 (1956), doi:https://doi.org/10.4271/560063.

15.16. El-Wakil, M.M. and Abdou, M.I., "The Self Ignition of Fuel Drops in Heated Air Streams," SAE Technical Paper 620284 (1962), doi:https://doi.org/10.4271/620284.

15.17. Adam, A., Inukai, N., Kidoguchi, Y., Miwa, K. et al., "A Study on Droplets Evaporation at Diesel Spray Boundary during Ignition Delay Period," SAE Technical Paper 2007-01-1893 (2007), doi:https://doi.org/10.4271/2007-01-1893.

15.18. Rosseel, E. and Sierens, R., "The Physical and the Chemical Part of the Ignition Delay in Diesel Engines," SAE Technical Paper 961123 (1996), doi:https://doi.org/10.4271/961123.

15.19. Ogasawara, M. and Sami, H., "A Study on the Behavior of a Fuel Droplet Injected into the Combustion Chamber of a Diesel Engine," SAE Technical Paper 670468 (1967), doi:https://doi.org/10.4271/670468.

15.20. Faeth, G.M. and Olson, D.R., "The Ignition of Hydrocarbon Fuel Droplets in Air," SAE Technical Paper 680465 (1968), doi:https://doi.org/10.4271/680465.

15.21. Temple-Pediani, R.W., "The Ignition Delay and Combustion of a Drop under Pressure on a Hot Surface," SAE Technical Paper 700502 (1970), doi:https://doi.org/10.4271/700502.

15.22. Henein, N.A., "A Mathematical Model for the Mass Transfer and Combustible Mixture Formation around Fuel Droplets," SAE Technical Paper 710221 (1971), doi:https://doi.org/10.4271/710221.

15.23. Chi, Y. and Kim, E., "Measurement of Droplet Size Distribution of Transient Diesel Spray," SAE Technical Paper 931949 (1993), doi:https://doi.org/10.4271/931949.

15.24. Adam, A., Yatsufusa, T., Gomi, T., Irie, N. et al., "Analysis of Droplets Evaporation Process of Diesel Spray at Ignition Delay Period Using Dual Nano-spark Shadowgraph Photography Method," SAE Technical Paper 2009-32-0017 (2009), doi:https://doi.org/10.4271/2009-32-0017.

15.25. Pedersen, P.S. and Qvale, B., "A Model for the Physical Part of the Ignition Delay in a Diesel Engine," SAE Technical Paper 740716 (1974), doi:https://doi.org/10.4271/740716.

15.26. Abraham, J. and Magi, V., "A Model for Multicomponent Droplet Vaporization in Sprays," SAE Technical Paper 980511 (1998), doi:https://doi.org/10.4271/980511.

15.27. Minagawa, T., Kosaka, H., and Kamimoto, T., "A Study on Ignition Delay of Diesel Fuel Spray via Numerical Simulation," SAE Technical Paper 2000-01-1892 (2000), doi:https://doi.org/10.4271/2000-01-1892.

15.28. Plee, S.L. and Ahmad, T., "Relative Roles of Premixed and Diffusion Burning in Diesel Combustion," SAE Technical Paper 831733 (1983), doi:https://doi.org/10.4271/831733.

15.29. Dixon, H.B., Bradshaw, L., and Campbell, C., "The Firing of Gases by Adiabatic Compression. Part I. Photographic Analysis of the Flame," *Journal of the Chemical Society, Transactions* 105 (1914): 2027-2036.

15.30. Becker, A.H. and Fischer, H.G.M., "A Suggested Index of Diesel Fuel Performance," SAE Technical Paper 340110 (1934), doi:https://doi.org/10.4271/340110.

15.31. Schweitzer, P.H. and Hetzel, T.B., "Cetane Rating of Diesel Fuels," SAE Technical Paper 360118 (1936), doi:https://doi.org/10.4271/360118.

15.32. Baxley, C.H. and Rendel, T.B., "Report of the Volunteer Group for Compression-Ignition Fuel Research," SAE Technical Paper 380121 (1938), doi:https://doi.org/10.4271/380121.

15.33. ASTM International, "Standard Test Method for Cetane Number of Diesel Fuel Oil," ASTM D613-23, ASTM International, 2023.

15.34. ASTM International, "Standard Test Method for Calculated Cetane Index of Distillate Fuels," ASTM D976-21e1, ASTM International, 2023.

15.35. ASTM International, "Standard Test Method for Determination of Derived Cetane Number (DCN) of Diesel Fuel Oils—Fixed Range Injection Period, Constant Volume Combustion Chamber Method," ASTM D7170-16, ASTM International, 2019 (Withdrawn 2019).

15.36. ASTM International, "Standard Test Method for Determination of Derived Cetane Number (DCN) of Diesel Fuel Oils—Ignition Delay and Combustion Delay Using a Constant Volume Combustion Chamber Method," ASTM D7668-17, ASTM International, 2017.

15.37. U.S. Department of Health and Human Services Public Health Service, Agency for Toxic Substances and Disease Registry, "Toxicological Profile for Naphthalene, 1-Methylnaphthalene, and 2-Methylnaphthalene," U.S. Department of Health and Human Services Public Health Service, Agency for Toxic Substances and Disease Registry, Washington, DC, 2005.

15.38. Hurn, R.W. and Hughes, K.J., "Combustion Characteristics of Diesel Fuels as Measured in a Constant-Volume Bomb—A Report of the Coordinating Research Council, Inc," SAE Technical Paper 520210 (1952), doi:https://doi.org/10.4271/520210.

15.39. Taracha, J.W. and Cliffe, J.O., "The Effects of Cetane Quality on the Performance of Diesel Engines," SAE Technical Paper 821232 (1982), doi:https://doi.org/10.4271/821232.

15.40. Needham, J.R. and Doyle, D.M., "The Combustion and Ignition Quality of Alternative Fuels in Light Duty Diesels," SAE Technical Paper 852101 (1985), doi:https://doi.org/10.4271/852101.

15.41. Siebers, D.L., "Ignition Delay Characteristics of Alternative Diesel Fuels: Implications on Cetane Number," SAE Technical Paper 852102 (1985), doi:https://doi.org/10.4271/852102.

15.42. Cellier, J., Mille, G., and Mottas, A., Process for measuring the cetane number of supply fuels for diesel engines and apparatus for performing this process. U.S. Patent 5,457,985, 1995.

15.43. Michailova, M.N. and Neumann, M.B., "The Cetene Scale and the Induction Period Preceding the Spontaneous Ignition of Diesel Fuels in Bombs," Technical Memorandum No. 813, National Advisory Committee for Aeronautics, Washington, DC, 1936.

15.44. Ryan, T.W. and Stapper, B., "Diesel Fuel Ignition Quality as Determined in a Constant Volume Combustion Bomb," SAE Technical Paper 870586 (1987), doi:https://doi.org/10.4271/870586.

15.45. Ryan, T.W. and Callahan, T.J., "Engine and Constant Volume Bomb Studies of Diesel ignition and Combustion," SAE Technical Paper 881626 (1988), doi:https://doi.org/10.4271/881626.

15.46. Datschefski, G. and Rickeard, D.J., "Diesel Fuel Ignition Quality Measurement by a Constant Volume Combustion Test," SAE Technical Paper 932743 (1993), doi:https://doi.org/10.4271/932743.

15.47. Aradi, A.A. and Ryan, T.W., "Cetane Effect on Diesel Ignition Delay Times Measured in a Constant Volume Combustion Apparatus," SAE Technical Paper 952352 (1995), doi:https://doi.org/10.4271/952352.

15.48. Allard, L.N., Webster, G.D., Hole, N.J., Ryan, T.W. et al., "Diesel Fuel Ignition Quality as Determined in the Ignition Quality Tester (IQT)," SAE Technical Paper 961182 (1996), doi:https://doi.org/10.4271/961182.

15.49. Allard, L.N., Hole, N.J., Webster, G.D., Ryan, T.W. et al., "Diesel Fuel Ignition Quality as Determined in the Ignition Quality Tester (IQT)—Part II," SAE Technical Paper 971636 (1997), doi:https://doi.org/10.4271/971636.

15.50. Allard, L.N., Webster, G.D., Ryan, T.W., Baker, G. et al., "Analysis of the Ignition Behaviour of the ASTM D-613 Primary Reference Fuels and Full Boiling Range Diesel Fuels in the Ignition Quality Tester (IQT™)—Part III," SAE Technical Paper 1999-01-3591 (1999), doi:https://doi.org/10.4271/1999-01-3591.

15.51. Allard, L.N., Webster, G.D., Ryan, T.W., Matheaus, A.C. et al., "Diesel Fuel Ignition Quality as Determined in the Ignition Quality Tester (IQT™)—Part IV," SAE Technical Paper 2001-01-3527 (2001), doi:https://doi.org/10.4271/2001-01-3527.

15.52. Yates, A.D.B., Viljoen, C.L., and Swarts, A., "Understanding the Relation between Cetane Number and Combustion Bomb Ignition Delay Measurements," SAE Technical Paper 2004-01-2017 (2004), doi:https://doi.org/10.4271/2004-01-2017.

15.53. Henein, N.A., Fragoulis, A.N., and Luo, L., "Correlation between Physical Properties and Autoignition Parameters of Alternate Fuels," SAE Technical Paper 850266 (1985), doi:https://doi.org/10.4271/850266.

15.54. ASTM International, "Standard Specification for Diesel Fuel Oils," ASTM D975-22a, ASTM International, 2022.

15.55. Murphy, M., "An Improved Cetane Number Predictor for Alternative Fuels," SAE Technical Paper 831746 (1983), doi:https://doi.org/10.4271/831746.

15.56. Ingham, M.C., Bert, J.A., and Painter, L.J., "Improved Predictive Equations for Cetane Number," SAE Technical Paper 860250 (1986), doi:https://doi.org/10.4271/860250.

15.57. Glavincevski, B., Gülder, Ö.L., and Gardner, L., "Cetane Number Estimation of Diesel Fuels from Carbon Type Structural Composition," SAE Technical Paper 841341 (1984), doi:https://doi.org/10.4271/841341.

15.58. Steere, D.E. and Nunn, T.J., "Diesel Fuel Quality Trends in Canada," SAE Technical Paper 790922 (1979), doi:https://doi.org/10.4271/790922.

15.59. Steere, D.E., "Development of the Canadian General Standards Board (CGBS) Cetane Index," SAE Technical Paper 841344 (1984), doi:https://doi.org/10.4271/841344.

15.60. ASTM International, "Standard Test Method for Calculated Cetane Index by Four Variable Equation," ASTM D4737-21, ASTM International, 2021.

15.61. Grunder, L., "West Coast Diesel Smoke and Odor Problems," SAE Technical Paper 490132 (1949), doi:https://doi.org/10.4271/490132.

15.62. Ladommatos, N., Balian, R., Horrocks, R., and Cooper, L., "The Effect of Exhaust Gas Recirculation on Soot Formation in a High-Speed Direct-Injection Diesel Engine," SAE Technical Paper 960841 (1996), doi:https://doi.org/10.4271/960841.

15.63. Ladommatos, N., Abdelhalim, S.M., Zhao, H., and Hu, Z., "The Dilution, Chemical, and Thermal Effects of Exhaust Gas Recirculation on Diesel Engine Emissions—Part 1: Effect of Reducing Inlet Charge Oxygen," SAE Technical Paper 961165 (1996), doi:https://doi.org/10.4271/961165.

15.64. Rao, K.K., Winterbone, D.E., and Clough, E., "Influence of Swirl on High Pressure Injection in Hydra Diesel Engine," SAE Technical Paper 930978 (1993), doi:https://doi.org/10.4271/930978.

15.65. Winterbone, D.E., Yates, D.A., Clough, E., Rao, K.K. et al., "Combustion in High-Speed Direct Injection Diesel Engines—A Comprehensive Study," *Proceedings of the Institution of Mechanical Engineers, Part C: Journal of Mechanical Engineering Science* 208, no. 4 (1994): 223-240.

15.66. Ladommatos, N., Abdelhalim, S., Zhao, H., and Hu, Z., "The Dilution, Chemical, and Thermal Effects of Exhaust Gas Recirculation on Diesel Engine Emissions—Part 4: Effects of Carbon Dioxide and Water Vapour," SAE Technical Paper 971660 (1997), doi:https://doi.org/10.4271/971660.

15.67. Ullman, T.L., Spreen, K.B., and Mason, R.L., "Effects of Cetane Number on Emissions From a Prototype 1998 Heavy-Duty Diesel Engine," SAE Technical Paper 950251 (1995), doi:https://doi.org/10.4271/950251.

15.68. Li, X., Chippior, W.L., and Gülder, Ö.L., "Effects of Cetane Enhancing Additives and Ignition Quality on Diesel Engine Emissions," SAE Technical Paper 972968 (1997), doi:https://doi.org/10.4271/972968.

15.69. Lavoie, G.A., Heywood, J.B., and Keck, J.C., "Experimental and Theoretical Investigation of Nitric Oxide Formation in Internal Combustion Engines," *Combustion Science and Technology* 1, no. 4 (1970): 313-326.

15.70. Fenimore, C.P., "Formation of Nitric Oxide in Premixed Hydrocarbon Flames," *Symposium (International) on Combustion* 13, no. 1 (1971): 373-380.

15.71. Mellor, A.M., Mello, J.P., Duffy, K.P., Easley, W.L. et al., "Skeletal Mechanism for NOx Chemistry in Diesel Engines," SAE Technical Paper 981450 (1998), doi:https://doi.org/10.4271/981450.

15.72. Blauwens, J., Smets, B., and Peeters, J., "Mechanism of 'Prompt' No Formation in Hydrocarbon Flames," *Symposium (International) on Combustion* 16, no. 1 (1977): 1055-1064.

15.73. Flynn, P.F., Durrett, R.P., Hunter, G.L., and zur Loye, A.O., "Diesel Combustion: An Integrated View Combining Laser Diagnostics, Chemical Kinetics, And Empirical Validation," SAE Technical Paper 1999-01-0509 (1999), doi:https://doi.org/10.4271/1999-01-0509.

15.74. Ladommatos, N., Balian, R., Horrocks, R., and Cooper, L., "The Effect of Exhaust Gas Recirculation on Combustion and NOx Emissions in a High-Speed Direct-injection Diesel Engine," SAE Technical Paper 960840 (1996), doi:https://doi.org/10.4271/960840.

15.75. Ladommatos, N., Abdelhalim, S.M., Zhao, H., and Hu, Z., "The Dilution, Chemical, and Thermal Effects of Exhaust Gas Recirculation on Diesel Engine Emissions—Part 2: Effects of Carbon Dioxide," SAE Technical Paper 961167 (1996), doi:https://doi.org/10.4271/961167.

15.76. Arcoumanis, C., Bae, C., Nagwaney, A., and Whitelaw, J.H., "Effect of EGR on Combustion Development in a 1.9L DI Diesel Optical Engine," SAE Technical Paper 950850 (1995), doi:https://doi.org/10.4271/950850.

Further Reading

Kasab, J. and Strzelec, A., *Automotive Emissions Regulations and Exhaust Aftertreatment Systems* (Warrendale, PA: SAE International, 2020).

Rogers, D., *Engine Combustion: Pressure Measurement and Analysis*, 2nd ed. (Warrendale, PA: SAE International, 2021).

Zhao, H., *Laser Diagnostics and Optical Measurement Techniques in Internal Combustion Engines* (Warrendale, PA: SAE International, 2012).

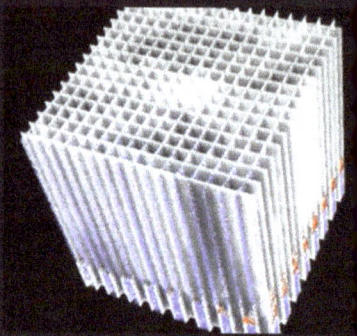

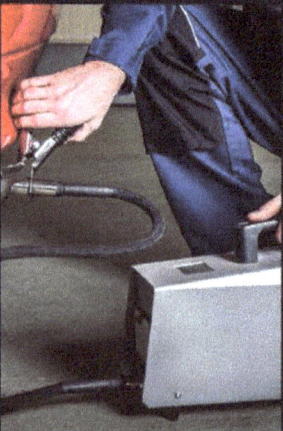

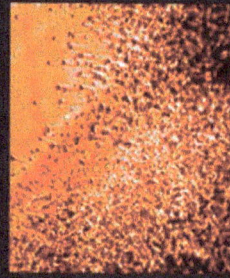

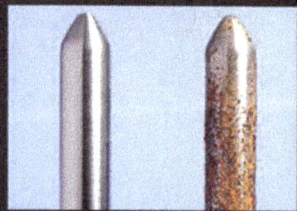

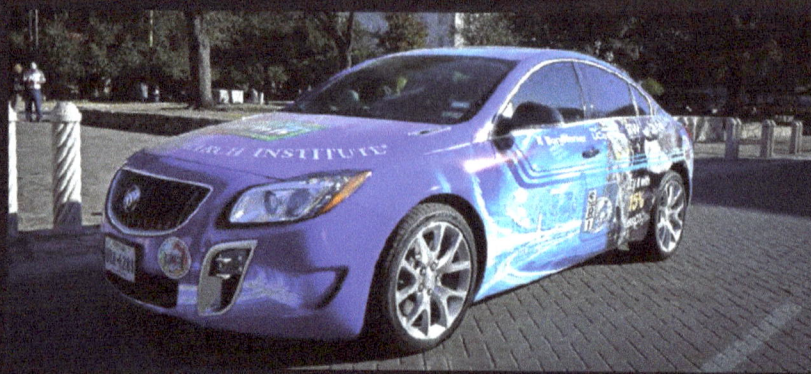

16

Diesel Engine Design and Influence of Fuel Characteristics

16.1. Introduction

Chapter 2, Section 2.2.1, describes how the compression ignition engine came into existence as the successful outcome of the independent efforts by Herbert Akroyd Stuart [16.1] in England and Rudolf Diesel [16.2] in Germany to overcome particular limitations of the spark-ignition engine. Akroyd Stuart's vaporizer engine was the first internal combustion engine to run without spark-induced ignition, and it was put on exhibition in 1891, two years before Diesel's first prototype was completed.

Both inventors designed their engines to breathe only air, with the fuel only being injected toward the end of the compression stroke. This avoided the problem of pre-ignition during compression of an explosive mixture. However, Akroyd Stuart's engine had a relatively low compression ratio and relied on the vaporizer to aid ignition. Diesel's approach was to achieve greater thermal efficiency by utilization of a higher expansion ratio that was achieved by increasing the compression ratio. This had the benefit of producing a temperature that was high enough to cause the fuel to ignite spontaneously when injected into the hot compressed air.

Inspection of the two patents reveals that Akroyd Stuart's vaporizer was in effect a pre-chamber, and Diesel's conception delivered the fuel directly into the combustion chamber. Akroyd Stuart's engine was thus an indirect injection (IDI) engine, whereas Diesel's original conception was a direct injection (DI) engine. For high-speed applications, there were problems associated with the DI approach that resulted in many mechanical designs in the direction of the IDI configuration used by Akroyd Stuart. These designs in turn had their particular disadvantages, and modern diesel engine design has reverted to the DI concept of Diesel's original patent. The design of the combustion chamber will be discussed in more detail in the following section.

Another difference between the initial designs of these two pioneers was in the fuel injection systems. Akroyd Stuart used a mechanical pump to spray a liquid fuel into the combustion space, and Diesel's conception relied on gravity to deliver pulverized coal into the combustion chamber. Diesel was unable to engineer a working system to this design, and it was abandoned in favor of a blast of compressed air system [16.3].

The compressed air blew the pulverized coal into the combustion chamber. Air blasting was subsequently abandoned in favor of the mechanical pump, liquid injection system similar to that used by Akroyd Stuart. Fluid injection now forms the basis of all modern automotive diesel engines. The fuel injection system is now a fundamental part of the diesel engine and will be discussed in more detail in Section 16.3.

16.2. The Diesel Compression Ignition Engine

By the turn of the millennium, there were two broad categories of automotive diesel engines based on their combustion chamber arrangement: DI and IDI. Figure 16.1 [16.4] illustrates examples of these two basic combustion chamber configurations.

FIGURE 16.1 The two basic combustion chamber configurations: DI (left) and IDI (right).

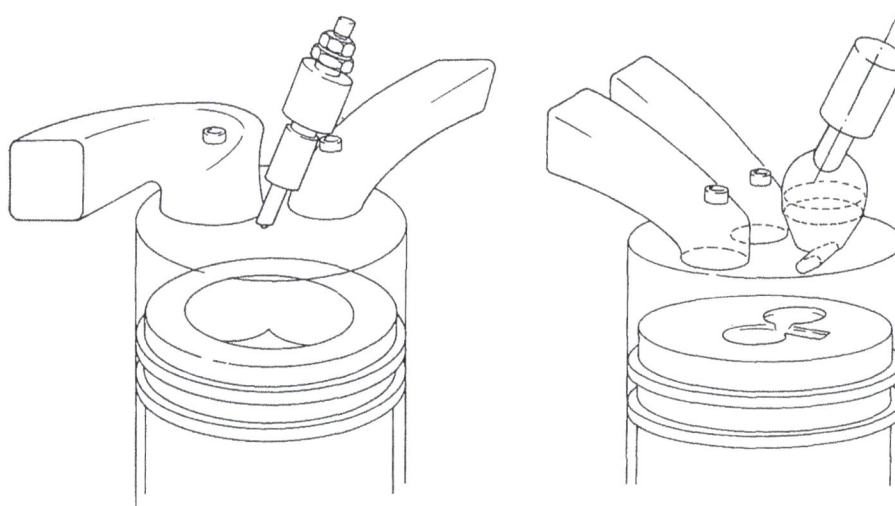

© SAE International.

16.2.1. Direct Injection

With DI engines, there is a combustion chamber formed by the cylinder, cylinder head, and piston top. The fuel is injected directly into the chamber. In Diesel's original design and in early examples of these engines, the combustion chamber was quiescent. No design elements were deliberately incorporated to induce any air motion, and the slow speed of the engines promoted very little motion during the compression stroke. With Diesel's air-blast, coal dust injection system, the air blast provided reasonable mixing of the dust and compressed air. However, with liquid injection, the low pressures available at the time produced poor atomization and hence poor mixing of fuel and air. Poor mixing extends the ignition delay period, which increases the amount of pre-mixed burning, resulting in high rates of cylinder pressure rise. This increased combustion noise and mechanical stresses. Various schemes were devised to try and promote better mixing. This could be achieved by designing the intake ports to induce a swirling motion to the air within the cylinder [16.5, 16.6]. This motion was then promoted in the vicinity of the fuel spray by means of differently shaped combustion chambers formed within the piston crown. Some examples of this are shown in Figure 16.2 [16.7, 16.8].

FIGURE 16.2   Various bowl-in-piston configurations for DI combustion chambers.

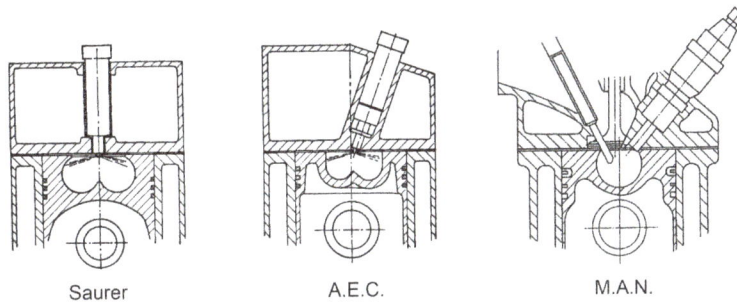

Saurer   A.E.C.   M.A.N.

The Saurer and A.E.C. designs date to the 1930s, and the M.A.N. design dates to the 1950s [16.9]. In the latter design, fuel impinges on the hot combustion chamber wall in order to aid evaporation. One of the drawbacks of these systems was that with the degree of swirl necessary to achieve optimum performance, it became a problem to start the engine under low-temperature conditions. An alternative approach was the use of indirect injection.

16.2.2. Indirect Injection

The IDI engines, which were developed to overcome these problems, have a divided combustion chamber. A pre-chamber or a swirl chamber within the cylinder head is connected to the cylinder by a narrow passage through which air is forced during the compression stroke. This configuration provides a high degree of air motion in the second chamber to promote good mixing without recourse to high-pressure fuel injection and without the complexity of multihole injection nozzles. The fuel injector, usually a single-hole pintle-type nozzle, sprays fuel into the mixing chamber, where it mixes with the incoming compressed air and ignites. The idea of the pintle is that it forms a hollow cone spray to improve mixing at high engine speeds. It is also to some extent self-cleaning. Figure 16.3 [16.10] shows examples of different divided chamber configurations.

FIGURE 16.3   Divided chamber configurations.

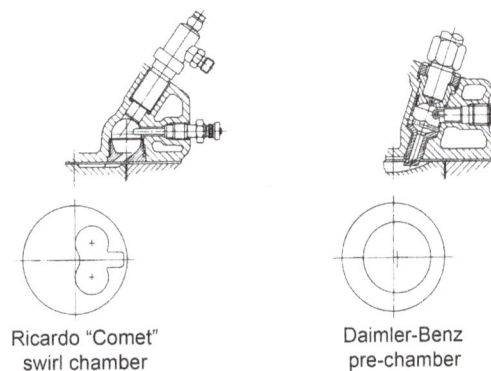

Ricardo "Comet"   Daimler-Benz
swirl chamber   pre-chamber

The ignition of fuel in the swirl chamber, of the Ricardo Comet design, increases the pressure, which forces the hot burning gases to expand and pass through the narrow connecting passage into the main cylinder. Mixing is further enhanced by combustion-driven mixing in the swirl-chamber throat, and as a further aid, a recess in the piston crown is shaped to give high air swirl. A further advantage of the IDI approach is that the offset pre-chamber enables the injector to be located to one side of the cylinder head. Within the constraints of a two-valve per cylinder arrangement, this provides space for larger valves, allowing better volumetric efficiency and hence a wider speed range. Compared with DI engines, the IDI system therefore allows operation at higher speeds and the use of simpler lower pressure pumps and less expensive injection equipment.

Despite these advances, there are weaknesses in the indirect injection approach. Combustion chambers have larger surface-to-volume ratios than those of DI engines, and therefore higher compression ratios are used to compensate for the greater heat losses during compression and to ensure an adequate ignition temperature. The combination of this surface-to-volume ratio deficiency, plus the pumping loss through the pre-chamber throat, endows the IDI engine with poorer thermal efficiency and lower specific power than an equivalent DI engine. For example, the fuel consumption of an IDI power unit is generally as much as 10% higher than a comparable DI diesel engine [16.11]. This meant that there was still incentive to revert to the original concept of direct injection.

16.2.3. Other Considerations

A key element in the return to the DI system was advances in the fuel injection equipment (FIE). Initially this was via the use of two-stage injection and latterly by full electronic control allowing multiple injections per combustion cycle and shaping of the rate of injection. This will be discussed further in the following section. Injecting only a small portion of the fuel required allowed this fuel to produce a small amount of pre-mixed fuel that had started the combustion process before the main fuel requirement was injected. This smoothed the combustion process, thus reducing noise and mechanical stresses.

Although the efficiency of the diesel engine has always been a major advantage when compared with a gasoline engine, its specific power output was a disadvantage. This had been overcome for industrial and heavy-duty vehicle applications by the use of pressure charging. Using a compressor to force the air into the engine at higher pressures, the effective capacity of the engine was increased and the amount of fuel that could be burned was correspondingly increased. By using a turbine in the exhaust pipe to drive a compressor, some of the heat in the exhaust could be recovered. The combination is then referred to as a turbocharger.

As with gasoline engines, diesels also exist in two- and four-stroke versions. The vast majority of the world's current automotive engines operate on the four-stroke principle, although two-strokes, with their power-to-weight ratio advantage, were historically well established in the US, where high engine power output was of greater consequence than fuel economy. Two- and four-stroke diesel engines operate on the same type of fuel and generally respond in a similar way to variations in fuel characteristics. However, as environmental controls tightened, two-stroke engines—which have difficulty in meeting some emissions legislation—have generally been replaced by their four-stroke counterparts.

16.3. Diesel Vehicle Fuel Systems

As already mentioned, Diesel's patent exemplified a fuel delivery system that relied on gravity to feed pulverized coal into the combustion chamber. Figure 16.4 [16.2] is taken from Diesel's patent and illustrates his mechanism.

FIGURE 16.4 Fuel "injection" system from Diesel's 1892 patent.

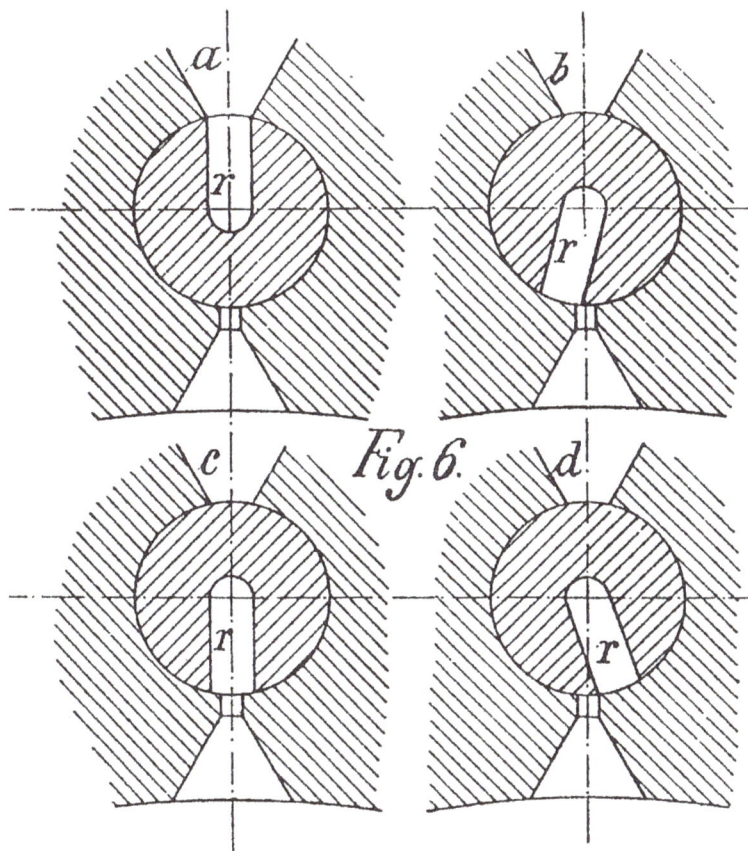

The cock illustrated in Figure 16.4 simply rotated at half crankshaft speed to deliver the coal. Beginning at point a in the diagram, it collected the coal from the hopper. Then at point b in the diagram, it started delivering the coal to the combustion chamber. Finally, at point d in the diagram it ceased delivery of any coal. This system was never successful and Diesel resorted to using compressed air to blow the pulverized coal into the combustion chamber [16.3]. In 1910 British engineer James McKechnie filed a patent [16.12] where a fuel pump was used to pump fuel into the injection system and an engine driven cam and spring mechanism was used to generate the high pressure necessary to actually inject the liquid fuel directly into the combustion chamber. Similarities between this and current mechanical unit injectors will become apparent.

Two years later, Thomas Gaff, a US engineer, proposed a system [16.13] that relied on a pump that was not synchronized to the engine to generate fuel pressure within an accumulator. The accumulator was connected to an individual injector for each cylinder. The fuel injection event was then governed by an electromagnetic valve for each cylinder. Figure 16.5 [16.13] shows Gaff's proposed system.

FIGURE 16.5 Gaff's electromagnetic fuel injection system.

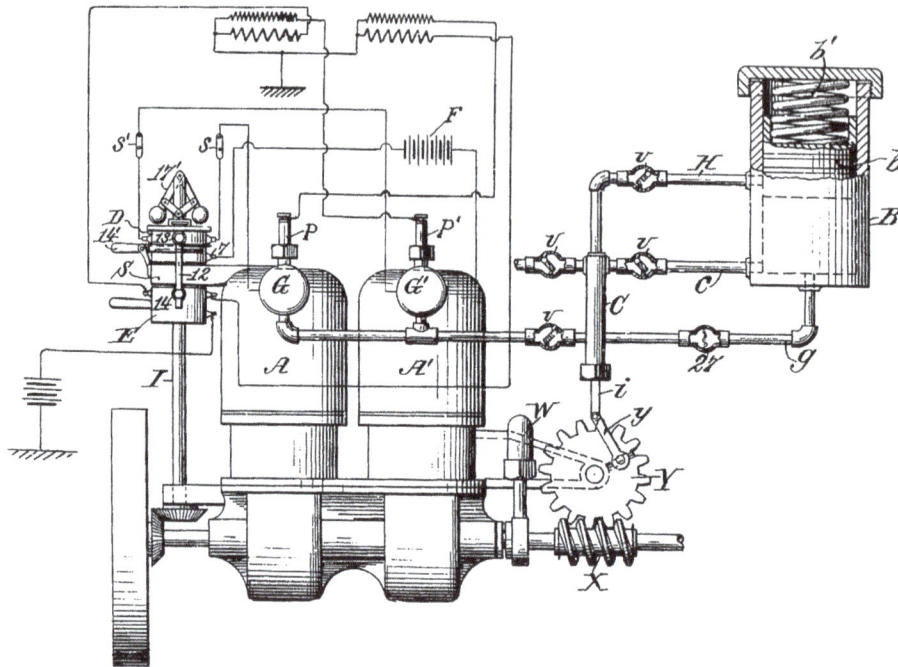

The similarities between Gaff's system and current high-pressure common-rail fuel injection systems will later become apparent.

What is clear from the history is that the fuel injection system was a major source of interest and development and that the diesel engine as we know it would not have existed without the efforts of these fuel injection pioneers. Since that time, different FIE manufacturers have tended to develop their own version of a few basic types of system, as will be described in more detail in Section 16.3.3.

The diesel vehicle fuel system will typically consist of both a low-pressure section and a high-pressure section. In the low-pressure section, a transfer or lift pump draws fuel from the tank and delivers it, at a relatively low pressure of typically less than 200 kPa, to the injection pump. The low-pressure section will also include one or more filters and a water separator. The injection pump along with the fuel injectors comprise the high-pressure section. The different aspects of the low- and high-pressure systems will be discussed in the following sections.

16.3.1. Strainers, Filters, and Separators

Some vehicle tanks are fitted with a coarse mesh strainer to prevent any large pieces of rust, dirt, or other foreign matter that gets into the tank from being drawn into the fuel lines. Drain plugs allow regular draining of water and sediment from fuel tanks, but these are becoming less common. A water sensor has been fitted in the fuel tanks of some diesel vehicles to give a visual signal on the instrument panel when a critical level has been reached.

Where a fuel transfer pump is fitted, it is sometimes protected by a pre-filter or a sedimenter fitted in the line from the tank. Woven wire or nylon mesh screen filters of about 250 μm (micron) aperture sizes are considered adequate for diaphragm and plunger pumps, but finer porosity material of cloth or paper is required for gear pumps.

The small clearances between the pumping elements of injection pumps necessitate the use of fine porosity filters to protect against damage or excessive wear. Paper, cloth, or felt main filters are capable of retaining particles larger than 5 or 10 µm. Unlike the simple woven mesh screens, these are depth filters that become more restrictive with use due to entrapment of dirt particles, so they need to be replaced periodically before becoming completely choked. To avoid high pressure drops because of their fine pore size, main filters generally have a large surface area, 3500 cm^2 being typical for the spirally wound paper element fitted in many automotive diesel systems. Some vehicles are fitted with two main filters, either connected in parallel to increase the total surface area or in series to provide better protection for the injection pump.

16.3.2. Transfer or Lift Pumps

Various types of transfer pumps, such as diaphragm, plunger-in-barrel, or gear type, are used in diesel systems to lift fuel from the tank and supply it to the injection pump. Transfer pumps have to be capable of delivering more fuel than the full load requirement of the engine, so the excess fuel is recycled back to the fuel tank. The exceptions are some diaphragm pumps fitted with an internal pressure relief valve that opens to prevent further delivery when the injection pump has taken the fuel it needs. In some designs, the recycled fuel is returned to the bottom of the tank near the suction line, and in others, it is returned through a short pipe at the top of the tank, sometimes remote from the suction line. Fuel recirculation ratios vary widely, depending on the requirements of the engine designer or the recommendations of the FIE manufacturer.

Transfer pumps may actually be placed in the fuel tank, from where they push the fuel to the injection pump. This eliminates any possibility of vapor generation in the suction line. However, placing an electric pump in a fuel tank does introduce safety concerns. Due to the high volatility of gasoline, the headspace in the fuel tank is usually too rich to be considered a flammable mixture. With diesel fuel, the low volatility usually implies a mixture that is too lean to be ignited. However, under adverse conditions, this can change; for example, misfueling is alarmingly common for diesel passenger cars. Additional care must therefore be exercised if an electric pump is placed in the fuel tank.

16.3.3. Injection Systems

In early fuel injection systems, the control of the fuel injection event was largely brought about by the injection pump. The fuel pump was remote from the fuel injector and connected by a pipe, hence the term pump-line-nozzle system. With low fuel injection pressures, the fuel can be treated as an incompressible fluid, and it would be acceptable to have long lengths of pipe between pump and injector. This was certainly the case for Akroyd Stuart. However, as injection pressures increase, the compressibility of the fuel becomes important, and in order to maintain good control the volume of high-pressure fuel needed to be minimized or the control of the injection event had to be incorporated into the injector. This favored unit injector designs in which the pump and injector were effectively a single unit. For large diesel engine applications, many engine manufacturers have traditionally utilized their own design of mechanically driven unit pump and injector units. However, for smaller automotive applications, it was not easy to downsize this technology. The power requirement for operating such systems was also quite high. Additional cams and followers were necessary with attendant frictional losses. Hydraulic unit injector systems attempted to combine the benefits of the unit injector and the pump-line-nozzle systems [16.14]. For automotive applications, the hydraulic pump did not really provide that much of a benefit compared with the pump-line-nozzle systems that could be bought and bolted-on. It is now common for engine manufacturers to install fuel pump and injection systems from specialist FIE manufacturers.

As injection pressures have risen and injector technology has advanced, these systems are giving way to common-rail systems. All these systems are described in more detail in the following sections.

16.3.3.1. Pump-Line-Nozzle Systems

For pump-line-nozzle systems, a separate injection pump is externally mounted on the side of the engine and connected by gears or chains to the crankshaft so that it runs at half engine speed for four-stroke engines. Operation of the pumping elements is affected by cams within the pump. The high-pressure fuel from the injection pump is delivered through rigid external pipes to an individual injector for each cylinder.

Pump-line-nozzle systems can be further subdivided according to whether there is a separate pumping element for each cylinder or whether a single pumping unit provides the pressurized fuel that is then distributed between the individual cylinders.

16.3.3.1.1. Inline Pumps. An early design of inline injection pump for multicylinder diesel engines was developed by Robert Bosch in 1926 [16.15], although a patent was also taken out in which the individual pumping elements were arranged radially around a single cam [16.16]. For the inline pump, each pumping element has its own cam to operate the plunger and barrel pumping elements supplying fuel to each injector in turn. Most inline block pumps have oil lubrication of these cams and followers, but the pumping elements rely on the diesel fuel being pumped to also lubricate the sliding surfaces of the plunger and the barrel. This can lead to cross-contamination of the fuel and lubricant [16.17], which can result in problems that are discussed in Chapter 18, Section 18.2.2.2.

High pressure is generated in the chamber above the plunger within the barrel. This chamber is connected by a central drilling to a helical groove cut into the plunger. There are two drillings in the barrel wall, one to allow fuel to flow into the chamber, known as the filling port, and the other to allow fuel out of the chamber, termed the spill port. With the cam and plunger at their lowest points, the filling port is open, allowing fuel to flow into the chamber under low pressure from the transfer pump. The bulk modulus of the fuel is quite high so the pressure rises rapidly as the plunger continues to rise. The delivery valve opens when the pressure exceeds its setting, and the motion of the plunger creates a high-pressure acoustic wave in the injector pipe. This travels along the pipe until it meets the seat of the injector needle valve. At this point, the acoustic wave is reflected, creating additional pressure at the seat. When the combined pressure of the incident and reflected waves exceeds the nozzle opening pressure, the needle valve lifts, exposing a larger area of the needle tip to the pressure. This ensures that the valve snaps open and remains that way, despite the fall in seat pressure as the needle lifts. The needle stays open until the wave pressure in the injector falls below its normal closing pressure.

While the acoustic pressure wave travels toward the injector needle, the pump plunger continues to travel upward until the helical groove in the plunger reaches the spill port. When this occurs, fuel pressure in the pump chamber collapses as fuel flows through the central drilling to the helical groove and out through the spill port. This drop in pressure generates an expansion wave in the injector pipe and injector, which in turn causes the pressure under the needle to collapse below the nozzle closing pressure. The injector spring returns the needle to its seat to end the injection.

Toward the end of injection, low pressure may remain across the nozzle holes, caused by the downward movement of the needle under the action of the spring. This can cause large droplets of fuel to be injected at low velocity into the diffusion flames. This is sometimes referred to as injector dribbling. Although this fuel may evaporate quickly, it is unlikely to penetrate far enough into the combustion chamber to find fresh air, resulting in incomplete combustion and unwanted hydrocarbon or particulate emissions. The heavier fractions of the fuel may not evaporate completely and can contribute to deposit build-up around the injection holes.

Control of the amount of fuel delivered to the engine is by variation of the effective length of the plunger stroke, which is achieved by changing the relative angular position of the plunger and the barrel. This determines at what stroke length the helical groove uncovers the spill port. An internal linkage ensures that changes in fuel delivery apply to all engine cylinders [16.18]. An example of an inline pump is shown in Figure 16.6.

FIGURE 16.6   Cut-away of Minimec inline fuel injection pump.

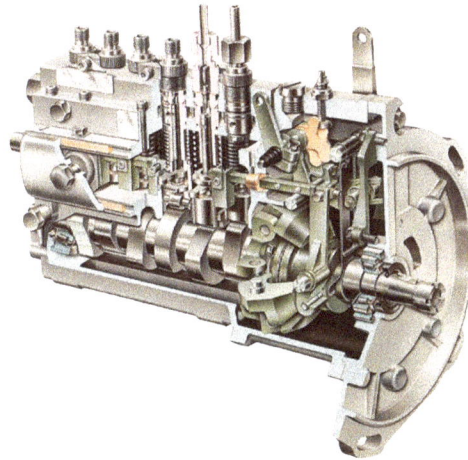

Courtesy of Delphi.

Inline block pump performance is reliable and repeatable provided the components are made to tight manufacturing tolerances. Historically, however, there have been limitations in their performance, particularly as emissions control became more critical. One problem was that injection timing could not be altered unless a helical spline arrangement was fitted in the pump drive to advance or retard the pump camshaft with respect to the engine camshaft [16.14]. Another approach has been to introduce a helical groove into the crown of the pump plunger, so that timing varies as the fueling varies. This approach was used in the 1980s in combination with changed cam profiles to give faster plunger movement toward the end of injection. As the timing was retarded by rotating the plunger, the delivery rate increased and so did the velocity of the plunger. This increased the amplitude of the acoustic wave, giving higher pressures at the nozzle and therefore higher fuel momentum as it was injected. This higher momentum translated into smaller droplet size, giving faster evaporation and, as a consequence, a lower propensity for soot formation.

16.3.3.1.2. Rotary Distributor Pumps. During the 1950s, considerable research was conducted on fuel injection and its effect on combustion in diesel engines. Despite advances in inline pumps, there was still a need for lower-cost injection equipment, without which smaller diesel engines for the light-duty vehicle market could not compete with their gasoline counterparts. The key was to be able to produce a unit with a single pumping element coupled to a "distributor" that sequentially directed the pressurized fuel to each injector. Although the original concepts dated back to the 1930s, the real breakthrough occurred during the 1950s and 1960s, and two major designs emerged: the single-axial-piston pump arrangement and the cam-ring acting on radial plungers.

The axial-piston pump approach is illustrated by the Bosch VE pump shown in Figure 16.7 [16.19].

FIGURE 16.7   Isometric section of axial-piston distributor pump.

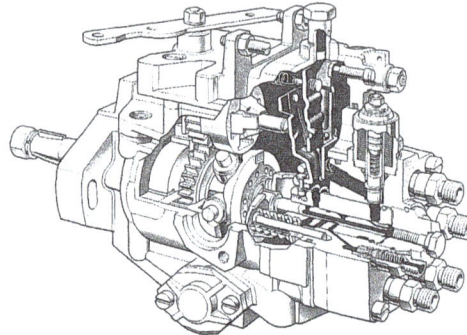

© SAE International.

The main pumping mechanism is in the center of the unit and employs a rotating face cam to impart reciprocating motion to a single, axially mounted pump plunger that rotates to distribute the fuel via a port situated halfway along the pumping element. The face cam rides on four rollers that are fixed to a cage that can also be rotated through a small angle to bring about changes to the injection phasing.

Control of the quantity of fuel injected is achieved by moving a sleeve, often referred to as a spill-muff, along the pumping element. A central drilling is extended beyond the delivery port and onto a spill port in the plunger. This port is covered by the spill-muff during pumping, but as the plunger moves, this port reaches the end of the spill-muff and the pressure in the pumping chamber collapses. The fall in pressure generates an expansion wave, which travels along the injection pipe and terminates injection. The spill-muff position is controlled by accelerator pedal position and the internal governor. This whole process provides a constant start, variable end of injection characteristic.

A peg attached to the roller cage engages with a piston that moves transversely in a housing underneath the cage. This allows the dynamic timing to be steadily increased with speed to counter the effects of wave travel time in the injector pipes. Additional mechanisms advance or retard injection timing during cold starting.

A nonreturn valve is fitted in each injector pipe outlet and maintains each line at its design residual pressure to minimize pumping work. These valves are sometimes manufactured to provide a controlled impedance to the acoustic wave reflected from the closing injector and are often termed snubber valves. The effect of this impedance is to dissipate energy from reverberant waves within the pipes.

The cam-ring type of pump is illustrated in Figure 16.8.

FIGURE 16.8   DPA cam-ring type rotary pump.

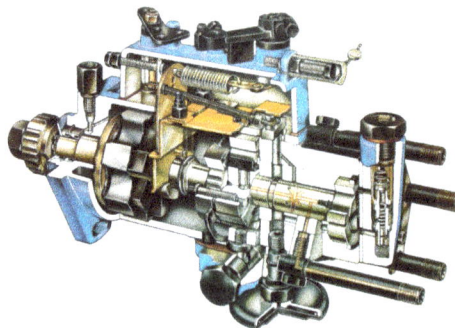

Courtesy of Delphi.

These pumps have an internally lobed cam ring to actuate radially mounted pumping plungers carried in a rotor, which also distributes the fuel to each injector in turn [16.20]. The pump relies entirely on the fuel for lubrication, and an incorporated vane pump draws fuel from the tank and floods the interior of the pump housing with fuel at a transfer pressure that varies with engine speed. The vane pump provides servo-pressure for variation of injection timing by rotating the cam ring as a whole and for the hydraulic governor version of the pump. Alteration of the plunger stroke to regulate the engine speed is achieved by limiting the quantity of fuel entering the pumping chamber by means of a metering valve controlled by the movement of the accelerator pedal and the governor. It is possible to tap off the metering valve pressure downstream to activate a light load advance system, or to use a two-port valve to achieve the same effect. Depending on whether the governor is hydraulic or mechanical, the metering valve is either lifted or rotated to vary the opening of the metering orifice. The amount of fuel admitted through the metering orifice will depend on the transfer pressure, the time available when the stator and rotor feed ports coincide, as well as the size of the metering orifice.

It is important to recognize the limited time available for the pump to function. For a six-cylinder engine, filling, pumping, and recovery all have to take place within a 60° rotation of the pump shaft, with some allowance for an 11° rotation of the cam ring to alter timing. The pumping event can take up to 20°, the recovery (to settle each injector pipe to a stable residual pressure) another 15°. After allowing 11° for cam-ring movement, a mere 14° is left to fill the pump. This is performed through a secondary porting system connected to a central drilling in the rotor. The secondary port is phased to be opposite the filling port in the rotor sleeve (stator) for just the rotational angle available to fill the pumping mechanism on each stroke. To achieve adequate filling, the transfer pressure must be raised by up to 600 kPa to move the fuel into the space between the plungers within the time available. As stated earlier, the transfer pressure increases with speed, a characteristic of the vane pump incorporated in the unit. This feature provides consistent filling across the speed range, simply because the time during which the fill ports are open reduces with increasing speed.

The progressive increase in transfer pressure is also used to provide speed advance. As the pressure rises, the advance piston (moving against the action of a return spring) moves the position of the cam between injections. When pumping starts, any cam reaction load is held by the hydraulic pressure and a nonreturn valve ensures that the cam is locked in position by that pressure. If the engine slows down, leakage past the advance piston allows the pressure to drop sufficiently to follow the reduction in engine speed.

The Lucas CAV, now Delphi Diesel Systems, DPA (Distributor Pump Type A) was designed to pump not only on the cam flank but also over the "nose" of the cam. This allows the plungers to retract at the end of injection and thereby unload the injector pipes, reducing both residual pressure and the potential for secondary injections. As a result, delivery valves were not needed, which significantly reduced manufacturing costs. The concept provided adequate rapid termination of injection at high engine speeds, but the hydraulic wave action in the injection pipes had to be carefully tuned to provide both clean termination and appropriate torque. The end of injection is constant and is a natural characteristic of the pump. As a consequence, the start of injection advances as load increases, simply because pumping starts earlier on the cam flank as load increases.

Development continued; excess fueling to aid cold starting was introduced, and the DPS pump from the 1970s featured higher rate cams to increase injection pressures. Later pumps incorporated snubber valves to control hydraulic waves in the injection pipes, which had become more difficult to contain as pumping rates increased. Such pumps were successfully matched to high-speed direct injection engines when they were first introduced in the late 1970s and early 1980s.

This type of fuel pump relied totally on the fuel for its lubrication and as a result was an early indication of a reduction in diesel fuel lubricating characteristics caused by a reduction in polar compounds brought about by the reduction in sulfur content [16.21].

16.3.3.1.3. Electronic Control. The advent of electronic engine management systems allowed these mechanical systems to be modified to provide better control. Sensors can monitor engine operating parameters such as engine speed, various fluid temperatures, manifold pressure, and pedal demand and feed this information into an electronic control unit (ECU). The ECU can then determine desired fueling parameters. Actuators, usually electro-hydraulic valves, are added to the fuel injection pumps to give more precise control of parameters that were previously controlled by mechanical systems. These servo valves can act on various parameters including fuel shut-off, the rate of fueling, and injection timing [16.15]. Electronic actuation can also take the place of the old mechanical governors on both inline pumps and rotary pumps [16.18, 16.19]. Further advances in electronic control of inline and rotary pumps have been overtaken by the development of the common-rail systems that will be described in Section 16.3.3.3.

16.3.3.2. Unit Injector Systems

To reduce emissions to meet ever lower legislative limits, it is necessary to improve fuel atomization and provide better control over the injection timing and quantity. In the pump-line-nozzle systems, there is a rise and fall of the pressure in the fuel line to each injector. To improve fuel atomization, it is necessary to increase the fuel injection pressure. This means a greater rate of pressure rise and a greater pressure in these fuel lines. For pump-line-nozzle systems, one of the performance-limiting problems is the elasticity of the injection lines, which are possibly of different lengths and incorporate changes in direction and bends that vary from cylinder to cylinder. Due to the volume of fuel within the lines, this can lead to imprecise timing control and acoustic oscillations that could affect the injection rate and may cause variation in injection-to-injection and pipe-to-pipe performance.

The unit injector provides one route to overcome these limitations. These injectors are driven from the engine camshaft and contain the pumping element and nozzle within the same assembly. The requirement for fully flexible control of injection timing and delivery has led to the development of the electronic unit injector in which the injection is controlled electromagnetically [16.22]. Initially the electronic control was brought about by use of a solenoid to open and close a spill port. As the cam starts the pumping process, the spill port is open, allowing the fuel to escape. When the solenoid is energized it closes the normally open spill control valve, allowing fuel pressure to build up to open the injector valve and initiate the high-pressure injection. When the solenoid is deactivated, it allows the spill port to open and the pressure to collapse and the spring force to close the injector.

By incorporating a second electronically controlled valve, the injection pressure and timing of the opening and closing of the nozzle needle can be determined electronically. This also allows for multiple injections. If the second valve is activated throughout the period when the spill port is closed, then the nozzle opens and closes according to the nozzle opening and closing pressures set by the nozzle needle return spring. The injector thus performs as for a single-actuator system. If the spill port is closed before the second valve is activated, the increasing fuel pressure is applied to the back of the nozzle needle, preventing it from opening. When the second valve is activated, there is a higher pressure in the system, resulting in a higher injection pressure. The single-actuator and double-actuator designs are shown in Figure 16.9 [16.23].

FIGURE 16.9 Single-actuator and double-actuator electronic unit injectors.

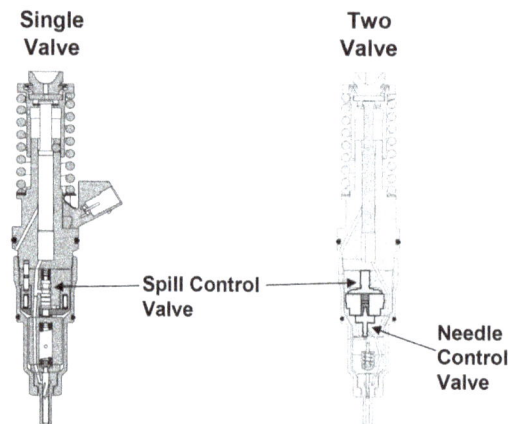

The multiple-injection capability allows for pilot injections that are useful for lowering combustion noise and NO_X emissions [16.24–16.27]. The capability can also be used to provide a late injection event that may be necessary for aftertreatment systems, which will be discussed in Section 16.4.

Unit injectors require a stiffer camshaft and add somewhat to engine height, so they were initially employed on larger, medium, and heavy-duty engines. In the 1990s, however, their application was extended to the passenger car market, where the combination of high injection pressures and flexible injection strategy allows high-speed DI engines to operate with low combustion noise and excellent emissions performance.

A variation to this approach is to use the pressure pulse from a distributor pump to actuate the injector rather than a cam. A differential piston within the fuel injector generates the high injection pressure when driven from the lower pressure produced by the distributor pump [16.14]. The distributor pump also provides the timing control. Other systems have been proposed [16.28] where the driving pressure is produced on a continuous basis and the supply to the injector is modulated electronically. Similar systems have been commercialized for heavy-duty application, where the lubricating oil system provides the driving pressure [16.29].

16.3.3.3. High-Pressure Common-Rail Systems

Despite the advances that have been made in pump-line-nozzle systems and the introduction of unit injectors to larger diesel engines, these systems all still rely on generating a high pressure for injection and then relaxing this pressure after the injection has taken place. Improvements in the electronic actuators, whether electromagnetic or piezo-electric, have allowed for better control of the injection event. However, higher engine speeds and the requirement for higher injection pressures have dictated a need for higher rates of pressure rise to achieve these higher pressures in such a short space of time.

To overcome this problem, the high-pressure common-rail (HPCR) system has been developed. This type of system is now commonplace for passenger car application and its use is spreading to heavier and heavier engines. The system can be described basically as a pump system, usually a low-pressure pump followed by a high-pressure pump, providing a continuous supply of fuel at fuel injection pressure to an accumulator to which each injector is connected. Each injector is then controlled electrically to provide the fuel injection event on demand. This is not dissimilar to the system proposed by Gaff a century ago; however, these days the pressures are a little higher! The modern common-rail fuel system is often credited to Fiat Auto and Centro Ricerche Fiat in the mid-1980s [16.30], who were certainly the first to implement such a system in passenger cars. The Alfa Romeo model 156 with a 1.9-L JTD engine equipped with a common-rail system was introduced in 1997. Work was also being conducted by Robert Bosch [16.31] and Nippondenso Co., Ltd. [16.32]. A typical system is illustrated schematically in Figure 16.10 [16.33].

FIGURE 16.10   Schematic of a high-pressure common-rail system.

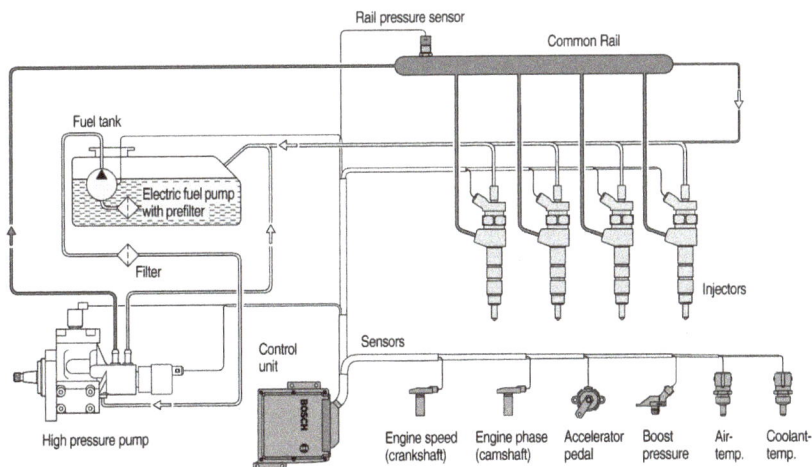

For packaging reasons, the accumulator is often a small-diameter cylinder that is placed along the engine block with short lengths of pipe between it and the injector. It has become known as the "rail," and the system has acquired its name. As the rail and the pipes to the injector are all ostensibly at injection pressure, the lengths of the pipes are not as important as they are in pump-line-nozzle systems. Although there will inevitably be some pressure drop along each pipe during the injection event, it is wise to keep pipe lengths matched and short.

For the injection systems described in the previous section, the pump generates the fuel pressure for the injection event to each injector, whereas with a HPCR system the pump has to provide the high pressure on a continuous basis. The axial-piston and cam-ring technologies that were used for distributor pumps can still be used but supply just one common outlet. Figure 16.11 [16.34] shows a longitudinal cross section of a high-pressure pump and a cross section of the cam ring and plungers.

FIGURE 16.11   Cross sections of high-pressure pump and cam-ring arrangement.

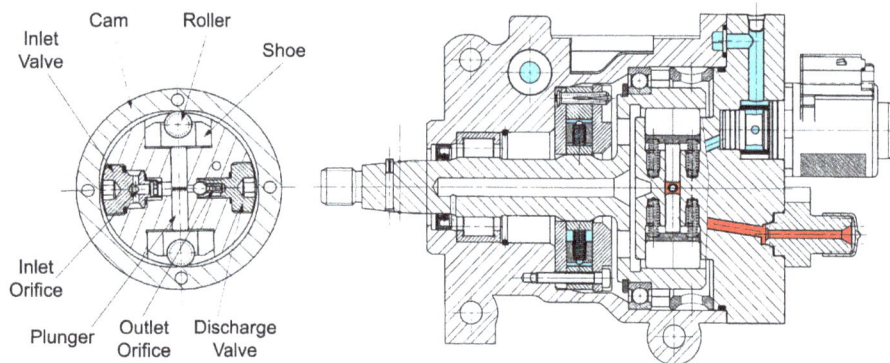

© SAE International.

The pressure in the rail can be regulated by an electronically controlled valve on the pump or on the rail. Generating these high pressures requires a fair amount of power, typically about 2 kW/L of engine capacity. There is thus a trade-off; increasing injection pressure can be used to increase combustion efficiency but it increases parasitic losses. The energy input to produce the high pressures also imparts a significant heat input to the fuel, and it is often necessary to include a heat exchanger in the fuel return line to prevent excessive heat accumulating in the fuel tank. The maximum injection pressure generated in unit injector systems thus tends to be higher than in HPCR systems [16.35–16.37].

An alternative approach is to employ an inlet metering valve to control the amount of fuel being pressurized. When maximum pressure is not required, less energy is expended and less heated high-pressure fuel is returned to the fuel tank.

From the standpoint of diesel combustion, one of the major advantages of an HPCR system is that it allows the injection event to take place at any point in the cycle. Current systems allow up to five injection events per combustion cycle [16.38, 16.39]. A pilot injection allows a small amount of fuel to be injected and undergo pre-flame reactions, thereby reducing the ignition delay for the main injection. This reduces combustion noise and NO_x formation. Pilot injection could be achieved by purely mechanical means such as twin spring injectors. Another benefit of the HPCR system is that it gives greater flexibility over this pilot injection. The pilot injection can even be split into multiple injections [16.40]. The main injection can also be split into more than one part in order to reduce emissions formation [16.41–16.43]. The rise and fall of the injection pressure during the main injection event can also be controlled, commonly known as rate shaping. Figure 16.12 shows some possible injection rates for a single injection at different speed/load scenarios [16.44].

FIGURE 16.12 Injection rate shaping for different speed/load scenarios.

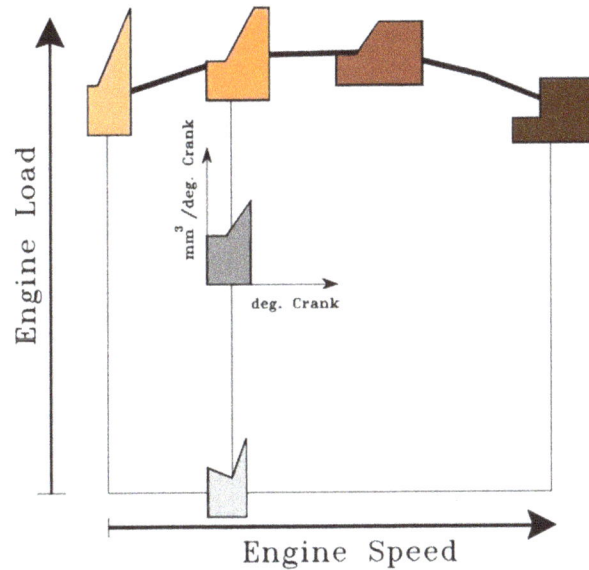

16.4. Exhaust Aftertreatment

The diesel engine has been favored for stationary applications and for use in heavy-duty vehicles due to its fuel economy. In these days of concern over CO_2 emissions, low fuel consumption not only equates to a financial benefit: it also equates to a perceived environmental benefit. Along with low CO_2 emissions, the diesel engine is also known for low hydrocarbon (HC) and carbon monoxide (CO) emissions. Unfortunately, these plus points are offset by relatively high emissions of NO_X, which cannot be controlled by a three-way catalyst, and by sooty exhaust stacks.

When Diesel filed his patent in 1892, he acknowledged that,

The residue of combustion, which is suspended in very small quantities as a very fine dust in the quickly moved and whirling air are evacuated together with the exhaust gases. [16.2]

This "very fine dust" was seen as visible smoke, which due to the widespread use of coal as a fuel source was all too common at the time. As discussed, the "pulverized dry burning coal" that Diesel proposed for his engine has been replaced with liquid petroleum-derived fuel, and to a greater extent these days, with renewable liquid hydrocarbons. The "very fine dust" emitted by more modern diesel engines is a complex mixture that varies in size from a few nanometers to agglomerates that are measured in micrometers.

As noted in Chapter 15, the fundamentals of the combustion process impose a trade-off between NO_X and soot or particulate matter (PM) emissions. Though engine designers, and particularly the FIE engineers, have made great strides in moving the NO_X v. PM curve toward zero, the trade-off still remains. Ever more stringent emissions regulation has thus made it virtually inevitable that the problem must be cured downstream of the engine.

As with the gasoline engine, the use of catalytic exhaust aftertreatment systems is now commonplace for modern automotive diesel applications. Initially, PM emissions could be met by reducing fuel sulfur content, controlling lubricating oil consumption, and by fitting a diesel oxidation catalyst (DOC) to reduce the amount of adsorbed volatile HCs from the soot. However, it was widely believed that this was only an interim solution, and work was under way to trap the soot before it reached the tailpipe. Soot traps were proved in the laboratory—all that remained was to make them a practical solution. Attention then turned to reducing the NO_X emissions.

These different technologies are discussed in more detail in the following section.

16.4.1. Diesel Oxidation Catalysts

The first commercial use of a DOC was on a standby diesel generator set, in 1967 [16.45]. The diesel engine has always produced very low levels of HC and CO emissions, and for automotive applications this did not require control. However, in workplace environments with low levels of air exchange, these emissions can become important. The DOC was originally developed to overcome these problems. At sufficiently high exhaust temperatures, DOCs can provide very effective control of HC and CO emissions, with reduction efficiencies in excess of 90%.

Stricter control of PM emissions, starting in the 1990s, resulted in the adaption of the DOC to reduce the PM emissions. This was done by developing the DOC to reduce the amount of liquid HCs that are adsorbed onto the carbonaceous core of the soot particles. The DOC has very little effect on the carbonaceous core of the soot particles and does not significantly alter the NO_X emissions. They have been widely fitted to automotive diesel engines since 1989 [16.46].

In general, a DOC uses a monolithic support for the catalyst coating. Metallic and ceramic substrates are used [16.47]. The metallic substrate usually has thinner walls, which results in a reduced resistance to exhaust gas flow for the same surface area of catalyst. However, metallic substrates tend to be more expensive. The catalytic elements are typically platinum or palladium based that are dispersed on a high surface area, base metal oxide washcoat such as aluminum oxide or silicon dioxide.

The noted very high conversion efficiencies for HC and CO are achieved through the use of highly active, platinum group metal (PGM)–based catalysts. Unfortunately, such high activity will also oxidize sulfur dioxide (SO_2) and nitrogen oxide (NO) that may be present in the exhaust gas. The SO_2 is oxidized to sulfur trioxide (SO_3) that reacts with the water vapor present in the exhaust gas to form sulfuric acid (H_2SO_4). As the exhaust gas cools, this gaseous H_2SO_4 will combine with more water and nucleates to form very fine particles of hydrated sulfuric acid. These fine particles can then agglomerate to form large sulfuric acid particles, or they will be readily adsorbed onto the soot particles. As the regulated tests for PM emissions require the collection of PM on a filter paper at 52°C, the hydrated sulfuric acid will also condense on this paper and add to the measured weight. If high-activity catalysts are used with high-sulfur fuels, then they can actually increase the measured PM emissions rather than reducing them.

Care must also be given to the washcoat formulation, which must not store sulfate because this will then be released at high temperatures [16.48]; the inclusion of silica [16.49], zirconia [16.50], and titania [16.51] into the washcoats has been suggested.

Oxidation of NO to nitrogen dioxide (NO_2) is another potentially undesirable effect of high-activity DOCs. Because NO_2 is known to be a poisonous gas, ambient air quality limits are imposed in many countries. It has also been recognized in the enclosed working environments where DOCs were first introduced [16.52]. However, NO_2 is also a very effective oxidizing agent for the oxidation of the carbonaceous core of the soot particles, and this can be utilized in particulate reduction systems (discussed in the following section).

When a DOC is used as the sole means of PM reduction, the catalysts are normally formulated to be less active toward the formation of sulfates and NO_2. This can be achieved by adding base metal oxides, such as vanadia, to suppress the sulfate formation [16.53]. It is desirable to maintain a low level of PGM catalyst to oxidize heavier HCs that exist as an aerosol at low exhaust temperatures. This aerosol will tend to adsorb or condense in the pores of the DOC before being catalytically oxidized when the exhaust heats up [16.54].

Provided that the catalyst is formulated to avoid production of NO_2, there is an added benefit in that a DOC is very effective at reducing polycyclic aromatic hydrocarbon (PAH) emissions. The PAH can be present in the fuel and survive the combustion process, or as discussed in Chapter 15, Section 15.3.1, they can be generated by the combustion process. The PAHs can be present in the exhaust in the gas phase or the liquid phase, which will readily be adsorbed onto the PM. The heavier PAHs will be adsorbed onto the PM and contribute to the soluble organic fraction (SOF) of the PM. The lighter PAH species may remain as part of the gas-phase HC emissions. The distribution between the gas phase and the liquid phase will depend on the gas temperature. At lower temperature, the PAH are more likely to be in the liquid phase absorbed into the particulate and are thus more likely to be in direct contact with the catalytic surface of the DOC, resulting in good conversion efficiency. High levels of PAH reduction can then be achieved, even at low temperatures [16.55]. However, the DOC is a chemical reactor, and it is possible for the oxidation of NO in the presence of water to produce nitric acid, which can nitrate PAHs to form nitro-PAHs. Nitro-PAHs are also carcinogenic. Sulfuric acid can also catalyze these reactions [16.56]. This again demonstrates the importance of careful formulation of the catalyst and rigorous testing.

16.4.2. Diesel Particulate Filters

As previously discussed, diesel engines make soot. This soot is a mixture of small, carbonaceous particles with adsorbed HC water sulfates. The irony is that these unwanted particles are not that different from the pulverized coal that Diesel intended his engines to run on! The amount and nature of the soot that is made is influenced by the fuel characteristics and the engine design. Engine design clearly includes the design of the FIE. The amount of this soot that reaches the exhaust is also dependent on these parameters. As we have seen in the previous section, the nature of these soot particles can be changed as they pass through the exhaust pipe, resulting in a much-reduced mass of material reaching the tailpipe. The DOC was a solution to the immediate problem of meeting emissions legislation that was proposed at the time. At that time, work was already being performed on methods of trapping the soot. The two major obstacles that needed to be overcome were the durability of the proposed systems and how to dispose of the trapped soot.

Some early developments aimed at trapping the soot particles included electrostatic precipitators [16.57], cyclone separators [16.58], and various barrier methods. The barrier methods are better termed as particulate filters. Various forms of filters have been developed that can be categorized as deep bed or surface filtration. As will be discussed next, surface filtration is usually accomplished using wall flow filters.

16.4.2.1. Deep Bed Filters

In deep bed filtration, the size of the pores in the surface of the material is much larger than the size of the particles that must be trapped. However, there is significant depth to the material so that eventually the particles will become deposited on the filter material; this can be likened to a foam dust filter. Examples of deep bed filtration systems include ceramic fibers [16.59–16.63], metal fibers [16.64, 16.65], ceramic [16.66], and metal foams [16.67, 16.68].

16.4.2.2. Wall Flow Filters

When using surface filtration, the pore sizes are similar to the size of the particles to be trapped so less depth is required; this can be likened to a filter paper. The most commonly used diesel particulate filter (DPF) system is the wall flow monolith. This type of filter uses surface filtration and is formed from an extruded ceramic monolith similar to that used for oxidation catalyst for both diesel and gasoline application. To make the monolith function as a filter, opposite ends of adjacent channels are closed to force the gas to flow through the wall of the DPF. The channel ends are usually closed by inserting a plug of ceramic, but other methods of closure have been developed [16.69]. The general arrangement is illustrated in Figure 16.13 [16.70].

FIGURE 16.13   Schematic representation of a wall flow DPF.

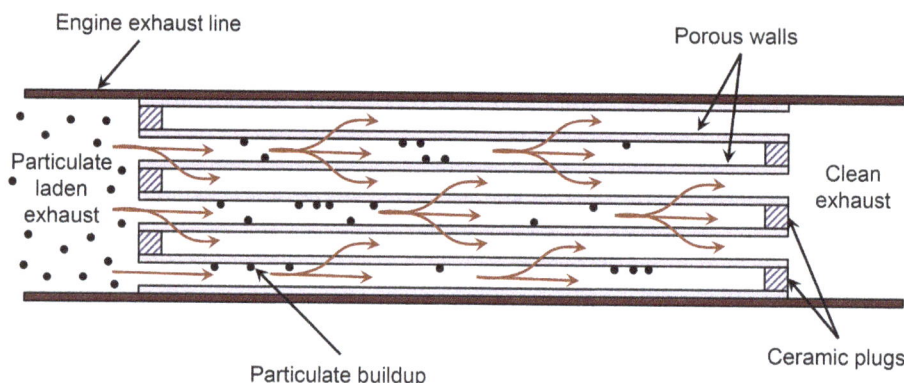

The exhaust gas enters the upstream open end of channels. Since the downstream end of the channel is blocked, the exhaust gas is forced through the porous wall to an adjacent channel, from where it can exit. The filter walls impose a restriction to the flow of exhaust gas that imposes an additional back pressure that the engine must overcome. It is therefore desirable to minimize this resistance while also maintain high filtration efficiency. This can be helped by maximizing the filtration area while maintaining a reasonable package volume. As with a flow-through catalyst, it is thus desirable to maximize the number of channels within a given frontal area. Careful design of the canning of the filter element can allow the DPF to replace a standard muffler assembly and provide equal noise attenuation.

The structure of the channel wall will influence the resistance to flow and determine how effective it is as a filter. A great deal of work is ongoing to develop DPF materials that will give a lower pressure drop while still giving high filtration efficiency with the necessary durability [16.71, 16.72]. As will be discussed in Section 16.4.2.4, a DPF can be subjected to far higher temperature gradients than automotive catalysts. This imposes greater demands on the integrity of the monolith. For this reason, it is often desirable to form the DPF from segments rather than as a single block of ceramic [16.73, 16.74].

Various materials have been developed to satisfy the demanding requirements of these substrates. Due to many years of using cordierite for catalyst supports, this material was an obvious choice as a material for DPFs [16.71, 16.75–16.77]. Other materials that have been developed and commercialized include silicon carbide (SiC) [16.73, 16.78, 16.79], silicon nitride [16.80, 16.81], and aluminum titanate [16.82, 16.83].

These ceramic wall flow DPFs have very high filtration efficiency, typically greater than 95%. This can pose problems regarding the lifetime of the DPF. Although the accumulated carbonaceous material can be burned off during regeneration, there is residual ash from the metals present in the lubricant, either as lubricant additives or as wear metals. This ash will accumulate in the DPF and gradually reduce its operating volume. To ensure satisfactory lifetime for the DPF, the overall volume must be increased or alternatively the trapping volume must be increased while maintaining the overall volume. For the simple system shown schematically in Figure 16.13, the inlet area, which is the trapping volume, is equal to the outlet volume. Various designs have been proposed that allow for different inlet and outlet volumes [16.83–16.86]; some of these are shown in Figure 16.14 [16.83–16.86]. The darker-colored cells represent the inlet channels, and the lighter-colored cells are the outlet channels.

FIGURE 16.14   Different configurations that provide increased inlet volume compared with outlet volume.

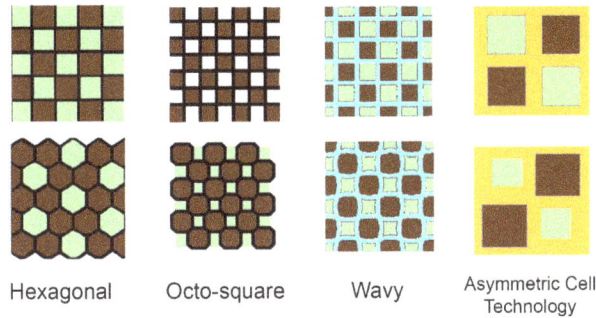

Hexagonal   Octo-square   Wavy   Asymmetric Cell
Technology

© SAE International.

16.4.2.3. Partial Filtration

The DPF systems described are designed to have filtration efficiencies greater than about 95% across the full particle size range that is to be expected of diesel particles. These systems will allow engines to meet current regulated homologation emissions tests. Proposed legislation that includes a limit on the number of particles emitted should also be met as a result of fitting one of these DPF systems. To encourage retrofitting of DPF systems, some countries have introduced schemes that will provide incentives to fit systems that provide at least 30% filtration efficiency. This is a similar situation to that which led to the introduction of the DOC as original equipment. This has led to the development of systems designed specifically to meet these low-filtration efficiency targets.

Systems designed to meet these less-stringent particulate reduction limits usually consist of an oxidation catalyst that has been designed to provide greater residence time for the soot particles [16.87–16.89]. A significant drawback of these systems is that there is a possibility that particles can pass through the system untrapped. Under certain conditions, it is also possible for trapped soot particles to be discharged in significant quantities [16.90].

16.4.2.4. DPF Regeneration

The more effective a DPF, then the less soot is emitted to the atmosphere but the more soot is trapped in the DPF. This soot must obviously be removed, otherwise the increase in exhaust back pressure would soon stall the engine. The trapped soot must therefore be removed periodically or continuously. In industrial applications, it may be acceptable to remove a DPF from the engine to allow it to be cleaned overnight. However, for automotive applications, this would clearly not be acceptable even if the cleaning frequency was the same as the fuel tank filling frequency.

To clean the DPF on the vehicle (or regenerate the DPF, as it is usually termed) requires the right combination of time, temperature, and oxidant. There is usually adequate oxygen in a diesel exhaust stream, but temperatures are usually too low to ensure high oxidation rates. The rates of the reaction could be increased by incorporating a catalyst into the DPF system. The catalyst could be incorporated into the DPF itself [16.91, 16.92] or into the fuel [16.93–16.95].

Placing the catalyst on the DPF substrate means that the catalyst is effectively downstream of the accumulated soot and is only in intimate contact with the thin layer of soot that is trapped on the DPF wall. If the exhaust gas temperature falls below that necessary for oxidation of the soot, then a thicker layer of soot will accumulate and it will thus be more difficult to burn off [16.96]. If the catalyst actively produces NO_2 then this will diffuse back through the catalyst and soot layer and aid the oxidation process, because NO_2 is a more active oxidant than oxygen. However, NO_2 emissions will be increased and the catalyst will be susceptible to sulfur poisoning unless the fuel-borne sulfur levels are very low.

Placing the catalyst directly into the accumulated soot eliminates the problem of intimate contact, and a less active catalyst can be used that eliminates the NO_2 issue. To obtain intimate mixing of the soot and the catalyst, the catalyst could be dosed directly into the exhaust gas [16.97] or a compound can be mixed with the fuel such that on combustion a catalyst is formed that is intimately mixed with the soot [16.98–16.101]. These compounds are usually referred to as fuel-borne catalyst (FBC) and should only be used with fuel that is to be burned in an engine equipped with a DPF. Without the use of a DPF, the use of FBCs would clearly increase the tailpipe emissions of metals used in the FBC. Careful design of a DPF system using FBC ensures that there is a decrease in total metal emissions and that there is no increase in secondary emissions [16.101]. Use of a FBC means that the catalytic substances are continually being replenished; this significantly reduces the susceptibility to sulfur poisoning.

Oxidation rates with NO_2 are higher than with oxygen, but relatively small quantities of NO_2 are normally present in the exhaust gas. One elegant solution is to place a flow-through-catalyst upstream of the DPF. As noted in Section 16.4.1, high PGM loadings will very effectively oxidize the NO to NO_2, which can then be used to oxidize the trapped soot at much lower temperatures. This technique has been patented by Johnson Matthey [16.102] and is commonly referred to as the Continuously Regenerating Trap (CRT®). One of the limitations of this system is that it requires the NO_X-to-PM ratio to be above a certain level to ensure adequate oxidation of the trapped soot [16.103, 16.104]. If the NO_X-to-PM ratio is too high, then there will be a high proportion of poisonous NO_2 in the exhaust. Another potential issue is that fuel-borne-sulfur can poison the PGM catalyst.

The methods of regenerating the DPF onboard the vehicle that have been discussed are all totally passive; there is no active control over what is happening in the DPF. Although passively regenerated DPF was being offered as an original equipment option by the turn of the century [16.105], this relied on the manufacturer and customer being confident that the required regeneration conditions would be met in service. The extremely wide variation in duty cycles experienced by passenger cars meant that the required conditions were unlikely to be met in many cases. The widespread adaption of common-rail FIE gave the engine manufacturer the scope to alter the engine operating conditions to actively regenerate a DPF. The first serial application of a DPF to passenger cars occurred at the turn of the century [16.106].

The first serial production system utilized a DOC upstream of the DPF, but unlike the CRT® system, the primary purpose of the DOC was not to produce NO_2 to continuously regenerate the DPF. The primary aim of the DOC was to generate heat to raise the temperature of the DPF. By utilizing the flexibility of the HPCR system, some of the fuel could be injected late in the combustion cycle. Some of this fuel would still be unburned or partially burn when it was exhausted from the engine. This would raise the exhaust temperature. The unburned fuel would then further oxidize over the DOC, further raising the exhaust gas temperature. By using an FBC, the ignition temperature of the trapped soot was lowered and by the combination of the above mechanisms reliable regeneration could be assured during normal operating conditions. The presence of the catalyst in the accumulated soot also ensured that the regeneration was quite rapid, thus minimizing the quantity of fuel that had to be injected late.

Similar systems are now routinely fitted to diesel passenger cars. Not all systems utilize an FBC. The basic system outlined will still function without the use of an FBC, but the regeneration time and the amount of fuel injected late in the cycle will be increased.

16.4.3. NO_X Reduction Systems

Reducing NO_X and PM emissions are the two major challenges for diesel engine technologists. As described, the PM issue can be addressed by fitting DPFs, although durability and regeneration strategies are still being improved. The control of NO_X emissions was initially addressed by the use of exhaust gas recirculation (EGR) to suppress the formation of NO_X during the combustion process [16.107, 16.108]. The use of EGR alone will not meet the most stringent legislated limits that are being imposed. Although attempts have been made to develop NO reduction catalysts, the oxidizing environment and temperature regime prevailing in a diesel exhaust make this extremely challenging.

Other techniques that are available for controlling NO_X emissions from the diesel engine are the NO_X absorber systems that are also being developed for lean-burn gasoline engines and the so-called selective catalytic reduction (SCR) systems. Catalytic reduction of NO_X can be achieved by adding a reducing agent into the exhaust system upstream of a catalyst. This approach has been utilized for other applications for a number of years [16.109, 16.110]. Either ammonia or urea is the most widely used reductant. The presence of the catalyst selectively promotes the oxidation of the ammonia with NO rather than oxygen. The reaction is given:

$$4NO + 4NH_3 + O_2 \rightarrow 4N_2 + 6H_2O$$

A high NO_X conversion efficiency can be achieved with this system, but it requires an additional fluid to be carried. A lot of research has therefore been directed at developing systems that would allow the use of diesel fuel to produce a HC reductant. The following sections briefly describe some of the available NO_X reduction technologies.

16.4.3.1. Hydrocarbon-Based NO_X Reduction

Early research efforts were directed at developing catalysts to control the NO_X by reducing it in the presence of HCs [16.111, 16.112]. The conversion efficiency was very low due to the low levels of HC normally present in the exhaust. This could be improved by adding HCs to the exhaust. This could be achieved by late injection utilizing the flexibility of the HPCR injection system, as discussed previously for regenerating DPFs [16.113]. However, this would be required on an almost continuous basis, which would have a negative impact on oil dilution and fuel economy. These systems were known as lean NO_X catalysts and never achieved the high NO_X conversion efficiencies that would be required to meet forthcoming legislation.

It was found that by formulating the catalyst washcoat to adsorb HCs at low temperatures, these could then be released at higher temperatures and allow reduction of the NO_X with the adsorbed HCs. Zeolites are often incorporated into a DOC to enhance the overall low-temperature HC reduction efficiency through storage of HCs in the catalyst washcoat [16.114, 16.115]. Careful design of the adsorber and catalyst combination are required to ensure that the adsorbed HC is not released until the catalytic compounds are active and able to oxidize the HC. Because these catalysts would oxidize HC and CO, provide a reduction in PM emissions by reducing the SOF content of the particulate, and provide some NO_X reduction, they were often termed four-way catalysts [16.116–16.118]. The overall NO_X reduction efficiency of these systems is limited by the fact that the overall stoichiometry of the engine remains unchanged; therefore, the overall availability of HC is low compared with the amount of NO_X produced.

An alternative approach to the problem is to store the NO_X and then reduce it periodically by introducing a higher amount of HC. This approach has been receiving a lot of attention in relation to gasoline engines, where the familiar three-way catalyst does not significantly reduce NO_X if the engine operates lean of stoichiometry. This is discussed in Chapter 8, Section 8.5.3.3.

The technology is variously referred to as a lean NO_X trap (LNT) or a NO_X storage reduction (NSR) catalyst, NOx adsorber catalyst (NAC), NOx storage catalyst (NSC) and passive NOx adsorber (PNA). These systems function by adsorbing NO_2 and forming nitrates, which under fuel-rich conditions become thermodynamically unstable, resulting in the release of NO or NO_2. This NO_X then oxidizes the HC and/or CO that exists in the exhaust gas during fuel-rich operation. The requirements for a diesel NSR and a gasoline NSR differ because the diesel engine would normally operate from overall lean to very lean, whereas the gasoline engine would normally operate from slightly rich to lean. The temperature of a diesel exhaust is also lower. A particular "problem" of an NSR is that, without careful calibration of the rich spike required to reduce the NO_X, there is a tendency to produce ammonia [16.119]. However, this problem can be turned to an advantage by placing the NSR upstream of an SCR and utilizing the ammonia produced in the NSR to reduce NO_X in the SCR [16.120]. This is discussed further in the following section.

16.4.3.2. **Selective Catalytic Reduction**

Due to handling problems with using ammonia as a reductant, most work has focused on the use of urea, usually in an aqueous form. In its application as a reductant for SCR systems, aqueous urea has now become known as diesel exhaust fluid (DEF) [16.121]. The DEF usually contains a 32.5 wt.% solution of urea. At this concentration, urea forms a eutectic solution characterized by the lowest crystallization point of −11°C [16.122]. It also means that if the DEF begins to freeze, the concentration of the urea in the liquid phase would remain the same. This would ensure that the correct concentration is fed to the SCR system. The DEF is injected into the exhaust gas stream, and when the temperature is above 160°C the urea starts to hydrolyze and decompose according to the following equations:

$$CO(NH_2)_2 + H_2O \rightarrow 2NH_3 + CO_2$$

$$CO(NH_2)_2 \rightarrow 2 \bullet NH_2 + CO$$

The ammonia will then reduce NO as explained earlier. The •NH$_2$ radical formed from the decomposition of urea will then reduce the NO according to the following reaction:

$$\bullet NH_2 + NO \rightarrow N_2 + H_2O$$

An additional catalyst may be incorporated in the system to promote the hydrolysis of the urea [16.123]. This can aid the performance but adds to the complexity of the system.

Ammonia slip is a potential problem with urea-SCR systems. During normal operation there can be significant quantities of ammonia stored on the surface of the SCR catalyst. During high-temperature excursions, this can result in the release of some of this ammonia. Without precise control over the dosing of the DEF, it is also possible to have excess ammonia in the exhaust. Ammonia slip can be counteracted by the use of an ammonia slip catalyst downstream of an SCR catalyst [16.124], as discussed in the following section.

Various catalysts have been investigated for their SCR performance. A platinum catalyst is very effective at low temperatures but is effective over a narrow range of temperatures. For a platinum catalyst to give high selectivity, the temperature window is usually between 200°C and 250°C; this is too narrow for automotive applications. In fact, platinum can be considered a poison of SCR catalysts due to the adverse reactions it promotes [16.125].

Base metal catalysts such as vanadia-based formulations have a higher temperature window [16.126]. Unfortunately, these catalysts have a tendency to oxidize sulfur to SO_2 to SO_3. This tendency can be reduced by adding other metals, and as a result vanadia-titania SCR catalysts have been developed [16.127]. Other base metal oxides, such as tungsten trioxide (WO_3) [16.128] can also be added to decrease the SO_3 formation and improve catalyst durability.

The most popular SCR technology currently utilizes zeolite-based catalysts. These are usually iron or copper exchange zeolites [16.129–16.132]. The advantage of these catalysts is that their selectivity toward NO_X reduction does not fall off at higher temperatures due to the formation of NO_X by the oxidation of ammonia.

16.4.3.3. **Ammonia Slip Catalysts**

As discussed in the previous section, the release of ammonia stored within an SCR catalyst can result in the release of some of this ammonia into the exhaust. Ammonia slip can be counteracted by the use of an ammonia slip catalyst (ASC) downstream of an SCR catalyst, the SCR and ASC can be incorporated into a single catalyst brick [16.133].

The primary function of the ASC is to remove ammonia from the exhaust gas by converting it to nitrogen and water as shown below.

$$4NH_3 + 3O_2 \rightarrow 2N_2 + 6H_2O$$

However, there is the potential for unwanted oxidation reactions to also take place; these include:

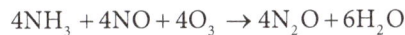

$$2NH_3 + 2O_2 \rightarrow N_2O + 3H_2O$$

$$4NH_3 + 3O_2 \rightarrow 2N_2 + 6H_2O$$

$$4NH_3 + 4NO + 4O_3 \rightarrow 4N_2O + 6H_2O$$

$$4NH_3 + 5O_2 \rightarrow 4NO + 6H_2O$$

It is worth noting that N_2O is a potent greenhouse gas and Tier 3 light-duty diesel vehicles are required to meet a US EPA tailpipe N_2O standard of 6.2 mg/km (10 mg/mi). It is thus important that the ASC has high activity but good selectivity toward N_2 across the range of operating temperatures and flow, emanating from the SCR [16.132]. The positioning of the SCR and ASC in the whole aftertreatment system must also be considered [16.134], as will be discussed further in the next section.

Because the primary function of the ASC is to oxidize the ammonia, PGM catalysts are favored. Platinum, or Pt along with aluminum oxide, is thus a favored material [16.135, 16.136], but can be combined with the reducing behavior (SCR) by the inclusion of Fe/zeolite or Cu/zeolite to improve the selectivity toward N_2. Other combinations have been proposed, including: one or more platinum group metals supported on titania, a silica-titania mixed oxide, a Ce-Zr mixed oxide, or mixtures thereof [16.137], a noble metal on an acidic tungsten-containing mixed oxide [16.138], and at least one precious metal; and at least one alkali metal or at least one alkaline earth metal [16.139]. Non-PGM systems have also been proposed [16.140].

The fact that the ASC is an oxidizing catalyst also presents the possibility that an SCR system followed by an ASC can negate the need for a separate DOC [16.141].

16.4.4. Integrated Systems

Diesel emissions control technologies have been described in isolation in the preceding three subsections. They can be designed to perform in isolation, but for a vehicle to meet all of the legislated emission limits, it is likely that a combination of these technologies will be necessary. Combining these technologies is not a straightforward exercise.

An uncatalyzed DPF will function perfectly well for a period of time at low temperatures but will then need high temperatures to regenerate. The NO_X storage devices will also function for a period of time at low temperatures, but then they will require higher temperatures to reduce the stored NO_X. All the other catalytic technologies described require higher temperatures before they become active.

A CRT-type DPF system requires NO_X in the exhaust gas to generate NO_2 to oxidize the trapped soot. A catalyzed DPF also regenerates with less heat input if there is NO_X in the exhaust gas stream. Placing these devices downstream of an effective NO_X reduction system would therefore make it harder to regenerate the DPF.

Placing a NO_X reduction system downstream of the DPF will rob the NO_X reduction system of important temperature.

The performance of SCR catalysts is also influenced by the composition of the NO_X; a ratio of 1:1 between NO and NO_2 is usually found to be best [16.142]. It is therefore advantageous for an SCR system if there is a DOC upstream that will provide conversion of NO to NO_2 to give approximately equal to this ideal ratio. Depending upon the position in the system of the SCR catalyst, it may also be necessary to include a further catalyst to cope with any ammonia slip.

It is clear that integrating the required emissions reduction devices into a functioning system is far from simple, and it is highly likely that there is no single solution. This is currently being demonstrated by different manufacturers adapting different strategies and the same manufacturer adopting different strategies for different types of vehicles.

Early attempts to solve this problem simply combined a CRT system with a urea-based SCR system [16.143], which was marketed as the SCRT®. This approach was successful for heavy-duty applications where it had been shown that the CRT would function well. This approach was also well suited to retrofit, as it did not require any alteration to the engine fueling system. It was not totally passive because it needed control of the urea dosing system. This system gives high levels of reduction of all four pollutants, namely HC, CO, PM, and NO_X. Unfortunately, the CRT system was not well suited to passenger car applications because the DPF would not regenerate reliably. The SCRT system is thus also unsuited to passenger car application.

Another early approach that was commercialized by Mercedes-Benz under the Bluetec name was to mount an NSR close to the engine, mount a DOC-DPF combination in the conventional under-floor position, and then a final NO_X reduction catalyst after the DPF [16.144]. This system is illustrated in Figure 16.15 [16.144].

FIGURE 16.15 Bluetec NO_X storage system.

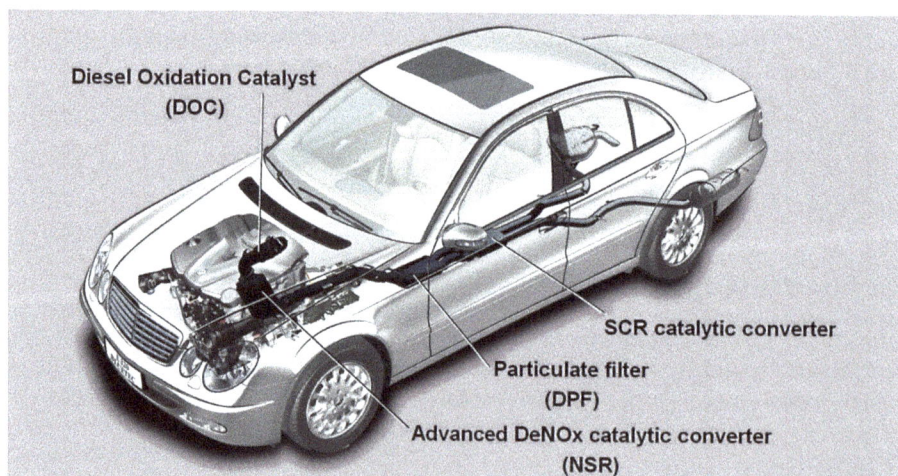

To meet stricter emissions regulation, the NSR catalyst had to be replaced by a urea-based SCR catalyst. It was then considered advantageous to move the DOC-DPF unit to the close-coupled position and move the SCR catalyst to the under-floor position [16.144]. This is shown in Figure 16.16 [16.144].

FIGURE 16.16   Bluetec urea-SCR system.

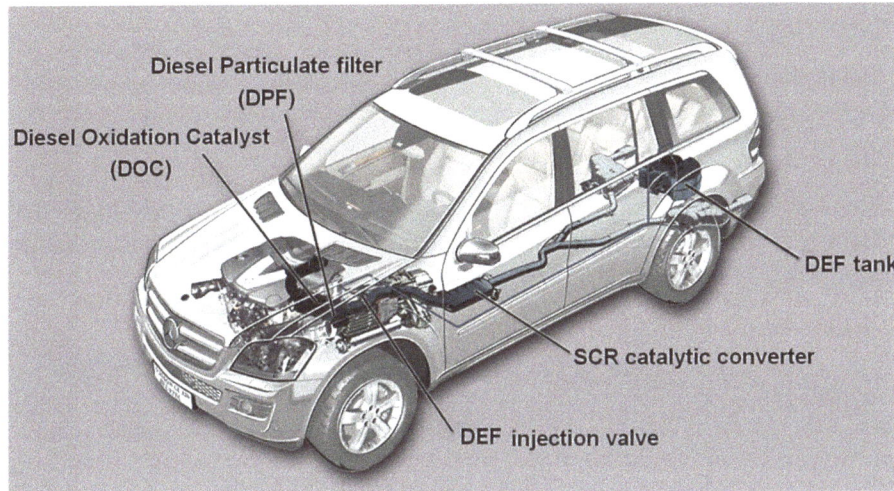

Another approach is to utilize the synergistic combination of an NSR and SCR. As noted in the previous section, there is a tendency for an NSR to produce ammonia during the rich NO_X reduction phase. This effect can be utilized to provide onboard generation of ammonia to act as the reductant for the SCR system [16.120, 16.145]. This approach has been demonstrated as part of an integrated system where a DOC is placed upstream of the NSR-SCR system with the DPF downstream [16.146, 16.147]. The DOC controls HC and CO emissions under normal operation and provides heat for regenerating the DPF. The NSR-SCR combination reduces the NO_X emissions.

16.5.   Influence of Fuel Properties on Engine Systems Performance

In his patent of 1892, Rudolf Diesel envisaged an engine that could run on any sort of fuel. This attribute has made the diesel engine popular in applications from small, portable generator sets, with an engine swept volume of less than 0.2 L, to large power plant generator sets where the engine swept volume is over 17.5 m^3 and more than one engine is used [16.148]. Large diesel engines are also used for marine propulsion applications were the engines swept volume can exceed 32.7 m^3. These larger engines run on heavy fuel oil that is taken from the heavier end of the crude oil, while the smaller engines run on more refined fuel that would be suitable for automotive application. Neither of these engines would be able to run on the fuel that is used by the other engine. The engine and the fuel have to be developed to work with each other.

A major fuel characteristic for determining combustion performance is the cetane number (discussed in Chapter 15, Section 15.2). Other key factors for cold starting and for long-term durability are cold flow characteristics and lubricity performance, respectively. These characteristics are discussed in Chapter 17 and in Chapter 20, Section 20.4. The influence of fuel characteristics on tailpipe emissions is discussed in Chapter 21. The following sections will consider how other fuel characteristics influence the combustion performance of diesel engines.

16.5.1. Influence of Fuel Density

Density, the weight of a unit volume of diesel fuel, can provide useful indications about its composition and performance-related characteristics, such as ignition quality, power, economy, low-temperature properties, and smoking tendency. There is a strong correlation between the density and the volumetric energy content of the fuel such that a denser fuel will tend to have greater energy content per unit volume. FIE usually dispenses fuel on a volumetric basis. Therefore, a greater mass of fuel will be injected for a higher density fuel. This will tend to produce more power, but at full load conditions this will tend to produce higher emissions. This is discussed further in Chapter 21, Section 21.3.1.

16.5.2. Influence of Diesel Fuel Volatility

The distillation range of a diesel fuel does not allow much flexibility to the refiner because of the interrelated and interdependent properties and constraints imposed by other specification items. Raising the back-end distillation temperature also raises the cloud point; lowering the temperature at the front-end lowers the flash point and also raises the vapor pressure, which might result in vapor lock in the fuel injection system and cause engine misfiring or failure to restart after a brief shutdown in hot conditions.

There are some aspects of the distillation test that can relate to the combustion performance of the fuel. A low 10% recovery temperature reflects the ease with which the fuel will start to vaporize, while the 90 or 95% recovery temperature gives an indication of the extent to which complete vaporization of the fuel may be expected in the combustion zone of the diesel engine. High boiling components may not burn completely, forming engine deposits and increasing smoke levels. Some diesel fuel specifications impose a limit on the back-end distillation temperature to avoid such problems. The influence of volatility on emissions is discussed in Chapter 21, Section 21.3.3.

Front-end volatility is not usually specified, and it will depend on the amount of low boiling material incorporated in the blend. This will be determined by the need to control cloud point by that means, the demand for kerosene for the jet fuel market and the availability of alternative light fractions. Also, the initial boiling point must be high enough to ensure that the flash point is at or above the specified legal limit.

The mid-volatility of a diesel fuel has been shown to relate to its smoking tendency, possibly through influence on the injection and mixing of the fuel, but there is also interest in mid-volatility for the calculation of Cetane Index by the various methods. The 50% recovery temperature is specified in some countries and was considered for inclusion in the pan-European diesel fuel specification.

Although engine tests comparing fuels differing widely in boiling range have indicated some differences attributable to volatility, the effects of variations in volatility that are likely with commercial diesel fuels are generally modest and usually far less significant than the influence of individual engine design features [16.149].

To comply with ever more stringent emissions regulation, there is an increasing use of exhaust aftertreatment system. As discussed in Section 16.4.2, periods of high temperature are required to regenerate the DPF, and NO_x storage systems require periodic rich fuel mixtures to reduce the stored NO_x. Both of these requirements are often achieved by the use of late injection or post-injection strategies [16.106, 16.150, 16.151]. These strategies inject fuel late in the expansion stroke when the piston is toward the bottom of the cylinder and much of the cylinder wall is exposed. The ambient in-cylinder gas densities are also relatively low. Both of these factors increase the probability that the liquid fuel jet will deposit some fuel on the cylinder wall, known as wall wetting. Liquid fuel on the cylinder wall can make its way past the piston rings and accumulate in the engine's lubricating oil.

The distillation characteristics of the fuel will influence the spray penetration and hence the amount of fuel that impinges on the cylinder wall. It will also influence how much fuel is vaporized due to residence on the cylinder wall and hence how much fuel enters the lubricating oil. The volatility of the fuel will also influence how rapidly the fuel is volatilized from the oil and how much fuel accumulates in the oil over time [16.152].

The inclusion of biodiesel components can distort the distillation curve due to the poor volatility and narrow boiling range of these components. For engines with a late injection strategy for the management of aftertreatment devices, the use of high biodiesel contents may be a problem. Due to the lower volatility of biodiesel, the fuel injection spray liquid length of biodiesel is approximately 20 to 30% longer than ultra-low sulfur diesel (ULSD) fuel [16.153]. This has been demonstrated to increase the accumulation of fuel within the lubricant. Figure 16.17 [16.154] shows the total fuel content in the lubricating oil of a European passenger car that had been operated on ULSD and the same car that had been operated on the ULSD with the addition of 20% of biodiesel (B20).

FIGURE 16.17 Total fuel content in the lubricating oil of a European passenger car operated over highway conditions with ULSD and B20.

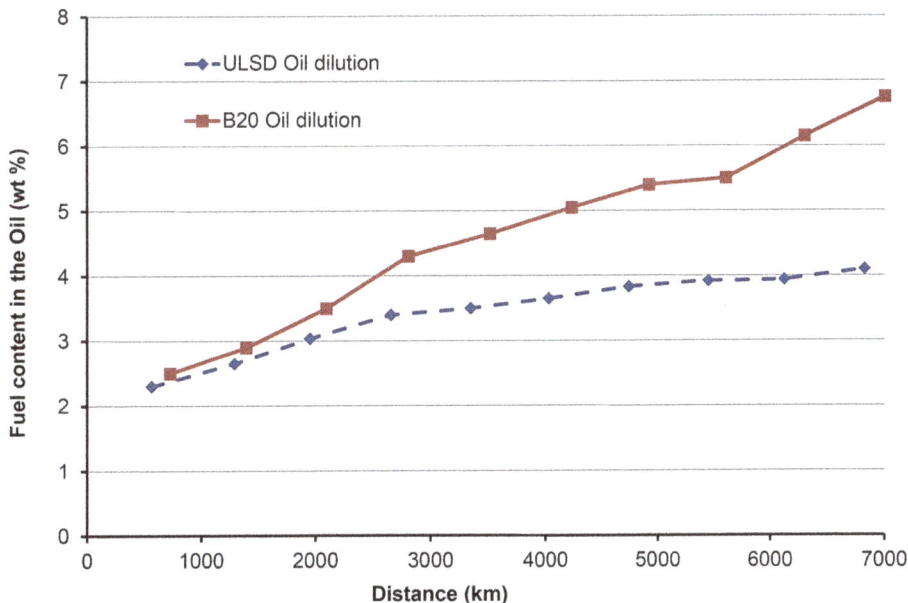

© SAE International.

Further analysis of these results shows that the proportion of this fuel that is biodiesel also increases with time. This is shown in Figure 16.18 [16.154].

FIGURE 16.18 Composition of the fuel within the lubricating oil for B20 test.

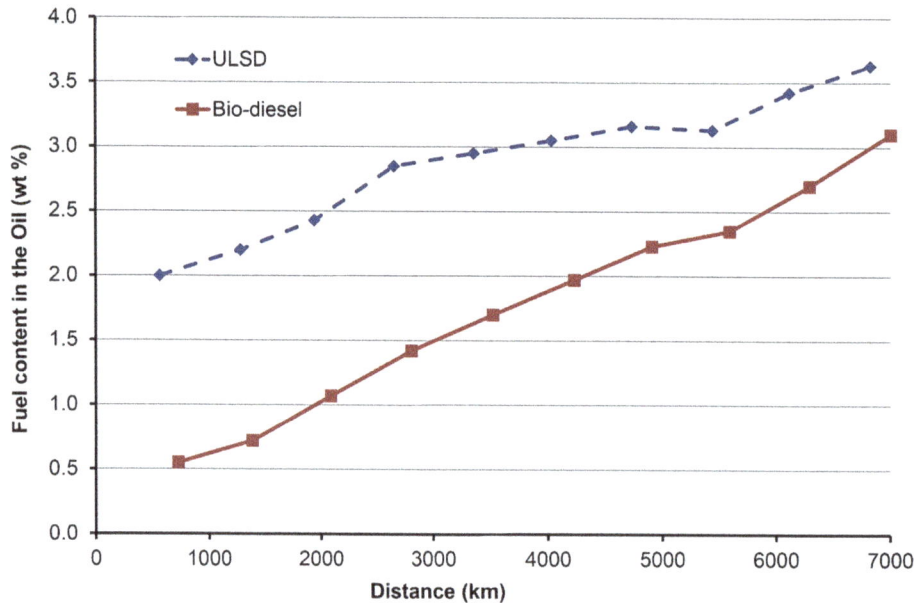

The accumulation of biodiesel within the lubricant can negatively impact the lubricant performance. Fuel dilution will reduce the lubricating oil viscosity [16.155], and because of the low volatility of the biodiesel component there is greater accumulation of fuel in the lubricant. The biodiesel can also accelerate the degradation of the lubricant [16.156–16.158].

16.5.3. Influence of Diesel Fuel Viscosity

Viscosity is an important property of a diesel fuel because of its relevance to the performance of the FIE, particularly at low temperatures when the increase in viscosity affects the fluidity of the fuel. The FIE has to meter, with great accuracy, the small quantity of fuel to be injected. As viscosity varies inversely with temperature, the tolerance band between maximum and minimum viscosity values should be kept as small as practicable. A very high viscosity at low temperature could reduce fuel flow rates and result in incomplete filling of the metering chamber in some types of injection equipment. This can result in a smaller volume of fuel being injected. If the fuel is very viscous, there is also a possibility of pump distortion due to heat generated by the shearing action in the small clearances.

Good sealing of the pumping chambers in an injection system depends on the close clearances of the principal elements. While a certain amount of leakage will occur normally, a low viscosity fuel could result in almost total leakage from the pumping elements, particularly at low speeds. In practice, such a situation can arise when attempting a hot restart after a brief shutdown, following a period of operation at high load. The temperature of the already hot FIE will be further increased by heat soak-back from the engine, reducing the fuel viscosity so much that leakage may make restarting impossible until the fuel system has cooled.

16.5.4. Influence of Diesel Fuel Composition

The chemical composition of the diesel fuel will affect the discussed physical properties. These affect the physical processes leading to the preparation of the fuel-air mixture within the combustion chamber. Beyond that, the chemical composition of the fuel will affect the combustion process. The carbon-to-hydrogen ratio of the fuel will affect the amount of oxygen required for complete combustion of the fuel. The structure of the individual fuel components will affect how readily they burn.

The major impact that the chemical composition (as opposed to the physical properties that are influenced) has on engine system performance is on the emissions production. With engines that are not equipped with exhaust aftertreatment, this will have a major effect on the tailpipe emissions. This is discussed in Chapter 21, Section 21.4. For engines that are fitted with aftertreatment systems, the effect on tailpipe emissions will be decreased due to the efficiency of the aftertreatment. However, the characteristics of the engine-out emissions will influence how hard the aftertreatment systems have to work. For example, if the fuel produces high soot emissions, the DPF will need regenerating more frequently. Because there is a fuel consumption penalty associated with regenerating the DPF, there is a fuel consumption penalty associated with a fuel that has a propensity to produce soot. However, there is often a correlation between the density of the fuel and its propensity to produce soot emissions. If fuel consumption is measured on a volumetric basis, these two factors may cancel each other out.

Nonhydrocarbon elements within the fuel, such as sulfur and alkali metals in biodiesel, can affect the durability of these aftertreatment devices.

References

16.1. Stuart, H.A. and Binney, C.R., Improvements in engines operated by the explosion of mixtures of combustible vapour of gas and air. G.B. Patent No. 7146, 1890.

16.2. Diesel, R., A process for producing motive work from the combustion of fuel. G.B. Patent No. 7241, 1892.

16.3. Diesel, R., Improvements in regulating fuel supply for slow combustion motors, and apparatus for that purpose. G.B. Patent No. 4243, 1895.

16.4. Monaghan, M.L., "The High Speed Direct Injection Diesel for Passenger Cars," SAE Technical Paper 810477 (1981), doi:https://doi.org/10.4271/810477.

16.5. Shimada, T., Sakai, K., and Kurihara, S., "Variable Swirl Inlet System and Its Effect on Diesel Performance and Emissions," SAE Technical Paper 861185 (1986), doi:https://doi.org/10.4271/861185.

16.6. Arcoumanis, C. and Tanabe, S., "Swirl Generation by Helical Ports," SAE Technical Paper 890790 (1989), doi:https://doi.org/10.4271/890790.

16.7. Wittek, H.L., Rosen, C.G.A., and List, H., "Diesel Engine Design—Past, Present, and Future," SAE Technical Paper 640456 (1964), doi:https://doi.org/10.4271/640456.

16.8. Collins, D., "Light Duty Power Plants for the Future," SAE Technical Paper 852190 (1985), doi:https://doi.org/10.4271/852190.

16.9. Meurer, J.S., "Evaluation of Reaction Kinetics Eliminates Diesel Knock—The M-Combustion System of MAN," SAE Technical Paper 560024 (1956), doi:https://doi.org/10.4271/560024.

16.10. Barnes-Moss, H.W. and Scott, W.M., "The Light Duty Diesel Engine for Private Transportation," SAE Technical Paper 750331 (1975), doi:https://doi.org/10.4271/750331.

16.11. Leipold, F.W. and Hardenberg, H.O., "Noise, Emissions and Performance of the Diesel Engine - A Comparison between DI and IDI Combustion Systems," SAE Technical Paper 750796 (1975), doi:https://doi.org/10.4271/750796.

16.12. McKechnie, J., Improvements in internal combustion engines. GB Patent No. 27,579, 1911.

16.13. Gaff, T.T., Explosion engine. U.S. Patent 1,059,604, 1913.

16.14. Schweimer, G.W. and Bader, T., "Hydraulic Unit Injector for Passenger Car DI-Diesel Engines," SAE Technical Paper 920628 (1992), doi:https://doi.org/10.4271/920628.

16.15. Zimmermann, K.D., "New Robert Bosch Developments for Diesel Fuel Injection," SAE Technical Paper 760127 (1976), doi:https://doi.org/10.4271/760127.

16.16. Robert Bosch Aktiengesellschaft, Improvements in or relating to pumping mechanism. GB Patent 336,126, 1930.

16.17. Mikkonen, S. and Tenhunen, E., "Deposits in Diesel Fuel-Injection Pumps Caused by Incompatibility of Fuel and Oil Additives," SAE Technical Paper 872119 (1987), doi:https://doi.org/10.4271/872119.

16.18. Glikin, P.E., "Fuel Injection in Automotive Diesel Engines—1985 SAE/I Mech E Exchange Lecture," SAE Technical Paper 851458 (1985), doi:https://doi.org/10.4271/851458.

16.19. Straubel, M. and Krieger, K., "Distributor Type Injection Pump for High-Speed Diesel Engines with Direct Injection," SAE Technical Paper 872222 (1987), doi:https://doi.org/10.4271/872222.

16.20. Mowbray, D.F. and Drori, M., "The CAV DP15 Fuel Injection Pump," SAE Technical Paper 780163 (1978), doi:https://doi.org/10.4271/780163.

16.21. Mitchell, K., "The Lubricity of Winter Diesel Fuels," SAE Technical Paper 952370 (1995), doi:https://doi.org/10.4271/952370.

16.22. Soteriou, C. and Smith, M., "From Concept to End Product—Computer Simulation in the Development of the EUI-200," SAE Technical Paper 960866 (1996), doi:https://doi.org/10.4271/960866.

16.23. Greeves, G., Tullis, S., and Barker, B., "Advanced Two-Actuator EUI and Emission Reduction for Heavy-Duty Diesel Engines," SAE Technical Paper 2003-01-0698 (2003), doi:https://doi.org/10.4271/2003-01-0698.

16.24. Aoyama, T., Mizuta, J., and Oshima, Y., "NOx Reduction by Injection Control," SAE Technical Paper 900637 (1990), doi:https://doi.org/10.4271/900637.

16.25. Shundoh, S., Komori, M., Tsujimura, K., and Kobayashi, S., "NOx Reduction from Diesel Combustion Using Pilot Injection with High Pressure Fuel Injection," SAE Technical Paper 920461 (1992), doi:https://doi.org/10.4271/920461.

16.26. Minami, T., Takeuchi, K., and Shimazaki, N., "Reduction of Diesel Engine NOx Using Pilot Injection," SAE Technical Paper 950611 (1995), doi:https://doi.org/10.4271/950611.

16.27. Tullis, S. and Greeves, G., "Improving NOx versus BSFC with EUI 200 Using EGR and Pilot Injection for Heavy-Duty Diesel Engines," SAE Technical Paper 960843 (1996), doi:https://doi.org/10.4271/960843.

16.28. Cross, R.K., Lakra, P., and O'Neill, C.G., "Electronic Fuel Injection Equipment for Controlled Combustion in Diesel Engines," SAE Technical Paper 810258 (1981), doi:https://doi.org/10.4271/810258.

16.29. Coldren, D.R., Schuricht, S.R., and Smith, R.A., "Hydraulic Electronic Unit Injector with Rate Shaping Capability," SAE Technical Paper 2003-01-1384 (2003), doi:https://doi.org/10.4271/2003-01-1384.

16.30. Buratti, R., Imarisio, R., and Peters, B., "Experiences with Common Rail, a Technology Changing the Image of Diesels in Europe," SAE Technical Paper 2004-28-0072 (2004), doi:https://doi.org/10.4271/2004-28-0072.

16.31. Stumpp, G. and Ricco, M., "Common Rail—An Attractive Fuel Injection System for Passenger Car DI Diesel Engines," SAE Technical Paper 960870 (1996), doi:https://doi.org/10.4271/960870.

16.32. Miyaki, M., Fujisawa, H., Masuda, A., and Yamamoto, Y., "Development of New Electronically Controlled Fuel Injection System ECD-U2 for Diesel Engines," SAE Technical Paper 910252 (1991), doi:https://doi.org/10.4271/910252.

16.33. Flaig, U., Polach, W., and Ziegler, G., "Common Rail System (CR-System) for Passenger Car DI Diesel Engines; Experiences with Applications for Series Production Projects," SAE Technical Paper 1999-01-0191 (1999), doi:https://doi.org/10.4271/1999-01-0191.

16.34. Russell, M.F., Greeves, G., and Guerrassi, N., "More Torque, Less Emissions and Less Noise," SAE Technical Paper 2000-01-0942 (2000), doi:https://doi.org/10.4271/2000-01-0942.

16.35. Gibson, D., Shinogle, R., and Moncelle, M., "Meeting the Customer's Needs - Defining the Next Generation Electronically Controlled Unit Injector Concept for Heavy Duty Diesel Engines," SAE Technical Paper 961285 (1996), doi:https://doi.org/10.4271/961285.

16.36. Mulemane, A., Han, J., Lu, P., Yoon, S. et al., "Modeling Dynamic Behavior of Diesel Fuel Injection Systems," SAE Technical Paper 2004-01-0536 (2004), doi:https://doi.org/10.4271/2004-01-0536.

16.37. Fisher, B. and Mueller, C., "Effects of Injection Pressure, Injection-Rate Shape, and Heat Release on Liquid Length," *SAE Int. J. Engines* 5, no. 2 (2012): 415-429, doi:https://doi.org/10.4271/2012-01-0463.

16.38. Oki, M., Matsumoto, S., Toyoshima, Y., Ishisaka, K. et al., "180MPa Piezo Common Rail System," SAE Technical Paper 2006-01-0274 (2006), doi:https://doi.org/10.4271/2006-01-0274.

16.39. Tomishima, H., Matsumoto, T., Oki, M., and Nagata, K., "The Advanced Diesel Common Rail System for Achieving a Good Balance between Ecology and Economy," SAE Technical Paper 2008-28-0017 (2008), doi:https://doi.org/10.4271/2008-28-0017.

16.40. Okude, K., Mori, K., Shiino, S., Yamada, K. et al., "Effects of Multiple Injections on Diesel Emission and Combustion Characteristics," SAE Technical Paper 2007-01-4178 (2007), doi:https://doi.org/10.4271/2007-01-4178.

16.41. Park, C., Kook, S., and Bae, C., "Effects of Multiple Injections in a HSDI Diesel Engine Equipped with Common Rail Injection System," SAE Technical Paper 2004-01-0127 (2004), doi:https://doi.org/10.4271/2004-01-0127.

16.42. Liu, Y. and Reitz, R.D., "Optimizing HSDI Diesel Combustion and Emissions Using Multiple Injection Strategies," SAE Technical Paper 2005-01-0212 (2005), doi:https://doi.org/10.4271/2005-01-0212.

16.43. Ehleskog, R. and Ochoterena, R.L., "Soot Evolution in Multiple Injection Diesel Flames," SAE Technical Paper 2008-01-2470 (2008), doi:https://doi.org/10.4271/2008-01-2470.

16.44. Hwang, J.W., Kal, H.J., Kim, M.H., Park, J.K. et al., "Effect of Fuel Injection Rate on Pollutant Emissions in DI Diesel Engine," SAE Technical Paper 1999-01-0195 (1999), doi:https://doi.org/10.4271/1999-01-0195.

16.45. Mooney, J. and Wolfgang, J., "Save the Diesel Fueled Engine: A Clean Diesel Engine with Catalytic Aftertreatment—The Alternative to Alternate Fuels," SAE Technical Paper 931182 (1993), doi:https://doi.org/10.4271/931182.

16.46. Beckmann, R., Engeler, W., Mueller, E., Engler, B.H. et al., "A New Generation of Diesel Oxidation Catalysts," SAE Technical Paper 922330 (1992), doi:https://doi.org/10.4271/922330.

16.47. Campbell, M.G. and Martin, E.P., "Substrate Selection for a Diesel Catalyst," SAE Technical Paper 950372 (1995), doi:https://doi.org/10.4271/950372.

16.48. Horiuchi, M., Saito, K., and Ichihara, S., "Sulfur Storage and Discharge Behavior on Flow-Through Type Oxidation Catalysts," SAE Technical Paper 910605 (1991), doi:https://doi.org/10.4271/910605.

16.49. Ball, D.J. and Stack, R.G., "Catalyst Considerations for Diesel Converters," SAE Technical Paper 902110 (1990), doi:https://doi.org/10.4271/902110.

16.50. Verdier, S., Harle, V., Huang, A., Rohart, E. et al., "Doped Zirconia with High Thermal Stability, for High Sulfur Resistance Diesel Oxidation Catalysts," SAE Technical Paper 2006-01-0031 (2006), doi:https://doi.org/10.4271/2006-01-0031.

16.51. Farrauto, R.J., Voss, K.E., and Heck, R.M., "A Base Metal Oxide Catalyst for Reduction of Diesel Particulates," SAE Technical Paper 932720 (1993), doi:https://doi.org/10.4271/932720.

16.52. Mine Safety and Health Administration, "Potential Health Hazard Caused by Platinum Based Catalyzed Diesel Particulate Matter Exhaust Filters," Program Information Bulletin No. P02-4, Mine Safety and Health Administration, 2002.

16.53. Wyatt, M., Manning, W.A., Roth, S.A., D'Aniello, M.J. et al., "The Design of Flow-Through Diesel Oxidation Catalysts," SAE Technical Paper 930130 (1993), doi:https://doi.org/10.4271/930130.

16.54. Horiuchi, M., Saito, K., and Ichihara, S., "The Effects of Flow-through Type Oxidation Catalysts on the Particulate Reduction of 1990's Diesel Engines," SAE Technical Paper 900600 (1990), doi:https://doi.org/10.4271/900600.

16.55. Andrews, G.E., Iheozor-Ejiofor, I.E., and Pang, S.W., "Diesel Particulate SOF Emissions Reduction Using an Exhaust Catalyst," SAE Technical Paper 870251 (1987), doi:https://doi.org/10.4271/870251.

16.56. Porter, B.C., Doyle, D.M., Faulkner, S.A., Lambert, P. et al., "Engine and Catalyst Strategies for 1994," SAE Technical Paper 910604 (1991), doi:https://doi.org/10.4271/910604.

16.57. Kittelson, D.B., Pui, D.Y.H., and Moon, K.C., "Electrostatic Collection of Diesel Particles," SAE Technical Paper 860009 (1986), doi:https://doi.org/10.4271/860009.

16.58. Akhter, M.S. and Nabi, M.N., "Design, Construction and Performance Testing of a Cyclonic Separator to Control Particulate Pollution from Diesel Engine Exhaust," SAE Technical Paper 2005-01-3695 (2005), doi:https://doi.org/10.4271/2005-01-3695.

16.59. Hardenberg, H.O., Daudel, H.L., and Erdmannsdörfer, H.J., "Experiences in the Development of Ceramic Fiber Coil Particulate Traps," SAE Technical Paper 870015 (1987), doi:https://doi.org/10.4271/870015.

16.60. Mayer, A. and Buck, A., "Knitted Ceramic Fibers—A New Concept for Particulate Traps," SAE Technical Paper 920146 (1992), doi:https://doi.org/10.4271/920146.

16.61. Child, D. and Cioffi, J., "An Electronically Controlled Diesel Particulate Filter System Employing Electrically Regenerated Ceramic Fiber Wound Cartridges," SAE Technical Paper 942265 (1994), doi:https://doi.org/10.4271/942265.

16.62. Bloom, R., "The Development of Fiber Wound Diesel Particulate Filter Cartridges," SAE Technical Paper 950152 (1995), doi:https://doi.org/10.4271/950152.

16.63. Sakaguchi, T., Ohgushi, A., Suzuki, S., Kita, H. et al., "Development of High Durability Diesel Particulate Filter by using SiC Fiber," SAE Technical Paper 1999-01-0463 (1999), doi:https://doi.org/10.4271/1999-01-0463.

16.64. McMahon, M.A., Faist, C.H., Virk, K.S., and Tierney, W.T., "Alumina Coated Metal Wool as a Particulate Filter for Diesel Powered Vehicles," SAE Technical Paper 820183 (1982), doi:https://doi.org/10.4271/820183.

16.65. Zarvalis, D., Vlachos, N., Buergler, L., Seewald, G. et al., "A Metal Fibrous Filter for Diesel Hybrid Vehicles," *SAE Int. J. Engines* 4, no. 1 (2011): 537-552, doi:https://doi.org/10.4271/2011-01-0604.

16.66. Mizrah, T., Maurer, A., Gauckler, L., and Gabathuler, J.-P., "Open-Pore Ceramic Foam as Diesel Particulate Filter," SAE Technical Paper 890172 (1989), doi:https://doi.org/10.4271/890172.

16.67. Yoro, K., Itsuaki, S., Saito, H., Nakajima, S. et al., "Diesel Particulate Filter Made of Porous Metal," SAE Technical Paper 980187 (1998), doi:https://doi.org/10.4271/980187.

16.68. Koltsakis, G.C., Katsaounis, D.K., Samaras, Z.C., Naumann, D. et al., "Filtration and Regeneration Performance of a Catalyzed Metal Foam Particulate Filterm," SAE Technical Paper 2006-01-1524 (2006), doi:https://doi.org/10.4271/2006-01-1524.

16.69. Stobbe, P., Petersen, H.G., Sorenson, S.C., and Høj, J.W., "A New Closing Method for Wall Flow Diesel Particulate Filters," SAE Technical Paper 960129 (1996), doi:https://doi.org/10.4271/960129.

16.70. Howitt, J.S. and Montierth, M.R., "Cellular Ceramic Diesel Particulate Filter," SAE Technical Paper 810114 (1981), doi:https://doi.org/10.4271/810114.

16.71. Boger, T., He, S., Collins, T., Heibel, A. et al., "A Next Generation Cordierite Diesel Particle Filter with Significantly Reduced Pressure Drop," *SAE Int. J. Engines* 4, no. 1 (2011): 902-912, doi:https://doi.org/10.4271/2011-01-0813.

16.72. Nakamura, K., Vlachos, N., Konstandopoulos, A., Iwata, H. et al., "Performance Improvement of Diesel Particulate Filter by Layer Coating," SAE Technical Paper 2012-01-0842 (2012), doi:https://doi.org/10.4271/2012-01-0842.

16.73. Ohno, K., Shimato, K., Taoka, N., Santae, H. et al., "Characterization of SiC-DPF for Passenger Car," SAE Technical Paper 2000-01-0185 (2000), doi:https://doi.org/10.4271/2000-01-0185.

16.74. Mizutani, T., Ito, M., Masukawa, N., Ichikawa, S. et al., "The Study for Structural Design of the Segmented SiC-DPF," SAE Technical Paper 2006-01-1527 (2006), doi:https://doi.org/10.4271/2006-01-1527.

16.75. Gulati, S.T. and Sherwood, D.L., "Dynamic Fatigue Data for Cordierite Ceramic Wall-Flow Diesel Filters," SAE Technical Paper 910135 (1991), doi:https://doi.org/10.4271/910135.

16.76. Murtagh, M.J., Sherwood, D.L., and Socha, L.S., "Development of a Diesel Particulate Filter Composition and Its Effect on Thermal Durability and Filtration Performance," SAE Technical Paper 940235 (1994), doi:https://doi.org/10.4271/940235.

16.77. Rao, V.D.N., Cikanek, H.A., and Horrocks, R.W., "Diesel Particulate Control System for Ford 1.8L Sierra Turbo-Diesel to Meet 1997-2003 Particulate Standards," SAE Technical Paper 940458 (1994), doi:https://doi.org/10.4271/940458.

16.78. Ohno, K., Taoka, N., Ninomiya, T., Sungtae, H. et al., "SiC Diesel Particulate Filter Application to Electric Heater System," SAE Technical Paper 1999-01-0464 (1999), doi:https://doi.org/10.4271/1999-01-0464.

16.79. Miwa, S., Abe, F., Hamanaka, T., Yamada, T. et al., "Diesel Particulate Filters Made of Newly Developed SiC," SAE Technical Paper 2001-01-0192 (2001), doi:https://doi.org/10.4271/2001-01-0192.

16.80. Miyakawa, N., Maeno, H., and Takahashi, H., "Characteristics and Evaluation of Porous Silicon Nitride DPF," SAE Technical Paper 2003-01-0386 (2003), doi:https://doi.org/10.4271/2003-01-0386.

16.81. Okano, H., Yamaguchi, H., Shigenobu, R., Obuchi, A. et al., "Porous Silicon Nitride Ceramics with High Performance for Diesel Exhaust After-Treatment System," SAE Technical Paper 2012-01-0849 (2012), doi:https://doi.org/10.4271/2012-01-0849.

16.82. Ogunwumi, S.B., Tepesch, P.D., Chapman, T., Warren, C.J. et al., "Aluminum Titanate Compositions for Diesel Particulate Filters," SAE Technical Paper 2005-01-0583 (2005), doi:https://doi.org/10.4271/2005-01-0583.

16.83. Iwasaki, K., "Innovative Aluminum Titanate-Based Diesel Particulate Filter Having Asymmetric Hexagonal Cell Geometry," SAE Technical Paper 2012-01-0838 (2012), doi:https://doi.org/10.4271/2012-01-0838.

16.84. Ogyu, K., Ohno, K., Hong, S., and Komori, T., "Ash Storage Capacity Enhancement of Diesel Particulate Filter," SAE Technical Paper 2004-01-0949 (2004), doi:https://doi.org/10.4271/2004-01-0949.

16.85. Bardon, S., Bouteiller, B., Bonnail, N., Girot, P. et al., "Asymmetrical Channels to Increase DPF Lifetime," SAE Technical Paper 2004-01-0950 (2004), doi:https://doi.org/10.4271/2004-01-0950.

16.86. Aravelli, K. and Heibel, A., "Improved Lifetime Pressure Drop Management for Robust Cordierite (RC) Filters with Asymmetric Cell Technology (ACT)," SAE Technical Paper 2007-01-0920 (2007), doi:https://doi.org/10.4271/2007-01-0920.

16.87. Pace, L., Konieczny, R., and Presti, M., "Metal Supported Particulate Matter-Cat, a Low Impact and Cost Effective Solution for a 1.3 Euro IV Diesel Engine," SAE Technical Paper 2005-01-0471 (2005), doi:https://doi.org/10.4271/2005-01-0471.

16.88. Müller-Haas, K., Rice, M., and Rodovanovic, R., "Low Back Pressure Metallic Substrates Technology for CI Engines Used in Vehicles and Machinery to Meet Future Emission Requirements," SAE Technical Paper 2006-01-3508 (2006), doi:https://doi.org/10.4271/2006-01-3508.

16.89. Yoon, Y., Lee, D., Blakeman, P., Matsuda, K. et al., "Development and In-Field Application of a New Type of Partial Filter System for Diesel Retrofit," SAE Technical Paper 2009-01-1264 (2009), doi:https://doi.org/10.4271/2009-01-1264.

16.90. Mayer, A., Czerwinski, J., Comte, P., and Jaussi, F., "Properties of Partial-Flow and Coarse Pore Deep Bed Filters Proposed to Reduce Particle Emission of Vehicle Engines," *SAE Int. J. Fuels Lubr.* 2, no. 1 (2009): 497-511, doi:https://doi.org/10.4271/2009-01-1087.

16.91. Koberstein, E., Pletka, H.D., and Völker, H., "Catalytically Activated Diesel Exhaust Filters—Engine Test Methods and Results," SAE Technical Paper 830081 (1983), doi:https://doi.org/10.4271/830081.

16.92. Romero-López, A.F., Gutirrez-Salinas, R., and García-Moreno, R., "Soot Combustion During Regeneration of Filter Ceramic Traps for Diesel Engines," SAE Technical Paper 960469 (1996), doi:https://doi.org/10.4271/960469.

16.93. Wade, W.R., White, J.E., Florek, J.J., and Cikanek, H.A., "Thermal and Catalytic Regeneration of Diesel Particulate Traps," SAE Technical Paper 830083 (1983), doi:https://doi.org/10.4271/830083.

16.94. Dementhon, J.B., Martin, B., Richards, P., Rush, M. et al., "Novel Additive for Particulate Trap Regeneration," SAE Technical Paper 952355 (1995), doi:https://doi.org/10.4271/952355.

16.95. Zelenka, P., Reczek, W., Mustel, W., and Rouveirolles, P., "Towards Securing the Particulate Trap Regeneration: A System Combining a Sintered Metal Filter and Cerium Fuel Additive," SAE Technical Paper 982598 (1998), doi:https://doi.org/10.4271/982598.

16.96. Richards, P. and Kalischewski, W., "Retrofitting of Diesel Particulate Filters—Particulate Matter and Nitrogen Dioxide," JSAE Paper Number 20030053, 2003.

16.97. Margraf, J., Schrewe, K., and Steigert, S., "Enhancement of Diesel Soot Combustion with Oxygen on Particulate Filters after Injection of Dicyclopentadienyl Iron (Ferrocene) in the Exhaust Pipe," *SAE Int. J. Engines* 4, no. 1 (2011): 126-142, doi:https://doi.org/10.4271/2011-01-0303.

16.98. Richards, P., Terry, B., Vincent, M.W., and Cook, S.L., "Assessment of the Performance of Diesel Particulate Filter Systems with Fuel Additives for Enhanced Regeneration Characteristics," SAE Technical Paper 1999-01-0112 (1999), doi:https://doi.org/10.4271/1999-01-0112.

16.99. Blanchard, G., Seguelong, T., Michelin, J., Schuerholz, S. et al., "Ceria-Based Fuel-Borne Catalysts for Series Diesel Particulate Filter Regeneration," SAE Technical Paper 2003-01-0378 (2003), doi:https://doi.org/10.4271/2003-01-0378.

16.100. Rocher, L., Seguelong, T., Harle, V., Lallemand, M. et al., "New Generation Fuel Borne Catalyst for Reliable DPF Operation in Globally Diverse Fuels," SAE Technical Paper 2011-01-0297 (2011), doi:https://doi.org/10.4271/2011-01-0297.

16.101. Richards, P., Vincent, M.W., Johansen, K., and Mogensen, G., "Metal Emissions, NO_2 and HC Reduction from a Base Metal Catalysed DPF/FBC System," SAE Technical Paper 2006-01-0420 (2006), doi:https://doi.org/10.4271/2006-01-0420.

16.102. Cooper, B.J., Radnor, H.J., Jung, W., and Thoss, J.E., Treatment of diesel exhaust gases. U.S. Patent 4,902,487, 1990.

16.103. Sumiya, S. and Yokota, H., "Diesel Oxidation Catalyst System for PM Control," SAE Technical Paper 2004-28-0069 (2004), doi:https://doi.org/10.4271/2004-28-0069.

16.104. Suresh, A., Yezerets, A., Currier, N., and Clerc, J., "Diesel Particulate Filter System - Effect of Critical Variables on the Regeneration Strategy Development and Optimization," *SAE Int. J. Fuels Lubr.* 1, no. 1 (2009): 173-183, doi:https://doi.org/10.4271/2008-01-0329.

16.105. Czerwinski, J., Jaussi, F., Wyser, M., and Mayer, A., "Particulate Traps for Construction Machines Properties and Field Experience," SAE Technical Paper 2000-01-1923 (2000), doi:https://doi.org/10.4271/2000-01-1923.

16.106. Salvat, O., Marez, P., and Belot, G., "Passenger Car Serial Application of a Particulate Filter System on a Common Rail Direct Injection Diesel Engine," SAE Technical Paper 2000-01-0473 (2000), doi:https://doi.org/10.4271/2000-01-0473.

16.107. Parker, R.F. and Walker, J.W., "Exhaust Emission Control in Medium Swirl Rate Direct Injection Diesel Engines," SAE Technical Paper 720755 (1972), doi:https://doi.org/10.4271/720755.

16.108. Plee, S.L., Ahmad, T., and Myers, J.P., "Flame Temperature Correlation for the Effects of Exhaust Gas Recirculation on Diesel Particulate and NOx Emissions," SAE Technical Paper 811195 (1981), doi:https://doi.org/10.4271/811195.

16.109. Hug, H.T., Mayer, A., and Hartenstein, A., "Off-Highway Exhaust Gas After-Treatment:Combining Urea-SCR, Oxidation Catalysis and Traps," SAE Technical Paper 930363 (1993), doi:https://doi.org/10.4271/930363.

16.110. Cooper, D.A., "Exhaust Emissions from High Speed Passenger Ferries," *Atmospheric Environment* 35, no. 24 (2001): 4189-4200.

16.111. Heimrich, M.J. and Deviney, M.L., "Lean NOx Catalyst Evaluation and Characterization," SAE Technical Paper 930736 (1993), doi:https://doi.org/10.4271/930736.

16.112. Litorell, M., Allansson, R., Andreasson, A., Fredholm, S. et al., "Development of Test Methods for Lean-NOx Catalyst Evaluation," SAE Technical Paper 952489 (1995), doi:https://doi.org/10.4271/952489.

16.113. Peters, A., Langer, H., Jokl, B., Müller, W. et al., "Catalytic NOx Reduction on a Passenger Car Diesel Common Rail Engine," SAE Technical Paper 980191 (1998), doi:https://doi.org/10.4271/980191.

16.114. Leyrer, J., Lox, E.S., Ostgathe, K., Strehlau, W. et al., "Advanced Studies on Diesel Aftertreatment Catalysts for Passenger Cars," SAE Technical Paper 960133 (1996), doi:https://doi.org/10.4271/960133.

16.115. Kharas, K.C.C., Bailey, O., and Vuichard, J., "Improvements in Intimately Coupled Diesel Hydrocarbon Adsorber/Lean NOx Catalysis Leading to Durable Euro 3 Performance," SAE Technical Paper 982603 (1998), doi:https://doi.org/10.4271/982603.

16.116. Kharas, K.C.C. and Theis, J.R., "Performance Demonstration of a Precious Metal Lean NOx Catalysts in Native Diesel Exhaust," SAE Technical Paper 950751 (1995), doi:https://doi.org/10.4271/950751.

16.117. Deeba, M., Feeley, J., Farrauto, R.J., Steinbock, N. et al., "Catalytic Abatement of NOx from Diesel Engines: Development of Four Way Catalyst," SAE Technical Paper 952491 (1995), doi:https://doi.org/10.4271/952491.

16.118. Katoh, K., Kosaki, Y., Watanabe, T., and Funabiki, M., "Compatibility of NOx and PM Abatement in Diesel Catalysts," SAE Technical Paper 980931 (1998), doi:https://doi.org/10.4271/980931.

16.119. Kodama, Y. and Wong, V.W., "Study of On-Board Ammonia (NH_3) Generation for SCR Operation," *SAE Int. J. Fuels Lubr.* 3, no. 1 (2010): 537-555, doi:https://doi.org/10.4271/2010-01-1071.

16.120. Gandhi, H.S., Cavataio, J.V., Hammerle, R.H., and Cheng, Y., Catalyst system for the reduction of NOx and NH_3 emissions. U.S. Patent 8,240,132, 2012.

16.121. Fulks, G., Fisher, G., Rahmoeller, K., Wu, M. et al., "A Review of Solid Materials as Alternative Ammonia Sources for Lean NOx Reduction with SCR," SAE Technical Paper 2009-01-0907 (2009), doi:https://doi.org/10.4271/2009-01-0907.

16.122. Koebel, M., Elsener, M., and Kleemann, M., "Urea-SCR: A Promising Technique to Reduce NOx Emissions from Automotive Diesel Engines," *Catalysis Today* 59, no. 3-4 (2000): 335-345.

16.123. Kowatari, T., Hamada, Y., Amou, K., Hamada, I. et al., "A Study of a New Aftertreatment System (1): A New Dosing Device for Enhancing Low Temperature Performance of Urea-SCR," SAE Technical Paper 2006-01-0642 (2006), doi:https://doi.org/10.4271/2006-01-0642.

16.124. Guo, G., Dobson, D., Warner, J., Ruona, W. et al., "Advanced Urea SCR System Study with a Light Duty Diesel Vehicle," SAE Technical Paper 2012-01-0371 (2012), doi:https://doi.org/10.4271/2012-01-0371.

16.125. Jen, H.-W., Girard, J.W., Cavataio, G., and Jagner, M.J., "Detection, Origin and Effect of Ultra-Low Platinum Contamination on Diesel-SCR Catalysts," *SAE Int. J. Fuels Lubr.* 1, no. 1 (2009): 1553-1559, doi:https://doi.org/10.4271/2008-01-2488.

16.126. Blakeman, P., Arnby, K., Marsh, P., Newman, C. et al., "Vanadia-Based SCR Systems to Achieve EUIV HDD Legislation," SAE Technical Paper 2009-26-0013 (2009), doi:https://doi.org/10.4271/2009-26-0013.

16.127. Chapman, D., Fu, G., Augustine, S., Crouse, J. et al., "New Titania Materials with Improved Stability and Activity for Vanadia-Based Selective Catalytic Reduction of NOx," *SAE Int. J. Fuels Lubr.* 3, no. 1 (2010): 643-653, doi:https://doi.org/10.4271/2010-01-1179.

16.128. Tao, T., Xie, Y., Dawes, S., Melscoet-Chauvel, I. et al., "Diesel SCR NOx Reduction and Performance on Washcoated SCR Catalysts," SAE Technical Paper 2004-01-1293 (2004), doi:https://doi.org/10.4271/2004-01-1293.

16.129. Gieshoff, J., Pfeifer, M., Schäfer-Sindlinger, A., Spurk, P.C. et al., "Advanced Urea SCR Catalysts for Automotive Applications," SAE Technical Paper 2001-01-0514 (2001), doi:https://doi.org/10.4271/2001-01-0514.

16.130. Cavataio, G., Jen, H.-W., Warner, J.R., Girard, J.W. et al., "Enhanced Durability of a Cu/Zeolite Based SCR Catalyst," *SAE Int. J. Fuels Lubr.* 1, no. 1 (2009): 477-487, doi:https://doi.org/10.4271/2008-01-1025.

16.131. Ido, T., Yoshimura, K., Kunieda, M., Tamura, Y. et al., "Volume Reduction of SCR Catalyst Using Zeolite-Base Honeycomb Substrate," *SAE Int. J. Fuels Lubr.* 3, no. 1 (2010): 614-624, doi:https://doi.org/10.4271/2010-01-1170.

16.132. Kamasamudram, K., Currier, N., Szailer, T., and Yezerets, A., "Why Cu- and Fe-Zeolite SCR Catalysts Behave Differently at Low Temperatures," *SAE Int. J. Fuels Lubr.* 3, no. 1 (2010): 664-672, doi:https://doi.org/10.4271/2010-01-1182.

16.133. Burgess, G.A., Chandler, G.R., Flanagan, K.A., Marvell, D. et al., Catalytic wall-flow filter with an ammonia slip catalyst. U.S. Patent 10,369,555, 2019.

16.134. Ahari, H., Smith, M., Zammit, M., Price, K. et al., "Impact of SCR Integration on N_2O Emissions in Diesel Application," *SAE Int. J. Passeng. Cars - Mech. Syst.* 8, no. 2 (2015): 526-530, doi:https://doi.org/10.4271/2015-01-1034.

16.135. Kamasamudram, K., Yezerets, A., Chen, X., Currier, N. et al., "New Insights into Reaction Mechanism of Selective Catalytic Ammonia Oxidation Technology for Diesel Aftertreatment Applications," *SAE Int. J. Engines* 4, no. 1 (2011): 1810-1821, doi:https://doi.org/10.4271/2011-01-1314.

16.136. Ottinger, N., Foley, B., Xi, Y., and Liu, Z., "Impact of Hydrocarbons on the Dual (Oxidation and SCR) Functions of Ammonia Oxidation Catalysts," *SAE Int. J. Engines* 7, no. 3 (2014): 1262-1268, doi:https://doi.org/10.4271/2014-01-1536.

16.137. Howells, H., Micallef, D., Newman, A., and Greenham, N., Catalyst article with SCR active substrate, ammonia slip catalyst layer and SCR layer for use in an emission treatment system. World Patent WO2018178627A1, 2018.

16.138. Reichinger, M., Maletz, G., Wanninger, K., Bentele, A. et al., Ammonia oxidation catalyst having low N_2O by-product formation. U.S. Patent 9,573,097. 2017.

16.139. Dotzel, R., Leppelt, R., Mountstevens, E.H., Münch, J.W. et al., NO_X absorber catalysts. U.S. Patent 8,641,993, 2014.

16.140. Andersen, P.J. and Doura, K., Non-PGM ammonia slip catalyst. U.S. Patent 11,198,094, 2021.

16.141. Larsson, M., Exhaust system without a DOC having an ASC acting as a DOC in a system with an SCR catalyst before the ASC. U.S. Patent 10,589,261, 2020.

16.142. Gieshoff, J., Schäfer-Sindlinger, A., Spurk, P.C., van den Tillaart, J.A.A. et al., "Improved SCR Systems for Heavy Duty Applications," SAE Technical Paper 2000-01-0189 (2000), doi:https://doi.org/10.4271/2000-01-0189.

16.143. Conway, R., Chatterjee, S., Beavan, A., Goersmann, C. et al., "NOx and PM Reduction Using Combined SCR and DPF Technology in Heavy Duty Diesel Applications," SAE Technical Paper 2005-01-3548 (2005), doi:https://doi.org/10.4271/2005-01-3548.

16.144. Enderle, C., Vent, G., and Paule, M., "BLUETEC Diesel Technology - Clean, Efficient and Powerful," SAE Technical Paper 2008-01-1182 (2008), doi:https://doi.org/10.4271/2008-01-1182.

16.145. Wan, C.-Z., Zheng, X., Stiebels, S., Wendt, C. et al., Emissions treatment system with amminia-generating and SCR catalyst. U.S. Patent 2011/0173950, 2011.

16.146. Xu, L., McCabe, R., Dearth, M., and Ruona, W., "Laboratory and Vehicle Demonstration of '2nd-Generation' LNT + In-Situ SCR Diesel NOx Emission Control Systems," *SAE Int. J. Fuels Lubr.* 3, no. 1 (2010): 37-49, doi:https://doi.org/10.4271/2010-01-0305.

16.147. Xu, L., McCabe, R., Tennison, P., and Jen, H., "Laboratory and Vehicle Demonstration of '2nd-Generation' LNT + In-Situ SCR Diesel Emission Control Systems," *SAE Int. J. Engines* 4, no. 1 (2011): 158-174, doi:https://doi.org/10.4271/2011-01-0308.

16.148. Cordeiro, V. and Jensen, J.B., "Two-Stroke Diesels Meet Macau Electric Power Needs," *Diesel and Gas Turbine Worldwide* 28, no. 6 (1996): 22-26.

16.149. Hills, F.J. and Schleyerbach, C.G., "Diesel Fuel Properties and Engine Performance," SAE Technical Paper 770316 (1977), doi:https://doi.org/10.4271/770316.

16.150. Hiranuma, S., Takeda, Y., Kawatani, T., Doumeki, R. et al., "Development of DPF System for Commercial Vehicle—Basic Characteristic and Active Regenerating Performance," SAE Technical Paper 2003-01-3182 (2003), doi:https://doi.org/10.4271/2003-01-3182.

16.151. Parks, J., Huff, S., Kass, M., and Storey, J., "Characterization of In-Cylinder Techniques for Thermal Management of Diesel Aftertreatment," SAE Technical Paper 2007-01-3997 (2007), doi:https://doi.org/10.4271/2007-01-3997.

16.152. Andreae, M., Fang, H., and Bhandary, K., "Biodiesel and Fuel Dilution of Engine Oil," SAE Technical Paper 2007-01-4036 (2007), doi:https://doi.org/10.4271/2007-01-4036.

16.153. Genzale, C.L., Pickett, L.M., and Kook, S., "Liquid Penetration of Diesel and Biodiesel Sprays at Late-Cycle Post-Injection Conditions," *SAE Int. J. Fuels Lubr.* 3, no. 1 (2010): 479-495, doi:https://doi.org/10.4271/2010-01-0610.

16.154. He, X., Williams, A., Christensen, E., Burton, J. et al., "Biodiesel Impact on Engine Lubricant Dilution during Active Regeneration of Aftertreatment Systems," *SAE Int. J. Fuels Lubr.* 4, no. 2 (2011): 158-178, doi:https://doi.org/10.4271/2011-01-2396.

16.155. Zdrodowski, R., Gangopadhyay, A., Anderson, J.E., Ruona, W.C. et al., "Effect of Biodiesel (B20) on Vehicle-Aged Engine Oil Properties," *SAE Int. J. Fuels Lubr.* 3, no. 2 (2010): 579-597, doi:https://doi.org/10.4271/2010-01-2103.

16.156. Blackburn, J.H., Pinchin, R., Nobre, J.L.T., Crichton, B.A.L. et al., "Performance of Lubricating Oils in Vegetable Oil Ester-Fuelled Diesel Engines," SAE Technical Paper 831355 (1983), doi:https://doi.org/10.4271/831355.

16.157. Devlin, C.C., Passut, C.A., Campbell, R.L., and Jao, T.-C., "Biodiesel Fuel Effect on Diesel Engine Lubrication," SAE Technical Paper 2008-01-2375 (2008), doi:https://doi.org/10.4271/2008-01-2375.

16.158. Fujimoto, K., Yamashita, M., Kaneko, T., and Ishikawa, M., "Influence of Bio Diesel Fuel on Engine Oil Performance," *SAE Int. J. Fuels Lubr.* 3, no. 2 (2010): 362-368, doi:https://doi.org/10.4271/2010-01-1543.

Further Reading

Kasab, J. and Strzelec, A., *Automotive Emissions Regulations and Exhaust Aftertreatment Systems* (Warrendale, PA: SAE International, 2020).

Lai, M.-C.D. and Dingle, P.J., *Diesel Common Rail and Advanced Fuel Injection Systems* (Warrendale, PA: SAE International, 2005).

Rogers, D., *Engine Combustion: Pressure Measurement and Analysis*, 2nd ed. (Warrendale, PA: SAE International, 2021).

Zhao, H., *Laser Diagnostics and Optical Measurement Techniques in Internal Combustion Engines* (Warrendale, PA: SAE International, 2012).

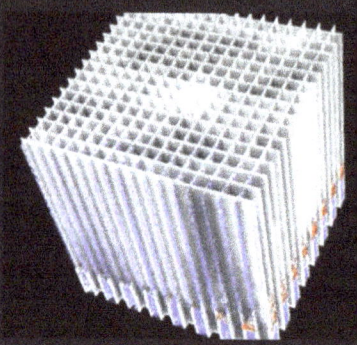

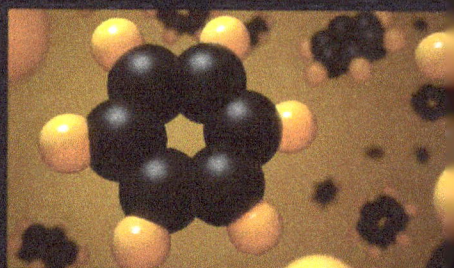

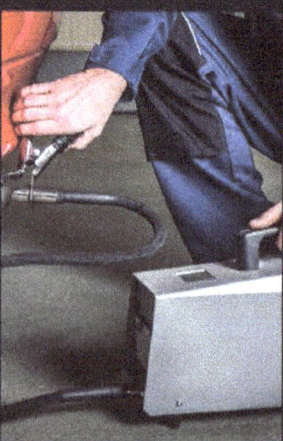

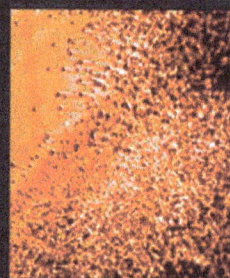

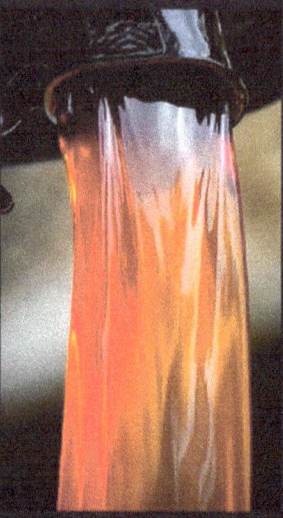

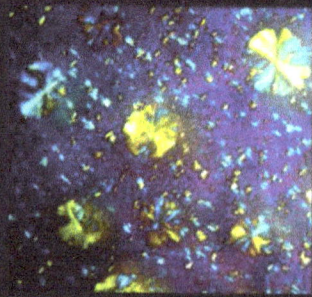

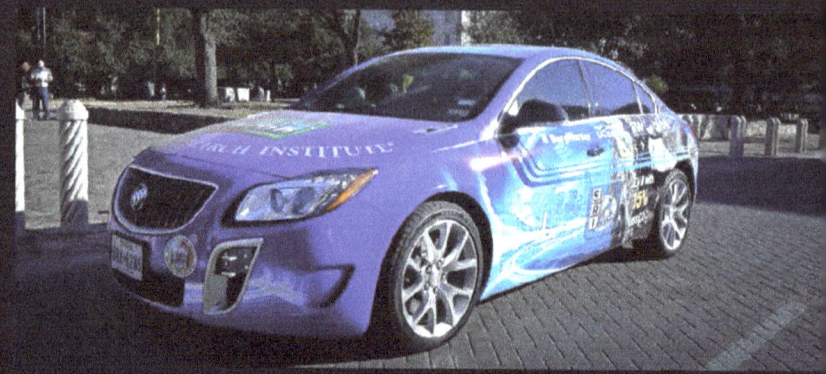

17

Diesel Fuel Low-Temperature Characteristics

Diesel fuels are complex mixtures of hydrocarbons that are usually blended from a number of refinery streams, as discussed in Chapter 3. As much as 20% of the diesel fuel can consist of relatively heavy paraffinic hydrocarbons, which, if cooled sufficiently, will form wax crystals. As a general rule, the longer the paraffin chain length, the higher the temperature at which it will begin to crystalize. In many regions, this picture is now complicated by the inclusion of fatty acid methyl esters (FAME), also referred to as biodiesel. Some of these ester molecules will also freeze at relatively high temperature.

Whether these initial crystal formations will result in well-dispersed small crystals, whether they will agglomerate, or whether they will grow into larger crystals will depend on the other components and characteristics of the fuel. Which of these routes is followed will determine how the engine or vehicle systems are affected. Engine and vehicle manufacturers are making advances to ensure their products are less susceptible to the effects of low temperatures, but the fuel producers must be cognizant of the legacy fleet and how it responds to fuel properties [17.1–17.8].

The following sections will examine the different ways of assessing the low-temperature characteristics of diesel fuels, how they can be improved by the use of fuel additives, and how the performance of fuels can be investigated.

17.1. Diesel Fuel Low-Temperature Properties

The paraffinic components of the petroleum diesel fuel will tend to form wax crystals as the temperature falls below the freezing point of that particular component. Wax in a vehicle fuel system is a potential source of operating problems, so the low-temperature properties of the fuel are defined by wax-related tests that measure the temperature at which

- The separation of wax out of solution is first observed.

- The amount of wax is considered sufficient to restrict flow in a vehicle fuel system.

- The amount of wax is sufficient to cause complete gelling of the fuel.

Similarly, the FAMEs in biodiesel will tend to solidify when the temperature drops below their freezing point. Due to their physical structure, the saturated FAMEs have higher melting points than their unsaturated counterparts. Some of the FAMEs that are commonly found in biodiesel have relatively high melting points. Figure 17.1 [17.9] shows the melting points of some saturated FAMEs.

FIGURE 17.1 Melting point of saturated FAMEs commonly found in biodiesel.

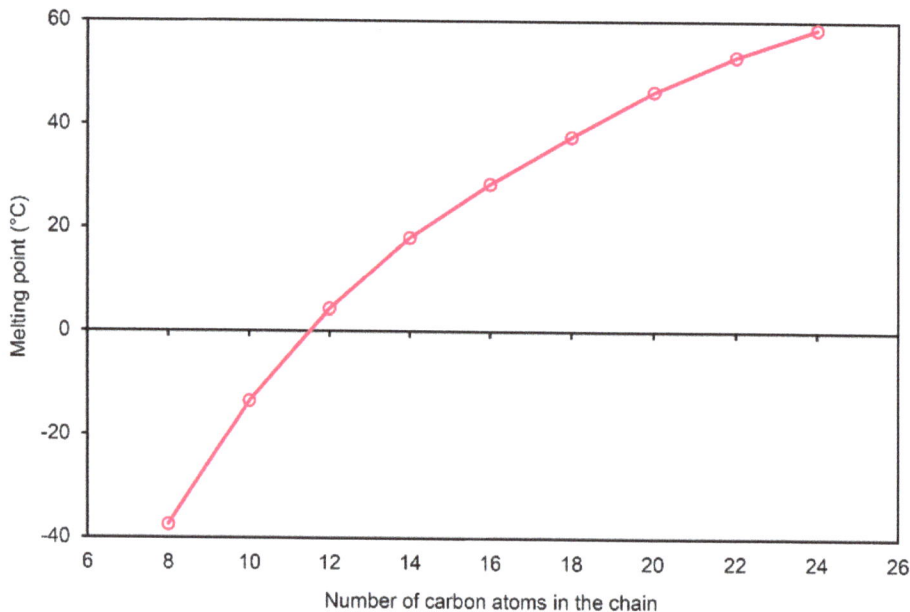

© SAE International.

It is therefore believed that a high level of saturated compound in the FAME will result in poorer cold flow properties [17.10].

The following sections look at how these different phenomena can be defined and quantified.

17.1.1. Cloud Point

One way of quantifying the temperature at which wax first becomes visible when the fuel is cooled is the cloud point (CP), and there is a standard test procedure, ASTM D2500 [17.11], for measuring the CP.

The test method is applicable to petroleum-derived diesel and biodiesel fuels. In the test, a small sample of the fuel is placed in a glass jar, subjected to a specific cooling profile, and examined at intervals of 1°C. A thermometer is immersed in the fuel to measure the temperature at the bottom of the jar, where the first waxes appear. It is a subjective test, depending on the operator's judgment of whether wax particles are visible.

Interlaboratory correlation tests to establish the precision of the test have shown that duplicate results on identical fuel samples, when tested by the same operator using the same apparatus, should not differ by more than 2°C. This value is referred to as the repeatability of the test. When comparing results obtained by different operators in different laboratories, the difference should not be more than 4°C. This is known as the reproducibility of the test. Both values represent the 95% limit of confidence that can be placed on results obtained from the test.

Two alternative methods allow the determination of CP to 0.1°C using either a linear cooling rate [17.12] or a constant cooling rate [17.13]. When the results from either of these two tests are rounded down to the nearest 1°C, the results are equivalent to the D2500 standard [17.11]. A more precise automated test method [17.14] is also available. Other standard methods are now available [17.15–17.17] that are quicker to perform, but the results have to be adjusted for bias.

17.1.2. Wax Appearance Point

Another procedure that was used to determine the temperature at which wax starts to come out of solution was the wax appearance point [17.18]. In this test, a 25 mL portion of the fuel is agitated while being cooled in a double-walled (Dewar-type), jacketed, clear glass tube. The cooling bath is an unsilvered vacuum flask, and mixing of the fuel is by means of a twin-helix stirrer operating at 110 strokes/minute through a 50 mm amplitude within the test portion. The test is terminated when a distinct swirl of wax crystals in the fuel makes the stirring pattern obvious. The thermometer is read immediately to the nearest 0.2°C.

With agitation, there is no temperature gradient, so the wax crystals will be dispersed throughout the fuel and the temperature at which this occurs is reported. In the CP test the jar is examined at intervals of 1°C to check for wax crystals in the coldest part of the fuel, at the bottom of the jar. The wax appearance test method has now been withdrawn by ASTM.

17.1.3. Pour Point

The pour point (PP) test is used to measure the temperature at which the amount of wax out of solution is sufficient to gel the fuel. It is traditionally carried out in a similar way and under the same controlled cooling conditions as CP, except that the thermometer is located with its bulb just below the surface of the fuel, and checks on the condition of the fuel are made at intervals of 3°C [17.19].

Checks are made by briefly removing the test jar from the cooling bath and tilting it to see whether the fuel flows. This procedure is continued until the fuel fails to move when the jar is held horizontally for 5 seconds. At this point, the fuel is not completely solidified; movement is prevented by the formation of an interlocking wax structure. When there is as little as 2% of wax coming out of solution, this can result in the remaining 98% of liquid fuel being locked in by the wax structure [17.20].

The PP is the lowest temperature at which the fuel was observed to flow. This will be 3°C above the temperature of the last check, when the fuel was seen to have gelled. The repeatability of the D97 [17.19] test procedure is 3°C, and its reproducibility, for comparisons between laboratories, is 6°C.

Alternative automated test procedures are available that apply tilting of the specimen jar to determine the PP [17.21, 17.22]. There are also standard methods that apply a pulse of air [17.23] or nitrogen [17.24] to the surface of the sample to determine whether it has reached the PP. There is also a standard method that rotates the sample and uses an immersed detector to determine the point at which the sample ceases to flow [17.25].

17.1.4. Significance of CP and PP

These tests measure two physical properties of diesel fuel. They are the temperatures at which wax starts to come out of solution as the fuel cools and when there is sufficient wax to prevent the fuel from flowing under the low-shear conditions. Neither test measures the intermediate temperature at which the amount of wax becomes sufficient to restrict flow in a vehicle fuel system.

Both tests have been used for a long time to define the low-temperature properties of diesel fuels, but neither indicates precisely how the fuel will perform in vehicle service. This is particularly true if the fuel contains an additive to improve its low-temperature properties.

In many instances, the CP underestimates the ability of the fuel to perform at low temperatures, while the PP tends to be overoptimistic. Data obtained by different researchers have been plotted in Figure 17.2 [17.26–17.28] to show how vehicle operability relates to the CP of the fuel.

FIGURE 17.2 Correlation of CP with diesel vehicle low-temperature operability.

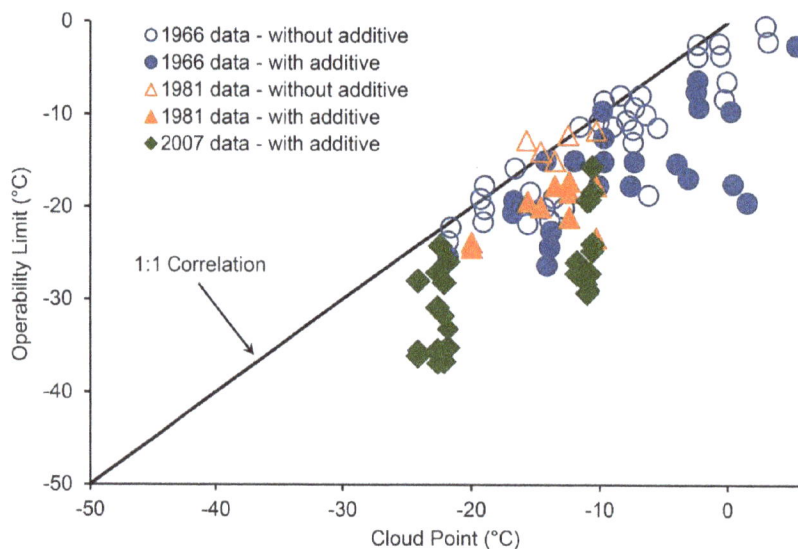

The diagonal line on the graph indicates where the points should lie if the laboratory test prediction is correct. A large number of fuels were found to be operating satisfactorily at temperatures well below their CP. These included base fuels as well as those containing additives designed to improve the low-temperature performance without affecting the CP.

A similar plot, showing the relationship between PP and operability, is given in Figure 17.3 [17.26–17.28]. In this case, most fuels reached their operability limit before cooling to their PP temperature, irrespective of whether or not they contained a cold flow improver additive.

FIGURE 17.3 Correlation of PP with diesel vehicle low-temperature operability.

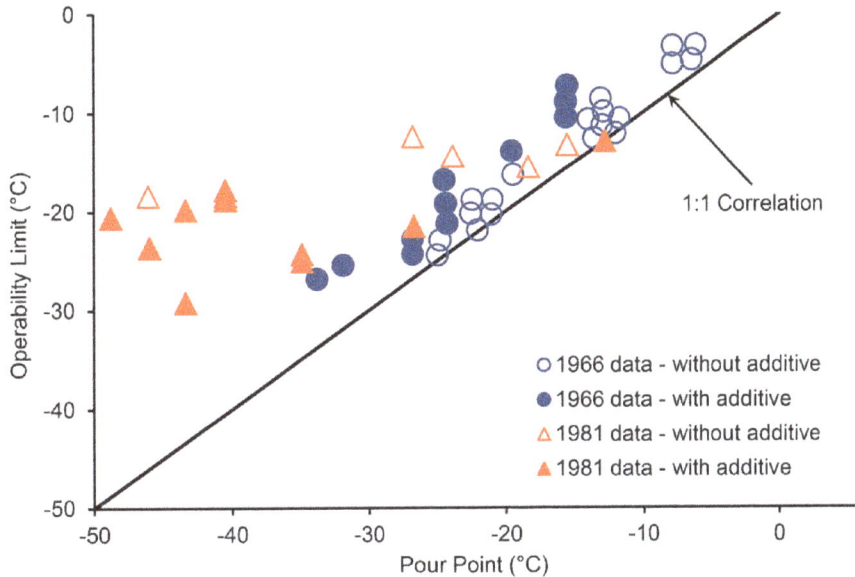

© SAE International.

The low-temperature operability limit is defined as the lowest temperature at which acceptable performance is possible. Because this is rarely a precisely measured value, it is often termed the estimated minimum operating temperature (EMOT). Interpretation of vehicle operability test results will be covered in more detail later.

To overcome the limitations of cloud and PP in defining diesel fuel quality, other laboratory tests have been developed to predict fuel performance in service. The other methods aim to simulate realistic operability criteria by assessing the ability of a waxy fuel to flow through a filter screen, as it has to do in a vehicle fuel system.

17.1.5. Cold Filter Plugging Point

A test that has become widely accepted in temperate regions of the world to predict low-temperature performance is the cold filter plugging point of distillate fuels (CFPP) [17.29]. The CFPP test was developed from vehicle operability data generated during a winter field trial on a fleet of European diesel vehicles [17.26, 17.30]. Fuels of different origins and base qualities were tested to observe how treatment with a PP depressant affected the low-temperature performance of the test vehicles. Additive treating levels were chosen to give good PP reduction, and tests were made when suitable below-CP temperatures were forecast.

The field trial showed that operation down to lower temperatures was possible when a fuel was treated with the additive. Failures were due to fuel starvation caused by wax accumulation on the filters in the vehicle system.

The field test results confirmed the inadequacy of the PP as a predictive test for vehicle operability at low temperatures and provided the stimulus for development of the CFPP as a more realistic laboratory procedure. The test was found to be relevant for untreated fuels as well as those containing a wax-modifying additive.

The CFPP test measures the lowest temperature at which 20 mL of the fuel will pass through a fine wire mesh screen of 45 µm nominal aperture in less than 60 seconds. Checks are made at 1°C intervals as the fuel cools under conditions similar to those for CP and PP. Figure 17.4 [17.31, 17.32] shows a schematic of the general arrangement of the test apparatus.

FIGURE 17.4 General arrangement of CFPP apparatus.

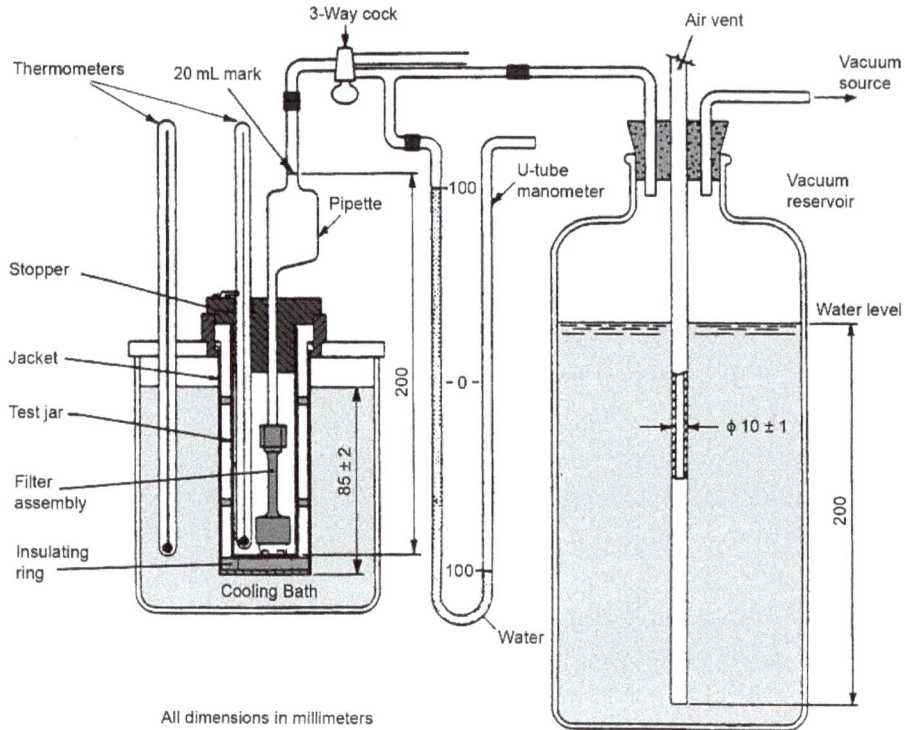

All dimensions in millimeters

© SAE International.

The closer correlation with operability given by the CFPP is illustrated in Figure 17.5 [17.26, 17.28].

FIGURE 17.5 Correlation of CFPP with diesel vehicle low-temperature operability.

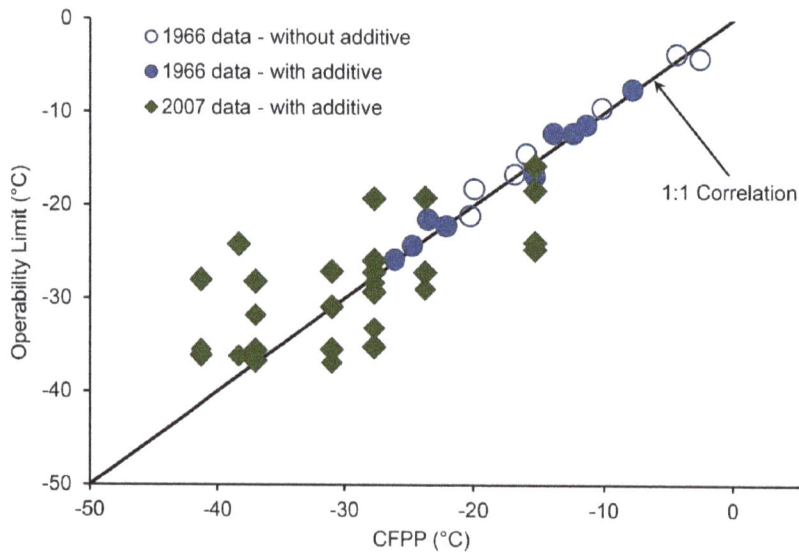

© SAE International.

While there is some scatter of results around the 1:1 correlation line due to differences between the fuels and the vehicle systems, this tends to be either side of the 1:1 line. There is not a consistent under-prediction or overestimation of the performance. The prediction is thus better than that was obtained using either CP or PP. These operability data were obtained from field and chassis dynamometer tests on a wide range of commercial vehicles and passenger cars and over a significant period of time, thus covering a variety of technology. The fuels, both with and without flow improver treatment, came from a variety of crude oil and refining sources.

17.1.6. Low-Temperature Flow Test

The low-temperature flow test (LTFT) was developed in the US to predict how a diesel fuel will perform at low temperatures [17.33]. This test uses the same concept as the CFPP test in that it requires a quantity of chilled fuel to pass through a fine mesh screen within a short period of time, but the detail features are different. The general arrangement of the test equipment is shown in Figure 17.6 [17.31].

FIGURE 17.6   General arrangement of LTFT apparatus.

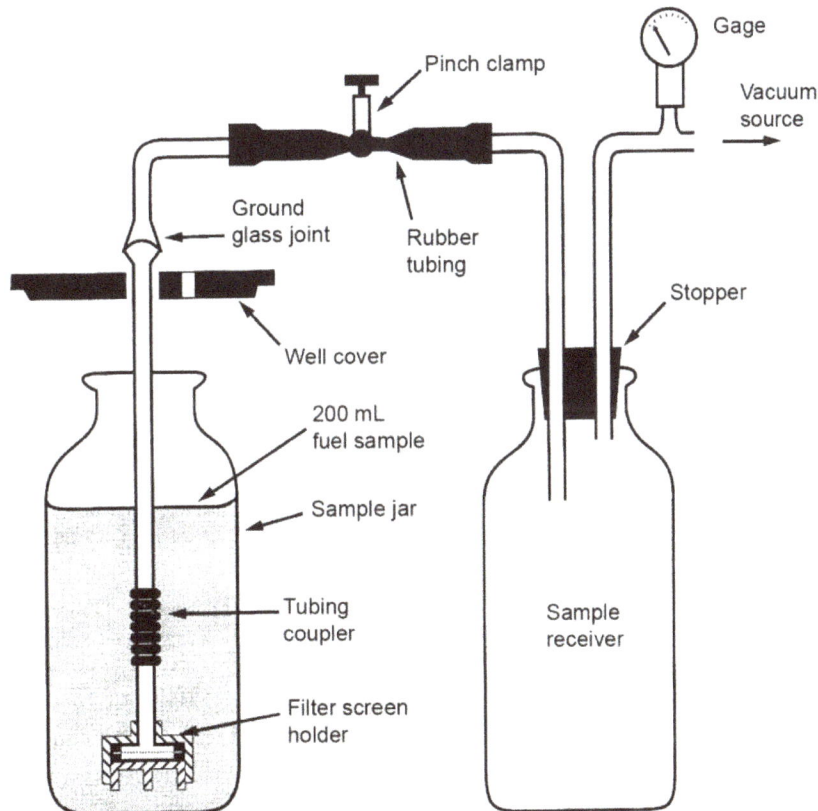

© SAE International.

A 200 mL sample of test fuel is cooled at 1°C/hour to the chosen test temperature, and then a 20 kPa vacuum is applied to draw the fuel through a 17 μm filter screen. Results are reported as the lowest temperature at which approximately 180 mL passes through the screen in less than 60 seconds. A study by CRC in

1981 showed the LTFT procedure to be a slightly better predictor than CFPP of the low-temperature operability limit of some US diesel vehicles [17.34]. A later CRC study using light-duty US vehicles showed a correlation between operability and CFPP and LTFT were similar [17.28]. These findings were in agreement with other independent field tests [17.35] and chassis dynamometer studies [17.36–17.38]. Figure 17.7 [17.27, 17.28, 17.39] shows a compilation of the results from three of these studies.

FIGURE 17.7 Correlation of LTFT with diesel vehicle low-temperature operability.

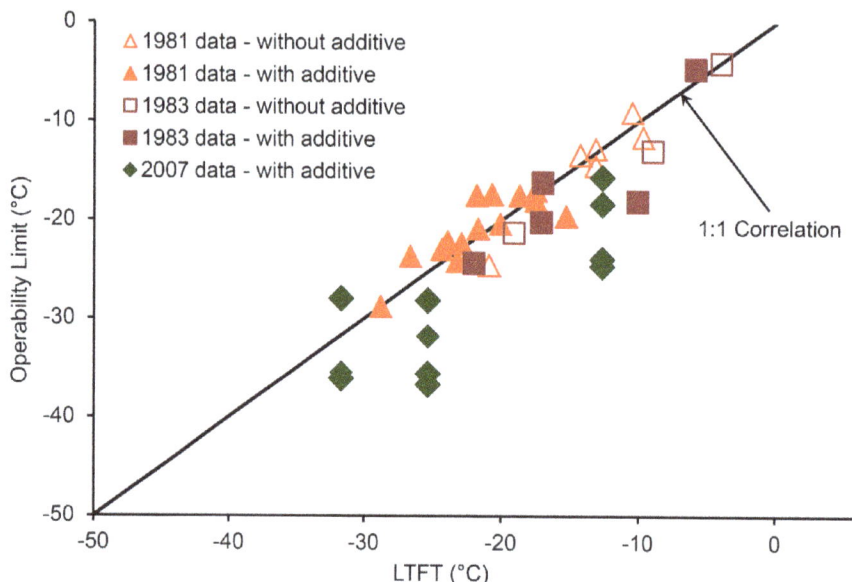

A test procedure has been used where the fuel is cooled at a rate of 2°C/hour and filtered at 1°C decrements below its CP [17.40]. Three different filter screens of 37, 125, and 420 μm were used. This demonstrated the importance of filter porosity in relation to crystal size as regards filter plugging.

It has been suggested that the wax precipitation index (WPI) could be used as a measure of vehicle performance. The WPI is given by the following formula:

$$WPI = CP - 1.3\left(CP - PP - 1.1\right)^{0.5}$$

It was found that this formula correlated with the average estimated minimum operating temperatures of vehicles in a CRC field test [17.39]. It should be noted that there was a high degree of scatter.

17.1.7. Simulated Filter Plugging Point (SFPP)

The CFPP test has been incorporated into many diesel fuel specifications as the principal criterion of low-temperature quality. The application of the test has enabled refiners to increase production to meet the demand for diesel fuel through the use of flow improver additives.

There was a trend in the late 1980s in Europe toward further improvement in the low-temperature quality of diesel fuels. This was stimulated partly by two consecutive severe winters in the mid-1980s and also by the introduction of the concept of product differentiation in the diesel fuel market.

Improved diesel fuel properties were achieved through changes in fuel formulation, increased use of flow improvers, and the application of more-advanced flow improver technology. As a result, fuels offering low-temperature operability at temperatures more than 10°C below the fuel CP became available. Over the years, it has been shown that CFPP becomes less reliable as an indication of vehicle operability when it has been depressed to more than 10 to 12°C below the fuel CP. Consequently, there has been an industry initiative to develop a new test that would be a more reliable predictor of operability for those fuels with a larger cloud–CFPP difference.

The SFPP [17.41] test was developed by a group of European oil companies in response to that need. The SFPP test was based on a CFPP apparatus, with a number of significant modifications. In particular, these include a controlled (and slower) rate of cooling; a finer mesh filter; a higher vacuum, progressively applied; and controlled convection within the test cell. The criterion for a "pass" result is for the 5 mL pipette to fill within 60 seconds of the vacuum being applied and for the fuel to flow back into the test jar within 40 seconds when the vacuum is switched off. The SFPP was standardized as the Energy Institute Method IP419, but this method is now obsolete.

The comparison with vehicle operability, based on fuel tank temperatures, was obtained for 129 fuels from various sources. The test vehicles are known to be sensitive with regard to cold weather waxing problems. Most of these fuels had large differences between CP and CFPP. They were deliberately selected to represent those cases where CFPP is known to give inadequate protection. The relationship between vehicle operability and SFPP is illustrated in Figure 17.8 [17.42].

FIGURE 17.8   Correlation of SFPP with vehicle operability.

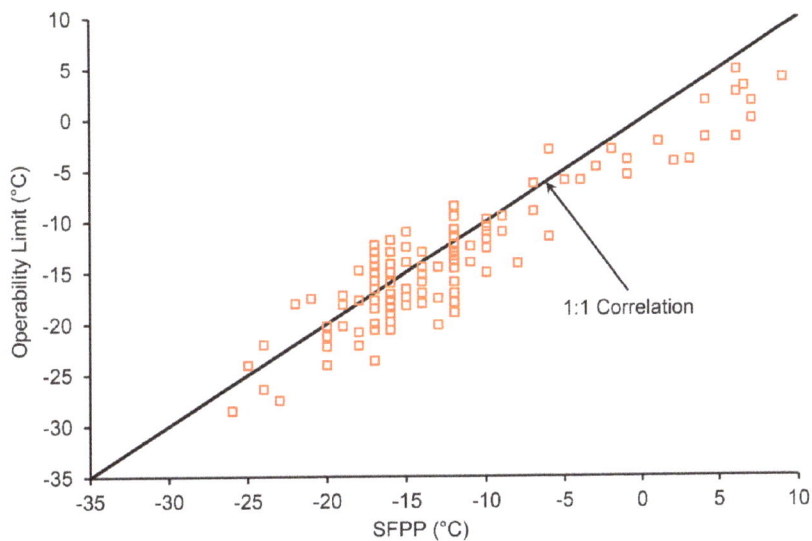

© SAE International.

SFPP is not currently used in any fuel specifications. However, the test is used by many researchers to provide additional information on the cold flow behavior of new fuels and additives [17.8, 17.43, 17.44].

17.1.8. Cold Soak Filtration Test

The test procedures discussed are applicable to petroleum diesel fuels and diesel fuels containing a proportion of biodiesel. However, there are concerns that these tests do not adequately represent the low-temperature performance of these low-level blends of biodiesel [17.45, 17.46]. This has resulted in the inclusion of the cold soak filtration test (CSFT) [17.47] into the biodiesel (B100) fuel specification [17.48].

The cold flow properties of biodiesel blending components are determined by the characteristics of the feedstock, including the fatty acid chain length, degree of unsaturation, and branching of the chains. Saturated fatty esters have higher melting points than unsaturated fatty esters of the same carbon number [17.49]. The shape of the crystals formed on cooling a biodiesel blend can be significantly different from those formed in petroleum diesel. This is shown in Figure 17.9 [17.50].

FIGURE 17.9   Crystals formed in petroleum diesel fuel (left) and 5% biodiesel blend (right) at −6°C.

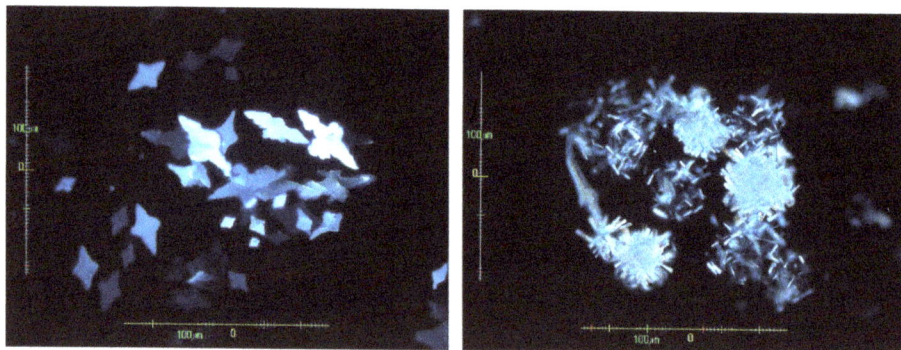

© SAE International.

The image on the left of Figure 17.9 shows the expected platelet crystal formation resulting from the paraffins in the petroleum diesel, and the image on the left shows a more three-dimensional crystal formation resulting from that particular sample of fuel that included 5% rape methyl ester (RME). This had a deleterious effect on engine performance.

The low-temperature performance can also be affected by the presence of trace amounts of impurities resulting from the manufacturing process. These can include total glycerin, especially saturated monoglyceride (SMG), and high molecular weight polar compounds such as sterol-glucosides (SG) [17.51–17.53]. As the biodiesel ages, there is also the possibility that oxidation products can be formed that may also affect the low-temperature performance.

The CSFT measures the time taken for a 300 mL sample of neat biodiesel to filter through a 7-μm glass fiber filter paper. Prior to the test, the sample is maintained at a temperature of 4.4°C for 16 hours before being allowed to warm to a temperature of between 20 and 22°C. The sample is then filtered with the aid of a vacuum of between 71.1 and 84.7 kPa. The time for the 300 mL to pass through the filter is recorded as the test time. If the total sample does not pass through the filter, then the amount that has passed after 6 minutes is noted.

Extensive testing conducted a few years ago demonstrated that high CSFT biodiesels can lead to fuel filter plugging at and above the CP [17.54]. The study also showed that for a fully blended fuel, the LTFT was a more conservative predictor of potential field problems than the CFPP.

17.2. Additives to Improve Cold Weather Performance

In the past, it was often necessary for an operator to blend lower boiling diesel fuels, kerosene, or even gasoline into their diesel fuel to ensure vehicle performance during the winter [17.55, 17.56]. However, these lower boiling point blends also tend to have a lower density that will result in reduced power [17.57, 17.58]. A fuel additive that enhances low-temperature operation is therefore a preferred alternative.

Additives to lower the PPs of engine lubricating oils have been in use for many years. The effective application of additives to improve the low-temperature properties of middle distillate fuels was a more recent development. Interest in their use for diesel fuels dates from around 1960, after the development of a PP depressant for domestic heating oil.

The selection of additives available today has evolved from PP depressants to include filterability or flow improvers, CP depressants, and additives that lessen the tendency for wax crystals to settle in the fuel.

The most commonly used unit for the treat rates of additives for diesel fuels is parts per million (ppm) by weight, thus mg/kg. As a percentage, 0.1% is equal to 1000 ppm.

17.2.1. Wax Crystal Modifiers (WCM)

Most diesel fuels contain a significant proportion of paraffinic components, which include waxes. Paraffins are important because they have good ignition characteristics, but due to their limited solubility, when the fuel cools, wax crystals form that can restrict flow in the vehicle fuel system. Additives that improve the low-temperature properties of diesel fuels by changing the way the wax crystals grow are known as wax crystal modifiers. The different types of wax modifier additive will be dealt with in the chronological order of their appearance on the scene.

17.2.2. PP Depressants

The use of pour point depressants (PPD) in lubricating oils has been long established [17.59], but a different type of additive was needed for distillate fuels, and these began to make their appearance in Europe and North America during the 1950s. PPDs work by interacting with the waxes in the fuel to modify their size and shape, making them more compact and less able to form interlocking structures that would prevent the fuel from flowing. The mechanisms of wax crystallization and PPD are described in detail by Holder and Winkler [17.20].

Some early attempts to lower fuel pour points were unsuccessful because the additives used were designed as lubricating oil PPDs. These had little or no effect on the fuel because they had been developed to interact with the waxes in lubricating oils, which differ from those normally found in middle distillate fuels. Extensive screening tests were carried out to identify materials capable of modifying the growth habits of the waxes in diesel fuel.

Natural wax crystals that precipitate when an untreated fuel is cooled below its CP form as thin plates that overlap and interlock. Eventually these will create a structure that gels the fuel. When a PPD has been blended into the fuel, the resulting crystals are smaller and thicker than those formed naturally, as shown in Figure 17.10.

FIGURE 17.10 Photomicrographs of wax crystals without (above) and with (below) a PPD additive.

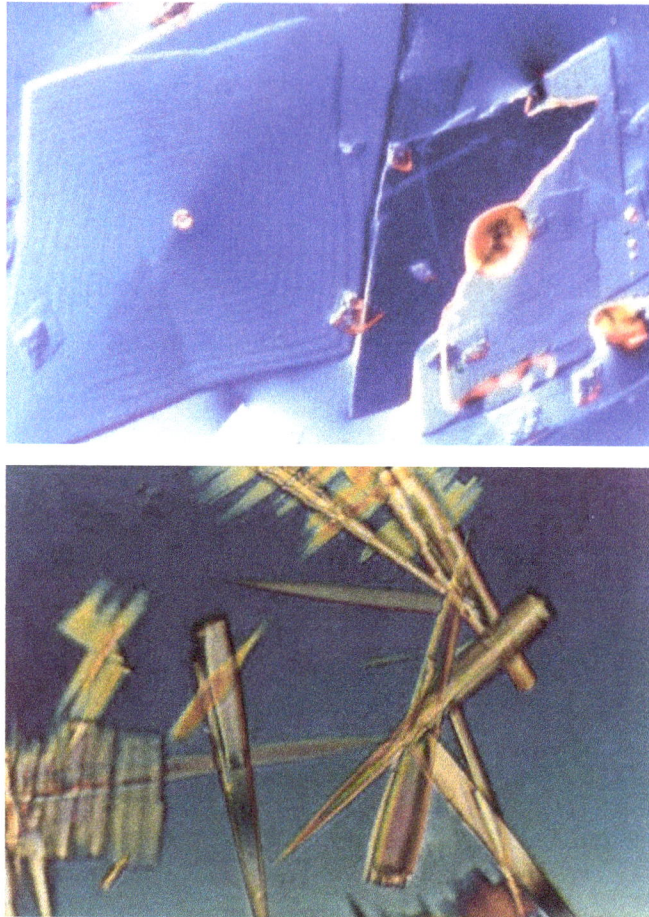

These changes in the size and shape also reduce the tendency for adjacent crystals to interlock, thereby lowering the PP of the fuel. The depressed PP will be reached after further cooling to a lower temperature, bringing more wax out of solution. Gelling or solidification of the fuel will occur when the entire additive dose has been adsorbed by the precipitated wax, allowing the formation of larger, unmodified crystals that readily interlock.

One of the first applications for a fuel PPD was for domestic heating oils in Canada, where the severe winter climate caused problems in the distribution and use of stove oil. PPDs are still used in countries where PP is the low-temperature specification. It is usually easy to modify wax crystals and achieve a substantial PPD [17.1].

Diesel vehicle road tests in North America and Europe during the 1960s had also confirmed the effectiveness of these additives in lowering the minimum temperature at which satisfactory operation was possible [17.60].

Unfortunately, the improvement in vehicle operability was smaller than that indicated by the reduction in PP, and a more satisfactory predictive test method was required. The development of such a test, the CFPP, is described in Section 17.1.5.

17.2.3. Flow Improvers

Flow improvers are the types of additives used to lower the CFPP and LTFT. They are sometimes referred to as middle distillate flow improvers (MDFI) because they can be applied to distillate heating oils as well as to diesel fuels. They function in a similar way to PPDs but to a greater extent. They interact with the waxes that separate out of the fuel as it cools, making small, three-dimensional crystals rather than the larger platelet crystals produced by an untreated fuel [17.61, 17.62]. Figure 17.11 shows the greater degree of wax modification given by the flow improver additive.

FIGURE 17.11   Photomicrographs of wax crystals without (left) and with (right) a MDFI.

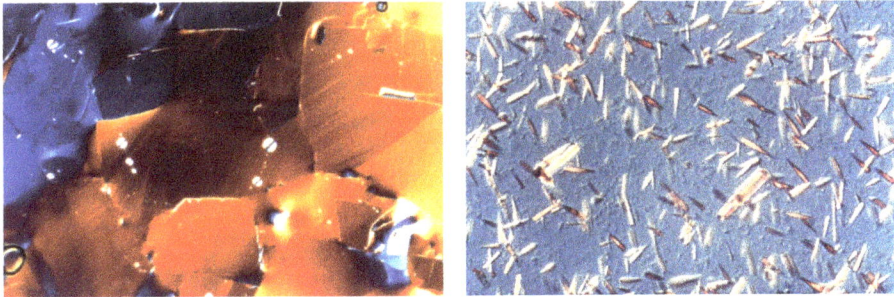

Due to the presence of the additive molecules, modified wax crystals are also less prone to attach themselves to each other and form agglomerates that could restrict the flow of liquid fuel through the lines and filters in the vehicle system. Modified wax crystals are able to pass through the fairly coarse prefilters and strainers that are fitted to hold back large items of foreign material, but they are too large to go through the main filter of paper, felt, or cloth protecting the closely machined clearances of the fuel injection equipment. Because of the shape of the modified crystals, the wax layer on the filter is permeable, allowing liquid fuel to pass through, whereas the large, unmodified platelet crystals readily interlock, making a structure that will impede flow through the waxy layer on the main filter or even the coarse, woven mesh strainers. This is illustrated schematically in Figure 17.12 [17.63].

FIGURE 17.12   Schematic of wax deposition on fuel filters.

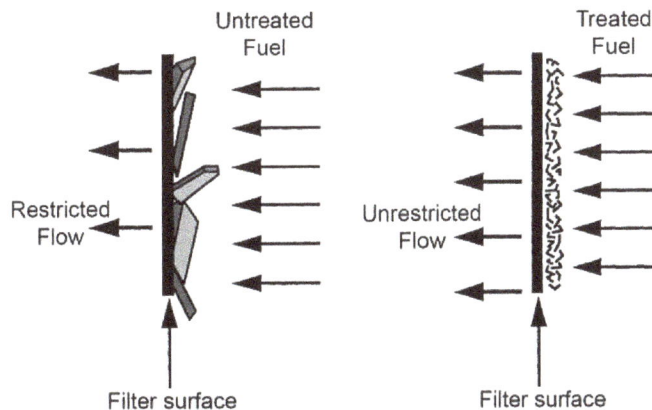

A number of different materials have been found to be effective at modifying the crystal growth and hence lowering the CFPP or LTFT. The response of a particular fuel to a particular additive is dependent upon the characteristics of the base fuel. Figure 17.13 [17.64] shows the response of two different fuels to two different flow improver additives.

Figure 17.13

FIGURE 17.13 Response of two different diesel fuels to two different flow improver additives.

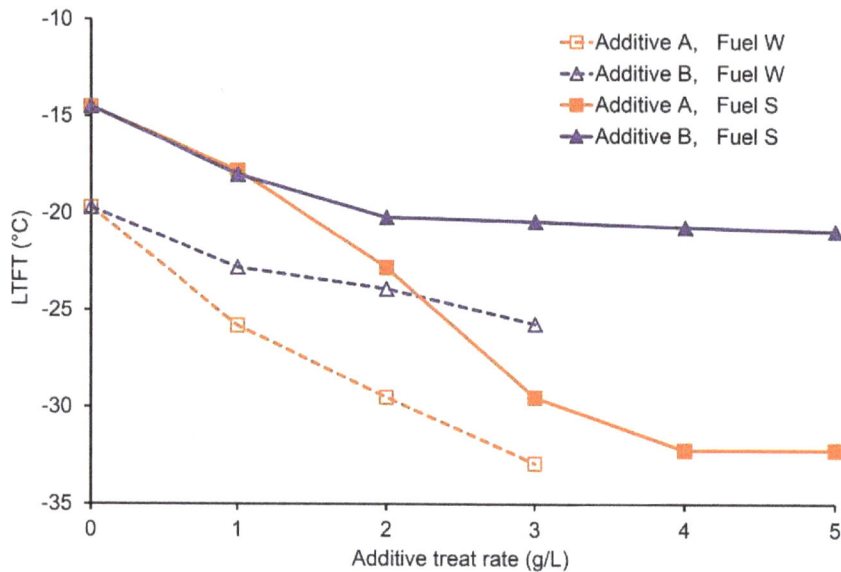

As for most additives, flow improvers are subject to the law of diminishing returns. The rate of LTFT improvement decreases as the treat rate is raised, effectively reducing the cost-effectiveness of additive use.

17.2.4. CP Depressants

Cloud point depressants (CPDs) are another type of wax-modifying additive developed for use in distillate fuels. Producing a diesel fuel with satisfactory properties for winter use is one of the major constraints on the refiner, and before flow improvers were used, winter quality was achieved by blending to a lower CP. This improved the low-temperature characteristics by eliminating some of the heavier distillate components, but it also reduced the amount of fuel produced. This is the reason why most diesel fuel specifications generally allow seasonal cold property grades.

CP is not always included in diesel fuel specifications, but when it is, as well as being a constraint on the refiner (who usually blends to CP), it is an important indicator of base fuel quality.

It was formerly considered impossible for a small amount of a chemical additive to influence the wax solubility of a diesel fuel sufficiently to give a reduction in its CP. However, detailed studies into the effect of wax modifiers have shown that some additives appear capable of suppressing wax crystallization by a few degrees [17.65].

The amount of CP depression obtainable, even with high treat rates, is relatively small, rarely exceeding 3 or 4°C, whereas flow improver additives provide CFPP improvements two or three times greater at much lower treat levels. CPD additives also lower the base CFPP by a similar amount, but they tend to be antagonistic to conventional flow improvers. Adding a CPD additive to a fuel treated containing flow improver may worsen its CFPP and give little CPD. Although some additive combinations can provide dual benefits, treat costs are high, and the use of CPDs may only be economically attractive in fuels without flow improver additives.

17.2.5. Wax Anti-Settling Additives (WASA)

Wax crystals have a higher density than the fuel from which they have crystallized; hence they will tend to settle to the bottom of a storage vessel. In nonadditive treated fuel, the wax platelets have a strong attraction for each other, and an interlocking matrix of crystals forms only 2 or 3°C below the fuel CP. This prevents the wax from settling. In additive treated fuels, the crystal-to-crystal attraction is reduced, the interlocking matrix does not form, and the crystals are free to settle under the influence of gravity.

Wax settling is not a new phenomenon. It has been observed ever since flow improvers were introduced. Layers of wax were reported in vehicle fuel tanks after overnight cooling during the very early vehicle operability trials [17.26]. The relationship of tests, such as CFPP, to vehicle operability therefore incorporates the fact of wax settlement, and wax settling in these circumstances is not a problem.

Wax settling becomes a problem if it takes place in storage tanks earlier in the fuel distribution chain. When this occurs, the quality of fuel withdrawn from the tank can be unpredictable and different from the original fuel [17.66]. Figure 17.14 [17.66] illustrates the channeling effect when fuel is drawn from a tank in which there is a settled wax layer.

FIGURE 17.14   Wax settling in storage and the channeling effect during fuel draw-off.

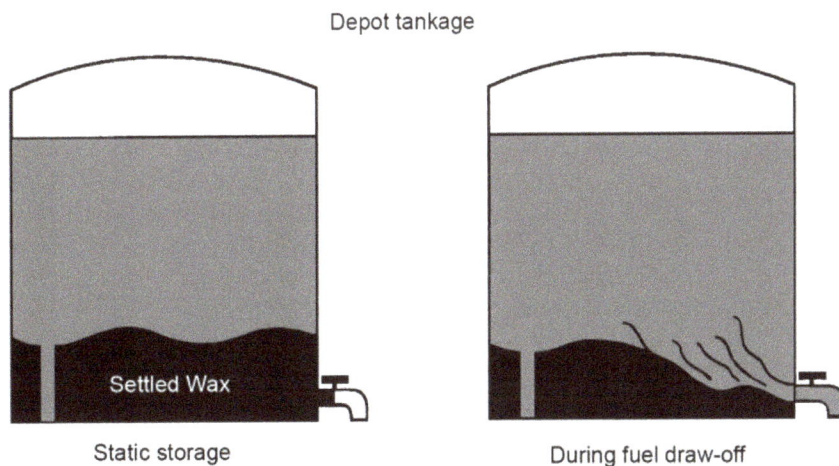

The very first portion withdrawn may contain some of the wax. However, a channel then quickly develops through the more viscous layer, which allows the clear fuel to flow out, probably drawing some of the wax with it. Eventually there is very little clear fuel left, and the final portion withdrawn from the tank will be highly wax enriched.

The risk of wax forming and settling in storage tanks depends upon

- Fuel CP (higher cloud increases the risk).
- Volume of fuel in the tank (larger volume reduces the risk).
- Ambient temperature.

As a general rule, refiners and fuel distributors place an upper limit on fuel CP, which minimizes the probability that wax settling will occur in the fuel distribution system. Nevertheless, there will be a proportion of higher risk situations, particularly where small-volume tanks are in use.

WASAs are used in combination with MDFI to reduce the tendency for wax settling during storage, thus reducing the likelihood of problems in high-risk situations. This benefit may be especially useful for heating oils, which tend to be handled in smaller volumes compared with diesel fuel, toward the end of the distribution chain. The effectiveness of a WASA can be assessed by measuring the ΔCP by what is often referred to as the short sedimentation test. In this test, the CP of a base fuel is measured. The fuel is then treated with the WASA and cooled to a temperature below the measured CP. After leaving the fuel to allow any wax to settle, the CP of the bottom 20% by volume of the sample is measured. The difference between this measurement and the value obtained for the base fuel is the ΔCP. The more effective the WASA, then the lower will be the value of ΔCP.

WASA used in combination with MDFI achieves control of wax crystals to much smaller sizes than are possible by use of MDFI alone. As well as reducing the rate of wax settling in storage, there is the additional benefit of improved diesel vehicle operability. This is discussed further in Section 17.2.6.

17.2.6. Mechanism of Wax Crystal Modification

Holder and Winkler concluded from their detailed studies on wax crystallization and the mechanism of PPD action [17.20] that polymeric additive molecules, by virtue of their chain length and structure, are able to incorporate themselves at the dislocation step on the growing crystal and stop the natural spiral growth pattern. Adsorption of the additive onto the wax crystal appeared not to hinder its precipitation from solution, but the resulting smaller and more compact wax crystals effectively lowered the PP of the treated fuel.

MDFI are wax modifier additives developed to give better cold filterability in the CFPP test rather than only depressing the PP. Like PPDs, MDFI have a polymeric active ingredient, but modification of the crystal growth is achieved through a dual action of nucleation and growth-arresting.

The additive provides polymeric nuclei as nucleation sites for the growth of the wax crystals. But as the wax crystals begin to grow, other additive molecules attach themselves to the crystal face, inhibiting further growth of that crystal. Other nucleation sites then become the target for further wax formation. This results in the formation of many smaller crystals that are less likely to restrict flow in a vehicle fuel system than a few larger crystals. The effect on wax crystal growth is illustrated by Figure 17.15 [17.62].

FIGURE 17.15   Wax crystallization in narrow and broad boiling distillates.

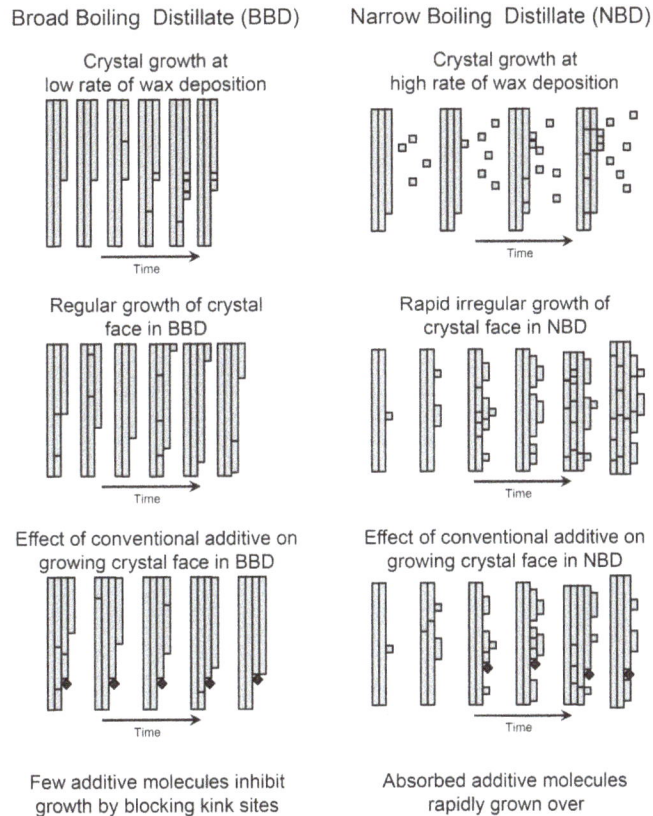

WASAs form crystal interactions in a similar way to conventional MDFI additives, but they make even smaller wax crystals, which has the effect of slowing down their rate of settling [17.67] and giving additional benefits in diesel vehicle operability. Photomicrographs of crystals in Figure 17.16 show the extent of size reduction achieved with the anti-settling additive.

FIGURE 17.16   Photomicrograph of wax crystals obtained with a WASA.

Full-scale vehicle chassis dynamometer and field tests comparing a simple MDFI with an additive giving anti-settling protection as well as CFPP improvement showed that dual-action additive provided better vehicle performance. Figure 17.17 [17.66] shows results obtained with two portions of the same −1°C CP fuel, one treated with conventional MDFI and the other with the dual-action, wax anti-settling flow improver (WAFI) additive. Each had the same CFPP value of −24°C, which was well below the usual level for commercial fuels.

FIGURE 17.17   Influence of WASA on diesel low-temperature operability.

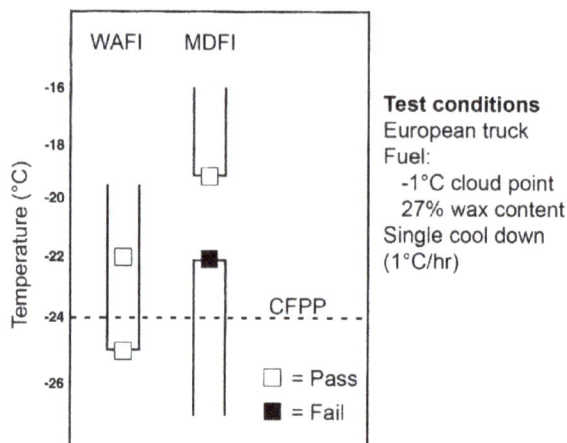

The fuel containing the MDFI reached its operability limit a few degrees above its CFPP, at about −21°C, while with the anti-settling treatment, the vehicle was still operating satisfactorily at −26°C.

17.2.7. Factors Influencing Choice of Wax Modifier Additive

There is a great variation in diesel fuel and heating oil quality across the world. This is driven by the very wide range of crude oils, different refinery processing schemes, and regional differences in product specifications. Climatic differences, which dictate local low-temperature specifications, are most influential. For example, even within Europe, winter CFPP specifications vary from maximum values of −5 to −32°C. It is not possible to effectively treat all these fuels with one single flow improver additive.

The need for different additives to give effective wax crystal modification in fuels from different sources is due to the influence of a number of factors on the distribution of wax types in the fuel. The factors include the crude being processed; the availability of blend components and the way they are fractionated; the boiling range of the final blend and the target CFPP temperature to be met. Raising the end point of a fuel will bring in some higher carbon number waxes, which may increase response to the original flow improver or require the use of a different flow improver type.

The fuel producer must also ensure that other diesel fuel properties conform to specification limits. These factors will influence not only the amount of wax to be modified but also the range and distribution of individual normal paraffins (n-alkanes) in the wax and the rate at which they come out of solution as the fuel cools below its CP. The optimum composition of the flow improver is determined by the type of wax in the fuel to be treated. The additive giving the most cost-effective treatment will therefore vary from one type of fuel to another.

Maximizing jet fuel production to meet a higher demand reduces the proportion of kerosene available as a blend component. In a simple hydroskimming refinery, with no alternative blend component to replace kerosene, the only way to keep within the CP specification is to lower the end point of the diesel fuel. A similar situation can arise in high-conversion refineries, when heavy streams from the atmospheric and vacuum distillation units are preferentially used as a feed for the cracker units to reduce fuel oil production or make more gasoline components. The result of lowering the end point is that the boiling range of the diesel fuel becomes narrower, which changes the distribution of paraffins present in the fuel. Changes in diesel fuel specifications have forced many refineries to change the distribution of blend components between diesel fuel and heating oil grades. This often results in lighter boiling diesel and heavier boiling heating oils, and both having a narrower boiling range than before. Figure 17.18 shows the n-alkane distribution in the waxes that precipitate from narrow boiling fuels.

FIGURE 17.18   Distribution of n-paraffins separating 10°C below the CP.

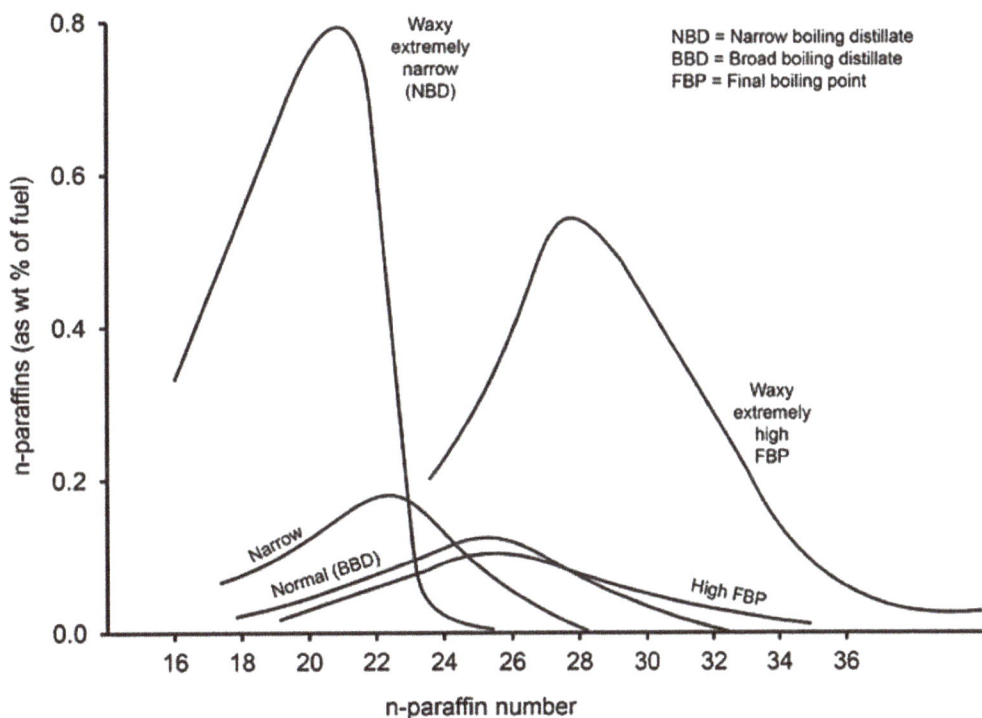

The steeper slope of the distribution pattern for the narrow boiling distillate also reflects the sharper fractionation typically used to maximize the yield from a narrower cut. Fuels having a narrow boiling range produce large, irregular-shaped wax crystals that are less easy to treat than the regular crystals formed by broader boiling fuels. The different appearance of the wax crystals from untreated broad boiling and narrow boiling distillates (BBD and NBD) is evident from the photomicrographs in Figure 17.19.

FIGURE 17.19   Photomicrographs of wax crystals formed by a BBD (above) and an NBD (below).

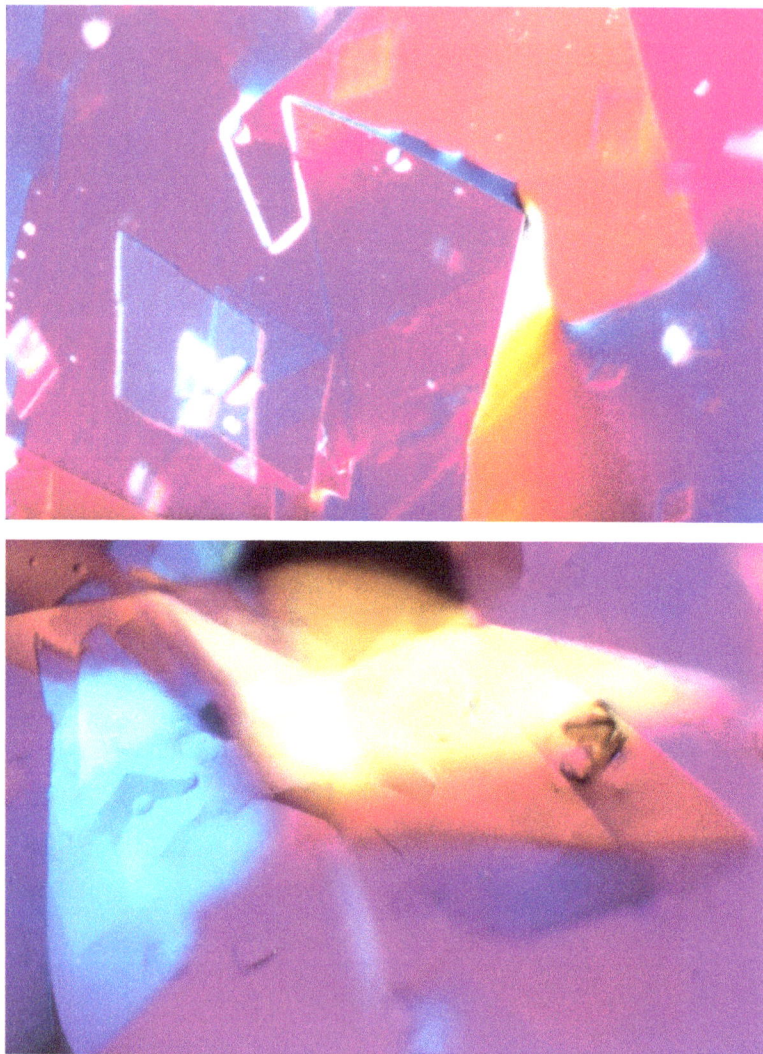

Orthorhombic crystals showing a regular growth pattern are typical of distillates having a broad boiling range and n-alkane distribution, whereas narrow cut fuels under identical cooling conditions produce large, amorphous wax crystals exhibiting no regularity in their growth [17.62].

Figure 17.20 demonstrates the growth-arresting action of the additives developed to suit the particular types of wax crystal.

FIGURE 17.20   Photomicrograph of additive-modified wax crystals from a narrow boiling distillate.

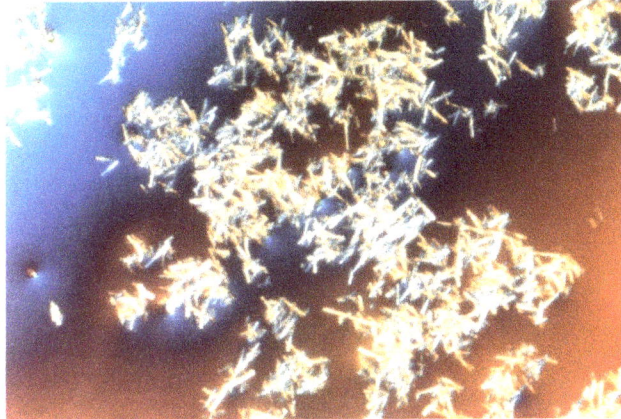

Narrow boiling distillates have a high initial rate of wax precipitation as the temperature goes below the CP. As a result, the formation of attachments on the crystal face is irregular, making it more difficult for a flow improver to arrest growth. Additives have been developed for these narrow boiling fuels, but higher treating levels are inevitable because of the large number of growth sites on the crystal and the higher quantity of wax to be treated.

Fuels produced in different parts of the world vary widely in boiling range and wax content as a result of different crude types, refinery operations, and climatic conditions. An indication of the range of variations in fuel distillation is illustrated by the chart in Figure 17.21 [17.68] and confirms the broad differences in various regions of the world.

FIGURE 17.21   Distillate types around the world.

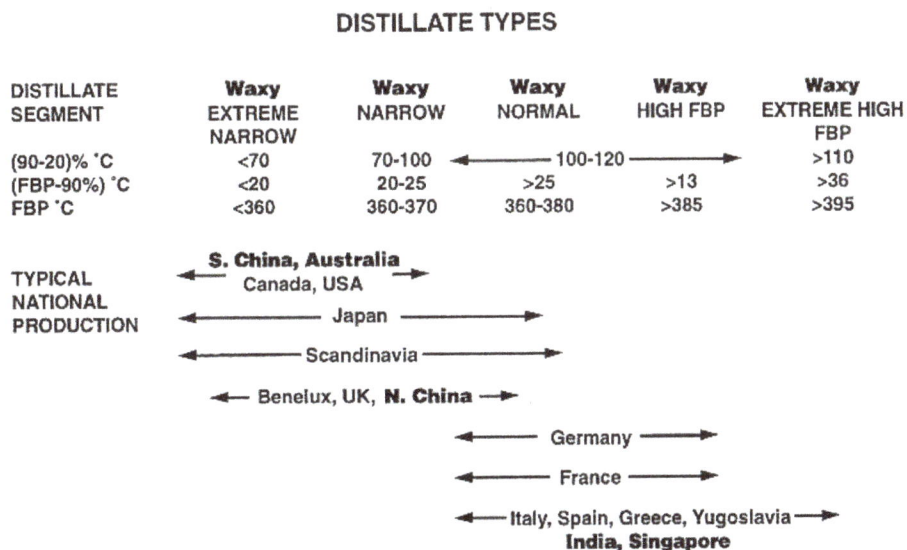

DISTILLATE TYPES

DISTILLATE SEGMENT	Waxy EXTREME NARROW	Waxy NARROW	Waxy NORMAL	Waxy HIGH FBP	Waxy EXTREME HIGH FBP
(90-20)% °C	<70	70-100	←——— 100-120 ———→		>110
(FBP-90%) °C	<20	20-25	>25	>13	>36
FBP °C	<360	360-370	360-380	>385	>395

TYPICAL NATIONAL PRODUCTION

← **S. China, Australia** →
   Canada, USA

←——————— Japan ———————→

←——— Scandinavia ———→

← Benelux, UK, **N. China** →

←——— Germany ———→

←——— France ———→

← Italy, Spain, Greece, Yugoslavia →
   India, Singapore

Fuels with higher end points tend to be associated with warmer climates requiring less stringent CFPP specifications. Countries such as Australia, India, and China process their own indigenous, waxy crudes, producing distillates with very high wax content. Many of these fuels cannot be successfully treated with flow improver or may require specially developed additives, used at high treat levels, to achieve acceptable CFPP benefits.

17.2.8. The Incorporation of Biodiesel

As discussed in Chapter 5, Section 5.4, the biodiesel feedstock has a significant effect on the distribution of different FAMEs within the resulting product. Without the added complication of any impurities that might be present, biodiesel is therefore still a complex mixture of different FAME compounds. With petroleum diesel fuel, the cold flow characteristics will be slightly different from batch to batch and with larger geographical differences due to the source of the crude oil feedstock. Similarly with biodiesel, there will be slight batch-to-batch variation and larger geographical differences reflecting the preferred feedstock, although with biodiesel the batch size is usually much smaller. It is thus difficult to predict the cold flow properties of mixtures of these different biodiesel and petroleum diesel fuels.

Figure 17.22 [17.69] shows the CFPP for four different petroleum diesel fuels (labeled as D1 to D4) when they are blended with 5 and 20% of biodiesel derived from either rapeseed oil or soy bean oil to produce RME and soy methyl ester (SME), respectively.

FIGURE 17.22   CFPP of 5 and 20% biodiesel blends using RME and SME.

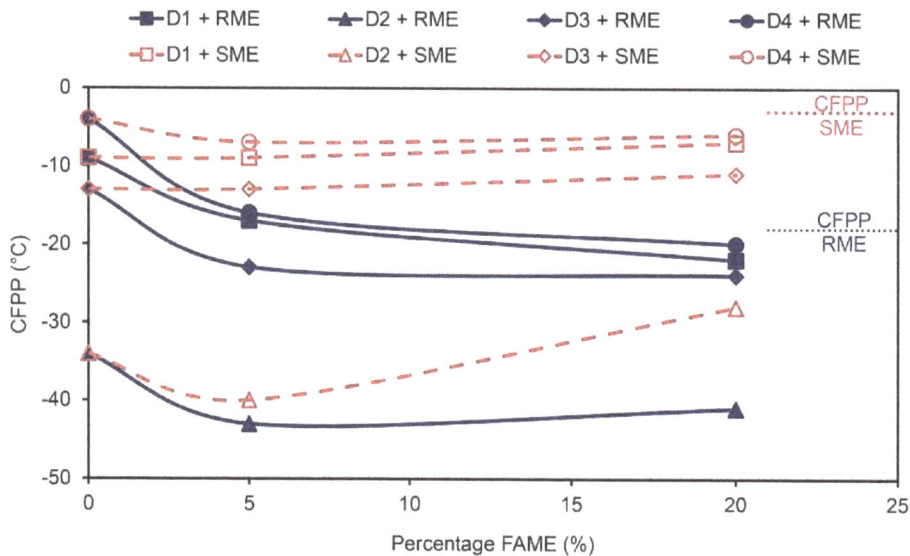

© SAE International.

The chart also shows the CFPP of the neat biodiesel components. The most noticeable point from this chart is that both the RME and the SME depress the CFPP of the D2 petroleum base diesel fuel when blended at low concentrations, despite the fact that both have a much higher CFPP when neat. At 20% addition, the RME also depresses the CFPP of all the petroleum diesel fuels. This behavior has also been observed in other studies [17.70].

The response of FAME and FAME blends to flow improvers is also difficult to predict. Different samples of FAME from the same type of feedstock behave differently to different flow improver additives. This is demonstrated in Figure 17.23 [17.69] where five different samples of RME and two samples of SME have each been treated with two different flow improver additives and show totally different responses. One sample of SME actually has a negative response!

FIGURE 17.23   Change in CFPP of different biodiesels when treated with flow improver additives.

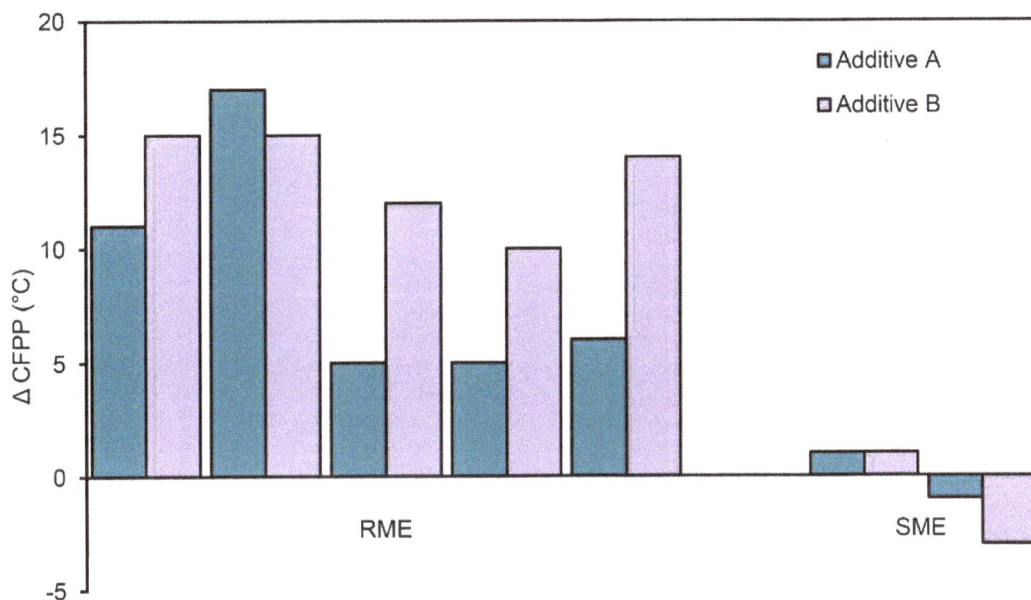

The upper part of Figure 17.24 shows the crystals formed in a sample of 100% RME that has been cooled at 0.5°C/minute until it has reached +5°C. The sample appears to be completely crystalized despite the relatively high temperature. The same RME fuel treated with an additive is shown in the lower part of the figure. In this case, the cooling at 0.5°C/minute has been continued down to −26°C. The degree of crystal growth can thus be reduced in the same way as it can with petroleum diesel.

FIGURE 17.24 Crystals formed in untreated RME (above) cooled to +5°C and treated RME (below) cooled to −26°C.

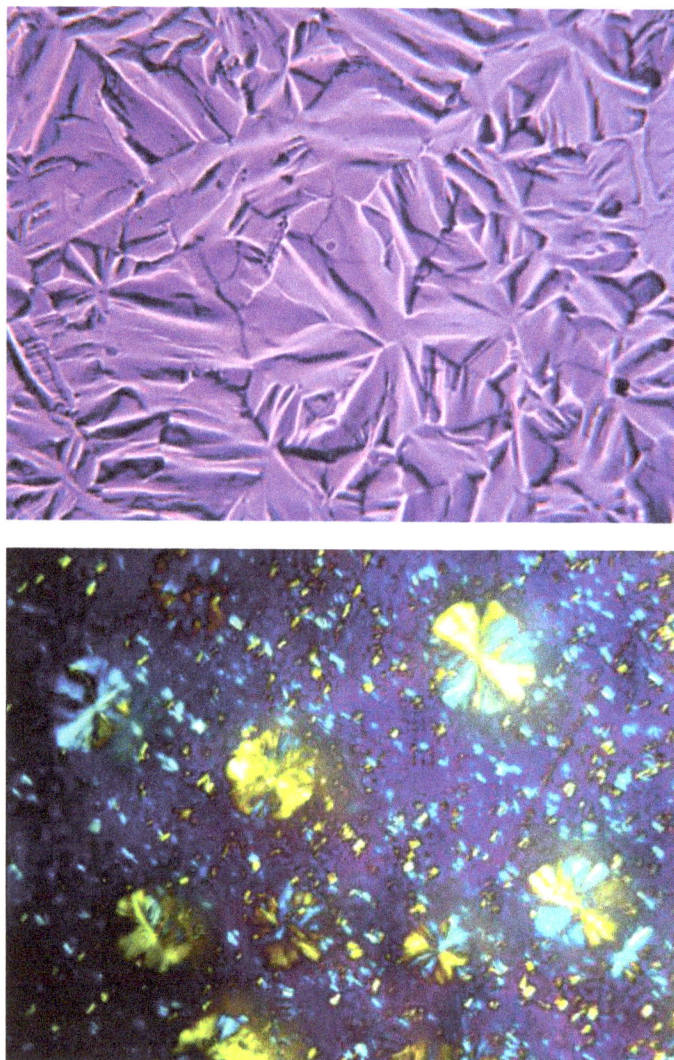

Other studies have shown that some flow improver additives are very effective in reducing the PP of biodiesel blends but have little effect on the CP [17.71]. Figure 17.25 [17.71] shows the results of using 1000 mg/kg of 12 different commercial flow improver additives in a 20% volume blend of SME in diesel fuel.

FIGURE 17.25   Improvements in PP and CP from using commercial additives in a 20% SME biodiesel blend.

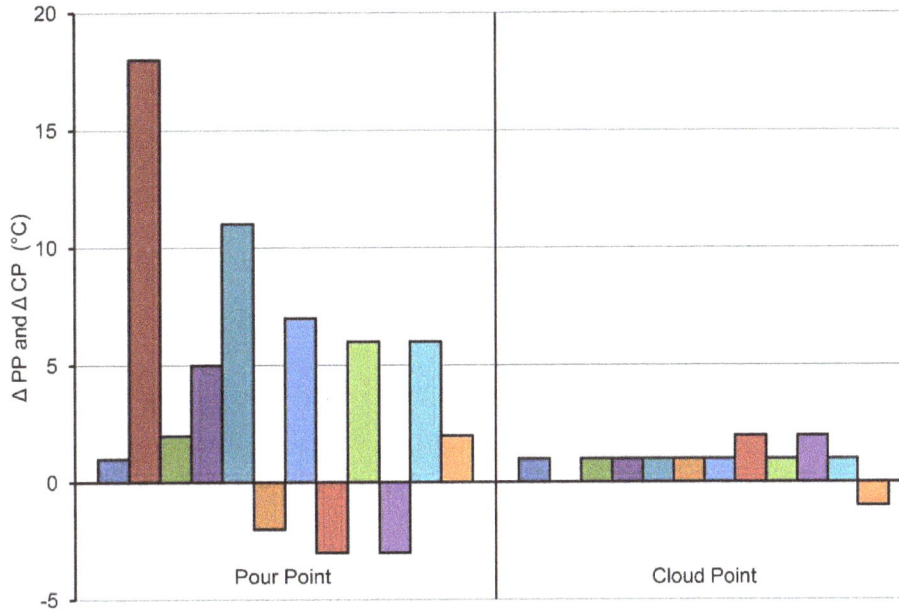

17.3. Measurement of Diesel Fuel Low-Temperature Performance

Specifications for diesel fuels define the minimum quality acceptable for low-temperature operation of diesel vehicles in terms of one or more of the rapid laboratory tests described in Section 17.1. Many variations are possible in the composition of diesel fuels and in vehicle fuel systems. Consequently, it is appropriate to consider the procedures to be followed when assessing the validity of specification methods used for defining vehicle performance in service. The justification is to ensure that tests are conducted in a manner that is realistic and acceptable to the manufacturer of the vehicle, the operator, and the fuel supplier.

It is standard practice for the motor industry to carry out extensive field testing of prototype and production models to ensure that they will perform acceptably in the different environments in which they are likely to be used. The tests are intended to evaluate the performance of the vehicle under both normal and severe climatic and road conditions. Critical attention has to be given to its various parts, which include the chassis, transmission, and braking systems; the bodywork; and lighting, heating, ventilation, soundproofing, and general comfort of the driver and passenger-carrying areas. In addition, it is necessary to assess the performance of the engine with its ancillary and auxiliary equipment in terms of cranking, fueling, ignition, speed control, and cooling. The need for such testing programs applies equally to gasoline and diesel vehicles.

The following sections cover the various aspects that need to be considered when carrying out low-temperature test programs in the field and under simulated conditions in a climate chamber.

17.3.1. Selection of Field Test Site

Siting of the field test is determined by the particular objectives of the tests and can range from tropical to arctic climates, low- and high-altitude areas, and a wide variety of road surfaces and grades. For low-temperature testing of diesel vehicles and fuels, the first requirement is a site where there is a high likelihood that temperatures within the required range will be encountered during much of the testing period.

Other important factors to be considered include the accessibility of the site for transportation of test vehicles, fuels, and all the other materials and personnel needed to carry out the work; facilities for working on the vehicles; and storage for equipment, spare parts, and fuels. Suitable routes will be required for realistic driving practices and hotel accommodations arranged for the test team. An additional consideration may be the availability of frequent and accurate forecasting of temperature conditions at the site to facilitate selection of the most appropriate quality fuel for each test run.

17.3.2. Procedure for Low-Temperature Testing

In the late 1980s, the CEC developed a Code of Practice for low-temperature performance tests on diesel fuels and vehicles. This was intended to provide a standard procedure that, if adopted by vehicle testing laboratories, would facilitate direct comparison of results obtained by different operators. The Code of Practice was never taken beyond that stage and is no longer supported by the CEC. However, the following briefly outlines some of the recommendations of that Code of Practice, which covered cold startability and cold operability testing of vehicles in the field and in cold climate chambers.

17.3.2.1. Cold Startability

The purpose of cold startability testing is to check the ease with which the engine will start on the test fuel under the prevailing temperature conditions. Factors influencing the result include battery performance, cranking speed, and fueling rate as well as the ignition quality and fluidity of the fuel.

Cold startability at a particular temperature is evaluated on the basis of the total time of starter motor cranking to obtain engine autorotation, the time at which engine speed starts to rise, the number of attempts to start, the number of stalls, and the idling speed.

17.3.2.2. Cold Operability

Cold operability tests are made to ascertain whether the engine will not only start on the test fuel but will also continue to run and perform satisfactorily while the vehicle is being driven under road load conditions. Once the engine has started, the outcome will be determined largely by two factors: the rate at which fuel can be delivered to the engine and the time taken for the fuel system to warm up and melt any wax that might accumulate and starve the engine of fuel. Serious restriction of the flow will reduce the amount of excess fuel recycled back to the tank and slow down the fuel system warm-up.

The objective of some diesel vehicle field tests is to ascertain whether satisfactory performance will be obtained from a particular fuel and vehicle at a particular temperature level. In this case, the likely outcome would be a simple pass or fail result. A more complex objective is to determine the lowest temperature at which satisfactory performance can be achieved. This temperature is often referred to as the operability limit, and several tests in the critical temperature range are usually needed for an acceptably precise value to be defined.

17.3.3. Fuel Storage

All diesel fuels contain a significant portion of waxy material that will separate out of solution when the fuel temperature falls below its CP. As field test fuels tend to be delivered and stored in barrels at the test site, care must be taken when refueling to avoid the use of a waxy, nonhomogeneous product. Such a fuel would not be representative of normal situations, where vehicles are refueled from large underground or over-ground tanks, in which temperatures are unlikely to go below the fuel CP. To minimize the risk of contamination with water or other foreign material, the fuel should be stored in closed containers.

Large quantities of drummed fuel may be left in the open if there is no suitable building, but provision must be made for warming the drums before use to ensure that a homogeneous, wax-free fuel is available for the test vehicles. This may require a heated room in which a few drums of fuel can be stored above the CP temperature before they are used. Alternatively, steam coils or electric heaters can be used to warm up individual drums. If the fuel has been stored below its CP for several days, the waxes may have settled, and sufficient time must be allowed for all the wax to be re-dissolved. It is advisable to mix the contents thoroughly, by rolling, shaking, or stirring the drums, before sampling to check that all the wax is back in solution.

17.3.4. Vehicle Instrumentation

Testing to determine the low-temperature operability limit of diesel fuels and vehicles requires instrumentation to monitor air, fuel system, and engine temperatures, pressure and vacuum in the fuel system, and vehicle speed. Figure 17.26 [17.62] indicates the recommended locations for temperature and pressure measurements to be made in the engine and fuel system.

FIGURE 17.26   Fuel system layout showing locations for temperature and pressure measurements.

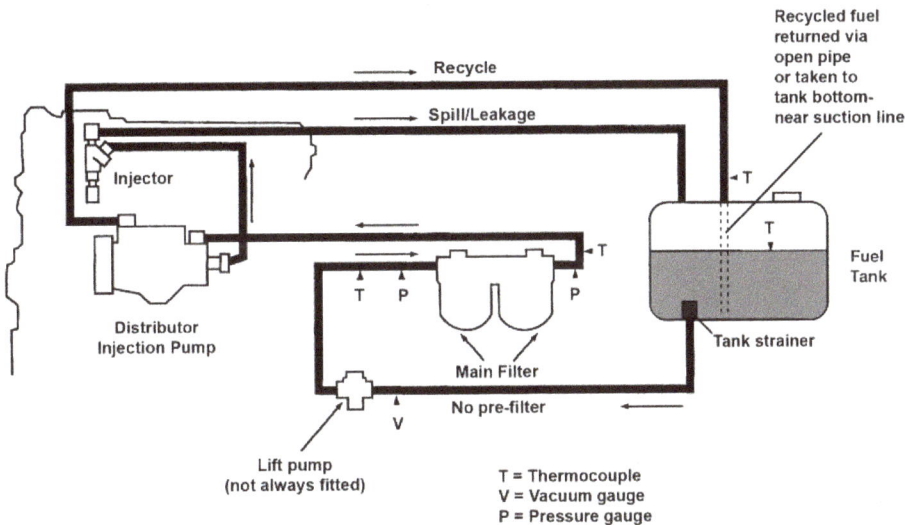

If it is not possible to use a speed-recording instrument, the vehicle speedometer can be used after calibration. Measurement of battery voltage and electrolyte temperature can also provide useful information during cold starting tests.

17.3.5. Preparation of Test Vehicles

Before starting test work, vehicle checks should be carried out to ensure that the correct grades of lubricant are used in the engine, gearbox transmission, and auxiliaries and that the coolant system has sufficient antifreeze for the expected minimum temperature. Drive belt tensions should be adjusted, if necessary, and a check made on the coolant thermostat and fan switch operating temperatures. Checks should also be made on the engine itself, electrical components, starting aids, and fuel injection equipment, making corrective adjustments as necessary to fuel leaks, valve clearances, and injector opening pressure. Safety considerations make it advisable to ensure that tires and inflation pressures are appropriate for the driving conditions. When testing for fuel-limited cold operability, normal practice is to fit a new fuel filter for each test. Also, to avoid starting difficulties due to inadequate cranking speed rather than restricted fuel flow, a fully charged battery is recommended for each test.

17.3.6. Operational Procedure

Once the preparatory checks have been completed, the operational procedure is typically based on a 24-hour cycle of events, starting with the selection of the fuel to be tested. If there is a choice, a fuel suitable for the expected overnight temperature should be used. A good starting point is to take a fuel having a CFPP about 3°C above the expected minimum temperature. After putting in the test fuel, the engine should be run for about 15 minutes to purge the fuel system completely and to verify that there are no leaks or operating faults. The vehicle can then be parked in the open to cool under the prevailing ambient conditions. The battery should be stored in a warm place and, if possible, left on charge to ensure that any failure to start is not due to poor battery condition. The recorder should be left running to monitor ambient and fuel system temperatures. The driving test to see if the engine will perform satisfactorily on waxy fuel is made after the vehicle has been standing overnight, ideally starting early, before ambient and fuel system temperatures start to rise. The overnight minimum temperature will determine the amount of wax separating in the lines and filters between the tank and the injection pump, which are usually the coldest parts of the fuel system.

After refitting the battery, the cold start procedure is carried out as recommended by the vehicle manufacturer. Only a short period of fast idle is recommended before starting the road test, driving at speeds appropriate to road conditions and traffic regulations. For vehicles with air brakes, longer idling will be needed to recharge the system.

Test durations of 30 minutes are generally sufficient to determine whether or not the vehicle will perform satisfactorily. Engine stalling due to wax restriction of the fuel flow generally occurs within the first 15 minutes of driving, before the engine and fuel system have warmed sufficiently to disperse the wax accumulation. Recording fuel system temperature and pressure variations and noting the time of hesitations or power reduction during the test run will provide useful information on the location and severity of any wax accumulation in the fuel system.

When there is sufficient time available and weather conditions permit, several tests should be made in the critical temperature range to provide more data for evaluation of results. Depending on how the vehicle performs on the first test, additional results at lower or higher temperatures may be needed to help define the operability limit of the particular fuel and vehicle combination. If the first test is satisfactory, tests should be made at lower temperatures until performance is affected by wax in the system. Tests at higher temperatures will be needed if wax plugging causes an engine stall on the first test.

17.3.7. Climate Chamber Testing

Cold climate chambers equipped with chassis dynamometers are used to simulate field conditions in a controlled environment. This method of testing overcomes the major problem of unsuitable weather conditions that often afflicts field programs, and it can be used at any time of the year. Also, as the test vehicle is stationary, there are fewer constraints on instrumentation and data gathering. The main limitation is that it is not usually possible to test more than one vehicle at a time.

The vehicle is located with its driving wheels on large rollers coupled to the dynamometer, which allows it to be driven under normal road load conditions. A large fan circulates the air over refrigerator coils to control the room temperature and to provide a cooling air flow at the equivalent road speed of the vehicle. A typical cold climate chassis dynamometer arrangement is shown in Figure 17.27 [17.72].

FIGURE 17.27   Layout of cold climate chassis dynamometer.

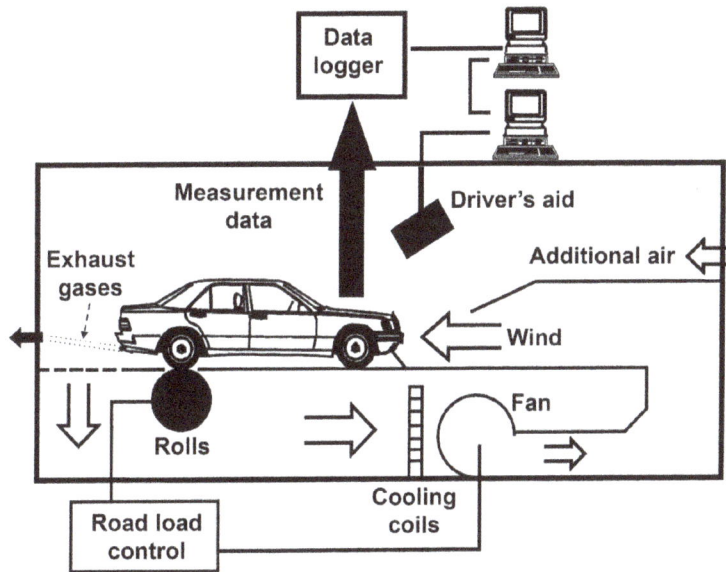

Preparation of the vehicle for cold climate chassis dynamometer (CCCD) testing is the same as prescribed for field testing. The starting procedure is also the same, but in the absence of road traffic constraints, a well-defined driving procedure can be followed. This will minimize the test-to-test variability that can arise and possibly affect road test results. The suggested driving pattern for passenger cars is shown in Figure 17.28 [17.72].

FIGURE 17.28   Drive cycles for passenger car low-temperature operability.

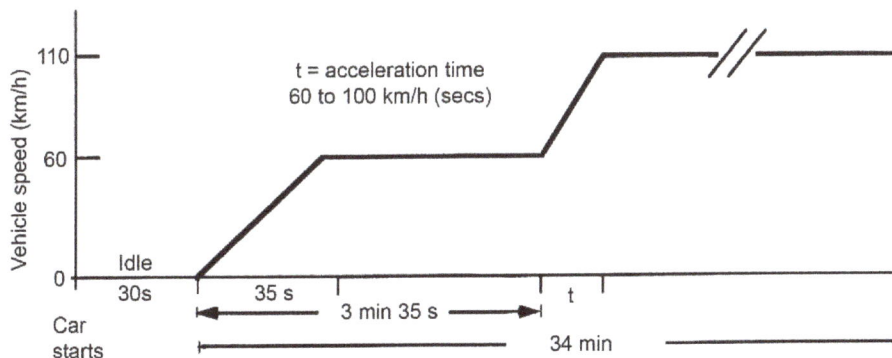

When testing in a climate chamber, the temperature programmer is set to give a controlled rate of cooling to simulate realistic field conditions. This is typically in the range of 1 to 2°C/hour during the overnight period, when vehicles are normally parked. Also the programmer can be set so that the temperature selected to suit the fuel being evaluated is reached at a convenient time in the working day. Taking an approach similar to that for field testing, a realistic starting point would be at a temperature 3°C below the CFPP, if the fuel contains a flow improver, or 3°C below the CP if it is untreated.

Depending on the first result, subsequent tests will be at lower or higher temperatures. Chassis dynamometer testing is expensive, so as only one test per vehicle per day is normally practicable, it is advisable to minimize the total number of tests needed to evaluate one fuel/vehicle pair. Steps of 4°C will usually give an early indication of the operability limit zone, and then 2°C steps will allow fine-tuning. Between each test it is advisable to move the vehicle to a warmer environment for complete draining of the fuel system and main filter replacement before adding a fresh batch of test fuel. Otherwise, the chamber has to be warmed to ambient temperature, or well above the fuel CP, and then cooled down again.

17.3.8. Interpretation of Results

During cold operability tests, partial or complete power loss may sometimes occur before the engine has warmed sufficiently to re-dissolve any wax accumulating in the critical part of the fuel system. For the researcher, whether working for the motor industry, a petroleum company, or for an additive supplier, such results are essential to help identify the lowest temperature at which the material under test will operate.

For realistic comparisons to be made of results derived from different test programs, a standardized procedure for interpretation of test data is required. This illustrates the meaning of some of the terminology that has been used to describe the low-temperature performance of diesel fuels.

Satisfactory operability, often expressed as a pass result, is reported when the vehicle performs as it would on clear, wax-free fuel, with no hesitations or reduction of power.

Unacceptable operability may be reported as a fail when an accumulation of wax in the fuel system restricts the flow of fuel to the engine, so that it is not possible to start the engine or the engine stalls during driving and fails to complete the test run.

Operability limit is a term that has been used to describe the lowest temperature at which satisfactory performance can be obtained. When testing at temperatures near the operability limit, there may be a partial loss of engine power as the system is warming up, after which the vehicle is able to perform normally and complete the test run. This is sometimes called a borderline result and has been used to define the operability limit. When no borderline result is obtained, and the lowest pass and highest fail are within about 2 or 3°C, the operability limit is sometimes taken as the midway temperature.

To accommodate borderline situations when the result is neither a clear pass nor a clear fail, the obsolete CEC Code of Practice proposed a system of demerit rating. A numerical value is assigned to all misfires, surges, pedal adjustments (to maintain speed), and driving stalls to provide a quantitative assessment of how the vehicle performed. As there are different opinions on what constitutes acceptable performance, no recommendation was given to indicate a limiting demerit. The choice is left to individual organizations to select a level to suit the particular criteria involved.

17.3.9. Low-Temperature Test Experience

Results obtained by companies and organizations that had carried out low-temperature operability tests were collated and incorporated in a CEC report [17.73]. In the commercial trials of buses and trucks in normal service, reported complaints were associated with periods of operation at temperatures below the CFPP of the fuel under test. No fuel-related problems were experienced when the temperature was above the CFPP of the fuel. Field trails and climatic chamber tests showed a consistent pattern when they were conducted at different times and places on various models of diesel equipment, including trucks, agricultural tractors, and passenger cars. Most of the borderline and fail results occurred when the fuel temperature was at or below its CFPP. Relatively few operating problems were experienced at temperatures higher than the CFPP [17.74]. Similar correlation data were obtained from tests on two Japanese diesel vehicles, a light truck and a passenger car tested on a chassis dynamometer [17.75].

In Canada, a mixed fleet of diesel trucks and passenger cars was field tested in a CRC cooperative program to investigate their low-temperature operability on different fuels and flow improver additives [17.39]. Significant differences in fuel system design were evident among the test fleet. Minimum operating temperatures derived from the test results indicated that, overall, the LTFT had provided the best available correlation for flow-improved fuels. The CFPP was only partially successful, being unable to give satisfactory predictions of the operability limits of some of the fuel/vehicle combinations.

17.3.10. Reducing Sensitivity to Waxing Problems by Vehicle Design

A number of the test results collated in the CEC report indicate that satisfactory performance was obtained at temperatures well below the CFPP. It is probable that these may have been due to some vehicles being less sensitive to the presence of fuel wax. Differences in cold performance between vehicles running on the same fuel have been a feature of many field tests and have led to investigations into the causes of sensitivity to wax. Understandably, there have also been studies into ways of improving vehicle operability at low temperatures.

Wax in the vehicle tank can be drawn into the fuel lines and restrict fuel flow if it accumulates on fine filters or in sharp bends in the lines, particularly if they are in exposed locations. Water in the fuel system is also a potential hazard at freezing temperatures. Care is needed to prevent water from getting into the tank, and the layout of fuel pipes should avoid U traps where free water could freeze while the vehicle is parked.

A vehicle fuel system designed to minimize low-temperature problems is illustrated in Figure 17.29 [17.62], which shows a typical European heavy-duty diesel layout with a separate injection pump mounted on the side of the engine.

FIGURE 17.29   Well-designed diesel fuel system to minimize risk of low-temperature problems.

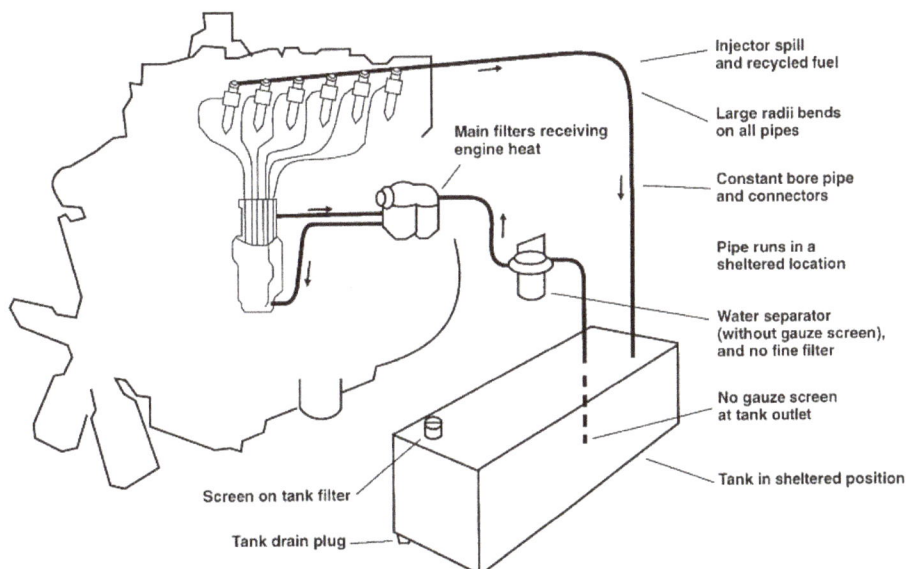

Injector spill
and recycled fuel

Large radii bends
on all pipes

Constant bore pipe
and connectors

Pipe runs in a
sheltered location

Water separator
(without gauze screen),
and no fine filter

No gauze screen
at tank outlet

Tank in sheltered position

Main filters receiving
engine heat

Screen on tank filter

Tank drain plug

© SAE International.

The design is equally applicable to engines fitted with unit injectors, because waxing problems usually occur in the low-pressure system bringing fuel from the vehicle tank to the injection pump.

Water separators should not contain any filter screen, and makers recommend that they are installed close to the tank. Constant bore pipes and connectors with large radius bends fitted in a sheltered location will minimize the likelihood of wax plugs building up. Main filters should be mounted where they can receive engine heat and be of adequate size for the maximum fueling rate. Series filters give better wear protection to the injection pump but offer more resistance to fuel flow than a parallel arrangement.

The fuel recycle line is shown close to the engine supply line, and in some models, it goes down to the bottom of the tank. If the recycled fuel is likely to become very hot, a more remote return position may be necessary to avoid vapor lock in the injection pump, but this will slow the rate of fuel system warming.

In addition to "rationalizing" the standard fuel system, new devices have been developed that are suitable for fitting to vehicles in service as well as being available either as original equipment or as an option for the purchaser of a new vehicle. These include heaters for the various parts of the fuel system (such as the tank, lines, or main filters) and bypass valves to divert some or all of the recycled fuel back to the injection pump during warm-up, rather than into the tank. Electric filter heaters may be built into modified replacement filter housings or supplied as a separate unit to be installed close to the main filter inlet of existing vehicle systems. Many new vehicles, especially passenger cars, are now fitted with a fuel heating device.

17.3.11. Experience with Modified Fuel Systems

An assessment of fuel system requirements for satisfactory low-temperature operability identified the use of a tank screen and/or prefilter, the size of the main filter, and its location as major factors [17.76]. The survey of 24 diesel passenger car models from manufacturers in the US, Japan, and several European countries revealed a range of differences. Half of the models had no tank or prefilter screen, whereas main filters were located in various positions around the engine compartment, some close enough to benefit readily from engine heat during the critical warming-up period.

Tank screens and/or prefilters have been found to be prone to blocking in cold weather, particularly when the fuel has not been treated with flow improver additives to give smaller wax crystals, so locating the prefilter where it will pick up engine heat can help. Although relocation to a warmer position is not possible for the tank screen, directing the recycled fuel toward it can be beneficial in lowering the rate of wax buildup.

The largest variable was main filter surface area and its influence on the specific fuel flow through the filter, expressed as the rate of flow per unit area. Flows were measured at three engine speeds and averaged. Filter surface areas ranged from 900 to 6400 cm^2, and specific flow rates varied widely between 0.05 and 2.29 $mL/s/cm^2$.

The main filter tends to be the most sensitive part of the fuel system when flow-improved fuel is used. Most of the modified wax crystals will pass through the mesh screens unless the temperature is well below the CFPP level. Positioning the filter to receive engine heat soon after the engine starts will contribute to melting the wax, but the size and flow rate through the filter are more critical features. When waxy fuel is being drawn from the tank, the filter size influences the amount of fuel that can be filtered before it plugs. A larger surface area will lower the specific flow through the filter and extend the time before it becomes plugged with wax, thereby allowing more time for the fuel system to warm up and melt the wax on the filter.

Chassis dynamometer tests on some of the passenger cars from the survey showed improvements in low-temperature operability as a result of fuel system modifications on the lines discussed [17.68]. Several of the vehicles were able to operate to temperatures well below the CFPP with their standard fuel system. Fitting a larger size filter gave about the same benefit as an electric filter heater, lowering the operability limit of the more critical vehicle models by 8 to 12°C. Similar levels of improvement were observed in studies on heavy-duty diesel vehicles [17.77].

Electric filter heaters must be used with care to avoid draining the vehicle battery. A recommended practice is to switch on the heater only when the engine has been started. An alternative source of heat for the filter is the engine coolant, which warms up rapidly after engine start-up. Some passenger cars have a water-heated filter connected to the engine coolant system. A thermostatic valve in the coolant line prevents the fuel from overheating.

Where high fuel recirculation rates are needed to avoid overheating in the fuel manifold, the fuel is normally returned to the tank at a point well away from the off-take. This is necessary at normal engine operating temperatures, but getting the engine running properly can be difficult if there is waxy fuel in the tank. The use of a thermostatically controlled bypass between the recycle and feed lines can overcome this problem. The bypass diverts some of the recycled fuel to the main filter inlet, reducing the amount of cold fuel drawn from the tank and allowing the heat in the recycled fuel to melt the wax on the filter. The temperature of fuel coming from the tank controls the opening and closing of the thermostat.

Despite the advances in vehicle design and the use of fuel additives, there are still occasional incidences of bouts of field problems due to poor cold flow performance. As a result, there is an ongoing need for exercises to assess the correlation between the cold flow properties discussed and the in-service operability of diesel vehicles.

References

17.1. Deen, H.E., Swanson, E.S., and Peltola, W.M., "Designing Diesel Fuels and Fuel Systems for Low Temperatures," SAE Technical Paper 640353 (1964), doi:https://doi.org/10.4271/640353.

17.2. Mickel, B.L. and Fergesen, L.D., "Dimensions of Diesel Fuel Performance: Design, Depressants, and Response," SAE Technical Paper 660371 (1966), doi:https://doi.org/10.4271/660371.

17.3. Miller, J.A., Porter, H.R., and Lewis, J.D., "Will It Run at 70 Below? A Progress Report on Arctic Winter Operation of Automotive Equipment," SAE Technical Paper 710717 (1971), doi:https://doi.org/10.4271/710717.

17.4. McMillan, M.L. and Reddy, S.R., "Fuel and Fuel System Effects on Low Temperature Operation of Diesel Vehicles," SAE Technical Paper 811180 (1981), doi:https://doi.org/10.4271/811180.

17.5. Berg, P.G., "Cold Weather Diesel Fuel Preparation with PTC Heaters," SAE Technical Paper 840539 (1984), doi:https://doi.org/10.4271/840539.

17.6. Lundberg, M., "Development of Cold Operability of Diesel Fuels and Vehicles," SAE Technical Paper 845012 (1984), doi:https://doi.org/10.4271/845012.

17.7. Polyakov, Y.T. and Valeev, D.K., "Diesel Fuel Heaters for Arctic Vehicles," SAE Technical Paper 920035 (1992), doi:https://doi.org/10.4271/920035.

17.8. Mikkonen, S., Kiiski, U., Saikkonen, P., and Sorvari, J., "Diesel Vehicle Cold Operability: Design of Fuel System Essential Besides Fuel Properties," *SAE Int. J. Fuels Lubr.* 5, no. 3 (2012): 977-989, doi:https://doi.org/10.4271/2012-01-1592.

17.9. Knothe, G. and Dunn, R.O., "A Comprehensive Evaluation of the Melting Points of Fatty Acids and Esters Determined by Differential Scanning Calorimetry," *Journal of the American Oil Chemists' Society* 86, no. 9 (2009): 843-856.

17.10. Alleman, T.L., McCormick, R.L., Christensen, E.D., Fioroni, G. et al., *Biodiesel Handling and Use Guide*, 5th ed. (Washington, DC: US Department of Energy, 2016).

17.11. ASTM International, "Standard Test Method for Cloud Point of Petroleum Products," ASTM D2500-23, ASTM International, 2023.

17.12. ASTM International, "Standard Test Method for Cloud Point of Petroleum Products (Linear Cooling Rate Method)," ASTM D5772-21, ASTM International, 2021.

17.13. ASTM International, "Standard Test Method for Cloud Point of Petroleum Products (Constant Cooling Rate Method)," ASTM D5773-21, ASTM International, 2021.

17.14. ASTM International, "Standard Test Method for Cloud Point of Petroleum Products (Optical Detection Stepped Cooling Method)," ASTM D5771-21, ASTM International, 2021.

17.15. ASTM International, "Standard Test Method for Cloud Point of Petroleum Products (Miniaturized Optical Method)," ASTM D7397-21, ASTM International, 2021.

17.16. ASTM International, "Standard Test Method for Cloud Point of Petroleum Products (Small Test Jar Method)," ASTM D7683-21, ASTM International, 2021.

17.17. ASTM International, "Standard Test Method for Cloud Point of Petroleum Products (Mini Method)," ASTM D7689-21, ASTM International, 2021.

17.18. ASTM International, "Standard Test Method for Wax Appearance Point of Distillate Fuels," ASTM D3117-03, ASTM International, 2003 (Withdrawn 2010).

17.19. ASTM International, "Standard Test Method for Pour Point of Petroleum Products," ASTM D97-17b(2022), ASTM International, 2022.

17.20. Holder, G.A. and Winkler, J., "Wax Crystallisation from Distillate Fuels," *Journal of the Institute of Petroleum* 51, no. 499 (1965): 228-234.

17.21. ASTM International, "Standard Test Method for Pour Point of Petroleum Products (Robotic Tilt Method)," ASTM D6892-03(2020), ASTM International, 2020.

17.22. ASTM International, "Standard Test Method for Pour Point of Petroleum Products (Automatic Tilt Method)," ASTM D5950-14(2020), ASTM International, 2020.

17.23. ASTM International, "Standard Test Method for Pour Point of Petroleum Products (Automatic Air Pressure Method)," ASTM D6749-02(2018), ASTM International, 2018.

17.24. ASTM International, "Standard Test Method for Pour Point of Petroleum Products (Automatic Pressure Pulsing Method)," ASTM D5949-16(2022), ASTM International, 2022.

17.25. ASTM International, "Standard Test Method for Pour Point of Petroleum Products (Rotational Method)," ASTM D5985-02(2020), ASTM International, 2020.

17.26. Coley, T., Rutishauser, L.F., and Ashton, H.M., "New Laboratory Test for Predicting Low-Temperature Operability of Diesel Fuels," *Journal of the Institute of Petroleum* 52, no. 510 (1966): 173-189.

17.27. Steere, D.E. and Marino, J.P., "Low Temperature Field Performance of Flow Improved Diesel Fuels," SAE Technical Paper 810024 (1981), doi:https://doi.org/10.4271/810024.

17.28. CRC, "Evaluation of Low Temperature Operability Performance of Light-Duty Diesel Vehicles for North America," CRC Report No. 649, 2007.

17.29. ASTM International, "Standard Test Method for Cold Filter Plugging Point of Diesel and Heating Fuels," ASTM D6371-17a, ASTM International, 2017.

17.30. CEC, "The Cold Filter Plugging Point of Distillate Fuels," A European Test Method, CEC Report No. P-O1-74, 1974.

17.31. Tupa, R.C. and Dorer, C.J., "Gasoline and Diesel Fuel Additives for Performance/Distribution Quality—II," SAE Technical Paper 861179 (1986), doi:https://doi.org/10.4271/861179.

17.32. Dale, D.L.E. and Williams, D., "The Influence of Fuel Additives on the Cold Climate Operation of Diesel Engines," SAE Technical Paper 920036 (1992), doi:https://doi.org/10.4271/920036.

17.33. ASTM International, "Standard Test Method for Filterability of Diesel Fuels by Low-Temperature Flow Test (LTFT)," ASTM D4539-22, ASTM International, 2022.

17.34. CRC, "1981 CRC Diesel Fuel Low-Temperature Operability Field Test," CRC Report No. 528, 9/83, 1981.

17.35. Tharby, R.D., "Experiences with Diesel Fuel Containing Cold Flow Improver Additives," SAE Technical Paper 831753 (1983), doi:https://doi.org/10.4271/831753.

17.36. Chandler, J.E., "Comparison of All Weather Chassis Dynamometer Low Temperature Operability Limits for Heavy and Light Duty Trucks with Standard Laboratory Test Methods," SAE Technical Paper 962197 (1996), doi:https://doi.org/10.4271/962197.

17.37. Mitchell, K. and Chandler, J., "The Use of Flow Improved Diesel Fuel at Extremely Low Temperatures," SAE Technical Paper 982576 (1998), doi:https://doi.org/10.4271/982576.

17.38. Chandler, J.E. and Zechman, J.A., "Low Temperature Operability Limits of Late Model Heavy Duty Diesel Trucks and the Effect Operability Additives and Changes to the Fuel Delivery System Have on Low Temperature Performance," SAE Technical Paper 2000-01-2883 (2000), doi:https://doi.org/10.4271/2000-01-2883.

17.39. McMillan, M.L. and Barry, E.G., "Fuel and Vehicle Effects on Low-Temperature Operation of Diesel Vehicles—The 1981 CRC Field Test," SAE Technical Paper 830594 (1983), doi:https://doi.org/10.4271/830594.

17.40. Reddy, S.R. and McMillan, M.L., "Understanding the Effectiveness of Diesel Fuel Flow Improvers," SAE Technical Paper 811181 (1981), doi:https://doi.org/10.4271/811181.

17.41. David, P., Brown, G.I., and Lehmann, E.W., "SFPP—A New Laboratory Test for Assessment of Low Temperature Operability of Modern Diesels Fuels," in *CEC 4th International Symposium on the Performance Evaluation of Automotive Fuels and Lubricants*, Paper No. CEC/93/EF15, Birmingham, UK, 1993.

17.42. Brown, G.I. and Chandler, J.E., "Evaluation of Faster LTFT and SFPP for Protection of Low Temperature Operability in North American Heavy Duty Diesel Trucks," SAE Technical Paper 932769 (1993), doi:https://doi.org/10.4271/932769.

17.43. Nylund, N.-O. and Aakko, P., "Characterization of New Fuel Qualities," SAE Technical Paper 2000-01-2009 (2000), doi:https://doi.org/10.4271/2000-01-2009.

17.44. Coultas, D.R., "The Effect of FAME Additive on Diesel Vehicle Operability and Harms," SAE Technical Paper 2009-01-2801 (2009), doi:https://doi.org/10.4271/2009-01-2801.

17.45. Poirier, M., Lai, P.K., and Lawlor, L., "The Effect of Biodiesel Fuels on the Low Temperature Operability of North American Heavy Duty Diesel Trucks," SAE Technical Paper 2008-01-2380 (2008), doi:https://doi.org/10.4271/2008-01-2380.

17.46. Dunn, R.O., "Effects of Minor Constituents on Cold Flow Properties and Performance of Biodiesel," *Progress in Energy and Combustion Science* 35, no. 6 (2009): 481-489.

17.47. ASTM International, "Standard Test Method for Determination of Fuel Filter Blocking Potential of Biodiesel (B100) Blend Stock by Cold Soak Filtration Test (CSFT)," ASTM D7501-22, ASTM International, 2022.

17.48. ASTM International, "Standard Specification for Biodiesel Fuel Blend Stock (B100) for Middle Distillate Fuels," ASTM D6751-23a, ASTM International, 2023.

17.49. Knothe, G., "'Designer' Biodiesel: Optimizing Fatty Ester Composition to Improve Fuel Properties," *Energy & Fuels* 22, no. 2 (2008): 1358-1364.

17.50. Ohshio, N., Saito, K., Kobayashi, S., and Tanaka, S., "Storage Stability of FAME Blended Diesel Fuels," SAE Technical Paper 2008-01-2505 (2008), doi:https://doi.org/10.4271/2008-01-2505.

17.51. Jolly, L., Kitano, K., Sakata, I., Strojek, W. et al., "A Study of Mixed-FAME and Trace Component Effects on the Filter Blocking Propensity of FAME and FAME Blends," SAE Technical Paper 2010-01-2116 (2010), doi:https://doi.org/10.4271/2010-01-2116.

17.52. Lopes, S., Conran, D., and Russell, M., "Effectiveness of Cold Soak Filtration Test to Predict Precipitate Formation in Biodiesel," SAE Technical Paper 2011-01-1201 (2011), doi:https://doi.org/10.4271/2011-01-1201.

17.53. Camerlynck, S., Chandler, J., Hornby, B., and van Zuylen, I., "FAME Filterability: Understanding and Solutions," *SAE Int. J. Fuels Lubr.* 5, no. 3 (2012): 968-976, doi:https://doi.org/10.4271/2012-01-1589.

17.54. CRC, "Biodiesel Blend Low-Temperature Performance Validation," CRC Report 650, 2008; CRC, "Biodiesel Blend Low-Temperature Performance Validation," CRC Report 656, 2010.

17.55. Beyreis, K.A., Catto, V.P., and Swanson, E.S., "The Role of Flow Improvers in Solving Autodiesel Winter Fuel Problems," SAE Technical Paper 660372 (1966), doi:https://doi.org/10.4271/660372.

17.56. Taniguchi, B.Y. and Benson, J.D., "Cold Weather Fuel Requirements of Oldsmobile Diesels," SAE Technical Paper 800223 (1980), doi:https://doi.org/10.4271/800223.

17.57. Mikkonen, S. and Juva, A., "Good Diesel Fuel Cold Properties Reduce Engine Performance," SAE Technical Paper 890016 (1989), doi:https://doi.org/10.4271/890016.

17.58. Rickeard, D.J., Cartwright, S.J., and Chandler, J.E., "The Impact of Ambient Conditions, Fuel Characteristics and Fuel Additives on Fuel Consumption of Diesel Vehicles," SAE Technical Paper 912332 (1991), doi:https://doi.org/10.4271/912332.

17.59. Wright, W.A., "A Survey of Past and Present Trends in Lubricating Oil Additives," SAE Technical Paper 440111 (1944), doi:https://doi.org/10.4271/440111.

17.60. Tiedje, J.L., "The Use of Pour Depressants in Middle Distillates," in *6th World Petroleum Congress*, Frankfurt am Main, Germany, 1963; Fallon, T.J., "Flow Improver Additives—Effective for Winterizing Diesel Fuels," SAE Technical Paper 680537, 1968, https://doi.org/10.4271/680537.

17.61. Feldman, N., "Operability of Automotive Diesel Equipment at Temperatures below Fuel Cloud Point," SAE Technical Paper 730677 (1973), doi:https://doi.org/10.4271/730677.

17.62. Zielinski, J. and Rossi, F., "Wax and Flow in Diesel Fuels," SAE Technical Paper 841352 (1984), doi:https://doi.org/10.4271/841352.

17.63. Chandler, J.E., Horneck, F.G., and Brown, G.I., "The Effect of Cold Flow Additives on Low Temperature Operability of Diesel Fuels," SAE Technical Paper 922186 (1992), doi:https://doi.org/10.4271/922186.

17.64. Fenstermaker, R.W. and Riley, R.K., "Additives for Improving the Low-Temperature Filterability of Diesel Fuel Oils," SAE Technical Paper 841350 (1984), doi:https://doi.org/10.4271/841350.

17.65. Damin, B., Faure, A., Denis, J., Sillion, B. et al., "New Additives for Diesel Fuels: Cloud-Point Depressants," SAE Technical Paper 861527 (1986), doi:https://doi.org/10.4271/861527.

17.66. Brown, G.I., Tack, R.D., and Chandler, J.E., "An Additive Solution to the Problem of Wax Settling in Diesel Fuels," SAE Technical Paper 881652 (1988), doi:https://doi.org/10.4271/881652.

17.67. Coley, T.R. "Diesel Fuel Additives Influencing Flow and Storage Properties," in *Gasoline and Diesel Fuel Additives*, Ed. Owen, K. (Chichester, UK: John Wiley & Sons, 1989).

17.68. Coley, T.R., Rossi, F., Taylor, M.G., and Chandler, J.E., "Diesel Fuel Quality and Performance Additives," SAE Technical Paper 861524 (1986), doi:https://doi.org/10.4271/861524.

17.69. Richards, P., Reid, J., Tok, L.-H., and MacMillan, I., "The Emerging Market for Biodiesel and the Role of Fuel Additives," SAE Technical Paper 2007-01-2033 (2007), doi:https://doi.org/10.4271/2007-01-2033.

17.70. Kono, N., Fukumoto, J., Iizuka, M., and Takeda, H., "Influence of FAME Blends in Diesel Fuel on Driveability Performance of Diesel Vehicles at Low Temperatures," SAE Technical Paper 2006-01-3306 (2006), doi:https://doi.org/10.4271/2006-01-3306.

17.71. Dunn, R.O., Shockley, M.W., and Bagby, M.O., "Improving the Low-Temperature Properties of Alternative Diesel Fuels: Vegetable Oil-Derived Methyl Esters," *Journal of the American Oil Chemists' Society* 73, no. 12 (1996): 1719-1728.

17.72. Kolhanen, A., Kokko, J., and Rautiola, A., "Neste's Solution to Computerize CEC Cold Driveability and CEC Cold Operability Tests Performed in Chassis Dynamometer," SAE Technical Paper 920007 (1992), doi:https://doi.org/10.4271/920007.

17.73. CEC, "Low Temperature Operability of Diesels," A Report by CEC Investigation Group IGF-3, CEC Report No. P-171-82, 1982.

17.74. Coley, T.R., "Low Temperature Operability of Diesels," SAE Technical Paper 830596 (1983), doi:https://doi.org/10.4271/830596.

17.75. Ise, H., Hirano, H., Nozaki, N., Takizawa, H. et al., "Cold Flow Properties of Automotive Diesel Fuels and Diesel Fuel Systems," SAE Technical Paper 885019 (1988), doi:https://doi.org/10.4271/885019.

17.76. Heinze, P., "Low Temperature Operability of Diesel Powered Passenger Cars," in *CEC 2nd International Symposium on Performance Evaluation of Automotive Fuels and Lubricants*, Wolfsburg, Germany, 1985.

17.77. Lanoë, M., "Operabilite a Froid des Poids Lourds Diesel," GFC Comite Technique Carburants Moteurs, Journees d'Etudes, Paris, France, October 1982.

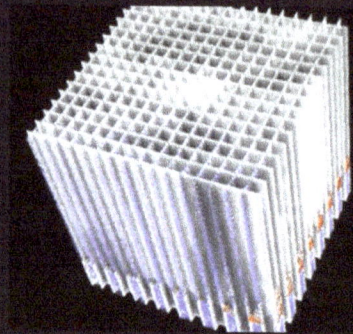

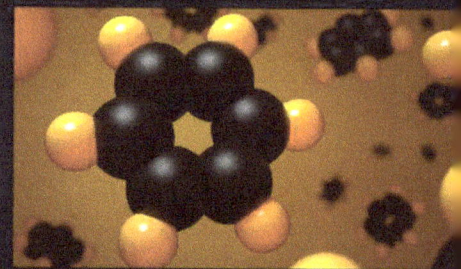

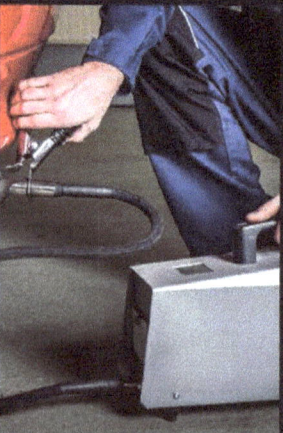

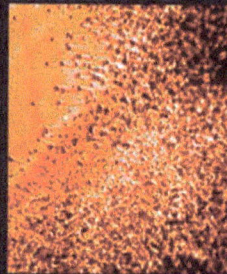

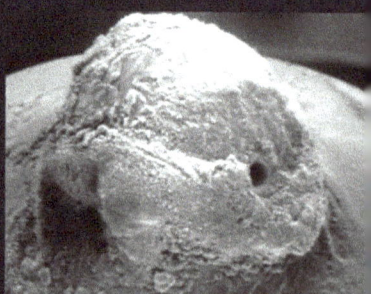

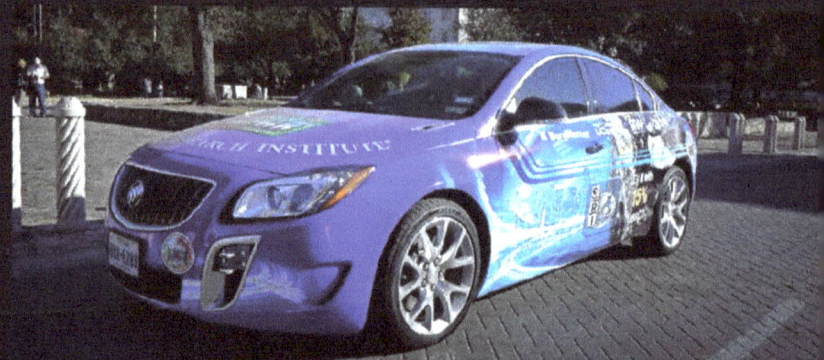

18

Influence of Diesel Fuel Composition on Stability and Engine Deposits

The formation of gum in fuels upon storage is a long-standing problem. The gum is formed in two physical states; one type is dissolved in the fuel and cannot be filtered from it but can be recovered by evaporation of the fuel. This dissolved non-volatile material is called "soluble gum." It is thought to be undesirable in fuels because it may be a precursor to formation of insoluble gum, and soluble gum as such may contribute to injector sticking and engine deposits.

The second type of gum exists as a precipitate in the fuel. The material is insoluble in the fuel and contributes to filter clogging and fouling of the fuel injection system on the engine. This material is referred to as "insoluble gum." It is believed that insoluble gum is the more harmful product of fuel deterioration in storage. [18.1]

The formation of gums and their effects on engine systems was clearly a long-standing issue over 50 years ago. Today, thanks to a better understanding and technology, it should no longer be a problem for the user. But it is still a serious issue that the fuel producers, blenders, and marketers must address in producing a finished fuel.

Diesel fuels marketed in many parts of the world have traditionally been prepared from predominantly straight-run components having satisfactory stability characteristics. However, the situation has changed as refiners have had to adjust their operations to produce a greater proportion of lighter products, such as jet and diesel fuel, and less residual fuel. Cracking processes are used to convert heavy streams into lighter fractions suitable for use as blend components for gasoline, jet, and diesel fuel.

Distillates from cracking operations are less paraffinic than those from atmospheric distillation and without further hydrotreating will contain more nitrogen compounds. As a result, they are less stable, being prone to oxidation by free-radical reactions, as discussed briefly in Section 6.6.5.

Studies on an unstable catalytically cracked diesel fuel have led to the identification of two classes of compounds that are produced during storage under ambient conditions [18.2, 18.3]. Their structure is believed to consist of linked indole and phenalene ring systems. Such compounds are soluble in diesel fuel, but if allowed to react with acids, they form insoluble sediment virtually indistinguishable from the polar, highly colored part of naturally formed sediment.

Generally, cracked material has been blended into heating oil, with the straight-run material being used to make diesel fuel, particularly where cetane numbers of 45 or higher are specified. However, with the heating oil market shrinking, many refiners in Europe needed to divert some cracked stocks to meet the growing demand for diesel fuel. With tightening diesel fuel specifications, the scope for introducing cracked material has been limited by, in particular, the cetane number, density, and polyaromatics limits. Diesel fuels are generally processed with hydrogen to bring their sulfur content down to the specification level. The traditional view is that this treatment has an additional benefit of providing improved stability, so unless either plant capacity or hydrogen availability is limited, the stability of fuels containing cracked components should be adequate. However, there is a possibility that severe hydrotreatment can create a potential for peroxide formation due to the removal of natural antioxidants. As the industry is also coming under increasing pressure to increase the renewable credentials of all fuels, the introduction into the refinery of plant and waste based blending products (see Chapter 3, Section 3.1) and as post-refinery blending products such as biodiesel and fatty acid methyl esters (FAME) has brought about significant changes in physical and chemical properties of diesel fuel. There is concern that the presence of FAME has a deleterious effect on storage stability and importantly the ability of fuels to solubilize other material; this can lead to operational difficulties, which will be discussed later [18.4, 18.5].

18.1. The Influence of Diesel Fuel Composition on Stability

The ability of a diesel fuel to remain unchanged during the period between its manufacture and its eventual use in an engine is an important quality requirement. Most deliveries of diesel fuel to retail forecourts are consumed within a few weeks of leaving the distribution terminal, and the likelihood of degradation should be small. However, it is the policy of many governments to lay down stocks of all types of fuel for use in emergencies. These strategic reserves are normally subject to periodic turnover. Quantities will be withdrawn for use and replaced by newer batches at intervals, but it is possible that some fuels may be in extended storage for periods of more than one year before turnover. Medical and nuclear facilities have similar long-term storage requirements [18.6].

Because of requirements to store fuel for extended periods, it was the long-term stability of diesel fuel that was considered an issue. The stability of jet fuel, a lighter distillate fuel, was more of a short-term issue due the tendency to recirculate fuel as a coolant for aircraft systems and lubricants [18.7, 18.8]. This imposed significant thermal stresses on the fuel. Advances in diesel fuel injection systems, necessary to improve efficiency and reduce emissions, have resulted in similar thermal stresses on diesel fuel. There are thus similarities in the mechanisms that can affect the stability of diesel fuel in a vehicle as those that affect the stability of jet fuel, and much of what has been learned about jet fuel can be applied to diesel fuel.

Changes that can occur with an unstable fuel are due to oxidative decomposition, resulting in the formation of sediments and gums. These processes are accelerated by increased temperature. Diesel fuel may contain up to about 70 mg/kg of oxygen [18.9], and this is readily replaced if the fuel is exposed to air [18.10]. The reaction products may degrade the color of the fuel but, more seriously, can also cause engine operating problems due to blocked filters or deposits within the fuel injection system and within the combustion chamber. The proposed degradation mechanisms are discussed in the following sections.

18.1.1. Petroleum Fuel Composition Effects

Petroleum diesel fuel will naturally contain small amounts of oxygen that can be replenished by exposure to air. Recirculation of fuel to a vehicle tank can increase the degree of aeration of the fuel. There is ample opportunity for reactions involving molecular oxygen, the stable form of which has two unpaired electrons. It is thus able to react rapidly with carbon-centered radicals while still retaining one unpaired electron. This is then capable of abstracting a hydrogen atom to form a new radical. This forms the basis of a widely accepted mechanism for the oxidative instability of distillate fuels [18.11]. The simplified mechanism is shown schematically in Figure 18.1, where R represents an alkyl group and Ar is an aryl moiety.

FIGURE 18.1   Schematic of oxidation mechanism within distillate fuels.

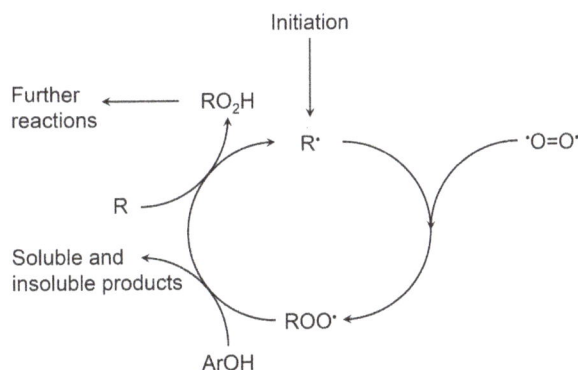

© SAE International.

The homolysis of existing hydroperoxide [18.9] is a possible starting point for this process. It has also been found that transition metals can catalyze single electron transfer reactions [18.12]. When biodiesel is present, the oxidation of highly vulnerable fuel components, such as those containing diallylic methylene groups [18.13], is another possible starting point.

The autoxidation of paraffinic fuel components at temperatures that could be encountered in modern fuel injection equipment can form keto-hydroperoxides [18.14]; aldehydes, olefins, and carboxylic acids [18.15]; as well as cyclic ether products [18.16]. In a less oxygen-rich environment, dimers may also be formed [18.17].

Some highly polar species, such as polycyclic aromatic materials, including fluorenes, phenalenes, and phenalenones, along with heteroaromatic materials including indoles and other nitrogen containing compounds, have been found in diesel fuel. The presence of oxygen-based [18.18] and nitrogen-based [18.19] aromatic and heteroatomic, polar species have each been shown to affect the stability of fuel under thermal stress. None-basic nitrogen species such as alkyl indoles have been found to be important constituents of fuel sediment [18.2]. Acid-catalyzed coupling reactions of phenalenones with indoles have also been implicated in deposit formation [18.20].

The high degree of hydrotreatment necessary to reduce fuel sulfur levels to meet many current specifications will tend to significantly reduce the quantity of these polar compounds. Although this will reduce the instability of the fuel and reduce the quantity of sedimentary particle formation, it will also tend to reduce the solvency of the fuel. The balance of these two competing processes will depend on the composition of the feedstock and how this has been processed.

18.1.2. Biodiesel Composition Effects

As discussed in Section 6.7.4, biodiesel can be prone to autoxidation, and this depends upon the feedstock and the quality of the finished product. The biodiesel will begin to oxidize as soon as it is produced. The oxidation process can be inhibited by treating the biodiesel with an antioxidant. The sooner this is done, then the more effective will be the long-term performance of the additive. In most of the markets using biodiesel, including the US, Europe, and Japan, the neat biodiesel has to meet an oxidation stability specification. The petroleum diesel/biodiesel blend will usually also have to meet an oxidation stability specification.

The primary reason for the differences in stability of different biodiesels is that different FAMEs have different levels of unsaturation. The biodiesel feedstock determines the degree of unsaturation in the finished biodiesel. Individual FAMEs can be monounsaturated or polyunsaturated. The greater the degree of unsaturation of a FAME then the more reactive it is [18.21, 18.22]. However, the oxidation stability does not correlate with the number of double bonds [18.23] but rather with the total number of *bis*-allylic sites. The *bis*-allylic site is the most vulnerable position for the formation of the free radicals required to initiate the oxidation process.

The radicals formed at these *bis*-allylic sites will readily isomerize to form a more stable conjugated structure, which can react directly with oxygen to form a peroxide. The peroxide species can form a cyclic intermediate that will cleave to form acids, alcohols, aldehydes, and ketones [18.24, 18.25]. The peroxide can also react with other fatty acid chains to form polymers, although these polymers rarely become larger than trimers or tetramers [18.26]. These polymers will increase the viscosity of the fuel [18.27, 18.28]. The polymerization can occur at ambient temperatures but is accelerated at higher temperatures [18.29]. At temperatures above 250°C, Diels-Alder reactions can take place to form cyclic compounds [18.27]. These temperatures may occur locally in high-pressure fuel injection equipment. Figure 18.2 [18.30] demonstrates some of the possible products that have been found to occur from the oxidation of three of the FAMEs commonly found in biodiesel.

FIGURE 18.2 Acids, ketone, aldehyde, and dimers found from the oxidative decomposition of three commonly biodiesel FAMEs.

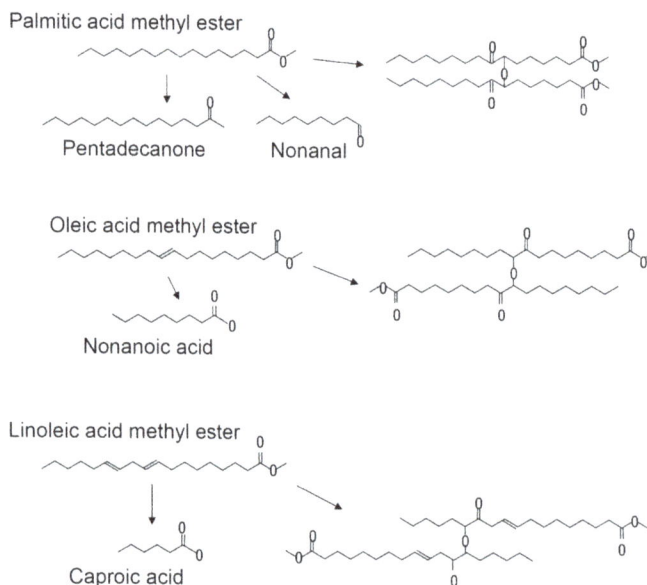

The presence of free fatty acids [18.31] and metals [18.32, 18.33] can increase the instability of the FAMEs. However, it has been found that oxidized biodiesel is a good solvent for some of the polymers formed [18.34] but that the solvency diminishes when petroleum diesel fuel is added. Figure 18.3 [18.35] shows the time for deposit precipitation at various FAME blending ratios; the shorter the precipitation time, the less stable the blend.

FIGURE 18.3   Precipitation time for different FAME blending ratios.

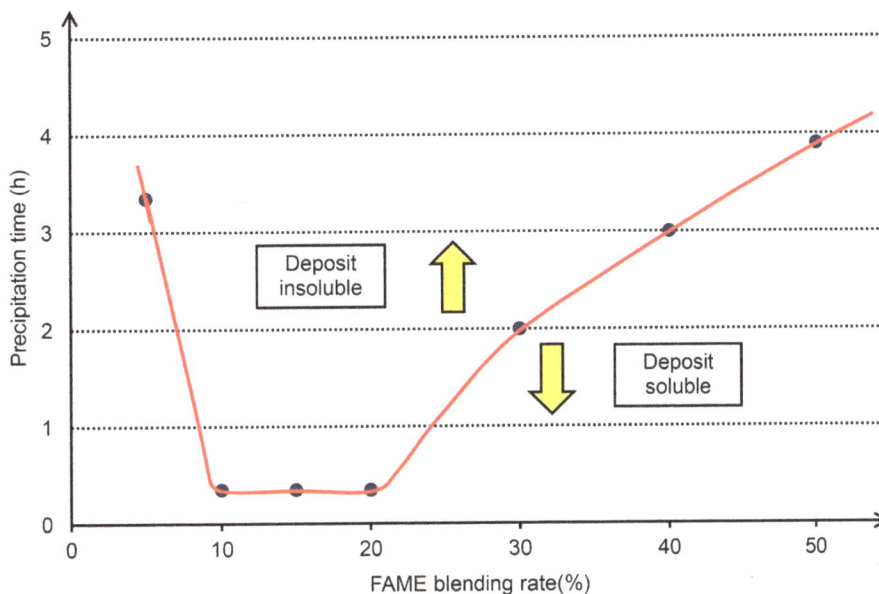

18.1.3. Measurement of Stability

The rate at which changes are likely to occur under representative storage conditions is usually slow. As noted, the stability of the fuel is determined by the temperature, its thermal stability, and the availability of oxygen, its oxidative stability. To obtain a quantitative value for the long-term stability of a diesel fuel, it is necessary to accelerate the degradation process by the use of abnormally severe test conditions. This can be done by raising the fuel temperature and/or by increasing oxygen availability.

The different conditions used in the various test methods may emphasize either the thermal stability or the oxidative stability, or they may be designed to simulate long-term storage conditions without particular emphasis on thermal or oxidative deterioration. Some of the most commonly used tests are discussed briefly in the following sections.

18.1.3.1. Measurement of Storage Stability

One test that has been used fairly widely for many years is the ASTM D2274 Standard Test Method for Oxidation Stability of Distillate Fuel Oil (Accelerated Method) [18.36]. This is an accelerated method for oxidation stability that contains the cautionary statement that any correlation between the test and field storage may vary significantly under different field conditions or with distillates from different sources. The fundamental difficulty with this test, as with all other accelerated methods, is that changing the storage conditions (in this case, the elevated temperature and the addition of oxygen) change the oxidation mechanisms. The use of this test is declining due to its poor correlation with real storage experience.

A sample of fuel is heated to a temperature of 95°C, which is maintained for 16 hours with continuous bubbling of oxygen, at a rate of 3 L/h. The fuel is then allowed to cool in the dark, prior to filtration and solvent washing, to recover sediment and gummy residues. The total amount of insoluble deposits is reported as mg/100 mL of fuel. Typical specification values are usually in the range of 0.02 to 0.03 mg/100 mL. Another long-term test is based on the Du Pont F8-81 procedure; the sample is placed into accelerated conditions at 50°C for 13 weeks. Samples ae stored in open bottles in the dark and tested for color and insolubles at the start, and at 6- and 13-week periods. The test is regarded as equivalent to one year's storage under normal conditions [18.37].

The color of the fuel before and after aging can be determined using the ASTM color scale [18.38]. A sample is placed in a container and, using a standard light source, is compared with colored glass discs ranging in value from 0.5 to 8.0, in steps of 0.5. The number of the matching glass, or the higher of the two numbers where there is no exact match, is reported as the ASTM color. Although it does not feature in many diesel fuel specifications, quoted levels tend to range between 2.5 and 5.0 ASTM color.

Some of the other tests, such as ISO 12205 [18.39], for predicting diesel fuel stability are variants on the ASTM D2274 method; the conditions are altered to modify the test severity. The combination of high temperature and continuous oxygen bubbling is considered by some researchers to induce reactions that would not occur under normal conditions. One approach to reduce severity is to replace the oxygen with air, using either continuous bubbling or presaturation to limit the amount of oxygen available for reaction. Changes to the temperature and duration of exposure to oxidative conditions are other factors influencing severity.

Another short-term test is the ASTM D5304 Storage Stability by Oxygen Overpressure [18.40]. In this test a 100 mL sample of fuel, in a glass jar, is placed in a preheated pressure vessel. The pressure vessel is kept at a temperature of 90°C and an absolute oxygen pressure of 800 kPa. The pressure vessel is then placed in an air-forced oven at 90°C for 16 hours. The amount of fuel insoluble produced is determined when the sample has cooled.

An alternative test method for predicting long-term storage stability uses more representative conditions, but this is at the expense of greater test time [18.41]. For this test, the fuel is presaturation with oxygen rather than being subjected to the continuous bubbling of air or oxygen. The saturated fuel is then held at a temperature of 43°C for a period of 100 days. The results are believed to correlate with storage at ambient conditions for one year. Storage stability can, and often is, assessed by measuring oxidation stability, which is discussed later in Section 18.1.3.3.

18.1.3.2. Measurement of Thermal Stability

The degree of fuel degradation at significantly elevated temperatures that may be encountered in recirculating fuel systems can be assessed using the ASTM D6468 High Temperature Stability test [18.42]. For this test, the 50 mL fuel sample is left exposed to air but with no forced aeration for 90 or 180 minutes. During this time the fuel is maintained at a temperature of 150°C. At the end of the test the insoluble deposits that are formed can be assessed either gravimetrically or by the reflectance of the deposits on the filter paper. The color of the deposits is dependent upon the nature of the insoluble formed; there is thus no correlation between the gravimetric and reflectance results.

18.1.3.3. Measurement of Oxidation Stability

The measurement of oxidative stability per se has become more important with the inclusion of biodiesel into the blending pool. The most commonly used test method for assessing the oxidation stability of biodiesel blends is the Rancimat test [18.43]. A general schematic of the apparatus is shown in Figure 18.4 [18.33] and described next.

FIGURE 18.4   Schematic of Rancimat test apparatus.

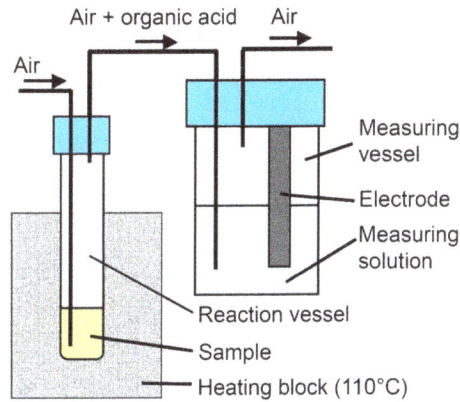

A 3 g sample of fuel is maintained at 110°C while air is bubbled through it at a rate of 10 L/h. The air, along with any volatile organic acid that is formed within the fuel, is carried over to the measuring vessel. The measuring vessel begins the test containing electrically neutral de-ionized water. As the off-gas is bubbled through it, any acids will be dissolved in this water, increasing its acidity and electrical conductivity. The conductivity in the measuring vessel will typically remain constant for some time before transitioning to a rapid rate of increase in conductivity. The time taken before this rapid increase begins is known as the induction period (IP). A typical output from a Rancimat test is shown in Figure 18.5 [18.44].

FIGURE 18.5   Typical Rancimat output.

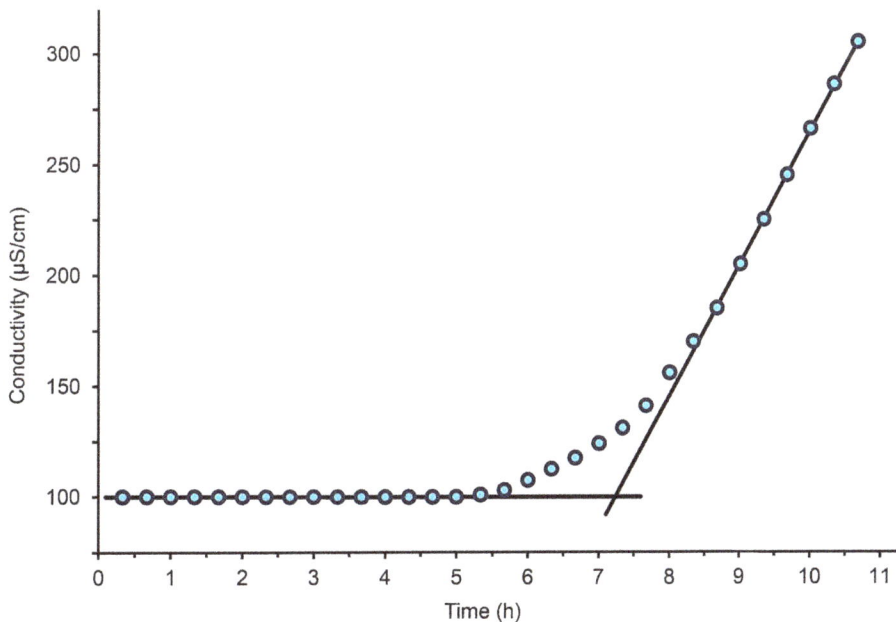

The IP can be determined graphically, as shown in Figure 18.5, by determining the intersection of the tangents to the initial stable values and the rapidly increasing values as the fuel oxidizes. As an alternative, the conductivity data can be differentiated twice, and the peak of the second derivative will yield the IP. It should be noted that the Rancimat test is intended for fuels containing at least 2% biodiesel because the degradation mechanism detected in this test is specific to polyunsaturated fatty acid chains.

A very similar method is recommended by the American Oil Chemists' Society [18.45] and uses a 5 g sample of material. The Oil Stability Index (OSI) method is also sometimes referred to as the Oxidation Stability Index method. Because the OSI specifies temperatures of 100, 110, 120, 130, and 140°C, the resultant IP must also specify at which temperature the test was performed.

Limitations to these methods are that they do not consider the formation of heavy acids remaining in the biodiesel and do not measure properties directly related to engine performance. Further storage life estimates may be unreliable as the method assumes the oxidation mechanism does not alter at the extrapolated lower temperatures. It is therefore not a substitute for real world storage conditions for biodiesel [18.13, 18.46].

Other methods for assessing the stability of biodiesel and biodiesel blends subject the sample to thermal and oxidative degradation and measure the amount of insoluble that are produced. These include methods specifically intended for biodiesel and biodiesel blends [18.47], this method was withdrawn in 2016. Another approach is the ASTM D7545 Rapid Small Scale Oxidation Test (RSSOT) [18.48] also known as the Petroxy method [18.49] where a 5 mL sample is sealed in a chamber with a fixed amount of oxygen at a pressure of 700 kPa. The temperature of the sample is then raised to 140°C. As the sample heats up, the pressure will rise until rapid oxidation begins. The pressure in the chamber is automatically monitored; the point at which the pressure drops by 10% of the maximum pressure attained is used as the break point. The time from the start of sample heating to the break point is defined as the IP. This test method is also used as a measure of fuel stability in the Top Tier™ Diesel Fuel Performance Standard as shown in Appendix 3.

18.2. Engine Deposits

As noted at the start of this chapter, it has long been recognized that there is a propensity of fuel to form deposits within the engine. Some early work was conducted to evaluate the effect of fuel characteristics on engine performance. One early performance criterion was combustion chamber deposits, and this included the deposits on fuel injection valves [18.50]. This work showed the relationship between certain fuel characteristics and the mass of combustion chamber deposits. These correlations were basically indicating the tendency of the fuels to produce soot and fuel impingement on the combustion chamber surfaces.

With the introduction of emissions legislation, researchers began to investigate possible links between these deposits and observed in-service deterioration of emissions performance, such as increased smoke emissions [18.51–18.53]. Combustion chamber deposits have been linked to engine wear and engine wear to emissions [18.54]. A more direct correlation between emissions and deposits has been attributed specifically to deposit formation on and within the fuel injectors.

The manner in which deposits affect emissions changes as the injector technology changes. For indirect injection (IDI) engines, the injection nozzles are specifically designed to control the rate of injection by the degree of needle lift. Deposit build-up on these injectors will restrict the fuel flow, particularly at the low needle lift conditions. This can alter the fuel injection rate profile and result in an increase in engine noise and emissions [18.55, 18.56]. The location of these deposits for different types of IDI injector nozzles are shown schematically in Figure 18.6 [18.57].

FIGURE 18.6 Location of deposit build-up that restricts the flow in IDI injector nozzles.

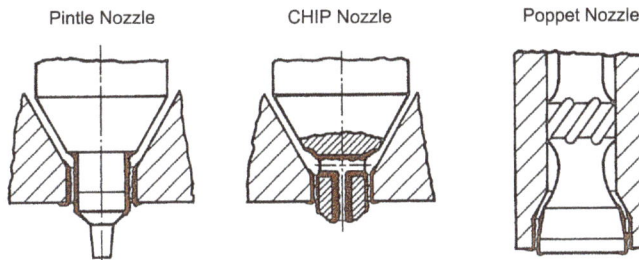

Note. Flow passages and deposits are not to scale

With direct injection (DI) engines, the nozzle geometry is different and the location of the deposits differs accordingly. With multihole DI injector nozzles, the deposit tends to build up around the injector hole and does not significantly reduce the flow of fuel. For DI engines the deleterious effect of deposits is a result of the way the deposits alter the fuel spray formation. The deposits have been shown to reduce fuel spray cone angle and penetration [18.58]. These deposits can be considered as external injector deposits, because they occur on the combustion chamber side of the needle seat.

The formation of deposits within the fuel injectors has been found to cause reduced precision in the control of the fuel injection event and even resulted in injector sticking [18.59, 18.60]. These can be considered as internal injector deposits (IDID).

There are many possible sources for the observed deposits, and their precursors, and this is shown graphically in Figure 18.7 [18.61].

FIGURE 18.7 Sources of diesel injector deposits.

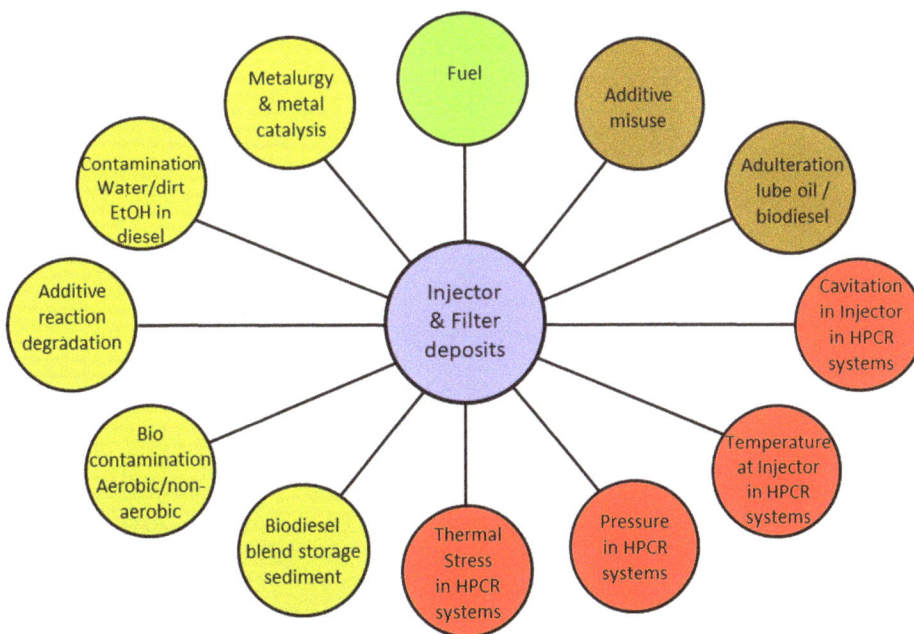

The figure shows in red those sources of deposit forming mechanisms associated with temperature and pressure rises resulting from the introduction of high-pressure common rail (HPCR) diesel injector systems. The sources colored green indicates changes in fuel components, the sources shown in yellow/green indicate fuel solubilization capability, housekeeping, poor manufacturing standards, materials of construction and contamination, the presence of water, dirt, and ethanol in diesel fuel. The sources shown in brown in the figure indicate adulteration of the fuel, e.g., with used lubricating oil, biodiesel, and additive misuse.

18.2.1. Deposit Formation in the Fuel Tank, Lines, and Filters

As discussed earlier, it is possible for deposit precursor and fine particles to be formed within the bulk of the fuel circulating within the fuel system. Therefore, the deposits do not necessarily occur at the same place as the chemical reactions that initiate them. The reduced aromatic content of many ULSD fuels is likely to reduce the solubility of high molecular weight and/or polar species. Fine carbon-based particles have been produced in fuel injection equipment (FIE) operated in the absence of combustion [18.62]. These particles can precipitate within the FIE as a result of reduced fuel solvency.

As discussed in Section 18.1.2, the inclusion of biodiesel components can result in the formation of acids within the fuel. These acids can result in corrosion of metal parts within the fuel system [18.63]. In turn, this corrosion can accelerate the oxidative degradation of the biodiesel [18.33, 18.64], which will result in greater deposit formation. The acids formed in the degradation of FAME are carboxylic acids, which can cause corrosion of iron surfaces leading to the formation of carboxylate iron salt. This can result in deposits elsewhere in the FIE [18.35].

As discussed in Chapter 6, Section 6.7.4, the oxidative stability of the biodiesel component is dependent upon the nature of the source fatty acids and upon the purity of the finished product. Extensive testing has shown that if the biodiesel is within specification, including its oxidative stability, the blends of up to 20% biodiesel in petroleum diesel do not cause problems in service [18.65–18.68]. It should be noted that many specifications do not request the reporting of numerous FAME impurities that have been implicated in filter blocking, such as oxidized biodiesel, steroyl glucosides and free fatty acids [18.69–18.72]. In cases where filter plugging problems were encountered, they were thought to be due to off-specification biodiesel or microbial contamination [18.65]. When testing has been conducted with biodiesel that has been deliberately oxidized, it has been confirmed that out-of-specification biodiesel will in fact cause these problems [18.73, 18.74]. Interestingly, it was also noted that there was evidence of sludge in the fuel return line, suggesting that this had either passed through the filter or had been formed downstream of the filter.

The application of D7501 has had some success with B100 for the above filter blocking compounds, but this method has no equivalent for petroleum diesel and biodiesel blends. The FAME impurities have ill-defined solubility limits in such fuels. Thus, filter blocking problems may still occur even though the blended fuel has met international specifications; usually occurring when the fuel has been exposed to unexpected low temperatures [18.70].

Microbial contamination has been an issue in diesel fuel tanks, lines, and filters [18.75]. This can also be an issue in bulk storage, as discussed in Chapter 6, Section 6.6.2. Despite recent measures taken to reduce the problems, the introduction of biofuels and ethanol to the industry has seen bulk storage and corrosion issues still being reported [18.76, 18.77]. The EPA has found that ethanol was present in 90% of 42 diesel storage facilities [18.78], this suggests that cross-contamination of diesel fuel with ethanol is likely to be the norm, and that the quality of diesel fuel stored in underground storage tanks is variable. The presence of ethanol has other connotations, in that its presence is often associated with acetic acid producing microbes and the concurrent issue of severe corrosion. In vehicle systems, corrosion has been noted in common-rail fuel systems.

Toward the end of the first decade of the 21st century, there was a rise in the number of reported field problems due to filter blocking. This appeared to be coincident with the widespread introduction of biodiesel blends, the more widespread use of high-pressure common-rail fuel injection systems, and the reduction of fuel sulfur levels. A typical example of an unacceptably fouled fuel filter is shown on the left of Figure 18.8 [18.79], while the right side of the figure shows a filter with biodiesel fouling, the center of the figure shows a used filter that still exhibits acceptable performance.

FIGURE 18.8   Used fuel filters showing unacceptable fouling on the left, absence of fouling in the center and biodiesel fouling on the right.

© SAE International.

Deposits in fuel filters were also found when using fuels that did not contain biodiesel components.

18.2.2. Internal Diesel Injector Deposits (IDID)

Fuel injection pressures have increased over the years [18.80–18.84], which results in increased temperatures within the fuel system. This puts additional stresses on the fuel, thus increasing the likelihood of the formation of deposit precursors. Internal diesel injector clearances have also become smaller, making the operation of the injector more prone to impaired performance from deposition of these materials. These deposits may be formed from the hydrocarbon fuel components via the fuel degradation processes already discussed. They may also be formed or compounded by additional components that have been deliberately included in the fuel; this can include biodiesel components and fuel additives. There may be other elements in the fuel that are not included intentionally. These can include overtreatment of certain fuel additives, trace elements or compounds that remain from the fuel feedstock, and compounds used in the production process that have not been completely removed.

Early work suggested deposits were "waxy" or "soap," while another type appeared carbonaceous or of a lacquer appearance with the latter being associated with the presence of amides [18.85]. Others [18.86] have categorized the groups as follows:

- Carbonaceous (black in color); thermal or pressure degradation of fuel.

- Inorganic salts (off-white in color, e.g., sulfates, chlorides).

- Carboxylate salts (white in color, often sodium- or calcium-based).

- Amides (brown in color polymeric, unstable FAME feedstock or non-commercial low-molecular-weight PIBSI DCAs).

- Lacquer based visualized on some injectors may be a carbonaceous precursor one molecule thick.

There is continued worldwide study of IDIDs because of their importance to the industry and a need to fully understand their nature and cause [18.87–18.93]. Studies have shown IDIDs to be complex and multilayer in their constitution [18.94–18.97] with advanced analytical techniques indicating highly ordered graphene-like structures [18.98, 18.99].

18.2.2.1. Carbonaceous Deposits

In many passenger car diesel fuel systems, including the XUD-9 and DW10 engine systems, the fuel pump is driven directly of the engine crankshaft. The fuel pump speed and the volume of fuel pumped are thus roughly proportional to the engine speed. The fuel volume demand will vary according to the power requirement of the engine and the excess fuel pumped will be returned. The proportioned of the fuel that is returned can thus be high at light loads. Parts of the fuel can thus be pumped around the system many times before being consumed in the combustion process. This subjects the fuel to repeated pressurization to the very high pressures of current systems, pressure relaxation, corresponding temperature cycling, and aeration.

Lower stability petroleum-based fuels have the potential to degrade, causing precipitates that result in filter plugging, as discussed in the previous section. However, as these precipitates are most likely formed in the high-pressure regions downstream of the filter, there is also a likelihood that they will precipitate in these high-pressure regions. These deposits have a very high carbon content and a more crystalline nature than soot particles formed in the combustion chamber [18.100]. Data has shown the presence of aromatic ring species, of more than six rings and such structure are postulated as intermediates between the initial fuel degradation products and graphene-like deposits [18.100–18.102]. Carbonaceous deposits can also be formed in a test stand in the absence of any combustion and using pure paraffinic diesel fuel. It is possible that they are formed by the very high temperatures created by the collapse of cavitation bubbles during steep reductions of fuel pressure within the system [18.62]. The presence of catalytically active metals such as zinc or copper have been shown to exacerbate the formation of these deposits.

The diesel fuel in many markets now contains some biodiesel. The IP requirement for these fuels varies from region to region. In Europe, biodiesel blended fuels must have an IP of at least 20 hours, whereas the US requirement is only 6 hours IP. It has also been found that the fuel available to the customer does not always meet these specifications [18.103, 18.104]. Deposits with high-pressure injectors have been shown to have a similar composition to the residue from Rancimat testing of the same fuel [18.62]. The use of biodiesel blends in high-pressure fuel injection systems also reflect Rancimat testing in that there is a period over which in-service biodiesel blends produce no change in deposit-forming tendency. Then after a period when the stability reserve is depleted, there is a noticeable increase in the deposit-forming tendency; this is similar to the way in which acid formation increases rapidly at the end of the IP in the Rancimat test.

This reinforces the point that the fuel formulator must ensure the stability of batches of biodiesel that are being used for fuel blending. A batch of biodiesel may have a very long IP when it has just been produced and tested, but if this fuel is then stored for a period of time before being blended, then some of that stability reserve may be depleted. The IP of the resultant blend can then fall outside of the specification.

18.2.2.2. Metal Salts

Metal salt deposits have previously been found in diesel fuel injection systems due to interactions between basic metal compounds found in lubricating oil and acidic fuel additives. This has been due to comingling of the fuel and lubricant in older types of diesel FIE [18.105] and as a result of "recycling" of lubricant [18.106]. In today's high-pressure common-rail systems, there is no possibility for comingling of the fuel and lubricant, and deliberate adulteration of the fuel is always prone to cause problems. However, problems are still being experienced.

18.2.2.3. Metal Carboxylates

Metal carboxylate salts have been found in high-pressure diesel fuel injection systems [18.59, 18.60]. The organic acids may originate from fuel contamination, microbiological contamination, biodiesel, degraded biodiesel, or organic acids added to the fuel as corrosion or lubricity additives [18.62, 18.107–18.110].

If alkaline metal compounds are present in the fuel, then they can readily react with acids present in the fuel to form salts. Metals that have been found in IDIDs include calcium, iron, sodium, tin, and zinc [18.111].

Sodium is the most frequently reported metal [18.59, 18.60, 18.112]. Possible sources of sodium in the fuel include salt (sodium chloride) driers and caustic (sodium hydroxide) wash used at petroleum refineries, sodium hydroxide used in the production of biodiesel, sodium nitrite salts used as corrosion inhibitors, storage tank water bottoms, and seawater used as ship ballast (sodium chloride) and bleaching agents (sodium hypochlorite) [18.62, 18.113]. Organic acids may be present in the fuel as impurities from the biodiesel esterification process or from decomposition of the biodiesel. Organic acids may also be added to the fuel as lubricity agents or as corrosion inhibitors. It should be noted that these weak acids will not react with sodium chloride to form other sodium salts [18.114].

The acid-base reactions that form metal carboxylate salts can occur during fuel manufacture and storage prior to filing the vehicle tank. The more severe environment of the high-pressure, high-temperature, recirculating vehicle fuel system is likely to further promote the formation of these salts. However, it has been found that these compounds can be present and circulate within the fuel system without forming deposits in the low-pressure, lower-temperature regions of the system. Deposit formation tends to occur only in the critical areas often associated with high pressure or temperature gradients [18.62].

It is clear that the fuel formulator must consider the quality and possible adverse interactions of different components in a finished fuel blend.

18.2.2.4. Amide Deposits

The formation of lacquer deposits on the internal components of fuel injectors is not new. Investigations performed in the 1980s into the use of plant oils as diesel fuel blending components highlighted an injector fouling problem [18.115]. The deleterious effect of these deposits was found to increase as the injector temperature increased and the clearance between the injector needle and its guide was decreased [18.116, 18.117]. In modern high-pressure common-rail FIE, the temperatures are higher and the clearances are lower than on the DI fuel system used in the 1980s. These lacquer deposits were considered to be due to polymerization of the plant oils and unstable FAME stocks were the source [18.85]. However, earlier work had shown that esterification of the plant oils to produce FAME reduced the lacquer formation [18.118].

There has recently been a resurgence in concern about lacquer type IDIDs due to the ever-diminishing clearances within the injector and the importance of the FIE in enabling engine manufacturers to meet emissions legislation. A great deal of research has been conducted to characterize deposits found in field service and to reproduce these deposits using controlled laboratory tests. Research has highlighted possible interactions of small quantities of compounds deliberately or accidently added to the fuel that can lead to such deposits.

Nitrogen-bearing deposit control additives (DCA) such as amines and succinimides have been suggested as precursors to the formation of lacquers within FIE systems [18.112]. Laboratory bench scale testing has been used to support this supposition, but engine testing has not always corroborated these findings [18.119]. A possible explanation for these differences is the composition of the polyisobutene (PIB) backbone of the additive. The PIB tail of the molecule is used to provide solubility of the polar additive within the fuel and will typically contain a range of molecular weights. Some testing has shown that the inclusion of low-molecular-weight compounds within the additive resulted in IDID formation, whereas without these compounds there were no deposits formed [18.62, 18.120]. However, it is not simply a question of whether such low-molecular-weight PIB is used to produce the DCA but also the manufacturing process and the resultant composition of the finished product [18.120, 18.121].

18.2.2.5. Lacquer Appearance

As noted above, early reports of deposits with a lacquer appearance have been found to be amide based and are discussed in the previous section, however, there are still reports of deposits with a lacquer appearance where the deposits are too thin to have been analyzed using current techniques. These are multicolored and appear similar to the patterns formed by gasoline on water, suggesting molecular monolayers, which may be precursors to carbonaceous deposits.

IDID Test Methods

There is currently no industry-recognized test method for assessing the propensity of a fuel to cause IDID. Due to the current concern over the problem, both the US Coordinating Research Council (CRC) and the Coordinating European Council (CEC) have set up groups to work on developing an appropriate test method.

18.2.2.6.1. CRC Test Development. The objective of the CRC work was to identify or develop a test rig for evaluating a fuel's tendency to cause IDID. This would also allow for the evaluation of the effectiveness of a fuel additive in preventing such deposit formation. An initial scoping study was undertaken using two in-house tests to determine if fuels that were expected to cause IDID could be differentiated from those that were not expected to form such deposits [18.122].

One approach used a test rig [18.123] where the test fuel and air were passed over a test specimen at elevated temperature and pressure to simulate the condition that may be encountered in the diesel fuel system. The test specimen is subsequently analyzed for deposit formation. This approach is similar to that employed in the jet fuel thermal oxidation tester (JFTOT) used primarily for testing aviation fuel according to the ASTM D3241 procedure [18.124] but which has also been used to assess the deposit-forming tendency of diesel fuels [18.120, 18.125].

The other approach utilized commercial fuel pump tests that had been modified to provide heating to the fuel injector tip. This approach is effectively the same as a full engine system except that the high-pressure fuel pump is driven by an electric motor and the injected fuel is collected and recirculated instead of being burned in the combustion chamber. This approach has previously been used to assess the results of thermal degradation of fuels [18.126, 18.127].

The results of this scoping study have unfortunately concluded that neither of these rigs, in their present state, could discriminate among deposit-forming and not deposit-forming fuels. Work is ongoing using an injector deposit rig with the application of a Variable Angle Spectroscopic Ellipsometry (VASE) [18.128].

18.2.2.6.2. CEC Test Development. In response to the formation of IDIDs in the field the CEC has attempted to develop a further test based on DW10 C engine, CEC F-110-16. The test aims to simulate sodium carboxylate deposits by dosing the fuel with sodium naphthenate and dodecenyl succinic acid, amide deposits are simulated by dosing the fuel with a "non-commercial" low-molecular-weight PIBSI. The "non-commercial" low-molecular-weight PIBSI selected was never established as representing the quality of the material, which resulted in field issues, thus making any results questionable. The test failed to produce reliable fouling and suitable precision, for this contaminant, and thus a method making use of it was never approved [18.129]. The carboxylate version of the test suffered from precision issues but a new sodium naphthenate contaminant solution has been produced and the cold soak operating conditions have been tightened. The test method is now under surveillance status. The test itself is being used worldwide to show the effectiveness of DCAs to combat the problem of IDIDs, from both a cleanup and keep clean standpoint. However, the F-098 (DW10B) surveillance group continues to try and extend the life of that method (see Section 18.2.3.4) while working on developing a replacement test. Many continue to use a modified DW10B test as an alternative to the DW10C [18.130].

18.2.3. External Injectors Deposits

It is now widely known that at the 1900 Exposition Universelle in Paris, Rudolf Diesel exhibited an engine running on peanut oil [18.131, 18.132]. It is not documented for how long Diesel's Paris expo engine ran on this fuel; nor is it documented what effect this had on the FIE. However, work conducted toward the end of the last century [18.133] and the beginning of this one [18.134] suggests that using vegetable oil can result in significant deposit formation on fuel injectors. It is therefore likely that the issue of fuel injector fouling is almost as old as the diesel engine itself. An example of the deposit formation after approximately 20 hours of running on peanut oil is shown in Figure 18.9 [18.133].

FIGURE 18.9 Deposit formation when running on peanut oil.

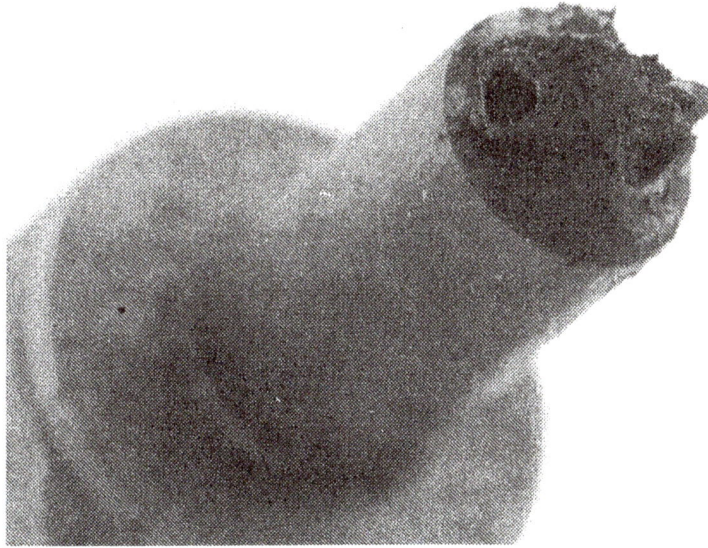

It is interesting to note that when Andre Attendu presented his development of a small diesel engine [18.135], he was questioned as to whether the fuel nozzle remained free of carbon. At the time, he replied, "The fuel nozzle is always absolutely free from any carbon formation," yet when the first high-speed diesel engine was introduced by General Motors Corporation in 1938, it was found to be susceptible to carbon formation on injector tips when running on regular diesel fuel [18.136].

Diesel designed his engine to run on any type of fuel, and over the years the diesel engine became known for its ability to run on a wide range of fuel qualities. However, as emissions legislation has tightened, and the diesel engine, and in particular the FIE, has been developed to meet the legislation, fuel quality has become more important. It has also become more important that the FIE is maintained in the state the manufacturers intended. The level of deposit shown in Figure 18.8 certainly could not be permitted today.

With successive steps in fuel and engine technology, there have been bouts of concern over the effect of deposit formation within the FIE. This has led to repeated attempts to define a test protocol that replicates field problems and thus allows fuel producers to assess the propensity of different fuels to form deposits and hence lower the incidence of poor fuels reaching the market.

The following sections present the work done to develop standardized test methods to assess fuel propensity to form injector deposits and current concerns that suggest further standard tests are still required.

18.2.3.1. Peugeot XUD-9 Test Method

Exhaust smoke was initially the only controlled diesel exhaust emission. One solution to this problem was found to be the use of barium-containing fuel additives [18.52]. These additives were also found to limit deposit build-up within the engine, including on fuel injectors [18.137]. Ashless additives were also developed specifically to reduce the build-up of deposits on the fuel injectors [18.57]. The performance of these additives was assessed by accumulating vehicle mileage under normal service conditions. This highlighted a need for a short-term test method that could be conducted under controlled laboratory conditions that would allow the assessment of different fuels and fuel additive combinations for their injector deposit-forming tendencies.

Along with the need for a recognized test procedure, there was a need for a quantitative method for assessing the degree of deposit formed. It was also important that this assessment reflected a metric that would influence the performance of the engine. For passenger car IDI diesel engines, the loss of flow through the fuel injector was considered a good parameter, and this could more easily be assessed by flowing air through a dry injector nozzle [18.138].

One early proposal to assess the deposit-forming tendency of fuels was a short screening procedure [18.53] that can briefly be described as follows. The procedure consisted of using slave injectors to warm up the engine at 2000 rev/min and 20% load for 20 minutes. The slave injectors were then removed, and the test injectors installed. Combustion chamber pressure and needle lift data were then acquired at an engine speed of 1200 rev/min and 10% load. The engine was then run for 2 hours at a deposit accumulating condition of 1500 rev/min and 40% load. At the end of this period a further set of measurements were taken at the 1200 rev/min, 10% load condition.

A longer test procedure was also proposed that was considered to simulate freeway driving. The longer test procedure differed from the proposed screening test only in the deposit accumulation phase. The operating conditions for the deposit accumulation phase were changed to an engine speed of 3000 rev/min with 75% load. The duration of the accumulation phase was also increased from 2 to 6 hours. This procedure became the basis for the CEC test procedure for IDI diesel engine nozzle coking assessment [18.139] often referred to in the literature of the time as the CEC PF26 procedure. The test cycle is illustrated schematically in Figure 18.10 [18.53].

FIGURE 18.10   Six-hour test procedure adapted for the CEC injector nozzle coking test.

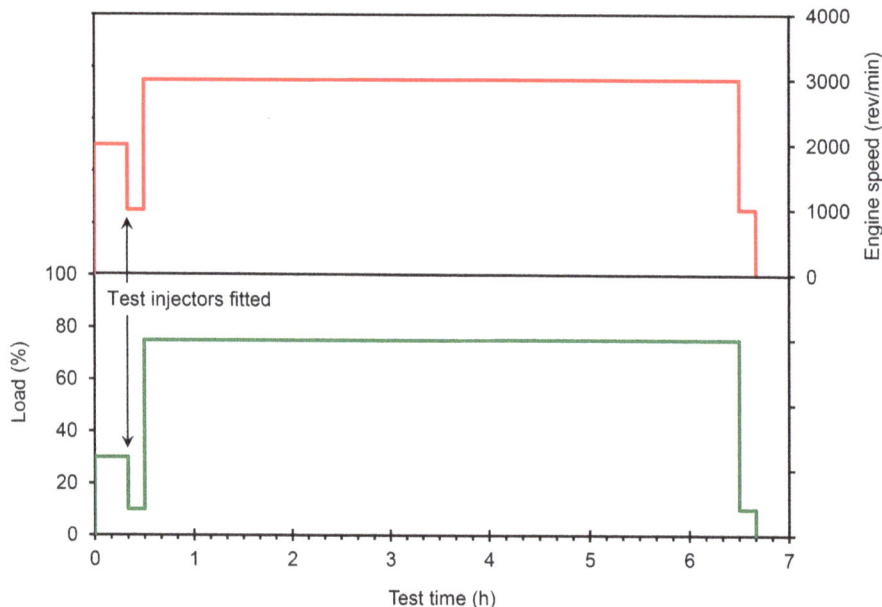

The CEC procedure became the industry standard for assessing the deposit-forming tendency of diesel fuels. Attempts were made to develop less-expensive tests by using a smaller engine that was part of an electrical generator set. The test duration remained the same. While this proposed procedure gave good correlation with the CEC test procedure it did require modification of the engine [18.140]. This added some complexity when compared to the CEC procedure that used an unmodified engine.

A disadvantage of using the unmodified engine was that the severity of the test procedure could not readily be altered, and this led to the eventual replacement of this procedure. The efficacy of the CEC test procedure had been maintained by running round-robin programs using a reference fuel and the same reference fuel with the addition of a known DCA. The reference fuel was blended to give a specified level of deposit. Unfortunately, variation in successive batches of reference fuel and the lack of engine adjustment resulted in the inability of this test method to provide the necessary discrimination. This led to the development of a replacement procedure that will be discussed in Section 18.2.2.3.

18.2.3.2. Cummins L10 Test Method

The test procedure described using the XUD-9 engine was at the time considered adequate to test the propensity of fuels to cause deposits in IDI engines. However, there was still concern over deposit formation in DI engines where the deposit formation mechanisms may be different. Various attempts were made to develop procedures for assessing fuel performance in DI engines. What were visually significant levels of deposit could be accumulated in a very short period of time, as is illustrated in Figure 18.11 [18.58].

FIGURE 18.11 Clean (above) and fouled (below) injector tip after 1.5 hours running.

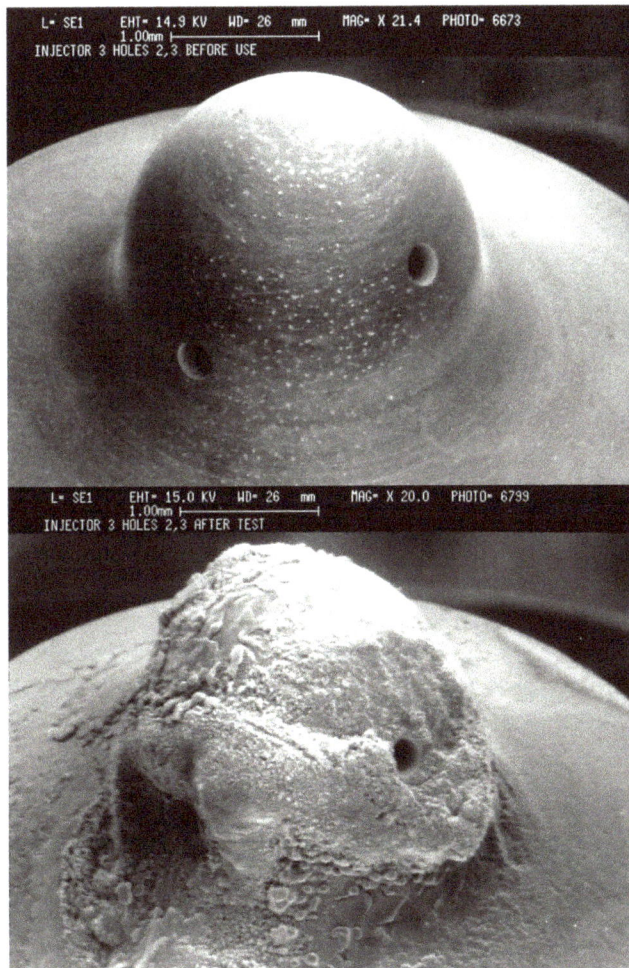

© SAE International.

These very short screening tests did not produce any measurable change in engine performance, and the only metric for assessing deposit-forming tendency was the weight of deposit. The weight of deposit did show a response to DCA treat rate as shown in Figure 18.12 [18.58], but as this could not be correlated with engine performance it was not considered a meaningful test procedure.

FIGURE 18.12   Effect of DCA treat rate on injector deposit weight.

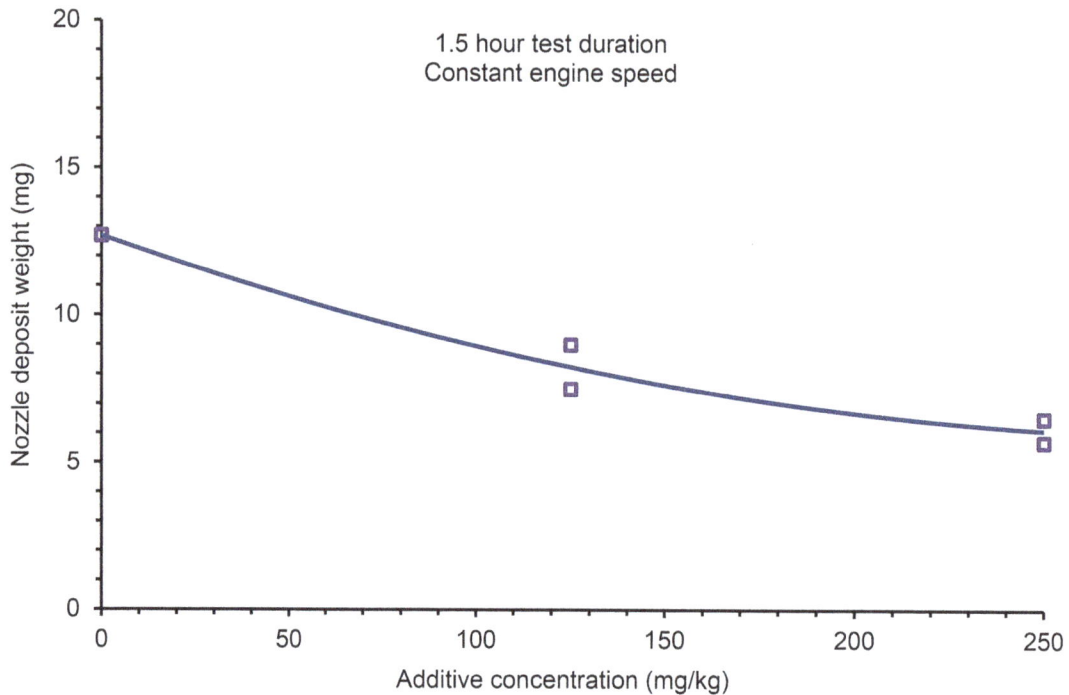

Other in-house test procedures were developed to assess the deposit-forming tendency of fuels by measuring the smoke emissions from various DI engines [18.58, 18.141]. These test procedures used DI engines with cylinder swept volumes of up to 1 L and in-line fuel injection pumps. Such engines could be considered as representative of the upper end of the size range of light-duty vehicles of the time. In one of these procedures, to obtain any measurable change in engine performance it was necessary to reduce the injector opening pressure to increase the possibility of combustion gases entering the injector [18.141].

At around the same time, a procedure was proposed using a multicylinder engine with individual cylinder swept volumes of 1667 cm^3. This gave the complete engine a swept volume of 10 L, which put it into the heavy-duty category. This test procedure also differed from those mentioned in that it employed the Cummins PT® fuel system, which was claimed to be a

Novel means for substantially preventing products of combustion from entering the injector and passing to critical parts thereof or to fuel lines connected to the injector

[18.142]. The design of the injector nozzle is shown schematically in Figure 18.13 [18.142].

FIGURE 18.13 Schematic of Cummins PT® injector nozzle showing the charging position (left) and the injection position (right).

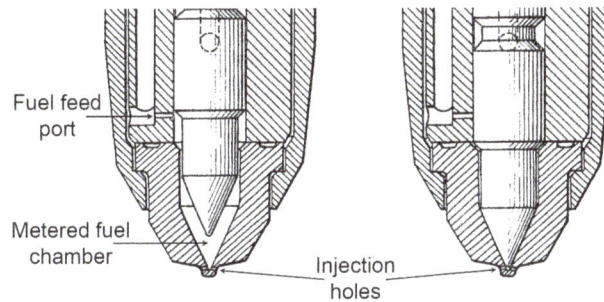

The injector needle valve is held in the charging position (as shown to the left of Figure 18.12) by a stiff spring (not shown in the figure). Fuel is fed into the metering chamber through the feed port. During the injection process, the injector needle is driven against the return spring by a cam and push-rod mechanism to the injection position (as shown to the right of Figure 18.12). This forces the fuel from the metering chamber and into the combustion chamber through the nozzle holes. After the injection event, the plunger returns under the action of the return spring to allow the chamber to refill with fuel. Although the injection system was designed to prevent combustion chamber gases entering the fuel system, in all probability during this return stroke some combustion chamber products would be drawn into the metering chamber. This would allow combustion products to mix with the incoming fuel for the next injection event. This would promote degradation of the fuel, given sufficient time.

In the proposed test procedure [18.143] two Cummins L10 engines were connected together, nose to tail, without a dynamometer. One engine was run under power to motor the other engine, and every 15 seconds the engines were switched from powering to motoring. Engine speed was controlled by adjusting the high idle governor on the fuel pumps, and the engines were then run at either full rack or closed rack. The load and speed cycle were thus as shown in Figure 18.14 [18.143], and this cycle was repeated to give a total test duration of 125 hours.

FIGURE 18.14 Cummins L10 test cycle.

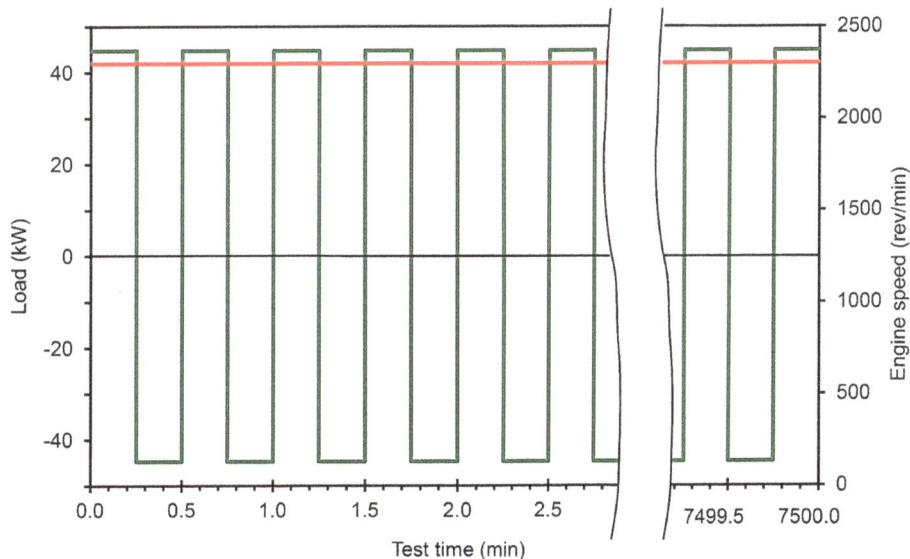

The degree of fouling was determined by measuring the loss of flow through the injector and by visual rating of the deposits on the injector needle. The tip of the needle is the portion that would be exposed to fuel that may have been mixed with a small amount of combustion products as described. To make the visual assessment of the deposits less subjective, an image analysis system was proposed and demonstrated [18.144].

Although the Cummins L10 test procedure was widely used and recognized as a heavy-duty diesel engine nozzle fouling test, the engine became obsolete before it was approved by any of the testing authorities. Other test procedures have been proposed [18.145], the most recent being in 2002 [18.146], but there is currently only one approved test procedure for nozzle fouling tendency in DI engine, and that is the Peugeot DW10 procedure discussed in Section 18.2.3.4.

18.2.3.3. Peugeot XUD-9A Test Method

As discussed in Section 18.2.3.1, the XUD-9 test procedure was not able to provide the required discrimination between fuels, and it was necessary to develop a new procedure. The replacement procedure used the later version of the XUD-9 engine designated as the XUD-9 A/L, and the six-hour constant speed load test was replaced with a ten-hour cyclic test, similar to that used for gasoline inlet valve deposit testing. The test cycle is shown schematically in Figure 18.15 [18.147].

FIGURE 18.15 Ten-hour XUD-9 A/L test cycle.

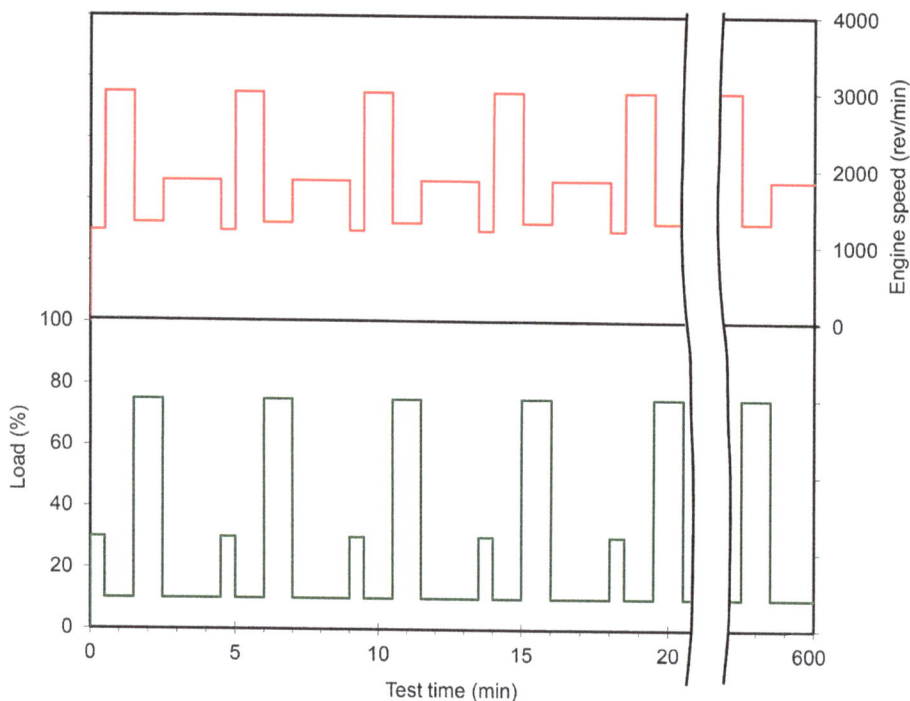

The revised procedure also differed in that the fuel injection timing was set to give a prescribed degree of injector fouling on an un-additised reference fuel. As the severity of different reference fuel batches changed, it was thus possible to adjust the engine calibration to suit. This reference fuel is referred to as the calibration fuel. To ensure the engine provides the necessary degree of fouling, the automatic timing adjustment mechanism within the fuel injection pump is locked, and the dynamic fuel injection timing is adjusted according to a well-defined set of criteria. The dynamic timing is then set within these criteria to produce a specified level of nozzle coking for the calibration fuel. The ability to adjust the engine calibration to match changes in fuels has allowed this test method to continue as the only industry standard for IDI injector deposit-forming tendency [18.148].

The nozzle fouling tendency is determined by measuring the air flow through the clean and fouled injector nozzles at needle lifts of 0.1, 0.2, and 0.3 mm. An average flow reduction at the 0.1 mm condition is used as the performance criteria.

18.2.3.4. Peugeot DW10 Test Method

The XUD-9A/L test procedure described is still a universally recognized test procedure for assessing the propensity of fuels to cause deposit formation on IDI fuel injectors. However, the use of IDI technology is rapidly being replaced by high-pressure common-rail fuel injection technology. It was believed necessary to develop a test procedure using an engine employing that technology.

Despite reported field problems, it proved extremely difficult to reproduce such problems under controlled laboratory conditions. A CEC group has developed a test procedure using a Peugeot DW10 ATED engine that uses fuel doped with zinc neodecanoate. It has been shown that the inclusion of the zinc dopant makes the fuel significantly more prone to deposit formation [18.149, 18.150]. The reason for this appears to be that the zinc in the fuel forms zinc compounds that deposit in the small spray holes of the injector. This reduces fuel flow and thus reduces the maximum power attainable from the engine. Maximum deliverable power is the metric for assessing injector fouling.

The mechanism by which the zinc is believed to contribute to deposit formation is described next and illustrated in Figure 18.16 [18.151].

FIGURE 18.16 Schematic of zinc-induced nozzle fouling mechanism.

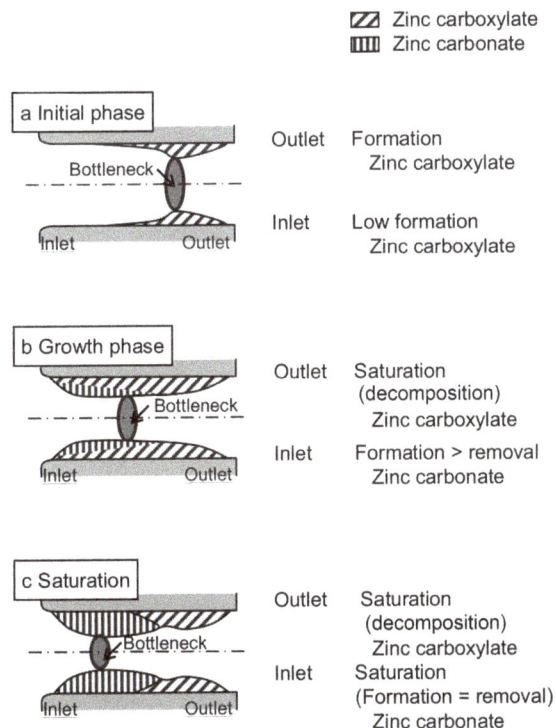

The accumulation of zinc compounds deposited in the fuel injector holes can be considered as three phases: the initial phase, the growth phase, and finally the saturation stage.

During the initial phase, zinc carboxylates form along the length of the nozzle hole as a result of the reaction between zinc and lower molecular weight carboxylic acids, such as formic or acetic acid, in the fuel or combustion gas. Larger amounts of deposit form toward the outlet of the nozzle hole due to higher temperatures toward the hot combustion gases. This forms a bottleneck for the injected fuel flow, as shown in the upper part of Figure 18.16.

During the growth phase, the deposits accumulate farther into the nozzle hole. The deposits accumulating close to the nozzle hole inlet increase and gradually react with water and carbon dioxide that are forced back into the injector hole to form zinc carbonate. The higher temperatures toward the nozzle hole outlet prevent the formation of the carbonate. The deposits close to the nozzle hole reach a saturation point where the formation and removal of deposits are equal. This is shown in the center of Figure 18.16.

During the saturation phase, the amount of zinc carbonate close to the nozzle hole inlet gradually accumulates, moving the bottleneck farther into the nozzle hole. Finally, a saturation condition is reached where the deposit formation is in equilibrium with the decomposition and removal. This is shown in the lower part of Figure 18.16.

The CEC test procedure [18.152] utilizes a severe 1-hour operating cycle that is repeated eight times, followed by a 4-hour shutdown and soak period before the whole procedure is repeated. The test is normally run to give a total of 32 hours of engine operating time. The 1-hour operating cycle is illustrated schematically in Figure 18.17 [18.153].

FIGURE 18.17 DW10 test cycle.

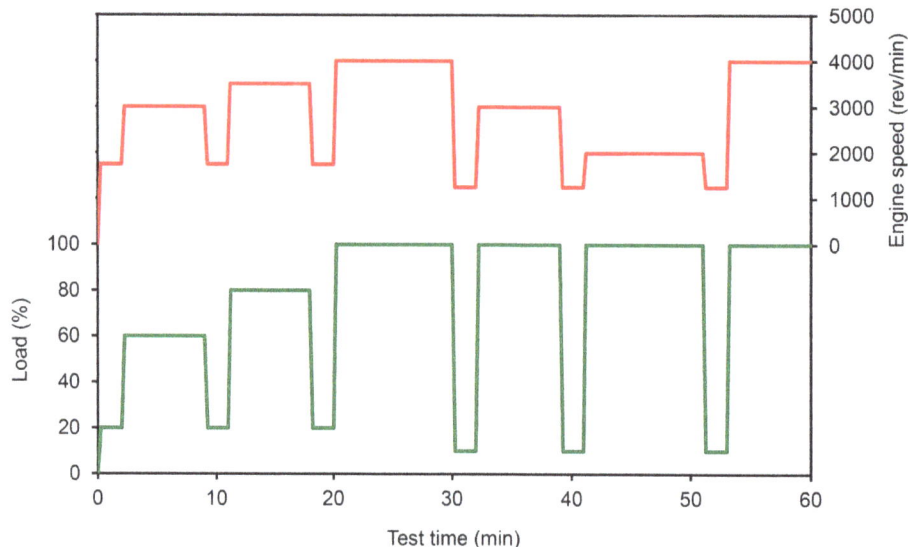

The fuel for the test is doped with zinc neodecanoate to give 1 mg/kg of zinc in the fuel. This is at least ten times the level that would commonly be found in market diesel fuels [18.153]. Due to the severity of the operating cycle, it has been estimated that 48 hours of operation on the cycle is "equivalent to what a vehicle would experience over a full life time" [18.154]. The efficacy of including zinc has also been called into question as the morphology of the deposits differs as a result of the inclusion of the zinc [18.155]. It has been shown that with a fully formulated fuel there is no measurable zinc pick-up during the operation of a vehicle over 50,000 km [18.156]. It has also been shown that different fuels have a different propensity to pick up zinc during storage, transportation, and use. These fuels also have different tendency to keep zinc compounds such as zinc neodecanoate in solution [18.157]. The DW10 test cannot therefore be considered applicable to all commercial diesel fuels but merely a test of a fuel additive's propensity to prevent deposits from the adulterated reference fuel.

The use of a fuel containing 10% by volume of biodiesel and no zinc adulteration has shown wide variations in performance. Some tests have resulted in no measurable injector fouling [18.153], while others showed levels of fouling below that of the zinc adulterated fuel [18.158] and others showed higher levels of fouling than the zinc adulterated fuel [18.154].

References

18.1.   Schwartz, F.G., Ward, C.C., and Smith, H.M., "Studies on the Storage Stability of Distillate Fuels Results of Storage Tests," SAE Technical Paper 530181 (1953), doi:https://doi.org/10.4271/530181.

18.2.   Pedley, J.F., Hiley, R.W., and Hancock, R.A., "Storage Stability of Petroleum-Derived Diesel Fuel: 1. Analysis of Sediment Produced during the Ambient Storage of Diesel Fuel," *Fuel* 66, no. 12 (1987): 1646-1651.

18.3.   Pedley, J.F. and Hiley, R.W., "Investigation of 'Sediment Precursors' Present in Cracked Gas Oil," in *3rd International Conference on Stability and Handling of Liquid Fuels*, London, UK, 1988.

18.4.   Batts, B.D. and Zuhdan, F., "A literature Review on Fuel Stability Studies with Particular Emphasis on Diesel Oil," *Energy and Fuels* 5 (1991): 2-21.

18.5.   Hartikka, T., Kisski, U., Kuronenen, M., and Mikkonen, S., "Diesel Fuel Oxidation Stability: A Comparative Study," SAE Technical Paper 2013-01-2768 (2013), doi:https://doi.org/10.4271/2013-01-2768.

18.6.   Bentley, J.R. and Schellhase, H.-U., "Fuel Stability and Storage Life of Middle Distillate Fuels," SAE Technical Paper 831205 (1983), doi:https://doi.org/10.4271/831205.

18.7.   Bachman, K.C., "Heat Transfer Unit Evaluates Performance of Jet Fuels for Supersonic Aircraft," SAE Technical Paper 650803 (1965), doi:https://doi.org/10.4271/650803.

18.8.   Harrison, W.E., Binns, K.E., Anderson, S.D., and Morris, R.W., "High Heat Sink Fuels for Improved Aircraft Thermal Management," SAE Technical Paper 932084 (1993), doi:https://doi.org/10.4271/932084.

18.9.   Pickard, J.M. and Jones, E.G., "Liquid Phase Oxidation Kinetics: Paraffin Blends," *Energy and Fuels* 12, no. 6 (1998): 1241-1244.

18.10. Morris, R.E., Hazlett, R.N., and McIlvaine, C.L. III, "The Effects of Stabilizer Additives on the Thermal Stability of Jet Fuel," *Industrial & Engineering Chemistry Research* 27, no. 8 (1988): 1524-1528.

18.11. Zabarnick, S. and Mick, M.S., "Inhibition of Jet Fuel Oxidation by Addition of Hydroperoxide-Decomposing Species," *Industrial & Engineering Chemistry Research* 38, no. 9 (1999): 3557-3563.

18.12. Beaver, B., "Long Term Storage Stability of Middle Distillate Fuels from a Chemical Mechanistic Point of View. Part 1," *Fuel Science and Technology International* 9, no. 10 (1991): 1287-1335.

18.13. Fang, H.L. and McCormick, R.L., "Spectroscopic Study of Biodiesel Degradation Pathways," SAE Technical Paper 2006-01-3300 (2006), doi:https://doi.org/10.4271/2006-01-3300.

18.14. Jensen, R.K., Korcek, S., Mahoney, L.R., and Zinbo, M., "Liquid-Phase Autoxidation of Organic Compounds at Elevated Temperatures. 1. The Stirred Flow Reactor Technique and Analysis of Primary Products from n-Hexadecane Autoxidation at 120-180°C," *Journal of the American Oil Chemists Society* 101, no. 25 (1979): 7574-7584.

18.15. Jensen, R.K., Korcek, S., Mahoney, L.R., and Zinbo, M., "Liquid-Phase Autoxidation of Organic Compounds at Elevated Temperatures. 2. Kinetics and Mechanisms of the Formation of Cleavage Products in n-Hexadecane Autoxidation," *Journal of the American Oil Chemists Society* 103, no. 7 (1981): 1742-1749.

18.16. Jensen, R.K., Korcek, S., and Zinbo, M., "Formation, Isomerization, and Cyclization Reactions of Hydroperoxyalkyl Radicals in Hexadecane Autoxidation at 160-190°C," *Journal of the American Oil Chemists Society* 114, no. 20 (1992): 7742-7748.

18.17. Mayo, F.R. and Lan, B.Y., "Gum and Deposit Formation from Jet Turbine and Diesel Fuels at 130°C," *Industrial & Engineering Chemistry Product Research and Development* 25, no. 2 (1986): 333-348.

18.18. Balster, L.M., Zabarnick, S., Striebich, R.C., Shafer, L.M. et al., "Analysis of Polar Species in Jet Fuel and Determination of Their Role in Autoxidative Deposit Formation," *Energy and Fuels* 20, no. 6 (2006): 2564-2572.

18.19. Link, D.D. and Baltrus, J.P., "Isolation and Identification of Nitrogen Species in Jet Fuel and Diesel Fuel," *Energy and Fuels* 21, no. 3 (2007): 1575-1581.

18.20. Pedley, J.F., Hiley, R.W., and Hancock, R.A., "Storage Stability of Petroleum-Derived Diesel Fuel: 3. Identification of Compounds Involved in Sediment Formation," *Fuel* 67, no. 8 (1988): 1124-1130.

18.21. Cosgrove, J.P., Church, D.F., and Pryor, W.A., "The Kinetics of the Autoxidation of Polyunsaturated Fatty Acids," *Lipids* 22, no. 5 (1987): 299-304.

18.22. Tang, H., De Guzman, R.C., Salley, S.O., and Ng, K.Y.S., "The Oxidative Stability of Biodiesel: Effects of FAME Composition and Antioxidant," *Lipid Technology* 20, no. 11 (2008): 249-252.

18.23. Knothe, G., "Structure Indices in FA Chemistry. How Relevant is the Iodine Value?" *Journal of the American Oil Chemists Society* 79, no. 9 (2002): 847-854.

18.24. McCormick, R.L. and Westbrook, S.R., "Empirical Study of the Stability of Biodiesel and Biodiesel Blends," NREL/TP-540-41619, 2007.

18.25. Ohshio, N., Saito, K., Kobayashi, S., and Tanaka, S., "Storage Stability of FAME Blended Diesel Fuels," SAE Technical Paper 2008-01-2505 (2008), doi:https://doi.org/10.4271/2008-01-2505.

18.26. Wexler, H., "Polymerization of Drying Oils," *Chemical Reviews* 64, no. 6 (1964): 591-611.

18.27. Johnson, O.C. and Kummerow, F.A., "Chemical Changes Which Take Place in an Edible Oil during Thermal Oxidation," *Journal of the American Oil Chemists Society* 34, no. 8 (1957): 407-409.

18.28. Miyata, I., Takei, Y., Tsurutani, K., and Okada, M., "Effects of Bio-Fuels on Vehicle Performance: Degradation Mechanism Analysis of Bio-Fuels," SAE Technical Paper 2004-01-3031 (2004), doi:https://doi.org/10.4271/2004-01-3031.

18.29. McCormick, R.L. and Westbrook, S.R., "Storage Stability of Biodiesel and Biodiesel Blends," *Energy Fuels* 24, no. 1 (2010): 690-698.

18.30. Ogawa, T., Kajiya, S., Kosaka, S., Tajima, I. et al., "Analysis of Oxidative Deterioration of Biodiesel Fuel," *SAE Int. J. Fuels Lubr.* 1, no. 1 (2009): 1571-1583, doi:https://doi.org/10.4271/2008-01-2502.

18.31. Miyashita, K. and Takagi, T., "Study of the Oxidative Rate and Prooxidant Activity of Free Fatty Acids," *Journal of the American Oil Chemists Society* 63, no. 10 (1986): 1380-1384.

18.32. Bondioli, P., Gasparoli, A., Lanzani, A., Fedeli, E. et al., "Storage Stability of Biodiesel," *Journal of the American Oil Chemists Society* 72, no. 6 (1995): 699-702.

18.33. Shiotani, H. and Goto, S., "Studies of Fuel Properties and Oxidation Stability of Biodiesel Fuel," SAE Technical Paper 2007-01-0073 (2007), doi:https://doi.org/10.4271/2007-01-0073.

18.34. Bondioli, P., Gasparoli, A.D., Bella, L., and Tagliabue, S., "Evaluation of Biodiesel Storage Stability Using Reference Methods," *European Journal of Lipid Science and Technology* 104, no. 12 (2002): 777-784.

18.35. Omori, T., Tanaka, A., Yamada, K., and Bunne, S., "Biodiesel Deposit Formation Mechanism and Improvement of Fuel Injection Equipment (FIE)," SAE Technical Paper 2011-01-1935 (2011), doi:https://doi.org/10.4271/2011-01-1935.

18.36. ASTM International, "Standard Test Method for Oxidation Stability of Distillate Fuel Oil (Accelerated Method)," ASTM D2274-14(2019), ASTM International, 2019.

18.37. Owen, K. (ed.) "Gasoline and Diesel Fuel Additives," in *Critical Reports on Applied Chemistry*, Vol. 25 (Chichester, UK: John Wiley & Sons Ltd, 1989).

18.38. ASTM International, "Standard Test Method for ASTM Color of Petroleum Products (ASTM Color Scale)," ASTM D1500-12(2017), ASTM International, 2018.

18.39. International Organization for Standardization, "Petroleum Products—Determination of the Oxidation Stability of Middle-Distillate Fuels," ISO 12205:1995, International Organization for Standardization, 1996.

18.40. ASTM International, "Standard Test Method for Assessing Middle Distillate Fuel Storage Stability by Oxygen Overpressure," ASTM D5304-20, ASTM International, 2020.

18.41. ASTM International, "Standard Test Method for Distillate Fuel Storage Stability at 43°C (110°F)," ASTM D4625-21, ASTM International, 2021.

18.42. ASTM International, "Standard Test Method for High Temperature Stability of Middle Distillate Fuels," ASTM D6468-22, ASTM International, 2022.

18.43. British Standards Institute, "Fat and Oil Derivatives. Fatty Acid Methyl Esters (FAME). Determination of Oxidation Stability (Accelerated Oxidation Test)," BS EN 14112:2003, British Standards Institute, 2003.

18.44. Conconi, C.C., Canale, L.C., and Totten, G.E., "Determination of Biodiesel Oxidation Stability of Biodiesel B100 with Optical Spectroscopies of Eletronic Absorption UV-Visible Correlation with Rancimat Method DIN EN 14112," SAE Technical Paper 2010-36-0144 (2010), doi:https://doi.org/10.4271/2010-36-0144.

18.45. Firestone, D. (Ed.), *Official Methods and Recommended Practices of the AOCS Cd 12b-92 Oil Stability Index (OSI)*, 6th ed., 2nd Printing (Champaign, IL: AOCS Press, 2009).

18.46. Pullen, J. and Saed, K., "An Overview of Biodiesel Oxidation Stability," *Renewable and Sustainable Energy Reviews* 16, no. 8 (2012): 5924-5950.

18.47. ASTM International, "Standard Test Method for Oxidation Stability of Biodiesel (B100) and Blends of Biodiesel with Middle Distillate Petroleum Fuel (Accelerated Method) (Withdrawn 2016)," ASTM D7462-11, ASTM International, 2016.

18.48. ASTM International, "Standard Test Method for Oxidation Stability of Middle Distillate Fuels—Rapid Small Scale Oxidation Test (RSSOT)," ASTM D7545-14(2019)e1, ASTM International, 2019.

18.49. British Standards Institute, "Liquid Petroleum Products—Middle Distillates and Fatty Acid Methyl Ester (FAME) Fuels and Blends—Determination of Oxidation Stability by Rapid Small Scale Oxidation Test (RSSOT)," BS EN 16091:2022, British Standards Institute, 2022.

18.50. Ainsley, W.G., "Evaluation of Diesel Fuels in Full-Scale Engines (Report of the Cooperative Fuel Research Committee)," SAE Technical Paper 410134 (1941), doi:https://doi.org/10.4271/410134.

18.51. Irish, G.E. and Mattson, R.W., "Cleaner Injectors and Less Smoke with Hydrogen-Treated Diesel Fuel," SAE Technical Paper 640459 (1964), doi:https://doi.org/10.4271/640459.

18.52. Brubacher, M.L., "Reduction of Diesel Smoke in California," SAE Technical Paper 660548 (1966), doi:https://doi.org/10.4271/660548.

18.53. Montagne, X., Herrier, D., and Guibet, J.-C., "Fouling of Automotive Diesel Injectors-Test Procedure, influence of Composition of Diesel Oil and Additives," SAE Technical Paper 872118 (1987), doi:https://doi.org/10.4271/872118.

18.54. Chaput, H., "Urban Transportation and Emissions," SAE Technical Paper 710559 (1971), doi:https://doi.org/10.4271/710559.

18.55. Sutton, D.L., "Investigation into Diesel Operation with Changing Fuel Property," SAE Technical Paper 860222 (1986), doi:https://doi.org/10.4271/860222.

18.56. Blanco, J.C., "Effect of Diesel Fuel Quality and Additives on Engine Performance—An Update," SAE Technical Paper 911710 (1991), doi:https://doi.org/10.4271/911710.

18.57. Olsen, R.E., Ingham, M.C., and Parsons, G.M., "A Fuel Additive Concentrate for Removal of Injector Deposits in Light-Duty Diesels," SAE Technical Paper 841349 (1984), doi:https://doi.org/10.4271/841349.

18.58. Winterbone, D.E., Clough, E., Rao, K.K., Richards, P. et al., "The Effect of DI Nozzle Fouling on Fuel Spray Characteristics," SAE Technical Paper 922232 (1992), doi:https://doi.org/10.4271/922232.

18.59. Schwab, S.D., Bennett, J.J., Dell, S.J., Galante-Fox, J.M. et al., "Internal Injector Deposits in High-Pressure Common Rail Diesel Engines," *SAE Int. J. Fuels Lubr.* 3, no. 2 (2010): 865-878, doi:https://doi.org/10.4271/2010-01-2242.

18.60. Lacey, P., Gail, S., Kientz, J., Milovanovic, N. et al., "Internal Fuel Injector Deposits," *SAE Int. J. Fuels Lubr.* 5, no. 1 (2012): 132-145, doi:https://doi.org/10.4271/2011-01-1925.

18.61. Barker, J., Reid, J., Wilmot, E., Mulqueen, S. et al., "Investigations of Diesel Injector Deposits Characterization and Testing," SAE Technical Paper 2020-01-2094 (2020), doi:https://doi.org/10.4271/2020-01-2094.

18.62. Lacey, P., Gail, S., Kientz, J.-M., Benoist, G. et al., "Fuel Quality and Diesel Injector Deposits," *SAE Int. J. Fuels Lubr.* 5, no. 3 (2012): 1187-1198, doi:https://doi.org/10.4271/2012-01-1693.

18.63. Tsuchiya, T., Shiotani, H., Goto, S., Sugiyama, G. et al., "Japanese Standards for Diesel Fuel Containing 5% FAME: Investigation of Acid Generation in FAME Blended Diesel Fuels and its Impact on Corrosion," SAE Technical Paper 2006-01-3303 (2006), doi:https://doi.org/10.4271/2006-01-3303.

18.64. Sugiyama, G., Maeda, A., and Nagai, K., "Oxidation Degradation and Acid Generation in Diesel Fuel Containing 5% FAME," SAE Technical Paper 2007-01-2027 (2007), doi:https://doi.org/10.4271/2007-01-2027.

18.65. Fraer, R., Dinh, H., Proc, K., McCormick, R.L. et al., "Operating Experience and Teardown Analysis for Engines Operated on Biodiesel Blends (B20)," SAE Technical Paper 2005-01-3641 (2005), doi:https://doi.org/10.4271/2005-01-3641.

18.66. Proc, K., Barnitt, R., Hayes, R., Ratcliff, M. et al., "100,000-Mile Evaluation of Transit Buses Operated on Biodiesel Blends (B20)," SAE Technical Paper 2006-01-3253 (2006), doi:https://doi.org/10.4271/2006-01-3253.

18.67. Lammert, M., Barnitt, R., and McCormick, R.L., "Field Evaluation of Biodiesel (B20) Use by Transit Buses," *SAE Int. J. Commer. Veh.* 2, no. 2 (2010): 209-221, doi:https://doi.org/10.4271/2009-01-2899.

18.68. Bartoli, Y., Lyford-Pike, E.J., Lucke, J.E., Khalek, I.A. et al., "1000-Hour Durability Evaluation of a Prototype 2007 Diesel Engine with Aftertreatment Using B20 Biodiesel Fuel," *SAE Int. J. Fuels Lubr.* 2, no. 2 (2010): 290-304, doi:https://doi.org/10.4271/2009-01-2803.

18.69. Barker, J., Langley, G., Carter, A., Herniman, J. et al., "Investigations Regarding the Causes of Filter Blocking in Diesel Powertrains," SAE Technical Paper 2022-01-1069 (2022), doi:https://doi.org/10.4271/2022-01-1069.

18.70. Heiden, R.W., Schober, S., and Mittelbach, M., "Solubility Limitations of Residual Steryl Glucosides, Saturated Monoglycerides and Glycerol in Commercial Biodiesel Fuels as Determinants of Filter Blockages," *Journal of the American Oil Chemists' Society* 98, no. 12 (2021): 1143-1165.

18.71. Csontos, B., Bernemyr, H., Erlandsson, A., Forsberg, O. et al., "Characterization of Deposits Collected from Plugged Fuel Filters," *SAE Int. J. Adv. & Curr. Prac. in Mobility* 2, no. 2 (2020): 672-680, doi:https://doi.org/10.4271/2019-24-0140.

18.72. Csontos, B., Shinkhede, S., Bernemyr, H., Pach, M. et al., "Comparison of Fuel Filters and Adsorption Filters for Metal Carboxylate Separation," SAE Technical Paper 2021-24-0064 (2021), doi:https://doi.org/10.4271/2021-24-0064.

18.73. Terry, B., McCormick, R.L., and Natarajan, M., "Impact of Biodiesel Blends on Fuel System Component Durability," SAE Technical Paper 2006-01-3279 (2006), doi:https://doi.org/10.4271/2006-01-3279.

18.74. Lopes, S.M. and Cushing, T., "The Influence of Biodiesel Fuel Quality on Modern Diesel Vehicle Performance," SAE Technical Paper 2012-01-0858 (2012), doi:https://doi.org/10.4271/2012-01-0858.

18.75. Wright, R.H. and Hostetler, H.F., "Microbiological Diesel Fuel Contamination," SAE Technical Paper 630120 (1963), doi:https://doi.org/10.4271/630120.

18.76. Williamson, C.H., Jain, L.A., Mishra, B., Olson, D.L. et al., "Microbially Influenced Corrosion Communities Associated with Fuel-Grade Ethanol Environments," *Applied Microbiology and Biotechnology* 99 (2015): 6945-6957.

18.77. Barker, J. and Reid, J., "Injector and Fuel System Deposits," in *10th International Colloquium Fuels Conventional and Future Energy for Automobiles*, Stuttgart, Germany, January 2015.

18.78. US Environmental Protection Agency, "Investigation of Corrosion Influencing Factors in Underground Storage Tanks with Diesel Service," EPA 510-R-16-001. 2016.

18.79. Barker, J., Richards, P., Goodwin, M., and Wooler, J., "Influence of High Injection Pressure on Diesel Fuel Stability: A Study of Resultant Deposits," *SAE Int. J. Fuels Lubr.* 2, no. 1 (2009): 877-884, doi:https://doi.org/10.4271/2009-01-1877.

18.80. Miyaki, M., Fujisaw, H., Masuda, A., and Yamamoto, Y., "Development of New Electronically Controlled Fuel Injection System ECD-U2 for Diesel Engines," SAE Technical Paper 910252 (1991), doi:https://doi.org/10.4271/910252.

18.81.   Ohishi, K., Maeda, T., and Hummel, K., "The New Common Rail Fuel System for Duramax 6600 V8 Diesel Engine," SAE Technical Paper 2001-01-2704 (2001), doi:https://doi.org/10.4271/2001-01-2704.

18.82.   Dober, G., Tullis, S., Greeves, G., Milovanovic, N. et al., "The Impact of Injection Strategies on Emissions Reduction and Power Output of Future Diesel Engines," SAE Technical Paper 2008-01-0941 (2008), doi:https://doi.org/10.4271/2008-01-0941.

18.83.   Matsumoto, S., Yamada, K., and Date, K., "Concepts and Evolution of Injector for Common Rail System," SAE Technical Paper 2012-01-1753 (2012), doi:https://doi.org/10.4271/2012-01-1753.

18.84.   Matsumoto, S., "4th Generation Common Rail System," SAE Technical Paper 2013-01-1590 (2013), doi:https://doi.org/10.4271/2013-01-1590.

18.85.   Quigley, R., Barbour, R., Arters, D., and Bush, J., "Understanding the Spectrum of Diesel Deposits" in *9th International Colloquium Fuels: Conventional and Future Energy for Automobiles*, Stuttgart Germany, 2013.

18.86.   Barker, J., Reid, J., Angel Smith, S., Snape, C. et al., "The Application of New Approaches to the Analysis of Deposits from the Jet Fuel Thermal Oxidation Tester (JFTOT)," *SAE Int. J. Fuels Lubr.* 10, no. 3 (2017), Erratum published in *SAE Int. J. Fuels Lubr.* 11, no. 1 (2018): 143.

18.87.   Risberg, P. and Alfredson, S., "The Effect of Zinc and other Metal Carboxylates on Nozzle Fouling," SAE Technical Paper 2016-01-0837 (2016), doi:https://doi.org/10.4271/2016-01-0837.

18.88.   Stępień, Z., "Investigations of Injector Deposits in Modern Diesel Engines," *Combustion Engines* 55, no. 2 (2016): 9-20.

18.89.   Stępień, Z., "Deterioration of the Fuel Injection Parameters as a Result of Common Rail Injectors Deposit Formation," *MATEC Web of Conferences* 118 (2017): 00002.

18.90.   Stępień, Z., "Zinc as a Catalyst Supports Processes of Diesel Injector Deposit Formation," *Nafta-Gaz* 74, no. 2 (2018): 130-137.

18.91.   Stępień, Z., Mazanek, A., and Suchecki, A., "Impact of Fuel on Real Diesel Injector Performance in Field Test," *Proceedings of the Institution of Mechanical Engineers, Part D: Journal of Automobile Engineering* 232, no. 8 (2018): 1047-1059.

18.92.   Sykes, D., de Sercey, G., Gold, M., Pearson, R. et al., "Visual Analyses of End Injection Liquid Structures and the Behaviour of Nozzle Surface Bound Fuel in Direct Injection Diesel Engine," SAE Technical Paper 2019-01-0059 (2019), doi:https://doi.org/10.4271/2019-01-0059.

18.93.   Bernemyr, H., Csontos, B., Hittig, H., and Forsberg, O., "Study of Nozzle Fouling: Deposit Build-Up and Removal," SAE Technical Paper 2019-01-2231 (2019), doi:https://doi.org/10.4271/2019-01-2231.

18.94.   Barker, J., Snape, C., and Scurr, D., "A Novel Technique for Investigating the Characteristics and History of Deposits Formed within High Pressure Fuel Injection Equipment," *SAE Int. J. Fuels Lubr.* 5, no. 3 (2012): 1155-1164, doi:https://doi.org/10.4271/2012-01-1685.

18.95.   Barker, J., Snape, C., and Scurr, D., "Diesel Deposits," in *9th International Colloquium Fuels: Conventional and Future Energy for Automobiles*, Stuttgart Germany, 2013.

18.96.   Dallanegra, R. and Caprotti, R., "Chemical Composition of Ashless Polymeric Internal Diesel Injector Deposits," SAE Technical Paper 2014-01-2728 (2014), doi:https://doi.org/10.4271/2014-01-2728.

18.97.   Feld, H. and Oberender, N., "Characterization of Damaging Biodiesel Deposits and Biodiesel Samples by Infrared Spectroscopy (ATR-FTIR) and Mass Spectrometry (TOF-SIMS)," *SAE Int. J. Fuels Lubr.* 9, no. 3 (2016): 717-724, doi:https://doi.org/10.4271/2016-01-9078.

18.98.   Barker, J., Reid, J., Piggott, M., Fay, M. et al., "The Characterisation of Diesel Internal Injector Deposits by Focussed Ion-Beam Scanning Electron Microscopy (FIB-SEM), Transmission Electron Microscopy (TEM), Atomic Force Microscopy (AFM) and Raman Spectroscopy," SAE Technical Paper 2015-01-1826 (2015), doi:https://doi.org/10.4271/2015-01-1826.

18.99.   Barker, J., Reid, J., Smith, S.A., Snape, C. et al., "Internal Injector Deposits (IDID)," in *11th International Colloquium Fuels Conventional and Future Energy for Automobiles*, Stuttgart Germany, 2017.

18.100. Barker, J., Richards, P., Pinch, D., and Cheeseman, B., "Temperature Programmed Oxidation as a Technique for Understanding Diesel Fuel System Deposits," *SAE Int. J. Fuels Lubr.* 3, no. 2 (2010): 85-99, doi:https://doi.org/10.4271/2010-01-1475.

18.101. Barker, J., Snape, C., and Scurr, D., "Information on the Aromatic Structure of Internal Diesel Injector Deposits from Time of Flight Secondary Ion Mass Spectrometry (ToF-SIMS)," SAE Technical Paper 2014-01-1387 (2014), doi:https://doi.org/10.4271/2014-01-1387.

18.102. Lau, K., Junk, R., Klingbeil, S., Schümann, U. et al., "Analysis of Internal Common Rail Injector Deposits via Thermodesorption Photon Ionization Time of Flight Mass Spectrometry," *Energy & Fuels* 29, no. 9 (2015): 5625-5632.

18.103. Geng, P.Y., Buczynsky, A.E., and Konzack, A., "US and EU Market Biodiesel Blends Quality Review—An OEM Perspective," *SAE Int. J. Fuels Lubr.* 2, no. 1 (2009): 860-869, doi:https://doi.org/10.4271/2009-01-1850.

18.104. Alleman, T.L., Fouts, L., and McCormick, R.L., "Quality Analysis of Wintertime B6–B20 Biodiesel Blend Samples Collected in the United States," *Fuel Processing Technology* 92, no. 7 (2011): 1297-1304.

18.105. Mikkonen, S. and Tenhunen, E., "Deposits in Diesel Fuel-Injection Pumps Caused by Incompatibility of Fuel and Oil Additives," SAE Technical Paper 872119 (1987), doi:https://doi.org/10.4271/872119.

18.106. Stehouwer, D.M., Fang, H.L., Wooton, D., and Martin, H., "Interaction between Fuel Additive and Oil Contaminant: (I) Field Experiences," SAE Technical Paper 2003-01-3139 (2003), doi:https://doi.org/10.4271/2003-01-3139.

18.107. CRC, "CRC Internal Diesel Injector Deposit (IDID) Test: Hardware, Fuel, and Additive Evaluations," accessed February 2023, https://crcao.org/wp-content/uploads/2021/06/CRC-DP-04-17_Final-Report_Rev-4-Combined.pdf.

18.108. Cardenas Almena, M., Lucio Esperilla, O., Martin Manzanero, F., Murillo Duarte, Y. et al., "Internal Diesel Injector Deposits: Sodium Carboxylates of C12 Succinic Acids and C16 and C18 Fatty Acids," SAE Technical Paper 2012-01-1689 (2012), doi:https://doi.org/10.4271/2012-01-1689.

18.109. Reid, J., Cook, S., and Barker, J., "Internal Injector Deposits from Sodium Sources," *SAE Int. J. Fuels Lubr.* 7, no. 2 (2014): 436-444, doi:https://doi.org/10.4271/2014-01-1388.

18.110. Trobaugh, C., Burbank, C., Zha, Y., Whitacre, S. et al., "Internal Diesel Injector Deposits: Theory and Investigations into Organic and Inorganic Based Deposits," SAE Technical Paper 2013-01-2670 (2013), doi:https://doi.org/10.4271/2013-01-2670.

18.111. Barker, J., Langley, G.J., and Richards, P., "Insights into Deposit Formation in High Pressure Diesel Fuel Injection Equipment," SAE Technical Paper 2010-01-2243 (2010), doi:https://doi.org/10.4271/2010-01-2243.

18.112. Ullmann, J., Geduldig, M., Stutzenberger, H., Caprotti, R. et al., "Investigation into the Formation and Prevention of Internal Diesel Injector Deposits," SAE Technical Paper 2008-01-0926 (2008), doi:https://doi.org/10.4271/2008-01-0926.

18.113. Cardenas Almena, M., Lucio Esperilla, O., Martin Manzanero, F., Murillo Duarte, Y. et al., "Internal Diesel Injector Deposits: Sodium Carboxylates of C12 Succinic Acids and C16 and C18 Fatty Acids," SAE Technical Paper 2012-01-1689 (2012), doi:https://doi.org/10.4271/2012-01-1689.

18.114. Cook, S., Barker, J., Reid, J., and Richards, P., "Possible Mechanism for Poor Diesel Fuel Lubricity in the Field," *SAE Int. J. Fuels Lubr.* 5, no. 2 (2012): 711-720, doi:https://doi.org/10.4271/2012-01-0867.

18.115. Baranescu, R.A. and Lusco, J.J., "Sunflower Oil as a Fuel Extender in Direct Injection Turbocharged Diesel Engines," SAE Technical Paper 820260 (1982), doi:https://doi.org/10.4271/820260.

18.116. Ziejewski, M. and Goettler, H.J., "Reduced Injection Needle Mobility Caused by Lacquer Deposits from Sunflower Oil," SAE Technical Paper 880493 (1988), doi:https://doi.org/10.4271/880493.

18.117. Ziejewski, M. and Goettler, H.J., "Effect of Lacquer Deposits from Sunflower Oil on Injection Needle Mobility for Different Needle Guide Clearances," SAE Technical Paper 881336 (1988), doi:https://doi.org/10.4271/881336.

18.118. Goettler, H.J., Harwood, R.F., Ziejewski, M., and Klosterman, H.J., "On the Thermal Decomposition and Residue Formation of Plant Oils," SAE Technical Paper 861582 (1986), doi:https://doi.org/10.4271/861582.

18.119. Quigley, R., Barbour, R., Fahey, E., Arters, D.C. et al., "A Study of the Internal Diesel Injector Deposit Phenomenon," in *8th International Colloquium Fuels—Conventional and Future Energy for Automobiles*, Technische Akademie Esslingen e.V., Stuttgart, Germany, 2011.

18.120. Reid, J. and Barker, J., "Understanding Polyisobutylene Succinimides (PIBSI) and Internal Diesel Injector Deposits," SAE Technical Paper 2013-01-2682 (2013), doi:https://doi.org/10.4271/2013-01-2682.

18.121. Barker, J., Reid, J., Snape, C., Scurr, D. et al., "Spectroscopic Studies of Internal Injector Deposits (IDID) Resulting from the Use of Non-Commercial Low Molecular Weight Polyisobutylenesuccinimide (PIBSI)," *SAE Int. J. Fuels Lubr.* 7, no. 3 (2014): 762-770, doi:https://doi.org/10.4271/2014-01-2720.

18.122. CRC, "Scoping Study to Evaluate Two Rig Tests for Internal Injector Sticking," CRC Report No. DP-04, 2012.

18.123. Altin, O., Carbon deposit simulation bench and method therefor. U.S. Patent 2012/0090384, 2012.

18.124. ASTM International, "Standard Test Method for Thermal Oxidation Stability of Aviation Turbine Fuels," ASTM D3241-20c, ASTM International, 2022.

18.125. Pidol, L., Lecointe, B., and Jeuland, N., "MicroCoking Test: An Accelerated Test Method for Predicting the Thermal Stability of Biodiesel," SAE Technical Paper 2008-01-1804 (2008), doi:https://doi.org/10.4271/2008-01-1804.

18.126. Osawa, M., Ebinuma, Y., Sasaki, S., Takashiba, T. et al., "Influence of Base Diesel Fuel Upon Biodiesel Sludge Formation Tendency," *SAE Int. J. Fuels Lubr.* 2, no. 1 (2009): 127-138, doi:https://doi.org/10.4271/2009-01-0482.

18.127. Bouilly, J., Mohammadi, A., Iida, Y., Hashimoto, H. et al., "Biodiesel Stability and its Effects on Diesel Fuel Injection Equipment," SAE Technical Paper 2012-01-0860 (2012), doi:https://doi.org/10.4271/2012-01-0860.

18.128. Coordinating Research Council, "CRC Internal Diesel Injector Deposit (IDID) Test: Hardware, Fuel, and Additive Evaluations," CRC Report No. DP-04-17, September 2019.

18.129. CEC, "CEC Activity Report July–December 2021," accessed February 2023, https://cectests.org/assets/CEC-Activity-Report_July---December-2021_.pdf.

18.130. Kolkowski, B., Williams, R., Gee, M., Rimmer, J. et al., "Development and Application of an Engine Test Method to Rate the Internal Injector Deposit Formation of Diesel Fuels and Additives," SAE Technical Paper 2022-01-1070 (2022), doi:https://doi.org/10.4271/2022-01-1070.

18.131. Altına, R., Çetinkayab, S., and Yücesu, H.S., "The Potential of Using Vegetable Oil Fuels as Fuel for Diesel Engines," *Energy Conversion and Management* 42, no. 5 (2001): 529-538.

18.132. Demirbaş, A., "Chemical and Fuel Properties of Seventeen Vegetable Oils," *Energy Sources* 25, no. 7 (2003): 721-728.

18.133. Barsic, N.J. and Humke, A.L., "Performance and Emissions Characteristics of a Naturally Aspirated Diesel Engine with Vegetable Oil Fuels," SAE Technical Paper 810262 (1981), doi:https://doi.org/10.4271/810262.

18.134. Li, H., Lea-Langton, A., Biller, P., Andrews, G.E. et al., "Effect of Multifunctional Fuel Additive Package on Fuel Injector Deposit, Combustion and Emissions Using Pure Rape Seed Oil for a DI Diesel," SAE Technical Paper 2009-01-2642 (2009), doi:https://doi.org/10.4271/2009-01-2642.

18.135. Attendu, A.C., "The Attendu Heavy-Oil Engine," SAE Technical Paper 260011 (1926), doi:https://doi.org/10.4271/260011.

18.136. Moore, C.C., Mahan, R.I., and Anderson, B.T., "Diesel Engine Fuel Developments," SAE Technical Paper 420016 (1942), doi:https://doi.org/10.4271/420016.

18.137. Norman, G.R., "A New Approach to Diesel Smoke Suppression," SAE Technical Paper 660339 (1966), doi:https://doi.org/10.4271/660339.

18.138. Reynolds, E.G., "A Procedure for the Assessment of Pintle Injector Nozzle Blockage (Nozzle Coking) in Indirect Injection Diesel Engines," SAE Technical Paper 861409 (1986), doi:https://doi.org/10.4271/861409.

18.139. CEC, "Procedure for IDI Diesel Injector Nozzle Coking Test," CEC F-23-X-93, 1993.

18.140. Mulard, P.P. and China, P.N., "Development of a Diesel Fuel Screening Test for Injector Nozzle Coking," SAE Technical Paper 922184 (1992), doi:https://doi.org/10.4271/922184.

18.141. Virk, K., Herbstman, S., and Rawdon, M., "Development of Direct Injection Diesel Engine Injector Keep Clean and Clean Up Tests," SAE Technical Paper 912329 (1991), doi:https://doi.org/10.4271/912329.

18.142. Perr, J.P., Fuel injector. U.S. Patent 3,351,288, 1967.

18.143. Gallant, T.R., Cusano, C.M., Gray, J.T., and Strete, N.M., "Cummins L10 Injector Depositing Test to Evaluate Diesel Fuel Quality," SAE Technical Paper 912331 (1991), doi:https://doi.org/10.4271/912331.

18.144. Blythe, G.H. and Flask, C.A., "Development of an Image Analysis System to Rate Injectors from the Cummins L10 Injector Depositing Test," SAE Technical Paper 972902 (1997), doi:https://doi.org/10.4271/972902.

18.145. Gutman, M., Tartakovsky, L., Kirzhner, Y., and Zvirin, Y., "Development of a Screening Test for Evaluating Detergent/Dispersant Additives to Diesel Fuels," SAE Technical Paper 961184 (1996), doi:https://doi.org/10.4271/961184.

18.146. Williams, R., "Development of a Nozzle Fouling Test for Additive Rating in Heavy Duty DI Diesel Engines," SAE Technical Paper 2002-01-2721 (2002), doi:https://doi.org/10.4271/2002-01-2721.

18.147. Panesar, A., Martens, A., Jansen, L., Lal, S. et al., "Development of a New Peugeot XUD9 10 Hour Cyclic Test to Evaluate the Nozzle Coking Propensity of Diesel Fuels," SAE Technical Paper 2000-01-1921 (2000), doi:https://doi.org/10.4271/2000-01-1921.

18.148. CEC, "Procedure for Diesel Engine Injector Nozzle Coking Test (PSA XUD9A/L 1.9 Litre 4 Cylinder Indirect Injection Diesel Engine)," CEC F-23-01, 2012.

18.149. Arpaia, A., Catania, A.E., d'Ambrosio, S., Ferrari, A. et al., "Injector Coking Effects on Engine Performance and Emissions," in *ASME Conference Proceedings*, ICEF2009-14094, Lucerne, Switzerland, 229-241, 2009.

18.150. Birgel, A., Ladommatos, N., Aleiferis, P., Milovanovic, N. et al., "Investigations on Deposit Formation in the Holes of Diesel Injector Nozzles," *SAE Int. J. Fuels Lubr.* 5, no. 1 (2012): 123-131, doi:https://doi.org/10.4271/2011-01-1924.

18.151. Ikemoto, M., Omae, K., Nakai, K., Ueda, R. et al., "Injection Nozzle Coking Mechanism in Common-Rail Diesel Engine," *SAE Int. J. Fuels Lubr.* 5, no. 1 (2012): 78-87, doi:https://doi.org/10.4271/2011-01-1818.

18.152. CEC, "Direct Injection, Common Rail Diesel Engine Nozzle Coking Test," CEC F-98-08, 2012.

18.153. Uitz, R., Brewer, M., and Williams, R., "Impact of FAME Quality on Injector Nozzle Fouling in a Common Rail Diesel Engine," SAE Technical Paper 2009-01-2640 (2009), doi:https://doi.org/10.4271/2009-01-2640.

18.154. Hawthorne, M., Roos, J.W., and Openshaw, M.J., "Use of Fuel Additives to Maintain Modern Diesel Engine Performance with Severe Test Conditions," SAE Technical Paper 2008-01-1806 (2008), doi:https://doi.org/10.4271/2008-01-1806.

18.155. Tang, J., Pischinger, S., Lamping, M., Körfer, T. et al., "Coking Phenomena in Nozzle Orifices of Dl-Diesel Engines," *SAE Int. J. Fuels Lubr.* 2, no. 1 (2009): 259-272, doi:https://doi.org/10.4271/2009-01-0837.

18.156. Caprotti, R., Takaharu, S., and Masahiro, D., "Impact of Diesel Fuel Additives on Vehicle Performance," SAE Technical Paper 2008-01-1600 (2008), doi:https://doi.org/10.4271/2008-01-1600.

18.157. Velaers, A.J., de Goede, S., Woolard, C., and Burnham, R., "Injector Fouling Performance and Solubility of GTL Diesel Dosed with Zinc," *SAE Int. J. Fuels Lubr.* 6, no. 1 (2013): 276-288, doi:https://doi.org/10.4271/2013-01-1697.

18.158. Barbour, R., Arters, D., Dietz, J., Macduff, M. et al., "Diesel Detergent Additive Responses in Modern, High-Speed, Direct-Injection, Light-Duty Engines," SAE Technical Paper 2007-01-2001 (2007), doi:https://doi.org/10.4271/2007-01-2001.

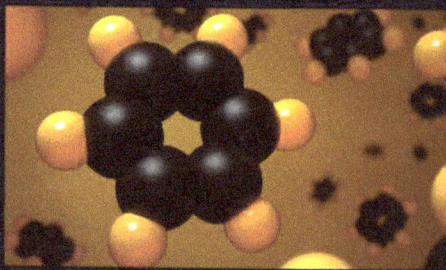

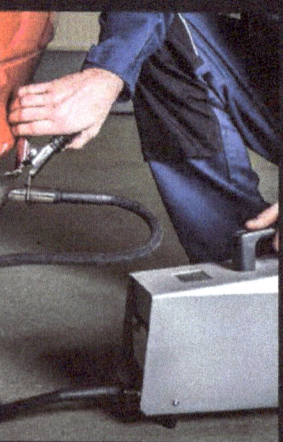

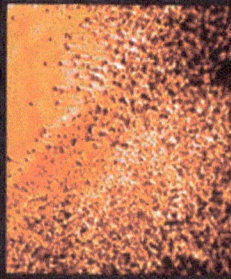

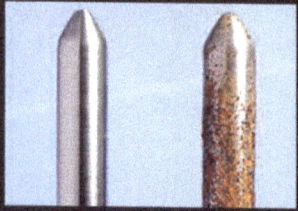

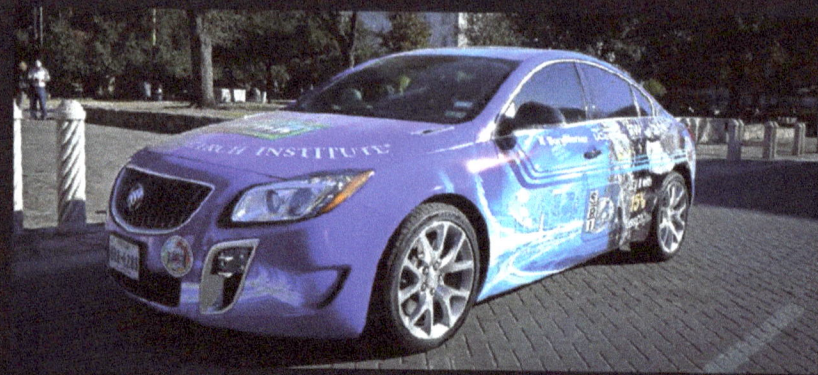

19

Diesel Fuel Additives

U ntil the late 1960s, there was little or no use of additives in automotive diesel fuel. The diesel fuel manufactured at most refineries around the world was traditionally a blend of straight-run atmospheric distillate components that allowed the refiner to meet the specification points without the need for further processing or the use of additives. In the US, where the market was severely biased toward gasoline production, this had necessitated a high level of downstream conversion to yield more gasoline components, with some cracked gas oils being added into diesel fuel blends. Depending upon the source of crude oil, some additional processing was required to meet reduced sulfur specifications. Low-temperature performance could be achieved by blending kerosene into the diesel fuel.

Routine use of diesel fuel additives effectively started in the late 1960s in Europe, with the introduction of cold flow improvers. With the largest proportion of diesel-powered road vehicles of any world region, the growth in demand for diesel fuel was starting to pose problems for the refining industry. The supply situation was further aggravated by the crude oil prices rising during the 1970s. Although total demand for petroleum products went down, refiners had to increase the yield of diesel fuel while reducing crude throughput. The use of flow improvers enabled the refiner to produce more diesel fuel by cutting deeper into the crude oil and using the additive to restore the low-temperature properties of the fuel. In the early 1990s the first of a series of comprehensive reports on the various types of diesel (and gasoline) fuel additives, together with background information on their application and the environmental implications, was published by the European Additives Technical Committee (ATC), updates have been issued [19.1–19.4].

Over the last 50 years the use of a number of different diesel fuel additives used to meet specifications has grown considerably. Cold weather performance additives are now widely used. Changes to the refinery processes required to meet the demand for automotive diesel fuel have in many cases led to potentially less stable fuels being produced, resulting in the use of fuel stabilizers. Cetane number specifications have risen, resulting in the wider use of ignition improver additives. As lower fuel sulfur levels have been legislated, the ability of the diesel fuel to lubricate the fuel injection equipment has diminished; this has necessitated the use of lubricity additives.

An additional factor influencing the trends in additive use is a growing awareness of the need for fuel product differentiation in the marketplace. It is common practice in many countries for oil refining companies to exchange and rebrand products to keep down the costs of fuel transportation, the exchanged product being accepted on the basis of an agreed specification and marketed as such. Nowadays, further additive treatment may be made before an exchanged fuel is sold, in order to support the marketing company's advertising claims for a superior quality product compared with those of its competitors. This practice has been widely adopted in Europe and other parts of the world.

The additives used to meet specifications and to impart marketable performance benefits are discussed in more detail in the following sections. Additive treatment of diesel fuels is usually by weight and expressed either in parts per million (ppm), which shall be expressed as mg/kg to make clear it is by weight, or as a percentage, where 0.1% w/w is equal to 1000 mg/kg. An exception to this is ignition improvers, which are often added on a percentage volume basis. Although there is no rigorous definition of what constitutes an additive, as opposed to a blending component, it is generally accepted that an additive is something added at less than 1% w/w (i.e., 10,000 mg/kg). In 2022 the German federal environmental agency—Umweltbundesamt (UBA)—released the final report of a study on the "Impact of fuel additives on exhaust aftertreatment systems, emissions, environment and health" [19.5]. This work found that a "premium additive" that goes beyond the so-called "refinery additive" can show positive effects on engine function, and thus the emission levels, in long-term use and that there is nothing to suggest that they pose a risk to the environment and health. These findings echo the findings of various stakeholders such as OEMs, members of Truck and Engine Manufacturers Association (EMA), fuel additive companies, producers and marketers in the US [19.6] that resulted in the development of the Top Tier™ Diesel Standard, see Appendix 3.

19.1. Additives to Improve Fuel Stability

Fuel stability improvers can include antioxidants, stabilizers, and metal deactivators. Antioxidant additives are used in diesel fuels considered to be prone to oxidative or thermal instability due to the components used in their preparation. The additives work by terminating free-radical chain reactions that would result in color degradation and the formation of sediment and insoluble gums [19.7]. If oxidation takes place, engine operation could be affected due to filter blocking or gummy deposits in the injection system and on fuel injector nozzles [19.8]. In some countries a fuel might be unacceptable for marketing as automotive diesel if the maximum color specification is exceeded. The color formation has been attributed to quinoid structures from minor fuel components such as pyrroles [19.9], and compounds such tertiary amines have been shown to be effective in reducing color formation [19.10]

Cracked gas oils were predominantly used as blend stocks for distillate and residual heating fuels, but with those markets declining, more cracked gas oil was diverted into diesel fuel. Distillates from cracking operations are more olefinic than those from atmospheric distillation and contain more nitrogen compounds. As a result, they are less stable, being prone to oxidation by free-radical reactions. This is the main reason why oxidation stability limits were introduced into more diesel fuel specifications. With drastic reductions in fuel sulfur limits in many countries, additional hydrotreating of diesel fuel has become necessary. This process also improves fuel stability by removing nitrogen- and oxygen-containing compounds and saturating the more reactive olefinic compounds that are typically present in catalytically cracked gas oils. As virtually all diesel fuel is hydro-desulfurized, there is normally no need for routine antioxidant treatment. However, when only a mild degree of hydrogen treating is required to meet the sulfur specification, as when refining a low-sulfur crude oil, or if the availability of hydrogen is limited, an antioxidant may be added to ensure that the fuel is adequately stabilized.

Despite the comments about the significant benefits of hydrotreating, potential problems have been identified at very low sulfur levels (50 mg/kg and less). These have been attributed to a tendency for very low-sulfur fuels to form peroxides. Investigations have shown that hydrotreating to achieve these very low sulfur levels results in the removal of some fuel species that behave as antioxidants [19.11]. This work concluded that additives currently used as antioxidants in gasoline and aviation fuel successfully control any peroxide-forming tendency in diesel fuels.

In addition to the likely need for the use of antioxidants in oil refinery streams, the increasing use of biodiesel to meet renewable fuel mandates has potentially introduced a new source of instability to the fuel. This is discussed in Chapter 18, Section 18.1.2.

Typical antioxidants are radical traps, such as hindered phenols [19.12, 19.13] that are intended to prevent gum-forming reactions. Stabilizers are basic compounds, such as long-chain paraffinic amines, cyclic amines [19.14], or other nitrogen-containing compounds, that interfere with acid-base sediment forming reactions, such as cyclohexylamine.

Metal deactivators are sometimes used in conjunction with other stability improver additives to prevent oxidation reactions from being catalyzed by heavy metal ions, particularly those of copper, that may be present in trace amounts in the fuel. The use of copper is usually avoided in vehicle fuel systems, but the possibility of fuel contamination with copper or its alloys cannot be excluded. One example of a metal deactivator is 2-(2-hydroxyphenyl)benzoxazole [19.15], which will chelate copper as shown in Figure 19.1.

FIGURE 19.1   Two molecules of 2-(2-hydroxyphenyl)benzoxazole chelating a copper atom.

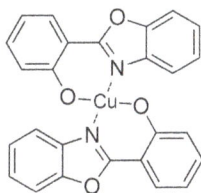

© SAE International.

Other examples of metal deactivators are N,N′-disalicylidene-1,2-propanediamine [19.16], tetra-(acetylacetone) pentraerythrityltetramine [19.17], and hydroxyl-oximes [19.18]. Treat rates are similar to those for gasolines, typically in the region of 10 mg/kg [19.19]. Chelating functionality can also be incorporated into antioxidants [19.20].

The effectiveness of additive treatment will depend very much on the dominant fuel characteristics that determine the degradation reactions. The choice of additive is generally decided by trial-and-error to find out which is best for the particular fuel. Treating levels are usually in the range of 25 to 200 mg/kg, and it is critical that the additive be injected into the fuel as soon as possible, preferably in the run-down streams of a refinery.

19.2. Additives Used to Aid Distribution and Handling of Diesel Fuel

19.2.1. Additives to Aid Cold Flow Performance

Cold flow improver additives or wax crystal modifiers are used to improve a number of characteristics associated with the low-temperature performance of the fuel. This is discussed in more detail in Chapter 17.

As discussed in Section 17.2, there are a number of areas of cold flow performance that can be targeted by different additives. The ability of an additive to affect one performance parameter does not necessarily mean that it will also affect the other cold weather performance parameters; this is very much dependent upon the base fuel characteristics.

The cloud point is the temperature at which wax crystals start to come out of solution and become visible, making the fuel cloudy. In many instances a vehicle can still operate satisfactorily at this temperature while vehicle operation warms the fuel. A cloud point depressant is therefore an additive that will delay the onset of wax crystallization as the temperature is decreased. Chemistries that have been used as cloud point depressants have included comb polymers utilizing a variety of polymer mixtures [19.21], maleic anhydride α-olefins [19.22–19.24], and polymers of hydrocarbyl esters of unsaturated acids [19.25].

The pour point is the temperature at which the amount of wax that is no longer in solution is sufficient to gel the fuel and prevent it from pouring. A vehicle will often reach its limit of cold weather operability at temperatures above the pour point. Pour point depressant chemistries include EVA copolymers [19.26, 19.27], polyacrylates [19.28], fumarates [19.29], and poly-α-olefins [19.30, 19.31].

Many commercial cold flow improvers for diesel fuel will include proprietary mixtures of different polymer types to impart specific cold flow properties dependent upon the characteristics of the base fuel. These flow improvers that have in the past been formulated to address problems encountered in paraffinic petroleum diesel fuels are now being used to solve problems encountered with biodiesel blends. Other additives are being developed specifically to address the problems found with biodiesel blends. These additives may employ similar technologies to the petroleum diesel additive but with revisions to the manufacturing process to afford better control of the molecular weight and polydispesancy [19.32–19.35] or by altering the type of parts of the comb polymer [19.36]. The biodiesel problem has been attributed to the degree of saturation of the FAME [19.37]. The use of branched chain fatty acid esters as additives has helped the problem [19.38]. There is increasing concern over the suitability of biodiesel for use in various geographical locations [19.39].

19.2.2. Corrosion Inhibitors

Corrosion inhibitors, sometimes referred to as antirust additives, can be added to diesel fuels being transported by pipeline to protect the pipeline and to avoid rust contaminating the fuel and blocking filters in the distribution system.

These additives are surfactant materials with a polar group at one end and an oleophilic/hydrophobic group at the other. The polar group attaches itself to metal surfaces in the system, and the other group repels water and provides an oily layer to prevent rust formation.

A wide range of chemical types are used as anticorrosion additives. They include esters or amine salts of alkenyl succinic acids. In many cases, treat rates are controlled to a low level that is just sufficient to protect the distribution systems, leaving only about 1 mg/kg in the emerging fuel. A typical treat rate for pipeline protection is in the region of 5 mg/kg. Alkyl orthophosphoric acids, alkyl phosphoric acids, and aryl sulfonic acids have been extensively used in the past, but phosphorous and sulfur-containing compounds are now rarely used due to the adverse effects that these elements can have on exhaust aftertreatment systems.

Protection of the vehicle fuel system is also a critical consideration, particularly for the latest technology precision fuel injection systems. A corrosion inhibitor will therefore be included in multifunctional diesel additive packages to ensure protection of the vehicle fuel system. Selection of additive type and treat rate is usually determined by using a standard rust-preventing test [19.40]. In this test, a polished steel spindle is immersed in a 9:1 mixture of fuel and water, which is continuously stirred and kept at a temperature of 60°C for 24 hours. The appearance of the steel spindle is compared with a control specimen from a portion of untreated fuel. Figure 19.2 shows an example of a spindle at the end of a test run with treated fuel (above) and a spindle at the end of a test with untreated fuel (below).

FIGURE 19.2 Corrosion test spindles with treated (above) and untreated (below) fuel.

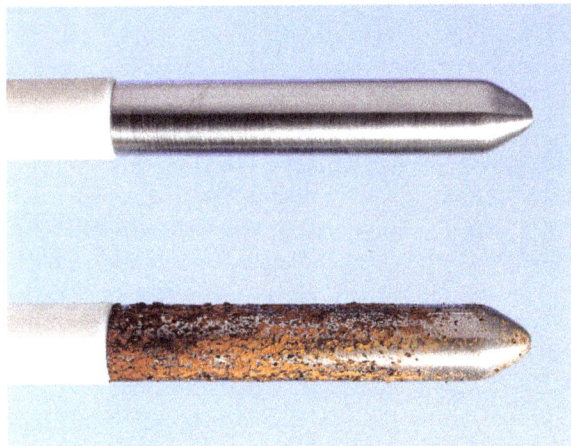

© SAE International.

19.2.3. Static Dissipater Additives

Antistatic or, more correctly, static dissipater additives are now becoming common in diesel fuel applications. This type of additive was first used in aviation kerosene to avoid the risk of an explosion due to sparks from the discharge of static electricity. The static electricity could build up as a result of high circulation rates within aircraft fuel system. Static electricity may also accumulate as a result of high pumping rates that may occur when loading or unloading large quantities of fuel from road tankers. The refinery processes now being used to remove the sulfur from distillate fuel, in order to produce ultra-low-sulfur diesel fuel, will also remove many of the other polar compounds in the fuel. This will tend to reduce any natural static dissipation properties of the fuel, thereby making the use of a static dissipater additive more desirable. It should be stressed that there is virtually no risk of an excessive charge building up when refueling diesel vehicles at a filling station.

The additives work by increasing the conductivity of the fuel, allowing any electrostatic charge generated during pumping to be dissipated. Chemistries that have been employed as static dissipaters include chromium salts [19.41], aliphatic amines-fluorinated polyolefins [19.42], and olefin-acrylonitrile copolymers [19.43]. The most commonly used static dissipaters are polysulfone copolymers [19.44]. Additive treat rates are low, typically only single-digit part per million by mass.

19.2.4. Dehazers and Demulsifiers

Dehazer treatment may occasionally be needed if the fuel becomes hazy due to the presence of finely dispersed droplets of water. Contamination with water can occur at almost any stage, as the fuel passes from the refinery and through the distribution network until it reaches the vehicle tank. As described in Chapter 20, Section 20.5, it can be the result of dissolved water coming out of solution or condensing from the air when the temperature falls; leakage of rain water into the tank; or entrainment of water accumulated in storage tank bottoms. The situation may be aggravated by the characteristics of the fuel or the type of additives it contains and by excessive turbulence in the pumping system.

If the haze persists after the normal one or two days of settling time, additive treatment may be necessary to accelerate clearance and meet the usual "clear and bright" requirement. Effective dehazer additives include quaternary ammonium salts, typically used at dose rates between 5 and 20 ppm.

As hazy fuel tends to be a spot problem, the practical approach is for alternative additives to be tested on-site, in cold samples drawn directly from the affected tank. Samples taken away for testing will usually have cleared by the time they reach the laboratory because of a temperature change or contact with the sample container. However, the haze-forming tendency of a diesel fuel can be assessed in the laboratory using a modified version of the Waring blender test for gasolines. A portion of diesel fuel, to which is added 10% distilled or synthetic seawater, is stirred at 10,000 rev/min, for 2 minutes, in the standard laboratory blender. The mixture is poured into a glass jar and its appearance is rated at intervals over a period of 24 hours, either visually or using a photometer, to see how long it takes for the haze to clear.

A demulsifier may be included when other surfactant-type additives are used to avoid problems due to pick-up of storage tank water bottoms. Entrainment of water and debris in pipelines and during product transfer might result in the formation of stable emulsions and suspended matter that could plug filters or otherwise make the fuel unacceptable. The ASTM D7451 test method [19.45] has been developed to allow additive manufacturers to evaluate new additive packages to ensure that the surfactant-type products in the package do not adversely affect the fuel-water separability due to the use of the package. The additive package can be tested at either its normal treat rate or at several times the intended treat rate to evaluate the impact of potential overtreatment. The amount and type of demulsifier can thus be assessed to ensure the performance of the total package. Typical examples are based on alkoxylate chemistry with for example phenolic resins with ethylene or propylene oxide.

Demulsifiers are highly surface-active chemicals selected for their limited solubility in oil and water. They are usually prepared by reacting a hydrophobic molecule such as a long-chain alkylphenol with ethylene or propylene oxide. Because the demulsifiers are themselves surfactants, their use must be carefully controlled. Too much can make matters worse, and, more importantly, they can interfere with other surfactants that may be present, reducing or eliminating their effectiveness.

The effectiveness of different additive types and treat rates can also be checked using the ten-cycle multiple contact [19.46] test in which a small amount of water is successively agitated with ten portions of fuel to represent repeated filling and emptying of a storage tank. The assessment is based on the amount of emulsion and suspended matter at the oil/water interface. Treat rates are generally less than 10 mg/kg.

19.2.5. Biocides

Biocidal treatment of diesel fuel is sometimes carried out to prevent the growth of bacteria and fungi in the bottom of fuel tanks. These aerobic or anaerobic organisms live in the water bottoms and feed at the fuel/water interface. They lead to accumulation of suspended matter that will obstruct flow through filters if drawn out with the fuel. Warm conditions can encourage growth of the organisms.

Regular draining of water bottoms will help minimize the risk of bacterial growths, but this is not always possible, especially if natural or man-made caverns are used for long-term storage of strategic reserves. Commercial biocides come from a wide range of chemical types, including boron compounds, amines, and imines, which need to be soluble in water and in the fuel to be properly effective. A problem with biocidal treatments is that the bacteria can develop resistance, so the additive type must be rotated, and must be on the approved active substance list for the geographical area or organization, e.g., NATO.

Evaluation of biocides is by incubation of treated and untreated nutrient media that have been inoculated with the relevant strain of bacteria. A treat level of 200 mg/kg is used for direct addition to water bottoms, but a smaller dose rate would be appropriate for injection into the fuel as it goes into storage.

19.2.6. Anti-Foamants

Anti-foamants are sometimes added to diesel fuel, often as a component of a multifunctional additive package, to help speed up or to allow more complete filling of vehicle tanks. Their use also minimizes the likelihood of fuel splashing on the ground or onto clothing, avoiding the nuisance of stains and unpleasant odor and reducing the risk of spills polluting the ground and the atmosphere.

As with lubricating oils, silicon additives are effective in suppressing the foaming tendency of diesel fuels, the choice of silicone, and cosolvent depending on the characteristics of the fuel to be treated. Tests used to assess the tendency of lubricating oils to form a stable foam in an engine are not suitable for diesel fuels. Selection of a diesel anti-foamant is generally decided by the speed with which the foam collapses after vigorous manual agitation to simulate the effect of air entrainment during tank filling. Treat levels are normally below 10 mg/kg and are kept to a minimum due to concerns over silicon compounds contributing to sediment formation, resulting in sludge, clogged filters, and injector deposits. Polydimethylsiloxane (PDMS) chemistry is used.

19.2.7. Odor Masks and Odorants

As the odor of diesel fuel is considered objectionable by many people, odor masks are occasionally employed to improve the market acceptability of some branded diesel fuels. Because diesel fuel is less volatile than gasoline, the stain and smell of spills will persist, which can be very annoying, particularly if clothing is contaminated.

Attitudes to smells vary widely and are subjective, but odor panel tests suggest that the market preference is for a neutral rather than a positive odor, which puts the emphasis mainly on odor-masking effects. Various products, with a choice of fragrances, are commercially available for use at treating rates of 10 to 20 mg/kg.

19.2.8. Dyes and Markers

Dyes are often used in petroleum fuels for identification of particular branded products or for legal reasons as a means of providing evidence, such as in cases of theft, tax evasion, or fuel adulteration. For example, a lower rate of tax is payable on diesel fuel for off-road use. This diesel fuel can be dyed to enable detection of its illegal use for on-road applications. The need for markers that are robust to laundering methods has seen the development of the Accutrace™ portfolio of markers one of which is used as a fiscal gas oil marker [19.47]. Azo dyes are normally used when red or orange colors are needed and anthraquinone dyes when greens or blues are required. Treat rates are usually less than 10 mg/kg.

19.2.9. Drag Reducers

Drag reducers are sometimes used in petroleum products when pipeline capacity is limited to increase throughput and postpone the need for investment to construct additional pipelines. They are high molecular weight, oil-soluble polymers that shear very rapidly and reduce drag. Treating levels in the region of 50 mg/kg gives significant reductions in frictional drag.

19.2.10. Anti-Icers

In some countries and localities, anti-icers may be added to road tankers making deliveries of diesel fuel. This is to prevent ice plugging of fuel lines by lowering the freezing point of small amounts of free water that may separate from the fuel. These additives are relatively low molecular weight alcohols or glycols that are soluble in diesel fuel and have a strong affinity for water, so they are effective at low dose rates in the region of 30 mg/kg.

19.3. Additives Used to Protect Engines and Fuel Systems

19.3.1. Deposit Control Additives

Successive pieces of legislation have significantly reduced the permissible level of emissions from diesel engines. To achieve these new limits, diesel engine technology, and particularly the fuel injection equipment (FIE), has evolved to enable more precise control of the combustion process. Whereas the diesel engine has always been prone to deposit formation, due to its inherent characteristic of producing soot, these deposits are becoming far less tolerable. Deposit accumulation is due to an imbalance of the processes of deposit lay down and deposit removal by the action of fuel flow or gas flow. The use of a deposit control additives (DCA) alters this balance to limit deposit accumulation and at high treat rates can even result in a reduction of accumulated deposits.

The use of DCAs is therefore becoming far more important to control the accumulation of deposits that can have a detrimental effect on combustion. Deposits that form within the fuel injection system can cause sticking of injector needles, resulting in misfires, power loss, and increased smoke. The build-up of lacquer and carbonaceous deposits on injector tips can affect the amount of fuel injected and the spray pattern, causing problems of reduced power and higher smoke. Starting the engine may also become more difficult. Besides being an annoyance to the operator, this will also increase emissions of unburned or partially burned fuel.

Over the years, a number of in-house and standardized test methods have been developed to assess the propensity of fuels to form deposits and the efficacy of DCAs to control this. The test methods were developed to cater for the prevailing engine technology. Although the technology may have become obsolete, some of these tests remain in operation and are well recognized. An example of this is the XUD-9A test [19.48], which uses an indirect injection (IDI) diesel engine that is no longer in mass production. During the development of some of these in-house tests, it has been demonstrated that the build-up of deposits in and on the injector can increase regulated emissions over a short period of operation [19.49, 19.50]. A test method using a more modern high-pressure common-rail type engine [19.51] relies on the loss of power as a metric for assessing the degree of nozzle fouling. A fuller discussion of nozzle fouling and the recognized test methods is given in Chapter 18, Section 18.2.

Figure 19.3 is a cross section of an IDI injector nozzle showing the deposit formation with and without the use of a DCA; the corresponding fuel spray patterns are also shown.

FIGURE 19.3 Pintle nozzle deposits and spray pattern with (left) and without (right) the use of a DCA.

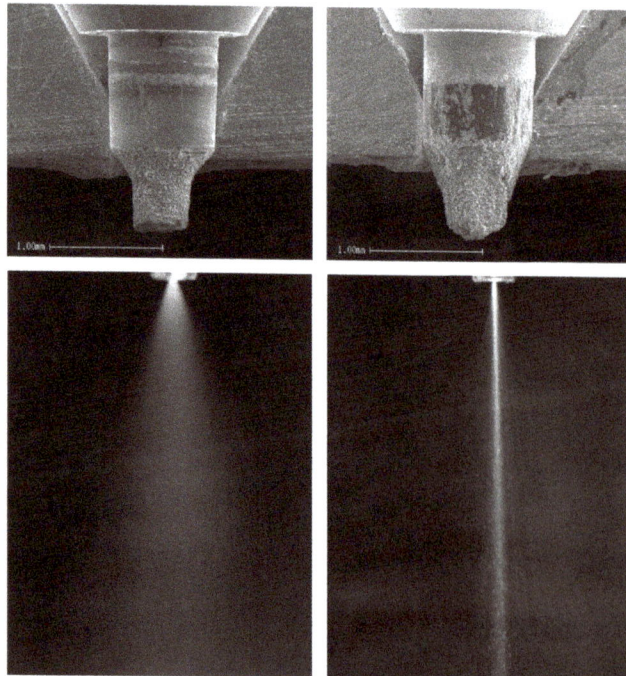

© SAE International.

The image on the top left shows a cross section of the nozzle with the nonsectioned needle in the 0.1 mm lift position with a limited amount of deposit on the needle tip and around the outside of the injector hole. The test had been run using a fuel containing a DCA. The image on the lower left shows the spray pattern formed using the standard fuel injection pressure. There is good atomization of the fuel spray, leading to good mixing of the fuel and air. The images on the right of the figure show the corresponding situation when the test was run using an untreated fuel. The fuel spray is more jet like, displaying very poor atomization that will result in poor mixing of the fuel. This will result in the increased emissions observed.

Figure 19.4 [19.52] shows the loss of power that results from deposit build-up within the injector holes of the high-pressure common-rail test procedure.

FIGURE 19.4   Power loss using Zn adulterated fuel and the same fuel plus a DCA.

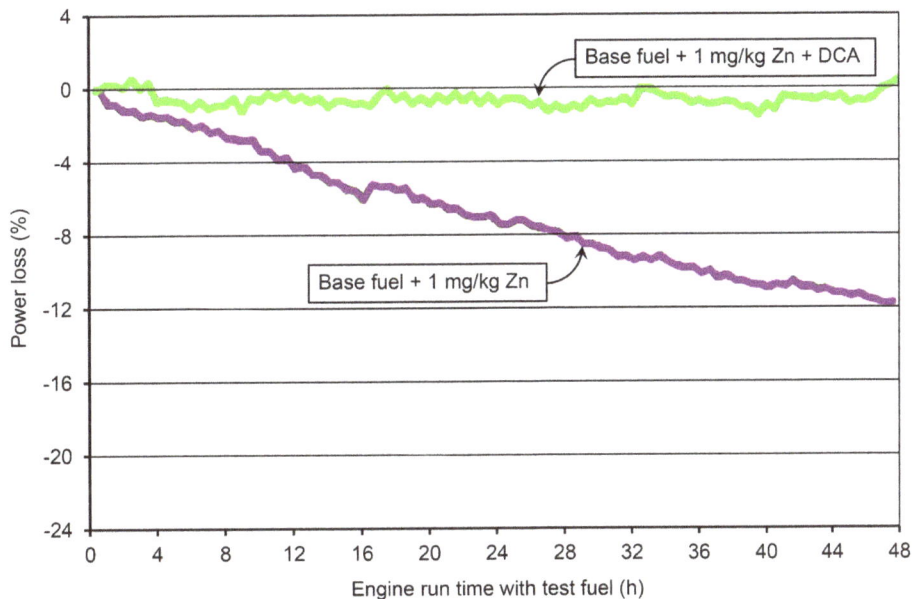

DCAs for diesel fuels perform by a similar mechanism to DCAs for gasoline applications as discussed in Chapter 11, Section 11.3.3. Due to the different conditions prevailing within the diesel engine fuel system, a gasoline DCA will not necessarily function well within a diesel system and vice versa. DCAs are surface-active molecules; they function by way of the polar group, at one end of the molecule, being attracted to a particle or a surface and the large nonpolar, oleophilic group at the other end being attracted to the fuel. This tends to form a monomolecular film around any particles, effectively solubilizing them by forming a micelle which prevents aggregation of the particles and allows the particulate matter to be carried along with the fuel. The deposit precursor is thus carried into the combustion chamber or returned with the fuel recirculation to be subsequently trapped in the fuel filter. Metal surfaces are protected against deposition in a similar way to that in which they are protected by surfactant corrosion inhibitors. The oleophilic group repels the deposit particle or the micelle in the same way it would repel a water droplet.

Some of the chemistries available as DCAs for diesel fuel include polyetheramides [19.53], polyamines [19.54, 19.55], Mannich reaction products [19.56, 19.57], succinimides [19.58, 19.59], and quaternized nitrogen compounds [19.60–19.62]. Figure 19.5 shows an example of a quaternized DCA where R_1, R_2, and R_7 are H or a low molecular weight hydrocarbyl radical; R_3 is a long-chain hydrocarbyl radical; R_4, R_5, and R_6 are low-molecular-weight hydrocarbyl or hydroxyl substituted hydrocarbyl radicals; and L_1 and L_2 are straight-chain or branched alkylene groups. These are purported to function by reducing cavitation and turbulence in injectors [19.63].

FIGURE 19.5   Example of a quaternized DCA.

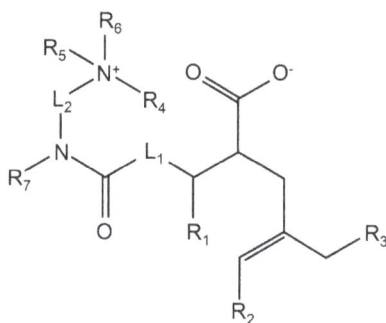

© SAE International.

The most appropriate additive type and treat rate will be determined by the characteristics of the fuel and the required performance targets. A lower treat rate of additive is required if the target is to prevent deposit accumulation on a clean injector nozzle than is required to reduce the level of deposit on an already fouled injector. Treat rates to control deposits on new or cleaned injectors are typically in the range of 100 to 200 mg/kg [19.64].

19.3.2. Lubricity Additives

In the automotive community, the issue of a lack of diesel fuel lubricity was highlighted in the early 1990s [19.65] when Sweden enacted legislation mandating the reduction of fuel sulfur and aromatic content to a maximum of 50 mg/kg and 20%, respectively, for Class 2 diesel, and to a maximum of 10 mg/kg and 5%, respectively, for Class 1 diesel. In the 1950s, the aircraft engine community had already been alerted to the potential problems of fuel lubricity. The high temperatures that prevailed in certain aircraft applications [19.66, 19.67] had driven the requirement for increased fuel stability, and in many cases, this was brought about by increased refining of the fuel. This increased refining commonly utilized hydrotreating, which had been shown to benefit stability [19.67].

Hydro-desulfurization had been introduced to "sweeten" the increasingly common high-sulfur fuels that were being produced. It was commonly thought that removing the sulfur compounds reduced the fuels lubricity. Independent studies had shown that high fuel sulfur concentrations resulted in increased engine deposits and wear [19.68–19.71], and it was initially supposed that removing sulfur from the fuel might reduce wear. However, the correlation between sulfur content and wear rates was due to the production of sulfur trioxide during the combustion process [19.70, 19.71], rather than any lubricity-enhancing properties of the sulfur compounds. It must be noted that at the time, diesel fuel was predominantly straight-run distillate and the low-sulfur fuel considered still had in excess of 500 mg/kg of sulfur.

The hydrotreatment not only removed sulfur compounds but also removed other polar compounds containing nitrogen and oxygen plus heavier aromatic compounds. Various studies showed that it was these polar compounds [19.72], heavy aromatic compounds [19.73], polyaromatics and particularly oxygen-containing components reduced wear [19.74]. The beneficial effect of sulfur compounds was contradicted by tests where sulfur compounds were added to the fuel and reduced lubricity was observed [19.75]. It was also found that certain polar fuel additives, such as corrosion inhibitors, would reduce friction and wear [19.76].

Following the identification of a lubricity problem in Northern Europe, a great deal of effort was expended on developing methods of assessing the lubricity of diesel fuels. These methods include the pump rig tests [19.77, 19.78], bench top tests including the Ball-On-Cylinder Lubricity Evaluator (BOCLE) [19.79], the roller on cylinder lubricity evaluator (ROCLE), the Scuffing Load Ball-On-Cylinder Lubricity Evaluator (SLBOCLE) [19.80], the Thornton Aviation Fuel Lubricity Evaluator (TAFLE) [19.81, 19.82], the Ball on Three Disks (BOTD) [19.83], and the High Frequency Reciprocating Rig (HFRR) [19.84, 19.85]. These test methods are discussed further in Chapter 20, Section 20.4.

As was noted, early investigations showed that replacing some of the naturally occurring polar fuel components with polar additives could improve the lubricity of the fuel. Commonly used lubricity-improving additives include carboxylic acids [19.86, 19.87] and esters [19.88–19.91]. Other surfactant materials have also been proposed [19.92]. The response of different additives varies according to the properties of the fuel. Figure 19.6 [19.93] shows the response of three different lubricity improver additives in a low-sulfur fuel when tested in the HFRR procedure with a fuel temperature of 60°C.

FIGURE 19.6 HFRR results at 60°C for three different additives in low-sulfur fuel.

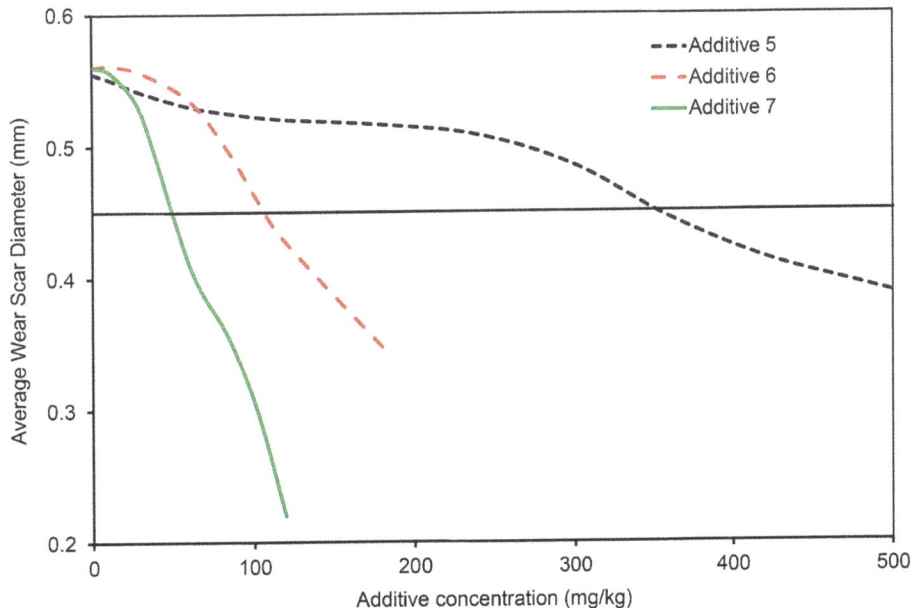

19.4. Additives That Influence Combustion

19.4.1. Ignition Improvers

Ignition improvers are used to decrease the time between when fuel is sprayed into the combustion chamber and ignition occurs. This is known as the ignition delay and is a measure of the fuel's ignition quality; a fuller discussion is given in Chapter 15, Section 15.2.1. The cetane number (CN) is the most widely accepted measure of ignition quality and it is determined using the Cooperative Fuels Research (CFR) cetane engine [19.94]. The CN of a fuel is determined by comparing its ignition delay, under standard operating conditions, with those of blends of two reference fuels, n-cetane and 2,2,4,4,6,8,8-heptamethylnonane (iso-cetane), having, by definition, a CN of 100 and 15, respectively. Because ignition improvers are characterized by the fact that they increase CN, they are often referred to as cetane number improvers (CNI).

Several types of chemicals have been identified as effective CNIs: alkyl nitrates, ether nitrates, nitroso compounds, and certain peroxides. They are all materials that at elevated temperatures decompose readily to form free-radicals that accelerate the oxidation of the fuel. This shortens the measured ignition delay period. Table 19.1 shows some possible CNIs that were tested in the early 1950s and the change in CN for high treat rates of typically 1.5% v/v. The additive treat rate is listed in the table; note also that the base fuels have a very low CN that was typical at the time.

TABLE 19.1 Possible CNIs and their observed CN improvement [19.95].

Additive	Fuel code	Treat (%)	Δ CN	Additive	Fuel code	Treat (%)	Δ CN
2-Chloroethyl nitrate	a	1.5	13.3	Nitroethane	a	1.5	1.7
2-Ethoxyethyl nitrate	a	1.5	21.0	1-Nitropropane	a	1.5	0.7
Isopropyl nitrate	a	1.5	17.9	2-Nitropropane	a	1.5	2.7
1,3-Dinitrate-2-dimethyl propane	a	1.5	9.2	Nitrobenzene	a	1.5	1.3
Trimethylal propane trinitrate	a	1.5	9.2	Diphenylamine	a	1.5	−4.3
Butyl nitrate	a	1.5	16.8	Pyridine	a	1.5	−2.5
Mixed amyl nitrates	a	1.5	12.1	Tetraethylene pentamine	a	1.5	−0.2
Primary amyl nitrate	a	1.5	13.0	Aniline	a	1.5	0.1
Secondary amyl nitrate	a	1.5	15.0	Triamylamine	a	1.5	0.9
Isoamyl nitrate	a	1.5	14.1	Triethanolamine	a	1.5	1.4
Secondary hexyl nitrate	a	1.5	15.1	Dioctylamine	a	1.5	0.9
Primary hexyl nitrate	a	1.5	17.6	Ethoxy ethyl morphollne	a	1.5	0.9
2-Ethylhexyl nitrate	a	1.5	12.1	Dimethyl glyoxlme	a	1.5	1.3
n-Heptyl nitrate	a	1.5	14.8	Isophorone	a	1.5	1.9
n-Octyl nitrate	a	1.5	20.3	Acetone	a	1.5	1.1
n-Nonyl nitrate	a	1.5	11.3	Acetomyl acetone	a	1.5	2.2
Cyclohexyl nitrate	a	1.5	21.6	Methyl acetate	a	1.5	1.7
Ethylene glycol dinitrate	a	1.5	2.1	Propyl acetate	a	1.5	0.3
Diethylene glycol dinitrate	k	1.5	18.0	Amyl acetate	a	1.5	3.8
Dipropylene glycol dinitrate	k	1.5	19.0	Butyl bromide	a	1.5	0.9
Propylene glycol dinitrate	a	1.5	2.5	Trichloroethane	a	1.5	−2.0
Ethylene propylene diglycol dinitrate	k	1.5	20.0	Trichlorethylene	a	1.5	3.0
Glycol dinitrate	a	1.5	2.1	Triglycol dichloride	a	1.5	1.7

(Continued)

TABLE 19.1 (Continued) Possible CNIs and their observed CN improvement [19.95].

Additive	Fuel code	Treat (%)	Δ CN	Additive	Fuel code	Treat (%)	Δ CN
Diglycol dinitrate	a	0.7	9.7	Tetraglycol dichloride	a	1.5	3.5
Triglycol dinitrate	b	1.5	24.0	Dimethoxy tetraethylene glycol	a	1.5	3.2
Tetraglycol dinitrate	b	1.1	19.0	Trimethoxy propane	a	1.5	2.9
Butyl peroxide	a	1.5	20.2	Triethoxy propane	a	1.5	2.1
Caproyl peroxide	a	1.5	24.7	Dimethoxy tetraglycol	a	1.5	8.0
Heptylyl peroxide	a	1.5	9.1	Ethyl ether	a	1.5	−0.2
Chaulmoogryl peroxide	a	1.5	5.4	n-Dibutyl ether	a	1.5	7.5
Oleyl peroxide	a	1.5	6.2	Octaethylene ether	a	1.5	9.3
Stearyl peroxide	a	1.5	6.2	Butyl carbitol	a	1.5	2.6
Lauroyl peroxide	a	1.5	12.1	Diethyl carbitol	a	1.5	8.0
Triacetone peroxide	a	1.5	16.3	Acetaldol	a	1.5	3.5
Acetyl benzoyl peroxide	a	1.5	16.8	Paraldehyde	a	1.5	4.1
Nitramethane	a	1.5	−0.7	Carbon disulfide	a	1.5	−0.6

Fuel codes: a—93% mixed kerosene distillate from coastal and midcontinent crudes, and 1% straight-run midcontinent gas oil. CN = 39.1; b—Straight-run untreated distillate from Elk Hill crude. CN = 25.6; k—No. 2 furnace oil. CN = 51.0.

Commercial and safe-handling considerations have resulted in most attention being given to primary alkyl nitrates, and various versions have been marketed. At present, low production costs and good response in a wide range of fuels have identified 2-ethyl-hexyl-nitrate (2EHN), also known as iso-octyl nitrate, as one of the most cost-effective CNIs [19.7], with peroxides being used where low nitrogen alternatives are required [19.96]. Figure 19.7 [19.97] shows the increase in CN with respect to the treat rate of a 2EHN additive.

FIGURE 19.7 Response to CNI additive.

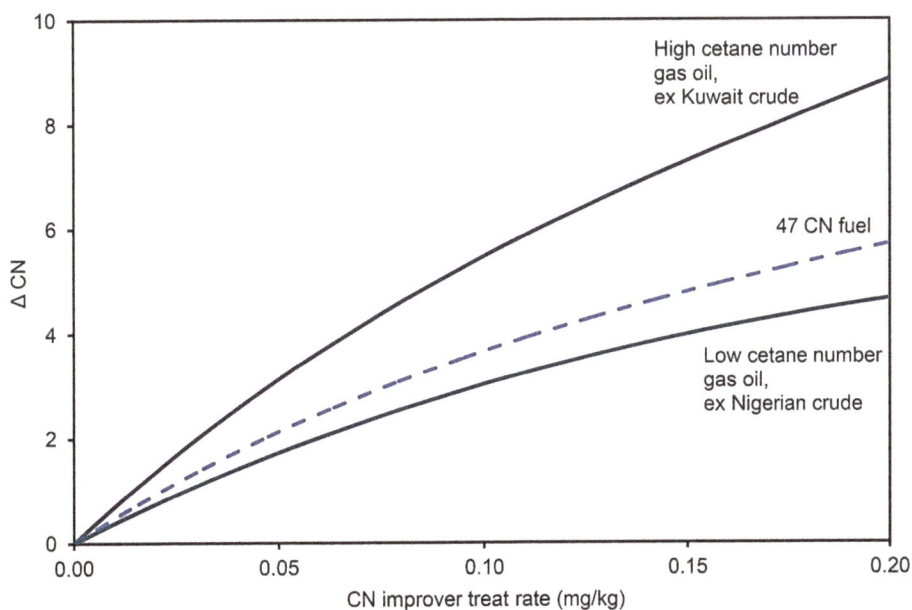

From Figure 19.7, it is clear that the response to CNI varies considerably from fuel to fuel. The response is dependent on individual fuel characteristics, including the base fuel CN. Early attempts were made to develop a means of predicting the amount of CNI additive required to deliver a given increase in CN. An early attempt produced an equation the related the CN of a treated fuel to the density of the base fuel (specified as the API gravity), the 50% distillation temperature, and the volume treat rate of the additive [19.98]. However, later studies showed that this equation tended to over-predict the CN of the fuels that were becoming available due to reductions in specified sulfur levels in particular [19.99, 19.100]. A later predictive equation was developed where the change in CN was a quadratic function of the weight addition of CNI additive [19.101]. The general form of the equation was

$$\Delta CN = B_0 + B_1 * X + B_2 * X^2$$

where X is the percentage weight addition of CNI additive and B_0, B_1, and B_2 are constants determined from the API gravity, the aromatic content, the flash point, pour point, and midrange distillation temperatures of the fuel. A different set of coefficients are used according to the type of CNI additive. The CN of the finished blend can be determined by adding this change in CN to the base fuel CN. This equation has been shown to provide a better correlation between predicted and measured CN values, and where the base fuel CN is not known the calculated cetane index [19.102] can be used with little loss in prediction accuracy [19.100].

Various studies have been conducted using engines and laboratory combustion chambers to investigate the relative performance of the two most commonly used CNIs, 2EHN and di-tertiary-butyl peroxide (DTBP), in different fuels [19.96, 19.100, 19.101, 19.103–19.106]. Most of these studies showed that DTBP was less effective than 2EHN, and in some cases as much as 30% less effective.

An increase in the CN of the fuel tends to reduce regulated emissions and improve efficiency independent of engine technology [19.107–19.110]. This is discussed further in Chapter 21, Section 21.2. There is no apparent difference in the beneficial effect of increased CN, whether the high CN is derived from the natural composition of the fuel or as a result of the use of a CNI additive [19.108, 19.111].

In refineries, CNI additives are used mainly to give fairly modest improvements of two or three numbers, to bring off-grade fuel blends in specification. This would require additive treat levels in the 500 to 1000 ppm range. This type of additive is used in some multifunctional packages, where the package formulation will give a CNI treat level of around 500 ppm.

19.4.2. Combustion Improvers

Combustion improvers are additives that enhance the oxidation process of the fuel in the diesel engine. Strictly speaking, the ignition improver additives discussed in the previous section are combustion improver; they improve to initial stages of the oxidation process. Traditionally, the term combustion improver is reserved for additives that influence the later stages of the combustion process. During the combustion process, it is likely that organic molecules will break down and oxidize. The most effective combustion improver additives are therefore likely to be organo-metallic compounds. During the combustion process these compounds are likely to break down and oxidize, leaving a metal oxide that will act as a catalyst for the oxidation of soot particles. It is interesting to note that toward the end of the first half of the 20th century organo-metallic fuel additives were being investigated as ignition improver additives [19.112].

During the 1960s, there was a great deal of interest in organo-metallic additives as a means of reducing visible smoke from diesel engines [19.113]. Barium was found to be particularly successful in reducing the level of visible black smoke emissions [19.114, 19.115]. Other metals, including calcium, iron, and manganese, have been proposed. Because these additives must produce catalytic products that are active during the combustion process; they will inevitably result in the some of these catalytic compounds being present in the exhaust gas. When the only emissions metric was visible smoke, this was not a concern. However, the remainder of the combustion products of the additive will remain in the engine as ash deposits, and these can be of concern to engine manufacturers [19.116]. As with metallic gasoline additives, notable tetra-ethyl lead antiknock additives, the health effects of metallic emissions were questioned. Investigations were conducted into the effects of barium emissions, resulting from the use of barium smoke suppressant additives [19.117–19.119] with some results showing increased mutagenicity as a result of using a barium additive [19.120]. Concern over the emissions of metal particles in general has resulted in the prohibition of metallic additives from automotive fuels in many jurisdictions including the US, Europe, and Japan. However, there are limited waivers in certain circumstances. Metallic fuel additives are therefore being promoted for reducing emissions and improving economy in a limited number of situations.

As emissions legislation moved from simply restricting visible smoke to the limiting of the mass of particulate matter emitted, the mass of metallic ash became an important consideration. Improvements in diesel fuel injection systems resulted in better fuel-air mixing and reduced smoke emissions. This also resulted in a lower mass of soot emissions; this in turn made the contribution from the mass of ash, from a smoke suppressant, more significant. It therefore became more difficult to balance the benefits of these combustion improver additives with the detrimental effects of increased metal ash emissions.

Successive emissions legislation has continued to reduce the permissible level of particulate mass (and potentially particle number) emissions; it has thus become increasingly necessary to trap the particulate to prevent it reaching the tailpipe. As discussed in Chapter 16, Section 16.4, the use of diesel exhaust aftertreatment systems is now commonplace to meet this ever more stringent emissions regulation. Diesel particulate filters (DPFs) are now fitted as standard equipment to most diesel passenger cars sold in Western Europe. Any DPF system that consistently performs well will by definition accumulate the soot particles that it filters out of the exhaust. This accumulated soot must be periodically removed to regenerate the DPF. Organometallic additives that are effective at reducing smoke emissions by aiding the oxidation of soot particles are also likely to assist the oxidation of soot particles trapped in a DPF system. This has proved to be the case with proposals to use a wide selection of different metals, either alone or in combination, as catalysts for the regeneration of DPFs [19.121, 19.122] and such additives have become known as fuel borne catalysts (FBCs). Due to the restrictions on the use of metallic fuel additives, the use of an FBC with a DPF is generally restricted to applications where the FBC is stored on the vehicle and dosed into the fuel onboard the vehicle. As such, FBCs should not be considered as a fuel additive but as part of the emissions control system. However, for completeness a brief discussion of the technology is included in the following section.

19.4.3. FBCs for DPF Regeneration

There are many examples of DPF/FBC systems that have been demonstrated where there is no active control over the regeneration process; these are known as passive systems. Some of these passive systems have been shown to perform even when fitted to vehicles that are operating over low duty operating cycles [19.123–19.125]. There will inevitably be some duty cycles where passive regeneration will not take place and active regeneration technologies will be necessary. An engine or vehicle manufacturer will therefore design an active system into the emissions compliant products. In these situations, the use of an FBC can still be useful in reducing the energy input required to bring about regeneration of the DPF.

The relative efficacy of such additives is often compared by using the balance point temperature. The balance point temperature is the temperature at which the rate of soot deposition within the DPF is matched by the rate at which the soot is being oxidized with the aid of the FBC. To achieve this balance, the DPF is loaded with soot and the temperature is raised by increasing the load on the engine in a stepwise fashion. This increase in engine duty increases the exhaust gas flow and thus the pressure drop across the DPF. At each step, the pressure drop increases more gradually as soot is deposited within the DPF. When the balance point is reached, this gradual increase in pressure no longer occurs, and the pressure begins to fall as the soot continues to oxidize. However, the balance point temperature is also dependent upon the mass of accumulated soot within the DPF, oxygen flow rate, and the ratio of catalyst-to-soot within the DPF. To be meaningful, the balance point temperature should also quote these other parameters. Figure 19.8 [19.126] shows how the balance point temperature can vary with these other parameters.

FIGURE 19.8   Variation in balance point temperature with changes in soot loading, oxygen flow rate, and metal to soot ratio.

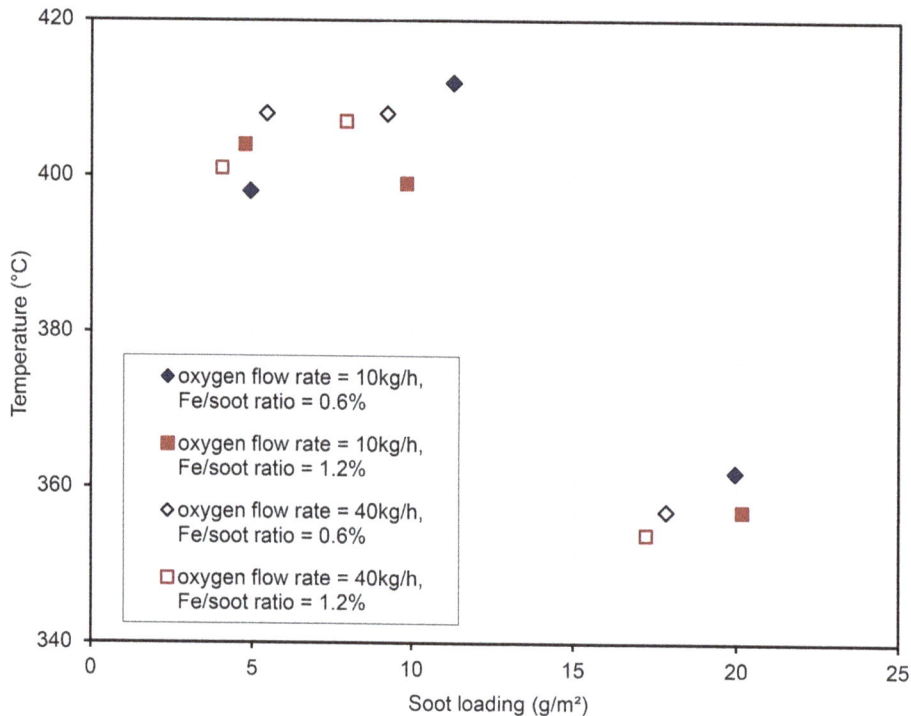

© SAE International.

To be acceptable, an FBC for the regeneration of a DPF must be developed, bearing in mind the following points:

- The combustion products of the additive must be as catalytically active toward the soot as is possible.

- The combustion products of the additive should be nontoxic. Although the ash will be trapped within the DPF, the end-of-life for the aftertreatment system must be considered.

- The FBC will have to be stored on the vehicle for onboard dosing into the fuel. The FBC must therefore be easily soluble and miscible with the fuel. Because the FBC must be stored on the vehicle, it is desirable that the metal concentration in the additive be as high as possible to minimize the quantity that must be carried.

- The FBC must be stable over the anticipated range of temperatures that it will encounter on the vehicle. It must also be stable under these conditions over the period of time for which it is likely to be stored on the vehicle.

- The FBC should not promote any unfavorable reactions within the fuel system, engine, or aftertreatment systems. A vehicle fitted with a DPF/FBC system could be re-fueled with a wide variety of fuels containing a wide variety of other fuel additives.

- The FBC itself and the combustion products it produces must not promote wear in any part of the engine systems.

These considerations have tended to limit the commercial compositions of FBC to compounds of cerium and iron [19.126–19.131]. Examples of the chemistries employed for these additives are colloidal dispersions [19.132–19.135], ferrocene (dicyclopentadienyl iron) derivatives [19.136, 19.137], and metal salts [19.138, 19.139].

19.5. Multifunctional Additive Packages

The preceding sections of this chapter make it clear that adequate diesel fuel performance can no longer be achieved by refinery processes, and the use of fuel additives is required. Some performance issues can be addressed by the use of a single additive, for example, a CNI may enable a refiner to meet the CN specification. In other cases, a single additive may provide functionality in more than one area, for example, a DCA, being surface active, may also act as a corrosion inhibitor. However, to meet all the requirements of a modern high-quality diesel fuel it is necessary to include more than one active component. Sometimes the inclusion of one component may necessitate the inclusion of another, for example, a surfactant additive may necessitate the inclusion of a dehazer or demulsifier.

Clearly, a fuel producer cannot simply buy individual additives off-the-shelf, add them to the base diesel fuel, and hope that the finished fuel achieves all the required performance attributes. It is thus desirable for the fuel producer to buy and add a single fuel additive package that contains all the required components to give the desired quality finished diesel fuel.

As with the gasoline additives that are discussed in Chapter 11, testing these additive packages is no longer a case of only running laboratory tests or even engine tests to look at various performance aspects. Testing these additive packages must involve not only looking at all the intended benefits, but must also consider the possible interactions of components added to differentiate marketed fuels and components added by the refiner or distributor to meet specified performance limits. It is also important to consider the influence on vehicle emissions and emissions control devices, and whether the performance can be achieved in a range of vehicles in the real world.

References

19.1.   Haycock, R.F. and Thatcher, R.G.F., "Fuel Additives and the Environment," ATC Document 52, 1994.

19.2.   Haycock, R.F. and Thatcher, R.G.F., "Fuel Additives and the Environment," ATC Document 52, 2004.

19.3.   Technical Committee of Petroleum Additive Manufacturers in Europe (ATC), "Fuel Additives; Use and Benefit," September 2013/ATC Document 113, 2013.

19.4.   Technical Committee of Petroleum Additive Manufacturers in Europe (ATC), "Fuel Additives; Use and Benefit," June 2020/ATC Document 113, 2020.

19.5.   Crusius, S., Müller, M., Seehack, S., Meyer, N.-S. et al., "Auswirkungen von Additiven für Kraftstoffe auf Abgasnachbehandlungssysteme, Emissionen sowie Umwelt und Gesundheit," Umweltbundesamtes, March 2022, accessed March 2023, https://www.umweltbundesamt.de/publikationen/auswirkungen-von-additiven-fuer-kraftstoffe-auf.

19.6.   Lopes, S., Sigelko, J., Monroe, R., Kozub, D. et al., "Development of the TOP TIERTM Diesel Standard," SAE Technical Paper 2019-01-0264 (2019), doi:https://doi.org/10.4271/2019-01-0264.

19.7. Coley, T.R., "Diesel Fuel Additives Influencing Flow and Storage; Critical Reports on Applied Chemistry," in *Gasoline and Diesel Fuel Additives*, Vol. 25, Ed. Owen, K. (New York: John Wiley & Sons, 1989).

19.8. Russell, T.J., "Petrol and Diesel Additives," *Institute of Petroleum Review* 42, no. 501 (1988): 35-42.

19.9. Bergeron, I., Charland, J.P., and Ternan, M., "Color Degradation of Hydrocracked Diesel Fuel," *Energy & Fuels* 13, no. 3 (1999): 686-693.

19.10. Banavali, R.M. and Chheda, B.D., Use of tertiary-alkyl primary amines in fuel compositions used as heat-transfer fluid. EP Patent 0 947 577 B1, 2004.

19.11. Batt, R.J., Henry, C.P., and Whitesmith, P.R. "Stabiliser Additive Performance in Diesel Fuels and Gas Oils Meeting New Environment Targets," Presented at in *5th International Conference on Stability and Handling of Liquid Fuels*, Rotterdam, the Netherlands, 1994.

19.12. Ethyl Corperation, Alkylated bis-phenols and hydroxy benzylamines and their method of manufacture. GB Patent 806,961, 1959.

19.13. Balfanz, U., Frohling, J.-C., Spieckermann, A., and Terschek, R., Fuel for diesel engines. U.S. Patent 2006/207167, 2006.

19.14. Poirier, M.-A. and Lang, A.S., Method of improving the oxidation stability of biodiesel as measured by the rancimat test. U.S. Patent 2011/0146139, 2011.

19.15. Chenicek, J.A., Metal deactivator. U.S. Patent 2,754,216, 1956.

19.16. Downing, F.B. and Pedersen, C.J., Stabilization of organic substances. U.S. Patent 2,181,121, 1939.

19.17. Pedrsen, C.J. and Downing F.B., Condensation products of pentra-erythrityletramine with beta-diketones and beta-hydroxy methylene ketones. U.S. Patent 2,462,668, 1949.

19.18. Roling, P.V. and Martin, J.E., Selected hydroxy-oximes as iron deactivators. U.S. Patent 5,100,532, 1992.

19.19. Pedley, J.F., Hiley, R.W., and Hancock, R.A., "Storage Stability of Petroleum-Derived Diesel Fuel: 3. Identification of Compounds Involved in Sediment Formation," *Fuel* 67, no. 8 (1988): 1124-1130.

19.20. Fisher, S.L. and Street, J.P., Stabilization of hydrocarbon fluids using metal deactivators. U.S. Patent 5,641,394, 1997.

19.21. Lewtas, K. and Bland, J.D., Flow improvers and cloud point depressants. U.S. Patent 5,011,505, 1991.

19.22. Damin, B., Faure, A., Denis, J., Sillion, B. et al., "New Additives for Diesel Fuels: Cloud-Point Depressants," SAE Technical Paper 861527 (1986), doi:https://doi.org/10.4271/861527.

19.23. Botros, M.G., Cloud point depresants for middle distillate fuels. U.S. Patent 6,143,043, 2000.

19.24. Botros, M.G., Cloud point depresants for middle distillate fuels. U.S. Patent 6,342,081, 2002.

19.25. Herbert, G.L., Cloud point depressant composition. EP Patent 0 654 526, 1995.

19.26. Ilnyckyj, S. and Rupar, C.B., Ethylene-vinyl ester pour point depressant for middle distillates. U.S. Patent 3,048,470, 1962.

19.27. Schield, J.A. and Weers, J.J., Methods and compositions for improvement of low temperature fluidity of fuel oils. U.S. Patent 5,857,287, 1999.

19.28. Handa, S. and Hodgson, P.K.G., Polyacrylate esters, their preparation and use as a low-temperature flow-improver in middle distillate oils. WO Patent 01/48032, 2001.

19.29. Davies, B.W., More, I., Costello, J.K., Brown, G.I. et al., Fuel oil additives. EP Patent 0 308 176, 1989.

19.30. More, I., Camarco, M.W., and Smith, D.R.T., Fuel oil additives and compositions. WO Patent 91/15562, 1991.

19.31. Wu, M.M., Baillargeon, D.J., and Jackson, A., Pour point depressant for hydrocarbon compositions. WO Patent 2009/148685, 2009.

19.32. Krull, M., Siggelkow, B., Hess, M., and Neuhaus, U., Cold flow improvers for fuel oils of vegetable or animal origin. U.S. Patent 2004/0010072, 2004.

19.33. Martyak, N.M., Macy, N.E., Gernon, M.D., Schmidt, S.C. et al., Acrylic polymer low temperature flow modifiers in bio-derived fuels. WO Patent 2008/154558, 2008.

19.34. Scanlon, E., DeSantis, K., Auschra, C., and Mock, A., Biodiesel cold flow improver. U.S. Patent 2009/0152235, 2009.

19.35.   Sondjaja, R., Koschabek, R., Weber, M., Nenito, J. et al., Composition to improve cold flow properties of fuel oils. U.S. Patent 2012/0174474, 2012.

19.36.   Westlake, A., McGlashan, S., Clarke, S., Fisher, M. et al., Biodiesel additive. WO Patent 2009/143566, 2009.

19.37.   Sbihi, H.M., Nehdi, I.A., Mokbli, S., Romdhani-Younes, M. et al., "Study of Oxidative Stability and Cold Flow Properties of Citrillus Colocynthis Oil and Camelus Dromedaries Fat Biodiesel Blends," *Industrial Crops and Products* 122 (2018): 133-141.

19.38.   Hazrat, M.A., Rasul, M.G., Mofijur, M., Khan, M.M.K. et al., "A Mini Review on the Cold Flow Properties of Biodiesel and its Blends," *Frontiers in Energy Research* 8 (2020): 598651.

19.39.   Zhang, X., Kushner, P.J., Saville, B.A., and Posen, I.D., "Cold Temperature Limits to Biodiesel Use under Present and Future Climates in North America," *Environmental Science & Technology* 56, no. 12 (2022): 8640-8649.

19.40.   ASTM International, "Standard Test Method for Rust-Preventing Characteristics of Inhibited Mineral Oil in the Presence of Water," ASTM D665-19, ASTM International, 2020.

19.41.   Matt, J.W., Conductivity additive for liquid hydrocarbons. U.S. Patent 3,758,283, 1973.

19.42.   Bialy, J.J., Siegart, W.R., Blackley, W.D., and Chafetz, H., AntiStatic fuel composition. U.S. Patent 3,652,238, 1972.

19.43.   Naiman, M.I. and Buriks, R.S., Olefin-acrylonitrile copolymers and uses thereof. U.S. Patent 4,333,741, 1982.

19.44.   Henry, C.P., Compositions of olefin-sulfur dioxide copolymers and polyamines as antistatic additives for hydrocarbon fules. U.S. Patent 3,917,466, 1975.

19.45.   ASTM International, "Standard Test Method for Water Separation Properties of Light and Middle Distillate, and Compression and Spark Ignition Fuels," ASTM D7451-21, ASTM International, 2021.

19.46.   Sheahan, T.J., Dorer, C.J., and Miller, C.O., "Detergent-Dispersant Fuel Performance and Handling," SAE Technical Paper 690516 (1969), doi:https://doi.org/10.4271/690516.

19.47.   Langley, G.J., Herniman, J., Carter, A., Wilmot, E. et al., "Detection and Quantitation of ACCUTRACE S10, a New Fiscal Marker Used in Low-Duty Fuels, Using a Novel Ultrahigh-Performance Supercritical Fluid Chromatography–Mass Spectrometry Approach," *Energy & Fuels* 32, no. 10 (2018): 10580-10585.

19.48.   CEC, "Procedure for Diesel Engine Injector Nozzle Coking Test (PSA XUD9A/L 1.9 Litre 4 Cylinder Indirect Injection Diesel Engine)," CEC F-23-01, 2012.

19.49.   Reading, K., Roberts, D.D., and Evans, T.M., "The Effects of Fuel Detergents on Nozzle Fouling and Emissions in IDI Diesel Engines," SAE Technical Paper 912328 (1991), doi:https://doi.org/10.4271/912328.

19.50.   Richards, P., Walker, R.D., and Williams, D., "Fouling of Two Stage Injectors—An Investigation into Some Causes and Effects," SAE Technical Paper 971619 (1997), doi:https://doi.org/10.4271/971619.

19.51.   CEC, "Direct Injection, Common Rail Diesel Engine Nozzle Coking Test," CEC F-98-08, 2012.

19.52.   Hawthorne, M., Roos, J.W., and Openshaw, M.J., "Use of Fuel Additives to Maintain Modern Diesel Engine Performance with Severe Test Conditions," SAE Technical Paper 2008-01-1806 (2008), doi:https://doi.org/10.4271/2008-01-1806.

19.53.   Su, W.-Y., Hydrocarbon compositions containing a polyetheramide additive. EP Patent 0732390, 1995.

19.54.   Courtney, R.L., Diesel fuel and method for deposit control in compression ignition engines. U.S. Patent 4,568,358, 1986.

19.55.   Cherpeck, R.E., Fuel additive compositions containing poly(oxyalkylene) amines and polyalkyl hydroxyaromatics. U.S. Patent 5,192,335, 1993.

19.56.   Malfer, D.J. and Thomas, M.D., Mannich detergent for hydrocarbon fuels. U.S. Patent 2009/0071065, 2009.

19.57.   Reid, J., Fuel compositions. U.S. Patent 2010/0258070, 2010.

19.58.   Decanio, E., Herbstman, S., Kaufman, B., and Parke, B., Detergent additive compositions for diesel fuels. WO Patent 98/12282, 1998.

19.59.   Hou, P.W., Colucci, W.J., and Nichols, T.W., Fuel additive to control deposit formation. U.S. Patent 2011/0010985, 2011.

19.60.   Reid, J., Method and uses relating to fuel compositions. WO Patent 2010/097624, 2010.

19.61.   Stevenson, P.R., Ray, J.C., Moreton, D.J., and Bush, J.H., Quternary ammonium amide and/or ester salts. WO Patent 2010/132259, 2010.

19.62.   Grabarse, W., Böhnke, H., Tock, C., Röger-Göpfert, C. et al., Acid-free quaternized nitrogen compounds and use thereof as additives in fuels and lubricants. U.S. Patent 2012/0010112, 2012.

19.63.   Naseri, H., Trickett, K., Mitroglou, N., Karathanassis, I. et al., "Turbulence and Cavitation Suppression by Quaternary Ammonium Salt Additives," *Scientific Reports* 8, no. 1 (2018): 1-15.

19.64.   Tupa, R. and Dorer, C., "Gasoline and Diesel Fuel Additives for Performance/Distribution Quality—II," SAE Technical Paper 861179 (1986), doi:https://doi.org/10.4271/861179.

19.65.   Booth, M. and Wolveridge, P.E., "Severe Hydrotreating of Diesel Fuel Can Cause Fuel-Injector Pump Failure," *Oil and Gas Journal* 91, no. 33 (1993): 71-76.

19.66.   Crampton, A.B., Gleason, W.W., and Wieland, E.R., "Thermal Stability—A New Frontier for Jet Fuels," SAE Technical Paper 550096 (1955), doi:https://doi.org/10.4271/550096.

19.67.   Barringer, C.M., "Stability of Jet Fuels at High Temperatures," SAE Technical Paper 550097 (1955), doi:https://doi.org/10.4271/550097.

19.68.   Cloud, G.H. and Blackwood, A.J., "The Influence of Diesel Fuel Properties on Engine Deposits and Wear," SAE Technical Paper 430164 (1943), doi:https://doi.org/10.4271/430164.

19.69.   Moore, C.C. and Kent, W.L., "Effect of Nitrogen and Sulfur Content of Fuels on Diesel-Engine Wear," SAE Technical Paper 470257 (1947), doi:https://doi.org/10.4271/470257.

19.70.   Broeze, J.J. and Wilson, A., "Sulphur in Diesel Fuels," *Proceedings of the Institute of Mechanical Engineers* 2, no. 1 (1948): 128-145.

19.71.   Fursund, K., "Wear in Cylinder Liners," *Wear* 1, no. 2 (1957): 104-118.

19.72.   Vere, R.A., "Lubricity of Aviation Turbine-Fuels," SAE Technical Paper 690667 (1969), doi:https://doi.org/10.4271/690667.

19.73.   Appeldoorn, J.K. and Tao, F.F., "The Lubricity Characteristics of Heavy Aromatics," *Wear* 12, no. 2 (1968): 117-130.

19.74.   Wei, D. and Spikes, H.A., "The Lubricity of Diesel Fuels," *Wear* 111, no. 2 (1986): 217-235.

19.75.   Nikanjam, M. and Henderson, P.T., "Lubricity of Low Sulfur Diesel Fuels," SAE Technical Paper 932740 (1993), doi:https://doi.org/10.4271/932740.

19.76.   Appeldoorn, J.K. and Dukek, W.G., "Lubricity of Jet Fuels," SAE Technical Paper 660712 (1966), doi:https://doi.org/10.4271/660712.

19.77.   Mitchell, K., "The Lubricity of Winter Diesel Fuels," SAE Technical Paper 952370 (1995), doi:https://doi.org/10.4271/952370.

19.78.   Caprotti, R., Jansen, E.B.M., Kraft, T., and Woodall, K., "Laboratory Test for Distributor Type Diesel Fuel Pumps - CEC PF032 Code of Practice," SAE Technical Paper 2004-01-016 (2004), doi:https://doi.org/10.4271/2004-01-016.

19.79.   Jenkins, S.R., Landells, R.G.M., and Hadley, J.W., "Diesel Fuel Lubricity Development of a Constant Load Scuffing Test Using the Ball on Cylinder Lubricity Evaluator (BOCLE)," SAE Technical Paper 932691 (1993), doi:https://doi.org/10.4271/932691.

19.80.   Mitchell, K., "Diesel Fuel Lubricity: A Survey of 1994/95 Canadian Winter Diesel Fuels," SAE Technical Paper 961181 (1996), doi:https://doi.org/10.4271/961181.

19.81.   Hadley, J.W., "A Method for the Evaluation of the Boundary Lubricating Properties of Aviation Turbine Fuels," *Wear* 101, no. 3 (1985): 219-253.

19.82.   Tucker, R.F., Stradling, R.J., Wolveridge, P.E., Rivers, K.J. et al., "The Lubricity of Deeply Hydrogenated Diesel Fuels–The Swedish Experience," SAE Technical Paper 942016 (1994), doi:https://doi.org/10.4271/942016.

19.83.   Voitik, R.M. and Ren, N., "Diesel Fuel Lubricity by Standard Four Ball Apparatus Utilizing Ball on Three Disks, BOTD," SAE Technical Paper 950247 (1995), doi:https://doi.org/10.4271/950247.

19.84.   Hadley, J.W., Owen, G.C., and Mills, B., "Evaluation of a High Frequency Reciprocating Wear Test for Measuring Diesel Fuel Lubricity," SAE Technical Paper 932692 (1993), doi:https://doi.org/10.4271/932692.

19.85.   Nikanjam, M., Crosby, T., Henderson, P., Gray, C. et al., "ISO Diesel Fuel Lubricity Round Robin Program," SAE Technical Paper 952372 (1995), doi:https://doi.org/10.4271/952372.

19.86.   Schwab, S.D., Lubricity additive for fuels. U.S. Patent 2006/0288638, 2006.

19.87.   Schield, J.A., and Biggerstaff, P.J., Branched carboxylic acids as fuel lubricity additives. U.S. Patent 7,867,295, 2011.

19.88.   Giavazzi, F. and Panarello, F., Gas oil composition, U.S. Patent 5,599,358, 1997.

19.89.   Vrahopoulou, E.P., Schlosberg, R.H., and Turner, D.W., Polyol ester distillate fuels additive. U.S. Patent 5,993,498, 1999.

19.90.   Eber, D., Germanaud, L., and Maldonado, P., Additive for fuel oiliness. U.S. Patent 6,511,520, 2003.

19.91.   Choo, Y.M., Cheng, S.F., Ma, A.N., and Basiron, Y., Fuel lubricity additive. U.S. Patent 2006/0117648, 2006.

19.92.   Schwab, S.D., Succinimide lubricity additive for diesel fuel and a method for reducing wear scarring in an engine. U.S. Patent 2009/0249683, 2009.

19.93.   Batt, R.J., McMillan, J.A., and Bradbury, I.P., "Lubricity Additives—Performance and No-Harm Effects in Low Sulfur Fuels," SAE Technical Paper 961943 (1996), doi:https://doi.org/10.4271/961943.

19.94.   ASTM International, "Standard Test Method for Cetane Number of Diesel Fuel Oil," ASTM D613-23, ASTM International, 2023.

19.95.   Robbins, W.E., Audette, R.R., and Reynolds, N.E., "Performance and Stability of Some Diesel Fuel Ignition Quality Improvers," SAE Technical Paper 510200 (1951), doi:https://doi.org/10.4271/510200.

19.96.   Goodrich, B.E., McDuff, P.J., Krupa, C.C., Alvarez, L. et al., "A No-Harm Test Matrix Investigating the Effect of Di-tert-butyl Peroxide (DTBP) Cetane Number Improver on Diesel Fuel Properties," SAE Technical Paper 982574 (1998), doi:https://doi.org/10.4271/982574.

19.97.   Sutton, D.L., Rush, M.W., and Richards, P., "Diesel Engine Performance and Emissions Using Different Fuel/Additive Combinations," SAE Technical Paper 880635 (1988), doi:https://doi.org/10.4271/880635.

19.98.   Collins, J.M. and Unzelman, G.H., "Diesel Trends Emphasize Cetane Quality, Economics, and Prediction," *Proceedings, American Petroleum Institute* 61 (1982): 7-42.

19.99.   Sobotowski, R.A., "Investigation of Cetane Response of US Diesel Fuels," SAE Technical Paper 950249 (1995), doi:https://doi.org/10.4271/950249.

19.100.   Thompson, A.A., Lambert, S.W., and Mulqueen, S.C., "Prediction and Precision of Cetane Number Improver Response Equations," SAE Technical Paper 972901 (1997), doi:https://doi.org/10.4271/972901.

19.101.   Nandi, M.K. and Jacobs, D.C., "Cetane Response of Di-tertiary-butyl Peroxide in Different Diesel Fuels," SAE Technical Paper 952368 (1995), doi:https://doi.org/10.4271/952368.

19.102.   ASTM International, "Standard Test Method for Calculated Cetane Index by Four Variable Equation," ASTM D4737-21, ASTM International, 2021.

19.103.   Liotta, F.J., "A Peroxide Based Cetane Improvement Additive with Favorable Fuel Blending Properties," SAE Technical Paper 932767 (1993), doi:https://doi.org/10.4271/932767.

19.104.   Nandi, M.K., Jacobs, D.C., Liotta, F.J., and Kesling, H.S., "The Performance of a Peroxide-Based Cetane Improvement Additive in Different Diesel Fuels," SAE Technical Paper 942019 (1994), doi:https://doi.org/10.4271/942019.

19.105.   Aradi, A.A. and Ryan, T.W., "Cetane Effect on Diesel Ignition Delay Times Measured in a Constant Volume Combustion Apparatus," SAE Technical Paper 952352 (1995), doi:https://doi.org/10.4271/952352.

19.106.   Schwab, S.D., Guinther, G.H., Henly, T.J., and Miller, K.T., "The Effects of 2-Ethylhexyl Nitrate and Di-Tertiary-Butyl Peroxide on the Exhaust Emissions from a Heavy-Duty Diesel Engine," SAE Technical Paper 1999-01-1478 (1999), doi:https://doi.org/10.4271/1999-01-1478.

19.107.   Ullman, T.L., Mason, R.L., and Montalvo, D.A., "Effects of Fuel Aromatics, Cetane Number, and Cetane Improver on Emissions from a 1991 Prototype Heavy-Duty Diesel Engine," SAE Technical Paper 902171 (1990), doi:https://doi.org/10.4271/902171.

19.108.   Ullman, T.L., Spreen, K., and Mason, R.L., "Effects of Cetane Number on Emissions from a Prototype 1998 Heavy-Duty Diesel Engine," SAE Technical Paper 950251 (1995), doi:https://doi.org/10.4271/950251.

19.109. Barbier, P., "Evaluation of Cetane Improver Effects on Regulated Emissions from a Passenger Car Equipped with a Common Rail Diesel Engine," SAE Technical Paper 2000-01-1853 (2000), doi:https://doi.org/10.4271/2000-01-1853.

19.110. Nanjundaswamy, H., Tatur, M., Tomazic, D., Koerfer, T. et al., "Fuel Property Effects on Emissions and Performance of a Light-Duty Diesel Engine," SAE Technical Paper 2009-01-0488 (2009), doi:https://doi.org/10.4271/2009-01-0488.

19.111. Lange, W.W., Cooke, J.A., Gadd, P.J., Zürner, H. et al., "Influence of Fuel Properties on Exhaust Emissions from Advanced Heavy-Duty Engines Considering the Effect of Natural and Additive Enhanced Cetane Number," SAE Technical Paper 972894 (1997), doi:https://doi.org/10.4271/972894.

19.112. Miller, P. and Cloud, G.H., Organo-metallic diesel fuel ignition promotors. U.S. Patent 2,258,297, 1941.

19.113. Miller, C.D., "Diesel Smoke Suppression by Fuel Additive Treatment," SAE Technical Paper 670093 (1967), doi:https://doi.org/10.4271/670093.

19.114. Norman, G.R., "A New Approach to Diesel Smoke Suppression," SAE Technical Paper 660339 (1966), doi:https://doi.org/10.4271/660339.

19.115. Miyamoto, N., Hou, Z., Harada, A., Ogawa, H. et al., "Characteristics of Diesel Soot Suppression with Soluble Fuel Additives," SAE Technical Paper 871612 (1987), doi:https://doi.org/10.4271/871612.

19.116. Brandes, J.G., "Diesel Fuel Specification and Smoke Suppressant Additive Evaluations," SAE Technical Paper 700522 (1970), doi:https://doi.org/10.4271/700522.

19.117. Golothan, D.W., "Diesel Engine Exhaust Smoke: The Influence of Fuel Properties and the Effects of Using Barium-Containing Fuel Additive," SAE Technical Paper 670092 (1967), doi:https://doi.org/10.4271/670092.

19.118. Saito, T. and Nabetani, M., "Surveying Tests of Diesel Smoke Suppression with Fuel Additives," SAE Technical Paper 730170 (1973), doi:https://doi.org/10.4271/730170.

19.119. Apostolescu, N.D., Matthew, R.D., and Sawyer, R.F., "Effects of a Barium-Based Fuel Additive on Particulate Emissions from Diesel Engines," SAE Technical Paper 770828 (1977), doi:https://doi.org/10.4271/770828.

19.120. Draper, W.M., Phillips, J., and Zeller, H.W., "Impact of a Barium Fuel Additive on the Mutagenicity and Polycyclic Aromatic Hydrocarbon Content of Diesel Exhaust Particulate Emissions," SAE Technical Paper 881651 (1988), doi:https://doi.org/10.4271/881651.

19.121. Pattas, K., Samaras, Z., Sherwood, D., Umehara, K. et al., "Cordierite Filter Durability with Cerium Fuel Additive: 100,000 km of Revenue Service in Athens," SAE Technical Paper 920363 (1992), doi:https://doi.org/10.4271/920363.

19.122. McKinnon, D.L., Pavlich, D.A., Tadrous, T., and Shephard, D., "Results of North American Field Trials Using Diesel Filters with a Copper Additive for Regeneration," SAE Technical Paper 940455 (1994), doi:https://doi.org/10.4271/940455.

19.123. Richards, P., Terry, B., and Pye, D., "Demonstration of the Benefits of DPF/FBC Systems on London Black Cabs," SAE Technical Paper 2003-01-0375 (2003), doi:https://doi.org/10.4271/2003-01-0375.

19.124. Richards, P., "Field Experience of DPF Systems Retrofitted to Vehicles with Low Duty Operating Cycles," SAE Technical Paper 2004-28-0013 (2004), doi:https://doi.org/10.4271/2004-28-0013.

19.125. Richards, P., "DPF Technology for Older Vehicles and High Sulphur Fuel," SAE Technical Paper 2005-26-020 (2005), doi:https://doi.org/10.4271/2005-26-020.

19.126. Schrewe, K., Belcour, C., and Richards, P., "A Study of the Parameters Ensuring Reliable Regeneration of a Sintered Metal Particulate Filter using a Fuel Borne Catalyst," SAE Technical Paper 2008-01-2485 (2008), doi:https://doi.org/10.4271/2008-01-2485.

19.127. Salvat, O., Marez, P., and Belot, G., "Passenger Car Serial Application of a Particulate Filter System on a Common Rail Direct Injection Diesel Engine," SAE Technical Paper 2000-01-0473 (2000), doi:https://doi.org/10.4271/2000-01-0473.

19.128. Blanchard, G., Colignon, C., Griard, C., Rigaudeau, C. et al., "Passenger Car Series Application of a New Diesel Particulate Filter System Using a New Ceria-Based Fuel-Borne Catalyst: From the Engine Test Bench to European Vehicle Certification," SAE Technical Paper 2002-01-2781 (2002), doi:https://doi.org/10.4271/2002-01-2781.

19.129. Jeuland, N., Dementhon, J.-B., Gagnepain, L., Plassat, G. et al., "Performances and Durability of DPF (Diesel Particulate Filter) Tested on a Fleet of Peugeot 607 Taxis: Final Results," SAE Technical Paper 2004-01-0073 (2004), doi:https://doi.org/10.4271/2004-01-0073.

19.130. Mayer, A., Nöthiger, P., Andreassen, L., Kany, S. et al., "Retrofitting TRU-Diesel Engines with DPF-Systems Using FBC and Intake Throttling for Active Regeneration," SAE Technical Paper 2005-01-0662 (2005), doi:https://doi.org/10.4271/2005-01-0662.

19.131. Rocher, L., Seguelong, T., Harle, V., Lallemand, M. et al., "New Generation Fuel Borne Catalyst for Reliable DPF Operation in Globally Diverse Fuels," SAE Technical Paper 2011-01-0297 (2011), doi:https://doi.org/10.4271/2011-01-0297.

19.132. Blanchard, G., Chane-Ching, J.-Y., and Tolla, B., Organic colloidal dispersion of iron particles, method for preparing same and use thereof as fuel additive for internal combustion engines. WO Patent 03/053560, 2003.

19.133. Caprotti, R., Fava, C.S., Thompson, R.M., Willis, M.J. et al., Fuel additive. U.S. Patent 2006/0000140, 2006.

19.134. Reed, K.J., Cerium dioxide nanoparticle-containing fuel additive. WO Patent 2008/030805, 2008.

19.135. Thompson, R.M., Method of supplying iron to the particulate trap of a diesel engine exhaust. WO Patnet 2008/116550, 2008.

19.136. Cook, S.L., Kalischewski, W., Lohmann, G., and Marchewski, A., Composition comprising dimeric or oligomeric ferrocenes. WO Patent 02/18398, 2002.

19.137. Cook, S.L., Kalischewski, W., Lohmann, G., and Marchewski, A., Composition. WO Patent 03/020733, 2003.

19.138. Caprotti, R. and Pilling R.J., Iron salt diesel fuel additive compositions for improvement of particulate traps. U.S. Patent 7,306,634, 2007.

19.139. Harle, V., and Rocher, L., Operation of diesel/lean-burn engines having easily regenerated particle filters in the exhaust system therefor. U.S. Patent 2010/0300079, 2010.

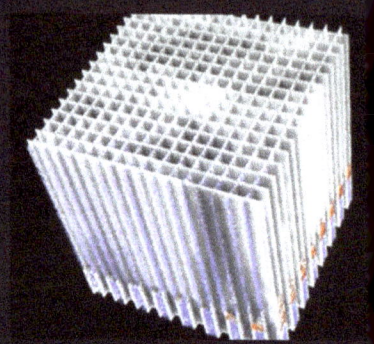

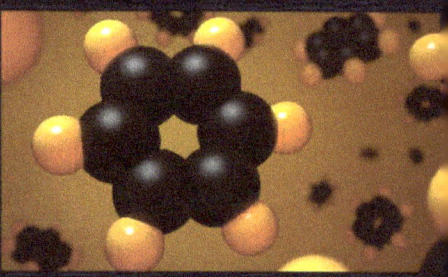

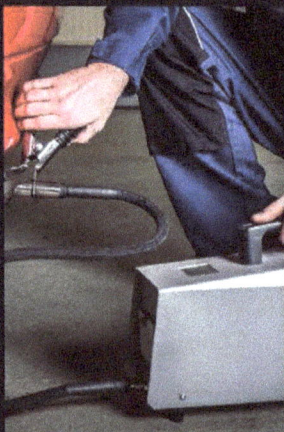

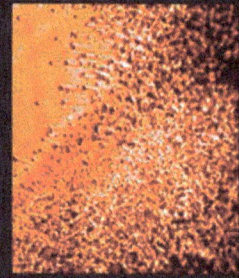

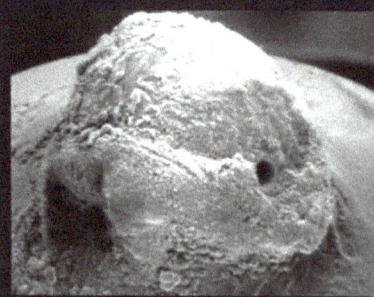

20

Other Diesel Specification and Nonspecification Properties

Two of the major diesel fuel properties that affect the performance of diesel engines are ignition quality and low-temperature flow characteristics; these have been discussed in Chapters 15 and 17, respectively. Ensuring that the diesel fuel quality remains as the fuel producer intended and also that it does not produce unwanted degradation products within the engine was discussed in Chapters 18 and 19. How fuel composition influences exhaust emissions will be dealt with in Chapter 21. This chapter will consider other diesel fuel properties that can influence vehicle performance and customer perception, and how various fuel properties are measured.

20.1. Diesel Fuel Stability

Diesel fuel users expect that when they fill their tank, they are doing so with fuel that meets the producer's quality specifications. The ability of the fuel to remain unchanged during the period between its manufacture and use is therefore a prime consideration to the producer.

In many parts of the world, diesel fuel is no longer prepared from predominantly straight-run components. Most diesel fuels now contain cracked components, usually further processed by some degree of hydrotreatment. Low levels of hydrotreatment can leave large quantities of olefinic material, which is less stable, whereas very high levels of hydrotreatment can remove naturally occurring antioxidants. This can then leave the product prone to oxidative deterioration. The desire or the requirement to include biodiesel in the final blend can also introduce potentially unstable product into the fuel dispensed at the pump.

Instability in the fuel can result in a degradation in the color of the product, blocked filters, or deposit formation within the fuel injection system and within the combustion chamber. The factors affecting stability are discussed in more detail in Chapter 18, Section 18.1. These issues can be address by the correct use of appropriate fuel additive technology as discussed in Chapter 19, Section 19.1. Various test methods for assessing the stability of the petroleum fuel, the bio-components, and the finished fuel are described in Chapter 18, Section 18.1.3.

20.2. Corrosivity

It is important to ensure that the fuel will not attack metals in the distribution and storage network or in the engine fuel system. Copper and copper alloys are vulnerable to attack by certain sulfur compounds, and many specifications call for a copper corrosion and tarnish test [20.1]. In this test, a polished copper strip is immersed in a portion of the fuel and heated for 3 hours at 50°C, after which it is removed, washed, and compared with ASTM standards for copper strip corrosion. The standards are color reproductions of test strips showing increasing degrees of tarnish and corrosion. Typical specification limits allow a measure of tarnishing (ratings of 1 to 3), but no corrosion (rating of 4) is permitted. A quicker test method can be used for screening purposes [20.2].

Another standard method for indicating whether the fuel has a tendency to corrode metals with which it may come into contact is the neutralization number by color-indicator titration [20.3]. A sample of the fuel is dissolved in a solvent mixture, a color indicator is added, and the mixture is titrated with a standard base, potassium hydroxide (KOH) or acid, hydrochloric acid, solution until a color change indicates that the end point has been reached. Different equations are used to calculate the result (in mg KOH/g) according to whether the specification requirement is total acid number, strong acid number, or strong base number. A similar European standard [20.4] is used for determining the neutralization number of biodiesel. In the US an automatic titrator [20.5] is used to assess the neutralization number of biodiesel to meet the US specification [20.6].

20.3. Ignition Quality

The cetane number (CN) of a fuel is a measure of its ignition quality. The significance of ignition quality regarding diesel engine performance is discussed in Chapter 15, Section 15.2. The CN is determined by comparing the ignition delay characteristics of the test fuel with those of two reference fuels in a Cooperative Fuels Research (CFR) engine under specified standard conditions [20.7]. The reference fuels that are usually used are secondary reference fuels designated U-11 and T-18; these fuels have CN of 20.5 and 75.0, respectively.

Due to the time taken to measure the CN using the engine, a constant volume combustion chamber determination has replaced the engine test in many applications. This method [20.8] yields a corresponding value known as the derived cetane number (DCN). The range of application of the DCN method is slightly narrower than the engine method, 30 DCN to 65 DCN as opposed to 20.5 CN to 75 CN for the engine method.

For routine monitoring of diesel ignition quality, it is usually more convenient to use a calculated value based on the fuel's physical properties. The Calculated Cetane Index (CCI), often referred to as simply the cetane index, can be determined according to two different formulae, depending upon the degree of agreement with the engine test that is required. The simpler method [20.9] uses only the fuel density and mid-boiling point temperature. The CCI is defined by the following equation:

$$CCI = 454.74 - 1641.416D + 774.74D^2 - 0.554T_{50} + 97.803 \left(\log T_{50} \right)^2$$

where

CCI is the calculated cetane index

D is the density at 15°C (kg/L)

T_{50} is the mid-boiling temperature (°C)

Closer correlation with the engine test procedure is usually achieved by using a four variables equation [20.10] based on the fuel density and the 10%, 50%, and 90% distillation points. In this case, the CCI is given by the following equation:

$$\begin{aligned} CCI = 45.2 &+ 0.0892 \left(T_{10} - 215 \right) + 0.131 \left(T_{50} - 260 \right) \\ &+ 0.0523 \left(T_{90} - 310 \right) + 0.901B \left(T_{50} - 260 \right) - 0.420B \left(T_{90} - 310 \right) \\ &+ 0.0049 \left(T_{10} - 215 \right)^2 - 0.0049 \left(T_{90} - 310 \right)^2 + 107.0B + 60.0B^2 \end{aligned}$$

where T_{10}, T_{50}, and T_{90} are the 10%, 50%, and 90% recovery temperatures in °C, corrected to standard barometric pressure, respectively, D is density at 15°C, and

$$B = e^{\left[-3.5 \left(D - 0.85 \right) \right]} - 1$$

The fuel density and distillation characteristics included in these equations are not significantly affected by the inclusion of cetane number improver (CNI) additives. The inclusion of CNIs can result in noticeable discrepancies between CCI and CN. A diesel fuel specification that includes a limit value for both CN and CCI can restrict the refiner ability to meet CN specifications by including large amounts of CNI. CNI additives are discussed in Chapter 19, Section 19.3.

20.4. Lubricity

The lubricity of a fuel, sometimes referred to as its oiliness, is the ability of the fuel to build-up and to maintain a fluid film that will prevent contact between two solid surfaces—in short, the ability of the fuel to lubricate. This is a particularly relevant property for diesel fuel as modern fuel injection equipment (FIE) relies on the fuel to lubricate many of the moving parts within the system. Poor lubricity was often associated with low-viscosity fuel [20.11]. This may be due to the practice of adding kerosene to diesel fuel to prevent cold weather problems in cold climates; the kerosene prevented wax formation but also reduced the viscosity of the fuel and increased wear problems. In such cases, a lubricity agent was sometimes added to the fuel to limit wear in the injection equipment. It is now known that lubricity is not provided by the fuel viscosity per se, but rather it is provided by components within the fuel that prevent wear of metal surfaces in contact with each other. To appreciate this, a rudimentary understanding of friction and wear is required. Figure 20.1 illustrates two types of lubrication, namely hydrodynamic lubrication and boundary lubrication.

FIGURE 20.1 Illustration of hydrodynamic and boundary lubrication.

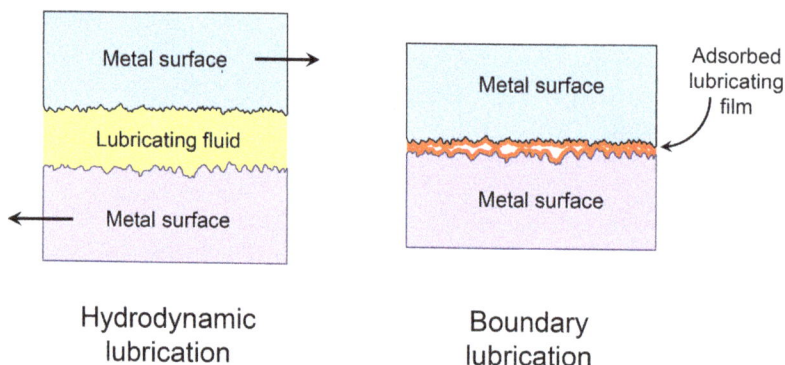

© SAE International.

Hydrodynamic lubrication exists when the two surfaces, moving relative to each other, are completely separated by a viscous fluid film. There is no surface-to-surface contact and no wear will occur. In this situation, the viscosity of the fluid is an important parameter, along with the relative speed of the two parts. This is illustrated on the left of Figure 20.1. If either of these parameters, usually the speed, is too low, then the fluid film will decrease in thickness. Boundary lubrication occurs when the dynamics of the system do not allow this fluid film to build up between the two surfaces, and the only thing that prevents surface-to-surface contact is an adsorbed layer of lubricant. This is illustrated on the right of Figure 20.1. In reality, the asperities of the surfaces will, in places, break through these protective layers, and some wear will result when relative movement occurs. At low relative speeds, the lubrication will be boundary lubrication, and at higher speed there will be hydrodynamic lubrication. The transition stage between the two is known as mixed film lubrication.

During boundary lubrication and mixed film lubrication regimes, it is vital that the fuel has the ability to form the adsorbed layer necessary to protect the mating surfaces. In straight-run diesel fuels, these protective components of the fuel are believed to be polycyclic aromatic types with sulfur, oxygen, and nitrogen content [20.12, 20.13]. With the severe hydrotreatment necessary to meet ultra-low sulfur limits, these polar compounds tend to be removed from the fuel. This fact was brought to the industry's attention in the early 1990s when Sweden introduced new environmental classifications for diesel fuels, with tax incentives to encourage their use. The fuels were intended to bring benefits of lower emissions levels but were found to cause premature injection pump wear due to its low lubricity characteristics. Vehicle endurance testing on Swedish Class I and Class II fuels demonstrated rotary injection pump failures in less than 30,000 km of operation [20.14]. Repeat testing demonstrated that the use of a lubricity additive could eliminate the problems. Lubricity is now included in diesel fuel specifications to ensure that FIE is not put at risk.

20.4.1. Evaluation of Different Test Methods

Bench tests that measure film strength or load-carrying capacity were studied as possible methods for assessing the lubricity or antiwear quality of different fuel types. After initial work by an SAE committee, an ISO working group, WG6, was set up to devise a method of measuring the lubricity of diesel fuel and subsequently setting a standard. A parallel activity in Europe, to assess the merits of alternative lubricity testing machines, was carried out by the Coordinating European Council (CEC) group CEC/PF 006. A subgroup of lubricity experts from ISO WG6 and CEC/PF 006 was set up to analyze the data and recommend a test. Support for the program was provided by oil companies, motor manufacturers, fuel additive companies, and all diesel injection pump manufacturers.

Four test procedures were evaluated [20.15]: the ball on three seats (BOTS) machine (later modified and renamed as the ball on three disks [BOTD]); two versions of the ball-on-cylinder lubricity evaluator (BOCLE), one being a Lubrizol modification and the other the US Army's scuffing load BOCLE; and the high-frequency reciprocating rig (HFRR). Figure 20.2 shows schematic arrangements of the wearing elements of the three lubricity test machines.

FIGURE 20.2 Schematic arrangements of three different types of lubricity test machine.

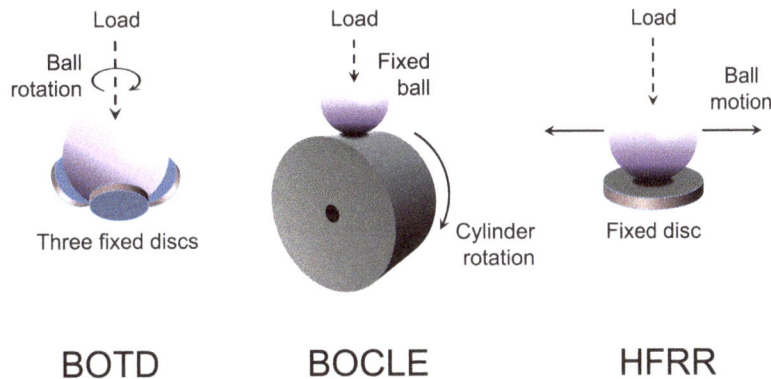

© SAE International.

Figure 20.2 shows that in the BOTD and BOCLE rigs there is constant sliding contact between the ball and the mating surfaces. Whether there is boundary lubrication or hydrodynamic lubrication will depend upon the relative speed and the load at the point of contact. In the HFRR procedure, because the ball reverses direction, it will come to rest at the end of each stroke, and there will always be a region of boundary lubrication. If the speed and load conditions are set appropriately, the contact may transition through mixed lubrication into hydrodynamic lubrication.

Data for the evaluation program were derived from round-robin tests carried out in 28 different laboratories, each running one or more of the four test techniques. Twelve fuels, including four treated with an additive, were evaluated on the laboratory test machines. To provide information regarding correlation with fuel injection system performance, the same fuels were tested by three FIE manufacturers using rotary fuel injection pumps and by one engine manufacturer using unit injectors. After eliminating data sets from laboratories that did not follow the correct test procedure, incorrect data points, and points identified as statistical outliers, a total of 931 results remained. The laboratory test methods were statistically analyzed for repeatability, reproducibility, and discrimination between fuels; they were also assessed for correlation with wear measurements conducted by the manufacturers on their fuel injection systems.

The outcome of this extensive test program was that both the BOCLE and HFRR tests correlated well with the FIE tests, and both tests had good repeatability and comparable reproducibility, although the HFRR showed better discrimination. However, later analysis [20.16] of this and other test data found that the BOCLE procedure gave the best correlation with full-scale pump tests while the HFRR at 60°C gave the worst correlation! In the ISO program, other factors such as ease of operation and cost were also included in the decision process, which led to the selection of the HFRR as the ISO and CEC test method. The HFRR test method is thus used for specifying lubricity limits in the US, Europe, Japan, and many other Asian countries that follow the European guidelines. The different tests are described in slightly more detail in the following sections.

20.4.2. Ball on Three Seats (BOTS) Method

The BOTS, and the very similar BOTD method [20.17, 20.18], were both variants of the four-ball machine [20.19] that was used for lubricants testing. In the BOTS method, the three fixed balls were replaced by three seats set at an angle to the axis of the rotating ball. In the BOTS equipment the three seats were machined to provide a spherical contact patch to match that of the rotating ball. The ball was rotated in loaded contact with the seats, which were immersed in the test fuel. The fuels lubricity was assessed by the average weight loss of the three seats due to wear. In the BOTD equipment, the three spherical seats are replaced with three-disc seats as shown in Figure 20.2. The assessment of wear was changed to the measurement of the average wear scar diameter. For the BOTD test, the load on the ball was selected as 3 kg to produce a Hertz stress on each of the three discs that was more closely matched to that of fuel pump components in field service. The rotational speed of the ball was set at 60 rev/min to avoid the contacts entering the hydrodynamic lubrication regime [20.17].

20.4.3. Ball-on-Cylinder Lubricity Evaluator (BOCLE)

Service problems with gas turbine fuel pumps running on hydrofined kerosene led to the development of the BOCLE [20.20], which became the most widely used test method to measure aviation fuel lubricity. The ball is loaded against a rotating cylinder, partly immersed in the test fuel, which continuously transfers a thin film of the fuel to the ball-to-cylinder interface. Fuel lubricity is assessed by measurement of the wear scar on the ball [20.21].

Initial use of the standard BOCLE procedure to assess diesel fuel lubricity gave poor correlation with field experience. Several variants of the test operation had been recommended to give a better representation of conditions inside the fuel injection pump and improve its suitability for short duration screening tests of diesel fuels and additives. Two methods, intended to represent the scuffing process, were studied by the ISO working group. One variation was developed by Lubrizol [20.22] and the second, known as the scuffing load BOCLE (SLBOCLE), was developed for the US Army by Southwest Research Institute (SwRI) [20.23].

In the Lubrizol procedure, the machine is operated at 300 rpm for a conditioning period of 15 minutes, and then a load of 7 kg is applied for 2 minutes. The test is carried out with the relative humidity of the air controlled at 50%. The wear scar measured at the end of the test represents the scuffing wear that has occurred when running on the test fuel. The SLBOCLE procedure determines the maximum load that can be applied without causing scuffing. The result is attained by running a series of tests at low, nonscuffing loads and high loads that cause scuffing, to bracket closely the highest nonscuffing load that the test fuel can carry. Attempts to correlate the lubricity of a fuel to its physical characteristics suggest both hydrodynamic and boundary lubrication are influential parameters in this test [20.24]. At the end of the last century the SLBOCLE was developed into a standardize procedure [20.25] but this has since been withdrawn.

20.4.4. High-Frequency Reciprocating Rig (HFRR)

The HFRR [20.26, 20.27] uses a hardened steel ball oscillating transversely in loaded contact with a hardened steel disc immersed in the test fuel. It was designed to give negligible hydrodynamic film formation and low frictional heating. Contact temperature can be controlled as an independent test variable, enabling assessment of temperature-dependent effects as well as the performance of chemical lubricity additives. Figure 20.3 shows the general arrangement of the HFRR apparatus, and the standard test conditions are given in Table 20.1.

FIGURE 20.3 General arrangement of high-frequency reciprocating rig.

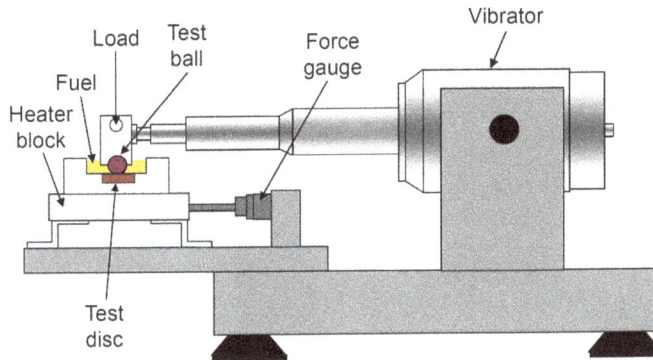

TABLE 20.1 HFRR test conditions.

Load	200 g (1.96 N)
Stroke	1 mm (0.5 mm amplitude)
Frequency	50 Hz
Temperature	25 and 60°C
Metallurgy	Ball: ANSI 52 100 (hardened bearing tool steel 60-65 HRC)
	(F.A.G. Pt. No. KV6.G28)
	Disc: ANSI 52 100 (bearing tool steel 200 HV 30)
Surface finish	Ball: <0.05 μm Ra
	Disc: <0.02 μm Ra
Duration	75 minutes

The results of the HFRR test are usually quoted as the mean wear scar diameter (WSD) [20.28, 20.29]. The instrument also allows measurement of the friction force and the film thickness. This additional information can provide an insight into how different additives are behaving. In the US a maximum WSD of 520 μm is specified, whereas in Europe a maximum of 460 μm is specified.

Whereas the HFRR test is well known, there have been concerns about a lack of precision the measurement range of 350–500 μm WSD and the ability of the test to differentiate low levels of lubricity additive. As a result of work in this area, a modification of the method has been proposed where the point contact, provided by the ball on plate, is changed to line contact, provided by a pin on plate, to reduce the Hertzian contact pressure and increase the active test surface area [20.30, 20.31]. The modified test is referred to as high-frequency reciprocating rig—line contact (HFRR-LC), but this is not yet a standardized method.

20.5. Water and Sediment Content

Water cannot be completely eliminated from diesel fuel. It can get into the fuel at various stages as it progresses through the distribution network from the oil refinery to the fuel system of the vehicle in which it is finally burned.

The earliest stage at which water can get in is during manufacture, when the hot fuel is in contact with process water. Much of this water will be removed in the stripping units at the refinery and more will separate as the fuel cools. The main risks of water contamination are during transportation and storage in tanks that have not been thoroughly emptied of water or into which water can enter and accumulate. This can be due to rain leaking past the roof seals or breathing in large volumes of humid air, which can happen during fuel withdrawals and as a result of changes in ambient air temperature. Moisture in the air will condense when the temperature falls, and it is a recommended practice to maintain tanks as full as possible to minimize water contamination due to breathing. Provision is usually made for regular draining of storage tank bottoms, but this may not be completely effective if the tank bottom is distorted. Where natural or man-made caverns are used for intermediate fuel storage, there is no possibility of eliminating contact with water.

The presence of water in storage tanks may encourage growth of fungi or bacteria that feed on the fuel and may block filters if drawn out with the fuel. This is particularly true with the inclusion of biodiesel, which is particularly hygroscopic. Also, entrainment of water bottoms when pumping from contaminated tanks can sometimes produce hazy fuel that may be slow to reach the clear and bright appearance required of a merchantable product.

Sediment likely to be found in diesel fuel tends to be mainly inorganic in origin: rust and metal particles from fuel tanks and lines and dirt entering from the atmosphere or because of poor housekeeping practices. Organic deposits may come from degradation of unstable fuel components, bacterial action at the oil–water interface, or wax from the fuel. Wax settling may occur if the fuel has cooled below its cloud point as a result of unexpectedly low temperatures or long storage in small, exposed tanks. Although settled waxes usually re-dissolve when the tank is refilled or the ambient temperature rises, tank cleaning is advisable if the amount of accumulated wax present could put the new batch of fuel out of specification. As noted in Chapter 17, Section 17.1.8, when biodiesel is included, these problems can occur above the cloud point, and some solids may not go back into solution [20.32].

Water and sediment can contribute to filter plugging in the distribution network or the vehicle and cause problems due to corrosion and wear in the engine and fuel injection system. A standard test for water and sediment content is a centrifuge method [20.33, 20.34]. A small sample of fuel is mixed with an equal volume of water-saturated solvent in each of two cone-shaped graduated tubes and centrifuged to concentrate the water and sediment in the bottom of the tube. The total volume of contaminant in each tube is noted, and the sum is reported as the percentage of water and sediment.

Some diesel specifications impose separate limits for water and sediment. In such cases, water content is determined by distillation [20.35] or titration with a chemical reagent [20.36]. Typical maximum limits tend to be 0.05% for water content and 0.01% for sediment, or 0.05% if combined water and sediment is specified; all percentages are by volume. In Europe, the water content is limited to 200 mg/kg [20.37].

20.6. Ash Content

Diesel fuels may contain small amounts of ash-forming materials, such as suspended solids and soluble organometallic compounds. These can cause damage in the close tolerances of FIE and contribute to increased deposit levels and abrasive wear of piston rings and other components in high-temperature zones of the engine. The ash content is not to be confused with the sediment content of the fuel. Organometallic compounds, such as methylcyclopentadienyl manganese tricarbonyl (MMT), will not be detected as sediment but would form ash.

In the standard quantitative method for determining ash in petroleum products [20.38], a small sample of the fuel is burned in a weighed dish until all combustible material has been consumed. The weight of the unburned residue is reported as a percentage of the original fuel sample. A typical specification limit for ash content is 100 mg/kg maximum.

20.7. Carbon Residue

A carbon residue test is included in many diesel fuel specifications to indicate the tendency of the fuel to form carbonaceous deposits, although it bears little relationship to fouling of fuel injector nozzles in service. Some entrainment of higher boiling material occurs during distillation to separate the diesel fraction, and the test measures the amount of these low-volatility components in the finished fuel.

Two tests widely used around the world to measure the carbon residue of automotive diesel fuels are the Conradson method [20.39, 20.40] and the Ramsbottom method [20.41]. As the level of carbon residue in automotive diesel fuel is low, often less than 0.1%, the test is carried out on the 10% residue from the laboratory distillation test [20.42] in order to increase the accuracy of the result. Both methods are prone to erroneous results due to the presence of certain types of additives such as alkyl nitrate CNIs; the test is therefore only meaningful when used on base fuels.

In the Conradson procedure, a larger sample, 10 g of distillation bottoms, is put in a tarred crucible and subjected to destructive distillation, undergoing cracking and coking reactions during a fixed period of severe heating. After cooling and reweighing to calculate the amount remaining, the result is reported as percent Conradson carbon residue on 10% distillation residue.

In the Ramsbottom procedure, a 4 g sample of the 10% bottoms is placed in a tarred glass coking bulb and heated at 550°C for 20 minutes. After cooling, the bulb is reweighed to determine the amount of residue, which is reported as percent Ramsbottom carbon residue on 10% distillation residue.

The two methods have similar precisions in the region of 0.1%. The 95% confidence level for repeatability is about 0.025%, while, for reproducibility, the figures are 0.036% for Ramsbottom and 0.058% for Conradson. There is no exact correlation between results obtained by the two methods, but an approximate correlation chart is included in each procedure.

20.8. Density

The density of the fuel is often expressed as "relative density" or "specific gravity" and is the ratio of the weight of a given volume of gasoline to the weight of the same volume of water, when both are at a temperature of 15°C and at a pressure of 101.325 kPa. To a small extent, the older term "degrees API" is also used and is based on an arbitrary hydrometer scale that is related to specific gravity as follows:

$$\text{Degrees API} = \frac{141.5}{\text{SG}} - 131.5$$

where SG is the specific gravity that is the ratio of the density of the gasoline to the density of water when both are measured at 15°C.

Fuel is usually sold by units of volume; because of this, motorists usually measure fuel economy or fuel consumption on a volumetric basis. Therefore, a denser fuel means that motorists are actually getting more fuel, by mass, every time they fill their tank. Diesel fuel is typically about 10% denser than gasoline, so this will account for some of the fuel economy benefits of a diesel engine vehicle when compared to the same vehicle with a gasoline engine.

Density tends to correlate well with the amount of carbon in the fuel, and this tends to correlate with the energy content of the fuel. For hydrocarbon components, generally speaking, aromatic compounds have the highest density, with the di-olefins and olefins being intermediate and the paraffins having the lowest density when compounds having the same number of carbon atoms are compared. A high-carbon-number compound will have a higher density than a low carbon number compound of the same type.

The aromatic content of the fuel can be determined by fluorescent indicator adsorption (FIA) [20.43] or by chromatography [20.44]. A chromatography procedure has the ability to separate the aromatics into monoaromatic and polynuclear aromatic hydrocarbon compounds, the latter of which is regulated in many jurisdictions.

Most diesel fuels have a relatively narrow range of density, typically between about 0.82 and 0.86 kg/L. Measurement is usually by a hydrometer method [20.45] or by an oscillating U-tube method [20.46].

Diesel FIE is designed to meter the fuel on a volume basis; therefore, changes in fuel density will affect the mass of fuel injected and hence the air-to-fuel ratio. Unlike a gasoline engine, which must operate within a narrow range of air-to-fuel ratios, in a diesel engine the air-to-fuel ratio varies over a wide range and is engine load dependent. The change in air-to-fuel ratio as a result of changes in fuel density will therefore simply mean the engine produces more or less power and the driver compensates by a change in accelerator pedal position. However, for modern diesel engines with electronic engine management systems, parameters such as exhaust gas recirculation (EGR) control valves and boost pressure control are mapped to the expected fuel delivery. Changes in fuel density and the driver's compensation can mean that inappropriate engine map settings are implemented; this can affect the emissions performance of the engine. For older technology engines with purely mechanical fuel injection systems, a change in fuel density can also affect the injection timing. This can affect emissions performance, noise, and fuel economy.

20.9. Heating Value

The heating value is an important fuel property as it is a measure of the energy available from a fuel when it is burned. This forms the basis for calculating the thermal efficiency of an engine using that fuel. The heating value is also referred to as the calorific value, thermal value, heat content, or heat of combustion.

The heating value of a diesel fuel is determined using a bomb calorimeter [20.47]. The test measures the amount of heat released by burning a known quantity of fuel, and this may be expressed in Joules per kilogram (J/kg) or per liter (J/L). This is reported as the higher or gross heat of combustion, as it includes the heat released when water vapor produced during combustion is condensed. However, as water from combustion normally escapes as steam in the engine's exhaust gases, it is usual to subtract the latent heat of condensation of water vapor to give the lower or net heating value, which is used for the calculation of thermal efficiency. The difference between the two values is a function of the hydrogen content of the fuel. A number of empirical formulas have been developed for estimating heating values based on the hydrogen, carbon, oxygen, nitrogen, sulfur, and ash composition of the fuel [20.48, 20.49] or from the fuel density, sulfur, water, and ash content [20.50].

Values derived from empirical formulas will generally be of sufficient accuracy for most routine test work, but the bomb calorimeter method should be used when an exact determination of heating value is required. The heating value is not controlled during manufacture except indirectly through other fuel properties, so it does not normally appear in specifications for automotive diesel fuel. Its application relates to the calculation of engine thermal efficiency.

20.10. Flash Point

The flash point is the temperature to which the fuel has to be heated to produce a vapor and air mixture that will ignite when a flame is applied. The most common test method used for diesel fuels is by Pensky–Martens closed cup [20.51]. In this test, a portion of the fuel is heated slowly in a covered cup, at a constant rate. At regular intervals the cover is opened, briefly, to admit a small flame into the cup. This procedure is continued until the fuel temperature is high enough for flash ignition to occur. Other methods [20.52–20.55] may be specified or permitted in some regulations.

The primary reason for including a flash point specification is to ensure safe handling of the product. If the flash point is too low, there could be a fire hazard, and for this reason, mandatory minimum limits on flash point have been set and storage criteria established by insurance companies and government agencies. For typical automotive diesel fuels the minimum flash point is normally between 52 and 55°C but can be as low as 38°C where ambient temperatures are known to be low.

The flash point relates to the front-end volatility of the fuel. Consequently, it will have an influence on the choice of light refinery streams that are included in the fuel blend, as its initial boiling point must be sufficiently high for the flash point to conform to the specification. Assuming that the correct blend components have been used, a low flash point can indicate contamination of the diesel fuel with a more volatile, lower flash point product such as gasoline.

As far as performance in an engine is concerned, the flash point of a diesel fuel has no direct significance. There is no direct correlation between flash point and autoignition temperature or other combustion characteristics.

20.11. Electrical Conductivity

Although conductivity is not currently a specification requirement for diesel fuel, there is a potential safety risk due to the build-up of a static electric charge during bulk handling of any flammable liquid at high pumping rates. Grounding leads are normally used to conduct away any charge when large quantities of fuel are being transferred into and out of storage tanks, but a volatile product like aviation kerosene is routinely treated with a static dissipator additive.

The electrical conductivity of distillate fuels can be determined by a direct current meter [20.56]. For fuels containing a static-dissipater the conductivity can be measured by portable meters, for the direct measurement, or by in-line meters for the continuous measurement, such as in a fuel distribution system [20.57]. In all test methods, a voltage is applied across two electrodes immersed in the fuel, and the current that flows is measured. The conductivity is then reported in pico-Siemens per meter (pS/m).

The conductivity of diesel fuels influences the likelihood of an excessive charge building up when refueling vehicles at service stations. However, faster filling rates have been suggested to speed up refueling stops for trucks having large-capacity fuel tanks, and as a precaution, some oil companies use an antistatic additive in diesel fuel.

20.12. Appearance and Color

A clear and bright appearance is generally a requirement for any diesel fuel, whether or not it is included in the specification, to demonstrate its freedom from water haze, suspended matter, and other contaminants that could result in the product being visually unacceptable. Color is sometimes specified to preclude incorporation of materials such as heavy cracked gas oils and residues, which would give a darker appearance to the fuel, whereas in some specifications a minimum color value is stipulated.

Color is usually assessed by comparison to standard color charts [20.58]. A sample of the fuel is placed in a cylindrical jar and compared, under a standard light source, with colored glass discs having numerical values ranging in regular steps from 0.5 for the palest to 8.0 for the darkest. The number of the matching glass or the higher of the two numbers where there is no exact match is reported as the ASTM Color. In specifications where color is included, the limiting values can vary widely.

References

20.1. ASTM International, "Standard Test Method for Corrosiveness to Copper from Petroleum Products by Copper Strip Test," ASTM D130-19, ASTM International, 2019.

20.2. ASTM International, "Standard Test Method for Rapid Determination of Corrosiveness to Copper from Petroleum Products Using a Disposable Copper Foil Strip," ASTM D7095-17, ASTM International, 2017.

20.3. ASTM International, "Standard Test Method for Acid and Base Number by Color-Indicator Titration," ASTM D974-22, ASTM International, 2023.

20.4. British Standards Institute, "Fat and Oil Derivatives—Fatty Acid Methyl Esters (FAME)—Determination of Acid Value," BS EN 14104:2021, 2021.

20.5. ASTM International, "Standard Test Method for Acid Number of Petroleum Products by Potentiometric Titration," ASTM D664-18e2, ASTM International, 2019.

20.6. ASTM International, "Standard Specification for Biodiesel Fuel Blend Stock (B100) for Middle Distillate Fuels," ASTM D6751-23a, ASTM International, 2023.

20.7. ASTM International, "Standard Test Method for Cetane Number of Diesel Fuel Oil," ASTM D613-23, ASTM International, 2023.

20.8. ASTM International, "Standard Test Method for Determination of Derived Cetane Number (DCN) of Diesel Fuel Oils—Ignition Delay and Combustion Delay Using a Constant Volume Combustion Chamber Method," ASTM D7668-17, ASTM International, 2017.

20.9. ASTM International, "Standard Test Method for Calculated Cetane Index of Distillate Fuels," ASTM D976-21e1, ASTM International, 2023.

20.10. ASTM International, "Standard Test Method for Calculated Cetane Index by Four Variable Equation," ASTM D4737-21, ASTM International, 2021.

20.11. Mitchell, K., "Diesel Fuel Lubricity: A Survey of 1994/95 Canadian Winter Diesel Fuels," SAE Technical Paper 961181 (1996), doi:https://doi.org/10.4271/961181.

20.12. Wei, D.P., "The Lubricity of Fuels I. Wear Studies on Diesel Fuel Components," *Acta Petrolei Sinica (Petroleum Processing Section)* 2 (1986): 79-87.

20.13. Wei, D.P., "The Lubricity of Fuels II. Wear Studies Using Model Compounds," *Acta Petrolei Sinica (Petroleum Processing Section)* 4 (1990): 90-99.

20.14. Tucker, R.F., Stradling, R.J., Wolveridge, P.E., Rivers, K.J. et al., "The Lubricity of Deeply Hydrogenated Diesel Fuels - The Swedish Experience," SAE Technical Paper 942016 (1994), doi:https://doi.org/10.4271/942016.

20.15. Nikanjam, M., Crosby, T., Henderson, P., Gray, C. et al., "ISO Diesel Fuel Lubricity Round Robin Program," SAE Technical Paper 952372 (1995), doi:https://doi.org/10.4271/952372.

20.16. Lacey, P.I. and Mason, R.L., "Fuel Lubricity: Statistical Analysis of Literature Data," SAE Technical Paper 2000-01-1917 (2000), doi:https://doi.org/10.4271/2000-01-1917.

20.17. Voitik, R. and Ren, N., "Diesel Fuel Lubricity by Standard Four Ball Apparatus Utilizing Ball on Three Disks, BOTD," SAE Technical Paper 950247 (1995), doi:https://doi.org/10.4271/950247.

20.18. Voitik, R., "Diesel Fuel Lubricity BOTD Status—1995," SAE Technical Paper 952371 (1995), doi:https://doi.org/10.4271/952371.

20.19. ASTM International, "Standard Test Method for Wear Preventive Characteristics of Lubricating Fluid (Four-Ball Method)," ASTM D4172-21, ASTM International, 2022.

20.20. ASTM International, "Standard Test Method for Measurement of Lubricity of Aviation Turbine Fuels by the Ball-on-Cylinder Lubricity Evaluator (BOCLE)," ASTM D500-95(2016), ASTM International, 2016.

20.21. Jenkins, S.R., Landells, R.G.M., and Hadley, J.W., "Diesel Fuel Lubricity Development of a Constant Load Scuffing Test Using the Ball on Cylinder Lubricity Evaluator (BOCLE)," SAE Technical Paper 932691 (1993), doi:https://doi.org/10.4271/932691.

20.22. Hadley, J.W. and Blackhurst, P., "An Appraisal of the Ball-on-Cylinder Technique for Measuring Aviation Turbine Fuel Lubricity," *Society of Tribologists and Lubrication Engineers* 47, no. 5 (1991): 404-411.

20.23. Lacey, P.I., "Development of a Lubricity Test Based on the Transition from Boundary Lubrication to Severe Adhesive Wear in Fuels," *Lubrication Engineering* 50, no. 10 (1994): 749-757.

20.24. Wang, J.C. and Cusano, C.M., "Predicting Lubricity of Low Sulfur Diesel Fuel," SAE Technical Paper 952564 (1995), doi:https://doi.org/10.4271/952564.

20.25. ASTM International, "Standard Test Method for Evaluating Lubricity of Diesel Fuels by the Scuffing Load Ball-on-Cylinder Lubricity Evaluator (SLBOCLE) (Withdrawn 2021)," ASTM D6078-04(2016), ASTM International, 2021.

20.26. Hadley, J.W., Owen, G.C., and Mills, B., "Evaluation of a High Frequency Reciprocating Wear Test for Measuring Diesel Fuel Lubricity," SAE Technical Paper 932692 (1993), doi:https://doi.org/10.4271/932692.

20.27. Nikanjam, M. and Rutherford, J., "Improving the Precision of the HFRR Lubricity Test," SAE Technical Paper 2006-01-3363 (2006), doi:https://doi.org/10.4271/2006-01-3363.

20.28. The International Organization for Standardization (ISO), "Diesel Fuel—Assessment of Lubricity Using the High-Frequency Reciprocating Rig (HFRR)—Part 1:Test Method," ISO 12156-1:2016, 2016.

20.29. ASTM International, "Standard Test Method for Evaluating Lubricity of Diesel Fuels by the High-Frequency Reciprocating Rig (HFRR)," ASTM D6079-22, ASTM International, 2023.

20.30. Hansen, G.A. and Sattler, E., "Military Ground Vehicle Fuel Lubricity Tester (Improving the Sensitivity of the HFRR)," US Army TARDEC Fuels and Lubricants Research Facility (SWRI), San Antonio, TX, 2018.

20.31. Henderson, P.T., Christison, K., Evers-McGregor, D., Martinez, J. et al., "ISO Paraffinic Diesel Fuel Lubricity Study," SAE Technical Paper 2022-01-5073 (2022), doi:https://doi.org/10.4271/2022-01-5073.

20.32. Jolly, L., Kitano, K., Sakata, I., Strojek, W. et al., "A Study of Mixed-FAME and Trace Component Effects on the Filter Blocking Propensity of FAME and FAME Blends," SAE Technical Paper 2010-01-2116 (2010), doi:https://doi.org/10.4271/2010-01-2116.

20.33. ASTM International, "Standard Test Method for Water and Sediment in Fuel Oils by the Centrifuge Method (Laboratory Procedure)," ASTM D1796-22, ASTM International, 2022.

20.34. ASTM International, "Standard Test Method for Water and Sediment in Middle Distillate Fuels by Centrifuge," ASTM D2709-22, ASTM International, 2022.

20.35. ASTM International, "Standard Specification for Apparatus for Determination of Water by Distillation," ASTM E123-02(2018), ASTM International, 2018.

20.36. ASTM International, "Standard Test Method for Determination of Water in Petroleum Products, Lubricating Oils, and Additives by Coulometric Karl Fischer Titration," ASTM D6304-20, ASTM International, 2021.

20.37. British Standards Institute, "Automotive Fuels. Diesel. Requirements and Test Methods," BS EN 590:2022, British Standards Institute, 2022.

20.38. ASTM International, "Standard Test Method for Ash from Petroleum Products," ASTM D482-19, ASTM International, 2019.

20.39. ASTM International, "Standard Test Method for Conradson Carbon Residue of Petroleum Products," ASTM D189-06(2019), ASTM International, 2019.

20.40. ASTM International, "Standard Test Method for Determination of Carbon Residue (Micro Method)," ASTM D4530-15(2020), ASTM International, 2020.

20.41. ASTM International, "Standard Test Method for Ramsbottom Carbon Residue of Petroleum Products," ASTM D524-15(2019), ASTM International, 2019.

20.42. ASTM International, "Standard Test Method for Distillation of Petroleum Products at Atmospheric Pressure," ASTM D86-23, ASTM International, 2023.

20.43. ASTM International, "Standard Test Method for Hydrocarbon Types in Liquid Petroleum Products by Fluorescent Indicator Adsorption," ASTM D1319-20a, ASTM International, 2020.

20.44. ASTM International, "Standard Test Method for Determination of Aromatic Content and Polynuclear Aromatic Content of Diesel Fuels and Aviation Turbine Fuels by Supercritical Fluid Chromatograph," ASTM D5186-22, ASTM International, 2022.

20.45. ASTM International, "Standard Test Method for Density, Relative Density (Specific Gravity), or API Gravity of Crude Petroleum and Liquid Petroleum Products by Hydrometer Method," ASTM D1298-12b(2017)e1, ASTM International, 2023.

20.46. ASTM International, "Standard Test Method for Density, Relative Density, and API Gravity of Liquids by Digital Density Meter," ASTM D4052-22, ASTM International, 2022.

20.47. ASTM International, "Standard Test Method for Heat of Combustion of Liquid Hydrocarbon Fuels by Bomb Calorimeter," ASTM D240-19, ASTM International, 2019.

20.48. Grummel, E.S. and Davis, I.A., "A New Method of Calculating the Calorific Value of a Fuel from Its Ultimate Analysis," *Fuel* 12 (1933): 199-203.

20.49. Boie, W., "Fuel Technology Calculations," *Energietechnik* 3 (1953): 309-316.

20.50. ASTM International, "Standard Test Method for Estimation of Net and Gross Heat of Combustion of Burner and Diesel Fuels," ASTM D4868-17 ASTM International, 2022.

20.51. ASTM International, "Standard Test Methods for Flash Point by Pensky-Martens Closed Cup Tester," ASTM D93-20, ASTM International, 2020.

20.52. ASTM International, "Standard Test Method for Flash Point and Fire Point of Liquids by Tag Open-Cup Apparatus," ASTM D1310-14(2021), ASTM International, 2021.

20.53. ASTM International, "Standard Test Methods for Flash Point of Liquids by Small Scale Closed-Cup Apparatus," ASTM D3278-21, ASTM International, 2021.

20.54. ASTM International, "Standard Test Method for Flash Point by Tag Closed Cup Tester," ASTM D56-22, ASTM International, 2022.

20.55. ASTM International, "Standard Test Methods for Flash Point by Small Scale Closed Cup Tester," ASTM D3828-16a(2021), ASTM International, 2021.

20.56. ASTM International, "Standard Test Method for Electrical Conductivity of Liquid Hydrocarbons by Precision Meter," ASTM D4308-21, ASTM International, 2022.

20.57. ASTM International, "Standard Test Methods for Electrical Conductivity of Aviation and Distillate Fuels," ASTM D2624-22, ASTM International, 2022.

20.58. ASTM International, "Standard Test Method for ASTM Color of Petroleum Products (ASTM Color Scale)," ASTM D1500-12(2017), ASTM International, 2018.

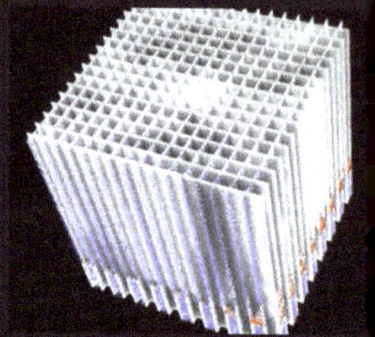

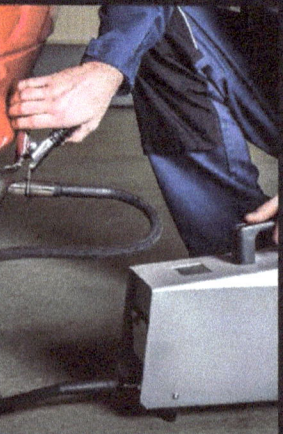

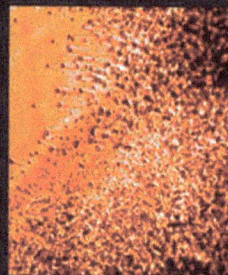

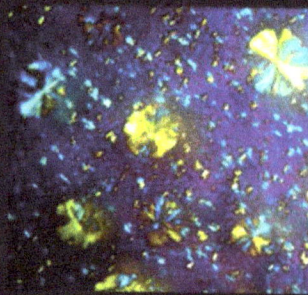

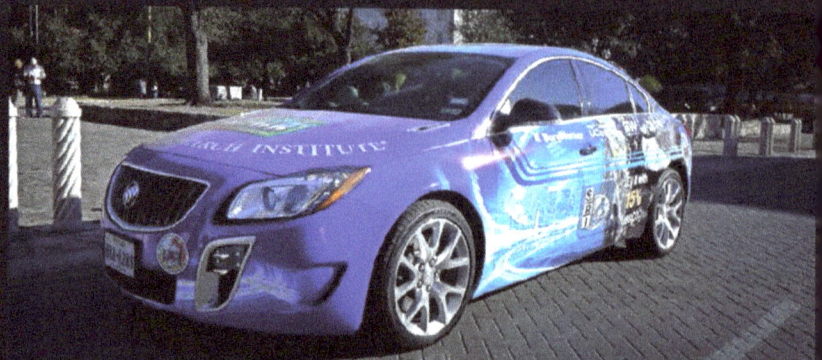

21

Influence of Diesel Fuel Characteristics on Emissions

A diesel engine running on hydrocarbon fuel will always produce soot particles. The nature of the combustion process discussed in Chapter 15, Section 15.1, means that there is always the opportunity for fuel to burn in a fuel-rich environment. This will inevitably result in some fuel not completely oxidizing. This fuel will also be subjected to high temperatures leading to dehydrogenation. Whether the resultant soot will be emitted from the tailpipe depends upon the ability of these soot particles to undergo subsequent oxidation. Thus, diesel engines will always have the potential to produce a visible emissions problem. It is the aim of the fuel injection equipment (FIE) system designer to maximize the mixing of the fuel and air in order to minimize the number and size of the soot particles produced. In turn, it is the aim of the engine designer to maximize the availability of air to provide a plentiful supply of oxygen and the opportunity for this oxygen to oxidize these soot particles. However, under heavy load, if the fuel delivered is not limited, then a diesel engine can produce exhaust soot, visible as a plume of black smoke. This is epitomized by Figure 21.1, which shows the exhaust stack of a piece of off-road diesel equipment.

FIGURE 21.1   Off-road equipment showing what was a typical diesel engine emissions trait.

Due to the visibility of these emissions, the diesel engine was the first automotive engine to be subject to legislative emissions control. In 1946, the Smoke Control Hearing Board was set up under the Los Angeles County Air Pollution Control Board (APCD). Laws were set to limit the emissions of visible smoke, and the official patrol car of the APCD was legally authorized to stop any highway vehicle violating the ordinance [21.1]. What was less visible at the time was the other Achilles heel of the diesel engine: the quantity of oxides of nitrogen that were produced. However, persistent smog problems in the Los Angeles basin spurred research that soon identified the oxides of nitrogen (NO_X) present in automotive exhaust emissions as ozone precursors [21.2]. It was estimated at the time that automotive exhaust emissions were adding 922 tons of hydrocarbons (HC) to the atmosphere every day [21.3]. This along with the ozone generated by the NO_X was the cause of the smog. Because the diesel engine produced low levels of HC emissions, the initial focus of (non-black smoke) emissions was thus the gasoline engine. The development of emissions regulations for gasoline engine vehicles is discussed in Chapter 13, Section 13.1. The development of legislation concerning diesel engine emissions tended to move in step with that for the gasoline engine, although initially the limits appeared less stringent. These will be outlined in the following section.

21.1. Development of Emissions Legislation and Fuel Quality Regulations

Los Angeles was not the only place to suffer from pollution episodes that could be related to road transport [21.4], but the strength of a body such as the California Air Resources Board (CARB) gave impetus to the formation of control measures. The following subsections will briefly outline how automotive emissions legislation and fuel quality standards have evolved globally in over half a century of development.

21.1.1. Development in the US

The Clean Air Act (CAA) of 1968 set the first nationally legislated automotive exhaust emission limits for the United States. The CAA was updated in 1970 to become the Clean Air Act Amendments (CAAA). The CAAA required a 90% reduction in exhaust emissions from those prevailing at that time. However, discussions between the industry and the Environmental Protection Agency (EPA) suggested the technology required to achieve these goals was not yet developed, and implementation was thus delayed beyond the planned 1975–1976 time frame.

The CAAA was revised in 1977 to incorporate National Ambient Air Quality Standards (NAAQS), proposed by the EPA, with a further update being signed into law in 1990. How these targets were to be implemented was decided following discussions between the EPA and the fuel and motor industries. The outcome was a set of detailed rules incorporating the Tier One and Tier Two exhaust emissions limits: the later was to take effect from 2004.

The Tier One standards applied to all new vehicles of less than 3855 kg (8500 lb) gross vehicle weight rating (GVWR), termed light-duty vehicles (LDV). The GVWR is the weight of the vehicle plus its rated cargo capacity. The LDV category included passenger cars, sport utility vehicles (SUV), minivans, pick-up trucks, and other light-duty trucks. LDVs were then further subdivided into the following subcategories:

- Passenger cars.
- Light light-duty trucks (LLDT), below 2721 kg (6000 lb) GVWR.
- Heavy light-duty trucks (HLDT), above 2721 kg GVWR.

The LLDTs were further subdivided into two groups according to their loaded vehicle weight (LVW); those with a LVW below 1700 kg (3750 lb) and those with an LVW above 1700 kg. The LVW is the vehicles curb weight plus 136 kg (300 lb). The HLDTs were similarly subdivided according to their averaged LVW (ALVW), which was defined as the numerical average of the vehicles GVWR and its curb weight. The division point was an ALVW of 2608 kg (5750 lb).

Considering that emissions limits for LDVs are set as a mass per unit distance traveled, it is clearly easier for smaller, lighter vehicles to meet the limits. Because the US LDV class covers a significant range of vehicle sizes, the emissions limits were set differently according to the different size classifications outlined; the heavier the vehicle classification, the higher the emissions limit. The limits for the lighter LLDTs were the same as those for passenger cars.

Legislated limits for emissions of total hydrocarbons (THC), non-methane hydrocarbons (NMHC), and carbon monoxide (CO) were the same regardless of the type of fuel. For passenger cars and the lighter LLDTs, the NO_X limits were less stringent for diesel-powered vehicles than for gasoline-powered vehicles. Beyond Tier One, the emissions limits apply equally to gasoline- and diesel-powered vehicles. For diesel-powered vehicles there was, of course, the additional limit for particulate matter (PM) emissions.

The state of California implemented its own stricter limits, and during the 1990s, the 12 eastern states that comprised the Ozone Transport Region (OTR) adopted some elements of the California legislation. To avoid the complexity of different emission limits applying in different federal states, a voluntary agreement was developed, known as the National Low Emissions Vehicles (NLEV) program. These vehicles operated with a NO_X standard that was 50% below the Tier One standards.

In 1998, the CARB identified diesel PM as a toxic air contaminant (TAC) that by law required the CARB to determine if there was a need for further control measures. This led to the development of the Risk Reduction Plan to Reduce Particulate Matter Emissions from Diesel-Fueled Engines and Vehicles, which was approved by the CARB in September 2000. This is discussed further in Section 21.1.1.1.

Also in 1998, as required by the CAAA of 1990, the EPA issued the Tier Two Report to Congress. This report put forward strong justification for tightening the tailpipe emission standards that had already been proposed to take effect in 2004. The new Tier Two standards were the same as for gasoline engine vehicles and would be phased in from 2004 to 2008. This is discussed in Chapter 13, Section 13.1.1, but is repeated here for completeness. The Tier Two regulations introduced the idea of fleet averaging and the "bin" structure. The bins are a series of increasingly severe limits as the bin number decreases; culminating in bin 1, which is zero emissions. The limits for each bin are shown in Table 21.1. Note that the US limits are set in mixed units of g/mile but have been converted to g/km for comparison with other regions.

TABLE 21.1 Tier to emissions limits converted to g/km.

| Bin no. | Emission limits (g/km) | | | | |
	NO_X	NMOG	CO	HCHO	PM
11	0.559	0.174	4.536	0.020	0.075
10	0.373	0.097 (0.143)	2.610 (3.977)	0.011 (0.017)	0.050
9	0.186	0.056 (0.112)	2.610	0.011	0.037
8	0.124	0.078 (0.097)	2.610	0.011	0.012
7	0.093	0.056	2.610	0.011	0.012
6	0.062	0.056	2.610	0.011	0.006
5	0.043	0.056	2.610	0.011	0.006
4	0.025	0.043	1.305	0.007	0.006
3	0.019	0.034	1.305	0.007	0.006
2	0.012	0.006	1.305	0.002	0.006
1	0.000	0.000	0.000	0.000	0.000

In Table 21.1, the figures in parenthesis applied only to HLDTs until 2008, when this relaxed limit expired. Bin 11 was only applicable to medium-duty passenger vehicles (MDPVs) and along with bins 9 and 10 were deleted in 2006, except for HLDTs, for which they expired in 2008.

Under the bins system, the motor manufacturer selects a bin to which they will produce a family of vehicles. A test group must meet all of the limits within that bin. The manufacturer must also ensure that the "corporate average" emissions meet the 0.043 g/km NO_X limit put in place to ensure air quality targets should be met. The number of vehicles produced in each family and complying with each bin must be controlled to meet this corporate average.

In 2014 the Tier 3 standards came into law, to be phased-in from 2017 through 2025. Tier 3 followed the idea of fleet averaging and the bin structure. For Tier 3 the limits were set in terms of NO_X + NMOG rather than for the individual pollutants and the bins were named according to the NO_X + NMOG value expressed in terms of mg/mile, all values have been converted to SI units of mg/km and are shown in Table 21.2. The Tier 3 Bin 160 and Bin 30 are equivalent to the old Tier 2 Bin 5 and Tiers 2 Bin 2, respectively.

TABLE 21.2 Tier 3 emissions limits converted to mg/km.

| Bin no. | Emission limits (mg/km) | | | |
	NO_X + NMOG	CO	HCHO	PM
Bin 160	99.4	2609.8	2.5	1.9
Bin 125	77.7	1304.9	2.5	1.9
Bin 70	43.5	1056.3	2.5	1.9
Bin 50	31.1	1056.3	2.5	1.9
Bin 30	18.6	621.4	2.5	1.9
Bin 20	12.4	621.4	2.5	1.9
Bin 0	0.0	0.0	0.0	0.0

Vehicles had to comply with these limits for a vehicle life of 150,000 miles (241,402 km) to demonstrate the durability of the emissions control system. The certification fuel had a maximum sulfur content of 15 ppm.

21.1.1.1. California Risk Reduction Program for Diesel PM

The aim of the Risk Reduction Plan to Reduce Particulate Matter Emissions from Diesel-Fueled Engines and Vehicles was to reduce diesel PM emissions in California by 75% by 2010 and by 85% by 2020. Californian regulations included the following steps:

- Low Emission Vehicle (LEV) standards to the year 2003, equivalent to Tier 1.

- LEV II regulations phased-in for model years 2004–2010.

- LEV III regulations phased-in for model years 2015–2025.

- LEV IV regulations to be phased-in for model years 2026–2030.

Though it was accepted that new, tighter emissions standards would reduce emissions in the future, in the shorter time frame other measures would be needed. These included the retrofitting of aftertreatment technology to reduce diesel PM directly and through in-use compliance programs to maintain the improvements achieved through cleaner, new engine standards and retrofit programs. Figure 21.2 shows a tractor retrofitted with a diesel particulate filter (DPF) as part of a CARB and South Coast Air Quality Management District (SCAQMD) showcase program conducted in 2008.

FIGURE 21.2   Tractor retrofitted with a DPF as part of a CARB and SCAQMD showcase.

© SAE International.

When the risk reduction plan was issued, it was estimated that in the year 2000 there were nearly 278,000 diesel-powered LDVs in California. The CARB estimated that this would fall to just over 175 thousand by the year 2010 due to the age of the vehicles and their being replaced by non-diesel-powered vehicles. They estimated that these vehicles would emit 477 tons of PM in the year 2000 and would emit 196 tons in 2010. The predicted decline in diesel PM emissions from LDVs was accounted for by the predicted population decrease and by the effects of existing regulations. This was only 4% of the projected diesel PM emissions, and no recommendations were made for retrofit programs for on-road LDVs. However, the CARB stated that the existing fleet of diesel-fueled LDVs would be examined in more detail to determine if retrofitting these vehicles would be a cost-effective diesel PM reduction strategy.

The LEV III regulations basically follow the national US Tier 3 limits but include more stringent fleet emission requirements, with NO_X + NMOG equivalent to Tier 3 Bin 30.

The LEV IV emission standards have now been finalized by CARB and approved by the Office of Administrative Law [21.5] and will be phased-in over the model years 2026–2030. The LEV IV standards again follow the Tier 3 standards but eliminate the Bin 160 and introduce a new Bin 25 and Bin 15 with corresponding NO_X + NMOG emissions limits (25 and 15 mg/mi), and with CO, HCHO and PM limits as for Bin 20.

21.1.2. Development in Europe

Through the 1970s and 1980s, European Emissions regulations in Europe were formulated by the United Nations Economic Commission for Europe (UN-ECE) via its technical advisory body the Group Rapporteurs Pollution and Energy (GRPE). The role of the GRPE was to produce model standards that could be voluntarily adopted by member nations. In what was then the European Union (EU), the member countries generally adopted regulations that were technically identical with the UN-ECE equivalents. Since then, an expert group within the European Commission has assumed the leading role in formulating automotive emissions standards. The GRPE is now unlikely to adopt any proposal that has not been put forward by this expert group, called the Motor Vehicle Emissions Group (MVEG).

The EU publishes its standards as directives that have the force of law within EU member states, under the provisions of the Treaty of Rome. The EU countries may not prohibit the marketing of vehicles that comply with the provisions of the directives but may prohibit vehicles that do not comply. With the introduction of the Consolidated Emissions Directive in June 1991 (91/441/EEC), implementation became mandatory for all EU member states and was no longer left to the discretion of individual national governments. With the later Directive 93/59/EEC, these became known as the Euro 1 limits. Vehicles produced before these limits are often referred to as Euro 0. The EU Council of Ministers adopted Directive 94/12/EC and 96/69/EC, applying more stringent emissions limits from 1996 that were known as the Euro 2 limits. Contrary to earlier standards, no conformity of production allowance was permitted over and above the type approval limits. Note that the corresponding standards for heavy-duty vehicles are referred to as Euro I, Euro II, and so on.

As a result of Directive 94/12/EC, the European Commission set up the European Programme on Emissions, Fuels and Engines (EPEFE). This initiative involved legislators, the European Parliament, academia, consumer groups, the fuel industry, and automotive manufacturers. This program resulted in the submission of proposals for the further regulation of vehicle emissions to take effect during the period 2000/2005/2010. The proposals also included the introduction of requirements for onboard diagnostic systems, more rigorous inspection, and maintenance programs. In addition, there were proposals concerning fuel quality specifications.

A common position was reached regarding the first step, and this became EU Directive 98/69/EC, which set the Euro 3 and Euro 4 limits that came into force from 2000 and 2005, respectively. The regulations were updated by Directive 2002/80/EC, which included new specifications for reference fuels for testing LDVs. These new emissions regulations had separate limits for diesel- and gasoline-powered vehicles.

Further reductions were set out in Regulation (EC) 715/2007, known as the Euro 5 and Euro 6 limits. This was later amended by Commission Regulation (EC) No. 692/2008. The evolution of the limits for the criteria pollutants is illustrated in Table 21.3.

TABLE 21.3 Evolution of EU emissions limits.

| | Year | Emissions limit (g/km) | | | | | | | PN (#/km) |
		CO	HC	NMHC	HC + NO$_X$	NO$_X$	PM	NH$_3$	
Euro 1	1992	2.72	—	—	0.97	—	0.140	—	—
Euro 2	1996	1.00	—	—	0.70	—	0.080	—	—
Euro 3	2000	0.64	—	—	0.56	0.50	0.050	—	—
Euro 4	2005	0.50	—	—	0.30	0.25	0.025	—	—
Euro 5	2009	0.50	—	—	0.23	0.18	0.005	—	—
Euro 6	2014	0.50	—	—	0.17	0.08	0.0045	—	6.0 E^{11}
Euro 7*	2025	0.50	0.10	0.068	—	0.06	0.0045	0.02	6.0 E^{11}

* Proposal yet to be ratified by the European Parliament and the Council.

© SAE International.

The Euro 1 limits applied equally to diesel- and gasoline-powered vehicles, from Euro 2 to Euro 6 the values stated in the table applied only to diesel-powered vehicles. For Euro 7 the standards applied equally irrespective of fuel type. The situation regarding gasoline-powered vehicles is discussed in Chapter 13. Reference to Chapter 13, Section 13.1.2, will show that from Euro 2 to Euro 6 the CO limits for diesel-powered vehicles are lower than for gasoline-powered vehicles, but the NO$_X$ limits are higher. This reflects the ability of the corresponding engine and aftertreatment technologies to meet increasingly strict limits and differs from the US, where the legislation does not differentiate between the different fuel types. As noted above, for Euro 7 the values are once again the same for gasoline and diesel fuel. However, the Euro 7 proposal is meeting a lot of opposition within the industry [21.6, 21.7] and has yet to be ratified.

In January 2001 the GRPE set up an informal group to develop type approval test protocols, with instrumentation, to assess and control nano-particle emissions from both LDVs and heavy-duty engines. This became known as the Particle Measurement Programme (PMP). Based on the work of this group, new measurement procedures were proposed and adopted for the measurement of these very fine and potentially health-damaging particles. The Euro 6 and Euro 7 standards also include a particle number (PN) limit, which is set at a maximum of 6×10^{11} particles per km, using this new protocol. The PM and PN emissions limits for Euro 5 onwards apply equally to diesel- and gasoline-powered vehicles. Measurement of PN is discussed in Section 21.1.2.1. Diesel passenger cars fitted with full-flow DPFs (see Chapter 16, Section 16.4.2.2) to reduce PM emissions should not have difficulty meeting the PN standard. Indeed it should be noted that these mitigation strategies have made non-tailpipe emissions (PM$_{10}$; PM$_{2.5}$) a larger proportion of road traffic emissions. Source such as brake dust, tire wear particles and road dust being the culprits. [21.8]

21.1.2.1. Measurement of PN

Toward the end of the last century, a cooperative piece of work was sponsored by the UK Department of the Environment Transport and the Regions (DETR), the Society of Motor Manufacturers and Traders (SMMT), and the European oil companies' organization for environment, health, and safety: CONCAWE. The program was run to investigate the sampling and measurement of particles emitted from LDVs and heavy-duty engines. This included a characterization exercise using conventional type approval full-flow dilution systems to enable simultaneous measurements of mass, mass weighted size distributions, and number weighted size distributions from diluted exhaust aerosols [21.9]. These typically provide dilution ratios in the range 3:1 to 60:1. Particle size distributions were determined in number terms by twin Scanning Mobility Particle Sizers (SMPS) and in mass terms by a Micro Orifice Uniform Deposit Impactor (MOUDI). The initial work was conducted mainly using heavy-duty engines that were tested on both steady-state and transient test cycles. This work highlighted shortcoming in the sampling and measurement techniques that would require further investigation. The program also considered the effects of different fuels and engine technologies on PN emissions for both LDVs [21.10] and heavy-duty engines [21.11].

In 2001, the GRPE set up the PMP to develop type approval test protocols, with instrumentation, to assess and control nano-particle emissions from both LDVs and heavy-duty engines. The project assessed both the solid "accumulation mode" and the volatile "nucleation mode" particles. This produced a methodology capable of measuring both modes of particles from a minimum size range of ca. 7 nm from all types of vehicles. Based on this methodology, a significant number of both gasoline and diesel LDVs were assessed and fuel and lubricant effects were also investigated [21.12, 21.13]. Besides providing valuable information regarding the performance of these vehicles, the exercise also demonstrated the efficacy of the proposed methodology. As noted, these procedures have now been adopted by the UN-ECE, and a PN limit has been set as part of the future European emissions legislation. The PN measuring system takes its sample from the standard emissions testing dilution tunnel. The sample then passes through a cyclone preclassifier that filters out larger particles, thus setting an upper size measured to a nominal 2.5 µm. The sample is then heated to 150°C and diluted to evaporate the volatile material and reduce its partial pressure to prevent recondensation. The sample is then further heated to 300°C before further dilution before entering the particle counter. This procedure ensures that volatile material does not condense to form nucleation particles, and the count is of only the solid-cored carbonaceous particles that are considered to be the most damaging to human health.

21.1.3. Development in Japan

The Japanese introduced vehicular CO emissions limits in 1966. These were followed by the first long-term plan put forward by the Ministry of Transport (MOT) in 1970 [21.14]. These proposals limited the emissions of the criteria pollutants (i.e., CO, HC, and NO_X) with separate limits for hot start and cold start cycles. Following proposals from the Central Council for Environmental Pollution Control (CCEPC), more stringent exhaust emission standards took effect in 1975. These limits were further tightened in 1978 by additional NO_X reductions. Changes to the test procedures effectively made these limits harder to achieve.

In December 1989, the CCEPC recommended new emission limits with short-term and long-term targets. The aim was to set the most stringent standards that were technologically feasible and to apply the same standards for both diesel- and gasoline-fueled vehicles. Based on this proposal, the MOT revised the emission regulations in May 1991 to meet the short-term limits. The 10-mode light-duty test cycle was modified to include a high-speed section. Now referred to as the 10–15 mode cycle, it applies to both diesel- and gasoline-powered LDV. Emission measurement results are now expressed as g/kWh.

A joint MOT/Ministry of International Trade and Industry (MITI)/Environmental Agency (EA) proposal to further reduce NO_X in urban areas took effect from December 1993. Toward the end of 1997, the Central Environment Council (CEC) issued a report, "Future policy for motor vehicle exhaust emissions reduction (Second Report)," and the EA embarked on a revision of the existing regulations for motor vehicles to take effect between 2000 and 2002. Amendments include an extension of durability requirements, the introduction of onboard diagnostics, and the adoption of a sealed housing for evaporative determination (SHED) procedure for evaporative emissions control.

In addition to the emissions legislation, the CEC recommended new long-term regulations for fuels. The proposed fuel quality requirements included a reduction in the diesel sulfur limit to less than 50 mg/kg by the end of 2004 and a further reduction to 10 mg/kg by 2007. However, by voluntary agreement these targets were generally met about two years before the mandatory requirements came into force.

Further amendments, New Long Term Regulations, and the Post New Long Term Regulations, which were phased in by 2010. This has been followed by the Japan 2018 Target Emission Regulations, which are illustrated in Table 21.4, testing is performed according to the World-Harmonized Light-Duty Transient Cycle (WLTC).

TABLE 21.4 Japanese future criteria pollutant emission standards.

Test	Emissions limit (g/km)			
	CO	NMHC	NO$_X$	PM
Passenger cars	1.15	0.10	0.05	0.005
Vehicles with GVW ≤ 1700 kg	1.15	0.10	0.05	0.005
Vehicles with GVW ≤ 3500 kg	2.25	0.15	0.07	0.007

© SAE International.

21.1.4. Development in the Rest of the World

Canada has standards and procedures that closely match those of the US. Central and South American countries also tend to follow the US procedures and limits, although different countries are at different stages of the development of the US standards.

Australia and New Zealand follow European procedures and are broadly in line with the EU limits.

China and India have procedures equivalent to the EU procedures, but implementation timescale lags Europe and is also different for the country as a whole and for the largest metropolitan areas. Other Asian countries are adopting the EU procedures and regulations, although the timing is different from country to country. Although these countries are currently some years behind Europe, many have a timetable for bringing their regulations in line with Europe. The exception is Taiwan, which is following the US procedures, with limits slightly behind the US stages.

Central and Eastern European countries are adopting the EU regulation. Many of these countries are joining the EU and are therefore obliged to not only adopt the EU procedures but also the EU regulatory limits. The Russian Federation currently has its own limits. These are less severe than the EU limits.

South Africa, the Middle East, and other African countries follow EU procedures, although different countries are at different stages of the reduction steps that have taken place in Europe. Many of the African countries having no indigenous vehicle manufacturing capability simply rely on the fact that their imported vehicles did comply with relevant standards, of the country of manufacture, when the vehicles were first put into production.

21.2. Diesel Fuel Cetane Number and Emissions

The cetane number (CN) of a fuel is defined as the percentage of n-cetane in a mixture with α-methyl naphthalene that produces the same ignition delay in a specific engine under defined operating conditions [21.15]. The CN of a fuel is determined by bracketing its ignition delay under these well-defined standard operating conditions with two reference fuel blends of known CN. The CN of the test fuel is then calculated by interpolation. The reference fuels are prepared by blending mixtures of primary reference fuels or secondary reference fuels. The primary reference fuels are n-cetane, having by definition a value of 100, and iso-cetane with an assigned value of 15. The secondary reference fuels are designated T-26 and U-19 and have CNs of 75.2 and 19.4, respectively. The CN is therefore not a measure of ignition delay per se but à measure of the corresponding reference fuel that produces the same ignition delay. There is, however, a relationship between CN and ignition delay with higher CN fuels having a shorter ignition delay. This is the basis for other test methods that can be used for determining CN equivalents; this is discussed in Chapter 15, Section 15.2.1.2.

The ignition delay is not only determined by the chemical kinetics of the fuel but also by the physical mixing processes. This is often referred to as the physical and chemical ignition delay. The standard test procedure only measures the sum of the two. The relative importance of the two parts of the ignition delay will vary according to engine type and engine operating condition. For conventional petroleum diesel fuels, it has generally been found that the CN of the fuel does correlate well with the ignition delay that is measured in commercial engines [21.16–21.19]. However, if the physical properties of the fuel differ significantly from conventional petroleum diesel fuel, then the relationship may be broken.

A classic example would be a vegetable oil that has a distillation range above that of petroleum diesel fuel and a viscosity that is an order of magnitude higher than petroleum diesel fuel. Such a fuel would have a low CN and under low-temperature operating conditions, such as cold start, would actually have a longer ignition delay than a petroleum fuel of the same CN. However, under fully warmed conditions of many current engines, the vegetable oil would exhibit an ignition delay comparable with a much higher CN petroleum fuel [21.20].

In general, a higher CN fuel will result in a shorter ignition delay for a given engine operating point. A shorter ignition delay will reduce the amount of fuel that has already mixed with the surrounding air prior to ignition and hence there will be a lower amount of premixed burn. This will reduce the peak rate of heat release and the peak rate of pressure rise. This will reduce engine noise [21.21]. Reducing the premixed burn reduces the spike in heat release, resulting in a more balanced burning of the fuel, closer to Rudolf Diesel's original concept. This will usually reduce the peak cylinder temperature, which will reduce the NO_x formation. Reducing the ignition delay will also mean less fuel has had a chance to disperse within the combustion chamber prior to burning. This will reduce the opportunity for fuel to impinge on cylinder walls, where it can be quenched and result in high HC emissions.

A high CN fuel reduces ignition delay by increasing the reaction rate of the low-temperature chemical reactions taking place early in the combustion process. This is because the high CN fuel produces more reactive radicals. This feature will also influence the kinetics of the whole combustion process. The burning of the fuel should therefore take place at a slightly higher rate. Thus, if sufficient oxygen is available, a higher CN fuel should result in more complete combustion and reduced emissions of CO. Because this relies upon sufficient oxygen being available, it would be expected that the benefits of a high CN fuel would diminish as the load on the engine increases. Engine strategies to reduce the amount of fuel burned during the premixed phase would also be expected to reduce the effects of changes in the CN of the fuel. However, if more fuel is burned during the diffusion burning phase, then this may produce more particulate emissions. The following sections discuss how these theoretical considerations manifest themselves in practical applications.

21.2.1. Effect of CN under Fully Warm Operating Conditions

Diesel fuel properties are inter-related, which makes it more difficult to determine the effects of a single parameter such as natural CN. Therefore, in most cases the effect of CN has to be determined by statistical models to try and eliminate the response from inter-related fuel properties. If small variations in CN are to be considered, then these changes can be brought about by the use of CN improver (CNI) additives. This is discussed in Section 21.2.3. The following sections outline the results of some previous studies, divided chronologically, which will tie-in with changes in engine technology.

Pre-1990 Studies

Some early work was conducted using fuels that had been blended to give what would today be considered extremely low CN values [21.22]. This demonstrated a very nonlinear response to CN as shown in Figure 21.3 [21.22].

FIGURE 21.3   Effect of CN on HC emissions from a turbocharged and a naturally aspirated engine.

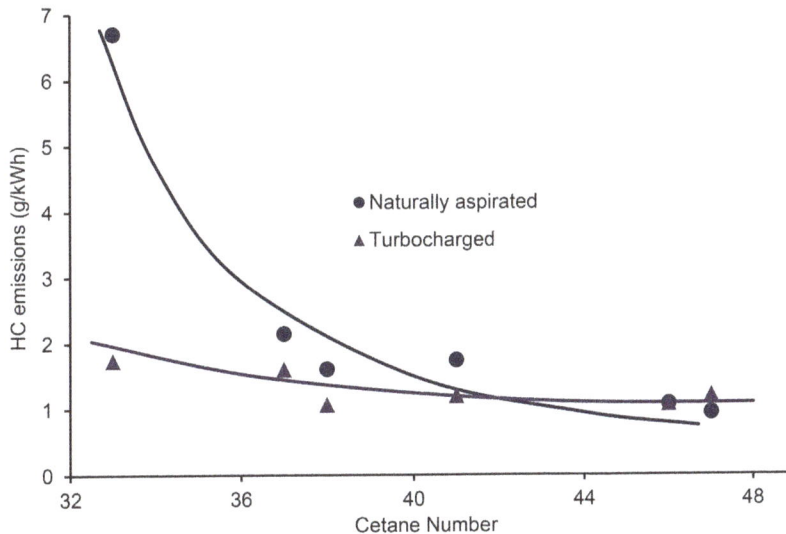

This figure shows the HC emissions from a naturally aspirated and a turbocharged production heavy-duty engine of the time, when tested over the 13-mode test cycle. The figure also shows that the naturally aspirated engine is more sensitive to changes in CN than the turbocharged engine. A similarly nonlinear trend was demonstrated on indirect injection (IDI) passenger cars [21.23], although there was a significant variation from vehicle to vehicle. Further work in the 1980s, on both large [21.17] and small [21.24] cylinder capacity IDI engines, consistently showed a strong correlation between increasing CN and reduced HC emissions. There was generally a reduction of CO emissions reported as a result of increasing CN, but there was little reported influence of CN on NO_X or smoke emissions.

21.2.1.2. **Studies during the 1990s**

In the 1990s, pending changes to emissions legislation in the US prompted research into fuel effects on prototype heavy-duty diesel engines. This work again showed a strong correlation between increasing CN and reduced emissions of HC and CO. However, there were significant differences in the magnitude of the reduction for a given increase in CN [21.25, 21.26]. It was found that the relationship was nonlinear and that there were thus diminishing benefits [21.27] to be achieved from very high CN fuels. Most of this work showed that increasing CN reduced NO_X emissions, but in some cases, there were no significant changes in NO_X due to changes in CN [21.28].

Also in the 1990s, the EPEFE was undertaken to investigate the gaps in the knowledge on the effect of fuels and engine technologies on exhaust emissions [21.29]. This program considered gasoline engines, but the major emphasis was on diesel engines with both heavy-duty and light-duty being considered [21.30]. As with the US studies, a carefully designed and formulated matrix of fuels was developed to allow subsequent statistical analysis to investigate the individual effects of density, polyaromatics, CN, and back-end volatility on emissions [21.31]. As with the US work, the effects of the different properties were determined by building regression models based on the data produced from the matrix of fuels with inter-related properties [21.32].

The results of the heavy-duty diesel program [21.33] were in broad agreement with the US research and showed that an increase in CN resulted in a reduction in emissions of CO and HC by 10.3 and 6.3%, respectively, when averaged over the fleet of engines tested. The directional trend was consistent across the fleet, although the sensitivity to changes in CN varied from engine to engine. There was also a small reduction in NO_X emissions of 0.6% when averaged over the fleet, although the directional trend differed from engine to engine. There was no significant effect on PM emissions attributable to changes in CN; again, there were directional variations from engine to engine.

The LDV study included 17 passenger cars and two light-duty trucks with engine technologies including DI and IDI, naturally aspirated and turbocharged, with and without exhaust gas recirculation (EGR), and with purely mechanically controlled fuel injection and with electronically controlled fuel injection. Some of the turbocharged engines had intercoolers and some did not, and some of the engines with EGR had open-loop control of the EGR and others had closed-loop control. All engines were fitted with diesel oxidation catalysts (DOC), and emissions measurements were taken downstream of the catalyst. Compared with the heavy-duty studies there was thus a far greater variation in engine technology [21.30].

Although there was a greater variety of engine technology, the light-duty study [21.34] also showed that there was consistent reduction in HC and CO emissions as a result of increasing CN. Across the fleet there was no significant effect of CN on the NO_X emissions. Here the trend was for an increase in CN to reduce the NO_X emissions from the DI fleet but to increase the NO_X emissions from the IDI fleet. The NO_X reductions in the DI fleet were about 2.5 times the increase in the IDI fleet. There was a small but statistically significant increase in the PM emissions. Again, there was a difference between the DI and IDI fleets, with the DI engine vehicle showing an increase in PM emissions as a result of an increase in CN, whereas there was an opposite trend for the IDI vehicles. The DI engines exhibited a larger PM emissions response to CN that resulted in the fleet averaged result showing the statistical increase in PM for an increase in CN. Due to the wide variety of engine technology, the variation in emissions was greater across the different engine technologies than it was across the fairly narrow range of CN values studied.

21.2.1.3. Studies in the 21st Century

A Japan Clean Air Program (JCAP) study conducted at the end of the 1990s did not specifically target CN effects, but a later Japanese study [21.19] included petroleum-based fuel that had been doped to decrease the CN and gas-to-liquid (GTL) based fuels that gave a spread of CN from 35.7 to 78.9. The engine technology variation was smaller, with all the engines being DI with common-rail fuel injection. The vehicle choice did include different aftertreatment technologies. This study did not statistically separate the influence of the different fuel parameters but also suggested that increasing CN decreased HC and CO emissions. The Japanese study also found a less significant effect of CN on NO_X and PM emissions. The results from two of the three vehicles studied are shown in Figure 21.4 [21.19].

FIGURE 21.4 Emissions results for two Japanese vehicles with different aftertreatment systems.

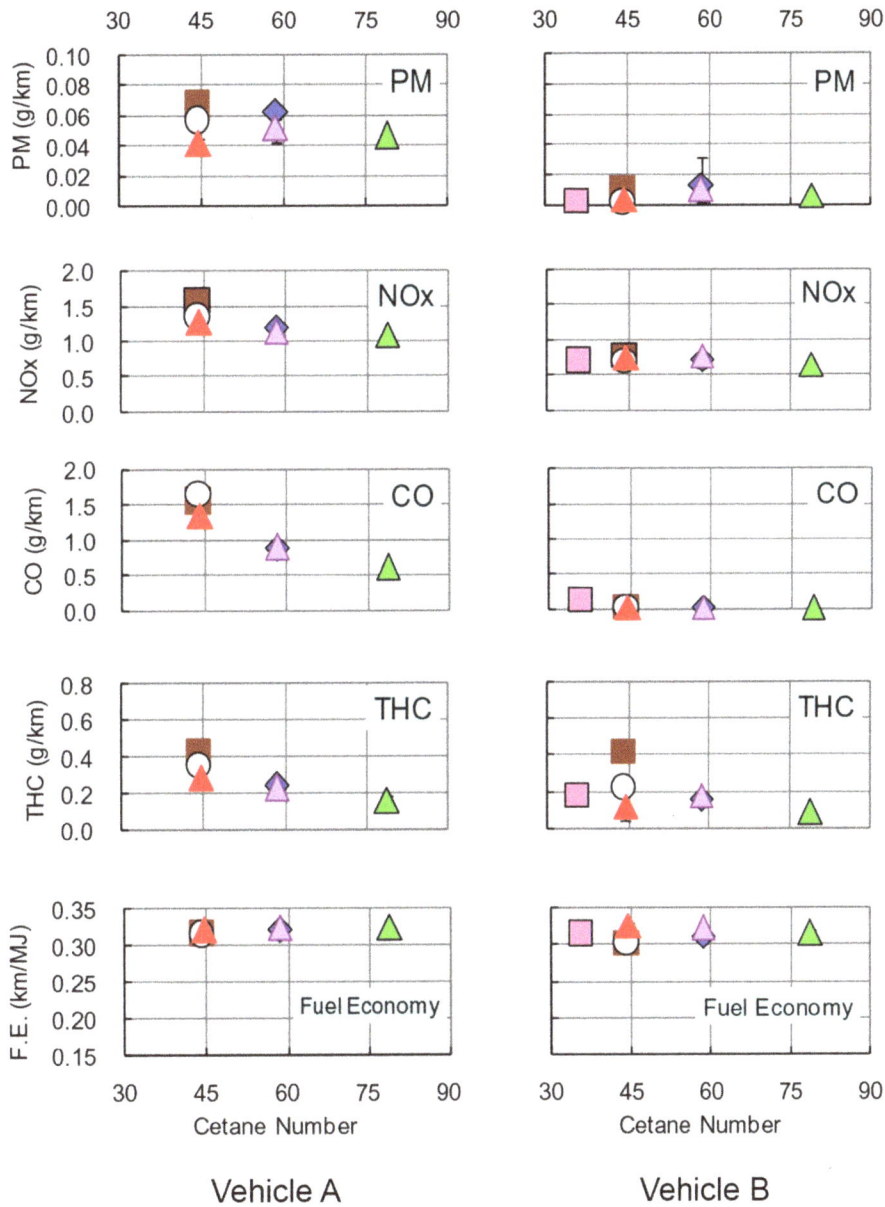

Vehicle A   Vehicle B

© SAE International.

The figure shows the results for each individual fuel, from Vehicle A, which was fitted with a DOC, and Vehicle B, which was fitted with a DOC and NO_X storage reduction (NSR) catalyst and a DPF. The emissions of CO, NO_X, and PM are all noticeably lower for Vehicle B with the more advanced aftertreatment system. The response to changes in CN is also reduced. Figure 21.5 [21.19] shows both the engine-out and the tailpipe emissions results from Vehicle C, which was fitted with a DOC and a DPF.

FIGURE 21.5 Engine-out and tailpipe emissions with different CN fuels.

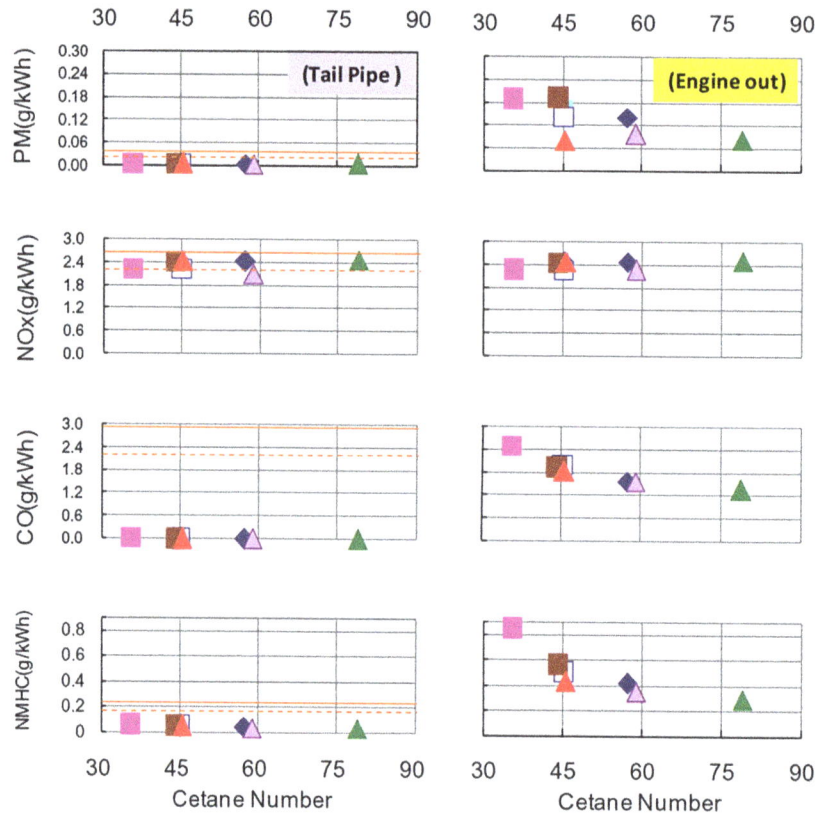

This figure shows the familiar trend of a reduction in CO and HC emissions with increasing CN. The effect of the catalytic aftertreatment is also clearly visible with the CO and HC emissions significantly reduced and the trend with increasing CN also significantly reduced. The DPF also significantly reduces the PM emissions. As this vehicle was not fitted with the NSR, there is little change in the NO_X emissions between the engine-out and the tailpipe. There is no trend for changes in CN to significantly change NO_X emissions. The absence of any effect of CN on NO_X emissions could be partly explained by the fact that these vehicles utilized sophisticated common-rail fuel injection systems. These systems employed pilot injection; the CN of the fuel affected the ignition delay from these pilot injections, but the energy released from these pilot injections resulted in an increased temperature and pressure at the start of the main injection. The effect of the CN on the chemical element of the ignition delay for the main injection was thus reduced.

A European study, using Euro 4, 5, and 6 passenger cars [21.35] demonstrated varying effects due to the emissions limits that the vehicle/engine technology was calibrated for, for example how the engine is calibrated to sit on the NO_X – PM trade-off curve. It also demonstrated that with advance exhaust aftertreatment technology the effect on tailpipe emissions is vastly reduced. A study looking at medium-duty, Euro V, truck and a heavy-duty, Euro VI, bus [21.36] also demonstrated that more advance engine and aftertreatment technologies mask the traditional benefits that might be associated with changes in fuel specification.

21.2.2. Effect of CN under Cold Start Condition

The issue of black smoke that can be generated by diesel engines under heavy load or hard acceleration has already been mentioned. The other highly visible and offensive emission from diesel engines is the white smoke that occurs at start-up, especially under cold weather conditions. A major element of this smoke plume is condensing droplets of unburned fuel. When a diesel engine is started cold, the charge air is at a relatively low temperature. Inlet air heaters or glow plugs can partially rectify this situation. Engine block heaters will also have a significant effect, but these are rarely fitted, except in very cold climates. When fuel is first injected into a cold engine, the physical and chemical processes that occur during the ignition delay period are slowed down by the low temperature. Due to the low temperatures, vaporization of the fuel is impeded and the fuel jet penetrates further into the combustion chamber, thus increasing the degree of wall-wetting. The cold combustion chamber walls do not help with vaporization of the fuel.

It is likely that the first injection will not result in ignition. Some of this fuel may then be blown out of the cylinder during the first exhaust stroke; much of it will remain as liquid fuel. When the first few firing cycles do occur, they will generate enough heat in the combustion chamber to vaporize this liquid fuel but probably not result in its oxidation. The fuel vapor is thus emitted during the exhaust stroke and later condenses as it leaves the tailpipe. It is this unburned fuel that is seen, and smelled, as white smoke. The CN of the fuel will influence the chemical part of the ignition delay period and will thus affect the probability of the fuel reaching full ignition during this cold start phase of engine operation.

The influence of CN on cold starting performance was clearly demonstrated in some work that was performed in the 1940s using a single cylinder laboratory engine [21.37]. Another set of work about 30 years later showed a similar set of results on heavy-duty, naturally aspirated, and turbocharged production engines [21.22]. The time for which the engine had to be cranked before the engine fired was measured for fuels of different CN. In the 1970s work, it was found that at 0°C, increasing the CN from 33 to 50 reduced the cranking time from 60 to 18 seconds. Clearly, these times would be unacceptable for a modern diesel engine vehicle, especially a passenger car. The variation in cranking time with changes in CN from the 1940s is illustrated on the left of Figure 21.6 [21.22], while the results from the 1970s are included on the right of the figure.

FIGURE 21.6 Variation in engine cranking times with different CN fuels.

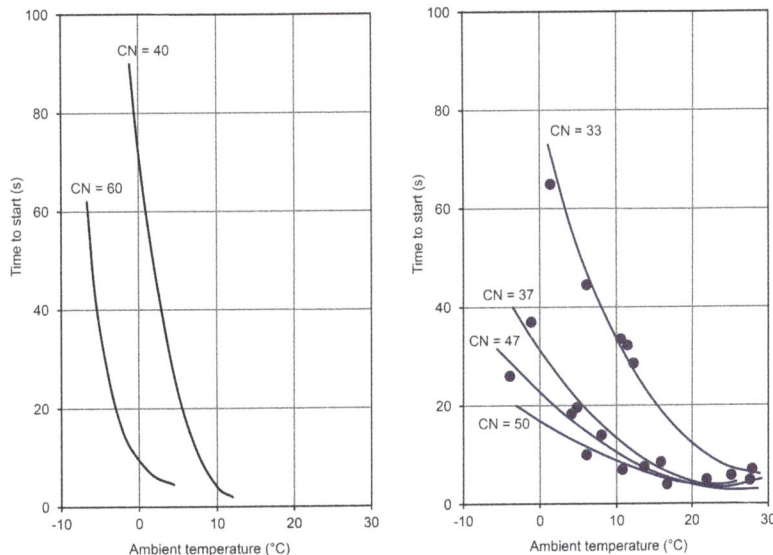

© SAE International.

Work in the 1980s using a DI light-duty truck engine and an IDI passenger car engine, using an inlet air heater and glow plugs, respectively, showed considerably shorter starting times [21.38]. The passenger car generally achieved a stable idle speed within 1 second irrespective of the fuel CN, but the truck engine needed a longer period of time and was therefore more sensitive to changes in CN. At an ambient temperature of

−10°C, with the use of the inlet air heater, this engine required between 30 and 50 seconds to reach a stable idle, depending upon the CN of the fuel. The general response to changes in CN followed the characteristics shown to the right of Figure 21.6. The time for the white smoke emissions to fall to 50% opacity followed a similar response to changes in CN as the time to start.

Work throughout the 1990s [21.39–21.41] continued to show that increased CN shortened starting time and white smoke clearance time. Although engine technology has had a significant effect on engine starting performance, the increasingly strict legislative emissions limits means that the emissions during start-up continue to be a relevant concern.

21.2.3. Natural and Improved CN

As noted, changes in CN are easy to achieve without making any other significant changes to fuel characteristics. Therefore, some of the work reported has shown CN effects based on adding CNI base fuels. This inevitably leads to the question of whether the performance of two fuels of equal CN will be comparable if one is of a lower CN base fuel treated with CNI additive and the other naturally has a high CN. The answer depends on what other characteristics of the fuel are different between the naturally lower CN fuel and the naturally higher CN fuel.

Because of the inter-relationship of fuel parameters, much of the research to determine the effects of fuel properties on emissions relies on statistical analysis to separate the effects of individual parameters. This then allows the researcher to include both natural CN fuels and fuels containing a CNI additive. By separating the data from these two groups of fuel, it is possible to determine whether the result is dependent or independent of whether a CNI additive is used. The general conclusion is that the use of a CNI additive does not significantly affect the relationship between CN and engine performance [21.27, 21.41–21.44]. An example of this is shown in Figure 21.7 [21.44].

FIGURE 21.7   Effects of natural and improved CN fuels on emissions.

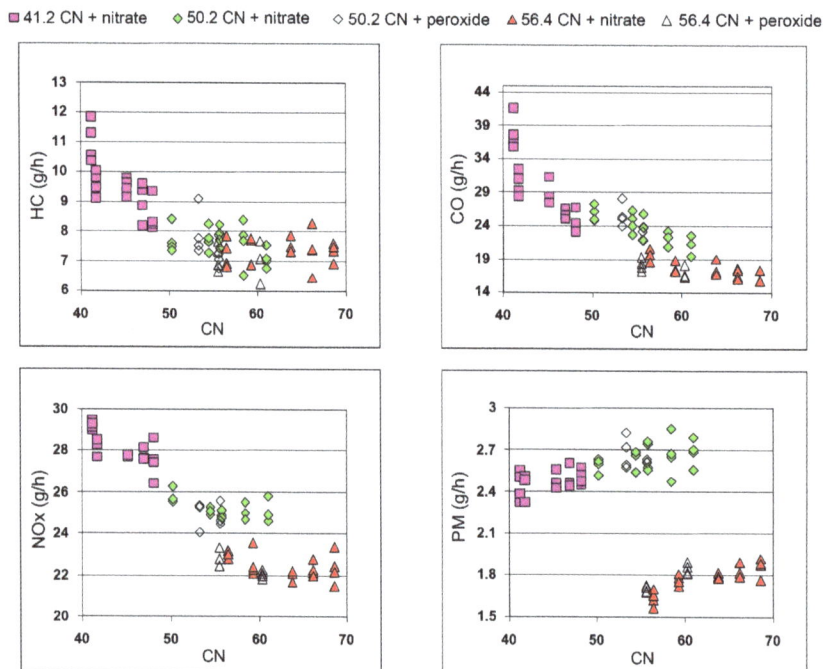

The figure shows the emissions results for three different CN base fuels, each of which has been treated with 0.5, 1, 2, and 3 g/kg of nitrate CNI, and two of the fuels have also been treated with 0.5 and 2 g/kg of peroxide CNI. The trends for HC and CO emissions to fall with increasing CN are clearly visible, with CNI treated fuels and the untreated fuels all conforming to the same trend. The lack of a strong correlation between CN and PM emissions is again clear for each set of base fuels. The PM emissions are in a different grouping due to the differences in the composition of the higher CN base fuel, which do significantly affect PM emissions. The conclusions regarding NO_X emissions are less clear. There is some suggestion that there is a downward trend for NO_X emissions with increasing CN of the base fuel, but that adding the CNI additive increases the CN without significantly reducing the NO_X emissions. However, there is no significant difference between the emissions when using the two different CNI additives.

Some research has indicated that peroxide based CNI will create a greater CN related NO_X reduction than the nitrate based CNIs [21.45, 21.46]. It has been suggested that this is because there are some additional NO_X emissions as a result of adding nitrogen radical forming components to the fuel. A specific investigation to consider this phenomenon did show that the use of a nitrate CNI, at a high treat rate of 3200 mg/kg, did produce additional NO_X emissions [21.47]. The additional NO_X at this high treat rate was not as much as would theoretically be produced if all of the nitrogen in the additive was released as NO_2. Figure 21.8 [21.47] illustrates the results of this work.

FIGURE 21.8 Effect of the high treat rate of CNI on NO_X emissions.

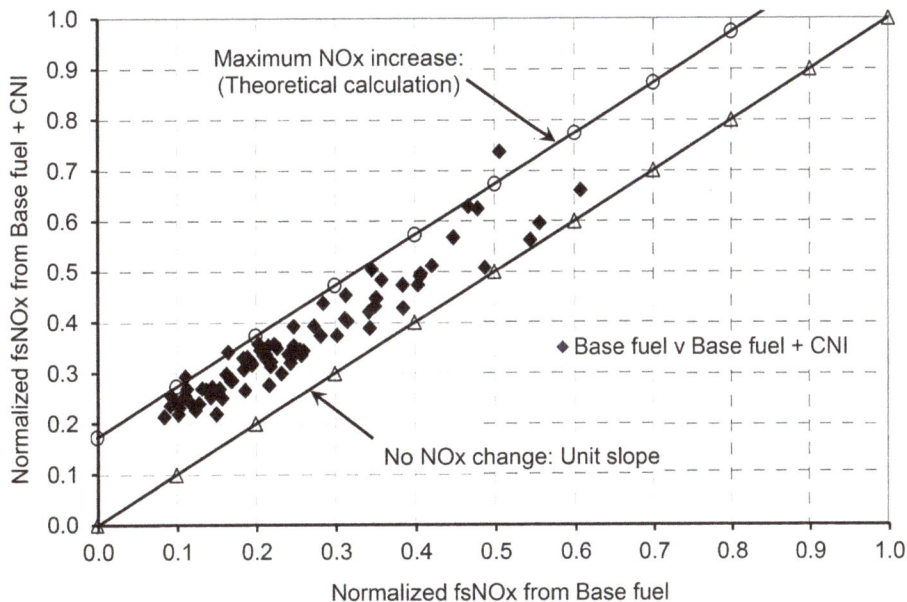

The figure compares the normalized experimental fuel specific NO_X (fsNO_X) (the amount of NO_X produced per unit of fuel consumed) measurements for the base fuel and that of the base fuel plus the high treat rate of CNI. The straight line, with triangular markers, indicates where the results would lie if there was no change in the NO_X emissions, due to the addition of the CNI. The maximum NO_X increase line shows the result of a theoretical calculation that assumes that all the nitrogen contained in the CNI is released as NO_2. The fact that the data points, comparing the two fuel sets, lie between the two lines shows that the treated fuel produces higher NO_X emissions than the untreated fuel.

A recent laboratory study, comparing "natural cetane number" fuels and CNI induced cetane number fuels, showed a strong positive correlation between increased ignition delay and increased CO engine-out emissions, with both sets of fuels, i.e., a CN effect rather than a CNI effect. This behavior was attributed to the production of over-lean mixtures and quenching of the mixture at cylinder walls, due to the longer ignition delay [21.48].

21.3. The Influence of Diesel Fuel Physical Characteristics on Emissions

The physical characteristics of a fuel are clearly determined by the chemical composition of that fuel. Theoretical considerations can suggest how changes in physical characteristics may influence the mechanical processes leading up to the combustion process. How the physical characteristics influence the mechanical processes during the combustion event is much harder to separate from the influence of the chemical effects on combustion. Therefore, trying to separate influence of physical and chemical characteristics on the emissions performance of real engines is a daunting task. Over the years there has understandably been a lot of effort devoted to investigating the link between fuel properties and emissions performance. The following sections try to elucidate how some of the physical characteristics of diesel fuel can influence the engine and vehicle emissions in the ever-changing environment of evolving engine technology.

21.3.1. Diesel Fuel Density Effects

The density of diesel fuel is very strongly correlated with its hydrocarbon composition. This is illustrated where researchers have presented results showing the correlation of different fuel properties that have been used in programs investigating fuel effects. The density of diesel fuel is usually very strongly correlated with the aromatic or polyaromatic content of the fuel [21.24, 21.47, 21.49]. Attempts to formulate test fuel matrices to decouple these parameters are therefore difficult and can limit the range of these normally correlated parameters [21.31]. Care must therefore be taken when interpreting the results to ensure that it is a density effect that is being observed rather than a fuel chemical composition effect.

Diesel fuel injection systems deliver fuel on a volumetric basis. From a fundamental point of view, the most obvious effect that the fuel density will have on the engine's performance is that a greater mass of fuel will be delivered during each injection event. If the testing technique compares fuel performance on a fixed rack position, for example, full-power testing, then the mass of fuel delivered will differ. This difference in fuel delivery can significantly affect the emissions performance; using the full-power example, an increase in fuel density will increase the fuel delivery, which will increase the fuel-to-air ratio. This can result in over-fueling and a significant increase in particulate or smoke emissions.

There is usually a strong correlation between the density of the fuel and its calorific value on a mass basis; there will usually be an increase in calorific value on a volume basis when the density of the fuel is increased. Therefore, an increase in fuel density will result in a greater amount of energy being injected into the combustion chamber for each injection. The engine technology will have a significant impact on how this is compensated for and the emissions response.

In a simple open-loop mechanical system, as used on older technology engines, the additional energy content, delivered at every injection, with a high-density fuel, would result in higher power output. Under most driving conditions, the driver demand, via the accelerator pedal, is for a given power output. The driver compensates for the additional power by using a lesser accelerator pedal position. Increased fuel density will therefore have no direct influence on the amount of energy injected on each cycle. However, on more modern engines with elaborate electronic control of many engine parameters, there are implications as to how the electronic engine management systems respond to changes in the amount of energy delivered at each fuel injection event. This is discussed further in Section 21.3.1.2.

The density of the fuel can also influence the processes within the diesel engine at a more detailed level. If the accelerator pedal position, or the rack position, is adjusted to give equal mass of fuel injected, then it follows that for a denser fuel there will be a smaller volume of fuel injected. This means that the duration of the injection event will be shortened; in the absence of other influencing parameters this would tend to increase the amount of premixed combustion and reduce the amount of diffusion combustion. This change to the balance of the combustion process would obviously affect the emissions.

The hydrodynamic performance of the fuel injection system can also be influenced by the physical properties of the fuel. Fuel density, viscosity, and compressibility will affect the hydraulic behavior of the fuel injection system. In older pump-line-nozzle systems, the injection timing is controlled at the pump. Variations in the hydraulic performance of the system can therefore lead to changes in the dynamic fuel injection timing. With modern designs like the high-pressure common-rail system, where the injection event is controlled electronically at the injector itself, these effects are significantly reduced but are still present.

21.3.1.1. Density Emissions Correlations

Work performed in the 1990s [21.34] showed that for a fleet of LDVs, reducing the fuel density would reduce emissions of CO, HC, and PM with a slight increase in NO_X emissions. However, the relationship with NO_X emissions was not consistent across the fleet, with some vehicles showing the opposite trend. Considering the DI and IDI engine vehicles separately suggested that the DI engine were more prone to an increase in NO_X emissions as a result of reducing fuel density. Further work conducted around this time also found a strong correlation between the fuel density and PM emissions [21.50, 21.51], although it must be noted that there were strong correlations between density and other fuel properties such as aromatic content and the high end of the distillation curve. Attempts were made to break this correlation by restricting the fuels analyzed to those where there was only a weak correlation. This showed that there was still a strong correlation between fuel density and PM emissions.

A parallel piece of work on heavy-duty vehicles [21.33] showed an opposite trend for density and gaseous emissions; a reduction in density increased both CO and HC but reduced NO_X emissions. In the heavy-duty vehicles there was no overall correlation between density and PM emissions. There were also differences in the way different fuel injection systems responded. The engine using a distributor type fuel pump was found to be the most sensitive to fuel changes. The in-line injection system also responded to changes in density. The engine using a closed-loop fuel injection timing system was the least sensitive. Unfortunately, these conclusions were based solely on a single engine of each type.

In these programs, the vehicles were tested with the engines in their standard state of tune, and it was noted that the changes in fuel properties was affecting the injection timing on some vehicles. Additional testing was performed on heavy-duty vehicles to try to eliminate any changes in emissions performance that might result from these changes in injection timing [21.52]. Adjusting the engine setup to restore the injection timing to a set value for each fuel significantly reduced variations in emissions performance, and it was concluded that the observed density effects were due to interactions with the fuel injection system and not the combustion process.

Work was also performed on a passenger car engine where the standard engine management unit was by-passed to allow independent control of the injection process [21.44]. This showed that when eliminating any effects due to changes in fuel injection timing, there was a significant decrease in the observed effects, but there was still an apparent density effect; although density and 90% distillation point could not be decoupled. It was again noted that reducing density reduced the emissions of CO, HC, and PM. In this case NO_X emissions were also reduced by a reduction in fuel density.

Work has also been performed on an engine with a common-rail fuel injection system [21.53]. A heavy-duty engine was tested according to the Japanese diesel 13-mode procedure using nine fuels with a range of density from 0.796 and 0.856 kg/m^3. Six of these fuels were also tested according to the Japanese ED12-mode procedure. The 13-mode procedure is a steady-state test procedure, whereas the ED12-mode is a transient test procedure. Regression analysis of the results showed a stronger correlation between density and PM emissions than between any of the other fuel properties and emissions characteristics considered. A reduction in density resulted in a reduction in PM emissions. The emissions results for the diesel 13-mode and ED12 procedures, relative to changes in density, are shown graphically in Figure 21.9 [21.53].

FIGURE 21.9 Relationship between fuel density and emissions on a heavy-duty common-rail engine.

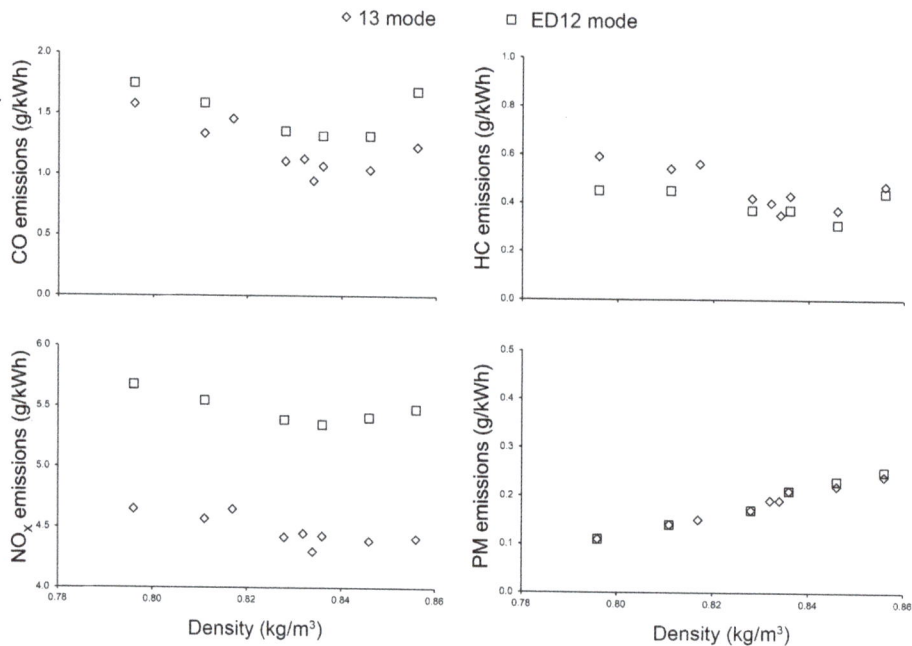

The results for the PM emissions are almost identical in both the steady-state and the transient cycles. The NO_X emissions are noticeably higher on the transient cycle. For this work, there was a strong correlation between the fuel density and its viscosity. Further work was performed to try to determine how these two properties were influencing the combustion process; this is discussed further in Section 21.3.2.

21.3.1.2. The Influence of Fuel Density on Emission Control Systems

Modern diesel engines that use a mapped engine management strategy assume a standard fuel density in order to determine the fuel injector pulse with settings or sleeve position in older electronically controlled rotary pumps. These fuel injection parameters are mapped against an accelerator pedal position to represent power demand and an engine speed signal. This should result in a given engine speed and load set point. Other parameters such as EGR valve demand are also mapped against these inputs. An increase in fuel density would result in a greater power output, but the engine would be using other set points to match the lower anticipated output. Changes in fuel density may therefore result in a mismatch between the EGR valve demand for the actual load and speed point and that where the map believes the engine is operating. This can then alter the exhaust emission characteristics, particularly the NO_X response.

Testing with a DI passenger car using an electronically controlled distributor pump, EGR and an oxidation catalyst showed the expected increase in PM emissions as a result of increased fuel density. Disabling the EGR and adjusting the management unit to compensate for the change in fuel density resulted in almost all of the density effect being eliminated. With the same density adjustment to the electronic control but with the EGR system active there was still about 50% of the density effect still present [21.54].

Further test work aimed at eliminating interactions between the engine management systems and fuel properties also showed that there remained an in-cylinder or combustion mechanism by which fuel density was having a small effect on emissions [21.55]. Thus, as electronic engine management systems continue to become more sophisticated and employ closed-loop control, there is likely to remain a fuel density effect on emissions performance. As exhaust aftertreatment systems reduce a greater proportion of the engine-out emissions, the tailpipe fuel effects will become smaller. While aftertreatment will reduce the tailpipe emissions, there are still implications of fuel property effects. Changes in engine-out PM emissions will influence the required regeneration frequency of a DPF, and changes in engine-out NO_X emissions can influence the reductant usage for selective catalytic reduction (SCR) systems for NO_X control. If regeneration of the DPF is accomplished through additional fuel injection, then the energy content of these additional injections becomes important. Again, if the injected quantity is open-loop and assume a given fuel density, then changes in fuel density will influence the regeneration performance.

21.3.2. The Influence of Viscosity

Fuel effects studies usually demonstrate a strong correlation between fuel viscosity and other physical properties, particularly density. Results are sometimes discussed in terms of density/viscosity [21.56] because it is hard to decouple these properties. However, density effects are likely to become more significant as the industry moves toward greater use of bio-derived fuels. Vegetable oils typically have a kinematic viscosity in the range of 20 to 45 mm^2/s [21.57]. This is reduced by about a factor of 10 by transesterification to the methyl ester. This is one of the main reasons for the transesterification process [21.58, 21.59]. This is still slightly higher than petroleum diesel fuel.

Work in the 1990s that tried to decouple the viscosity from other fuel parameters found that NO_X and smoke emissions could be described in terms on the viscosity, ignition delay, and aromatic content of the fuel [21.60]. The smoke emissions were increased by increasing fuel viscosity and the NO_X emissions were reduced. Further work showed a positive correlation between viscosity and PM emissions [21.61]. It should be noted that in both these studies there was a correlation between viscosity and the distillation characteristics of the test fuels.

A later piece of work that again showed a strong correlation between density and viscosity tried to explain the performance at a more fundamental level and thus determine how these two properties may be influencing engine performance [21.53]. Nonevaporating spray studies were carried out using a matrix of fuels that included three trial production fuels that were formulated to deliberately break the correlation between density and viscosity that existed for the other fuels in the matrix. Examination of the spray characteristics included the spray penetration, the cone angle, and the Sauter mean diameter (SMD) of the fuel droplets. This showed that there was no correlation between the spray penetration or cone angle and either density or viscosity. However, there was a correlation between density and viscosity and the SMD of the droplets. The trial fuels that had a significant variation in viscosity over a very narrow range of density still showed this strong correlation, indicating that it was the viscosity that was the main factor in the correlation. The increased droplet size is then likely to result in the higher levels of particulate production and hence PM emissions that had previously been observed.

Other fundamental studies have also shown that an increase in viscosity results in an increase in SMD [21.62, 21.63]. For a pintle nozzle there is also an inverse correlation between viscosity and rate of spray penetration; that is, an increase in viscosity results in a decrease in the rate of penetration [21.62]. These studies also confirmed that there is no clear correlation with viscosity and cone angle.

21.3.3. Diesel Fuel Distillation Characteristics and Emissions

The distillation characteristics of the fuel are an indication of how readily given percentages of the fuel will volatilize. This will have an effect on how readily the fuel droplets, produced during fuel injection, will vaporize. This in turn will influence the mixing of the air and fuel and the subsequent combustion. The lighter components in the fuel, then the more likely it is for this fuel to mix well and result in more complete combustion and hence lower HC, CO, and PM emissions; the effect on emissions is less straightforward. The less volatile fuel components will volatilize less readily and can result in wall-wetting and high emissions of unburned fuel.

The effect of fuel volatility will be most apparent during cold start operation. As noted in Section 21.2.2, when a diesel engine is cold started the charge air is at a relatively low temperature. Starting aid will help negate this situation. The physical and chemical processes that occur during the ignition delay period are slowed down by the low temperature. The physical delay is determined by the fuel injection characteristics and the distillation characteristics of the fuel. Due to the low temperatures, vaporization of the fuel is impeded and the fuel jet penetrates farther into the combustion chamber, thus increasing the degree of wall-wetting. It is usually the less volatile fractions of the fuel that are most related to this problem. The 90% distillation point is commonly shown to correlate with white smoke, odor, and high HC emissions [21.64, 21.65].

With respect to warmed-up operation, there was found to be a correlation between increasing top end distillation and HC emissions. In the EPEFE program, the 95% distillation point (T95) was used as the metric and was shown to correlate with HC emissions in both LDVs [21.34] and heavy-duty engines [21.33] such that increasing T95 resulted in increased HC emissions. The effect was far greater on the heavy-duty engines than in the LDVs.

21.4. The Influence of Chemical Composition of Diesel Fuel on Emissions

The chemical composition of the fuel is intrinsically linked to its physical properties. Consideration of some of the fundamental physical processes and the related fuel physical characteristics can suggest why some physical changes in the fuel can result in changes to emissions. Some changes to the chemical nature of the fuel can be accomplished without significantly changing the physical properties. A clear example of this is the CN as discussed in Section 21.2. Another chemical compositional characteristic that is easy to change to investigate its effects on emissions is the fuel sulfur content. Other parameters that can also be changed over quite a wide range with relatively small changes in physical parameters are the aromatic and oxygen content within the fuel. The effect that these parameters have on emissions will be discussed in the following sections.

21.4.1. Diesel Fuel Sulfur Effects on Emissions

The influence of fuel sulfur concentration on engine emissions has been a topic of research for many years now. The interaction of fuel sulfur and exhaust aftertreatment became a topic of interest with the introduction of aftertreatment systems on gasoline engines that occurred before aftertreatment was fitted to diesel engines. As a result of this work, the fuel sulfur concentration of both gasoline and diesel fuels have been legislated to very low levels in many countries throughout the world. Most other countries already have plans to follow this lead. The following sections look at this past work and shed light on the rationale for reducing the level of sulfur in fuel to near zero levels.

21.4.1.1. The Effects of Fuel Sulfur on Tailpipe Emissions

Most of the sulfur in diesel fuel will burn to produce sulfur dioxide (SO_2), most of which, in the absence of after-treatment, will be emitted in the diesel exhaust. The SO_2 is in itself a hazardous pollutant. Most of the sulfur that is not converted to SO_2 is converted to sulfur trioxide (SO_3). The SO_3 readily combines with water to form sulfuric acid (H_2SO_4), which may react with metals to form metal sulfates. These metals may be present in the lubricating oil or they may be present as part of the engine structure. The sulfuric acid and the metal sulfates that are emitted in the exhaust will add to the collected mass of PM. The effect of the sulfate particles can be significantly increased by their hygroscopic nature. This can be considered as a direct contribution to emitted particulates.

Sulfuric acid formed in the combustion chamber can also significantly increase corrosive wear of piston rings and cylinder liners [21.66, 21.67]. It is the sulfuric acid entering the lubricating oil that necessitates the inclusion of alkaline metal additives. It is these additives that can then contribute sulfates to the PM emissions.

Sulfuric acid that is not adsorbed onto the PM can also nucleate to form a sulfuric acid aerosol [21.68, 21.69]. The SO_2 that is emitted can undergo further reactions within the atmosphere to produce other compounds, such as ammonium sulfate. This can be considered as an indirect contribution to the atmospheric particulate burden. With high sulfur fuels, this can be a far greater burden than the direct contribution [21.70]. Numerous studies have been published that demonstrate the effect of fuel sulfur on the direct contribution to PM emissions [21.71–21.73].

21.4.1.2. Effect of Sulfur on Diesel Aftertreatment Systems

With the tighter emissions regulations that were introduced in the early 1990s, the major challenge for the diesel engine manufacturer was to meet the new PM emissions standards. The PM emissions are composed primarily of carbonaceous particles, adsorbed organic matter, and inorganic ash. The adsorbed organic matter is either volatile or soluble, the volatile component is referred to as the volatile organic fraction (VOF), and the soluble material as the soluble organic fraction (SOF)—in most cases these are a similar thing. The SOF is composed of unburned or partially burned fuel and lubricant. The ash is usually composed of sulfates, phosphates, and metal compounds from the burning of lubricant.

Emissions of gaseous pollutants were not as much of a challenge for the diesel engineer as for the gasoline engine manufacturers. However, the use of a DOC would not only significantly reduce the emissions of HC and CO, but it would also greatly reduce the amount of SOF, thereby reducing the total mass of PM emissions. The SOF is adsorbed by and adhered to the DOC at low temperatures (100 to 300°C) and is then oxidized at higher temperatures (200 to 500°C) resulting in a 40 to 90% reduction in SOF [21.74].

Platinum group metals (PGM), particularly Platinum (Pt) and palladium (Pd), are known to be particularly active toward the oxidation of SOF, as well as HC and CO emissions. Unfortunately, Pt and Pd are also active toward the oxidation of SO_2 to SO_3, especially at temperatures in excess of 350 to 400°C. This can then have a negative impact on the PM emissions. This produces a challenge to the system designer: to try and ensue that the operating window of the DOC is close to that dictated by favorable chemistry. Base metal catalysts are less active in oxidizing the SO_2 but are more readily poisoned by sulfur [21.75].

To comply with the latest emissions standards, automotive diesel engines will also have to be fitted with NO_X reduction strategies and far more effective PM reduction devises. In general, to meet the PM standards, a DPF will be required. To control NO_X, it was initially thought that SCR technology would be required along with the necessity to carry the SCR reductant, often referred to as diesel exhaust fluid (DEF). Advances in NO_X aftertreatment make NSR a more attractive solution, particularly for LDVs. Diesel NO_X aftertreatment systems are discussed further in Chapter 16, Section 16.4.3.

For DPF systems, the effects of sulfur have some dependence on the method chosen for regenerating the DPF. The use of PGM catalyst on the DPF substrate [21.76] or as part of a system where a DOC is placed in front of an uncatalyzed DPF [21.77] can lead to the same SO_3 generation problems as discussed for the DOC [21.78]. For fuel sulfur levels ranging from 3 to 350 mg/kg, there was still a linear dependence of PM emissions to sulfur content. Increased sulfur content increased PM emissions even when the DPF was fitted [21.78].

The operation of NSR systems relies on the formation and storage of nitrogen dioxide (NO_2). The NO_2 is subsequently released and reduced by HC and CO present in the exhaust during the regeneration phase. Oxides of sulfur in the exhaust will not only poison the precious metal catalyst that oxidizes the nitric oxide (NO) to NO_2 but it also preferentially reacts with the storage components to form sulfates rather than nitrates [21.79], thus reducing the storage capacity of the devise. The NSR must therefore go through a desulfation cycle to restore its efficacy. The desulfation cycle requires raising the NSR to a high temperature; this has a fuel consumption penalty and ages the devise [21.80–21.82]. The higher the fuel sulfur levels, then the more frequently this process must be implemented and the more rapidly the NSR will lose its efficiency. Figure 21.10 [21.83] shows the tailpipe emissions from a passenger car fitted with an NSR catalyst for mileage accumulation with different sulfur content fuels.

FIGURE 21.10 Tailpipe NO$_X$ emissions during a distance accumulation test on a passenger car with NSR.

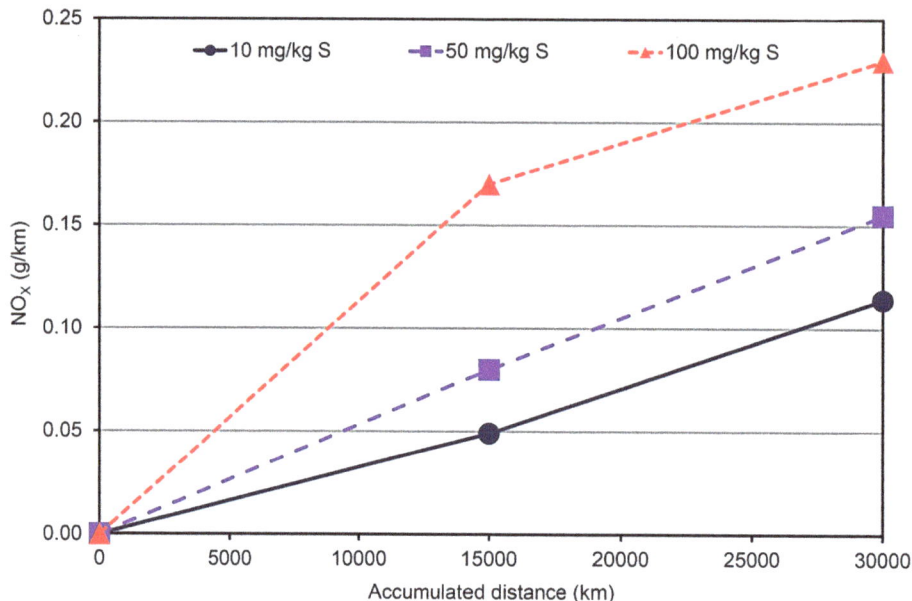

21.4.2. The Influence of Aromatic Content on Emissions

The aromatic content of diesel fuel has for a long time been considered an influential fuel parameter regarding diesel engine performance [21.84, 21.85]. Due to the recognition of the carcinogenic aspects of polycyclic aromatic hydrocarbons (PAH), there was also concern over the emissions of these compounds either in volatile form or adsorbed onto the PM emissions.

An early piece of work on the influence of different fuel components on emissions used a carbon-14 tracer to investigate the relationship between these components and the PM emissions [21.86]. Three paraffinic components, one olefinic component, and three aromatic components were tagged using the radioactive isotope, carbon-14. The paraffins decane, hexadecane, and octadecane were selected to give an indication of the effect of boiling point on particulates. The olefin, decene, was selected because of its low boiling point. Xylene was selected as a low boiling point single-ring aromatic component, and 1-methyl naphthalene was chosen as a two-ringed aromatic with a methyl branch. Two differently tagged naphthalenes were used: one with the carbon-14 tag on the methyl branch and the other with the carbon-14 tag in the ring opposite the methyl radical. The results showed that each of these compounds contributed to the PM emissions; the PM emissions emanating from the paraffins were independent of the boiling point of the ones tested; and that as a group the aromatic compounds were the greatest contributors to PM emissions. The ring labeled naphthalene made a greater contribution to PM emissions than did the methyl radical labeled compound.

To try and relate emissions performance to fuel compositional effects, a great many studies have been conducted to try and break the correlation between aromatic content and other parameters such as CN and distillation characteristics. Some of these studies indicate that a greater aromatic content results in greater PM emissions [21.86–21.89], while others have found the effect of aromatic content to be small and not statistically significant [21.24, 21.25, 21.71, 21.90, 21.91]. In most of these studies the commercial fuel components were used, and the aromatic structures were not considered. Other studies have specifically looked at the effects of the different structure of individual compounds such as decalin, 1-2-3-4 tetrahydronaphthalene (tetralin), naphthalene, and 1-methylnaphthalene to elucidate specific processes [21.92, 21.93]. However, it is also possible for aromatic compound to be formed by pyrolysis of non-aromatic compounds or be derived from the lubricating oil [21.94, 21.95]. It is also possible for aromatic fuel components to form PAH compounds within the combustion chamber [21.96, 21.97].

The diesel fuel aromatic content also influences the flame temperature during combustion; this can affect the NO_X formation process as well as the particulate formation and oxidation processes [21.98]. The aromatics present in the fuel can thus also directly influence the NO_X emissions. The following sections discuss the observed influence of the aromatic component and the PAH content, of the fuel, on the regulated and unregulated emissions with changes in engine and aftertreatment technologies.

21.4.2.1. Effect of Aromatics Free Fuel

With the more wide-spread use of "aromatic-free" fuels, there is growing interest in the effect of these fuels on emissions performance. Aromatic-free fuels or paraffinic fuels, include those synthesized directly from hydrogen, from water, and carbon, from CO_2, those produced from fossil hydrocarbons using the Fischer-Tropsch (F-T) process, such as CTL and GTL fuels, and those produced by hydrogenating vegetable oils (HVO). Because these fuels are ostensibly free of aromatic compounds, and sulfur, they are thought to have the potential to produce far fewer emissions, particularly of particulate matter.

Many studies have been performed to investigate the effect on emissions behavior of such paraffinic fuels, in comparison with petroleum-derived fuels with varying aromatic content [21.99–21.103]. These studies showed that the aromatic-free fuel reduced the engine-out and tailpipe HC, CO, NO_X, PM and PN emissions. Unfortunately, these studies were considering the aromatic-free fuels as drop-in replacements for petroleum diesel fuel without any engine recalibration. The aromatic-free fuels tend to have a higher CN, and this factor can itself lead to some of the observed effects. However, it has been found that too high a CN fuel, which results in a decreased ignition delay, can initiate combustion before sufficient fuel-air mixing has occurred, thus producing higher particulate emissions [21.18, 21.104]. Work has also been conducted to consider recalibration of the engine to take advantage of the increased CN of the paraffinic fuels [21.105], this has demonstrated the potential for greater reductions in regulated emissions and improved efficiency [21.106].

21.4.2.2. Effect of Total Aromatics Content on Emissions

Some early testing of heavy-duty US diesel engines showed opposing trends for aromatic content on smoke emissions [21.107]. A 10.4 L, eight-cylinder engine exhibited a reduction in smoke when using a fuel with increased aromatic content while a 13.7 L, six-cylinder engine showed an increase in smoke when using the same fuel. Both engines were naturally aspirated and of an open chamber design. Apart from the obvious difference in individual cylinder volume, 1.3 L per cylinder for the smaller engine against 2.28 L per cylinder for the larger engine, these engines displayed similar technology but opposite responses to changes in aromatic content. Of course, these were real fuel blends, and it was thus impossible to decouple some fuel parameters. These engines showed similarly opposite emission responses to CN.

A program of work in the 1980s, using a single type of heavy-duty US engine, showed that the aromatic content of the fuel had no HC, CO, or PM emissions under steady-state testing [21.108]. There was an increase in NO_X emissions associated with increasing the fuel aromatic content. However, under transient testing of the same engine, there was an increase in HC, NO_X, PM, and smoke emissions associated with increasing fuel aromatic content. Because this engine was turbocharged, it is likely that the changes in response to fuel compositional changes were due to over-fueling brought about by a lag in the turbocharger response during transients. This study did also note that at high aromatic contents there was a sharp increase in the emissions of the one measured PAH compound.

A more recent piece of work, on an engine with common-rail fuel injection and exhaust aftertreatment, using blends of petroleum diesel fuel and a GTL fuel that contained no aromatics, showed that reducing aromatic content reduced HC, CO, and PM emissions [21.109]. The GTL fuel has a high CN, probably as a consequence of the zero aromatic composition, and it was therefore not possible to determine whether this effect was due to the aromatic content of the CN of the fuel. This work also showed that downstream of the DOC and DPF fitted to the engine, there was no appreciable difference in any of these emissions. The NO_X emissions were unchanged by the aftertreatment and showed a slight increase with increasing aromatic content on driving cycles with a high average speed.

21.4.2.3. Effect of Fuel Di- and Tri-Aromatic Content on Emissions

In general, the work discussed used fuels where the nature of the aromatic components was not classified. Work in the early 1990s showed that although there was no strong correlation between total aromatics and emissions, there was a strong correlation between two- and three-ringed aromatics and PM emissions [21.110]. This work used a DI, turbocharged engine with a rotary pump fuel injection system and showed that increasing di- and tri-aromatics increased PM emissions. Later studies have also tended to concentrate on these polyaromatic fuel constituents rather than the total aromatic content.

In the LDV part of the EPEFE program, it was also found that increasing polyaromatics in the fuel increased PM and NO_X emissions but reduced HC and CO emissions. Both DI and IDI vehicles produced the same directional response [21.34]. In the heavy-duty engine part of the program there was also an increase in PM and NO_X as a result of increasing polyaromatic content, but in this case, it was accompanied by an increase in HC emissions [21.33]. There was no significant effect on CO emissions.

A more detailed study considered the effects on PAH emissions of doping a paraffinic fuel with specific chemical species [21.111]. The species were naphthenes (50% mono-naphthene and 50% di-naphthene), mono-aromatics (80% alkyl-benzene and 20% naphthene-benzene), di-aromatics (methyl naphthalene), and tri-aromatics (phenanthelene). The addition of the naphthenes did not affect the PAH emissions, but the addition of di- and tri-aromatic compounds produced a significant increase in PAH emissions. The effect of phenanthelene on the PAH emissions was about an order of magnitude greater than that of the mono-aromatics. There was clear hierarchy of effect on PAH emissions; tri-aromatics produced the greatest effect, followed by di-aromatics, then mono-aromatics, and finally naphthenes. Additional testing was conducted with more than one class of compound added to the base fuel. This showed that the combination of the tri-aromatic compound and the mono-aromatics or naphthenes increased the effect disproportionately. Other combinations produced only an additive effect. This suggests that the influence of tri-aromatics on PAH emissions is promoted by the coexistence of mono-aromatic or naphthene compounds. The influence of mono-aromatics and naphthenes was smaller than that of the di-aromatic compound.

This is in agreement with the theory that the carbon-to-hydrogen ratio is an important chemical property in the determination of flame temperature and soot formation [21.98, 21.112]. The carbon-to-hydrogen ratio can also affect the growth of soot particles [21.113]. However, the subsequent oxidation of this soot within the combustion chamber and hence the exhaust emissions will be very dependent upon engine design and operating condition.

21.4.3. The Influence of Oxygenate Content on Emissions

The use of oxygenated compounds as diesel fuel additives for soot suppression has been of interest for many years. Previous work has included the performance of many oxygenated compounds, including dimethoxymethane [21.114], dimethoxyethane (monoglyme) [21.26], diethylene glycol dimethyl ether (diglyme) [21.26, 21.115, 21.116], dimethyl carbonate [21.117–21.119], diethyl carbonate [21.119, 21.120], diethyl ether [21.121, 21.122], acetates [21.123–21.125], maleates [21.124, 21.126–21.128], and alcohols [21.125, 21.129, 21.130]. Many studies have shown that the PM emissions are reduced linearly with increasing oxygen content [21.131–21.134] but that the structure of the oxygenate can have a significant effect on its performance [21.124, 21.130].

The precise mechanism by which oxygenated fuels reduce PM emissions and why different types of oxygenates produce different benefits is still being researched. Oxygenated fuels are in themselves low-sooting fuels; therefore, replacing an inherently soot-producing component with a low-soot-producing component will have a beneficial effect [21.135]. This can explain the almost linear relationship between oxygen addition to the fuel and the observed PM reduction.

During the early part of the diesel combustion process, the incoming fuel jet draws hot gas from the combustion chamber into the fuel jet. This produces a rise in fuel temperature and resultant evaporation. The evaporating fuel mixes with the air while preflame chemical reactions occur at temperatures of about 550°C. In this reaction zone the partially burned fragments of fuel that are believed to be soot precursors are formed. This is discussed more fully in Chapter 15, Section 15.3.1. The rate of growth of these soot particles is dependent upon the local air-to-fuel ratio. The inclusion of oxygen within the fuel will increase the local air-to-fuel ratio, thus limiting soot particle growth [21.136].

In many of the oxygenated molecules considered, a single oxygen atom is bonded to one or more carbon atoms. When these molecules react, they undergo a series of reactions, including hydrogen abstraction and bond breaking, to form smaller oxygen containing species until ultimately CO is formed. An increase in the local concentration of CO can result in a decrease in PAH production [21.137] and also an increase in OH radicals [21.138]. The increase in OH radicals has been proposed as a mechanism for reduced soot emissions [21.139–21.141]. If the oxygenated compound contains two oxygen atoms bonded to a single carbon atom, then the hydrogen abstraction and bond-breaking process can result in the formation of CO_2 rather than CO. However, under the high-temperature fuel-rich conditions of the combustion zone, this CO_2 can react with radicals to produce CO and further radicals.

In many countries the desire to replace petroleum-derived fuels had resulted in the use of plant or animal-derived oils and fats. Due to the high viscosities of many of these products, transesterification has been used to yield viscosities closer to diesel fuel. The resulting esters, usually methyl esters known as biodiesel, have of course resulted in an oxygenated blending component, which would be expected to reduce soot emissions. This has been shown to be the case [21.142–21.145], although at low biodiesel blending ratios the benefit can be small or reversed, especially on LDVs [21.146]. However, the use of biodiesel has also widely been reported to increase the emissions of NO_X [21.147–21.149].

The increase NO_X emissions can in some cases be attributed to the higher bulk modulus of the biodiesel, causing an advance in injection timing, particularly on older pump-line-nozzle injection systems. There can also be a decrease in ignition delay times due to the higher CN of most biodiesel fuels. Both of these mechanisms will result in an earlier start of combustion, yielding longer residence times and/or higher in-cylinder temperatures. This will result in the observed increase in NO_X emissions. These effects will be less apparent on newer technology engines employing a common-rail fuel system. It has also been found that the combustion process generally progresses faster with biodiesel fuels than petroleum fuels. This can result in higher in-cylinder temperatures and longer residence times at high temperature. This will lead to greater NO_X formation, regardless of engine technology [21.150]. The lower soot production can also result in reduced radiative heat transfer to the combustion chamber walls, which will again result in higher local temperatures and increased thermal NO_X formation [21.151].

21.4.3.1. The Influence of Biodiesel on Exhaust Aftertreatment Systems

As noted, the introduction of biodiesel is mainly to satisfy the desire or legislation to replace some petroleum diesel fuel in order to reduce petroleum imports and/or greenhouse gas emissions. The introduction of biodiesel will not in itself allow engines to meet stricter emissions legislation. Biodiesel will therefore be used in vehicles equipped with all levels of emissions control technology. The fact that biodiesel has some significantly different chemical and physical characteristics has implications for the functioning of systems that were designed around the assumption that petroleum diesel will be used. The higher boiling point of biodiesel can result in higher levels of condensation downstream of the DPF, resulting in slightly lower measured filtration efficiency, particularly at low loads [21.152]. However, there is no measurable change in the toxicity of the emissions when using biodiesel [21.152].

To meet tighter PM emissions regulations, vehicles will have to be fitted with a DPF. In many applications the regeneration of the DPF is accomplished by introducing unburned fuel upstream of an oxidation catalyst. The fuel is oxidized across this catalyst to generate heat and/or NO_2 as an oxidant for the trapped PM. This is discussed further in Chapter 16, Section 16.4.2.4. The unburned fuel, upstream of the catalyst, can be generated by either late injection into the combustion chamber or by injecting it directly into the exhaust system.

Differences in viscosity and distillation characteristics of the biodiesel become important when considering the strategy of late fuel injection. When fuel is injected late during the expansions stroke, the cylinder pressure is lower than for normal injection timings; the temperatures are also higher than normal. As a result of this, the fuel jet will penetrate farther into the combustion chamber, and there is a high probability of liquid fuel hitting the combustion chamber walls. This wall-wetting can result in liquid fuel getting past the piston rings and into the lubricant. For late injection of biodiesel, the penetration of the liquid core of the fuel spray has been found to increase by about 15% [21.153], which will exacerbate the problem of wall-wetting and oil dilution.

In addition to the greater probability of fuel entering the lubricant, the fact that the biodiesel has a high and narrow distillation characteristic, means that less of the fuel will subsequently evaporate out of the lubricant. This will result in even higher long-term lubricant dilution. On a heavy-duty engine equipped with a DPF and NSR aftertreatment system, the oil dilution level, using petroleum diesel fuel, has been found to stabilize at about 4%. The amount of fuel evaporating matches the amount that is being added as a result of the late injection strategy. However, when using fuel containing 20% biodiesel, the oil dilution does not appear to stabilize, and it is estimated that the dilution rate could reach 12% at the end of the manufacturer-recommended oil drain interval [21.154].

To avoid these oil dilution issues, some manufacturers have elected to inject the fuel directly into the exhaust system. To ensure reliable regeneration of the DPF, which may contain accumulated soot from petroleum diesel or biodiesel blends, a DPF inlet temperature of about 600°C is required. This requires a slightly higher temperature exiting the DOC. For a given aftertreatment system, it has been found that to achieve a temperature of 625°C downstream of the DOC, an inlet temperature of 270°C is required with petroleum diesel. With 100% biodiesel, this temperature rises to approximately 340°C, with intermediate biodiesel blend requiring intermediate temperatures, although the relationship is not linear [21.155]. This higher exhaust temperature requirement will restrict the opportunity to regenerate the DPF, which could obviously lead to operational problems.

The fact that biodiesel fuel may contain alkali metals from its production is also a potential concern for aftertreatment systems. These metals can obviously add to the ash that will accumulate within the DPF [21.156]. These metals may also have a detrimental effect on the integrity of certain DPF substrate materials [21.156, 21.157]. The introduction of LM (Low Metal) grades of B100 in the ASTM standard [21.158] is designed to address this issue. Biodiesel meeting the ASTM LM grades has a maximum combined sodium, potassium, calcium, and magnesium concentration of 4 µg/g. Testing of a B20 blend fuel, based on these limits for B100, demonstrated no deleterious physical effects on a DPF substrate nor negative effect on DPF pressure drop or regeneration rates [21.159, 21.160].

There are also concerns regarding the response of NO_X aftertreatment systems to the use of biodiesel blends. As discussed, it is commonly found that the use of biodiesel increases the engine-out NO_X emissions. It has also been found that the NO_2 emissions fall; therefore, there is a significant change in the NO/NO_2 ratio in the exhaust [21.161]. The NO/NO_2 ratio has an effect on the conversion efficiency of urea-based SCR systems. It has been found that for some vehicles equipped with urea-based SCR systems, operation on biodiesel can produce tailpipe NO_X emissions that will no longer meet the legislated targets [21.162].

References

21.1.   Grunder, L., "West Coast Diesel Smoke and Odor Problems," SAE Technical Paper 490132 (1949), doi:https://doi.org/10.4271/490132.

21.2.   Haagen-Smit, A.J. and Fox, M.M., "Automobile Exhaust and Ozone Formation," SAE Technical Paper 550277 (1955), doi:https://doi.org/10.4271/550277.

21.3.   Larson, G.P., Fischer, G.I., and Hamming, W.J., "Evaluating Sources of Air Pollution," *Industrial & Engineering Chemistry* **45**, no. 5 (1953): 1070-1074.

21.4.   Wilkins, E.T., "Exhaust Gases from Motor Vehicles: Some Measurements of Carbon Monoxide in the Air of London," *Journal of the Royal Society of Health* (4) (1956): 677-687.

21.5.   "Final Regulation Order Adoption of New Section 1961.4, Title 13, California Code of Regulations, Exhaust Emission Standards and Test Procedures - 2026 and Subsequent Model Year Passenger Cars, Light-Duty Trucks, and Medium-Duty Vehicles," accessed January 2023, https://ww2.arb.ca.gov/sites/default/files/barcu/regact/2022/accii/acciifro1961.4.pdf.

21.6.   Automotive News Europe, "EU, Germany Reach Deal to Allow e-Fuels after 2035 Law Ends Sales of CO_2-Emitting Cars," March 25, 2023, accessed June 5, 2023, https://europe.autonews.com/environmentemissions/eu-germany-reach-car-emissions-deal-includes-e-fuels.

21.7.   Automotive News Europe, "Europe's Euro 7 Emission Limits Face Pushback from Eight Countries," May 22, 2023, accessed June 5, 2023, https://europe.autonews.com/environmentemissions/eight-eu-nations-call-scrapping-euro-7-emission-rules.

21.8.   Harrison, R.M., "Airborne Particulate Matter," *Philosophical Transactions of the Royal Society A* 378 (2020): 20190319.

21.9.   Andersson, J.D., Wedekind, B.G.A., Hall, D., Stradling, R. et al., "DETR/SMMT/CONCAWE Particle Research Programme: Sampling and Measurement Experiences," SAE Technical Paper 2000-01-2850 (2000), doi:https://doi.org/10.4271/2000-01-2850.

21.10. Andersson, J.D., Wedekind, B.G.A., Hall, D., Stradling, R. et al., "DETR/SMMT/CONCAWE Particulate Research Programme: Light Duty Results," SAE Technical Paper 2001-01-3577 (2001), doi:https://doi.org/10.4271/2001-01-3577.

21.11. Wedekind, B.G.A., Andersson, J.D., Hall, D., Stradling, R. et al., "DETR/SMMT/ CONCAWE Particle Research Programme: Heavy Duty Results," SAE Technical Paper 2000-01-2851 (2000), doi:https://doi.org/10.4271/2000-01-2851.

21.12. Ntziachristos, L., Mamakos, A., Samaras, Z., Mathis, U. et al., "Overview of the European 'Particulates' Project on the Characterization of Exhaust Particulate Emissions from Road Vehicles: Results for Light-Duty Vehicles," SAE Technical Paper 2004-01-1985 (2004), doi:https://doi.org/10.4271/2004-01-1985.

21.13. Andersson, J., Preston, H., Warrens, C., and Brett, P., "Fuel and Lubricant Effects on Nucleation Mode Particle Emissions from a Euro III Light Duty Diesel Vehicle," SAE Technical Paper 2004-01-1989 (2004), doi:https://doi.org/10.4271/2004-01-1989.

21.14. Miyazaki, T. and Nishimoto, T., "Motor Vehicle Emission Control Measures of Japan," SAE Technical Paper 922178 (1992), doi:https://doi.org/10.4271/922178.

21.15. ASTM International, "Standard Test Method for Cetane Number of Diesel Fuel Oil," ASTM D613-23, ASTM International, 2023.

21.16. Wong, C.L. and Steere, D.E., "The Effects of Diesel Fuel Properties and Engine Operating Conditions on Ignition Delay," SAE Technical Paper 821231 (1982), doi:https://doi.org/10.4271/821231.

21.17. Olree, R.M. and Lenane, D.L., "Diesel Combustion Cetane Number Effects," SAE Technical Paper 840108 (1984), doi:https://doi.org/10.4271/840108.

21.18. Nishiumi, R., Yasuda, A., Tsukasaki, Y., and Tanaka, T., "Effects of Cetane Number and Distillation Characteristics of Paraffinic Diesel Fuels on PM Emission from a DI Diesel Engine," SAE Technical Paper 2004-01-2960 (2004), doi:https://doi.org/10.4271/2004-01-2960.

21.19. Takahashi, K., Sakurai, Y., Furuse, T., Sakuraba, T. et al., "Effects of Cetane Number and Chemical Components on Diesel Emissions and Vehicle Performance," SAE Technical Paper 2009-01-2692 (2009), doi:https://doi.org/10.4271/2009-01-2692.

21.20. Siebers, D.L., "Ignition Delay Characteristics of Alternative Diesel Fuels: Implications on Cetane Number," SAE Technical Paper 852102 (1985), doi:https://doi.org/10.4271/852102.

21.21. Anderton, D. and Waters, P.E., "Effect of Fuel Composition on Diesel Engine Noise and Performance," SAE Technical Paper 820235 (1982), doi:https://doi.org/10.4271/820235.

21.22. Broering, L.C. and Holtman, L.W., "Effect of Diesel Fuel Properties on Emissions and Performance," SAE Technical Paper 740692 (1974), doi:https://doi.org/10.4271/740692.

21.23. Kagami, M., Akasaka, Y., Date, K., and Maeda, T., "The Influence of Fuel Properties on the Performance of Japanese Automotive Diesels," SAE Technical Paper 841082 (1984), doi:https://doi.org/10.4271/841082.

21.24. Weidmann, K., Menrad, H., Reders, K., and Hutcheson, R.C., "Diesel Fuel Quality Effects on Exhaust Emissions," SAE Technical Paper 881649 (1988), doi:https://doi.org/10.4271/881649.

21.25. Ullman, T.L., Spreen, K.B., and Mason, R.L., "Effects of Cetane Number, Cetane Improver, Aromatics, and Oxygenates on 1994 Heavy-Duty Diesel Engine Emissions," SAE Technical Paper 941020 (1994), doi:https://doi.org/10.4271/941020.

21.26.   Spreen, K.B., Ullman, T.L., and Mason, R.L., "Effects of Cetane Number, Aromatics, and Oxygenates on Emissions from a 1994 Heavy-Duty Diesel Engine with Exhaust Catalyst," SAE Technical Paper 950250 (1995), doi:https://doi.org/10.4271/950250.

21.27.   Ullman, T.L., Spreen, K.B., and Mason, R.L., "Effects of Cetane Number on Emissions From a Prototype 1998 Heavy-Duty Diesel Engine," SAE Technical Paper 950251 (1995), doi:https://doi.org/10.4271/950251.

21.28.   Ullman, T.L., Mason, R.L., and Montalvo, D.A., "Effects of Fuel Aromatics, Cetane Number, and Cetane Improver on Emissions from a 1991 Prototype Heavy-Duty Diesel Engine," SAE Technical Paper 902171 (1990), doi:https://doi.org/10.4271/902171.

21.29.   MacKinven, R. and Hublin, M., "European Programme on Emissions, Fuels and Engine Technologies—Objectives and Design," SAE Technical Paper 961065 (1996), doi:https://doi.org/10.4271/961065.

21.30.   Steinbrink, R., Cahill, G.F., Signer, M., and Smith, G., "European Programmes on Emissions, Fuels and Engine Technologies (EPEFE)—Vehicle/Engine Technology," SAE Technical Paper 961067 (1996), doi:https://doi.org/10.4271/961067.

21.31.   Rainbow, L.J., Le Jeune, A., Lang, G.J., and McDonald, C.R., "European Programme on Emissions, Fuels and Engine Technologies (EPEFE)—Gasoline and Diesel Test Fuels Blending and Analytical Data," SAE Technical Paper 961066 (1996), doi:https://doi.org/10.4271/961066.

21.32.   Zemroch, P.J., Schimmerling, P., Sado, G., Gray, C.T. et al., "European Programme on Emissions, Fuels and Engine Technologies (EPEFE)—Statistical Design and Analysis Techniques," SAE Technical Paper 961069 (1996), doi:https://doi.org/10.4271/961069.

21.33.   Signer, M., Heinze, P., Mercogliano, R., and Stein, H.J., "European Programme on Emissions, Fuels and Engine Technologies (EPEFE)—Heavy Duty Diesel Study," SAE Technical Paper 961074 (1996), doi:https://doi.org/10.4271/961074.

21.34.   Hublin, M., Gadd, P.G., Hall, D.E., and Schindler, K.P., "European Programmes on Emissions, Fuels and Engine Technologies (EPEFE)—Light Duty Diesel Study," SAE Technical Paper 961073 (1996), doi:https://doi.org/10.4271/961073.

21.35.   Williams, R., Hamje, H., Rickeard, D., Bartsch, T. et al., "Effect of Diesel Properties on Emissions and Fuel Consumption from Euro 4, 5 and 6 European Passenger Cars," SAE Technical Paper 2016-01-2246 (2016), doi:https://doi.org/10.4271/2016-01-2246.

21.36.   Williams, R., Pettinen, R., Ziman, P., Kar, K. et al., "Fuel Effects on Regulated and Unregulated Emissions from Two Commercial Euro V and Euro VI Road Transport Vehicles," *Sustainability* 13, no. 14 (2021): 7985.

21.37.   Porter, H.R., "Cold-Starting Tests on Diesel Engines," SAE Technical Paper 430157 (1943), doi:https://doi.org/10.4271/430157.

21.38.   Sutton, D.L., "Investigation into Diesel Operation with Changing Fuel Property," SAE Technical Paper 860222 (1986), doi:https://doi.org/10.4271/860222.

21.39.   Buchsbaum, A. and Zeiner, W., "Cold Start Properties of Diesel Fuel Blends with Varying Low End Volatility and Cetane Number," SAE Technical Paper 920034 (1992), doi:https://doi.org/10.4271/920034.

21.40.   Bickerton, R.A., Such, C.H., Sørum, P.A., and Fløysand, S.A., "The Development of a Method for Evaluating the Effect of Fuel Quality on the Cold Starting of a Range of Diesel Engines," SAE Technical Paper 921748 (1992), doi:https://doi.org/10.4271/921748.

21.41.   Hara, H., Itoh, Y., Henein, N., and Bryzik, W., "Effect of Cetane Number with and without Additive on Cold Startability and White Smoke Emissions in a Diesel Engine," SAE Technical Paper 1999-01-1476 (1999), doi:https://doi.org/10.4271/1999-01-1476.

21.42.   Sienicki, E.J., Jass, R.E., Slodowske, W.J., McCarthy, C.I. et al., "Diesel Fuel Aromatic and Cetane Number Effects on Combustion and Emissions from a Prototype 1991 Diesel Engine," SAE Technical Paper 902172 (1990), doi:https://doi.org/10.4271/902172.

21.43.   Lange, W., Cooke, J.A., Gadd, P., Zürner, H. et al., "Influence of Fuel Properties on Exhaust Emissions from Advanced Heavy-Duty Engines Considering the Effect of Natural and Additive Enhanced Cetane Number," SAE Technical Paper 972894 (1997), doi:https://doi.org/10.4271/972894.

21.44.   Kwon, Y., Mann, N., Rickeard, D.J., Haugland, R. et al., "Fuel Effects on Diesel Emissions—A New Understanding," SAE Technical Paper 2001-01-3522 (2001), doi:https://doi.org/10.4271/2001-01-3522.

21.45.   Liotta, F.J., "A Peroxide Based Cetane Improvement Additive with Favorable Fuel Blending Properties," SAE Technical Paper 932767 (1993), doi:https://doi.org/10.4271/932767.

21.46. Nandi, M.K., Jacobs, D.C., Liotta, F.J., and Kesling, H.S., "The Performance of a Peroxide-Based Cetane Improvement Additive in Different Diesel Fuels," SAE Technical Paper 942019 (1994), doi:https://doi.org/10.4271/942019.

21.47. Kumar, S., Stanton, D.W., Fang, H., Gustafson, R.J. et al., "The Effect of Diesel Fuel Properties on Engine-Out Emissions and Fuel Efficiency at Mid-Load Conditions," SAE Technical Paper 2009-01-2697 (2009), doi:https://doi.org/10.4271/2009-01-2697.

21.48. Erman, A.G., Hellier, P., and Ladommatos, N., "The Impact of Ignition Delay and Further Fuel Properties on Combustion and Emissions in a Compression Ignition Engine," *Fuel* 262 (2020): 116155.

21.49. Zannis, T.C., Hountalas, D.T., Papagiannakis, R.G., and Levendis, Y.A., "Effect of Fuel Chemical Structure and Properties on Diesel Engine Performance and Pollutant Emissions: Review of the Results of Four European Research Programs," *SAE Int. J. Fuels Lubr.* 1, no. 1 (2009): 384-419, doi:https://doi.org/10.4271/2008-01-0838.

21.50. Betts, W.E., Fløysand, S.Å., and Kvinge, F., "The Influence of Diesel Fuel Properties on Particulate Emissions in European Cars," SAE Technical Paper 922190 (1992), doi:https://doi.org/10.4271/922190.

21.51. Den Ouden, C.J.J., Clark, R.H., Cowley, L.T., Stradling, R.J. et al., "Fuel Quality Effects on Particulate Matter Emissions from Light- and Heavy-Duty Diesel Engines," SAE Technical Paper 942022 (1994), doi:https://doi.org/10.4271/942022.

21.52. Rickeard, D.J., Bonetto, R., and Signer, M., "European Programme on Emissions, Fuels and Engine Technologies (EPEFE)—Comparison of Light and Heavy Duty Diesel Studiesm," SAE Technical Paper 961075 (1996), doi:https://doi.org/10.4271/961075.

21.53. Morita, A. and Sugiyama, G., "Influence of Density and Viscosity of Diesel Fuel on Exhaust Emissions," SAE Technical Paper 2003-01-1869 (2003), doi:https://doi.org/10.4271/2003-01-1869.

21.54. Heinze, P., Hutcheson, R., Kapus, P., and Cartellieri, W., "The Interaction between Diesel Fuel Density and Electronic Engine Management Systems," SAE Technical Paper 961975 (1996), doi:https://doi.org/10.4271/961975.

21.55. Mann, N., Kvinge, F., and Wilson, G., "Diesel Fuel Effects on Emissions: Towards a Better Understanding," SAE Technical Paper 982486 (1998), doi:https://doi.org/10.4271/982486.

21.56. Stradling, R., Gadd, P., Signer, M., and Operti, C., "The Influence of Fuel Properties and Injection Timing on the Exhaust Emissions and Fuel Consumption of an Iveco Heavy-Duty Diesel Engine," SAE Technical Paper 971635 (1997), doi:https://doi.org/10.4271/971635.

21.57. Demirbas, A., "Relationships Derived from Physical Properties of Vegetable Oil and Biodiesel Fuels," *Fuel* 87, no. 8-9 (2008): 1743-1748.

21.58. Ma, F. and Hanna, M.A., "Biodiesel Production: A Review," *Bioresource Technology* 70, no. 1 (1999): 1-15.

21.59. Knothe, G. and Steidley, K.R., "Kinematic Viscosity of Biodiesel Fuel Components and Related Compounds. Influence of Compound Structure and Comparison to Petrodiesel Fuel Components," *Fuel* 84, no. 9 (2005): 1059-1065.

21.60. Miyamoto, N., Ogawa, H., Shibuya, M., and Suda, T., "Description of Diesel Emissions by Individual Fuel Properties," SAE Technical Paper 922221 (1992), doi:https://doi.org/10.4271/922221.

21.61. Gerini, A. and Montagne, X., "Automotive Direct Injection Diesel Engine Sensitivity to Diesel Fuel Characteristics," SAE Technical Paper 972963 (1997), doi:https://doi.org/10.4271/972963.

21.62. Callahan, T.J., Ryan, T.W., Dodge, L.G., and Schwalb, J.A., "Effects of Fuel Properties on Diesel Spray Characteristics," SAE Technical Paper 870533 (1987), doi:https://doi.org/10.4271/870533.

21.63. Chang, C.T. and Farrell, P.V., "A Study on the Effects of Fuel Viscosity and Nozzle Geometry on High Injection Pressure Diesel Spray Characteristics," SAE Technical Paper 970353 (1997), doi:https://doi.org/10.4271/970353.

21.64. Tahara, Y. and Akasaka, Y., "Effects of Fuel Properties on White Smoke Emission from the Latest Heavy-Duty DI Diesel Engine," SAE Technical Paper 952354 (1995), doi:https://doi.org/10.4271/952354.

21.65. Ogawa, H., Raihanl, K., Iizuka, K., and Miyamoto, N., "Cycle-to-Cycle Transient Characteristics of Diesel Emissions during Starting," SAE Technical Paper 1999-01-3495 (1999), doi:https://doi.org/10.4271/1999-01-3495.

21.66. Blanc, L.A., "Effect of Diesel Fuel on Deposits and Wear," SAE Technical Paper 480197 (1948), doi:https://doi.org/10.4271/480197.

21.67. Furstoss, R.J., "Field Experience with High Sulfur Diesel Fuels," SAE Technical Paper 490221 (1949), doi:https://doi.org/10.4271/490221.

21.68. Opris, C.N., Gratz, L.D., Bagley, S.T., Baumgard, K.J. et al., "The Effects of Fuel Sulfur Concentration on Regulated and Unregulated Heavy-Duty Diesel Emissions," SAE Technical Paper 930730 (1993), doi:https://doi.org/10.4271/930730.

21.69. Liu, Z.G., Vasys, V.N., Swor, T.A., and Kittelson, D.B., "Significance of Fuel Sulfur Content and Dilution Conditions on Particle Emissions from a Heavily-Used Diesel Engine during Transient Operation," SAE Technical Paper 2007-01-0319 (2007), doi:https://doi.org/10.4271/2007-01-0319.

21.70. Wall, J.C., Shimpi, S.A., and Yu, M.L., "Fuel Sulfur Reduction for Control of Diesel Particulate Emissions," SAE Technical Paper 872139 (1987), doi:https://doi.org/10.4271/872139.

21.71. Wall, J.C. and Hoekman, S.K., "Fuel Composition Effects on Heavy-Duty Diesel Particulate Emissions," SAE Technical Paper 841364 (1984), doi:https://doi.org/10.4271/841364.

21.72. Baranescu, R.A., "Influence of Fuel Sulfur on Diesel Particulate Emissions," SAE Technical Paper 881174 (1988), doi:https://doi.org/10.4271/881174.

21.73. Hori, S. and Narusawa, K., "The Influence of Fuel Components on PM and PAH Exhaust Emissions from a DI Diesel Engine—Effects of Pyrene and Sulfur Contents," SAE Technical Paper 2001-01-3693 (2001), doi:https://doi.org/10.4271/2001-01-3693.

21.74. Horiuchi, M., Saito, K., and Ichihara, S., "The Effects of Flow-through Type Oxidation Catalysts on the Particulate Reduction of 1990's Diesel Engines," SAE Technical Paper 900600 (1990), doi:https://doi.org/10.4271/900600.

21.75. Farrauto, R.J. and Mooney, J.J., "Effects of Sulfur on Performance of Catalytic Aftertreatment Devices," SAE Technical Paper 920557 (1992), doi:https://doi.org/10.4271/920557.

21.76. Koberstein, E., Pletka, H.D., and Völker, H., "Catalytically Activated Diesel Exhaust Filters - Engine Test Methods and Results," SAE Technical Paper 830081 (1983), doi:https://doi.org/10.4271/830081.

21.77. Cooper B.J., Radnor H.J., Jung W., and Thoss J.E., Treatment of diesel exhaust gases. U.S. Patent 4,902,487, 1990.

21.78. Liang, C.J., Baumgard, K.A., Gorse, R.E., Orban, J.M.E. et al., "Effects of Diesel Fuel Sulfur Level on Performance of a Continuously Regenerating Diesel Particulate Filter and a Catalyzed Particulate Filter," SAE Technical Paper 2000-01-1876 (2000), doi:https://doi.org/10.4271/2000-01-1876.

21.79. Shoji, A., Kamoshita, S., Watanabe, T., Tanaka, T. et al., "Development of a Simultaneous Reduction System of NOx and Particulate Matter for Light-Duty Truck," SAE Technical Paper 2004-01-0579 (2004), doi:https://doi.org/10.4271/2004-01-0579.

21.80. Molinier, M., "NOx Adsorber Desulfurization under Conditions Compatible with Diesel Applications," SAE Technical Paper 2001-01-0508 (2001), doi:https://doi.org/10.4271/2001-01-0508.

21.81. Parks, J., Watson, A., Campbell, G., and Epling, B., "Durability of NOx Absorbers: Effects of Repetitive Sulfur Loading and Desulfation," SAE Technical Paper 2002-01-2880 (2002), doi:https://doi.org/10.4271/2002-01-2880.

21.82. Thornton, M., Webb, C.C., Weber, P.A., Orban, J. et al., "Fuel Sulfur Effects on a Medium-Duty Diesel Pick-Up with a NOX Adsorber, Diesel Particle Filter Emissions Control System: 2000-Hour Aging Results," SAE Technical Paper 2006-01-0425 (2006), doi:https://doi.org/10.4271/2006-01-0425.

21.83. Oyama, K. and Kakegawa, T., "Evaluation of Diesel Exhaust Emission of Advanced Emission Control Technologies Using Various Diesel Fuels, and Sulfur Effect on Performance after Mileage Accumulation. JCAP Diesel WG (fuel) Report for Step II Study," SAE Technical Paper 2003-01-1907 (2003), doi:https://doi.org/10.4271/2003-01-1907.

21.84. Hurn, R.W. and Hughes, K.J., "Combustion Characteristics of Diesel Fuels as Measured in a Constant-Volume Bomb—A Report of the Coordinating Research Council, Inc," SAE Technical Paper 520210 (1952), doi:https://doi.org/10.4271/520210.

21.85. Marshall, W.F. and Hum, R.W., "Factors Influencing Diesel Emissions," SAE Technical Paper 680528 (1968), doi:https://doi.org/10.4271/680528.

21.86. Burley, H.L. and Rosebrock, T.A., "Automotive Diesel Engines-Fuel Composition vs Particulates," SAE Technical Paper 790923 (1979), doi:https://doi.org/10.4271/790923.

21.87. Bykowski, B.B. and Baines, T.M., "Effects of Alternate Source Diesel Fuels on Light-Duty Diesel Emissions," SAE Technical Paper 820771 (1982), doi:https://doi.org/10.4271/820771.

21.88. Barry, E.G., Axelrod, J.C., McCabe, L.J., Inoue, T. et al., "Effects of Fuel Properties and Engine Design Features on the Performance of a Light-Duty Diesel Truck—A Cooperative Study," SAE Technical Paper 861526 (1986), doi:https://doi.org/10.4271/861526.

21.89. Ullman, T.L. and Human, D.M., "Fuel and Maladjustment Effects on Emissions from a Diesel Bus Engine," SAE Technical Paper 910735 (1991), doi:https://doi.org/10.4271/910735.

21.90. Cunningham, L.J., Henly, T.J., and Kulinowski, A.M., "The Effects of Diesel Ignition Improvers in Low-Sulfur Fuels on Heavy-Duty Diesel Emissions," SAE Technical Paper 902173 (1990), doi:https://doi.org/10.4271/902173.

21.91. van Beckhoven, L.C., "Effects of Fuel Properties on Diesel Engine Emissions—A Review of Information Available to the EEC-MVEG Group," SAE Technical Paper 910608 (1991), doi:https://doi.org/10.4271/910608.

21.92. Flanigan, C.T., Litzinger, T.A., and Graves, R.L., "The Effect of Aromatics and Cycloparaffins on DI Diesel Emissions," SAE Technical Paper 892130 (1989), doi:https://doi.org/10.4271/892130.

21.93. Fukuda, M., Tree, D.R., Foster, D.E., and Suhre, B.R., "The Effect of Fuel Aromatic Structure and Content on Direct Injection Diesel Engine Particulates," SAE Technical Paper 920110 (1992), doi:https://doi.org/10.4271/920110.

21.94. Abbass, M.K., Andrews, G.E., Williams, P.T., Bartle, K.D. et al., "Diesel Particulate Emissions: Pyrosynthesis of PAH from Hexadecane," SAE Technical Paper 880345 (1988), doi:https://doi.org/10.4271/880345.

21.95. Andrews, G.E., Abdelhalim, S.M., and Williams, P.T., "Pyrosynthesis of PAH in a Modern IDI Diesel Engine," SAE Technical Paper 961230 (1996), doi:https://doi.org/10.4271/961230.

21.96. Richter, H., Granata, S., Green, W.H., and Howard, J.B., "Detailed Modeling of PAH and Soot Formation in a Laminar Premixed Benzene/Oxygen/Argon Low-Pressure Flame," *Proceedings of the Combustion Institute* 30, no. 1 (2005): 1397-1405.

21.97. Yamada, H. and Goto, Y., "Formation Process of Soot Precursors in a Laminar Flow Reactor," SAE Technical Paper 2007-01-0061 (2007), doi:https://doi.org/10.4271/2007-01-0061.

21.98. Azetsu, A., Sato, Y., and Wakisaka, Y., "Effects of Aromatic Components in Fuel on Flame Temperature and Soot Formation in Intermittent Spray Combustion," SAE Technical Paper 2003-01-1913 (2003), doi:https://doi.org/10.4271/2003-01-1913.

21.99. Bays, J., Gieleciak, R., Viola, M., Lewis, R. et al., "Detailed Compositional Comparison of Hydrogenated Vegetable Oil with Several Diesel Fuels and Their Effects on Engine-Out Emissions," *SAE Int. J. Fuels Lubr.* 16, no. 3 (2023), doi:https://doi.org/10.4271/04-16-03-0015.

21.100. McCaffery, C., Karavalakis, G., Durbin, T., Jung, H. et al., "Engine-Out Emissions Characteristics of a Light Duty Vehicle Operating on a Hydrogenated Vegetable Oil Renewable Diesel," SAE Technical Paper 2020-01-0337 (2020), doi:https://doi.org/10.4271/2020-01-0337.

21.101. Shukla, P., Shamun, S., Gren, L., Malmborg, V. et al., "Investigation of Particle Number Emission Characteristics in a Heavy-Duty Compression Ignition Engine Fueled with Hydrotreated Vegetable Oil (HVO)," *SAE Int. J. Fuels Lubr.* 11, no. 4 (2018): 495-505, doi:https://doi.org/10.4271/2018-01-0909.

21.102. Wu, Y., Ferns, J., Li, H., and Andrews, G., "Investigation of Combustion and Emission Performance of Hydrogenated Vegetable Oil (HVO) Diesel," SAE Technical Paper 2017-01-2400 (2017), doi:https://doi.org/10.4271/2017-01-2400.

21.103. Erkkilä, K., Nylund, N., Hulkkonen, T., Tilli, A. et al., "Emission Performance of Paraffinic HVO Diesel Fuel in Heavy Duty Vehicles," SAE Technical Paper 2011-01-1966 (2011), doi:https://doi.org/10.4271/2011-01-1966.

21.104. Nakakita, K., Ban, H., Takasu, S., Hotta, Y. et al., "Effect of Hydrocarbon Molecular Structure in Diesel Fuel on In-Cylinder Soot Formation and Exhaust Emissions," SAE Technical Paper 2003-01-1914 (2003), doi:https://doi.org/10.4271/2003-01-1914.

21.105. Dimitriadis, A., Natsios, I., Dimaratos, A., Katsaounis, D. et al., "Evaluation of a Hydrotreated Vegetable Oil (HVO) and Effects on Emissions of a Passenger Car Diesel Engine," *Frontiers in Mechanical Engineering* 4 (2018): 7.

21.106. Omari, A., Pischinger, S., Bhardwaj, O., Holderbaum, B. et al., "Improving Engine Efficiency and Emission Reduction Potential of HVO by Fuel-Specific Engine Calibration in Modern Passenger Car Diesel Applications," *SAE Int. J. Fuels Lubr.* 10, no. 3 (2017): 756-767, doi:https://doi.org/10.4271/2017-01-2295.

21.107. Hills, F.J. and Schleyerbach, C.G., "Diesel Fuel Properties and Engine Performance," SAE Technical Paper 770316 (1977), doi:https://doi.org/10.4271/770316.

21.108. Barry, E.G., McCabe, L.J., Gerke, D.H., and Perez, J.M., "Heavy-Duty Diesel Engine/Fuels Combustion Performance and Emissions—A Cooperative Research Program," SAE Technical Paper 852078 (1985), doi:https://doi.org/10.4271/852078.

21.109. Hasegawa, M., Sakurai, Y., Kobayashi, Y., Oyama, N. et al., "Effects of Fuel Properties (Content of FAME or GTL) on Diesel Emissions under Various Driving Modes," SAE Technical Paper 2007-01-4041 (2007), doi:https://doi.org/10.4271/2007-01-4041.

21.110. Bertoli, C., Del Giacomo, N., Iorio, B., and Prati, M.V., "The Influence of Fuel Composition on Particulate Emissions of DI Diesel Engines," SAE Technical Paper 932733 (1993), doi:https://doi.org/10.4271/932733.

21.111. Tanaka, S., Takizawa, H., Shimizu, T., and Sanse, K., "Effect of Fuel Compositions on PAH in Particulate Matter from DI Diesel Engine," SAE Technical Paper 982648 (1998), doi:https://doi.org/10.4271/982648.

21.112. Tosaka, S., Fujiwara, Y., and Murayama, T., "The Effect of Fuel Properties on Particulate Formation (The Effect of Molecular Structure and Carbon Number)," SAE Technical Paper 891881 (1989), doi:https://doi.org/10.4271/891881.

21.113. Furutani, H., Tsuge, S., and Goto, S., "The Dependence of Carbon/Hydrogen Ratio on Soot Particle Size," SAE Technical Paper 920689 (1992), doi:https://doi.org/10.4271/920689.

21.114. Maricq, M.M., Chase, R.E., Podsiadlik, D.H., Siegl, W.O. et al., "The Effect of Dimethoxy Methane Additive on Diesel Vehicle Particulate Emissions," SAE Technical Paper 982572 (1998), doi:https://doi.org/10.4271/982572.

21.115. Bennethum, J.E. and Winsor, R.E., "Toward Improved Diesel Fuel," SAE Technical Paper 912325 (1991), doi:https://doi.org/10.4271/912325.

21.116. Beatrice, C.B., Bertoli, C.B., Del Giacomo, N., and Migliaccio, M., "Potentiality of Oxygenated Synthetic Fuel and Reformulated Fuel on Emissions from a Modern DI Diesel Engine," SAE Technical Paper 1999-01-3595 (1999), doi:https://doi.org/10.4271/1999-01-3595.

21.117. Murayama, T., Zheng, M., Chikahisa, T., Oh, Y. et al., "Simultaneous Reductions of Smoke and NOx from a DI Diesel Engine with EGR and Dimethyl Carbonate," SAE Technical Paper 952518 (1995), doi:https://doi.org/10.4271/952518.

21.118. Kitagawa, H., Murayama, T., Tosaka, S., and Fujiwara, Y., "The Effect of Oxygenated Fuel Additive on the Reduction of Diesel Exhaust Particulates," SAE Technical Paper 2001-01-2020 (2001), doi:https://doi.org/10.4271/2001-01-2020.

21.119. Kozak, M., Merkisz, J., Bielaczyc, P., and Szczotka, A., "The Influence of Oxygenated Diesel Fuels on a Diesel Vehicle PM/NOx Emission Trade-Off," SAE Technical Paper 2009-01-2696 (2009), doi:https://doi.org/10.4271/2009-01-2696.

21.120. Kozak, M., Merkisz, J., Bielaczyc, P., and Szczotka, A., "The Influence of Synthetic Oxygenates on Euro IV Diesel Passenger Car Exhaust Emissions—Part 3," SAE Technical Paper 2008-01-2387 (2008), doi:https://doi.org/10.4271/2008-01-2387.

21.121. Bailey, B., Eberhardt, J., Goguen, S., and Erwin, J., "Diethyl Ether (DEE) as a Renewable Diesel Fuel," SAE Technical Paper 972978 (1997), doi:https://doi.org/10.4271/972978.

21.122. Kapilan, N., Mohanan, P., and Reddy, R., "Performance and Emission Studies of Diesel Engine Using Diethyl Ether as Oxygenated Fuel Additive," SAE Technical Paper 2008-01-2466 (2008), doi:https://doi.org/10.4271/2008-01-2466.

21.123. Nikanjam, M., "Development of the First CARB Certified California Alternative Diesel Fuel," SAE Technical Paper 930728 (1993), doi:https://doi.org/10.4271/930728.

21.124. Hallgren, B.E. and Heywood, J.B., "Effects of Oxygenated Fuels on DI Diesel Combustion and Emissions," SAE Technical Paper 2001-01-0648 (2001), doi:https://doi.org/10.4271/2001-01-0648.

21.125. Srinivasan, P. and Devaradjane, G., "Experimental Investigations on Performance and Emission Characteristics of Diesel Fuel Blended with 2-Ethoxy Ethyl Acetate and 2-Butoxy Ethanol," SAE Technical Paper 2008-01-1681 (2008), doi:https://doi.org/10.4271/2008-01-1681.

21.126. Stoner, M. and Litzinger, T., "Effects of Structure and Boiling Point of Oxygenated Blending Compounds in Reducing Diesel Emissions," SAE Technical Paper 1999-01-1475 (1999), doi:https://doi.org/10.4271/1999-01-1475.

21.127. Buchholz, B.A., Mueller, C.J., Upatnieks, A., Martin, G.C. et al., "Using Carbon-14 Isotope Tracing to Investigate Molecular Structure Effects of the Oxygenate Dibutyl Maleate on Soot Emissions from a DI Diesel Engine," SAE Technical Paper 2004-01-1849 (2004), doi:https://doi.org/10.4271/2004-01-1849.

21.128. Kozak, M., Merkisz, J., Bielaczyc, P., and Szczotka, A., "The Influence of Synthetic Oxygenates on Euro IV Diesel Passenger Car Exhaust Emissions—Part 2," SAE Technical Paper 2008-01-1813 (2008), doi:https://doi.org/10.4271/2008-01-1813.

21.129. Myburgh, I.S., "Performance and Durability Testing of a Diesel Engine Fuelled with a Propanol-Plus/Diesel Blend," SAE Technical Paper 861583 (1986), doi:https://doi.org/10.4271/861583.

21.130. Yeh, L.I., Rickeard, D.J., Duff, J.L.C., Bateman, J.R. et al., "Oxygenates: An Evaluation of their Effects on Diesel Emissions," SAE Technical Paper 2001-01-2019 (2001), doi:https://doi.org/10.4271/2001-01-2019.

21.131. Liotta, F.J. and Montalvo, D.M., "The Effect of Oxygenated Fuels on Emissions from a Modern Heavy-Duty Diesel Engine," SAE Technical Paper 932734 (1993), doi:https://doi.org/10.4271/932734.

21.132. Tsurutani, K., Takei, Y., Fujimoto, Y., Matsudaira, J. et al., "The Effects of Fuel Properties and Oxygenates on Diesel Exhaust Emissions," SAE Technical Paper 952349 (1995), doi:https://doi.org/10.4271/952349.

21.133. Miyamoto, N., Ogawa, H., Arima, T., and Miyakawa, K., "Improvement of Diesel Combustion and Emissions with Addition of Various Oxygenated Agents to Diesel Fuels," SAE Technical Paper 962115 (1996), doi:https://doi.org/10.4271/962115.

21.134. González, D.M.A., Piel, W., Asmus, T., Clark, W. et al., "Oxygenates Screening for Advanced Petroleum-Based Diesel Fuels: Part 2. The Effect of Oxygenate Blending Compounds on Exhaust Emissions," SAE Technical Paper 2001-01-3632 (2001), doi:https://doi.org/10.4271/2001-01-3632.

21.135. Curran, H.J., Fisher, E.M., Glaude, P.-A., Marinov, N.M. et al., "Detailed Chemical Kinetic Modeling of Diesel, Combustion with Oxygenated Fuels," SAE Technical Paper 2001-01-0653 (2001), doi:https://doi.org/10.4271/2001-01-0653.

21.136. Westbrook, C.K., Pitz, W.J., and Curran, H.J., "A Chemical Kinetic Modeling Study of the Effects of Oxygenated Hydrocarbons on Soot Emissions from Diesel Engines," *The Journal of Physical Chemistry* 110, no. 21 (2006): 6912-6922.

21.137. Litzinger, T., Stoner, M., Hess, H., and Boehman, A., "Effects of Oxygenated Blending Compounds on Emissions from a Turbo-Charged Direct Injection Diesel Engine," *International Journal of Engine Research* 1, no. 1 (2000): 57-70.

21.138. Nag, P., Song, K.-H., Litzinger, T.A., and Haworth, D.C., "A Chemical Kinetic Modelling Study of the Mechanism of Soot Reduction by Oxygenated Additives in Diesel Engines," *International Journal of Engine Research* 2, no. 3 (2001): 163-175.

21.139. Ni, T., Gupta, S., and Santoro, R.J., "Suppression of Soot Formation in Ethene Laminar Diffusion Flames by Chemical Additives," *Symposium (International) on Combustion* 25, no. 1 (1994): 585-592.

21.140. Puri, R., Santoro, R.J., and Smyth, K.C., "The Oxidation of Soot and Carbon Monoxide in Hydrocarbon Diffusion Flames," *Combustion and Flame* 97, no. 2 (1994): 125-144.

21.141. Cheng, A.S., Dibble, R.W., and Buchholz, B.A., "The Effect of Oxygenates on Diesel Engine Particulate Matter," SAE Technical Paper 2002-01-1705 (2002), doi:https://doi.org/10.4271/2002-01-1705.

21.142. Graboski, M.S., Ross, J.D., and McCormick, R.L., "Transient Emissions from No. 2 Diesel and Biodiesel Blends in a DDC Series 60 Engine," SAE Technical Paper 961166 (1996), doi:https://doi.org/10.4271/961166.

21.143. Clark, N.N., Atkinson, C.M., Thompson, G.J., and Nine, R.D., "Transient Emissions Comparisons of Alternative Compression Ignition Fuels," SAE Technical Paper 1999-01-1117 (1999), doi:https://doi.org/10.4271/1999-01-1117.

21.144. Sharp, C.A., Howell, S.A., and Jobe, J., "The Effect of Biodiesel Fuels on Transient Emissions from Modern Diesel Engines, Part I Regulated Emissions and Performance," SAE Technical Paper 2000-01-1967 (2000), doi:https://doi.org/10.4271/2000-01-1967.

21.145. Sze, C., Whinihan, J.K., Olson, B.A., Schenk, C.R. et al., "Impact of Test Cycle and Biodiesel Concentration on Emissions," SAE Technical Paper 2007-01-4040 (2007), doi:https://doi.org/10.4271/2007-01-4040.

21.146. Karavalakis, G., Alvanou, F., Stournas, S., and Bakeas, E., "Regulated and Unregulated Emissions of a Light Duty Vehicle Operated on Diesel/Palm-Based Methyl Ester Blends over NEDC and a Non-Legislated Driving Cycle," *Fuel* 88, no. 6 (2009): 1078-1085.

21.147. Hansen, K.F. and Jensen, M.G., "Chemical and Biological Characteristics of Exhaust Emissions from a DI Diesel Engine Fuelled with Rapeseed Oil Methyl Ester (RME)," SAE Technical Paper 971689 (1997), doi:https://doi.org/10.4271/971689.

21.148. Bouché, T., Hinz, M., Pittermann, R., and Herrmann, M., "Optimising Tractor CI Engines for Biodiesel Operation," SAE Technical Paper 2000-01-1969 (2000), doi:https://doi.org/10.4271/2000-01-1969.

21.149. Lance, D.L., Goodfellow, C.L., Williams, J., Bunting, W. et al., "The Impact of Diesel and Biodiesel Fuel Composition on a Euro V HSDI Engine with Advanced DPNR Emissions Control," *SAE Int. J. Fuels Lubr.* 2, no. 1 (2009): 885-894, doi:https://doi.org/10.4271/2009-01-1903.

21.150. Mueller, C.J., Boehman, A.L., and Martin, G.C., "An Experimental Investigation of the Origin of Increased NOx Emissions When Fueling a Heavy-Duty Compression-Ignition Engine with Soy Biodiesel," *SAE Int. J. Fuels Lubr.* 2, no. 1 (2009): 789-816, doi:https://doi.org/10.4271/2009-01-1792.

21.151. Cheng, A.S., Upatnieks, A., and Mueller, C.J., "Investigation of the Impact of Biodiesel Fuelling on NOx Emissions Using an Optical Direct Injection Diesel Engine," *International Journal of Engine Research* 7, no. 4 (2006): 297-318.

21.152. Czerwinski, J., Eggenschwiler, P.D., Heeb, N., Astorga-Ilorens, C. et al., "Diesel Emissions with DPF & SCR and Toxic Potentials with BioDiesel (RME) Blend Fuels," SAE Technical Paper 2013-01-0523 (2013), doi:https://doi.org/10.4271/2013-01-0523.

21.153. Genzale, C.L., Pickett, L.M., and Kook, S., "Liquid Penetration of Diesel and Biodiesel Sprays at Late-Cycle Post-Injection Conditions," *SAE Int. J. Engines* 3, no. 1 (2010): 479-495, doi:https://doi.org/10.4271/2010-01-0610.

21.154. He, X., Williams, A., Christensen, E., Burton, J. et al., "Biodiesel Impact on Engine Lubricant Dilution during Active Regeneration of Aftertreatment Systems," *SAE Int. J. Fuels Lubr.* 4, no. 2 (2011): 158-178, doi:https://doi.org/10.4271/2011-01-2396.

21.155. Kattwinkel, P., Reith, C., and Petersson, M., "Different Properties of Biodiesel in Comparison with Standard Diesel Fuel and Their Impact on EURO VI Exhaust Aftertreatment Systems," SAE Technical Paper 2012-01-1733 (2012), doi:https://doi.org/10.4271/2012-01-1733.

21.156. Williams, A., McCormick, R., Luecke, J., Brezny, R. et al., "Impact of Biodiesel Impurities on the Performance and Durability of DOC, DPF and SCR Technologies," *SAE Int. J. Fuels Lubr.* 4, no. 1 (2011): 110-124, doi:https://doi.org/10.4271/2011-01-1136.

21.157. Vertin, K., He, S., and Heibel, A., "Impacts of B20 Biodiesel on Cordierite Diesel Particulate Filter Performance," SAE Technical Paper 2009-01-2736 (2009), doi:https://doi.org/10.4271/2009-01-2736.

21.158. ASTM International, "Standard Specification for Biodiesel Fuel Blendstock (B100) for Middle Distillate Fuels," ASTM D6751-23a, ASTM International, 2023.

21.159. Lakkireddy, V., McCormick, R.L., Weber, P., and Howell, S., "Diesel Particulate Filter Durability Performance Comparison Using Metals Doped B20 vs. Conventional Diesel Part I: Accelerated Ash Loading and DPF Performance Evaluation," SAE Technical Paper 2023-01-0297 (2023), doi:https://doi.org/10.4271/2023-01-0297.

21.160. Lakkireddy, V., McCormick, R.L., Weber, P., and Howell, S., "Diesel Particulate Filter Durability Performance Comparison Using Metals Doped B20 vs. Conventional Diesel Part II: Chemical and Microscopic Characterization of Aged DPFs," SAE Technical Paper 2023-01-0296 (2023), doi:https://doi.org/10.4271/2023-01-0296.

21.161. Mizushima, N., Murata, Y., Suzuki, H., Ishii, H. et al., "Effect of Biodiesel on NOx Reduction Performance of Urea-SCR System," *SAE Int. J. Fuels Lubr.* 3, no. 2 (2010): 1012-1020, doi:https://doi.org/10.4271/2010-01-2278.

21.162. Kawano, D., Mizushima, N., Ishii, H., Goto, Y. et al., "Exhaust Emission Characteristics of Commercial Vehicles Fuelled with Biodiesel," SAE Technical Paper 2010-01-2276 (2010), doi:https://doi.org/10.4271/2010-01-2276.

Further Reading

CONCAWE, "Comparison of Particle Emissions from Advanced Vehicles Using DG TREN and PMP Measurement Protocols," Report No. 2/09, CONCAWE, 2009.

Kasab, J. and Strzelec, A., *Automotive Emissions Regulations and Exhaust Aftertreatment Systems* (Warrendale, PA: SAE International, 2020).

Khair, M.K. and Majewsky, W.A., "Diesel Emissions and Their Control," SAE International, 2006.

Konrad, R., *Diesel Engine Management–Systems and Components* (Braunschweig: Springer Vieweg, 2014).

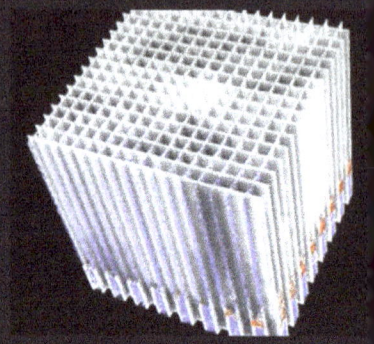

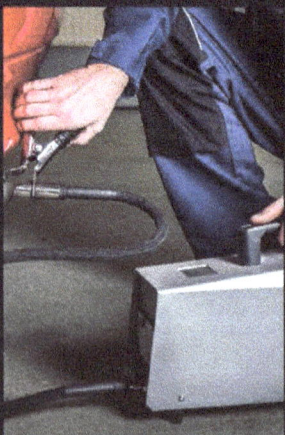

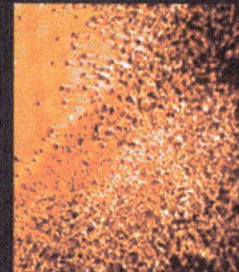

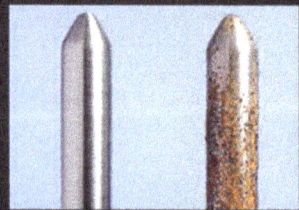

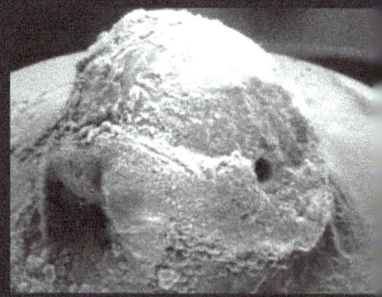

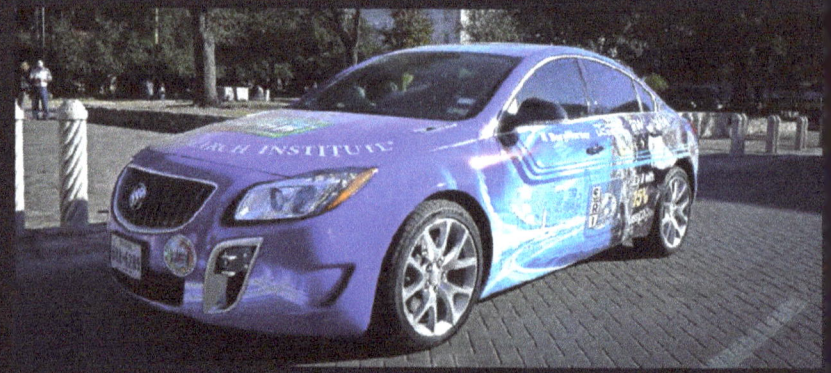

22

The Kinetically Controlled Compression Ignition Engine and Combustion Process

Chapter 7 discusses the combustion process, and Chapter 8 discusses design and fuel influences, on the positive ignition engine often referred to as the spark ignition (SI) engine. The charge is positively ignited at one or more discrete points by a spark plug. This initiates a flame front that propagates through the fuel and air mixture. The flame front is a thin zone of intense chemical reaction that is essentially uncontrolled and self-propagating. There are some significant disadvantages to the process from the point of view of the combustion chemistry and the engine design to utilize it.

From the point of view of the chemistry, the first limiting factor is that the mixture around the point of ignition must be within a fairly narrow range of air-to-fuel mixture ratios to guarantee reliable ignition. The mixture that the flame will propagate through must also be within a fairly narrow band of air-to-fuel ratios. Emissions control technology may significantly restrict this even further. This brings about a restriction on engine design; controlling the power output of the engine entails controlling the amount of air going through the engine as well as the amount of fuel. Without the mechanical complexities of variable displacement engine, this imposes efficiency losses. Charge stratification reduces this problem and is now becoming more common. This is discussed in Chapter 8, Section 8.3.1.4.1.

Another issue relating to the chemistry is that as the flame propagates, the temperature at the flame front is high enough to generate large quantities of oxides of nitrogen (NO_X), which is an undesirable and difficult to reduce pollutant. The generation of NO_X can be limited by retarding the combustion process, waiting until the piston is in the expansion stroke when the gas temperatures are slightly lower. This will reduce the engine efficiency. Another approach is to introduce an inert diluent in with the air and fuel. This is usually achieved by exhaust gas recirculation (EGR), but this requires careful control as charge dilution must be kept within certain limits.

If the whole of the combustion chamber was filled with a well-mixed, ostensibly homogeneous, mixture of air and fuel, then relying on the propagation of the flame front to consume all of the mixture imposes another serious mechanical constraint. The hot, burned gas behind the flame front expands rapidly. It is this expansion that produces the power out of the engine, but this expansion increases the pressure on the unburned gas in front of the flame. If this pressure rises too high, then the mixture in front of the flame can spontaneously ignite, leading to detonation. This is discussed more fully in Chapter 7, Section 7.2. Ensuring this situation does not arise requires two things: The fuel must be highly resistant to auto-ignition (this may be difficult or very expensive to produce), or the in-cylinder pressures and temperatures must be kept low at the start of the combustion process. This usually means designing the engine with a low compression ratio, which will reduce the engine's efficiency. Charge stratification can make this less of a problem. The flame front passing through a homogeneous charge therefore imposes mechanical inefficiencies and produces NO_X emissions.

Chapter 15 discusses the combustion process, and Chapter 16 discusses the design and fuel influences on the diesel engine. This type of engine is sometimes referred to as a compression ignition (CI) engine or a mixing (or diffusion) controlled CI engine. In these engines the combustion chamber is filled with gas not containing fuel. The amount of charge does not have to be regulated to match the fuel usage. This eliminates one of the drawbacks of the SI engine. Fuel is injected directly into the hot compressed charge toward the end of the compression stroke; this results in the auto-ignition of the fuel. The rate of combustion is controlled by the rate at which the fuel can mix with air; hence it is referred to as mixing controlled or diffusion-controlled combustion. The engine designer does have some control over this rate by controlling the rate of fuel addition. However, from a thermodynamic point of view, reducing the rate of combustion will reduce the engine efficiency.

In diffusion-controlled combustion processes, the combustion takes place in the mixing zone where the air-to-fuel ratio is within a narrow window close to the stoichiometric mixture. This results in very high temperatures analogous to the flame front in the SI engine. The difference between the flame front of the SI engine and the reaction zone in the diesel engine is that the mixture is far from homogeneous. The high NO_X formation occurs on the fuel-lean side of stoichiometry, and soot is formed on the fuel-rich side of stoichiometry. This is discussed further in Chapter 15, Section 15.3. For this type of combustion, the fuel requirement is for a fuel that will very easily auto-ignite. The diffusion combustion reduces the mechanical inefficiencies but produces soot emissions in addition to the NO_X emissions.

Both combustion processes suffer from the fact that the combustion process takes place in a restricted region of the combustion chamber; the temperature at the reaction zone is high, leading to NO_X production, although the bulk temperature is relatively low.

An alternate combustion process is therefore being investigated in which a well-mixed fuel and air mixture auto-ignites throughout the mixture. The combustion process is then controlled by the kinetics of the auto-ignition process. This avoids the locally high temperature that results in NO_X formation and also eliminates the fuel-rich zones that result in soot formation. This overcomes a significant drawback of diffusion-controlled combustion. Because this process relies on auto-ignition, it eliminates the need for SI and the resultant flame front. This eliminates NO_X production and concerns over knock, and the low compression ratio that imposes. It also allows a wide range of mixture ratios, thus reducing the other inefficiency imposed by the SI combustion process.

The following sections will discuss the background to the process, its own limitations, how these are currently being investigated, and what the fuel requirements would be for engines employing this type of combustion process.

22.1. Brief History of Kinetically Controlled Combustion

The implementation of the concept of kinetically controlled combustion is usually attributed to Shigeru Ohnishi. Ohnishi had set out in the late 1960s to solve the problem of simultaneously reducing the emissions of unburned hydrocarbons (HC), carbon monoxide (CO), and NO_X from small two-stroke gasoline engines. The two-stroke engine is discussed briefly in Chapter 8, Section 8.2.1, but has generally disappeared from automotive applications due to high emissions. Ohnishi's original solution to the problem was a stratified charge system that went by the name of the Yui and Ohnishi Combustion Process (YOCP) [22.1]. Because, in a two-stroke engine the exhaust stroke and the inlet stroke of the four-stroke cycle are condensed into the later fraction of the power-stroke and the initial fraction of the compression stroke, good scavenging of the burned gas is always a concern. This poor scavenging of residual exhaust gas could cause the equivalent of high levels of EGR. This was used to advantage in the YOCP to produce charge stratification that allowed the combustion of overall fuel-lean mixtures but required direct fuel injection. This was further refined to produce the Nippon Clean Engine (NiCE) [22.2].

Further research into the part load instabilities of lean mixtures showed that high levels of radicals were present during the compression stroke and that these could auto-ignite. This auto-ignition was a problem for conventional gasoline two-stroke engines because it generated roughness and noise. However, it was realized that if this auto-ignition could be controlled then the instability would be eliminated and smoother operation would ensue. This led to the development in the mid-1970s of the Active Thermo-Atmosphere Combustion (ATAC) process [22.3] where the controlled auto-ignition (CAI) of gasoline was utilized. This concept was also investigated, within the same conglomerate, under the name of TS combustion [22.4]; TS stood for Toyota-Soken after Toyota Motor Co. Ltd. and Nippon Soken Inc. This work was again performed on a two-stroke engine, in this case an opposed piston design, using standard gasoline as the fuel.

Work in the US to investigate the response of the process to changes in operating parameters, and the phenomena involved, utilized a four-stroke engine; this introduced the acronym CIHC, standing for compression ignition homogeneous charge combustion [22.5]. To achieve the high levels of residuals in a four-stroke engine, an external EGR loop was employed and auto-ignition was helped by the use of low octane number fuels of clearly defined chemical composition. Later work used full boiling range fuel in a four-stroke engine and introduced the term homogeneous charge compression ignition (HCCI) to describe the process [22.6]. This work showed that by utilizing HCCI at part load conditions, the inefficiencies of intake throttling could be reduced and diesel-like fuel economy could be achieved for a gasoline engine. It also demonstrated that HCCI was difficult to achieve at very low loads, including idle, because charge temperatures were too low. HCCI combustion was also problematic at high loads because the necessary high EGR rates reduced the amount of available fresh air, and hence fuel, that was available to produce power. Subsequent work in Japan introduced the term premixed charge compression ignition (PCCI) and the idea of heating the incoming air by either electrical heating or supercharging. This extended the range of operation in the low-load region [22.7].

The use of PCCI with diesel fuel demonstrated that both low NO_X and smoke emissions could be achieved. The diesel fuel could be injected very early in the compression stroke to produce a PREmixed lean DIesel Combustion (PREDIC) engine [22.8]. Other implementations of this concept included the Homogeneous charge intelligent Multiple Injection Combustion System (HiMICS) [22.9] and MULtiple stage DIesel Combustion (MULDIC) [22.10]. Both of these systems used multiple injection strategies to produce some premixed combustion as well as late injection in order to minimize NO_X and soot formation, but they did suffer from increased HC and CO emissions. The injection of diesel fuel into the intake ports, as in a port injected gasoline engine, was also investigated [22.11].

The challenges for HCCI combustion using diesel fuel are different from when using gasoline. Gasoline has been developed to have a high resistance to auto-ignition and measures must be taken to overcome this, such as raising the temperature of the trapped charge. When using diesel fuel that has been developed to readily auto-ignite the challenge is to control the rate of heat release, to prevent high rates of pressure rise, as the mixture becomes more fuel-rich and approaches stoichiometry [22.12].

22.2. **The Low-Temperature Combustion Process**

In kinetically controlled CI combustion, the fuel is mixed with the air and the resulting mixture is compressed. The compression process increases the temperature and pressure, causing the mixture to ignite. This is true CI: in a diesel engine the air is already at high temperature and pressure when the fuel is injected; it is the mixing of the hot air with the fuel that causes the ignition. To reduce the rate of reaction, the mixture is diluted with excess air or another diluent, usually EGR. For a naturally aspirated engine, running on iso-octane fuel, an air-to-fuel ratio greater than three times stoichiometric has been used [22.13]. With pressure charging, this mixture can be made almost twice as dilute [22.13]. If EGR is used as a diluent, then the actual air-to-fuel ration can be equal to stoichiometric. These mixtures are too dilute to support a flame; a spark-ignited engine would not operate with such dilute mixtures. However, if the temperature is high enough, oxidation reactions will take place, but the rate of heat release will be low enough to avoid increasing the in-cylinder temperature to the level that will promote NO_X formation. Because the mixture is so fuel-lean, it will not create soot. Figure 22.1 is a map of fuel-to-air equivalence ratio against combustion temperature showing the regions where soot formation and NO_X formation occurs.

FIGURE 22.1 Map of fuel-to-air equivalence ratio against temperature.

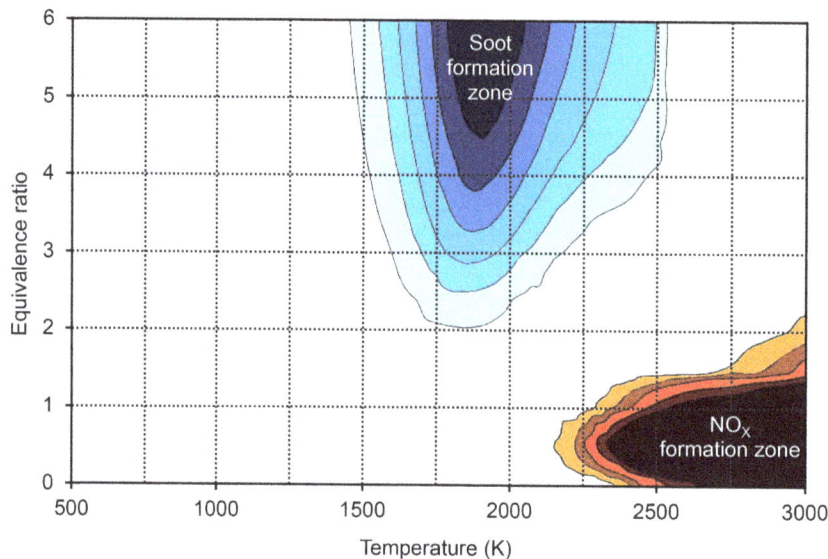

© SAE International.

Inspection of Figure 22.1 shows that for traditional spark-ignited gasoline combustion where the mixture is approximately stoichiometric, an equivalence ratio of one, soot will not be formed, but at high temperatures NO_X will be formed. In a diesel engine, because neat fuel is injected, the equivalence ratio will start of infinitely high but combustion will normally occur in the mixing zone when the equivalence ratio has been reduced to between 4 and 2 [22.14]. This will pass down through the soot formation zone. The aim of a kinetically controlled CI process is clearly to operate in the low-temperature combustion (LTC) region toward the low-left corner of the map.

To achieve true HCCI combustion, the fuel must be completely mixed with the air and EGR, if present, before the combustion process starts. If fuel is injected directly into the combustion chamber, then this will obviously not be the initial case. However, if the injection takes place early enough in the compression stroke and the temperature is not too high, then LTC will still take place and remain outside the soot formation zone (i.e., it will be to the left in Figure 22.1).

The LTC is thought to be a result of the auto-oxidation of paraffins, predominantly the n-paraffins, and branched paraffins to a lesser extent. The reactions taking place during the LTC process can be simplified as follows.

An initiation step abstracts a hydrogen atom from a paraffin molecule; this is followed by the addition of an oxygen molecule to the alkyl radical to form an alkylperoxy radical. The alkylperoxy radical isomerizes to form an alkylhydroperoxy radical. The alkylhydroperoxy radical can then undergo decomposition to form a carbonyl, an olefin, and a hydroxyl radical. Alternatively, the alkylhydroperoxy radical can undergo further oxidation, followed by hydrogen abstraction and isomerization to yield another hydroxyl radical. This is followed by further hydrogen abstraction to form a ketone group and another hydroxyl radical [22.15]. This is shown below:

$$RH + O_2 \rightarrow R\bullet + HO_2$$

$$R\bullet + O_2 \rightarrow ROO\bullet$$

$$ROO\bullet \rightarrow \bullet ROOH$$

$$\bullet ROOH \rightarrow Carbonyl + Olefin + OH\bullet$$

$$\bullet ROOH + O_2 \rightarrow \bullet OOROOH$$

$$\bullet OOROOH \rightarrow HOOR{=}O + OH\bullet$$

$$HOOR{=}O \rightarrow \bullet OR{=}O + OH\bullet$$

The hydroxyl radicals formed in these reactions continue to participate in other reactions, such as abstracting a hydrogen atom to form water and releasing energy. As the temperature rises due to heat released by LTC and the continued compression, then dissociation of alkyl radicals is favored over oxidation; this will limit further LTC. When this occurs, the rate of reaction decreases with a rise in temperature and is consequently known as the negative temperature coefficient (NTC) regime. The preceding equations are a simplification of the reactions taking place, and more detail sets of equations have been developed to model the auto-ignition processes of different paraffins [22.16–22.20], initially due to their relevance to knock in gasoline engines. Note that the simplest carbonyl is formaldehyde which can be used as an indication of the start of the oxidation/combustion process.

As the in-cylinder temperatures continue to rise, as a combination of LTC and compression, the combustion will transition to high-temperature combustion.

22.3. LTC Engines

To design an engine to operate in the LTC mode throughout the required operating range will present a challenge. In relation to a conventional gasoline SI engine, a kinetically controlled combustion engine may be able to overcome the part load inefficiencies of the conventional engine, but it would struggle to meet the maximum power performance. However, the emissions from a conventional gasoline engine operating with a stoichiometric mixture can be addressed by a three-way catalyst. Operating on gasoline provides a challenge in generating sufficient heat input to promote ignition at very low-load conditions. Gasoline engines using kinetically controlled combustion are often referred to as CAI engines or gasoline CI (GCI) engines.

In relation to a conventional diesel engine, the kinetically controlled combustion concept would struggle to match the maximum power range with such a diluted charge. Modern diesel engines already utilize pressure charging to allow them to operate with overall fuel-lean mixtures. Due to the low volatility of diesel fuel, the complete vaporization and mixing of the fuel presents a challenge, and diesel engines using kinetically controlled combustion are often referred to as PCCI engines.

Current research is investigating a combination of the characteristics of gasoline and diesel fuel to control the reactivity of the fuel to produce a reactivity-controlled compression ignition (RCCI) engine. There is also a lot of research being performed on the concept of on-board reforming of part of the fuel to produce a second fuel with significantly different reactivity.

22.3.1. LTC Engines Using Gasoline-Type Fuels

The implementation of CAI in a gasoline engine is currently restricted to the mid-load operating range. At idle and light loads, the engine is liable to misfire due to low charge temperatures and excessive fuel-lean mixtures. At the high loads, the auto-ignition of large amounts of fuel in a near stoichiometric mixture generates high rates of heat release that can result in very high rates of pressure rise; this produces high noise levels and possibly engine damage.

Various approaches can be used to enable the auto-ignition of gasoline-type fuels within this mid-load operating range and to extend this range as far as possible. The use of hot combustion gases was of course how the whole idea of CAI started [22.3]. This was an inherent feature of two-stroke engines at light loads but is not necessarily true of conventional four-stroke engines. A relatively simple approach to ensuring high levels of combustion products within the trapped charge is to use an external EGR loop [22.6, 22.21–22.23]. The alternative approach, which more closely replicates the situation in two-stroke engine, is to use different valve timings within the four-stroke engine. This can take different forms that can be classified as trapping or rebreathing of the exhaust gas [22.24]. Trapping of the exhaust gas is accomplished by changing the valve timing to produce negative valve overlap (NVO). In a conventional four-stroke engine the exhaust valve would close just after the top of the exhaust stroke and just after the inlet valve opens, producing valve overlap (i.e., both valves being open at the same time). To produce NVO, the exhaust valve is closed before the top of the exhaust stroke and before the inlet valve opens. Some of the exhaust gas is thus trapped and remains in the cylinder as the fresh charge is drawn in. To accomplish rebreathing, some of the already exhausted gas is drawn directly back into the cylinder. This can be achieved by delaying the closure of the exhaust valve until well into the induction stroke [22.25–22.27], sometimes referred to as positive valve overlap (PVO), or by opening the exhaust valve for a second time during the induction stroke [22.28]. Both of these strategies allow some of the exhaust gas to be drawn back in from the exhaust manifold during the induction stroke. An example of valve timings for NVO and PVO are shown in comparison to conventional SI engine timings in Figure 22.2.

FIGURE 22.2 Valve timings for conventional, NVO, and PVO configurations.

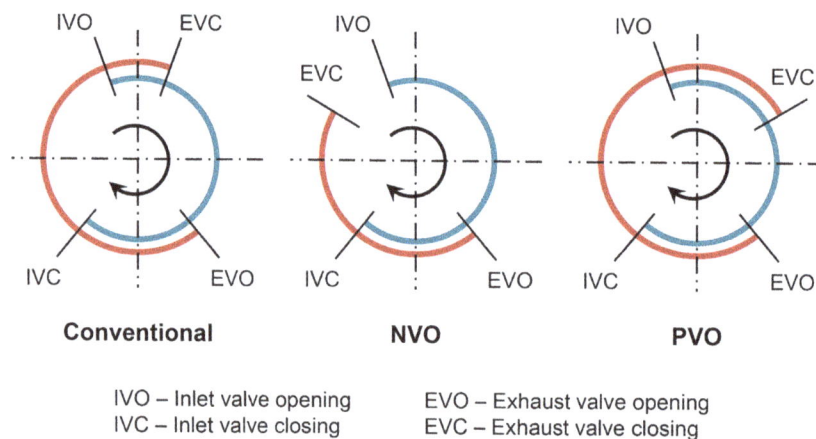

Conventional   NVO   PVO

IVO – Inlet valve opening   EVO – Exhaust valve opening
IVC – Inlet valve closing   EVC – Exhaust valve closing

© SAE International.

The use of these burned gases was initially to aid and control the auto-ignition of a fuel that is designed not to ignite. The exhaust gas raises the temperature of the charge during the compression stroke, but the burned gas also contains reactive species that will promote auto-ignition [22.29]. However, the burned gas also has a controlling effect on the rate of reactions once auto-ignition has occurred, in the same way that EGR can be used to suppress NO_X formation. This is a beneficial factor in extending the high-load operating range for a CAI engine in that it limits the maximum rate of pressure rise. The burned gas reduces the oxygen concentration and has a higher specific-heat capacity, which slows the oxidation process.

Heating of the inlet air has also been employed to promote CAI [22.30, 22.31], but due to the lack of reactive species this is not as effective as using exhaust gas. For practical applications of this strategy, the parasitic losses would be high unless waste heat could be utilized [22.32]. The temperature of the incoming charge can also be controlled by changing the engine coolant temperature [22.33]. However, the time response of this approach does not lend itself to practical application where transitions in operating conditions are much faster.

A combination of GCI within the attainable operating window and conventional stoichiometric SI combustion at low loads and at high loads would appear to be the solution if the transition between the two can be achieved. Early investigations with variable valve timings to achieve HCCI with SI [22.34] did indicate that in the transitional phase the spark did trigger auto ignition. The spark produces a propagating flame that initially consumes some of the charge, releasing some of the fuel energy. The heat released by this flame increases the temperature and pressure of the remaining charge, promoting auto-ignition [22.35, 22.36]. At higher loads there is a conventional rate of pressure rise, due to the propagation of the spark-ignited flame front, followed by a much more rapid rate of pressure rise due to the homogeneous combustion of the remainder of the charge [22.37]. This is demonstrated by the cylinder pressure diagram and in-cylinder images shown in Figure 22.3 [22.38]. After the SI there is only a low and steadily increasing rate of heat release before the auto-ignition occurs leading to the very rapid increase in the rate of heat release and the corresponding rapid increase in cylinder pressure.

FIGURE 22.3   Spark assisted gasoline ignition process.

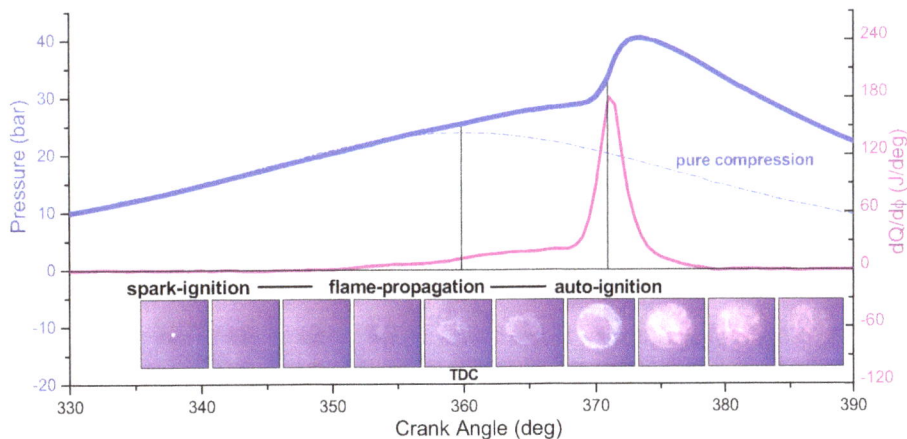

22.3.2. LTC Engines Using Diesel Type Fuels

Taking a conventional diesel engine as the starting point for achieving PCCI, an obvious route is to inject the fuel during the induction stroke or early in the compression stroke. This was the route chosen for the PREDIC system mentioned previously [22.8]. However, with no other changes from conventional diesel engine operation, then at very light loads with a homogeneous mixture the mixture would effectively be too lean, leading to a significant increase in HC and CO emissions and possibly even misfire [22.39–22.41]. Some mixture inhomogeneity was necessary to produce multiple ignition sources. Pockets of fuel-lean mixture are likely to remain, resulting in increased emissions of HC and CO. Another potential problem with early injection is the impingement or condensation of liquid fuel on the cylinder walls [22.42–22.44]; this can result in lubricating oil dilution. Changes in fuel injector geometry could reduce wall-wetting but increase piston bowl impingement. This could result in pools of fuel in the piston bowl and potential pool fires, leading to increased NO_x and soot missions [22.45].

Some of the limitations of these single early injection strategies can be alleviated by the use of multiple injection strategies. Some early examples were the HiMICS [22.9] and MULDIC [22.10] systems. These systems injected part of the required fuel very early in the compression stroke, which resulted is some power being produced from the premixed charge via LTC. The remaining part of the required fuel was injected just after the top of the compression stroke. This fuel is injected into hotter gases than would be the case without the LTC, and these gases are also partially depleted of oxygen. In the HiMICS, this fuel was actually injected as two distinct injections, one just after the top of the compression stroke and a smaller amount about 30 degrees of crank rotation into the expansion stroke. These strategies relied on the kinetically controlled combustion for only part of the energy release, and while they achieved some NO_X and PM reduction, this was usually at the expense of increased HC and CO emissions. Timing of the early injection was critical to avoid too much energy being released too early in the cycle, leading to diesel knock.

A better understanding of the processes taking place with these split injection strategies has been developed using advanced imaging techniques [22.46]. Work has been performed by injecting n-dodecane ($C_{12}H_{26}$), as diesel fuel surrogate, into a high-pressure, high-temperature preburn combustion chamber using a pilot-main injection strategy. With a fairly long time between the two injections, the combustion products of the pilot injection have sufficiently mixed with the in-cylinder gases, so that the ignition delay before the main injection is virtually the same as for a single injection. If time between the two injections is reduced then the combustion products of the pilot injection may not have mixed so well with remaining in-cylinder gases, and a local region of hot gases can lead to a much shorter ignition delay for the main injection and subsequent combustion. This makes the time between the two injections (dwell) a critical parameter [22.47]. Mapping the formaldehyde distribution using planar laser-induced fluorescence (PLIF) imaging provides an insight to this situation. The evolution of the ignition process, determined from formaldehyde PLIF, for the pilot-main injection strategy is illustrated in Figure 22.4 for bulk gas temperatures of 750, 800, and 900 K [22.47].

FIGURE 22.4 Evolution of ignition process by formaldehyde PLIF at 700 K (left), 800 K (center), and 900 K (right).

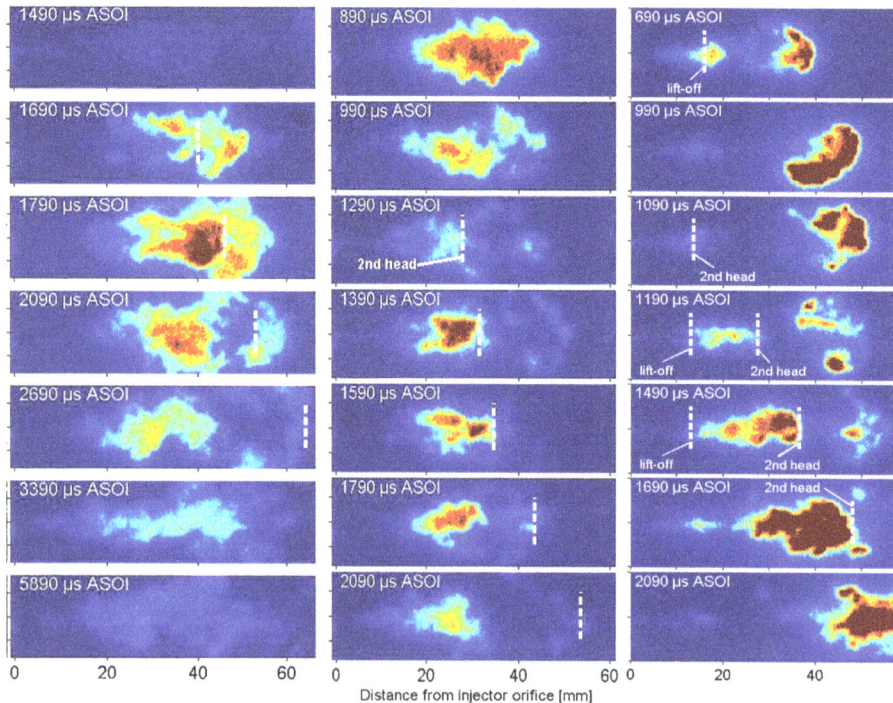

Due to a long ignition delay, at the 750 K bulk gas temperature, the first injection is still undergoing cool-flame combustion when the second injection starts to penetrate, allowing the second injection to interact with the cool-flame products of the first injection. The prolonged ignition delay also allows for a longer mixing time, leading to a more homogeneous fuel-lean mixtures with temperatures close to the 750 K ambient, resulting in a moderate, first-stage, heat release. The interaction of the second injection with the combustion products aids cool-flame ignition, this is then followed by the onset of high-temperature combustion. Because much of the upstream and downstream mixture remain fuel-lean, high-temperature combustion appears to be confined to isolated regions as indicated by the large regions of formaldehyde, signifying broad regions of incomplete combustion leading to low combustion efficiencies [22.48].

At 900 K bulk gas temperature the first injection, due to its shorter ignition delay, the formaldehyde that begins to form downstream of the spray is quickly consumed by the high-temperature combustion [22.49] whereby, the flame advances from the characteristic lift-off length toward the injector. Due to the rapid penetration of the second injection into the high-temperature reactive products, from the first injection, this second injection undergoes rapid transition into high-temperature combustion without much indication of the cool-flame reactions previously observed.

As an alternative to early/pilot injection, late fuel injection was also utilized in the modulated kinetics (MK) system [22.50–22.52], which was incorporated into a production diesel engine. At light loads, the engine operated with a high swirl ratio to enhance mixing and a high level of EGR. Under these conditions, the time required to achieve a high degree of mixing was shorter than the ignition delay period. Therefore, combustion occurs in an almost homogeneous mixture. This strategy could only be used at light loads, as depicted in Figure 22.5 [22.51].

FIGURE 22.5 Operating region for the MK system.

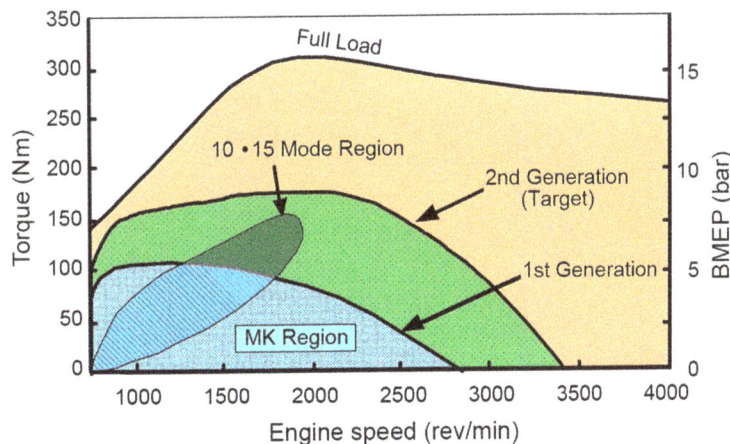

A commercial version of this multiple injection strategy was introduced in 2000 [22.53]. This system, known as Uniform Bulky Combustion System (UNIBUS), used the early injection and fuel mixing to allow the early, low-temperature, radical forming, and fractionation reactions to take place before the second injection [22.54]. The second injection was used as an ignition trigger for all the fuel. This differs from a dual fuel engine in that energy is released from the LTC prior to the second injection, which releases further energy. Precise control of the premixed burn was achieved by using a variable geometry turbocharger to control the boost pressure; this will alter the fuel-to-air ratio, as well as the compression pressure and temperature, thus affecting the LTC.

An extension of this multiple injection strategy that tries to overcome the problems of wall-wetting yet still maximizes the amount of premixed LTC can use up to six early injections [22.55]. The greater number of injections means that lower quantities of fuel have to be injected during each injection. This minimizes the penetration of each injection spray, thus reducing wall-wetting. The earliest injections will be into the cooler, lower pressure air and will have to be shorter. As the compression stroke advances, the pressure and temperature will increase, which increases the rate of evaporation of the injected fuel. Therefore, less-advanced injections can be of a greater quantity of fuel. The later injection pulses have less time to allow complete mixing before combustion takes place. The timing of these multiple injections and the quantity of fuel delivered during each injection is therefore critical to obtaining a smooth combustion event and preventing diffusion burning.

Due to the ready ignitability of diesel fuel, the quantity of fuel that can be injected early and combusted via LTC is limited in a conventional diesel engine. This therefore limits the maximum power that can be derived under the LTC regime. This limitation can be tackled by reducing the compression ratio of the engine by means such as changing the intake valve timing [22.56, 22.57]. However, this will inevitably reduce the thermodynamic efficiency. The challenge for diesel HCCI therefore remains the high-load operating points.

22.3.3. LTC Engines Using More Than One Fuel

As discussed at the start of this chapter, the automotive engine market has been dominated by the gasoline engine and the diesel engine, both of which have their drawbacks. Developments in LTC have therefore tended to concentrate on using this technique to reduce the drawbacks of these alternative engine types. The previous two sections have examined the use of LTC in a gasoline engine using gasoline as the fuel and in a diesel engine using diesel fuel. Each fuel has its own benefits and disadvantages. The benefits of gasoline are that its high volatility produces ready volatilization and mixing [22.58] and its high resistance to auto-ignition makes the onset of ignition easier to control. The disadvantage is that it is harder to achieve LTC at low loads. Diesel fuel has the benefit of easier auto-ignition, which makes LTC easier to achieve at low-load conditions. The disadvantage is that the heat release is harder to control at high loads. Diesel fuel also has the disadvantage of low volatility, which makes it more difficult to ensure good mixing and low emissions of HC and CO [22.59].

This situation has prompted much research into the ideal fuel for LTC operation of either a gasoline or diesel engine; this will be discussed further in the next section. It became clear from this work that the ideal fuel for LTC was one that had properties somewhere between those of gasoline and diesel fuel, and that the ideal varied according to operating condition [22.60, 22.61]. The use of two different fuels that could be blended within the engine had also been utilized as a means of testing LTC with different fuel properties [22.62–22.64]. These early investigations utilized volatile fuels that were injected into the inlet manifold, ensuring adequate mixing of the incoming charge. By inducting a mixture of air and a volatile gasoline-type fuel and the directly injecting diesel fuel, it is possible to achieve charge stratification to give control over the LTC process [22.65].

Stratification of the fuel is an effective means of controlling the rate of heat release to increase the higher load operating range of LTC [22.66–22.69].

The use of two distinct fuels thus allows the reactivity of the mixture of air and fuel to be adjusted to meet the engine requirement, hence RCCI. Current research is investigating the combination of gasoline injected into the inlet manifold and diesel fuel injected directly into the combustion chamber [22.69]. Mechanically, an RCCI engine can be achieved by combining a direct injection diesel engine with a port fuel injection system from a gasoline engine; such an arrangement is shown on Figure 22.6 [22.70].

FIGURE 22.6   Multicylinder RCCI engine.

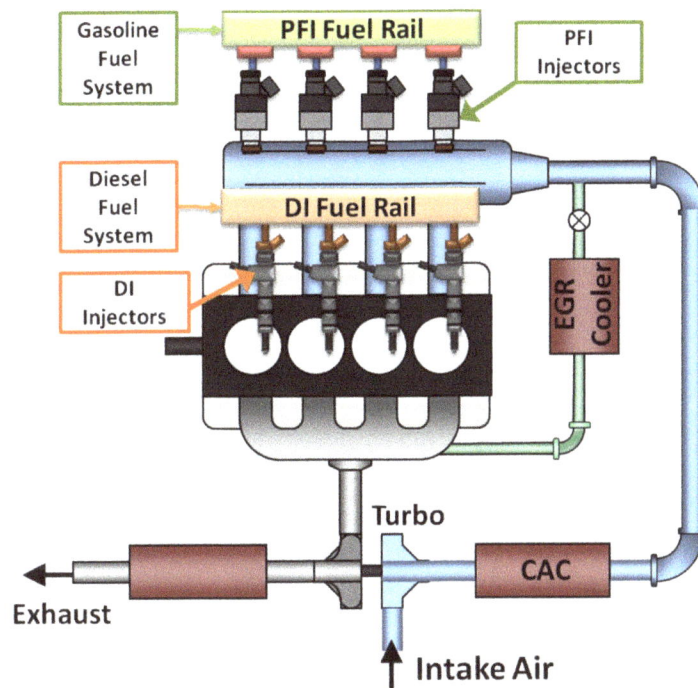

The idea of ingesting a less reactive fuel into the cylinder and using a more reactive fuel to cause ignition and hance combust the less reactive fuel is not a new idea, as Rudolf Diesel's patent of 1901 demonstrates [22.71] but it is only in the later half of that time frame that it has become widely used with natural gas, or compressed natural gas, as the common low reactivity fuel. Investigations into RCCI using dual fuel gas engines is thus becoming common [22.72, 22.73] and combinations of natural gas and hydrogen [22.74, 22.75]. As with PCCI, discussed in the previous section, in-cylinder observations are adding to the understanding of the processes. In an example presented in Figure 22.7, work was conducted in a single-cylinder optical engine [22.76] with natural luminosity images presented alongside images based on color image segmentation

[22.77] to show non-sooting premixed flame and luminous non-premixed flame. The left-most pair of columns show conventional diesel combustion which as expected shows that the natural luminosity is predominantly generated by the non-premixed flame. The central pair of columns show the equivalent sequence with 60% natural gas substitution ratio (NGSR), on a fuel energy flow rate basis, where there is noticeably less non-premixed flame and hence less soot production. The right-most pair of columns show the equivalent images with the start of injection (SOI) advanced from −10 to −30°ATDC. This again shows a reduction in soot production.

FIGURE 22.7 Evolution of Diesel/Natural Gas combustion for different fuel ratios and start of injection.

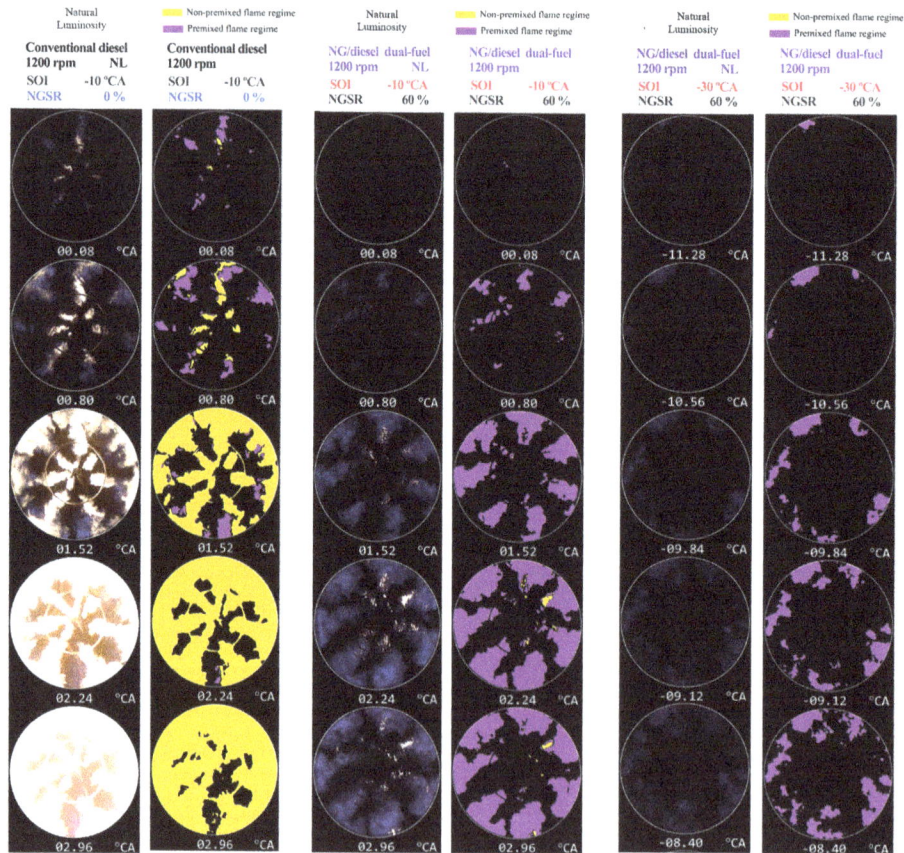

Adapted from "Combustion Phenomena and Emissions in a Dual-Fuel Optical Engine Fueled with Diesel and Natural Gas," SAE Int. J. Advances & Curr. Prac. in Mobility 4(2):502–513, 2022. © SAE International.

The use of RCCI combustion can result in high efficiency and controllability over a broad range of speed and load conditions, in both light-duty and heavy-duty applications [22.78–22.81]. Without careful control over the diesel fuel injection, there is still a possibility for increased emissions of HC and CO. It may thus be necessary to incorporate techniques such as the multiple injection strategies discussed in the previous section [22.70].

22.3.4. LTC Engines with a Fuel Reformer

The dual fuel approach discussed above does however present challenges including the requirement to have two fuel tanks which will undoubtedly have different fuel consumption rates leading to different refueling intervals, the possibility of misfuelling, etc. An alternative approach to the challenge of changing the reactivity of the fuel is to use a single fuel but to reform a portion of this fuel within the vehicle systems, thereby effectively creating a dual fuel system from a single fuel source. The concept of on-board hydrogen generations to provide a hydrogen rich fuel/air mixture had been demonstrated about half a century ago [22.82]. In those days the idea behind using a modified fuel was simply to reduce NO_X emissions, which could be achieved by operating the engine very lean, which required greatly increasing the flammability range of the fuel by the inclusion of hydrogen. The hydrogen was produced by catalytic reformation of the gasoline and the "syn-gas" (a mixture of H_2, CO, and some remaining hydrocarbons), which was mixed with the incoming air. Interestingly, it was concluded that initial work with leaded gasoline had shown promise that catalyst poisoning may not be a problem! An alternative approach was to reform all of the fuel and feed the reformed fuel to the engine, either by using a catalyst [22.83] or by simply relying on heat generated by burning part of the fuel [22.84]. The latter approach "resulted in extremely fine road performance from the test vehicle, comparable to the production model prior to modification," however, carbon deposits in the reformer were a problem. The "Boston Car," [22.85], which used the steam reformer approach, did however rely on storing a quantity of the reformed fuel for future use, such as start-up.

As an alternative the idea of using the engine to perform partial oxidation of an air/fuel mixture to produce "syn-gas" has been considered, either through the use of EGR [22.86, 22.87] or directly in the cylinder [22.88]. Using the former approach, the EGR, or part thereof is blended with a small quantity of fresh fuel and passed to the reformer where the exhaust gas temperature and the CO_2 present bring about the partial oxidation of the fuel to produce the syn-gas, which is then fed back into the engine with fresh air and more fuel. With developments in fuel injection equipment, it became possible to fuel different cylinders of an engine differently; thus, it was possible to use one or more cylinders as a fuel reformer and the other cylinders to produce work from a combination of reformed fuel and unreformed fuel. This became the basis of the dedicated EGR engine [22.89, 22.90] which was successfully use in a demonstration vehicle [22.91].

In addition to this, advances in inlet and exhaust valve timing control have enabled the implementation of NVO, as discussed in Section 22.3.1. By injecting a small quantity of fuel into the hot residual gas present during the NVO period it is possible to produce a similar effect to the use of EGR for reforming the fuel, thus the incoming air charge is mixed with H_2 and CO [22.92] to control the reactivity of the mixture [22.93].

Alternative fuels have been considered as the basis for fuel reforming, methanol, ethanol, and dimethyl ether (DME) being probably the most studied. Using waste heat form the exhaust to power a reformer methanol can be converted to DME, H_2, CO, and water. By controlling the proportions of DME and H_2 plus CO, or "syn-gas" an HCCI engine can be controlled [22.94, 22.95], however to control these proportions two reformers were required [22.96]. By careful design of the reformer both dehydration of methanol, to produce DME and water, and steam reforming of methanol, to produce hydrogen, can be accomplished in a single unit [22.97].

22.4. Fuel Effects on Kinetically Controlled CI

Fuel effects studies are usually performed on production type engines and vehicles. As kinetically controlled CI engines are still at various stages of development, it is not possible to determine how these engines will respond to changes in fuel properties when installed in a vehicle and driven over a test cycle. However, as part of the engine development process, it has become clear that neither current gasolines nor current diesel fuels are ideal for the CAI or PCCI engines that are being developed. How gasoline-type fuels affect CAI engines and how different fuels affect PCCI engines will be discussed briefly in the following sections. Developments around RCCI engines aim to make the best use of a combination of gasoline and diesel fuel properties according to operating conditions. However, fuel characteristics can still influence engine performance, and this is discussed briefly in the final section of this chapter.

22.4.1. Fuel Effects on CAI Engine Performance

For conventional gasoline engines, a prime fuel consideration is its antiknock quality. Knock is an undesirable characteristic, and gasoline specifications have been designed to provide gasoline that helps engine manufacturers to avoid knock. The antiknock quality of a gasoline can be balanced by ignition timing, but this has a penalty regarding engine efficiency and emissions performance. For engines operating with CAI, the poor antiknock characteristic of the fuel may be considered desirable as similar preflame reactions, similar to those leading to knock, are required to produce the LTC. As with conventional SI operation, the antiknock quality can to some degree be compensated for by adjusting the engine operating conditions, such as the amount of exhaust gas trapped within the combustion chamber during the compression stroke. However, this again carries a price of reduced efficiency or increased emissions.

The antiknock quality of gasoline is usually defined by its research octane number (RON) and its motor octane number (MON). Both of these values are determined in standard engine tests as discussed in Chapter 7, Section 7.3. The difference between these two values is known as the sensitivity (S). Because the RON and MON test procedures use the same iso-paraffin reference fuels, then by definition these two iso-paraffins will have S = 0, and iso-paraffins in general will lower the S value of a gasoline. Aromatic, olefinic, cycloparaffinic, and oxygenated components will tend to increase S [22.98].

As discussed in Chapter 7, Section 7.3.3, another parameter known as the octane index (OI) can be defined as

$$OI = RON - K\,S$$

where K is a constant that is dependent upon the engine and operating conditions.

For a given pressure within the cylinder, the value of K decreases as the compression temperature decreases and can be negative if this temperature is lower than in the RON test [22.58]—this has been termed "beyond RON" [22.99]. Conversely, as the compression temperature increases, then the value of K increases and can become greater than 1—this can be termed "beyond MON" [22.70]. For a gasoline engine operating in CAI mode, it is likely that the pressure will be low and the temperature high to promote auto-ignition [22.100]. This is shown in Figure 22.8 [22.101].

FIGURE 22.8   Temperature and pressure history during the compression stroke for CAI operation compared with the RON and MON tests.

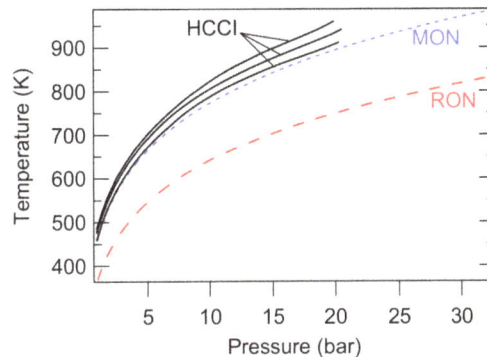

Figure 22.8 shows the temperature against pressure histories for the compression stroke of gasoline engine operating with NVO at three different speed and load points, plus the equivalent histories for a RON and MON test engine set up for iso-octane. A fuel with a high value of S results in a low effective octane number for CAI conditions. The value of S is thus an important parameter in determining the suitability of a gasoline for CAI combustion when using NVO [22.102].

The values of RON and MON are determined by engine testing; the auto-ignition performance can also be estimated from the MON value and the chemical composition of the fuel. In this case the auto-ignition index (HCCI Index (abs)) is given by the following equation [22.103]:

$$\text{HCCI Index}\left(\text{abs}\right) = m\text{MON} + a\left(\text{n-P}\right) + b\left(\text{i-P}\right) + c\left(\text{O}\right) + d\left(\text{A}\right) + e\left(\text{Ox}\right) + Y$$

where

n-P, i-P, O, A, and Ox are the percentage volume of n-paraffins, iso-paraffins, olefins, aromatics, and oxygenates, respectively

m, a, b, c, d, e, and Y are constants

As with the value of K in the OI equation, the constants in the HCCI Index (abs) equation will vary according to the in-cylinder pressure and temperature histories. If engine parameters, such as exhaust valve closing, are adjusted to compensate for changes in the ignitability of the fuels, then the chemical composition of the fuels make little difference to the engine performance [22.101, 22.104, 22.105].

Due to the developmental nature of CAI engines, there is little long-term testing data or field experience to indicate potential durability issues. However, it has been observed that the formation of combustion chamber deposits can have an influence on the timing of auto-ignition [22.52, 22.106], although the deposits may stabilize in a relatively short period of time [22.106].

22.4.2. Fuel Effects on PCCI Engine Performance

Traditionally, the major advantage of the diesel engine over the gasoline engine has been its higher efficiency. This has mainly been due to the higher compression ratios that can be used and the reduced pumping losses at part load conditions. For light-duty applications this has been tempered by the need to meet increasingly strict emissions regulation, particularly for NO_X and PM. With the latest round of emissions reduction legislation, the use of exhaust aftertreatment is adding significant cost and complexity to the diesel engine package. The prospect of using LTC to significantly reduce the production of NO_X and PM, and hence eliminate the exhaust aftertreatment, is obviously very attractive. However, this must not be done at the expense of efficiency, because advances in gasoline engine technology are already narrowing the gap between gasoline and diesel engine efficiencies. The use of PCCI at part load conditions and reversion to conventional diesel combustion at high loads offers an opportunity until more advanced systems such as RCCI are fully developed. This section looks at how diesel fuel characteristics influence PCCI operation.

To obtain a premixed charge, the distillation characteristics of the fuel will obviously have an influence, but as with fuel atomization in a conventional diesel spray, characteristics such as density, viscosity, and surface tension will also play a role [22.107, 22.108]. Fuels with high back-end distillation characteristics are likely to increase HC emissions due to fuel droplet impingement and condensation on the cylinder walls [22.108, 22.109] and can also increase CO emissions [22.110]. The distribution of boiling points is also important because this can result in a degree of fuel stratification. This can assist ignition through richer areas of the cylinder charge [22.111]. The back-end distillation characteristics can influence the physical part of the ignition delay and hence the phasing of PCCI combustion. A reduction in back-end distillation will tend to advance the combustion phasing [22.110] due to better vaporization and mixing of the fuel and air.

The ignitability and oxidation rate of the fuel are major factors in determining when and how rapidly the energy will be released from the fuel. For diesel type fuels, the ignitability is usually measured by the cetane number (CN). Because the CN is a measure of the auto-ignition quality of the fuel in a diesel engine, it incorporates the physical delays (atomization, vaporization, mixing) and chemical delays (reactivity). For PCCI operation there is a considerable difference in fuel injection timing from the conventional diesel engine; thus, the physical delay may play a bigger role than is indicated by the CN. However, without engine adjustment the CN will still have a strong influence on PCCI combustion phasing [22.15, 22.112–22.115].

Due to their higher volatility, gasoline-type fuels have been investigated in PCCI applications, and derived cetane numbers (DCN) [22.116] can be used as an alternative measure of ignitability [22.60, 22.117]. It has been found that with the correct engine settings, using either gasoline or diesel boiling range fuels with DCN values in the range of about 27 to 32, full load can be achieved with PCCI conditions [22.60]. The thermal efficiency for the engine was measured to be similar to that obtained with a conventional diesel engine.

The chemical composition of the fuel has also been found to effect PCCI exhaust emissions. An increased fuel aromatic content has been found to increase NO_X [22.118] and PM [22.60] emissions but reduces both HC and CO emissions [22.118].

22.4.3. Fuel Effects on RCCI Engine Performance

As discussed in the previous sections, there does not appear to be an ideal fuel to allow the operation of an engine using kinetically controlled combustion over the entire operating range. An engine with an operating range that will satisfy customer demand will need to operate in conventional SI or diesel operating mode for part of the operating range. For CAI operation the high knock resistance of gasoline-type fuels is an impediment for CAI operation; for PCCI operation the propensity for auto-ignition (i.e., a low knock resistance) is an impediment. A blend of gasoline and diesel fuel, referred to as dieseline, has been proposed as a compromise [22.61, 22.119, 22.120], but this still has limitations at certain operating points. The concept of RCCI allows the use of two fuels with different characteristics in order to achieve optimum combustion behavior by the use of the correct mixture of the two fuels. The mixture is achieved in the cylinder rather than in the fuel system.

Most studies on RCCI engine operation have been conducted with gasoline and diesel fuels meeting the prevailing standards for commercial fuels. Testing has been conducted using purely petroleum-based gasoline and diesel fuel and also using gasoline containing ethanol and diesel fuel–containing biodiesel [22.121]. The inclusion of ethanol, in the port injected gasoline, increases volumetric efficiency and increases overall brake thermal efficiency. The use of an ethanol gasoline mix also extends the RCCI operating range at high loads. However, there was a small negative impact of NO_X emissions. The use of biodiesel also resulted in higher brake thermal efficiency but slightly increased CO emissions.

As noted, the impediment of standard gasoline for CAI is its high resistance to knock. It has a high octane number, whereas for auto-ignition it requires a low octane number and a high CN. As discussed in Chapter 19, Section 19.4.1, the CN of a fuel can be increased by the addition of a small amount of a CN improver (CNI) additive. It is therefore not unreasonable to consider an RCCI engine where standard gasoline is used as one fuel and gasoline plus a CNI used as the other fuel. This is shown schematically in Figure 22.9 [22.81].

FIGURE 22.9   Schematic of a RCCI concept using gasoline and a CNI.

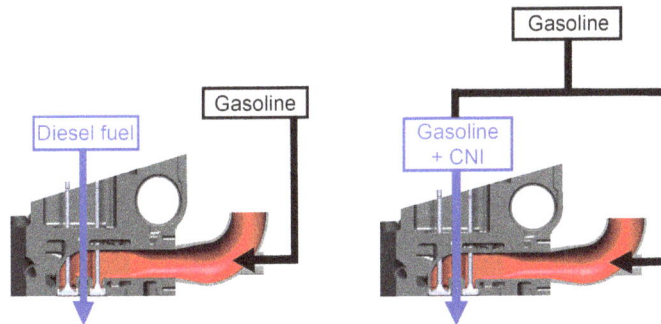

This approach has been tested using the two commonly used CNIs, di-tertiary-butyl-peroxide (DTBP) [22.122] and 2-ethyl-hexyl-nitrate (2EHN) [22.81]. The performance of this single-fuel RCCI concept has been compared with the dual fuel approach and is found to give better performance. The fact that the gasoline and the gasoline plus CNI both have gasoline distillation characteristics helps in producing a well-mixed charge. As noted in Chapter 19, Section 19.4.1, 2EHN is more effective as a CNI than is DTBP, which would reduce the amount of CNI required. However, as noted in Chapter 21, Section 21.2.3, there is a tendency for 2EHN to have a slight negative impact on NO_X emissions. This is also true when using a gasoline plus 2EHN mixture for an RCCI engine [22.123], however, the results are heavily dependent on the combustion phasing [22.124].

Interestingly, work with fuel doped with a reactivity enhancing additive such as DTBP, in comparison to the neat fuel, shows that it is not simply a question of the fuel's reactivity. As outlined above, using gasoline as the low reactivity fuel to form a premixed charge that is made more reactive by the addition of the same gasoline plus the reactivity enhancing additive, such as DTBP, the reactivity of the whole mixture can be controlled to control the CI. Repeating this exercise with methanol as the low reactivity fuel and methanol in combination with DTBP did not allow the same degree of combustion control as was achieved with the gasoline mixtures. It was hypothesized that this was due to charge cooling brought about by the direct injection of the methanol, i.e., inducing thermal stratification [22.125] and brought about the idea of inverted RCCI (iRCCI) [22.125].

References

22.1. Yui, S. and Ohnishi, S., "A New Concept of Stratified Charge Two Stroke Engine Yui and Ohnishi Combustion Process (YOCP)," SAE Technical Paper 690468 (1969), doi:https://doi.org/10.4271/690468.

22.2. Jo, S.-H., Jo, P.-D., Gomi, T., and Ohnishi, S., "Development of a Low-Emission and High-Performance 2-Stroke Gasoline Engine (NiCE)," SAE Technical Paper 730463 (1973), doi:https://doi.org/10.4271/730463.

22.3. Onishi, S., Jo, S.-H., Shoda, K., Jo, P.-D. et al., "Active Thermo-Atmosphere Combustion (ATAC)—A New Combustion Process for Internal Combustion Engines," SAE Technical Paper 790501 (1979), doi:https://doi.org/10.4271/790501.

22.4. Noguchi, M., Tanaka, Y., Tanaka, T., and Takeuchi, Y., "A Study on Gasoline Engine Combustion by Observation of Intermediate Reactive Products during Combustion," SAE Technical Paper 790840 (1979), doi:https://doi.org/10.4271/790840.

22.5. Najt, P.M. and Foster, D.E., "Compression-Ignited Homogeneous Charge Combustion," SAE Technical Paper 830264 (1983), doi:https://doi.org/10.4271/830264.

22.6. Thring, R.H., "Homogeneous-Charge Compression-Ignition (HCCI) Engines," SAE Technical Paper 892068 (1989), doi:https://doi.org/10.4271/892068.

22.7. Aoyama, T., Hattori, Y., Mizuta, J., and Sato, Y., "An Experimental Study on Premixed-Charge Compression Ignition Gasoline Engine," SAE Technical Paper 960081 (1996), doi:https://doi.org/10.4271/960081.

22.8. Takeda, Y., Keiichi, N., and Keiichi, N., "Emission Characteristics of Premixed Lean Diesel Combustion with Extremely Early Staged Fuel Injection," SAE Technical Paper 961163 (1996), doi:https://doi.org/10.4271/961163.

22.9. Yokota, H., Kudo, Y., Nakajima, H., Kakegawa, T. et al., "A New Concept for Low Emission Diesel Combustion," SAE Technical Paper 970891 (1997), doi:https://doi.org/10.4271/970891.

22.10. Hashizume, T., Miyamoto, T., Hisashi, A., and Tsujimura, K., "Combustion and Emission Characteristics of Multiple Stage Diesel Combustion," SAE Technical Paper 980505 (1998), doi:https://doi.org/10.4271/980505.

22.11. Ryan, T.W. and Callahan, T.J., "Homogeneous Charge Compression Ignition of Diesel Fuel," SAE Technical Paper 961160 (1996), doi:https://doi.org/10.4271/961160.

22.12. Walter, B. and Gatellier, B., "Development of the High Power NADI™ Concept Using Dual Mode Diesel Combustion to Achieve Zero NOx and Particulate Emissions," SAE Technical Paper 2002-01-1744 (2002), doi:https://doi.org/10.4271/2002-01-1744.

22.13. Christensen, M., Johansson, B., Amnéus, P., and Mauss, F., "Supercharged Homogeneous Charge Compression Ignition," SAE Technical Paper 980787 (1998), doi:https://doi.org/10.4271/980787.

22.14. Dec, J.E., "A Conceptual Model of DI Diesel Combustion Based on Laser-Sheet Imaging," SAE Technical Paper 970873 (1997), doi:https://doi.org/10.4271/970873.

22.15. Szybist, J.P. and Bunting, B.G., "Cetane Number and Engine Speed Effects on Diesel HCCI Performance and Emissions," SAE Technical Paper 2005-01-3723 (2005), doi:https://doi.org/10.4271/2005-01-3723.

22.16. Pitz, W.J., Westbrook, C.K., Proscia, W.M., and Dryer, F.L., "A Comprehensive Chemical Kinetic Reaction Mechanism for the Oxidation of N-Butane," *Symposium (International) on Combustion* 20, no. 1 (1985): 831-843.

22.17. Green, R.M., Cernansky, N.P., Pitz, W.J., and Westbrook, C.K., "The Role of Low Temperature Chemistry in the Autoignition of N-Butane," SAE Technical Paper 872108 (1987), doi:https://doi.org/10.4271/872108.

22.18. Gaffuri, P., Faravelli, T., Ranzi, E., Cernansky, N.P. et al., "Comprehensive Kinetic Model for the Low Temperature Oxidation of Hydrocarbons," *AIChE Journal* 43, no. 5 (1997): 1278-1286.

22.19. Curran, H.J., Gaffuri, P., Pitz, W.J., and Westbrook, C.K., "A Comprehensive Modeling Study of n-Heptane Oxidation," *Combustion and Flame* 114, no. 1-2 (1998): 149-177.

22.20. Curran, H.J., Gaffuri, P., Pitz, W.J., and Westbrook, C.K., "A Comprehensive Modeling Study of Iso-Octane Oxidation," *Combustion and Flame* 129, no. 3 (2002): 253-280.

22.21.   Oakley, A., Zhao, H., Ladommatos, N., and Ma, T., "Experimental Studies on Controlled Auto-ignition (CAI) Combustion of Gasoline in a 4-Stroke Engine," SAE Technical Paper 2001-01-1030 (2001), doi:https://doi.org/10.4271/2001-01-1030.

22.22.   Kuboyama, T., Moriyoshi, Y., Hatamura, K., Yamada, T. et al., "An Experimental Study of a Gasoline HCCI Engine Using the Blow-Down Super Charge System," SAE Technical Paper 2009-01-0496 (2009), doi:https://doi.org/10.4271/2009-01-0496.

22.23.   Kuboyama, T., Moriyoshi, Y., Hatamura, K., Takanashi, J. et al., "Extension of Operating Range of a Multi-Cylinder Gasoline HCCI Engine using the Blowdown Supercharging System," *SAE Int. J. Engines* 4, no. 1 (2011): 1150-1168, doi:https://doi.org/10.4271/2011-01-0896.

22.24.   Borgqvist, P., Tunestål, P., and Johansson, B., "Investigation and Comparison of Residual Gas Enhanced HCCI Using Trapping (NVO HCCI) or Rebreathing of Residual Gases," SAE Technical Paper 2011-01-1772 (2011), doi:https://doi.org/10.4271/2011-01-1772.

22.25.   Kaahaaina, N.B., Simon, A.J., Caton, P.A., and Edwards, C.F., "Use of Dynamic Valving to Achieve Residual-Affected Combustion," SAE Technical Paper 2001-01-0549 (2001), doi:https://doi.org/10.4271/2001-01-0549.

22.26.   Yang, C., Zhao, H., and Megaritis, T., "Investigation of CAI Combustion with Positive Valve Overlap and Enlargement of CAI Operating Range," SAE Technical Paper 2009-01-1104 (2009), doi:https://doi.org/10.4271/2009-01-1104.

22.27.   Li, L., Xie, H., Chen, T., Yu, W. et al., "Experimental Study on Spark Assisted Compression Ignition (SACI) Combustion with Positive Valve Overlap in a HCCI Gasoline Engine," SAE Technical Paper 2012-01-1126 (2012), doi:https://doi.org/10.4271/2012-01-1126.

22.28.   Duffour, F., Vangraefschèpe, F., Knop, V., and de Francqueville, L., "Influence of the Valve-Lift Strategy in a CAI™ Engine using Exhaust Gas Re-Breathing—Part 1: Experimental Results and 0D Analysis," SAE Technical Paper 2009-01-0299 (2009), doi:https://doi.org/10.4271/2009-01-0299.

22.29.   Law, D., Allen, J., and Chen, R., "On the Mechanism of Controlled Auto Ignition," SAE Technical Paper 2002-01-0421 (2002), doi:https://doi.org/10.4271/2002-01-0421.

22.30.   Olsson, J.-O., Tunestål, P., and Johansson, B., "Closed-Loop Control of an HCCI Engine," SAE Technical Paper 2001-01-1031 (2001), doi:https://doi.org/10.4271/2001-01-1031.

22.31.   Hiraya, K., Hasegawa, K., Urushihara, T., Iiyama, A. et al., "A Study on Gasoline Fueled Compression Ignition Engine ~ A Trial of Operation Region Expansion," SAE Technical Paper 2002-01-0416 (2002), doi:https://doi.org/10.4271/2002-01-0416.

22.32.   Yang, J., Culp, T., and Kenney, T., "Development of a Gasoline Engine System Using HCCI Technology—The Concept and the Test Results," SAE Technical Paper 2002-01-2832 (2002), doi:https://doi.org/10.4271/2002-01-2832.

22.33.   Milovanovic, N., Blundell, D., Pearson, R., Turner, J. et al., "Enlarging the Operational Range of a Gasoline HCCI Engine By Controlling the Coolant Temperature," SAE Technical Paper 2005-01-0157 (2005), doi:https://doi.org/10.4271/2005-01-0157.

22.34.   Kalian, N., Standing, R., and Zhao, H., "Effects of Ignition Timing on CAI Combustion in a Multi-Cylinder DI Gasoline Engine," SAE Technical Paper 2005-01-3720 (2005), doi:https://doi.org/10.4271/2005-01-3720.

22.35.   Persson, H., Hultqvist, A., Johansson, B., and Remón, A., "Investigation of the Early Flame Development in Spark Assisted HCCI Combustion Using High Speed Chemiluminescence Imaging," SAE Technical Paper 2007-01-0212 (2007), doi:https://doi.org/10.4271/2007-01-0212.

22.36.   Fuerhapter, A., Piock, W.F., and Fraidl, G.K., "CSI - Controlled Auto Ignition - The Best Solution for the Fuel Consumption - Versus Emission Trade-Off?" SAE Technical Paper 2003-01-0754 (2003), doi:https://doi.org/10.4271/2003-01-0754.

22.37.   Hyvönen, J., Haraldsson, G., and Johansson, B., "Operating Conditions Using Spark Assisted HCCI Combustion During Combustion Mode Transfer to SI in a Multi-Cylinder VCR-HCCI Engine," SAE Technical Paper 2005-01-0109 (2005), doi:https://doi.org/10.4271/2005-01-0109.

22.38.   Wang, Z., Wang, J., Shuai, S., He, X. et al., "Research on Spark Induced Compression Ignition (SICI)," SAE Technical Paper 2009-01-0132 (2009), doi:https://doi.org/10.4271/2009-01-0132.

22.39. Iwabuchi, Y., Kawai, K., Shoji, T., and Takeda, Y., "Trial of New Concept Diesel Combustion System—Premixed Compression-Ignited Combustion," SAE Technical Paper 1999-01-0185 (1999), doi:https://doi.org/10.4271/1999-01-0185.

22.40. Kook, S., Bae, C., Miles, P.C., Choi, D. et al., "The Effect of Swirl Ratio and Fuel Injection Parameters on CO Emission and Fuel Conversion Efficiency for High-Dilution, Low-Temperature Combustion in an Automotive Diesel Engine," SAE Technical Paper 2006-01-0197 (2006), doi:https://doi.org/10.4271/2006-01-0197.

22.41. Mendez, S., Kashdan, J.T., Bruneaux, G., Thirouard, B. et al., "Formation of Unburned Hydrocarbons in Low Temperature Diesel Combustion," *SAE Int. J. Engines* 2, no. 2 (2010): 205-225, doi:https://doi.org/10.4271/2009-01-2729.

22.42. Boot, M.D., Luijten, C.C.M., Somers, L.M.T., van Erp, D.D.T.M. et al., "Uncooled EGR as a Means of Limiting Wall-Wetting under Early Direct Injection Conditions," SAE Technical Paper 2009-01-0665 (2009), doi:https://doi.org/10.4271/2009-01-0665.

22.43. Boot, M.D., Luijten, C.C.M., Rijk, E.P., Albrecht, B.A. et al., "Optimization of Operating Conditions in the Early Direct Injection Premixed Charge Compression Ignition Regime," SAE Technical Paper 2009-24-0048 (2009), doi:https://doi.org/10.4271/2009-24-0048.

22.44. Boot, M., Rijk, E., Luijten, C., Somers, B. et al., "Spray Impingement in the Early Direct Injection Premixed Charge Compression Ignition Regime," SAE Technical Paper 2010-01-1501 (2010), doi:https://doi.org/10.4271/2010-01-1501.

22.45. Martin, G.C., Mueller, C.J., Milam, D.M., Radovanovic, M.S. et al., "Early Direct-Injection, Low-Temperature Combustion of Diesel Fuel in an Optical Engine Utilizing a 15-Hole, Dual-Row, Narrow-Included-Angle Nozzle," *SAE Int. J. Engines* 1, no. 1 (2009): 1057-1082, doi:https://doi.org/10.4271/2008-01-2400.

22.46. Rajasegar, R. and Srna, A., "A Review of Current Understanding of the Underlying Physics Governing the Interaction, Ignition and Combustion Dynamics of Multiple-Injections in Diesel Engines," SAE Technical Paper 2022-01-0445 (2022), doi:https://doi.org/10.4271/2022-01-0445.

22.47. Skeen, S., Manin, J., and Pickett, L., "Visualization of Ignition Processes in High-Pressure Sprays with Multiple Injections of n-Dodecane," *SAE Int. J. Engines* 8, no. 2 (2015): 696-715, doi:https://doi.org/10.4271/2015-01-0799.

22.48. Maes, N., Sim, H.S., Weiss, L., and Pickett, L., "Simultaneous High-Speed Formaldehyde PLIF and Schlieren Imaging of Multiple Injections from an ECN Spray D Injector," in *Internal Combustion Engine Division Fall Technical Conference*, Vol. 84034, V001T05A001, November 2020, American Society of Mechanical Engineers.

22.49. Knox, B., Genzale, C., Pickett, L., Garcia-Oliver, J. et al., "Combustion Recession after End of Injection in Diesel Sprays," *SAE Int. J. Engines* 8, no. 2 (2015): 679-695, doi:https://doi.org/10.4271/2015-01-0797.

22.50. Kimura, S., Aoki, O., Ogawa, H., Muranaka, S. et al., "New Combustion Concept for Ultra-Clean and High-Efficiency Small DI Diesel Engines," SAE Technical Paper 1999-01-3681 (1999), doi:https://doi.org/10.4271/1999-01-3681.

22.51. Kimura, S., Aoki, O., Kitahara, Y., and Aiyoshizawa, E., "Ultra-Clean Combustion Technology Combining a Low-Temperature and Premixed Combustion Concept for Meeting Future Emission Standards," SAE Technical Paper 2001-01-0200 (2001), doi:https://doi.org/10.4271/2001-01-0200.

22.52. Shirawaka, T., Miura, M., Itoyama, H., Aiyoshizawa, E. et al., "Study of Model-Based Cooperative Control of EGR and VGT for a Low-Temperature, Premixed Combustion Diesel Engine," SAE Technical Paper 2001-01-2006 (2001), doi:https://doi.org/10.4271/2001-01-2006.

22.53. Hasegawa, R. and Yanagihara, H., "HCCI Combustion in DI Diesel Engine," SAE Technical Paper 2003-01-0745 (2003), doi:https://doi.org/10.4271/2003-01-0745.

22.54. Yanagihara, H., "A Simultaneous Reduction of NOx and Soot in Diesel Engines under a New Combustion System (Uniform Bulky Combustion System-UNIBUS)," in *17th International Vienna Motor Symposium*, Vienna, Austria, 303-314, 1996.

22.55. Su, W., Wang, H., and Liu, B., "Injection Mode Modulation for HCCI Diesel Combustion," SAE Technical Paper 2005-01-0117 (2005), doi:https://doi.org/10.4271/2005-01-0117.

22.56. Duret, P., Gatellier, B., Monteiro, L., Miche, M. et al., "Progress in Diesel HCCI Combustion within the European SPACE LIGHT Project," SAE Technical Paper 2004-01-1904 (2004), doi:https://doi.org/10.4271/2004-01-1904.

22.57. Kawano, D., Suzuki, H., Ishii, H., Goto, Y. et al., "Ignition and Combustion Control of Diesel HCCI," SAE Technical Paper 2005-01-2132 (2005), doi:https://doi.org/10.4271/2005-01-2132.

22.58. Kalghatgi, G.T., "Auto-Ignition Quality of Practical Fuels and Implications for Fuel Requirements of Future SI and HCCI Engines," SAE Technical Paper 2005-01-0239 (2005), doi:https://doi.org/10.4271/2005-01-0239.

22.59. Opat, R., Ra, Y., Gonzalez, D.M.A., Krieger, R. et al., "Investigation of Mixing and Temperature Effects on HC/CO Emissions for Highly Dilute Low Temperature Combustion in a Light Duty Diesel Engine," SAE Technical Paper 2007-01-0193 (2007), doi:https://doi.org/10.4271/2007-01-0193.

22.60. Bessonette, P.W., Schleyer, C.H., Duffy, K.P., Hardy, W.L. et al., "Effects of Fuel Property Changes on Heavy-Duty HCCI Combustion," SAE Technical Paper 2007-01-0191 (2007), doi:https://doi.org/10.4271/2007-01-0191.

22.61. Turner, D., Tian, G., Xu, H., Wyszynski, M.L. et al., "An Experimental Study of Dieseline Combustion in a Direct Injection Engine," SAE Technical Paper 2009-01-1101 (2009), doi:https://doi.org/10.4271/2009-01-1101.

22.62. Olsson, J.-O., Tunestål, P., Haraldsson, G., and Johansson, B., "A Turbo Charged Dual Fuel HCCI Engine," SAE Technical Paper 2001-01-1896 (2001), doi:https://doi.org/10.4271/2001-01-1896.

22.63. Strandh, P., Bengtsson, J., Johansson, R., Tunestål, P. et al., "Cycle-to-Cycle Control of a Dual-Fuel HCCI Engine," SAE Technical Paper 2004-01-0941 (2004), doi:https://doi.org/10.4271/2004-01-0941.

22.64. Yao, M., Chen, Z., Zheng, Z., Zhang, B. et al., "Effect of EGR on HCCI Combustion Fuelled with Dimethyl Ether (DME) and Methanol Dual-Fuels," SAE Technical Paper 2005-01-3730 (2005), doi:https://doi.org/10.4271/2005-01-3730.

22.65. Inagaki, K., Fuyuto, T., Nishikawa, K., Nakakita, K. et al., "Dual-Fuel PCI Combustion Controlled by In-Cylinder Stratification of Ignitability," SAE Technical Paper 2006-01-0028 (2006), doi:https://doi.org/10.4271/2006-01-0028.

22.66. Sjöberg, M. and Dec, J.E., "Smoothing HCCI Heat-Release Rates Using Partial Fuel Stratification with Two-Stage Ignition Fuels," SAE Technical Paper 2006-01-0629 (2006), doi:https://doi.org/10.4271/2006-01-0629.

22.67. Dec, J.E., Hwang, W., and Sjöberg, M., "An Investigation of Thermal Stratification in HCCI Engines Using Chemiluminescence Imaging," SAE Technical Paper 2006-01-1518 (2006), doi:https://doi.org/10.4271/2006-01-1518.

22.68. Dec, J.E. and Hwang, W., "Characterizing the Development of Thermal Stratification in an HCCI Engine Using Planar-Imaging Thermometry," *SAE Int. J. Engines* 2, no. 1 (2009): 421-438, doi:https://doi.org/10.4271/2009-01-0650.

22.69. Kokjohn, S., Hanson, R., Splitter, D., Kaddatz, J. et al., "Fuel Reactivity Controlled Compression Ignition (RCCI) Combustion in Light- and Heavy-Duty Engines," *SAE Int. J. Engines* 4, no. 1 (2011): 360-374, doi:https://doi.org/10.4271/2011-01-0357.

22.70. Curran, S., Hanson, R., Wagner, R., and Reitz, R., "Efficiency and Emissions Mapping of RCCI in a Light-Duty Diesel Engine," SAE Technical Paper 2013-01-0289 (2013), doi:https://doi.org/10.4271/2013-01-0289.

22.71. Diesel, R., Method of igniting and regulating combustion for internal-combustion engines. U.S. Patent 673,160, 1901.

22.72. Konigsson, F., Stalhammar, P., and Ångström, H., "Combustion Modes in a Diesel-CNG Dual Fuel Engine," SAE Technical Paper 2011-01-1962 (2011), doi:https://doi.org/10.4271/2011-01-1962.

22.73. Di Blasio, G., Belgiorno, G., Beatrice, C., Fraioli, V. et al., "Experimental Evaluation of Compression Ratio Influence on the Performance of a Dual-Fuel Methane-Diesel Light-Duty Engine," *SAE Int. J. Engines* 8, no. 5 (2015): 2253-2267, doi:https://doi.org/10.4271/2015-24-2460.

22.74. Kobashi, Y., Inagaki, R., Shibata, G., and Ogawa, H., "Improvements of Combustion and Emissions in a Natural Gas Fueled Engine with Hydrogen Enrichment and Optimized Injection Timings of the Diesel Fuel," SAE Technical Paper 2022-32-0095 (2022), doi:https://doi.org/10.4271/2022-32-0095.

22.75. Mancaruso, E., De Robbio, R., and Vaglieco, B., "Hydrogen/Diesel Combustion Analysis in a Single Cylinder Research Engine," SAE Technical Paper 2022-24-0012 (2022), doi:https://doi.org/10.4271/2022-24-0012.

22.76. Kim, W., Park, C., and Bae, C., "Combustion Phenomena and Emissions in a Dual-Fuel Optical Engine Fueled with Diesel and Natural Gas," *SAE Int. J. Adv. & Curr. Prac. in Mobility* 4, no. 2 (2022): 502-513, doi:https://doi.org/10.4271/2021-01-1175.

22.77. Kim, W., Park, C., and Bae, C., "Characterization of Combustion Process and Emissions in a Natural Gas/Diesel Dual-Fuel Compression-Ignition Engine," *Fuel* 291 (2021): 120043.

22.78. Kokjohn, S.L., Hanson, R.M., Splitter, D.A., and Reitz, R.D., "Experiments and Modeling of Dual-Fuel HCCI and PCCI Combustion Using In-Cylinder Fuel Blending," *SAE Int. J. Engines* 2, no. 2 (2010): 24-39, doi:https://doi.org/10.4271/2009-01-2647.

22.79. Hanson, R., Kokjohn, S., Splitter, D., and Reitz, R., "Fuel Effects on Reactivity Controlled Compression Ignition (RCCI) Combustion at Low Load," *SAE Int. J. Engines* 4, no. 1 (2011): 394-411, doi:https://doi.org/10.4271/2011-01-0361.

22.80. Splitter, D., Hanson, R., Kokjohn, S., and Reitz, R., "Reactivity Controlled Compression Ignition (RCCI) Heavy-Duty Engine Operation at Mid-and High-Loads with Conventional and Alternative Fuels," SAE Technical Paper 2011-01-0363 (2011), doi:https://doi.org/10.4271/2011-01-0363.

22.81. Hanson, R., Curran, S., Wagner, R., Kokjohn, S. et al., "Piston Bowl Optimization for RCCI Combustion in a Light-Duty Multi-Cylinder Engine," *SAE Int. J. Engines* 5, no. 2 (2012): 286-299, doi:https://doi.org/10.4271/2012-01-0380.

22.82. Houseman, J. and Cerini, D., "On-Board Hydrogen Generator for a Partial Hydrogen Injection Internal Combustion Engine," SAE Technical Paper 740600 (1974), doi:https://doi.org/10.4271/740600.

22.83. Lindstrom, O.B., Fuel treatment for combustion engines. U.S. Patent 3,918,412, 1975.

22.84. Newkirk, M.S., Method and means for generating hydrogen and a motive source incorporating same. U.S. Patent 3,682,142, 1972.

22.85. Newkirk, M. and Abel, J., "The Boston Reformed Fuel Car," SAE Technical Paper 720670 (1972), doi:https://doi.org/10.4271/720670.

22.86. Tsolakis, A. and Megaritis, A., "Exhaust Gas Fuel Reforming for Diesel Engines - A Way to Reduce Smoke and NOX Emissions Simultaneously," SAE Technical Paper 2004-01-1844 (2004), doi:https://doi.org/10.4271/2004-01-1844.

22.87. Yap, D., Peucheret, S.M., Megaritis, A., Wyszynski, M.L. et al., "Natural Gas HCCI Engine Operation with Exhaust Gas Fuel Reforming," *International Journal of Hydrogen Energy* 31, no. 5 (2006): 587-595.

22.88. Shibata, G., Asai, G., Ishiguro, S., Watanabe, Y. et al., "Fuel Reformation by Piston Compression of Rich Air-Fuel Mixture," *International Journal of Engine Research* 24, no. 1 (2023): 240-261.

22.89. Alger, T. and Mangold, B., "Dedicated EGR: A New Concept in High Efficiency Engines," *SAE Int. J. Engines* 2, no. 1 (2009): 620-631, doi:https://doi.org/10.4271/2009-01-0694.

22.90. Chadwell, C., Alger, T., Zuehl, J., and Gukelberger, R., "A Demonstration of Dedicated EGR on a 2.0 L GDI Engine," *SAE Int. J. Engines* 7, no. 1 (2014): 434-447, doi:https://doi.org/10.4271/2014-01-1190.

22.91. Robertson, D., Chadwell, C., Alger, T., Zuehl, J. et al., "Dedicated EGR Vehicle Demonstration," *SAE Int. J. Engines* 10, no. 3 (2017): 898-907, doi:https://doi.org/10.4271/2017-01-0648.

22.92. Yang, S., "Generation of Reactive Chemical Species/Radicals through Pilot Fuel Injection in Negative Valve Overlap and Its Effects on Engine Performances," SAE Technical Paper 2022-01-1002 (2022), doi:https://doi.org/10.4271/2022-01-1002.

22.93. Szybist, J., Steeper, R., Splitter, D., Kalaskar, V. et al., "Negative Valve Overlap Reforming Chemistry in Low-Oxygen Environments," *SAE Int. J. Engines* 7, no. 1 (2014): 418-433, doi:https://doi.org/10.4271/2014-01-1188.

22.94. Shudo, T. and Ono, Y., "HCCI Combustion of Hydrogen, Carbon Monoxide and Dimethyl Ether," SAE Technical Paper 2002-01-0112 (2002), doi:https://doi.org/10.4271/2002-01-0112.

22.95. Shudo, T., Ono, Y., and Takahashi, T., "Influence of Hydrogen and Carbon Monoxide on HCCI Combustion of Dimethyl Ether," SAE Technical Paper 2002-01-2828 (2002), doi:https://doi.org/10.4271/2002-01-2828.

22.96. Shudo, T. and Takahashi, T., "Influence of Reformed Gas Composition on HCCI Combustion of Onboard Methanol-Reformed Gases," SAE Technical Paper 2004-01-1908 (2004), doi:https://doi.org/10.4271/2004-01-1908.

22.97. Eyal, A. and Tartakovsky, L., "Reforming Controlled Homogenous Charge Compression Ignition-Simulation Results," SAE Technical Paper 2016-32-0014 (2016), doi:https://doi.org/10.4271/2016-32-0014.

22.98. Kalghatgi, G., Risberg, P., and Ångstrom, H.-E., "A Method of Defining Ignition Quality of Fuels in HCCI Engines," SAE Technical Paper 2003-01-1816 (2003), doi:https://doi.org/10.4271/2003-01-1816.

22.99. Kalghatgi, G.T., "Fuel Anti-Knock Quality - Part I. Engine Studies," SAE Technical Paper 2001-01-3584 (2001), doi:https://doi.org/10.4271/2001-01-3584.

22.100. Kalghatgi, G.T. and Head, R.A., "The Available and Required Autoignition Quality of Gasoline - Like Fuels in HCCI Engines at High Temperatures," SAE Technical Paper 2004-01-1969 (2004), doi:https://doi.org/10.4271/2004-01-1969.

22.101. Farrell, J.T. and Bunting, B.G., "Fuel Composition Effects at Constant RON and MON in an HCCI Engine Operated with Negative Valve Overlap," SAE Technical Paper 2006-01-3275 (2006), doi:https://doi.org/10.4271/2006-01-3275.

22.102. Bunting, B.G., "Combustion, Control, and Fuel Effects in a Spark Assisted HCCI Engine Equipped with Variable Valve Timing," SAE Technical Paper 2006-01-0872 (2006), doi:https://doi.org/10.4271/2006-01-0872.

22.103. Shibata, G. and Urushihara, T., "Auto-Ignition Characteristics of Hydrocarbons and Development of HCCI Fuel Index," SAE Technical Paper 2007-01-0220 (2007), doi:https://doi.org/10.4271/2007-01-0220.

22.104. Angelos, J.P., Andreae, M.M., Green, W.H., Cheng, W.K. et al., "Effects of Variations in Market Gasoline Properties on HCCI Load Limits," SAE Technical Paper 2007-01-1859 (2007), doi:https://doi.org/10.4271/2007-01-1859.

22.105. Zuehl, J.R., Ghandhi, J., Hagen, C., and Cannella, W., "Fuel Effects on HCCI Combustion Using Negative Valve Overlap," SAE Technical Paper 2010-01-0161 (2010), doi:https://doi.org/10.4271/2010-01-0161.

22.106. Güralp, O., Hoffman, M., Assanis, D., Filipi, Z. et al., "Characterizing the Effect of Combustion Chamber Deposits on a Gasoline HCCI Engine," SAE Technical Paper 2006-01-3277 (2006), doi:https://doi.org/10.4271/2006-01-3277.

22.107. Butts, R.T., Foster, D., Krieger, R., Andrie, M. et al., "Investigation of the Effects of Cetane Number, Volatility, and Total Aromatic Content on Highly-Dilute Low Temperature Diesel Combustion," SAE Technical Paper 2010-01-0337 (2010), doi:https://doi.org/10.4271/2010-01-0337.

22.108. Hosseini, V., Neill, W., Guo, H., Dumitrescu, C.E. et al., "Effects of Cetane Number, Aromatic Content and 90% Distillation Temperature on HCCI Combustion of Diesel Fuels," SAE Technical Paper 2010-01-2168 (2010), doi:https://doi.org/10.4271/2010-01-2168.

22.109. Yang, B., Li, S., Zheng, Z., Yao, M. et al., "A Comparative Study on Different Dual-Fuel Combustion Modes Fuelled with Gasoline and Diesel," SAE Technical Paper 2012-01-0694 (2012), doi:https://doi.org/10.4271/2012-01-0694.

22.110. Bunting, B.G., Eaton, S.J., and Crawford, R.W., "Performance Evaluation and Optimization of Diesel Fuel Properties and Chemistry in an HCCI Engine," SAE Technical Paper 2009-01-2645 (2009), doi:https://doi.org/10.4271/2009-01-2645.

22.111. Bunting, B.G., Crawford, R.W., Wolf, L.R., and Xu, Y., "The Relationships of Diesel Fuel Properties, Chemistry, and HCCI Engine Performance as Determined by Principal Components Analysis," SAE Technical Paper 2007-01-4059 (2007), doi:https://doi.org/10.4271/2007-01-4059.

22.112. Shibata, G., Oyama, K., Urushihara, T., and Nakano, T., "The Effect of Fuel Properties on Low and High Temperature Heat Release and Resulting Performance of an HCCI Engine," SAE Technical Paper 2004-01-0553 (2004), doi:https://doi.org/10.4271/2004-01-0553.

22.113. Shibata, G., Oyama, K., Urushihara, T., and Nakano, T., "Correlation of Low Temperature Heat Release with Fuel Composition and HCCI Engine Combustion," SAE Technical Paper 2005-01-0138 (2005), doi:https://doi.org/10.4271/2005-01-0138.

22.114. Risberg, P., Kalghatgi, G., Ångstrom, H.-A., and Wåhlin, F., "Auto-Ignition Quality of Diesel-Like Fuels in HCCI Engines," SAE Technical Paper 2005-01-2127 (2005), doi:https://doi.org/10.4271/2005-01-2127.

22.115. Shibata, G. and Urushihara, T., "Dual Phase High Temperature Heat Release Combustion," *SAE Int. J. Engines* 1, no. 1 (2009): 1-12, doi:https://doi.org/10.4271/2008-01-0007.

22.116. ASTM International, "Standard Test Method for Determination of Derived Cetane Number (DCN) of Diesel Fuel Oils—Ignition Delay and Combustion Delay Using a Constant Volume Combustion Chamber Method," ASTM D7668-17, ASTM International, 2017.

22.117. Hildingsson, L., Johansson, B., Kalghatgi, G.T., and Harrison, A.J., "Some Effects of Fuel Autoignition Quality and Volatility in Premixed Compression Ignition Engines," *SAE Int. J. Engines* 3, no. 1 (2010): 440-460, doi:https://doi.org/10.4271/2010-01-0607.

22.118. Cho, K., Han, M., Sluder, C.S., Wagner, R.M. et al., "Experimental Investigation of the Effects of Fuel Characteristics on High Efficiency Clean Combustion in a Light-Duty Diesel Engine," SAE Technical Paper 2009-01-2669 (2009), doi:https://doi.org/10.4271/2009-01-2669.

22.119. Zhang, F., Xu, H., Zhang, J., Tian, G. et al., "Investigation into Light Duty Dieseline Fuelled Partially-Premixed Compression Ignition Engine," *SAE Int. J. Engines* 4, no. 1 (2011): 2124-2134, doi:https://doi.org/10.4271/2011-01-1411.

22.120. Zhang, F., Xu, H., Zeraati Rezaei, S., Kalghatgi, G. et al., "Combustion and Emission Characteristics of a PPCI Engine Fuelled with Dieseline," SAE Technical Paper 2012-01-1138 (2012), doi:https://doi.org/10.4271/2012-01-1138.

22.121. Hanson, R., Curran, S., Wagner, R., and Reitz, R., "Effects of Biofuel Blends on Light-Duty Multi-Cylinder RCCI Operation," SAE Technical Paper 2013-01-0289 (2013), doi:https://doi.org/10.4271/2013-01-0289.

22.122. Splitter, D., Reitz, R., and Hanson, R., "High Efficiency, Low Emissions RCCI Combustion by Use of a Fuel Additive," *SAE Int. J. Fuels Lubr.* 3, no. 2 (2010): 742-756, doi:https://doi.org/10.4271/2010-01-2167.

22.123. Kaddatz, J., Andrie, M., Reitz, R., and Kokjohn, S., "Light-Duty Reactivity Controlled Compression Ignition Combustion Using a Cetane Improver," SAE Technical Paper 2012-01-1110 (2012), doi:https://doi.org/10.4271/2012-01-1110.

22.124. Dempsey, A., Walker, N., and Reitz, R., "Effect of Cetane Improvers on Gasoline, Ethanol, and Methanol Reactivity and the Implications for RCCI Combustion," *SAE Int. J. Fuels Lubr.* 6, no. 1 (2013): 170-187, doi:https://doi.org/10.4271/2013-01-1678.

22.125. Lawler, B., Splitter, D., Szybist, J., and Kaul, B., "Thermally Stratified Compression Ignition: A New Advanced Low Temperature Combustion Mode with Load Flexibility," *Applied Energy* 189 (2017): 122-132.

Further Reading

Kalghatgi, G., *Fuel/Engine Interactions* (Warrendale, PA: SAE International, 2014).

Zhao, H., *HCCI and CAI Engines for the Automotive Industry* (Cambridge, UK: Woodhead Publishing Ltd., 2007).

Zhao, F., Asmus, T.N., Assanis, D.N., Dec, J.E. et al., *Homogeneous Charge Compression Ignition (HCCI) Engines* (Warrendale, PA: SAE International, 2003).

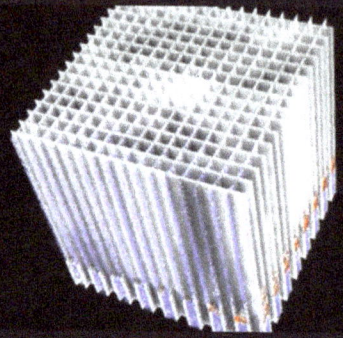

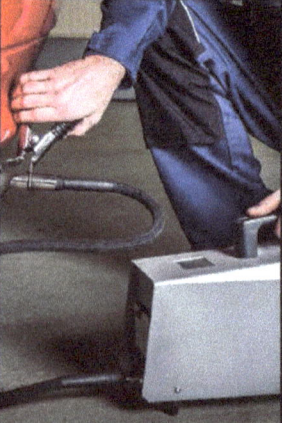

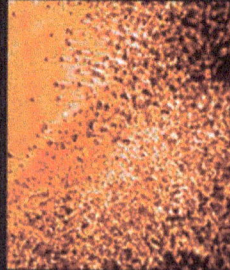

23

Future Trends and Alternative Fuels

Over the latter half of the twentieth century, the main driver for changes in the automotive sector, both engines and fuels, has been legislation to curtail pollutant emissions. The formation of the Intergovernmental Panel on Climate Change (IPCC) in 1988, and its subsequent Assessment Reports (FAR, SAR, TAR, 4AR, 5AR, and 6AR) have driven global thinking, and government policy, in the direction of decarbonization in order to limit CO_2 emissions to counter global warming. This is leading many countries to promote the battery electric vehicle (BEV) as the solution for future transport. The widespread adoption of BEVs can make a reduction in CO_2 emissions, assuming the electricity is generated in a carbon-neutral way, but the manufacture of BEVs has a negative impact due to the large use of metals, chemicals, and energy requirements of the powertrain construction process [23.1].

Combustion research, particularly of fossil fuels and in internal combustion engines is currently seen as unnecessary in many countries. However, it will be absolutely necessary, along with the development of the alternatives in order to ensure that energy use is improved since combustion will continue to be central to supplying global energy and driving transport for decades to come. The gap between current policies and reality will perhaps be bridged as energy security concerns come to the fore [23.2].

Against this background, we discuss the changes that can be envisaged for gasoline and diesel fuel, and the combustion engines that use them, as well as what might be the alternatives in the longer term. We try to anticipate how the market for automotive fuels will develop in the short (5 years) and medium (15 years) term and to what extent this will include alternatives.

In the short term and medium term, the internal combustion engine will continue to dominate as the power plant of choice for automotive transport. These engines will continue to be fueled predominantly by what we have discussed as gasoline and diesel fuel. The use of gaseous fuels, from fossil sources as discussed in Chapter 4, Section 4.4, will continue in niche markets.

A number of factors will ensure the continuing dominance of gasoline and diesel fuel:

- There is an established infrastructure for vehicle fueling and supply.

- These liquid fuels have a high energy density and ease of handling, which is vital for satisfying consumer expectations for vehicle driving range and refueling time.

- The existing vehicle fleet cannot quickly be replaced, and vehicle design and development programs are based on known fuel technologies.

It is clear that in the short term, it will not be possible to make large changes in the characteristics of gasolines and diesel fuels. To market additional types of fuel is possible, but it still takes time to phase out the incumbent. This has already been demonstrated by the phase-out of leaded gasolines. Unleaded gasoline was introduced to satisfy vehicles with catalytic converters, but leaded gasoline remained on sale in parallel until the legacy fleet was small enough and demand had fallen to the point where leaded gasoline could be removed from general sale. A slightly different scenario is currently playing out with the introduction of high ethanol gasoline blends. Vehicles do not have to use these fuels and can operate on "normal" gasoline, so there has to be some incentive for the consumer to use these fuels. There are a small number of consumers that will use these fuels for environmental reasons, but in general the incentive has to be financial. The market penetration of these fuels remains low. It should be noted that "normal" gasoline in many jurisdictions now contains low levels of ethanol, typically 10% by volume.

The pressures for change will be the following:

- Environmental considerations and legislation.

- Fossil fuel price and security of supply.

- Price and availability of renewable fuel sources.

- Advances in engine and vehicle technology.

These topics will be addressed in more detail in the following sections, and finally we will look at what might be further down the line in terms of alternative fuels.

23.1. Environmental Considerations and Legislation

Environmental pressures will continue to be the major factor in influencing vehicle technology and in turn, fuel quality and engine technology. Following the UN Climate Change Conference (COP21) in Paris, France, in December 2015, the Paris Agreement was adopted by 196 Parties [23.3]. Many countries have now made policy commitments to reach carbon neutrality by 2050, which is beyond the 15-year medium-term window we are considering here. However, some countries have already introduced policies that will take effect during this 15-year window, for example the UK was intending to end of sale of new gasoline and diesel cars (and light vans) by 2030 [23.4]. This would not ban the sale of vehicles using gasoline or diesel engines because plug-in hybrids or full hybrids vehicles would still be allowed until 2035, which would then be in line with the rest of the European Union [23.5]. This will obviously promote the sale of hybrid electric vehicles (HEVs) and the continued pressure to develop improved gasoline and diesel engine technology as discussed in Section 23.4 and demonstrated by the newly developed Mazda 3.3-litre in-line six-cylinder e-Skyactiv D diesel engine [23.6].

There will be continued pressure to reduce the amount of sulfur in fuel; this will not only apply to automotive fuels but also to aviation fuel and marine diesel fuel. Whereas many regions have already reduced the level of sulfur in automotive fuels to below 10 mg/kg, some of the fastest growing markets are still aiming to achieve this target in the medium term. The exhaust aftertreatment technology that can be utilized in the absence of these ultra-low sulfur fuels will have significant implications for the ability to meet air quality standards. Refinery processes to reduce sulfur to these low levels also eliminates some compounds that impart antioxidant and lubricity performance to the fuel; these properties must be restored by the use of fuel additives.

There will also be continued pressure, and legislation, to increase the renewable component of the input to fuel refineries, as mentioned in Chapter 3 and Chapter 5, Section 5.5. The development of e-fuels that can produce drop-in replacements for gasoline and diesel fuel via the hydrolysis of water and sequestration of carbon dioxide, as discussed in Chapter 5, Section 5.7; this will gain importance to generate not only carbon-neutral, but even carbon-negative, fuels. In 2023 this was demonstrated by Porche who fueled a 911 with e-fuel produced at a plant in Chile [23.7] discussed in Section 5.7.

In the longer term there will be continued desire to move from hydrocarbon-based fuels toward a hydrogen-based economy. Some of the technical challenges that are being addressed to achieve this are discussed in Section 23.4.4.

23.2. Fossil Fuel Price and Security of Supply

Probably one of the major consequences of the Paris Agreement will be changes in the fuel demands for the power generating industry. While it is projected [23.8] that the world power generating demand will rise by almost a quarter between 2020 and 2050 the demand for coal to satisfy this demand will fall by over a third, and the demand for oil will fall by about two-thirds. This is shown in Figure 23.1. This figure also clearly shows that oil is now only a small fraction of the power generation demand.

FIGURE 23.1 Forecast power generating demand from 2000 to 2050.

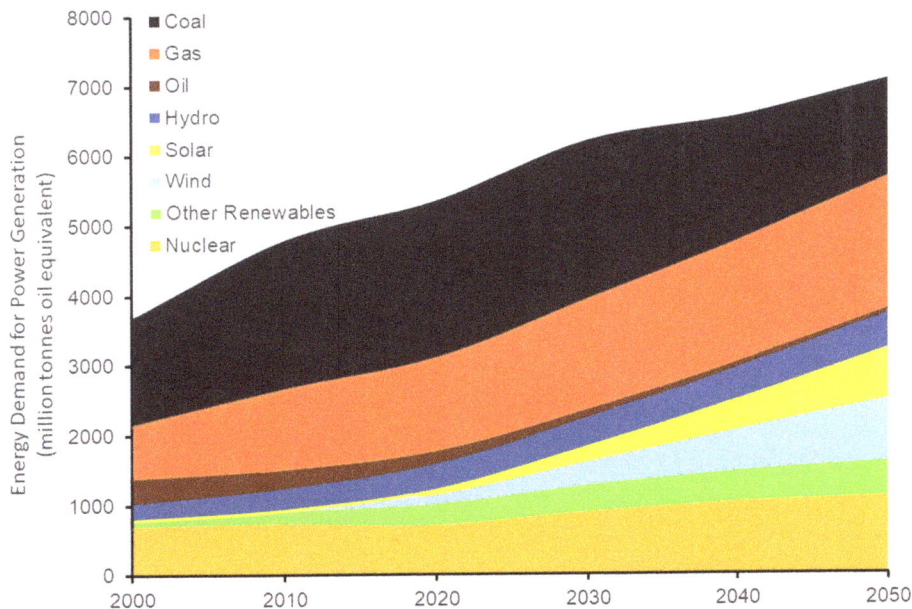

The other thing that is clear is, that for electric vehicles to fulfill their promise then a far greater proportion of the electricity must come from renewable sources. Although the amount of the demand being met from renewable sources, hydro, wind, solar, and other renewable sources, is clearly anticipated to rise significantly by 2050, it will still only make up just over a third of the demand.

How changes in price and availability of different energy sources might affect gasoline and diesel fuel quality will be discussed briefly in the following sections.

23.2.1. Crude Oil

High crude oil prices might favor the use of heavier sour crudes if they are somewhat cheaper but would depress the market for the residual fuels resulting from the normal refining processes. This may then make it attractive to crack these heavier components for use as automotive fuels, which would give rise to changes in gasoline and diesel fuel composition. Increased cracking could lead to reduced oxidation stability as more and more cracked naphthas and gas oils are used. On the other hand, because of the limitations in terms of octane quality and the cetane number of such cracked stocks, a portion of them will be upgraded by reprocessing, and this may also improve their stability. Deficiencies in octane number (RON and MON), if necessary, would have to be made up by such processes as isomerization and alkylation.

Any increase in the use of cracked stocks will lead to a rise in the use of additives both for gasoline and diesel fuel. These additives will be antioxidants, particularly for gasoline, to prevent degradation of olefinic compounds; cetane number improvers to improve diesel ignition quality; and additives to keep fuel systems free of deposits that will adversely influence performance.

23.2.2. Other Oil Sources

As the security of crude oil supply becomes more of a concern and prices rise, the use of non-crude sources becomes more attractive. As discussed in Chapter 4, Section 4.5, oil sands are already being commercially exploited on a large scale. Canada is estimated to have oil sands reserves equivalent to 175 billion barrels of oil [23.9], using current technology.

As outlined in Chapter 4, Section 4.6, there are also significant reserves of oil trapped within oil shale. The exploitation of these reserves generates significant environmental concerns. Because of China's growing need for energy and its current reliance on imported oil, China is increasing its exploitation of non-crude oil resources. China is estimated to have almost 12 billion tons of recoverable oil from shale deposits, and the exploitation of this resource is expanding. Shale oil production is expected to reach 6 million tons in 2030 [23.10]. The quality of these oils is variable, and how much will be used to produce transport fuels is unknown; most will probably be used for power generation.

23.2.3. Coal

Coal is the most abundant fossil fuel and has been commercially processed to produce automotive fuels since 1955; the coal-to-liquids (CTL) process is described in Chapter 4, Section 4.2. For economic reasons the use of GTL technology is generally the favored approach for producing fuel. However, due to vast reserves of coal and a desire to reduce dependence on imported oil, China embarked on a rapid program of CTL development and currently has the most active CTL program in the world [23.11, 23.12]. The quality of the fuels produced by CTL technology is a known parameter and is anticipated to be able to meet proposed specifications.

23.2.4. Natural Gas

Known reserves of natural gas have increased significantly in recent years due to increased potential from shale gas in North America and coal bed methane in Australia. As discussed in Chapter 3, Section 4.3, natural gas can be used to produce fuels via GTL processes. The installation of a new plant is obviously expensive and time consuming, in addition it makes little difference to the overall carbon balance. In 2022 the failed attempt by the Russian Federation to invade Ukraine, and sanctions imposed by most of the countries in the rest of the world, had a drastic effect on the price of gas which has made many countries reassess their reliance on gas. The medium-term effect this will have on the market remains to be seen.

23.3. Renewable Fuel Price and Availability

Figure 23.2 shows the energy demand for transport, from 2000 and projected up to 2050 [23.8]. Biofuels are clearly a growing area, although still a relatively small proportion by 2050. The 21st century has seen a significant increase in the use of fuel derived from renewable sources and this is projected to continue in the long term. This is predominantly driven by legislation, as such fuels are more expensive to produce than gasoline and diesel fuel from fossil sources. It is projected that the use of renewables will result in a more than 40-fold increase in the first half of the 21st century. The following sections look briefly at future possibilities for new feedstocks for bio-ethanol and biodiesel and other renewables.

FIGURE 23.2 Forecast energy demand for transport from 2000 to 2050.

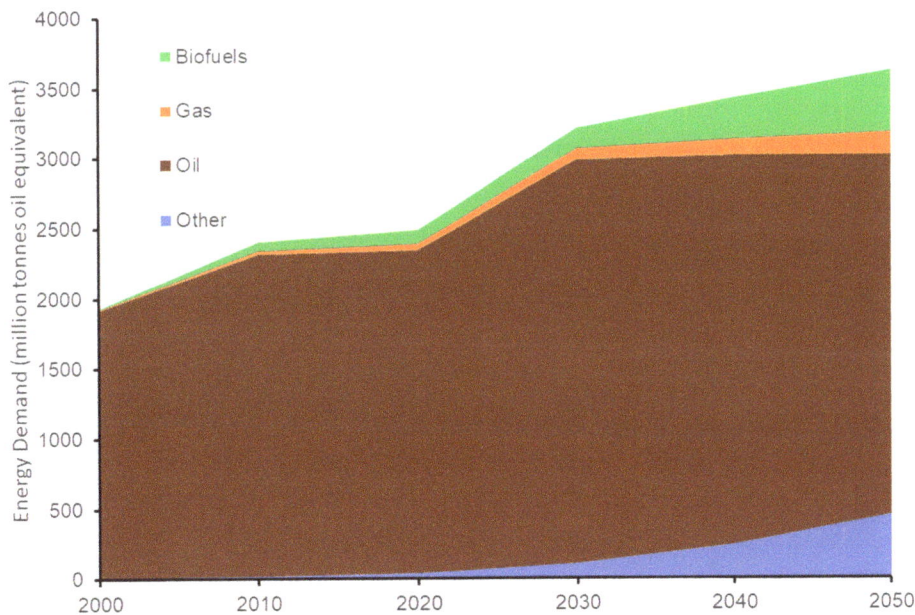

© SAE International.

23.3.1. Bio-Ethanol Feedstocks

The production of ethanol from non-edible feedstocks is obviously to be preferred over the current situation, where the fuel ethanol is produced from food crops. However, the non-edible crops must be grown on land that was not previously used for food. Fuel crops that can be grown on poor agricultural land are the focus of attention for researchers looking for alternative feedstocks. The production of methanol from such materials as wood chippings, corn stover, corn cobs, and straw or specifically grown crops such as switch grass and miscanthus is commonly referred to as cellulosic ethanol. This is discussed in more detail in Chapter 5, Section 5.3.3. Research will continue into more economical and less land intensive methods of producing cellulosic ethanol. Research will also continue into alternative fuels to ethanol, such as di-methyl-furan (DMF), as discussed in Section 23.5.

23.3.2. Biodiesel Feedstocks

Biodiesel can be produced from a vast array of different animal and plant oils; this is discussed further in Chapter 5, Section 5.4. The most commonly used feedstocks can also be used for food production. A great deal of attention is thus focused on producing biodiesel from non-edible plant oils. Two widely investigate plants families are *Jatropha curcas* [23.13–23.18], known simply as Jatropha and *Pongamia pinnata* [23.19–23.23], variously known as Pongamia, Pungai, Honge, or Karanja. Not only are the seeds from these plants inedible but the plants themselves are also very tolerant of poor habitat [23.24], although their oil yields are not exceptionally high [23.25].

An alternative approach to using seed oils is to cultivate algae, which also contain triglycerides that can be processed to produce biodiesel [23.26–23.29]. Algae, like seed-bearing plants, produce carbon-based compounds by photosynthesis and can fix larger amounts of CO_2 per unit area of land than other plants [23.30]. Certain algal strains can also use wastewater treatment effluents for nutrients [23.31–23.36].

Some of the other most commonly cited factors for favoring algae as a biofuel feedstock include the following:

- High growth rate.
- High oil content.
- Use of non-arable land.
- A tolerance to poor quality water.
- Produce valuable coproducts.

There are still significant challenges to be overcome in developing large-scale commercial algae to biodiesel facilities; these include temperature control, effective light dispersion, maintenance of healthy algal growth, avoidance of invasive native algae, reliable harvesting methods, effective extraction of oils, and of course the economics [23.37, 23.38]. Small-scale operations are already in place to grow algae for protein and food supplements The common procedures involve open "raceway ponds," with sparging of CO_2 into the ponds and closed photobioreactor designs. Photobioreactors help to maintain the purity of the algal strains and offer better control of operating conditions, but at the expense of higher capital and operating costs [23.39].

23.3.3. Other Renewable Feedstocks

Feedstocks for hydrotreated vegetable oil (HVO) are not limited to vegetable oils; the general term HVO also covers the products of hydrotreating of animal waste, including, tallow, yellow grease, choice white grease, poultry fat, fish oil, etc. used cooking oil (UCO), and with some preprocessing, agricultural residues, including corn stover, wheat straw, rice straw, etc., and possibly forestry residues. Supply of any of these individual products can inevitably be subject to unforeseen circumstances such as was demonstrated by the pandemic that struck in 2019–2020, when social distancing regulations resulted in restaurant closures and a big decrease in the availability of UCO [23.40], but the flexibility of the hydrotreatment process allows for the substitution of other products such as soybean oil. In some countries there is also ongoing interest in developing the use of forestry waste [23.41, 23.42]. These bio-derived products can also be used as feedstocks for coprocessing within a petroleum refinery [23.43].

23.4. Changes in Engine and Vehicle Technology

The big driver for changes in engine and vehicle technology in the short and medium terms will be to increase fuel economy and reduce CO_2 emissions while meeting proposed emissions legislation and still providing the customers with the comfort, safety, and performance that they have grown accustomed to. Ten governments, including Brazil, Canada, China, the European Union, India, Japan, Mexico, Saudi Arabia, South Korea, and the US, have already established fuel economy or GHG emission standards for LDVs.

In August 2012 the US Environmental Protection Agency (EPA) and the National Highway Traffic Safety Administration (NHTSA) issued a joint Final Rulemaking to further tighten the corporate average fuel economy IE) standards for passenger cars and light trucks for the years 2017 to 2025. By 2020 the passenger car fuel economy target was equivalent to 113.1 gCO_2/km, and for light trucks the target will be equivalent to 167.1 gCO_2/km. By 2025 the targets will be tightened to the equivalent of 88.9 and 126.1 gCO_2/km for cars and light trucks, respectively, with a combined light-duty fleet target equivalent to 101.3 gCO_2/km.

Based on a European Commission strategy to meet its greenhouse gas emission reduction targets under the Kyoto Protocol, legislation has been introduced setting targets for new passenger car and light truck fleets.

In Japan, at the end of March 2020, METI (Minister of Economy, Trade and Industry) issued new fuel efficiency standards for passenger vehicles for 2030. The aim of the standard was to require manufacturers to improve the fuel efficiency of their vehicles by 32.4% compared with the actual fuel efficiency in 2016. The 2030 standards are expanded to include BEVs and plug-in HEVs (PHEVs). According to METI's electric vehicle development target, it is estimated that the combined market penetration of BEVs and PHEVs will be 20% by 2030. The 2030 standards are defined in terms of the vehicles curb weight using the following formula:

$$FE = 30.65 - 8.52 \times 10^{-4} \times CW - 2.47 \times 10^{-6} \times CW^2$$

where FE is the fuel economy (in km/L) according to the WLTP (World harmonized Light vehicle Test Procedure) and CW is curb weight (in kg). This applies to vehicles weighing less than 2759 kg, for vehicles weighing more than 2759 kg there is a fixed standard of 9.50 km/L (approximately 22.35 miles/gallon (US)). The targets are for CAFE, gasoline equivalent. As the standards also apply to diesel, liquid petroleum gas (LPG), and electric vehicles, adjustments are made for diesel fueled and LPG fueled vehicles using a simple factor for each fuel, while for BEVs the energy consumption (Wh/km) for that vehicle is used, for PHEVs an average utility factor is incorporated.

These targets can be achieved by the evolution of existing technology rather than revolutionary approaches. The following sections suggest how the gasoline and diesel engines burning carbon-based fuels may evolve to meet these challenges and what will be necessary to move from this carbon-based economy to a hydrogen-based economy.

23.4.1. Gasoline Engine Development

Recent developments in gasoline engines have made significant improvements in engine efficiency. The widespread introduction of direct injection technology as discussed in Chapter 8, Section 8.3.1.4, have resulted in economy improvements while the stratified charge versions with advanced exhaust aftertreatment offer even greater benefits.

The ability to change inlet and exhaust valve timing, referred to as variable valve actuation (VVA), has also resulted in fuel economy benefits. The VVA can be accomplished by various means, including mechanically actuated cams such as multiple-cam-profile mechanisms [23.44] or fully controlled electro-hydraulic [23.45], or pneumatic systems [23.46]. VVA systems also allow the possibility to provide cylinder deactivation. Such technologies were first developed over 30 years ago [23.47, 23.48] but were not generally accepted due to the difficulty in providing smooth transitions between the different displacements. Modern control systems have overcome some of the original issues, allowing the capacity of the engine to be matched to the power requirements [23.49–23.52]. It is expected that this sort of technology will find wider application on larger engines. Further advances in VVA and control systems will allow the implementation of controlled autoignition (CAI) over parts of the engine operating range. The benefits of CAI were discussed in the previous chapter.

The use of VVA to control the engine displacement is one approach to providing an effective engine capacity to suit the power demand. Another approach is to reduce the size of the engine but then supplement the maximum power by pressure charging the engine. The smaller engine has lower frictional losses and has reduced pumping losses during low-load operation. This is commonly referred to as downsizing and is already taking place. For the same power output, a gasoline engine can be downsized by about 30%, resulting in a fuel economy improvement of about 10% [23.53]. Using a turbocharger to provide this engine boosting has the advantage of utilizing some waste energy in the exhaust to drive the compressor. However, to meet driver demand in terms of engine response requires more sophisticated turbo machinery such as variable geometry turbochargers (VGT) [23.54]. Two-stage pressure charging may also be necessary to ensure good transient response [23.55].

A turbocharger can recover some kinetic energy from the flow of exhaust gas, but most of the energy in the exhaust gas is in the form of thermal energy; very little of this energy is recovered in current automotive engines. This will be an area of future development, initially for gasoline engines, which currently have a higher waste heat output. Not all of this waste heat is dissipated through the exhaust system; the cooling system also dissipates a significant amount of fuel energy. Proposals were made 30 years ago to utilize this waste heat to provide useful work [23.56]. The current drive to improve average fleet fuel economy is generating new impetus to develop waste heat recovery systems [23.57–23.59]. While some of the early systems were directed at producing useful work that fed directly back into the powertrain, the spread of hybridization makes it attractive to convert waste heat into electrical energy [23.60, 23.61].

23.4.2. Diesel Engine Development

The diesel engine currently has a fuel economy advantage over its gasoline counterpart of between 20 and 30% [23.62, 23.63]. This makes it attractive to motor manufacturers aiming to meet fleet average CO_2 emissions targets. Advances in gasoline engine technology, as discussed in the previous section, are reducing this difference. Increasing diesel engine efficiency is usually achieved at the expense of increased NO_X emissions. Exhaust aftertreatment to reduce tailpipe NO_X emissions requires the use of an additional reducing agent or a lean NO_X trap (LNT). These technologies are discussed in more detail in Chapter 16, Section 16.4.3. The use of LNTs has a fuel economy penalty that can negate some of the advantages of calibrating the engine for improved fuel economy. In-cylinder control of NO_X by better exhaust gas recirculation (EGR) control and enhanced mixture control will increasingly be employed. High-pressure fuel injection with increased use of multiple injection strategies will allow diesel engine manufacturers to move toward the use of low-temperature combustion during part of the operating cycle.

The use of VVA in gasoline engines is promoted because it can reduce part-load pumping losses. One of the factors contributing to the fuel economy benefits of the diesel engine, when compared to the gasoline engine, is the absence of these part-load pumping losses. However, VVA still has a part to play in improving the fuel economy of the diesel engine [23.64, 23.65] as it influences the in-cylinder charge motion and the effective compression ratio. This in turn influences the in-cylinder pressures and temperatures, the dynamics of the EGR system, and will also affect the turbocharger operating point. Fully flexible VVA has also been proposed as a means of switching from a four-stroke to two-stroke operating regime [23.66]. However, the scope for VVA in a diesel engine is also limited by the reduced clearance between the valves and the piston when it is at the top of the exhaust stroke; this is due to the higher compression ratio of the diesel engine [23.67]. Most passenger car diesel engines already employ pressure charging to allow them to produce similar power outputs to their gasoline counterpart. Therefore, the scope for boosting and downsizing of diesel engines is more limited than for gasoline engines.

Although the diesel engine produces less waste heat than its gasoline counterpart, there is still an economic benefit to be obtained from the use of waste heat recovery systems, particularly for heavy-duty applications [23.68, 23.69]. Studies have also been undertaken to assess the benefits to be gained by applying waste heat recovery systems to the less conducive diesel passenger car engine [23.70]. Figure 23.3 [23.71] shows exergy (i.e., the available useful energy) contour plots for post-turbine exhaust and high-pressure EGR streams of a 1.9 L diesel engine.

FIGURE 23.3 Exergy contour plots for post-turbine exhaust and high-pressure EGR streams of a 1.9 L diesel engine.

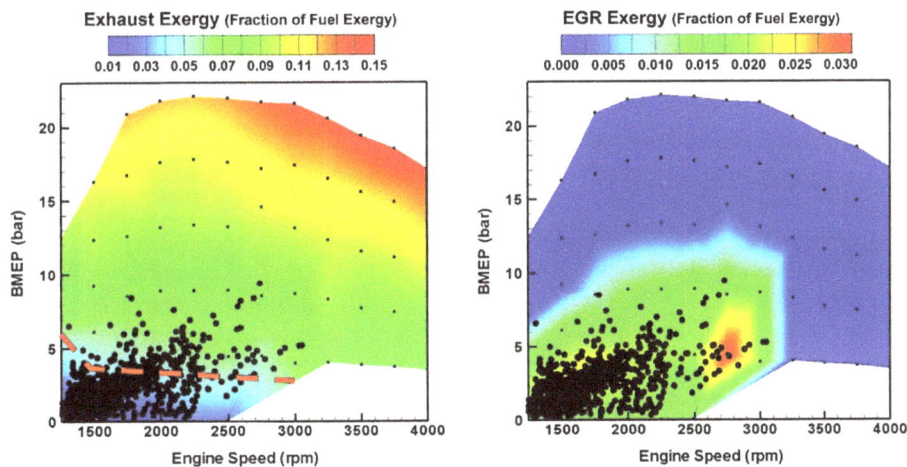

The left chart in Figure 23.3 shows the exergy of the post-turbine exhaust stream, with the larger black dots indicating the engine operating points corresponding to the US EPA Urban Dynamometer Driving Schedule (UDDS) and the thick dashed-line indicating a 250°C isotherm. As can be seen from the chart, the amount of available energy during the UDDS is relatively low. If the exhaust gas temperature falls below the 250°C line, this will reduce the performance of aftertreatment systems. Putting the heat recovery systems downstream of the aftertreatment systems will further reduce the amount of exergy reaching the recovery systems. The right-hand chart in Figure 23.3 shows the exergy in the high-pressure EGR circuit. During the UDDS there is a greater proportion of the total fuel exergy to be recovered. While fuel economy can be improved by utilizing some of this available energy, it is clear that it must consider the engine system as a whole, and the added complexity will be a deterrent. There are also opportunities to employ hybridization to improve the overall diesel vehicle efficiency [23.72–23.74].

23.4.3. Hybrid Electric Vehicles

Electric vehicles have been considered as the answer to the vehicle emissions problem, as the vehicle itself is perceived to produce zero emissions; this assumes that generating the electricity produces no emissions! However, there is a middle ground between the internal combustion engine running on gasoline or diesel fuel and the electric vehicle; these are termed HEVs. These HEVs can be configured in many forms, and some forms have been in commercial production since the turn of the 21st century. These HEVs can be divided into three general categories, full hybrid, mild hybrid, and micro hybrid, as indicated in Table 23.1.

TABLE 23.1 Comparison of hybrid categories [23.75].

		Full hybrid	Mild hybrid	Micro hybrid
Functions	Electric drive	Yes	No	No
	Torque assist	Yes	Yes	No
	Stop/start	Yes	Yes	Yes
	Regenerative braking	Yes	Yes	Yes (limited)
Hardware	Electric machine	High power electric motor	Crankshaft driven integrated starter generator or Belt driven integrated starter generator	Belt driven integrated starter generator or Enhanced starter motor
	Power source	High-voltage NiMH or Li-Ion battery pack	High-voltage NiMH or Li-Ion battery pack	Lead/acid batteries Ultracapacitors
Hybridization cost		High	Medium	Low

© SAE International.

The degree of hybridization determines the degree of functionality that can be provided by the electric motor and the cost of the system. Most of these vehicles have little or no effect on the fuel requirements in comparison with non-hybrid gasoline or diesel engine vehicles, other than that they are all designed to use less fuel. Because of existing or forthcoming legislation there are probably more HEV cars sold now than non-HEVs.

Full hybrid systems can be further subdivided into parallel hybrid and series hybrid configurations. The majority of HEVs are currently parallel hybrid configurations. In parallel hybrid design, the driven wheels are connected to the engine and the electric motor and can be driven by either or both power sources. In a series hybrid configuration, the driven wheels are not connected to the IC engine. The IC engine is connected to an electrical generator that charges a battery or supplies power directly to an electric motor, or motors, which drive the wheels. If the IC engine is sized only to recharge the battery, then this is known as a range-extended vehicle (REV). The vehicle runs as an electric vehicle, but the IC engine can then be used to recharge the battery to extend the range of the vehicle. For short journeys the battery can be recharged at an electrical charging point. Because the IC engine is only used to recharge the batteries, the normally wide range of operating conditions of the IC engine can be reduced and the engine can be designed to operate under more fuel-efficient conditions [23.76–23.78].

The fuel implication of REVs is that if the vehicle is operated in purely electric mode for a long period of time, the same batch of fuel may stand in the vehicle tank for a long period of time. If the fuel is gasoline, or more probably a petroleum/ethanol blend, then over an extended period of time the lighter fractions of the gasoline can be lost by evaporation. Because ethanol has a low boiling point and a high octane number, the octane number of the gasoline in the REV tank could fall over a prolonged period. If the REV is operating on a diesel fuel containing biodiesel, then the oxidation stability of the biodiesel will become more important than it is currently.

23.4.4. Fuel Cell Vehicles

A fuel cell is an electrochemical device in which chemical energy is converted directly into electrical energy. In December 1838, William Robert Grove, a British lawyer, sent a letter to the editors of the *Philosophical Magazine Journal*, describing a new voltaic cell [23.79] and four years later Gove sent another letter describing what was in effect the first hydrogen fuel cell [23.80]. Twenty-four years later General Motors described probably the first example of a hydrogen fuel cell vehicle (HFCV), the Electrovan [23.81, 23.82]. Figure 23.4 [23.81] shows the layout and the exterior view of the vehicle. What is of particular interest is that the vehicle not only carried a supply of hydrogen but also a supply of oxygen, an electrolyte top-up supply was also required.

FIGURE 23.4   Layout (left) and exterior view (right) of the GM Electrovan.

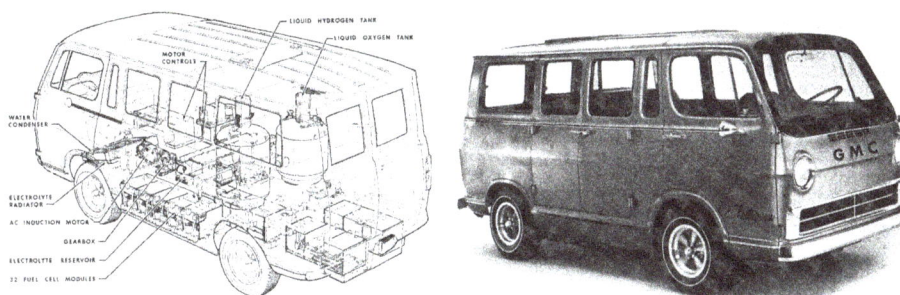

Over the last century many different types of fuel cells have been developed, with the majority of the work being done during the final quarter of that century, each type of cell has its own advantages and disadvantages [23.83]. One important characteristic of these different fuel cell technologies is their operating temperature. Some of the important types of fuel cell technology are listed below in terms of decreasing operating temperature:

- Solid oxide fuel cells (SOFC)
- Molten carbonate fuel cells (MCFC)
- Phosphoric acid fuel cells (PAFC) and alkaline fuel cells (AFC)
- Proton exchange membrane (PEM)

The low-temperature fuel cells are favored for automotive applications because they have the ability to start from cold. The medium- and high-temperature fuel cells must be artificially heated before they will start to operate [23.84]. The higher operating temperatures also necessitate greater thermal shielding, making low-temperature fuel cells preferable for vehicle applications.

At the start of the second half of the twentieth century, the major obstacles to the idea of hydrogen fuel-cell-powered vehicles were onboard storage of hydrogen and the hydrogen infrastructure. This led to the idea of onboard generation of hydrogen, primarily from hydrocarbon fuels that were readily available, methanol [23.85–23.87], ethanol [23.88], gasoline [23.89, 23.90], multifuel [23.91, 23.92], although ammonia has also been considered as a carbon-free hydrogen carrier [23.93, 23.94]. The idea of direct methanol fuel cells was also investigated [23.95–23.97].

Since GM trialed the Electrovan many motor manufacturers have produced development and prototype vehicles that are powered by hydrogen fuel cells [23.98]. In 2008 Honda began limited production of the Honda FCX Clarity [23.99–23.101] and in 2009 Mercedes-Benz presented its B-Class F-Cell, again for limited release [23.102]. In 2013, Hyundai Motor Co. announced that it has become the first company to begin mass production of a hydrogen fuel-cell-powered vehicle. The company planned to have 1000 of its vehicles leased to private companies and governments by 2015 and to start selling vehicles to customers shortly after that. Figure 23.5 [23.103] shows the layout and the exterior view of the vehicle.

Reprinted from "Development of Hyundai's Tucson FCEV," SAE Technical Paper 2005-01-0005, 2005 © SAE International.

FIGURE 23.5 Layout (left) and exterior view (right) of Hyundai fuel cell vehicle.

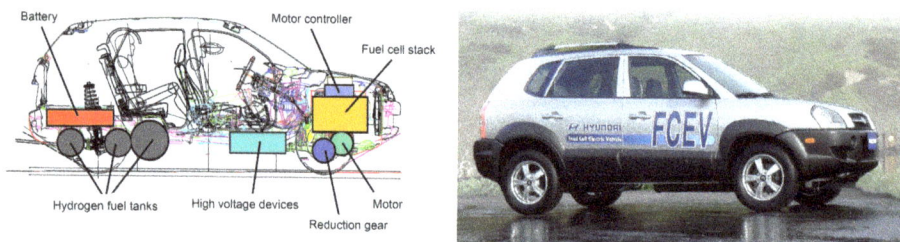

There are still considerable challenges to be overcome before there is a significant uptake of fuel cell vehicles. Although the Hyundai vehicle is capable of starting from an ambient condition of −20°C [23.104], the cold starting ability of fuel cell vehicles is still an issue. The water generated in a PEM fuel cell must be purged on shutdown, otherwise at subzero temperatures this will freeze. The ice formed will expand in the sandwiched structure of the PEM, causing mechanical stress and impeding the transport of reactants within the cell. The range of the vehicle is also a hurdle for widespread acceptance, although the use of multiple tanks and careful packaging to ensure safety has made the range of current models acceptable. The major challenge remains the hydrogen infrastructure, but this is changing as described in Section 23.5.1.

23.5. Alternative Fuels

The use of what were considered alternative fuels in gasoline and diesel engines is now common place, to some extent, in many areas. Niche markets already use LPG, liquefied natural gas (LNG), and compressed natural gas (CNG). Ethanol is commonly used in many countries as a low-volume blending component in the majority of commercial gasoline. In countries such as the US and many European countries, gasoline with a high blend of ethanol is also available for dedicated flex-fuel vehicles. In Brazil, high ethanol blend gasolines have been available for many years. The use of fatty acid methyl esters (FAME) as a diesel blending component is also now common. Some vehicles are capable of running on neat FAME. The use of HVO is also now common place.

As discussed in Section 23.2, the vast majority of renewable source ethanol is produced by fermentation of material that could be used as food. Similarly, the majority of FAME feedstock could also be used for foodstuff production. There are also issues with the use of FAME in diesel engines that have been discussed elsewhere in this book. While work continues to identify methods of producing ethanol from cellulosic feedstocks, other products have been identified that can be derived from such feedstocks.

Probably the most important "alternative fuel" these days is hydrogen which can be generated by electrolysis of water, as discussed in Chapter 5, Section 5.7.1. and its use id discussed in the following section. This is then followed by a brief discussion of ammonia which is another carbon-free product that is being investigated as a potential alternative fuel. There is then a brief discussion of hydrocarbon fuels that can be produced renewably.

23.5.1. Hydrogen

The concept of using hydrogen as a fuel for powered mobility is not new [23.105]. During the last century there have been significant bouts of interest in using hydrogen to replace conventional fossil fuels; the interest was often spurred by the knowledge that fossil fuels were finite, and their exhaustion was seen as being fairly soon. Before concerns regarding greenhouse gas (GHG) emissions, it was also appreciated that the use of hydrogen as opposed to hydrocarbons would eliminate unwanted emissions of CO and HC. In the late 1960s it was written:

> *Hydrogen, as a fuel, has unique properties; however, the difficulties of cryogenic or compressed gas storage have limited its field of application. A study performed by the Allis-Chalmers Mfg. Co. [23.106] indicates a projected cost for electrolytic hydrogen produced by off-peak nuclear power from large breeder-type reactors of 20¢/thousand standard cu ft. (0.7¢/m³). This cost corresponds, on the basis of energy content, to a cost of 8.5¢/gal. (2.2¢/L) of gasoline. Thus, the economic advantage of producing large blocks of electricity in advanced nuclear plants can be applied to small power package applications through the use of hydrogen as a fuel.*

> *Of particular interest, because of the absence of hydrocarbon pollutants in the combustion of hydrogen, is the use of hydrogen, stored in a metal hydride, as a fuel for an internal combustion engine for vehicular propulsion. Hydrogen could also be used to power a fuel cell for propulsion and in the longer term, as the cost of fuel cells is decreased, this may prove to be a more attractive application; however, as a near-term alternative to the bulky and limited range battery-driven vehicles now being developed, the hydrogen fueled internal-combustion engine appears promising. This approach also represents a minimum departure from current practice as compared with the fuel cell or battery-powered vehicle. The exhaust of such a vehicle would be free of the hydrocarbon and carbon monoxide emissions that now account for >50% of the air pollution in our cities. [23.107].*

The answer to dwindling fossil fuel supplies and the reduction of emissions gave rise to the idea of the hydrogen economy [23.108].

This was before concern regarding a carbon footprint and the vast majority of hydrogen was derived from methane [23.109], which is now known as blue hydrogen as opposed to green hydrogen [23.110], which is generated from carbon-neutral sources.

23.5.1.1. Hydrogen as a Fuel for Spark Ignition Engines

Hydrogen-fueled IC engines were seen as a means of transitioning from a liquid hydrocarbon to a hydrogen economy. Nobody would buy a hydrogen-powered car if there were no hydrogen filling stations, and nobody would install a hydrogen infrastructure if there were no customers. By developing passenger cars like the BMW 7 Series that could run on hydrogen or gasoline, a demand for hydrogen could be established with the ultimate goal that people would then buy hydrogen fuel cell vehicles (see Section 23.4.4) that would eliminate the need for hydrocarbon fuels.

The replacement of gasoline with hydrogen was initially seen as an easy alternative to fuel cell vehicles. Simply replacing gasoline with hydrogen is not a trivial matter. Hydrogen has a very high flame speed and a wide flammability range; this makes it attractive in that it can be used at extremely lean air-to-fuel ratios. This means that for most of the engine operating range the engine can be run unthrottled, thus reducing pumping losses and hence improving efficiency. However, these characteristics will also result in undesirable behavior that was initially a major problem when using hydrogen in a gasoline engine. The wide ignitability range meant that engines were prone to backfire, and the high flame speed, while good for thermodynamic efficiency, made the engine prone to knock.

There is also the issue of safe storage and handling of hydrogen. Initial enthusiasm was for metal hydride storage [23.107, 23.111] although this has not yet become a reality and is still an area of much research [23.112–23.115]. The use of pressurized hydrogen was not the preferred solution, but it was not an insurmountable hurdle. In the early 1970s as part of the Urban Vehicle Design Competition [23.116] undergraduate students at the University of California, Los Angeles designed and built a hydrogen-power subcompact car that was declared the overall winner in Class I, Internal Combustion Engines, and from a practical point of view its only major limitation was its driving range [23.117].

Since that time some of the major motor manufacturers have made fully functional vehicles that can be powered by hydrogen. In the US one example was the Ford P2000 [23.118–23.120], which was developed as the first production viable hydrogen-powered vehicle. In building the car, great care was taken regarding the safety aspects and a triple redundant hydrogen safety system consisting of gas sensing, and active and passive elements. A schematic of the fuel and safety systems is shown in Figure 23.6 [23.118].

FIGURE 23.6   Vehicle wire frame showing locations of fuel and safety system within the vehicle.

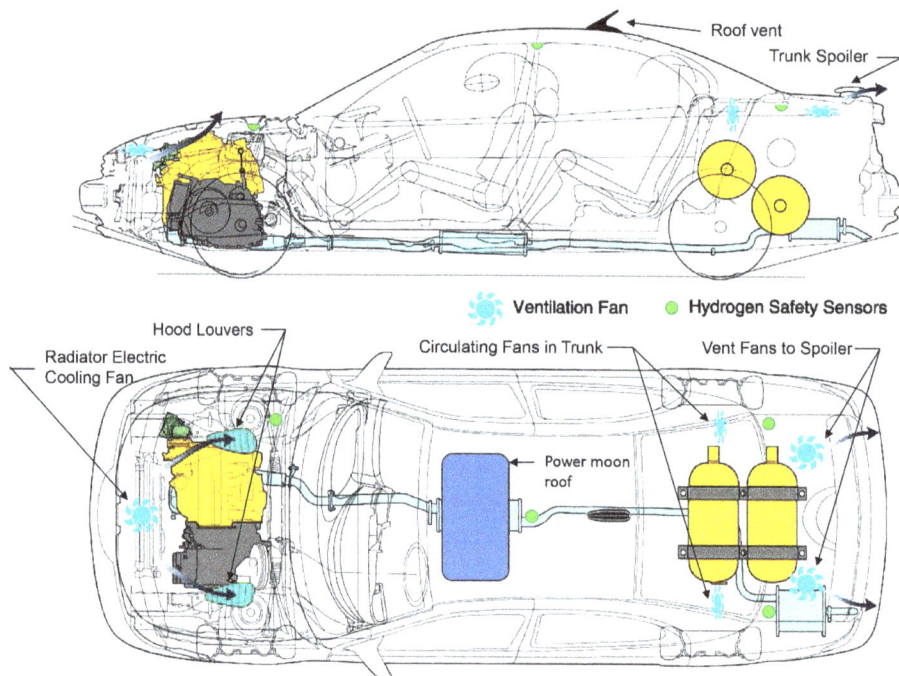

© SAE International.

The P2000 had an engine based on a naturally aspirated production engine. The maximum power was less than the stock gasoline engine but still gave acceptable in-city driving feel. The HC and CO emissions were below the SULEV targets but the NO_X emissions would require further changes to the engine design. Pressure charging would increase the torque and power output while the use of EGR or aftertreatment would reduce the tailpipe NO_X emissions.

Ford also produced a demonstration fleet of 30 E-450 shuttle buses with a 6.8 L engine that runs on hydrogen [23.121]. The 8- to 12-passenger shuttle bus with a 4.47 m wheelbase and an estimated gross vehicle weight of 6373 kg is equipped with a compressed hydrogen onboard storage system that holds up to 29.6 kg of hydrogen at a pressure of 34.5 MPa with a resulting vehicle range of 240 to 320 km [23.122, 23.123]. In this case, the engine was fitted with a supercharger to compensate for the air that was displaced by the manifold injection of hydrogen.

In Japan Mazda developed a bi-fuel version of their RX-8 sports car. This used the twin-rotor Wankel engine which could be fueled with either hydrogen or conventional gasoline [23.124]. The car was made available to a limited market from 2006 [23.125].

In Europe the BMW group developed a bi-fuel version of their 7 Series sedan. The group developed their 12-cylinder, 5.9 L gasoline engine so that it could run as a direct injection gasoline engine or as a manifold injected hydrogen engine [23.126]. This engine operated with an approximately stoichiometric hydrogen-to-air mixture that allowed the three-way catalytic converter to function normally when the engine was running on gasoline and also to function when running on hydrogen. The engine was able to operate with NO_X emissions well below the regulated levels, although the catalyst formulation had to be adapted to favor NO_X reduction. Due to the manifold mixing of air and hydrogen, some air is displaced in a naturally aspirated engine; the amount of mixture trapped each cycle is thus reduced when operating on hydrogen. The maximum power is thus also reduced. A hydrogen-only vehicle was also developed, which again exhibited extremely low emissions values that were well below regulated values. The fuel economy on the FTP-75 test cycle was 3.7 kg of hydrogen per 100 km, which, on an energy basis, is equivalent to a gasoline fuel consumption of 13.8 L/100 km [23.127]. Of course, there are no CO_2 emissions. None of the vehicle projects discussed above have progressed and whether the idea of hydrogen fueled IC engines will materialize is unknown.

A stumbling block for these projects may have been the sparsity of hydrogen filling stations, but this situation is changing with many producers setting up their own infrastructure or teaming up with well know fuel producers [23.128–23.131]. As a result, the number of hydrogen filling stations is growing rapidly, Figure 23.7 shows a filling station in Germany.

FIGURE 23.7 Hydrogen filling station in Germany.

H2 MOBILITY Deutschland.

Filling stations typically dispense hydrogen at a higher pressure for light-duty vehicles and a lower pressure for all other vehicles, the higher-pressure system dispense hydrogen at 70 MPa (10,000 psi) which is labeled as H70 with the lower pressure system at 35 MPa (5,000 psi) labeled H35, this is the pressure used by Ford for their E-450 shuttle buses mentioned above. By the end of 2022, there were 814 hydrogen refueling stations in operation around the world [23.132].

23.5.2. Ammonia

Over half a century ago the US military realized that:

> *Because of increased mechanization, petroleum supply has become one of the major problems of military logistics, especially in Army operations where small, dispersed energy demands often necessitate an extensive, vulnerable fuel supply complex. The nuclear powered energy depot, conceived as a potential solution to the problem, will utilize a nuclear reactor to produce a chemical fuel for vehicle and aircraft engines. The energy depot, logistically independent for a year, would operate with or near the consumer in the field and considerably broaden Army capabilities [23.133].*

While the term e-fuel had not been invented at that time, this was in effect the conception of the e-fuel phenomenon. The concept of the Energy Depot was illustrated as shown in Figure 23.8 [23.134]. As can be seen from the figure the concept intended that water be electrolyzed to produce hydrogen as the basis of the fuel, but to increase its storage density, consideration was given to compounding the hydrogen with nitrogen and/or oxygen. The favored solution was to use ammonia (NH_3) as the energy carrier for use in conventional engines. Ammonia has a better volumetric energy density than hydrogen, with ammonia, more hydrogen can be stored in liquid form without the need for cryogenic storage, ammonia vaporizes at −33.4°C as opposed to hydrogen that vaporizes at −252.9°C.

FIGURE 23.8 Energy Depot—mobile fuel manufacturing plant.

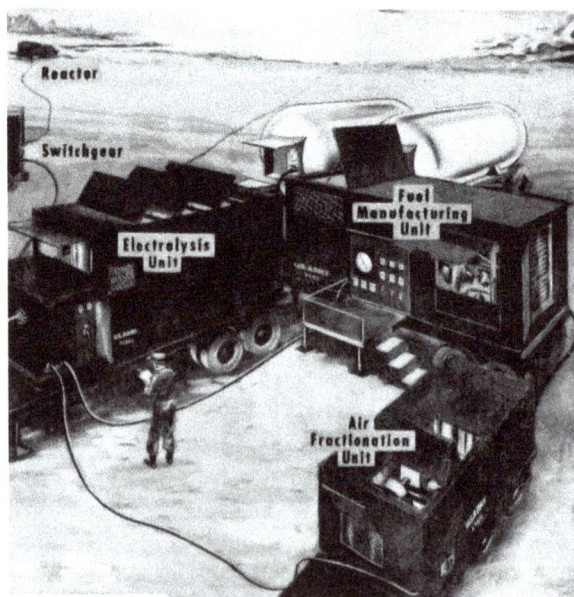

At the time work, was done considering the use of ammonia as a fuel for both spark ignition [23.135, 23.136] and compression ignition engines [23.137]. Although, at that time the logistical need for local fuel production has probably diminishing, the environmental incentive to move away from a carbon economy to a hydrogen economy has renewed interest in ammonia as a hydrogen carrier, as an enabler of the move to a hydrogen economy [23.138]. As a potential fuel for IC engines, the use of ammonia in combination with hydrogen is also under investigation [23.139–23.142]. The combustion of ammonia can inevitably lead to some NO_X emissions [23.143, 23.144] but these can be controlled as with other IC engines.

As in the 1960s the use of a fuel cell was the preferred option for converting the chemical energy stored in the ammonia into a more usable form [23.145–23.149]. Alkaline and alkaline membrane direct ammonia fuel cells have been investigated. SOFCs are the favored type of cell for use with ammonia as the fuel source [23.150]. This is because a zirconia based solid electrolyte is stable in the presence of the ammonia which can be catalytically decomposed on conventional Ni-based anodes; the heat released from the oxidation of hydrogen in the cells can be used for the endothermic decomposition of ammonia.

23.5.3. Dimethyl Ether

Di-methyl-ether (DME), also known as methoxymethane, has the formula CH_3-O-CH_3 and is produced as a side-stream of methanol production from synthesis gas. The yield of DME can be improved by selectivity in the catalyst [23.151]. At room temperature it is a colorless gas; its boiling point is –24.9°C. For automotive fuel applications, it was first used to enhance the autoignition quality of methanol [23.152–23.155]. Neat DME has a cetane number (CN) of about 60, but fuel grade DME contains some methanol and water and has a slightly lower CN. Because the DME molecule does not contain any carbon-carbon bonds, it has an extremely low tendency to form soot. The high CN and the absence of a carbon-carbon bond within the DME molecule suggests that it would make a good fuel for diesel engine application. This was first seriously investigated in the mid-1990s [23.156–23.159]. Unfortunately, DME has very low inherent lubricity and therefore requires the use of lubricity additives. The evaluation of DMEs lubricity performance presents its own challenge due to the low boiling point of the DME [23.160].

Because DME is gaseous at normal ambient temperatures, it obviously cannot be stored, handled, and used like diesel fuel. However, with a vapor pressure of 0.51 MPa it can be handled in a similar fashion to LPG, which is discussed further in Chapter 4, Section 4.4. The need for a pressurized fuel tank usually limits the shape of these tanks to being cylindrical. Cylindrical tanks can easily be mounted along the chassis rails of commercial vehicles as shown in Figure 23.9 [23.161] but is less easy to accommodate within the tight packaging constraints of a passenger car.

FIGURE 23.9 Layout of medium-duty truck with two DME fuel tanks.

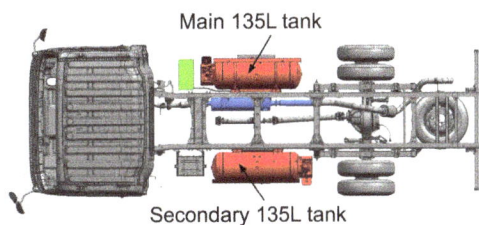

Main 135L tank

Secondary 135L tank

The lower heating value of DME is 28.43 MJ/kg, which is only about 65% of that of diesel fuel, which means that far more fuel has to be injected. This fact, along with the requirement to keep the DME liquid, means that specialized fuel injection equipment needed to be developed [23.162, 23.163]. In the 21st century,

Japanese researchers have made a great deal of progress in investigating the development of diesel fuel injection equipment to handle DME [23.164–23.167]. If the DME is produced from fossil sources, namely natural gas, then the well-to-wheels GHG emissions are not significantly different from petroleum diesel fuel, if produced from non-fossil sources such as manure waste, then there are more significant reductions in GHG emissions [23.168].

23.5.4. Polyoxymethylene Dimethyl Ethers (OME$_X$)

Polyoxymethylene dimethyl ethers (PODE or POMDME), also known simply as oxymethylene dimethyl ethers (OME$_X$), have the formula CH$_3$-O-(CH$_2$-O)$_X$-CH$_3$ where x is typically from 3 to 8, with smaller quantities formed with x being 1 or 2. Like DME, which can be classed as OME$_0$, OME$_X$ can be produced as an e-fuel from the electrolysis of water and captured CO$_2$, making it an attractive alternative fuel. Due to the large number of oxygen atoms and the absence of carbon-carbon bonds within the molecule, OME$_X$ has an extremely low sooting tendency. However, unlike DME, OME$_X$ is liquid at normal ambient temperatures which makes it more attractive as an alternative fuel. Some of the properties of the different OME$_X$ fuels are given in Table 23.2 [23.169].

TABLE 23.2 Properties of different of OME$_X$ fuels [23.169].

Name	Formula	Oxygen content [wt.%]	LHV [MJ/kg]	Density [kg/ dm^3]	Cetane number	Melting point [°C]	Boiling point [°C]
OME$_0$ (DME)	C$_2$H$_6$O	34.7	28.6	0.66	>55	−142	−25
OME$_1$	C$_3$H$_8$O$_2$	42.1	23.3	0.86	28	−105	42
OME$_2$	C$_4$H$_{10}$O$_3$	45.2	21.0	0.98	68	−70	105
OME$_3$	C$_5$H$_{12}$O$_4$	47.0	19.7	1.03	72	−43	156
OME$_4$	C$_6$H$_{14}$O$_5$	48.1	19.0	1.07	84	−10	202
OME$_5$	C$_7$H$_{16}$O$_6$	48.9	18.5	1.11	93	18	242

© SAE International.

As can be seen from Table 23.2, as the number of oxymethylene groups increases so does the melting and boiling temperatures. OME1 has a low cetene number and has been found to have poor lubricity performance when blended with diesel fuel and also result in high methene emissions [23.170]. Other OME$_X$ tend to have an increasing CN as the number of oxymethylene groups increases. OME$_X$ has been studied in a diesel engine, as a substitute for diesel fuel [23.170] and as a blend with diesel fuel [23.171–23.174] as a means of reducing criteria emissions.

With the further development of advanced engine technologies such as RCCI (see Chapter 22, Section 22.3.3) the idea of using OME$_X$ with gasoline become an interesting proposition for dual-fuel strategies. The OME$_X$ is a low sooting, high reactivity fuel and the gasoline is a low reactivity fuel, so controlling the ratio of these two fuels can be used to control the reactivity of the fuel for compression ignition [23.175, 23.176]. The use of OMEX and diesel fuel in dual-fuel engines has also been investigated [23.177].

23.5.5. Furans

Another group of chemicals based on furan have been investigated as possible fuels from renewable sources. A furan consists of an aromatic ring composed of four carbon atoms and one oxygen atom. In its simplest form, each of the four carbon atoms has one hydrogen atom attached, giving it the chemical formula C$_4$H$_4$O.

Furans can be produced from sugars such as cellulose, fructose, and glucose. Recent research has suggested that lignocellulosic biomass can also be converted directly into 5-hydroxy-methyl-furfural (HMF) [23.178,

23.179]. The HMF can then be converted to DMF over a copper-based catalyst [23.180] or without isolation using a carbon supported palladium catalyst [23.181]. Furan derivatives, such as the esters and ethers, can also be made directly from sugars [23.182, 23.183]. This makes these compounds attractive as the catalytic process is faster than the fermentation process that is currently used to produce alcohols, such as ethanol.

A limited amount of testing has been conducted using the ether derivatives as blending components in diesel fuel [23.184]. These furan derivatives have a high energy density, high CN, and do not have the low-temperature operability problems that have been associated with biodiesel. While the addition of these compounds produces emissions reductions, there can be phase separation when blended at high concentrations into petroleum diesel fuel.

The compounds 2-methyl-furan (2-MF) and DMF have been investigated as alternative gasoline blending components. Both of these compounds have octane numbers slightly higher than regular gasoline [23.185]. Both compounds also have properties that make them more attractive than ethanol as gasoline blending components. Ethanol has a high oxygen content that is partly responsible for its low volumetric energy content. The lower heating value of 2-MF and DMF are 27.6 and 30.1 MJ/L, respectively. This compares favorably with gasoline at about 32.9 MJ/L and is significantly higher than the 21.4 MJ/L for ethanol. Neither of these furan derivatives are readily soluble in water and will thus not have the problem encountered with ethanol/gasoline blends regarding the pickup of water bottoms.

Due to the higher energy density, DMF is possibly more attractive as a gasoline blending component, and this compound has been more thoroughly investigated [23.186–23.190]. These studies show that the behavior of DMF is closer to gasoline than is the behavior of ethanol. DMF does not have a deleterious effect on the emissions performance of gasoline. Due to its lower oxygen content, DMF does not reduce the emissions to the same extent as ethanol [23.191]. On the other hand, although 2-MF reduced the emissions of HC and CO, it was found to increase NO_X emissions [23.192].

One note of caution about these compounds: they both have far higher oral toxicity than gasoline or ethanol [23.193], and there are studies that suggest DMF may be genotoxic [23.194] and that its combustion products may also be toxic [23.195, 23.196].

References

23.1.   Del Pero, F., Delogu, M., and Pierini, M., "Life Cycle Assessment in the Automotive Sector: A Comparative Case Study of Internal Combustion Engine (ICE) and Electric Car," *Procedia Structural Integrity* 12 (2018): 521-537.

23.2.   Kalghatgi, G., "Is It the End of Combustion and Engine Combustion Research? Should It Be?" *Transportation Engineering* 10 (2022): 100142.

23.3.   UN, "Paris Agreement," in *Report of the Conference of the Parties to the United Nations Framework Convention on Climate Change (21st Session)*, Paris, France, 2015.

23.4.   UK Government, "Government Takes Historic Step towards Net-Zero with End of Sale of New Petrol and Diesel Cars by 2030," November 2020, accessed March 2023, https://www.gov.uk/government/news/government-takes-historic-step-towards-net-zero-with-end-of-sale-of-new-petrol-and-diesel-cars-by-2030.

23.5.   European Parliament, "EU Ban on the Sale of New Petrol and Diesel Cars from 2035 Explained," June 2023, accessed March 2023, https://www.europarl.europa.eu/news/en/headlines/economy/20221019STO44572/eu-ban-on-sale-of-new-petrol-and-diesel-cars-from-2035-explained.

23.6.   Mildenhall, O., "Mazda Announces Low Emission in-Line Six-Cylinder Diesel Engine for Mazda CX-60," *Mazda News*, July 2022, acessed August 2023, https://www.insidemazda.co.uk/2022/07/05/mazda-announces-low-emission-in-line-six-cylinder-diesel-engine-for-mazda-cx-60-2/.

23.7.   The Sunday Times, "Porsche 911 Sport Classic Review—Powered by e-Fuel," August 2023, accessed August 2023, https://www.thetimes.co.uk/article/porsche-911-sport-classic-review-powered-by-e-fuel-p9ms7qxv0.

23.8.   Exxon Mobil Corporation, "The Outlook for Energy Is ExxonMobil's Latest View of Energy Demand and Supply through 2050," 2023, accessed March 2023, https://corporate.exxonmobil.com/energy-and-innovation/outlook-for-energy.

23.9.   Pasqualetti, M.J., "The Alberta Oil Sands from Both Sides of the Border," *Geographical Review* 99, no. 2 (2009): 248-267.

23.10. Liu, C.L., Che, C., Zhu, J., and Yang, H., "China's Endowment 2: China Assesses Unconventional Land Oil Shale, Oil Sands, Coal Gas Resources," *Oil & Gas Journal* 108, no. 15 (2010): 36-39.

23.11. Nolan, P., Shipman, A., and Rui, H., "Coal Liquefaction, Shenhua Group, and China's Energy Security," *European Management Journal* 22, no. 2 (2004): 150-164.

23.12. Liu, Z., Shi, S., and Li, Y., "Coal Liquefaction Technologies—Development in China and Challenges in Chemical Reaction Engineering," *Chemical Engineering Science* 65, no. 1 (2010, 2010): 12-17.

23.13. Mandpe, S., Kadlaskar, S., Degen, W., and Keppeler, S., "On Road Testing of Advanced Common Rail Diesel Vehicles with Biodiesel from the *Jatropha curcas* Plant," SAE Technical Paper 2005-26-356 (2005), doi:https://doi.org/10.4271/2005-26-356.

23.14. Choudhury, S. and Bose, P.K., "Jatropha Derived Biodiesel – Its Suitability as CI Engine Fuel," SAE Technical Paper 2008-28-0040 (2008), doi:https://doi.org/10.4271/2008-28-0040.

23.15. Che Hamzah, N.H., Khairuddin, N., Siddique, B.M., and Hassan, M.A., "Potential of *Jatropha curcas* L. as Biodiesel Feedstock in Malaysia: A Concise Review," *Processes* 8, no. 7 (2020): 786.

23.16. Neupane, D., Bhattarai, D., Ahmed, Z., Das, B. et al., "Growing Jatropha (*Jatropha curcas* L.) as a Potential Second-Generation Biodiesel Feedstock," *Inventions* 6, no. 4 (2021): 60.

23.17. Yaqoob, H., Teoh, Y.H., Sher, F., Ashraf, M.U. et al., "*Jatropha curcas* Biodiesel: A Lucrative Recipe for Pakistan's Energy Sector," *Processes* 9, no. 7 (2021): 1129.

23.18. Riayatsyah, T.M.I., Sebayang, A.H., Silitonga, A.S., Padli, Y. et al., "Current Progress of *Jatropha curcas* Commoditisation as Biodiesel Feedstock: A Comprehensive Review," *Frontiers in Energy Research* 9 (2022): 1019.

23.19. Muralidharan, M., Thariyan, M.P., Roy, S., Subrahmanyam, J.P. et al., "Use of Pongamia Biodiesel in CI Engines for Rural Application," SAE Technical Paper 2004-28-0030 (2004), doi:https://doi.org/10.4271/2004-28-0030.

23.20. Kumar, R., Sharma, M., Ray, S.S., Sarpal, A.S. et al., "Biodiesel from *Jatropha curcas* and *Pongamia pinnata*," SAE Technical Paper 2004-28-0087 (2004), doi:https://doi.org/10.4271/2004-28-0087.

23.21. Aminah, A. and Syamsuwida, D., "Natural Growing Site and Cultivation of Pongamia (*Pongamia pinnata* (L.) Pierre) as a Source of Biodiesel Raw Materials," *IOP Conference Series: Earth and Environmental Science* 308, no. 1 (2019): 012050.

23.22. Hasnah, T., Leksono, B., Sumedi, N., Windyarini, E. et al., "Pongamia as a Potential Biofuel Crop: Oil Content of *Pongamia pinnata* from the Best Provenance in Java, Indonesia," in *International Conference and Utility Exhibition on Energy, Environment and Climate Change (ICUE)*, 1-6, 2021, IEEE.

23.23. Degani, E., Prasad, M.V.R., Paradkar, A., Pena, R. et al., "A Critical Review of *Pongamia pinnata* Multiple Applications: From Land Remediation and Carbon Sequestration to Socioeconomic Benefits," *Journal of Environmental Management* 324 (2022): 116297.

23.24. Scott, P.T., Pregelj, L., Chen, N., Hadler, J.S. et al., "*Pongamia pinnata*: An Untapped Resource for the Biofuels Industry of the Future," *BioEnergy Research* 1, no. 1 (2008): 2-11.

23.25. Divakara, B.N., Upadhyaya, H.D., Wani, S.P., and Laxmipathi Gowda, C.L., "Biology and Genetic Improvement of *Jatropha curcas* L.: A Review," *Applied Energy* 87, no. 3 (2010): 732-742.

23.26. Singh, A., Nigam, P.S., and Murphy, J.D., "Renewable Fuels from Algae: An Answer to Debatable Land Based Fuels," *Bioresource Technology* 102, no. 1 (2011): 10-16.

23.27. Adeniyi, O.M., Azimov, U., and Burluka, A., "Algae Biofuel: Current Status and Future Applications," *Renewable and Sustainable Energy Reviews* 90 (2018): 316-335.

23.28. Saad, M.G., Dosoky, N.S., Zoromba, M.S., and Shafik, H.M., "Algal Biofuels: Current Status and Key Challenges," *Energies* 12, no. 10 (2019): 1920.

23.29. Bošnjaković, M. and Sinaga, N., "The Perspective of Large-Scale Production of Algae Biodiesel," *Applied Sciences* 10, no. 22 (2020): 8181.

23.30. Brown, L.M. and Zeiler, K.G., "Aquatic Biomass and Carbon Dioxide Trapping," *Energy Conversion and Management* 34, no. 9-11 (1993): 1005-1013.

23.31. Woertz, I., Feffer, A., Lundquist, T., and Nelson, Y., "Algae Grown on Dairy and Municipal Wastewater for Simultaneous Nutrient Removal and Lipid Production for Biofuel Feedstock," *Journal of Environmental Engineering* 135, no. 11 (2009): 1115-1122.

23.32. Chinnasamy, S., Bhatnagar, A., Hunt, R.W., and Das, K.C., "Microalgae Cultivation in a Wastewater Dominated by Carpet Mill Effluents for Biofuel Applications," *Bioresource Technology* 101, no. 9 (2010): 3097-3105.

23.33. Pittman, J.K., Dean, A.P., and Osundeko, O., "The Potential of Sustainable Algal Biofuel Production Using Wastewater Resources," *Bioresource Technology* 102, no. 1 (2011): 17-25.

23.34. Zhou, W., Chen, P., Min, M., Ma, X. et al., "Environment-Enhancing Algal Biofuel Production Using Wastewaters," *Renewable and Sustainable Energy Reviews* 36 (2014): 256-269.

23.35. Marella, T.K., Datta, A., Patil, M.D., Dixit, S. et al., "Biodiesel Production through Algal Cultivation in Urban Wastewater Using Algal Floway," *Bioresource Technology* 280 (2019): 222-228.

23.36. Ahmad, A., Banat, F., Alsafar, H., and Hasan, S.W., "Algae Biotechnology for Industrial Wastewater Treatment, Bioenergy Production, and High-Value Bioproducts," *Science of The Total Environment* 806 (2022): 150585.

23.37. Benemann, J.R., "CO_2 Mitigation with Microalgae Systems," *Energy Conversion and Management* 38, no. Supplement (1997): S475-S479.

23.38. Jacob, A., Ashok, B., Alagumalai, A., Chyuan, O.H. et al., "Critical Review on Third Generation Micro Algae Biodiesel Production and Its Feasibility as Future Bioenergy for IC Engine Applications," *Energy Conversion and Management* 228 (2021): 113655.

23.39. Chisti, Y., "Biodiesel from Microalgae," *Biotechnology Advances* 25, no. 3 (2007): 294-306.

23.40. Media, A., "Renewable Feedstocks Present New and Complex Opportunities," *Industrial Biotechnology* 17, no. 3 (2021): 166-169.

23.41. Börjesson, P., Hansson, J., and Berndes, G., "Future Demand for Forest-Based Biomass for Energy Purposes in Sweden," *Forest Ecology and Management* 383 (2017): 17-26.

23.42. Soam, S. and Börjesson, P., "Considerations on Potentials, Greenhouse Gas, and Energy Performance of Biofuels Based on Forest Residues for Heavy-Duty Road Transport in Sweden," *Energies* 13, no. 24 (2020): 6701.

23.43. Han, X., Wang, H., Zeng, Y., and Liu, J., "Advancing the Application of Bio-Oils by Co-Processing with Petroleum Intermediates: A Review," *Energy Conversion and Management* 10 (2021): 100069.

23.44. Sellnau, M. and Rask, E., "Two-Step Variable Valve Actuation for Fuel Economy, Emissions, and Performance," SAE Technical Paper 2003-01-0029 (2003), doi:https://doi.org/10.4271/2003-01-0029.

23.45. Turner, C.W., Babbitt, G.R., Balton, C.S., Raimao, M.A. et al., "Design and Control of a Two-Stage Electro-Hydraulic Valve Actuation System," SAE Technical Paper 2004-01-1265 (2004), doi:https://doi.org/10.4271/2004-01-1265.

23.46. Trajkovic, S., Milosavljevic, A., Tunestål, P., and Johansson, B., "FPGA Controlled Pneumatic Variable Valve Actuation," SAE Technical Paper 2006-01-0041 (2006), doi:https://doi.org/10.4271/2006-01-0041.

23.47. Bates, B., Dosdall, J.M., and Smith, D.H., "Variable Displacement by Engine Valve Control," SAE Technical Paper 780145 (1978), doi:https://doi.org/10.4271/780145.

23.48. Fukui, T., Nakagami, T., Endo, H., Katsumoto, T. et al., "Mitsubishi Orion-MD—A New Variable Displacement Engine," SAE Technical Paper 831007 (1983), doi:https://doi.org/10.4271/831007.

23.49. Leone, T.G. and Pozar, M., "Fuel Economy Benefit of Cylinder Deactivation—Sensitivity to Vehicle Application and Operating Constraints," SAE Technical Paper 2001-01-3591 (2001), doi:https://doi.org/10.4271/2001-01-3591.

23.50. Falkowski, A., McElwee, M., and Bonne, M., "Design and Development of the DaimlerChrysler 5.7L HEMI® Engine Multi-Displacement Cylinder Deactivation System," SAE Technical Paper 2004-01-2106 (2004), doi:https://doi.org/10.4271/2004-01-2106.

23.51. Rebbert, M., Kreusen, G., and Lauer, S., "A New Cylinder Deactivation by FEV and Mahle," SAE Technical Paper 2008-01-1354 (2008), doi:https://doi.org/10.4271/2008-01-1354.

23.52. Senapati, U., McDevitt, I., and Hankinson, A., "Vehicle Refinement Challenges for a Large Displacement Engine with Cylinder Deactivation Capability," SAE Technical Paper 2011-01-1678 (2011), doi:https://doi.org/10.4271/2011-01-1678.

23.53. Petitjean, D., Bernardini, L., Middlemass, C., and Shahed, S.M., "Advanced Gasoline Engine Turbocharging Technology for Fuel Economy Improvements," SAE Technical Paper 2004-01-0988 (2004), doi:https://doi.org/10.4271/2004-01-0988.

23.54. Buckland, J., Cook, J., Kolmanovsky, I., and Sun, J., "Technology Assessment of Boosted Direct Injection Stratified Charge Gasoline Engines," SAE Technical Paper 2000-01-0249 (2000), doi:https://doi.org/10.4271/2000-01-0249.

23.55. Fraser, N., Blaxill, H., Lumsden, G., and Bassett, M., "Challenges for Increased Efficiency through Gasoline Engine Downsizing," *SAE Int. J. Engines* 2, no. 1 (2009): 991-1008, doi:https://doi.org/10.4271/2009-01-1053.

23.56. Oomori, H. and Ogino, S., "Waste Heat Recovery of Passenger Car Using a Combination of Rankine Bottoming Cycle and Evaporative Engine Cooling System," SAE Technical Paper 930880 (1993), doi:https://doi.org/10.4271/930880.

23.57. Boretti, A., "Improving the Efficiency of Turbocharged Spark Ignition Engines for Passenger Cars through Waste Heat Recovery," SAE Technical Paper 2012-01-0388 (2012), doi:https://doi.org/10.4271/2012-01-0388.

23.58. Boretti, A., Osman, A., and Aris, I., "Design of Rankine Cycle Systems to Deliver Fuel Economy Benefits over Cold Start Driving Cycles," SAE Technical Paper 2012-01-1713 (2012), doi:https://doi.org/10.4271/2012-01-1713.

23.59. Bae, S., Heo, H., Park, J., Lee, H.-Y. et al., "Performance Design of Low Temperature Condenser for Waste Heat Recovery System," SAE Technical Paper 2013-01-0046 (2013), doi:https://doi.org/10.4271/2013-01-0046.

23.60. Lee, J., Ohn, H., Choi, J., Kim, S.J. et al., "Development of Effective Exhaust Gas Heat Recovery System for a Hybrid Electric Vehicle," SAE Technical Paper 2011-01-1171 (2011), doi:https://doi.org/10.4271/2011-01-1171.

23.61. Mori, M., Yamagami, T., Sorazawa, M., Miyabe, T. et al., "Simulation of Fuel Economy Effectiveness of Exhaust Heat Recovery System Using Thermoelectric Generator in a Series Hybrid," *SAE Int. J. Mater. Manuf.* 4, no. 1 (2011): 1268-1276, doi:https://doi.org/10.4271/2011-01-1335.

23.62. Körfer, T., Lamping, M., Kolbeck, A., Pischinger, S. et al., "Potential of Modern Diesel Engines with Lowest Raw Emissions—A Key Factor for Future CO_2 Reductionm," SAE Technical Paper 2009-26-0025 (2009), doi:https://doi.org/10.4271/2009-26-0025.

23.63. Johnson, T.V., "Diesel Emissions in Reviewm," *SAE Int. J. Engines* 4, no. 1 (2011): 143-157, doi:https://doi.org/10.4271/2011-01-0304.

23.64. Lancefield, T., Methley, I., Räse, U., and Kuhn, T., "The Application of Variable Event Valve Timing to a Modern Diesel Engine," SAE Technical Paper 2000-01-1229 (2000), doi:https://doi.org/10.4271/2000-01-1229.

23.65. Gurney, D., Mitcalf, J., Warth, M., Schneider, S. et al., "Integrated Simulation, Analysis and Testing of a Variable Valve Train for Passenger Car Diesel Engines," SAE Technical Paper 2012-01-0829 (2012), doi:https://doi.org/10.4271/2012-01-0829.

23.66. Dittrich, P., Peter, F., Huber, G., and Kuehn, M., "Thermodynamic Potentials of a Fully Variable Valve Actuation System for Passenger-Car Diesel Engines," SAE Technical Paper 2010-01-1199 (2010), doi:https://doi.org/10.4271/2010-01-1199.

23.67. Parvate-Patil, G.B., Hong, H., and Gordon, B., "Analysis of Variable Valve Timing Events and Their Effects on Single Cylinder Diesel Engine," SAE Technical Paper 2004-01-2965 (2004), doi:https://doi.org/10.4271/2004-01-2965.

23.68. Teng, H., "Waste Heat Recovery Concept to Reduce Fuel Consumption and Heat Rejection from a Diesel Engine," *SAE Int. J. Commer. Veh.* 3, no. 1 (2010): 60-68, doi:https://doi.org/10.4271/2010-01-1928.

23.69. Edwards, S., Eitel, J., Pantow, E., Geskes, P. et al., "Waste Heat Recovery: The Next Challenge for Commercial Vehicle Thermomanagement," *SAE Int. J. Commer. Veh.* 5, no. 1 (2012): 395-406, doi:https://doi.org/10.4271/2012-01-1205.

23.70. Briggs, T.E., Wagner, R., Edwards, K.D., Curran, S. et al., "A Waste Heat Recovery System for Light Duty Diesel Engines," SAE Technical Paper 2010-01-2205 (2010), doi:https://doi.org/10.4271/2010-01-2205.

23.71. Edwards, K.D., Wagner, R., and Briggs, T., "Investigating Potential Light-Duty Efficiency Improvements through Simulation of Turbo-Compounding and Waste-Heat Recovery Systems," SAE Technical Paper 2010-01-2209 (2010), doi:https://doi.org/10.4271/2010-01-2209.

23.72. Srinivas, A., Prasad, T., Satish, T., Dhande, S. et al., "Diesel Hybrids - The Logical Path towards Hybridisation," SAE Technical Paper 2009-28-0046 (2009), doi:https://doi.org/10.4271/2009-28-0046.

23.73. Millo, F., Badami, M., Ferraro, C.V., Lavarino, G. et al., "A Comparison between Different Hybrid Powertrain Solutions for an European Mid-Size Passenger Car," SAE Technical Paper 2010-01-0818 (2010), doi:https://doi.org/10.4271/2010-01-0818.

23.74. Johnson, T.V., "Review of CO_2 Emissions and Technologies in the Road Transportation Sector," *SAE Int. J. Engines* 3, no. 1 (2010): 1079-1098, doi:https://doi.org/10.4271/2010-01-1276.

23.75. Gao, B., Svancara, K., Walker, A., Kok, D. et al., "Development of a BISG Micro-Hybrid System," SAE Technical Paper 2009-01-1330 (2009), doi:https://doi.org/10.4271/2009-01-1330.

23.76. Turner, J., Blake, D., Moore, J., Burke, P. et al., "The Lotus Range Extender Engine," *SAE Int. J. Engines* 3, no. 2 (2010): 318-351, doi:https://doi.org/10.4271/2010-01-2208.

23.77. Bassett, M., Thatcher, I., Bisordi, A., Hall, J. et al., "Design of a Dedicated Range Extender Engine," SAE Technical Paper 2011-01-0862 (2011), doi:https://doi.org/10.4271/2011-01-0862.

23.78. Mattarelli, E., Rinaldini, C., Cantore, G., and Baldini, P., "2-Stroke Externally Scavenged Engines for Range Extender Applications," SAE Technical Paper 2012-01-1022 (2012), doi:https://doi.org/10.4271/2012-01-1022.

23.79. Grove, W.R., "On a New Voltaic Combination," *The London, Edinburgh, and Dublin Philosophical Magazine and Journal of Science* 13, no. 84 (1938): 430-431.

23.80. Grove, W.R., "On a Gaseous Voltaic Battery," *The London, Edinburgh, and Dublin Philosophical Magazine and Journal of Science* 21, no. 140 (1942): 417-420.

23.81. Marks, C., Rishavy, E., and Wyczalek, F., "Electrovan-A Fuel Cell Powered Vehicle," SAE Technical Paper 670176 (1967), doi:https://doi.org/10.4271/670176.

23.82. Rajashekara, K., "History of Electric Vehicles in General Motors," *IEEE Transactions on Industry Applications* 30, no. 4 (1994): 897-904.

23.83. Perry, M.L. and Fuller, T.F., "A Historical Perspective of Fuel Cell Technology in the 20th Century," *Journal of the Electrochemical Society* 149, no. 7 (2002): 59-67.

23.84. Gay, S.E. and Ehsani, M., "High-Temperature Fuel Cell Warm-Up Influence on Vehicle Fuel Consumption," SAE Technical Paper 2003-01-2269 (2003), doi:https://doi.org/10.4271/2003-01-2269.

23.85. Rothfleisch, J., "Hydrogen from Methanol for Fuel Cells," SAE Technical Paper 640377 (1964), doi:https://doi.org/10.4271/640377.

23.86. Lewis, R. and Dolan, G., "Looking Beyond the Internal Combustion Engine: The Promise of Methanol Fuel Cell Vehicles," SAE Technical Paper 1999-01-0531 (1999), doi:https://doi.org/10.4271/1999-01-0531.

23.87. Ramaswamy, S., Sundaresan, M., Hauer, K., Freidman, D. et al., "Efficiency, Dynamic Performance and System Interactions for a Compact Fuel Processor for Indirect Methanol Fuel Cell Vehicle," SAE Technical Paper 2003-01-0810 (2003), doi:https://doi.org/10.4271/2003-01-0810.

23.88. Franco, E., Linardi, M., Colosio, M., and Barboza, J., "Fuel Cells and Ethanol: A Technological Advantage," SAE Technical Paper 2003-01-3623 (2003), doi:https://doi.org/10.4271/2003-01-3623.

23.89. Dircks, K., "Recent Advances in Fuel Cells for Transportation Applications," SAE Technical Paper 1999-01-0534 (1999), doi:https://doi.org/10.4271/1999-01-0534.

23.90. Ernst, W., Boyer, J., Ho, D., and Podolski, W., "Fuel-Flexible Automotive Fuel Cell Power System," SAE Technical Paper 2000-01-1530 (2000), doi:https://doi.org/10.4271/2000-01-1530.

23.91. Wagner, A., Wagner, J., Krause, T., and Carter, J., "Autothermal Reforming Catalyst Development for Fuel Cell Applications," SAE Technical Paper 2002-01-1884 (2002), doi:https://doi.org/10.4271/2002-01-1884.

23.92. Bowers, B., Zhao, J., Dattatraya, D., Rizzo, V. et al., "Development of an Onboard Fuel Processor for PEM Fuel Cell Vehicles," SAE Technical Paper 2004-01-1473 (2004), doi:https://doi.org/10.4271/2004-01-1473.

23.93. Gay, S. and Ehsani, M., "Ammonia Hydrogen Carrier for Fuel Cell Based Transportation," SAE Technical Paper 2003-01-2251 (2003), doi:https://doi.org/10.4271/2003-01-2251.

23.94. Ishimatsu, S., Saika, T., and Nohara, T., "Ammonia Fueled Fuel Cell Vehicle: The New Concept of a Hydrogen Supply System," SAE Technical Paper 2004-01-1925 (2004), doi:https://doi.org/10.4271/2004-01-1925.

23.95. Muramatsu, Y., Furukawa, K., and Adachi, S., "Evaluation of Direct Methanol Fuel Cell Systems for Two-Wheeled Vehicles," SAE Technical Paper 2007-32-0112 (2007), doi:https://doi.org/10.4271/2007-32-0112.

23.96. Furukawa, K., Muramatsu, Y., and Adachi, S., "Development and Operation of a 1 kW Direct Methanol Fuel Cell Stack," SAE Technical Paper 2009-32-0031 (2009), doi:https://doi.org/10.4271/2009-32-0031.

23.97. Shudo, T. and Naganuma, S., "Performance Improvement in Direct Methanol Fuel Cell by Using a Novel Porous Flow Field Made of Sintered Metal Powder," SAE Technical Paper 2011-39-7261 (2011), doi:https://doi.org/10.4271/2011-39-7261.

23.98. Jain, A. and Agrawal, S., "An Analysis of Trends in Vehicle Technologies Based on Alternative Fuels: Battery Electric Vehicles and Fuel Cell Electric Vehicles," SAE Technical Paper 2011-01-1743 (2011), doi:https://doi.org/10.4271/2011-01-1743.

23.99. Matsunaga, M., Fukushima, T., and Ojima, K., "Powertrain System of Honda FCX Clarity Fuel Cell Vehicle," *World Electric Vehicle Journal* 3, no. 4 (2009): 820-829.

23.100. Morikawa, H., Kikuchi, H., and Saito, N., "Development and Advances of a V-Flow FC Stack for FCX Clarity," *SAE Int. J. Engines* 2, no. 1 (2009): 955-959, doi:https://doi.org/10.4271/2009-01-1010.

23.101. Matsunaga, M., Fukushima, T., and Ojima, K., "Advances in the Power Train System of Honda FCX Clarity Fuel Cell Vehicle," SAE Technical Paper 2009-01-1012 (2009), doi:https://doi.org/10.4271/2009-01-1012.

23.102. Mohrdieck, D., "Next Generation Fuel Cell Technology for Passenger Cars and Buses," *World Electric Vehicle Journal* 3, no. 2 (2009): 209-213.

23.103. Kim, S.H., Kum, Y.B., Lee, K.C., Lim, T.W. et al., "Development of Hyundai's Tucson FCEV," SAE Technical Paper 2005-01-0005 (2005), doi:https://doi.org/10.4271/2005-01-0005.

23.104. Sung, W., Song, Y.-I., Yu, K.-H., and Lim, T.-W., "Recent Advances in the Development of Hyundai • Kia's Fuel Cell Electric Vehicles," *SAE Int. J. Engines* 3, no. 1 (2010): 768-772, doi:https://doi.org/10.4271/2010-01-1089.

23.105. Verhelst, S. and Wallner, T., "Hydrogen-Fueled Internal Combustion Engines," *Progress in Energy and Combustion Science* 35, no. 6 (2009): 490-527.

23.106. Costa, R.L. and Grimes, P.G., "Electrolysis as a Source of Hydrogen and Oxygen," *Chemical Engineering Progress* 63, no. 4 (1967): 56-58.

23.107. Hoffman, K.C., Winsche, W.E., Wiswall, R.H., Reilly, J.J. et al., "Metal Hydrides as a Source of Fuel for Vehicular Propulsion," SAE Technical Paper 690232 (1969), doi:https://doi.org/10.4271/690232.

23.108. Gregory, D.P., "The Hydrogen Economy," *Scientific American* 228, no. 1 (1973): 13-21.

23.109. Mittal, V., Shah, R., Aragon, N., and Douglas, N., "A Perspective on the Challenges and Future of Hydrogen Fuel," *SAE J. STEEP* 3, no. 1 (2022): 31-39, doi:https://doi.org/10.4271/13-03-01-0003.

23.110. Brandon, N.P. and Kurban, Z., "Clean Energy and the Hydrogen Economy," *Philosophical Transactions of the Royal Society A: Mathematical, Physical and Engineering Sciences* 375, no. 2098 (2017): 20160400.

23.111. Billings, R.E., "Hydrogen Fuel in the Subcompact Automobile," SAE Technical Paper 760572 (1976), doi:https://doi.org/10.4271/760572.

23.112. Young, R.C., Chao, B., Li, Y., Myasnikov, V. et al., "A Hydrogen ICE Vehicle Powered by Ovonic Metal Hydride Storage," SAE Technical Paper 2004-01-0699 (2004), doi:https://doi.org/10.4271/2004-01-0699.

23.113. Diwan, M., Hwang, H.T., Al-Kukhun, A., and Varma, A., "Hydrogen Generation from Noncatalytic Hydrothermolysis of Ammonia Borane for Vehicle Applications," *AIChE Journal* 57, no. 1 (2011): 259-264.

23.114. Devarakonda, M., Brooks, K., Ronnebro, E., Rassat, S. et al., "Chemical Hydrides for Hydrogen Storage in Fuel Cell Applications," SAE Technical Paper 2012-01-1229 (2012), doi:https://doi.org/10.4271/2012-01-1229.

23.115. Tarasov, B.P., Fursikov, P.V., Volodin, A.A., Bocharnikov, M.S. et al., "Metal Hydride Hydrogen Storage and Compression Systems for Energy Storage Technologies," *International Journal of Hydrogen Energy* 46, no. 25 (2021): 13647-13657.

23.116. Darago, V.S. and McCuen, C.M., "Urban Vehicle Design Competition—History, Progress, Development," SAE Technical Paper 720497 (1972), doi:https://doi.org/10.4271/720497.

23.117. Finegold, J.G., Lynch, F.E., Baker, N.R., Takahashi, R. et al., "The UCLA Hydrogen Car: Design, Construction, and Performance," SAE Technical Paper 730507 (1973), doi:https://doi.org/10.4271/730507.

23.118. Stockhausen, W.F., Natkin, R.J., Kabat, D.M., Reams, L. et al., "Ford P2000 Hydrogen Engine Design and Vehicle Development Program," SAE Technical Paper 2002-01-0240 (2002), doi:https://doi.org/10.4271/2002-01-0240.

23.119. Tang, X., Kabat, D.M., Natkin, R.J., Stockhausen, W.F. et al., "Ford P2000 Hydrogen Engine Dynamometer Development," SAE Technical Paper 2002-01-0242 (2002), doi:https://doi.org/10.4271/2002-01-0242.

23.120. Szwabowski, S.J., Hashemi, S., Stockhausen, W.F., Natkin, R.J. et al., "Ford Hydrogen Engine Powered P2000 Vehicle," SAE Technical Paper 2002-01-0243 (2002), doi:https://doi.org/10.4271/2002-01-0243.

23.121. Natkin, R.J., Denlinger, A.R., Younkins, M.A., Weimer, A.Z. et al., "Ford 6.8L Hydrogen IC Engine for the E-450 Shuttle Van," SAE Technical Paper 2007-01-4096 (2007), doi:https://doi.org/10.4271/2007-01-4096.

23.122. Gopalakrishnan, R., Throop, M.J., Richardson, A., and Lapetz, J., "Engineering the Ford H2 IC Engine Powered E-450 Shuttle Bus," SAE Technical Paper 2007-01-4095 (2007), doi:https://doi.org/10.4271/2007-01-4095.

23.123. Richardson, A., Gopalakrishnan, R., Chhaya, T., Deasy, S. et al., "Design Considerations for Hydrogen Management System on Ford Hydrogen Fueled E-450 Shuttle Bus," *SAE Int. J. Commer. Veh.* 2, no. 1 (2009): 101-109, doi:https://doi.org/10.4271/2009-01-1422.

23.124. Morimoto, K., Teramoto, T., and Takamori, Y., "Combustion Characteristics in Hydrogen Fueled Rotary Engine," SAE Technical Paper 920302 (1992), doi:https://doi.org/10.4271/920302.

23.125. Wakayama, N., Morimoto, K., Kashiwagi, A., and Saito, T., "Development of Hydrogen Rotary Engine Vehicle," in *16th World Hydrogen Energy Conference*, Lyon, France, 2006.

23.126. Kiesgen, G., Klüting, M., Bock, C., and Fischer, H., "The New 12-Cylinder Hydrogen Engine in the 7 Series: The H2 ICE Age Has Begun," SAE Technical Paper 2006-01-0431 (2006), doi:https://doi.org/10.4271/2006-01-0431.

23.127. Wallner, T., Lohse-Busch, H., Gurski, S., Duoba, M. et al., "Fuel Economy and Emissions Evaluation of BMW Hydrogen 7 Mono-Fuel Demonstration Vehicles," *International Journal of Hydrogen Energy* 33, no. 24 (2008): 7607-7618.

23.128. Phillips 66, "Phillips 66 and H2 Energy Europe Close on Joint Venture to Create European Network of Hydrogen Refueling Stations," July 2022, accessed March 2023. https://investor.phillips66.com/financial-information/news-releases/news-release-details/2022/Phillips-66-and-H2-Energy-Europe-close-on-joint-venture-to-create-European-network-of-hydrogen-refueling-stations/default.aspx.

23.129. Chevron Corporation, "Chevron, Iwatani Announce Agreement to Build 30 Hydrogen Fueling Stations in California," February 2022, accessed March 2023, https://chevroncorp.gcs-web.com/news-releases/news-release-details/chevron-iwatani-announce-agreement-build-30-hydrogen-fueling.

23.130. TotalEnergies SE, "Total Energies and Air Liquide Join Forces to Develop a Network of over 100 Hydrogen Stations for Heavy Duty Vehicles in Europe," February, 2023, accessed March 2023, https://totalenergies.com/media/news/press-releases/totalenergies-and-air-liquide-join-forces-develop-network-over-100.

23.131. Robert Bosch GmbH, "Bosch Hydrogen Offensive: Technology for Climate-Neutral Factories and Zero-Carbon Traffic," May 2022, accessed March 2023, https://www.bosch-presse.de/pressportal/de/en/bosch-hydrogen-offensive-technology-for-climate-neutral-factories-and-zero-carbon-traffic-241154.html.

23.132. Ludwig-Bölkow-Systemtechnik GmbH, "Another Record Addition of European Hydrogen Refuelling Stations in 2022," February 1, 2023, accessed March 2023, https://www.h2stations.org/press-release-2023-another-record-addition-of-european-hydrogen-refuelling-stations-in-2022/.

23.133. Lindell, L., "An Introduction to the Nuclear Powered Energy Depot Concept," SAE Technical Paper 650049 (1965), doi:https://doi.org/10.4271/650049.

23.134. Rosenthal, A., "Energy Depot - A Concept for Reducing the Military Supply Burden," SAE Technical Paper 650050 (1965), doi:https://doi.org/10.4271/650050.

23.135. Starkman, E., Newhall, H., Sutton, R., Maguire, T. et al., "Ammonia as a Spark Ignition Engine Fuel: Theory and Application," SAE Technical Paper 660155 (1966), doi:https://doi.org/10.4271/660155.

23.136. Gray, J., Dimitroff, E., Meckel, N., and Quillian, R., "Ammonia Fuel - Engine Compatibility and Combustion," SAE Technical Paper 660156 (1966), doi:https://doi.org/10.4271/660156.

23.137. Starkman, E., James, G., and Newhall, H., "Ammonia as a Diesel Engine Fuel: Theory and Application," SAE Technical Paper 670946 (1967), doi:https://doi.org/10.4271/670946.

23.138. Zamfirescu, C. and Dincer, I., "Using Ammonia as a Sustainable Fuel," *Journal of Power Sources* 185, no. 1 (2008): 459-465.

23.139. Frigo, S., Gentili, R., and Doveri, N., "Ammonia Plus Hydrogen as Fuel in a S.I. Engine: Experimental Results," SAE Technical Paper 2012-32-0019 (2012), doi:https://doi.org/10.4271/2012-32-0019.

23.140. Pozzana, G., Bonfanti, N., Frigo, S., Doveri, N. et al., "A Hybrid Vehicle Powered by Hydrogen and Ammonia," SAE Technical Paper 2012-32-0085 (2012), doi:https://doi.org/10.4271/2012-32-0085.

23.141. Frigo, S., Gentili, R., and De Angelis, F., "Further Insight into the Possibility to Fuel a SI Engine with Ammonia Plus Hydrogen," SAE Technical Paper 2014-32-0082 (2014), doi:https://doi.org/10.4271/2014-32-0082.

23.142. Lhuillier, C., Brequigny, P., Contino, F., and Rousselle, C., "Performance and Emissions of an Ammonia-Fueled SI Engine with Hydrogen Enrichment," SAE Technical Paper 2019-24-0137 (2019), doi:https://doi.org/10.4271/2019-24-0137.

23.143. Kobayashi, H., Hayakawa, A., Somarathne, K.K.A., and Okafor, E.C., "Science and Technology of Ammonia Combustion," *Proceedings of the Combustion Institute* 37, no. 1 (2019): 109-133.

23.144. Elbaz, A.M., Wang, S., Guiberti, T.F., and Roberts, W.L., "Review on the Recent Advances on Ammonia Combustion from the Fundamentals to the Applications," *Fuel Communications* 10 (2022): 100053.

23.145. Grimes, P., "Energy Depot Fuel Production and Utilization," SAE Technical Paper 650051 (1965), doi:https://doi.org/10.4271/650051.

23.146. Arai, A., Kanzaki, Y., Saito, Y., Nohara, T. et al., "A Fuel-Cell Electric Vehicle with Cracking and Electrolysis of Ammonia," SAE Technical Paper 2010-01-1791 (2010), doi:https://doi.org/10.4271/2010-01-1791.

23.147. Lan, R. and Tao, S., "Ammonia as a Suitable Fuel for Fuel Cells," *Frontiers in Energy Research* 2 (2014): 35.

23.148. Afif, A., Radenahmad, N., Cheok, Q., Shams, S. et al., "Ammonia-Fed Fuel Cells: A Comprehensive Review," *Renewable and Sustainable Energy Reviews* 60 (2016): 822-835.

23.149. Jeerh, G., Zhang, M., and Tao, S., "Recent Progress in Ammonia Fuel Cells and Their Potential Applications," *Journal of Materials Chemistry A* 9, no. 2 (2021): 727-752.

23.150. Alemu, M.A. and Ilbas, M., "Direct Ammonia Powered Solid Oxide Fuel Cells: Challenges, Opportunities, and Future Outlooks," *Journal of Engineering and Applied Sciences Technology* 2, no. 3 (2020): 2-11.

23.151. Voss, B., Joensen, F., and Hansen, J.B., Preparation of fuel grade dimethyl ether. U.S. Patent 5,908,963, 1999.

23.152. Green, C.J., Cockshutt, N.A., and King, L., "Dimethyl Ether as a Methanol Ignition Improver: Substitution Requirements and Exhaust Emissions Impact," SAE Technical Paper 902155 (1990), doi:https://doi.org/10.4271/902155.

23.153. Karpuk, M.E., Wright, J.D., Dippo, J.L., and Jantzen, D.E., "Dimethyl Ether as an Ignition Enhancer for Methanol-Fueled Diesel Engines," SAE Technical Paper 912420 (1991), doi:https://doi.org/10.4271/912420.

23.154. Murayama, T., Chikahisa, T., Guo, J., and Miyano, M., "A Study of a Compression Ignition Methanol Engine with Converted Dimethyl Ether as an Ignition Improver," SAE Technical Paper 922212 (1992), doi:https://doi.org/10.4271/922212.

23.155. Guo, J., Chikahisa, T., Murayama, T., and Miyano, M., "Improvement of Performance and Emissions of a Compression Ignition Methanol Engine with Dimethyl Ether," SAE Technical Paper 941908 (1994), doi:https://doi.org/10.4271/941908.

23.156. Fleisch, T., McCarthy, C., Basu, A., Udovich, C. et al., "A New Clean Diesel Technology: Demonstration of ULEV Emissions on a Navistar Diesel Engine Fueled with Dimethyl Ether," SAE Technical Paper 950061 (1995), doi:https://doi.org/10.4271/950061.

23.157. Kapus, P. and Ofner, H., "Development of Fuel Injection Equipment and Combustion System for DI Diesels Operated on Dimethyl Ether," SAE Technical Paper 950062 (1995), doi:https://doi.org/10.4271/950062.

23.158. Hansen, J.B., Voss, B., Joensen, F., and Siguroardóttir, I.D., "Large Scale Manufacture of Dimethyl Ether—A New Alternative Diesel Fuel from Natural Gas," SAE Technical Paper 950063 (1995), doi:https://doi.org/10.4271/950063.

23.159. Sorenson, S.C. and Mikkelsen, S.-E., "Performance and Emissions of a 0.273 Liter Direct Injection Diesel Engine Fuelled with Neat Dimethyl Ether," SAE Technical Paper 950064 (1995), doi:https://doi.org/10.4271/950064.

23.160. Oguma, M., Goto, S., Yanai, T., and Mikita, Y., "Methodology of Lubricity Evaluation for DME Fuel Based on HFRR," SAE Technical Paper 2011-32-0651 (2011), doi:https://doi.org/10.4271/2011-32-0651.

23.161. Hara, T., Shimazaki, N., Yanagisawa, N., Seto, T. et al., "Study of DME Diesel Engine for Low NOx and CO_2 Emission and Development of DME Trucks for Commercial Use," *SAE Int. J. Fuels Lubr.* 5, no. 1 (2012): 233-242, doi:https://doi.org/10.4271/2011-01-1961.

23.162. Ofner, H., Gill, D.W., and Krotscheck, C., "Dimethyl Ether as Fuel for CI Engines—A New Technology and Its Environmental Potential," SAE Technical Paper 981158 (1998), doi:https://doi.org/10.4271/981158.

23.163. Gill, D. and Ofner, H., "DiMethyl Ether - A Clean Fuel for Transportation," SAE Technical Paper 990059 (1999), doi:https://doi.org/10.4271/990059.

23.164. Kinoshita, K., Oguma, M., Goto, S., Sugiyama, K. et al., "Development of Retrofit DME Diesel Engine Operating with Rotary Distributor Fuel Injection Pump," SAE Technical Paper 2003-01-0758 (2003), doi:https://doi.org/10.4271/2003-01-0758.

23.165. Kato, M., Takeuchi, H., Koie, K., Sekijima, H. et al., "A Study of Dimethyl Ether(DME) Flow in Diesel Nozzle," SAE Technical Paper 2004-01-0081 (2004), doi:https://doi.org/10.4271/2004-01-0081.

23.166. Sato, Y., Nozaki, S., and Noda, T., "The Performance of a Diesel Engine for Light Duty Truck Using a Jerk Type In-Line DME Injection System," SAE Technical Paper 2004-01-1862 (2004), doi:https://doi.org/10.4271/2004-01-1862.

23.167. Oguma, M., Goto, S., and Watanabe, T., "Engine Performance and Emission Characteristics of DME Diesel Engine with Inline Injection Pump Developed for DME," SAE Technical Paper 2004-01-1863 (2004), doi:https://doi.org/10.4271/2004-01-1863.

23.168. Lee, U., Han, J., Wang, M., Ward, J. et al., "Well-to-Wheels Emissions of Greenhouse Gases and Air Pollutants of Dimethyl Ether from Natural Gas and Renewable Feedstocks in Comparison with Petroleum Gasoline and Diesel in the United States and Europe," *SAE Int. J. Fuels Lubr.* 9, no. 3 (2016): 546-557, doi:https://doi.org/10.4271/2016-01-2209.

23.169. Pélerin, D., Gaukel, K., Härtl, M., and Wachtmeister, G., "Nitrogen Oxide Reduction Potentials Using Dimethyl Ether and Oxymethylene Ether in a Heavy-Duty Diesel Engine," SAE Technical Paper 2020-01-5084 (2020), doi:https://doi.org/10.4271/2020-01-5084.

23.170. Härtl, M., Seidenspinner, P., Jacob, E., and Wachtmeister, G., "Oxygenate Screening on a Heavy-Duty Diesel Engine and Emission Characteristics of Highly Oxygenated Oxymethylene Ether Fuel OME1," *Fuel* 153 (2015): 328-335.

23.171. Gelner, A.D., Beck, H.A., Pastoetter, C., Härtl, M. et al., "Ultra-Low Emissions of a Heavy-Duty Engine Powered with Oxymethylene Ethers (OME) under Stationary and Transient Driving Conditions," *International Journal of Engine Research* 23, no. 5 (2022): 738-753.

23.172. Pellegrini, L., Marchionna, M., Patrini, R., Beatrice, C. et al., "Combustion Behaviour and Emission Performance of Neat and Blended Polyoxymethylene Dimethyl Ethers in a Light-Duty Diesel Engine," SAE Technical Paper 2012-01-1053 (2012), doi:https://doi.org/10.4271/2012-01-1053.

23.173. Pellegrini, L., Marchionna, M., Patrini, R., and Florio, S., "Emission Performance of Neat and Blended Polyoxymethylene Dimethyl Ethers in an Old Light-Duty Diesel Car," SAE Technical Paper 2013-01-1035 (2013), doi:https://doi.org/10.4271/2013-01-1035.

23.174. Zacherl, F., Wopper, C., Schwanzer, P., and Rabl, H.P., "Potential of the Synthetic Fuel Oxymethylene Ether (OME) for the Usage in a Single-Cylinder Non-Road Diesel Engine: Thermodynamics and Emissions," *Energies* 15, no. 21 (2022): 7932.

23.175. Pastor, J.V., García, A., Micó, C., and Lewiski, F., "Simultaneous High-Speed Spectroscopy and 2-Color Pyrometry Analysis in an Optical Compression Ignition Engine Fueled with OMEX-Diesel Blends," *Combustion and Flame* 230 (2021): 111437.

23.176. Benajes, J., Garcia, A., Monsalve-Serrano, J., and Sari, R., "Clean and Efficient Dual-Fuel Combustion Using OMEx as High Reactivity Fuel: Comparison to Diesel-Gasoline Calibration," *Energy Conversion and Management* 216 (2020): 112953.

23.177. Garcia, A., Monsalve-Serrano, J., Villalta, D., and Guzmán Mendoza, M., "OMEx Fuel and RCCI Combustion to Reach Engine-Out Emissions Beyond the Current EURO VI Legislation," SAE Technical Paper 2021-24-0043 (2021), doi:https://doi.org/10.4271/2021-24-0043.

23.178. Garcia, A., Gil, A., Monsalve-Serrano, J., and Sari, R.L., "OMEx-Diesel Blends as High Reactivity Fuel for Ultra-Low NOx and Soot Emissions in the Dual-Mode Dual-Fuel Combustion Strategy," *Fuel* 275 (2020): 117898.

23.179. Binder, J.B. and Raines, R.T., "Simple Chemical Transformation of Lignocellulosic Biomass into Furans for Fuels and Chemicals," *Journal of the American Chemical Society* 131, no. 5 (2009): 1979-1985.

23.180. Yang, Y., Hu, C., and Abu-Omar, M.M., "Conversion of Carbohydrates and Lignocellulosic Biomass into 5-Hydroxymethylfurfural Using AlCl$_3$•6H$_2$O Catalyst in a Biphasic Solvent System," *Green Chemistry* 14, no. 2 (2012): 509-513.

23.181. Roman-Leshkov, Y., Barrett, C.J., Liu, Z.Y., and Dumesic, J.A., "Production of Dimethylfuran for Liquid Fuels from Biomass-Derived Carbohydrates," *Nature* 447 (2007): 982-985.

23.182. Chidambaram, M. and Bell, A.T., "A Two-Step Approach for the Catalytic Conversion of Glucose to 2,5-Dimethylfuran in Ionic Liquids," *Green Chemistry* 12, no. 7 (2010): 1253-1262.

23.183. Gruter, G.J.M. and Dautzenberg, F., Method for synthesis of organic acid esters of 5-hydroxymethylfufural and their use. U.S. Patent 2009/0306415, 2009.

23.184. Gruter, G.J.M. and Dautzenberg, F., Method for synthesis of 5-alkoxymethyl fufural ethers and their use. U.S. Patent 2012/0083610, 2012.

23.185. de Jong, E. and Gruter, G.-J., "Furanics: A Novel Diesel Fuel with Superior Characteristics," SAE Technical Paper 2009-01-2767 (2009), doi:https://doi.org/10.4271/2009-01-2767.

23.186. Ohtomo, M., Nishikawa, K., Suzuoki, T., Miyagawa, H. et al., "Auto-Ignition Characteristics of Biofuel Blends for SI Engines," SAE Technical Paper 2011-01-1989 (2011), doi:https://doi.org/10.4271/2011-01-1989.

23.187. Tian, G., Li, H., Xu, H., Li, Y. et al., "Spray Characteristics Study of DMF Using Phase Doppler Particle Analyzer," *SAE Int. J. Passeng. Cars – Mech. Syst.* 3, no. 1 (2010): 948-958, doi:https://doi.org/10.4271/2010-01-1505.

23.188. Tian, G., Daniel, R., Li, H., Xu, H. et al., "Laminar Burning Velocities of 2,5-Dimethylfuran Compared with Ethanol and Gasoline," *Energy and Fuels* 24, no. 7 (2010): 3898-3905.

23.189. Wu, X., Huang, Z., Wang, X., Jin, C. et al., "Laminar Burning Velocities and Flame Instabilities of 2,5- Dimethylfuran-Air Mixtures at Elevated Pressures," *Combustion and Flame* 158, no. 3 (2011): 539-546.

23.190. Simmie, J.M. and Metcalfe, W.K., "Ab Initio Study of the Decomposition of 2,5-Dimethylfuran," *The Journal of Physical Chemistry A* 115, no. 32 (2011): 8877-8888.

23.191. Daniel, R., Wang, C., Xu, H., and Tian, G., "Effects of Combustion Phasing, Injection Timing, Relative Air-Fuel Ratio and Variable Valve Timing on SI Engine Performance and Emissions Using 2,5-Dimethylfuran," *SAE Int. J. Fuels Lubr.* 5, no. 2 (2012): 855-866, doi:https://doi.org/10.4271/2012-01-1285.

23.192. Zhong, S., Daniel, R., Xu, H., Zhang, J. et al., "Combustion and Emissions of 2,5-Dimethylfuran in a Direct-Injection Spark-Ignition Engine," *Energy and Fuels* 24, no. 5 (2010): 2891-2899.

23.193. Thewes, M., Muether, M., Pischinger, S., Budde, M. et al., "Analysis of the Impact of 2-Methylfuran on Mixture Formation and Combustion in a Direct-Injection Spark-Ignition Engine," *Energy Fuels* 25, no. 12 (2011): 5549-5561.

23.194. Lewis, R.J. Sr., *Sax's Dangerous Properties of Industrial Materials*, 11th ed. (New York: John Wiley & Sons, 2004).

23.195. Fromowitz, M., Shuga, J., Wlassowsky, A.Y., Ji, Z. et al., "Bone Marrow Genotoxicity of 2,5-Dimethylfuran, a Green Biofuel Candidate," *Environmental and Molecular Mutagenesis* 53, no. 6 (2012): 488-491.

23.196. Phuong, J., Kim, S., Thomas, R., and Zhang, L., "Predicted Toxicity of the Biofuel Candidate 2,5-Dimethylfuran in Environmental and Biological Systems," *Environmental and Molecular Mutagenesis* 53, no. 6 (2012): 478-487.

Further Reading

Corbo, P., Migliardini, F., and Veneri, O., *Hydrogen Fuel Cells for Road Vehicles* (London: Springer, 2011).

German, J., *Hybrid-Powered Vehicles*, 2nd ed. (Warrendale, PA: SAE International, 2011).

Holt, D.J., *Fuel Cell Powered Vehicles—Automotive Technology of the Future—Update* (Warrendale, PA: SAE International, 2003).

Morey, B., *Automotive 2030—North America* (Warrendale, PA: SAE International, 2011).

Morey, B., *Future Automotive Fuels and Energy Technology Profile* (Warrendale, PA: SAE International, 2013).

Senecal, K. and Leach, F., *Racing toward Zero: The Untold Story of Driving Green* (Warrendale, PA: SAE International, 2021).

Wakefield, E.H., *History of the Electric Automobile* (Warrendale, PA: SAE International, 1998).

Appendix 1: Introduction to Fuel Chemistry

This appendix gives an introduction to those aspects of organic chemistry (i.e., the chemistry of carbon compounds) that are important to the understanding of automotive fuels.

A1.1 Hydrocarbons

As the name indicates, these compounds contain carbon and hydrogen only, but there are many thousands of different possibilities depending on how the individual atoms are arranged. Carbon is quadrivalent, which means that in almost all cases it has four chemical bonds that allow it to combine with other atoms. Hydrogen has one chemical bond and so has a valency of one.

The simplest hydrocarbon is methane (CH_4), which can be written as:

$$
\begin{array}{c}
H \\
| \\
H - C - H \\
| \\
H
\end{array}
$$

Although the formula above is shown in two dimensions, in fact the carbon atom behaves as if it were in the center of a tetrahedron with carbon bonds going out to each point, so that the angle between the bonds is about 109°.

The valency or combining power of an atom is determined by the number of electrons in its outer shells. When the outer shells are full, either by sharing electrons with other atoms (giving a covalent bond, as in most organic compounds) or by having electrons donated by another atom (ionic bond), a relatively stable chemical compound result. The stability will depend on the strength of the chemical bonds and this, in turn, depends on the nature and structure of the various groupings present.

Carbon can combine with itself with single, double, or triple bonds, as shown below:

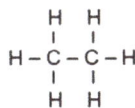

ethane   ethene or ethylene   ethyne or acetylene and these are more normally written:

$$CH_3 \bullet CH_3 \text{ or } C_2H_6 \qquad CH_2 = CH_2 \text{ or } C_2H_4 \qquad CH \equiv CH \text{ or } C_2H_2$$

Hydrocarbons with only single bonds are described as saturated whereas those with double or triple bonds are unsaturated. The unsaturated hydrocarbons, which contain a lower proportion of hydrogen, are less stable than the saturated compounds since other compounds such as oxygen can react quite readily with them across the double bond to form new materials. Although unsaturated compounds with double bonds are commonly present in automotive fuels, compounds with triple bonds are not. Hydrocarbons can also occur as ring compounds in which several carbon atoms are joined together to form a ring.

A1.2 The Alkanes or Paraffins, C_nH_{2n+2}

This class of compound consists of a homologous series of saturated hydrocarbons. Methane is the simplest member, then ethane having two carbon atoms, propane with three carbon atoms, and so on. The carbon atoms can be arranged as a straight chain or as branch chain compounds and the alternative forms of compounds having the same basic formula are known as isomers. Thus, there are two forms of butane, which is an alkane with four carbon atoms, as follows:

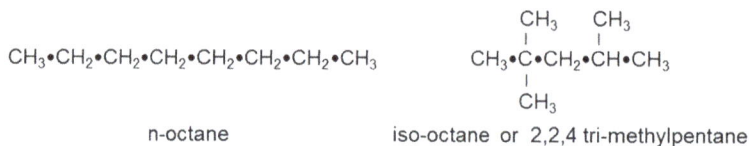

$$CH_3 \bullet CH_2 \bullet CH_2 \bullet CH_2 \bullet CH_2 \bullet CH_2 \bullet CH_2 \bullet CH_3$$

n-octane

$$\begin{array}{ccc} & CH_3 & CH_3 \\ & | & | \\ CH_3 \bullet C & \bullet CH_2 \bullet CH \bullet CH_3 \\ & | & \\ & CH_3 & \end{array}$$

iso-octane or 2,2,4 tri-methylpentane

The "n" in n-butane stands for normal; this means that all the carbon atoms are arranged in a straight line. Although both butanes have the same simple formula of C_4H_{10}, they are two entirely separate compounds which differ in boiling point, octane quality, and many other characteristics. The more carbon atoms there are in a compound, the more isomers are possible.

The group $CH_3 \bullet$, derived from methane, is called a methyl group or radical so that it is easy to see why iso-butane, which is propane ($CH_3 \bullet CH_2 \bullet CH_3$) with one of its hydrogen atoms on the second carbon atom (counting from the left) replaced by a methyl group, can also be called 2-methylpropane. There are no less than 18 different isomers of octane (having eight carbon atoms), two of which are n-octane and iso-octane:

$$CH_3 \bullet CH_2 \bullet CH_2 \bullet CH_3$$

n-butane

$$\begin{array}{c} CH_3 \bullet CH \bullet CH_3 \\ | \\ CH_3 \end{array}$$

iso-butane or 2-methylpropane

The boiling point of n-octane is 126°C and its octane quality is very low; it blends as if it had an octane number of about −17. The highly branched compound iso-octane, on the other hand, has a boiling point of 99°C and an octane value of 100.

Other groups or radicals that can be attached to carbon atoms are: the ethyl group, C_2H_5; the propyl group, C_3H_7; the butyl group, C_4H_9; and so on. In the case of groups with more than two carbon atoms, different arrangements of the atoms mean that there are: two propyl groups, isopropyl and n-propyl; three butyl groups, n-butyl, isobutyl, and tertiary butyl; and so on through the series. Such groups derived from alkanes are called alkyl groups.

A1.3 The Cycloparaffins or Naphthenes, C_nH_{2n}

As the name suggests, these are cyclic compounds consisting, in their simplest form, of CH, groups arranged in a circle. The most stable structures, which give the minimum distortion of the carbon bond angles, have either five or six carbon atoms (i.e., cyclopentane and cyclohexane), and these form more readily than any other ring structures. The hydrogen atoms attached to each carbon atom can be substituted by methyl or other groups.

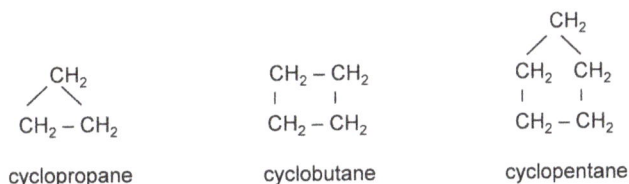

$$
\begin{array}{ccc}
& & & CH_2 \\
& & & \diagup \quad \diagdown \\
CH_2 & CH_2 - CH_2 & CH_2 \quad CH_2 \\
\diagup \quad \diagdown & | \qquad | & | \qquad | \\
CH_2 - CH_2 & CH_2 - CH_2 & CH_2 - CH_2 \\
\\
\text{cyclopropane} & \text{cyclobutane} & \text{cyclopentane}
\end{array}
$$

A1.4 The Alkenes or Olefins, C_nH_{2n}

Although the alkenes have the same carbon-to-hydrogen ratio and the same general formula as the cyclo-paraffins, their behavior and characteristics are entirely different. They are straight or branch chain compounds with one or more double bond. The position of the double bond is indicated by the number of the first carbon atom to which it is attached, i.e., the compound having the formula:

$$CH_2 = CH \cdot CH_2 \cdot CH_2 \cdot CH_3$$

can be called pentene-1 or 1-pentene or even pent-1-ene. Similarly, the compound:

$$CH_3 \cdot CH = CH \cdot CH_3$$

is called either butene-2, 2-butene or but-2-ene.

The double bond is a very reactive group so that the alkenes, or olefins as they are frequently called, can react readily with many other compounds. For this reason, the olefinic compounds are easily oxidized and so tend to have a poor oxidation stability. Again, since the double bond can occur in different parts of the molecule, different compounds having the same general formula are possible.

More than one double bond can occur in a compound as, for example, butadiene:

Dienes are particularly undesirable in fuel due to their high reactivity and consequent poor resistance

$$CH_2 = CH \cdot CH = CH_2$$

to oxidation. When the double bonds are arranged alternatively with single bonds, they are known as conjugated double bonds, and compounds having such structures are even more prone to oxidation and other reactions.

Olefins can be present as groups or radicals in the same way as alkyl groups, and examples are the vinyl group, $CH_2=CH-$, and the allyl group, $CH_2=CH-CH_2-$.

A1.5 **The Aromatic Hydrocarbons, C_nH_{2n-6}**

Aromatic hydrocarbons were so called because some of these compounds have a pleasant "aromatic" odor. However, the term is now applied to a class of hydrocarbons based on a six-membered ring having three apparently conjugated double bonds. The simplest member this group is benzene, C_6H_6, as follows:

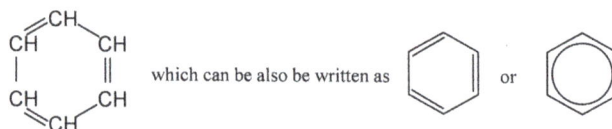

The double bonds do not behave in the way that would be expected if they were a noncyclic molecule in that the aromatic ring itself is not very reactive. This is because the double bonds behave as if they were not in fixed positions but "resonate" between the two possible structures:

The hydrogen atoms can be substituted by other groups, the positions of which are denoted by numbering the carbon atoms or by the use of names for the relative positions, as follows:

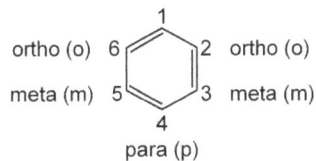

The following are examples of aromatic compounds:

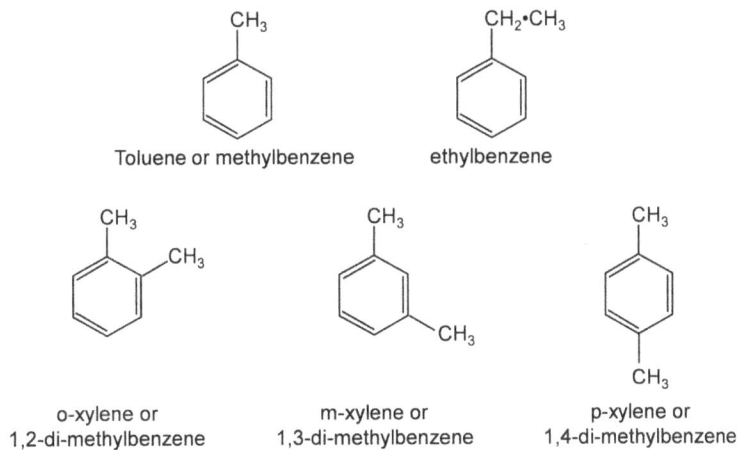

Toluene or methylbenzene   ethylbenzene

o-xylene or
1,2-di-methylbenzene

m-xylene or
1,3-di-methylbenzene

p-xylene or
1,4-di-methylbenzene

The aromatic rings can be fused together to give polynuclear aromatics such as naphthalene and anthracene:

naphthalene   anthracene

or much more complex compounds:

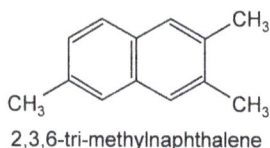

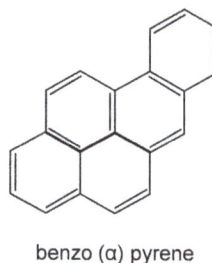

2,3,6-tri-methylnaphthalene   benzo (α) pyrene

Benzo(α)pyrene is of concern because it can be present in small amounts in exhaust gases and it is highly carcinogenic. These polynuclear aromatics (PNAs) are also sometimes called polycyclic aromatic hydrocarbons (PCAs or PAHs).

A1.6 Combustion of Hydrocarbons

Under ideal conditions all hydrocarbons, no matter what type, combust when ignited in the presence of oxygen to form water and carbon dioxide, as illustrated below for n-heptane:

$$C_7H_{16} + 11O_2 \rightarrow 7CO_2 + 8H_2O$$

This equation shows that n-heptane reacts with oxygen in the proportion of 1 molecule of heptane to 11 molecules of oxygen to form 7 molecules of carbon dioxide and 8 molecules of water. Carbon has a molecular weight of 12, hydrogen of 1 and oxygen of 16 so that it can be calculated that 100 g of n-heptane reacts with 352 g of oxygen to give 308 g of carbon dioxide and 144 g of water. This ratio of oxygen to hydrocarbon to give theoretically complete combustion is known as the stoichiometric ratio.

In practice, air is used instead of oxygen so that an oxygen-to-fuel ratio of 3.52:1 translates to approximately 15:1 for air on a weight basis for the combustion of n-heptane. For gasoline, one part by weight requires about 14.5 parts of air by weight, although the exact stoichiometric amount of air will depend on the precise composition of the fuel.

Using air instead of pure oxygen allows some of the oxygen to combine with nitrogen at the temperatures reached during fuel combustion, so that small amounts of oxides of nitrogen are formed. There are three common oxides of nitrogen and a mixture of them is usually formed so that they are not given a precise formula, but simply NO_X since the relative proportions of each will depend on the combustion conditions. Sometimes there is less than the stoichiometric amount of air present and, in this case, a mixture of carbon monoxide (CO) and carbon dioxide is formed as well as water. If there is an excess of air, the amount of carbon monoxide formed will be very low but the oxides of nitrogen may be high. In a gasoline engine, because the air-to-fuel mixture is not entirely homogeneous and because the metal of the cylinder head quenches the flame near it, combustion does not proceed exactly as indicated by the theoretical equation. In addition, there are various impurities present in the fuel in small amounts and some of them, such as the polynuclear aromatics, are quite difficult to combust, and a percentage of them will survive the combustion process. Finally, partially combusted materials can undergo reactions after the combustion chamber to form compounds that were not present in or before the combustion chamber.

A1.7 Monohydric Alcohols, $C_nH_{2n+1}OH$

These compounds are used as automotive fuels and fuel components, and include methanol (methyl alcohol), ethanol (ethyl alcohol), propanol (propyl alcohol), and butanol (butyl alcohol), as the compounds of most interest in automotive fuels. The OH group, which replaces one of the hydrogen atoms in an alkane, gives these compounds their characteristic properties such as solubility in water, which gets progressively less as the number of carbon atoms increases.

Methanol (CH_3OH, sometimes written as MeOH) is the simplest of the monohydric (i.e., having one OH group) alcohols.

Ethanol (C_2H_5OH, sometimes written as EtOH) is the next in the series, followed by propyl alcohol (C_3H_7OH) of which there are two isomers:

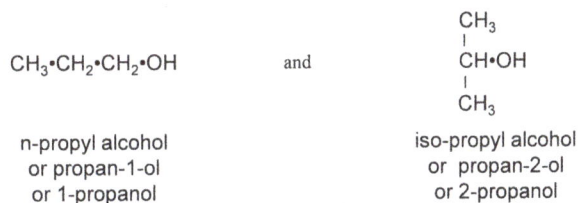

$$CH_3 \cdot CH_2 \cdot CH_2 \cdot OH \qquad \text{and} \qquad \begin{array}{c} CH_3 \\ | \\ CH \cdot OH \\ | \\ CH_3 \end{array}$$

n-propyl alcohol	iso-propyl alcohol
or propan-1-ol	or propan-2-ol
or 1-propanol	or 2-propanol

There are three different butyl alcohols:

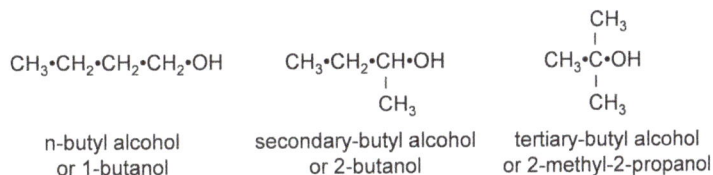

$$CH_3 \cdot CH_2 \cdot CH_2 \cdot CH_2 \cdot OH \qquad \begin{array}{c} CH_3 \cdot CH_2 \cdot CH \cdot OH \\ | \\ CH_3 \end{array} \qquad \begin{array}{c} CH_3 \\ | \\ CH_3 \cdot C \cdot OH \\ | \\ CH_3 \end{array}$$

n-butyl alcohol	secondary-butyl alcohol	tertiary-butyl alcohol
or 1-butanol	or 2-butanol	or 2-methyl-2-propanol

A1.8 Ethers, R′–O–R

Ethers are a group of compounds that contain an oxygen atom connected to the two carbon atoms of two alkyl or aryl groups. The simplest forms are alkyl ethers where the oxygen atom is link to two similar, or different, low carbon number alkyl groups, i.e.:

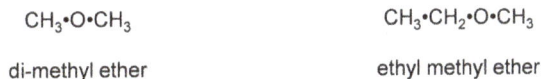

$$CH_3 \cdot O \cdot CH_3 \qquad\qquad CH_3 \cdot CH_2 \cdot O \cdot CH_3$$

di-methyl ether　　　　　　　　ethyl methyl ether

These simple ethers are too volatile to be readily used in automotive fuels, although dimethyl ether (DME) has been used as a fuel for diesel engines, other simple ethers of interest are:

$$\begin{matrix} & CH_3 & \\ & | & \\ CH_3 \cdot C \cdot O \cdot CH_3 & & \\ & | & \\ & CH_3 & \end{matrix}$$ methyl tertiary-butyl ether (MTBE)

$$\begin{matrix} & CH_3 & \\ & | & \\ CH_3 \cdot C \cdot O \cdot CH_3 & & \\ & | & \\ & C_2H_5 & \end{matrix}$$ tertiary-amyl methyl ether   (TAME)

$$\begin{matrix} & CH_3 & \\ & | & \\ CH_3 \cdot C \cdot O \cdot C_2H_5 & & \\ & | & \\ & CH_3 & \end{matrix}$$ ethyl tertiary-butyl ether   (ETBE)

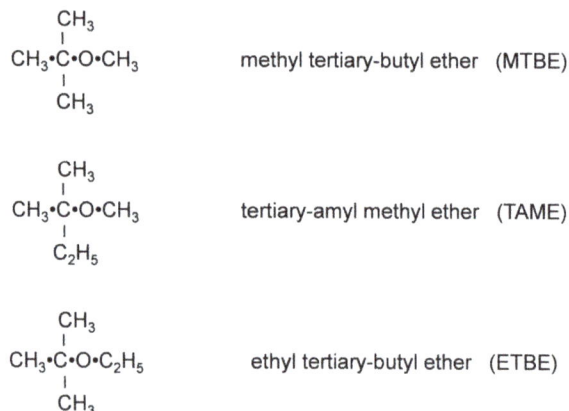

These ethers have been used as gasoline blending components.

Other ethers of interest are the group polyoxymethylene dimethyl ethers $CH_3\text{-}O\text{-}(CH_2\text{-}O)_X\text{-}CH_3$

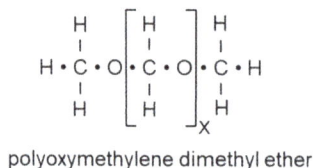

polyoxymethylene dimethyl ether

These compounds are of interest because they have no carbon–carbon bonds and a high oxygen content.

A1.9 Combustion of Oxygenates

Oxygenates require proportionately less oxygen for complete combustion as illustrated by the following equations for the combustion of methanol and MTBE:

$$2CH_3OH + 3O_2 \rightarrow 2CO_2 + 4H_2O$$

i.e., 64 g of methanol require 96 g of oxygen for complete combustion of methanol and so 100 g of methanol requires 150 g of oxygen.

$$2CH_3OC(CH_3)_3 + 15O_2 \rightarrow 10CO_2 + 12H_2O$$

i.e., 176 g MTBE requires 480 g of oxygen and so 100 g of MTBE would require 273 g oxygen.

For the hydrocarbon heptane, which is similar to gasoline in terms of stoichiometric air-to-fuel ratio, 100 g of heptane requires 352 g oxygen, as outlined in Section A1.6, which is approximately 2.3 times as much as for methanol and 1.3 times as much as for MTBE. It is clear that substituting or blending an oxygenated fuel component can significantly alter the air-to-fuel ratio unless there is some compensation.

Appendix 2: Worldwide Fuel Charter Recommendations

The Worldwide Fuel Charter provides fuel quality recommendations published by the members of the Worldwide Fuel Charter Committee as a service to worldwide legislators, fuel users, and producers. It contains information from sources believed to be reliable; however, the Committee makes no warranty, guarantee, or other representation, express or implied, with respect to the Charter's sufficiency or fitness for any particular purpose. The Charter imposes no obligation on any users or producers of fuel, and it does not prohibit use of any engine or vehicle technology or design, fuel, or fuel quality specification. It is not intended to, nor does it, replace engine and vehicle manufacturers' fueling recommendations.

The Worldwide Fuel Charter Committee is composed of representatives from the European Automobile Manufacturers Association (ACEA), the Alliance of Automobile Manufacturers (Alliance), the Truck and Engine Manufacturers Association (EMA), the Japan Automobile Manufacturers Association (JAMA), and associated organizations from around the world.

A full copy of the Worldwide Fuel Charter can be obtained from the members listed above. The following pages summarize the major points from the full document. The Charter recommends specification limits for gasoline and diesel fuel in five categories as described below.

- Category 1: Markets with no or first level of emission control; based primarily on fundamental vehicle/engine performance and protection of emission control systems, for example, markets requiring US Tier 0, Euro 1 or equivalent emission standards. For gasoline this category is being retired as obsolete, given the significant fuel quality progress particularly regarding lead and sulfur content.

- Category 2: Markets with requirements for emission control or other market demands, for example, markets requiring US Tier 1, US 1998 and 2004 Heavy-Duty On-Highway, Euro 2/II, Euro 3/III or equivalent emission standards.

- Category 3: Markets with more stringent requirements for emission control or other market demands, for example, markets requiring US LEV, California LEV or ULEV, US 1998 and 2004 Heavy-Duty On-Highway, Euro 4/IV (except lean burn gasoline engines), JP 2005 or equivalent emission standards.

- Category 4: Markets with advanced requirements for emission control, for example, markets requiring US Tier 2, US Tier 3, US 2007/2010 Heavy-Duty On-Highway, US Non-Road Tier 4, California LEV II, Euro 4/IV, Euro 5/V, Euro 6/6b/VI, JP 2009 or equivalent emission standards. Category 4 fuels enable sophisticated NO_X and particulate matter after-treatment technologies.

- Category 5: Markets with highly advanced requirements for emission control (including GHG) and fuel efficiency, such as US Tier 3 Bin 5, US light-duty vehicle fuel economy, California LEV III and as amended, Euro 6c, Euro 6dTEMP, Euro 6d, current EU CO_2 target, China 6a, China 6b or equivalent emission control and fuel efficiency standards, or other market demands. This category aims to minimize real driving emissions (RDE) required for Euro 6dTEMP (from 2017), Euro 6d (from 2020) and China 6b (from 2023).

- Category 6: Markets with potential emission control and fuel efficiency targets more stringent than those in Category 5, such as future or anticipated US light-duty vehicle GHG and fuel economy standards and future EU targets for CO_2. This category aims to minimize RDE required for Euro 6dTEMP (from 2017), Euro 6d (from 2020) and China 6b (from 2023). This category is intended to enable the introduction of engines and vehicles with higher fuel efficiency and lower exhaust emissions and to specify a gasoline with lower carbon intensity, thereby enabling greater efficiency and GHG benefits than possible with lower category fuels.

Requirements for all markets:

Fuel in the market will meet the quality specifications only if blendstock quality is monitored and good management practices are used. The following recommendations apply broadly to fuel systems in all markets:

- Additives must be compatible with engine oils, to prevent any increase in engine sludge or deposits of varnish.

- Ash-forming components must not be added.

- Good housekeeping practices must be used throughout distribution to minimize contamination from dust, water, different fuels, and other sources of foreign matter.

- Pipeline corrosion inhibitors must not interfere with fuel quality, whether through formulation or reaction with sodium.

- Dispenser pumps must be labeled adequately to help customers identify the appropriate fuels for their vehicles.

- Fuel should be dispensed through nozzles meeting SAE J285, "Dispenser Nozzle Spouts for Liquid Fuels Intended for Use with Spark Ignition and Compression Ignition Engines."

- Ethanol used for blending with gasoline, and biodiesel (FAME) used for blending with diesel fuel, should adhere to the E100 Guidelines and the B100 Guidelines, respectively, published by the Worldwide Fuel Charter Committee.

The following pages show the test methods and recommended limits for the specified fuel properties for the five different categories.

TABLE A2.1 Test methods for gasoline properties.

Property	Units	ISO	ASTM	JIS	Others
RON		EN 5164	D2699	K 2280	
MON		EN 5163	D2700	K 2280-96	
Oxidation stability	Minutes	7536	D525	K 2287	
Sulfur content	mg/kg		D2622	K 2541	
		20846	D5453		
		20844			
Lead content	mg/L		D3237	K 2255	EN 237
Potassium content	mg/L				NF M 07065
					EN 14538
Trace metal content	mg/kg				ICP; ASTM D7111 modified
Phosphorous content	mg/L		D3231		
Silicon content	mg/kg				ICP-AES (Reference in-house methods with detection limit = 1 mg/kg)
Oxygen content	% m/m		D4815	K 2536	EN 13132
Olefin content[1]	% v/v	3837	D1319	K 2536	
Aromatic content[1]	% v/v	3837	D1319	K 2536	EN 14517
Benzene content	% v/v		D5580	K 2536	EN 238
			D3806		EN 14517
Vapor pressure	kPa		D5191	K 2258	EN 13016/1 DVPE
Distillation: T10/T50/T90, E70/E100/E180. End point, residue		3405	D86	K 2254	
Vapor/Liquid ratio (V/L)	°C		D5188		
Sediment	mg/L		D5452		
Unwashed gums	mg/100 mL	6246	D381	K 2261	May be replaced with CCD test
Washed gums	mg/100 mL	6246	D381	K 2261	
Density	kg/m³	3675	D4052	K 2249	
		12185			
Copper corrosion	Rating	2160	D130	K2513	
Silver corrosion	Rating		D7671		
Appearance			D4176		Visual inspection
Fuel injector cleanliness, Method 1	% Flow loss		D5598		
Fuel injector cleanliness, Method 2	% Flow loss		D6421		
Particulate contamination	Code rating	4406			Intake valve sticking
Size distribution	No. of particles/mL	4407 and 11500			
Pass/fail				CEC F-16-T	
Intake valve cleanliness II					
Method 1 (CEC F-05-A-93)	Avg. mg/valve				CEC F-05-A
Method 3 (ASTM D6201)	Avg. mg/valve		D6201		
Combustion chamber deposits					
Method 1	% of base fuel		D6201		
Method 2	mg/engine				CEC F-20-A
Method 3	% Mass at 450°C				FLTM-BZ154[2]

Notes:

[1] Some methods for olefin and aromatic content are used in legal documents; more precise methods are available and may be used.

[2] This method is available at http://global.ihs.com.

TABLE A2.2 Gasoline specifications—recommended limits.

Property	Units		Category 2	Category 3	Category 4	Category 5	Category 6
"91 RON"[1]	RON	Min	91	91	91		
	MON	Min	82.5	82.5	82.5		
"95 RON"[1]	RON	Min	95	95	95	95	
	MON	Min	85	85	85	85	
"98 RON"[1]	RON	Min	98	98	98	98	98
	MON	Min	88	88	88	88	88
"102 RON"[1]	RON	Min					102
	MON	Min					88
Oxidation stability	Minutes	Min	480	480	480	480	480
Sulfur	mg/kg	Max	150	30	10	10	10
Trace metals[2]	mg/kg	Max	1 or non-detectable, whichever is lower				
Oxygen[3]	% m/m	Max	2.7[4]	2.7[4]	2.7[4]	3.7[4]	3.7[4]
Olefins	% v/v	Max	18.0	10.0	10.0	10.0	10.0
Aromatics	% v/v	Max	40.0	35.0	35.0	35.0	35.0
Benzene	% v/v	Max	2.5	1.0	1.0	1.0	1.0
Sediment	mg/L	Max	1	1	1	1	1
Unwashed gums	mg/100 mL	Max	70	30	30	30	30
Washed gums	mg/100 mL	Max	5	5	5	5	5
Density	kg/m³	Min	715	715	715	720	720
Density	kg/m³	Max	770	770	770	775	775
Copper corrosion	Rating	Min	Class 1	Class 1	Class 1	Class 1	Class 1
Silver corrosion	Rating	Min	Class 1		Class 1	Class 1	Class 1
Appearance			Clear and bright; no free water or particulates				
Fuel injector cleanliness, Method 1	% Flow loss	Max	5	5	5	5	5
Fuel injector cleanliness, Method 2	% Flow loss	Max	10	10	10	10	10
Particulate contamination	Code rating		18/16/13 per ISO 4406				
Size distribution	No. of particles/mL						
Intake valve sticking	Pass/fail		Pass	Pass	Pass	Pass	Pass
Intake valve cleanliness II							
Method 1 (CEC F-05-A-93)	Avg. mg/valve	Max	50	30	30	30	30
Method 2 (ASTM D6201)	Avg. mg/valve	Max	90	50	50	50	50

(Continued)

TABLE A2.2 (Continued) Gasoline specifications—recommended limits.

Property	Units		Category 2	Category 3	Category 4	Category 5	Category 6
Combustion chamber deposits			(5)	(5)	(5)		
Method 1 (ASTM D6201)	% of base fuel	Max	140	140	140	140	140
Method 2 (TGA - FLTM BZ154-01)	% Mass at 450°C	Max	20	20	20	20	20
Method 3 (DISI)[6]							
PMI[7]							

Notes:

(1) Three octane grades are defined for maximum market flexibility; availability of all three is not needed.

(2) Examples of trace metals (and other undesirable elements) include, but are not limited to, Cl, Cu, Fe, Mn, Na, P, Pb, and Si. Metal-containing additives are acceptable only for valve seat protection in non-catalyst cars; in this case, potassium-based additives are recommended. No intentional addition of metal-based additives is allowed.

(3) Where oxygenates are used, ethers are preferred. Methanol is not permitted.

(4) By exception, up to 10% by volume ethanol content is allowed if permitted by existing regulation. In this case, the blendstock ethanol should meet the E100 Guidelines published by the Worldwide Fuel Charter Committee. Fuel pump labeling is recommended to enable customers to determine if their vehicles can use gasoline containing 10% ethanol.

(5) To provide flexibility (for example, to enable the use of detergency additives that increase unwashed gum levels), the fuel may comply with either the Unwashed Gum limit or the Combustion Chamber Deposits limit. This does not apply in category 5 and category 6.

(6) Method and limit are in development.

(7) A PMI limit will be adopted when improved methodology is available.

© SAE International.

TABLE A2.3 Test methods for diesel fuel properties.

Property	Units	ISO	ASTM	JIS	Others
Cetane number		5165	D613	K 2280	D6890, D7170[1]
Cetane index		4264	D4737	K 2280	
Density at 15°C	kg/m^3	3675	D4052	K 2249	
		12185			
Viscosity at 40°C	mm^2/s	3104	D445	K2283	
Sulfur content	mg/kg	20846	D5453	K 2541	
		20884	D2622		
Trace metal content	mg/kg				ICP, D7111 modified
Total aromatic content	% m/m		D5186		EN 12916
PAH (di+, tri+)	% m/m		D5186		EN 12916, D2425
T90, T95, Final boiling point	°C	3405, 3924	D86	K 2254	D2287
Flash point	°C	2719	D93	K 2265	D56
Carbon residue	% m/m	10370	D4530	K 2270	
Cold Filter Plugging Point (CFPP)	°C		D6371	K 2288	EN 116, IP 309
Low Temperature Flow Test (LTFT)	°C		D4539		
Cloud Point (CP)	°C	3015	D2500	K 2269	D5771, D5772, D5773
Water content	mg/kg	12937	D6304	K 2275	
Oxidation stability, Method 1	g/m^3	12205	D2274		
Oxidation stability, Method 2 (modified Rancimat)	Hours				EN 15751
Oxidation stability, Method 3 (Delta TAN)	mg KOH/g		D664 and D2274 (modified)		
Foam volume	mL				NF M 07-075
Foam vanishing time	Seconds				NF M 07-075
Biological growth					NF M 07-070
FAME content	% v/v		D7371		EN 14078
Ethanol/Methanol	% v/v		D4815 (modified)		
Total acid number	mg KOH/g	6618	D664 and D2274 (modified)		
Ferrous corrosion			D665[2]		
Copper corrosion	Rating	2160	D130	K 2513	
Ash content	% m/m	6245	D482[3]	K 2272	
Particulate contamination, total	mg/kg		D5452		EN 12662[4]
Particulate contamination	Code rating	4406			
Size distribution	No. of particles/mL	4407 and 11500			
Appearance			D4176		Visual inspection
Injector cleanliness (Method 1)	% Air flow loss				CEC (PF-023) TBA[5]
Injector cleanliness (Method 2)	% Power loss				CEC F-098
Lubricity (HFRR wear scar dia. at 60°C)	μm	12156-1.3	D6079		CEC F-06-A, D7688

Notes:
[1] ASTM D6890 and D7170 measure Derived Cetane Number (DCN) and are being widely used as alternatives to D613.
[2] Procedure A, run at 38°C for 5 hours.
[3] Minimum 100 g sample size.
[4] Method under review.
[5] CEC has initiated test development for Internal Diesel Injector Deposits (IDID).

© SAE International.

TABLE A2.4 Diesel fuel specifications—recommended limits.

Property	Units		Category 1	Category 2	Category 3	Category 4	Category 5
Cetane number		Min	48.0	51.0	53.0	55.0	55.0
Cetane index[1]		Min	48.0 (45.0)	51.0 (48.0)	53.0 (50.0)	55.0 (52.0)	55.0 (52.0)
Density at 15°C	kg/m³	Min	820[2]	820[2]	820[2]	820[2]	820[2]
		Max	860	850	840	840	840
Viscosity at 40°C	mm²/s	Min	2.0[3]	2.0[3]	2.0[3]	2.0[3]	2.0[3]
		Max	4.5	4.0	4.0	4.0	4.0
Sulfur content	mg/kg	Max	2000	300	50	10	10
Trace metal content[4]	mg/kg	Max	1 or non-detectable, whichever is lower				
Total aromatic content	% m/m	Max		25	20	15	15
PAH (di+, tri+)	% m/m	Max		5	3.0	2.0	2.0
T90	°C	Max		340[5]	320[5]	320[5]	320[5]
T95	°C	Max	370	355[5]	340[5]	340[5]	340[5]
Final boiling point	°C	Max		365	350	350	350
Flash point	°C	Min	55[6]	55	55	55	55
Carbon residue	% m/m	Max	0.30	0.30	0.20	0.20	0.20
CFPP or LFTT or CP	°C	Max	At least 5°C lower than the lowest expected ambient temperature.[7]				
Water content	mg/kg	Max	500	200	200	200	200
Oxidation stability, Method 1	g/m³	Max	25	25	25	25	25
Oxidation stability, Method 2	Hours	Min	30	35	35	35	
Oxidation stability, Method 3 (Delta TAN)	mg KOH/g	Max	0.12[8]	0.12[8]	0.12[8]	0.12[8]	0.12
Foam volume	mL		100		100	100	100
Foam vanishing time	Seconds		25		15	15	15
Biological growth				"Zero" content			
FAME content	% v/v	Max	5[9]	5[9]	5[9]	5[9]	Non-detectable
Other biofuels[10]	% v/v	Max				5[9]	
Ethanol/Methanol	% v/v	Max			Non-detectable[11]		

(Continued)

TABLE A2.4 (Continued) Diesel fuel specifications—recommended limits.

Property	Units		Category 1	Category 2	Category 3	Category 4	Category 5
Total acid number	mg KOH/g			0.08	0.08	0.08	0.08
Ferrous corrosion						Light rusting	
Copper corrosion	Rating	Max	Class 1	Class 1	Class 1	Class 1	Class 1
Ash content	% m/m	Max	0.01	0.01	0.01	0.001(12)	0.001(12)
Particulate contamination, total(13)	mg/kg	Max	10	10	10	10	10
Particulate contamination	Code rating				18/16/13 per ISO 4406		
Size distribution	No. of particles/mL						
Appearance			Clear and bright; no free water or particulates				
Injector cleanliness (Method 1)(14)	% Air flow loss			85	85	85	85
Injector cleanliness (Method 2)	% Power loss				2	2	2
Lubricity (HFRR wear scar dia. at 60°C)	µm	Max	460	460	460	400	400

Notes:

(1) Cetane Index is acceptable instead of Cetane Number if a standardized engine to determine the Cetane Number is unavailable and Cetane improvers are not used. When Cetane improvers are used, the estimated Cetane Number must be greater than or equal to the specified value and the Cetane Index must be greater than or equal to the number in parenthesis.

(2) May relax the minimum limit to 800 kg/m³ when ambient temperatures are below −30°C. For environmental purposes, a minimum of 815 kg/m³ can be adopted.

(3) May relax the minimum limit to 1.5 mm²/s when ambient temperatures are below −30°C or to 1.3 mm²/s when ambient temperatures are below −40°C.

(4) Examples of trace metals (and other undesirable elements) include, but are not limited to, Cl, Cu, Fe, Mn, Na, P, Pb, and Si. No trace metal should exceed 1 mg/kg. No intentional addition of metal-based additives is allowed.

(5) Compliance with either T90 or T95 is required.

(6) The minimum limit can be relaxed to 38°C when ambient temperatures are below −30°C.

(7) If compliance is demonstrated by meeting CFPP, then it must be no more than 10°C less than cloud point.

(8) Change in total acid number (delta TAN) can be used as an alternative to Oxidation Stability Method 2.

(9) For FAME, both EN14214 and ASTM D6751, or equivalent standards, should be considered. Where FAME is used, the blendstock should meet the B100 Guidelines published by the Worldwide Fuel Charter Committee, and fuel pumps should be labeled accordingly.

(10) Other biofuels include HVO and BTL. Blending level must allow the finished fuel to meet all the required specifications.

(11) At or below detection limit of the test method used.

(12) Limit and test method are under review to assure DPF endurance.

(13) Limit and test method are under review.

(14) CEC is developing a new method for Internal Diesel Injector Deposits (IDID).

© SAE International.

Appendix 3: TOP TIER™ Fuel Standards

The following pages contain copies of the performance standards for gasoline and diesel fuel promoted by the Top Tier™ program.

- Top Tier Detergent Gasoline Deposit Control Performance Standard Revision F—December 2019
- Top Tier Diesel Fuel Performance Standard Revision A—February 2017

These are the latest versions of the performance standards; updates are available on the website at https://www.toptiergas.com/performance-standards/

TOP TIER DETERGENT GASOLINE
DEPOSIT CONTROL PERFORMANCE STANDARD
Revision F – December 2019

1. Scope

1.1 Performance Description. This document describes the deposit control performance of an unleaded gasoline at the retail level that minimizes deposits on fuel injectors, intake valves, and combustion chambers.

2. References

Note: Only the latest versions of standards are applicable or as indicated.

2.1 ASTM International (American Society for Testing and Materials)
D 86 D 381 D 1319 D 2622 D 4806 D 4814 D 4815 D 5453 D 5580 D 5845 D 6201 D 6550 D 6729

2.2 California **Air Resources Board (CARB)**
Advisory letter (April 19, 2001).
Test Method for Evaluating Intake Valve and Combustion Chamber Deposits in Vehicle Engines (March 12, 1999).

2.3 Coordinating European Council (CEC)
CEC F-16-T-96

2.4 Code of Federal Regulations (CFR)
Parts 79 and 80

2.5 International Organization for Standardization (ISO)
ISO 17025, General Requirements for the Competence of Testing and Calibration Laboratories

2.5 General Motors (GM)
TOP TIER fuel injector fouling vehicle test

2.6 Southwest Research Institute (SwRI)
Intake Valve Sticking Test in GM 5.0L V-8

3. Definitions

3.1 "Independent Laboratory" - a mechanical and/or chemical testing organization which is accredited by a national or international accreditation agency such as the American Association for Laboratory Accreditation for testing competence in mechanical and chemical testing or ISO 17025 "General Requirements for the Competence of Testing and Calibration Laboratories"; provided, however, that an Independent Laboratory cannot be affiliated with a TOP TIER Licensee, a fuel marketer or the manufacturer of any additive package approved for use in TOP TIER Detergent Gasoline.

4. Standards

4.1 Retail Gasoline Performance Standards. The deposit control performance of an additive package conforming to section 4 of this document shall be included at the retail level in all grades of gasoline sold by a fuel company in all marketing areas of a selected nation. In addition, conformance to the standards shall mean gasoline sold in the selected nation shall not contain metallic additives, including methylcyclopentadienyl manganese tricarbonyl (MMT).

4.2 Deposit Control Additive Requirements. The deposit control additive used to meet the performance Standards described in 4.3 shall meet the substantially similar definition under Section 211(f) of the Clean Air Act. Also, the additive shall be certified to have met the minimum deposit control requirements established by the U.S. Environmental Protection Agency (EPA) in 40 CFR Part 80. Lastly, the additive shall be registered with the EPA in accordance with 40 CFR Part 79.

TOP TIER DETERGENT GASOLINE
DEPOSIT CONTROL PERFORMANCE STANDARD
Revision F – December 2019

4.3 Deposit Control Initial Performance Standard. All performance testing and fuel composition analysis shall be conducted by an Independent Laboratory. Initial deposit control performance shall be demonstrated using the tests shown below.

4.3.1 Intake Valve Keep Clean Initial Performance Standard

4.3.1.1 Test Method. Intake valve deposit (IVD) keep clean performance shall be demonstrated using ASTM D 6201, *Standard Test Method for Dynamometer Evaluation of Unleaded Spark-Ignition Engine Fuel for Intake Valve Deposit Formation.* Tests demonstrating base fuel minimum deposit level (4.3.1.2) and additive performance (4.3.1.3) shall be conducted using the same engine block and cylinder head. All results shall be derived from operationally valid tests in accordance with the test validation criteria of ASTM D 6201. IVD results shall be reported for individual valves and as an average of all valves.

4.3.1.2 Base Fuel. The base fuel shall conform to ASTM D 4814 and shall contain commercial fuel grade ethanol conforming to ASTM D 4806. All gasoline blend stocks used to formulate the base fuel shall be representative of normal territory refinery operations and shall be derived from conversion units downstream of distillation. Butanes and pentanes are allowed for vapor pressure adjustment. The use of chemical streams is prohibited. The base fuel shall have the following specific properties after the addition of ethanol:

1. Contain enough denatured ethanol such that the ethanol content is no less than 8.0 and no more than 10.0 volume percent as measured by ASTM D 4815 or D 5845. In markets with lower fuel ethanol content, fuel matching the market conditions of fuel ethanol content can be used up on approval.
2. Contain no less than 8 volume percent olefins as measured by ASTM D 1319 or D 6729.
3. Contain no less than 15 volume percent aromatics as measured by ASTM D 1319 or D 6729.
4. Contain no more than 80 mg/kg sulfur as measured by ASTM D 2622 or D 5453.
5. Produce a 90% evaporated distillation temperature no less than 290°F as measured by ASTM D86.
6. Produce IVD no less than 500 mg averaged over all intake valves.
7. A Certificate of Analysis showing both the detailed test fuel composition results and source should accompany the additive results package. This certificate should also contain the unwashed and washed gum level of the base fuel according to ASTM D381.

4.3.1.3 Demonstration of Performance. The base fuel from 4.3.1.2 shall contain enough deposit control additive such that IVD is no more than 50 mg averaged over all intake valves. Results for individual valves and an average shall be reported. The unwashed gum level of the fuel containing deposit control additive shall be determined according to ASTM D 381 and reported.

4.3.2 Combustion Chamber Deposit Initial Performance Standard

4.3.2.1 Test Method. Combustion chamber deposits (CCD) shall be collected and weighed along with IVD using ASTM D 6201, *Standard Test Method for Dynamometer Evaluation of Unleaded Spark-Ignition Engine Fuel for Intake Valve Deposit Formation.* ASTM D 6201 does not contain a procedure for collecting and measuring CCD. Adapting a scrape and weigh procedure developed by CARB is recommended (see referenced test method dated March 12, 1999). Results for individual cylinders and an average shall be reported.

4.3.2.2 Base Fuel. Combustion chamber deposits shall be measured for the base fuel from 4.3.1.2.

4.3.2.3 Demonstration of Performance. The base fuel from 4.3.1.2 treated with additive at the concentration meeting the standard found in 4.3.1.3 shall not result in more than 140% of the average CCD weight for the base fuel without additive.

TOP TIER DETERGENT GASOLINE
DEPOSIT CONTROL PERFORMANCE STANDARD
Revision F – December 2019

4.3.3 Intake Valve Sticking Initial Performance Standard

4.3.3.1 Test Method. Intake valve sticking tendency shall be determined using either the 1.9 L Volkswagen engine (Wasserboxer according to CEC F-16-T-96) or the 5.0 L 1990-95 General Motors V-8 engine (SWRI IVS test). Two options are available for demonstrating intake valve sticking tendency.

4.3.3.2 Option 1. The valve-sticking tendency of the test fuel by itself will not have to be demonstrated prior to testing the candidate additive. The following shall be required of all tests:

1. Test fuel shall be either the same as in 4.3.1.2 or CEC valve sticking reference fuel.
2. Concentration of deposit control additive in the test fuel shall be at three times the amount determined in 4.3.1.3.
3. Test temperature shall be -20°C.
4. Three 16-hr cold soak cycles, each followed by a compression pressure check, shall constitute a complete test.

4.3.3.2.1 Demonstration of Performance. A pass shall result in no stuck valves during any of the three cold starts. A stuck valve is defined as one in which the cylinder pressure is less than 80% of the normal average cylinder compression pressure.

4.3.3.3 Option 2. The valve-sticking tendency of the test fuel together with an additive known to cause valve sticking shall be demonstrated prior to testing the candidate additive. The following shall be required of all tests:

1. Test fuel shall be either the same as in 4.3.1.2 or CEC valve sticking test reference fuel.
2. An additive known to cause valve sticking shall be selected, and, when blended into test fuel, shall demonstrate valve sticking tendency as follows: (a) for the Volkswagen engine, at least two valves shall be stuck; (b) for the GM engine, at least three valves shall be stuck.
3. Tests demonstrating performance of the candidate additive shall be conducted at a concentration that is at least three times the amount determined in 4.3.1.3
4. Test temperature shall be -20°C.
5. One 16-hr cold soak cycle followed by a compression pressure check shall constitute a complete test.

5. Process to Register Additive Packages for use in Licensed TOP TIER™ Detergent Gasoline

5.1 Submission of Test Results. An additive or fuel company desiring to register an additive package for use in licensed TOP TIER™ Detergent Gasoline products shall forward the test results issued by the Independent Laboratory ("Test Results") to the following address:

Center for Quality Assurance
Attn: TOP TIER™ Licensing Program
4800 James Savage Rd.
Midland, MI 48642 USA
Telephone: +1 989-496-2399
Facsimile: +1 989-496-3438
Email: TopTier@CenterForQA.com

5.2 Notification of receipt. The Test Results shall be reviewed by the TOP TIER Detergent Gasoline programand, if deemed acceptable in its sole discretion, the additive company will be notified the additive package will be registered for use in licensed TOP TIER Detergent Gasoline products. Such notification does not allow the use of the TOP TIER™ trademarks or logos.

3

TOP TIER DETERGENT GASOLINE
DEPOSIT CONTROL PERFORMANCE STANDARD
Revision F – December 2019

6. Release, Effective Date, and Revisions

6.1 Release. This document was first released in April 2004.

6.2 Revisions. Revision B is effective May 2008.
Revision C is effective February 2015.
Revision D is effective May 2016.
Revision E is effective March 2017.
Revision F is effective December 2019.

Revision	Date	Description of change
A	March 2008	Modified document to reflect new license agreement; removed attestation forms.
B	May 2008	Changed minimum sulfur limit to 24 mg/kg.
C	February 2015	Removed Fuel Injector Fouling Initial Performance Standard requirement; Changed ethanol limit from 8.0-10.0% up to 10% ± 1%; Changed sulfur to a maximum limit of 80 mg/kg; Removed base fuel requirement for proof of 75% olefin and 60% sulfur content from FCC streams; Added requirement to provide full base fuel composition analytical results and gum values
D	May 2016	Changed contact to CQA or submission of results Changed reference in option 1 valve-sticking to 4.3.1.3 (TOP TIER Concentration) from Injector Fouling concentration reference.
E	March 2017	3.1 Modified from A2LA to Independent Laboratory with further definition to accommodate international laboratory accreditation. 4.3.1.2 Base Fuel: ethanol changed to 8-10% to align with CARB and to market ethanol outside US.
F	October 2019	4.1 An approved additive package at the correct treat rate must be used at all grades of gasoline at a station vs. all gasoline must meet the standard. 4.3.1.2 Minimum aromatics changed to 15% from 28% as it is difficult to obtain fuel with over 28% aromatics. 4.3.3.2 Valve sticking test concentration changed from 2 times to 3 times to align with allowable treat rate range. 5 Updated language to reflect the process to obtain TOP TIER approval. Added additive company as a submitter and also removed reference to licensing- which is handled separately.

4

TOP TIER DIESEL FUEL PERFORMANCE STANDARD
Revision A - February 2017

1. Scope.

1.1 Performance Description. This document describes the performance standard for diesel fuel at a retail level or private fueling site that has detergents to prevent deposit formation on injectors, enhanced stability, improved lubricity, and lower water & particulate contamination.

2. References.
Note: Only the latest versions of standards are applicable or as indicated.

2.1 ASTM International (American Society for Testing and Materials)
ASTM D975, ASTM D6751, ASTM D7467, ASTM D6079, ASTM D6304, ASTM D6079, ASTM D6217, ASTM D7545, ASTM D7501, ASTM D2274, ASTM D524, ASTM D2500, ASTM D23015, ASTM D6371, ASTM D130, ASTM D664, ASTM D93, ASTM D971, ASTM D471, ASTM D6201

2.2 Coordinating European Council (CEC)
CEC F-98-08 (DW10B) and CEC F-110-16 (DW-10C), EN590, EN 15751, EN 14214, DIN EN 12662, DIN EN 116

2.5 International Organization for Standardization (ISO)
ISO 17025, General Requirements for the Competence of Testing and Calibration Laboratories, ISO 4406, ISO 16889, ISO 12205, ISO 10370, ISO 2160, ISO 2719, ISO 1817, ISO 37

3. Definitions

3.1 Independent Laboratory – a mechanical and/or chemical testing organization which has been approved by General Motors for testing the performance of diesel fuel to the TOP TIER Diesel Fuel Performance Standard; provided, however, that the Independent Laboratory is not affiliated with the entity seeking TOP TIER Diesel Fuel status or with the manufacturer of any additive package approved for use in TOP TIER Diesel Fuel.

4. Standards.

4.1 Fuel Additive Requirements. The additive(s) to be used as part of TOP TIER diesel requirements described in the sections below must meet all the requirements of the regulatory body in country of sales. Documentation demonstrating such compliance is required. For example, in the United States, the additive must be registered as per regulations of Title 40 CFR Part 79 of Section 211 as stipulated by Environmental Protection Agency (EPA) to assess impact of the product on emissions.

4.2 Diesel Performance Standards. The performance standards of diesel fuel conforming to section 4 of this document shall be met at the point of delivery to the vehicle at the retail level or private fueling site of diesel sold by a fuel company in all or selected marketing areas of a nation. In addition, conformance to the standards shall mean that the diesel fuel sold in the selected nation will meet the latest existing standards for diesel/biodiesel blends such as ASTM D975 and ASTM D7467 standards in the United States.

4.2.1. Fuel Stability. Biofuels such as fatty acid esters, commonly called "biodiesel," are increasingly being used in the marketplace. In the United States, the quality of common fatty acid alkyl esters is defined in the ASTM D6751 standard for biodiesel fuel blend stock (B100) intended for use in blends at concentrations of up to 20 volume percent. There are two additional standards that define the quality of the blended fuels, ASTM D975 which applies to blended fuels up to B5 and ASTM D7467 which applies to blended fuels from B6 to B20.

Field warranty and validation tests have shown significant concerns with stability of biodiesel fuels including fuel system deposits, engine oil deterioration, and efficiency loss of the after treatment system. Maintaining high stability of biodiesel blended fuels is critical for its optimal performance as a vehicle fuel.

In the United States, the ASTM D6751 standard for FAME has a stability requirement as measured by an induction period of 3 hours minimum, which is significantly lower than the induction period of a minimum of 8 hours that is required by the European EN 14214 standard. There are no stability requirements in ASTM D975, while the stability

1

TOP TIER DIESEL FUEL PERFORMANCE STANDARD
Revision A - February 2017

requirements for B6 to B20 blends as per ASTM D7467 is a minimum of 6 hours. In comparison, the EN590 standard which allows biodiesel content up to 7 volume percent has a limit for stability of a minimum 20 hours.

Fuel injection equipment and after-treatment systems could be significantly damaged from the use of oxidized fuel. Diesel OEMs supporting the TOP TIER diesel program would like to ensure that the level of oxidation stability in the diesel as well as biodiesel blended fuel is sufficient to prevent damage due to oxidized fuel under all normal scenarios in which the fuel is stored and used.

Fuel injection equipment and other components could have significant maintenance and/or operating issues from use of oxidized fuel. Diesel OEMs supporting TOP TIER diesel program would like to ensure that the level of oxidation stability in the diesel as well as biodiesel blended fuel is sufficient to prevent any issues due to oxidized fuel in all scenarios of fuel storage and use.

4.2.1.1 Test Method. Fuel stability of biodiesel and biodiesel blended fuels (≥B2) is best described by the parameter 'ageing reserve' determined as Rancimat induction period (IP) according to the EN 15751 test method.

Fuel stability of diesel fuels without any biodiesel (B0) or up to B2 blend must be determined by Rapid Small Scale Oxidation Test (RSSOT) - PetroOXY test as per ASTM D7545 method which measures the induction period under specified conditions and can be used as an indication of the oxidation and storage stability of diesel fuels.

4.2.1.2 Acceptable Performance Limit. The fuel should have following stability properties to meet the TOP TIER diesel performance:
1. The FAME Biodiesel (B100) to be used to make biodiesel blends must have induction period of minimum of 8 hours as measured by EN15751 test method
2. If it is not possible to qualify the stability performance of B100 used to make a biodiesel blend then following stability requirements must be fulfilled at the blend levels:
   a. For biodiesel blends up to 5 volume percent as allowed by ASTM D975 standard, the induction period should be minimum of 24 hours as measured by EN15751 test method.
   b. For biodiesel blends from B6 to B20 as per ASTM D7467 or for blended fuels above B20 biodiesel blends, the induction period should be minimum of 20 hours as measured by EN15751 test method.
3. For petroleum diesel without any biodiesel the expected stability as measured by petro-oxy test (ASTM D7545) should have an induction period of > 60 minutes.

4.2.2 Lubricity. Good lubricity of diesel fuel is important to minimize friction and damage in diesel fuel system components, especially in the fuel pumps. A position statement by the Diesel Fuel Injection Manufacturers of 2012 suggests that the lubricity of the fuel as measured by the HFRR according to ASTM D6079 must have a lubricity value of 460 microns maximum. Fuel with lubricity exceeding 460 microns can adversely affect the lifetime of some of the fuel lubricated injection system components.

4.2.2.1 Test Method. The lubricity of the diesel fuel should be measured by the *High Frequency Reciprocating Rig* (HFRR) method which measures the wear scar on the test specimen ball as per the procedure described in ASTM D6079.

4.2.2.2 Demonstration of Performance. To meet TOP TIER diesel performance the fuels must have an HFRR at 60°C wear scar diameter of less than 460 µm to provide sufficient lubricity.

4.2.3 Particulate Contamination in Fuel. Particulates and water are the most common contaminants in diesel fuel. Modern diesel engines have fine clearances and high operating pressures which makes them highly sensitive to particulate contamination in the fuel. High level of particulate contamination in the diesel fuel can lead to premature clogging of vehicle fuel filters or if the hard particles are carried past the fuel filters, they can cause premature wear of fuel injection systems as well as engine failures. For effective control of particulate contamination in the fuel, it is important to control both the number of particles as well as size of the particles.

4.2.3.1 Test Method. The ISO 4406 test standard provides a measure of particulate contamination based on size and number of particles. The standard defines three code numbers corresponding to the numbers of particles of size greater than 4, 6 and 14 µm per milliliter, respectively.

TOP TIER DIESEL FUEL PERFORMANCE STANDARD
Revision A - February 2017

4.2.3.2 Demonstration of Performance. There are numerous ways particulate contamination can enter the fuel during storage and transportation some of which are beyond the control of the fuel station operators. Also, there is increased risk that improper fuel sampling procedures during audits will artificially increase particulates in the fuel. Thus, for the TOP TIER diesel program, there is no specific requirement for particulate contamination but it is expected that the sediment content be within the specifications defined in latest version of ASTM D975 or ASTM D7467 standards. Although there is no specification for particulate contamination, it is preferable that the fuel cleanliness for particulates meets the typical engine manufacturer requirements identified as per an ISO 4406 Code18/16/13 fuel cleanliness rating at the dispenser nozzle. This rating reflects scale #18 (1300-2500 particles greater than 4 microns), scale #16 (320-640 particles greater than 6 microns), and scale #13 (40-80 particles greater than 14 microns) in any given milliliter sample of fuel.

4.2.4 Water Contamination in Fuel. Water contamination, if not controlled, can led to many problems in the fuel system such as corrosion, fuel degradation, microbial growth, fuel pump wear and ice formation during cold weather. Water is also a good solvent for inorganic salts and can contain dissolved acids or other contaminants that can harm the fuel system and engine either directly or indirectly by deposits formation due to interactions of these contaminants with additives in the fuel. Water may enter the fuel in various ways, such as part of the refining process, as rain, as ship ballast water or as condensation in storage tanks and equipment. Good water management will help minimize the contamination of the fuel during storage and transportation. It is recommended to follow the guidelines and procedures provided in CRC Report 667 – Diesel Fuel Storage and Handling Guide to lower risks for water and particulate contamination in fuel.

Good housekeeping practices and routine maintenance are the best ways to keep water accumulation and particulate contamination in control during storage. In addition, removal of free water from the fuel before the fuel is filled into vehicle fuel tank is also desirable.

4.2.4.1 Test Method. Total water content in diesel fuel shall be measured by procedure described in ASTM D6304 "Standard Test Method for Determination of Water in Petroleum Products, Lubricating Oils, and Additives by Coulometric Titration."

4.2.4.2 Demonstration of Performance. There are numerous ways water can enter the fuel during storage at the fuel station, some of which are beyond the control of the station operators. Thus, for the TOP TIER diesel program there is no specific requirement for absolute water content of the fuel but it is expected that the water content be within the specifications defined in latest version of ASTM D975 or ASTM D7467 standards Additionally, it is preferable that the water content be less than 200 mg/kg at 25°C as measured by ASTM D6304. Audit samples will be tested for water content as per ASTM D6304.

4.2.5 Dispenser Filters. Diesel dispenser filters are last line of defense before particulate or water contamination in the fuel can be delivered into a vehicle's fuel tank. Thus, better requirements on the fuel dispenser filters can help to significantly reduce concerns associated with fuel contaminations like particulates and water. Generally, the dispenser filters are of the 30 microns to 10 microns in size. A finer size filter is expected to be more efficient in cleaning up the particulate contamination and reduce vehicle concerns related to these contamination including increased filter plugging within defined vehicle filter service intervals.

4.2.5.1 Test Method. The filtration efficiency shall be as per the multipass test procedure defined in ISO 16889 method which describes a procedure for determining the contaminant capacity, particulate removal and differential pressure characteristics and a test using ISO medium test dust contaminant and a test fluid. For water absorption capacity testing, there is no official test due to variety of applications and conditions in the field.

4.2.5.2 Demonstration of Performance (Particulate Filter). Dispenser stations with low speed pumps (fuel flow of less than 15 gallons per minute) shall be equipped with a filter having 10 micron or smaller pore size with a filtration efficiency of greater than 90% at removing 10 micron size particles as demonstrated by ISO 16889 procedure. For dispensers with flow rates greater than 15 gallons per minute, it is recommended to have a 10-micron or smaller nominal pore-sized filter; however, it is a requirement to install a dispenser filter of at least 30-micron or smaller nominal pore size filter with filtration efficiency greater than 50% as demonstrated by ISO 16889 procedure.

4.2.5.3 Demonstration of Performance (Water Absorbing Media). For the water absorption capacity, the filter or a separate water absorption unit shall act as detection devices to alert the station operator of excess water conditions

TOP TIER DIESEL FUEL PERFORMANCE STANDARD
Revision A - February 2017

by restricting or stopping flow when excessive water contamination is present. Alternately, it would be acceptable if the retail station or private fueling site uses fuel monitoring technology with robust process to monitor and control excess water in the fuel tanks. Fueling sites using monitoring or detection technology should review their process and quality procedure with the TOP TIER Diesel Fuel program to obtain written approval.

4.3 Detergent Performance Standard. All performance testing and fuel composition analysis shall be conducted by an Independent Laboratory. Initial deposit control performance shall be demonstrated using the tests shown below.

4.3.1 Diesel Injector Nozzle Deposit Test

4.3.1.1 Test Method. Diesel Injector nozzle keep clean performance shall be demonstrated using CEC F-98-08 (DW10B) *Diesel injector nozzle fouling*. In the above test, the base reference fuel minimum deposit level and additive performance shall be conducted using the same engine block and cylinder head. For the CEC-F-98-08 (DW10B), performance data shall be reported in terms of torque loss for each cycle of the test. The final result shall be reported as loss of engine power as measure of injector fouling performance. Reported results should be within the stated reproducibility of the test procedure.

4.3.1.2 Base Fuel. For the CEC F-98-08 (DW10B) test, the base fuel shall conform to the specifications as per RF-06-03 (European Certification test fuel) reference fuel to break in the fuel injectors. For the key part of the CEC F-98-08 (DW10B) test to induce external nozzle deposits, one part per million (ppm) of zinc (Zn) in form of zinc neodecanoate is added to the reference fuel as defined in CEC F-98-08 test procedure.

4.3.1.3 Demonstration of Performance. The base fuel from 4.3.1.2 treated with additive at the concentration meeting the standard found in 4.3.1.1 shall not result in more than 2% power loss while the power loss for the base fuel with Zn neodecanoate and without detergent additive should exceed 6% power loss.

4.3.2 Internal Diesel Injector Deposits Test

4.3.2.1 Test Method and Demonstration of Performance. Engine test method using a common rail fuel system engine demonstrating dirty-up without use of additive and clean-up performance with use of the detergent additives for internal diesel injector deposits caused by carboxylates salts of calcium and sodium. Additional demonstration of keep clean performance of additive in presence of any other contaminants that can cause internal injector deposits is optional but would be valuable for demonstrating effectiveness of additive formulation. In addition, any test data demonstrating effectiveness of the detergent additive to prevent deposit formations in vehicle tests or fleet tests should be submitted for evaluation. The decision on the approval for additives performance for IDID deposits shall be made after review of data by the OEM sponsors. Reported results should be within the stated reproducibility of the test procedure.

4.3.2.2 DW-10 C Test Method (As of January 2017, this test method is a placeholder until formally approved by CEC). Internal Diesel Injector deposit test keep clean performance shall be demonstrated using CEC F-110-16 (DW10C), *Internal Diesel Injector Deposits*. In the above test, the base reference fuel minimum deposit level and additive performance shall be conducted using the same engine block and cylinder head. For CEC F-110-16 (DW10C) test performance data shall be reported in terms of global rating system derived from complex demerit scale for operational issues. Reported results should be within the stated reproducibility of the test procedure.

4.3.1.2 Base Fuel. For the CEC F-110-16 (DW10C) test, the base fuel shall conform to the specifications as per RF-06-03 (European Certification test fuel) reference fuel to break in the fuel injectors. For the key part of the CEC F-98-08 (DW10B) test to induce external nozzle deposits, one part per million (ppm) of zinc (Zn) in form of zinc neodecanoate is added to the reference fuel as defined in CEC F-98-08 test procedure.

4.3.1.3 Demonstration of Performance. The base fuel from 4.3.1.2 treated with additive at the concentration meeting the standard found in 4.3.1.1 shall not result in more than 2% power loss while the power loss for the base fuel with Zn neodecanoate and without detergent additive should exceed 6% power loss.

4.4 No-Harm Testing Requirements. The requirements for no-harm testing should demonstrate suitable interaction of the candidate detergent deposit control additive used in the fuel marketer final blend package. The additive

4

TOP TIER DIESEL FUEL PERFORMANCE STANDARD
Revision A - February 2017

package will have to pass 'no-harms' testing with fuel containing biodiesel at highest blend rate along with lubricity, conductivity, stability additives tested with each detergent deposit control additive.

4.4.1 Demonstration of Performance. The acceptable reference fuels shall be the representative retail pump fuel meeting the ASTM D975 or ASTM D7467 specifications to which this package is expected to be added. The additive package will have to pass the 'no- harms' testing listed in Table 1 and section 4.4.2 that shall be the representative of the retail pump fuel meeting the ASTM D975 or ASTM D7467 specifications. The fuel for 'no-harms' testing must contain biodiesel at highest blend rate with additives for lubricity, conductivity, and stability required to meet TOP TIER diesel performance tested with the detergent deposit control additive. If there is more than one type of detergent deposit control additive that will be offered to meet the TOP TIER diesel performance than each of the formulation must be evaluated in market representative fuel. For each test, the entire TOP TIER diesel package shall be mixed at 3x the recommended concentration of the additives in the reference fuel and its performance will be compared to that of the relevant reference fuel. Tests defined in Table 1 shall be used to determine the candidate additive final package compatibility. All the final additive package blend combinations must be tested under these conditions.

Table 1: Tests for Fuel Compatibility

Properties	Standard	Unit	Pass/Fail Criteria
Fuel Contamination by Additive Precipitation	ASTM D6217 or	mg/L	</= 20
	DIN EN 12662	mg/kg	</=24
Cold Soak Filterability	ASTM D7501	Seconds	< 360
Gum Forming Potential	ASTM D2274 or ISO 12205 or ASTM D7462 (for > B6 fuels)	mg/L	< 25
Carbon Residue (on 10% Distillation Residue), mass%	ASTM D524 or ISO 10370	%	0.35
Cloud Point	ASTM D2500 or ASTM D23015	°C	No change relative to reference fuel unless additive used to effect cold properties of fuel [Note 1]
Cold Filter Plugging Point	ASTM D6371 or DIN EN 116	°C	No change relative to reference fuel unless additive used to effect cold properties of fuel [Note 1]
Copper Strip Corrosion (3 h at 50 °C), Maximum	ASTM D130 or ISO 2160	Rating	No. 3
National Association of Corrosion Engineers (NACE International)	NACE TM0172	Rating	Same or Better than reference fuel
Acid Number, mg KOH/g, Maximum	ASTM D664	mg KOH/g	0.3 Maximum
Flash Point	ASTM D93 or ISO 2719	°C	55 Minimum (Summer Grade) 38 Minimum (Winter Grade)
Water Interfacial Tension	ASTM D971	Dynes/cm	To Report

5

TOP TIER DIESEL FUEL PERFORMANCE STANDARD
Revision A - February 2017

Fuel Additive Resistance Diesel			
Elastomer Testing Note 2	A minimum of three (3) specimens to be tested per elastomer type. Required for fuels that have additives		
SAE J2643 Fluorocarbon (FKM-1)	ASTM D471 or ISO 1817, 168 h -2 h, at 23 °C ± 3 °C. Test ratio fuel to test pieces = (40 ± 5) to 1 All tests performed 1 minute after removal from test liquid Elongation at break, Tensile strength, Change in Mass, Properties after additional drying at +150 °C ± 3 °C for 24 h -2 h Elongation at break, Tensile strength, Change in Mass,	% MPa % % MPa %	120 Minimum 3 Minimum 0 to +30 150 Minimum 6 Minimum -10 to +5
SAE J2643 Hydrogenated Nitrile Butadiene Rubber (HNBR-1)	ASTM D471 or ISO 1817, 168 h -2 at 23 ± 3 °C Test ratio fuel to test pieces = (40 ± 5) to 1 All tests performed 1 minute after removal from test liquid Elongation at break, ASTM D412 or ISO 37 Tensile strength, ASTM D412 or ISO 37 Change in Mass, ASTM D471 or ISO 1817 Properties after additional drying at +150 °C ± 3 °C for 24 h -2 h Elongation at break, Tensile strength, Change in Mass,	% MPa % % MPa %	150 Minimum 5 Minimum 0 to +35 140 Minimum 5 minimum 0 to +25

Note 1: Values within the precision of the described test methods will be acceptable as no change.
Note 2: Purchase elastomer samples from Akron Rubber Development Laboratories at www.ardl.com.

4.4.2 Compatibility with Engine Oil.

4.4.2.1 Test Procedure. Samples must be tested per modified ASTM D2274 or ISO 10370 test procedure. The standard preparation and test procedures in accordance with ASTM D2274 specifications must be followed, except that bubbling oxygen during 16 hours heating period shall not be performed. The following two (2) samples shall be evaluated side by side:
a. Reference fuel without candidate additive but with 0.5% of engine oil which meets service fill specification for dexos2™.
b. Reference fuel with 3x the recommended treat rate of the candidate additive and with 0.5% of engine oil which meets the service fill specification for dexos2™.
The amount of insoluble material generated during the test shall be measured gravimetrically as defined in the test procedure for ASTM D2274. The amount of insoluble generated per 100 mL of the test fuel can provide measure of potential of fuel additive to react adversely with lubricant additive.

4.4.2.2 Demonstration of Performance. The amount of insoluble generated per 100 mL as determined by the modified ASTM D2274 method for sample (**a**) in 3.4.1, (reference fuel with engine oil but no candidate additive) and sample (**b**) in 3.4.1 (reference fuel with engine oil and with 3x candidate additive) shall be ≤ 2.5 mg/100 mL.

6

TOP TIER DIESEL FUEL PERFORMANCE STANDARD
Revision A - February 2017

Note: Values within the precision of the described test methods will be acceptable as no change. Detection limit of test method ASTM D2274 is 1.0 mg/100 mL. Values lower than the detection limit of the test method shall be reported as < 1.0 mg/100 mL.

For a list of dexos2™ approved brands for diesel engine oil, visit: www.gmdexos.com.

It is recognized that no-harm compatibility testing can take significant resources, and, therefore, it is recommended that compatibility data generation test plan be reviewed by the TOP TIER coordinator in advance of conducting the tests.

The above no-harms tests should be regarded as minimum requirements to qualify the additive. Any additional information in terms of vehicle demonstration or field experiences or engine dyno tests would form a favorable basis to confirm consistent performance of fuel additive package.

5. Process to Attain TOP TIER Diesel Fuel Status.

5.1 Submission of Test Results. A fuel company desiring TOP TIER Diesel status shall forward the test results issued by the Independent Laboratory ("Test Results") to the following address:

> Center for Quality Assurance
> Attn: TOP TIER™ Licensing Program
> 4800 James Savage Rd.
> Midland, MI 48642, USA
> Telephone: +1 989-496-2399
> Facsimile: +1 989-496-3438
> Email: TopTier@CenterForQA.com

5.2 Notification of receipt. The Test Results shall be reviewed by representatives of General Motors & other OEM sponsors and, if deemed acceptable in its sole discretion, the fuel company will be provided a TOP TIER License Agreement for their execution. Only upon complete execution of the TOP TIER License Agreement by both the Fuel Company and General Motors shall the fuel company be entitled to begin use the TOP TIER name in connection with the distribution, promotion and sale of their gasoline, pursuant to the terms and conditions of the TOP TIER License Agreement.

6. Release, Effective Date, and Revisions.

6.1 Release. This document was first released in January 2017.

6.2 Revisions. Revision A is effective February 2017.

Revision	Date	Description of change
Original	January 2017	✓ Initial Release
A	February2017	✓ Remove reference to ASTM D6201 in section 4.3 ✓ Update language in section 4.3 to clarify keep clean performance requirement for detergent for external coking and IDID. ✓ Update section 4.4 to clarify requirements for no harms testing

7

Appendix 4: Composition of Biodiesel from Different Feedstocks

The following tables indicate typical fatty acid compositions for some of the feedstocks that have been used to produce fatty acid methyl esters (FAME) as biodiesel. The different degrees of saturation of the fatty acids are indicated as follows.

Notation	Common name	Degree of saturation
C8:0	Caprylic	These are saturated fatty acids
C10:0	Capric	These are more oxidatively stable than the other fatty acids
C12:0	Lauric	
C14:0	Myristic	
C15:0	Pentadecanoic	
C16:0	Palmitic	
C17:0	Margaric	
C18:0	Stearic	
C20:0	Arachidic	
C22:0	Behenic	
C24:0	Lignoceric	
C16:1	Palmitoleic	These are unsaturated fatty acids
C17:1	Margaroleic	These are less oxidatively stable than the saturated fatty acids
C18:1	Oleic	
C18:1-9c, 12(OH)	Ricinoleic	but more stable than the polyunsaturated fatty acids
C20:1	Gadoleic	
C20:1-11c, 14(OH)	Lesquerolic	
C22:1	Erucic	
C24:1	Nervonic	
C18:2	Linoleic	These are polyunsaturated fatty acids
C18:3	Linolenic	These fatty acids are more prone to oxidation than the other fatty acids.
C18:3-9c,11t,13t	α-Eleostearic	
C20:2	Eicosadienoic	
C20:5	Timnodonic	

© 2023 SAE International

Feedstock	Macauba pulp oil *Acrocomia aculeata*	Almond oil *Amygdalus communis*	Peanut oil *Arachis hypogaea*	Silkweed *Asclepias syriaca*	Babassu *Attalea speciosa*	Neem *Azadirachta indica*	Desert date *Balanites aegyptiaca*	Borage *Borago officinalis*	Abyssinian mustard *Brassica carinata*	Mustard *Brassica juncea*	Canola/ Rape oil *Brassica napus*	Indian doomba *Calophyllum inophyllum*
Fatty acid components												
C8:0	0-1				0-1							
C10:0	0-1				3-4							
C12:0	1-3				48-49							
C14:0	0-1				17-18							
C15:0												
C16:0	18-24	6-7	11-12	5-6	9-10	14-15	13-14	9-10	5-6	2-3	3-4	12-13
C16:1	3-5			6-7		0-1				0-1	0-1	
C17:0												
C17:1												
C18:0	3-4	1-2	2-3	2-3	4-5	20-21	11-12	3-4	0-1	1-2	0-2	13-14
C18:1	53-56	69-71	48-49	34-35	14-15	43-44	43-44	17-18	10-44	20-21	61-65	34-35
C18:1-9c, 12(OH)												
C18:2	12-18	17-20	32-33	48-49	1-2	17-18	31-32	38-39	24-36	20-21	19-23	38-39
C18:3	0-2		0-1	1-2		0-1		26-27	15-17	13-14	7-10	0-1
C18:3-9c,11t,13t												
C20:0	0-1			0-1		1-2				0-1	0-1	
C20:1										10-11	0-1	
C20:1-11c, 14(OH)												
C20:2										0-1		
C20:5												
C22:0						0-1		0-1		0-1	0-1	
C22:1								2-3	0-44	25-26		
C24:0						0-1				0-1	0-1	
C24:1								1-2		1-2	0-1	

Fatty acid components	Feedstock	Camelina oil *Camelina sativa*	Hemp *Cannabis sativa*	Safflower oil *Carthamus tinctorius*	Melon seed *Citrullus colocynthis L.*	Coconut *Cocos nucifera*	Coffee *Coffea canephora*	Hazelnut kernel *Corylus sp. L.*	Pumpkin *Cucurbita pepo*	Blue waxweed *Cuphea viscosissima*	Cardoon *Cynara cardunculus*	Palm *Elaeis guineensis*	Soybean *Glycine max L.*
	C8:0					0–7							
	C10:0					5–7						0–4	
	C12:0					47–50						0–48	
	C14:0					18–19			0–1	4–5		0–16	
	C15:0												
	C16:0	5–6	5–6	7–9	12–13	9–10	11–12	4–5	12–13	18–19	14–15	43–45	9–14
	C16:1			0–2	0–1	0–1	0–1	0–1	0–1			0–1	0–1
	C17:0												
	C17:1												
	C18:0	2–3	2–3	1–14	6–7	2–3	3–4	2–3	5–6	3–4	3–4	4–5	4–5
	C18:1	14–18	13–14	11–78	13–14	6–7	70–71	82–84	37–38	46–47	25–26	39–46	22–24
	C18:1-9c,12(OH)												
	C18:2	14–18	57–58	73–78	62–63	1–3	12–13	8–9	43–44	22–23	56–57	8–11	52–57
	C18:3	37–39	19–21		2–3		0–1	0–1	0–1	2–3		0–1	4–9
	C18:3-9c,11t,13t												
	C20:0	0–2	0–1				0–1		0–1	0–1		0–1	
	C20:1	9–17					0–1		0–1				
	C20:1-11c,14(OH)												
	C20:2	0–2											
	C20:5												
	C22:0	0–2	0–1				0–1		0–1	0–1		0–1	0–1
	C22:1	2–5	0–1										
	C24:0	0–1	0–1				0–1			0–1			
	C24:1	0–1											

Feedstock	Soybean soapstock *Glycine max L.*	Cottonseed oil *Gossypium hirsutum*	Sunflower oil *Helianthus annuus*	Rubber tree *Hevea brasiliensis*	African bush mango *Irvingia gabonensis*	Jatropha *Jatropha curcas*	Walnut oil *Juglans regia L.*	Bay laurel leaf *Laurus nobilis*	Yellowtop *Lesquerella Fendleri* (A. Gray) S. Watson	Linseed *Linum usitatissimum*	Mahwa/ Illipe/ butter tree *Madhuca indica*	Bead-tree or Cape lilac *Melia azedarach*
Fatty acid components												
C8:0												
C10:0												
C12:0								26-27				
C14:0		0-1				1-2		4-5	0-1		0-1	
C15:0												
C16:0	17-18	11-29	6-7	10-11	0-1	11-13	7-11	25-26	0-1	4-6	17-18	10-11
C16:1			0-1			0-1	0-1	0-1	0-1	0-1		
C17:0												
C17:1												
C18:0	4-5	0-3	3-5	8-9	0-4	5-17	1-5	3-4	1-2	2-7	14-45	3-4
C18:1	15-16	13-28	17-29	24-25	33-34	12-40	18-28	10-11	13-14	18-21	35-46	21-22
C18:1-9c, 12(OH)												
C18:2	55-56	54-58	58-74	39-40	2-3	41-48	51-56	11-12	5-6	15-19	1-18	64-65
C18:3	7-8	0-1	0-1	16-17	58-60	0-1	5-17	17-18	10-11	53-56		0-1
C18:3-9c,11t,13t												
C20:0					0-1	0-4			0-1	0-1	0-3	0-1
C20:1												0-1
C20:1-11c, 14(OH)									66-67			
C20:2												
C20:5												
C22:0		0-1								0-1		
C22:1			0-1									
C24:0										0-1		
C24:1									0-1			

Fatty acid components

Feedstock	Drumstick tree *Moringa oleifera*	Tabaco *Nicotiana tabacum*	Evening primrose *Oenothera caespitosa*	Olive oil *Olea europaea*	Rice bran *Oryza sativa*	Poppy seed *Papaver somniferum*	Perilla seed *Perilla frutescens*	Karanja *Pongamia pinnata*	Radish *Raphanus sativus*	Castor *Ricinus communis*	Marula *Sclerocarya birrea*	Sesame oil *Sesamum indicum*
C8:0												
C10:0												
C12:0												
C14:0		0-1			0-1							
C15:0												
C16:0	5-7	11-12	6-7	5-13	12-19	12-13	5-6	10-11	5-7	0-2	14-15	9-14
C16:1	1-2			0-1		0-1	0-1				0-1	
C17:0												
C17:1												
C18:0	5-6	3-4	1-2	1-3	2-3	4-5	2-3	6-7	2-3	1-4	8-9	4-53
C18:1	72-77	14-15	6-7	71-75	43-48	22-23	16-17	49-50	34-35	3-5	67-68	30-43
C18:1-9c, 12(OH)										90-91		
C18:2	0-1	69-70	76-77	10-18	33-36	60-61	13-14	19-20	17-18	1-4	5-6	0-45
C18:3		0-1	9-10	0-1	0-2	0-1	62-63		12-13	0-1		
C18:3-9c,11t,13t												
C20:0	3-4		0-1		0-1			4-5	0-1			
C20:1	1-3							2-3	10-11			
C20:1-11c, 14(OH)												
C20:2												
C20:5												
C22:0	4-8				0-1			5-6				
C22:1									16-17			
C24:0	0-1				0-1			2-3				
C24:1												

Fatty acid components Feedstock	Jojoba Simmondsia chinensis	Stillingia Stillingia linearifolia	Beach almond Terminalia catappa	Cocoa butter Theobroma cacao	Wheat grain Triticum sp.	Tung Vernicia fordii	Shea nut Vitellaria paradoxa	Grape seed OilVitis vinifera	Chinese prickly-ash Zanthoxylum bungeanum	Corn Zea mays
C8:0					1.4					
C10:0										
C12:0	0–1						0–2			
C14:0	0–1				0–1					
C15:0										
C16:0	0–1	7–8	35–36	25–26	20–21	3–5	3–4	8–9	10–11	11–12
C16:1	0–3				0–1				5–6	0–1
C17:0	0–1									
C17:1										0–1
C18:0	0–1	2–3	5–6	38–39	1–2	2–4	39–40	4–5	1–2	1–2
C18:1	5–15	16–17	32–33	32–33	16–17	50–64	44–45	15–16	32–33	27–28
C18:1-9c,12(OH)										
C18:2	0–5	31–35	28–29	3–4	56–57	11–28	5–6	73–74	25–26	56–58
C18:3	0–1	41–42			3–4	0–1			24–25	1–2
C18:3-9c,11t,13t										
C20:0	0–1									0–1
C20:1	65–80									
C20:1-11c,14(OH)										
C20:2						0–9				
C20:5										
C22:0	0–1									0–2
C22:1	10–20									
C24:0	0–5					0–1				
C24:1										

Feedstock	Algae	Beef tallow	Poultry fat	Butterfat (cow)	Butterfat (goat)	Butterfat (human)	Pork fat	Fish oil
Fatty acid components								
C8:0								
C10:0				3-4	6-8	1-3		0-1
C12:0		0-1	0-1	3-4	2-4	4-6		
C14:0	0-1	2-6	0-2	11-12	8-10	7-9	1-2	6-8
C15:0		0-1						
C16:0	6-7	24-32	19-25	27-28	24-26	24-26	26-30	17-19
C16:1	0-1	0-5	3-8					0-10
C17:0		0-2	0-1					0-1
C17:1		0-1	0-1					0-1
C18:0	3-4	12-25	5-8	12-13	11-13	7-9	12-18	0-4
C18:1	75-76	28-47	36-41	29-30	26-28	34-36	40-50	15-22
C18:1-9c, 12(OH)								
C18:2	12-13	2-4	18-29	2-3	2-4	8-10	7-13	2-5
C18:3	1-2	0-1	0-2	1-2	0-2	0-2	0-1	0-12
C18:3-9c,11t,13t								
C20:0	0-1	0-1	0-1					0-1
C20:1		0-1						0-2
C20:1-11c, 14(OH)								
C20:2			0-1					
C20:5								0-26
C22:0	0-1		0-1					0-1
C22:1			0-1					0-2
C24:0								
C24:1								0-1

Feedstock	Choice white grease	Yellow grease	Brown grease	Used cooking oil	Sorghum bug *Agonoscelis pubescens*	Melon bug *Aspongobus viduatus*
Fatty acid components						
C8:0						
C10:0		0–1				
C12:0	1–2	0–3				
C14:0			1–2			
C15:0				0–17		
C16:0	21–22	11–24	22–23	1–30	12–13	30–31
C16:1	2–3	0–4	3–4	0–5	0–2	10–11
C17:0	0–1	0–1				
C17:1	0–1	0–1				
C18:0	8–10	4–13	12–13	2–6	7–8	3–4
C18:1	50–51	25–45	42–43	5–53	40–41	46–47
C18:1-9c, 12(OH)						
C18:2	12–13	7–50	12–13	2–51	34–35	3–4
C18:3	0–2	0–8	0–1	0–1		
C18:3-9c,11t,13t				0–73		
C20:0	0–1	0–1		0–1		
C20:1	0–1			0–1		
C20:1-11c, 14(OH)						
C20:2				0–1		
C20:5						
C22:0		0–1		0–1		
C22:1	0–1	0–1				
C24:0		0–1		0–11		
C24:1						

Appendix 5: Material Safety Data Sheets

The following text contains copies of material safety data sheets (MSDS) for:

- A typical non-oxygenated European gasoline
- A typical European gasoline containing up to 10% ethanol
- A typical non-oxygenated European diesel fuel
- A typical European diesel fuel containing up to 10% fatty acid methyl ester (FAME)

All the data sheets were kindly supplied by and are reproduced with the kind permission of Coryton Advanced Fuels Ltd.

CORYTON
FOR A CLEANER FUTURE

EN228 E0 Gasoline
Safety Data Sheet
according to Regulation (EC) No. 1907/2006 (REACH) with its amendment Regulation (EU) 2015/830
Issue date: 20/06/2022 Version: 1.0

SECTION 1: Identification of the substance/mixture and of the company/undertaking

1.1. Product identifier

Product form   : Mixture
Product name   : EN228 E0 Gasoline

1.2. Relevant identified uses of the substance or mixture and uses advised against

1.2.1. Relevant identified uses

Use of the substance/mixture   : Fuels

1.2.2. Uses advised against

No additional information available

1.3. Details of the supplier of the safety data sheet

Coryton Advanced Fuels Ltd.
The Manorway
Stanford-le-Hope, SS17 9LN - United Kingdom
T +44 (0) 1375 511668
lab@coryton.com - www.coryton.com

1.4. Emergency telephone number

Emergency number   : +44 (0)2031 307667

Country	Organisation/Company	Address	Emergency number	Comment
United Kingdom	Guy's & St Thomas' Poisons Unit Medical Toxicology Unit, Guy's & St Thomas' Hospital Trust	Avonley Road SE14 5ER London	+44 20 7188 7188	
United Kingdom	National Poisons Information Service (Birmingham Centre) City Hospital	Dudley Road B18 7QH Birmingham	0344 892 0111 .	Only for healthcare professionals

SECTION 2: Hazards identification

2.1. Classification of the substance or mixture

Classification according to Regulation (EC) No. 1272/2008 [CLP]

Flammable liquids, Category 1	H224
Skin corrosion/irritation, Category 2	H315
Germ cell mutagenicity, Category 1B	H340
Carcinogenicity, Category 1B	H350
Reproductive toxicity, Category 2	H361
Specific target organ toxicity – Single exposure, Category 3, Narcosis	H336
Aspiration hazard, Category 1	H304
Hazardous to the aquatic environment – Chronic Hazard, Category 2	H411

Full text of H- and EUH-statements: see section 16

Adverse physicochemical, human health and environmental effects

Extremely flammable liquid and vapour. May cause cancer. May cause genetic defects. Suspected of damaging fertility or the unborn child. May cause drowsiness or dizziness. Causes skin irritation. May be fatal if swallowed and enters airways. Toxic to aquatic life with long lasting effects.

EN228 E0 Gasoline
Safety Data Sheet
according to Regulation (EC) No. 1907/2006 (REACH) with its amendment Regulation (EU) 2015/830

2.2. Label elements

Labelling according to Regulation (EC) No. 1272/2008 [CLP]

Hazard pictograms (CLP)   :

GHS02   GHS07   GHS08   GHS09

Signal word (CLP)	: Danger
Contains	: Gasoline
Hazard statements (CLP)	: H224 - Extremely flammable liquid and vapour.
	H304 - May be fatal if swallowed and enters airways.
	H315 - Causes skin irritation.
	H336 - May cause drowsiness or dizziness.
	H340 - May cause genetic defects.
	H350 - May cause cancer.
	H361 - Suspected of damaging fertility or the unborn child.
	H411 - Toxic to aquatic life with long lasting effects.
Precautionary statements (CLP)	: P201 - Obtain special instructions before use.
	P210 - Keep away from heat, hot surfaces, sparks, open flames and other ignition sources. No smoking.
	P280 - Wear protective gloves/protective clothing/eye protection/face protection/hearing protection.
	P301+P310+P331 - IF SWALLOWED: Immediately call a POISON CENTER or doctor. Do NOT induce vomiting.
	P308+P313 - IF exposed or concerned: Get medical advice/attention.
	P403+P235 - Store in a well-ventilated place. Keep cool.

2.3. Other hazards

No additional information available

SECTION 3: Composition/information on ingredients

3.1. Substances

Not applicable

3.2. Mixtures

Name	Product identifier	%	Classification according to Regulation (EC) No. 1272/2008 [CLP]
Gasoline	(CAS-No.) 86290-81-5 (EC-No.) 289-220-8 (EC Index-No.) 649-378-00-4 (REACH-no) 01-2119471335-39	100	Flam. Liq. 1, H224 Skin Irrit. 2, H315 Muta. 1B, H340 Carc. 1B, H350 Repr. 2, H361 STOT SE 3, H336 Asp. Tox. 1, H304 Aquatic Chronic 2, H411

Full text of H- and EUH-statements: see section 16

SECTION 4: First aid measures

4.1. Description of first aid measures

First-aid measures general	: Call a physician immediately.
First-aid measures after inhalation	: Remove person to fresh air and keep comfortable for breathing.
First-aid measures after skin contact	: Rinse skin with water/shower. Take off immediately all contaminated clothing. If skin irritation occurs: Get medical advice/attention.

EN228 E0 Gasoline
Safety Data Sheet
according to Regulation (EC) No. 1907/2006 (REACH) with its amendment Regulation (EU) 2015/830

First-aid measures after eye contact	: Rinse eyes with water as a precaution.
First-aid measures after ingestion	: Do not induce vomiting. Call a physician immediately.

4.2. Most important symptoms and effects, both acute and delayed

Symptoms/effects	: May cause drowsiness or dizziness.
Symptoms/effects after skin contact	: Irritation.
Symptoms/effects after ingestion	: Risk of lung oedema.

4.3. Indication of any immediate medical attention and special treatment needed

Treat symptomatically.

SECTION 5: Firefighting measures

5.1. Extinguishing media

Suitable extinguishing media	: Water spray. Dry powder. Foam. Carbon dioxide.

5.2. Special hazards arising from the substance or mixture

Fire hazard	: Extremely flammable liquid and vapour.
Hazardous decomposition products in case of fire	: Toxic fumes may be released.

5.3. Advice for firefighters

Protection during firefighting	: Do not attempt to take action without suitable protective equipment. Self-contained breathing apparatus. Complete protective clothing.

SECTION 6: Accidental release measures

6.1. Personal precautions, protective equipment and emergency procedures

6.1.1. For non-emergency personnel

Emergency procedures	: No open flames, no sparks, and no smoking. Only qualified personnel equipped with suitable protective equipment may intervene. Avoid breathing dust/fume/gas/mist/vapours/spray.

6.1.2. For emergency responders

Protective equipment	: Do not attempt to take action without suitable protective equipment. For further information refer to section 8: "Exposure controls/personal protection".

6.2. Environmental precautions

Avoid release to the environment. Notify authorities if product enters sewers or public waters.

6.3. Methods and material for containment and cleaning up

For containment	: Collect spillage.
Methods for cleaning up	: Take up liquid spill into absorbent material. Notify authorities if product enters sewers or public waters.
Other information	: Dispose of materials or solid residues at an authorized site.

6.4. Reference to other sections

For further information refer to section 13.

EN228 E0 Gasoline
Safety Data Sheet
according to Regulation (EC) No. 1907/2006 (REACH) with its amendment Regulation (EU) 2015/830

SECTION 7: Handling and storage

7.1. Precautions for safe handling

Precautions for safe handling	: Ensure good ventilation of the work station. Keep away from heat, hot surfaces, sparks, open flames and other ignition sources. No smoking. Ground/bond container and receiving equipment. Use only non-sparking tools. Take precautionary measures against static discharge. Flammable vapours may accumulate in the container. Use explosion-proof equipment. Wear personal protective equipment. Obtain special instructions before use. Do not handle until all safety precautions have been read and understood. Take all necessary technical measures to avoid or minimize the release of the product on the workplace. Limit quantities of product at the minimum necessary for handling and limit the number of exposed workers. Provide local exhaust or general room ventilation. Floors, walls and other surfaces in the hazard area must be cleaned regularly. Avoid breathing dust/fume/gas/mist/vapours/spray. Avoid contact with skin and eyes.
Hygiene measures	: Separate working clothes from town clothes. Launder separately. Wash contaminated clothing before reuse. Do not eat, drink or smoke when using this product. Always wash hands after handling the product.

7.2. Conditions for safe storage, including any incompatibilities

Technical measures	: Ground/bond container and receiving equipment.
Storage conditions	: Store in a well-ventilated place. Keep cool. Keep container tightly closed. Store locked up.

7.3. Specific end use(s)

No additional information available

SECTION 8: Exposure controls/personal protection

8.1. Control parameters

Exposure limit values for the other components

Benzene (71-43-2)		
EU - Biological Limit Value (BLV)		
Local name	Benzene	
BLV	28 µg/l Parameter: benzene - Medium: blood - Sampling time: immediately end of shift 46 µg/g creatinine Parameter: phenylmercapturic - Medium: urine - Sampling time: end of exposure/shift	
Regulatory reference	SCOEL List of recommended health-based BLVs and BGVs	
United Kingdom - Occupational Exposure Limits		
Local name	Benzene	
WEL TWA (OEL TWA) [1]	3.25 mg/m³	
WEL TWA (OEL TWA) [2]	1 ppm	
Remark	Carc (Capable of causing cancer and/or heritable genetic damage), Sk (Can be absorbed through the skin. The assigned substances are those for which there are concerns that dermal absorption will lead to systemic toxicity)	
Regulatory reference	EH40/2005 (Fourth edition, 2020). HSE	

EN228 E0 Gasoline
Safety Data Sheet
according to Regulation (EC) No. 1907/2006 (REACH) with its amendment Regulation (EU) 2015/830

Toluene (108-88-3)

EU - Indicative Occupational Exposure Limit (IOEL)		
Local name	Toluene	
IOEL TWA	192 mg/m³	
IOEL TWA [ppm]	50 ppm	
IOEL STEL	384 mg/m³	
IOEL STEL [ppm]	100 ppm	
Remark	Skin	
Regulatory reference	COMMISSION DIRECTIVE 2006/15/EC	

United Kingdom - Occupational Exposure Limits		
Local name	Toluene	
WEL TWA (OEL TWA) [1]	191 mg/m³	
WEL TWA (OEL TWA) [2]	50 ppm	
WEL STEL (OEL STEL)	384 mg/m³	
WEL STEL (OEL STEL) [ppm]	100 ppm	
Remark	Sk (Can be absorbed through the skin. The assigned substances are those for which there are concerns that dermal absorption will lead to systemic toxicity)	
Regulatory reference	EH40/2005 (Fourth edition, 2020). HSE	

n-Hexane (110-54-3)

United Kingdom - Occupational Exposure Limits		
Local name	n-Hexane	
WEL TWA (OEL TWA) [1]	72 mg/m³	
WEL TWA (OEL TWA) [2]	20 ppm	
Regulatory reference	EH40/2005 (Fourth edition, 2020). HSE	

EN228 E0 Gasoline

DNEL/DMEL (Workers)	
Acute - systemic effects, inhalation	1286.4 mg/m³
Acute - local effects, inhalation	1066.67 mg/m³
Long-term - local effects, inhalation	837.5 mg/m³
DNEL/DMEL (General population)	
Acute - systemic effects, inhalation	1152 mg/m³
Acute - local effects, inhalation	640 mg/m³
Long-term - local effects, inhalation	178.57 mg/m³

8.2. Exposure controls

Appropriate engineering controls:
Ensure good ventilation of the work station.

Hand protection:
Protective gloves

EN228 E0 Gasoline
Safety Data Sheet
according to Regulation (EC) No. 1907/2006 (REACH) with its amendment Regulation (EU) 2015/830

Eye protection:

Safety glasses

Skin and body protection:

Wear suitable protective clothing

Respiratory protection:

[In case of inadequate ventilation] wear respiratory protection.

Personal protective equipment symbol(s):

Environmental exposure controls:
Avoid release to the environment.

SECTION 9: Physical and chemical properties

9.1. Information on basic physical and chemical properties

Physical state	:	Liquid
Colour	:	Colourless.
Odour	:	Hydrocarbon.
Odour threshold	:	No data available
pH	:	No data available
Relative evaporation rate (butylacetate=1)	:	No data available
Melting point	:	No data available
Freezing point	:	No data available
Boiling point	:	25 – 210 °C
Flash point	:	≤ -40 °C
Auto-ignition temperature	:	> 250 °C
Decomposition temperature	:	No data available
Flammability (solid, gas)	:	Not applicable
Vapour pressure	:	45 – 100 kPa (37.8°C)
Relative vapour density at 20 °C	:	No data available
Relative density	:	No data available
Density	:	0.72 – 0.775 kg/l (15°C)
Solubility	:	Material insoluble in water.
Partition coefficient n-octanol/water (Log Pow)	:	No data available
Partition coefficient n-octanol/water (Log Kow)	:	> 3
Viscosity, kinematic	:	< 1 mm²/s (40°C)
Viscosity, dynamic	:	No data available
Explosive properties	:	No data available
Oxidising properties	:	No data available
Lower explosive limit (LEL)	:	1.4 vol %
Upper explosive limit (UEL)	:	7.6 vol %

9.2. Other information

No additional information available

EN228 E0 Gasoline
Safety Data Sheet
according to Regulation (EC) No. 1907/2006 (REACH) with its amendment Regulation (EU) 2015/830

10.1. Reactivity

Extremely flammable liquid and vapour.

10.2. Chemical stability

Stable under normal conditions.

10.3. Possibility of hazardous reactions

No dangerous reactions known under normal conditions of use.

10.4. Conditions to avoid

Avoid contact with hot surfaces. Heat. No flames, no sparks. Eliminate all sources of ignition.

10.5. Incompatible materials

No additional information available

10.6. Hazardous decomposition products

Under normal conditions of storage and use, hazardous decomposition products should not be produced.

SECTION 11: Toxicological information

11.1. Information on toxicological effects

Acute toxicity (oral)   : Not classified
Acute toxicity (dermal)   : Not classified
Acute toxicity (inhalation) : Not classified

Gasoline (86290-81-5)	
LD50 oral rat	> 5000 mg/kg bodyweight Animal: rat, Guideline: OECD Guideline 401 (Acute Oral Toxicity)

Skin corrosion/irritation   : Causes skin irritation.
Serious eye damage/irritation   : Not classified
Respiratory or skin sensitisation : Not classified
Germ cell mutagenicity   : May cause genetic defects.
Carcinogenicity   : May cause cancer.

Reproductive toxicity   : Suspected of damaging fertility or the unborn child.

STOT-single exposure   : May cause drowsiness or dizziness.

STOT-repeated exposure   : Not classified

Aspiration hazard   : May be fatal if swallowed and enters airways.

EN228 E0 Gasoline	
Viscosity, kinematic	< 1 mm²/s (40°C)

SECTION 12: Ecological information

12.1. Toxicity

Ecology - general : Toxic to aquatic life with long lasting effects.
Hazardous to the aquatic environment, short–term (acute) : Not classified

EN228 E0 Gasoline
Safety Data Sheet
according to Regulation (EC) No. 1907/2006 (REACH) with its amendment Regulation (EU) 2015/830

Hazardous to the aquatic environment, long–term (chronic) Not rapidly degradable	: Toxic to aquatic life with long lasting effects.

Gasoline (86290-81-5)	
LC50 - Fish [1]	8 – 10 mg/l
EC50 - Crustacea [1]	4.5 mg/l (Daphnia)
EC50 72h - Algae [1]	3.1 mg/l
NOEC chronic crustacea	2.6 mg/l (Daphnia)

12.2. Persistence and degradability

Gasoline (86290-81-5)	
Persistence and degradability	Not readily biodegradable.

12.3. Bioaccumulative potential

EN228 E0 Gasoline	
Partition coefficient n-octanol/water (Log Kow)	> 3

Gasoline (86290-81-5)	
Bioaccumulative potential	Potentially bioaccumulable.

12.4. Mobility in soil

No additional information available

12.5. Results of PBT and vPvB assessment

Component	
Gasoline (86290-81-5)	PBT: not relevant – no registration required vPvB: not relevant – no registration required

12.6. Other adverse effects

No additional information available

SECTION 13: Disposal considerations

13.1. Waste treatment methods

Waste treatment methods	: Dispose of contents/container in accordance with licensed collector's sorting instructions.
Additional information	: Flammable vapours may accumulate in the container.
European List of Waste (LoW) code	: 13 07 02* - petrol

SECTION 14: Transport information

In accordance with ADR / IMDG / IATA / ADN / RID

ADR	IMDG	IATA	ADN	RID
14.1. UN number				
UN 1203	UN 1203	UN 1203	UN 1203	UN 1203
14.2. UN proper shipping name				
GASOLINE	GASOLINE	Gasoline	GASOLINE	GASOLINE

EN228 E0 Gasoline
Safety Data Sheet
according to Regulation (EC) No. 1907/2006 (REACH) with its amendment Regulation (EU) 2015/830

Transport document description

UN 1203 GASOLINE, 3, II, (D/E), ENVIRONMENTALLY HAZARDOUS	UN 1203 GASOLINE, 3, II, MARINE POLLUTANT/ENVIRONMENTALLY HAZARDOUS	UN 1203 Gasoline, 3, II, ENVIRONMENTALLY HAZARDOUS	UN 1203 GASOLINE, 3, II, ENVIRONMENTALLY HAZARDOUS	UN 1203 GASOLINE, 3, II, ENVIRONMENTALLY HAZARDOUS

14.3. Transport hazard class(es)

3	3	3	3	3

14.4. Packing group

II	II	II	II	II

14.5. Environmental hazards

Dangerous for the environment : Yes	Dangerous for the environment : Yes Marine pollutant : Yes	Dangerous for the environment : Yes	Dangerous for the environment : Yes	Dangerous for the environment : Yes

No supplementary information available

14.6. Special precautions for user

Overland transport

Classification code (ADR)	: F1
Special provisions (ADR)	: 243, 534, 664
Portable tank and bulk container instructions (ADR)	: T4
Transport category (ADR)	: 2
Hazard identification number (Kemler No.)	: 33
Orange plates	:

33
1203

Tunnel restriction code (ADR)	: D/E
EAC code	: 3YE

Transport by sea

Special provisions (IMDG)	: 243
Tank instructions (IMDG)	: T4
EmS-No. (Fire)	: F-E
EmS-No. (Spillage)	: S-E

Air transport

Special provisions (IATA)	: A100
ERG code (IATA)	: 3H

Inland waterway transport

Classification code (ADN)	: F1
Special provisions (ADN)	: 243, 534
Carriage permitted (ADN)	: T

Rail transport

Classification code (RID)	: F1
Special provisions (RID)	: 243, 534
Portable tank and bulk container instructions (RID)	: T4
Transport category (RID)	: 2
Hazard identification number (RID)	: 33

14.7. Transport in bulk according to Annex II of Marpol and the IBC Code

Not applicable

EN228 E0 Gasoline
Safety Data Sheet
according to Regulation (EC) No. 1907/2006 (REACH) with its amendment Regulation (EU) 2015/830

SECTION 15: Regulatory information

15.1. Safety, health and environmental regulations/legislation specific for the substance or mixture

15.1.1. EU-Regulations

Contains no REACH substances with Annex XVII restrictions
Contains no substance on the REACH candidate list
Contains no REACH Annex XIV substances
Contains no substance subject to Regulation (EU) No 649/2012 of the European Parliament and of the Council of 4 July 2012 concerning the export and import of hazardous chemicals.
Contains no substance subject to Regulation (EU) No 2019/1021 of the European Parliament and of the Council of 20 June 2019 on persistent organic pollutants

Directive 2012/18/EU (SEVESO III)

Seveso Additional information	:	Seveso III: Directive 2012/18/EU of the European Parliament and of the Council on the control of major-accident hazards involving dangerous substances.
		34a Petroleum products: (a) gasolines and naphthas, (b) kerosenes (including jet fuels), (c) gas oils (including diesel fuels, home heating oils and gas oil blending streams),(d) heavy fuel oils (e) alternative fuels serving the same purposes and with similar properties as regards flammability and environmental hazards as the products referred to in points (a) to (d)

15.1.2. National regulations

No additional information available

15.2. Chemical safety assessment

No chemical safety assessment has been carried out

SECTION 16: Other information

Abbreviations and acronyms:	
ADN	European Agreement concerning the International Carriage of Dangerous Goods by Inland Waterways
ADR	European Agreement concerning the International Carriage of Dangerous Goods by Road
ATE	Acute Toxicity Estimate
BCF	Bioconcentration factor
BLV	Biological limit value
BOD	Biochemical oxygen demand (BOD)
COD	Chemical oxygen demand (COD)
DMEL	Derived Minimal Effect level
DNEL	Derived-No Effect Level
EC-No.	European Community number
EC50	Median effective concentration
EN	European Standard
IARC	International Agency for Research on Cancer
IATA	International Air Transport Association
IMDG	International Maritime Dangerous Goods
LC50	Median lethal concentration
LD50	Median lethal dose
LOAEL	Lowest Observed Adverse Effect Level
NOAEC	No-Observed Adverse Effect Concentration

EN228 E0 Gasoline
Safety Data Sheet
according to Regulation (EC) No. 1907/2006 (REACH) with its amendment Regulation (EU) 2015/830

NOAEL	No-Observed Adverse Effect Level
NOEC	No-Observed Effect Concentration
OECD	Organisation for Economic Co-operation and Development
OEL	Occupational Exposure Limit
PBT	Persistent Bioaccumulative Toxic
PNEC	Predicted No-Effect Concentration
RID	Regulations concerning the International Carriage of Dangerous Goods by Rail
SDS	Safety Data Sheet
STP	Sewage treatment plant
ThOD	Theoretical oxygen demand (ThOD)
TLM	Median Tolerance Limit
VOC	Volatile Organic Compounds
CAS-No.	Chemical Abstract Service number
N.O.S.	Not Otherwise Specified
vPvB	Very Persistent and Very Bioaccumulative
ED	Endocrine disrupting properties

Full text of H- and EUH-statements:

Aquatic Chronic 2	Hazardous to the aquatic environment – Chronic Hazard, Category 2
Asp. Tox. 1	Aspiration hazard, Category 1
Carc. 1B	Carcinogenicity, Category 1B
Flam. Liq. 1	Flammable liquids, Category 1
H224	Extremely flammable liquid and vapour.
H304	May be fatal if swallowed and enters airways.
H315	Causes skin irritation.
H336	May cause drowsiness or dizziness.
H340	May cause genetic defects.
H350	May cause cancer.
H361	Suspected of damaging fertility or the unborn child.
H411	Toxic to aquatic life with long lasting effects.
Muta. 1B	Germ cell mutagenicity, Category 1B
Repr. 2	Reproductive toxicity, Category 2
Skin Irrit. 2	Skin corrosion/irritation, Category 2
STOT SE 3	Specific target organ toxicity – Single exposure, Category 3, Narcosis

SDS CORYTON (REACH Annex II)

This information is based on our current knowledge and is intended to describe the product for the purposes of health, safety and environmental requirements only. It should not therefore be construed as guaranteeing any specific property of the product.

EN228 E0 Gasoline
Safety Data Sheet
according to Regulation (EC) No. 1907/2006 (REACH) with its amendment Regulation (EU) 2015/830

CORYTON
FOR A CLEANER FUTURE

EN228 E10 Gasoline
Safety Data Sheet
according to Regulation (EC) No. 1907/2006 (REACH) with its amendment Regulation (EU) 2015/830
Issue date: 20/06/2022   Version: 1.0

SECTION 1: Identification of the substance/mixture and of the company/undertaking

1.1. Product identifier

Product form   : Mixture
Product name   : EN228 E10 Gasoline

1.2. Relevant identified uses of the substance or mixture and uses advised against

1.2.1. Relevant identified uses

Use of the substance/mixture   : Fuels

1.2.2. Uses advised against

No additional information available

1.3. Details of the supplier of the safety data sheet

Coryton Advanced Fuels Ltd.
The Manorway
Stanford-le-Hope, SS17 9LN - United Kingdom
T +44 (0) 1375 511668
lab@coryton.com - www.coryton.com

1.4. Emergency telephone number

Emergency number   : +44 (0)2031 307667

Country	Organisation/Company	Address	Emergency number	Comment
United Kingdom	Guy's & St Thomas' Poisons Unit Medical Toxicology Unit, Guy's & St Thomas' Hospital Trust	Avonley Road SE14 5ER London	+44 20 7188 7188	
United Kingdom	National Poisons Information Service (Birmingham Centre) City Hospital	Dudley Road B18 7QH Birmingham	0344 892 0111	Only for healthcare professionals

SECTION 2: Hazards identification

2.1. Classification of the substance or mixture

Classification according to Regulation (EC) No. 1272/2008 [CLP]

Flammable liquids, Category 1	H224
Skin corrosion/irritation, Category 2	H315
Germ cell mutagenicity, Category 1B	H340
Carcinogenicity, Category 1B	H350
Reproductive toxicity, Category 2	H361
Specific target organ toxicity – Single exposure, Category 3, Narcosis	H336
Aspiration hazard, Category 1	H304
Hazardous to the aquatic environment – Chronic Hazard, Category 2	H411

Full text of H- and EUH-statements: see section 16

Adverse physicochemical, human health and environmental effects

Extremely flammable liquid and vapour. May cause cancer. May cause genetic defects. Suspected of damaging fertility or the unborn child. May cause drowsiness or dizziness. Causes skin irritation. May be fatal if swallowed and enters airways. Toxic to aquatic life with long lasting effects.

EN228 E10 Gasoline
Safety Data Sheet
according to Regulation (EC) No. 1907/2006 (REACH) with its amendment Regulation (EU) 2015/830

2.2. Label elements

Labelling according to Regulation (EC) No. 1272/2008 [CLP]

Hazard pictograms (CLP)   :

| GHS02 | GHS07 | GHS08 | GHS09 |

Signal word (CLP)   : Danger
Contains   : Gasoline
Hazard statements (CLP)   : H224 - Extremely flammable liquid and vapour.
                                  H304 - May be fatal if swallowed and enters airways.
                                  H315 - Causes skin irritation.
                                  H336 - May cause drowsiness or dizziness.
                                  H340 - May cause genetic defects.
                                  H350 - May cause cancer.
                                  H361 - Suspected of damaging fertility or the unborn child.
                                  H411 - Toxic to aquatic life with long lasting effects.
Precautionary statements (CLP) : P201 - Obtain special instructions before use.
                                  P210 - Keep away from heat, hot surfaces, sparks, open flames and other ignition sources. No smoking.
                                  P280 - Wear protective gloves/protective clothing/eye protection/face protection/hearing protection.
                                  P301+P310+P331 - IF SWALLOWED: Immediately call a POISON CENTER or doctor. Do NOT induce vomiting.
                                  P308+P313 - IF exposed or concerned: Get medical advice/attention.
                                  P403+P235 - Store in a well-ventilated place. Keep cool.

2.3. Other hazards

No additional information available

SECTION 3: Composition/information on ingredients

3.1. Substances

Not applicable

3.2. Mixtures

Name	Product identifier	%	Classification according to Regulation (EC) No. 1272/2008 [CLP]
Gasoline	(CAS-No.) 86290-81-5 (EC-No.) 289-220-8 (EC Index-No.) 649-378-00-4 (REACH-no) 01-2119471335-39	90 – 100	Flam. Liq. 1, H224 Skin Irrit. 2, H315 Muta. 1B, H340 Carc. 1B, H350 Repr. 2, H361 STOT SE 3, H336 Asp. Tox. 1, H304 Aquatic Chronic 2, H411
Ethanol	(CAS-No.) 64-17-5 (EC-No.) 200-578-6 (EC Index-No.) 603-002-00-5 (REACH-no) 01-2119457610-43	0 – 10	Flam. Liq. 2, H225

Full text of H- and EUH-statements: see section 16

EN228 E10 Gasoline
Safety Data Sheet
according to Regulation (EC) No. 1907/2006 (REACH) with its amendment Regulation (EU) 2015/830

SECTION 4: First aid measures

4.1. Description of first aid measures

First-aid measures general	: Call a physician immediately.
First-aid measures after inhalation	: Remove person to fresh air and keep comfortable for breathing.
First-aid measures after skin contact	: Rinse skin with water/shower. Take off immediately all contaminated clothing. If skin irritation occurs: Get medical advice/attention.
First-aid measures after eye contact	: Rinse eyes with water as a precaution.
First-aid measures after ingestion	: Do not induce vomiting. Call a physician immediately.

4.2. Most important symptoms and effects, both acute and delayed

Symptoms/effects	: May cause drowsiness or dizziness.
Symptoms/effects after skin contact	: Irritation.
Symptoms/effects after ingestion	: Risk of lung oedema.

4.3. Indication of any immediate medical attention and special treatment needed

Treat symptomatically.

SECTION 5: Firefighting measures

5.1. Extinguishing media

Suitable extinguishing media	: Water spray. Dry powder. Foam. Carbon dioxide.

5.2. Special hazards arising from the substance or mixture

Fire hazard	: Extremely flammable liquid and vapour.
Hazardous decomposition products in case of fire	: Toxic fumes may be released.

5.3. Advice for firefighters

Protection during firefighting	: Do not attempt to take action without suitable protective equipment. Self-contained breathing apparatus. Complete protective clothing.

SECTION 6: Accidental release measures

6.1. Personal precautions, protective equipment and emergency procedures

6.1.1. For non-emergency personnel

Emergency procedures	: No open flames, no sparks, and no smoking. Only qualified personnel equipped with suitable protective equipment may intervene. Avoid breathing dust/fume/gas/mist/vapours/spray.

6.1.2. For emergency responders

Protective equipment	: Do not attempt to take action without suitable protective equipment. For further information refer to section 8: "Exposure controls/personal protection".

6.2. Environmental precautions

Avoid release to the environment. Notify authorities if product enters sewers or public waters.

6.3. Methods and material for containment and cleaning up

For containment	: Collect spillage.
Methods for cleaning up	: Take up liquid spill into absorbent material. Notify authorities if product enters sewers or public waters.
Other information	: Dispose of materials or solid residues at an authorized site.

6.4. Reference to other sections

For further information refer to section 13.

EN228 E10 Gasoline
Safety Data Sheet
according to Regulation (EC) No. 1907/2006 (REACH) with its amendment Regulation (EU) 2015/830

SECTION 7: Handling and storage

7.1. Precautions for safe handling

Precautions for safe handling	:	Ensure good ventilation of the work station. Keep away from heat, hot surfaces, sparks, open flames and other ignition sources. No smoking. Ground/bond container and receiving equipment. Use only non-sparking tools. Take precautionary measures against static discharge. Flammable vapours may accumulate in the container. Use explosion-proof equipment. Wear personal protective equipment. Obtain special instructions before use. Do not handle until all safety precautions have been read and understood. Take all necessary technical measures to avoid or minimize the release of the product on the workplace. Limit quantities of product at the minimum necessary for handling and limit the number of exposed workers. Provide local exhaust or general room ventilation. Floors, walls and other surfaces in the hazard area must be cleaned regularly. Avoid breathing dust/fume/gas/mist/vapours/spray. Avoid contact with skin and eyes.
Hygiene measures	:	Separate working clothes from town clothes. Launder separately. Wash contaminated clothing before reuse. Do not eat, drink or smoke when using this product. Always wash hands after handling the product.

7.2. Conditions for safe storage, including any incompatibilities

Technical measures	:	Ground/bond container and receiving equipment.
Storage conditions	:	Store in a well-ventilated place. Keep cool. Keep container tightly closed. Store locked up.

7.3. Specific end use(s)

No additional information available

SECTION 8: Exposure controls/personal protection

8.1. Control parameters

Ethanol (64-17-5)	
United Kingdom - Occupational Exposure Limits	
Local name	Ethanol
WEL TWA (OEL TWA) [1]	1920 mg/m³
WEL TWA (OEL TWA) [2]	1000 ppm
Regulatory reference	EH40/2005 (Fourth edition, 2020). HSE

Exposure limit values for the other components

Benzene (71-43-2)		
EU - Biological Limit Value (BLV)		
Local name	Benzene	
BLV	28 µg/l Parameter: benzene - Medium: blood - Sampling time: immediately end of shift 46 µg/g creatinine Parameter: phenylmercapturic - Medium: urine - Sampling time: end of exposure/shift	
Regulatory reference	SCOEL List of recommended health-based BLVs and BGVs	
United Kingdom - Occupational Exposure Limits		
Local name	Benzene	
WEL TWA (OEL TWA) [1]	3.25 mg/m³	
WEL TWA (OEL TWA) [2]	1 ppm	

EN228 E10 Gasoline
Safety Data Sheet
according to Regulation (EC) No. 1907/2006 (REACH) with its amendment Regulation (EU) 2015/830

Benzene (71-43-2)

Remark	Carc (Capable of causing cancer and/or heritable genetic damage), Sk (Can be absorbed through the skin. The assigned substances are those for which there are concerns that dermal absorption will lead to systemic toxicity)
Regulatory reference	EH40/2005 (Fourth edition, 2020). HSE

Toluene (108-88-3)

EU - Indicative Occupational Exposure Limit (IOEL)

Local name	Toluene
IOEL TWA	192 mg/m³
IOEL TWA [ppm]	50 ppm
IOEL STEL	384 mg/m³
IOEL STEL [ppm]	100 ppm
Remark	Skin
Regulatory reference	COMMISSION DIRECTIVE 2006/15/EC

United Kingdom - Occupational Exposure Limits

Local name	Toluene
WEL TWA (OEL TWA) [1]	191 mg/m³
WEL TWA (OEL TWA) [2]	50 ppm
WEL STEL (OEL STEL)	384 mg/m³
WEL STEL (OEL STEL) [ppm]	100 ppm
Remark	Sk (Can be absorbed through the skin. The assigned substances are those for which there are concerns that dermal absorption will lead to systemic toxicity)
Regulatory reference	EH40/2005 (Fourth edition, 2020). HSE

n-Hexane (110-54-3)

United Kingdom - Occupational Exposure Limits

Local name	n-Hexane
WEL TWA (OEL TWA) [1]	72 mg/m³
WEL TWA (OEL TWA) [2]	20 ppm
Regulatory reference	EH40/2005 (Fourth edition, 2020). HSE

EN228 E10 Gasoline

DNEL/DMEL (Workers)

Acute - systemic effects, inhalation	1286.4 mg/m³
Acute - local effects, inhalation	1066.67 mg/m³
Long-term - local effects, inhalation	837.5 mg/m³

DNEL/DMEL (General population)

Acute - systemic effects, inhalation	1152 mg/m³
Acute - local effects, inhalation	640 mg/m³

EN228 E10 Gasoline

Safety Data Sheet

according to Regulation (EC) No. 1907/2006 (REACH) with its amendment Regulation (EU) 2015/830

Long-term - local effects, inhalation	178.57 mg/m³

8.2. Exposure controls

Appropriate engineering controls:
Ensure good ventilation of the work station.

Hand protection:
Protective gloves

Eye protection:
Safety glasses

Skin and body protection:
Wear suitable protective clothing

Respiratory protection:
[In case of inadequate ventilation] wear respiratory protection.

Personal protective equipment symbol(s):

Environmental exposure controls:
Avoid release to the environment.

SECTION 9: Physical and chemical properties

9.1. Information on basic physical and chemical properties

Physical state	:	Liquid
Colour	:	Colourless.
Odour	:	Hydrocarbon.
Odour threshold	:	No data available
pH	:	No data available
Relative evaporation rate (butylacetate=1)	:	No data available
Melting point	:	No data available
Freezing point	:	No data available
Boiling point	:	25 – 210 °C
Flash point	:	≤ -40 °C
Auto-ignition temperature	:	> 250 °C
Decomposition temperature	:	No data available
Flammability (solid, gas)	:	Not applicable
Vapour pressure	:	45 – 100 kPa (37.8°C)
Relative vapour density at 20 °C	:	No data available
Relative density	:	No data available
Density	:	0.72 – 0.775 kg/l (15°C)
Solubility	:	No data available
Partition coefficient n-octanol/water (Log Pow)	:	No data available
Partition coefficient n-octanol/water (Log Kow)	:	> 3
Viscosity, kinematic	:	< 1 mm²/s (40°C)
Viscosity, dynamic	:	No data available
Explosive properties	:	No data available
Oxidising properties	:	No data available
Lower explosive limit (LEL)	:	1.4 vol %

EN228 E10 Gasoline
Safety Data Sheet
according to Regulation (EC) No. 1907/2006 (REACH) with its amendment Regulation (EU) 2015/830

Upper explosive limit (UEL)	: 7.6 vol %

9.2. Other information

No additional information available

SECTION 10: Stability and reactivity

10.1. Reactivity

Extremely flammable liquid and vapour.

10.2. Chemical stability

Stable under normal conditions.

10.3. Possibility of hazardous reactions

No dangerous reactions known under normal conditions of use.

10.4. Conditions to avoid

Avoid contact with hot surfaces. Heat. No flames, no sparks. Eliminate all sources of ignition.

10.5. Incompatible materials

No additional information available

10.6. Hazardous decomposition products

Under normal conditions of storage and use, hazardous decomposition products should not be produced.

SECTION 11: Toxicological information

11.1. Information on toxicological effects

Acute toxicity (oral)	: Not classified
Acute toxicity (dermal)	: Not classified
Acute toxicity (inhalation)	: Not classified

Gasoline (86290-81-5)	
LD50 oral rat	> 5000 mg/kg bodyweight Animal: rat, Guideline: OECD Guideline 401 (Acute Oral Toxicity)

Ethanol (64-17-5)	
LD50 oral rat	10470 mg/kg bodyweight Animal: rat, Guideline: OECD Guideline 401 (Acute Oral Toxicity), 95% CL: 9720 - 11380

Skin corrosion/irritation	: Causes skin irritation.
Serious eye damage/irritation	: Not classified
Respiratory or skin sensitisation	: Not classified
Germ cell mutagenicity	: May cause genetic defects.
Carcinogenicity	: May cause cancer.
Reproductive toxicity	: Suspected of damaging fertility or the unborn child.
STOT-single exposure	: May cause drowsiness or dizziness.
STOT-repeated exposure	: Not classified

EN228 E10 Gasoline
Safety Data Sheet
according to Regulation (EC) No. 1907/2006 (REACH) with its amendment Regulation (EU) 2015/830

Ethanol (64-17-5)	
LOAEL (oral, rat, 90 days)	3200 mg/kg bodyweight Animal: rat, Animal sex: male, Guideline: OECD Guideline 408 (Repeated Dose 90-Day Oral Toxicity Study in Rodents)
NOAEL (oral, rat, 90 days)	1730 mg/kg bodyweight Animal: rat, Animal sex: male, Guideline: OECD Guideline 408 (Repeated Dose 90-Day Oral Toxicity Study in Rodents), Remarks on results: other:

Aspiration hazard : May be fatal if swallowed and enters airways.

EN228 E10 Gasoline	
Viscosity, kinematic	< 1 mm²/s (40°C)

SECTION 12: Ecological Information

12.1. Toxicity

Ecology - general	: Toxic to aquatic life with long lasting effects.
Hazardous to the aquatic environment, short–term (acute)	: Not classified
Hazardous to the aquatic environment, long–term (chronic) Not rapidly degradable	: Toxic to aquatic life with long lasting effects.

Gasoline (86290-81-5)	
LC50 - Fish [1]	8 – 10 mg/l
EC50 - Crustacea [1]	4.5 mg/l (Daphnia)
EC50 72h - Algae [1]	3.1 mg/l
NOEC chronic crustacea	2.6 mg/l (Daphnia)

Ethanol (64-17-5)	
EC50 - Crustacea [1]	> 10000 mg/l Test organisms (species): Daphnia magna
EC50 96h - Algae [1]	≈ 22000 mg/l Test organisms (species): Pseudokirchneriella subcapitata (previous names: Raphidocelis subcapitata, Selenastrum capricornutum)

12.2. Persistence and degradability

EN228 E10 Gasoline	
Persistence and degradability	Biodegradability in water: no data available.

Gasoline (86290-81-5)	
Persistence and degradability	Not readily biodegradable.

12.3. Bioaccumulative potential

EN228 E10 Gasoline	
Partition coefficient n-octanol/water (Log Kow)	> 3
Bioaccumulative potential	Potentially bioaccumulable.

Gasoline (86290-81-5)	
Bioaccumulative potential	Potentially bioaccumulable.

EN228 E10 Gasoline
Safety Data Sheet
according to Regulation (EC) No. 1907/2006 (REACH) with its amendment Regulation (EU) 2015/830

12.4. Mobility in soil

EN228 E10 Gasoline	
Ecology - soil	Product has only a limited biodegradability in soil and water.

12.5. Results of PBT and vPvB assessment

Component	
Gasoline (86290-81-5)	PBT: not relevant – no registration required vPvB: not relevant – no registration required

12.6. Other adverse effects

Other adverse effects	: Forms thin oil film on surface of water.

SECTION 13: Disposal considerations

13.1. Waste treatment methods

Waste treatment methods	: Dispose of contents/container in accordance with licensed collector's sorting instructions.
Additional information	: Flammable vapours may accumulate in the container.
European List of Waste (LoW) code	: 13 07 02* - petrol

SECTION 14: Transport information

In accordance with ADR / IMDG / IATA / ADN / RID

ADR	IMDG	IATA	ADN	RID
14.1. UN number				
UN 1203	UN 1203	UN 1203	UN 1203	UN 1203
14.2. UN proper shipping name				
GASOLINE	GASOLINE	Gasoline	GASOLINE	GASOLINE
Transport document description				
UN 1203 GASOLINE, 3, II, (D/E), ENVIRONMENTALLY HAZARDOUS	UN 1203 GASOLINE, 3, II, MARINE POLLUTANT/ENVIRONMENTALLY HAZARDOUS	UN 1203 Gasoline, 3, II, ENVIRONMENTALLY HAZARDOUS	UN 1203 GASOLINE, 3, II, ENVIRONMENTALLY HAZARDOUS	UN 1203 GASOLINE, 3, II, ENVIRONMENTALLY HAZARDOUS
14.3. Transport hazard class(es)				
3	3	3	3	3
14.4. Packing group				
II	II	II	II	II
14.5. Environmental hazards				
Dangerous for the environment : Yes	Dangerous for the environment : Yes Marine pollutant : Yes	Dangerous for the environment : Yes	Dangerous for the environment : Yes	Dangerous for the environment : Yes
No supplementary information available				

EN228 E10 Gasoline
Safety Data Sheet
according to Regulation (EC) No. 1907/2006 (REACH) with its amendment Regulation (EU) 2015/830

14.6. Special precautions for user

Overland transport

Classification code (ADR)	: F1
Special provisions (ADR)	: 243, 534, 664
Portable tank and bulk container instructions (ADR)	: T4
Transport category (ADR)	: 2
Hazard identification number (Kemler No.)	: 33
Orange plates	:

33

1203

Tunnel restriction code (ADR)	: D/E
EAC code	: 3YE

Transport by sea

Special provisions (IMDG)	: 243
Tank instructions (IMDG)	: T4
EmS-No. (Fire)	: F-E
EmS-No. (Spillage)	: S-E

Air transport

Special provisions (IATA)	: A100
ERG code (IATA)	: 3H

Inland waterway transport

Classification code (ADN)	: F1
Special provisions (ADN)	: 243, 534
Carriage permitted (ADN)	: T

Rail transport

Classification code (RID)	: F1
Special provisions (RID)	: 243, 534
Portable tank and bulk container instructions (RID)	: T4
Transport category (RID)	: 2
Hazard identification number (RID)	: 33

14.7. Transport in bulk according to Annex II of Marpol and the IBC Code

Not applicable

SECTION 15: Regulatory information

15.1. Safety, health and environmental regulations/legislation specific for the substance or mixture

15.1.1. EU-Regulations

Contains no REACH substances with Annex XVII restrictions
Contains no substance on the REACH candidate list
Contains no REACH Annex XIV substances
Contains no substance subject to Regulation (EU) No 649/2012 of the European Parliament and of the Council of 4 July 2012 concerning the export and import of hazardous chemicals.
Contains no substance subject to Regulation (EU) No 2019/1021 of the European Parliament and of the Council of 20 June 2019 on persistent organic pollutants

Directive 2012/18/EU (SEVESO III)

Seveso Additional information	: Seveso III: Directive 2012/18/EU of the European Parliament and of the Council on the control of major-accident hazards involving dangerous substances.
	34a Petroleum products: (a) gasolines and naphthas, (b) kerosenes (including jet fuels), (c) gas oils (including diesel fuels, home heating oils and gas oil blending streams),(d) heavy fuel oils (e) alternative fuels serving the same purposes and with similar properties as regards flammability and environmental hazards as the products referred to in points (a) to (d)

15.1.2. National regulations

No additional information available

EN228 E10 Gasoline
Safety Data Sheet
according to Regulation (EC) No. 1907/2006 (REACH) with its amendment Regulation (EU) 2015/830

15.2. Chemical safety assessment

No chemical safety assessment has been carried out

SECTION 16: Other information

Abbreviations and acronyms:

ADN	European Agreement concerning the International Carriage of Dangerous Goods by Inland Waterways
ADR	European Agreement concerning the International Carriage of Dangerous Goods by Road
ATE	Acute Toxicity Estimate
BCF	Bioconcentration factor
BLV	Biological limit value
BOD	Biochemical oxygen demand (BOD)
COD	Chemical oxygen demand (COD)
DMEL	Derived Minimal Effect level
DNEL	Derived-No Effect Level
EC-No.	European Community number
EC50	Median effective concentration
EN	European Standard
IARC	International Agency for Research on Cancer
IATA	International Air Transport Association
IMDG	International Maritime Dangerous Goods
LC50	Median lethal concentration
LD50	Median lethal dose
LOAEL	Lowest Observed Adverse Effect Level
NOAEC	No-Observed Adverse Effect Concentration
NOAEL	No-Observed Adverse Effect Level
NOEC	No-Observed Effect Concentration
OECD	Organisation for Economic Co-operation and Development
OEL	Occupational Exposure Limit
PBT	Persistent Bioaccumulative Toxic
PNEC	Predicted No-Effect Concentration
RID	Regulations concerning the International Carriage of Dangerous Goods by Rail
SDS	Safety Data Sheet
STP	Sewage treatment plant
ThOD	Theoretical oxygen demand (ThOD)
TLM	Median Tolerance Limit
VOC	Volatile Organic Compounds
CAS-No.	Chemical Abstract Service number
N.O.S.	Not Otherwise Specified
vPvB	Very Persistent and Very Bioaccumulative
ED	Endocrine disrupting properties

EN228 E10 Gasoline
Safety Data Sheet
according to Regulation (EC) No. 1907/2006 (REACH) with its amendment Regulation (EU) 2015/830

Full text of H- and EUH-statements:	
Aquatic Chronic 2	Hazardous to the aquatic environment – Chronic Hazard, Category 2
Asp. Tox. 1	Aspiration hazard, Category 1
Carc. 1B	Carcinogenicity, Category 1B
Flam. Liq. 1	Flammable liquids, Category 1
Flam. Liq. 2	Flammable liquids, Category 2
H224	Extremely flammable liquid and vapour.
H225	Highly flammable liquid and vapour.
H304	May be fatal if swallowed and enters airways.
H315	Causes skin irritation.
H336	May cause drowsiness or dizziness.
H340	May cause genetic defects.
H350	May cause cancer.
H361	Suspected of damaging fertility or the unborn child.
H411	Toxic to aquatic life with long lasting effects.
Muta. 1B	Germ cell mutagenicity, Category 1B
Repr. 2	Reproductive toxicity, Category 2
Skin Irrit. 2	Skin corrosion/irritation, Category 2
STOT SE 3	Specific target organ toxicity – Single exposure, Category 3, Narcosis

SDS CORYTON (REACH Annex II)

This information is based on our current knowledge and is intended to describe the product for the purposes of health, safety and environmental requirements only. It should not therefore be construed as guaranteeing any specific property of the product.

EN228 E10 Gasoline
Safety Data Sheet
according to Regulation (EC) No. 1907/2006 (REACH) with its amendment Regulation (EU) 2015/830

EN590 B0
Safety Data Sheet
according to Regulation (EC) No. 1907/2006 (REACH) with its amendment Regulation (EU) 2015/830
Issue date: 22/06/2022 Version: 1.0

SECTION 1: Identification of the substance/mixture and of the company/undertaking

1.1. Product identifier

Product form   : Mixture
Product name   : EN590 B0

1.2. Relevant identified uses of the substance or mixture and uses advised against

1.2.1. Relevant identified uses

Use of the substance/mixture   : Fuels

1.2.2. Uses advised against

No additional information available

1.3. Details of the supplier of the safety data sheet

Coryton Advanced Fuels Ltd.
The Manorway
Stanford-le-Hope, SS17 9LN - United Kingdom
T +44 (0) 1375 511668
lab@coryton.com - www.coryton.com

1.4. Emergency telephone number

Emergency number   : +44 (0)2031 307667

Country	Organisation/Company	Address	Emergency number	Comment
United Kingdom	National Poisons Information Service (Birmingham Centre) City Hospital	Dudley Road B18 7QH Birmingham	0344 892 0111	Only for healthcare professionals
United Kingdom	NHS 111/NHS 24/NHS Direct		111 0845 4647	or call a doctor

SECTION 2: Hazards identification

2.1. Classification of the substance or mixture

Classification according to Regulation (EC) No. 1272/2008 [CLP]

Flammable liquids, Category 3	H226
Acute toxicity (inhal.), Category 4	H332
Skin corrosion/irritation, Category 2	H315
Carcinogenicity, Category 2	H351
Specific target organ toxicity – Repeated exposure, Category 2	H373
Aspiration hazard, Category 1	H304
Hazardous to the aquatic environment – Chronic Hazard, Category 2	H411

Full text of H- and EUH-statements: see section 16

Adverse physicochemical, human health and environmental effects

Flammable liquid and vapour. Suspected of causing cancer. May cause damage to organs through prolonged or repeated exposure. Harmful if inhaled. Causes skin irritation. May be fatal if swallowed and enters airways. Toxic to aquatic life with long lasting effects.

EN590 B0

Safety Data Sheet

according to Regulation (EC) No. 1907/2006 (REACH) with its amendment Regulation (EU) 2015/830

2.2. Label elements

Labelling according to Regulation (EC) No. 1272/2008 [CLP]

Hazard pictograms (CLP)   :

| GHS02 | GHS07 | GHS08 | GHS09 |

Signal word (CLP)   : Danger
Contains   : Fuels, Diesel
Hazard statements (CLP)   : H226 - Flammable liquid and vapour.
                                       H332 - Harmful if inhaled.
                                       H315 - Causes skin irritation.
                                       H351 - Suspected of causing cancer.
                                       H373 - May cause damage to organs through prolonged or repeated exposure.
                                       H304 - May be fatal if swallowed and enters airways.
                                       H411 - Toxic to aquatic life with long lasting effects.

2.3. Other hazards

Other hazards which do not result in classification   : In use, may form flammable/explosive vapour-air mixture. Flammable.
PBT: not relevant – no registration required
vPvB: not relevant – no registration required

SECTION 3: Composition/information on ingredients

3.1. Substances

Not applicable

3.2. Mixtures

Name	Product identifier	%	Classification according to Regulation (EC) No. 1272/2008 [CLP]
Fuels, Diesel	(CAS-No.) 68334-30-5 (EC-No.) 269-822-7 (EC Index-No.) 649-224-00-6 (REACH-no) 01-2119484664-27	100	Flam. Liq. 3, H226 Acute Tox. 4 (Inhalation), H332 Skin Irrit. 2, H315 Carc. 2, H351 STOT RE 2, H373 Asp. Tox. 1, H304 Aquatic Chronic 2, H411

Full text of H- and EUH-statements: see section 16

SECTION 4: First aid measures

4.1. Description of first aid measures

First-aid measures general   : Call a physician immediately.
First-aid measures after inhalation   : Remove person to fresh air and keep comfortable for breathing. Call a poison center or a doctor if you feel unwell.
First-aid measures after skin contact   : Take off immediately all contaminated clothing. Wash immediately with plenty of soap and water. If skin irritation occurs: Get medical advice/attention. Rinse skin with water/shower.
First-aid measures after eye contact   : Remove contact lenses, if present and easy to do. Continue rinsing. Rinse cautiously with water for several minutes. If eye irritation persists: Get medical advice/attention.
First-aid measures after ingestion   : Do not induce vomiting. Rinse mouth. Call a physician immediately.

4.2. Most important symptoms and effects, both acute and delayed

Symptoms/effects after inhalation   : May cause headache, nausea and irritation of respiratory tract.
Symptoms/effects after skin contact   : Causes mild skin irritation. Irritation.
Symptoms/effects after eye contact   : May cause eye irritation.

EN590 B0

Safety Data Sheet

according to Regulation (EC) No. 1907/2006 (REACH) with its amendment Regulation (EU) 2015/830

Symptoms/effects after ingestion	: There may be soreness and redness of the mouth and throat. Harmful if swallowed. Risk of lung oedema.

4.3. Indication of any immediate medical attention and special treatment needed

Treat symptomatically.

SECTION 5: Firefighting measures

5.1. Extinguishing media

Suitable extinguishing media	: Dry chemical powder, alcohol-resistant foam, carbon dioxide (CO2). Cool down the containers exposed to heat with a water spray. Water spray. Dry powder. Foam. Carbon dioxide.

5.2. Special hazards arising from the substance or mixture

Fire hazard	: Flammable liquid and vapour.
Explosion hazard	: May form flammable/explosive vapour-air mixture.
Hazardous decomposition products in case of fire	: Toxic fumes may be released. Carbon dioxide. Carbon monoxide.

5.3. Advice for firefighters

Precautionary measures fire	: Keep container tightly closed and away from heat, sparks and flame. Stop leak if safe to do so. Eliminate all ignition sources if safe to do so. Evacuate area.
Protection during firefighting	: Self-contained breathing apparatus. Do not attempt to take action without suitable protective equipment. Complete protective clothing.

SECTION 6: Accidental release measures

6.1. Personal precautions, protective equipment and emergency procedures

General measures	: Clean up any spills as soon as possible, using an absorbent material to collect it. No flames, no sparks. Eliminate all sources of ignition. Prevent from entering sewers, basements and workpits, or any place where its accumulation can be dangerous. All equipment used when handling the product must be grounded.

6.1.1. For non-emergency personnel

Protective equipment	: Wear recommended personal protective equipment. For further information refer to section 8: "Exposure controls/personal protection".
Emergency procedures	: No open flames, no sparks, and no smoking. Only qualified personnel equipped with suitable protective equipment may intervene. Do not breathe dust/fume/gas/mist/vapours/spray.

6.1.2. For emergency responders

Protective equipment	: Do not attempt to take action without suitable protective equipment. For further information refer to section 8: "Exposure controls/personal protection".
Emergency procedures	: Evacuate unnecessary personnel.

6.2. Environmental precautions

Avoid release to the environment. Notify authorities if product enters sewers or public waters. For a large spillage, contain the spillage by bunding.

6.3. Methods and material for containment and cleaning up

For containment	: Collect spillage.
Methods for cleaning up	: Take up liquid spill into absorbent material. Collect leaking and spilled liquid in sealable containers as far as possible. Notify authorities if product enters sewers or public waters.
Other information	: Dispose of materials or solid residues at an authorized site.

6.4. Reference to other sections

For further information refer to section 8: "Exposure controls/personal protection". For disposal of contaminated materials refer to section 13 : "Disposal considerations". For further information refer to section 13.

EN590 B0
Safety Data Sheet
according to Regulation (EC) No. 1907/2006 (REACH) with its amendment Regulation (EU) 2015/830

SECTION 7: Handling and storage

7.1. Precautions for safe handling

Additional hazards when processed	:	Keep container tightly closed. Keep away from heat, hot surfaces, sparks, open flames and other ignition sources. No smoking.
Precautions for safe handling	:	Ensure good ventilation of the work station. Use only non-sparking tools. Wear personal protective equipment. Do not eat, drink or smoke when using this product. Avoid contact with skin and eyes. Keep away from heat, hot surfaces, sparks, open flames and other ignition sources. No smoking. Ground/bond container and receiving equipment. Take precautionary measures against static discharge. Flammable vapours may accumulate in the container. Use explosion-proof equipment. Obtain special instructions before use. Do not handle until all safety precautions have been read and understood. Do not breathe dust/fume/gas/mist/vapours/spray. Use only outdoors or in a well-ventilated area.
Hygiene measures	:	Separate working clothes from town clothes. Launder separately. Wash contaminated clothing before reuse. Do not eat, drink or smoke when using this product. Always wash hands after handling the product.

7.2. Conditions for safe storage, including any incompatibilities

Technical measures	:	Ground/bond container and receiving equipment.
Storage conditions	:	Keep container tightly closed. Keep away from ignition sources. Prevent the build-up of electrostatic charge. Store in a well-ventilated place. Keep cool. Store locked up.
Incompatible materials	:	Sources of ignition. Heat sources.
Heat and ignition sources	:	flames or sparks.
Packaging materials	:	Aluminium. Steel.

7.3. Specific end use(s)

(No data available specific to the product).

SECTION 8: Exposure controls/personal protection

8.1. Control parameters

No additional information available

8.2. Exposure controls

Appropriate engineering controls:
Ensure that there is a suitable ventilation system. Use spark-/explosionproof appliances and lighting system. Ensure good ventilation of the work station.

Personal protective equipment:
Gloves. Protective clothing. Protective goggles.

Materials for protective clothing:		
Condition	Material	Standard
Antistatic clothing, Chemical resistant clothing, Fire-resistant protective clothing		

Hand protection:					
Protective gloves					
Type	Material	Permeation	Thickness (mm)	Penetration	Standard
Disposable gloves	Nitrile rubber (NBR)				
Reusable gloves	Butyl rubber				

EN590 B0
Safety Data Sheet
according to Regulation (EC) No. 1907/2006 (REACH) with its amendment Regulation (EU) 2015/830

Eye protection:			
Safety glasses			
Type	**Field of application**	**Characteristics**	**Standard**
Safety glasses	Droplet	With side shields	EN 166

Skin and body protection:	
Wear suitable protective clothing	
Type	**Standard**
Safety shoes	
Chemically resistant protective gloves, Impermeable clothing	

Respiratory protection:			
Insufficient ventilation: wear respiratory protection. [In case of inadequate ventilation] wear respiratory protection.			
Device	**Filter type**	**Condition**	**Standard**
Self-contained breathing apparatus (SCBA)			

Personal protective equipment symbol(s):

Environmental exposure controls:
Avoid release to the environment.

SECTION 9: Physical and chemical properties

9.1. Information on basic physical and chemical properties

Physical state	:	Liquid
Colour	:	No data available
Odour	:	Characteristic.
Odour threshold	:	No data available
pH	:	No data available
Relative evaporation rate (butylacetate=1)	:	No data available
Melting point	:	No data available
Freezing point	:	No data available
Boiling point	:	160 – 370 °C
Flash point	:	> 60 °C
Auto-ignition temperature	:	> 250 °C
Decomposition temperature	:	No data available
Flammability (solid, gas)	:	Not applicable
Vapour pressure	:	No data available
Relative vapour density at 20 °C	:	No data available
Relative density	:	No data available
Density	:	0.833 – 0.837 kg/l (15°C)
Solubility	:	Insoluble in water.
Partition coefficient n-octanol/water (Log Pow)	:	No data available
Viscosity, kinematic	:	2 – 4.5 mm²/s (40°C)
Viscosity, dynamic	:	No data available
Explosive properties	:	No data available
Oxidising properties	:	No data available
Lower explosive limit (LEL)	:	0.5 vol %

EN590 B0
Safety Data Sheet
according to Regulation (EC) No. 1907/2006 (REACH) with its amendment Regulation (EU) 2015/830

Upper explosive limit (UEL)	: 5 vol %

9.2. Other information

Additional information	: This material is a static accumulator. Even with proper grounding and bonding, this material can still accumulate an electrostatic charge. If sufficient charge is allowed to accumulate, electrostatic discharge and ignition of flammable air-vapour mixtures can occur.

SECTION 10: Stability and reactivity

10.1. Reactivity

Stable under recommended handling and storage conditions (see section 7). Flammable liquid and vapour.

10.2. Chemical stability

Stable under normal conditions. Stable at room temperature.

10.3. Possibility of hazardous reactions

No dangerous reactions known under normal conditions of use.

10.4. Conditions to avoid

Open flame. Heat. Sources of ignition. No contact with hot surfaces. Avoid contact with hot surfaces. No flames, no sparks. Eliminate all sources of ignition.

10.5. Incompatible materials

Strong acids. Oxidizing agent.

10.6. Hazardous decomposition products

On combustion releases : Carbon dioxide. Carbon monoxide. Toxic fumes.

SECTION 11: Toxicological information

11.1. Information on toxicological effects

Acute toxicity (oral)	: Not classified
Acute toxicity (dermal)	: Not classified
Acute toxicity (inhalation)	: Harmful if inhaled.

EN590 B0	
ATE CLP (gases)	4500 ppmv/4h
ATE CLP (vapours)	11 mg/l/4h
ATE CLP (dust,mist)	1.5 mg/l/4h

Fuels, Diesel (68334-30-5)	
LD50 dermal rabbit	> 2000 mg/kg bodyweight Animal: rabbit, Remarks on results: other:

Skin corrosion/irritation	: Causes skin irritation.
Serious eye damage/irritation	: Not classified
Respiratory or skin sensitisation	: Not classified
Germ cell mutagenicity	: Not classified
Carcinogenicity	: Suspected of causing cancer.
Reproductive toxicity	: Not classified
STOT-single exposure	: Not classified

EN590 B0

Safety Data Sheet

according to Regulation (EC) No. 1907/2006 (REACH) with its amendment Regulation (EU) 2015/830

| STOT-repeated exposure | : May cause damage to organs through prolonged or repeated exposure. |

Fuels, Diesel (68334-30-5)	
NOAEL (dermal, rat/rabbit, 90 days)	≈ 2000 mg/kg bodyweight Animal: rabbit

| Aspiration hazard | : May be fatal if swallowed and enters airways. |

EN590 B0	
Viscosity, kinematic	2 – 4.5 mm²/s (40°C)

SECTION 12: Ecological Information

12.1. Toxicity

Ecology - general	: Toxic to aquatic life. Toxic to soil organisms. Toxic to aquatic life with long lasting effects.
Hazardous to the aquatic environment, short–term (acute)	: Not classified
Hazardous to the aquatic environment, long–term (chronic) Not rapidly degradable	: Toxic to aquatic life with long lasting effects.

Fuels, Diesel (68334-30-5)	
LC50 - Fish [1]	> 3.2 mg/l
EC50 - Crustacea [1]	> 5.3 ml/l
ErC50 algae	> 2.9 mg/l

12.2. Persistence and degradability

No additional information available

12.3. Bioaccumulative potential

No additional information available

12.4. Mobility in soil

Fuels, Diesel (68334-30-5)	
Ecology - soil	Product has only a limited biodegradability in soil and water.

12.5. Results of PBT and vPvB assessment

EN590 B0
PBT: not relevant – no registration required
vPvB: not relevant – no registration required

12.6. Other adverse effects

No additional information available

SECTION 13: Disposal considerations

13.1. Waste treatment methods

Waste treatment methods	: Dispose of contents/container in accordance with licensed collector's sorting instructions.
Sewage disposal recommendations	: Disposal must be done according to official regulations.
Additional information	: Flammable vapours may accumulate in the container.
Ecology - waste materials	: Avoid release to the environment.

EN590 B0

Safety Data Sheet

according to Regulation (EC) No. 1907/2006 (REACH) with its amendment Regulation (EU) 2015/830

European List of Waste (LoW) code   :   13 07 01* - fuel oil and diesel

SECTION 14: Transport information

In accordance with ADR / IMDG / IATA / ADN / RID

ADR	IMDG	IATA	ADN	RID
14.1. UN number				
UN 1202	UN 1202	UN 1202	UN 1202	UN 1202
14.2. UN proper shipping name				
DIESEL FUEL	DIESEL FUEL	Diesel fuel	DIESEL FUEL	DIESEL FUEL
Transport document description				
UN 1202 DIESEL FUEL, 3, III, (D/E), ENVIRONMENTALLY HAZARDOUS	UN 1202 DIESEL FUEL, 3, III, MARINE POLLUTANT/ENVIRONME NTALLY HAZARDOUS	UN 1202 Diesel fuel, 3, III, ENVIRONMENTALLY HAZARDOUS	UN 1202 DIESEL FUEL, 3, III, ENVIRONMENTALLY HAZARDOUS	UN 1202 DIESEL FUEL, 3, III, ENVIRONMENTALLY HAZARDOUS
14.3. Transport hazard class(es)				
3	3	3	3	3
14.4. Packing group				
III	III	III	III	III
14.5. Environmental hazards				
Dangerous for the environment : Yes	Dangerous for the environment : Yes Marine pollutant : Yes	Dangerous for the environment : Yes	Dangerous for the environment : Yes	Dangerous for the environment : Yes
No supplementary information available				

14.6. Special precautions for user

Overland transport

Classification code (ADR)	:	F1
Special provisions (ADR)	:	640M, 664
Portable tank and bulk container instructions (ADR)	:	T2
Transport category (ADR)	:	3
Hazard identification number (Kemler No.)	:	30
Orange plates	:	

Tunnel restriction code (ADR)	:	D/E
EAC code	:	3Y

Transport by sea

Special provisions (IMDG)	:	363
Tank instructions (IMDG)	:	T2
EmS-No. (Fire)	:	F-E
EmS-No. (Spillage)	:	S-E

Air transport

Special provisions (IATA)	:	A3
ERG code (IATA)	:	3L

Inland waterway transport

Classification code (ADN)	:	F1

EN590 B0

Safety Data Sheet

according to Regulation (EC) No. 1907/2006 (REACH) with its amendment Regulation (EU) 2015/830

Special provisions (ADN)	: 640M
Carriage permitted (ADN)	: T
Rail transport	
Classification code (RID)	: F1
Special provisions (RID)	: 640M
Portable tank and bulk container instructions (RID)	: T2
Transport category (RID)	: 3
Hazard identification number (RID)	: 30

14.7. Transport in bulk according to Annex II of Marpol and the IBC Code

Not applicable

SECTION 15: Regulatory information

15.1. Safety, health and environmental regulations/legislation specific for the substance or mixture

15.1.1. EU-Regulations

Contains no REACH substances with Annex XVII restrictions
Contains no substance on the REACH candidate list
Contains no REACH Annex XIV substances
Contains no substance subject to Regulation (EU) No 649/2012 of the European Parliament and of the Council of 4 July 2012 concerning the export and import of hazardous chemicals.
Contains no substance subject to Regulation (EU) No 2019/1021 of the European Parliament and of the Council of 20 June 2019 on persistent organic pollutants

Directive 2012/18/EU (SEVESO III)

Seveso Additional information	:	Seveso III: Directive 2012/18/EU of the European Parliament and of the Council on the control of major-accident hazards involving dangerous substances.
		34c Petroleum products: (a) gasolines and naphthas, (b) kerosenes (including jet fuels), (c) gas oils (including diesel fuels, home heating oils and gas oil blending streams),(d) heavy fuel oils (e) alternative fuels serving the same purposes and with similar properties as regards flammability and environmental hazards as the products referred to in points (a) to (d)

15.1.2. National regulations

No additional information available

15.2. Chemical safety assessment

No chemical safety assessment has been carried out

SECTION 16: Other information

Abbreviations and acronyms:	
ADN	European Agreement concerning the International Carriage of Dangerous Goods by Inland Waterways
ADR	European Agreement concerning the International Carriage of Dangerous Goods by Road
ATE	Acute Toxicity Estimate
BCF	Bioconcentration factor
BLV	Biological limit value
BOD	Biochemical oxygen demand (BOD)
COD	Chemical oxygen demand (COD)
DMEL	Derived Minimal Effect level
DNEL	Derived-No Effect Level
EC-No.	European Community number

EN590 B0

Safety Data Sheet

according to Regulation (EC) No. 1907/2006 (REACH) with its amendment Regulation (EU) 2015/830

EC50	Median effective concentration
EN	European Standard
IARC	International Agency for Research on Cancer
IATA	International Air Transport Association
IMDG	International Maritime Dangerous Goods
LC50	Median lethal concentration
LD50	Median lethal dose
LOAEL	Lowest Observed Adverse Effect Level
NOAEC	No-Observed Adverse Effect Concentration
NOAEL	No-Observed Adverse Effect Level
NOEC	No-Observed Effect Concentration
OECD	Organisation for Economic Co-operation and Development
OEL	Occupational Exposure Limit
PBT	Persistent Bioaccumulative Toxic
PNEC	Predicted No-Effect Concentration
RID	Regulations concerning the International Carriage of Dangerous Goods by Rail
SDS	Safety Data Sheet
STP	Sewage treatment plant
ThOD	Theoretical oxygen demand (ThOD)
TLM	Median Tolerance Limit
VOC	Volatile Organic Compounds
CAS-No.	Chemical Abstract Service number
N.O.S.	Not Otherwise Specified
vPvB	Very Persistent and Very Bioaccumulative
ED	Endocrine disrupting properties

Full text of H- and EUH-statements:

Acute Tox. 4 (Inhalation)	Acute toxicity (inhal.), Category 4
Aquatic Chronic 2	Hazardous to the aquatic environment – Chronic Hazard, Category 2
Asp. Tox. 1	Aspiration hazard, Category 1
Carc. 2	Carcinogenicity, Category 2
Flam. Liq. 3	Flammable liquids, Category 3
H226	Flammable liquid and vapour.
H304	May be fatal if swallowed and enters airways.
H315	Causes skin irritation.
H332	Harmful if inhaled.
H351	Suspected of causing cancer.
H373	May cause damage to organs through prolonged or repeated exposure.
H411	Toxic to aquatic life with long lasting effects.
Skin Irrit. 2	Skin corrosion/irritation, Category 2

EN590 B0

Safety Data Sheet

according to Regulation (EC) No. 1907/2006 (REACH) with its amendment Regulation (EU) 2015/830

| STOT RE 2 | Specific target organ toxicity – Repeated exposure, Category 2 |

SDS CORYTON (REACH Annex II)

This information is based on our current knowledge and is intended to describe the product for the purposes of health, safety and environmental requirements only. It should not therefore be construed as guaranteeing any specific property of the product.

EN590 B7

Safety Data Sheet

according to Regulation (EC) No. 1907/2006 (REACH) with its amendment Regulation (EU) 2015/830
Issue date: 27/06/2022 Version: 1.0

CORYTON
FOR A CLEANER FUTURE

SECTION 1: Identification of the substance/mixture and of the company/undertaking

1.1. Product identifier

Product form : Mixture
Product name : EN590 B7

1.2. Relevant identified uses of the substance or mixture and uses advised against

1.2.1. Relevant identified uses

Use of the substance/mixture : Fuels

1.2.2. Uses advised against

No additional information available

1.3. Details of the supplier of the safety data sheet

Coryton Advanced Fuels Ltd.
The Manorway
Stanford-le-Hope, SS17 9LN - United Kingdom
T +44 (0) 1375 511668
lab@coryton.com - www.coryton.com

1.4. Emergency telephone number

Emergency number : +44 (0)2031 307667

Country	Organisation/Company	Address	Emergency number	Comment
United Kingdom	National Poisons Information Service (Birmingham Centre) City Hospital	Dudley Road B18 7QH Birmingham	0344 892 0111	Only for healthcare professionals
United Kingdom	NHS 111/NHS 24/NHS Direct		111 0845 4647	or call a doctor

SECTION 2: Hazards identification

2.1. Classification of the substance or mixture

Classification according to Regulation (EC) No. 1272/2008 [CLP]

Flammable liquids, Category 3	H226
Acute toxicity (inhal.), Category 4	H332
Skin corrosion/irritation, Category 2	H315
Carcinogenicity, Category 2	H351
Specific target organ toxicity – Repeated exposure, Category 2	H373
Aspiration hazard, Category 1	H304
Hazardous to the aquatic environment – Chronic Hazard, Category 2	H411

Full text of H- and EUH-statements: see section 16

Adverse physicochemical, human health and environmental effects

Flammable liquid and vapour. Suspected of causing cancer. May cause damage to organs through prolonged or repeated exposure. Harmful if inhaled. Causes skin irritation. May be fatal if swallowed and enters airways. Toxic to aquatic life with long lasting effects.

EN590 B7

Safety Data Sheet

according to Regulation (EC) No. 1907/2006 (REACH) with its amendment Regulation (EU) 2015/830

2.2. Label elements

Labelling according to Regulation (EC) No. 1272/2008 [CLP]

Hazard pictograms (CLP)   :

GHS02	GHS07	GHS08	GHS09

Signal word (CLP)   : Danger
Contains   : Fuels, Diesel
Hazard statements (CLP)   : H226 - Flammable liquid and vapour.
H332 - Harmful if inhaled.
H315 - Causes skin irritation.
H351 - Suspected of causing cancer.
H373 - May cause damage to organs through prolonged or repeated exposure.
H304 - May be fatal if swallowed and enters airways.
H411 - Toxic to aquatic life with long lasting effects.

2.3. Other hazards

Other hazards which do not result in classification   : In use, may form flammable/explosive vapour-air mixture. Flammable.
PBT: not relevant – no registration required
vPvB: not relevant – no registration required

SECTION 3: Composition/information on ingredients

3.1. Substances

Not applicable

3.2. Mixtures

Name	Product identifier	%	Classification according to Regulation (EC) No. 1272/2008 [CLP]
Fuels, Diesel	(CAS-No.) 68334-30-5 (EC-No.) 269-822-7 (EC Index-No.) 649-224-00-6 (REACH-no) 01-2119484664-27	≥ 85	Flam. Liq. 3, H226 Acute Tox. 4 (Inhalation), H332 Skin Irrit. 2, H315 Carc. 2, H351 STOT RE 2, H373 Asp. Tox. 1, H304 Aquatic Chronic 2, H411
Fatty acids, Me Esters (FAME)	(CAS-No.) 67762-38-3 (EC-No.) 267-015-4 (REACH-no) 01-2119471664-32	1 – 10	Not classified

Full text of H- and EUH-statements: see section 16

SECTION 4: First aid measures

4.1. Description of first aid measures

First-aid measures general   : Call a physician immediately.
First-aid measures after inhalation   : Remove person to fresh air and keep comfortable for breathing. Call a poison center or a doctor if you feel unwell.
First-aid measures after skin contact   : Wash immediately with plenty of soap and water. Rinse skin with water/shower. Take off immediately all contaminated clothing. If skin irritation occurs: Get medical advice/attention.
First-aid measures after eye contact   : Remove contact lenses, if present and easy to do. Continue rinsing. Rinse cautiously with water for several minutes. If eye irritation persists: Get medical advice/attention.
First-aid measures after ingestion   : Rinse mouth. Do not induce vomiting. Call a physician immediately.

EN590 B7
Safety Data Sheet
according to Regulation (EC) No. 1907/2006 (REACH) with its amendment Regulation (EU) 2015/830

4.2. Most important symptoms and effects, both acute and delayed

Symptoms/effects after inhalation	: May cause headache, nausea and irritation of respiratory tract.
Symptoms/effects after skin contact	: Causes mild skin irritation. Irritation.
Symptoms/effects after eye contact	: May cause eye irritation.
Symptoms/effects after ingestion	: There may be soreness and redness of the mouth and throat. Harmful if swallowed. Risk of lung oedema.

4.3. Indication of any immediate medical attention and special treatment needed

Treat symptomatically.

SECTION 5: Firefighting measures

5.1. Extinguishing media

Suitable extinguishing media	: Dry chemical powder, alcohol-resistant foam, carbon dioxide (CO2). Cool down the containers exposed to heat with a water spray. Water spray. Dry powder. Foam. Carbon dioxide.

5.2. Special hazards arising from the substance or mixture

Fire hazard	: Flammable liquid and vapour.
Explosion hazard	: May form flammable/explosive vapour-air mixture.
Hazardous decomposition products in case of fire	: Toxic fumes may be released. Carbon dioxide. Carbon monoxide.

5.3. Advice for firefighters

Precautionary measures fire	: Keep container tightly closed and away from heat, sparks and flame. Stop leak if safe to do so. Eliminate all ignition sources if safe to do so. Evacuate area.
Protection during firefighting	: Self-contained breathing apparatus. Do not attempt to take action without suitable protective equipment. Complete protective clothing.

SECTION 6: Accidental release measures

6.1. Personal precautions, protective equipment and emergency procedures

General measures	: Clean up any spills as soon as possible, using an absorbent material to collect it. No flames, no sparks. Eliminate all sources of ignition. Prevent from entering sewers, basements and workpits, or any place where its accumulation can be dangerous. All equipment used when handling the product must be grounded.

6.1.1. For non-emergency personnel

Protective equipment	: Wear recommended personal protective equipment. For further information refer to section 8: "Exposure controls/personal protection".
Emergency procedures	: Only qualified personnel equipped with suitable protective equipment may intervene. No open flames, no sparks, and no smoking. Do not breathe dust/fume/gas/mist/vapours/spray.

6.1.2. For emergency responders

Protective equipment	: Do not attempt to take action without suitable protective equipment. For further information refer to section 8: "Exposure controls/personal protection".
Emergency procedures	: Evacuate unnecessary personnel.

6.2. Environmental precautions

Avoid release to the environment. Notify authorities if product enters sewers or public waters. For a large spillage, contain the spillage by bunding.

6.3. Methods and material for containment and cleaning up

For containment	: Collect spillage.
Methods for cleaning up	: Take up liquid spill into absorbent material. Collect leaking and spilled liquid in sealable containers as far as possible. Notify authorities if product enters sewers or public waters.
Other information	: Dispose of materials or solid residues at an authorized site.

EN590 B7
Safety Data Sheet
according to Regulation (EC) No. 1907/2006 (REACH) with its amendment Regulation (EU) 2015/830

6.4. Reference to other sections

For further information refer to section 8: "Exposure controls/personal protection". For disposal of contaminated materials refer to section 13 : "Disposal considerations". For further information refer to section 13.

SECTION 7: Handling and storage

7.1. Precautions for safe handling

Additional hazards when processed	: Keep container tightly closed. Keep away from heat, hot surfaces, sparks, open flames and other ignition sources. No smoking.
Precautions for safe handling	: Ensure good ventilation of the work station. Do not eat, drink or smoke when using this product. Keep away from heat, hot surfaces, sparks, open flames and other ignition sources. No smoking. Ground/bond container and receiving equipment. Use only non-sparking tools. Take precautionary measures against static discharge. Flammable vapours may accumulate in the container. Use explosion-proof equipment. Wear personal protective equipment. Obtain special instructions before use. Do not handle until all safety precautions have been read and understood. Do not breathe dust/fume/gas/mist/vapours/spray. Use only outdoors or in a well-ventilated area. Avoid contact with skin and eyes.
Hygiene measures	: Separate working clothes from town clothes. Launder separately. Wash contaminated clothing before reuse. Do not eat, drink or smoke when using this product. Always wash hands after handling the product.

7.2. Conditions for safe storage, including any incompatibilities

Technical measures	: Ground/bond container and receiving equipment.
Storage conditions	: Keep away from ignition sources. Prevent the build-up of electrostatic charge. Store in a well-ventilated place. Keep cool. Keep container tightly closed. Store locked up.
Incompatible materials	: Sources of ignition. Heat sources.
Heat and ignition sources	: flames or sparks.
Packaging materials	: Aluminium. Steel.

7.3. Specific end use(s)

(No data available specific to the product).

SECTION 8: Exposure controls/personal protection

8.1. Control parameters

No additional information available

8.2. Exposure controls

Appropriate engineering controls:
Ensure that there is a suitable ventilation system. Use spark-/explosionproof appliances and lighting system. Ensure good ventilation of the work station.

Personal protective equipment:
Gloves. Protective clothing. Protective goggles.

Materials for protective clothing:		
Condition	**Material**	**Standard**
Antistatic clothing, Chemical resistant clothing, Fire-resistant protective clothing		

Hand protection:					
Protective gloves					
Type	**Material**	**Permeation**	**Thickness (mm)**	**Penetration**	**Standard**
Disposable gloves	Nitrile rubber (NBR)				

EN590 B7

Safety Data Sheet

according to Regulation (EC) No. 1907/2006 (REACH) with its amendment Regulation (EU) 2015/830

Reusable gloves	Butyl rubber				

Eye protection:

Safety glasses

Type	Field of application	Characteristics	Standard
Safety glasses	Droplet	With side shields	EN 166

Skin and body protection:

Wear suitable protective clothing

Type	Standard
Safety shoes	
Chemically resistant protective gloves, Impermeable clothing	

Respiratory protection:

Insufficient ventilation: wear respiratory protection. [In case of inadequate ventilation] wear respiratory protection.

Device	Filter type	Condition	Standard
Self-contained breathing apparatus (SCBA)			

Personal protective equipment symbol(s):

Environmental exposure controls:
Avoid release to the environment.

SECTION 9: Physical and chemical properties

9.1. Information on basic physical and chemical properties

Physical state	:	Liquid
Colour	:	No data available
Odour	:	Characteristic.
Odour threshold	:	No data available
pH	:	No data available
Relative evaporation rate (butylacetate=1)	:	No data available
Melting point	:	No data available
Freezing point	:	No data available
Boiling point	:	160 – 370 °C
Flash point	:	> 60 °C
Auto-ignition temperature	:	> 250 °C
Decomposition temperature	:	No data available
Flammability (solid, gas)	:	Not applicable
Vapour pressure	:	No data available
Relative vapour density at 20 °C	:	No data available
Relative density	:	No data available
Density	:	0.833 – 0.837 kg/l (15°C)
Solubility	:	Insoluble in water.
Partition coefficient n-octanol/water (Log Pow)	:	No data available
Viscosity, kinematic	:	2 – 4.5 mm²/s (40°C)
Viscosity, dynamic	:	No data available

EN590 B7

Safety Data Sheet

according to Regulation (EC) No. 1907/2006 (REACH) with its amendment Regulation (EU) 2015/830

Explosive properties	:	No data available
Oxidising properties	:	No data available
Lower explosive limit (LEL)	:	0.5 vol %
Upper explosive limit (UEL)	:	5 vol %

9.2. Other information

Additional information	:	This material is a static accumulator. Even with proper grounding and bonding, this material can still accumulate an electrostatic charge. If sufficient charge is allowed to accumulate, electrostatic discharge and ignition of flammable air-vapour mixtures can occur.

SECTION 10: Stability and reactivity

10.1. Reactivity

Stable under recommended handling and storage conditions (see section 7). Flammable liquid and vapour.

10.2. Chemical stability

Stable under normal conditions. Stable at room temperature.

10.3. Possibility of hazardous reactions

No dangerous reactions known under normal conditions of use.

10.4. Conditions to avoid

Open flame. Sources of ignition. No contact with hot surfaces. Avoid contact with hot surfaces. Heat. No flames, no sparks. Eliminate all sources of ignition.

10.5. Incompatible materials

Strong acids. Oxidizing agent.

10.6. Hazardous decomposition products

On combustion releases : Carbon dioxide. Carbon monoxide. Toxic fumes.

SECTION 11: Toxicological information

11.1. Information on toxicological effects

Acute toxicity (oral)	:	Not classified
Acute toxicity (dermal)	:	Not classified
Acute toxicity (inhalation)	:	Harmful if inhaled.

EN590 B7	
ATE CLP (gases)	4500 ppmv/4h
ATE CLP (vapours)	11 mg/l/4h
ATE CLP (dust,mist)	1.5 mg/l/4h

Fuels, Diesel (68334-30-5)	
LD50 dermal rabbit	> 2000 mg/kg bodyweight Animal: rabbit, Remarks on results: other:

Skin corrosion/irritation	:	Causes skin irritation.
Serious eye damage/irritation	:	Not classified
Respiratory or skin sensitisation	:	Not classified
Germ cell mutagenicity	:	Not classified
Carcinogenicity	:	Suspected of causing cancer.
Reproductive toxicity	:	Not classified

EN590 B7

Safety Data Sheet

according to Regulation (EC) No. 1907/2006 (REACH) with its amendment Regulation (EU) 2015/830

STOT-single exposure	: Not classified
STOT-repeated exposure	: May cause damage to organs through prolonged or repeated exposure.

Fuels, Diesel (68334-30-5)	
NOAEL (dermal, rat/rabbit, 90 days)	≈ 2000 mg/kg bodyweight Animal: rabbit

Aspiration hazard	: May be fatal if swallowed and enters airways.

EN590 B7	
Viscosity, kinematic	2 – 4.5 mm²/s (40°C)

SECTION 12: Ecological information

12.1. Toxicity

Ecology - general	: Toxic to aquatic life. Toxic to soil organisms. Toxic to aquatic life with long lasting effects.
Hazardous to the aquatic environment, short–term (acute)	: Not classified
Hazardous to the aquatic environment, long–term (chronic) Not rapidly degradable	: Toxic to aquatic life with long lasting effects.

Fuels, Diesel (68334-30-5)	
LC50 - Fish [1]	> 3.2 mg/l
EC50 - Crustacea [1]	> 5.3 ml/l
ErC50 algae	> 2.9 mg/l

12.2. Persistence and degradability

No additional information available

12.3. Bioaccumulative potential

No additional information available

12.4. Mobility in soil

Fuels, Diesel (68334-30-5)	
Ecology - soil	Product has only a limited biodegradability in soil and water.

12.5. Results of PBT and vPvB assessment

EN590 B7
PBT: not relevant – no registration required
vPvB: not relevant – no registration required

12.6. Other adverse effects

No additional information available

SECTION 13: Disposal considerations

13.1. Waste treatment methods

Waste treatment methods	: Dispose of contents/container in accordance with licensed collector's sorting instructions.

EN590 B7

Safety Data Sheet

according to Regulation (EC) No. 1907/2006 (REACH) with its amendment Regulation (EU) 2015/830

Sewage disposal recommendations	:	Disposal must be done according to official regulations.
Additional information	:	Flammable vapours may accumulate in the container.
Ecology - waste materials	:	Avoid release to the environment.
European List of Waste (LoW) code	:	13 07 01* - fuel oil and diesel

SECTION 14: Transport information

In accordance with ADR / IMDG / IATA / ADN / RID

ADR	IMDG	IATA	ADN	RID
14.1. UN number				
UN 1202	UN 1202	UN 1202	UN 1202	UN 1202
14.2. UN proper shipping name				
DIESEL FUEL	DIESEL FUEL	Diesel fuel	DIESEL FUEL	DIESEL FUEL
Transport document description				
UN 1202 DIESEL FUEL, 3, III, (D/E), ENVIRONMENTALLY HAZARDOUS	UN 1202 DIESEL FUEL, 3, III, MARINE POLLUTANT/ENVIRONMENTALLY HAZARDOUS	UN 1202 Diesel fuel, 3, III, ENVIRONMENTALLY HAZARDOUS	UN 1202 DIESEL FUEL, 3, III, ENVIRONMENTALLY HAZARDOUS	UN 1202 DIESEL FUEL, 3, III, ENVIRONMENTALLY HAZARDOUS
14.3. Transport hazard class(es)				
3	3	3	3	3
14.4. Packing group				
III	III	III	III	III
14.5. Environmental hazards				
Dangerous for the environment : Yes	Dangerous for the environment : Yes Marine pollutant : Yes	Dangerous for the environment : Yes	Dangerous for the environment : Yes	Dangerous for the environment : Yes
No supplementary information available				

14.6. Special precautions for user

Overland transport

Classification code (ADR)	:	F1
Special provisions (ADR)	:	640M, 664
Portable tank and bulk container instructions (ADR)	:	T2
Transport category (ADR)	:	3
Hazard identification number (Kemler No.)	:	30
Orange plates	:	

Tunnel restriction code (ADR)	:	D/E
EAC code	:	3Y

Transport by sea

Special provisions (IMDG)	:	363
Tank instructions (IMDG)	:	T2
EmS-No. (Fire)	:	F-E
EmS-No. (Spillage)	:	S-E

Air transport

Special provisions (IATA)	:	A3

EN590 B7

Safety Data Sheet

according to Regulation (EC) No. 1907/2006 (REACH) with its amendment Regulation (EU) 2015/830

ERG code (IATA)	: 3L
Inland waterway transport	
Classification code (ADN)	: F1
Special provisions (ADN)	: 640M
Carriage permitted (ADN)	: T
Rail transport	
Classification code (RID)	: F1
Special provisions (RID)	: 640M
Portable tank and bulk container instructions (RID)	: T2
Transport category (RID)	: 3
Hazard identification number (RID)	: 30

14.7. Transport in bulk according to Annex II of Marpol and the IBC Code

Not applicable

SECTION 15: Regulatory information

15.1. Safety, health and environmental regulations/legislation specific for the substance or mixture

15.1.1. EU-Regulations

Contains no REACH substances with Annex XVII restrictions
Contains no substance on the REACH candidate list
Contains no REACH Annex XIV substances
Contains no substance subject to Regulation (EU) No 649/2012 of the European Parliament and of the Council of 4 July 2012 concerning the export and import of hazardous chemicals.
Contains no substance subject to Regulation (EU) No 2019/1021 of the European Parliament and of the Council of 20 June 2019 on persistent organic pollutants

Directive 2012/18/EU (SEVESO III)

Seveso Additional information	:	Seveso III: Directive 2012/18/EU of the European Parliament and of the Council on the control of major-accident hazards involving dangerous substances.
		34c Petroleum products: (a) gasolines and naphthas, (b) kerosenes (including jet fuels), (c) gas oils (including diesel fuels, home heating oils and gas oil blending streams),(d) heavy fuel oils (e) alternative fuels serving the same purposes and with similar properties as regards flammability and environmental hazards as the products referred to in points (a) to (d)

15.1.2. National regulations

No additional information available

15.2. Chemical safety assessment

No chemical safety assessment has been carried out

SECTION 16: Other information

Abbreviations and acronyms:	
ADN	European Agreement concerning the International Carriage of Dangerous Goods by Inland Waterways
ADR	European Agreement concerning the International Carriage of Dangerous Goods by Road
ATE	Acute Toxicity Estimate
BCF	Bioconcentration factor
BLV	Biological limit value
BOD	Biochemical oxygen demand (BOD)
COD	Chemical oxygen demand (COD)
DMEL	Derived Minimal Effect level

EN590 B7
Safety Data Sheet
according to Regulation (EC) No. 1907/2006 (REACH) with its amendment Regulation (EU) 2015/830

DNEL	Derived-No Effect Level
EC-No.	European Community number
EC50	Median effective concentration
EN	European Standard
IARC	International Agency for Research on Cancer
IATA	International Air Transport Association
IMDG	International Maritime Dangerous Goods
LC50	Median lethal concentration
LD50	Median lethal dose
LOAEL	Lowest Observed Adverse Effect Level
NOAEC	No-Observed Adverse Effect Concentration
NOAEL	No-Observed Adverse Effect Level
NOEC	No-Observed Effect Concentration
OECD	Organisation for Economic Co-operation and Development
OEL	Occupational Exposure Limit
PBT	Persistent Bioaccumulative Toxic
PNEC	Predicted No-Effect Concentration
RID	Regulations concerning the International Carriage of Dangerous Goods by Rail
SDS	Safety Data Sheet
STP	Sewage treatment plant
ThOD	Theoretical oxygen demand (ThOD)
TLM	Median Tolerance Limit
VOC	Volatile Organic Compounds
CAS-No.	Chemical Abstract Service number
N.O.S.	Not Otherwise Specified
vPvB	Very Persistent and Very Bioaccumulative
ED	Endocrine disrupting properties

Full text of H- and EUH-statements:	
Acute Tox. 4 (Inhalation)	Acute toxicity (inhal.), Category 4
Aquatic Chronic 2	Hazardous to the aquatic environment – Chronic Hazard, Category 2
Asp. Tox. 1	Aspiration hazard, Category 1
Carc. 2	Carcinogenicity, Category 2
Flam. Liq. 3	Flammable liquids, Category 3
H226	Flammable liquid and vapour.
H304	May be fatal if swallowed and enters airways.
H315	Causes skin irritation.
H332	Harmful if inhaled.
H351	Suspected of causing cancer.
H373	May cause damage to organs through prolonged or repeated exposure.

EN590 B7

Safety Data Sheet

according to Regulation (EC) No. 1907/2006 (REACH) with its amendment Regulation (EU) 2015/830

H411	Toxic to aquatic life with long lasting effects.
Skin Irrit. 2	Skin corrosion/irritation, Category 2
STOT RE 2	Specific target organ toxicity – Repeated exposure, Category 2

SDS CORYTON (REACH Annex II)

This information is based on our current knowledge and is intended to describe the product for the purposes of health, safety and environmental requirements only. It should not therefore be construed as guaranteeing any specific property of the product.

Appendix 6: Lead Alkyls

As mentioned in Chapter 2, Thomas Midgley, Jr., working for General Motor's Corporation, screened thousands of compounds for their anti-knock potential. By the end of 1921, they had selected tetra-ethyl-lead (TEL) as having the greatest effectiveness with the best potential for commercial development and filed a patent accordingly, the patent was granted in 1926 [A6.1]. This became the most widely used anti-knock additive but due to health and environmental concerns and the adverse effects of lead on catalyst the use of lead antiknock additives is no longer allowed in any motor gasoline. It is still allowed in aviation gasoline [A6.2] and is used in the RON [A6.3] and MON [A6.4] test procedures in order to achieve reference fuels of octane numbers greater than 100, i.e., pure iso-octane. With unleaded gasoline the required octane levels are generally obtained by more severe refining, but this reduces yield and increases operating costs.

Lead alkyls function, as do all organometallic antiknocks, by decomposing at the appropriate temperature in the combustion cycle to form a cloud of catalytically active metal-oxide particles [A6.5–A6.7]. These particles interrupt the chain branching reactions that lead to the rapid combustion known as knock.

Commercial lead alkyl compounds also contain 1,2-dibromoethane which acts as a lead scavenger which prevents a buildup of lead compounds in the combustion chamber. Volatile lead halides are formed, which are exhausted from the engine. Without scavengers, hard deposits can build up in the combustion chamber. These deposits can flake off and cause valve burning by holding valves off their seats, thus allowing the hot combustion gases to escape past the valves. The relative effectiveness of lead alkyls depends on their volatility, when they decompose in the cycle, and the type of base gasoline being used. The response of gasoline components to lead alkyls is illustrated in Figure A6.1. The curves show that the greatest octane benefits are when they are blended into low-octane quality components, and that the octane improvements reduce as lead concentration increases. Paraffinic components tend to have a greater response than aromatic components.

FIGURE A6.1   Lead response curves for different gasoline components.

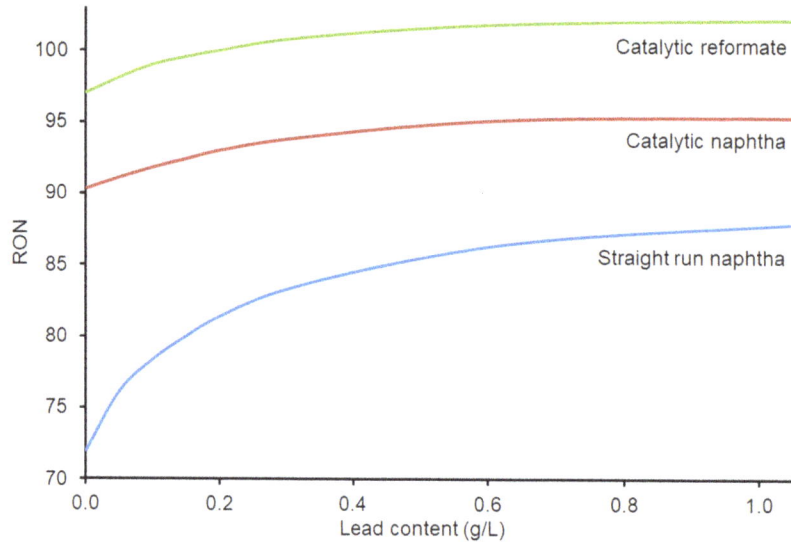

In finished gasolines, selection of the optimum concentration of lead alkyl is extremely important, particularly in view of the limitations on the maximum amount of lead that can be used. Figure A6.2 [A6.8] shows lead response curves for gasolines produced by hydroskimming refineries, that contain no olefins, and conversion refineries (see Chapter 3, Section 3.3) which have cracking processes, and do contain olefins.

FIGURE A6.2   Lead response curves for non-olefinic gasolines and for olefinic gasolines.

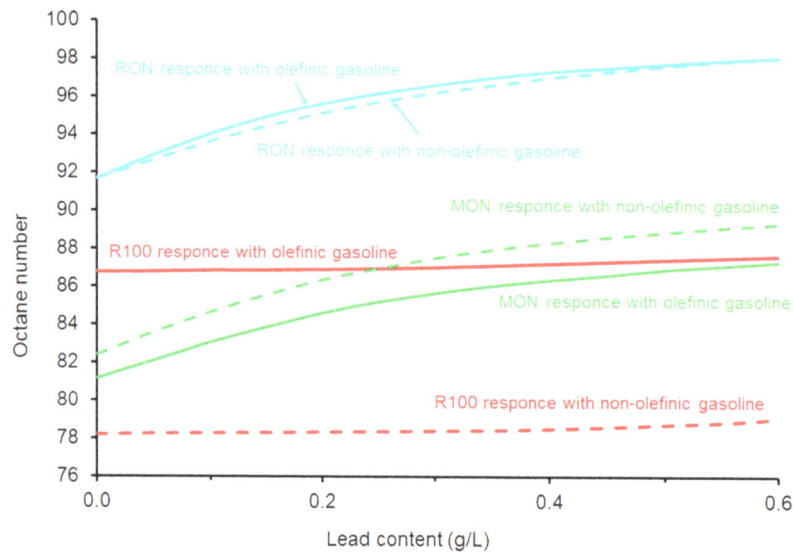

As lead is removed from the gasoline pool, it requires more and more energy to make the same volume of gasoline at the same octane quality. Although maintaining a high octane level enables engine manufacturers to use higher compression ratios and, therefore, achieve a better efficiency, this is pointless if the benefits are more than offset by efficiency losses at the refinery. A number of studies have been carried out in the past, to find the balance between these two conflicting factors. One of these, called RUFIT (rational utilization of fuels in transport), was a joint study by the oil and motor manufacturers with the objective of identifying the optimum octane quality to use at different lead levels. Input data used included the improvement in vehicle fuel economy with increasing compression ratio and the quantity of crude oil consumed to make a given volume of gasoline at different octane qualities and lead levels.

The combination of this vehicle and refinery fuel consumption data [A6.9] is given in Figure A6.3, which shows that at 0.4 g/L the optimum RON level is 96, at 0.15 g/L it is 95.5, and at zero lead it is about 92. This was re-evaluated in 1983 [A6.10] in the light of the availability of improved refinery equipment, and the optimum octane quality for unleaded gasoline moved to 94.5.

FIGURE A6.3 The effect of lead content on crude consumption/optimum octane number.

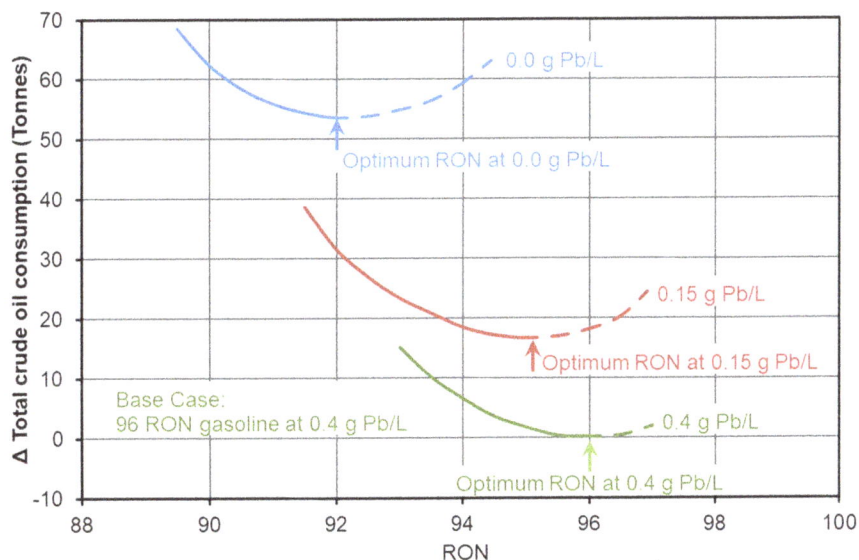

It should be noted that with the phase-out of lead antiknocks, in motor gasoline, these studies have not been repeated in recent years. Changes in refinery process technologies since that date mean that this data is only indicative of the situation at that particular time.

The presence of sulfur in the gasoline reduces the effectiveness of lead alkyls by promoting lead alkyl decomposition and by deactivating the lead-oxide species formed [A6.11–A6.13]. The order of antagonism of different sulfur compounds to lead alkyls, which is the same for all gasolines, is as follows:

Polysulfides > thiols > alkyl disulfides > alkyl sulfides
> elemental sulfur > aryl disulfides > aryl sulfides > thiophenes

Two other factors are important when considering the benefits and disadvantages of lead. One is that it prevents wear of exhaust valve seats by its lubricating action (this is discussed in Chapter 11, Section 11.4.6), and the other is that it may give lower levels of octane requirement increase (ORI) as discussed in Chapter 7, Section 7.7.

As noted at the start of this Appendix, lead alkyls are now only used in commercial gasoline for aviation gasoline. In the past there were different grades of avgas, as there were different grades of motor gasoline, but due to falling demand there is now only one grade of avgas available, this is Avgas 100. Avgas 100 has a lean mixture octane rating of 100 and can contain up to 1.12 g/L of lead [A6.2] in the US, but only up to 0.85 g/L in Europe [A6.14]. Avgas 100 is also colored green for identification purposes.

References

A6.1.   Midgley, T., Method and means for using motor fuels. U.S. Patent 1,573,846, issued February 23, 1926.

A6.2.   ASTM International, "Standard Specification for Leaded Aviation Gasolines," ASTM D910-21, ASTM International, 2022.

A6.3.   ASTM International, "Standard Test Method for Research Octane Number of Spark-Ignition Engine Fuel," ASTM D2699-12, ASTM International, 2012.

A6.4.   ASTM International, "Standard Test Method for Motor Octane Number of Spark-Ignition Engine Fuel," ASTM D2700-12, ASTM International, 2012.

A6.5.   Walsh, A.D. "The Mode of Action of Tetraethyl Lead as an Antiknock," Lectures on the Basic Combustion Process, Ethyl Corp., Detroit, MI, 1954.

A6.6.   Robinson, I.C.H. "Knock in the Gasoline Engine," in *Critical Reports on Applied Chemistry*, Vol. 10: Gasoline Technology, Ed. Hancock, E.G. (Oxford, UK: Society of Chemical Industry, Blackwell Scientific Publications, 1983), 81.

A6.7.   Rifkin, E.B. and Walcutt, C., "Decomposition of TEL in an Engine," *Industrial and Engineering Chemistry* 48, no. 9 (1956): 1532-1539.

A6.8.   Russell, T.J., "Motor Gasoline Antiknock Additives," Associated Octel Company Report 87/10, 1987.

A6.9.   CONCAWE, "Report on the Rational Utilisation of Fuels in Private Transport (RUFIT)," private communication, 1980.

A6.10. CONCAWE's Ad Hoc Group Automotive Emissions Fuel Characteristics; Kahsnitz, R. et al., "Assessment of the Energy Balances and Economic Consequences of the Reduction and Elimination of Lead in Gasoline," Report 11/83, CONCAWE, Den Haag, 1983.

A6.11. Livingstone, H.K., "Sulfur-Tetraethyl-Lead Interaction in Motor Fuels," *Industrial and Engineering Chemistry* 41, no. 5 (1949): 888.

A6.12. Livingstone, H.K., "The Effect of Sulfur Compounds on the Octane Number of Leaded Fuels," *Oil and Gas Journal* 46, no. 45 (1948): 81.

A6.13. Mieville, R.L. and Megeurian, G.H., "Mechanism of Sulfur Alkyl-Lead Antagonism," *Industrial and Engineering Chemistry* 6, no. 4 (1967): 253-257.

A6.14. Ministry of Defence, "DEF STAN 91-090 Issue 5," 2019.

Appendix 7: Physical Properties of Hydrocarbons

	Formula	Mole wt	Boiling pt °C	Melting pt	API	SG	Gross (MJ/kg)	Net (MJ/kg)	Gross (MJ/L)	Net (MJ/L)
Normal paraffins C_nH_{2n+2}										
Methane	CH_4	16	−161.6	−182.5	340.0	0.300	55.50	50.01	16.65	15.00
Ethane	C_2H_6	30.1	−88.9	−183.2	247.0	0.374	51.87	47.50	19.40	17.76
Propane	C_3H_8	44.1	−42.1	−187.6	147.0	0.508	50.36	46.36	25.58	23.55
Butane	C_4H_{10}	58.1	−0.5	−138.3	111.0	0.584	49.52	45.75	28.92	26.72
Pentane	C_5H_{12}	72.1	36.1	−129.7	92.7	0.631	49.01	45.36	30.92	28.62
Hexane	C_6H_{14}	86.2	68.7	−95.3	81.6	0.664	48.33	44.75	32.09	29.72
Heptane	C_7H_{16}	100.2	98.4	−90.6	74.2	0.688	48.08	44.57	33.08	30.66
Octane	C_8H_{18}	114.2	125.7	−56.8	68.6	0.707	47.89	44.43	33.86	31.41
Nonane	C_9H_{20}	128.2	150.8	−53.6	64.5	0.722	47.75	44.31	34.48	31.99
Decane	C_1OH_{22}	142.3	174.0	−29.7	61.3	0.734	47.64	44.24	34.97	32.47
Undecane	$C_{11}H_{24}$	156.3	195.8	−25.6	58.7	0.744	47.57	44.19	35.39	32.88
Dodecane	$C_{12}H_{26}$	170.3	216.3	−9.6	56.4	0.753	47.50	44.15	35.77	33.24
Iso-Paraffins C_nH_{2n+2}										
Iso-butane	C_4H_{10}	58.1	−11.7	−159.4	120.0	0.563	49.40	45.61	27.81	25.68
2-Methylbutane (Iso-pentane)	C_5H_{12}	72.1	27.9	−159.7	94.9	0.625	48.92	45.24	30.57	28.28
2,2-Dimethylpropane (Neo-pentane)	C_5H_{12}	72.1	9.4	−16.6	105.0	0.597	48.75	44.96	29.11	26.84
2-Methylpentane (Iso-hexane)	C_6H_{14}	86.2	60.3	−153.9	83.5	0.658	48.26	44.68	31.76	29.40
3-Methylpentane	C_6H_{14}	86.2	63.3	−117.8	80.0	0.669	48.29	44.71	32.30	29.91
2,2-Dimethylbutane (Neo-hexane)	C_6H_{14}	86.2	49.7	−99.8	84.9	0.654	48.15	44.57	31.49	29.15
2,3-Dimelhylbutane (Di-isoprpyl)	C_6H_{14}	86.2	58.0	−128.2	81.0	0.666	48.24	44.66	32.13	29.74
2-Methylhexane (Iso-heptane)	C_7H_{16}	100.2	90.1	−118.2	75.7	0.683	48.03	44.52	32.81	30.41
3-Methylhexane	C_7H_{16}	100.2	91.9	−119.4	73.0	0.692	48.06	44.54	33.25	30.82
3-Ethylpentane	C_7H_{16}	100.2	93.4	−118.6	69.8	0.703	48.08	44.57	33.80	31.33
2,2-Dimethylpentane	C_7H_{16}	100.2	79.2	−123.8	77.2	0.678	47.92	44.40	32.49	30.11
2,3-Dimethylpentane	C_7H_{16}	100.2	89.8	−17.8	70.6	0.700	48.01	44.50	33.61	31.15
2,4-Dimethylpentane	C_7H_{16}	100.2	80.5	−119.5	77.2	0.678	47.96	44.45	32.52	30.14

	Formula	Mole wt	Boiling pt °C	Melting pt	API	SG	Gross (MJ/kg)	Net (MJ/kg)	Gross (MJ/L)	Net (MJ/L)
3,3-Dimethylpentane	C_7H_{16}	100.2	86.1	−135.0	71.2	0.698	47.96	44.45	33.48	31.03
2,2,3-Trimelhylbutane (Triptane)	C_7H_{16}	100.2	80.9	−25.0	72.1	0.695	47.96	44.45	33.33	30.89
2-Methylheptane	C_8H_{18}	114.2	117.7	−109.5	71.2	0.698	47.85	44.38	33.59	31.15
3-Ethylhexane	C_8H_{18}	114.2	118.6	−17.8	65.6	0.718	47.85	44.38	34.35	31.86
2,5-Dimethylhexane (Di-isobutyl)	C_8H_{18}	114.2	109.1	−90.0	71.2	0.698	47.80	44.33	33.36	30.94
2,2,4-Trimethylpentane (Iso-octane)	C_8H_{18}	114.2	99.2	−107.3	71.8	0.696	47.78	44.31	33.25	30.84
Alcohols $C_nH_{2n+1}OH$										
Methanol	CH_3OH	32	65.0	−97.0		0.792	22.7	19.9		
Ethanol	C_2H_5OH	46.1	78.4	−114.0		0.789	29.7	28.9		
1-Propanol (n-propanol)	C_3H_7OH	60.1	97.0	−126.0		0.803	33.6	30.7		
2-Propanol (Iso-propanol)	C_3H_7OH	60.1	82.5	−89.0		0.786		30.4		
1-Butanol (n-butanol)	C_4H_9OH	74.1	118.0	−90.0		0.810		33.1		
2-Butanol (sec-butanol)	C_4H_9OH	74.1	99.0	−115.0		0.808				
2-Methyl-1-propanol (Iso-butanol)	C_4H_9OH	74.1	108.0	−102.0		0.802		33		
2-Methyl-2-propanol (tert-butanol)	C_4H_9OH	74.1	82.0	25.0		0.775		32.6		
Olefins C_nH_{2n}										
Ethylene	C_2H_4	28.0	−103.7	−169.2	273.0	0.35	50.10	47.19	17.54	16.52
Propylene	C_3H_6	42.1	−47.7	−185.2	140.0	0.522	48.94	45.80	25.55	23.91
Butene-I	C_4H_8	56.1	−6.3	−17.8	104.0	0.601	48.47	45.33	29.13	27.25
cis-Butene-2	C_4H_8	56.1	3.7	−138.9	94.2	0.627	48.33	45.19	30.31	28.34
trans-Butene-2	C_4H_8	56.1	0.9	−105.4	100.0	0.61	48.26	45.12	29.44	27.53
Iso-butene (Iso-butylene)	C_4H_8	56.1	−6.9	−140.3	104.0	0.6	48.19	45.05	28.92	27.03
Pentene-1 (Amylene)	C_5H_{10}	70.1	30.1	−138.0	87.2	0.647	48.17	45.03	31.17	29.14
cis-Pentene-2	C_5H_{10}	70.1	30.1	−179.0	87.2	0.661	48.06	44.92	31.76	29.69
trans-Pentene-2	C_5H_{10}	70.1	37.0	−135.0	84.9	0.654	48.01	44.87	31.40	29.34
2-Methylbutene-1	C_5H_{10}	70.1	31.1	−17.8	84.5	0.655	47.94	44.80	31.40	29.34
3-Methylbutene-1 (Iso-amylene)	C_5H_{10}	70.1	20.2	−180.0	92.0	0.633	48.06	44.92	30.42	28.43
2-Methylbutene-2	C_5H_{10}	70.1	38.4	−132.8	80.6	0.667	47.85	44.71	31.91	29.82
Hexene-1	C_6H_{12}	84.2	63.6	−138.9	77.2	0.678	47.57	44.43	32.25	30.12
cis-Hexene-2	C_6H_{12}	84.2	68.6	−146.1	73.9	0.689	47.50	44.36	32.73	30.56
trans-Hexene-2	$C6H_{12}$	84.2	67.9	−132.8	75.7	0.683	47.45	44.31	32.41	30.26
cis-Hexene-3	C_6H_{12}	84.2	67.6	−135.0	75.4	0.684	47.50	44.36	32.49	30.34
trans-Hexene-3	C_6H_{12}	84.2	68.1	−112.8	76.0	0.682	47.45	44.31	32.36	30.22
Diolefins C_nH_{2n-2}										
Propadiene	C_3H_4	40.1	−34.5	−136.1	106.0	0.595	48.57	46.36	28.90	27.58
Butadiene-1,2	C_4H_6	54.1	10.3		83.5	0.658				
Butadiene-1,3	C_4H_6	54.1	−4.4	−108.9	94.2	0.627	47.05	44.61	29.50	27.97
Pentadlene-1,2	C_5H_8	68.1	44.9	−65.0	71.5	0.697				
cis-Pentadiene-1,3	C_5H_8	68.1	44.2		71.8	0.696	46.87	44.29	32.62	30.82

	Formula	Mole wt	Boiling pt °C	Melting pt	API	SG	Gross (MJ/kg)	Net (MJ/kg)	Gross (MJ/L)	Net (MJ/L)
trans-Pentadiene-1,3	C_5H_8	68.1	42.3		76.0	0.682	46.87	44.29	31.96	30.20
Pentadiene-1,4	C_5H_8	68.1	26.1	−147.8	81.3	0.665	47.26	44.68	31.43	29.71
3-Methylbutadiene-1,2	C_5H_8	68.1	40.0	−120.0	82.9	0.685				
2-Methylbutadiene-1,3 (Isoprene)	C_5H_8	68.1	34.1	−146.1	74.8	0.686	46.66	44.08	32.01	30.24
Hexadiene-1,2	C_6H_{10}	82.1	77.8		64.5	0.722				
Hexadiene-1,3	C_6H_{10}	82.1	72.8		67.8	0.710				
Hexadiene-1,4	C_6H_{10}	82.1	65.0		70.6	0.700				
Hexadiene-1,5	C_6H_{10}	82.1	59.6	−140.8	71.8	0.696	46.82	44.15	32.59	30.73
Hexadiene-2,3	C_6H_{10}	82.1	68.0		75.1	0.685				
Hexadiene-2,4	C_6H_{10}	82.1	80.0		63.7	0.725				
3-Methylpentadiene-1,2	C_6H_{10}	82.1	70.0		65.0	0.720				
4-Methylpentadiene-1,2	C_6H_{10}	82.1	70.0		67.0	0.713				
2-Methylpentadiene-1,3	C_6H_{10}	82.1	76.1		63.9	0.724				
3-Methylpentadiene-1,3	C_6H_{10}	82.1	77.2		59.7	0.740				
4-Methylpentadiene-1,3	C_6H_{10}	82.1	76.3	−70.0	63.9	0.724				
2-Methylpentadiene-1,4	C_6H_{10}	82.1	56.1		70.9	0.699				
2-Methylpentadiene-2,3	C_6H_{10}	82.1	72.2		66.1	0.716				
2,3-Dimethylbutadiene-1,3	C_6H_{10}	82.1	68.7	−76.1	62.1	0.731	46.24	43.57	33.80	31.85
2-Ethylbutadiene-1,3	C_6H_{10}	82.1	75.0		61.0	0.735				
Acetylenes C_nH_{2n-2}										
Acetylene	C_2H_2	26	−84.0	−81.1	209.0	0.416	49.94	48.24	20.77	20.07
Methylacetylene	C_3H_4	40.1	−23.2	−102.8	94.9	0.625	48.40	46.19	30.25	28.87
Butyne-1 (Ethylacetylene)	C_4H_6	54.1	8.7	−122.5	86.2	0.650	48.03	45.59	31.22	29.63
Butyne-2 (Dimethylacetylene)	C_4H_6	54.1	26.9	−32.2	71.2	0.698	47.71	45.26	33.30	31.59
Pentyne-1 (Propylacetylene)	C_5H_8	68.1	40.2	−106.1	71.8	0.696	47.80	45.22	33.27	31.47
Pentyne-2	C_5H_8	68.1	56.0	−100.0	66.1	0.716	47.57	44.98	34.06	32.21
3-Methylbutyne-1 (Iso-propylacetylene)	C_5H_8	68.1	27.8		79.7	0.670	47.68	45.10	31.95	30.22
Heyne-1 (Butylacetylene)	C_6H_{10}	82.1	71.6	−132.0	65.0	0.720				
Hexyne-2	C_6H_{10}	82.1	84.5	−88.0	60.8	0.736				
Hexyne-3	C_6H_{10}	82.1	81.8	−101.0	63.1	0.727				
4-Methylpentyne-1	C_6H_{10}	82.1	61.2	−105.1	67.5	0.711				
4-Methylpentyne-2	C_6H_{10}	82.1	72.2		65.3	0.719				
3,3-Dimethylbutyne-1	C_6H_{10}	82.1	37.8	−81.2	78.7	0.673				
Olefins·Acetylenes C_nH_{2n-4}										
Buten-3-yne-1 (Vinylacetylene)	C_4H_4	52.1	5.6		73.9	0.689				
Penten-1-yne-3	C_5H_6	66.1	59.2		58.7	0.744				
Penten-1-yne-4 (Allylacetylene)	C_5H_6	66.1	41.7		49.4	0.782				
2-Methylbuten-1-yne-3	C_5H_6	66.1	32.2							
Hexen-1-yne-3	C_6H_8	80.1	85.0		56.4	0.753				
Hexen-1-yne-5	C_6H_8	80.1	70.0		32.8	0.861				

	Formula	Mole wt	Boiling pt °C	Melting pt	API	SG	Gross (MJ/kg)	Net (MJ/kg)	Gross (MJ/L)	Net (MJ/L)
2-Methylpenten-1-yne-3	C_6H_8	80.1	76.1							
3-Methylpenten-3-yne-1	C_6H_8	80.1	68.9							
Aromatics C_nH_{2n-6}										
Benzene	C_6H_6	78.1	80.1	5.5	28.6	0.884	41.84	40.17	36.99	35.51
Toluene	C_7H_8	92.1	110.6	−95.0	30.8	0.872	42.50	40.59	37.06	35.39
o-xylene	C_8H_{10}	106.2	144.4	−25.2	28.4	0.885	43.03	40.96	38.08	36.25
m-xylene	C_8H_{10}	106.2	139.1	−47.9	31.3	0.869	43.03	40.96	37.39	35.59
p-xylene	C_8H_{10}	106.2	138.3	13.3	31.9	0.866	42.87	40.80	37.12	35.33
Ethylbenzene	C_8H_{10}	106.2	136.2	−94.9	30.8	0.872	43.01	40.94	37.50	35.70
1,2,3-Trimethylbenzene	C_9H_{12}	120.2	176.1	−25.4	25.7	0.9				
1,2,4-Trimethylbenzene (Pseudocumene)	C_9H_{12}	120.2	169.2	−44.1	29.1	0.81	43.19	40.98	34.99	33.20
1,3,5-Trimethylbenzene (Mesitylene)	C_9H_{12}	120.2	164.6	−44.8	31.1	0.87	43.31	41.10	37.68	35.76
Propylbenzene	C_9H_{12}	120.2	159.2	−99.5	31.9	0.866	43.40	41.19	37.59	35.67
Lso-propylbenzene (Cumene)	C_9H_{12}	120.2	152.4	−96.0	31.9	0.866	43.43	41.22	37.61	35.69
1-Methyl-2-ethylbenzene	C_9H_{12}	120.2	165.1	−88.1	28.7	0.883				
1-Methyl-3-ethylbenzene	C_9H_{12}	120.2	161.5		31.1	0.87				
1-Methyl-4-ethylbenzene	C_9H_{12}	120.2	162.5	−63.7	31.5	0.868				
Cycloparaffins C_nH_{2n}										
Cyclopropane	C_3H_6	42.1	−32.8	−127.0	98.6	0.615				
Cyclobutane	C_4H_8	56.1	12.6	−50.0	74.8	0.686				
Cyclopentane	C_5H_{10}	70.1	49.3	−93.7	56.9	0.751	47.33	44.19	35.55	33.19
Methylcyclopentane	C_6H_{12}	84.2	71.8	−142.4	56.2	0.754	46.78	43.64	35.27	32.90
1,1-Dimethylcyclopentane	C_7H_{14}	98.2	87.5	−76.1	54.7	0.760				
1,2-Dimethylcyclopentane-cis	C_7H_{14}	98.2	99.3	−52.2	50.4	0.778	46.57	43.43	36.23	33.79
1,2-Dimethylcyclopentane-trans	C_7H_{14}	98.2	91.9	−118.9	55.4	0.757	46.57	43.43	35.25	32.87
1,3-Dimethylcyclopentane-trans	C_7H_{14}	98.2	90.8	−136.1	57.2	0.750				
Ethylcyclopentane	C_7H_{14}	98.2	103.4	−138.3	52.0	0.771	46.78	43.64	36.06	33.64
Cyclohexane	C_6H_{12}	84.2	80.7	6.7	49.0	0.784	46.59	43.45	36.53	34.06
Methylcyclohexane	C_7H_{14}	98.2	100.9	−126.4	51.3	0.774	46.52	43.38	36.01	33.58

Appendix 8: Abbreviations and Acronyms

The following is a list of abbreviations and acronyms used throughout this book. Unfortunately due to the breadth of topics covered in the book, there are a number of abbreviations or acronyms that have more than one meaning. However, which meaning applies on which occasion should be clear from the context.

^{1}H NMR - Proton nuclear magnetic resonance

2EHN - 2-ethyl-hexyl-nitrate

2-MF - 2-Methyl-furan

ΔR100°C - Difference between RON and R100°C

A.P.C.D. - Los Angeles County Air Pollution Control Board

ACI - Advanced compression ignition

ACO - Automobile Club de l'Ouest

ACORC - Australian Cooperative Octane Requirement Council

AFC - Alkaline fuel cells

AFEX - Ammonia fiber explosion

ALMS - American Le Mans Series

ALVW - Averaged loaded vehicle weight

APPI - Atmospheric pressure photo-ionization

AQIRP - Air Quality Improvement Program

ARCO - Atlantic Richfield Company

ASI - After the start of injection

ATAC - Active thermo-atmosphere combustion

ATC - Additives technical committee

Avgas - Aviation gasoline

AWE - Alkaline water electrolysis

B100 - 100% biodiesel fuel

B20 - Diesel fuel containing 20% biodiesel

BBD - Broad boiling distillate

BDC - Bottom dead center

BEV - Battery electric vehicle

BHT - Butylated hydroxytoluene

BMEP - Brake mean effective pressure

BOB - Blendstock for oxygenated blends

BOCLE - Ball-on-cylinder lubricity evaluator

BOTD - Ball on three disks

BOTS - Ball on three seats

CAA - Clean Air Act

CAAA - Clean Air Act Amendments

CAFE - Corporate average fuel economy

CAI - Controlled auto-ignition

CAMS - Confederation of Australian Motor Sport

CARB - California Air Resources Board

CCCD - Cold climate chassis dynamometer

CCD - Combustion chamber deposit

CCDI - Combustion chamber deposit interference

CCEPC - Central Council for Environmental Pollution Control

CCI - Calculated Cetane Index

CCS - Carbon capture and sequestration (storage)

CCUS - Carbon capture utilization and storage

CEC - Central Environment Council

CEC - Coordinating European Council

CEN - European Committee for Standardization

CFPP - Cold filter plugging point

CFR - Cooperative fuels research

CI - Compression ignition

CIHC - Compression ignition homogeneous charge

CME - Coconut methyl ester

CN - Cetane number

CNG - Compressed natural gas

CNI - Cetane number improver

COP21 - 21st Conference of the Parties

CORC - Cooperative Octane Requirement Committee

CP - Cloud point

CPD - Cloud point depressants

CRC - Coordinating Research Council

CRT® - Continuously regenerating trap

CSFT - Cold soak filtration test

CSS - Cyclic steam stimulation

CSTR - Continuous stirred tank reactor

CTL - Coal-to-liquids

CU - Conductivity unit

DAC - Direct Air Capture

DBPC - 2,6-di-tert-butyl-p-cresol

DCA - Deposit control additive

DCL - Direct coal liquefaction

DCN - Derived cetane number

DDFS - Direct dual-fuel stratification

DEA - Di-ethanol-amine

DEF - Diesel exhaust fluid

DEPG - Di-methyl-ether of poly-ethylene-glycol

DETR - Department of the Environment Transport and the Regions

DI - Diesel index

DI - Direct injection

DI - Driveability Index

DIPE - Di-iso-propyl-ether

DISC - Direct injection stratified charge

DISI - Direct injection spark ignition

DME - Di-methyl-ether

DMF - Di-methyl-furan

DOC - Diesel oxidation catalyst

DON - Distribution octane number

DPA - Distributor pump type A

DPF - Diesel particulate filter

DPG - Di-propylene glycol

DRA - Drag reducing agents

DTBP - Di-tertiary-butyl peroxide

DVPE - Dry vapor pressure equivalent

E0 - Gasoline containing 0% ethanol

E10 - Gasoline containing 10% ethanol

E70 - The amount that evaporates at 70°C

EA - Environmental agency

ECU - Electronic control unit

EDS - Exxon donor solvent

EFE - Early fuel evaporation

EGO - Exhaust gas oxygen sensor

EGR - Exhaust gas recirculation

EMA - Engine Manufacturers Association

EMOT - Estimated minimum operating temperature

EOR - Enhanced oil recovery

EPA - Environmental Protection Agency

EPEFE - European Programme on Emissions, Fuels and Engines

ETBE - Ethyl-tertiary-butyl-ether

EtOH - Ethanol

EU - European Union

FAME - Fatty acid methyl ester

FAR - First Assessment Report

FBC - Fuel-borne catalyst

FBP - Final boiling point

FBR - Full boiling range

FCC - Fluidized catalytic cracker

FEVI - Front-end volatility index

FFA - Free fatty acids

FIA - Fédération Internationale de l'Automobile

FIA - Fluorescent indicator adsorption

FIE - Fuel injection equipment

FIP - Federal Implementation Plan

FIT™ - Fuel Ignition Tester

FSN - Filter smoke number

fsNOx - Fuel-specific NOx

F-T - Fischer-Tropsch

GC - Gas chromatography

GCI - Gasoline Compression Ignition

GHG - Green-house gas

GHS - Globally Harmonized System of Classification and Labelling of Chemicals

GLC - Gas/liquid chromatography

GPF - Gasoline particulate filters

GRPE - Group Rapporteurs Pollution and Energy

GTBA - Gasoline-grade t-butanol

GTL - Gas-to-liquid

GVWR - Gross vehicle weight rating

HCCI - Homogeneous charge compression ignition

HEFA - Hydrotreated esters and fatty acids

HEGO - Heated exhaust oxygen sensor

HER - Hydrogen evolution reaction

HEV - Hybrid electric vehicle

HFCV - Hydrogen fuel cell vehicle

HFRR - High-frequency reciprocating rig

HiMICS - Homogeneous charge intelligent multiple injection combustion system

HLDT - Heavy light-duty truck

HMF - Hydroxy-methyl-furfural

HPCR - High-pressure common-rail

HPS - Heavy paraffin synthesis

HSDI - High-speed direct injection

HTFT - High Temperature Fischer-Tropsch

HUCR - Highest useful compression ratio

HVO - Hydrotreated vegetable oil

IARC - International Agency for Research on Cancer

IBP - Initial boiling point

IC - Internal-combustion

ICCS - Integrated carbon capture and storage

ICE - Internal-combustion engine

ICL - Indirect Coal Liquefaction

IDI - Indirect injection

IDID - Internal diesel injector deposits

IFPEN - IFP Energies Nouvelles

IGT - Institute of Gas Technology

IIEC - Inter-Industry Emission Control Program

IMSA - International Motor Sport Association

IP - Induction Period

IP - Institute of Petroleum

IPA - Iso-propanol

IPCC - Intergovernmental Panel on Climate Change

IQT™ - Ignition Quality Tester

iRCCI - Inverted Reactivity-Controlled Compression Ignition

IVD - Intake valve deposits

IVDA - Intake valve deposits apparatus

JCAP - Japan Clean Air Program

JFTOT - Jet Fuel Thermal Oxidation Tester

JPI - Japanese Petroleum Institute

L-DAC - Liquid direct air capture

LDV - Light-duty vehicles

LH2 - Liquid hydrogen

LLDT - Light light-duty trucks

LNG - Liquefied Natural Gas

LNT - Lean-NOx trap

LP - Linear programming

LPG - Liquefied petroleum gas

LTC - Low-temperature combustion

LTFT - Low-temperature Fischer-Tropsch

LTFT - Low-temperature flow test

LVW - Loaded vehicle weight

MBT - Minimum for best torque

MCFC - Molten carbonate fuel cells

MDA - Metal deactivators

MDDW - Mobil distillate dewaxing

MDEA - Methyl-di-ethanol-amine

MDFI - Middle distillate flow improvers

MDPV - Medium-duty passenger vehicles

MEA - Monoethanol-amine

MeOH - Methanol

METI - Minister of Economy, Trade and Industry

MIIT - Ministry of Industry and Information Technology

MITI - Ministry of International Trade and Industry

MK - Modulated Kinetics

MMT - Methylcyclopentadienyl manganese tricarbonyl

MN - Methane number

MON - Motor Octane Number

MOT - Ministry of Transport

MOUDI - Micro Orifice Uniform Deposit Impactor

MPI - Multipoint injection

MSDS - Material Safety Data Sheets

MTBE - Methyl-tertiary-butyl-ether

MTF - Methanol to fuel

MTG - Methanol to gasoline

MTO - Methanol to olefins

MULDIC - MULtiple stage DIesel Combustion

MVEG - Motor Vehicle Emissions Group

NAAQS - National Ambient Air Quality Standards

NACE - National Association of Corrosion Engineers

NASCAR - National Association for Stock Car Auto Racing

NBD - Narrow boiling distillate

NEDO - New Energy and Industrial Technology Development Organization

NET - 'Negative emissions' technologies

NG - Natural gas

NGSR - Natural gas substitution ratio

NHRA - National Hot Rod Association

NHTSA - National Highway Traffic Safety Administration

NiCE - Nippon Clean Engine

NLEV - National Low Emissions Vehicle

NLP - Nonlinear programming

NMA - N-methyl-aniline

NMHC - Non-methane hydrocarbons

NMOG - Non-methane organic gas

NSR - NOx storage reduction

NTC - Negative temperature coefficient

NVO - Negative valve overlap

OBD - On-board diagnostics

OER - Oxygen evolution reaction

OI - Octane index

OME$_x$ - Oxymethylene dimethyl ethers

ORI - Octane requirement increase

ORVR - On-Board Refueling Vapor Recovery

OSC - Oxygen storage capacity

OSI - Oil Stability Index

OTF - Over the fence

OTR - Ozone Transport Region

P/V - Pressure/vacuum

PAFC - Phosphoric Acid Fuel Cells

PAH - Polycyclic aromatics hydrocarbons

PCCI - Premixed charge compression ignition

PCI - Premixed compression ignition

PCV - Positive crankcase ventilation

PDMS - Polydimethylsiloxane

PEM - Proton Exchange Membrane

PEMWE - Proton exchange membrane water electrolysis

PFI - Port Fuel Injection

PFR - Plug flow reactor

PGM - Platinum group metal

PHEV - Plug-in hybrid electric vehicles

PIB - Poly-iso-butylene

PIBA - Poly-iso-butylene-amine

PLIF - Planar laser-induced fluorescence

PM - Particulate matter

PME - Palm Methyl Ester

PMP - Particle Measurement Programme

PN - Particle number

PNA - Polynuclear aromatic hydrocarbon

PODE - Polyoxymethylene dimethyl ethers

POMDME - Polyoxymethylene dimethyl ethers

PONA - Paraffins, olefins, naphthenes, and aromatics

PPC - Partially premixed combustion

PPD - Pour Point Depressants

ppm - Parts per million

PREDIC - PREmixed lean DIesel Combustion

PRF - Primary Reference Fuels

PROCO - Programmed combustion control

PSA - Pressure swing adsorption

ptb - Pounds per thousand barrels

PVO - Positive valve overlap

R100°C - RON of the gasoline distilled off up to 100°C

R75% - RON of the first 75% of the gasoline to distill off

RCCI - Reactivity-controlled compression ignition

REV - Range extended vehicle

RFG - Reformulated gasoline

RME - Rape Methyl Ester

ROCLE - Roller on cylinder lubricity evaluator

RON - Research Octane Number

RSSOT - Rapid Small Scale Oxidation Test

RUFIT - Rational Utilization of Fuels In Transport

RVP - Reid Vapor Pressure

RWGS - Reversed water gas shift

S - Sensitivity

SACI - Spark Assisted Compression Ignition

SAGD - Steam-assisted gravity drainage

SAR - Second Assessment Report

SCAQMD - South Coast Air Quality Management District

SCR - Selective catalytic reduction

S-DAC - Solid direct air capture

SDS - Safety Data Sheet

SE - Specific Energy

SFC - Supercritical chromatography

SFPP - Simulated filter plugging point

SG - Sterol-glucosides

SHED - Sealed housing for evaporative determination

SI - Spark ignition

SiC - Silicon carbide

SIP - State implementation plan

SLBOCLE - Scuffing load ball-on-cylinder lubricity evaluator

SLP - Sequential linear programming

SMD - Sauter mean diameter

SMDS - Shell Middle Distillate Synthesis

SME - Soy Methyl Ester

SMG - Saturated monoglyceride

SMMT - Society of Motor Manufacturers and Traders

SMPS - Scanning Mobility Particle Sizer

SNG - Synthetic natural gas

SOE - Solid oxide electrolysis

SOF - Soluble organic fraction

SOFC - Solid Oxide Fuel Cell

SOI - Start of injection

SPD - Slurry Phase Distillate

SRC - Solvent Refined Coal

SSF - Simultaneous saccharification and fermentation

SUV - Sport utility vehicles

SwRI - Southwest Research Institute

T10 - The temperature at which 10% evaporates

T50 - The temperature at which 50% evaporates

T90 - The temperature at which 90% evaporates

T95 - 95% distillation point

TAC - Toxic air contaminant

TAFLE - Thornton Aviation Fuel Lubricity Evaluator

TAME - Tertiary-amyl-methyl-ether

TAN - Total acid number

TAR - Third assessment report

TBA - Tertiary-butanol

TBHP - Tertiary-butyl-hydro-peroxide

TBI - Throttle body injection

TCCS - Texaco controlled combustion system

TDC - Top dead center

TEL - Tetra-ethyl-lead

THC - Total hydrocarbons

TLV-STEL - Short-term exposure limit threshold limit value

TLV-TWA - Time-weighted average threshold limit value

TML - Tetra-methyl-lead

TS - Toyota-Soken

TWC - Three-way catalyst

TWD - Total weighted demerits

UBA - Umweltbundesamt

UCO - Used cooking oil

UDDS - Urban dynamometer driving schedule

UGC - Underground coal gasification

UHC - Unburned hydrocarbons

ULSD - Ultra-low sulfur diesel

UN-ECE - United Nations Economic Commission for Europe

UNIBUS - Uniform bulky combustion system

UOP - Universal oil products

V/L - Vapor–liquid ratio

VASE - Variable angle spectroscopic ellipsometry

VGT - Variable geometry turbochargers

VI - Viscosity index

VLI - Vapor lock index

VOC - Volatile organic compounds

VOF - Volatile organic fraction

VOLFE - Volumetric fuel economy

VSR - Valve-seat recession

VVA - Variable valve actuation

WAFI - Wax anti-settling flow improver

WASA - Wax anti-settling additives

WCM - Wax crystal modifiers

WLTC - World-harmonized light-duty transient cycle

WLTP - World-harmonized light vehicle test procedure

WOT - Wide open throttle

WPI - Wax precipitation index

WSD - Wear scar diameter

YOCP - Yui and Ohnishi Combustion Process

Appendix 9: Glossary of Terms

Abnormal combustion
This applies to gasoline engine combustion. Normal combustion is where the flame front is initiated by the spark and proceeds smoothly throughout the combustion chamber. Abnormal combustion is where there are other unintended sources of ignition; this includes knock, pre-ignition, run-on or surface ignition.

Additive
The material added in small amounts to petroleum products to improve certain properties or characteristics. Small amounts are usually considered to be less than 1%.

Adsorption
The adhesion by weak forces of materials to the surface of solid bodies in which they are in contact often as a monomolecular film, although the layer can sometimes be two or more molecules thick.

Air-to-fuel ratio
The proportions, by weight, of air and fuel supplied for combustion.

Alcohols
A group of colorless organic compounds; each contains a hydroxyl (OH) group. The simplest alcohol is methanol, CH_3OH.

Alkane
A hydrocarbon having the general formula C_nH_{2n+2}. Also called a paraffin.

Alkylation
A refinery process for producing high-octane components consisting mainly of branched chain paraffins. The process involves combining light olefins with iso-paraffins in the presence of a strong acid catalyst such as sulfuric acid or hydrofluoric acid.

Alkyl group
A group of atoms, derived from an alkane (paraffin), having the general formula C_nH_{2n+1}, which forms part of a molecule. Examples are the methyl group (CH_3), the ethyl group (C_2H_5), etc.

Alternative fuel
An alternative to gasoline or diesel fuel which is not produced in a conventional way from crude oil.

Amide
A compound containing the group $CONH_2$. The hydrogen atoms on the nitrogen can also be substituted by other groups.

Amine

There are three types of amine. Primary amines are compounds containing the group NH_2 attached to an alkyl or aryl radical. Secondary amines have the NH group attached to two alkyl and/or aryl groups, and tertiary amines have three alkyl or aryl groups attached to the nitrogen atom.

Aniline point

The minimum temperature for complete mixing of equal volumes of aniline and the test fuel sample when evaluated by the ASTM D611 procedure. Often used to provide an estimate of the aromatic hydrocarbon content of a mixture.

Antiknock additive

An additive which, when added in small amounts to a gasoline, improves the octane quality of the fuel by suppressing knock.

Antiknock index

The average of the RON and MON for a fuel. Used as a measure of the octane quality of a gasoline, particularly in North America.

API gravity

An arbitrary scale, representing the gravity or density of liquid petroleum products in terms of API degrees, in accordance with the formula: API Gravity (degrees) = (141.5/Specific Gravity 15/15°C) − 131.5 The higher the API gravity, the lighter the material.

Aromatic

A hydrocarbon based on a six-membered benzenoid ring. See Appendix 1.

Aryl

A hydrocarbon group containing a benzene ring where the benzene ring is directly attached to the rest of the molecule.

Auto-ignition

The spontaneous ignition of a mixture of air and fuel without an ignition source. The auto-ignition temperature is the minimum temperature at which this takes place.

Aviation gasoline

Special grades of gasoline produced for aircraft reciprocating engines, often referred to simply as Avgas. The antiknock quality is defined by a lean mixture rating (ASTM D2700) and a supercharged rating (ASTM D909). The Vapor Pressure is generally somewhat lower than for motor gasoline (Mogas) and the distillation range can be narrower ideally from about 30°C to about 150°C.

Azeotrope

A mixture of liquids whose distillation characteristics do not conform to Raoult's law, i.e., one that boils at a higher or lower temperature than that of any of its constituents.

Benzole

A mixture of aromatic hydrocarbons containing a high proportion of benzene, obtained from the distillation of coal tar.

Biocides
Additives used for killing microbiological growths, which often occur in the water interface at the bottom of storage tanks, particularly those containing middle distillates.

Biodiesel
The term describes mono-alkyl esters of long-chain fatty acids, hence fatty acid methyl esters (FAME). The fatty-acids are generally derived from vegetable oils or animal fats. The product is used as a diesel fuel blending component or substitute.

Bio-fuel
A fuel or fuel component that is derived from a biological source, e.g., biodiesel or bio-ethanol.

Black products
A general term for any refinery stream or product containing residuum.

Black smoke
Exhaust gas containing a visible dispersion of carbonaceous particulate matter.

Blending number
A value assigned to a compound or component that will enable it to be blended linearly with other fuel components so that the value of the finished blend can be predicted. They can refer to a number of different properties such as octane number, vapor pressure, etc.

Blow-by
Combustion gases, often containing unburned or partially burned fuel, that escape from the combustion chamber, past the pistons, into the crankcase.

BOB grade fuel
Blendstock for Oxygenated Blends is a grade of fuel that is produced at the refinery for the later incorporation of a fixed level of oxygenated component such that the finished fuel meets all the required specifications.

Boiling range
The spread of temperature over which a fuel, or other mixture of compounds, distills.

Bosch smoke number
A measure of diesel smoke determined by passing the exhaust gas through a white filter paper. The darkening of the paper is determined using a reflectance meter. Bosch Smoke numbers are reported on a scale from 0 (clear) to 10 (black). Also referred to as the Filter Smoke Number (FSN).

Branched chain
A description of a paraffinic hydrocarbon in which the carbon atoms are arranged in a branched form and not in a straight line.

Brightstock
Heavy oil to be used as a lubricant that is obtained by refining residuum.

Carbonium ion
A positively charged carbon radical in which the charge is due to the loss of one electron from the carbon atom.

Carbonyl group
A carbon atom double bonded to an oxygen atom.

Carburetor
The device in some engine fuel systems that mixes fuel with air in the correct proportions and delivers this mixture to the intake manifold.

Carburetor foaming
The formation of a foam in a carburetor caused by the rapid boiling of fuel as it enters a hot carburetor. The foam cannot support the weight of the float bowl so that more and more fuel enters the carburetor causing an increase in pressure and forcing excess fuel out through the vent and/or metering jets, so that an over-rich mixture is obtained.

Carburetor percolation
This occurs when the fuel in a carburetor bowl starts to boil, either during or after a hot soak, forcing excess fuel into the inlet manifold via the vent or metering jet, so that an over-rich mixture results.

Catalyst
A substance that influences the speed and direction of a chemical reaction without itself undergoing any significant change.

Catalytic converter
A device fitted in the exhaust system of an engine containing a catalyst. The purpose is to catalyze reactions which convert undesirable compounds in the exhaust gas into harmless gases.

Catalytic cracking
A refinery process in which heavy hydrocarbon streams are broken down into lighter streams by the use of a catalyst and high temperatures.

Catalytic desulfurization
A refinery process in which sulfur is removed from a hydrocarbon stream by combining it with hydrogen in the presence of a catalyst and then stripping out the hydrogen sulfide thus formed.

Catalytic reforming
A refinery process which converts low-octane quality naphtha to a high-octane blendstock (catalytic reformate) in the presence of a catalyst, mainly by converting naphthenes and paraffins into aromatics. There are many commercially licensed versions of this process.

Caustic soda
Sodium hydroxide (NaOH), a strongly alkaline chemical.

Cetane
A paraffinic hydrocarbon, hexadecane ($C_{16}H_{34}$). The straight chain isomer, n-cetane or n-hexadecane, is a primary reference fuel on which the cetane number scale for measuring the ignition quality of diesel fuels is based. It has a cetane number of 100. The other primary reference fuel is iso-cetane or 2,2,4,4,6,8,8-heptamethyl nonane which has a cetane number of 15.

Cetane index

An approximation of cetane number based on an empirical relationship with density and volatility parameters such as the mid-boiling point.

Cetane number

A measure of the ignition quality of diesel fuel based on ignition delay in a specific engine. The higher the cetane number, the shorter the ignition delay and the better the ignition quality.

Chassis dynamometer

Equipment used to measure the power output of a vehicle at the driven wheels.

Climatic chamber

A room or chamber, usually containing a chassis dynamometer, in which various climatic conditions can be reproduced. Temperature control is most commonly used but humidity, air pressure, sunshine, and rain can also be reproduced in a repeatable manner.

Cloud point

The temperature at which a sample of a petroleum product just shows a cloud or haze of wax crystals when it is cooled under standardized test conditions.

CNG

Compressed natural gas.

Coal tar

One of the products from the destructive distillation of coal, the other main products being gas and coke.

Coking

A refinery process that is an extreme form of thermal cracking in which fuel oil is converted to lighter boiling liquids and coke.

Cold Filter Plugging Point (CFPP)

A measure of the ability of a diesel fuel to operate satisfactorily under cold weather conditions. A standardized test measures the highest temperature at which wax separating out of a sample can stop or seriously reduce the flow of fuel through a standard filter under standard test conditions.

Common-rail fuel injection

A type of diesel fuel injection system where the fuel pump maintains high pressure in a fuel reservoir and the fuel injection process is controlled by the fuel injects connected to this common reservoir.

Compression ignition

The use of the rapid compression of air within the cylinders to generate the heat required to ignite fuel. Often used to describe the combustion process within a diesel engine but is also used to describe some advanced combustion processes.

Compression ratio

The volume of the cylinder and combustion chamber when the piston is at Bottom Dead Center (BDC) divided by the volume when the piston is at Top Dead Centre (TDC).

Conjugated olefin
An organic non-cyclic hydrocarbon with alternate double and single bonds, e.g., 1,3-butadiene $CH_2=CH-CH=CH_2$.

Controlled auto-ignition (CAI) engine
An engine design where the air and fuel are pre-mixed and this mixture auto-ignites during the compression stroke, in a controlled manner.

Conversion process
A process which converts heavy products to lighter products.

CFR (Cooperative Fuel Research) engine
A single-cylinder, overhead valve, variable compression ratio engine used for measuring octane number or cetane number.

Corrosion inhibitor
An additive used in a fuel or other liquid that protects metal surfaces from corrosion.

Cracking
A type of refinery process that involves converting large, heavy molecules into lighter, lower boiling point ones.

Crude oil
Naturally occurring hydrocarbon fluid containing small amounts of nitrogen, sulfur, oxygen and other materials. Crude oils from different areas can vary enormously.

Cyclic dispersion
The cycle-to-cycle variation in cylinder pressure that occurs when an engine is running under otherwise constant conditions.

Dehydrogenation
A chemical reaction that involves removing hydrogen atoms from alkanes or naphthenes to give olefins or aromatics.

Delay period
See ignition delay.

Demerit rating
A numerical rating system in which increasingly high numbers represent increasingly poor performance. It is often used in evaluating vehicle driveability or the cleanliness of engine parts.

Demulsifier
An additive used for breaking oil-in-water emulsions.

Density
Mass of a substance per unit volume.

Deposit Control Additive
An additive or additive package designed to limit or reverse the accumulation of deposits within the engine and its fuel systems.

Derived Cetane Number
A measure of ignition quality determined using a constant volume combustion chamber as opposed to an engine.

Detergent
A term that was commonly used to describe an oil-soluble surfactant additive that is used to control deposit accumulation within an engine.

Detonation
A term used to describe the uncontrolled explosion of the last portion of combustible mixture in the combustion chamber. See also Knock.

Diesel exhaust fluid (DEF)
Aqueous urea solution used as the reductant in selective catalytic reduction (SCR) systems.

Diesel Index
An obsolescent measure of ignition quality in a diesel engine, defined as:

$$\text{Diesel Index} = 0.01 * (\text{Aniline point}) * (\text{API gravity})$$

Diesel knock
The unpleasant sound resulting from the spontaneous combustion of too much premixed fuel within a diesel engine. This can occur when the engine is cold.

Diesel particulate filter (DPF)
An exhaust aftertreatment device primarily intended to trap the solid cored particles, commonly referred to as soot, that are present in diesel exhaust.

Diffusion controlled combustion
See mixing controlled combustion.

Di-iso-propyl ether (DIPE)
An oxygenated compound having the formula C_3H_7-O-C_3H_7 suitable for use as a high octane gasoline blend component.

Direct injection (DI)
Where the fuel is injected directly into the cylinder which forms the main combustion chamber. The term can be applied to both gasoline and diesel engines. The former are usually referred to as DI spark ignition (DISI) whilst the later are simply DI diesel engines.

Dispersant
A surfactant additive designed to hold particulate matter dispersed in a liquid.

Distillation
The general process of vaporizing liquids in a closed vessel, condensing the vapors, and collecting the condensed liquids. Since liquids vaporize generally in order of their boiling points, it provides a method of separating materials according to their volatility.

Distribution Octane Number (DON)
A measure of the way octane quality is distributed across the boiling range of a gasoline, as measured using a modified CFR engine.

Drag coefficient
A measure of the air resistance of a vehicle as it is being driven.

Drag reducing agent (DRA)
A fuel additive which can reduce resistance to fluid flow so that the capacity of a pipeline can be increased.

Driveability
The response of a vehicle to the accelerator pedal inputs. Good driveability requires such characteristics as smoothness of idle, ease of starting when hot or cold, smoothness during acceleration without hesitations or stumbles, and absence of surging at constant pedal position when cruising. Separate tests are used for hot weather and cold weather driveability, in which numerical assessments of performance are assigned to each type of malfunction.

Dry Vapor Pressure
The dry vapor pressure and the dry vapor pressure equivalent (DVPE) are used to determine the vapor pressure in the absence of water vapor. The use of alcohol in gasoline can result in the pick-up of water which will influence the measured vapor pressure.

Dual-Fuel (Compression Ignition) engine
A type of diesel engine where a mixture of air and vaporized fuel (or gaseous fuel) is ignited by injection of a small amount of diesel fuel, which self-ignites due to high temperature and pressure.

Elastomer
Synthetic rubber-type materials frequently used in vehicle fuel systems.

Electro-fuels (e-fuels)
These fuels, which are also known as Power-to-Liquids (PtL or P2L), Power-to-X (PtX), Power-to-Gas (PtG) and Powerfuels, are a combination of hydrogen, produced by the electrolysis of water using renewably sourced power/electricity, and CO_2 captured from the atmosphere, again using renewably generated power/electricity.

Emulsification
The formation of a dispersion of one liquid in another where the liquids are not miscible with each other, such as oil in water. Emulsifying agents, which are surfactant materials, will stabilize such emulsions and prevent them from separating into two layers.

Ester
A class of organic compounds containing a carbonyl group adjacent to an ether linkage.

Ethers
A class of organic compounds containing an oxygen atom linked to two groups which can be alkyl and/or aryl.

Ethyl tertiary butyl ether (ETBE)
An oxygenated compound having the formula C_2H_5-O-C_4H_9 suitable for use as a high octane gasoline blend compound.

Evaporative loss controls
Devices used on vehicles, service station pumps, tanks, etc., to prevent losses of light hydrocarbons by evaporation. Such controls on vehicles usually involve the use of a canister filled with activated charcoal into which vents from the fuel tank, carburetor, etc., are fed so that the vapors are adsorbed onto the charcoal. The canister is regenerated by drawing intake air through the canister while the engine is running so that the hydrocarbons are desorbed and burned in the engine.

Exhaust gas recirculation (EGR)
The recycling of some exhaust gas back into the inlet manifold so as to lower the combustion temperature and hence reduce the formation of oxides of nitrogen.

Fatty acid methyl ester (FAME)
The term is used to describe mono-alkyl esters of long-chain fatty acids, hence fatty acid methyl esters (FAME). These are produced from vegetable oil or animal fats and used as a diesel blend component. These products are commonly referred to as biodiesel.

A group of compounds formed by the transesterification of long-chain fatty acids plant oils and animal fats using methanol. The fatty acids are often derived from plant oils and animal fats and hence products are commonly referred to as biodiesel.

FIA hydrocarbon analysis
A fluorescent indicator adsorption test procedure that can be used for the determination of hydrocarbon types in terms of aromatics, olefins and saturates. The sample is adsorbed on silica gel containing a mixture of fluorescent dyes and is then desorbed down an activated silica gel column. The hydrocarbons are separated by types and their positions indicated under ultraviolet light.

Flammability limits
Mixtures of air and petroleum vapors will only burn or explode within a certain range of concentrations. The lean limit (or lower explosive limit) is where the mixture has just enough hydrocarbons to burn and the rich limit (or upper explosive limit) is where it is almost too rich to burn.

Flash Point
The lowest temperature at which vapors from a petroleum product will ignite on application of a small flame, under standard test conditions.

Fluidizer
A high-boiling-point, thermally stable organic liquid used as part of some fuel additive packages to increase the fluidity of the active component.

Fractionation
The separation by boiling point, of mixtures of compounds by a distillation process. The degree of separation, i.e., the fractionation efficiency, will depend on the design of the fractionating tower.

Free radical

A radical is a group of atoms such as the methyl group (CH_3) that is part of a larger molecule. Such a group will not normally exist on its own since it has a free electron, but when it does it is called a free radical and it will rapidly react with other materials such as oxygen, sometimes forming further free radicals.

Freezing point

The temperature, determined under standard conditions, at which crystals of hydrocarbons formed on cooling disappear when the temperature of the fuel is allowed to rise.

Fuel injector

A device for injecting fuel into a piston engine. They are used in all diesel engines and most gasoline engines where they replace the carburetor(s) used in older engines.

Fuel oil

A term usually applied to a heavy residual fuel although it can also be applied to heavy distillates.

Full boiling range (FBR) reference fuels

A number of series of fuels made from refinery streams with an incremental change, usually one, in octane number. Unlike Primary Reference Fuels the RON, MON, and R100°C values are not the same for a given fuel and different FBR series are required to test the different requirements of an engine.

Fungibility

The ability to interchange or mix products from different sources or of different compositions without interactions occurring.

Gasohol

A blend of 90% gasoline and 10% ethanol used as an automotive fuel in all states in the US This blend of gasoline and ethanol is now standard in many countries but is now commonly referred to as simply E10 gasoline.

Gasoline particulate filter (GPF)

An exhaust aftertreatment device primarily intended to trap the solid cored particles, commonly referred to as soot, that are present in gasoline exhaust for direct injection engines.

Glow plug

A plug-type electrical heater used as a cold starting aid in indirect injection diesel engines.

Gum

The oxidation product arising from the storage of automotive fuel such as gasoline. It is barely soluble in gasoline and so may separate out and form a sludge. In an engine it will form sticky deposits which can cause malfunctions.

Hartridge unit

A measurement of the black smoke emitted from a diesel engine in which the opacity of the smoke is determined.

Heat of combustion

Also called the thermal, calorific or heating value and refers to the heat liberated when a fuel is burned. The Upper or Gross heating value includes the latent heat of water from combustion which is condensed in the test procedure. In an engine the water is exhausted in the vapor form and so a correction is made to give the net or lower heating value by subtracting the latent heat of condensation of any water produced.

Heat of vaporization
Also called Latent Heat of Vaporization. The heat associated with the change of phase from liquid to vapor at constant temperature.

Heptane
An alkane or paraffin having seven carbon atoms with the formula C_7H_{16}. Normal heptane, in which the carbons are arranged in a straight chain, is a primary reference fuel with Research and Motor octane values of zero. There are 9 isomers of heptane altogether.

Homogeneous Charge Compression Ignition (HCCI) engine
An engine design wherein it is intended that the air and fuel form a homogeneous mixture that then ignites due to an increase in temperature and pressure brought about during the compression stroke.

Hydrocarbon
A compound made up of hydrogen and carbon only.

Hydrocracking
A refinery process in which heavy streams are cracked in the presence of hydrogen to yield high-quality middle distillates and gasoline streams.

Hydrodesulfurization
A refinery process in which sulfur is removed from petroleum streams by treating it with hydrogen to form hydrogen sulfide which can be removed from the oil as a gas by stripping.

Hydrofining
A proprietary name for one version of the hydrodesulfurization process.

Hydrogenation
Treatment of a stream with hydrogen, usually to remove sulfur or to stabilize it by saturating double bonds.

Hydrophilic group
An organic group which has an affinity for water, such as the hydroxyl group (OH) or an acid group.

Hydrophobic group
The opposite of Hydrophilic. Most hydrocarbon groups come into this category.

Hydroskimming refinery
A simple refinery consisting only of process units for distilling, catalytically reforming and hydrotreating.

Hydrotreated vegetable oil (HVO)
Vegetable oils that have been broken into smaller branched molecules and de-oxygenated by replacing the oxygen atoms with hydrogen atoms. This can be used as a renewable diesel fuel or diesel fuel blending component.

Ignition advance
Used in relation to the timing of a spark for initiating combustion, it is the time before the fixed or optimum setting. It is usually expressed as degrees of crankshaft rotation.

Ignition delay
This is the time between the start of injection and the ignition of a fuel, in a diesel engine.

Ignition timing

Used in relation to the timing of a spark for initiating combustion, it is the time before the end of the compression stroke. It is often expressed as degrees of crankshaft rotation Before Top Dead Center (°BTDC).

Indirect injection (IDI)

A type of diesel engine in which the fuel is sprayed into a pre-chamber to initiate combustion, rather than directly into the cylinder, as in a Direct Injection engine.

Induction period

The time taken before the rapid increase in the formation of decomposition products during oxidation stability testing.

Injection timing

The timing of fuel injection, it is usually expressed as the number of degrees of crankshaft rotation Before Top Dead Center (°BTDC).

Injectors

See fuel injectors.

Intake system icing

The formation of ice in the carburetor or parts of some injector systems of spark ignition engines that can cause vehicle malfunctions such as stalling and loss of power. It occurs only during cool humid weather when there is enough moisture in the intake air to condense and then freeze in the carburetor due to the temperature drop caused by the evaporation of the gasoline. If the ambient temperature is too low (below about 0°C) there is not enough water in the atmosphere to give icing, and if the temperature is too high (above about 15°C) then the temperature depression is not enough to freeze the water.

Isomers

Compounds which have the same composition in terms of the elements present but which have the individual atoms arranged in different ways. Thus there are two isomers of butane (C_4H_{10}):

$$CH_3 \bullet CH_2 \bullet CH_2 \bullet CH_3$$

n-butane

$$CH_3 \bullet CH_2 \bullet CH_3$$
$$|$$
$$CH_2$$

iso-butane or 2-methylpropane

Isomerization

A refinery process which converts normal or straight chain hydrocarbons that have a poor octane quality into high-octane branch chain isomers. Thus n-butane is converted into isobutane, etc.

Isooctane

The hydrocarbon 2,2,4-trimethylpentane which has 8 carbon atoms and is used as a primary reference fuel with assigned values of RON and MON of 100.

Jerk Pump

A term used to describe the cam-operated plunger-type diesel fuel injector pump.

Kerosene (or Kerosine)
A refined petroleum distillate of which different grades are used as lamp oil, as heating oil, and as fuel for aviation turbine engines.

Kinetically controlled combustion
Combustion process that is controlled by the rate of chemical reaction.

Knock
In a spark ignition engine it is the auto-ignition (sometimes called detonation) of the end gas in the combustion chamber and causes the characteristic knocking or pinging sound. It can cause damage to the engine and can be overcome by increasing the octane quality of the fuel or by engine modifications. In diesel engines it is caused by excessive pressures in the combustion chamber and is avoided by the use of higher cetane number fuels.

Knock sensor
A detector, usually fixed to the cylinder head, that detects when knock is occurring in a spark ignition engine and actuates a mechanism such as one which retards the ignition to overcome the knock.

Latent heat
The heat associated with a change of phase such as going from a solid to a liquid at constant temperature (latent heat of fusion) or from a liquid to a gas at constant temperature (latent heat of vaporization or, simply, heat of vaporization).

Lead alkyl
A class of lead compounds, most commonly with methyl and/or ethyl groups attached to the lead atom, which are used as antiknock compounds in gasoline. See also TEL and TML.

Lead antagonism
Some compounds, and particularly those containing sulfur, are antagonistic to lead in that when they are present in a gasoline they reduce the antiknock benefit given by lead alkyls. The sulfur compounds themselves vary, according to their chemical composition, in the degree to which they are antagonistic.

Lead response
The extent to which the octane quality of a stream or component is improved by the addition of lead alkyls.

Lean mixture
An air-to-fuel mixture which has an excess of air over the amount required to completely combust all the fuel.

Lean NO_X trap (LNT)
An exhaust aftertreatment device where the catalytic coating reacts with the NO_X in the exhaust stream under fuel lean operating conditions. The NO_X is later released and reduced during a fuel rich operating regime.

Light ends
The lower-boiling components of a mixture of hydrocarbons. Those that are more easily evaporated or distilled off.

Low Temperature Flow Test (LTFT)
A test to predict the low-temperature performance of a diesel fuel; gives a better correlation with field performance for US vehicles and fuels than the Cold Filter Plugging Point (CFPP) test used in Europe and elsewhere.

LPG
Liquefied Petroleum Gas which consists mainly of propane and/or butane, and which can be stored as a liquid under relatively low pressure for use as a fuel.

Lubricity
The ability of a fuel (or oil or grease) to lubricate. It has particular relevance to low viscosity, low sulfur and low aromatic diesel fuels, made to meet stringent exhaust emissions legislation.

Markers
Chemicals that are added to fuel and can be detected by a color reaction with another chemical. Used to detect theft, contamination, tax evasion, etc.

Mercaptans
Compounds also known as thiols, having the group -SH. They have an extremely unpleasant odor and are removed from automotive fuel components to avoid customer complaints. Very low concentrations are often added to LPG and CNG to give it a distinctive warning odor.

Merit rating
A numerical scale in which high numbers represent good performance. See also Demerit rating.

Merox treating
A proprietary refinery process for removing mercaptans from petroleum streams.

Metal deactivator
A fuel additive that deactivates the catalytic oxidizing action of dissolved metals, notably copper, on hydrocarbons during storage.

Methanol
Methyl alcohol, CH_3OH, the simplest of the alcohols. It has been used, together with some of the higher alcohols, as a high-octane gasoline component and is a useful automotive fuel in its own right.

Methyl tertiary butyl ether (MTBE)
An oxygenated compound having the formula $CH_3OC_4H_9$, used widely as a high-octane component of gasoline.

Misfire
Failure to ignite the air-to-fuel mixture in one or more cylinders without stalling the engine.

Mixing controlled combustion
Combustion process that is controlled by the rate at which the fuel and air are mixed to produce a combustible mixture.

MMT
Methylcyclopentadienyl Manganese Tricarbonyl, an antiknock additive sometimes used in conjunction with lead and sometimes used on its own in unleaded gasolines. Also known as Hitech 3000.

Motor Octane Number (MON)
A measure of the antiknock quality of a fuel as determined by the ASTM D2700 method. It is a guide to the antiknock performance of a fuel under relatively severe driving conditions as can occur under full throttle, i.e., when the inlet mixture temperature and the engine speed are both relatively high.

Multifunctional additive
An additive or blend of additives having more than one function.

Naphtha
Loosely defined term covering a range of light petroleum distillates, used as chemical and reformer feedstocks, gasoline blend components, solvents, etc.

Naphthenes
A group of hydrocarbons having a cyclic structure with the general formula C_nH_{2n}. Examples are cyclohexane and cyclopentane.

Natural gas
A naturally occurring gas, consisting mainly of methane.

Naturally aspirated engine
An engine in which the intake air entering the system is at atmospheric pressure.

Negative valve overlap (NVO)
When the exhaust valve(s) closes before the inlet valve(s) opens. This limits the amount of burned gas that is pushed into the exhaust system and results in an increased amount of residual gas mixed with the next charge of air and fuel.

Neutralization Number
An indication of the acidity or alkalinity of an oil or fuel.

NO_X storage reduction (NSR) catalyst
See lean NO_X trap.

Nozzle coking
Deposit formation in or on the nozzle of a fuel injector.

Nucleation (of wax crystals)
A function of wax crystal modifier additives, that provides nuclei onto which wax molecules attach themselves when they come out of solution from a diesel fuel as it cools below its cloud point.

Nucleator (of wax crystals)
Component of a cold flow improver additive which creates nuclei onto which wax molecules attach themselves as they come out of solution, when a middle distillate fuel is cooled to below its cloud point.

Octane number
A measure of the antiknock performance of a gasoline or gasoline component; the higher the octane number, the greater the fuel's resistance to knock. There are two main types of octane number, the Research Octane Number (RON) and the Motor Octane Number (MON), which are based on different engine operating

conditions and, therefore, relate to different types of driving mode. Both are based on the knocking tendencies of pure hydrocarbons; n-heptane has an assigned value of zero and isooctane a value of 100. The octane number of a fuel is the percentage of isooctane in a blend with n-heptane that gives the same knock intensity as the fuel under test when evaluated under standard conditions in a standard engine.

Above a level of 100, the octane rating is based on the number of milliliters of tetraethyl lead per gallon which is added to isooctane to give the same knock intensity as the fuel under test.

Octane requirement

The octane number of a reference fuel (which can be a primary reference fuel or a full boiling range fuel) that gives a trace knock level in an engine on a test bed or a vehicle on the road when being driven under specified conditions. As many modern vehicles are fitted with a knock detector and the engine management system adjusts engine calibration accordingly, the octane requirement can be redefined as the point at which engine calibration begins to change.

Octane Requirement Increase (ORI)

The increase in octane requirement that occurs in an engine over the first several thousand miles of its life, due to buildup of carbonaceous and other deposits in the combustion chamber. It is influenced by driving mode, gasoline composition and the presence of lead and other additives. In modern vehicles the use of knock detectors and engine management systems make this extremely difficult to measure.

Oil dilution

The dilution of the lubricating oil by fuel or partially combusted fuel, which can find its way past the piston rings into the crankcase. This can be prevalent during cold starting and with late injection in diesel engines. Because the fuel lowers the viscosity of the lubricant, it increases wear.

Oil sands

A mixture of sand, clay, water, and bitumen that can be processed to produce oil.

Olefin

An unsaturated hydrocarbon, that is, one containing one or more double or triple bonds.

Oleophilic group

A chemical group attached to a molecule having an affinity for oily materials.

Oleophobic group

The opposite of oleophilic.

Operability limit

The lowest ambient temperature at which a diesel fuel will just function satisfactorily without wax separation causing filter plugging problems.

Otto cycle

The 4-stroke cycle of most piston engines, i.e., intake, compression, power and exhaust.

Oxidation

Loosely it is the chemical combination of oxygen to a molecule, although strictly it has a much broader meaning. It can be part of a manufacturing process or it can represent the deterioration of organic materials such as gasoline or diesel fuel due to the slow combination of oxygen from the air.

Oxygenate, Oxygenated compound
Terms which have come to mean compounds of hydrogen, carbon and oxygen that are blended into fuel to enhance combustion, for example to increase the octane number of gasoline and/or to reduce emissions.

Negative temperature coefficient (NTC)
When the rate of reaction decreases as the temperature increases.

Paraffin
A hydrocarbon having the general formula CH_{2n+2}. Also called an Alkane.

Particulates
Particles, as opposed to gases, emitted from the exhaust systems of vehicles.

Pintle
A type of fuel injector nozzle in which a shaped extension at the end of the injector needle controls the initial rate of fuel injection through the orifice of the nozzle. When the needle is fully lifted, there is full flow.

Pipestill
The primary distillation equipment used in a refinery in which the crude oil is heated in a furnace and passed into a fractionating tower, where it is split into different boiling range fractions.

Polymerization
The combination of two or more molecules of the same type to form a single molecule having the same elements in the same proportion as in the original molecule, but with a higher molecular weight. The product is a polymer. The product of two or more dissimilar molecules is known as copolymerization.

It is also a refinery process in which propenes and butenes from cracking processes are combined to form heavier olefins having a boiling range that makes it suitable for use as a gasoline blend component (polymer).

Port fuel injection
Fuel injectors which inject into the intake port rather than directly into the cylinders. They can be electronically or mechanically operated.

Pour point
The lowest temperature at which a petroleum product will just flow when tested under standard conditions, as defined in ASTM D97.

Pre-flame reaction
The chemical reactions which take place in an air-to-fuel mixture in an engine prior to ignition.

Pre-ignition
The premature ignition of the fuel-air mixture in a combustion chamber, i.e., before the spark from the plug. It can be caused by glowing deposits, hot surfaces or the auto-ignition of the fuel itself.

Pre-mixed Charge Compression Ignition (PCCI) engine
An engine design wherein the air and fuel are pre-mixed before ignition occurs due to an increase in temperature and pressure brought about during the compression stroke.

Pressure charged
Where a mechanical device is used to provide air to the engine at above-atmospheric pressure. This enables more air and thus more fuel to be processed and hence produce more power.

Primary Reference Fuel (PRF)
For use in spark ignition engines, it is a blend of n-heptane and isooctane used as a primary standard for knock evaluations. The octane value (both RON and MON) of a primary reference fuel is the percentage of isooctane in the blend with n-heptane.

For compression ignition engines, primary reference fuels are used to define the cetane quality of a fuel and are usually n-cetane (cetane number of 100) and heptamethyl nonane (cetane number of 15).

Pump-line-nozzle system
A type of diesel fuel injection system where the fuel pump provides the primary control over the injection process.

Pyrolysis gasoline (pygas)
Terms frequently used to describe naphtha produced by steam cracking and used as a gasoline component or a gasoline.

Quench
The removal of heat during combustion from the end gas or outside layers of the air-to-fuel mixture by the cooler walls of the combustion chamber.

R100°C
The Research Octane number of the part of a gasoline distilling up to 100°C using a standard distillation apparatus (ASTM D86). The difference between the RON of the whole fuel and the R100°C is the ΔR100°C.

Reactivity-controlled compression-ignition (RCCI) engine
An internal combustion engine where the ignition of the air and fuel is brought about by the increase in temperature and pressure during the compression stroke and the process is controlled by changing the reactivity of the fuel. This is currently accomplished by combining two fuels of different reactivities such as gasoline and diesel fuel.

Reflux ratio
A term used in connection with distillation and refers to the ratio of condensed side stream returned to a distillation column to the amount taken off as a side stream.

Reforming
Sometimes used as a short form of Catalytic Reforming which is a process for making high-octane components from naphtha. Also, it is a mild thermal cracking process for naphtha.

Reformulated Gasoline (RFG)
A gasoline blended to minimize undesirable exhaust and evaporative emissions.

Refutas Chart
A temperature-viscosity chart devised to show a linear relationship for Newtonian fluids.

Reid Vapor Pressure (RVP)
A measure of the vapor pressure of a liquid as measured by the ASTM D323 procedure; usually applied to gasoline or gasoline components.

Renewable Fuels
These are fuels produced from renewable resources such as from bio mass or using renewably sourced electricity to produce hydrocarbons from electrolysis of water and captured carbon dioxide.

Repeatability
The maximum difference between duplicate test results carried out on the same sample by the same operator using the same test equipment, above which the test is considered suspect.

Reproducibility
The maximum difference between test results carried out on the same sample by different operators in different laboratories, above which the test is considered suspect.

Research Octane Number (RON)
A measure of the antiknock quality of a gasoline as determined by the ASTM D2699 method. It is a guide to the antiknock performance of a fuel when vehicles are operated under mild conditions such as at low speeds and low loads.

Residue, Residuum
The non-volatile portion of a crude oil resulting from distillation.

Response
The way a fuel responds to treatment with additives such as lead alkyls, cold flow improvers, cetane improvers, etc., which are used to improve particular properties of the fuel.

Reynolds Number
Proportional to the ratio of inertial force to viscous force in a flow system. The critical Reynolds number corresponds to the transition from turbulent flow to laminar flow as the velocity is reduced.

Rich mixture
A fuel-air mixture that has more fuel than the stoichiometric ratio.

Road Octane Number
Usually the octane number of a Primary Reference Fuel (PRF) that just gives trace knock in a vehicle on the road or chassis dynamometer when tested under specified conditions.

Rundown tank
A tank into which the product from a still or other processing plant is received and from which the product is pumped to the storage tanks.

Run-on
A condition in which a spark ignition engine continues to run after the ignition has been switched off. Also known as "after-running" or "dieseling."

Saturated compound
A paraffinic hydrocarbon (alkane), i.e., a hydrocarbon with only single bonds and no double or triple bonds.

Scavenger
Term applied to the halogen compounds (usually dibromoethane and/or dichloroethane) present in a lead antiknock compound to prevent lead compounds such as lead oxides and sulfates from building up in the combustion chamber.

Selective catalytic reduction (SCR)
An exhaust aftertratement device where a specially formulated catalyst is used to reduce NO_X back to nitrogen using and additional reductant.

Sensitivity
The difference between the Research Octane Number and the Motor Octane Number of a gasoline. It is a measure of the sensitivity of the fuel to changes in the severity of operation of the engine.

Shale gas
A mixture of light hydrocarbon that can be found and extracted from some shale deposits.

Shale oil
A largely hydrocarbon mixture derived from oil shale (a naturally occurring deposit) by distillation.

SHED
An acronym for "Sealed Housing for Evaporative Determination," a sealed chamber in which a vehicle is placed in order to determine the amount of evaporative losses that occur from its fuel system.

Silicone
Organic compounds containing silicon, often used as antifoaming agents.

Solvent oil
An alternative term for fluidizer.

Spark knock
The most common form of knock, so-called because it is influenced by the spark timing.

Squish
The squeezing of part of the air-to-fuel mixture out of the end gas region in certain types of cylinder head as the piston reaches the end of the compression stroke. It promotes turbulence and further mixing of the air and fuel and minimizes the tendency to knock.

Steam cracking
A petrochemical process for the production of ethylene in which naphtha is cracked in the presence of steam.

Stoichiometric air-to-fuel ratio
The exact air-to-fuel ratio required to completely combust a fuel to water and carbon dioxide.

Stoke
A unit of kinematic viscosity (the quotient of the dynamic viscosity and the density). One stoke is I cm^2/s. A more convenient unit is the centistoke where 1 cSt = 0.01 St or 1 mm^2/s.

Storage stability
The ability of a fuel to resist deterioration on storage due to oxidation.

Straight chain
A descriptive term applied to a hydrocarbon in which all the carbon atoms are arranged consecutively in a straight line.

Supercharger
A mechanical device that pressurizes the intake air or air-to-fuel mixture to an engine and so increases the amount delivered to the cylinders and, hence, the power output.

Surface ignition
The ignition of the air-to-fuel mixture in a combustion chamber by a hot surface rather than by the spark.

Surfactant additive
When applied to fuels it is an organic compound with oleophobic and oleophilic groups that will form a coating on metal and other surfaces with the oleophilic group sticking into the hydrocarbon and the oleophobic group attaching itself to the surface. In this way it can protect surfaces and act as a dispersant to help control deposits.

Susceptibility
The extent to which the octane quality of a gasoline stream is improved by the addition of lead alkyls.

Sweetening process
A refinery process for converting mercaptans (thiols) into nonodorous compounds.

Synthesis gas
A mixture of carbon monoxide and hydrogen obtained from coke or natural gas by partial oxidation or steam reforming.

Tar sands
See oil sands.

TEL
See tetraethyl lead.

Terne plate
Steel sheet coated with a lead/tin alloy and often used for the construction of vehicle fuel tanks.

Tertiary amyl methyl ether (TAME)
An oxygenate used as a gasoline blend component.

Tetraethyl Lead (TEL)
A volatile lead compound, $Pb(C_2H_5)_4$, used as an antiknock additive in gasoline. No longer used in commercial motor gasoline but only in aviation gasoline, which can be used as a blending component for racing gasoline.

Tetramethyl Lead (TML)
A volatile lead compound, $Pb(CH_3)_4$, that was, but is no-longer, used as an antiknock additive in gasoline and having a higher volatility than TEL. It was particularly useful for improving the octane quality of the front end of a gasoline.

Thermal cracking
A refinery process for converting heavy streams into lighter ones by heat treatment.

Thermostat
A device used for the automatic regulation of temperature.

Three-way-catalyst (TWC)
A catalytic converter using a specially formulated catalyst to simultaneously decrease the amount of all three regulate gaseous emissions, CO, HC, and NO_X.

Toluene
A relatively volatile aromatic compound, $CH_3 \cdot C_6H_5$, present in catalytic reformate and widely used as a solvent. It has excellent octane qualities and can be used as a gasoline blend component.

Turbocharger
A device consisting of an exhaust driven turbine and an air compressor used for pressure charging. Energy is recovered from the exhaust gas to drive the turbine which is usually directly coupled to the compressor.

Unsaturated compounds
Hydrocarbons having one or more double or triple bonds. Such compounds are reactive and will combine with other elements such as oxygen or hydrogen. Olefins are unsaturated compounds.

Valve overlap
The period during the engine cycle when the inlet valve(s) and the exhaust valve(s) are both open at the same time. Also known as positive valve overlap (PVO).

Valve seat recession
The wearing away of a valve seat in the cylinder head of an engine. Not usually a problem when lead is present in gasoline because the lead combustion products act as a lubricant. Engines designed for unleaded gasoline have hardened valve seats.

Vapor-liquid ratio
The ratio, at a specified temperature and pressure, of the volume of vapor in equilibrium with liquid to the volume of liquid charged, at a temperature of 0°C. It can be measured using test procedure ASTM D2533 and is used to define the tendency for a gasoline to vaporize in the fuel system of a vehicle.

Vapor Lock
A gasoline supply failure to the engine of a vehicle due to vaporization of the fuel preventing the pump from delivering an adequate supply of fuel. Factors favoring vapor lock are high ambient temperatures, low ambient pressure, volatile gasoline and vehicle designs where heat from the engine can give high fuel line temperatures.

Vapor Lock Index (VLI)
An index that combines the Reid Vapor Pressure and the Percentage Evaporated at 70°C of a gasoline, and is a measure of the likelihood of a gasoline to cause vapor lock in vehicles on the road.

Vapor pressure
The pressure exerted by the vapors derived from a liquid at a given temperature and pressure.

Venturi
In a carburetor it is the narrowing of the air passageway. This increases the velocity of the air moving through it and induces a vacuum which is responsible for the discharge of fuel through the jets into the air stream.

Visbreaking
A refinery process for thermally cracking residual fuel oil, originally to reduce its viscosity but now to produce cracked streams.

Viscosity
A measure of the resistance to flow of a liquid.

Volatility
The property of a liquid that defines its evaporation characteristics. Highly volatile liquids boil at low temperatures and evaporate rapidly.

Wankel engine
A type of rotary, as opposed to a reciprocating, internal combustion engine. It uses an eccentric rotor which rotates within a epitrochoid-shaped housing.

Wax
High-molecular-weight, generally straight chain paraffins having limited solubility in diesel fuel.

Wax anti-settiing additive (WASA)
An additive that reduces the tendency for wax crystals to settle out on storage of diesel fuel.

Wax anti-settling flow improver (WAFI)
An additive that improves the cold flow characteristics of a diesel fuel and also reduces the tendency for wax crystals to settle out during storage.

Wax appearance point
A measure of the likelihood of wax coming out of solution as temperature is reduced. It is the temperature at which separated wax just becomes visible when tested under standard conditions, as defined by ASTM D3117, and gives a similar result to cloud point.

Wax Precipitation Index (WPI)
A function of Cloud Point (CP) and the spread between cloud point and pour point (CP-PP), used to correlate with the average estimated minimum operating temperature of vehicles in a CRC field test.

White product
Products which do not contain any residue from distillation.

White smoke
The smoke emitted during a cold start from a diesel engine, consisting largely of unburned fuel and particulate matter.

Wide cut
A fraction from a distillation column which has a wide boiling range. Wide cut fuels are of interest for both diesel and spark ignition engines, particularly for military use, when a normal fuel might not be available.

Wobbe Index

The Wobbe Index of a gas is proportional to the heating value of the quantity of gas that will flow sub-sonically through an orifice in response to a given pressure drop. It is calculated from the formula:

$$W = H / d^{1/2}$$

where

W is the Wobbe Index
H is the volumetric heating value of the gas
d is the specific gravity

Xylene

An aromatic hydrocarbon, $(CH_3)_2C_6H_4$, present in catalytic reformate and widely used as a solvent.

Index

About the Authors

Paul Richards was educated at the University of Manchester, UK; obtaining bachelor's and master's degrees in mechanical engineering, and finally a PhD for research on the transient performance of a gasoline engine. He worked briefly at the University of Bath, UK, studying Diesel Combustion, before joining the Associated Octel Company Ltd, now known as Innospec Ltd.

Paul worked at Innospec Ltd for over 27 years, where he was involved in the development and testing of many of their automotive fuel additive products. He served on a number CEC (Coordinating European Council for the development of performance tests for transportation fuels, lubricants, and other fluids) groups. He was responsible for obtaining Energy Saving Trust and Vehicle Certification Agency recognition for the Diesel Particulate Filter and Fuel Borne Catalyst systems marketed under the name Adastra™.

Paul became a Research Fellow at Innospec and an Honorary Senior Research Associate at University College London, UK. He was also a member of the Engineering and Physical Sciences Research Council's peer review college, a member of The Combustion Institute (British Section) and he was a member of the Universities Internal Combustion Engine Group (UnICEG) for over 20 years. In 1983 he received the Richard Way Memorial Prize, awarded annually by UnICEG for the best PhD thesis in the field of internal combustion engine research. In 2006 and 2013, he received the SAE Forest R. McFarland Award, in 2016 he was awarded the International Leadership Citation. In 2011 was elected an SAE Fellow.

Paul has authored or co-authored over 40 SAE papers and over 30 other publications. He has lectured regularly as part of Continuing Professional Development courses at the University of Leeds, UK. He has also presented at the CARB, SCAQMD, MECA, and ADPF "Course On Ultrafine Diesel Particles And Retrofit Technologies For Diesel Engines" in Diamond Bar, CA and at the "Technical Workshop on DPF-Retrofit for On-Road and Off-Road Applications" held in Beijing, China. He co-organized the 2008 SAE Bio-fuels: Specifications and Performance Symposium that was held in Paris, France and was a co-organizer of the 2013 SAE Fuels, Lubricants, and Aftertreatment Symposium held in Long Beach, CA.

Paul is also an active participant in SAE International activities, where he has acted as Strategic Advisor—Fuels, Vice Chair and subsequently Chair of the Exhaust Aftertreatment and Emissions Committee and Vice Chair and subsequently chair of the Powertrain, Fuels and Lubricants Activity. He is currently a member of the Fuels and Lubricants Committee, the Mobile Emissions Committee, and the Small Engines Technology Conference committee. He is also an Associate Editor of the *SAE International Journal of Fuels and Lubricants*.

James (Jim) Barker was born in Hartlepool, UK; he obtained a BSc(hons) in applied chemistry from Newcastle-upon-Tyne Polytechnic, followed by a PhD at The University of Durham in "Amidino Complexes of the Platinum Group Metals." He then went on to take up the position of Senior Research Fellow at The University of Warwick, before joining The Associated Octel Company Limited, now known as Innospec Ltd.

Jim has been with Innospec Ltd for over 36 years. He has undertaken a number of analytical, research and management roles, with The Associated Octel Co. Limited; Great Lakes Chemical Co. Ltd; Octel Corporation and Innospec Ltd. He is now the Senior Research Fellow at Innospec Ltd.

Jim is a member of the Engineering and Physical Sciences Research Council's peer review college. He is also an active member of The Energy Institute London (EI), where he is Chair of TMS; TMS—Test Methods Standardization Committee, Chair of SC-G 4 Molecular Spectroscopy, SC-G7 Hydrogen for Mobility, SC-G8 Ammonia for Mobility and SC-G9 Methanol for Mobility, and is Secretary of SC-G2 Organic Analysis, and a member of numerous other EI committees.

Jim is presently the Chair of Heavy Fuel Oils Working Group and vice chair CS93-International Standards and Related Activities at The American Society for Technical Measurement (ASTM), he is also a member of many other D02-Sub-committees.

Jim is also a Member of British Standards Institute, Chair of PTI/13-Petroleum Testing and Terminology; Member PTI/15 Natural Gas and Gas Analysis. Convenor of TC28/WG25 Hydrocarbon Analysis. He is a member of the Engineering Equipment and Materials Users Association (EEMUA) On-Line Analysers Group and a previous member of the Royal Society of Chemistry Dalton Committee. Jim is a Fellow of Royal Society of Chemistry (RSC), a Chartered Chemist, a Chartered Scientist, a member of the American Chemical Society (ACS), a member of Society of Automotive Engineers and a Member of ASTM.

Jim has published over one hundred papers, standards, guides, and patents. He has been a visiting lecturer at The University of Warwick and the University of Nottingham, he was a co-organizer of the "Fuel System Deposits, Formation and Effects" Workshop Baltimore, 2016 and of the "Gasoline Direct Injection Deposits" Workshop Heidelberg 2018. He is the recipient of the following awards: SAE Forest R. McFarland Award 2019. ASTM Award of Appreciation 2014 and 2019, ASTM Award of Excellence 2018; ASTM Eagle Award 2021; Energy Institute London Certificate of Appreciation 2010.

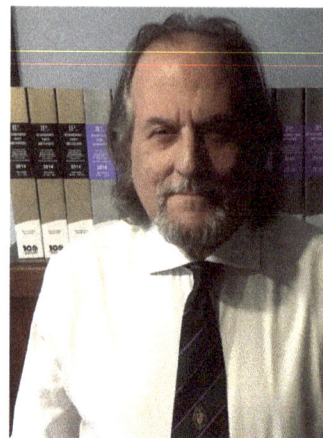